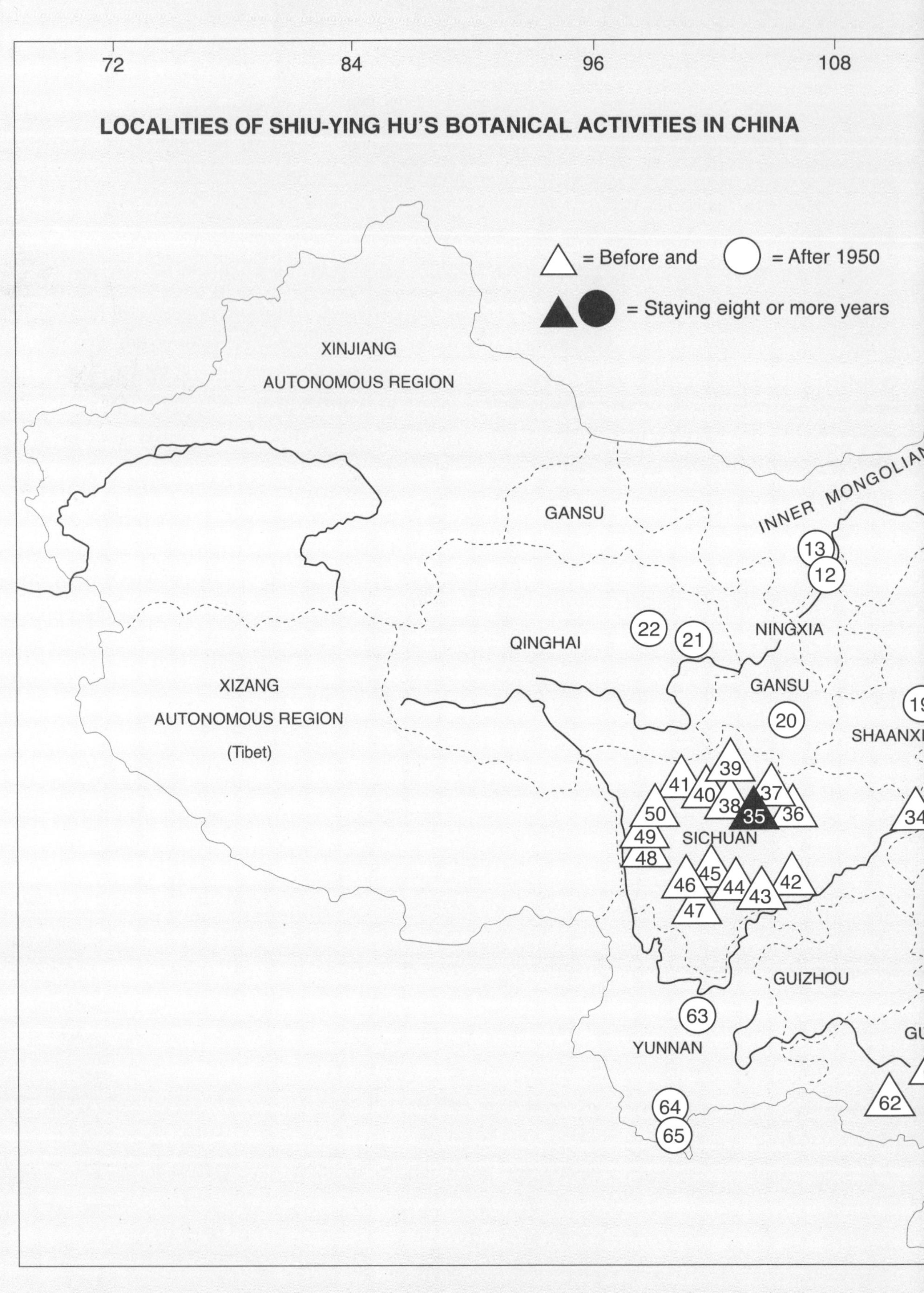

72
84
96
108
LOCALITIES OF SHIU-YING HU'S BOTANICAL ACTIVITIES IN CHINA
= Before and
= After 1950
= Staying eight or more years
XINJIANG
AUTONOMOUS REGION
GANSU
INNER MONGOLIAN
13
12
22
21
NINGXIA
QINGHAI
GANSU
20
19
XIZANG
AUTONOMOUS REGION
(Tibet)
SHAANXI
39
41
40
37
38
35
36
50
34
49
SICHUAN
48
45
46
44
42
43
47
GUIZHOU
63
YUNNAN
GU
64
62
65

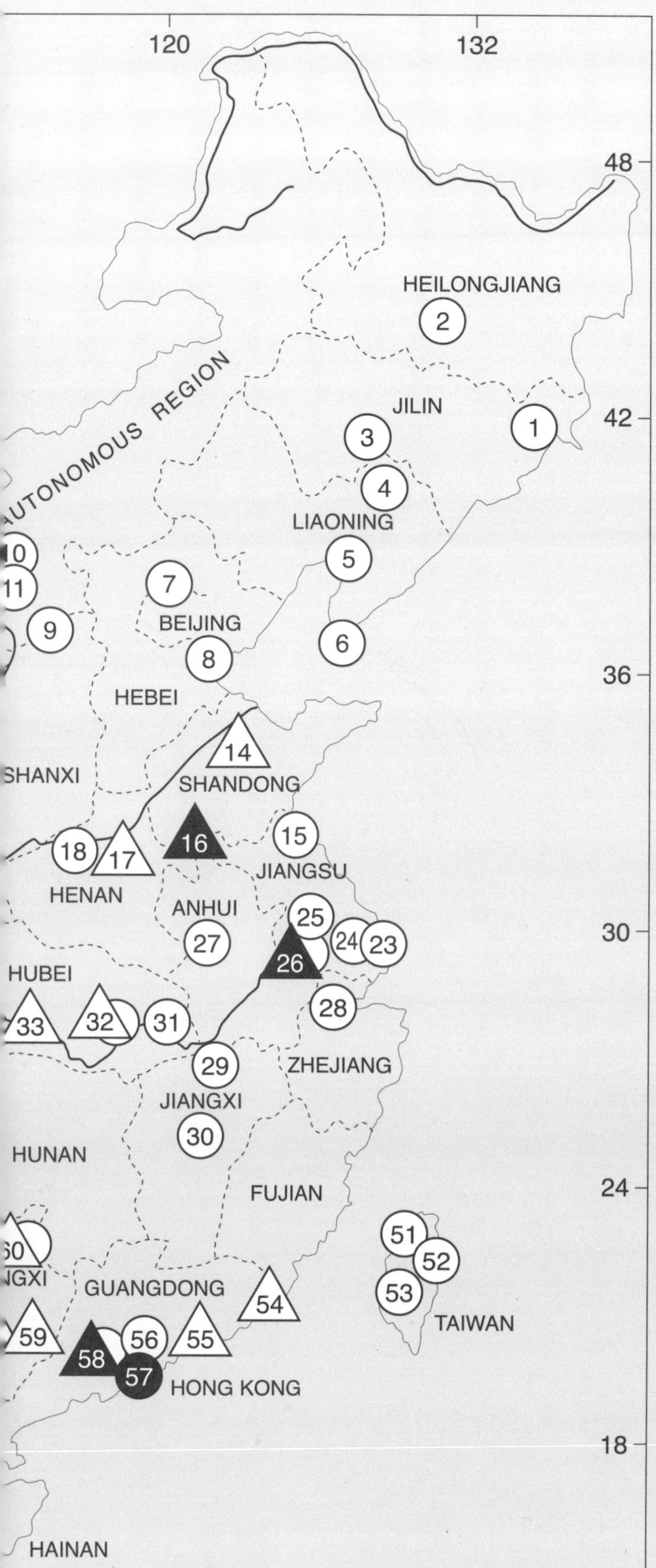

Map. A map of China showing the localities from where the author carried out her botanical activites in China. The localities are indicated by numbers designed to be read from north to south and east to west. The localities are:

(1) Long White Mountain
(2) Harbin
(3) Changchun
(4) Shenyang
(5) An Shan
(6) Dalian
(7) Beijing
(8) Tianjin
(9) Datong
(10) Yin Shan
(11) Hohhot
(12) Yinchuan
(13) Ala Shan
(14) Jinan
(15) Lianyun Harbor
(16) Xuzhou
(17) Zhengzhou
(18) Luoyang
(19) Xi-an
(20) Lanzhou
(21) Xining
(22) Koko Nor
(23) Shanghai
(24) Suzhou
(25) Yangzhou
(26) Nanjing
(27) Hefei
(28) Hangzhou
(29) Lu Shan
(30) Nanchang
(31) Qizhou
(32) Wuhan
(33) Yichang
(34) Wanxian
(35) Chengdu
(36) Xindu and Pixian (small towns too close to be seperated)
(37) Pengxian
(38) Guanxian
(39) Maoxian
(40) Wenchuan
(41) Lifan
(42) Chongqing
(43) Beipei
(44) Ziliujing
(45) Luo Shan
(46) Omei Mountains (Emei Mountains)
(47) Dragon Pool
(48) Ya-an
(49) Tianchuan
(50) Muping (Baoxing)
(51) Taipei
(52) Yilan
(53) Sun Moon Lake
(54) Shantou (Swatow)
(55) Loh Fau Shan (Luofu Shan)
(56) Conghua
(57) Hong Kong
(58) Guangzhou (Canton)
(59) Wuzhou
(60) Guilin
(61) Liuzhou
(62) Nanning
(63) Kunming
(64) Simao
(65) Xishuangbanna

Black triangles and dots represent areas of intensive studies with residence lasted for eight or more years. The white triangles and dots represent localities for shorter botanizations. The names of mountains (Ala Shan, An Shan, Loh Fau Shan, Long White Mountain, Lu Shan, Luo Shan, and Omei Mountains) and lakes (Sun Moon Lake in Taiwan, Koko Nor in Qinghai, and Dragon Pool in Sichuan) are kept as they appeared in earlier botanical literatures, and "Hong Kong" is used as by its local government (see text for more information).

FOOD PLANTS OF CHINA

Food Plants of CHINA

Shiu-ying Hu

Botanist, Arnold Arboretum, Harvard University
Honorary Professor of Chinese Medicine, The Chinese University of Hong Kong

The Chinese University Press

Acknowledgement: To Ms. Hsia-fei Wang for her painting *Peony* as the illustration on the cover.

Food Plants of China
  By Shiu-ying Hu

ISBN 962–201–860–2 (Hardcover)
      962–996–229–2 (Paperback)

THE CHINESE UNIVERSITY PRESS
The Chinese University of Hong Kong
SHA TIN, N.T., HONG KONG
Fax: +852 2603 6692
      +852 2603 7355
E-mail: cup@cuhk.edu.hk
Web-site: www.chineseupress.com

Printed in Hong Kong

CONTENTS

Foreword by Richard Evans Schultes vii

Foreword by George W. Staples, III ix

Introduction: Why the Book and to Whom It Is Addressed xi

Illustrations xv

PART ONE: CULTURAL ASPECTS OF CHINESE FOOD PLANTS

I. **The Sources and Nature of Information** 3
    Sources of Information 3
    Nature of Information 11

II. **The Production and Preparation of Chinese Plant Food** 19
    Kitchen Gardens 19
    Different Farming Systems 21
    Utilization of Suburban Land for Vegetable Gardens 22
    Plant Food Storage and Preservation 25
    The Use of Microbes in Chinese Plant Food 32
    Detoxification, Extraction and Limited Utilization 41

III. **Selected Chinese Food Plants with Instruction for Preparation** ... 47
    Chinese Cooking 48
    Leafy Shoots 50
    Flowers 66
    Fruits 81
    Seeds 98
    Special Subterranean Plant Food 126

IV. **Spices and Flavoring Materials** 147
    Cultural and Historical Background 148
    Common Spices Used in Chinese Food 150
    Combined Spices and Spicy Liquids for Making Cold Cuts 151

Natural Colors and Flavors 153

Table Dips 154

Samples for the Application of Spices 154

V. Health Food and Herbal Tea 161

Bupin 補品: Plant Esculents for the Conservation of Health 162

Chinese Herbal Teas: *Liangcha* and Parcelled Medicated Tea 230

Conclusions 249

PART TWO: BOTANICAL ASPECTS OF CHINESE FOOD PLANTS

VI. Nonvascular Plants: Monerans, Algae, Fungi and Lichens 257

Monerans 257

Nonvascular Eukaryotes — Algae 258

Nonvascular Eukaryotes — Fungi 262

VII. Vascular Plants: Pteridophytes and Spermatophytes 271

Pteridophytes — Filicopsids: True Ferns 271

Spermatophytes — Seed-bearing Plants 275

Gymnospermae: Naked-seed Plants 275

Angiospermae: Flowering Plants 280

Monocotyledoneae 280

Dicotyledoneae 332

Bibliography 753

Latin Name Index 763

Chinese Name Index 781

English Name Index 809

F O R E W O R D

With its very ancient cultural heritage, China naturally has a rich background in the utilitarian value of its many different vegetative environments. It is so deeply rooted in today's China that one would not expect this heritage to die out with competition from modern aspects of China's new development; quite on the contrary, there is no country where the value of useful plants is more thoroughly appreciated than in contemporary China — appreciation restricted not only to the general population but also in technical and scientific spheres. Country folk and city dwellers are still using herbal remedies, and the chemical, pharmacological and medical investigation of many aspects of traditional medicines is well under way and receives strong official encouragement. The same is true of other categories of economic plants, including in a place of prime importance, the food plants.

It is not only the worldwide love of the sundry Chinese cuisines that makes the study of food plants of this enormous and culturally very diverse nation of such wide attraction. Even the nutritional value of some of the lesser known foods is now receiving the attention of food scientists in China and in other countries as well.

This intense and growing interest consequently has both academic and practical value to mankind. The ethnobotanical and the health-nutritional and even commercial aspects, oftentimes working together, bode well for continued maintenance over many years of this interest in the wealth of plant species that the Chinese through millennia have found to be useful as foods.

In comparison with the plethora of popular and technical literature that has appeared on Chinese medicinal plants, little comprehensive material on food species has been easily available in one volume in Western publications. It is partly for this reason that this volume will be gratefully welcomed. But it will likewise be widely accepted as the dedicated labour of one of today's leading experts on the flora and economic botany of China — Dr. Shiu-ying Hu.

Dr. Hu, for many years connected with the Arnold Arboretum of Harvard University, a leading institution in the exploration and investigation of Asiatic floras, has published many technical and popular papers on the useful plants of China. This book by her on Chinese medicinal plants is a masterful contribution. Now we are offered a comprehensive survey, prepared from an interdisciplinary point of view, of the Chinese use of 1,430 species, varieties and cultivars of plants that nourish the extensive population of this enormous country.

Dr. Hu's encyclopedic contribution is destined to fill a void that has long been obvious and will remain for many years a vade-mecum not only for botanists but for scientists in related nutritional fields and for the interested public.

Richard Evans Schultes
Jeffrey Professor of Biology and
Director, Botanical Museum of Harvard University (Emeritus)

F O R E W O R D

I first met Shiu-ying Hu in 1981. Her reputation had preceded my meeting her in person, however, because long before I arrived at Harvard, friends and colleagues had recommended her to me as a person I must get to know. In the years that have elapsed since our first meeting, I have come to appreciate Dr. Hu's capabilities as a taxonomist and an economic botanist, and as an endless source of information about China and all things Chinese. Particularly in the latter regard she has been a benefactor to many people seeking guidance and information.

Throughout my years in graduate school Dr. Hu has generously and patiently shared with me, and others, her knowledge of her native land, its peoples, their customs, and the plants they use. When she began working on this book I encouraged her to press ahead with a project I believed to be very worthwhile. I am delighted that she is willing to share her knowledge with an even larger audience by setting it down in book form.

Dr. Hu asked me to read the manuscript. I was pleased to oblige. What I found in its pages was a fascinating discourse on the plants that are eaten by the Chinese, including answers to some of the very questions I had asked her over the years. A great deal of scientific information is presented here; Dr. Hu draws on the fields of plant taxonomy, morphology, economic botany, pharmacognosy, and phytochemistry to name a few, yet the text is far from a dry recitation of facts and statistics. What brings the work to life are the numerous anecdotes and personal experiences she shares. Indeed, this latter element gives the book the aspect of a personal memoir.

The subject of this book comes under the discipline of economic botany. More precisely, it covers the edible plants of one of the oldest continuous cultures on earth. And edible is applied in a broad connotation here; every plant consumed for food or for its therapeutic effects is treated. Because it incorporates information drawn from many fields the book will be of interest to scientists and laypersons alike.

Dr. Hu has organized her material into two parts: a text that introduces the cultural context of the work and presents detailed examples of the food plants and their preparation, followed by an annotated checklist of over 1,400 species. These two parts are extensively cross-referenced.

Most readers will probably focus attention on the text of the book, in which are found chapters discussing the cultural context of plant use in China, the agricultural practices used to produce plants, and the methods used to prepare them for consumption. The next three sections cover plants used for food, grouped by the plant parts (stems, roots, flowers, etc.) employed, plants used as spices and flavorings, and herbal health preparations, where the concepts of *liangcha* and *bupin* are introduced. Each of these sections is liberally provided with recipes that invite readers to sample the plants for themselves.

Of special interest to botanists is the annotated checklist of plants including algae and fungi, consumed in China. Major divisions of the list are taxonomic, and within each division the plants are arranged alphabetically according to scientific name. Also provided are the names in Chinese characters, the transliteration in English of both the Putonghua and Cantonese pronunciations for the Chinese names, common names in English, and a brief commentary. The commentary presents information such as the part of the plant used, its preparation, its use, and its availability in Boston's Chinatown, which may serve as a guide for those wishing to determine whether a particular product is imported into America at this time.

Not many botanists could write a book like this: a bilingual one, fluent in Chinese, could tap the enormous literature available in that language; a meticulous one could assemble the information and organize it into a manuscript; a botanist who enjoys cooking could develop recipes utilizing them so that readers might try the plants for themselves. Fortunately, Dr. Hu combines all these qualities, giving her a unique ability to write a book of this nature. Her work is further enriched because she has spent a lifetime using the plants firsthand, and consulting others who do so to gain their knowledge as well. The readers of the book are indeed fortunate, for they can enjoy the distillation of this experience and knowledge here assembled for the first time on such a broad scale.

George W. Staples, III
Bernice P. Bishop Museum
Honolulu, Hawaii

WHY THE BOOK AND TO WHOM IT IS ADDRESSED?

The food plants of an area provide the material basis for the survival of its population and furnish inspiring stimuli for its cultural development. In China, 1,156 species and 274 varieties and/or cultivars of food plants have been recorded and used by the peoples living between Taiwan in the Pacific East and Xining on the plateau of the arid west, and between Harbin in the frigid north and Xishuangbanna and Hong Kong in the humid south (see the map on the front endpaper). In association with these plants a reticulum of cultural patterns, both for survival and for enjoyment, has been developed by dozens of ethnic groups collectively called "Chinese". How was the botanical and cultural information acquired? Which plants have been selected by the Chinese people for food? How are these foodstuffs produced, preserved, detoxified (when poisonous) and prepared? What lessons can the people in the West learn from Chinese practices for preparing a healthy nutritious diet by using some of the Chinese plants? Answers to these questions are addressed within this book.

The preparation of this work began in 1957 at the request of the chairman of the Committee for the Vernacular Names of Pacific Plants, Pacific Science Association. I was invited to join the committee as a regular member and to provide a list of Chinese food plants with local names and their botanical identifications. Subsequently, *An Enumeration of the Food Plants of China*, containing 529 species, with Chinese names in hand-written ideograms, Wade-Giles system of romanization, botanical identifications, and parts of the plant used was compiled and distributed at the Ninth Pacific Science Congress, which took place in November 1957, in Bangkok, Thailand. Professor C. E. Wood of the Arnold Arboretum, Harvard University, has been keenly interested in

Chinese cuisine and the source plants. A complimentary copy of the *Enumeration* was presented to him. In late 1970s, he verified the listings with *Hortus III*, updated the nomenclature, added many English common names and suggested its publication as a special issue of *Arnoldia* (a quarterly publication of the Arnold Arboretum, Harvard University).

The original *Enumeration* was compiled at a time when China was closed to the West. Its content was limited to information I gathered for the Flora of China Project and from *List of Economic Plants in Taiwan* (Liu, 1942). Following the publication of the *Iconographia Cormophytorum Sinicorum* (Anonymous, 1972–83), all five volumes and two supplements were carefully checked for additional species of edible plants. Meanwhile, during my repeated trips to Hong Kong for field work toward the preparation of a flora of the area, I taught at Chung Chi College, The Chinese University of Hong Kong (CUHK). Economic botany, with special emphasis on the useful local species, was one of the courses taught there. The president of Chung Chi College, Dr. C. T. Yung, a botanist by training, was a connoisseur of genuine Chinese food. He delivered a lecture to my class on Chinese delicacies. With his help, numerous species of non-vascular plants were added to the *Enumeration*. As the director of the Institute of Science and Technology of CUHK, he always inspired and encouraged the staff members and students of the Department of Biology of Chung Chi College to carry out research projects on edible mushrooms and provided funds for their research programs. My office was adjacent to the research laboratory of Professor S. T. Chang, chairman of the department and a world-renowned specialist in edible fungi (Chang, 1978). From him I learned my first and most valuable lessons about mushrooms.

In July 1975, I returned to China for the first time after a sojourn of 29 years (Hu, 1975). In Guangzhou, Hangzhou, Shanghai, Nanjing, and Beijing, while other members of the tour went shopping for souvenirs or to the theater, I spent my time in the Xinhua Book Company, buying all the botanical publications I could find. Since the normalization of diplomatic relations between the United States of America (USA) and China in 1978, exchanges of botanists and botanical publications between these two countries have been promoted and accelerated. I personally made repeated trips and travelled in China extensively, making numerous new friends from whom I learned new uses of plants for food. All additional material available to me on Chinese food plants has been added to the *Enumeration*.

On July 19, 1985, after spending two months computerizing the revised list and adding the requested Pinyin to the original Wade-Giles system of romanization to all the Chinese names, I took the manuscript to Jamaica Plain to discuss its publication with the editor of *Arnoldia*, Edmund A. Schofield. After seeing the manuscript he suggested the addition of a good introduction, including some cultural aspects of Chinese food plants with illustrations and discussions of some species with recipes. These suggestions were accepted gladly and the work involved taken seriously.

In recent years, responding to a growing number of requests for information regarding the immunostimulating effects of special Chinese plant products received from people apprehensive about debilitating or life-threatening diseases, I have had to deal with the material on *bupin* (補品, revitalizing material), a subject included in a thesis prepared in 1937, for fulfilling a partial requirement of a Master of Science degree in Lingnan University, Guangzhou, China. In trying to investigate the availabilty of *bupin* in American market, I was indeed surprised to find that the Chinese businessmen had already imported numerous types of *bupin* and many kinds of medicated tea, *liangcha* (涼茶, cooling tea) into American stores. Before 1980 these articles were not in Chinese groceries. Subsequently, a chapter on *bupin* and *liangcha* was prepared and added to the manuscript.

Information on Chinese food plants grew steadily and by June 1988, documents on the cultural aspects on the subject covered more pages than the systematic enumeration of the species and varieties. By this time, there were some personnel change in the Arnold Arboretum, and the editor of *Arnoldia* resigned. Meanwhile, Dr. Theodore R. Dudley, research botanist, National Arboretum, Agricultural Research Service, United States Department of Agriculture (USDA), visited Harvard University Herbaria, examining the specimens of the genus *Ilex* L., which has been my area of research for thirty years (Hu, 1945; 1949–51; 1957). Mutual appreciation of works done on hollies has led us to share experiences in other areas of botanical investigations. Consequently, he saw the entire manuscript, took Part I, the cultural aspects, back to Washington, D.C. and read it and provided many critical notes. Moreover, he suggested the preparation of a short description for each species covered by the *Enumeration*, and being the general editor of Dioscorides Press (Plant Sciences) he advised to have the work published there. In complying with his suggestions and advice the description of the species was prepared. Descriptions of species available in common manuals and floras, including the *Iconographia Cormophytorum Sinicorum* were, abstracted and added to the manuscript. Descriptions of species not available in these general references were prepared from specimens deposited in Harvard University Herbaria and/or from plants raised in the Organismic Evolutionary Biology (OEB) experimental garden from seeds purchased in Chinese groceries. As these constitute the only descriptions of those species available in the English language, and some of them are given in detail, the volume of the work increased to four times that of the 1985 manuscript.

My personal involvement in the acquisition of the information is shown geographically by the map on the front endpaper indicating the localities where the data was collected, and socially by citing several cases showing the importance of harmonious human relations and friendship in learning practical cooking and in sharing genuine culinary art. To my knowledge, the cultural aspects involved in the producing, preserving and preparing plant food reported here are made available to my readers for the first time in the English language.

In the process of completing the *Enumeration*, Geradine C. Kaye, the librarian of the Farlow Reference Library and Herbarium of Cryptogamic Botany, Harvard University, has provided me with numerous references and helped me with the scientific names of the edible non-vascular plants. Beverly Huckins has voluntarily and systematically checked and rechecked the computerized list. In preparing the final manuscript she has given me invaluable editorial assistance with many helpful suggestions, and tested the recipes to prove their practicality. Elizabeth B. Schmidt, former managing editor of the *Journal of the Arnold Arboretum*, and an expert in Chinese cuisine, has readily given me assistance in the preparation of many recipes. Dr. Paul P. H. But, former director of the Chinese Medicinal Research Centre of The Chinese University of Hong Kong, is responsible for correctly placing the ideograms of Chinese names. Professor Paul D. Sorensen, Department of Biological Sciences, Northern Illinois University; Dr. Calvin R. Sperling, botanist, Germplasm Introduction and Evaluation Laboratory, Agricultural Research Service, USDA; and Miriam Ezust, a volunteer in Harvard University Herbaria, have kindly read the manuscript and given me culinary advice. I am deeply indebted to all of them. My special thanks are due to Hsia-fei Wang Chang, Enzyme Research Laboratory, Massachusetts General Hospital, for checking the chemical terms of the known compositions of articles used in *bupin* and *liangcha*; Professor Richard E. Schultes, former director, Botanical Museum, Harvard University, for the foreword; and Dr. George M. Staples, III, botanist, Bernice P. Bishop Museum, Honolulu, Hawaii, for painstakingly editing the final manuscript with many helpful suggestions and for the preface. I should like to take this opportunity to express my hearty appreciation to them all. Finally, to my friends Mrs. John Walden, Mrs. Linda Ooi, Miss Alice Tang and my assistant Miss Wong Yan for their valuable advice and final proof reading; and Mr. Fung Yat-kong and Mr. Tse Wai-keung of The Chinese University Press for their help in publishing this book.

Figures

1. Field investigation in western China
2. A communal kitchen garden situated near a retirement home in Chinatown at Boston
3. Suburban gardening in China
4. Planting, harvesting and marketing the vegetables produced in suburban gardens
5. Leafy shoots used for Chinese vegetables
6. Subtropical vegetables planted in kitchen gardens in Boston
7. Chinese herbs in a crowded kitchen garden in Boston
8. Two unusual potherbs in Chinese kitchen gardens of Boston
9. Leafy shoots for the special winter dish, *huo-guo* (火鍋, Fire Pot) in South China and Hong Kong
10. The Festival Gourd (*Benincasa hispida* [Thunb.] Cogn. var. *chiehqua* How) cultivated in a very small kitchen garden in Brookline, Massachuetts
11. Domesticated and semi-cultivated Chinese food plants
12. Market material of subterranean monocotyledonous plant food available in American Chinese stores
13. Market material of subterranean dicotyledonous plant food available in American Chinese stores
14. Chinese spices, material and marketing
15. *Bupin* (補品, immunostimulating plant material) sold in half- or one-pound packages in Chinese groceries in Boston
16. Fungal and algal food
17. A Chinese gymnosperm imported for horticultural purposes, producing valuable edible seed (*Ginkgo biloba* L.)
18. A perennial aquatic monocot (*Aponogeton natans* Engler et Krause) bearing large edible rootstock
19. A perennial aquatic plant (*Sagittaria sagittifolia* L.) cultivated as an annual crop in warmer regions of China for the edible corms
20. The true yam from the Chinese species of *Dioscorea* L.
21. Mulberry (*Morus alba* L.)
22. New Zealand Spinach (*Tetragonia tetragonoides* [Pallas] O. Ktze)
23. Peony (*Paeonia lactiflora* Pallas)
24. White flower forms of Chinese magnolias producing fleshy edible petals
25. Cruciferous Chinese vegetables introduced from southern China
26. Market material of Chinese cabbage and mustard
27. Five species of *Prunus* L. producing edible fruits and/or seeds available in China
28. Peanut (*Arachis hypogaea* L.)

29. Adsuki bean (*Azukia angularis* [Willd.] Ohwi)
30. Mungbean (*Cadelium radiatum* [L.] S. Y. Hu)
31. Traditional simple devices for extracting mungbean starch
32. Pegeon Pea (*Cajanus cajan* [L.] Millsp)
33. Scurgy pea (*Cullen corylifolium* [L.] Medikus, commonly known as *Psoralea corylifolia* L.)
34. Soybean (*Glycine max* [L.] Merr.)
35. Simple traditional devices for preparing soybean products
36. Garden pea (*Pisum sativum* L.)
37. Tamarind (*Tamarindus indicus* L.)
38. Yard-long bean (*Vigna sinensis* [L.] Savi ex Hasskarl)
39. Dragon's tongue (龍脷葉, *Sauropus changianus* S. Y. Hu)
40. Scarlet Sterculia (*Sterculia lanceolata* Cav.)
41. South China Mangosteen (*Garcinia oblongifolia* Champion ex Bentham)
42. Passion fruit (*Passiflora edulis* Sims)
43. Night-blooming cereus (*Hylocereus undatus* [Haw.] Britton et Rose)
44. Downy myrtle (逃軍糧, *Rhodomyrtus tomentosa* [Aiton] Hasskarl)
45. Natal plum (*Carissa macrocarpa* [Ecklon] A. DC. formerly known as C. *grandiflora* A. DC.)
46. Lycium berry (枸杞子, *Lycium barbarum* L.)
47. Husk tomato (*Physalis angulata* L.)
48. Cultivars of the eggplant (*Solanum melongena* L. var. *esculentum* Nees) from southern China
49. Tomato (*Lycopersicon lycopersicum* [L.] Karsten ex Farwell, generally known as *L. esculentum* Miller)
50. Witchweed (*Striga asiatica* [L.] O. Kuntze)
51. Cistanche, Mongolia Broom-rape (*Cistanche salsa* [C. A. Meyer] Beck.)
52. Gardenia, Huang-zhi (*Gardenia jasminoides* Ellis)
53. Honeysuckle (金銀花, *Lonicera japonica* Thunb.)
54. Winter melon (冬瓜, *Benincasa hispida* [Thunb.] Cogn.)
55. The common cultivars of 2 varieties of *Benincasa hispida* [Thunb.] Cogn.
56. Watermelon (*Citrullus battich* Forskal)
57. Oriental Picking Melon (*Cucumis melo* L. var. *conomon* [Thunb.] Nakai)
58. Cucumber (黃瓜, *Cucumis sativus* L.)
59. Bottle Gourd, Hulu (*Lagenaria siceraria* [Molina] Standley var. *siceraria*)
60. Bitter Melon (*Momordica charantia* L.)
61. Chayote (佛掌瓜, *Sechium edule* [Jacq.] Sw.)
62. Luohanguo (羅漢果, *Siraitia grosvenorii* [Swingle] C. Jeffrey)
63. Safflower (*Carthamus tinctorius* L.)
64. Chicory (*Cichorium intybus* L.)
65. Sunflower (*Helianthus annuus* L.)
66. Field Sow-thistle (*Sonchus arvensis* L.)
67. Dandelion (*Taraxacum officinale* Weber ex Wiggers)

Figure 1. Field investigation in western China: **a.** The author in Tibetan attire when she lived in different Gia-rong villages in the Sino-Tibetan mountains, formerly called Eastern Tibet (taken in summer of 1941). Courtesy of the Office of Registrar, Radcliffe College. **b.** A local field assistant and informant, a young Gia-rong medicine gatherer (summer of 1940).

Figure 2. A communal kitchen garden situated near a retirement home in Chinatown at Boston, built on the site of a condemned tenement. **a.** A fenced-in area divided into many small lots, each assigned to one gardener who uses the land very economically and builds frames for climbers. The Festival Gourd (*Benincasa hispida* [Thunb.] Cogn. cv. "Chiehqua" How) is a popular planting in this garden. **b.** The close-up of a plot in the garden, showing the crowded growth of different kinds of vegetables, particularly the Garden Amaranth (*Amaranthus inamoenus* Willd., in the foreground) and the Crown Daisy (*Chrysanthemum coronarium* L. var. *spatiosum* Bailey).

Figure 3. Suburban gardening in China. **a.** A garden outside Guangzhou (Map, Loc. 58), showing intensive cultivation of vegetables in beds between irrigation ditches. **b.** A cement tank for fertilizer, showing a simple roof for protecting the decomposing material from rain, and two wheelbarrows with large wooden containers for carrying human waste from the city to be biologically decomposed in the tank. **c.** A gardener working barefooted in an irrigation ditch on two raised beds simultaneously. In the background is a bed of Aquatic Morning Glory (*Ipomoea aquatica* Forskal) in flower, reserved for the production of mature seed. **d.** A gardener carrying the properly decomposed organic fertilizer in two buckets from the tank to the vegetable beds for application.

Figure 4. Planting, harvesting and marketing the vegetables produced in suburban gardens. **a.** Intensive planting in a vegetable garden outside Guangzhou (Map, Loc. 58), showing Sweet Kudzu (*Pueraria thomsonii* Bentham) supported by poles, planted for its fusiform-tuberous root, taro, and various species of *Brassica* L. **b.** The close up of the bed showing the cauliflower (*Brassica oleracea* L. var. *botrytis* L., left front), interplanted with and set in the bed just before the Vegetable Heart (*B. parachinensis* Bailey cv. 'Caixin' hort, right center and black), raised for repeated picking of the tender flowering shoots, is ready for the first harvest. **c.** Marketing Egg Plant on a street corner at Beijing, showing the fruits being unloaded from a truck carrying the products from a commune while the buyers are waiting. **d.** An itinerant vendor marketing Celery Cabbage (*B. pekinensis* [Lour.] Rupr.) in the street of Lanzhou (Map, Loc. 20), showing the dealer carrying his products and a scale on a cart, stopping at street corners for the convenience of customers. (**a** and **b** were taken in February 1980, **c** in late August 1984, and **d** in October 1979.)

Figure 5. Leafy shoots used for Chinese vegetables. **a.** A wild population of Field Amaranth (*Amaranthus paniculatus* L.) growing in the experimental garden near the Harvard University Herbaria. Young shoots have been gathered from this garden repeatedly and cooked as a potherb in potluck parties for staffs and graduate students of the herbaria; the dish has always been left empty. **b.** A small and healthy crop of garlic growing in the limited space of a kitchen garden in Chinatown at Boston. At this stage of growth, the plants are fully grown but the fleshy cloves are not developed. They are thinned and used for garlic green.

Figure 6. Subtropical vegetables planted in kitchen gardens in Boston. **a.** Aquatic Morning-glory (*Ipomoea aquatica* Forskal), a perennial climber, often seen floating on canals in Southeast Asia, grows in clumps as a summer crop in many gardens of Chinese Americans in eastern USA. When moved indoor in October and kept alive, the stock can be planted next June to start another crop. **b.** Bitter Melon (*Momordica charantia* L.) showing the deeply lobed leaves and a male flower.

Figure 7. Chinese herbs in a crowded kitchen garden in Boston. **a.** The closeup of a lot showing Chinese Leek (*Allium tuberosum* Rottler ex Sprengel), and the Coriander (*Coriandrum sativum* L.) in forefront. On the right, the Chinese Leek is the most popular species in gardens of Chinese Americans throughout USA. The leaves with the pseudostems are cut repeatedly. In Boston, the species flowers in August. The scapes at the stage as shown should be nipped off from the base with the flower buds attached and used for vegetable. The plant seeds readily and eventually becomes a weed and messes up the garden. The picking of the scapes for food and the umbels for spice eventually keeps the species under control and the garden clean. On the left, the coriander is kept for mature fruits, used for starting the next crop. **b.** Two Chinese herbs introduced and used for home remedies in the form of tea, the South China Motherwort (*Leonurus artemisia* [Lour.] S. Y. Hu) with red flowers in the foreground and the Cock's Comb Perilla (*Perilla crispa* [Thunb.] S. Y. Hu), with maroon-red herbage in the background. These species grow semispontaneously at the corner and near a fence of the garden, not observed elsewhere in Boston.

Figure 8. Two unusual potherbs in Chinese kitchen gardens of Boston. **a.** Herb-of-the-Goddess (神仙草, *Murdannia nudiflora* [L.] Brenan), a subtropical species brought from Hong Kong and propagated vegetatively among gardeners sharing the divisions of the crown. **b.** Edible Crown Daisy (*Chrysanthemum coronarium* L. var. *spatiosum* Bailey) raised from seed for the tender shoots in the foreground and Amaranth in the upper right hand corner.

a

b

c

Figure 9. Leafy shoots for the special winter dish, *huo-guo* (火鍋, Fire Pot) in South China and Hong Kong. **a** and **b** Chinese Matrimony-vine (*Lycium chinense* Miller), **c.** pea shoots (*Pisum sativum* L.). **a.** A wild plant growing by the fence of a playground of a grade school in Cambridge, Massachusetts. It has been a personal source of supply for tender shoots for many years. The stem from the plant has been gathered in every winter, chopped, dried and used for herbal tea. **b.** Sections from different stems of the same specimen, showing variation of leaf-sizes. The one on the right with spines and large leaves is from a sucker. Notice the positions of the flowers on the middle sample and of a fruit on the left one (photographs taken in August, 1985 by Thomas Hu). **c.** Pea shoots, the market material bought from Hong Kong on January 10, 1986, showing the exact stage of development for picking (either without tendril as the two central samples or with the leaf tendrils just begin to merge as shown by the two lateral samples).

Figure 10. The Festival Gourd (*Benincasa hispida* [Thunb.] Cogn. var. *chiehqua* How) cultivated in a very small kitchen garden in Brookline, Massachuetts. **a.** A flower with its pollination agent. **b.** A young fruit, very hairy, and dark green with lighter spots.

Figure 11. Domesticated and semi-cultivated Chinese food plants. **a.** Yard-long Bean (*Vigna sinensis* [L.] Savi ex Hasskarl) in a kitchen garden in Boston showing paired fruits, measuring 40–65 cm long and tasting best when 7–10 mm across. The three fruits on the right are saved for seeds. They have increased in thickness, indicating that seeds are approaching maturity. Eventually these fruits will attain a diameter of 1.5–2 cm across, and the pericarp become spongy and light green in color. **b.** Farmers harvesting Water Hyacinth (*Eichhornia crassipes* Solms) in a protected area in the Hong Kong countryside, using the outer leaves for feeding pigs and the inner tender portion for food (photo taken in 1972). **c.** A protected semi-wild population of Chinese Rhubarb (*Rheum officinale* Baillon) in the alpine meadow of "Eastern Tibet" at 3,500–4000m above sea level, where petioles of the plants are eaten locally (photo courtesy of the Arnold Arboretum).

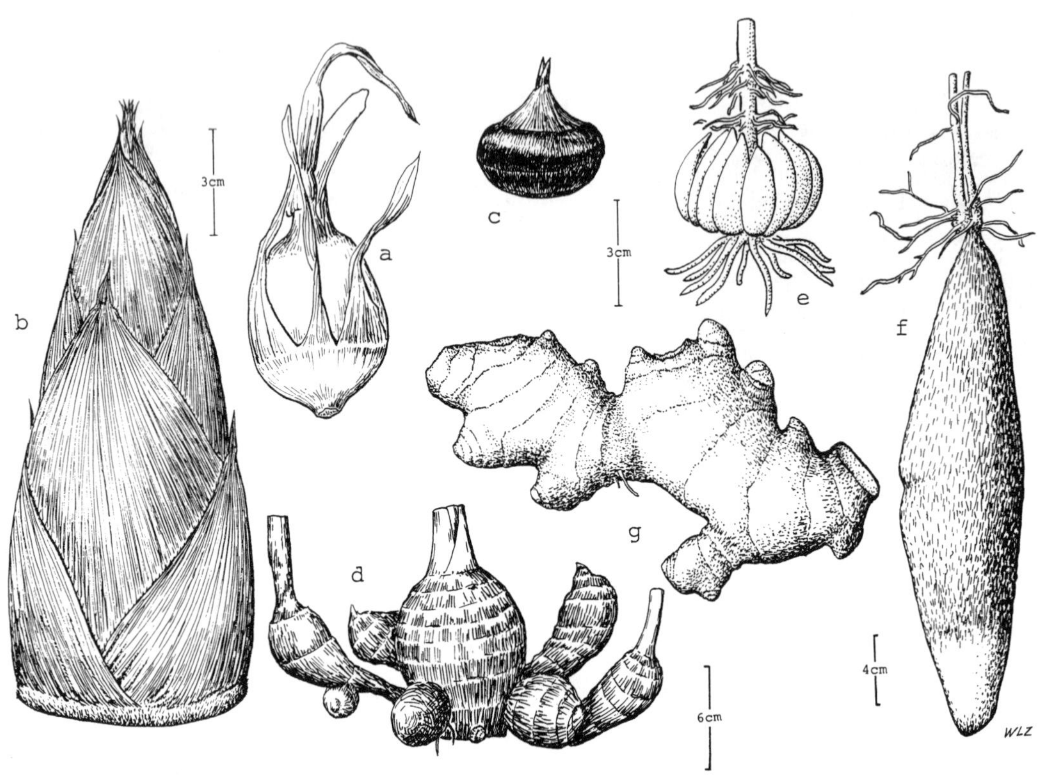

Figure 12. Market material of subterranean monocotyledonous plant food available in American Chinese stores. **a.** A white corm of fresh Old World Arrowhead (*Sagittaria sagittifolia* L.), imported from southern China. **b.** A fresh bamboo shoot (*Phyllostachys pubescens* Mazel ex H. de Lehaie). **c.** A black corm of fresh water chestnut, one of the most common ingredients of dishes in American Chinese restaurants. **d.** Brown fresh taro (*Colocasia esculenta* [L.] Schott), showing the central corm with several secondary cormlets and one tertiary comlet. **e.** A fresh white lily bulb, rarely available; dried scales in one-pound packages very common in Chinese stores. **f.** A fresh yam (*Dioscorea alata* L.). **g.** A piece of fresh ginger, the most important spice in Chinese cuisine. Morphologically, **a**, **c-e** and **g** are modified underground stems; **b** is a winter bud; and **f** a root tuber.

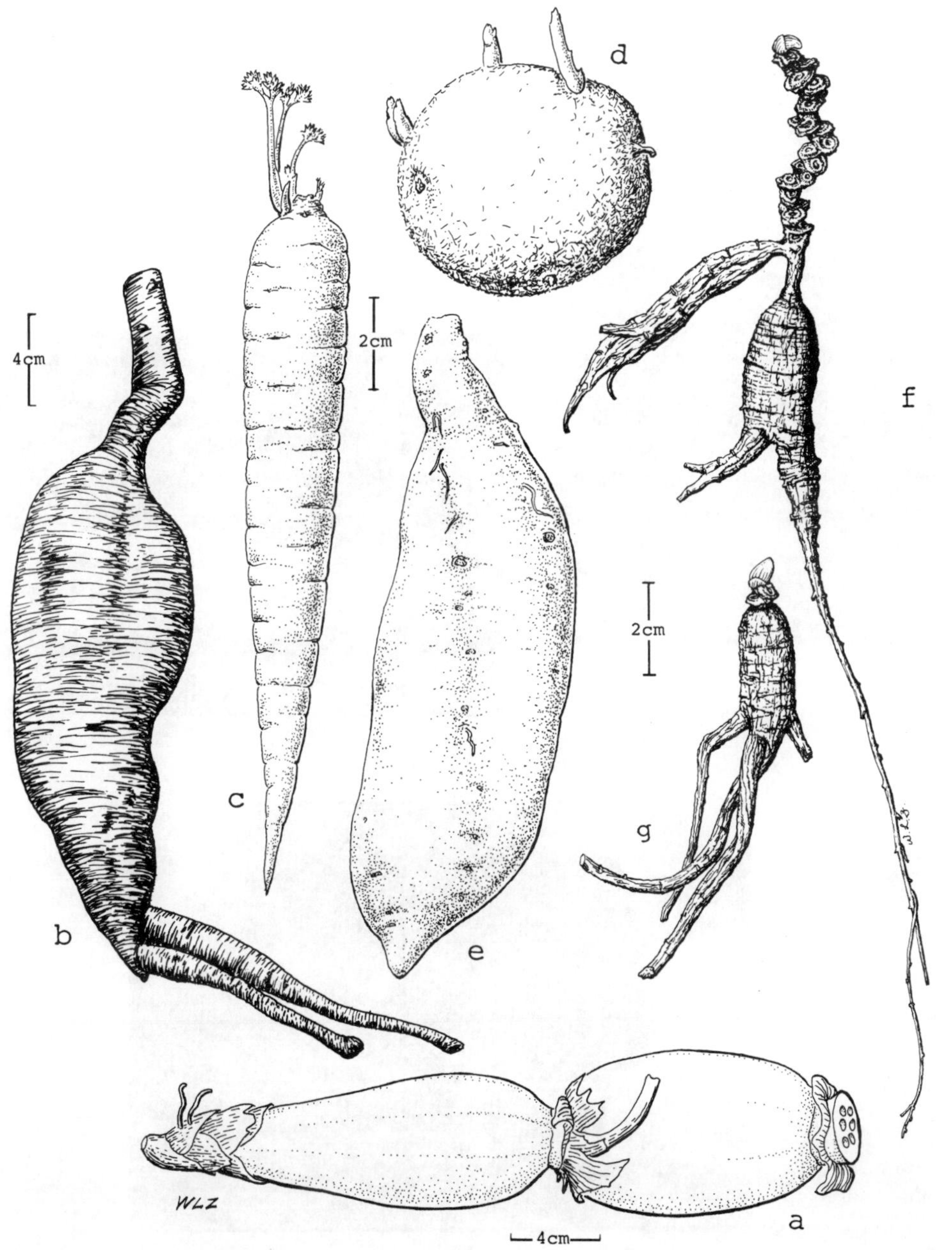

Figure 13. Market material of subterranean dicotyledonous plant food available in American Chinese stores. **a.** A piece of the fleshy rhizome of sacred lotus (*Nelumbo nucifera* Gaertner). **b.** A fresh sweet kudzu root tuber (*Pueraria thomsonii* Bentham). **c.** A carrot. **d.** An Irish potato with buds developed from several eyes. **e.** A sweet potato. **f.** The root of a wild ginseng about 25 years old, showing the slender zigzag and relatively long rhizome with a terminal bud, covered by numerous shoot scars, and an adventitious root emerging from the lower portion of the rhizome. **g.** The root of a cultivated 4-year old ginseng with short and stout rhizome bearing a terminal bud.

Figure 14. Chinese spices, material and marketing. **a.** Source species of the earliest recorded Chinese seasoning material *hua-jiao* (*Zanthoxylum simulans* Hance) showing the compound leaves and clusters of fruit. **b.** A spices store in Lanzhou, an important city on the ancient silkroute (Map, Loc. 20), showing the owner of the shop with various commodities exhibited in open bags and containers on a temporary stand over the sidewalk in front of his one-roomed, windowless store, where the stocks in numerous closed bags are stacked on shelves. **c.** A close up of the same store, showing the owner sitting behind a bronze grinder for pulverizing any material as requested. The grinder consists of a heavy boat-shaped tray fitted with a grinding wheel operated by his feet.

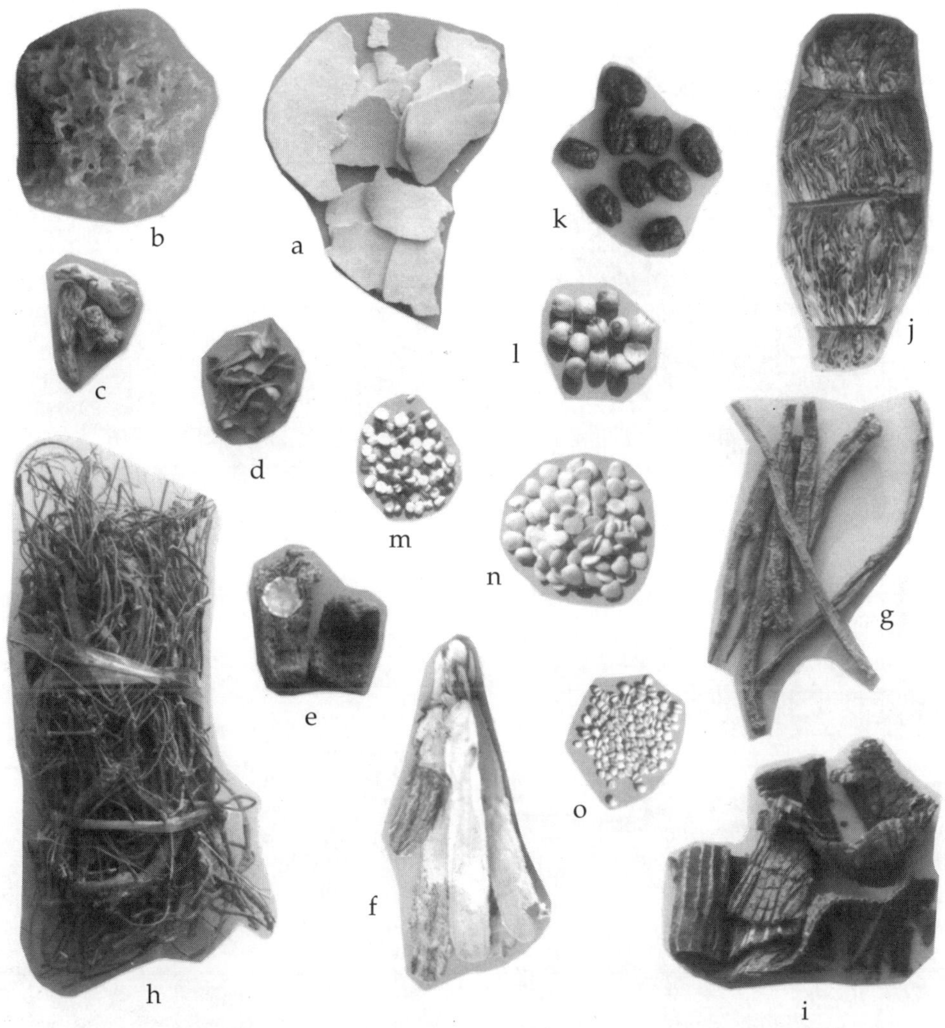

Figure 15. *Bupin* (補品, immunostimulating plant material) sold in half- or one-pound packages in Chinese groceries in Boston. **a.** China Root slices (茯苓, *Poria cocos* [Schw.] Wolf). **b.** Whole fruiting body of Silver Ear (銀耳, *Tremella fuciformis* Berk) **c.** Three Chinese Angelica (當歸, *Angelica sinensis* [Oliver] Diels). **d.** Dried scales of Lily Bulb (百合, *Lilium brownii* F. E. Br.) **e.** Whole rhizomes of Tian-ma (天麻, *Gastrodia elata* Bl.). **f.** Slices of Astragalus Root (黃耆, *Astragalus membranaceus* [Fischer] Bunge). **g.** Sections of Codonopsis (黨參, *Codonopsis pilosula* [Fr.] Nannf., or *C. tangshen* Oliver). **h.** A bundle of Abrus (雞骨草, *Abrus cantoniensis* Hance). **i.** Eucommia bark (杜仲, *Eucommia ulmoides* Oliver). **j.** A bundle of leaves of Dragon's Tongue (龍脷葉, *Sauropus changianus* S. Y. Hu). **k.** Dried fruit of Jujube (紅棗, *Zizyphus jujuba* Miller). **l.** Lotus seed with testa removed (蓮子, *Nelumbo nucifera* Gaertner). **m.** Euryale Seed (芡實, *Euryale ferox* [Lour.] Stend.). **n.** Detoxified Apricot Seed (*Prunus armeniaca* L. 白杏仁). **o.** Job's Tears (*Coix lachryma-jobi* L., photo courtesy of Crandell Huckins).

Figure 16. Fungal and algal food. **a.** Lingzhi (*Ganoderma lucidum* [Leysser ex Fries] Karsten), a widespread species in northern temperate regions, growing on decaying root or tree trunk. This specimen was collected in the shade of a garden in Harvard University, identified by Professor D. H. Pfister, Mycologist, Director of Harvard University Herbaria. (photo, courtesy of Professor Matthew Meselson, Department of Biochemistry, Harvard University.) **b.** Phoenix Mushroom (*Pleurotus sajor-caju* [Fries] Singer) cultivated in Hong Kong. (Photo, courtesy of Professor S. T. Chang, Chinese University of Hong Kong.) **c.** Collecting Sea lettuce (*Ulva lactuca* L.) at low tide by the seashore in Taipo, New Territories, Hong Kong. **d.** A close up of the same operation showing the population density of the alga and the amount of the material gathered by one family (photo taken in 1970).

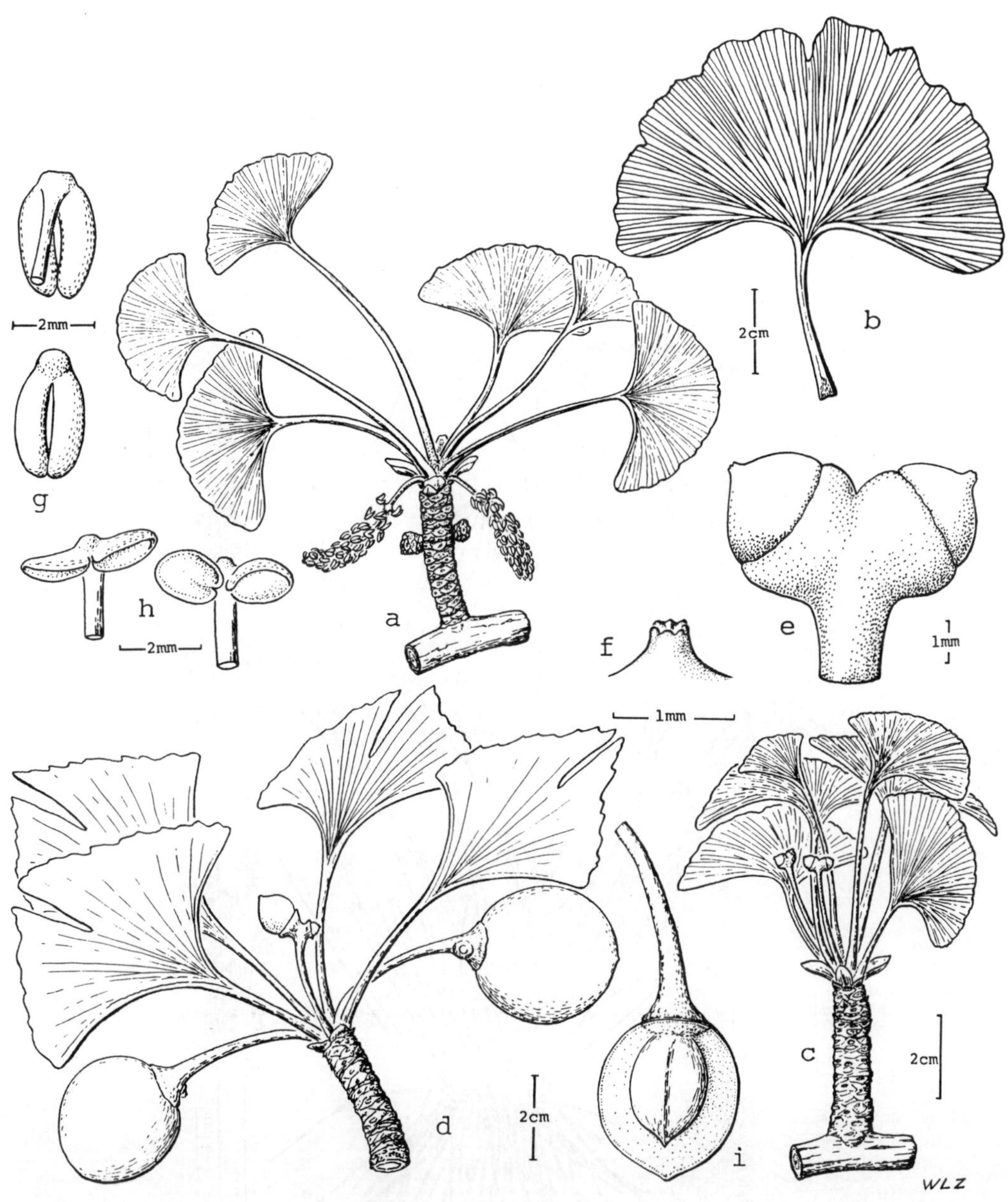

Figure 17. A Chinese gymnosperm imported for horticultural purposes, producing valuable edible seed (*Ginkgo biloba* L.). **a.** An abbreviate branch with catkin-like male strobiles, each bearing numerous stalked microsporangia (anthers). **b.** A leaf showing the dichotomous veins. **c.** An abbreviate ovulate branch bearing two axillary fertile brachlets (megasporophylls) slightly enlarged at the distal end, each bearing two ovules. **d.** An ovulate branch with some fertile ovules developed into globose apricot-like seeds, and others being abortive. **e.** The distal end of an ovulate branchlet showing the position of two naked ovules (megasporangia). **f.** The apical portion of an ovule. **g.** The adaxial and abaxial views of young male microsporangia (anthers). **h.** The same of older anthers after pollination. **i.** The vertical section of a seed showing the fleshy outer seed coat and the cartilaginous white inner seed coat. Ginkgoaceae 銀杏科.

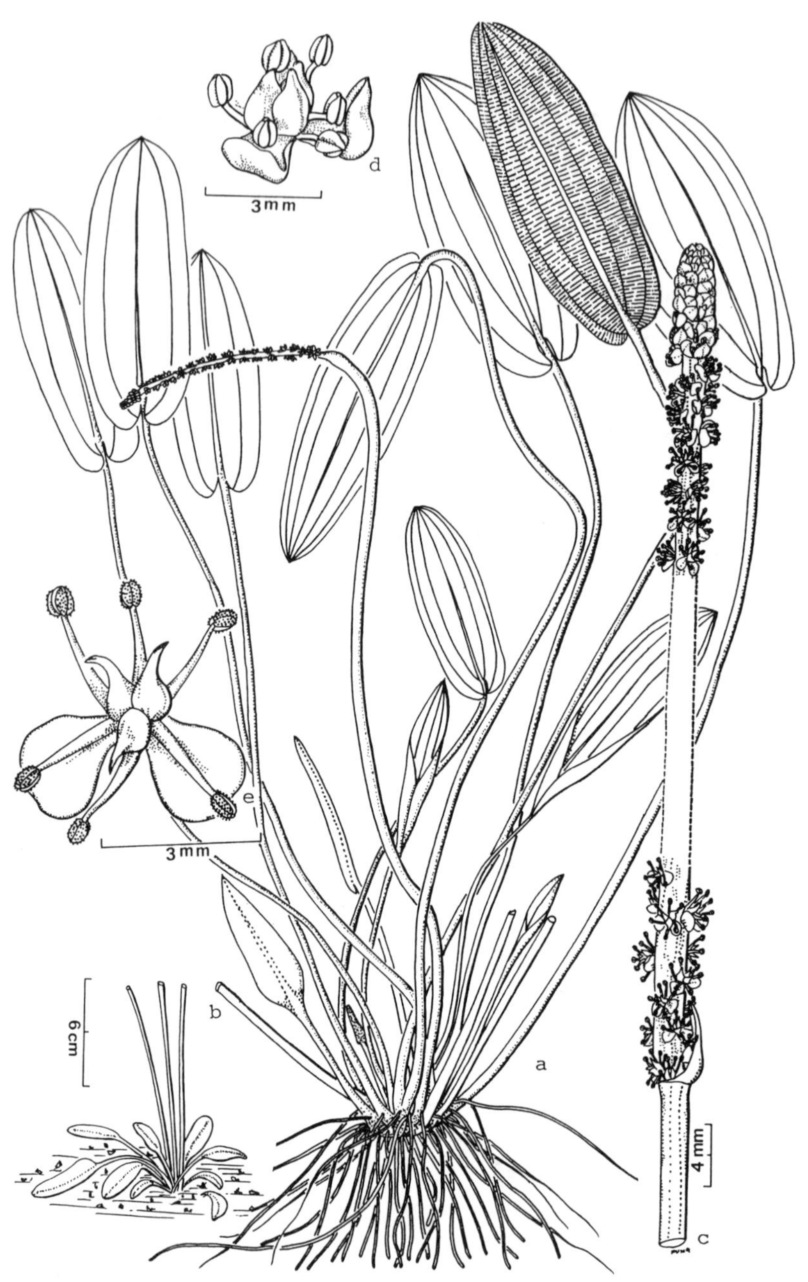

Figure 18. A perennial aquatic monocot (*Aponogeton natans* Engler et Krause) bearing large edible rootstock. **a.** A flowering plant showing the acaulescent habit and the scapose inflorescence. **b.** A juvenile leafy plant emerging from the rootstock with remnants of old scapes. **c.** The distal portion of the inflorescence showing a spike subtended by a small spathiform bract and numberous flowers at different stags of development (the middle portion of the rachis omitted). **d.** A young flower showing 2 tepals, 6 stamens, and 3 apocarpous ovaries. **e.** An older flower with the expanded, spathiform tepals, mature anthers full of pollen, and ovaries with oblique stigmas. Aponogetonaceae 水蕹科.

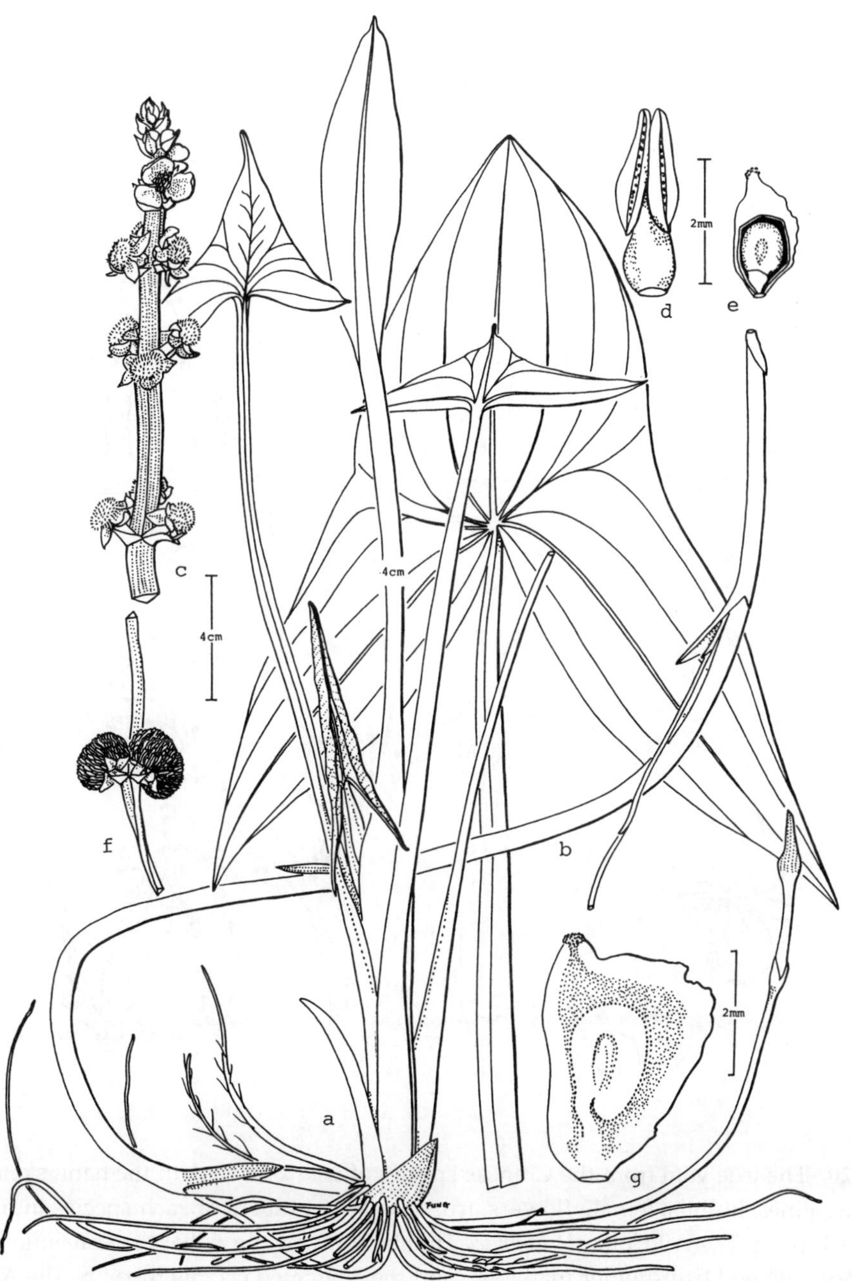

Figure 19. A perennial aquatic plant (*Sagittaria sagittifolia* L.) cultivated as an annual crop in warmer regions of China for the edible corms. **a.** An acaulescent plant emerging from a corm buried in mud. **b.** A stolon with six elongated internodes, penetrating and spreading in the mud showing a terminal bud, enlarged into the edible corm in cultivation (see Fig. 12a). **c.** A portion of the inflorescence showing the arrangement of flowers, their appearances at different stages of development, numerous stamens and apocarpous ovaries. **d.** A stamen showing the short stout filament and oblong anther. **e.** An apocarpous ovary with very short style and one ovule. **f.** Two fruits each bearing numerous achenes. Alismataceae 澤瀉科.

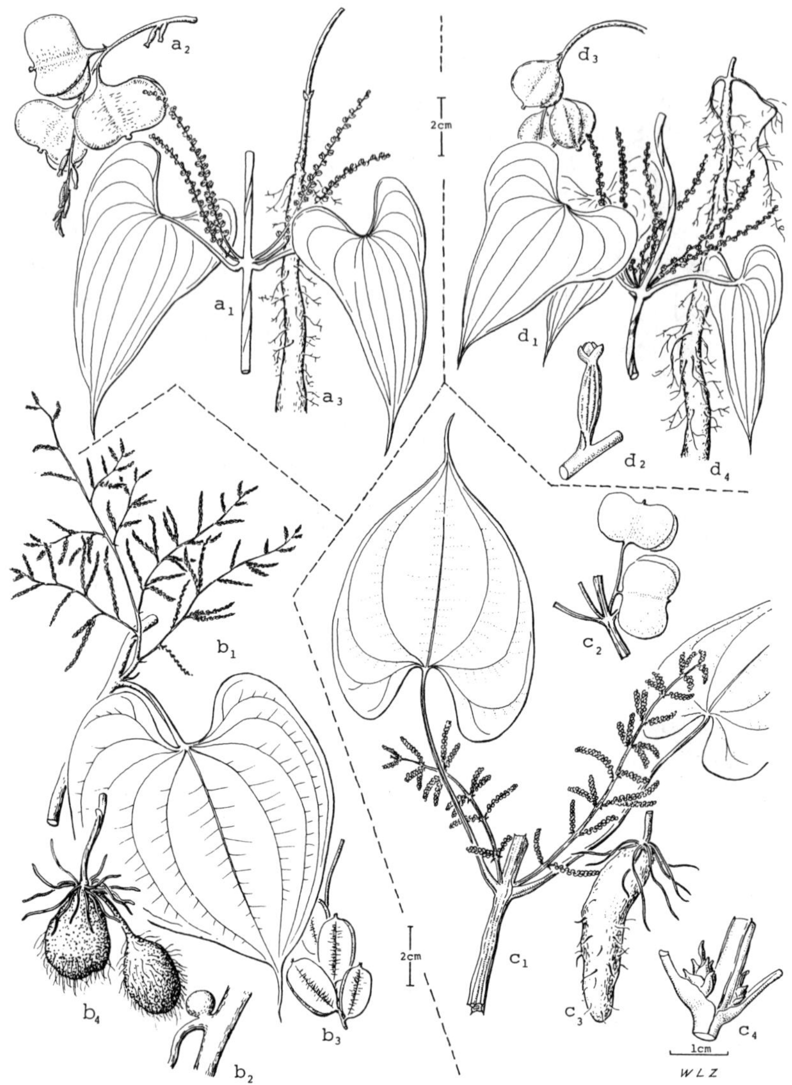

Figure 20. The true yam from the Chinese species of *Dioscorea* L. with the habit sketch of the flowering vines, bulbils, female flowers, fruits, and root tubers of each species indicated by numbers 1 to 4. **a.** The Wild Yam (*D. japonica* Thunb.) showing the axillary spicate inflorescences, the broadly winged fruits wider than long, and the elongated clavate tuber. **b.** The Air Potato (*D. bulbifera* L.) showing the paniculate inflorescence, axillary bulbil, oblong fruits with the wings longer than wide, and subglobose tubers. **c.** The White Yam (*D. alata* L.), a species of hybrid origin from the mountains of the Eastern Himalayan region, distributed by man eastward to the Pacific islands, westward to Europe and to the New World by the Portuguese in the sixteenth century, showing the angular and winged stems, branched verticils of spicate inflorescences, broadly winged fruits wider than long, large oblong tuber, and rugose bulbils. **d.** The Chinese Yam (*D. opposita* Thunb.), recorded in the earliest ancient Chinese herbals, cultivated extensively in northern China, showing the axillary fascicles of spicate inflorescences, a pistillate flower, fruits with wings longer than wide, and a long, cylindrical-clavate tuber. Dioscoreaceae 薯蕷科.

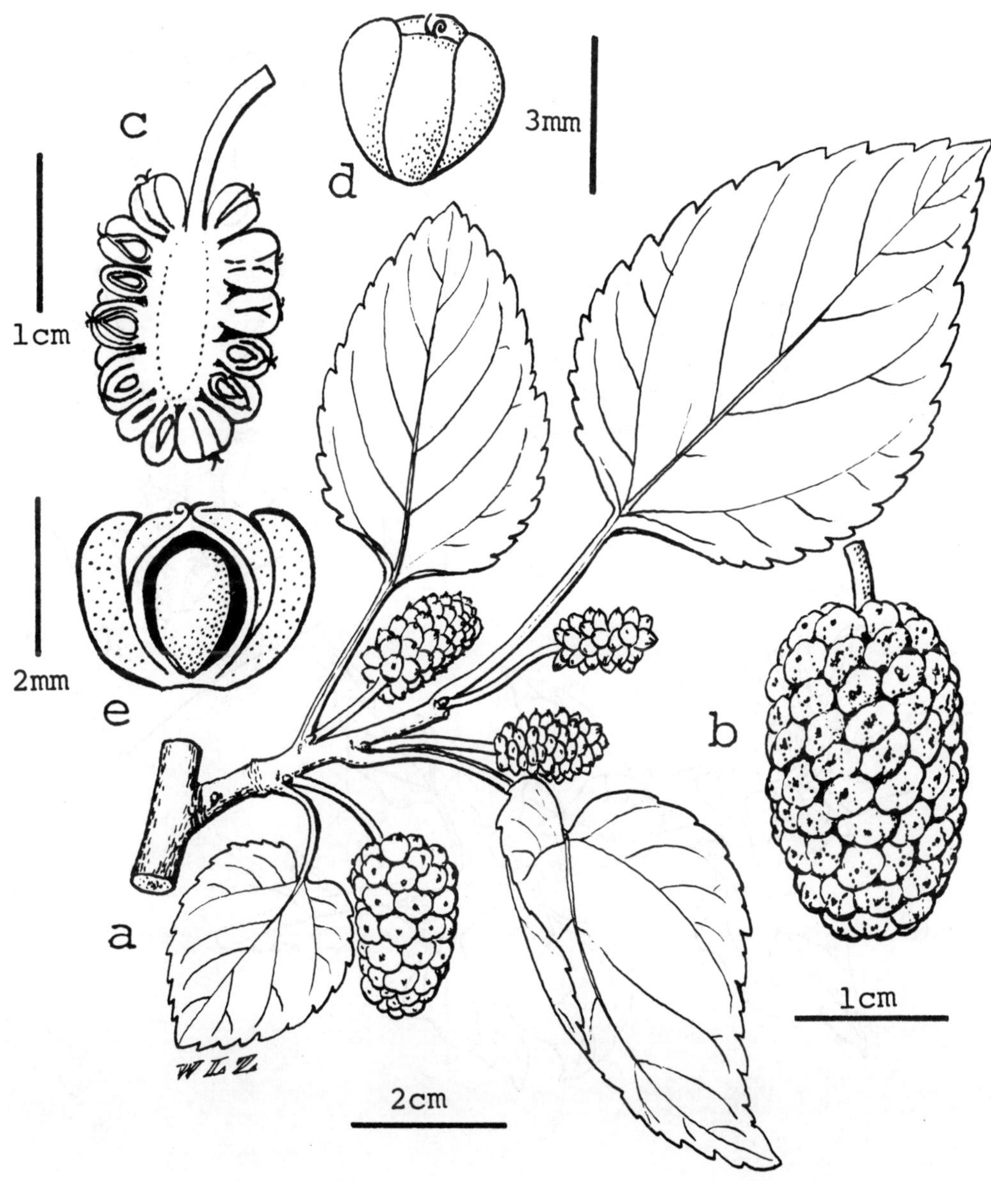

Figure 21. Mulberry (*Morus alba* L.). **a.** A fruiting branch. **b.** A multiple fruit. **c.** The vertical section of the same showing the central axis (rachis) and many individual fruits. **d.** A fruitlet with persistent stylar branches, enclosed in fleshy perianth. **e.** Vertical section of the same, showing the fleshy perianth segments, pericarp with attached stylar branches, and seed. Moraceae 桑科.

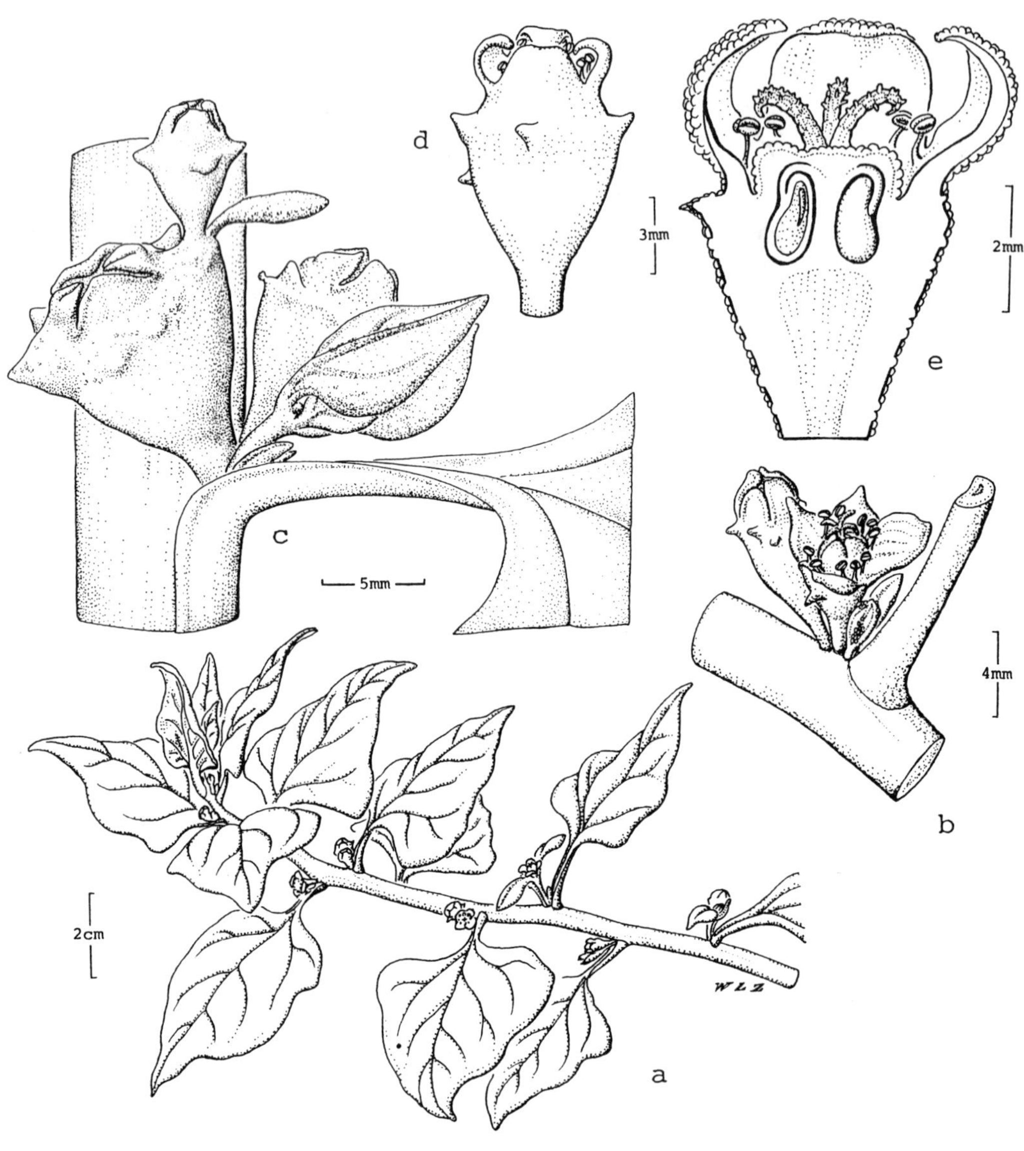

Figure 22. New Zealand Spinach (*Tetragonia tetragonoides* [Pallas] O. Ktze). **a.** A flowering branch showing the axillary position of flowers and new buds. **b.** A portion of the stem including one node with the lamina of the leaf removed, showing a fascicle of flowers and buds with the newer ones developed progressively nearer to the leaf. **c.** An older node showing two fruits, with the older one bearing a flower bud in the axil of a prophyll. A branch has emerged from the axil of the leaf. **d.** The lateral view of a flower showing the horn-like projection on the hypanthium. **e.** A vertical section of a flower showing the perianth, stamen, ovary, 3 styles and stigmas, and two hanging ovules. Aizoaceae 番杏科.

Figure 23. Peony (*Paeonia lactiflora* Pallas). **a.** A flowering branch showing the first flower that terminates the elongation of that branch, numerous petals and stamens. **b.** A flower with the petals and stamens removed, showing the sepals, spiral arrangement of the stamens, the annula, and the apocarpous ovaries. **c.** The front (adaxial) and lateral views of two stamens showing the basifixed anthers. **d.** A staminode. **e.** A pistil with portion of the ovary all removed, showing the orientation of numerous ovules. **f.** A transverse section of the ovary showing two ovules. **g.** The style showing the oblique stigma. Paeoniaceae 芍藥科.

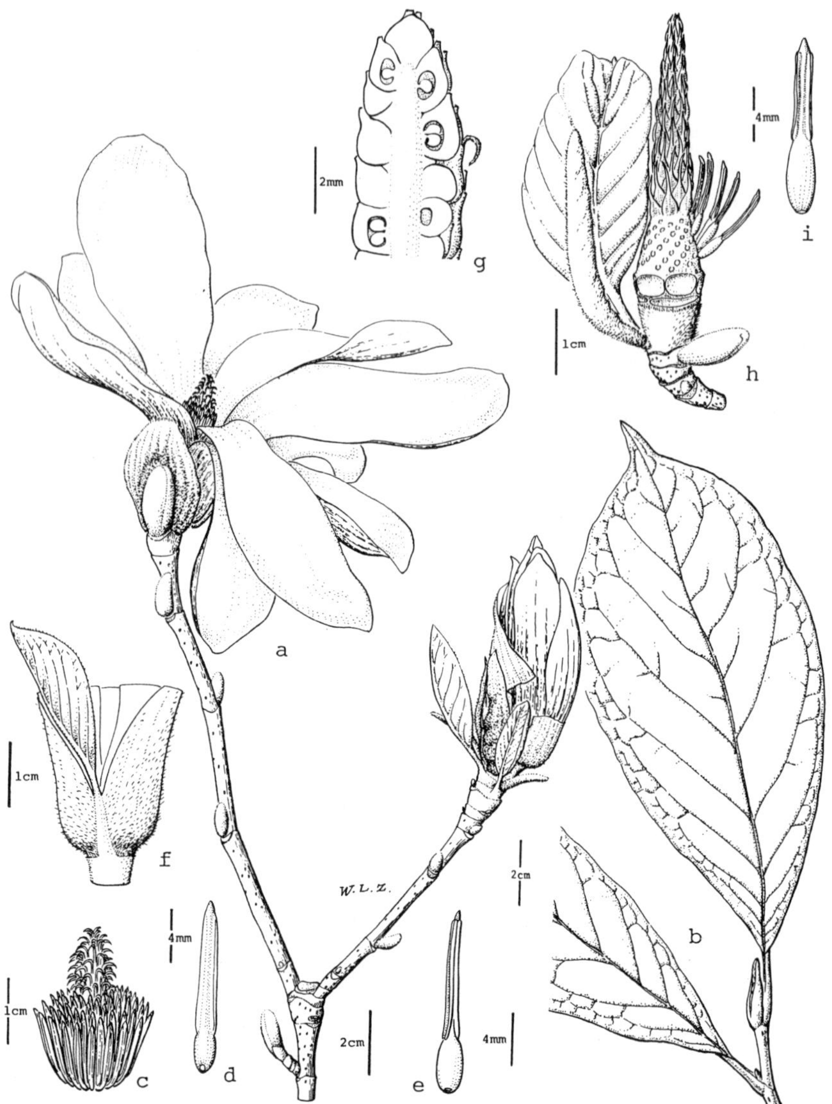

Figure 24. White flower forms of Chinese magnolias producing fleshy edible petals. **a-g.** *Magnolia solangiana* Soulange-Bodin, and **h-i.** *M. denudata* Desr. **a.** A flowering branch showing the solitary terminal flower or a flower bud and the small lateral vegetative buds, leaf scars, circular stipular scars, and conspicuous lenticels. **b.** A leafy branch showing the stipules of the distal leaf adnate to the basal portion of the petiole and enclosing the successive younger leaf. **c.** A young flower with the perianth removed showing numerous stamens and apocarpous pistils. **d-e.** The abaxial and adaxial views of two young stamens showing the elongated anthers apiculate at the apex. **f.** Portion of a bud showing the spathiform stipules adnated to the base of the petiole of a developing leaf. **g.** The vertical section of a gynoecium showing the apocarpous ovaries on the torus and two ovules in some of them. **h.** A terminal flower and its two associated vegetative buds showing the positions of the falling perianth and spiral arrangement of the stamens and pistils on the torus of *M. denudata* Desr. **i.** A stamen of the same flower showing a broad filament. Magnoliaceae 木蘭科.

Figure 25. Cruciferous Chinese vegetables introduced from southern China, available in the groceries in Boston. **a.** Cai-xin (菜心, *Brassica parachinensis* Bailey cv. 'Caixin'. **b.** Bai-cai (白菜, *B. chinensis* L.), the most common vegetable used by Chinese restaurants in USA. **c.** Chinese Kale (*B. alboglabra* Bailey). **d.** The habit sketch of a flowering plant of radish (*Raphanus sativus* L.) escaped, growing spontaneously in the experimental garden near Harvard Herbaria. **e.** A white juicy root of Luo-bo from the market with the larger ones attaining 50 cm long, 12 cm in diameter at the apical end. **f.** The lateral view of a flower showing the sepals, 4 petals and stamens. **g.** The same with a portion of the sepals and petals removed, showing the long and short stamens and the superior ovary. **h.** The same with the perianth and the upper portions of the stamens removed, showing the inner 4 stamens and the outer two stamens, and the glands near the base of the outer filaments. **i.** Two fruits showing their shape, and the position of the seeds embedded in the spongy tissue. Brassicaceae 十字花科.

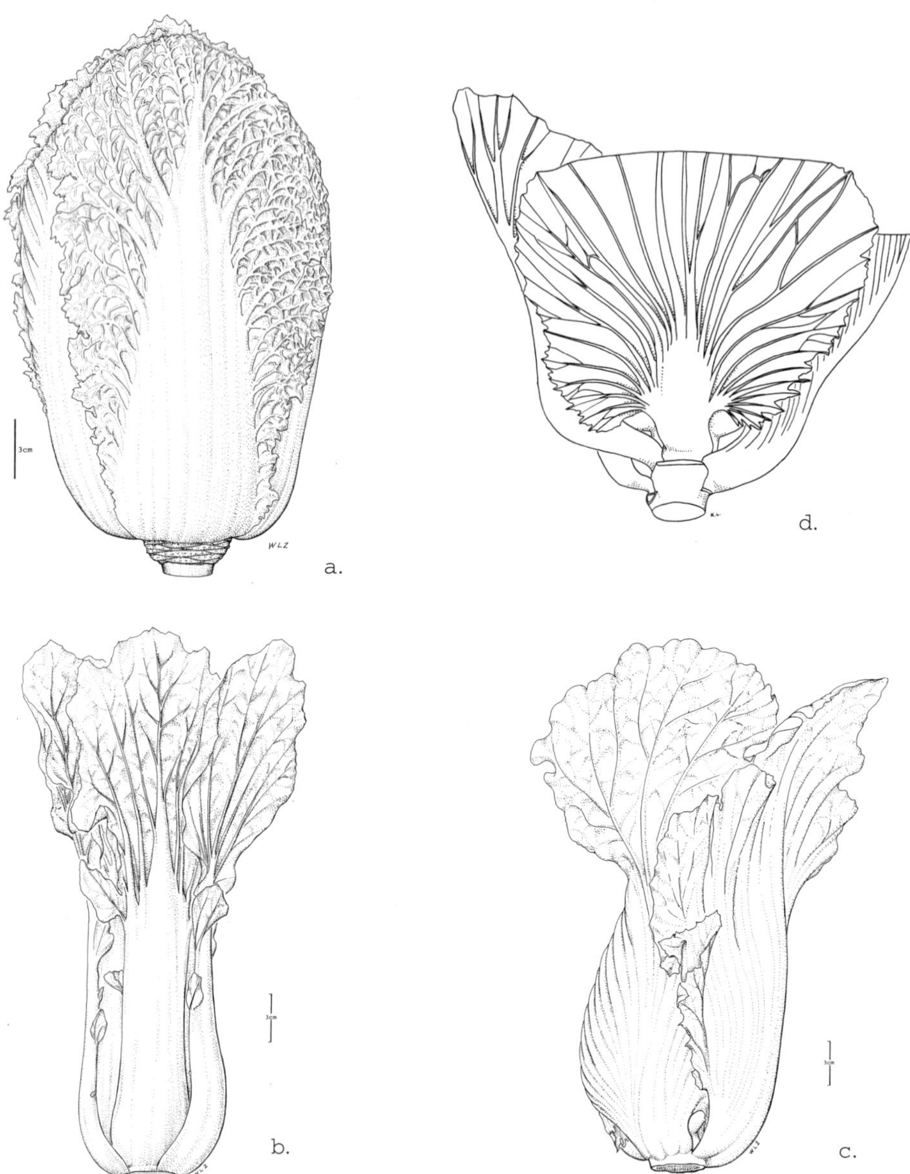

Figure 26. Market material of Chinese cabbage and mustard. **a.** An oblong, tight head of *Da-bai-cai* (*Brassica pekinensis* [Lour.] Rupr.), a selection of northern China, showing the leaf-scars of the outer green leaves and a proportion of the fleshy petioles merging into the broad fleshy midrib with wrinkled and folded lamina crinkled along the margin. This cultivar is called *Shao-cai* (紹菜) in Cantonese stores and *Song* (菘) in ancient herbals. **b.** The Celery Cabbage common in many supermarkets, showing longer fleshy petioles with auricles and shorter lamina. **c.** Swatow Mustard (*B. juncea* [L.] Czernajew var. *foliosa* Bailey) showing the large leaves with short petioles merging into the large fleshy midribs of the lamina with folded margin, very common in Chinese groceries. **d.** The same species from stocks stored too long and the outer leaves turning yellow; the laminas of the outer leaves are removed, and the trimmed material with tender hearts sold for higher prices. Brassicaceae 十字花科.

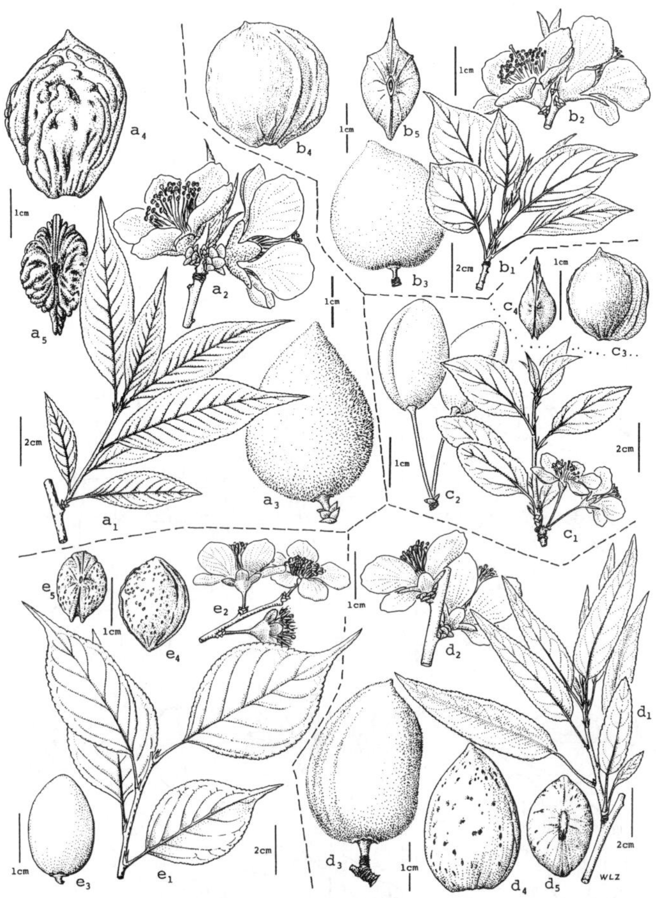

Figure 27. Five species of *Prunus* L. producing edible fruits and/or seeds available in China. The habit sketches of vegetative shoots, flowering branches, immature fruits as shown on herbarium specimens, lateral surface of the stone, and basal view the same with the marginal suture above, are indicated by numbers 1 to 5 for each species. **a.** Peach (桃, *P. persica* [L.] Batsch) showing the lanceolate leaves, sessile flowers developed from two lateral winter buds with the middle bud unfolding into a vegetative shoot, the densely hairy young fruit, and the deeply pitted and furrowed stone with roundish outline repeatedly grooved on one side of the marginal suture and a high broken ridge on the opposite side. **b.** Apricot (杏, *P. armeniaca* L.) showing ovate-suborbicular or oblong leaves, subsessile flowers developed from solitary buds, rather smooth stone with sharp marginal suture and rather blunt on the opposite side. **c1-2.** *Yu-li* (郁李, *P. japonica* Thunb.) showing ovate to broad-lanceolate leaves, paired or fasciculate, long-pedicellate flowers developed from solitary lateral bud, with the terminal bud emerging into a vegetative shoot. **c3-4.** *Yang-li* (洋李, *P. domestica* L.) with rather smooth stone, sharp outline on the side of the marginal suture and blunt but deeply grooved on the opposite side. **d.** Almond (扁桃, *P. dulcis* [Miller] D. A. Webb) showing the lanceolate leaves, sessile solitary flowers, smooth and slightly pitted stone with roundish obscure sutures. **e.** *Mei* (梅, *P. mume* Sieb. et Zucc.) showing broadly ovate or elliptic leaves, shortly pedicellate flowers developed from solitary lateral buds, smooth and shallowly pitted stone, roundish on the side of the marginal suture and broadly furrowed on the opposite side. Rosaceae 薔薇科.

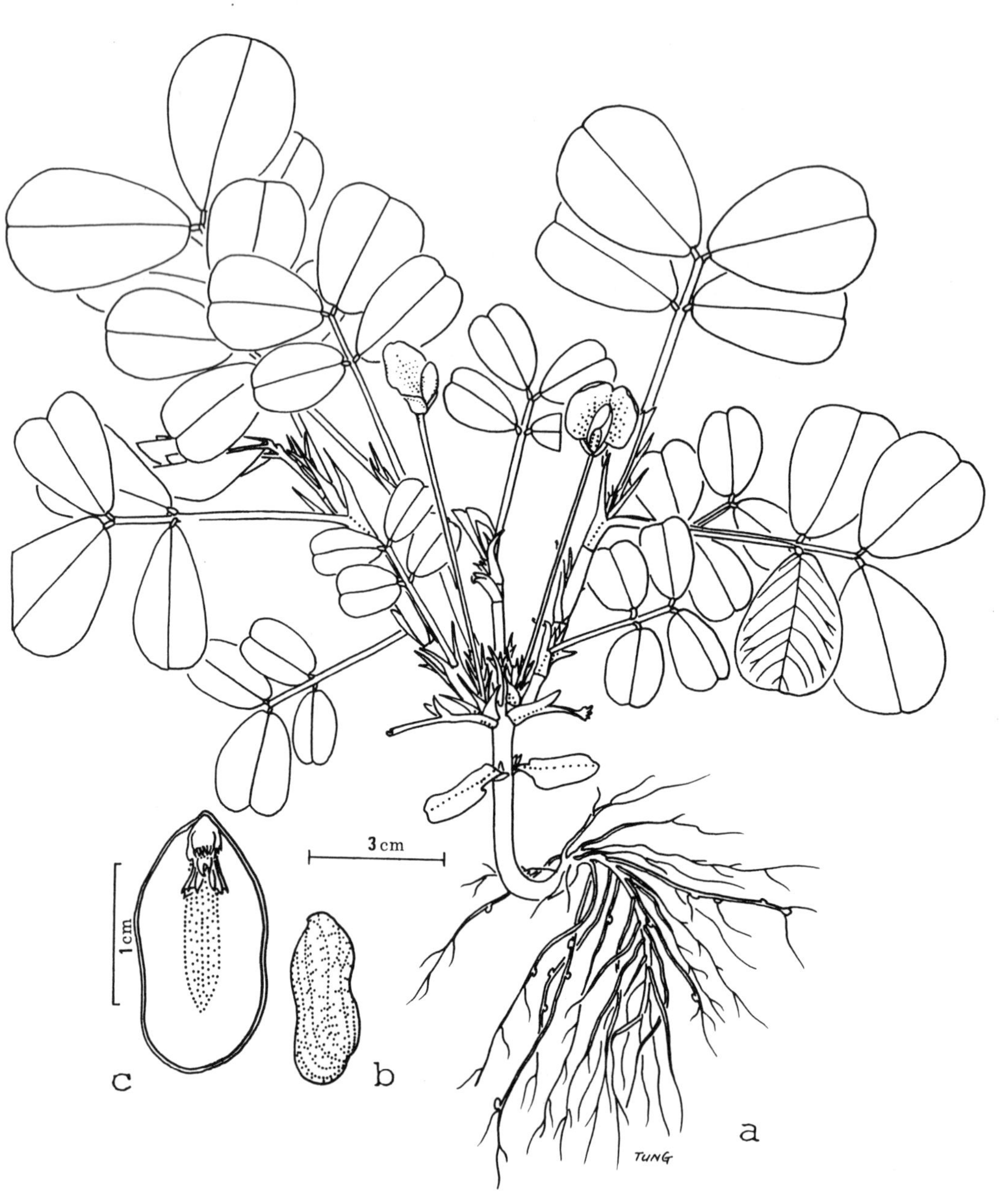

Figure 28. Peanut (*Arachis hypogaea* L.). **a.** The habit sketch of a flowering plant showing the even pinnate leaves each with 4 leaflets, and solitary flowers with elongated pedicels bending downward after anthesis hiding the fruit in the soil. **b.** A fruit, peanut of the market. **c.** A seed with one cotyledon removed showing the conspicuous plumule and a very small hypocotyl. Leguminosae 豆科.

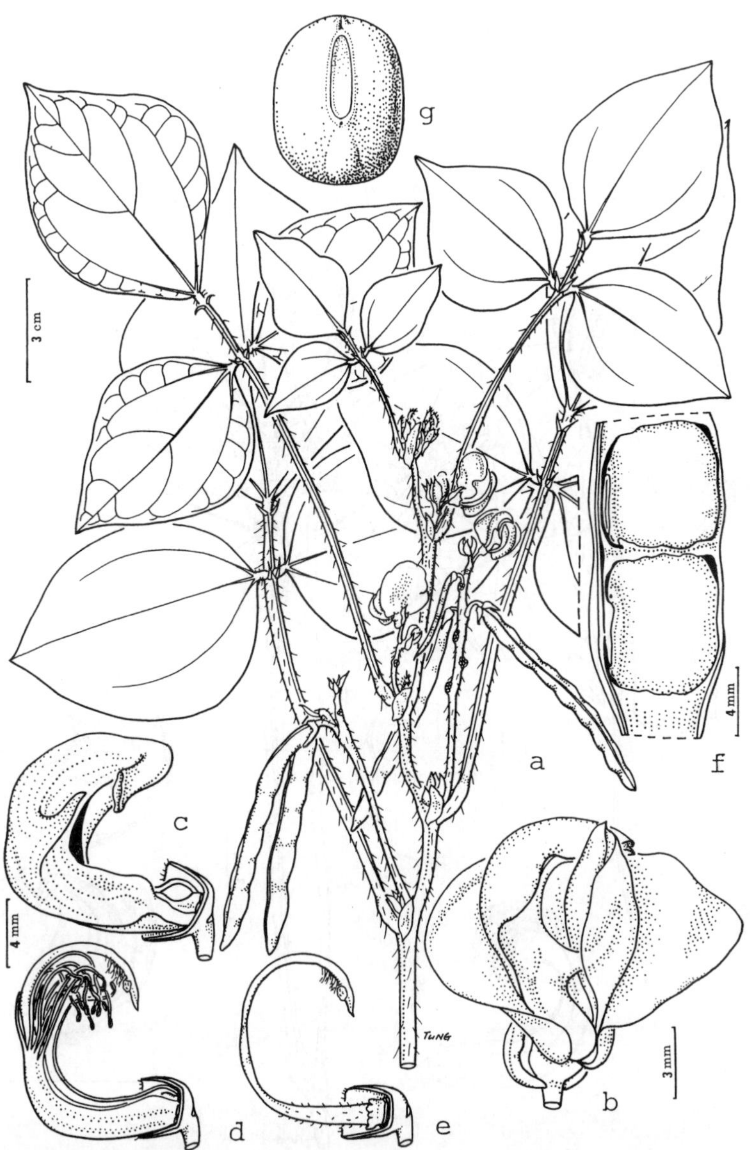

Figure 29. Adsuki bean (*Azukia angularis* [Willd.] Ohwi. **a.** The habit sketch of a flowering and fruiting branch showing the trifoliolate leaves with peltate (attached near the middle), elliptic stipules, and infrutescence with short nodose rachis. **b.** The front view of a flower showing the standard, left wing petal folding and embracing the spur on the lower left side of the keel petal, and the right wing folding over the strongly curved distal portion of the keel. **c.** The same with portion of the calyx, standard and wings removed, showing the keel with a strong spur on the left side and the folding and curving of its distal portion to approximately 300 degree to the left. **d.** The same with the keel removed showing the curved and barbate style, the lateral stigma and the subulate stylar appendage. **e.** The same with the stamens removed, showing the nectar at the base of the ovary. **f.** Portion of a septate fruit with half of the pericarp removed, showing the seeds and the septa. **g.** A short and broad seed with acentric hilum slightly or inconspicuously carunculate and an elevated short parahilum. Leguminosae 豆科.

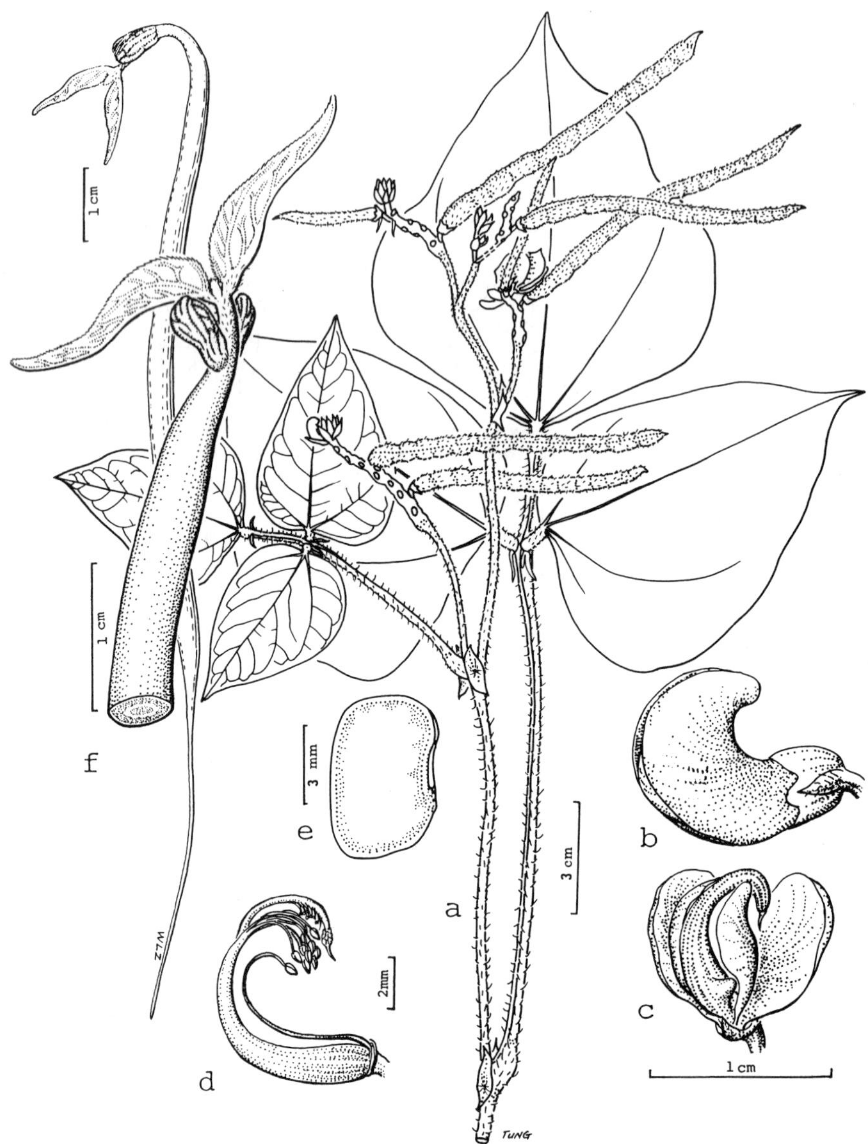

Figure 30. Mungbean (*Cadelium radiatum* [L.] S. Y. Hu). **a.** The habit sketch of a flowering and fruiting branch showing one trifoliolate leaf with elliptic stipules attached at the middle, stout and nodose rachis, and hispid fruits in a horizontal position. **b.** The lateral view of a mature flower bud showing the small ovate bracteole, the campanulate calyx with ovate-obtuse lobes, and the strongly recurved medium line of the standard. **c.** The front view of the same with the perianth segments slightly pushed opened mechanically, showing the spur on the left side of the keel and the curving of its distal portion to approximately 200 degree to the left. **d.** The same with the perianth segments removed, showing the stamens and the curved style, barbate below the lateral stigma and the subulate stylar appendage. **e.** The lateral view of a seed (the mungbean), showing the acentric hilum covered by white spongy tissue, and a smooth long parahilum opposite the micropyle. **f.** Market material of mungbean sprouts showing the shriveled cotyledons, the bending hypocotyl and the well-developed primary leaves. Leguminosae 豆科.

Figure 31. Traditional simple devices for extracting mungbean starch in the manufacturing of *fengsi* (粉絲), and the production of mungbean sprout in Hong Kong. **a.** A cement tank for soaking mungbean. **b.** Containers of ground mungbean for the precipitation of starch at the bottom. **c.** A tank of cold water to which the cooked starch paste falls from a sieve, forming soft thin strings, and in which the strings solidify and separate. **d.** A cold chamber with newly prepared *fengsi* hanging on bamboo sticks to be stiffened and dried. **e.** Packing *fengsi* into weighed bundles for local market and for exportation. **f.** Large earthware containers designed for growing mungbean sprouts in a dark room. **g.** Mungbean sprouts in large bamboo baskets ready for transporting to the market.

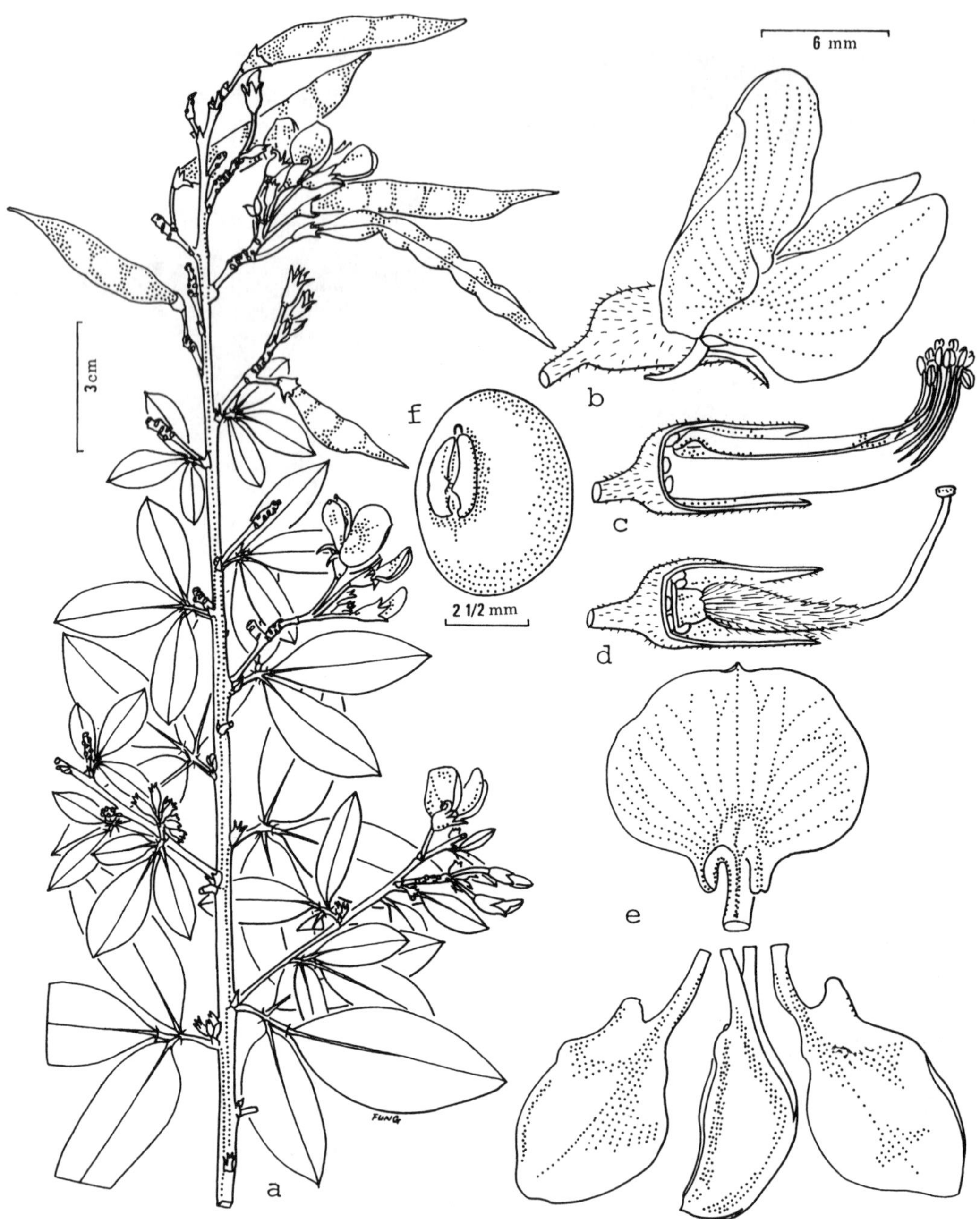

Figure 32. Pegeon Pea (*Cajanus cajan* [L.] Millsp). **a.** The habit sketch of a flowering and fruiting branch showing that the early flowers terminating the apical elongation of that branch, and the later flowering branches developed from lower lateral buds. **b.** The lateral view of a papilionaceous flower showing the campanulate hairy calyx with lanceolate lobes, the standard and wings. **c.** A flower with petals and half of calyx removed, showing the stamens, the filaments of 9 united, one free. **d.** The same with the petals and portion of the calyx removed, showing the nectar and the ovary. **e.** The standard, wings and keel petals. **f.** A seed showing the conspicuous caruncle and the hilum in a deep groove. Leguminosae 豆科.

Figure 33. Scurgy pea (*Cullen corylifolium* [L.] Medikus, commonly known as *Psoralea corylifolia* L.). **a.** The habit sketch of a flowering and fruiting branch showing punctate leaves, and axillary dense pseudoracemes. **b.** A unit of the inflorescence showing paired flowers, each subtended by a bract. **c.** The lateral view of a papilionaceous flower showing the pubescent, punctate, and deeply lobed calyx. **d.** The calyx mechanically opened, showing the unequal lobes with the lower one the larger. **e.** The standard, wing and keel petal. **f.** A flower with the perianth removed, showing the diadelphous stamens and the pistil. **g.** An one-seeded oblong fruit with portion of the persistent calyx removed, showing the glandular and rugose pericarp adhering to the cartilaginous testa, and the remnants of the perianth and stamens. **h.** An oblong seed, slightly compressed laterally, showing the acentric, suborbicular, carunculate hilum deeply sunk in a cavity. Leguminosae 豆科.

Figure 34. Soybean (*Glycine max* [L.] Merr.). **a.** The habit sketch of a fruiting plant showing few root nodules, trifoliolate leaves, subsessile racemose clusters of flowers and hispid fruits. **b.** A flowering branch. **c.** The front view of a flower showing a basal bract, 2 bracteoles, the campanulate, hispid calyx with large lobes, the standard, two wings and very small keel petals. **d.** The lateral view of the same. **e.** A fruit with one side of the pericarp removed, showing two large seeds. **f.** A seed (the soybean), showing the acentric hilum, the micropyle and the hypocotyl ridge. **g.** Market material of soybean sprouts showing the smooth fleshy cotyledons, the benting hypocotyl and the very small plumule. Leguminosae 豆科.

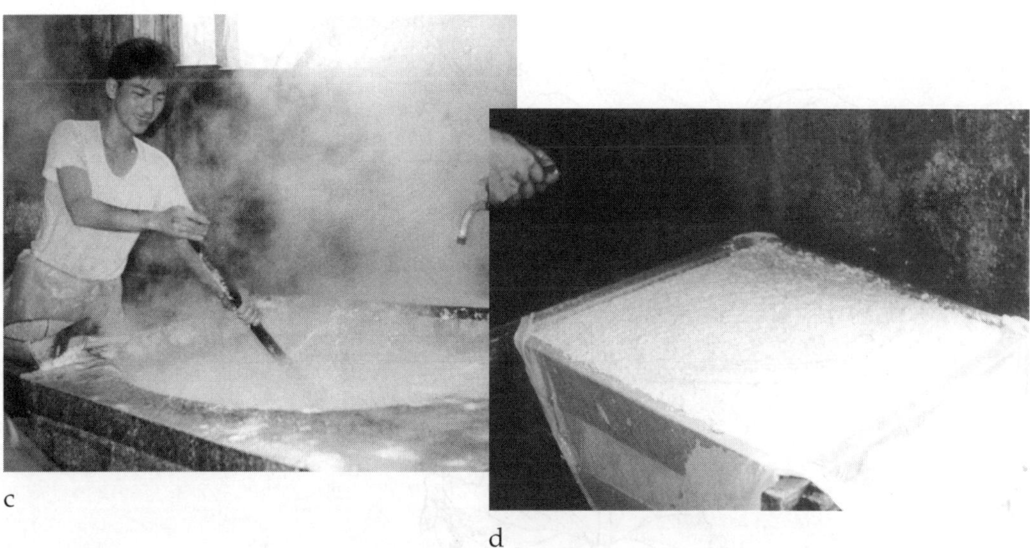

Figure 35. Simple traditional devices for preparing soybean products to be used as food in Hong Kong. **a.** For the manufacture of soy sauce, bamboo trays containing a mixture of cooked soybean, roasted wheat flour and fungal spawn, are set on shelves in a closed room to wait for fermentation. **b.** Large earthen jars with handy hat-like covers are set in the yard of the *jiang-yuan* (醬園), containing the fermented mixture, salt, and water, exposed to the sun with occasional stirs during the day and to dew at night, covered only when it rains. **c.** In the preparation of soybean curd, the filtrate from the ground bean is cooked in an open boiler before stirring in the pulverized gypsum (or calcium salt) for the precipitation of soybean protein. **d.** A wooden frame lined with cloth for filtering the coagulated protein and for solidifying (under slight pressure) the mass into bean curd.

Figure 36. Garden pea (*Pisum sativum* L.). **a.** The habit sketch of a flowering branch showing the pinnate leaves with foliaceous stipules and the terminal leaflets modified into tendrils, solitary and rarely paired flowers, and a young fruit. **b.** The standard, wings and keel petals of a papilionaceous pea flower. **c.** A flower with petals and portion of the calyx removed, showing the diadelphous stamens. **d.** The same with portion of the stamens removed showing the ovary, style and stigma. **e.** A seed showing the hilum, the micropyle and the hypocotyl ridge. Leguminosae 豆科.

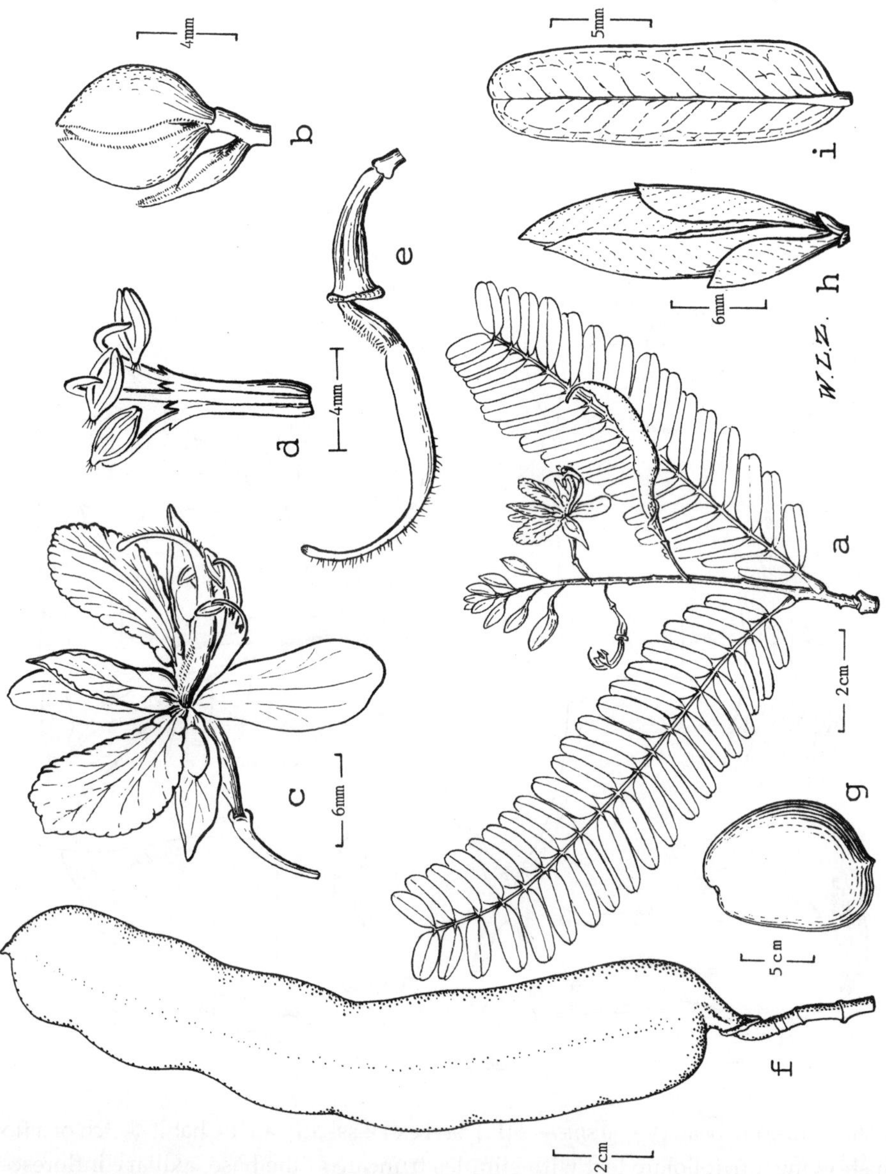

Figure 37. Tamarind (*Tamarindus indicus* L.). **a.** The habit sketch of flowering branch showing two even-pinnate leaves, a terminal raceme with a young fruit, a flower, and well-developed buds. **b.** The distal portion of a raceme with several very young flower buds covered by bracts at the apex and a lower bud subtended by a bract. **c.** A flower showing the slightly curved pedicel, an elongated tubular hypanthium, 4 sepals, 3 petals with ruffled margin, 3 fertile stamens, a stipitate ovary, and a horn-like style. **d.** The fertile stamens with the filaments connate at the base, and the basifixed ellipsoid anthers. **e.** A flower after anthesis, showing the point of the disartificulation of the falling sepals, petals and stamens on the rim of a tubular hypanthium, the ovary on a hairy stalk (gynophore), the style and punctiform stigma. **f.** A subcylindrical mature indehiscent fruit (the commercial tamarind). **g.** A seed. **h.** A vegetative bud. **i.** A leaflet. Leguminosae 豆科.

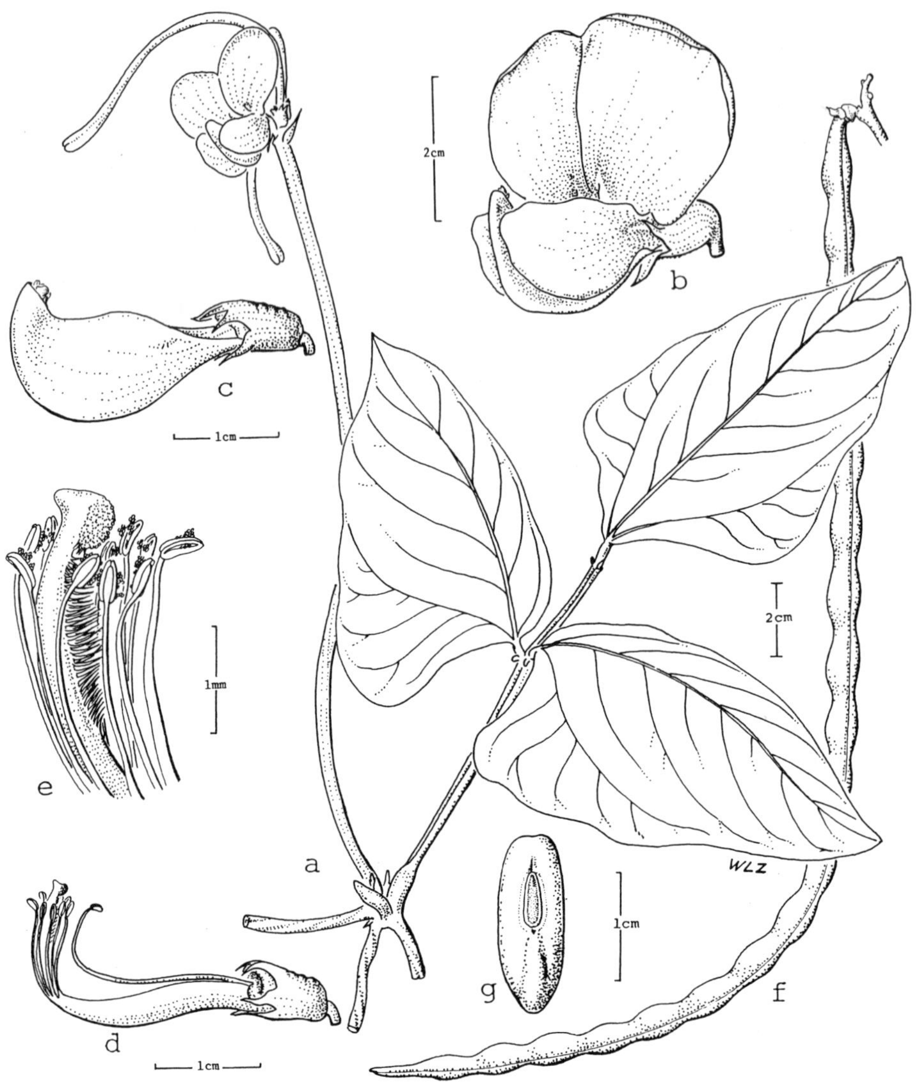

Figure 38. Yard-long bean (*Vigna sinensis* [L.] Savi ex Hasskarl). **a.** The habit sketch of a flowering branch showing a trifoliolate leaf with stipules truncate at the base, axillary inflorescence on elongated peducles, short nodose rachis and subsessile paired flowers. **b.** The lateral view of a papilionaceous flower showing the standard, two large wings and the keel petals equal to the length of the wings but with the shortly curved portion slightly exposed from them. **c.** The same with the standard and wings removed showing the campanulate calyx rugose on the upper side, and the smooth, obovate keel petals slightly curved along the upper margin toward the apex. **d.** The same with the keel petals removed showing the stamens, style, and stigma. **e.** The upper portion of the stamens and style showing the lateral hairs, the subterminal stigmatic surface and the minute cone-like stylar appendage. **f.** A fruit containing 20 seeds, fully grown but not yet mature, indehiscent, becoming greatly thickened and wrinkled approaching maturity, with the pericarp broken into fragments to release the seeds. **g.** The face view of an oblique-oblong slightly compressed seed showing the acentric hilum surrounded by an elevated gelatinized olive ring, and with obvious hypocotyl ridge and very faint parahilum. Leguminosae 豆科.

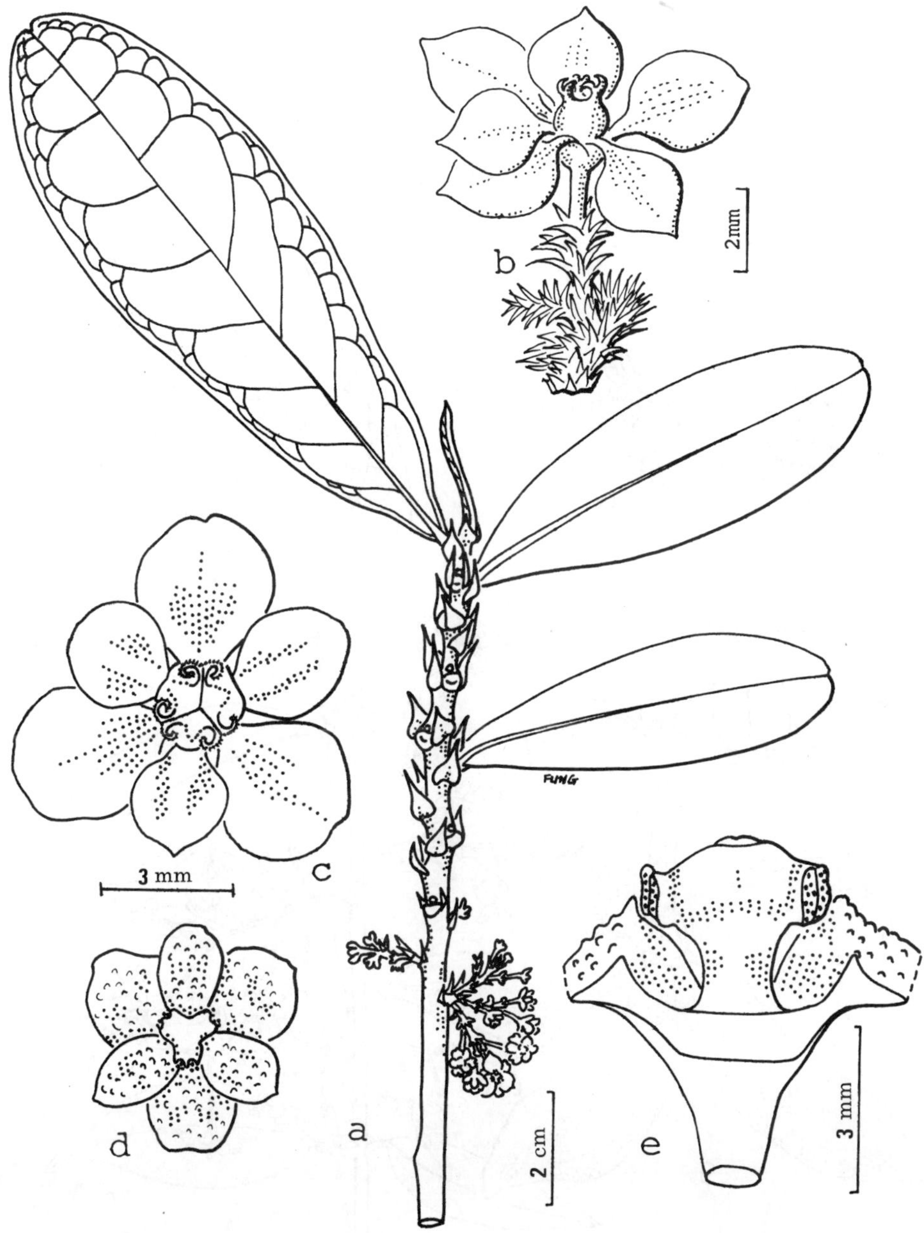

Figure 39. Dragon's tongue (龍脷葉, *Sauropus changianus* S. Y. Hu). **a.** A flowering branch showing the fascicles of staminate flower on leafless old wood and numerous persistent stipules remaining on the stem after the leaf-fall. **b.** The lateral view of a fascicle of abbreviate branches covered by subulate bracts, with one bearing a terminal solitary pistillate flower with maroon perianth segments in two series, and an ovary with 3 bifid styles. **c.** The face view of a pistillate flower showing the outer and inner series of perianth segments, the 3-carpellate ovary with 3 distinct bifid styles and 6 stigmas. **d.** The face view of a staminate flower showing the arrangement of the perianth segments and the central 3-lobed androphore. **e.** The lateral view of the same with the perianth segment complete or partially removed. Euphorbiaceae 大戟科.

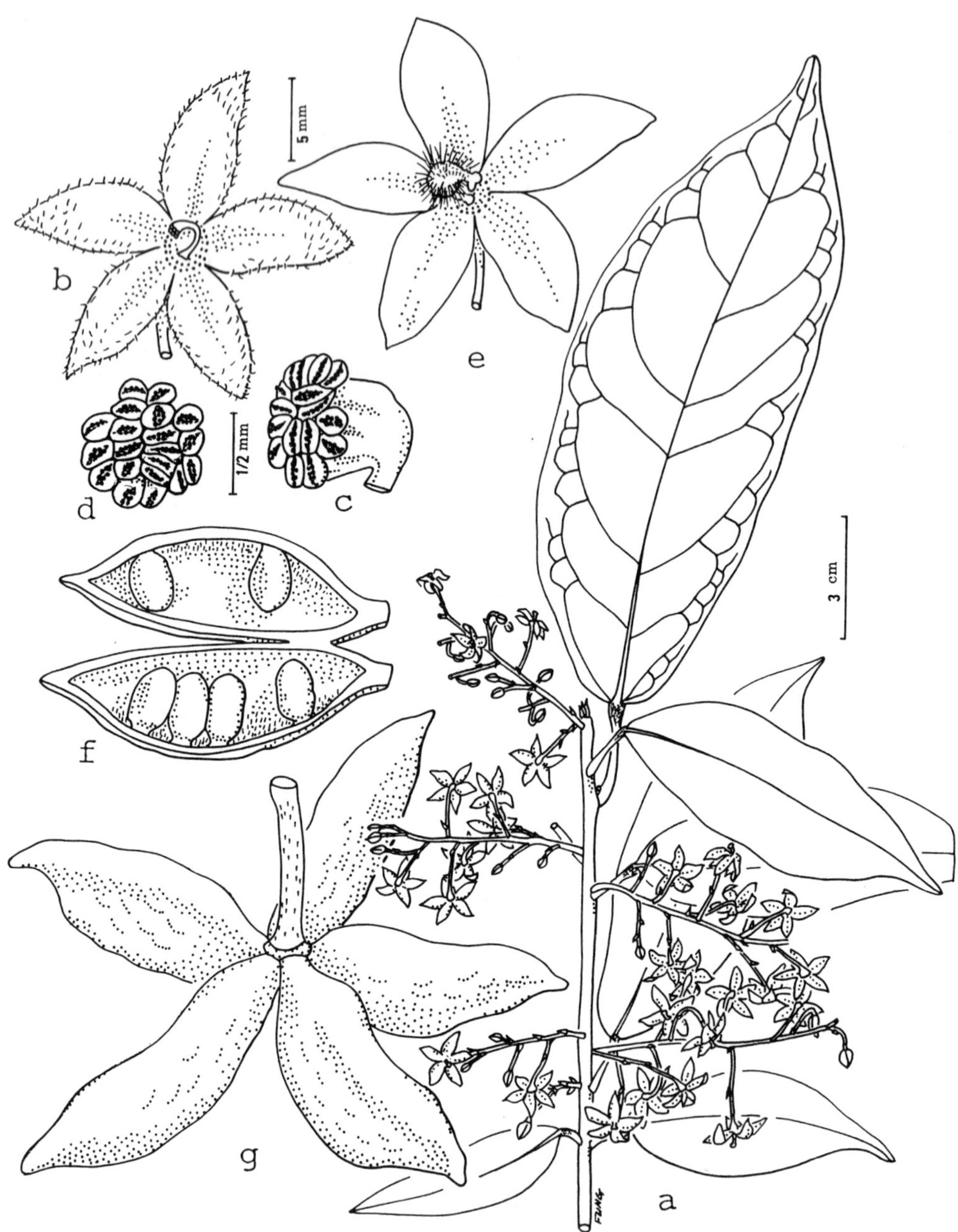

Figure 40. Scarlet Sterculia (*Sterculia lanceolata* Cav.). **a.** The habit sketch of a flowering branch showing the unifoliolate leaves each with an articulation at the distal end of the petiole, and the axillary subpaniculate polygamous flowers. **b.** A staminate flower showing the curved staminate column (androphore). **c.** The lateral view of the apical portion of the androphore showing the position and arrangement of the anthers. **d.** The front view of the same. **e.** A pistillate flower showing the stipitate pubescent ovary. **f.** A young fruit mechanically opened showing the marginal position of the seeds. **g.** A mature fruit in a hanging position, with the pericarps open, exposing the shiny black seeds along the margin, in contrast to the yellow and red iridescent velvety on the back. Sterculiaceae 梧桐科.

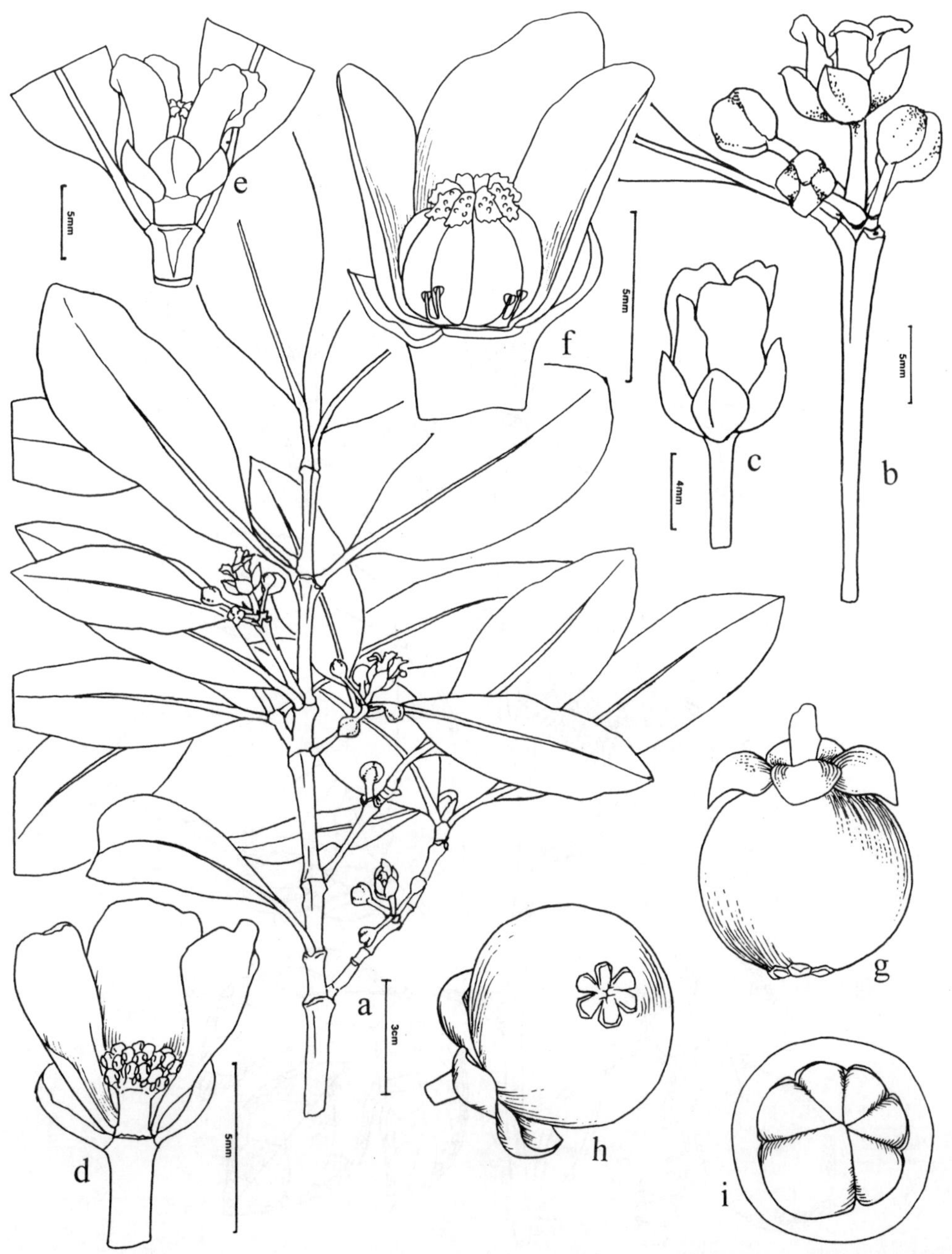

Figure 41. South China Mangosteen (*Garcinia oblongifolia* Champion ex Bentham). **a.** A flowering branch showing fascicles of staminate flowers. **b.** Face view of a staminate flower showing the sepals, petals and numerous stamens. **c.** A mature staminate flower bud with portion of the perianth removed showing the sessile anthers on a 4-angled column (androphore). **d.** A mature pistillate flower bud with portion of the perianth removed showing staminodes, and the ovary with discoid and recurved, lobed and crested stigma. **e.** A green fruit showing the persisting sepals and persistent crested stigmas. Clusiaceae 藤黃科.

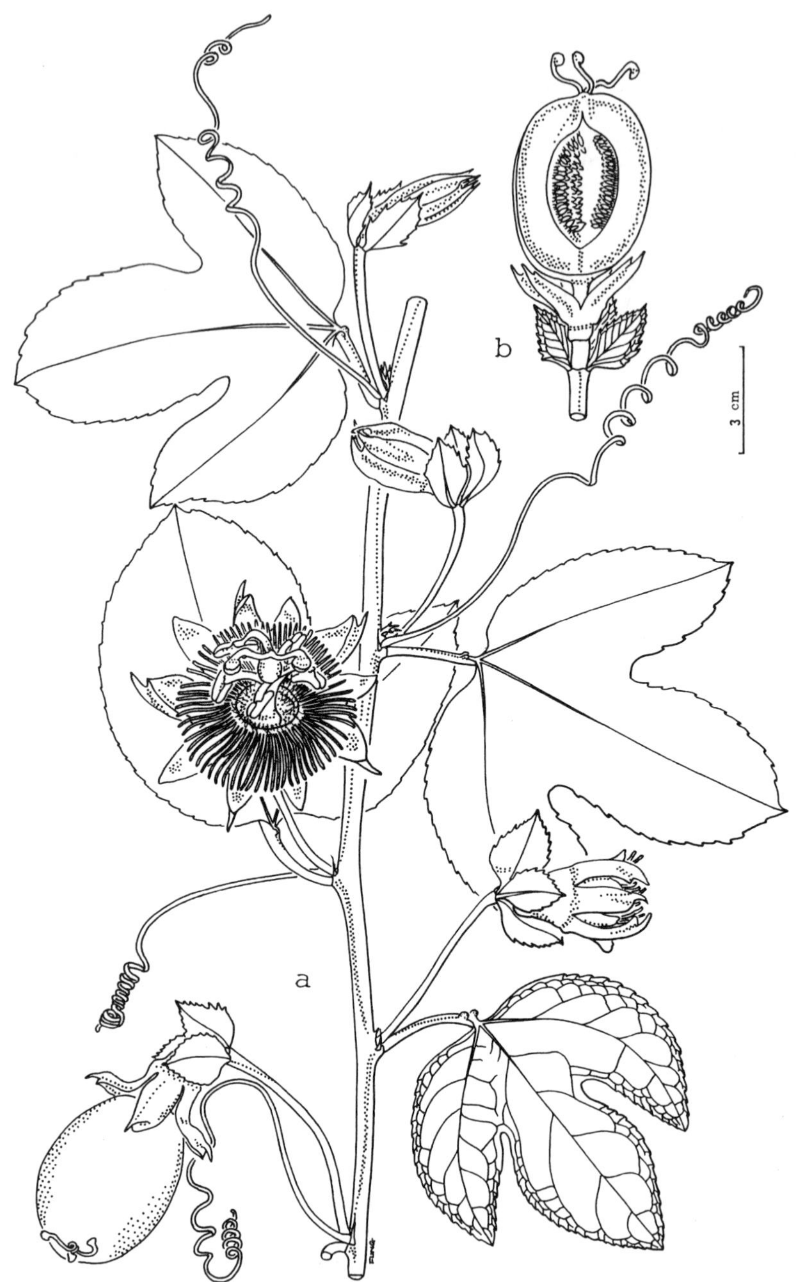

Figure 42. Passion fruit (*Passiflora edulis* Sims). **a.** A habit sketch of a flowering and fruiting portion of the vine showing simple tendrils, trilobed leaves, 2 large glands at the distal end of the petiole, axillary solitary flowers subtended by foliaceous bracts, sepals corniculate at the apex, filiform coronas on the rim of the cup-shaped hypanthium, a conspicuous column (androgynohpore) bearing 5 stamens with hanging anthers, and the stipitate ovary with 3 styles and capitate stigmas. **b.** The vertical section of a fruit showing short pedicel, persistent sepals connate at the base, parietal placentation, numerous seeds and the persistent styles and stigmas. Passifloraceae 西蕃蓮科.

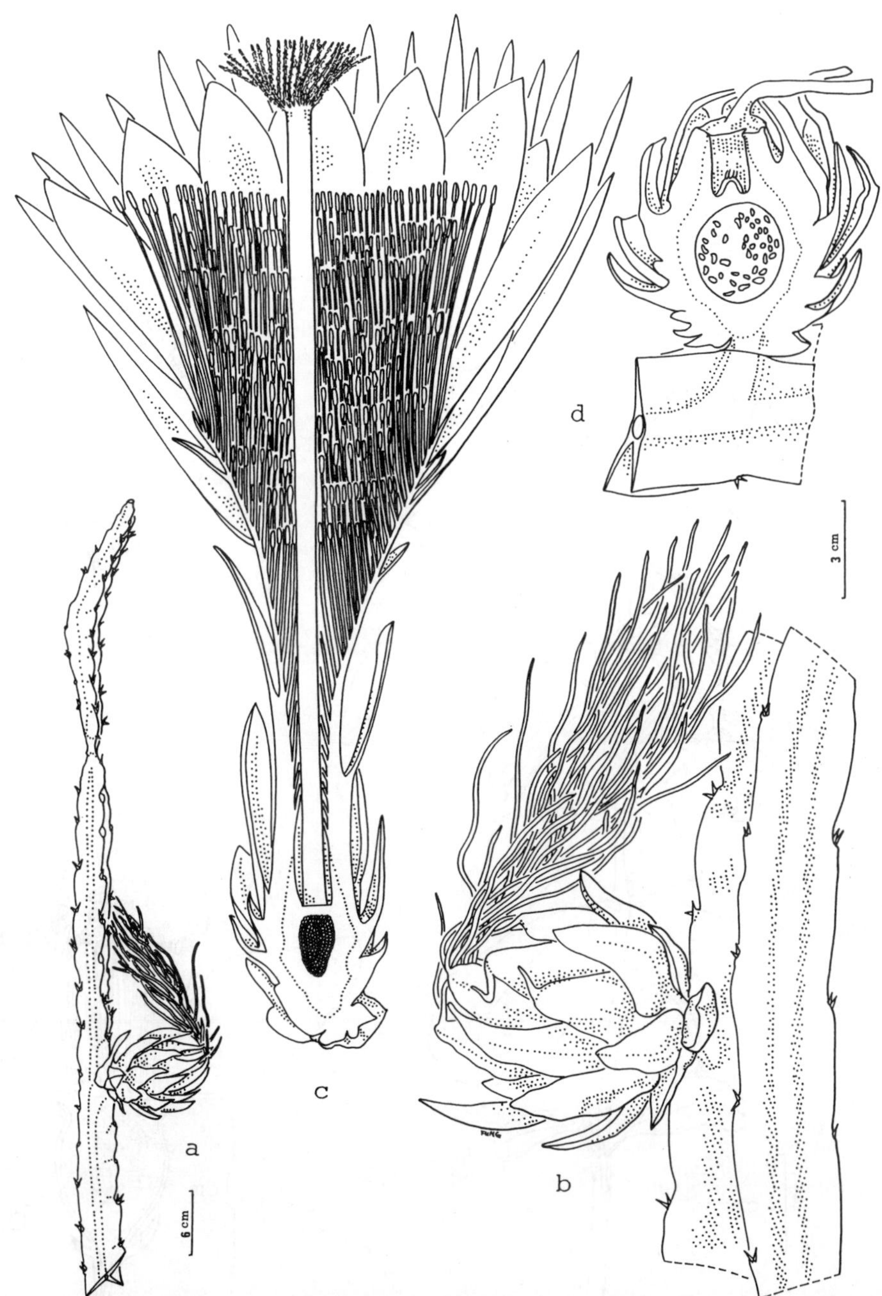

Figure 43. Night-blooming cereus (*Hylocereus undatus* [Haw.] Britton et Rose). **a.** The habit sketch of a leafless flowering section of the succulent stem showing the angular and winged features, many areoles furnished with short and stout spines, and the position of a solitary flower. **b.** A section of the same showing the spirally arranged fleshy scales gradually changing to linear ones on the hypanthium. **c.** The vertical section of a flower showing the hypanthium adnate to the inferior ovary, the gradual changes of the petaloid segments, the attachment of the stamens, and numerous ovules. **d.** A section of the fruit-bearing stem showing its 3 winged ridges, the shape of the fruit (vertical section) and the numerous seeds in the pulp. Cactaceae 仙人掌科.

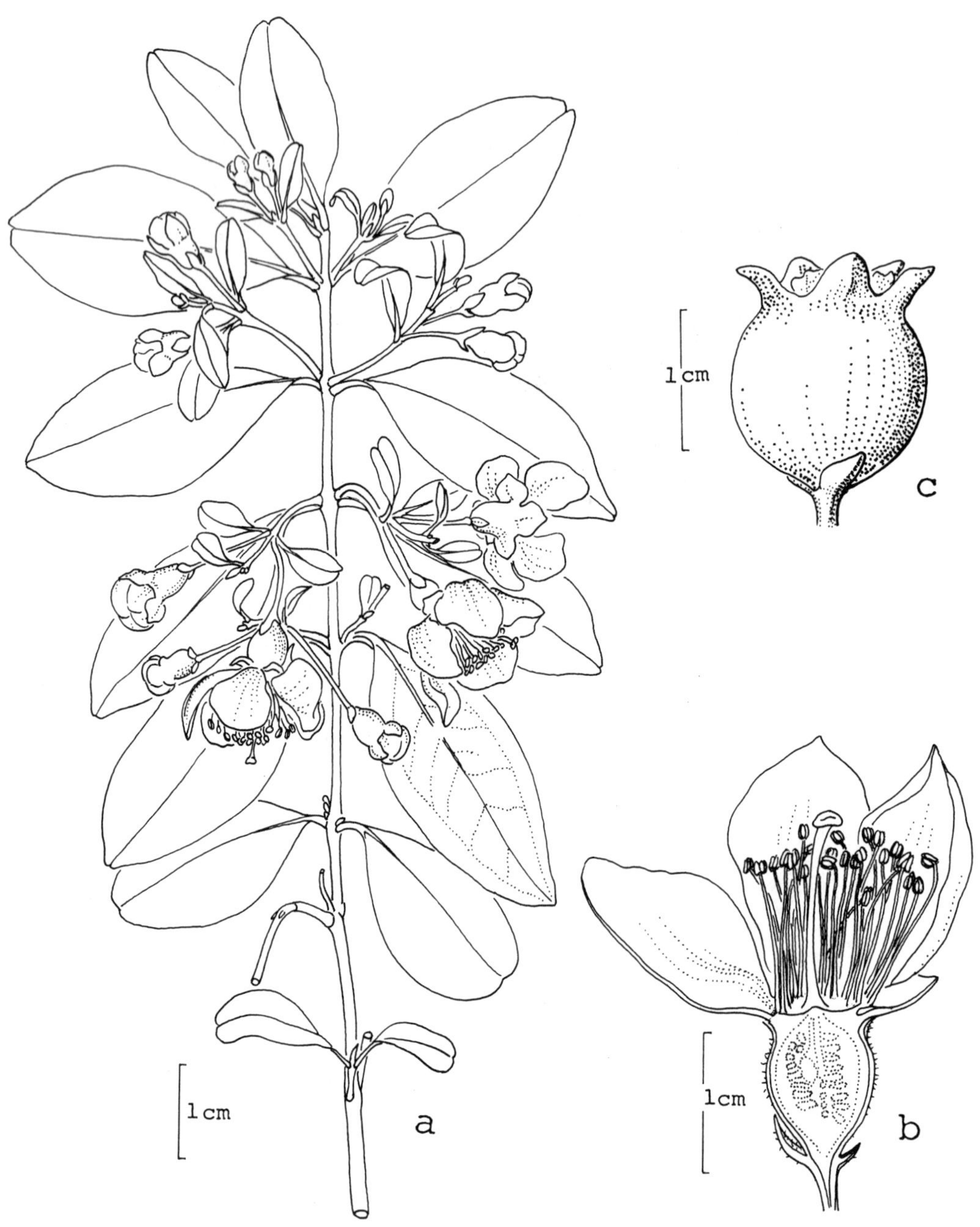

Figure 44. Downy myrtle (逃軍糧, *Rhodomyrtus tomentosa* [Aiton] Hasskarl). **a.** The habit sketch of a flowering branch showing opposite, simple, entire leaves and the axillary clusters of flowers. **b.** The vertical section of a flower showing low bracteoles, the tomentose hypanthium adnate to the inferior ovary, a sepal, 3 petals, numerous stamens, the style and stigma. **c.** The lateral view of a berry-like fruit (edible portion of the plant), showing the persistent bracteoles and sepals. Myrtaceae 桃金孃科.

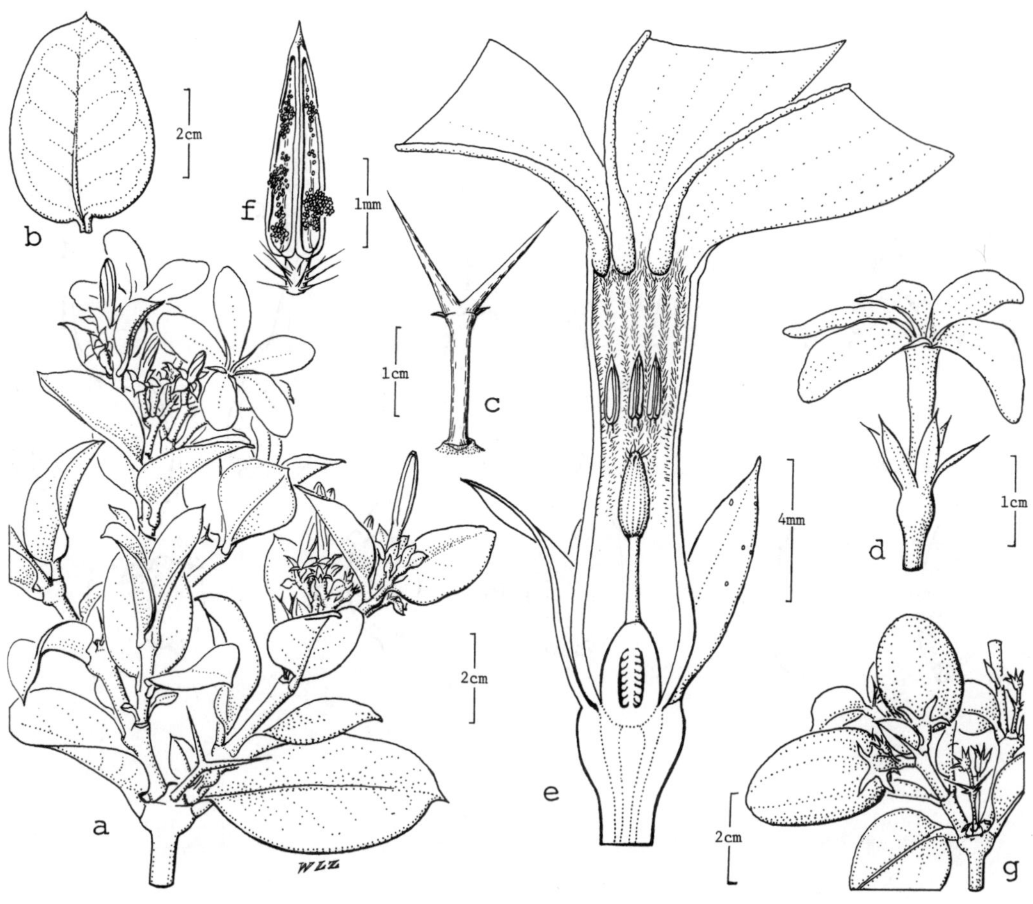

Figure 45. Natal plum (*Carissa macrocarpa* [Ecklon] A. DC. formerly known as *C. grandiflora* A. DC.). **a.** The habit sketch of a flowering branch showing the dense foliage, forked spines and cymose flower clusters. **b.** An ovate leaf. **c.** A sharp spine. **d.** The lateral view of a flower. **e.** The vertical section of a flower showing the hairy feature on the inside of the corolla tube, the superior ovary, style, and the clavuncle with the stigmatic surface on the ridges below the acute apex furnished with penicillate hairs. **f.** A stamen showing the short filament with long hairs locked with those at the apex of the stigma. **g.** A cluster of red, fleshy, juicy, very sour fruits with persistent calyx. Apocynaceae 夾竹桃科.

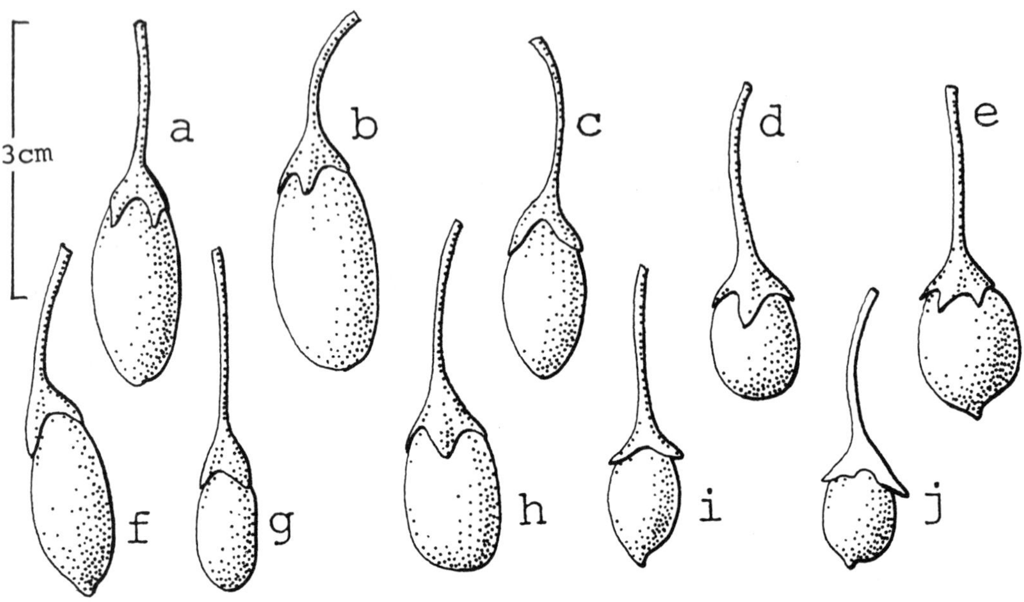

Figure 46. Lycium berry (枸杞子, *Lycium barbarum* L.). Selected cultivars of Ningxia lycium berry. **a.** cv. 'Hemp Leaf', with large red berry, very high protein content. **b.** cv. 'Greater Hemp Leaf', with very large red berry, rather low protein, and very high sugar content. **c.** cv 'White Branch', with medium-large fruit, containing less protein than 'Hemp Leaf', and tasting less sweet than 'Greater Hemp Leaf'. **d.** cv. 'Spherical Berry', having medium-sized sweet fruit and with rather high protein content for its size. **e.** cv. 'Spherical Acute Head', rarely kept in stock now. **f.** cv. 'Yellow Acute Head', with large and yellow fruit, very sweet, and low protein content. **g.** cv. 'Yellow Round Head', with rather small yellow fruit, rarely kept in stock. **h.** cv. 'Yellow Berry', with medium-large fruit, rarely kept in stock. **i.** cv. 'Curly Leaf', with small fruit, rarely cultivated. **j.** cv. 'Little Yellow', with small but very sweet fruit, having the highest ratio of sugar content. Solanaceae 茄科.

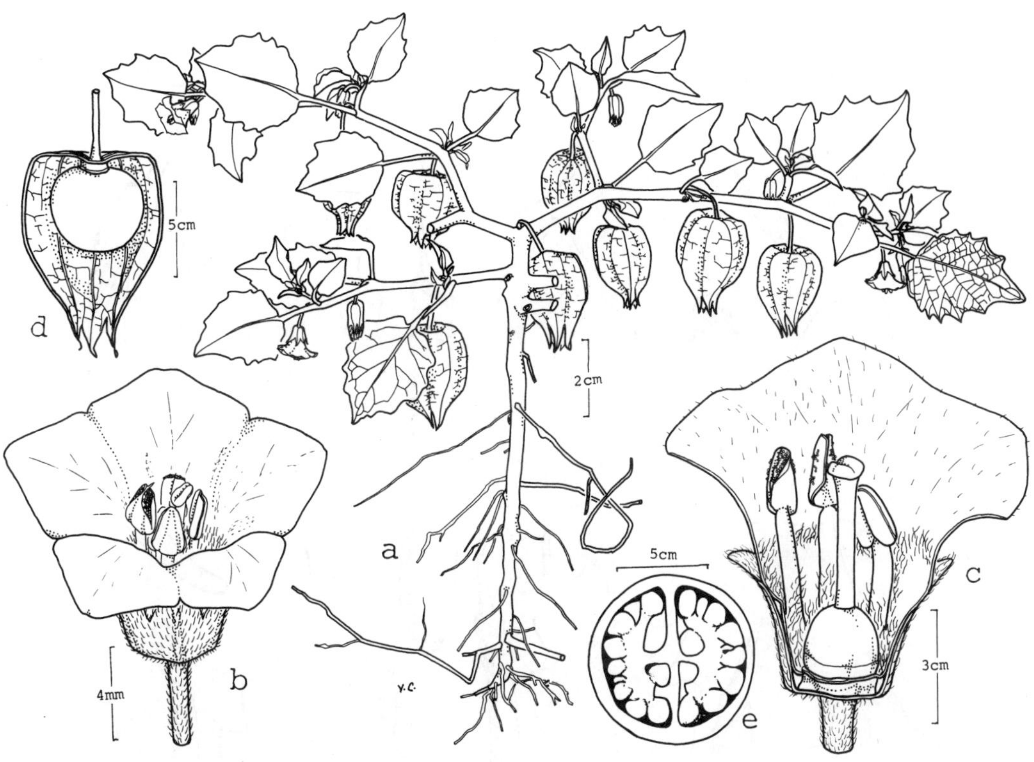

Figure 47. Husk tomato (*Physalis angulata* L.). **a.** The habit sketch of a flowering and fruiting plant showing the spreading branches over the surface of the soil, and the axillary position of individual flowers and fruits. **b.** The lateral view of a flower showing the broadly campanulate hairy calyx, funnelform corolla, stamens, and style. **c.** The same with portion of the calyx and corolla removed showing hairs on the inside of the corolla tube and filaments, the superior ovary, style and stigma. **d.** A fruit with portion of the bladder-like persistent calyx removed showing the globose berry. **e.** The transverse section of a berry showing the placentas and seeds. Solanaceae 茄科.

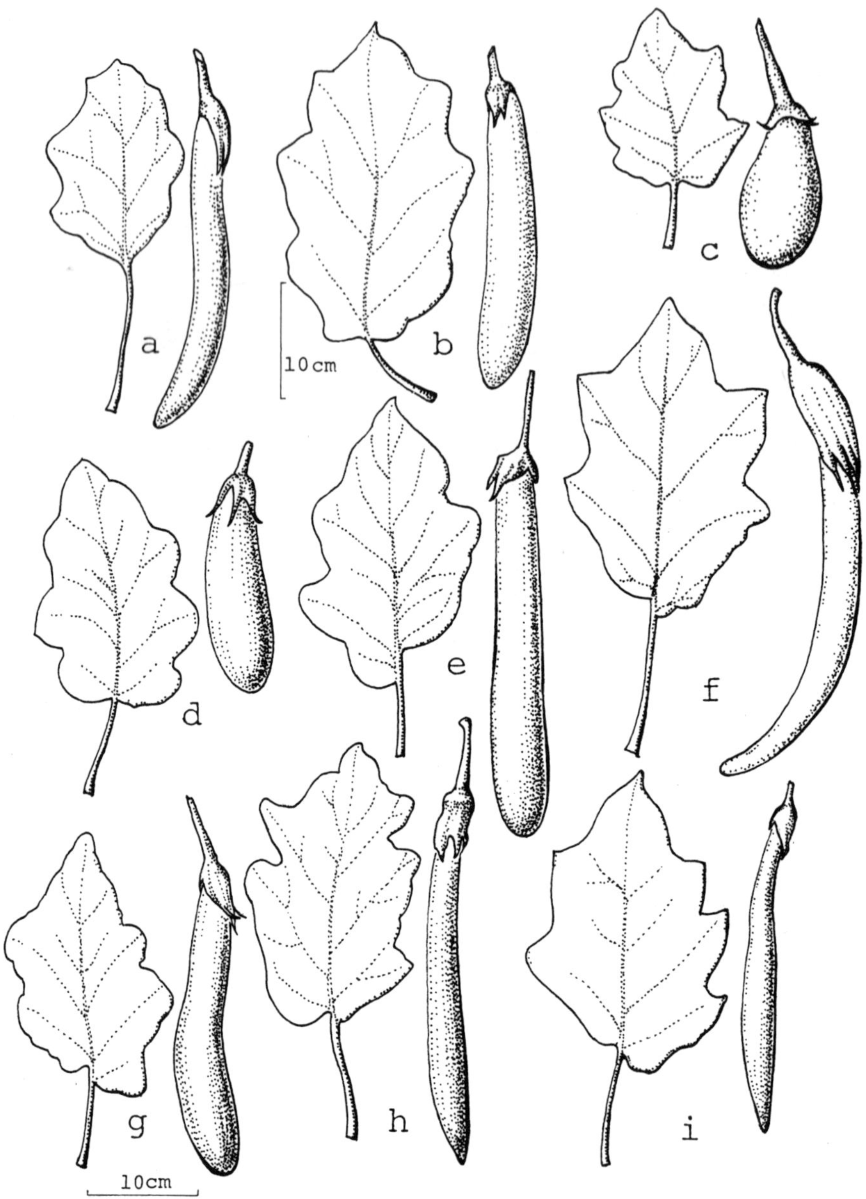

Figure 48. Cultivars of the eggplant (*Solanum melongena* L. var. *esculentum* Nees) from southern China. Fruits reddish-purple unless special mention of color. **a.** cv. 'Early Red', fruits 27–31 cm long, 3–4 cm across. **b.** cv. 'Early Green', fruits 28–35 cm long, 4 cm across, olive green. **c.** cv. 'Light Bulb', fruits 15 cm long, 7 cm across. **d.** cv. 'Hanging Purse', fruits 25–30 cm long, 4–5 cm across. **e.** cv. 'Extended Vision', fruits 20–30 cm long, 4–6 cm across. **f.** cv. 'White Dragon', fruits slender and curved, 30–36 cm long, 3 cm across, white. **g.** cv. 'Great Delight', fruits clavate, 26–30 cm long, 4–6 cm across. **h.** cv. 'Goddess-of-Mercy Finger', fruits 30–37 cm long, 2 cm across, light green. **i.** cv. 'Crane Cave Fall', fruits slender, 28–33 cm long, 3–4 cm across, good flavor, grayish green. A great majority of these cultivars bear slender and long eggplants, an indication of the preference of Chinese consumers. Solanaceae 茄科.

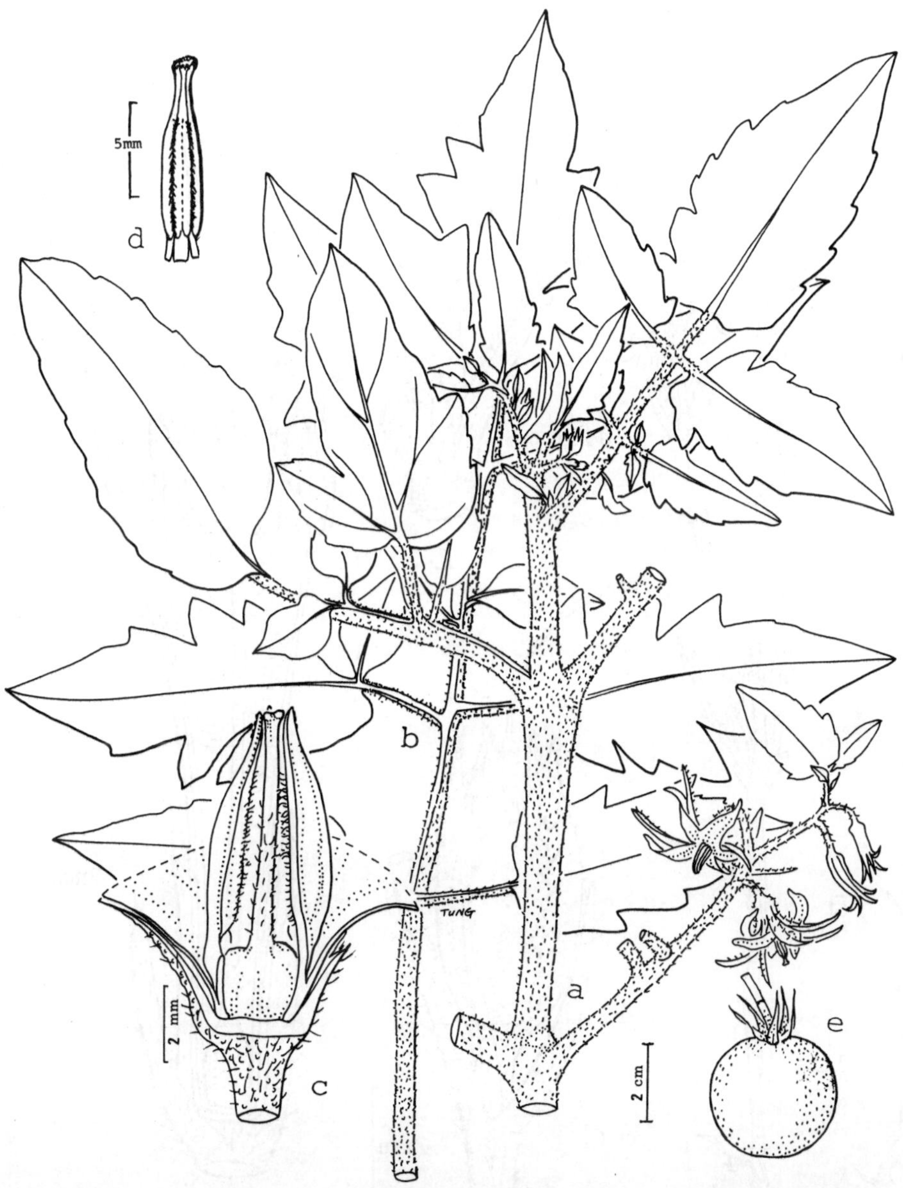

Figure 49. Tomato (*Lycopersicon lycopersicum* [L.] Karsten ex Farwell, generally known as *L. esculentum* Miller). **a.** The habit sketch of a flowering branch showing the mixture of glandular and villose hairs, irregularly pinnate or pinnatified leaves, extra-axillary orientation of the inflorescence, rotate corolla and the coherent anthers forming a central cone. **b.** A large cauline leaf. **c.** A flower with portion of the perianth removed showing the lateral view of stamens densely villose along the raised connectives, and the pistil with superior ovary and hairy style. **d.** The lateral view of the adroecium showing the abaxial aspect of a mature anther with dense hairs along the outer walls of its thecae locking with similar hairs of the neighboring thecae forming a coherent tube with the sterile apical portion. **e.** A fully grown tomato of a small-fruited cultivar showing the point of disarticulation on the stalk, and the reflexed persistent sepals. Solanaceae 茄科.

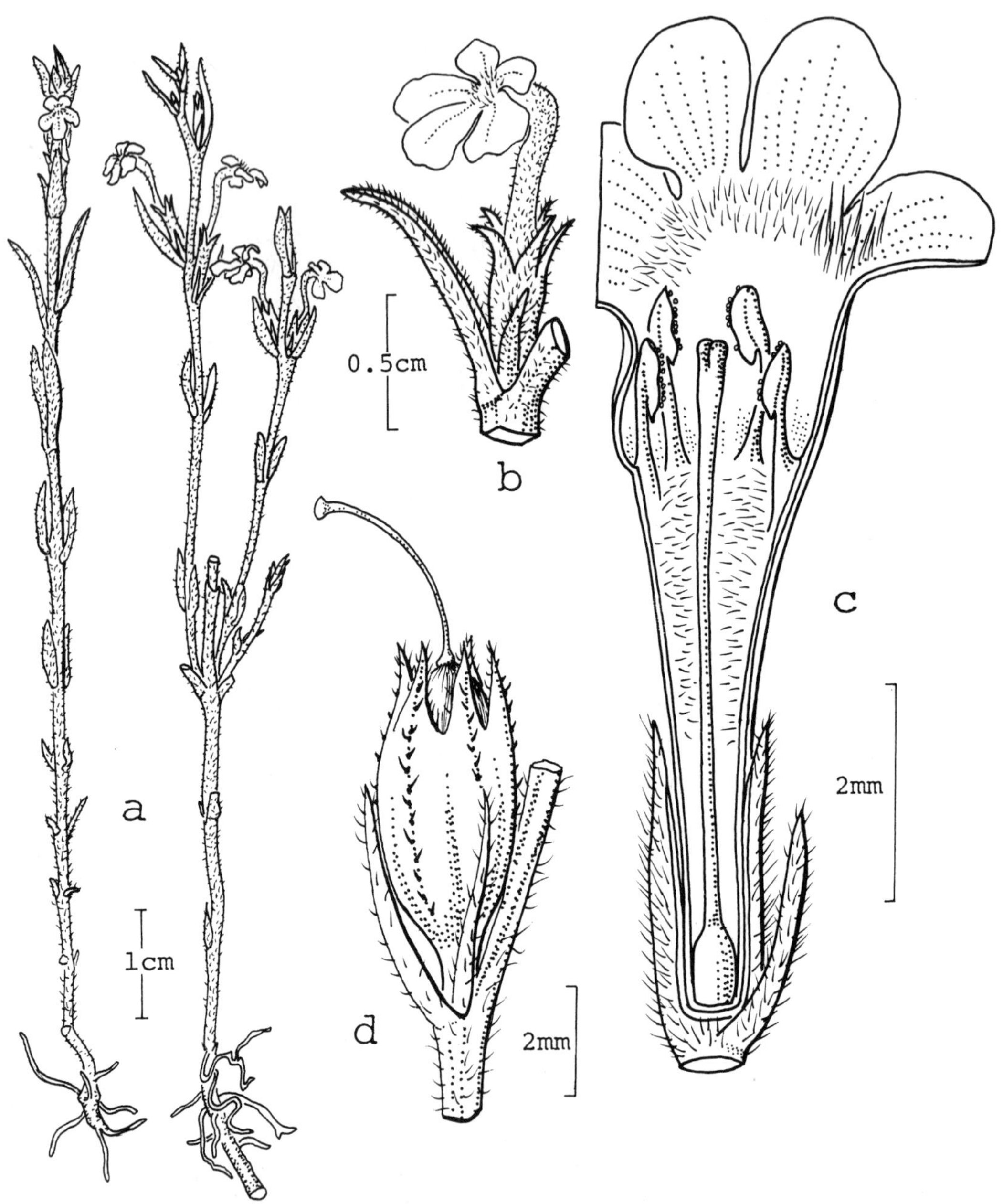

Figure 50. Witchweed (*Striga asiatica* [L.] O. Kuntze). **a.** The habit sketch of 2 flowering strigose plants. **b.** A portion of the stem showing the axillary flower with a bracteole, 5-toothed tubular calyx, and the corolla curved and gibbous below the throat. **c.** The vertical section of a flower showing the insertion of 4 stamens with unithecous anthers, different types of hairs on the inside of the corolla tube and throat, and the pistil with a superior ovary. **d.** A node from the fruiting section of the stem showing the oblong-ellipsoid capsule with persistent style and strigose calyx. Scrophulariaceae 玄參科.

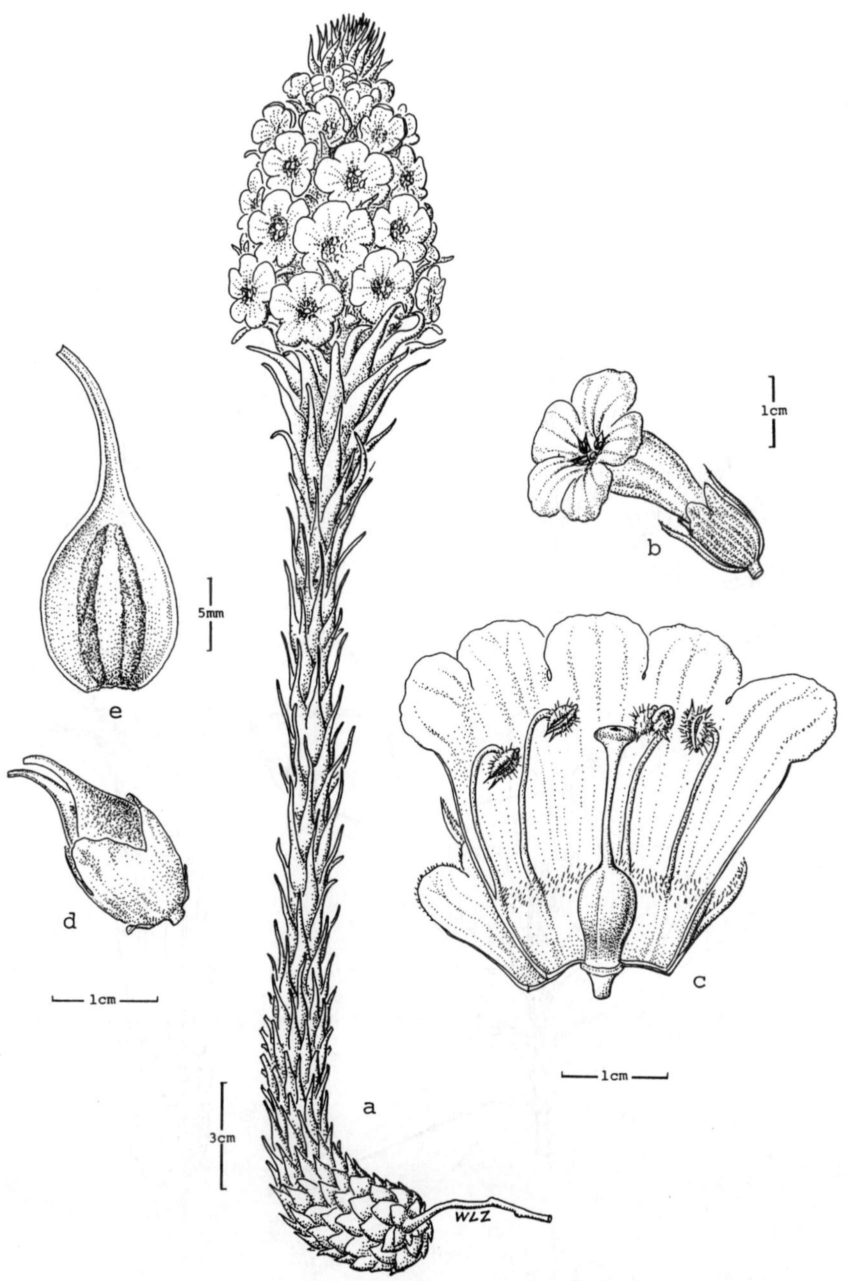

Figure 51. Cistanche, Mongolia Broom-rape (*Cistanche salsa* [C. A. Meyer] Beck.). **a.** The habit sketch of the flowering plant showing its connection with the host species, the imbricate fleshy scales on the stem terminated by a large spike with many flowers. **b.** The lateral view of a flower showing 2 bracteoles, the campanulate calyx, and the zygomorphic corolla. **c.** The same with the perianth spread mechanically showing the hairy area of the corolla tube, the insertion of the paired stamens with dorsifixed haired anthers, the pistil with superior ovary, style and captitate-discoid stigma. **d.** A 2-carpellate capsule with persistent calyx. **e.** The vertical section of a capsule showing numerous dust-like seeds on parietal placentas. Orobanchaceae 列當科.

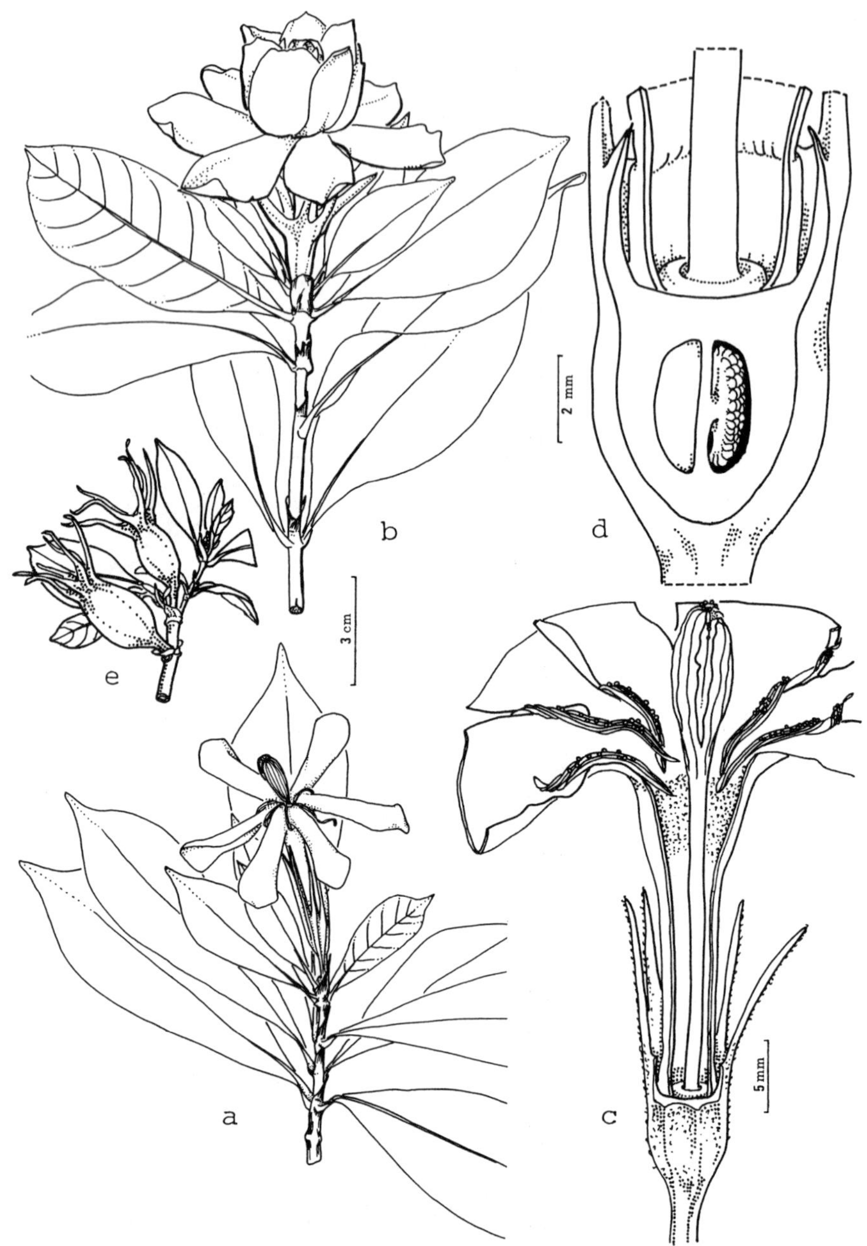

Figure 52. Gardenia, Huang-zhi (*Gardenia jasminoides* Ellis), a Chinese natural food color. **a.** The habit sketch of a flowering branch from a plant of wild population growing on Hong Kong hillsides showing the solitary flower with 6 reflexed petals and very large oblong stigma. **b.** The same from a plant cultivated in the campus of The Chinese University of Hong Kong. **c.** A flower with portion of the perianth removed showing the inferior ovary and the insertion of the stamens at the throat of the corolla. **d.** A vertical section of the lower portion of a flower showing the disk, and the placenta with numerous ovules. **e.** A fruiting branch bearing 2 capsule-like berries, golden-orange at maturity, with carnose pericarp and rather juicy pulp (gathered at this stage, dried, used for dying beancurd). Rubiaceae 茜草科.

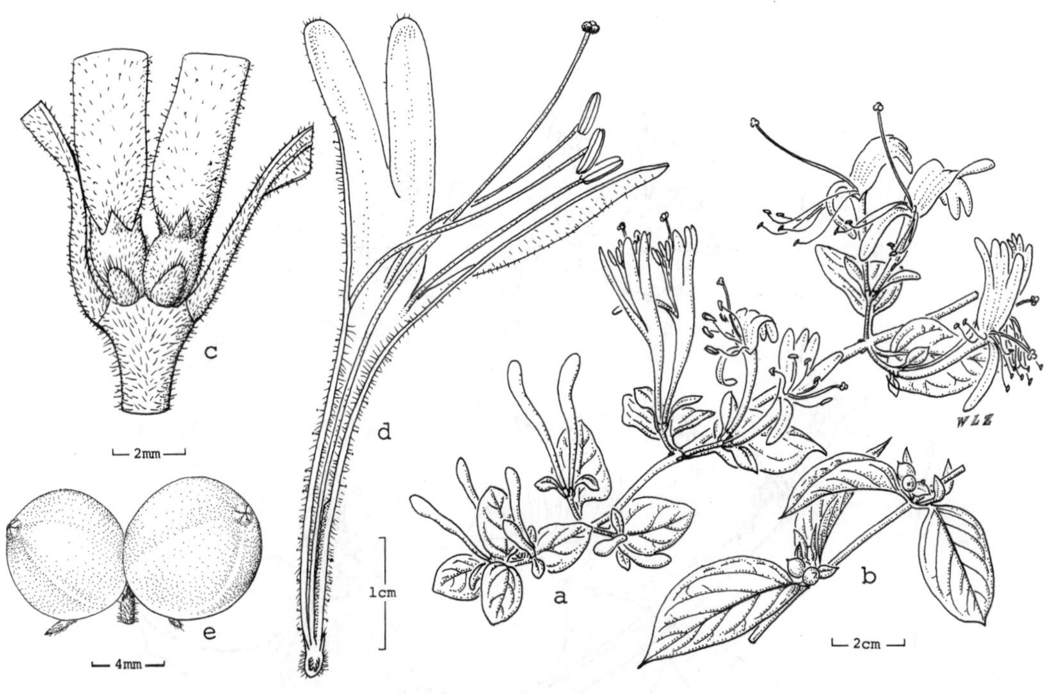

Figure 53. Honeysuckle (金銀花, *Lonicera japonica* Thunb.), a Chinese flower tea. **a.** The habit sketch of a branch of the vine showing paired flowers at different stages of development. **b.** A fruiting branch showing paired berries, turning black at maturity. **c.** The basal portion of a paired, sessile, and softly pilose flowers showing the petiolate foliaceous bracts, the sessile bracteoles, and the inferior ovaries. **d.** The vertical section of a flower showing the insertion of the stamens at the throat of the corolla, the elongated nectar on one side at the base of the zygomorphic corolla. **e.** A pair of fully grown berries with persistent calyx. Caprifoliaceae 忍冬科.

Figure 54. Winter melon (冬瓜, *Benincasa hispida* [Thunb.] Cogn.). **a.** The habit sketch of a flowering portion of the vine with the laminas removed, showing the branched tendrils, the relative positions of a young staminate flower and the emergence of a still younger branch bearing a pistillate flower (upper node), a staminate flower with lobate sepals on an elongated pedicel (middle node), and a pistillate flower with inferior ovary on a short pedicel (lower node). **b.** A leaf with reniform shallowly palmate-lobed lamina and a broad sinus at the base. **c.** The androecium with 3 stamens on a hispid disk, showing the short fleshy filaments and sigmoid thecae. **d.** The apical portion of the inferior ovary of the pistillate flower showing the hairy disk, the staminodes, styles and stigmas. Cucurbitaceae 葫蘆科.

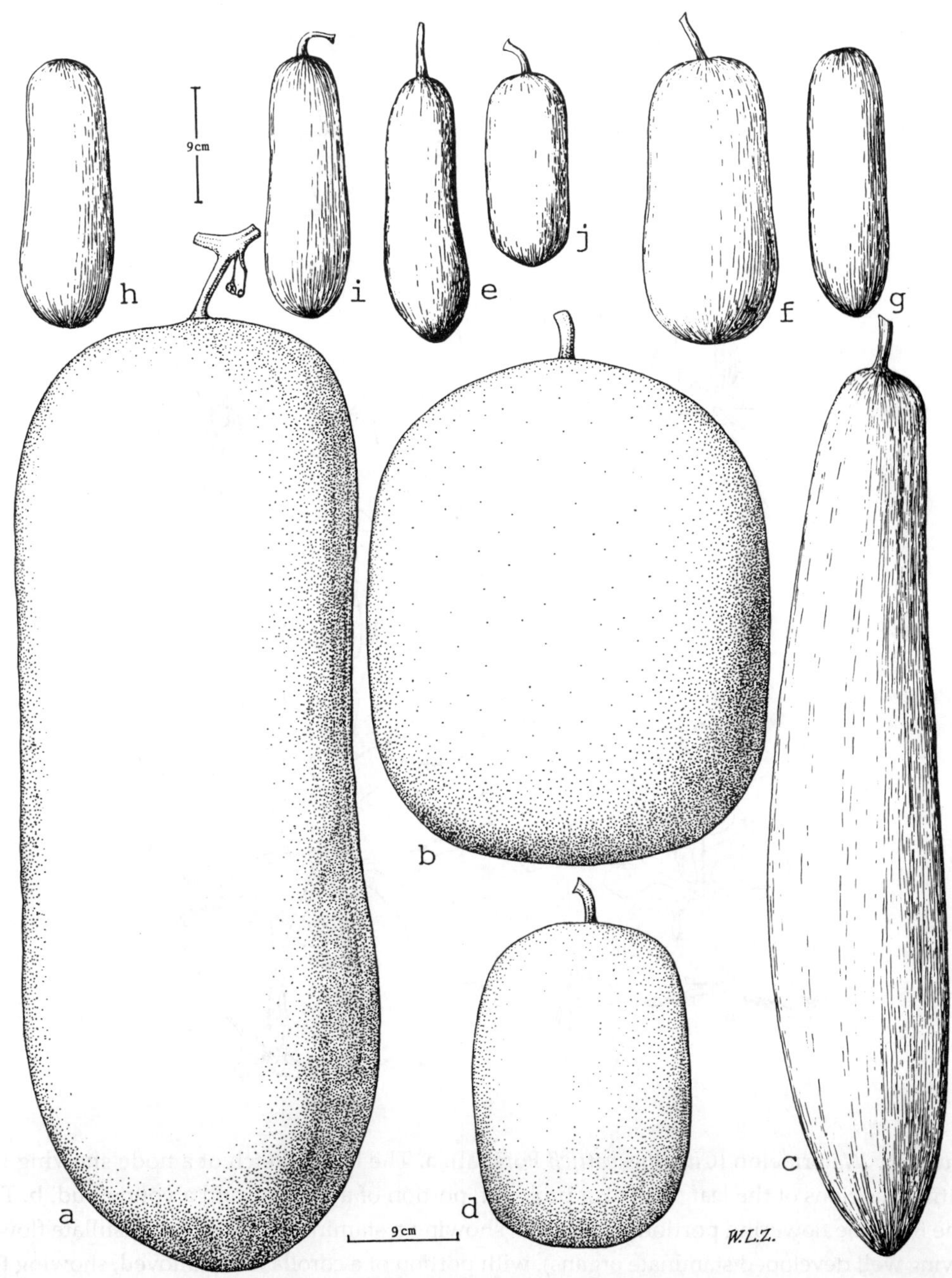

Figure 55. The common cultivars of 2 varieties of *Benincasa hispida* [Thunb.] Cogn. Recorded from southern China. **a-d.** Fruits of cultivars of var. *hispida*. **a.** cv. 'Big Green Rind', fruits 95 cm long, 28 cm across. **b.** cv. 'Gray Bushel', fruits 40 cm long, 33 cm across. **c.** cv. 'Ox Spleen', fruits 70 cm long, 14 cm across. **d.** cv. 'Yellow Climber', fruits 26 cm long, 17 cm across. **e-j.** Fruits of cultivars of var. *chiehqua* How. **e.** cv. 'Big Vine', fruits 21 cm long, 6 cm across. **f.** cv. 'Pineapple', fruits 23 cm long, 11 cm across. **g.** cv. 'Twin Carps', fruits 21 cm long, 6 cm across. **h.** cv. 'Patched Olive-green', fruits variable, 18–21 cm long, 6–7 cm across. **i.** cv. 'Seven Stars', fruits 21 cm long, 6 cm across. **j.** cv. 'Spotted Pineapple', fruits 15 cm long, 7 cm across. Cucurbitaceae 葫蘆科.

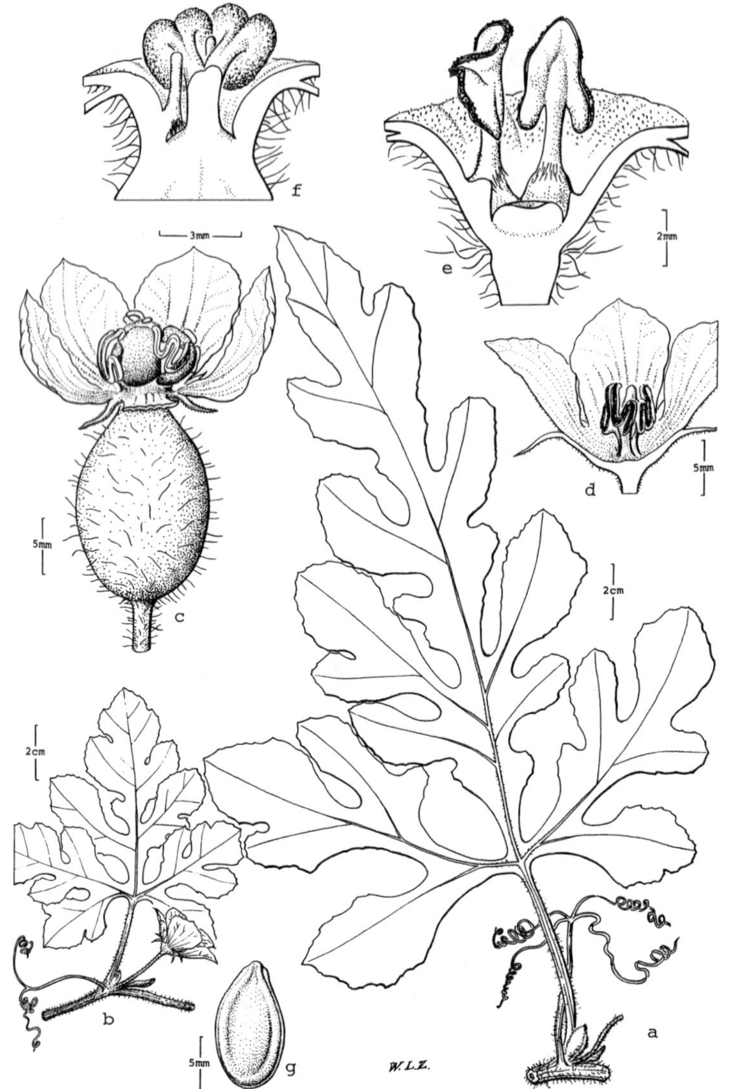

Figure 56. Watermelon (*Citrullus battich* Forskal): **a.** The habit sketch of a node showing the relative positions of the leaf, tendril, a probract, portion of a pedicel, and a young bud. **b.** The same from the flowering portion of the vine, showing a staminate flower. **c.** A pistillate flower (having well developed staminate organs), with portion of a corolla lobe removed, showing the calyx, the staminodes or stamens, hairy inferior ovary, short style and fleshy stigmas. **d.** A staminate flower with portion of the perianth removed showing the cup-shaped receptacle and 3 distinct stamens with filaments free from each other, 2 anthers 2-thecous, and 1 anther 1-thecous. **e.** The vertical section of a staminate flower cut through the middle, with portion of the sepals and corolla removed, showing the trichomes at the base of a filament and the cushion-shaped pistillode. **f.** The vertical section of the apical portion of a pistillate flower showing a rod-like staminode barbate at the base, the short style, and 2 fleshy stigmas. **g.** A seed (specimens used: **a.** V. Cory 51175, **b, d, e.** Shandong Univ. Oct. 1928, **c.** Wedermann & Oberdieck 1782, all in Gray Herbarium, Harvard University). Cucurbitaceae 葫蘆科.

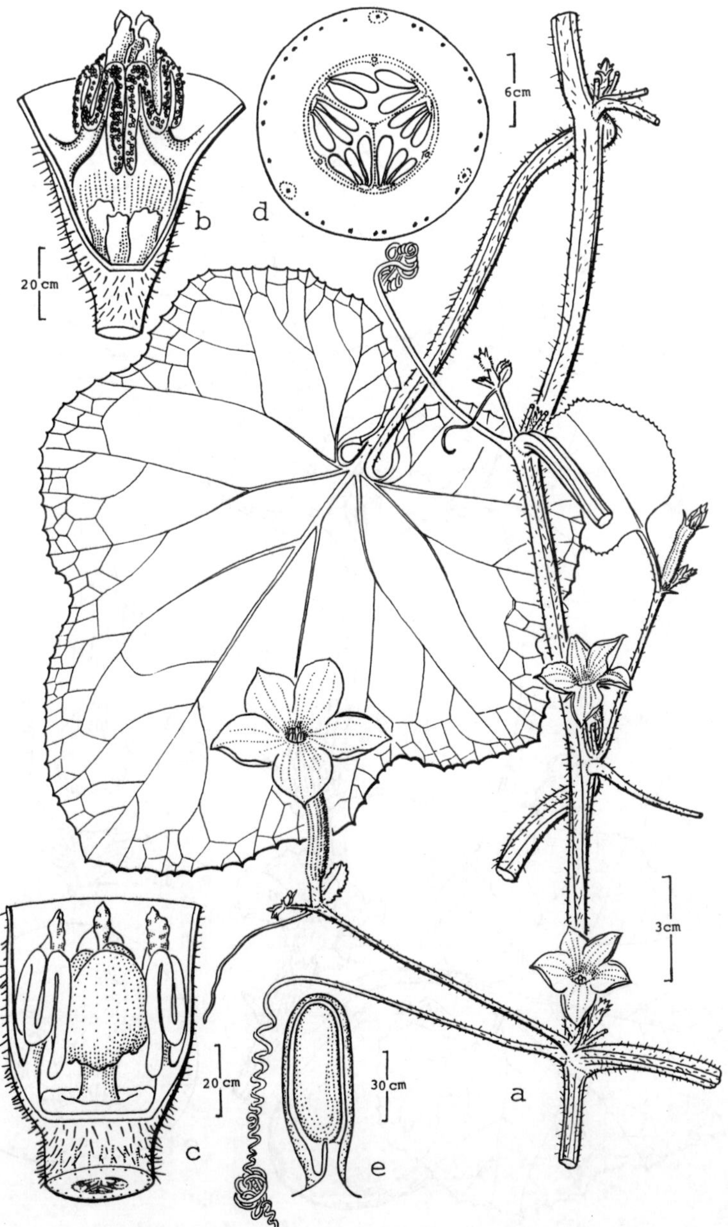

Figure 57. Oriental Picking Melon (*Cucumis melo* L. var. *conomon* [Thunb.] Nakai). **a.** The habit sketch of a flowering branch showing the relative orientation of the leaves, fasciculate staminate flowers and a short shoot bearing a solitary pistillate flower. **b.** A vertical section of the lower portion of a staminate flower showing the pilose hypanthium, the insertion of the stamens with sigmoid anthers and produced connectives. **c.** The same of a pistillate flower showing the very well developed staminodes, disk, transverse section of the ovary, short style and fleshy lobed stigma. **d.** The transverse section of a young fruit showing 3 major bundles of the vascular tissue supplying the hypanthium and another 3 bundles supplying the ovary wall, and 3 plancentas with young seeds. **e.** A section of the placenta showing an ovule, the funicle and related portion of the placenta. Cucurbitaceae 葫蘆科.

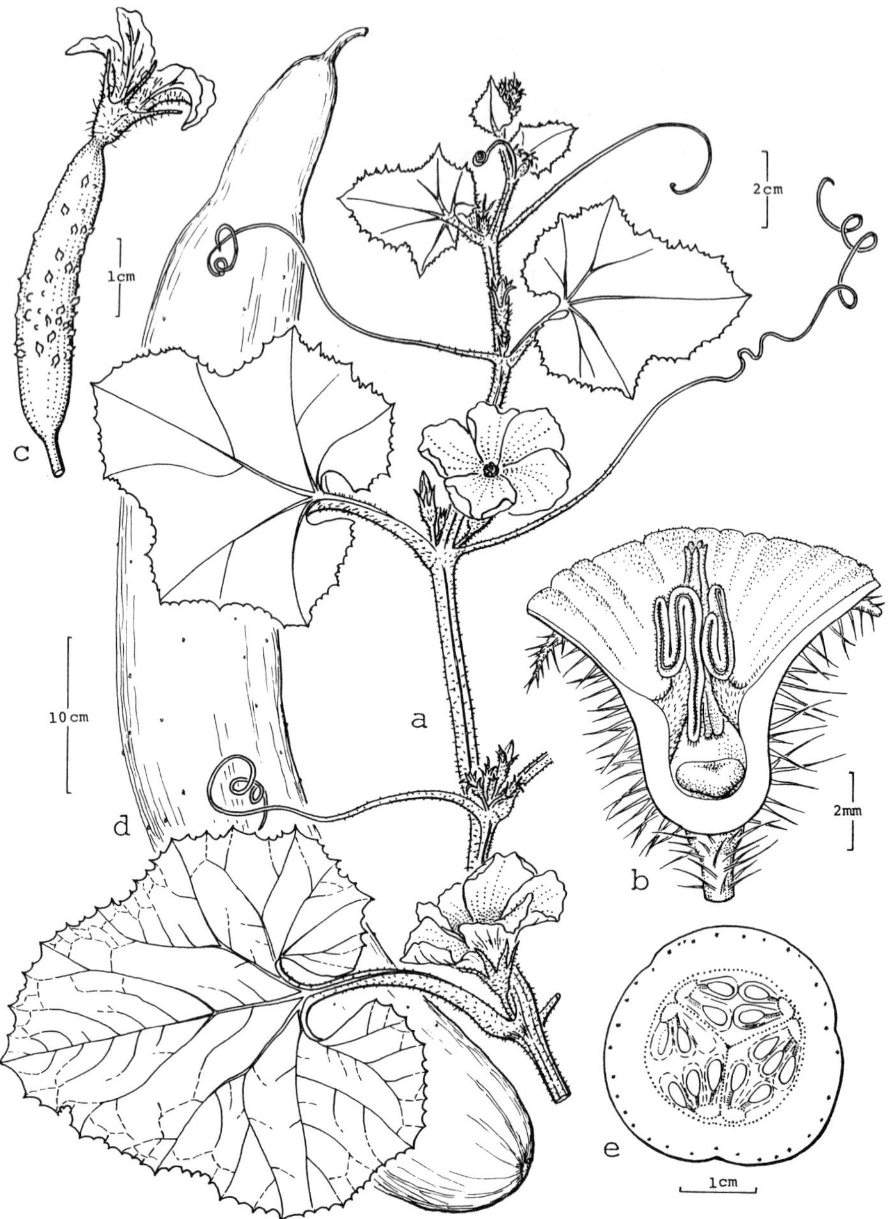

Figure 58. Cucumber (黃瓜, *Cucumis sativus* L.). **a.** The habit sketch of a branch from a young flowering plant bearing staminate flowers only, showing the strigose trichomes and prickles, simple tendrils, ovate-cordate leaves and funnelform hypanthium. **b.** The vertical section of a staminate flower with portion of the perianth removed, showing the insertion of stamens with pilose fleshy filaments and sigmoid anthers, prominently produced connectives, and discoid pistillode. **c.** The lateral view of a pistillate showing the oblong-fusiform spinescent ovary, funnelform calyx with linear lobes, and pilose corolla. **d.** A mature fruit of an elongated fruit form introduced from northern China to the experiment garden near the Harvard University Herbaria (harvested on September 30, 1991; 68 cm long, 7 cm across, weighing 1,800 kg fresh). **e.** A transverse section of the same, showing the placentas and seeds. Cucurbitaceae 葫蘆科.

Figure 59. Bottle Gourd, Hulu (*Lagenaria siceraria* [Molina] Standley var. *siceraria*). **a.** The habit sketch of a flowering branch showing the glandular trichomes, bifid tendrils and a solitary staminate flower on slender elongated peduncles. **b.** A lageniform (pyriform and narrowed at the middle), young fruit from a grocery in Hong Kong. **c.** Transverse section of the same showing the seeds embedded in fleshy placenta tissue. Cucurbitaceae 葫蘆科.

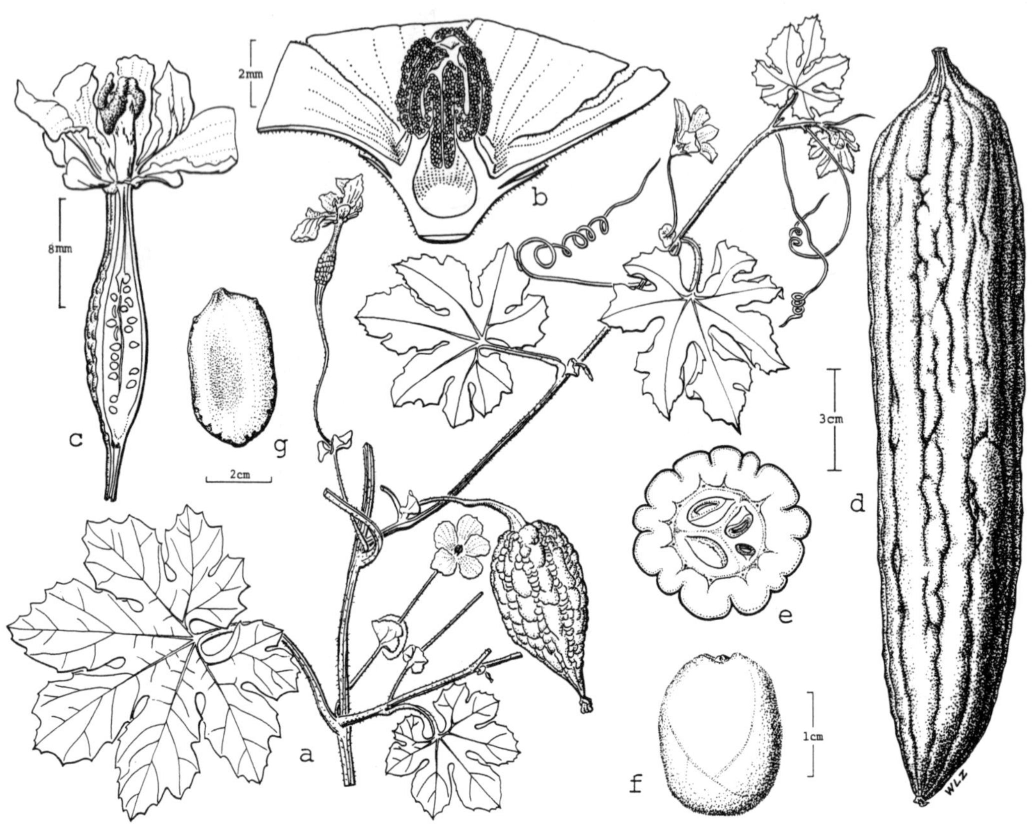

Figure 60. Bitter Melon (*Momordica charantia* L.). **a.** The habit sketch of a flowering branch showing the deeply lobate leaves, simple tendrils, solitary staminate and pistillate flowers in slender elongated peduncles each bearing a foliaceous bract below the middle. **b.** The vertical section of a staminate flower showing a shallow funnelform hypanthium, deeply lobed corolla, and stamens with free filaments and sigmoid anthers. **c.** The lateral view of a pistillate flower with a sepal and portion of the hypanthium removed showing the deeply lobed corolla, stylar column, fleshy stigmas, and numerous ovules embedded in the placental tissue. **d.** A subcylindrical, deeply furrowed and tubercled fruit from a Chinese grocery in Boston. **e.** A tranverse section of the same showing the seeds embedded in fleshy pulp. **f.** A seed wrapped in coriaceous placental material. **g.** The same with the fleshy cover removed showing a rectangular shape, and an irregular and grooved margin. Cucurbitaceae 葫蘆科.

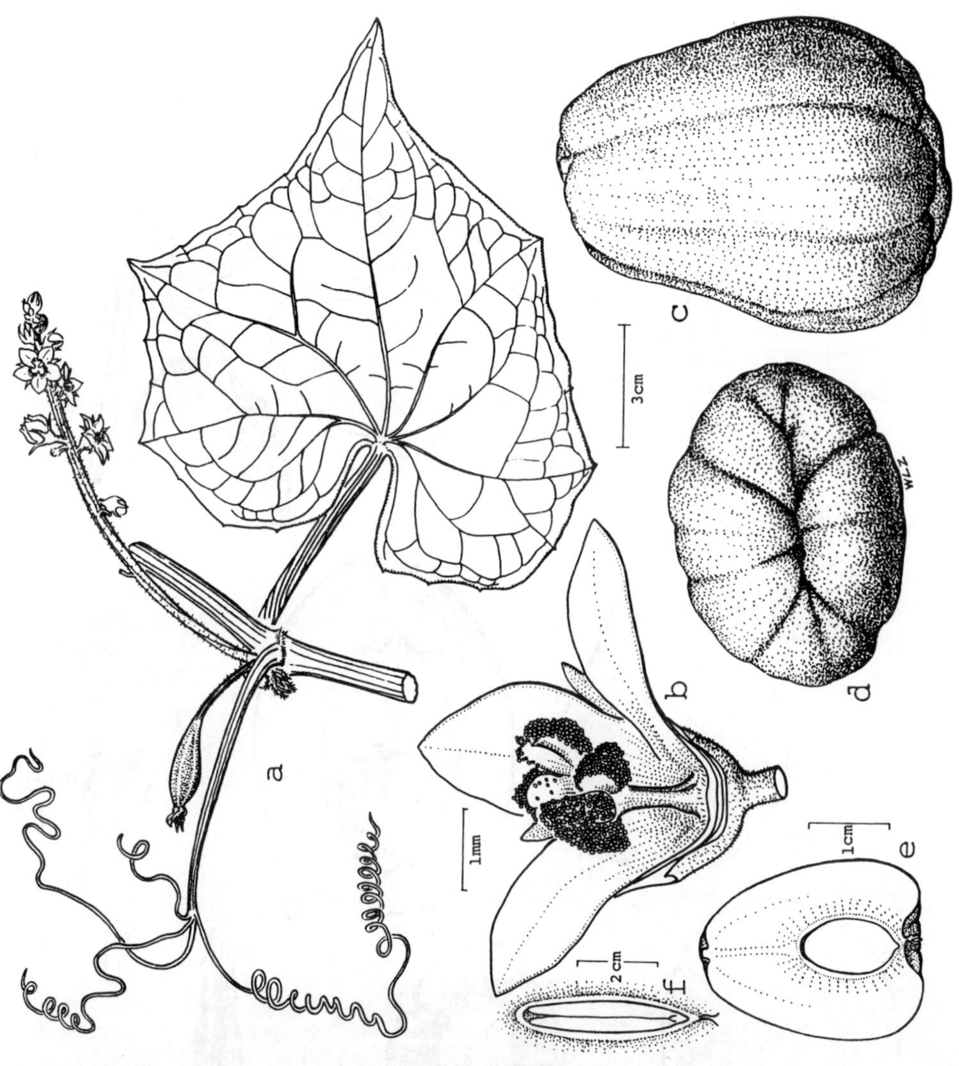

Figure 61. Chayote (佛掌瓜, *Sechium edule* [Jacq.] Sw.). **a.** The habit sketch of a node of a flowering vine showing the relative orientation of a leaf, a branched tendril, a pseudoracemose staminate inflorescence, and a solitary pistillate flower. **b.** The vertical section of a staminate flower showing the cupular hypanthium, the connate filaments and the separated sigmoid unithecous anthers. **c-d.** The lateral and apical views of a fruit from a grocery in Boston. **e.** The vertical section of a fruit showing the apical position of a solitary seed. **f.** A vertical section of the seed showing 2 large cotyledons attached to the minute plumule and hypocotyl, and the associated tissue of the pericarp. Cucurbitaceae 葫蘆科.

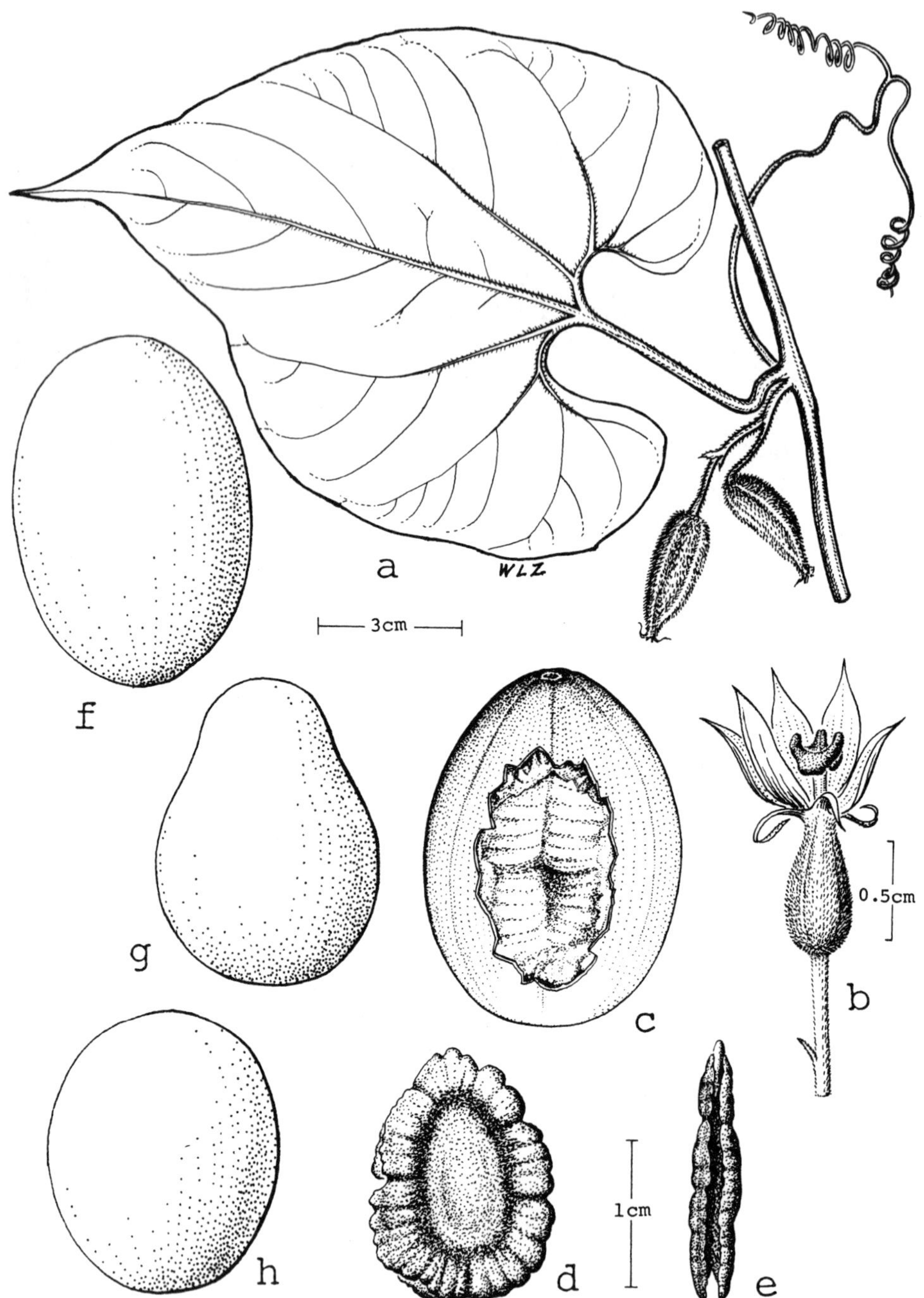

Figure 62. Luohanguo (羅漢果, *Siraitia grosvenorii* [Swingle] C. Jeffrey). **a.** A section of the vine showing an entire leaf with deep sinus at the base, a bifid tendril, and a cluster of young fruit. **b.** The lateral view of a pistillate flower showing the linear-lanceolate sepals, lanceolate corolla lobes, pilose hypanthium and style with lobate stigma. **c.** A fruit from a Boston store with the rind partially removed showing numerous seeds embedded in the delicate spongy placental material. **d.** A seed with the spongy material mechanically removed. **e.** The marginal view of the same showing a deep grove. Cucurbitaceae 葫蘆科.

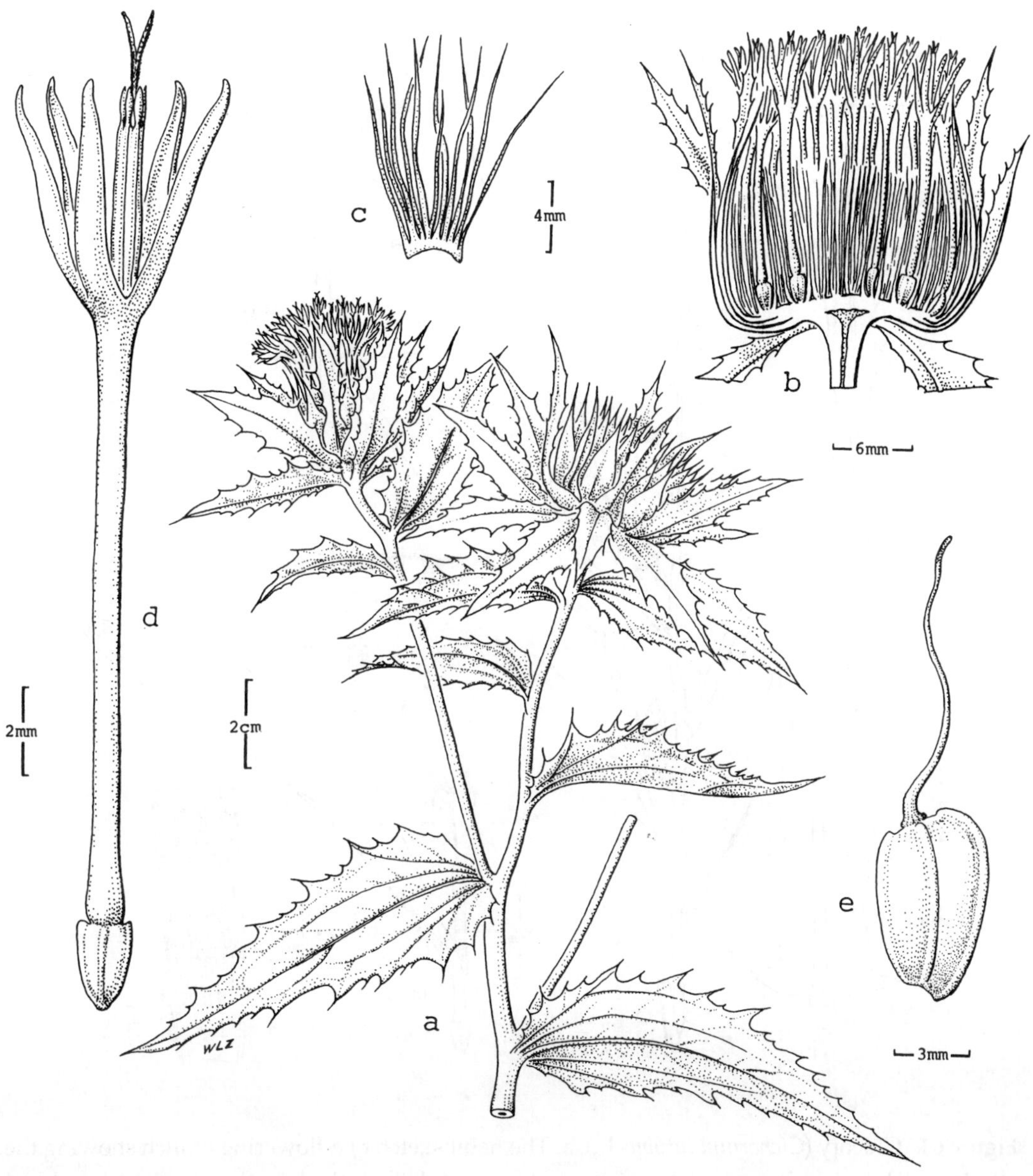

Figure 63. Safflower (*Carthamus tinctorius* L.): **a.** The habit sketch of a flowering branch showing the ovate-oblong, sinuate and spinose upper cauline leaves, and 2 solitary terminal capitula with prickly, foliaceous, coriaceous, and elliptic outer phyllaries. **b.** The vertical section of a capitulum showing the linear prickly inner phyllaries, flat paleate receptacle, and tubular flowers. **c.** Filiform paleae slightly connate at the base. **d.** A flower showing the slender tubular corolla suddenly enlarged and immediately divided into 5 linear lobes, synantherous stamens with conspicuous apical appendages, short stigmatic lobes, and obovoid ovary. **e.** An oblique-obovoid striate achene devoid of pappus. Asteraceae 菊科.

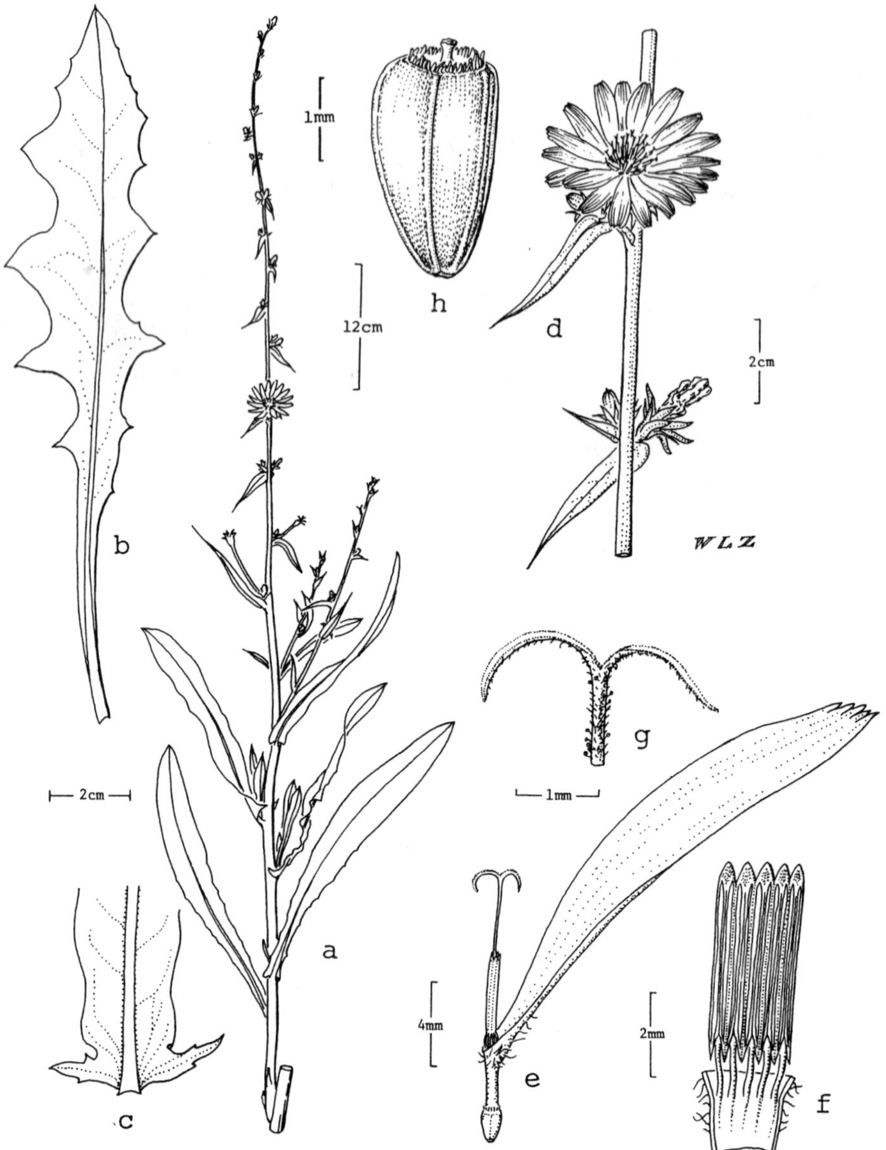

Figure 64. Chicory (*Cichorium intybus* L.). **a.** The habit sketch of a flowering branch showing the amplexicaul cauline leaves, and position of the capitulum with open flowers. **b.** A basal leaf showing the runcinate lamina, cuneate base and very short petiole. **c.** The basal portion of a cauline leaf showing the lobes clasping the stem. **d.** Portion of the stem showing a capitulum with withered ligulate corollas after anthesis, and another one with full blooming flowers. **e.** A flower with ligulate corolla and a short basal tube, the synantherous stamens on distinct filaments, inferior ovary, style and stigmatic branches. **f.** Portion of a coralla showing the insertion of the staments, and the adaxial aspect of the sagittate anthers. **g.** Apical portion of the style showing the setaceous trichome carrying pollen grains as the elongating style pushes through the anther tube, and the stigmatic surface. **h.** An angular-obovoid striate achene with the pappus modified into scuffy scales. Asteraceae 菊科.

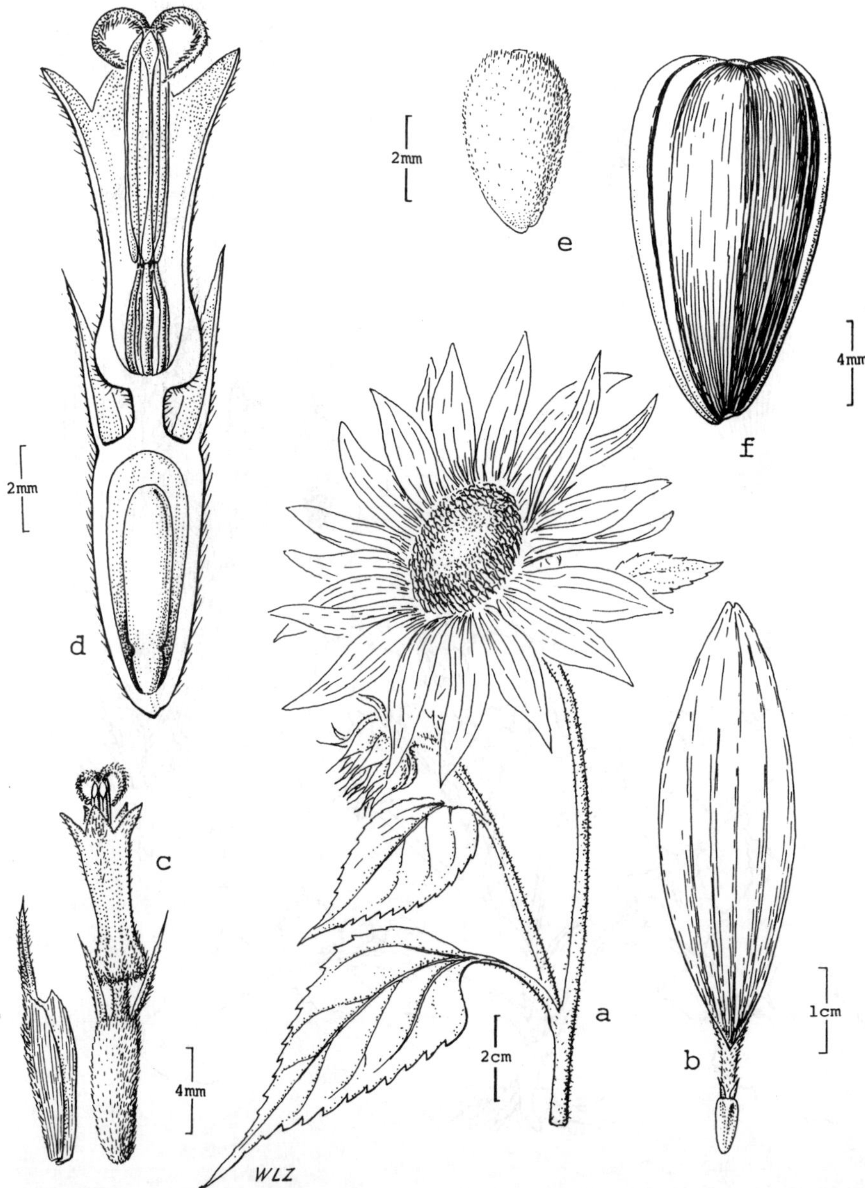

Figure 65. Sunflower (*Helianthus annuus* L.). **a.** The habit sketch of a small side shoot of a garden plant (with the central capitulum 20 cm in diameter terminating a bending stem), showing the herbaceous outer phyllaries (as shown by the tertiary bud), ligulate ray flowers and tubular disk flowers. **b.** A ligulate flower showing the corolla with a lanceolate large distal portion and a short hairy tubular basal portion. **c.** A tubular flower, showing the subtending tricuspidate bract (left), the goblet-shaped corolla tube suddenly narrowed into a slender portion at the base, and 2 pappi on the inferior ovary. **d.** Vertical section of the same showing the insertion of the filaments, the synandrium with anthers caudate at the base and apiculate at the apex, and an uniovulate, inferious ovary with its pappi. **e.** An obovoid hairy ovary. **f.** A strongly compressed obovoid achenes, gray with white stripes. Asteraceae 菊科.

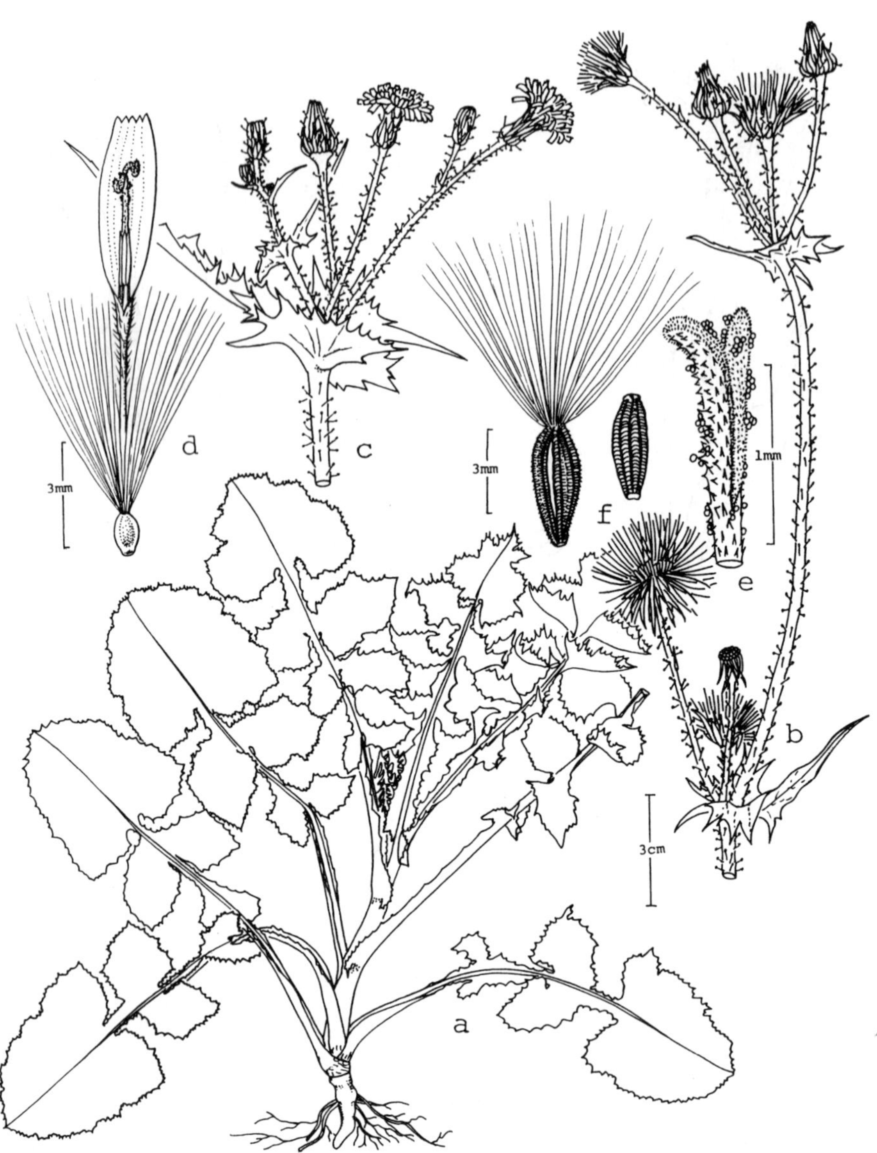

Figure 66. Field Sow-thistle (*Sonchus arvensis* L.). **a.** The habit sketch of a young plant (the edible stage) showing the lanceolate outline of the pinnatipartite leaves with the basal portion of the lamina narrowed and wing-like. **b.** The distal portion of a flowering branch showing the fascicles of capitula after anthesis and their effect on terminating the elongation of that branch, orientation of the secondary flowering branches, amplexicaul cauline leaves or foliaceous bracts and the glandular trichomes on the peduncles. **c.** A younger flowering branch showing the various stages of development, orientation of the secondary or tertiary capitula, and ligulate flowers. **d.** A flower (floret) showing the ligulate corolla on a slender hairy tube, the synantherous stamens and the capillary pappus, inferior ovary, style and stylar branches. **e.** The distal portion of the style of a newly open flower showing the spicular trichome on the stylar branches carrying pollen grains while pushing through the anther tube, and the stigmatic surfaces. **f.** Two truncate achenes showing the transversely wrinkled ribs and capillary pappus. Asteraceae 菊科.

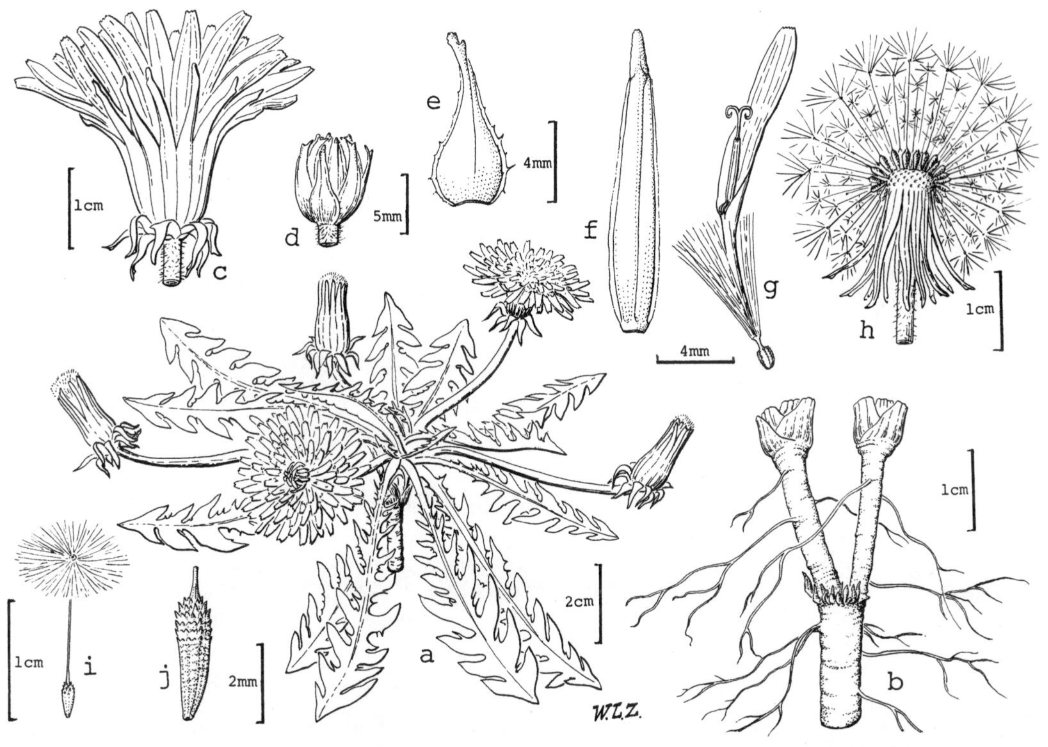

Figure 67. Dandelion (*Taraxacum officinale* Weber ex Wiggers), the first flowering species in lawn and gardens of New England, USA. **a.** The habit sketch of an acaulescent flowering plant with a rosette of leaves and several scapes (much elongated after anthesis) each terminated by a capitulum. **b.** The subterranean portion of this perennial species showing the branch-system. **c.** A capitulum showing the involucre with reflexed outer phyllaries and suberect inner ones. **d.** A flower bud protected by the outer phyllaries in cold weather. **e.** An ovate, herbaceous outer phyllary with membranous margin and slightly crested apex. **f.** A linear-lanceolate inner phyllary with hyaline margin and herbaceous apex. **g.** A flower showing the corolla with distal ligulate portion almost equal to the basal tube, stamens inserted at the corolla throat, anther tube, and ovary suddenly narrowed at the apical one-third, and enlarged into a disk carrying the capillary pappus. **h.** The distal portion of a scape with capitulum bearing mature fruits showing the reflexed phyllaries, rugose discoid receptacle, and many achenes and associated pappus. **i.** An achene with the distal portion specialized into a filiform beak, enlarged into a disk at the tip, and carrying the capillary white pappus spreading like an umbrella. **j.** The basal portion of the achene showing the rugose and spinulose surface. Asteraceae 菊科.

CULTURAL ASPECTS OF CHINESE FOOD PLANTS

THE SOURCES AND NATURE OF INFORMATION

The plants eaten by the Chinese people for survival and for pleasure cover a wide range in the plant kingdom, from a thalloid and a filamentous nostocs, the lowest organisms with prokaryotic cells (Monera), to the highly evolved species of orchids. This work covers two species of monera, 25 species of fungi, 27 algae, 7 lichens, 11 pteridophytes, 16 species and 5 varieties of gymnosperms, 156 species and 31 varieties and/or cultivars of monocots, and 912 species and 238 varieties and/or cultivars of dicots. An explanation is given below of the acquisition of the information for such an involved subject in such an extensive area of a country, where ancient cultural patterns coexist with modern practices.

Sources of Information

Personal experience and published Chinese records are the primary sources for the information reported here. Indeed, I am one of the few fortunate botanists to have had the opportunity to live with and to botanize among various ethnic groups of the Chinese population, as well as to study and to teach in several colleges and universities where I could exchange culinary information with students and faculty members. I am further favored by having been raised on a North China farm where I shared the heritage of surviving in famines and poverty. I have learned various dialects, and am able to read and write the ancient as well as the modern written language, to enjoy the genuine cooking in homes, banquets in restaurants, the beverages and vegetarian food prepared by Buddhist and Taoist monks in famous temples of large cities as well as in high mountain retreats such as Luofu 羅浮 in Guangdong, and Emei 峨嵋 and Qingcheng 青城 in Sichuan. Moreover, I was exposed to life in the rough primitive sheds of hunters in the wilderness of the Sino-Tibetan mountains, and in the yurts of Mongolian nomads. My close contact with the people is widespread in China, and the friendships I have made last for a life-time.

In order to explain the extent of the geographical area and the depth of the influence of friendly people on the source of the information gathered for this work, I would like to present some facts; for as the Chinese say, "Facts speak louder than arguments."

Geographical Aspects

First, to show the areas where information was acquired, a map has been printed on the front endpaper. The localities in which I have lived for eight or more years are indicated by black triangles and dots, and places where I have botanized for shorter periods are shown by white triangles and dots. These localities are indicated by numerals assigned arbitrarily by latitude from north to south into roughly three zones. Within each zone, the numbers read from east to west. Many of these localities are classical collecting sites of former botanical explorers; for example, Muping 牟平 by Armand David, Emei Mountains 峨嵋山 by E. H. Wilson, Luofu Shan 羅浮山 by Charles Ford, Yichang 宜昌 (Ichang) by A. Henry, Guangzhou (Canton) by Peter Osbeck, and Hong Kong by Charles Wright. Other localities are important not only botanically but also politically and culturally. For example, Beijing, Nanjing, Xi-an, Luoyang and Chongqing have all been sites of the central government at different periods in Chinese history, and Yangzhou is the city where Marco Polo was a civil officer. A short note for each locality is provided. After each note, few phrases about my activities in the area and the number of specimens collected are given in parenthesis.

1. Long White Mountain 長白山 — First explored by Henry E. M. James and Francis Younghusband in 1886, before the latter started his famous Central Asian tour (specimens collected).

2. Harbin 哈爾濱 — Capital of Heilongjiang Province (visited herbarium, Northeastern College of Forestry, Arboretum, lectured).

3. Changchun 長春 — Capital of Jilin Province (visited Ginseng Centers, Laboratories, field stations, lectured).

4. Shenyang 瀋陽 — Capital of Liaoning Province (visited several times, Medicinal Plant Gardens, Laboratories of the Institute of Forestry and Soil Sciences, Academia Sinica).

5. An Shan 鞍山 — Type locality of the fossil species of Metasequoia (descended into the mine).

6. Dalian 大連 — Location and commercial importance in China as Boston in USA, one of the early ports opened for foreign trade (visited Laminaria Culture Station).

7. Beijing 北京 — Capital of China (visited many times, worked for one month in the Institute of Botany, Academia Sinica, lectured, visited the Beijing Botanical Garden, Medicinal Plant Gardens at Xibaiwan, specimens 19738–19745).

8. Tianjin 天津 — Important port in North China (visited local gardens).

9. Datong 大同 — Coal capital of China (vegetation study).

10. Yin Shan 陰山 — A mountain range in Inner Mongolia (collected specimens 19565–19737).

11. Hohhot 呼和浩特 (Huhehaote) — Capital of Inner Mongolia (lectured at Inner Mongolia University, specimens 19560a–19565a).

12. Yinchuan 銀川 — Capital of Ningxia, Moslem Autonomous District (specimens 19530–19559).

13. Ala Shan 賀蘭山 — Desert and mountain vegetation studies (visited lycium berry orchard, Desert Reclamation Project, lectured, studied herbarium specimens of the Medicinal Institution Laboratory, specimens 19470–19529a).

14. Jinan 濟南 — Capital of Shandong Province (observation of the vegetation of North China, particularly of Qianfo Shan 千佛山).

15. Lianyun Harbor 連雲港 — A newly developed free port with a well preserved natural area; the vegetation shows the northern limit of some southern genera and species (specimens 19790–19830a).

16. Xuzhou 徐州 — Hometown of the First Emperor of Han Dynasty, a historical town for all important wars in China, for whichever army conquered the area ruled the nation (my home where basic knowledge in Chinese ethnobotany was acquired, and where the techniques were learned for preparing food from wheat, sorghum, soybean and many emergency foods).

17. Zhengzhou 鄭州 — An ancient city on the bank of the Yellow River (vegetation observation).

18. Luoyang 洛陽 — The ancient eastern capital of Zhou Dynasty (1122–255 B.C., vegetation observation).

19. Xi-an 西安 — The western capital of Zhou Dynasty and that of many subsequent dynasties, capital of Shaanxi Province (visited the Botanical Garden, lectured at the Shaanxi University).

20. Lanzhou 蘭州 — Capital of Gansu, important for the ancient silk-route (visited local vegetable gardens and orchards, vegetation studies, lectured at Northwest Teachers' University).

21. Xining 西寧 — Capital of Qinghai Province, seat of the Northwest Plateau Biological Institute, Academia Sinica (visited botanical garden, an arboretum, public park, vegetation studies).

22. Koko Nor 青海 (Qinghai) — A famous salt lake at 3,000 meter elevation, the largest lake of China (vegetation studies, specimens 19378–19467).

23. Shanghai 上海 — A commercial and industrial metropolis of China (visited botanical gardens, medicinal factories, many laboratories of the Academia Sinica, First Shanghai Medical School, lectured).

24. Suzhou 蘇州 — A famous city of unique gardens established in limited areas (visited orchards and gardens of communes, famous ancient gardens, observation of the vegetation of Tai Lake 太湖 and the hills in and around it, including the factory which produces the famous tea known as *Bi-luo-chun* 碧螺春).

25. Yangzhou 揚州 and its vicinity — An ancient city where Marco Polo was a civil officer (having a three-day study tour on the vegetation of the flat areas of Jiangsu seashores with special reference of the transplantation of American and European species of Spartina Schreber (marsh-grass) and land reclamation, specimens 19767–19780).

26. Nanjing 南京 — A former capital of many generations, a city of famous universities and colleges including my alma mater (Ginling College) where I began to study botany with Professors Cora D. Reeves and Albert N. Steward, and dendrology with Professor Ye Peizhong 葉培忠, and where I first learned to eat a meal of rice.

27. Hefei 合肥 — Capital of Anhui Province (vegetation study and lectured at Anhui University).

28. Hangzhou 杭州 — A garden city of China (visited Hangzhou Botanical Garden, famous temple gardens, special garden for bonsai).

29. Lu Shan 廬山 — Famous summer resort in central China with one of the largest and oldest botanical gardens (specimens 14014–14027).

30. Nanchang 南昌 — Capital of Jiangxi Province (visited medicinal plant gardens, lectured).

31. Qizhou 蘄州 — Birth place of Li Shizhen 李時珍, author of *Ben-cao-gang-mu* 本草綱目 (observation of lowland and swampy area vegetation, and the crops of central China).

32. Wuhan 武漢 — Capital of Hubei Province (visited Wuhan Botanical Garden, Wuhan University, lectured, specimens 19757–19766).

33. Yichang 宜昌 — Famous collecting site of A. Henry and W. Y. Chun, the first

port in the interior of China opened to foreign traders in the nineteenth century, the homeland of kiwi fruit (observation of the hillside area vegetation of central China).

34. Wanxian 萬縣 — The gateway into and out of Sichuan, homeland of the living type of *Metasequoia glyptostroboides* Hu et Cheng (first taste of Sichuan food in January 1938).

35. Chengdu 成都 — Capital of Sichuan Province, headquarters of Ernest H. Wilson's botanical expeditions in western China (taught in West China Union University for eight years, studies of the vegetation of Chengdu Plain, particularly of the medicinal and food plants of Chengdu, specimens 1–2700).

36. Xindu 新都 and Pixian 郫縣 — Quiet agricultural communities, with Xindu famous for its Osmanthus Lake, and Pixian for its *dou-ban-jiang* 豆瓣醬 made of *Vicia faba* L. (basic studies of the economic plants of the agricultural communities in western China).

37. Pengxian 彭縣 — The district in Sichuan which the famous White River Copper Mine is located, via the White River 白水河 (botanical investigation of the forest of White River Area, specimens 2147–2252).

38. Guanxian 灌縣 — The beginning of the ascent to the mountains of western Sichuan, famous ancient hydrologic construction, suspension bridge made of bamboo cables (numerous visits, first experienced a wine made of *Actinidia chinensis* Planchon [Kiwi Fruit] in a Taoist temple in March 1938, vegetation observation, economic plant studies among the people in the mountains, particularly in the temples of Qingcheng Shan 青城山 and Zhaogong Shan 趙公山).

39. Maoxian 茂縣 — Botanical collection and economic botany studies of the Qiang 羌 and Rong 戎 tribes.

40. Wenchuan 汶川 — Land of the giant panda (botanized in the area, studied the vegetation of giant panda habitat, specimens 2254–2722).

41. Lifan 理番 — Desert vegetation along the valley of the Min River (a tributary of Yangtze River); secondary growth in side valleys, due to disturbance made by the inhabitants of the Rong Tribe; virgin coniferous forest in the undisturbed valleys; alpine meadows above 3,500 meter altitude; (studied vegetation, specimens 1625–2129).

42. Chongqing 重慶 — Wartime capital of China between 1938 and 1945, stopover (visited herbarium, Central University wartime campus at Sapingpa 沙坪壩).

43. Beipei 北碚 — A quiet town on the shore of Jialing River where the Herbarium of the Biological Laboratories, Science Society of China, was situated, with the

specimens and staff moved from Nanjing, housed in new buildings on a hillside, the best facilities in wartime China (identification of specimens collected in western Sichuan, completion of a 300-page manuscipt on *Ilex* L. of China, studied the vegetation of Jinyun Shan 縉雲山 and the economic botany of the area, drank a sweet tea made of young leaves of a local species of *Quercus* L.).

44. Ziliujing 自流井 (Natural Flow Well) — Salt wells and natural gas production area (observation of the vegetation and the uses of native plants).

45. Luo Shan 樂山 — A communication center in southwestern Sichuan (study of local uses of native plants).

46. Emei Mountains 峨嵋山 — A famous mountain range with numerous Buddhist temples (stayed in a summer resort near Xinkai Temple 新開廟, visited many temples, climbed the summit about 4,000 meter altitude, learned the economic botany of the Buddhists and the farmers who supplied the necessities to the monks and the pilgrims).

47. Dragon Pool 龍溪 — An earthquake lake 30 miles south of Emei Mountains (studied the agriculture and the vegetation of the area).

48. Ya-an 雅安 — The beginning of our expedition, the end of the bus road in the 1940s, further travel all on foot (botanized Zhangjia Shan 張家山, experienced eating a large salamander caught in the river and a scaly ant-eater from the hillside).

49. Tianchuan 天全 — Quiet town in the mountains, on way to Muping (specimens 700–800).

50. Muping 穆坪 (= Baoxing 寶興) — Historical collecting site of Armand David where he discovered the giant panda (my first experience of collecting in alpine meadows and in studying the life of musk-deer hunters and the fritillary bulb collectors, and in seeing the cultivation and harvest of opium, lived in homes of the Rong Tribes and made ethnobotanical observations, specimens 800–1600).

51. Taipei 台北 — Capital of Taiwan, good forests with tree ferns and *Trochodendron aralioides* Sieb. et Zucc. (observation of vegetation).

52. Yilan 宜蘭 — Agriculture College (observation of vegetation across the island).

53. Sun Moon Lake 日月潭 — Good facilities for tourists (explored tropical forest with local botanists, impressive experience).

54. Shantou 汕頭 (Swatow) — A seaport, homeland of many Chinese in Southeastern Asia, particularly those settled in Vietnam and Singapore (special study on the cultivation and uses of taro and *ku-ding-cha* 苦丁茶, a holly tea).

55. Luofu Shan 羅浮山 — Famous collecting site of Charles Ford, many temples of

Chinese Taoists (conducted many field trips, lived in temples and studied local uses of plants).

56. Conghua 從化 — A resort center, hot springs, man-made lakes (vegetation observation, particularly on reforestation).

57. Hong Kong 香港 — An international free port with extreme differences in the vegetation of government protected islands and large areas in the New Territories which are burned repeatedly (over ten years of field work towards a flora of the area, specimens 5000–23000).

58. Guangzhou 廣州 (Canton) — Capital of Guangdong Province, with famous universities, including South China Agricultural University and Sun Yat-sen University (now on the campus of the former Lingnan University), and South China Institute of Botany, Academia Sinica (basic botanical studies and research on tropical economic botany, made many field trips).

59. Wuzhou 梧州 — Entrance to Guangxi Province (vegetation and economic botany observations).

60. Guilin 桂林 — Well known for the scenery, very good botanical garden with interesting experiments on tropical economic plants such as *Luo-han-guo* 羅漢果 (Buddha's Disciple Fruit), and *Actinidia chinensis* Planchon (studied the vegetation and economic plants with botanists of Guangxi Botanical Garden).

61. Liuzhou 柳州 — Famous for the production of coffins (vegetation and economic plants studied, visited coffin shops).

62. Nanning 南寧 — Capital of Guangxi Province, place for the First Six Chinese Scientific Societies Joint Meeting in 1936 (attended scientific meetings, observed local vegetation).

63. Kunming 昆明 — Capital of Yunnan Province, former headquarters of the Botanical Collecting Team organized by the Fan Memorial Institute of Biology, in cooperation with and financially supported by the Arnold Arboretum; now seat of Yunnan Botanical Institute, Academia Sinica, and Yunnan University (visited many former collectors who are now professors of various institutions, general observation of the vegetation of Kunming area, particularly of the Stone Forest, the severely eroded hillsides).

64. Simao 思茅 (Szemao) — Headquarters of A. Henry who collected in southern Yunnan in the first decade of 1900 (studied the economic plants of the area, especially the food plants in the market and cooked dishes, one of which I experienced for the first time was a soup made of *Cynanchum forrestii* Schlechtendal).

65. Xishuangbanna 西雙版納 — A recently established Tropical Botanical Garden,

under the direct auspices of the Academia Sinica, with a protected natural area showing the vegetation and composition of tropical flora of southern Yunnan (visited the tropical forest, the village of Tai Tribe 傣族, made general observation of the vegetation of the deforested areas newly opened up by the government for rubber plantations).

Many of the names of the 65 localities share the prefixes *dong* 東 (east), *xi* 西 (west), *nan* 南 (south), and *bei* 北 (north), which mean the four directions, or the suffixes *jing* 京 (national capital), *zhou* 州 (regional administrative site), *xian* 縣 (county seat), *shan* 山 (hill or mountain), *hai* 海 (ocean), *hu* 湖 (lake), *he* 河 (stream), and *jiang* 江 (river). *Jing*, *zhou*, and *xian* indicate the size and political importance of places. In Chinese history, four cities had the *jing* title. Beijing (north capital) and Nanjing (south capital) are still in use, while Xijing (west capital) is now called Xi-an and Dongjing (east capital) is now Luoyang. Cities with *zhou* status are now all modern metropolises: Guangzhou, Hangzhou, Lanzhou, Xuzhou and Yangzhou are some examples. *Shan, hai, hu, he* and *jiang* show topographical positions. Xishuangbanna is a name used by the Tai Tribe living in southern Yunnan, meaning 12 communities.

Social Aspects

The map on the front endpaper shows the spatial extent of the information gathered for this work. China is a developing country, and its means of communication are poor today, but were even worse four or five decades ago. Consequently, every area has developed its special food and eating habits. The knowledge of the food of any area depends upon the intimacy one may have with its people. My language facility, rural background, and my interest in and knowledge of the plants growing around the people, have always helped to draw them close to me. Many of them have subsequently become my friends and informants. In all the places I visited, I have made many intimate friends. Here are four examples of how my friends remember me with specific, simple Chinese food products.

Twenty years ago, a friend from China visited me in Brookline, Massachusetts and brought me one half of a black salt-pickled *Da-tou-cai* 大頭菜 from Shandong Province (Map, Loc. 14). A portion of this present is still in a closed jar in my pantry. I treasure it more than a ginseng root and occasionally put a pea-sized piece in my mouth. Its aroma always reminds me of life in a North China village.

In July 1984, Dr. Paul P. H. But went to Chengdu (Map, Loc. 35) on business. There he met my former colleague, Professor C. K. Xie, then Head of the Department of Pharmacology, Sichuan Medical University, who asked Dr. But to bring me 2 ounces of sesame oil in a glass bottle and 4 ounces of Chengdu *Hu-dou-ban* 胡豆瓣 (fermented broad bean sauce). From Hong Kong I carried back these presents to the US in my handbag, and they are kept for special occasions.

In October 1985, a scientist from the Zoological Institute, Academia Sinica, Beijing, visited me in my office at Cambridge, Massachusetts. Before leaving China, she met a grade school friend of mine who asked her to bring a handful of homemade fermented soybean from Xuzhou (Map, Loc. 16).

On January 10, 1986, as I worked on this manuscript, the receptionist at the Harvard Univerisity Herbaria paged me, asking me to go down to the front desk. There stood a 77-year-old friend who had just returned from China via Hong Kong. He handed me a plastic bag with one fresh bamboo shoot and a dozen withered garden pea shoots which he brought back to Boston in a handbag. I was deeply moved because the weather was very cold and travelling to Cambridge from Watertown, he had to change buses. I took the gifts to the Biological Laboratories, to be photographed as illustrations to be used in this work.

Nature of Information

Approximately 19 percent of the 1,156 species and 274 varieties and/or cultivars of the Chinese food plants recorded in this work were introduced from abroad. The introduced items, such as wheat, corn, and sweet potato, are common foodstuffs in the markets. They are not included in further discussions.

Ten percent of the recorded native Chinese species are cultivated. Products of some of these species, such as the Chinese yam (*Dioscorea opposita* Thunb.), longan, and lychee, are available in the Chinese stores of large American cities. However, the uses of the majority of these species are unknown outside of China. Good examples are the young shoots of *Murdannia nudiflora* (L.) Brenan and the half-opened fragrant flowers and mature buds of *Telosma cordata* (Burm. f.) Merr.. Some of the wild species are listed here as edible plants for the first time in botanical literature. The outstanding examples are the fleshy winter tubers of *Scilla japonica* Baker, the white rhizomes with tender shoots of *Convolvulus chinensis* Ker-Gawler (*C. arvensis* L., sensu lato), the roasted shoots of *Gaultheria veitchiana* Craib, the juicy-sweet, ripe fruit of *Ficus esquiroliana* Léveillé, and the snow white delicious small fruit of *Artocarpus hypargyreus* Hance ex Bentham. I record them as food plants of China because I have tasted mardannia and telosma as delicacies of the season in friends' houses, eaten scilla and convolvulus as famine food in a North China village, drunk Veitch's gaultheria tea prepared by lumberjacks in the homeland of the giant panda, and have been refreshed by the wild artocarpus and

Esquirol's figs in the humid, subtropical forests during the vegetal investigations in Hong Kong. The use of most of the native species is restricted in area. For example, the root tubers of *Potentilla anserina* L. and the drupes of *Elaeagnus angustifolia* L. are only found in local market places in the arid west of Lanzhou and Xining (Map, Loc. 20–21), whereas *Begonia fimbristipula* Hance is collected only in the monasteries of Dinghu Shan 鼎湖山 in Guangdong Province and in Lantau Island 大嶼山 in Hong Kong. Approximately 50 percent of the lesser known native species were recorded in the literature published after the 1950s, when university professors and high school teachers were mobilized and persuaded to investigate and record the uses of edible wild plants from farmers.

In regard to personal experiences, the first quarter of my life was spent on a communal homestead shared by descendants of my paternal grandparents. The village was situated on the southern bank of the ancient course of the Yellow River, with the Grand Canal running along the meridian to the east (Map, Loc. 16). There were no hills in sight, for the farmland was originally reclaimed from a swamp drained by several canals, all of which were in poor repair. The major crops were winter wheat, sorghum, soybean and mung bean. The minor crops were sesame, cotton, sweet potatoes, peanuts and carrots. Vegetables were from the kitchen gardens. In the summer, floods followed storms and destroyed all the crops; famines were a natural consequence. Prior to leaving home for college in Nanjing, I had never eaten a bowl of rice. First hand knowledge of many famine foods as well as the practices in gardening and farming was acquired during this period.

A tuition scholarship enabled me to enter a boarding school in Xuzhou (Map, Loc. 16), Mary Stevens Girls' High School, run by Presbyterian Missionaries. In like manner I went to Ginling College, a small liberal college of arts and sciences in Nanjing (Map, Loc. 26), and a sister institution of Smith College in Northampton, Massachusetts. My interest was agriculture, but since this subject was not available in a liberal arts college, I took biology for a major and sociology as a minor, hoping these subject would prepare me to serve the people in rural China. After college, an opportunity for a work-study program toward a Master's Degree in botany at Lingnan University, Guangzhou (Map, Loc. 58) was offered to me. As part of the degree requirements, I completed a thesis entitled "The Chinese Esculent Plants Used for the Conservation of Health". My teachers included Dr. F. A. McClure, who was studying bamboos in South China, drug collectors, Taoist monks and nuns residing at Luofu Shan (Map, Loc. 55), and shopkeepers and homemakers in Guangzhou.

After receiving a master degree of science in June 1937, the Sino-Japanese War broke out at Lugou Qiao 蘆溝橋 (Marco Polo Bridge), south of Beijing (Map, Loc. 7).

Very soon the maritime provinces of China were under Japanese occupation. As a refugee, I reached Chengdu (Map, Loc. 35), Sichuan in January 1938 and accepted a teaching position in the Department of Biology, West China Union University, where I stayed until 1946. During summer vacations, I took students to botanize the localities in Muping (Map, Loc. 49–50) where Armand David resided and many areas explored by E. H. Wilson of the Arnold Arboretum (Map, Loc. 38–41) . In July and August 1938, I lived on Emei Mountains with my former teacher, Dr. Cora D. Reeves, Chairman, Department of Biology, Ginling College. Taking a house at the Xinkai Temple Summer Resort for our headquarters, we climbed the Golden Summit, walked to the little known area of Hen-on-the-Nest Water Falls 雞婆蛋瀑布, and down to the Dragon Pool 龍溪, an earthquake lake (Map, Loc. 45–47). During the summers between 1939 and 1942, I stayed in various villages of the two Sino-Tibetan ethnic groups, the Qiang 羌 and the Jiarong 嘉戎, the latter lived in the homeland of the giant panda. My hosts and local assistants were my informants (Figure 1a–b). In 1942, I wrote an article on "The Ethnobotany of the Gia (Jia)-rong Tribes", described their life in relation to their botanical environment, especially the edible and the medicinal plants. The field numbers for my collections were 500–4000.

In March 1946, I received a telegram from the Graduate School of Radcliffe College, informing me about the grant of a graduate fellowship for a doctorate program under Professor E. D. Merrill of Harvard University. On August 2, 1946, I left China for graduate work in Radcliffe College. The change of life styles from China to America was to me like a complete metamorphosis is to insects. Everything was new.

The Marco Polo Bridge Incident of July 7, 1937 forced China to declare war on Japan. This Sino-Japanese War eventually merged with World War II. It did not end until after the bombing of Hiroshima. China was war-torn, whereas all the maritime provinces and the adjacent regions in northern, central and southern China were under Japanese occupation. Inflation fluctuated violently with the exchange value of the Chinese *yuan* dropping from three to two thousand for one US dollar, and even the necessities for daily life were extremely limited. Arriving in the USA, everything was new. All necessities were available, even sugar could be used freely at the table! Not having had any ice cream for eight years, a ten-cent cone was indeed a treat! It only took sixteen cents to get a loaf of Pepperige Farm bread. For a whole year I did no cooking. One day, walking with Professor E. D. Merrill on the path in front of the Administrative Building of the Arnold Arboretum we saw the common mallow (cheeses, *Malva neglecta* Wallroth) on the edge of the lawn. He asked me if weeds were used as food in China. I gathered some young shoots and leaves and prepared a "malva soup" for lunch (Hu, 1948). I was the last student of Professor E. D. Merrill and was in the last class of Professor N. L. Fernald, author of the eighth edition of *Gray's Manual of Botany* (Fernald, 1950). From these professors, I learned to eat American wild plants. Merrill used to gather dandelions before the plants flowered and leafless pokeweed shoots and ate them as people do with asparagus. Fernald took great pride in telling the class his story about "Professor Fernald's spinach". The story goes like this: One day in his summer home in Maine, his milk man spoke to him saying, "Professor Fernald, people say that you are eating spinach. I do not see any spinach around. May I see your spinach?" Fernald took the man to the back yard and pointed out to him the weed *Chenopodium album* L.. In a low voice the discouraged man said, "So, Professor Fernald's spinach is PIGWEED." The examples of these leading American botanists and kind professors have encouraged me to continue to eat the above mentioned weeds as well as the tender shoots of naturalized species such as amaranth in gardens and fields, and the matrimony vine found by the fences and walls of many school yards.

After completing the requirements and finishing the dissertation for a doctorate degree in April 1949, I became the herbarium assistant in the Arnold Arboretum, filing the Chinese specimens stored for many years during World War II in hundreds of the "Merrill boxes". This work gave me an opportunity to see samples of Chinese plants from most of the phytogeographical regions of China, including the types of species published by A. Rehder, E. H. Wilson, E. D. Merrill, C. Schneider, etc. At the time, the Chinese specimens in the genus *Philadelphus* L. were unnamed. Trying to identify this material with the aid of existing publications and herbarium specimens, I found that a deeper understanding of the characteristics for distinguishing species must be acquired first. This type of work plunged me into deep despair. In order not to be overwhelmed by the project, I made an all-out-effort, spending my research time, weekends and holidays tackling the problems. I used the herbarium and library facilities and the living collections in the Arnold Arboretum at Jamaica Plain and Weston, and examined specimens borrowed from New York Botanical Garden, the US National Herbarium and other outstanding herbaria. Subsequently, *A Monograph of the Genus Philadelphus* was published (Hu, 1954–56).

In the early 1950s, there was a predominant question in the minds of the majority of Chinese students and scholars residing in the USA. The question was: "What can we do here that would be of permanent value in the reconstruction of future China?" A group of young engineerers and businessmen formed the Continental Development Foundation, a privately endowed, non-profit, non-political, educational organization. In early 1952, this Foundation, in cooperation with the China Institiute in America, sent out a circular to all Chinese students and scholars, asking them to participate in a Project Suggestion Contest with the prospect that the project of the winning essay would be supported by the Foundation. In October of the same year, I participated in the contest and submitted an essay on "The Problem of the Preparation of A Flora of China". It won the first prize and by December, the Foundation allocated funds to support three botanists with secretarial assistance to work on a Flora of China Project in the Arnold Arboretum. In July 1953, with the authorization of the Administrative Committee of Harvard University, the Flora of China Project was initiated and I became a Research Associate of the Arboretum working full time for the Project. From this time onward, until I was asked to prepare a list of Chinese food plant for the Pacific Science Association, I concentrated on achieving the goals of the Flora of China Project.

According to the original proposal, the goals of the Project were to be completed in three stages: (1) the compilation of an index to the Chinese flora, (2) the publication of an enumeration of the flora of China, following the model of E. D. Merrill's *Enumeration of Philippine Plants*, and (3) the preparation of the flora by families with descriptions, keys, citations of important literature and specimens examined. The first two years were devoted to the indexing of the species of Chinese flowering plants published in English, French, German, Japanese, Italian and Russian books and periodicals. By the end of 1955, we had an index card catalogue with approximately 89,000 cards covering 248 families, 3,230 genera and 16,150 species of the flowering plants of China (now available as Hu card on the internet). Photocopying machines had not been invented and all cards were hand-written. My job was to check the species from the literature, and mark the family numbers for the typists, and then check each entry for spelling and accuracy. This work gave me a sound knowledge of the literature on the Chinese flora. In addition, models for the work to be done in the second and third stages of the Project were prepared. The first model consisted of two manuscripts on the enumeration of Chinese plants, the *Compositae of China* (Hu, 1965–69) and the other on a monocot, the *Orchidaceae of Chinese* (Hu, 1971–75). The second model was the publication of *Flora of China, Family 153, MALVACEAE* (Hu, 1955). It will suffice to say that the 1950s were uneasy years for me in the Arnold Arboretum and by early 1957 the director of the Arboretum wrote to the supporting Foundation declaring a moratorium of the Flora of China Project and to say that my service was needed in the Arnold Arboretum.

For the next ten years my work as a Chinese botanist in the Arnold Arboretum was to put the unidentified specimens of the herbarium in order so that they could be found and used in botanical research, to answer problems concerning Chinese plants sent from a worldwide botanical community, and to serve American horticulture concerning plants of Chinese origin. Much of this work called for a high degree of concentration. For example, in answering the questions on paulownia sent from a forester of the University of Maryland, I had to spend months in fundamental research of all species of the genus *Paulownia* Siebold et Zuccarini. In resolving his problems, I had gathered enough martterial for a *Monograph of the Genus Paulownia* (Hu, 1959). Many questions from individuals and/or business concerns about uses of Chinese plants took me back to food plants and medicinal plants of China. Between 1957 and 1970, the American Horticultural Society published three handbooks, two on Hollies: *Handbook of Hollies* (Hu, 1957, 1970), and one on the Daylily: *Daylily Handbook* (Hu, 1968). I was the only member of the Arboretum staff who contributed to these horticultural classics. Actually, it was in "Uses of Daylily as Food and Medicine" in the *Daylily Handbook* (Hu, 1968) that I first wrote six recipes on the use of Chinese plant material for food.

In March 1968, I returned to Hong Kong to begin field work on the Flora of Hong Kong Project and to teach in Chung Chi College of the Chinese University of Hong Kong (CUHK). During the next eight years, I made six round trips between Boston and Hong Kong, each time staying a minimum of three months and a maximum of one year. My field numbers for collections made in this area ran between 5,000 and 14,009.* This project was designed and operated as a university work-study program. I taught various courses in botany in Chung Chi College, including a very popular course on the local flora, in which every two students undertook a research project on the uses of Hong Kong plants. In such activities, I learned about the food plants, *bupin* and *liangcha* with many students. From the CUHK, I received all my expenses including international travel, an apartment, all the equipment to preserve the specimens collected, and some assistance for field work. My classes were scheduled on three days, thus I had long weekends for climbing the high mountains and camping out on various islands. A great advantage of botanizing an area connected with a well-known and much respected educational institution such as the CUHK, is that collection of specimens in many otherwise restricted areas was permitted. Lady M. Noel MacLehose, wife of the then Governor, not only helped to arrange for police boats to take the collecting party to islands not open to the public, but went with us to gather specimens along seashores and on mountain-tops under military control.

The quality of my research and the quantity of my performance was greatly enhanced by the friendship of two Hong Kong residents, Mrs. Gloria Barretto, an orchidophile of the native orchids, and Mrs. Beryl M. Walden, an accomplished artist specializing in mural and wild flowers painting. Constantly during these years, Gloria would bring a cup of tea, some refreshment and a flowering specimen from her orchid collection to the University Herbarium at 4:30 p.m.. With a ruler, two dissecting needles,

---

* By the time of final proof reading of this book, the field numbers are 5,000–23,988. January 10, 2004.

a binocular, and her plant, I dictated my observations and she took down the notes. Such work ususally took us two or more hours each time. Then, the specimen was handed over to a trained Botanical Artist, Teresa Fong Wang, for a habit sketch of the species and detailed analytical drawings of the flower parts. The data on the species of Hong Kong orchids, illustrated by line drawings and colored photographs was published as *The Genera of Orchidaceae in Hong Kong* (Hu, 1977).

Members of the Walden family are real naturalists who have made a beautiful collection of local butterflies. They are expert hikers and mountain climbers who have left more footprints in Hong Kong countryside, hilltops, deep gorges, forest floors, sandy beaches, steep seashores and isolated islands than any persons I have known. In their expeditions, wherever Beryl saw a wild flower of interest, she brought a piece with flowering buds back, raised it till the flower fully opened, and painted it in water color. No floristic botanist can match my blessings in having the living forms and colors of 555 species of the plants in an area like Hong Kong side by side with the herbarium specimens (Walden et Hu, 1977, 1988).

Regarding food plants of the humid tropical areas of China, I was most fortunate to have T. K. Woo, Technician, Department of Biology, CUHK, as my able field assistant and S. S. Chow as my friend and cook while I lived in Hong Kong. Woo had the sharpest eyes for locating unusual specimens and he never failed to climb the steepest cliffs and the tallest trees to get the materials he discovered. When he was in grade school, his father's properties in Guangdong were confiscated and distributed. He was deprived of the rights of a citizen to continue school, was abused by adults and ridiculed by children, and lived in poverty and hunger until he was old enough to escape by swimming to Hong Kong. During his youth, he had learned all the techniques for survival in the humid subtropics, techniques passed down from the ancient neolithic people who resided in Guangdong. In his escape, he walked with older companions at night and hid in woods during daytime, picked up roots, leaves, young shoots and berries to assuage hunger. It was from him that I learned to eat the fruits of *Artocarpus hypargyreus* Hance ex Bentham, and *Rhodomyrtus tomentosa* (Aiton) Hasskarl, and to understand the real meaning of the vernacular name of the latter species and to choose the correct Chinese ideograms " 逃軍糧 " for it.

Chow joined the Seven Sisters Club of South China rural community in her early teens. According to an ancient custom, members of such a club do not marry and they should be treated by the family and the community as boys. They were supposed to do heavy work and to support the family as men do. In the 1920–30s, many of them went to Guangzhou and Hong Kong to be domestic servants in these developing areas. They were women who could provide some cash to support their parents and younger sisters and brothers in the villages. Chow went to Guangzhou when she was eighteen, and she had worked for both Chinese and foreign families before World War II. When Guangzhou was occupied by the Japanese, she went with her British host family to Hong Kong. She was a good housekeeper and was very skilful in cooking both Chinese and Western dishes. On the retirement of Dr. A. T. Roy (Chinese name 芮陶菴), Vice President, Chung Chi College, she came to work for me, and was very eager to accompany me for short-distance field trips. As we walked in the countryside, she would tell me which species were edible, and she could make the most delicious dishes from very ordinary plants. For example, the bitter melon (*Momordica charantia* L.) was never used in North China as a vegetable, but from her I learned to eat and like it and many other plant foods of South China, particularly those for vegetarian dishes.

The extensive area as depicted by the localities visited on the map, the lengthy factual review of personal experiences, and over three decades of preparation for a manuscript all point to the fact that this work has been compiled with unusual thoroughness. The survey is comprehensive but not exhaustive. China is large, its topography varied, and its people ethnically complex. Future scientists are sure to discover many additions, particularly from little known and remote ethnic groups.

THE PRODUCTION AND PREPARATION OF CHINESE PLANT FOOD

Approximately 80 percent of the species of edible plants in China are gathered from the wild. Yet, no less than 70 percent of the life span of its population is spent in the production and preparation of plant food for the consumption of the people within the country. The methods and procedures vary by region, depending on the climate, ethnic background, social status, and cultural development of the consumers. However, certain general patterns prevail. These are: (a) the predominance of kitchen gardens; (b) dry-land agriculture vs paddy field cultivation; (c) utilization of suburban land for vegetable gardens; (d) the preservation of lesser crop products by drying, burying and pickling; and (e) the use of microbes (fermented plant food). Each of these topics involves extensive research and could be a good thesis topic. Simple explanations with some examples are provided here.

Kitchen Gardens

Kitchen gardens were the forerunners of agriculture in all major civilizations. Indeed, the practice has been going on in China for at least 5,000 years. Up until the 1920s, the Chinese farmers had been self-sufficient. Each farm produced everything the family needed for food and clothing. My village is 140 km (90 miles) south of Confucius' home. As for the life of the farming community, 2,500 years and 100 miles have not made much difference, although some alteration may have been made when people lived in communes. With the current reforms, whereby land is assigned to each farming family, the kitchen garden system has returned. For this practice, a small area near the house is dug and fenced-in to prevent disturbance by children and animals. The area is managed by the cook (the mother or grandmother) of the household, for she would know what is needed in the kitchen. Most of the plantings are annuals. The following description of my grandmother's garden is similar to kitchen gardens which prevailed throughout North China.

North China Kitchen Garden

This garden was established next to the threshing ground, fenced-in with tree branches and stems of sorghum, with necessary repairs made in the spring. It was fertilized with the chicken manure mixed with plant ashes, reserved specially for kitchen gardens. The Chinese leek was the only perennial in this garden. The other species were purse amaranth, field amaranth, chili pepper, Welsh onion, garlic, coriander, fennel, egg plant, cucumber, *hu-lu*, sponge gourd, summer squash, and asparagus lettuce. She had a tall shrub of zanthoxylum which seldom bore fruits because there were no male plants in the village. She shared its spiny leaves with neighbors who used fresh leaves in cooking fish, and for making their fermented soybean sauces.

Kitchen Gardens in South China

Kitchen gardens vary in complexity. The boat people of Guangzhou and the apartment dwellers in Hong Kong keep a few pots of fresh spices (such as common rue and sweet basil) and handy remedies (such as Barbados aloe) for burns and grape ginger for acute cystitis and whooping cough in children.

The mother of a colleague in Chung Chi College lived on the first floor of the same apartment building as mine. She dug a space two meters by one meter on the side of the lawn, and kept a few pots along the cement sidewalk. In such limited space, she supplied the family with a remarkable assortment of fresh spices and unusual vegetables of the seasons: Chinese leek, Chinese kale, vegetable heart, common rue, herb-of-the-goddess, coriander, common mint, green pepper, and white-flowered mugwort.

The South China Institute of Botany, Academia Sinica, in Guangzhou, is inconveniently located on the outskirts of the city. All the staff members living in apartments provided by the Institute keep small kitchen gardens in the very limited spaces at the back of the buildings. In addition to the species mentioned above, many of them plant different species of *Luffa* Miller, *Lagenaria* Seringe, and passion-flower for shade, and the edible young fruits for making a refreshing drink.

Kitchen Gardens of Chinese Americans

Obviously, tending a kitchen garden is a way of life for an average Chinese homemaker. When she moves from place to place, she carries the seeds of her favorite plants and raises them at the new residence. Examples of a communal kitchen garden of residents in a retirement home in the Chinatown of Boston, and one of an individual home in Brookline, Massachusetts (MA) are described below. The gardeners are all immigrants from Guangdong and unable to speak English.

Kitchen gardens in China, whether in the past or at present, are important ways and means for low income creative homemakers to contribute to the family economy and to supply healthy food for its members.

(1) *A communal garden*: A small communal garden has been developed on the site of a former building on the southern side of Washington Street, where the elevated subway trains pass a corner of Boston's Chinatown. In this enclosed area among tall buildings (Figure 2a), each gardener has a plot about 3.5 meters long and one meter wide. Various potherbs (Figures 7 and 8), yard-long beans, cucumber, sponge gourd, festival gourd (Figure 10), and rare Chinese home remedies (Figure 7b) are cultivated in it. In a Chinese kitchen garden, a gardener usually selects a couple of good, strong plants and leaves them to flower and fruit for seeds (Figure 7a).

(2) *An apartment garden*: Near St. Paul Street in Brookline (MA) lives a new immigrant with her daughter, taking care of her grandchildren when her daughter leaves for work. The apartment buildings in the neighborhood are very crowded, and the land in the fenced front yard is about 2-meters square, with a tall privet hedge (*Ligustrum ovalifolium* Sieb. et Zucc.) along the sidewalk. The grandmother digs up the arable land, plants a quick growing cultivar of field mustard, and along the wire fence, she puts *jie-gua* 節瓜 (Figure 10), letting the vines grow over the fence and guiding them in such a way that the gourds are concealed from children and pedestrians on the street. The pleasure of working with the soil, the happiness of sharing a handful of the Chinese mustard flowering stalks and/or a *jie-gua* with her friends means much more to her than just having a supply of vegetables.

Different Farming Systems

It does not take long for visitors in China to observe the striking differences between the paddy fields of the south and the dry-land agriculture of the north. If we draw a line over a map of China along the 30th north parallel and another one along the 107th meridian, these lines would meet in Shaanxi, southwest of Xi-an (Map, Loc. 19). The vast area on the upper right side marked by these lines supports dry-land agriculture. Until the 1920s, the farmers here did not have a rice diet. Many of them never travelled five miles from home. They did not know what rice looked like. Their food consisted of products raised on the farm: wheat, millet, sorghum, sweet potato, carrot, some soybean and mung bean. Approximately 90 percent of the urban population of the same area lived on wheat, millet, and/or sorghum. Animal protein was reduced to a minimum. In my village, located in this region (Map, Loc. 16), the per capita annual consumption of eggs was one or two and that of meat was less than five ounces. Fish was rare and very seasonal. Milk was never available, not even for children. Men died young, usually in their thirties, as did my grandfathers and father. Women lost their teeth after the birth of the first one or two children. Infant mortality in my village was 50–90 percent. In winter, no fresh food or vegetables were available. People craved for some green in early spring and they went out to search for wild herbs such as *Allium macrostemon* Bunge, Shepherd's Purse, *Lepidium sativum* L., and *Orychophragmus violaceus* (L.) O.E. Schulz. Even the young, leafy shoots of the field bindweed (*Convolvulus chinensis* Ker-Gawler), a laxative when taken in large enough dosage, were gathered and mixed with cracked wheat for porridge.

The region on the upper left side marked by these lines is barren desert, except for the irrigated areas along the rivers. Population density is very low, and the agricultural system, under irrigation, is similar to that of North China, except the very limited area irrigated by the Yellow River near Yinchuan (Map, Loc. 12).

The river plains south of the 34th parallel support the cultivation of rice. The climate is mild and the irrigation system well developed. People of this region have more plentiful food. However, those living in the hilly environment depend on rain for farming. The soil is relatively poor and the growing season comparatively short. The minority ethnic people rely heavily on wild plants.

Utilization of Suburban Land for Vegetable Gardens

In America, the suburbs are residential areas for high income people who belong to the affluent community. In China, immediately outside the city limits begins the vegetable gardening community. The transportation of human waste out of the town for fertilizer, and of the products into the town to market are the primary contributing factors leading to this phenomenon. All Chinese suburban vegetable gardening depends upon irrigation, more so in South China where temperature changes are not a limiting factor for plant growth, and where there is an alternation of dry and wet seasons.

In South China, suburban vegetable gardening is a very intensive type of agriculture which involves continuous planting, weeding, watering, fertilizing, and harvesting. Interplanting is a general practice. Often one vegetable is planted before the previous one is harvested, especially when the crop is cultivated for tender flower-shoots such as broccoli, which can be harvested several times a year.

In North China, the growing period corresponds to the rainy season. Suburban gardeners depend primarily on the rain. They build ponds to conserve some surface water and in some areas also dig wells to supplement the water supply. The gardeners in the suburbs of Lanzhou have devised the most ingenious practice for supplying soil moisture. They cover the vegetable beds with egg-sized pebbles. When I was shown such vegetable beds I asked the agricultural scientists for an explanation. Their answer was: "We can only tell you that the people have found it good. Perhaps the pebbles protect the soil water from evaporation in summer, and help to raise the soil temperature at the beginning of the growing season. Lanzhou is not only dry, its growing season is short due to its high altitude. The pebbles can raise the soil temperature in early summer." True as these explanations may be, another point was mentioned by Moses to the Isrealites in the desert, that rocks can "… distill … the dew … sweet dew over the vegetables." (Deut. 32: 2–3) That the suburban gardeners of Lanzhou employ pebbles to condense the moisture in the atmosphere into dew for increasing the soil moisture for their crops is a rare practice, and an advanced agricultural technique.

The Gardeners of the Chinese Suburban Community

Gardeners engaged in Chinese suburban vegetable production are middle-aged men and women who can withstand long-hours of hard labor, and who can carry heavy loads of liquid fertilizer (Figure 3d). In South China where vegetable beds are raised between irrigation ditches, gardeners always work in the mud and water, except when turning the soil with a heavy, long-handled mattock. In most gardens, during the harvest season, a bed is saved for seeds (Figure 3c).

Fertilizers

Until the 1950s, all Chinese suburban vegetable gardens used organic fertilizers, chiefly human waste acquired from nearby towns. Even now (1970–80), such wastes are collected at night and transported to the suburbs in large wooden containers on wheels, pulled by domestic animals or by men. In North China, this manure is mixed with leaves, garden wastes, barnyard manure, and garbage, then piled up for fermentation, turned over once or twice, and finally passed through a net to screen off rocks and rough material. The fine soft damp material is then applied to individual plants alone or mixed with soybean cake.

In South China, people use liquid manure. Human, animal and plant wastes are all dumped into a simple roofed cement pit where fermentation takes place (Figure 3b). In smaller gardens, open pits or large deep urns partially buried in the earth are used as liquid fertilizer tanks. Such liquid organic fertilizer containers are conveniently located at the corner of each garden, not far from the village or a house. They serve many purposes: a receptacle for wastes gathered from town, an open-air toilet for the workers, a place for the decay and fermentation of all the wastes accumulated in the garden, and a container of liquid manure. For the application of the fertilizer, a strong gardener carries the mixture directly from the pit to the beds in two wooden buckets on a pole, walks into the ditch, and skillfully spreads the liquid onto the crops of the two beds simultaneously.

Harvesting and Marketing

The harvest of Chinese suburban vegetable crops is all done by hand. Formerly, when the gardens were managed by individual families, the products were marketed by itinerant vendors and/or by a member of the owner's family. During the harvest season, vendors came to a garden, bargained with the producer, bought what fresh vegetables he could carry, and sold them from door to door to individual consumers, who recognized the shouting voices of the dealers. Before the commune system was instituted in the suburban vegetable gardening community, most gardening families had small pieces of land where many of them did the planting and tending, and from which their products were carried to the market-place to sell to consumers or dealers. In those days, vegetables were harvested in late afternoon so they could reach the consumers the next morning. They were very fresh.

After the suburban areas were communalized, the products in most suburban gardens were loaded onto carts, pulled to town by the less skilled laborers of the commune, usually women and teenagers, then on to the distribution centers in town, as I observed in Nanjing (Map, Loc. 26) in 1975, and in Nanchang (Map, Loc. 30) in 1979. In the 1970s, the varieties of vegetables were few, and were limited to egg-plants, squashes, yard-long beans and green onions. As I observed in the distribution centers from Guangzhou to Beijing, the products were exposed to sun and the heat of July and August, waiting for consumers to buy on their way home from work. The summer heat and the weight of the material spoiled the vegetables and reduced the amount of the edible portion. Green material deteriorated first. The hollow leaves of Welsh onions were yellow and partially rotten. Only in the suburbs of Beijing and Nanjing did I observe that some communes could transport products by truck to the distribution centers (Figure 4c).

Since the return of the free markets, gardeners again bring their crops fresh to the market, each carrying a weighing scale, and sell their own products (Figure 4d). This practice naturally assures the freshness of the plants and enhances the food value of the vegetables.

Plant Food Storage and Preservation

The staple food of the Chinese people consists of grains which are stored by the farmers for their family consumption. Surplus grains are sold to grain merchants who own storage facilities. Here the discussion is limited to perishable vegetables. In South China, fresh vegetables are available throughout the year, and their storage does not pose serious problems. In North China where the growing season is short and where modern facilities for freezing and canning are not available to the general public, keeping some vegetables for winter use becomes an important issue. As necessity is the mother of invention, many simple and ingenious methods have been devised for preserving vegetables. These measures may be grouped into temporary storage, burying, drying and pickling.

Temporary Storage of Fresh Vegetables

When the Chinese cabbage is in season in October, it is very reasonable in price, costing only a few cents a pound. People usually buy many heads, and stack them in a cool, dry place. The supply may last a couple of months. The outer leaves may become dry, but the major portion of a head is fresh and crisp. The center of a head may be used for salad and the outer portion for cooking.

In North China, farmers have developed a technique for growing large, white portion of the Welsh onion (locally called big scallion 大葱) into a cylinder 40 cm long and 2–3 cm in diameter, for winter use only. As a vegetable crop, the seed of this species is sown in late May, and the seedlings transplanted in June into parallel rows, which are continuously hilled up as the plants grow. By harvest time, in late October, most of the plant is covered by the soil, leaving only the slightest aerial portion exposed. On harvesting, the plants are dug, laid flat in rows, tied into bundles with the withered yellow leaves, and are ready for the market. People buy and store this products as they do with big Chinese cabbage heads.

In June when garlic is harvested, people also buy large quantities of it, for a great deal of raw garlic is consumed by rural people in North China. The stem and leaves are plaited with the heads all facing one direction. The braids are then hung in a convenient place for winter use. The outer scales become dry and membraneous, but most of the cloves remain fleshy for many months.

Burying Surplus Material

In North China, sweet potato is used as a staple food for breakfast by the low-income people and by farmers. Farmers build simple underground storage pits of 3 m long, 2 m wide, and 3 m deep. The sweet potatoes are carefully selected so that all broken pieces are kept out to be used first. The wholesome ones are stacked systematically, then a low roof made of sorghum straw is erected, leaving a small opening at one end, which allows for only a small teenager to crawl in and out. The entire structure is sealed with earth and is opened only once every two or three weeks to bring the potatoes out. Stored this way, sweet potatoes can last from harvest time in November until March of the next year.

Large Chinese red radish is also stored by burying. Compared with sweet potato, the amount of radish stored by each family is small, and it is kept expressly to be used for the Chinese New Year festivity, which comes in early February. For this crop, people usually dig a hole one meter deep and one-third to one-half meter in diameter near the kitchen, and the radishes are usually kept buried in sand.

Drying

Drying the surplus products for winter use is a widespread practice in North China. Baranov (1965) published an illustrated report of how vegetable marrow squash (*Cucurbita pepo* L.) is preserved. Elongated strips are cut in a spiral around the squash. These strips are hung in the air to dry, then plaited and hung over walls for storage. In my village, people slice young squash crosswise, and buried the slices in plant ashes to reduce the water content of the fleshy material before exposing them to dry in the sun. These slices are reserved for the New Year Festival. Other fleshy fruits preserved by drying are jujube, egg plant (sliced), lychee, longan, persimmons, red chili peppers and yard-long beans (parboiled first). Zanthoxylum is dried and stored to be used as a spice.

A few flowers are preserved by drying. The best known are the mature flower buds of the daylily (with the market name golden needle 金針菜), the whole heads of a small flowered cultivar of chrysanthemum, and jasmine. All the three products are available in Chinese stores in America. In preparing the daylily flower buds, the mature buds are gathered at daybreak, parboiled or steamed to kill the cells and then spread on a mat to dry in the sun. In the case of jasmine, fresh flowers are picked early in the morning, and used to perfume the highest grade of the scented green tea 茉莉花茶. They are then dried and placed with the less expensive tea.

Preserving young leaves and tender shoots of edible plants for special occasions and festivities in winter is a prevailing practice of all ethnic groups in China, particularly of those living in the less populated areas of the high mountains, where the climate is harsh and the growing season short. Young adults of the Sino-Tibetan mountains start their climbing trips to the alpine meadows in late April and May to collect fern fiddle heads and wild onion. The tender emerging leaves are parboiled, then spread in the sun to dry. Fiddle heads are rubbed repeatedly while drying to remove their tough scales. The dried young fronds are kept for special occasions and for entertaining guests. They are sometimes taken as presents to relatives living in lower elevations. Other tender, leafy shoots of cultivated species dried for special uses in winter include those of field amaranth, used for the Lantern Festival (the 15th day of the first month after the Chinese New Year). Young shoots of wild cock's comb are gathered and dried in summer for winter months in a similar manner.

Before 1950, vetch (*Vicia cracca* L.) was used for green manure by farmers in the Chengdu Plain (Map, Loc. 35). Before the plants flower, young shoots are collected and used for potgreen locally. Some farmers gathered and dried the young shoots and sold them to lumber companies. In my investigation of the vegetation in the homeland of the giant panda, I used to camp where the lumberjacks worked in the spruce and fir forests of the Sino-Tibetan mountains (Map, Loc. 40–41; Hu, 1956). These workmen obtained their food supplies from the lumber company which managed to bring food material up the mountains by carriers. The only vegetable in the rations of the lumberjacks were dried vetch shoots which were boiled with salt pork for oil and flavor.

Pickling

The Chinese people pickle plant material for two purposes: to preserve the perishable farm products and to produce aroma. Using pickled and fermented plant material has an important place in Chinese cuisine. The Chinese people like flavored and spicy food. Pickling and the fermentation process that accompany it generally develop a desirable aroma, and enhance the food value through microbial actions. In this repect, the production of soysauce is an excellent example. Learning the methods for preparing pickles and mastering the technique of using fermented plant material in cooking are important lessons for a teenage girl in her preparation for marriage. In northern and eastern China, pickles are made predominently from cruciferous plants, i.e., the leaves or fleshy roots of the various species and cultivars of mustards and radishes. In western

China, whole chili pepper is an indispensable pickle for giving color and flavor characteristic to Sichuan food. In southern China, special pickles are prepared from ginger, onion, lemon, canarium and various other fruits.

(1) *Cruciferous Pickles*

Several cultivars have been selected and developed by Chinese farmers and gardeners from the genera *Brassica* L. and *Raphanus* L. for fresh vegetables and for pickles. Some cultivars are planted for both purposes, the best examples are the root of the Chinese radish (*Raphanus sativus* L.) and the leaves of Swatow mustard (*Brassica juncea* [L.] Czernajew var. *foliosa* L. H. Bailey). Other cultivars are grown exclusively for pickles. The leaves of *Brassica juncea* (L.) Czernajew var. *megarrhiza* Tsen et Lee and var. *multisecta* L. H. Bailey, and the enlarged stems of var. *tamida* Tsen et Lee are good examples. Although the cultivars, the parts used, and the methods of preparation may be different, salt and pressure are the common factors to prepare good cruciferous pickles. All the cruciferous pickles discussed below are now available in American Chinese stores.

(a) *Brassica pickles.* In China, there is a distinct geographical variation in the kinds of the cultivars used to make *Brassica* pickles.

Da-tou-cai 大頭菜 (big-head vegetable) is a root crop developed from the enlarged root of *B. juncea* (L.) Czernajew var. *napiformis* Baillon et Bois in northern China. This cultivar is planted in early July and harvested in late October in the Yellow River Region and also in Yunnan. The white fleshy roots may weigh one to three pounds. After being cleaned, the root is cut vertically into two to several pieces, partially dehydrated by exposure to the sun, and then mixed with salt, and placed in large earthenware urns with pressure applied from the top. Gradually the material turns black and is then ready for food. This pickle has a very pleasing aroma, and can be kept for years. The climates of Shandong and Yunnan are ideal for the production of the best Da-tou-cai.

Xue-li-hong 雪裡紅 (red-in-snow). In Northern China a cut-leaf cultivar *B. juncea* (L.) Czernajew var. *multisecta* Tsen et Lee is raised for the leaves. This cultivar is hardy, and the leaves turn reddish-brown after the early snow, hence its name "red-in-snow". The crop is harvested in November, cleaned, the rosettes cut into several vertical portions, exposed to wind and sun to be dehydrated partially, kneaded with salt and then placed in large urns with pressure. As the ingredients turn black, they may be kept in the urn for family consumption or dried in the sun for packing and transportation to the market. It has a very pleasing aroma, and is used specially for cooking pork shoulder or duck, to absorb the grease and to enhance the flavor of the dish.

Zha-cai 榨菜 (pressed vegetable). This product is the best known Chinese pickle in

America, for it is used in many dishes in good Chinese restaurants. It is prepared from the fleshy stem of of *Brassica juncea* (L.) Czernajew var. *tamida* Tsen et Lee planted along the Yangtze River in Sichuan, with Fuling 涪陵 (Long. 107°30′ E, Lat. 29°45′ N) being the center of production, 180 km (113 miles) west of the Metasequoia Grove (Hu, 1980). The climate of the area is mild. The crop is harvested in late winter, at the conclusion of its vegetative growth and just before the elongation of the stem to produce the flowering stalks. After harvesting, the leaves are removed. The enlarged, tender, fleshy stem and the attached basal portions of the petioles are cut vertically into desired portions, and spread over mats to dry. The withered sections are mixed with a liberal amount of hot red peper flakes, crushed follicles of zanthoxylum, and salt. This mixture is placed in large earthenware urns and pressure is applied, hence the name, *zha-cai* (pressed vegetable). The finished product is olive green and very spicy. It is available in Chinese stores outside China, and is becoming very popular with fans of American Chinese cuisine.

Pao-cai 泡菜 (Nanjing pickle). In Jiangsu, the spring crop of a cultivar of *Brassica parachinensis* L. H. Bailey is used to prepare the special pickle very popular in Nanjing and its vicinity. The leafy rosettes are harvested in May. After each plant is cleaned, drained, cut into four sections, and exposed to wind to remove the excess moisture, the material is tightly packed with salt in a container with pressure applied from the top. The finished product appears yellowish-green and it has a very unusual and pleasing aroma. Most families prepare their own supply. In Nanjing this pickle is also available in market-places and small groceries. I have not seen this type of pickled cabbage elsewhere.

Suan-cai 酸菜 (sour mustard). Sour mustard in large quantity is available in American Chinese groceries. This product is prepared from a foliaceous cultivar of Swatow mustard, *Brassica juncea* (L.) Czernajew var. *foliosa* L. H. Bailey, which is grown for its leaves, and used fresh and pickled. This cultivar has been introduced into Taiwan and America. The pickled green leaves are imported from both Guangdong via Hong Kong and directly from Taiwan. It is found in large barrels at Chinese groceries in Boston under the Cantonese name *ham choy* (= pickled vegetable) or *shuen choy* (= sour vegetable).

Yan-cai 醃菜 (home-made pickle). When the celery cabbage, *Brassica pekinensis* (Lour.) Ruprecht, is in season, every family in North China prepares its own pickles with salt, some zanthoxylum, cloves of garlic, a few large pieces of hot, red pepper, and a little sorghum gin. The early Chinese immigrants to Korea brought seeds of celery cabbage and the recipes with them from North China, particularly from Shandong. Now the Koreans have perfected this type of pickle. Recently some Chinese restaurants in America have served pickled celery cabbage as an appetizer. In appearance and taste the preparation resemble what we used to have in North China, but it lacks the aroma and texture. Appearance and taste can be produced by combining various chemical ingredients, but the aroma comes only from proper microbial growth, and the texture requires correct timing and environmental conditions.

(b) *Chinese radish pickle.* Seeing this term "Chinese radish", the reader is advised to change his/her image of the small, round, red or cylindrical, white radish of the American market to forms like the European turnip or sugar beet, large oval or cylindrical roots, weighing one to six pounds, white, red, or green on the outside, and juicy, fleshy, pungent or slightly sweet. Among the species of food plants, *Raphanus sativus* L. is the species which shows the most successful achievements of Chinese gardeners in their selection of cultivars, for color, size, taste, eating quality, and adaptability to diverse environmental conditions. From the rich plain in Harbin in the north (Map, Loc. 2) to the salty, white sand of the seashores of Hong Kong in the south, numerous cultivars will grow in the varying conditions. In North China, cultivars of Chinese radish are summer crops, harvested in September or October, while in South China, winter crops give the best result. Just as gardeners have made the best selections to produce timely crops of Chinese radishes, so too have housewives developed various ways to serve them. The pickling of Chinese radishes begins in homes, and has become an item of international trade. Various pickles of Chinese Radish are available in large American cities.

Chinese radishes, like the celery cabbage, are vegetables of the common people, particularly the low income population. During the harvest season, farmers pickle their surplus to extend the supply, and city dwellers buy extra loads for pickles. The fleshy roots of the cultivars planted in North China for common use have red skins and the inside is firmer and more pungent than the white-skinned forms. They are good for pickles. In pickling, the small lateral roots are removed. The roots are washed, then cut into vertical sections about 5 cm long and 1.5 cm across. These sections are then exposed to the sun and wind to remove excess moisture. The withered sections are mixed with salt and spices and kept in closed urns. The choice of spices makes a difference in the finished product; some are very hot and spicy, while others are bland. The degree of drying is determined by the kind of pickle one plans to make. For liquid preserves, more water is retained, and for dry pickles, the sections are dried just short of remaining flexible. In order to minimize the drying period, the roots are shredded. In southern and western China where the white, large-rooted cultivars are planted, people often slice the roots and hang them on strings to dry. Pickled radishes are commonly eaten raw, or they may be chopped into small pieces and stir-fried with meat. Pickled radish is available in Boston in cans, bottles, and plastic bags.

(2) *Pickled Seasoning Material*

Several kinds of pickles are prepared from bulbs, rhizomes, and fruits, which are used as seasonings and/or appetizers. Among the wet pickles are the sour bulbs of *Allium chinense* G. Don, the sweet, young vegetable melon (*cai-gua* 菜瓜, = *Cucumis melo* L. var. *conomon* [Thunb.] Nakai), and the young ginger rhizomes. Two kinds of partially fermented and pickled fruits are prepared for seasoning food: the very sour fruit of *Citrus limonia* Osbeck used particularly for chicken dishes, and the oleoresinous flesh of the fruit of *Canarium pimela* Koenig. All of the above mentioned seasoning materials are used for special dishes in Cantonese cooking and are available in America. Pickled hot red pepper is an important seasoning material in Sichuan. It is used for color and flavor. In Chengdu, fancy restaurants use chopped pickled fresh red pepper to decorate and to flavor a dish called *cui-pi-yu* 脆皮魚 (crisp-skin fish).

(3) *Special Pickles*

In North China two special popular pickles are available in sauce stores (*jiang-yuan* 醬園). These are *Lu-jiao-cai* 鹿角菜 (deer-horn pickle) and *Bao-ta-cai* 寶塔菜 (pagoda pickle). The deer-horn pickle is prepared from the many-branched plant body of a red alga, *Chondrus elatus* Holmes. The finished product looks brown and slightly transparent and crunchy. The pagoda pickle is prepared from the fusiform, repeatedly constricted tubers of *Stachys affinis* Bunge. The finished product is black and crisp. These pickles are served as appetizers at banquets. They are occasionally seen at parties in America.

According to social custom in China, people entertain visitors with tea and titbits 點心 (touching-the-heart), particularly around the New Year. Among the favorite titbits are three pickled and dried fruits. These are *hua-mei* 話梅 (chatting mei), *chen-pi-mei* 陳皮梅 (orange peel mei), and *gan-lan* 橄欖 (canarium), all prepared in South China.

Hua-mei: There is no English equivalent for "mei", because the cultivar that bears the fruit has not been introduced into the botanical world outside China. The flowering cultivar developed from the prototype were introduced to Japan, hence to Europe, and are now planted in the southern United States and in Australia for ornamental purposes. It has been called "Japanese apricot" by some authors, and "flowering plum" by others. Actually *mei* is neither an apricot (for the flesh of the fruit clings to the stone like a prune) nor a plum. A mature fruit is as sour as a green one (Figure 27e). The fruit of *mei* retains its acidity and therefore can never be eaten without pickling. Edible *mei* is prepared by pickling the mature fruit in a solution of salt and licorice. In Guangzhou and Hong Kong, pickled *mei* in solution at refreshment stands or carried by itinerant vendors is available in railway stations and market-places. *Hua-mei*, the dried product of such pickles appears whitish gray and irregularly wrinkled. It is very hard and strongly salty-sour-sweet. One fruit goes a long way. People usually take tiny bits at first, then suck on the whole stone for hours, as it has a pleasant taste. Young people usually carry *hua-mei* on their outings, as the stone keeps the mouth from having a dry feeling.

Chen-pi-mei: This product is prepared by boiling sectioned *mei* in a weak salt solution with tangerine peels, sugar, and licorice. The pickled mixture is kept in the solution until it absorbs sufficient sweet and flavor ingredients. Then it is removed and dried. The finished product is soft, black, very tasty but on the strong side. Sick people and pregnant mothers like to keep a piece in the mouth. At social gatherings, people usually share small pieces when a wrapping is opened.

Gan-lan: This product is the ripe fruit of *Canarium album* (Lour.) Raeusch. The mature fruit is ellipsoid and olive-green, appearing like a fully grown European olive in shape and size, hence early European visitors called it Chinese-olive. Actually it does not even belong to the Olive Family (Oleaceae), and it is not used as an olive. It is edible raw and is easily transported. Consequently, it appears in large cities such as Chengdu, Beijing, Nanjing, and Harbin, north of the Yangtze River where it is called *qing-guo* 青果 (green fruit) and is used as an after-dinner-mint to freshen the mouth. For pickles, the fruit is crushed, soaked in a weak salt solution with sugar and licorice added to soften the flesh and to give flavor. Then the fruit is wrapped individually. The pickle is black, easy to break, and it has a very pleasant taste. Pickled *gan-lan* is available in Chinese stores in America.

In connection with pickling and drying fruits to preserve them for better taste and easier transportation, it is worth of note that Chinese material preserved in sugar rather than salt and spices are becoming increasingly popular in America. The outstanding ones are the Beijing honey date, kumquat, lotus seed, water chestnut, sliced lotus rhizome, coconut flakes and candied ginger.

The Use of Microbes in Chinese Plant Food

Fermentation plays a much more important role in Chinese plant food material than previously realized. Fermented plant products are used as food or as spices in food preparation throughout China. Paradoxically, the lower a family's income, the greater the use of fermented food. That is not to say that the lower income population consumes more fermented plant food than the higher income people, but rather, that the lower income people have no other palatable food to help them to eat the tasteless cooked rice or steamed bread than a tiny bit of fermented bean and/or hot pepper sauce, thus proportionally they eat more fermented food. Approximately 85 percent of the rural population and the low income city dwellers depend upon fermented material to give flavor and to add food value to their rice or wheat diet supplemented by millet or sorghum. Normally, a person eats large quantities of cooked rice, steamed bread, or noodles with a little addition of fermented soybean products, in various forms of *jiang* 醬. The Chinese term *"jiang"* is very close in sound and texture to the English "jam". It is a generic term used for all fermented bean and grain products, particularly in reference to the mushy, spicy mass. Authors of modern cookbooks on Chinese cuisine use the term "sauce" for *jiang*. For example, *la-dou-ban-jiang* 辣豆瓣醬 is translated into hot bean sauce, *la-jiao-jiang* 辣椒醬 to hot pepper sauce, *suan-mei-jiang* 酸梅醬 to duck sauce and *tian-mian-jiang* 甜麵醬 to sweet flour sauce. In this discussion, the fermented Chinese plant foods are grouped by their source plants into the leguminous, the gramineous and the solanaceous types. Most of them are called *jiang* in Chinese.

Fermented Leguminous Products

The primary sources of fermented Chinese plant food are from the soybean genus, *Glycine* L., actually from various cultivars of *Glycine max* (L.) Merr.. The outstanding ones used throughout China are all available in America. These are soy sauce 醬油, fermented black soybean 豆豉, soybean sauce 豆瓣醬 and bean curd cheese 豆腐乳. In West China, the broad bean sauce 胡豆瓣 is prepared from *Vicia faba* L.

(1) *Soy Sauce*

Soy sauce is the only liquid product of fermented Chinese food, and its manufacture is one of the most important food industries in China. Before the 1930s, soy sauce was prepared from fermented soy bean. Now, many bottled products, especially those available in American supermarkets, are chemical combinations with artificial colors. In China, soy sauce is made in every city and in all the large, rural market-places. Formerly the production of soy sauce was a family business run by the owner of a store. This particular store was called a *jiang yuan* 醬園 (sauce garden) and was devoted entirely to the production and sale of fermented and pickled foods. In small towns and rural market-places, soy sauce is made in the back yard of a store, while in large cities, it is manufactured in the outskirts; for example, in Hong Kong soy sauce is made in the New Territories.

In making soy sauce, the soy bean is soaked overnight, cooked with a small amount of cracked wheat, cooled, inoculated purposely or naturally with airborn spores, and heaped up to allow fermentation. A person experienced in the business can tell by the color of the molds and the texture of the beans when the process of microbial action is complete. Then the fermented material is transferred into large, top-shaped, thick-walled, earthenware urns about one meter tall and one meter in diameter across the top. Salt, water and spices are added and stirred to make a uniform mass. All the urns are arranged in rows and each one is provided with a roof-like cover which is used only when it rains. The urns are kept open day and night so the contents are exposed to fresh air, direct sunshine, and dew at night. On clear days the material in the urn is stirred regularly so that the bottom portion can be exposed to the sun and dew. Under these conditions, natural physical, chemical, and microbial actions gradually change the color of the mixture to a hue between chestnut-brown and black, and convert the ingredients to a homogenous mushy consistency. As soon as the color and texture of the mass become uniform inside an urn, a well is made in the center. Soy sauce gradually fills the well and is collected. The first collection is called *yuan-you* 原油 (prime sauce). It is premier in excellence, in aroma, and in nutrient value, and it is said to be used for table dipping in high class restaurants and/or the homes of officials and rich families.

Water and salt are then added to the urn, and the process of exposure to the sun and dew is repeated. The soy sauce thus obtained is called *tou-chou* 頭抽 (= *t'ou-chou*, first extract). This material is good sauce and is used for table dipping in restaurants and in the homes of affluent families. After the first extract is removed, the space in the urn is larger so that more water, salt, and spices can be added. Soy sauce extracted henceforth is called *sheng-chou* 生抽 (= *sheng-chou*, raw extract). It has a much lighter color and is the soy sauce used by most people for cooking. This type is available in American Chinese stores.

To the final puree left in the bottom of the urn, water, salt, and partially charred brown sugar are added, homogenized and bottled. This material is called *lao-tou-chou* 老頭抽 (= *lau-t'ou-chou*, old man extract, shortened to *lao-chou* 老抽), and is also available in America. The several Chinese cookbooks published in English available to me all mention two types of soy sauce, one light and the other dark. The raw extract and the old man extract are the sauces to which they refer. The old man extract, reddish-brown in color, is used to cook pork or chicken dishes to give the desired color.

Extensive research on fermentation agents has been carried out by microbiologists in China and Japan. Among the organisms identified in the manufacture of soy sauce are *Aspergillus soyae* Sakaguchi et Yamada, and *A. oryzae* (Ahlburg) Cohn.

(2) *Fermented Black Soybean*

To my knowledge, three types of Chinese black soybean are available in American Chinese stores. The very large and the medium-sized beans are used as special health food, primarily for medicinal purposes and are sold in one-pound packages. The small, oblong, black soybeans appear only in form as *dou-chi* 豆豉 (= *tou-ch'ih*, fermented black bean), used as a condiment and/or in prescriptions. The large and intermediate black soybeans are the seeds of erect plants. The small black soybean is the seed of a semiclimber, which matures one or two weeks later than the erect varieties.

Dou-chi 豆豉 (fermented black soybean) is very popular in the warmer regions of China, used as a seasoning for steamed fish and as an ingredient in many vegetable dishes. It is listed in the cookbooks published in America as an ingredient in different types of Chinese foods: Mandarin, Cantonese, and Sichuan styles. In southern China, the product is sold in open containers in the market-places or in groceries. *Dou-chi* is imported to America in plastic bags or in cans. It appears as a small, oval, shrivelled, charcoal black bean with a spicy, piquant flavor.

Dou-chi is prepared from the seed of the semiclimbing plants. In making *dou-chi* the seed is cleaned and soaked in a decoction prepared with the various herbs, including green mugwort (*Artemisia apiacea* Hance), licorice, mulberry leaves, perilla shoots, and water smartweed. The soaked bean is steamed, cooled, and placed in a container covered with the residue of the boiled herbs and allowed to ferment until a layer of yellow mold appears. Then the bean is mixed with salt and dried in the sun. The material keeps well at room temperature.

(3) *Fermented Soybean Pastes*

Several kinds of fermented soybean products look like pastes, with large or small pieces of soybean. In Chinese cookbooks published in America, they are called *dou-ban-jiang* 豆瓣醬, meaning sauce with cracked soybeans. The source plant for these pastes is the common, yellow soybean with erect and branched stems. The principle and procedure for making pastes are alike. The differences stem from the amount of spices added to the bean mass after it is fermented. In preparing the pastes, the beans are cleaned, soaked, steamed, inoculated with a leaven and allowed to ferment, as in the manufacture of soy sauce. The soybean pastes are made in the *jiang-yuans* throughout the country. According to consumer demands, most *jiang-yuans* make both the mild and the hot types, by adding water and a little or a lot of hot, red pepper flakes to the fermented mass before the mixture is exposed to the sunshine and dew. The mild type is *dou-ban-jiang*, and the hot type is *la-dou-ban-jiang* 辣豆瓣醬, meaning hot, cracked, soybean sauce. In making soy sauce, when the aroma has developed and sauce has matured after having been subjected to the sunshine and dew, the liquid is extracted repeatedly with additions of salt, water and spices. In making soybean pastes, water, salt, and spices are added only once at the outset. When the desired aroma has developed by exposure to sunshine and dew, the entire paste is collected for marketing. Both mild *dou-ban-jiang* and the hot *la-dou-ban-jiang* are available in Chinese stores in America. The latter has become increasingly popular among the Chinese gourmets.

(4) *Bean Curd Cheeses*

Bean curd is pressed protein extracted from soybean. The technique for the extraction, the art of cooking and the habit of eating bean curd spread from China to its neighboring countries and then worldwide as did soybean. Now, it is used as an ingredient in cooking many dishes and is available in supermarkets as well as in Asian stores. However, making cheese out of bean curd has not spread as fast as did bean curd. Presently, various types of bean curd cheese in American Chinese stores are imported largely from China. Unlike bean curd, bean curd cheese is soft and aromatic, used as a dip, a spread or an ingredient in cooking. For its soft texture, the ancient Chinese people called it *dou-fu-ru* 豆腐乳 (= *tou-fu-ru*, bean curd milk).

Bean curd cheese is made of small squares of bean curd, about 3 cm long and wide and 3 mm thick. These pieces are placed single file on flat trays which are stacked on shelves in a closed room to ferment. When white mold grows up to 2–3 cm thick over the bean curd, salt and spices and/or colored material are added to the mass and the mixture is placed in an earthenware urn for aging until it becomes very soft and has a savory taste and aroma.

The common type of bean curd cheese is *chou-dou-fu-ru* 臭豆腐乳 (rotten bean curd cheese) which appears gray and soft, each square covered with a thin, slimy film. The Chinese people eat it as a relish, and in America it has been served on crackers as an *hors d'oeuvre*. A similar type is *la-dou-fu-ru* 辣豆腐乳 (hot rotten bean curd cheese), which is seasoned with red hot pepper flakes or powder before aging. A milder type appears brown. It is prepared by adding powdered, red fermented rice, salt and spices to the bean curd before aging. The finished product is firmer than the common gray type, and it has a special aroma. It is called *nan-ru* 南乳 (= *nan-ru*, southern bean curd cheese), and is used not only as a relish, but also for cooking a special chicken dish, *nan-ru* Chicken. All three types of bean curd cheese are available in American Chinese stores, either in bottles or in cans.

(5) *Broad Bean Sauce*

In Sichuan, an extremely hot sauce used as a relish and/or for cooking is made from fermented broad bean 蠶豆. On account of the high Tsinglin Mountain Range to the north, protecting it from the cold Siberian current, the climate of Sichuan is much milder than the provinces to its east at the same latitudes. Consequently, while these provinces are covered with snow, Sichuan has a healthy growth of winter crops including various mustards, cabbages, and broad beans. In the broad bean sauce, soybean used in preparing *la-dou-ban-jiang* is replaced by broad bean and the product is locally called *hu-dou-ban* 胡豆瓣. In appearance, broad bean sauce is similar to *la-dou-ban-jiang*, but its aroma, taste and texture is slightly superior.

Fermented Gramineous Products

Just as many of the fermented bean products are called *jiang*, many of the fermented grain products are called *zao* 糟 (= *tsao*, spoilt grain), hence we have *hong-zao* 紅糟 (red *zao*) and *nuo-zao* 糯糟 (glutinous rice *zao*). In the food industry, the brewery in which fermentation and distillation of grains for gin takes place is called *zao-fang* 糟坊 (= *tsao-fang*, brewhouse). A *zao-fang* uses rice for spirit in South China and sorghum and barley for gin in North China. The commercial uses of fermented grains, be they rice, sorghum and/or barley, for the newly developed industry of making beer, are excluded from this account; these belong to different fields of research. This account is concerned only with fermented grains used in food production, and includes a homemade, non-distilled drink made of grain. It covers the fermented products of rice, wheat with a little soybean, and corn. In all cases, fermented grain production is rather restricted geographically.

(1) *Fermented Rice (FR)*

There are two types of fermented rice. They are, by their color, red FR and white FR; and by their texture, wet FR and dried FR. In all cases, the source material is glutinous rice, which is washed, soaked, steamed, and cooled before being inoculated with fungal material called *qu* 麴 (= *ch'ü*, brewer's yeast).

(a) *Red fermented rice*: The use of filamentous fungus in food, the discovery of culturing and drying techniques, and the perfection of the art of cooking meat, poultry, fish, and bean curd with the fermented red rice, originated in Fujian, a maritime province in eastern China, opposite Taiwan. Emigrants of this province introduced the uses of *hong zao* 紅糟 (red fermented rice) to the islands and the mainland of Southeast Asia, particularly to Taiwan, the Philippines, Indonesia, Malaysia, Thailand, Vietnam, Cambodia, and Burma. Recently, with the immigrants from these areas to America, Fermented rice has become available in Chinese stores in the USA.

According to my records, the preparation of fermented red rice had been a trade secret in Fujian, and the supplies in various countries in Southeast Asia have been imported from there. In the 1950s, as recorded by Palo (1960), the supply in Manila was very low due to the Chinese closed-door policy. Scientists of the National Institute of Science and Technology of the Philippines had to undertake a special research project on the biology and production of the Chinese red rice (*ang-khak*), so that the country could produce it independently.

Our botanical knowledge of the red rice began in 1890s, when a Dutch scientist at Utrecht, F.S.F.C. Went, received some Chinese red rice from Java. From this material he isolated a filamentous fungus and named it *Monascus purpureus* Went, the earliest scientific name for the organism which produces the red color, the aroma, and the flavor of its substrate. In botanical literature, fermented red rice is known as the Chinese red rice, *ang-khak* in the Philippines, and *ang-quac* in Indonesia. In China, the dried fermented red rice is called *hong-qu* 紅麴 (= *hung-ch'ü*, red yeast), and the wet fermented red rice is called *hong-zao* 紅糟 (= *hung-tsao*, red spoilt).

The dried fermented red rice is a starter. It consists of dull, red rice grains easily broken when pressed between two fingers. In the Lanzhou (Map, Loc. 20) spice market, pulverized red rice is sold in open containers. People buy it to fill the layer of yellow-green-red steamed pastry prepared for the New Year Festival. In Fujian cuisine, pulverized *hong-qu* is mixed with soaked, steamed, and cooled glutinous rice, in proportions of 20 to 1 by dry weight. Some sherry-like rice gin is added to the mixture (approximately 1 cc for each gram of starter used) and is mixed thoroughly and placed in an urn. The urn is then sealed with paper and kept undisturbed for 30 days to let the fungus multiply. The fungal growth softens the rice, and the pigments (one red, = monascorubin, and the other yellow, = monascoflavin) secreted by the mycelia color the substrate. The accumulated liquid is ladled off and filtered. The solid portion is returned to the urn which is again sealed, this time with bamboo leaves covered by mud, and is left undisturbed to age for one year. When it is re-opened, the contents look rosy-red, smell delightfully aromatic, and taste sweet-sour. This is the real *hong-zao* 紅糟 (used for preparing genuine Fujian dishes. The amount of wet fermented red rice used varies from two tablespoons to four cups, depending on the dish cooked, the color desired, and the style of cooking.

(b) *White fermented rice*: A homemade, white fermented rice is very popular in the Upper Yangtze River Region, in the mountains as well as the plains, where it is called *nuo-zao* 糯糟 (glutinous rice *zao*). In the Lower Yangtze River Region, it has several local names, such as *jiu-niang* 酒釀 (mother-of-gin) in Nanjing, and *tian-jiu* 甜酒 (sweet liquor) in Shanghai. As most people in China do not have refrigerators, the use of the white fermented rice is quite seasonal, mostly a winter food, eaten as a special treat, served sweet with a poached egg, or with small balls made with glutinous rice flour. Personally, I often take a cup of this preparation before going to bed.

Nuo-zao is prepared with a starter, a white, irregular yeast ball about 2 cm in diameter. It is available in large cities in China and also in American Chinese stores. This starter is made of pulverized, malted barley and rice, several herbs, and is cemented with rice flour. The proportion of the glutinous rice to be fermented to the starter is approximately 350:1 by dry weight. In making *nuo-zao*, glutinous rice is washed and soaked for two hours. Then it is drained, steamed for one hour, and cooled by spreading over a flat, bamboo basket for one hour, or by pouring cold water over the steamer until the cooked rice is about 30°C. After the inoculum is thoroughly mixed with the cooled rice, the mixture is placed in an earthenware urn, leaving a hollow space in the center. The container is covered with a piece of wood, burlap bags, and straw mats for one to two days. When the central space is filled with liquid, the contents are ready to eat. In Boston, most of my friends merely place the inoculated rice in a clean bottle, leave it at room temperature for two days and put the bottle in the refrigerator. It keeps well.

(2) *Fermented Wheat*

In North China *tian-mian-jiang* 甜麵醬 (= *tien-mian-chiang*, sweet flour sauce) is the common sauce served with the popular dish, Peking duck. Actually, it is a source of protein and vitamins in the diet of the rural population, where a flat, thin bread cooked quickly over a hot iron (locally called *bing* = *ping* 餅) is a staple food. *Tian-mian-jiang* is generally eaten with raw Welsh onion (*da-cong* 大葱). Placing a fresh, whole, cleaned scallion, roots removed, across the diameter of the thin bread, a line of sauce is dribbled down the length of each side of the scallion and all of this is rolled tightly in the bread. The sauce appears smooth and chestnut-brown. It has a fine texture, a mild, sweetish taste, and a very pleasing aroma. Children especially like it very much.

Tian-mian-jiang is a homemade product in a rural areas. Every summer a housewife first prepares her own starter by wrapping dried, moldy bread with fresh melon, including rind and flesh of a mealy melon (*Cucumis melo* L. var. *chinensis* Pangalo cv. 'Miangua', *mian-gua* = *mien-kua* 麵瓜), bound together with sorghum leaves. The wrapped mixture is hung in the shade with good ventilation until it becomes dry. In preparing *tian-mian-jiang*, approximately 95% wheat and 5% soybean are mixed, ground, and the meal used for steamed bread which is inoculated with pulverized, homemade starter. The mixture is placed in a covered basket until greenish mold appears on the surface. Then the bread is broken into pieces about 1 to 2 cm, long and wide. This fermented material is placed in an earthenware container about 45 cm in diameter and

20 cm deep. Water, salt, and whole leaves of zanthoxylum are added, whereupon the mixture is exposed to direct sunshine and dew, and only brought into the house when it rains. The mass gradually changes its consistency and color, becoming soft, increasingly brown, and thickened. As the ingredients take on a uniformity, oily droplets well up, a pleasant aroma develops, and the annual supply of sauce is ready to be collected and stored.

In cities and towns *tian-mian-jiang*, manufactured by the shop owners, is available in all *jiang-yuan* 醬園 (sauce gardens). Due to the increasing number of Chinese restaurants and Chinese cooking classes, and particularly the popularity of the Peking duck, *tian-mian-jiang* is now available in American Chinese stores. Genuine *tian-mian-jiang* is very different from some market-imitated duck sauce used in many oriental restaurants in America. The authors of Chinese cookbooks published in English have erred in assuming *tian-mian-jiang* is a fermented sauce made of soybean, because they do not understand that the word *"mian"* (麵) means wheat flour and the sweetish taste is a natural flavor produced by the microbial conversion of starch to sugar. The active micro-organisms that participate in the production of *tian-mian-jiang* must be many. The Japanese *miso* is a product of wheat, rice, and soybean in a 1:1:2 proportion. *Aspergillus oryzae* (Ahlburg) Cohn, *A. soyae* Sakaguchi et Yamada, *Saccharomyces rouxii* Boutroux, and *Lactobacillus delbrueckii* (Leichmann) Beijerinck have been isolated from *miso*, and most likely they are present in the Chinese *tian-mian-jiang* as well.

Fermented Solanaceous Products

When mature Cayenne pepper is in season, Chinese farmers use the red fruit to make *la-jiao-jiang* 辣椒醬 (= *la-chiao-chiang*, hot pepper sauce). This is the only fermented solanaceous product. Due to the introduction of Sichuan and Hunan cuisines by various restaurants, people request hot Chinese food, and many brands of *la-jiao-jiang* are available in Chinese stores. It is used either as a relish or for cooking.

La-jiao-jiang is made of fresh red hot pepper, which is washed, ground, mixed with water, salt, and spices such as the pericarp or leaves of zanthoxylum, and exposed to sunshine and dew for the development of a special aroma. The finished product is red, hot, and aromatic, with seeds and fragments of the ground pepper visible. *La-jiao-jiang* should not be confused with hot, fermented soybean paste (*la-dou-ban-jiang* 辣豆瓣醬), or fermented broad bean sauce (*hu-dou-ban* 胡豆瓣 or *pi-xian-dou-ban* 郫縣豆瓣), although in color and appearance they may look quite similar. In making the hot bean sauces, the soybean or broad bean must be cooked, inoculated with a fungal starter, and left in a dark place for fermentation. To the fermented mass, ground, hot pepper, salt, water and spices are added and the mixture exposed to the elements for aging and for the development of the aroma. In the preparation of the hot pepper sauce, the fungal fermentation process is omitted. The procedure is much simpler. I have made it many times myself at home in the USA. In the preparation of *la-jiao-jiang*, it is the airborne yeast spores, bacteria, and other microbial organisms attached to the pepper that change the consistency and produce the aroma. Good homemade *la-jiao-jiang* has the brilliant red color of the ripe fruit.

Detoxification, Extraction and Limited Utilization

Many edible plants contain toxic principles. People have tried to avoid the harmful effects of these plants by removing or converting their toxic elements and eating them for their beneficial effects. In my second year for graduate work at Harvard University, I moved to live in the Arnold Arboretum, to be near the supervising professor. That spring, I learned from Professor E. D. Merrill to eat a poisonous plant, Pokeweed (*Phytolacca americana* L.), by boiling the young shoots first and washing the material before cooking.

A large number of the 1,156 species of Chinese food plants enumerated here, especially those which are fleshy or aquatic species, are poisonous by nature. Through trial and error experiments on themselves and through their observations, the ancient and/or contemporary ethnic peoples in China have invented many ingenious devices to make use of the otherwise poisonous species of plants by detoxification, extraction, and limited utilization.

Detoxification

Many poisonous plants are rendered harmless after being cooked or dried. Ethnic peoples learned this principle independntly with the edible plant species in their areas. The Chinese people have not only employed this principle in using some poisonous species, but have also developed various ways of cooking them and have selected many superior horticultural forms to make it possible to advance these culinary arts, particularly with the species of the Arum Family (Araceae). Plants of the Spiderwort Family (Commelinaceae) contain calcium oxalate crystals. Young shoots of *Commelina* L., *Murdannia* Royle, and *Rhoeo* Hance are edible after boiling. Even in the Chinese stores in Boston, different forms of taro (*Colocasia esculenta* [L.] Schott) are available for Chinese home-makers to make cakes, soups, or sweet and/or salty dishes for the New Year Festival. Like the taros, the root tubers of the Yam Family (Dioscoreaceae) can be made into delicious dishes, but only after cooking. These corms and tubers cannot be eaten raw. The mucilaginous root tubers of many terrestrial orchids and the potato-like rhizomes of *tian-ma* 天麻 (*Gastrodia elata* Blume) are used in food, usually after prolonged boiling.

The nature of the toxic contents and/or the unpleasant tastes vary greatly. Some plant parts are not made edible by mere cooking alone. They may require repeated boiling and/or washing after being boiled. The young shoots of pokeweed (*Phytolacca americana* L.) in America and eucommia in China (*Eucommia ulmoides* Oliver) must be boiled, washed and/or boiled again before they are palatable. The tender portions of pigweed (*Chenopodium album* L.) and *Hemistepta carthamodes* (Buch.-Ham.) O. Kuntze need to be boiled, washed, and the waxy or woolly surfaces rubbed off before seasoning for food. The young bamboo shoots of *Phyllostachys pubescens* Mazel ex H. de Lehaie are bitter. They must be boiled and washed to remove the bitterness before being brought to the market fresh, dried, or canned.

An unusual procedure was devised by farmers of northern Sichuan for making the tubers of a well-known toxic species into a prized food product. A market material called *bai-fu-pian* 白附片, consisting of bony, dried, nearly translucent slices, 2–3 cm in diameter and 3–4 mm thick, is a prized gift to senior citizens who use it in winter to improve their circulation so that they do not suffer from cold hands and feet. *Bai-fu-pian* is prepared from *Aconitum carmichaeli* Debeaux by (1) only using very young tubers, and by (2) removing the toxic elements through repeated soaking and boiling.

Normally *A. carmichaeli* Debeaux is cultivated for the medicinal use of the tubers known as *Chuan-wu* 川烏 (Sichuan Aconite), which is harvested in November. For the preparation of *bai-fu-pian*, however, young, lateral tubers are cut from the mother plant between June 21 and July 7. These tubers are sorted into three grades by size. In preparing *bai-fu-pian*, only the grade of small size (less than 2.5–4 cm long and 1.5–2.5 cm in diameter) is used. The selected tubers are soaked in natural salt solution for several days, then boiled in brine until the center of each tuber is well cooked. Each tuber is then sliced vertically about 3 mm thick. The slices are soaked again, with repeated water changes, until all the acrid taste is removed. Then, they are steamed, partially dried, and bleached with sulfur dioxide gas. When completely dried, they are ivory-white, slightly translucent, and glossy. The product is available in some stores in Boston.

Extraction

For several reasons the Chinese people take the time and trouble to make plant extracts. Extractions are made from the edible tubers and rhizomes of some species for better storage and for more digestable products. The best examples are found in the extraction of starches from the water chestnut (*Eleocharis dulcis* [N. Burman] Trinius) and the rhizomes of lotus (*Nelumbo nucifera* Gaertn.). Both of these underground products are fleshy. When they are in season, the farmers merely grind up the surplus, wash off the starch, filter the mass through bamboo baskets, wait for the starch grains to settle to the bottom, recover and dry them. The dried starch is placed into small packets for the market, used by the affluent people.

In a similar manner, the starch is extracted from the poisonous roots of *Cynanchum bungei* Decaisne and related species, and from *Ipomoea mauritiana* Jacquin to remove the toxic elements. The starches thus extracted are dried and sold, to be used as a source of quick energy for the elderly and for sick people in convalescence.

Extracts from the seed of *Ficus pumila* L., from the whole plant of *Mesona chinensis* Bentham, and from the tubers of the devil's tongue (*Amorphophallus rivieri* Durieu) are used immediately after the extraction, and used to prepare a cool jelly as a summer treat for the entire family.

For the benefit of the readers who would like to cook Chinese dishes, three important extractions which have been profitably adopted by China's Asian neighbors are described in more detail here: the preparation of aromatic sesame oil as a seasoning agent, and the extractions of wheat gluten and of mung bean starch for enriching the diet of vegetarians.

(1) *Sesame Products*

Both sesame oil and sesame butter are available in American groceries and in health food stores. These products, however, are prepared from raw sesame seed, and consequently, have no aroma. The Chinese sesame oil and sesame butter are prepared from roasted seed. They are available in American Chinese stores. A few drops of sesame oil added to many cooked dishes completely change their aroma and taste! Many people use this natural seasoning in place of monosodium glutamate.

Chinese sesame oil is extracted by the hot water system. The procedure consists of roasting sesame seed in a large, iron wok (over 1 meter in diameter), grinding the roasted seed into paste, adding and vigorously stirring boiling water to the ground paste, and then by shaking the wok continuously while making slight strokes here and there over the surface of the mass with a mature cylinder gourd (*Lagenaria siceraria* [Molina] Standley var. *hispida* [Thunb.] Hara). The oil that gradually appears on the surface is skimmed off. It has a wonderful aroma, and is good for Chinese salad. The Chinese sesame butter is the ground mass containing the oil.

(2) *Wheat Gluten*

Wheat gluten imported from China is available in cans in Chinese stores. Home-made wheat gluten is prepared in rural areas in the summer. The procedure is simple but it requires patience and skill. Wheat flour is mixed with water until it reaches the consistency of a thick paste which is then stirred with a stick in one direction only. It is allowed to stand covered for an hour. Then, water is added to the container. Piece by piece a handful of the mass is held in the palms of two hands submerged in a pot of water. With the thumbs pressing the mass lightly, each stroke releases the starch and retains the gluten, and gradually the two are separated, with the starch diffused in the water and the gluten kept in the palms. Beginners often lose a large portion of the gluten; recovering it takes practice.

Due to the large demands of modern cities, both the extraction of sesame oil and of wheat gluten have been industrialized. The by-products of oil extraction are used as animal feed and as fertilizer. The starch from the extraction of wheat gluten is immediately converted into a white powder sold for pastries in South China (from personal observation in Hong Kong), for export to restaurants serving Chinese pastries abroad, and for other industrial uses in North China.

(3) *Mung Bean Starch*

Fen-si (粉絲) and *fen-pi* (粉皮), the two products prepared from mung bean starch, are important ingredients of Chinese dishes. The extraction of starch from mung bean and the preparation of *fen-si* and *fen-pi* are the earliest Chinese agricultural industries. Mun bean starch is extracted by first soaking the seed. The swollen seed are then ground, separating the starch from the liquid, and transforming the wet starch into *fen-si* (Figure 31), and/or rarely into *fen-pi*. A small amount of mung bean starch is also dried and used as corn starch in Western kitchens. All the three products, *fen-si*, *fen-pi* and mung bean starch are available in American Chinese groceries.

Limited Utilization

Many plant products are beneficial when used rationally in small amounts, but are toxic when taken in excessive quantity. The best example is ginseng. For thousands of years, among the affluent families in East Asia particularly, the fresh or dried root of ginseng has been used in the form of tea or a broth. Currently in Seoul, Korea, fresh ginseng is sold side by side with carrots in grocery stores. There, restaurants specializing in ginseng dishes are very popular. With rational use, adverse reactions are unknown. Nevertheless, concentrated ginseng powder and dried ginseng root have been introduced into America for over a decade. Young adults desiring a short-term psychoactive, stimulant effect, take large doses, and even put ginseng into coffee for synergistic action. Siegal in Los Angeles reported his observations of ginseng abuse syndrome, with patients who suffered nervousness, sleeplessness, morning diarrhea, skin eruptions, and hypertension (Siegal, 1979, 1980).

Some of the species included in the *Enumeration of Chinese Food Plants, Part II* are known to be poisonous. For example, the Black Nightshade is listed in every standard reference on poisonous plants in the Harvard University Herbaria Library. Yet, in North China it is eaten by all the children in the villages. Personally, I have eaten handfuls of the sweet, black, mature berries every summer. This may be due either to the berries in each cluster maturing one at a time, so a child could never gather a large enough quantity to produce a toxic effect, or the toxicity of a species varies by the place where it grows.

Other species in the *Enumeration* listed as common food plants have had toxic records. Ginkgo is a good example. In China, no seed is left lying around the tree in November, for it is gathered as a prized food. In Suzhou 蘇州 of Jiangsu Province, a selected cultivar with very large seed is grown in orchards by grafting. Fresh and canned seeds are available in American Chinese stores. Sweet puddings, meat dishes, and soup with chicken or duck are prepared using the soft kernel of ginkgo seed. Yet, I have witnessed a case where a seven-year-old boy who ate too many seeds had such a toxic reaction that he slept continuously for two days. A graduate student in the Department of Organismic and Evolutionary Biology at Harvard University had severe dermitits due to direct contact with the fleshy portion of the seedcoat. It is advisable to wear rubber gloves when gathering and cleaning the ginkgo seeds.

Recently, the American Medical Association (AMA) published a *Handbook of Poisonous and Injurious Plants* (Lampe and McCann, 1985). The purpose of the handbook is to "provide physicians and other health care professionals with … a reference for the management of plant intoxication …" and to supply "… a field guide for the recognition of dangerous and injurious plants." In this book, the common onion, garlic, chili pepper, tomato, and many edible Chinese plants, particularly wisteria and black locust flowers, and apricot seeds are treated as dangerous plants. In China almond is very rare. The apricot kernel after being soaked, the cyanogenic glycosides hydrolized, and the cyanide acid released, is dried and used in Chinese cooking as almonds are used in Western cooking. The apricot seed moon cake is a special pastry prepared from such treated seed for the Mid-autumn Full Moon Festival (= the Chinese Thanksgiving). Actually, the dividing line between "the edible" and "the poisonous" plants is very fine with some species, and is only crossed by the amount used. As many common food stuffs are listed by the AMA's *Handbook of Poisonous and Injurious Plants,* many so-called poisonous plants, such as the aroids, are used by various ethnic peoples. Careful choice and rational use of small amounts is the safest guide. In the struggle with natural forces for survival, the ancient and present-day people of China have learned and practiced this principle with their food plants.

SELECTED CHINESE FOOD PLANTS WITH INSTRUCTION FOR PREPARATION

Recently, various Chinese cookbooks have been published in America by experienced home-makers, cooks, and restaurant owners. Each of these publications provides an introduction to the material, methods, and special equipment necessary for cooking Chinese food. Readers are advised to consult these books for ideas and elementary techniques for preparing Chinese food. My primary concern as a botanist is the presentation of the ethnobotanical aspects of the Chinese food plants and the associated information not yet available in English. Such information represents the observations I made in intimate association with many ethnic groups in China, and the recipes which have been tried by myself and friends in America.

Mankind first experienced the tastes and benefits of plant food by eating fresh berries, tender leafy shoots, and/or fleshy rhizomes gathered in his environs. People living in rural and isolated, mountainous areas still eat many species of raw plants. In my village people enjoyed cooked vegetables only during the harvest season and at special festivals. The little they had in addition to steamed bread and/or noodles, usually included raw onions, radishes, carrots, and cabbages mixed with some seasonings.

Chinese culinary art was developed by people of leisure, most of whom lived in cultural centers. The secret of good cooking is to keep the natural flavor, color, and texture of the major ingredients by not overcooking.

People interested in Chinese food plants naturally want to know how to use them to add more variety to their diet and to prepare more tasty and nutritious foods. This subject will be discussed in three parts: (1) trying new plant foods, (2) cooking with a variety of plants, and (3) preparing immunostimulant and other health foods. Before turning to these topics, however, it would be helpful to know the criteria of good Chinese food and the fundamental principles involved in attaining its highest standards. Most Chinese eat bland food on a daily basis, while their good, home-made foods are prepared for festivals and special occasions. Proportionally, fewer Chinese than Americans eat in restaurants. For special occasions, certain villagers can help to prepare a banquet. In cities and towns, connoisseurs often ask top cooks from special restaurants to cook in their homes.

Chinese Cooking

A connoisseur of good Chinese food judges a dish by its *huo-hou* 火候 (timing of cooking), and its *se-xiang-wei* 色香味 (color, aroma, and taste). In rural China, instead of using *huo-hou*, people say *huo-kou* 火口 (fire mouth), which means to apply or to stop the fire according to the taste-buds of the cook (= judgement). A good Chinese cook never uses a cookbook, but samples a tiny piece and stops the fire to judge by his or her taste. This ability calls for experience, and it is from such experience that a cook chooses a method of preparation, such as quick stir-frying, slow simmering, or steaming.

Color, aroma, and taste are qualities of a well-prepared dish. In expressing appreciation of the skill of the cook who prepares the dish, diners usually praise: "What a perfect dish! It has just the right color, aroma, and taste!" The color of a dish refers to the harmonious combination of the ingredients, the aroma means the pleasant smell created by using the correct seasonings, and the taste indicates the dominant flavor, whether sour, sweet, bitter, hot, salty, or just its natural savor. Color, aroma, and taste are influenced by careful selection of ingredients, their preparation, cutting, and the order in which they enter the cooking process. Equally important is the skillful application of seasonings which include ginger, onion, garlic, fermented sauces of beans and grains, and various pickles. The addition of a few drops of sesame oil just before serving always enhances the aroma.

By the above standards, food prepared in many American Chinese restaurants falls short of good Chinese cooking. Actually, many dishes are not Chinese. They are prepared with no thought of *huo-hou*, or *se-xiang-wei*. The fundamental principles and the important techniques of good Chinese cooking are acquired either through family training or by strict apprenticeship in restaurants. Few of the cooks in American Chinese restaurants have any such disciplines. They neglect the principle "one dish, one taste and one hundred dishes, one hundred tastes" (*yi-cai-yi-wei, bai-cai-bai-wei* 一菜一味，百菜百味). Consequently, in many American restaurants, a chicken, fish, or meat dish is prepared with the same cutting style and the same combination of accessory vegetables, plus the same amount of monosodium glutamate and cornstarch. Subsequently, the result is that different dishes have similar appearances and tastes. Recently, many Sichuan and Hunan restaurants have been opened in large American cities, and have introduced hot food which has become very popular, particularly among college students. Riding the tide of popularity, a lot of cookbooks on hot dishes have been published, but many of them are written by people who have never been to Sichuan and/or any other areas famous for hot dishes. That "Chinese food is spicy" is a distorted impression. Actually, 80 percent or more of the Chinese population lives on bland food. Having lived for eight years in Sichuan, I know that food in middle-class homes and dishes of the outstanding restaurants are mild. The hot sauces are merely relishes on the table, used by individuals that like spicy food. It is true that small eating places that serve quick lunches at corners near schools have very hot noodles and a few other simple dishes. On my last trip to Sichuan, I bought a cookbook entitled *Popular Sichuan Dishes* (Liu et al., 1979). It contains 148 recipes, 53 percent of which do not use any hot pepper. A random selection of 25 dishes (20 with meat and 5 with only vegetable) from this book and another selection of equal number and similar nature from a cookbook on Sichuan food published in America shows the variation in the application of seasonings both in kind and in amount (the seasonings are arranged alphabetically and the numbers indicate the percentage of dishes using that particular seasoning, with that of the American cookbook in parenthesis): chili pepper 28 (56), cornstarch 36 (52), garlic 20 (20), ginger 60 (80), scallion 72 (80), soy sauce 68 (84), sugar 32 (68), vinegar 25 (48), wine 8 (76), and zanthoxylum (Sichuan peppercorns) 48 (52).

What a connoisseur thinks of Chinese food in American restaurants is well expressed by an eighty-year old legislator from Guangxi Province who visited his daughter and her husband in Los Angeles. The young couple spent their two week summer vacation showing him famous cities and parks of California and the adjacent states. They stopped in Monterey and dined at a restaurant called "Hongkong". In a reminiscence he recorded, "I have never had so poorly prepared Cantonese food in my life." Two days later, he was entertained by a family. The mother prepared a banquet about which he wrote: "No food is better than home-made dishes." With this introduction I present the following deliberations with recipes for those who would like to try new ideas for good food using Chinese plants and products.

All the plants discussed here are available in America. They are listed under new plants because they are not commonly sold in supermarkets and are not used as food by many American families. Some of them are native species, but most of them are exotic plants introduced into America by the Asian immigrants as food, or by horticulturalists as ornamentals. Forty-one items have been selected and these are subdivided into five categories according to the parts used: (a) leafy shoots, (b) flowers, (c) fruits, (d) seeds, and (e) special subterranean plant food. Within each group the entries are arranged alphabetically by their trivial names. In recipes where the best ham such as "Yunnan ham" or "Ginhua ham" is used in China, Virginia ham is given instead.

Leafy Shoots

Nine species are included for the discussion in this category. Several popular Chinese vegetables called *bai-cai* 白菜 are not included because they are available in most supermarkets. In North China *bai-cai* (*Brassica pekinensis* [Lour.] Rruprecht) is well cooked, pickled or raw for salad. The South China *bai-cai* (*Brassica chinensis* L.) is used in American Chinese restaurants, both with meat or in soups. If some readers grow these species in their kitchen gardens and the plants begin to flower early, rather than pulling them out and throwing them onto the compost pile, use the tender portion of the flowering shoots with a few leaves, one to three opened flowers and many flower buds to make a very good stir-fried pot-herb.

Amaranths (*Xian-cai* 莧菜)

In *The Herbalist*, I published a short article: "Shien … some noteworthy edible herbs of China" (Hu, 1948). In this article, six pot-herbs were described, five of which belong to the genus *Amaranthus* L. species in this genus lead the list of selected edible leafy shoots here, because of its nutritive value, tenderness, quick way to be cooked and easiness to be cultivated. Amaranths are a popular vegetable crop throughout China. Their use can be traced back to the third century of the Christian Era. The cultivars used for vegetables are fast growing plants with low stature and ovate green or red-green leaves, obtuse at the apex (Figure 2b). In Guangzhou and Hong Kong seeds are sown in early April, and up north in the Yangtze River and Yellow River Regions they are sown at the same time as cotton. Many farmers sow some amaranth seeds with cotton for family use. As a vegetable crop, the plants are harvested before they flower or branch. They are pulled from the ground and tied into one-pound bundles for the market. As a pot-herb, the tender stems are broken into 5–6 cm sections, each bearing 2 to 4 leaves. The tough portion of the stem and the root are not used. In kitchen gardens, the grower usually pinches off the top portion of the plant when it is 12 to 15 cm tall, and lets new branches grow from the basal portion. Naturally, the pinched tops are used as a pot-herb, and they can be harvested repeatedly.

Amaranth Greens (*Chao xian-cai* 炒莧菜)

Ingredients:

1 lb fresh amaranth (washed and drained)

1/4 medium-sized onion (sliced)

2 cloves garlic (slightly crushed)

1/4 tsp salt

3 Tbsp vegetable oil

3 Tbsp water

Brown the onion and garlic on medium heat. Raise the heat and add amaranth. Stir for 1 minute to avoid burning. Add water and salt. Simmer 1 or 2 minutes, stirring occasionally. Serve hot. Makes 4 servings.

How good fresh amaranth greens are, was illustrated by an American missionary who used to be stationed in Yichang, Hubei. On New Year's Day of 1947 she asked me: "Can you get me some *xian-cai* seeds? The thing that I missed most in Chinese dishes is *xian-cai*." Different cultivars have been introduced into America since that time (Figures 2b and 8b).

The field amaranth (*A. paniculatus* L.) attains the height of a man in China, hence it is called *ren-xian* 人莧 (= *jen-hsien*, man amaranth, Figure 5a). Recently, a similar species has become the focus of intensive nutritional and agronomic research by the government, private institutions, and universities in the United States, and it is now on the verge of commercial production. The leafy shoots are high in Vitamins A and C, riboflavin and folic acid; its grain contains twice the lysine of wheat and three times that of corn (Tucker, 1986). Here in Cambridge, Massachusetts I have pinched off the tender shoots when the plants are 15 cm tall, and continued to gather the later ones for several weeks. I have used the above recipe and cooked 4 or 5 pounds of the shoots for garden parties of staff and students in the Harvard University Herbaria. The dish was always widely praised and devoured with relish. Everyone commented on its being unusual and good. Garlic, by the way, is an important addition to the amaranth greens dish, as a condiment.

In China, seeds are gathered from field amaranth, heated, popped, and used in making candy. The wild plants in the experimental garden at Cambridge (Figure 5a) yielded very little seed, because while the upper flowers of a branch were still in bloom, the lower fruits had released their seeds. Moreover, the seeds gathered did not pop to give a fluffy mass as we were accustomed to in China. Consequently, the candy I prepared was full of tiny black specks. Nevertheless, the batch I made for the staff tea at the Harvard University Herbaria was received with good comments, and not a bit was left.

Amaranth Candy (*Xian-zi dian-xin* 莧籽點心)

Ingredients:

    1/2 cup popped black amaranth seed

    1/2 cup sesame seed (roasted to light brown, partially crushed by a roller)

    1/4 pound marshmallow

    1/2 cup flour (toasted to light brown)

Place 1/4 cup flour, amaranth seed and sesame seed in a warm skillet; melt the marshmallow and knead the amaranth and sesame seed into the marshmallow. Place the remaining flour on a board and transfer the amaranth-sesame-marshmallow mass to the floured board, making a sheet about 1 cm thick. Cut the sheet into pieces while it is still warm. The candy is ready when cool.

Aquatic Morning Glory (*Weng-cai* 蕹菜)

In southeastern Asia this species is a perennial, floating in ponds or stagnant streams or creeping on wetland. People eat the tender portion of the stem and use the tougher portion for animal feed. It flowers abundantly in late autumn (Figure 3c). In Boston it is planted in kitchen gardens as an annual. Its tender shoots can be harvested repeatedly (Figure 6a). Each time a section of 15–20 cm of the stem is pinched off the plant. Each section is further broken into 4–5 cm pieces, with at least one leaf attached. A few American Chinese restaurants are beginning to serve aquatic morning glory under the name "water spinach".

Aquatic Morning Glory Greens (*Chao weng-cai* 炒蕹菜)

Ingredients:

    1 lb fresh Aquatic Morning Glory tops (washed and drained)

    1/4 medium-sized onion (sliced)

    2 thin slices ginger (*ca.* 2 mm thick, shredded)

    1/3 tsp salt

    3 Tbsp vegetable oil

    1/2 cup water

Brown the onion on medium-high heat. Add ginger, salt, and tender shoots. Stir to avoid burning. When withered, add water, cover and stir occasionally. After 2 minutes, try a segment. If it is crispy and tender, the dish is done, otherwise, continue to cook for another minute or two. Serves 4.

Chinese Leek (*Jiu-cai* 韭菜)

This is the most common vegetable used throughout China, and the most widely cultivated species in the kitchen gardens of Chinese Americans (Figure 7a). It is a perennial which requires little care. The leafy shoots can be cut to ground level when they are 20–25 cm tall, and they can be harvested repeatedly during the growing season. A delicacy cultivated in darkness called *jiu-cai-huang* 韭菜黃 (= *chiu-tsai-huang*, yellow jiu-cai) is in high demand during the Chinese New Year season. People are willing to pay a high price for just a little bit. In southern China, the farmers invented a crude, hand-made, hollow earthenware cylinder with a top which is removed at night to expose the plants to dew or rain. In Chengdu farmers bank up the plants with sand as they grow, remove the tips and sell the yellow portions. In Beijing the root-stocks are brought indoors and forced in darkness for the market. Chinese leek can be cooked by numerous ways, three of which are described below.

Chinese Leek Omelette (*Jiu-cai chao-dan* 韭菜炒蛋)

Ingredients:

    1 lb Chinese leek (washed, drained, cut into 1.5 cm sections)

    4 eggs

    1/4 tsp salt

    3 Tbsp vegetable oil

Beat the eggs. Add leek segments and salt. Mix well. Heat oil on medium-high heat. Throw in a couple of leek segments to test the oil. When it is very hot, add the mixture and turn to avoid burning. Chinese omelettes should come out in pieces, both sides yellow. As the leek has its own flavor, no seasonings are needed. Serves 4.

Chinese Leek Salad (*Liang-ban jiu-cai* 涼拌韭菜)

Ingredients:

    1 lb Chinese leek (cleaned, washed, drained, cut into 1.5 cm sections)

    4 cups water

    3 eggs

    1 Tbsp vegetable oil

    1/2 tsp salt

    1 Tbsp soysauce

    1 tsp vinegar

    1 Tbsp sesame oil

    2 oz mung bean silk (*fen-si* 粉絲)

2 cloves garlic (crushed)

Boil water in a saucepan. Add sections of leek to the boiling water. Remove from heat and drain immediately. Rinse leek with cold water until it is at room temperature. Set aside for later use. Heat the vegetable oil in a skillet on medium heat. Beat the eggs with half of the salt. Make 2 or 3 thin egg sheets. Shred the egg sheets into 4–5 cm long pieces. Boil 2 cups of water. Add the mung bean silk and continue to cook until the contents become transparent and flexible. Rinse with cold water, drain and cool. Cut into 3–5 cm sections. Combine the leek, egg shreds, mung bean silk, sesame oil, vinegar, soy sauce, garlic and the remaining portion of salt for a tossed salad. This is a very handy dish for farmers to prepare during harvest season when they must have lunch in the field.

In Boston, the Chinese leek begins to flower in August. When the flowering stalk (scape) first emerges, the fresh apical portion bends. As soon as the stipe becomes straight (Figure 7a), it must be picked to let the leafy portion grow, and to prevent the plant from seeding itself freely throughout the garden. The tender stipes are the best part of the Chinese leek, and although it does take time to pinch off each one, it is worth the effort. The following recipe is a delicacy which I usually reserve for my most respected Chinese friends.

Stir-fried Chinese Leek Scapes (*Chao jiu-cai-tai* 炒韭菜苔)

Ingredients:

    4 oz Chinese leek flowering stalks (washed, drained, cut into 1.5 cm sections)

    4 oz lean pork chop (shredded)

    1 tsp cornstarch

    1 Tbsp soy sauce

    1 Tbsp sherry

    2 Tbsp vegetable oil

    1/2 tsp salt

    1 tsp sesame oil

Marinate the shredded pork in sherry, soy sauce, 1/4 tablespoon vegetable oil, and cornstarch. Heat the remainder of oil in a pan on high heat. Add marinated pork, stir and cook for 2 minutes. Add leek stalk and stir. Add water and cover for 2 minutes. Serves 2.

Chinese Mustard Greens (*Jie-cai* 芥菜)

Different cultivars of *Brassica juncea* (L.) Czern. are used as mustard greens in China.

Two of them have been introduced successfully into eastern United States. These are Swatow mustard (*jie-cai* 芥菜) from the warm regions of China, and red-in-snow (*xue-li-hong* 雪裡紅) from temperate and arid regions of China. Swatow mustard is available in American Chinese stores, both fresh and pickled. In stature it resembles the celery cabbage (*da-bai-cai* 大白菜) as it has large, obovate-oblong leaves 40–50 cm long with fleshy, broad petioles and coarsely wrinkled leaf-blades reaching to the lower one-fifth or one-third of the petiole. It differs from the celery cabbage in its uniformly jade green color and its unheading habit. Chinese restaurants normally do not use this vegetable as much as they use *bai-cai* 白菜 (= *pak-choi*) and the celery cabbage, hence it has not become popular to the American public and is not available in supermarkets. It is the choice, however, of all Chinese Americans even though its price is one-third higher than other forms of the genus *Brassica* L.. It has been successfully cultivated at Chang's Farm in Amherst, Massachusetts and it also grows well in New Jersey.

Red-in-snow has been introduced to private gardens. Mrs. Lucy Sie introduced it from Nanjing and raised it for decades in her garden at Hightstown, New Jersey. I personally had seeds sent to me from Lanzhou (Map, Loc. 20) and planted them in the Experimental Garden near the Harvard University Herbaria. There they have overwintered repeatedly. In the spring of 1985, the maple tree outside the garden was felled. Not realizing that the elimination of the tree would lead to other problems, I had hoped to see bigger plants because of the improved lighting. Evidently, however, the exposure of the yellow flowering clusters attracted purple finches (*Carpodacus purpureus* Gmelin) to the plants. At first I suspected they were searching for caterpillars. As they continued to visit the plants, I opened the garden gate to investigate the situation. Alas! The birds had stripped the seeds of all the fully grown but yet green fruits! Fortunately, it was not too late to net the cultivar as more flowers were on the stalk, but the season was getting late. The quality of the red-in-snow improves after a light snowfall, at which time some leaves turn reddish, particularly along the margin. Canned red-in-snow is available in American Chinese groceries.

Bean Curd with Swatow Mustard (*Jie-cai dou-fu* 芥菜豆腐)

Ingredients:

    1 lb Swatow mustard (cut into 2 x 1 cm pieces, washed and drained)

    1 pkg (supermarket style) bean curd

    4 oz baked ham (sliced into 2 x 1 mm pieces)

    1/2 medium-sized onion (sliced)

    3 Tbsp vegetable oil

1/2 tsp salt

1 Tbsp sesame oil

1/2 cup stock or chicken broth

Saute the onion on medium-high heat. Add baked ham, stir, and saute for 1 minute. Add Swatow mustard greens and salt, stir thoroughly. Cut the bean curd cross-wise into three portions. Hold one piece in the palm and slice, letting the pieces fall flat over the cooking mustard. Try to avoid overlapping. Add the stock and cover. Cook for 5 minutes on high heat, turning with care so that the bean curd pieces do not break. Cover and continue to cook for another 5 minutes. Stir, add the sesame oil. The dish is ready to serve. Serves 4. This is a dish which can be said to have color (red, green, and white), aroma (from the ham and sesame oil), and taste (unusual mustard flavor and tenderness). Eliminating the ham, the dish is just as delightful but less colorful and aromatic, and is a good vegetarian food.

Farmers' Mustard Salad (*Hua-bei jie-cai-dou* 華北芥菜豆)

Ingredients:

　　1 lb red-in-snow (carefully washed, leaving no sand particles lodged in wrinkles
　　　　of the leaf-blades, drained, cut into 5 mm sections, particularly the petiole and
　　　　midrib portions)

　　1/2 cup soybean (cleaned, soaked overnight)

　　2 medium-sized carrots (pared, diagonally sliced, and then shredded)

　　1 Tbsp soy sauce

　　1 Tbsp sesame oil

　　1 tsp salt

　　2 qts water

　　(Optional) 1 tsp vinegar

Wash and clean the soaked soybean and boil it in 1 quart of clean water for 20 minutes, adjusting the heat to prevent boiling over. Drain (keep water for soup, if desired), and cool. Boil the remaining 1 quart of water, add red-in-snow, and stir. Cover and bring to a boil. Drain (keep water, if desired) and cool under running water. Drain. Combine the soybean, carrot and mustard green, mix well. Place in a lightly covered container, and refrigerate. Before serving, add salt, soy sauce, vinegar, and sesame oil to taste. The recipe serves 6.

　　Chinese farmers with large families usually prepare 10 to 30 pounds of the preliminary mixture, adding salt only. This keeps for weeks. The other ingredients can be added at the time the salad is served.

Red-in-snow Bamboo Pork (*Xue-li-hong sun-si chao-rou* 雪裡紅筍絲炒肉)

Ingredients:

   10 oz fresh red-in-snow (cleaned, cut, boiled, cooled; as in the preceding dish)

   6 oz lean pork chop (shredded)

   1/4 small onion (chopped)

   1 Tbsp cornstarch

   1/3 cup water

   1 Tbsp soy sauce

   1 Tbsp sherry

   1/4 cup vegetable oil

   1 Tbsp sesame oil

   1 small can bamboo shoots (sliced 5 mm thick, and again into 5 mm shreds)

Marinate shredded pork in sherry, soy sauce, 1/4 Tbsp oil, onion, and cornstarch for 10 minutes. Heat the remaining oil on high heat. Add marinated pork and stir. Cook for 2 minutes. Add red-in-snow, bamboo shoots, and water. Stir, cover, and cook for 5 minutes on medium heat. Add sesame oil. Serves 4.

In Chinese Americans' homes and in cities of northern China, "Red-in-snow and Bamboo Shoots" is served as a delicacy. In such cases, 4 oz of pickled red-in-snow (washed and cut) is used. Home gardeners may pickle their surplus for this purpose. Actually, canned red-in-snow with shredded bamboo shoots is available in Chinese groceries. Opening a can and adding the marinated pork as described above makes a quick and very welcome dish for busy home-makers.

Edible Crown Daisy (*Tong-hao* 茼蒿)

Missionaries of the mid-nineteenth century have written about the cultivation of a crown daisy for use as a vegetable in China. It was not until the summer of 1917 when L. H. Bailey observed the crop in central China, that the edible crown daisy was thought to be a variety of the Mediterranean species. In the report *A Collection of Plants from China*, he named the edible form *Chrysanthemum coronarium* L. var. *spatiosum* Bailey (Bailey, 1920). Later, he even raised it to species level as *Chrysanthemum spatiosum* Bailey (Bailey 1949). In China, the area of cultivating edible crown daisy parallels that of the aquatic morning glory, and its use is not as widespread as amaranth. The tender shoot has a very strong smell. However, when one learns to eat it, one would like it very much. Here is a good example: In 1949 I was invited to visit the family of Richard E. Price in Concord, Massachusetts, a staff member of the Waltham Field Station. He had taught at Lingnan University, Guangzhou, China, where he learned to eat the edible crown daisy soup and the stir-fried tops. He had some young plants of this variety in his kitchen garden and served it in soup that evening. He had three daughters aged between five and nine. The children had learned to eat "Crown Daisy Soup" and liked the tender shoots so much that they fought over them by counting the number of the green shoots in their respective bowls and making sure they were having fair shares. The cooked shoot has a peculiar flavor which may be too strong for beginners, but after a few trials, it quickly becomes the favorite of many people.

As a crop, young plants are harvested four or five weeks after sowing the seed, by pulling from the root. In kitchen gardens, the shoots are pinched off to encourage the growth of lateral branches (Figure 8b), which can be harvested repeatedly.

Viable seed recently imported from Japan, raised and distributed in USA as a culinary vegetable and herb by the name 'Shungiku' is available on seed shelves in supermarkets.

Crown Daisy Soup (*Tong-hao ji-tang* 茼蒿雞湯)

Ingredients:

>6 oz crown daisy tender shoots (4 cm stem with attached leaves; washed and drained)
>
>2 eggs (slightly beaten)
>
>1 can chicken broth (diluted with 1 1/2 cans of water)
>
>1/4 medium-sized onion (sliced)
>
>1 Tbsp vegetable oil

Saute the onion with vegetable oil in a saucepan. Add the chicken broth and water, and bring to a boil. Add the crown daisy shoots and bring to a boil again, slowly pouring the egg around and over the surface of the boiling soup. Cook for 1 minute. Makes 6 servings.

Crown Daisy Greens (*Chao tong-hao* 炒茼蒿)

Ingredients:

>1 1/2 lb crown daisy tender shoots (washed, drained)
>
>1/4 medium-sized onion (sliced)
>
>3 Tbsp vegetable oil
>
>2 Tbsp water
>
>1/2 tsp salt

Brown the onion on medium-high heat. Then increase to high heat, add the crown daisy shoots and stir. Add salt and water. Cook for 1–2 minutes. Serve hot. Makes 4 servings.

Garden Pea Shoots (*Wan-dou-miao* 豌豆苗)

The fresh garden pea shoot is an elegant and expensive winter favorite in Guangzhou, Hong Kong, and other large cities in the warmer regions of China. The crop is planted only for the shoots which are individually hand-picked when a shoot attains leaf-bearing maturity and the foliaceous stipules are 2 to 4 cm long. The plant continues to produce lateral shoots for future harvesting. When pea shoots first appear in the market, they are used for a special winter combination dish cooked on the table-top. Later in the season, larger shoots with one fully developed leaf and another unfolding one become available. These are used for soups or as potherbs (Figure 9c).

Pea Shoot Soup (*Dou-miao-tang* 豆苗湯)

Ingredients:

> 6 oz pea shoots (cleaned, terminal tendrils removed, drained, kept fresh in a wet towel)
>
> 1 can chicken broth (thinned with 1 can of water)
>
> 2 eggs
>
> 1/2 medium-sized onion (sliced)
>
> 5 mm piece of ginger (minced)
>
> 1 tsp vegetable oil
>
> 1 tsp sesame oil

Saute the onion and ginger in a saucepan on medium heat. Add chicken broth and bring to a boil on high heat. Beat the eggs and stir in. Bring to a boil. Add the pea shoots. Stir till boiled. Add sesame oil. Serves 6.

Stir-fried Pea Shoot (*Chao dou-miao* 炒豆苗)

Ingredients:

> 1 1/2 lb pea shoots (terminal tendrils removed, shoots washed, drained, kept fresh)
>
> 1 medium-sized onion (sliced)
>
> 5 mm piece of ginger (thinly sliced, then shredded)
>
> 3 tsp vegetable oil
>
> 2/3 Tbsp sesame oil
>
> 1/3 tsp salt
>
> 1/3 cup water

Brown the onion and ginger in a frying pan. Add the pea shoots. Stir to avoid burning. Add water and stir again. Add salt and sesame oil. Serve immediately. Makes 6 servings.

Garlic Greens (*Suan-miao* 蒜苗)

Although relatively unknown to Americans, garlic greens are important items in Chinese cooking. During the warm season, garlic greens are always available in markets throughout the country, for they are used not merely as a condiment, but also as an essential ingredient in many dishes, such as double-cooked pork (回鍋肉). Normally, growers set the seedling cloves close together, and they harvest garlic greens by thinning the plantings (Figure 5b).

The flowering stalk (scape) of garlic is a highly preferred vegetable. As the plants approach maturity, each one sends up a scape terminated by an elongated spathe. A gentle pull can break the scape at the base without hurting the plant. The scapes are thus gathered and tied into bundles for the market.

Sichuan Double-cooked Pork (*Hui-guo-rou* 回鍋肉)

Ingredients:

> 1 lb fresh bacon with rind (parboiled with ginger and salt, cut into 4–5 cm squares, each portion sliced crosswise into 2–3 mm thin pieces)
>
> 6 oz garlic green (washed, cleaned, cut into 2.5 cm sections)
>
> 1 tsp salt (use half for boiling meat per above)
>
> 1 Tbsp *tian-mian* sauce 甜麵醬 (fermented flour sauce)
>
> 2 Tbsp hot soybean sauce (*la-dou-ban-jiang* 辣豆瓣醬)
>
> 1 tsp soy sauce
>
> 1 cm ginger (slice half for parboiling bacon, mince the rest)
>
> 1 1/2 medium-sized onion (sliced, use part for parboiling the bacon)
>
> 10 zanthoxylum (Sichuan peppercorns, use 5 for boiling bacon)
>
> 1/2 piece Syrian bread (cut into 4 x 1 cm pieces)

Saute the remaining onion, the zanthoxylum, and the minced ginger on medium-high heat. Add the sliced bacon, salt and stir until the slices begin to appear wavy. Add *tian-mian* sauce and continue to cook for 1 minute. Stir in the garlic greens, hot soybean sauce and Syrian bread. Stir and cook for 2 more minutes. Add soy sauce and sesame oil, mix well. The dish is ready. Serve with plain rice. Makes 6 servings.

Garlic Scape Sliced Pork (*Suan-tai chao rou-si* 蒜苔炒肉絲)

Ingredients:

> 3 pork chops (thinly sliced and shredded, each portion containing some fat, marinated for 15 minutes in soy sauce, minced ginger, sherry, and cornstarch)
>
> 8 oz garlic scapes (cut into 2–3 cm sections)

2 1/2 Tbsp vegetable oil

1 Tbsp sherry

1 Tbsp soy sauce

1/2 tsp salt

3 mm slice of ginger (minced)

1 Tbsp cornstarch

1 tsp sesame oil

1/2 cup water

Heat the oil and test with 1 or 2 sections of the garlic scape. When the test pieces turn brown, add the marinated meat to the very hot oil, stirring quickly. As the meat changes color, add the garlic scape and mix thoroughly. While continuing to stir, add water, mix, cover and cook for 3 minutes. Add sesame oil. Makes 4 servings with plain rice.

Mallow Green (*Kui-cai* 葵菜)

One day in the 1950s, while walking with Professor E. D. Merrill, the former director of the Arnold Arboretum, we talked about the common mallow (*Malva neglecta* Wallroth) that grew along the sidewalk. I told him, "Hundreds of acres of *Malva verticillata* L. are cultivated as a winter crop in West China. I had countless meals of it before coming to the United States." "So, the Chinese people are ahead of us in recognizing the value of this very nutritious herb." Dr. Merrill commented and continued: "Malva is the richest food ever analyzed in any food laboratory in the world." (Hu, 1956). Forty years have gone by during which Malva as a food of high nutritive value remains an academic subject in the United States. Perhaps the time for change has come. Forty years ago, mung bean sprouts were a commodity sold only in Chinese stores. Now they are available in all supermarkets. Perhaps Malva will have a place on the vegetable shelves soon. It will when people know how to use it for food.

Stir-fried Mallow Greens (*Chao kui-cai* 炒葵菜)

Ingredients:

> 1 1/2 lbs mallow greens (use the leaf-blades and tender shoots with developing leaves only, washed, and drained)

3 Tbsp vegetable oil

1/2 medium-sized onion (sliced)

5 zanthoxylum (Sichuan peppercorns)

1/10 tsp salt

3 Tbsp water

1 tsp sesame oil

Saute the sliced onion and the zanthoxylum. Add the mallow greens and stir. Add salt and water. Stir and cover to cook for 2 minutes. Add sesame oil and serve. Makes 6 servings.

Mallow Green Soup (*Kui-cai-tang* 葵菜湯)

Ingredients:

2 cups tender mallow shoots and leaf-blades (carefully washed, leaving no sand
   particles, drained)

1/4 medium-sized onion (sliced)

2 mm ginger slice (minced)

1/3 can chicken broth mixed with 1 can of water

2 eggs

1 tsp sesame oil

Saute the onion and mince the ginger. Add the chicken broth and water, and bring to a boil. Add the mallow greens and stir. Beat the eggs. As the soup boils again, drop in the beaten egg in circles, without stirring. Add the sesame oil and serve the clear soup with a garnish of white and green. Makes 6 servings. Home-made stock gives better results than canned. This can be prepared with chicken bones, pork or beef bones cooked slowly in crockpots with slices of ginger, some onion, and salt.

Matrimony-vine (*Gou-qi* 枸杞)

Over 100 species of matrimony-vine are known worldwide, particularly in the semi-arid and arid areas of the Americas. Two species in China are planted and used economically. The Chinese species are scandant, spiny shrubs which require supports to assume an upright habit (Figure 9). The species which yields edible leaves has a wide range of tolerance to environmental conditions. Normally it grows close to a wall or hedge, by neglected waysides, in graveyards and in saline or highly base soil with hardly any organic matter where other species cannot thrive. Chinese farmers have capitalized on the characteristic ecological plasticity of the species, and have planted annual winter crops for the leafy shoots in southern China and in orchards for the fruit in the deserts of northwestern China.

In the warmer areas of China, such as Guangzhou (Canton) and Hong Kong, the plants lose their leaves in summer. The healthy shoots are cut off, tied in bundles, and stacked in shady, wet places to allow them to form calluses and to produce roots and buds. As soon as the garden plots are free of the autumn crops and have been prepared for the next planting, the stacked branches are taken out of storage and planted. By the time of the Chinese New Year, the healthy new shoots are ready to be harvested for the market.

In the Boston area the species has been grown in kitchen gardens for at least 100 years, and has been naturalized as an escapee (Figure 9a). It may be found in the gardens of many Chinese Americans, particularly of those who immigrated from South China. Only the young shoots and large tender leaves are eaten. As soon as leaf miners appear in the leaves in summer, they are no longer used. Leaves on newly developed offshoots are two or more times larger than those on older branches, particularly those of the flowering and fruiting branches (Figure 9b). Growers often cut off the branches in early spring in order to induce the production of stout shoots with large leaves. These branches are chopped into small pieces and used as herb tea.

More acreage of matrimony-vine is cultivated in northern China for dried berries than is cultivated for the fresh, edible shoots and leaves. The production of these berries is a multimillion dollar business both for domestic distribution and for international trade.

Stir-fried Matrimony-vine Shoot (*Chao gou-qi-tou* 炒枸杞頭)

Ingredients:
>   4 oz tender matrimony-vine shoots (shoot apices with immature leaves)
>   2 Tbsp vegetable oil
>   1/2 medium-sized onion (vertically sliced)
>   2 Tbsp water
>   1/10 tsp salt

Saute the onion slices. Add the tender shoots. Stir for 1 minute. Add water and stir. Cover for 1 minute. Makes 2 servings. This is the simplest way to prepare a delicious and highly nutritious dish. When a cup of chicken broth and a cup and half of water replace the salt and water in this recipe, the dish becomes an equally good soup for two people.

The fruit of matrimony-vine is a special health food used particularly by the elderly (recipes for using the fruit will be described later). The leafy shoots and leaves are seasonal delicacies used particularly in connection with a fondu-type dish known as *Huo-guo* 火鍋 (*Huo-kuo*) by many people inside and/or outside of China, as *Da-bian-lu* 打邊爐 in Guangzhou and Hong Kong, and as *Mao-du-zi* 毛肚子 in Sichuan. This is a complex dish with raw meats, fish, and vegetables surrounding a sizzling hot, centrally placed container filled with clear stock.

In China the container is a portable, metal stove (usually copper) similar in shape to an angel-food cake pan, with the central cylinder rigged for burning charcoal, and the groove as the container for boiling the stock. A deep electric skillet can be an excellent substitute. *Huo-guo* is a winter dish designed for having fun with friends, hence the Cantonese name, *Da-bian-lu* 打邊爐 (everybody cooking-by-table-side). The ingredients and the seasonings all vary with the localities, with personal preferences and with market availability of products. In Chengdu, tripe and hot sauces are essential for this dish, hence it is called *Mao-du-zi* 毛肚子 (fussy tripe!). In Lanzhou, people prefer to put petals of white chrysanthemum in the dish, hence it is called *Ju-hua-guo* 菊花鍋 (chrysanthemum pot). Although the ingredients are flexible, the basic requirements remain the same: (1) a good home-made stock, and plenty of it, (2) two or three top grade cuts of meats, (3) chicken breasts, (4) some seafood, (5) several vegetables in season, (6) mung bean silk (softened), (7) egg noodles (precooked), (8) bean curd, (9) ginger, onion, and garlic, and (10) soy sauce, vinegar, sesame oil, and hot pepper sauce. Some people like to use tripe, liver, pork kidney, sesame butter, hot soybean sauce (*la-dou-ban-jiang* 辣豆瓣醬), and a hot peanut sauce (*sha-cha-jiang* 沙茶醬). The last item should improve the taste, but it is optional.

The stock can be prepared ahead of time by boiling and simmering some soup bone, with slices of ginger, some peppercorns, salt, all the trimmings from the meat, the skin from the chicken breast, and the odds and ends of all the vegetables in two gallons of water. A sample is given below.

Huo-guo 火鍋 (Fire Pot)

Ingredients:

    2 chicken breasts (thinly sliced, meat only, in 3 dishes)

    1 lb beef (trimmed, thinly sliced, evenly divided into 3 portions, placed in the dishes by the chicken slices)

    2 pork chops (lean meat only, sliced into bite-size, placed in the same 3 dishes)

    1 lb fish fillet (sliced into bite-size, divided evenly into a second set of 3 dishes)

    1 lb scallops (horizontally cut 2/3 through, apportioned into the fish dishes)

    3 squares bean curd (sliced 5 x 3 x 1 cm, put into the seafood dishes)

    2 lbs celery cabbage (preferrably *Brassica pekinensis*, use the outside 2 or 3 leaves in the stock pot, cut the petioles and midrib portions crosswise into 3 cm long pieces, and the blade portion twice as big. Divide into a third set of 3 dishes)

    1 pkg spinach (cleaned, washed, drained, use the petioles in the stock, divide and place into the cabbage dishes)

1 pkg broccoli or cauliflower (cut into bite-size pieces, divide into the cabbage
dishes)

4 oz mung bean silk (*fen-si* 粉絲 , softened by soaking in boiling water, place in
another set of dishes)

1 lb thin egg-noodles (precooked, liquid drained into the stock, rinsed once, mixed
with 2 tablespoons of vegetable oil, share the mung bean silk dishes)

9 Tbsp soy sauce (place in 3 separate sauce dishes)

3 tsp vinegar (add to the soy sauce dishes)

3 cm piece of ginger (slice half for the stock, mince the remainder, mix with the soy
sauce and vinegar)

5 cloves of garlic (crushed, mix with the above sauce)

4 tsp sesame oil (add to the above sauce dishes)

2 medium-sized onions (quartered, mince the central portions and add to the sauce
dishes, use the outer portions in the stock)

Optional Ingredients:

1 lb tripe (precooked, sliced into bite-size)

1 pkg chicken liver (cleaned and sectioned to large pieces)

4 pork kidneys (cut lengthwise with a very sharp knife and cut out all inside tough
tissues, soak the outer portion in cold water, change the water once or twice,
drain, and core diagonally on the smooth side)

3 Tbsp hot soybean sauce (*la-dou-ban-jiang* 辣豆瓣醬, Sichuan style)

3 Tbsp hot peanut sauce (*sha-cha-jiang* 沙茶醬, Swatow style)

1 lb Swatow mustard (cut bite-size)

3 heads of white chrysanthemum (fresh from the florist, break heads into 4 to 6
pieces, free the petals, wash and drain. Be sure to wash off any small insects.)

1 lb garden pea shoots

1/2 lb tender shoot of matrimony-vine (stem and leaves, 5 to 10 cm long, use whole)

3 Tbsp sesame butter

Have all the ingredients arranged in three sets of dishes, one set between two
persons. Since each person eats what he cooks, it is advisable for him to be able to
maneuver a pair of long bamboo chopsticks. A fork or spoon may do, but it is not very
handy for dipping pieces of food into the boiling stock and eating hot. The table is set
with a pair of chopsticks, a rice bowl and a Chinese porcelain spoon. From the sauce
dish each person takes some mixture into his bowl. He starts with the chosen meat,
holds it in the boiling stock until it is cooked, places it in a bowl to cool and to season it
before eating. After working on the meat for 10 to 15 minutes, the participants can try
the seafood. Here, it is advisable to hold the fish filets in a perforated ladle to keep

them from disintegrating in the stock. After another 10 minutes, one can put in some vegetables, and after another quarter of an hour, when most people have almost finished, the host can start to add mung bean silk, some egg-noodles, and more stock. When these are cooked, each person can hold his bowl close to the cooking device and remove the slippery mung bean silk and noodles, add some sauce and enjoy them. By this time, the best part of the dish is the soup! There is just nothing like it in this world! One can keep eating until nothing more remains to cook or to eat.

This dish can easily be converted into the Beijing or Mongolian rinsed (dipped) lamb by replacing meats, chicken breast, seafood, and the optional liver, tripe, and kidneys, with lamb or beef, and adding a raw egg to each person's sauce bowl as a cooling agent. The hot food eventually solidifies the egg portion attached to it.

Flowers

Various flowers are gathered and used locally as food. For example the male flowers of summer squash are used by kitchen gardeners for food. Few such flowers are brought to the market, although in Hong Kong, in Shatin market in summer, I have seen clusters of flowers and buds of *Telosma cordata* (Burm. f.) Merr. and have had soup made from it at my friends' houses. In North China, flowers of *Wisteria sinensis* (Sims) Sweet are sold in spring as a delicacy of the season. However, these species are cultivated essentially as ornamentals, not for food specially. Gardeners gather them for themselves and/or for some extra cash.

In general, cooking with flowers is a delicate practice which busy home-makers often avoid. Many of the recipes given here are translated from an illustrated book entitled, *Selected Royal Dishes* (Anonymous, 1985). According to this book, in the preparation of the royal dishes with delicate flowers, a special stock made with chicken wings, pork bones, carrots, celery, onion, ginger, salt, pepper, and a few follicle of zanthoxylum (peppercorns) is often used.

Daylily Flower Buds (*Jin-zhen-cai* 金針菜)

The only flower cultivated for food, to my knowledge, is the yellow daylily. An extensive review of this subject was published in *Daylily Handbook* (Hu, 1968). It has been over 30 years since that manuscript was completed. My present understanding of the distribution of the species of daylily, their physiological activities, and their blooming habits enable me to appreciate the wisdom of the ancient people in their timing for gathering the flower buds, and in their technique of curing them by the application of heat prior to drying in order to preserve their nutritive value. The following points were not included in that handbook.

Daylilies are native of temperate eastern Asia. The species with orange and/or fulvous-red flowers are diurnal (day blooming), and those with pale yellow flowers are nocturnal (night blooming) and fragrant. The flowering period for some daylilies (*Hemerocallis lilioasphodelus* L., *H. minor* Miller) are extended. They may last for one to three days. Flowers of these species begin to bloom in early afternoon, allowing farmers to pick the mature buds in the morning. They grow in North China, particularly on the grassy slopes of the eastern Qinling 秦嶺 mountains, where cultivars with slender, yellow, elongated buds have been selected.

An investigation into the physiological activities of *Hemerocallis lilioasphodelus* L. made in the 1950s shows that gathering the fully grown flower buds just before they open and curing them by applying heat to kill the cells quickly is the most effective way of getting the best out of the daylily flowers. In daylilies, 25 days before the flower opens, the bud begins gradually to build up water and nutrients. On the twelfth day before anthesis, the metabolic curve increases sharply. On the fourth day before anthesis a sudden steep increase of water, nitrogen, phosphorus, and sulfur compounds occurs. At the inception of anthesis the inorganic compounds drop sharply, while at the completion of blooming the organic compounds drop. Another experiment shows that the drop in inorganic and organic compounds from a mature flower bud to a fully open flower is due to a return of such elements to the plant body. Flowers withered on the flowering stalk lose 90% of their glucosides, and those cut and let withered in water lose only 22% of the same compound. Evidently, in nature, nutrients on the flowering branch return to the parent plant during anthesis. It takes keen observations of astute connoisseurs to detect the differences in quality and taste between a mature bud and the flower developed from it, and it takes brilliant minds to make use of such observations and experiences for practical purposes, i.e., designing the time of the harvests so that they correspond to the development of the flower buds and of applying heat to cure the buds in such a way as to preserve their fullest nutrient value for the most effective drying.

It is amusing that the keen observations and the ingenious designs are the creations of men of leisure rather than those with empty stomachs! Such were the Taoist and Buddhist monks living in temples situated in the homeland of the yellow daylily. The use of the dried daylily flowers developed by these monks for preparing their vegetarian dishes spread to their devotees in urban areas, especially to the monasteries and cities in the southern China where daylilies do not grow. Farmers plant daylilies as a labor demanding cash crop, therefore, they generally do not eat the fresh flower buds.

The establishment and growth of Chinese restaurants abroad created an international trade for *jin-zhen* 金針 (= *chin-chen*, golden needle), dried daylily flower buds. It was reported that in the 1930s the annual consumption of daylily flower buds in New York was 1,716 kg (4,000 pounds). In *Food Composition Table for Use in East Asia* (published by the US Department of Health, Education, and Welfare), the edible portion of *jin-zhen* contains six times of protein, four times of fiber, nine times of ash, 14 times of calcium, six times of potassium, four times of Vitamin B-complex (riboflavin, thiamins) and five times of niacin more than those of egg-plant (Leung, et al, 1972). Dried daylily flower buds are now available in American Chinese stores, and one package can be used for 6–10 times. In using the market product, the desired amount must be rehydrated in boiling water, washed, and the pedicel of each flower bud removed. My friends often keep the first water for enriching the stock or soup.

Daylily Egg-shred Pork or ***Mu-xu* Pork** (*Mu-xu rou* 木樨肉)

Ingredients:

> 3 pork chops (shredded, marinated in half of the salt, 1/3 of the ginger (minced), 1/2 of the onion (chopped), garlic, sherry, soysauce, and cornstarch)
>
> 2 eggs (beaten with 1/5 tsp salt, spread on a greased pan to make 3 thin sheets, then cut into 4 cm by 3 mm shreds)
>
> 3 oz dried daylily flower buds (revived by pouring boiling water over it in a container, then cooled and cleaned, with the stalk removed, and cut once through the middle)
>
> 2 oz wood ear (*mu-er* 木耳 (Auricularia; revived in another container with boiling water, washed, cleaned, with the hard portion removed, broken into irregular pieces about 2–3 cm across)
>
> 1 Tbsp soy sauce
>
> 1 tsp salt
>
> 1/2 cup water
>
> 1 large onion (sliced 3/4 lengthwise, 3–4 cm by 5 mm)
>
> 1 clove of garlic (mashed)
>
> 5 mm slice ginger (thinly shredded)
>
> 4 pieces of Syrian bread (warmed in the oven)
>
> 3 Tbsp vegetable oil

Brown the onion. Add the marinated pork, stir fry. Add 1/4 cup of water and cook for 2 minutes. Add the softened and cleaned daylily, wood ear, shredded egg, salt, 1/4 cup of water, and cook for 3 minutes. Wrapped in Syrian bread to serve 4. This dish is genuine *Mu-xu* Pork served with the addition of Syrian bread. It lacks cabbage as in the *Mu-xu* Pork served in most American Chinese restaurants. It is advisable that after reviving and cleaning the daylily and wood ear, keep them covered with water to render soft texture.

Buddha Disciples' Delight (*Luo-han-zhai* 羅漢齋)

Ingredients:

   4 oz daylily flower buds (revived, cleaned, stalk removed)

   1 oz hair vegetable (*fa-cai* 髮菜, Filamentous Nostoc; revived by soaking in warm water, washed, cleaned, separated)

   2 oz black mushroom (revived in hot water, save water, sliced lengthwise once)

   2 oz wood ear (revived, cleaned, separated)

   3 pieces of bean curd (sliced into 4 cm square, then cut into 1 cm thick pieces, brown with 1 tablespoon oil in a skillet)

   1 small can bamboo shoots (thinly sliced into 4 cm by 2 cm x 2 mm pieces)

   1 can wheat gluten (cut into 3 cm square by 1 cm pieces)

   1 small can water chestnuts (sliced horizontally once)

   1 cup fresh ginkgo seeds (or a small can if fresh seed is not available)

   1 section of fresh lotus rhizome (pared, sliced crosswise into 5 mm thick pieces, washed, drained; if not available, substitute with 1 cup revived, dried slices)

   2 cm long ginger (thinly sliced crosswise, then shredded)

   2 oz Laminaria (revived, cut into 4 by 3 cm pieces, washed)

   1 can straw mushroom

   1 can small corn-flower buds (from Taiwan)

   1 lb fried bean curd squares (cut lengthwise once)

   4 oz mung bean silk (revived in boiling water, washed, kept in cold water, cut into irregular sections)

   4 oz bean curd bamboo (soaked in cold water to revive it, cut into 4 cm sections)

   4 medium-sized onions (longitudinally sliced into 1 cm wide pieces)

   4 cloves garlic (cut lengthwise once or twice)

   1/4 tsp black pepper

   1/4 cup vegetable oil

   1 Tbsp sesame oil

   1 tsp salt

   1 cup water

This is a vegetarian dish, served often in Buddhist temples. In Chinese tradition, it is said that Buddha had 18 close disciples called *Shi-ba luo-han* 十八羅漢, therefore this dish calls for 18 principle ingredients. The amount of each ingredient depends on the number of people in the party. This recipe is planned for 6 people and is generally eaten with white rice. It is adviced that all the ingredients revived from dried material be kept under water after cleaning in order to keep soft.

Saute the onions in a large saucepan, then add all the ingredients and seasonings. Bring to a boil, reduce heat to medium and continue cooking for 30 minutes. Check the gravy, add some water if needed. Turn to low heat, and keep simmering for an hour. Cool and store in the refrigerator. Before the arrival of the guests, warm up on medium-low heat. It is a handy dish which can be prepared in advance.

Vegetarians' Shark-fins （素魚翅）

Ingredients:

> 4 oz daylily buds (revived in boiling water, cleaned, pedicels removed; push a needle through the middle of the swollen sepals and petals, and run it to the distal end, repeat once or twice)
>
> 1/2 oz black mushrooms (revived, shredded across the cap)
>
> 1/2 of a small can of bamboo shoots (sliced and shredded)
>
> 1/2 a medium-sized carrot (pared, diagonally sliced and shredded)
>
> 4 oz fresh coriander (cleaned, washed, drained dry)
>
> 1 cup vegetable oil
>
> 1 oz wheat gluten (shredded, mixed with 1 tsp cornstarch)
>
> 1/10 tsp Ac'cent
>
> 2 Tbsp sherry
>
> 2 cups chicken broth
>
> 2 Tbsp cornstarch
>
> 1 Tbsp flour

Use 2 tablespoons vegetable oil to stir-fry the coriander. Discard the brown coriander and keep the oil. Mix the cornstarch and flour to make a smooth batter. Dip the tube of the daylily bud into the batter, then deep-fry the daylily in a medium hot pan. Place the deep-fried daylily in a bowl, adding half of the Ac'cent, sherry, salt, and 1 cup of chicken broth, then steam the mixture for 15 minutes. Boil the shredded mushroom and bamboo shoots. Stir-fry the shredded gluten. Combine the mushroom, bamboo, and gluten. Add 1/2 cup of chicken broth, sherry, Ac'cent and salt. Cook for 2 minutes. Discard the liquid, and place the mixture on a platter. On top of the mixture, put the steamed daylily. Add stir-fried carrot. Prepare a sauce with the remaining chicken broth and coriander oil and ladle over the prepared platter. Dried daylily flower was used to prepare imitation shark's fin, a dish for the Empress Dowager on days scheduled for vegetarian food.

Sour and Hot Soup (*Suan-la-tang* 酸辣湯)

Ingredients:

> 2 oz daylily (revived, washed, stalk removed, cut once)
>
> 1 oz wood ear (revived, washed, torn into 2 cm pieces)
>
> 2 eggs
>
> 1 piece bean curd (sliced into 3 cm x 5 mm x 5 mm shreds)
>
> 1 pork chop (shredded, marinated)
>
> 1 Tbsp lemon juice (or 1 tsp vinegar)
>
> 1/2 tsp black pepper
>
> 1 Tbsp soy sauce
>
> 1/2 tsp sesame oil
>
> 2 1/2 cups water
>
> 1/2 cup chicken stock (or 1 boullion cube)
>
> 1/2 medium-sized onion (minced)
>
> 1 slice ginger (minced)
>
> 1 Tbsp flour
>
> 1/5 tsp salt

Saute the onion and ginger. Add the marinated pork and stir-fry. Add water, daylily, wood ear, and bring to a boil. Slice and shred the bean curd, letting the pieces fall into the boiling mixture below. Add lemon juice, black pepper, soy sauce, and salt. Use 4 tablespoons water to mix with flour to make a smooth, thin batter. Pour the batter slowly into the soup while stirring. Beat the eggs in a bowl. Holding the bowl with one hand and a pair of chopsticks in the other, slowly pour the beaten egg along the side of the pan, and stir in one direction. The movement of the liquid makes fine shreds of the egg. Add sesame oil and the soup is ready. A person feeling the onslaught of the flu will be helped by eating a bowl of this soup just before getting into a warm bed. A couple of aspirin in addition may help to relieve pains and to insure a good sleep. This is the origin of "Hot and Sour Soup" which started in North China, and it has recently become popular in Chinese restaurants in America.

In the sixteenth and seventeenth centuries, Chinese fleets visited the important ports in southeastern Asia and along the Indian Ocean, reaching the eastern coastal cities of northern Africa. Daylily and wood ear were among the food material taken on board.

Chrysanthemum (*Ju-hua* 菊花)

The wild species of chrysanthemum are known only in eastern Asia. Most of them have small yellow flowers, hence the generic names, *chrys* (o) = golden, *anthemon* = flower. Ancient people in China first used wild chrysanthemum for food (*Dendranthema lavandulifolium* [Fischer ex Trautv.] Ling et Shih), and for drink (*Dendranthema vestitum* [Hemsley] Ling). Later, the horticultural forms were developed from a hexaploid hybrid (*Dendranthema* x *grandiflorum* [Ramat.] Kitam.). In the fourth century A.D., Ge Hong 葛洪, a Taoist, gathered the flowering shoots of wild chrysanthemum and mixed them with grains to make fermented drink. Presently, the people of Nanjing still eat the tender young leafy shoots of a wild chrysanthemum and many people drink chrysanthemum tea which is available in Boston. The fresh petals of a white horticultural form is used as a delicacy in banquets by the Chinese people both in China and abroad.

Twin Flower Tea (*Shuang-hua-cha* 雙花茶)

Ingredients:

> 4 oz dried chrysanthemum flowers (available in Chinese stores, keep in closed glass jar)
>
> 4 oz dried honeysuckle flowers (available dried in Chinese stores, mix thoroughly with chrysanthemum flowers and keep them together in a closed jar; fresh honeysuckle flowers picked prior to opening in the early morning may be a substitute)

Boil 6 cups of water, then pour into a thermos bottle containing 2 tablespoons of the above flower mixture. Drink with or without sugar. Designed as a tonifying beverage for the elderly, and to dispel dizziness and ringing in the ears.

Chrysanthemum flowers mixed with green tea is a common combination used as a summer beverage in the Lower Yangtze River Region. Large white florist chrysanthemum is used in banquets.

Chrysanthemum and Rice Crust (*Ju-hua guo-ba* 菊花鍋粑)

Ingredients:

> 2 fresh, white chrysanthemums from a florist (break each head into pieces, pull off the individual flowers; soak, wash, and drain)
>
> 3 pork chops (shred meat, then marinate in onion, ginger, soy sauce, and sherry; make 1 cup of stock with the bones)
>
> 1 Tbsp sherry
>
> 4 tender leaves of celery cabbage (cut across into pieces of 1 1/2 cm long)
>
> 1 oz dried daylily (revived, cleaned, with stalk removed; keep in water for tenderness)

1 oz wood ear (revived, cleaned, and pulled into 2–3 cm sections; keep soaked)

1 small can bamboo shoots (cut into pieces of 3.5 cm by 1.5 cm, thinly slice the
    pieces)

1 small can water chestnuts (transversally sliced once)

1 cup glutinous rice (mixed with 1/2 cup common rice, washed, drained, add 3
    cups of water, 1/5 tsp salt, and 1/5 cup of margarine, cook on high heat. After
    water is absorbed, leave heat on low for 15 minutes and then turn to medium-
    low, and add 1/2 cup common rice (washed, drained) to be cooked with above
    mixture.

1 tsp salt

3 cups water

1 medium-sized onion (sliced lengthwise, 5 mm wide)

2 garlic cloves (minced)

1 cm ginger (thinly sliced, then shredded)

2 Tbsp soy sauce

1 tsp sesame oil

1/5 cup melted margarine

5 cups vegetable oil

To prepare the crust, carefully scrape all the soft rice away from the sides of the
wok; save the soft rice to make a dessert, using only the hardened crust. Take the crust
off the wok, and break it into bite-sized pieces.

Saute the garlic and onion in 2 tablespoons of vegetable oil. Stir-fry the marinated
pork. Add celery cabbage, daylily, wood ear, bamboo shoots, and water chestnuts. Mix
well. Add soy sauce, the stock from pork bones and chicken broth. Cover and cook on
medium heat.

In another pan heat the remaining oil and deep-fry the rice crust to pale brown.
Drain. Place the fried crust in a large, hot bowl and put it in the center of the table.

Pour the cooked vegetables and meat over the fried hot crust, making a sizzling
noise and much steam. Sprinkle with chrysanthemum and add sesame oil. Eat while
the fried rice crust is crispy. Makes 6 servings.

The secret to serving chrysanthemum flowers is to cook them as little as possible.
As mentioned above, fresh chrysanthemum can be used as an ingredient for *Huo-
guo.*

Jasmine (*Mo-li-hua* 茉莉花)

In Guongdong and Fujian jasmine is extensively cultivated for its flowers, formerly
used for seasoning tea and now also for the extraction of essential oil. Jasmine tea is

common in China and it is available in Chinese stores in America. Using jasmine in food is rare. The following recipe is adapted from *100-year Old Royal Recipes* (Anonymous, 1984).

Jasmine with Silver Ear (*Yin-er mo-li* 銀耳茉莉)

Ingredients:

    1 oz silver ear (*yin-er* 銀耳, Tremella; revived in a bowl of 4 cups of boiling water, covered to cool. Separated into 2 cm pieces, clean, remove yellow portions, and transfer to 2 cups of clean water. Bring to a boil, and keep simmering for 2 hours)

    2 dozen fresh jasmine flowers partially open (washed, then drained)

    3 cups home-made stock

    1 1/4 medium-sized onion (1/4 minced, 1 quartered)

    1 1/4 tsp salt

    4 mm section of ginger (thinly sliced)

    1 Tbsp sherry

    1/10 tsp Ac'cent

    1 1/2 Tbsp vegetable oil

    10 cups water

    1 lb chicken wings (washed and drained)

    1/2 lb pork bone (washed and drained)

    2 stalks of celery (sliced into 5 mm sections)

    2 carrots (pared, cut into 2 cm sections)

    5 zanthoxylum (peppercorns)

Home-made stock: Saute the four inner and center pieces of the quartered onion in 1 tablespoon of oil. Add the ginger, chicken wings and pork bone and stir for 1 minute. Add 10 cups of water, the outer portions of the quartered onion, celery, carrots, 1 teaspoon of salt and peppercorns. Bring to a boil, then reduce heat and simmer for 2 hours. Take out the solid material and keep the liquid to be used as stock.

Saute the minced onion in 1/2 tablespoon of vegetable oil. Add 3 cups of the stock and the silver ear with liquid. Bring to a boil. Continue to simmer for 30 minutes. Add sherry, Ac'cent, the jasmine flowers, and a little salt to taste. Stir and serve. Makes 4 servings. Keep the remaining stock in the refrigerator for future use.

Sacred Lotus Flower (*He-hua* 荷花)

The sacred lotus has an important place in Chinese culture. Thousands of hectares are cultivated throughout the country. Every portion of the plant is used in Chinese food and/or medicine. The following formula is adapted from a Chinese royal recipe.

Chicken with Lotus Flowers (*He-hua ji* 荷花雞)

Ingredients:

> 1 large chicken breast (thinly sliced to 4 x 2 cm size, marinated for 10 minutes in minced ginger, salt, sherry, cornstarch, and 1 egg white)
>
> 2 large, fresh flowers of sacred lotus (pull off the petals, wash, drain, cut diagonally once, dip in boiling water and cool immediately under cold, running water)
>
> 5 medium-sized black mushrooms (revive in boiling water, and keep soaked in cold water)
>
> 1 small can bamboo shoots (thinly sliced longitudinally to match the size of the chicken pieces)
>
> 2 oz Virginia ham (or Yunnan ham from China; washed, dried, sliced thinly to match the chicken pieces)
>
> 10 garden pea shoots (washed, drained, and kept fresh by wrapping in a damp paper towel)
>
> 1 egg white
>
> 1/4 cup home-made stock
>
> 1/10 tsp Ac'cent
>
> 1 Tbsp sherry
>
> 1 Tbsp melted chicken fat
>
> 1 tsp cornstarch
>
> 1/4 medium-sized onion (minced)
>
> 1/5 tsp salt
>
> 2 Tbsp vegetable oil
>
> 1/4 cm piece ginger (minced)

Saute the minced onion in vegetable oil. Stir-fry the marinated chicken slices, ham, black mushrooms, and bamboo shoots. Add the stock and cook for 1 minute, stirring constantly. Add the chicken fat, lotus flowers, and pea shoots. Makes 2 servings. Sesame oil (1 teaspoon) may be used as a substitute for chicken fat.

Magnolia Flower (*Yu-lan-hua* 玉蘭花)

The petals of fresh flowers of *Magnolia denudata* Desr., gathered soon after they open, are eaten as a delicacy. Because this species does not grow well in subtropical and tropical areas, this practice is limited geographically to those people who have this species in their gardens, which is rather rare. The Chinese name for Magnolia petals as they are used in food is *yu-lan-hua-pian* 玉蘭花片. It must not be confused by the term *yu-lan-pian* 玉蘭片, or *lan-pian* 蘭片 used by the royal cooks in Beijing, where they call dried winter bamboo shoot *lan-pian*, and the revived dried bamboo shoot *yu-lan-pian*. Actually, on careful examination of the 65 recipes of the former food served to the royal family and now available in a special restaurant called *Fang-shan* 仿膳, no magnolia flower petals are used now. However, in the early 1940s I did have deep-fried magnolia petals in Chengdu where the white magnolia trees were cultivated in gardens (Figure 24).

Deep-fried Magnolia Petals (*You-zha yu-lan-hua-pian* 油炸玉蘭花片)

Ingredients:

> 10 fresh, white magnolia flowers (carefully pull off the petals, wash, drain, and
> cover with damp cloth)
> 2 cups vegetable oil
> 1/2 cup flour
> 1 Tbsp cornstarch
> 2 Tbsp sugar
> 1/5 tsp baking powder

Make a smooth batter with flour, cornstarch, sugar, water, and baking powder. Heat the oil first on high heat, then reduce to medium-high. Holding each petal individually with a pair of chopsticks, dip it into the batter, then place it in the oil. Deep-fry the battered petal until it turns slightly yellow. Eat while hot. Makes 6 servings.

Osmanthus (*Gui-hua* 桂花)

The osmanthus are evergreen trees or shrubs having small, yellow or golden flowers which sweeten the autumn air in southern China, particularly in Guilin 桂林 (City of Osmanthus Woods). The cultivars of *Osmanthus fragrans* (Thunb.) Lour. are grown throughout China in large pots which can be moved indoors in northern China during the winter. Individual growers usually pick the sweet smelling flowers, mix them with brown sugar and lard, and use the mixture as fillings for pasteries. The following recipes are adapted from food served to the royal family in Beijing.

Imitation Dried Clam (*Su-yao-zhu* 素瑤柱)

Ingredients:

> 4 oz bean curd bamboo (*fu-zhu* 腐竹; revitalized by soaking in 2 cups of warm

water overnight, cut into 2.5 cm sections and seasoned with 4 tablespoons of home-made stock, soy sauce, sherry, and star anise in a warm covered pan until the liquid is completely absorbed)

3 egg whites

1 large stalk of coriander (about 25 cm long, cut into 4 cm pieces)

1/4 carrot (finely grated)

6 fresh water chestnuts (sliced)

1 Tbsp sherry

1/10 tsp Ac'cent

1 Tbsp soy sauce

2 sections of star anise

4 Tbsp home-made stock (see that for Jasmine)

2 cups warm water

2 Tbsp peanut oil

In a hot skillet over medium heat, brown the coriander until it becomes dry. Discard the coriander and keep the oil. Beat the egg whites, add the marinated bean curd bamboo sections, water chestnuts, and carrots. Keep half of the coriander oil in the skillet. Add and stir-fry the bean curd bamboo mixture, and gradually add the remaining oil while stirring. To the mixture add the remaining sherry and Ac'cent. Finally, stir in the osmanthus flowers. Serves 4.

Osmanthus Cake (*Gui-hua-gao* 桂花糕)

Ingredients:

1 cup sugar

4 eggs

2 cups flour (cooked by steaming)

1/2 cup osmanthus flowers (washed and drained)

2 Tbsp sherry

Beat the eggs. Add sugar, sherry, osmanthus flowers, and the steamed flour. Mix well and pour into an 11 inch generously greased pan. Steam for 20 minutes. Cool. Makes 6–8 servings.

Rose (*Mei-gui* 玫瑰)

The prickly and bristly rose, *Rosa rugosa* Thunb. growing on the less fertile hillsides, is cultivated for its fragrant flowers which are harvested between late April and July and dried in the air. The market demands for rose flowers are for flavoring tea and pastry and for medicinal uses. In China dried rose petals are available in tea stores. Farmers plant this rose as a cash crop. They seldom keep the flowers for their own use. The gentry who cultivate this fragrant rose usually make a sugar preserve of the petals and use it for pastries. In making a sugar preserve, the flowers must be gathered as soon as they open, spread thin in a tray allowing the petals to shrink. The withered petals are slightly kneaded in the hand with an equal amount of sugar, i.e. a cup of sugar to 1 cup of tightly packed petals. The mixture is kept in a closed jar for future use.

Rose Sesame Bread (*Mei-gui zhi-ma-bing* 玫瑰芝麻餅)

Ingredients:

> 2 cups all purpose flour
>
> 1 pkg yeast
>
> 1/2 tsp salt
>
> 1/2 cup warm water
>
> 2 Tbsp sugar
>
> 1 stick margarine (softened)
>
> 3/4 cup sugar-preserved rose petals
>
> 3/4 cup sesame seeds

Set aside 1/4 cup of flour to use for kneading. Mix the remaining flour, salt and yeast. Stir in water and margarine. Knead lightly to make a smooth dough. Cover with a wet towel and let it rise until it has doubled in size. Knead again and let it rise again. Meanwhile, roast the sesame seed in a skillet on medium heat. Stir until brown. Place the sesame seeds on a board and crush with a rolling pin to make a coarse aromatic powder. Mix the rose-flavored sugar and the sesame powder.

After the dough has risen a second time, knead it into a smooth cylinder about 30 cm long. Slice the dough into eight equal pieces, turning the cylinder so the cut surfaces of each section are not exactly parallel. Scatter some flour over the sections, particularly over the cut surfaces to avoid sticking. Roll a piece of the dough, 20 cm in diameter (about 8 inches). Spread 1/6 of the rose-sugar and sesame mixture evenly over it, leaving a 2 cm (1 inch) edge free. Repeat twice, each time placing the dough sheet over the preceding one and spreading 1/6 of the mixture over it. The fourth dough sheet serves as a pie crust. Pull the top and bottom layers slightly and seal them with two fingers. This makes one bread. Make another one with the remaining ingredients. Serve hot. Makes 6 servings.

Place a piece of damp cloth on the bottom of the steamer and carefully place your round bread over it. A Chinese steamer has two or more layers, so the two breads can easily be cooked together on high heat. However, a home-made steamer can be fashioned by placing two empty, cleaned, and hollow tuna cans in a two-gallon pot with a small rack to support a thin plate. In this instance, one can use only half of the ingredients and make one steamed rose sesame bread. This recipe is adapted from the recipe for home-made moon-cakes of North China farmers who cook with large woks (*guo* = *kuo* 鍋) one meter or more in diameter. For the Mid-autumn Festival (the full moon day in September) each family makes a large (30–40 cm in diameter) moon-cake for every member to share, and many small cakes so that every one can be satisfied. To fit the large woks, the farmers improvise the steamers using the long, slender peduncles of sorghum for the bottom and the cover, and wheat straw to build the surrounding wall. In such a device, one can steam many small cakes or one large cake in the center surrounded by small ones.

Rose-sesame *Tang-yuan* (*Mei-gui tang-yuan* 玫瑰湯圓)

Ingredients:
>   2/3 cup glutinous rice flour (available in American Chinese stores)
>   1/3 cup sugar-preserved rose petals
>   1/3 cup sesame seeds (roasted brown just before use)
>   1/2 stick of margarine (softened)
>   1/2 cup water
>   1/3 cup pancake syrup (maple syrup or fruit syrup as individual preference
>   dictates)

Roast the sesame seeds until golden brown and make a powder as described above for making rose sesame bread. Mix it well with the rose sugar and margarine. Roll the mixture into a 1 cm diameter stick. Cut it into 1 cm pieces which will be used as filling for the individual *tang-yuan* (湯圓). To prepare the body of this pastry, gradually add water to 1/2 cup of glutinous rice flour to make a smooth dough. Set some flour aside in a large bowl for rolling *tang-yuan*. Don't worry if some water remains. If the dough is too dry, add a little more water. Take a 1.5 cm piece of wet dough, roll it between the palms and press it into a 3.5 cm cake. Place a piece of the filling in the center and seal the *tang-yuan* into a ball. Place some reserved flour on a plate and roll the just made *tang-yuan* in the flour. Put the rolled *tang-yuan* on a plate and when all have been rolled, keep covered in a refrigerator.

Just before serving, boil one quart of water in a medium-sized saucepan. Into the boiling water drop the *tang-yuan* one by one, and with a wooden spoon moving them slightly so they will not stick to the bottom of the pan. When half of the *tang-yuan* are in the water, reduce the heat to medium, for too high heat may break the first ones, which by now are floating. After all the *tang-yuan* are floating on the surface, simmer for 2 or

3 minutes. If the first ones show signs of breaking, remove them carefully. Makes 4 or 5 servings.

To serve the *tang-yuan*, some people like syrup drizzled over the hot *tang-yuan*, while others prefer the water in which they were cooked. *Warning*: *Tang-yuan* is extremely hot inside! Allow sufficient time to cool and/or avoid eating whole!

Tang-yuan is a common winter dessert, usually home-made throughout China. The most famous brand name is *Lai-tang-yuan*, sold at a sidewalk stand in Chengdu. In the spring of 1938, I was introduced to this shabby stand where several people sat at a table and others stood by, some eating, others waiting. People liked *tang-yuan* because it was "sweet, aromatic, white, and soft yet chewy". The fillings vary. In place of rose petals, osmanthus flowers are used. Often no flowers are used at all, and black sesame is used in place of the ordinary white type. In China lard and chicken fat are used in place of margarine.

Wisteria (*Zi-teng-hua* 紫藤花)

The Chinese wisteria (*Wisteria sinensis* [Sims] Sweet), is a garden favorite throughout China. It is only used as food in northern China where the flower of an introduced tree, the Black Locust (*Robinia pseudoacacia* L.), is used in a similar manner. The flower clusters are gathered when the basal 1/10 of the flowers in the panicle have fallen, the next 1/10 of the flowers are in bloom, the middle flower buds are fully developed, and the remaining buds are in various stages of development, including some which are very young. Suburban gardeners having the plant near their houses gather the flower clusters to sell to people in town as a means for extra cash. Farmers living farther from town eat the flowers themselves or share them with their neighbors. Wisteria is not a common commodity, nor is it ever eaten by people living in warmer regions of China. The flower of the Chinese wisteria is not fragrant.

In the spring of 1948 when I first moved to the top floor of the former Bussey Institute dormitory, the wisteria plantings of the Arnold Arboretum were on the old Bussey grounds. The flowers were of various colors and in very large clusters. I took half a dozen clusters and cooked the flowers in the manner of the North China farmers. When I ate them, however, they had a peculiar aroma which I did not like. My room-mate thought they were delicious, and she ate them all. The next morning I went to her room and woke her up. As she stirred, I said: "Thank God!" When she awakened fully, she asked: "What is the matter?" "Last night you ate so much of the wisteria flower, it scared me. I was afraid that you might have been poisoned, for the flowers did not have the same taste of those which I was used to in North China." My room-mate was from the Yangtze River Region where people do not eat wisteria flowers, and that was her first experience. Later, with Rehder's *Manual of Cultivated Trees and Shrubs*, I learned that the flowers I cooked were *Wisteria floribunda* (Willd.) DC., a native of Japan.

Steamed Wisteria Flowers (*Zheng Zi-hua* 蒸紫花)

Ingredients:

> 4 cups wisteria flowers (some fully open, others in various stages of development; washed and drained)
>
> 3/4 cup all purpose flour
>
> 3 cloves of garlic
>
> 1/2 tsp salt
>
> 1 Tbsp soy sauce
>
> 1 tsp vinegar
>
> 1 tsp hot soybean sauce
>
> 1 Tbsp sesame oil

While the flowers and buds are damp, mix them with flour. Be sure that all of them are partially or entirely covered. Steam the mixture for 20 minutes. Meanwhile, crush and mince the garlic, mix it with salt, soy sauce, vinegar, soybean sauce and sesame oil. Take out the steamed flower and flour mixture. Use a pair of chopsticks to loosen and cool it. Stir in the mixed sauce and serve warm. Makes 2 to 4 servings.

Fruits

The term "fruit" has slightly different meanings in the market and in a botanical classroom. In the market, apple, apricot, avocado, papaya, peach, plum, pomegranate, squash, strawberry, tomato, and watermelon are all fruits, although some of them are used as vegetables. In the botanical sense, anything that is developed from a true ovary is a fruit. Technically, barley, corn, millet, rye, sun-flower seed, and wheat are all fruits. All fruits common in the market and the grains used as staple foods are omitted here. Only selected items of unusual Chinese fruits used for vegetables that are available in American Chinese groceries and/or in natural food stores are discussed here. They are arranged alphabetically by their common names.

Bitter Melon (*Ku-gua* 苦瓜)

The bitter melon is a summer vegetable in the warmer areas of China. Here in Boston, it is available throughout the year. The market product is a shiny green cucumber-like cylindrical fruit with a bumpy surface, 17–20 cm long, 3–4 cm in diameter, and tastes bitter. Like cucumber, it grows on a slender vine bearing unisexual, yellow flowers (Figure 60). In North China, the species is cultivated as an ornamental vine for its colorful, bitter-sweet-like, ripe fruit which opens with orange reflexed pericarps and scarlet red contents. In South China, the fruits are harvested as soon as they attain full size. People in Guangzhou and Hong Kong like bitter melon for its cooling effect and cook it with meat or bean curd, with or without fish.

Bitter-melon Pork (*Ku-gua rou-si* 苦瓜肉絲)

Ingredients:
> 4 pork chops (slice the meat and marinate it for 10 minutes in soy sauce, sherry, pepper, ginger, and cornstarch)
> 2 bitter melons (halve the melons vertically, dispose of the seeds, slice diagonally into 2–3 mm sections, and parboiled if desired but not advised)
> 2 cloves of garlic (minced)
> 1 1/2 Tbsp *dou-chi* 豆豉
> 1/3 tsp salt
> 1 tsp cornstarch
> 1 Tbsp sherry
> dash of pepper
> 5 mm piece of ginger (minced)
> 3 Tbsp vegetable oil
> 2/3 cup home-made stock (prepared by bringing the bones of the chops to a boil in 2 cups of water with sliced onion, celery tops, and salt, and then simmering for 45 minutes)
> 1/2 medium-sized onion (sliced)
> 2 cups water

Saute the garlic. Stir in the marinated meat and cook until its color all changed. Add the bitter melon and *dou-chi*. Pour in the home-made stock, stir and bring to a boil. Cover for 5 minutes on medium heat. Makes 4 servings.

Bitter melon is like papaya and some other tropical fruits in that, on first trial, one may not like it, but when one learns to eat it, the taste can be acquired. I did not like bitter melon before having a dish prepared by S. S. Chow 周賽茜. After eating her "Bitter Melon Bean Curd Fish", I have never refused a bitter melon dish no matter how it is prepared. Some people suggest parboiling the melon to make it less bitter. To preserve the food value, I always cook the melon fresh (Figure 60).

Bitter Melon Bean Curd Fish (*Ku-gua dou-fu-yu* 苦瓜豆腐魚)

Ingredients:

 1 lb filet of scrod or sole (cut horizontally into 3 cm sections)

 1 pkg bean curd (cut into 4 x 3 x 1 cm sections)

 2 bitter melons (vertically halved, seeds removed and sliced across into 3–4 mm thick pieces)

 3 cloves of garlic (smashed)

 2 Tbsp *dou-chi* (washed, drained, and minced)

 1/2 medium-sized onion (vertically sliced)

 5 mm piece of ginger (shredded)

 3 Tbsp vegetable oil

 1/2 cup chicken broth

 1/2 cup water

 1/2 tsp salt

 1 tsp sesame oil

Saute the onion and garlic. Spread the bean curd evenly and turn heat to medium until one side of the slices turns brown. Carefully turn over the bean curd sections and brown the other side. Place the melon and fish on top of the bean curd and pour the chicken broth and water over all the ingredients. Carefully move the bean curd so that no section is burned. Scatter the *dou-chi*, ginger, and salt over the mixture. Position the fish, melon, and bean curd pieces so that they can all be cooked in the liquid. Cover for 2 minutes. Turn the ingredients over and cover until the liquid has almost disappeared. Add sesame oil. Makes 6 servings.

Bitter Melon Steak (*Ku-gua niu-rou* 苦瓜牛肉)

Ingredients:

 8 oz lean beef (Sirloin steak advised; cut into 5 x 3 cm sections, each section sliced into 1 cm thick pieces, pounded slightly with the back of a chopping knife, then marinate for 15 minutes in soy sauce, ginger, cornstarch, salt, and pepper)

 1 bitter melon (washed, deseeded, sliced to 2 mm thick, parboiled if desired but not advised)

 3 Tbsp vegetable oil

 1 tsp sesame oil

 1/2 medium-sized onion (longitudinally sliced)

 2 cloves of garlic

1 Tbsp cornstarch

1 Tbsp soy sauce

5 mm pieces of fresh ginger (thinly sliced, then shredded)

1/4 cup water

1/3 tsp salt

a dash of pepper

Use 1 tablespoon oil to brown the onion and garlic. Stir in the sliced bitter melon. Add 1/3 teaspoon salt and the water. Stir and cover to cook for 3 minutes. Remove the melon and clean the pot. Heat the remaining oil on high heat and stir-fry the marinated beef. As soon as all the pieces are coated with hot oil, add the cooked melon. Mix well. Add sesame oil. The dish is ready to serve. Eat hot. Makes 2 to 4 servings.

Chinese Hawthorn (*Shan-zha* 山楂)

In North America many people know the hawthorns as ornamental plants, favored particularly for their red fruits and shiny green leaves. Rehder, in the *Manual of Cultivated Trees and Shrubs*, treated 42 species and discussed many related species, varieties, and hybrids. In China people know the hawthorns only as edible fruits and their preserves, such as *tang-hu-lu* 糖葫蘆 in Beijing, *shan-zha-gao* 山楂糕 and *shan-zha-pian* 山楂片 in other cities. When the wild hawthorns are in season, the red and yellow fruits are strung into leis like those made of flowers in Hawaii. Children like to buy them, first to wear and then to eat.

Tang-hu-lu or *tang-qiu* 糖球 (= *t'ang-ch'iu*, candied hawthorn) is a string of 5 or 10 candied hawthorn, strung on a bamboo stick. The fruit is fresh inside and crunchy outside. *Tang-hu-lu* is always a treat to children for good behavior or special accomplishment. It appears in autumn and winter in northern China, in small villages as well as in large cities such as Beijing.

Formerly, about 15% of the Chinese hawthorn production was consumed fresh and 85% used as medicine or canned. Between 1960 and 1970 eight scientific articles were published on the chemical composition of the Chinese hawthorn, ten on the biochemical and physiological activities, and six on the clinical observations of the efficacy of hawthorn for lowering blood pressure and cholesterol, particularly among the elderly. Then, much of the Chinese hawthorn crop was diverted to the pharmaceutical industry and the price rose sharply. The Chinese hawthorn, *Crataegus pinnatifida* Bunge grows very well in the Boston area, and in October the ground underneath the trees is red with the fallen fruits. These can be gathered, dried and used.

Hawthorn Tea (*Shan-zha-yin* 山楂飲)

Ingredients:

4 oz dried Chinese hawthorn (roasted and pulverized)

4 oz germinated barley (dried, roasted, and ground)

1 1/2 cups cooked rice (dried, roasted, and ground)

1/2 cup sugar

8 cups water

Boil the water in a pot, add all the ingredients and bring to a boil. Simmer until the ingredients appear as a thin paste. Bottle. After meals take 3 teaspoons of the paste to make a cup of tea with boiling water. Take this drink in place of tea or coffee. It improves the appetite and lowers blood pressure and cholesterol.

Hawthorn Dessert (*Shan-zha* 山楂)

Ingredients:

1 1/2 cups canned hawthorn preserve (the finest quality)

2 oz rock sugar

1 1/2 cup water

2 Tbsp cornstarch (softened in 2 Tbsp water)

Warm 1/2 cup of water and add the canned hawthorn preserve. With a spoon press the hawthorn to dissolve it over medium heat. Add sugar, stir slowly, then add the remaining water to make a even solution. Bring to a boil, and stir in the cornstarch. Makes 4 servings.

Festival Gourd (*Jie-gua* 節瓜)

This type of fruit vegetable is known only in the Pearl River Delta in Guangdong Province where it is extensively cultivated as a summer crop. It was first recorded in 1954, when F. C. How of the South China Institute of Botany described it as a new variety of *Benincasa hispida* (Thunb.) Cogniaut. It differs from the winter Gourd in size, color, and the texture of the fruit. The fully developed ones are 15–20 cm long, 4–8 cm in diameter, green and very hairy. In Guangzhou (Map, Loc. 58) the crop is ready for the Dragon Boat Festival (the fifth day of the fifth month by lunar calendar), hence the local name *jie-gua* (= *chieh-qua* in Cantonese pronunciation; festival gourd). It tastes like very tender zucchini, and it can be cooked with or without meat, stir-fried, or in soup. For seeds, the selected fruits are left on the vine to mature. The rind of the mature fruit hardens, the hairs gradually disappear and are replaced by a white bloom. The seeds inside are then ripe. They can be taken out and dried, and the flesh can be cooked as in the following recipe. It breeds true from seeds, and has been introduced into Boston kitchen gardens (Figure 10). Because the growing season in New England is so short, the plants do not bear fully mature seeds before the first frost. Therefore, gardeners usually select one good fruit, keep it in a cool, dry place for the winter, and then use the seeds for the next crop.

Festival Gourd Pork (*Jie-gua chao-rou* 節瓜炒肉)

Ingredients:

    3 pork chops (cut the meat into 3 or 4 cm pieces, trying to leave some meat on the bone)

    2 festival gourds (cut longitudinally into fourths, then cut each pieces into bite-size)

    1 medium-sized onion (sliced longitudinally about 1 cm wide on the back)

    1 Tbsp soy sauce

    4 or 5 zanthoxylum (Sichuan peppercorns)

    2 Tbsp vegetable oil

    1/3 tsp salt

    1/2 cup water

Saute the onion slices. As the onion turns color, add zanthoxylum after breaking the pericarps between the fingers and further reducing to smaller pieces. Add the pork, meat, and bones. Stir and add the soy sauce. As the meat turns color, add water and bring to a boil. Add salt, gourd pieces and stir. Turn the heat to medium-low, cover and cook for 20 minutes. Stir once or twice. Adjust the heat to keep the ingredients cooking but not burning. As soon as the gourd is soft and the liquid is almost cooked away, the dish is ready to be served. Makes 3 or 4 servings.

Stir-fried Festival Gourd (*Su-chao jie-gua* 素炒節瓜)

Ingredients:

    2 or 3 festival gourds (washed, cut into angular 3 cm pieces)

    1 medium-sized onion (sliced 1 cm wide on the back)

    1/2 tsp salt

    2 Tbsp vegetable oil

    1 sheet *fen-pi* 粉皮 (break into 3–4 cm irregular pieces, pour boiling water over the pieces and keep them soaked in water for perfect softness)

    1 Tbsp soy sauce

4 or 5 zanthoxylum (Sichuan peppercorns)

1/4 cup water

1 tsp sesame oil

Brown the onion. Add zanthoxylum, breaking them by pressing between the fingers before throwing them in. Add the gourds, water, soy sauce, and *fen-pi* while stirring. Bring to a boil, cover and simmer for 5 minutes. Stir so that all the pieces are cooked evenly. Add a little water if the mixture becomes too dry. When the pieces of the gourd are all soft, add the sesame oil and serve. Makes 4–6 servings.

Jujube (*Zao* 棗, the Chinese jujube)

Hill, in his *Economic Botany* (1952), wrote, "The jujube ... a native of China ... cultivated for at least 4000 years ... is still one of the five chief fruits of China ... since 1900 [it] has been grown to an increasing extent in California, Texas and Mexico ... It promises well as a fruit tree ... is remarkably free from pests. It is used fresh, dried or preserved, and is useful in cooking and candymaking." Since all these statements are true, why is there no American-produced jujube in local markets? All jujube products available in American Chinese stores are imported from China. The answer to the above question is very simple. The American people are not informed about the food value of jujube, nor do they know how to use it. Raw jujube is not very tasty; it is rather small with a hard stone, pointed at both ends. Most of the Chinese woody plants introduced to America have attractive flowers which can be used for landscape and ornamental purposes. The flowers of jujube, on the contrary, are very small and greenish. Moreover, on account of the Indian jujube, which has woolly leaves and larger fruit, most references treat the jujubes as "subtropical ... thriving in hot, dry climate" (*Encyclopedia Britannica*, 1962). This is not so with the Chinese jujube (*Ziziphus jujuba* Miller). In China the wild population of jujube is a dominant element in the desert of Ala Shan as well as on the neglected, dry slopes of large cities such as Jinan (Map, Loc. 13 and 14). It seldom reaches 1 m in height, and is multi-branched and very spiny. The cultivated plants are trees 5–10 m high growing best in temperate northern China, particularly in Shandong, Henan, and Hebei. The cultivar that produces the largest fruits grows in the vicinity of Tianjin and Luoyang (Map, Loc. 8 and 18). In Shanghai such large fruits are called *hong-zao-wang* 紅棗王 (= *hung-tzao-wang*, red jujube king).

The cultivated Chinese jujube has two types of branches, the woody elongated branches with paired, strong spines, the long one pointing forward and the short one recurved; and the herbaceous, deciduous branches with inconspicuous spines or without spines. The deciduous branches generally emerge from spurs on old woody branches, and only they can bear flowers and fruits. The immature fruits of jujube are as green as the leaves, hard, tasteless, and slimy. Approaching maturity, they turn to greenish ivory. As soon as a maroon ring appears on the basal end of most fruits, they are harvested. At this stage they are crisp, and slightly sweet. They turn maroon on storing, and become sweeter than when they were on the tree. Approximately 20% of the crop is consumed fresh and the remainder is preserved by drying or with sugar. All types of preserved Chinese jujube are available in American Chinese stores.

The famous Peking honey date, *Bei-jing-mi-zao* 北京蜜棗 (= *Pei-ching-mi-tsao*, Beijing honey jujube) is a brand name of the sugar preserve of Chinese jujube. It is prepared by incising the skin of the fresh fruit with heavy needles and then soaking in strong, brown sugar solution. After partial drying, the procedure is repeated and then the jujube is dried and packaged for the market.

Dried jujube in the market usualy falls into two types, the *hong-zao* or the *hei-zao*. *Hong-zao* 紅棗 (= *hung-tsao*, red jujube) is maroon in color. It is also called *da-zao* 大棗 (= *ta-tsao*, big jujube). It is prepared by drying the fruits directly in the sun. *Hei-zao* 黑棗 (= *hei-tsao*, black jujube) appears black. It is also called *nan-zao* 南棗 (= *nan-tsao*, southern jujube) in prescriptions. It is prepared by parboiling the fruits prior to drying in the sun. In general, *hong-zao* is used more often in food because of the cheerful, red color. It is used particularly at weddings when it is mixed with peanuts, sorghum grains, sunflower seeds, and walnuts to throw into the air at the end of the ceremony, just as rice and confetti are thrown over the bride and groom in America. The pronunciation of these fruits and seeds used in weddings contains the sounds of *"zao"*, *"sheng"*, *"zi"*, and *"he-tao"*, which are homonyms to the ideograms standing for early (*zao* 早), born (*sheng* 生), son (*zi* 子), and perfect union (*he* 合), all pertaining to goodwill and well wishes for weddings. Of course, when used in confections, a red fruit gives a more pleasant feeling than a black one. A paste made of dried jujube sweetened with sugar, preserved rose petals, or osmanthus flowers is used in pastries. Canned jujube paste, as well as dried jujube, are available in American Chinese stores.

Formerly, through empirical practices, the dried jujube was highly esteemed as a delicacy and a health food. Now, numerous reports from Chinese phytochemists show that jujube fruit is rich in proteins, simple sugars, and polysaccharides, glycosides, Vitamins A, B2, and C, organic acids, especially ascorbic acid, mucilage, riboflavine, carotene, and a little fat. The pharmacologists and physiologists have demonstrated its efficacy to improve strength with animal assays. Many physicians have clinical evidence of its merits. Jujube should, therefore, have its rightful place in the diet of the American people.

Stuffed Jujube Rolls (*Hong-zao-bu-tou* 紅棗餔頭)

Ingredients:

   1 1/2 cup red jujube (revived by soaking in an equal amount of boiling water for
      10 minutes, cover until the liquid is mostly absorbed, and save the juice for mak-
      ing a cool drink)

   10 buttermilk biscuits

In a two-gallon pot fashion a home-made steamer by placing a small food steamer flat on a tuna can with both ends removed. Cover the food steamer with wet cloth. Stuff 2 jujubes in each biscuit, seal it, and place the 10 stuffed biscuits on top of the food steamer. Add 5 cups of water, taking care that the water does not touch the cloth. Cover the pot and steam the contents on high heat for 10 minutes, reduce the heat to medium and cook for another 10 minutes. Makes 10 servings as an item for afternoon tea. Sugar-preserved jujube (market name: Peking honey date) can be used as a substitute for the red jujube, and it does not require reviving. Canned jujube can also be used and the name of the finished product is *zao-ni bao-zi* 棗泥包子 (mashed jujube patty).

Living Buddha Jujube Drink (*Huo-fo-yin* 活佛飲)

Ingredients:

   10 black jujubes (washed and drained)

   2 cm medium-sized American ginseng (washed, then pounded into irregular pieces)

   2 Tbsp sugar

   4 cups water

Boil the water in an enamel or Corningware pan. Add the jujubes, ginseng and sugar. As the mixture boils, reduce heat to medium-low, and continue the boiling until 2 cups of liquid are left. Serve hot. Makes 2–4 servings.

This recipe is adapted from a case report of a friend, C.Y. Chen (1975). He records that the Panchan Lama, when exiled from Tibet in the 1930s, was housed in Ling-yin Temple 靈隱寺 in Hangzhou 杭州. Unable to adjust to the sea-level climate and foods of the low latitudes of Hangzhou, he soon developed gastro-intestinal problems, and became severely dehydrated, weak, and listless. He was unable to sleep. Many medicines were recommended to him as his condition steadily worsened, but to no avail. One day, hearing of the plight of the poor Lama, Chen asked to visit him. On doing so, he advised the Lama to drink some of the above mentioned tea. The Lama drank the tea for a week, whereupon his condition began to improve. Contiuous use of the tea restored his health and appetite.

Stewed Jujube (*An-mian-zao* 安眠棗)

Ingredients:

1 lb sugar-preserved jujube (or dried jujube)

3 cups water

Juice of 1 lemon

Cook the ingredients as if stewing prunes. Keep cold and serve for breakfast or before going to bed. This recipe is adapted from one in which the Chinese use preserved jujube for people who suffer from insomnia due to anxiety and tension.

Luffa (*Si-gua* 絲瓜)

The tough fiber of the mature fruit of *Luffa aegyptiaca* Miller (= *L. cylindrica* auctt, non L.) is a familiar bathroom article in America. The very young fruits of the species are sold as vegetables in Chinese groceries as are zucchinis in American supermarkets. In China *L. aegyptiaca* Miller is cultivated throughout the country for the same purposes and also for medicinal uses. *L. acutangula* (L.) Roxb. is also cultivated in South China for the edible young fruit. It is distinguished by shallowly lobed leaves, fewer (2 or 3) stamens and deeply furrowed and strongly ridged young fruits. The edible stage of the fruits is when they are 3–4 cm in diameter. Bigger than this size, the inside becomes spongy and the outside tough. Chinese phytochemists have reported their findings on the food value of luffa: rich in proteins, glycosides, luffein, citrulline, mucilage, polysaccharides, and Vitamins C and B. Luffa is frequently served in American Chinese restaurants, particularly to the Asian customers. It can be served with or without meat.

Vegetarians' Delight (*Su-shi-jin* 素什錦)

Ingredients:

6–8 pieces fried bean curd cubes (diagonally quartered, available in Chinese groceries)

35 daylily buds (revived in boiling water, covered for 1 hour, washed with pedicel portion removed, petals separated by running a toothpick from the bottom up)

20 wood ears (revived in boiling water, cleaned, drained, and the large ones divided)

10 black mushrooms (revived in boiling water, covered for 1 hour, the caps sliced, stalk and water kept for home-made stock)

4 young fruits of luffa (with a paring knife, thinly strip off the rough portions, particularly near the base and apex; wash, quarter longitudinally, and cut

diagonally into irregular pieces, pointed at both ends, 4 cm long, 2 cm across the middle)

1 small can bamboo shoots (thinly sliced)

1 medium-sized carrot (peeled, sliced, with fancy pattern if preferred)

10 fresh mushrooms (cut large ones once or twice, keep small ones whole)

1 red pepper (plain or hot as preferred, cut into pieces to match the other ingredients)

3 celery cabbage leaves (cut to match the other ingredients)

1 medium-sized onion (sliced longitudinally)

2 cloves garlic (sliced)

1 tsp salt

3 Tbsp vegetable oil

Dash of pepper

1/2 cup water

1 Tbsp sesame oil

6 zanthoxylum (Sichuan peppercorns)

(Optional) 1/4 cup chicken broth

Saute the garlic and onion. Add zanthoxylum while stirring. Add luffa, carrot, celery cabbage and stir. Add the remaining ingredients except sesame oil. Stir, cover and cook for 2 minutes. Turn over and continue to cook until the liquid is almost entirely absorbed. Add sesame oil and serve immediately. Makes 6–8 servings.

The Chinese name of this dish is *Su-shi-jin* 素什錦 (Ten Plain Elegances). It is nutritious, colorful, and its aroma comes from all the natural ingredients. When the timing is judged correctly (through experience), it should be a delightful dish.

Luffa Ham Soup (*Si-gua-tang* 絲瓜湯)

Ingredients:

2 tender luffa (peeled, cut as described above)

2 oz Virginia ham (thinly sliced to match the size of the luffa)

1 medium-sized onion (chopped)

1/2 *fen-pi* 粉皮 (available in Chinese groceries, break into irregular pieces to match the luffa)

4 cups home-made stock (or 1 cup chicken broth and 3 cups water)

1 tsp sesame oil

Dash of pepper

Saute the onion. Add the ham and stir. After 1 minute stir in the luffa. Add stock and *fen-pi*. Cover and cook for 5 minutes. Add sesame oil and pepper (may be some salt to taste). Serve hot. Makes 4 servings.

White-flowered Gourd (*Hu-lu* 葫蘆)

The white-flowered gourd (*Lagenaria siceraria* [Molina] Standley) is one of the first cucurbitaceous species to enter the life and culture of the Chinese people. Long before the invention of the written language, it had become part of the folklore and songs of the people, and it was recorded in the *Classics of Songs* 詩經, edited by Confucius (551–479 B.C.). Presently, it is widely cultivated throughout the country. The varieties with large fruits are cultivated in gardens or along walls or fences for the edible, pear-shaped and/or cylindrical, young fruits, for the light and useful mature gourd with hard rind, and also for the valuable seeds used in medicine. The varieties with small fruits are cultivated for curios, often artistically carved. In the following recipies the short Chinese name, *hu-lu*, is used in place of the longer common name, white-flowered gourd. If the young fruit is from a kitchen garden, pierce an almost fully grown fruit with the finger nail. If the skin gives no resistance, it is at the best stage to be picked for food. Fruit harvested at this stage requires no peeling. Gourds from groceries may need peeling.

Hu-lu Soup (*Hu-lu-tang* 葫蘆湯)

Ingredients:

> 1 1/2 lb young *hu-lu* (washed, longitudinally sectioned, seeds removed from pulp and saved for making tea, meat sliced 4 cm x 5 mm)
>
> 2 oz Virginia ham (thinly sliced 2 x 1 cm)
>
> 1 medium-sized onion (sliced lengthwise)
>
> 1/2 *fen-pi* 粉皮 (broken into irregular pieces about 2 cm square, revived in boiling water)
>
> 1 Tbsp vegetable oil
>
> 1 tsp sesame oil
>
> 4 mm piece ginger (minced)
>
> 4 cups homemade stock (or 1 cup chicken broth, 3 cups water)
>
> Dash of pepper
>
> 1/2 tsp salt

Brown the onion and ginger. Stir in the sliced ham. Add *hu-lu*, the stock (or broth and water) and bring to a boil. Add *fen-pi* and continue to cook for 10 minutes on medium-low heat. Add sesame oil, salt and pepper (to taste), and serve hot. Makes 4 servings.

Hu-lu Butterfly Shrimp (*Hu-lu xia-ren* 葫蘆蝦仁)

Ingredients:

> 1 lb medium-large shrimps (wash, remove skin, cut the apical 4/5 along the dorsal

groove and remove intestine and veins, wash again, marinate with minced
ginger, onion, garlic, soy sauce, oil, and 1/2 tsp salt)

8 oz young *hu-lu* (washed, sliced 4 x 1 cm)

5 mm ginger (mince half, slice the remainder)

3 garlic cloves (mince 1, crush the other two)

1 medium-sized onion (mince 1/4, slice the remaining portion)

1 tsp soy sauce

4 Tbsp vegetable oil

1/4 cup water (or home-made stock)

2/3 tsp salt

1 tsp sesame oil

Brown the sliced ginger, onion, crushed garlic, and 1/4 tsp salt in 1 tablespoon of
oil. Stir in the sliced *hu-lu*. Add water (or stock). Stir, cover, and cook for 10 minutes on
medium-high heat (reduce to medium to prevent burning if necessary.) Empty the pot
and clean it. Turn the heat to high and heat the remaining vegetable oil to steaming. Stir
in the shrimp. As soon as it turns color, add the cooked *hu-lu* mixture. Stir and add
sesame oil. Serve immediately. Makes 4 servings.

Winter Melon (*Dong-gua* 冬瓜)

The winter melon (*Benincasa hispida* [Thunb.] Cogn.) is another plant which has been
associated with the Chinese people since ancient times. It appeared in the first record
of *Chinese Materia Medica* (*Shen-nong ben-cao jing* 神農本草經, *ca.* 202–260 A.D.; Hu, 1990,
pp. 495 and 523). The name is a direct translation of the Chinese term *"dong-gua"* .
Actually, it is the product of a summer crop which has such a good keeping quality
that it can last into winter. Although it is widely cultivated in China, the best varieties
were selected in the vicinity of Guangzhou, and the most ingenious methods of cooking
it were designed there also.

Many varieties grow in Guangzhou. The late varieties bear huge, oblong, cylindrical
fruits 40–45 cm long, 28–33 cm in diameter. It takes five or six months after sowing the
seed for such crops to be ready for the market. When the fruits are half grown, farmers
put a loose net around the individual fruits and hang them on a supporting framework.
With some varieties, as the fruits attain mature size, a white, waxy powder begins to
accumulate on the rind, giving it a grayish-green or white color. The edible portion of
the winter melon is snow white, hence the alternate name "white gourd" (*bai-gua* 白瓜
= *pai-kua*, Figures 54 and 55a–d).

Every part of the winter melon is used in China, either as food or for medicine.

Chinese phytochemists have worked out its chemical composition. Every 500 grams (*ca.* 1 lb) of fresh winter melon contains 1.5 g protein, 8 g polysaccharides, 15 g fiber, 1.5 g ash containing compounds of sulfur, phosphorous, and iron, 0.4 g carotene and also thiamine, 0.08 g riboflavin, 1.1 g nicotinic acid, and Vitamin C.

Winter melon can be prepared as food in many different ways, and every dish is good once a person develops a taste for it. It is always good in soup and with meat or poultry. Actually, the winter melon is the fruit of a large cultivar of the festival gourd. As such, it can be used as a substitute for the festival gourd in the recipes for festival gourd. In fact, I have used a mature festival gourd to replace winter melon in one of the following recipes.

Grated Winter Melon Egg (*Dong-gua-mo dan* 冬瓜末蛋)

Ingredients:

    1 lb winter melon (remove rind and seed, and grate finely)
    3 egg whites
    1/2 tsp salt
    5 mm ginger (thinly sliced)
    1/2 medium-sized onion (minced)
    2 tsp sherry
    1 Tbsp butter or margarine
    1/10 tsp Ac'cent
    2 cups home-made stock
    2 Tbsp cornstarch (mixed with 3 Tbsp water)
    1 tsp sesame oil

Mix in a large bowl the grated melon with salt, ginger slices, and 1 teaspoon of sherry, and steam the mixture for 30 minutes. Remove the ginger, keeping the mixture hot. In a sauce pan brown the minced onion with butter. Add the steamed melon, the remaining sherry and the stock and bring to a boil. Stir in the cornstarch and cook for 1–2 minutes. Meanwhile, beat the egg white with 1 teaspoon of water. Remove the sauce pan from the stove. Stir in the egg white, sesame oil, and Ac'cent. Serve hot. Makes 3 servings.

Modified Winter Melon Bell (*Fang dong-gua-zhong* 仿冬瓜盅)

Ingredients:

    3 lbs winter melon (seeds removed, cut into 3 cm squares)
    4 oz Virginia ham (diced into 1 cm cubes)
    6 oz pork chops (free the meat, cubed as above, use the bones for stock)

4 oz chicken breast (cubed as above, marinate in salt and sherry for 15 minutes, beat in one egg white)

4 oz cooked duck breast (cubed as above, available in Chinese cold-cut stores)

1 can crab meat (drained, cubed if needed)

1 can lotus seed (drained, or 3/4 cup dried skinned lotus seed, revived and boiled till soft)

1 cup fresh mushrooms

2 oz black mushrooms (revived in hot water, cut twice)

1 egg

3 cups water

1 can chicken broth

1 tsp salt

5 mm piece of ginger (sliced 1 mm thick)

1 Tbsp sherry

1 Tbsp sesame oil

1 large onion (diced into 1 cm cubes)

1 Tbsp cornstarch

2 Tbsp margarine

Melt the margarine and slightly brown the onion. Stir in the diced pork and cook for 1 minute. Add the winter melon, water, ginger, pork bones, and lotus seed. Cook for 40 minutes, first on high heat, then, after 5 minutes, on medium-low. Take out bones and turn heat to high and add diced ham, duck breast, crab meat, mushrooms, and chicken broth. Bring to a boil and cook for 5 minutes on medium-low. Turn to high, and stir in the marinated chicken so that the pieces are separated from one another. Thicken the liquid with cornstarch if desired. Add sesame oil. Serve hot. Makes 6 servings.

This recipe is a modified form of the Cantonese dish called *Dong-gua-zhong* 冬瓜盅. It contains all the ingredients, but the cooked material is not placed and steamed in a bell-shaped *dong-gua*.

To visitors of Guangzhou, China's most outstanding city for good food (there is a saying "食在廣州" (Eating in Guangzhou), the most enjoyable and impressive dish of the city is *Dong-gua-zhong*. In this dish the cooked *dong-gua* is a small, whole, melon about 3–4 kg (7–9 lbs) in weight. Approximately 20 cm of the apical portion of this melon is cut with a dentate margin, and the pulp and seeds removed. It is cooked for 10 minutes in boiling water, immediately changed to cold water. This parboiled melon is used as a bowl, and all the ingredients, prepared as described above, are placed in the hollowed portion of the melon and steamed for 30 minutes.

During the summers of 1947 to 1950, I lived in the Bussey Institute situated at the corner of South Street and Arborway on the ground of the Arboretum, as did all graduate students working in the summer projects of Professors Karl Sax and Paul Mangelstoff. Each of us was given a plot for a kitchen garden. I planted winter melon from seeds taken from fresh melon bought in Chinatown. The short growing season dwarfed the growth and consequently my harvest of melons was just big enough for *Dong-gua-zhong*, which I prepared several times. Not having the small melon to be used as a bowl, the above description is a good substitute, which I have prepared several times since.

Yard-long Beans (*Dou-jiao* 豆角)

The yard-long bean is a Chinese horticultural creation planted for the tender, young fruits which are 25–70 cm long and 7–8 mm in diameter, each containing 16–26 seeds. It is a climber supported by bamboo sticks in the garden. It is extensively cultivated in China, and is an important summer vegetable, particularly for students living on dormitory food and for the low income populace. It is a time- and labor-saving crop which continues to produce paired, string-like beans, generally for 30–40 days from the first harvest to the last. Eighteen cultivars have been reported, differing in the size of the fruits, color, and weight of the seeds. Fully grown, tender beans are eaten cooked (with or without meat or as a salad), or raw after being pickled in salt (Figures 11a and 38).

Yard-long Bean Salad (*Liang-ban dou-jiao* 凉拌豆角)

Ingredients:

    1 lb yard-long beans (broken and strung into 3–4 cm pieces, washed, and drained)
    2 Tbsp soy sauce
    2 garlic cloves (crushed and chopped)
    1 scallion (minced)
    1 Tbsp sesame oil
    1/4 tsp salt
    1 tsp vinegar (or lemon juice)
    4 cups water

Boil the water in a sauce pan. Stir in the yard-long beans. Cover and continue to cook for 5 minutes. Pour off the liquid and save for the home-made stock. Cool the beans with cold, running water, and then drain. Add the soy sauce, garlic, scallion, sesame oil, vinegar, and mix well. Makes 4 servings.

Yard-long Bean Pork (*Dou-jiao chao-rou* 豆角炒肉)

Ingredients:

> 8 oz boneless pork (thinly sliced, marinated in 1/2 tsp salt, soy sauce, shredded ginger, cornstarch, and minced onion)
>
> 6 oz tender yard-long beans (strung and broken into 3 cm pieces)
>
> 3/4 tsp salt
>
> 1/2 cup water
>
> 2 Tbsps soy sauce
>
> 4 mm ginger (sliced and shredded)
>
> 1 Tbsp cornstarch
>
> 1 large onion (mince one quarter, and slice the remainder)
>
> 1 tsp sesame oil
>
> 3 Tbsp vegetable oil

Brown the sliced onion in vegetable oil. Stir in the marinated meat and the yard-long beans. Turn several times. Add water and stir. Cover and cook for 8 minutes on medium heat. Add sesame oil and serve. Makes 3 servings.

Vegetarian's *Jiao-zi* (*Su-jiao-zi* 素餃子)

Ingredients:

> 4 oz very tender yard-long beans (washed, with both ends cut off, then cut them crosswise with a sharp knife into 2–4 mm sections)
>
> 2 eggs
>
> 1 medium-sized onion (minced)
>
> 4 mm ginger (minced)
>
> 3 Tbsp soy sauce (*sheng-chou* 生抽)
>
> 3/4 tsp salt
>
> 2 cloves garlic (crushed and chopped)
>
> 3 tsp sesame oil
>
> 1 tsp vinegar
>
> 2 Tbsp vegetable oil
>
> 1 1/2 cups all purpose flour
>
> 9 cups water
>
> (Optional) 2 Tbsp hot soybean sauce

In a medium-sized bowl mix 1 1/4 cups flour and 1/2 cup water to make a smooth dough. Cover the dough with a wet towel. In a skillet lightly saute 4/5 of the minced onion and ginger in 2 tablespoonful of oil, add 1/2 teaspoonful salt and the yard-long beans. Stir to prevent burning and cook for 3 minutes. Take out and place the bean in a mixing bowl. Clean and heat the skillet with medium heat. Spread the remaining oil evenly in it and make thin sheets of the slightly beaten eggs with the remaining salt. Control the heat so that the egg sheets are cooked evenly. Cut and chop the egg sheets finely to match the size of beans and mix the two in the mixing bowl. To the mixture, add 1 tablespoon each of soy sauce and sesame oil. Mix thoroughly (salt may be added to taste).

With the help of the remaining flour, roll the dough into a cylinder about 2 cm in diameter. Cut the cylinder into 2 cm disks. Roll each piece into an oblong sheet about 6 x 4.5 cm. Holding the little dough-sheet flat in the palm of the left hand, add 1 tablespoon of the yard-long bean mixture into the center. Fold the sheet on the long axis and seal the margin uniformly tight. Be sure no bean or egg pieces are on the sealing margin and no air will pass in and out. This pastry is called *jiao-zi* 餃子. The dough and filling should be enough to make 16–20 *jiao-zi*. Line them up on paper towels over a tray separately. Be sure none of them touch one another. Boil 8 cups of water, and add the *jiao-zi* to the boiling water (important!). With the back of a wooden spoon slightly move the *jiao-zi* so they will not stick to the bottom of the pan. Cover and bring to a boil. Add 1/2 cup of water and bring to a boil again. Repeat another time. Dish out and serve hot. Combine the remaining soy sauce, onion, ginger and sesame oil. Add garlic and vinegar to make a sauce. Put a spoonful of the sauce on the plate and dip the *jiao-zi* before eating. The number of servings may vary between 2 and 4 persons, depending on the kind of the other material served with it. I have seen students eating 30–40 *jiao-zi* easily. Many people like to dip *jiao-zi* in hot soybean sauce.

Seeds

Botanically a seed is a "ripened ovule, consisting of the embryo and its proper coats". The seeds of apple, pea, and squash are concealed within the fruits (angiosperms), and those of the ginkgo, pine, and yew are exposed (gymnosperms). The unusual seeds used in Chinese food selected for discussion here belong to both the gymnosperms and the angiosperms.

(1) *Gymnosperms*: Most gymnosperms are trees and shrubs. The cycads, pines, spruces, and yews are good examples. Many species of gymnosperms have been introduced from China to eastern North America for ornamental purposes. The seeds of some of them are eaten in China either as delicacies or a special health food . The best examples are *Ginkgo biloba* L., which was introduced to America in 1784, and the two species of the five-needle pines (*Pinus armandii* Franchet, introduced in 1861, and *P. koraiensis* Sieb. et Zucc., in 1895).

Ginkgo (*Bai-guo* 白果)

Ginkgo trees are planted extensively in eastern North America. Many large specimens are found in botanical gardens and city parks, while smaller trees are familiar along central avenues of cities, such as Washington D.C.. In the City Point area of Boston, ginkgo trees are planted along the seashore. Apparently, the species is tough, resistant to fungal diseases and insect attacks.

One hundred and fifty million years ago, in the Jurassic Period, the Ginkgoales was widespread throughout the world. Now the only living species is *Ginkgo biloba* L.. Even this species is known only in association with man, though a clump of several trees in Tian-mu-shan, Zhejiang, China, has been identified as "natural". This is not true, for these specimens all grow in temple grounds as do other ancient tree elsewhere. Horticulturists often ask, "What has caused the extinction of the natural population of ginkgo?" My personal observations may shed some light on this problem. I have observed: (1) It is very hard to raise ginkgo trees in Hong Kong and Guangzhou in the humid tropics; (2) although many ginkgo seeds germinate around the mother trees planted in association with yews at the Arnold Arboretum in Jamaica Plain and with rhododendrons at the Forest Hill Cemetery (in Boston), the seedlings do not survive the dense shade formed by the evergreen shrubs; (3) all the large trees I have seen grow in a mesic environment north of 30° latitude (including those growing on Tian-mu-shan, Zhejiang), and with no other trees near them. In short, the disappearance of the natural population of *Ginkgo biloba* L. is an intrinsic consequence of a clash between the biological nature of the species and the cultural pressure exerted by man in its homeland. Biologically, ginkgo is a heliophilous, mesic, and temperate species. Culturally, the homeland of ginkgo is the seat of an ancient Chinese civilization which began by the use of fire to clear the land and to drive away harmful wild animals (an assignment of Tang-di Yao — 2365 B.C. — to Shun who subsequently became Yu-di Shun — 2255 B.C.). The use of fire as a weapon to destroy the enemy seems to be the inherited nature of Chinese warfare. The best known instance is recorded for the Epoch of the Three Kingdoms (A.D. 200–260). The use of unlimited burning of the natural vegetation for clearing the land for agriculture is still practiced in some mountainous areas. At the time of the Tang Dynasty (A.D. 618–905), the natural vegetation of the homeland of the ginkgo had been completely destroyed and altered except for isolated properties of the Taoist temples and Buddhist monasteries. Ginkgo has been, however, one protected species that survived human destruction. During the Tang Dynasty, communication between Chinese and Japanese religious leaders was on the increase. Many Chinese economic plants were introduced to Japan, with ginkgo among them. It went by its monastic name, silver apricot (*yin-xing* 銀杏). In the early 1700s, a Dutchman, E. Kaempfer, residing in Japan, saw the species, made a drawing of it, inscribed its name in Chinese ideogram and gave the Japanese pronunciation, ginkgo. So, the origin of the name ginkgo is the Japanese pronunciation of the Chinese monastic name, silver apricot.

Many people wonder why a mature ginkgo seed is not a fruit like the apricot which it so closely resembles. The answer lies in the origin of its principle structures. The apricot is an angiosperm in which the ovules are enclosed in the ovary wall. The yellow flesh (mesocarp) and brown stone (endocarp) are developed from the ovary wall and they are parts of a botanical fruit (Figure 27b). The kernel inside the apricot stone represents its embryo which is covered by a brown membrane, the seed coat.

The ginkgo, on the other hand, is a gymnosperm which has naked ovules exposed at the end of a stalk from which the mature ginkgo seed is developed (Figures 17c, e–f). The yellow flesh and white "stone" in this instance are two layers of the seed coats (Figure 17i). When the seed matures and falls to the ground with the deciduous, yellow leaves, it contains no visible embryo, which develops in the following spring.

In China ginkgo grows only in limited areas, on temple grounds, or in special orchards. The seeds are sold in the market as a delicacy or for medicine. Most people only know the product with the white inner seed coat and call it *bai-guo* (白果, white fruit). In American Chinese stores, both dried seeds and canned kernels are available. The material imported from Shanghai, produced on a hill overlooking the Tai Lake (太湖) near Suzhou (Map, Loc. 24), has unusually large obovoid seeds. Here the farmers propagate the special large-seeded cultivar by grafting. As in China, the market name in Boston for ginkgo is *bai-guo,* and it is also known by Chinese Americans who cook with it as *bai-guo.*

In the 1950s the ginkgo seeds in the Boston gardens and parks were left on the ground. At that time the Chinese residents bought imported material in stores. They did not know that ginkgo trees grew in the area and there were seeds to be picked. I have shown some friends the large trees in the former Gray Herbarium Botanical Garden (Cambridge, Massachusetts), in the Arnold Arboretum and in the Forest Hills Cemetery in the suburb of Boston. Good news spreads fast. Now, there are no ginkgo seeds left under these trees. Often, I arrived too late to enjoy a good harvest!

The best time to gather ginkgo seeds is around Thanksgiving. At this time the weather has softened the fleshy outer seed coat to such a degree that a slight pressure applied to the seed would separate the juicy smelly portion from the inner white hard cover. This fleshy portion does not only carry an offensive odor, but it also contains ginkgolic acid and ginkgotoxin which may cause contact dermatitis in some people. It is advisable to wear rubber gloves for handling fresh ginkgo seeds.

The kernel of fresh ginkgo is greenish. It is rich in protein, carotene, riboflavin, calcium, phosphorous and iron compounds. Newly harvested ginkgo tastes good when it is cooked. Exposed seeds dry up after a month or so, the kernal becomes ivory and slightly bitter. If kept in a plastic bag or a glass jar and stored in the refrigerator, the seeds retain their freshness longer, and by February a green embryo will appear in the center of each kernel. For food the green cotyledons are usually removed. Seeds kept in a closed jar in a refrigerator may mold outside, but it won't hurt the edible portion. They can be washed, dried and returned to the jar. Seeds kept outside and dried in the air can be revived by boiling, and can be used for food.

The kernel of ginkgo may be toxic to some people who eat more than 20 seeds at a time. In China ginkgo is eaten as a delicacy and there has no documented record of its toxic effects. However, I know of several cases of ginkgo toxicosis. In Chengdu, a nine-year-old boy who stole seeds from a neighbor's garden roasted them in hot ash. He ate so many of the roasted seeds in one day that he slept 24 hours straight. A friend who has lived in Long Island, New York, for 40 years before she was introduced to a local ginkgo tree on Thanksgiving Day, 1987. She collected, washed, and roasted the seeds, and ate them like peanuts. She had too many and was poisoned, broke out in a rash, and became sensitized to ginkgo. She had never experienced ginkgo poisoning previously because she did not have the opportunity to eat so many seeds. I always suggest to people that one should eat no more than 15 ginkgo seeds at a time.

Roasted Fresh Ginkgo (*Chao-bai-guo* 炒白果)

Ingredients:

　1 cup fresh, clean, dried, white ginkgo

Heat the ginkgo seeds on medium heat in a frying pan. Stir or shake to avoid too much browning on one side. Cook for 4–5 minutes. When the seeds are hot, the white shell can be broken by applying pressure with the thumb-nails. Try one. If the center of the kernel is cooked, it appears light green, shiny, smooth and slightly transparent. These seeds are ready to be served. Place the roasted ginkgo in a bowl next to the other nuts of the Thanksgiving season. They are unusual, taste good, and they furnish a subject for conversation at any party.

Roasted, fresh ginkgo is a popular delicacy of the season in Shanghai where itinerant vendors carry their equipment and products and chant on the streets of residential areas to attract children, as do the ice-cream men in American cities. Their rhythmical chant, "*Tang shou re bai-guo; san chian mai wu ge.*" (= hot-to-touch, fresh ginkgo, three pennies, five seeds to go.). Hearing the chanting, many children soon gathered around while the ginkgo seeds are roasted in a wire basket over a small, charcoal burner. The children who get their three pennies' worth of ginkgo, break the white shell with their teeth and enjoy the kernels.

Mixed Ginkgo Dish (*Bai-guo za-hui* 白果雜燴)

Ingredients:

> 3 pork chops (parboiled in 1 1/2 cup of water, then diced to 1 cm cubes; keep the liquid for stock)
>
> 1 1/2 cup ginkgo seeds (shelled, boiled, then simmered for 10 minutes in 2 cups of water, 1 Tsp sugar, take out; dilute the water to taste and drink it 3 or 4 times as the cook's treat, or share with family)
>
> 2 carrots (pared, cut vertically into 4 sections, then cut into cubes to match the ginkgo's size)
>
> 3 celery stalks (cut to match the sizes of meat, carrot, and ginkgo)
>
> 2 Tbsp vegetable oil
>
> 1 large onion (cut into 1 cm squares)
>
> 6 zanthoxylum (Sichuan peppercorns)
>
> 1/3 cup fermented black soybean (*dou-chi* 豆豉)
>
> 1/2 tsp salt
>
> 3/4 cup raw peanuts (with or without the red, membranous seed cover, soaked in cold water for 2 hours)

Heat the oil, and brown the onion. Add salt, the zanthoxylum, and all the remaining ingredients including the liquid in which the meat was boiled, stirring continually. Mix well, and bring to a boil. Turn the heat to medium-low and cook for 30 minutes. Check to see that it does not burn. If the liquid is absorbed too soon, add a little more water and cook slowly until all the liquid is absorbed. This is my favorite dish because it is tasty, nutritious, and can be warmed when needed. Serves 4–5.

Eight Precious Pudding (*Ba-bao-fan* 八寶飯)

Ingredients:

> 1 1/2 cup of glutinous rice (washed and drained; cooked with 4 cups of water, 1/2 cup of sugar, 1/2 stick of margarine, and 2/3 tsp of salt, bring to a boil and turn heat to low to cook for 1/2 hour)
>
> 1 1/2 cup fresh ginkgo (shelled, boiled in 1 Tbsp of sugar and 3 cups of water for 1/2 hour, then drained; keep liquid for preparing a cool drink)
>
> 8 red jujube (soaked in 1 cup of water overnight)
>
> 5 red cherries

8 green cherries

1/2 cup palm dates

1/2 cup raisins (rinse with boiling water, and drain)

1/2 cup chestnuts (soak dried nuts overnight, cut into *ca.* 1 cm irregular pieces)

Some water, sugar, margarine, and salt (as asked above)

Use 1/4 of a stick of margarine to grease generously a 6 or 7 inch bowl. Place 1 red cherry in the middle of the bottom of the bowl. Use 4 green cherries to surround the red one, forming a cross design. Fill the space between the green cherries with the cooked ginkgo kernels, with the long axis pointing toward the red cherry. Arrange the raisins in a ring on the outside. Place the eight jujubes an equal distance out from the raisin ring and fill the spaces between them with ginkgo kernels. Cut 4 red and 4 green cherries vertically into halves, and arrange them in pairs with the cut surface face-up. Fill the space with longitudinally cut palm dates, rough surface up. Carefully ladle the cooked rice into the bowl, trying not to disturb the colorful design, and not to leave any air spaces at the bottom. When the bowl is one-third full, spread the chestnuts in the center. With the remaining ginkgos, dates, and raisins make designs against the sides of the bowl, and fill the bowl with the remaining rice.

The principles of preparing this dessert are: (1) to use 8 primary ingredients, all of which, except for the glutinous rice, are fruits; (2) to have a colorful and meaningful pattern to put before the company; and (3) to present a delicious dish. The red center, 4 green cherries and eight jujubes combine to show four directions and eight corners of the earth, signifying "all men are brothers". Ginkgo kernels are very good fillers for making the pattern. After the pattern has been prepared, all the left-over fruits can be placed in the center of the rice. The dish can be prepared in advance and kept cool. Just before serving, the bowl is put in a steamer and steamed while the party is eating the main course.

To serve, remove the bowl from the steamer, place the serving dish over it and quickly flip the bowl with the support of the cold dish, positioning the bowl bottom up on the dish. Now, insert a knife down the sides of the bowl and lightly tap them. If the bowl is greased sufficiently, the pudding will be freed. It will rest on the serving dish with the red, green, white, black and brown patterns signifying friends from all corners of the earth.

Many restaurants that serve "Eight Precious Pudding" do not use different kinds of fruits. Instead, they use the red bean paste to fill the center and prepare a sweet sauce covering the rice. A genuine "Eight Precious Pudding" should contain 8 kinds of fruits, with ginkgo being a special treat.

Ginkgo Duck Soup (*Bai-guo dun-ya* 白果燉鴨)

Ingredients:

    1 duckling (defrosted, washed, rub with salt and wash again)

    2 cups ginkgo kernels

    2 large onions (sliced)

    1 1/2 cm ginger (thinly sliced and then shredded)

    4 oz bean curd bamboo (*fu-zhu* 腐竹; soaked in cold water overnight)

    1 tsp salt

    8 whole pepper

    2 1/2 gallons water

Boil the water in a large pot. Add all ingredients and bring to a boil. Turn heat to medium-low and skim off any foam floating on the surface. After 1 hour, turn the heat to low and cook for 2 more hours. By this time the natural composition of the ginkgo kernels has reacted with the nutrient elements of the duck and forms a milky, white soup. Makes 6 servings of both soup and meat. This is a very famous Cantonese home-made dish for the cold season.

(Optional) If a sauce for the duck meat and skin is desired, the following ingredients is advised: 2 cloves of garlic smashed, the inner portion of an onion minced, 1 cm of ginger finely copped, 2 tablespoon of soy sauce and l teaspoon sesame oil, all mixed in a bowl.

Actually, ginkgo is a common ingredient for many vegetarian dishes. "Buddha's Disciples' Delight" is a good example.

Pine Seed (*Song-ren* 松仁)

In contrast to the voluminous information available on ginkgo as a food, the Chinese records regarding the supply and uses of pine seeds (piñons) are very scanty. With ginkgo, the supply comes from cultivated trees and the uses are widespread and various. With pines, however, the supply comes from wild population, and the uses are very limited, locally more as a confection than a food. Moreover, the farmers who gather pine seeds are more inclined to exchange the products for cash than to eat it. The distribution of the species presents another limitation. The species of pines that yield edible seeds (*Pinus armandii* Franchet and *P. koraiensis* Sieb. et Zucc.) grow in the mountainous areas of Yunnan, Sichuan, Helongjiang and Jilin. When the seeds are shipped to communication centers, it is more likely that they are bought by or used for the affluent people. This has been exactly the situation. Pine seeds used to be rare in China, and when available, it was served in the form of a tea or in making pastry.

Pine Seed Tea (*Song-zi-cha* 松子茶)

Ingredients:

2 oz shelled pine seeds

4 oz shelled walnuts

5 cups water

2 Tbsp honey

Very slightly brown the pine seed and walnut on medium-low heat. Cool. Roll the nuts into a coarse powder. Place the powder in a pan of water and bring to a boil. Turn the heat to medium-low to boil the water down to 4 cups. Add honey. Makes 4 servings. This is a nutritious drink for elderly affluent people, to be served before bed time.

Pine Seed Gruel (*Song-zi yu-zhou* 松子魚粥)

Ingredients:

1/4 cup pine seeds

1/3 cup glutinous rice

1/2 cup short grain rice

4 oz scrod or fillet of sole (sliced and pounded with the back of a heavy knife or cleaver, mixed with salt and slices of the white portion of the scallion, and keep covered in the refrigerator)

2 scallions (cut both the white and green portions into very thin cross-sections, use the white to cover the slices of the fish, keep the remainder)

1/2 tsp salt

6 cups water

Mix the rice, pine seeds, salt, and water and bring the mixture to a boil. Turn heat to low and continue to cook for 1 hour, stir occasionally to prevent burning on the bottom. Before serving, turn the heat to medium-high, stir in the fish and green scallion slices. Cook for 2 more minutes. Making 6 servings.

This dish is nutritious and easy to digest. It is good for an elderly person as a light supper. It can be kept in the refrigerator after the first serving, and reheated in a microwave oven. This recipe is designed at a low salt level. If desired, one can season with salt and some pepper to taste.

After the re-establishment of diplomatic relations between China and the United States, Chinese pine seeds are imported to America, shelled or unshelled. In the spring of 1985, R. A. Howard, Former Director of the Arnold Arboretum, brought some unshelled pine seeds from a Chinese store to the Harvard University Herbaria for identification. The Chinese government then encouraged the people to gather unusual plant products for exports. In 1984, I met farmers in the Long White Mountain (Map, Loc. 1) area in Jilin Province traveling on bicycles to the red pine forests to collect the cones for the seeds in August. They got a good price for the harvest. In a well-organized society, and supported by government agency, it does not take long for the pine seeds of Jilin to appear in American markets. Now, in Chinese groceries shelled Chinese pine seed is a common products, and in several restaurants, a special Shanghai dish of pine seed minced-chicken served with lettuce wrapping, has become increasing popular.

(2) *Angiosperms*: Every great civilization has developed its special angiosperms for food. Accounts are given below of the seeds of the seven species characteristic of Chinese culture. These are the Euryale seed, Job's tears, lotus seeds, mung bean, red bean (Adsuki), soybean, and watermelon seeds. Due to the geographical proximity of certain species and the ancient channels of communication among Southeast Asian countries, it is natural that techniques for cultivation and food preparation were introduced into neighboring countries from China. Historical records show that from these neighboring countries some of the unusual food products were first made known to Westerners. Nevertheless, wherever these seeds are used, one may trace the influence of Chinese culture.

In my investigation of the flora of China, I have never seen the wild progenitors of the mung bean or the red bean. Perhaps, like the ginkgo, the wild forms have been destroyed by man in his clearing for agricultural land, and now all the known progenies are cultivars. Whatever the native land may have been, the methods of using the seeds for food were developed by the ancient Chinese people.

Euryale (*Qian-shi* 芡實)

Euryale (*Euryale ferox* Salisb.) is a perennial, aquatic herb growing extensively in ponds and lakes throughout China. The entire plant is covered with numerous sharp, stinging prickles. The rope-like rhizomes are deeply buried in the mud where they pass the winter. In late April, new leaves emerge, gradually expand, and float over the surface of the water. Flowers of Euryale are inconspicuous, solitary, and terminal to the cylindrical scrapes. The purple-pink petals are concealed by the grayish, green prickly sepals. The ripe fruit enclosed in the spiny receptacle has the shape of the head of a hen, hence the vernacular name of the plant, "chicken head". This structure, like the fruit of the sacred lotus, is an aggregate fruit consisting of numerous bullet-like nuts embedded in the spongy, mucilaginous tissue of the receptacle.

The market material consists of shelled, broken seeds filled with a snow white perisperm, covered by a two-tone membrane, the dark lower 2/3 portion chestnut brown and the lighter upper portion nearly white. Occasionally an unbroken seed appears, which shows a tiny depression, central to the white portion. Buried in the depression is a tiny underdeveloped embryo wrapped in the fleshy endosperm. Unlike all known Chinese food products, the edible portion of Euryale is the perisperm (tissue derived from the nucellus) of the seed (Figure 15m).

Euryale is one of the most ancient useful plants recorded in Chinese history. It first appeared in *The Book of Rites* (禮記), the authorship of which is credited to Chou Gong 周公 (*ca*. 1115 B.C.). It was recorded as a grain used in official ceremonies. By the time of the Han Dynasty, as shown in *Shen-nong ben-cao jing* 神農本草經 (*ca*. 200 A.D.) it was treated as one of the food products which might lead to immortality when taken continuously. At present, the leading production centers of Euryale are in the Chinese Great Lakes areas in Jiangsu, Hubei, Hunan, and Shandong. Information published by the Pharmaceutical Institute, Chinese Academy of Medical Sciences between 1959 and 1966 give the food value of Euryale as: carbohydrates 32%, protein 4.4%, fats 0.2%, fibers 0.4%, ashes 0.5% (including compounds of calcium 0.009%, phosphor 0.11%, and iron 0.0004%), thiamine 0.04%, riboflavine 0.00008%, nicotinic acid 0.008%, ascorbic acid 0.006%, and carotene.

Some of my friends living in Boston area, mix Euryale with rice, Job's tears, and Greater Red Bean (Adsuki) for an enriched staple food, or use it with sacred lotus seed or rhizome, Chinese yam, lily bulb and jujube to make a broth with chicken or spareribs.

A Simple *Bupin* Beverage (*Chen-cun-ren-cha* 陳存仁茶)

Ingredients:

    1 cup Euryale

    1 cup rice

    1/2 cup Job's tears

    1/2 cup peanuts (with or without seed coat)

    1 cup jujube

    2 gallons water

    2 Tbsp brown sugar or honey

Combine, wash and drain the ingredients. Bring the mixture to a boil and simmer for 3 hours. Serves a family of 8, hot or cold, sweetened with honey or sugar if preferred. This recipe is adapted from one described by the most successful practitioner of traditional Chinese medicine in the twentieth century, Dr. C.Y. Chen, who wrote, "This food appears commonplace, but it is the most nutritious *bupin* with unusual carbohydrates, high amounts of protein, vitamins and other tonifying elements. It is economical and tasty, good for the populace." (Chen, 1975).

Revitalizing Euryale Drink (*Bu-xu-cha* 補虛茶)

Ingredients:

4 oz Euryale from the market

4 oz Chinese yam (山葯, dried slices)

4 oz China Root (*fu-ling* 茯苓, from the market)

4 oz sacred lotus seed (shelled, white)

4 oz Job's tears

4 oz cow peas (finely ground)

1 oz ginseng

Mix the ingredients and grind the mixture. Roast the powder on medium-low heat until it turns light brown. Pulverize it while still warm. Cool and bottle for later use.

Prepare a weak solution with 1/2 cup of rice cooked slowly in 5 cups of water for 20 minutes. Place one tablespoon of the above Euryale-mix in a rice bowl and pour the boiling rice water onto it, stirring all the while. Separate the remaining liquid from the rice and keep it for future use. Leave the parboiled rice on low heat for 20 minutes and use it for supper or lunch. Drink the Euryale-rice water on an empty stomach, preferably in the morning before rising in the case of weak or elderly people.

Job's Tears (*Yi-ren* 薏仁)

Job's tears is a tough, tall, perennial grass growing in tropical Asia. For cash the native people gather the fruits and string them to make necklaces, handbags, and other ornaments. Through these items, Job's tears was introduced to the American people. In temperate China the species is cultivated annually as a cereal. Its use can be traced back to the beginning of Chinese civilization. It is recorded in the earliest herbal of traditional Chinese medicine, *Shen-nong ben-cao jing*, as a remedy for muscle cramps. In a herbal composed in the sixth century, it is recommended as a diuratic. It has been prescribed for regulating the functions of the spleen, lungs, liver, intestine, and kidneys for two thousand years.

Now the Chinese scientists have ascertained the content of the grain: 16.2% protein, 4.65% fat, 79.17% carbohydrates, several amino acids (leucine, lysine, arginine, and tyrosine), coixol, coixenolide, and triterpenes. In the 1960s, Japanese scientists working on tissue culture reported that extracts of Job's tears inhibited the growth of carcinomatous cells. As each grain is enclosed in a bony involucre, the cost of freeing it is too high to be of value for everyday food. Therefore, in China Job's Tear is used more for medicine and less as a costly delicacy. Due to foreign exchange, the grain is exported and America has no short supply. Many elderly Chinese Americans are buying the grain and mixing it with rice and beans in their daily diet. The first of the following recipes is used by two senior citizens, one of whom has rheumatic pains of old age. The second recipe is adapted from a dish I had in Chengdu forty years ago.

Job's Tears Rice (*Yi-ren mi-fan* 薏仁米飯)

Ingredients:

    2 cup brown rice (washed and drained)

    1 cup Job's tears (washed, cleaned, and drained)

    1 cup greater red bean (*hong-dou* 紅豆, Adsuki), or soybean (washed, cleaned and soaked overnight)

    5 cups water

Boil the bean in 2 cups of water and then turn the heat to medium-low for 30 minutes. Add Job's tears and 1 1/2 cups of water and bring to a boil. Turn the heat to medium-low and continue to cook for 30 minutes. Add the remaining water and stir in the rice. Bring to a boil, turn to low heat and continue to cook for 45 minutes. Makes 8 to 10 servings. Keep in the refrigerator and warm up for the next meal or the next day. This recipe is designed by Hsiung Chieh, a ninety-year old gentleman who lives in a home for senior citizens and cooks for himself and his 85-year old wife.

Job's Tears with Duck (*Yi-ren ya* 薏仁鴨)

Ingredients:

    1 medium-sized duck (defrost, wash, drain, and then rub inside and outside with salt; wash off the salt and boil the liver and gizzard)

    1 1/2 cups Job's tears (washed, cleaned, soaked in cold water for 2 hours, and drained)

    1/2 cup glutinous rice (washed and soaked with Job's tears)

    2 medium-sized onions (mince one and slice the other)

    5 mm piece of ginger (mince 2 mm and slice 3 mm)

    2 Tbsp salt (use 1 1/2 Tbsp to rub the duck and 1/2 Tbsp for cooking)

    8 cups water

    1/2 cup red jujube (washed and drained)

    6 black mushrooms (washed)

    1 Tbsp soy sauce

    (Optional) 1 Tbsp *tian-mian sauce* 甜麵醬 (used as relish)

Use a cup of water and 1/2 teaspoon of salt to boil the liver and gizzard. Cut these into 5 mm cubes. Mix with Job's tears, rice, minced onion, ginger, and 1/2 teaspoon of salt to make a stuffing for the duck. Fill the duck and place it breast-side up on 2 chopsticks in a 2-gallon pot. Add the sliced onion and ginger, the neck, red jujube, black mushrooms, stock from the boiled liver and gizzard, and the remaining water, making 8 cups of liquid in total. Cook for 30 minutes on medium-high heat, then reduce to low heat for 1 1/2 hours, until the duck is well done. Serve hot as two dishes: a cup of soup with mushrooms, jujube and, for fun and for the meat attached, a piece of neck-bone for each person; and the duck, with soy sauce or *tian-mian* sauce if desired. Makes 6 servings.

Lotus Seed (*Lian-zi* 蓮子)

The sacred lotus is cultivated as a vegetable crop throughout China, particularly in the lowlands and lake areas of Hubei Province. Many varieties and cultivars have been developed, some recognizable by the color and shape of the flower petals and others by the texture of the edible rhizomes. The lotus seed is a by-product of the "root" crop. For this reason the immature fruits are often picked and marketed for young seeds which are eaten raw as fresh "fruit". This practice limits the annual supply of mature lotus seed, hence it is always expensive. Lotus seeds are available in American Chinese stores both dried and canned. The dried seed appears white when the seed coat is removed, and brown when the coat is intact. The edible portion of the lotus seed consists of the mealy cotyledons. In Chinese food the green embryo of the seed is generally removed. The canned product may be whole, white cotyledons or paste. Large amounts of lotus seed paste is used by Chinese bakeries in America for making the more expensive kind of moon-cake for the Mid-autumn Festival, the full moon day in September which is celebrated as a Chinese Thanksgiving Day. Such moon-cakes are prepared by specialty stores. The recipes given here are limited to home-made food.

Sweet Lotus Bowl (*Lian-zi-geng* 蓮子羹)

Ingredients:

    2 cups dried, white lotus seed (soaked in cold water overnight)

    6 oz rock sugar (or 1/2 cup regular sugar)

    10 cups water

    1 Tbsp preserved Osmanthus flower (see the section on flower)

Boil the ingredients together. Reduce heat, but keep the contents boiling for 3 hours or until the liquid is reduced to 8 cups. Just before serving, stir in the preserved Osmanthus flowers. Makes 8 hot servings.

At Chinese banquets, in the middle of the meal, a light, sweet dish is served to break the monotony of greasy meat or seafood courses. The "Sweet Lotus Bowl" is a very popular and welcome delicacy for such occasions.

Lotus Seed Refreshment (*Lian-zi dian-xin* 蓮子點心)

Ingredients:

1 1/2 cups dried, white lotus seed (washed, drained, then soaked overnight)

1 1/2 cups short grain rice (washed and drained)

1 cup millet (cleaned, washed until water is clear and all floating particles removed, drained)

1 cup red beans (soaked overnight; replaced with mung bean in summer)

1 cup Chinese red jujube (washed and soaked in cold water; save the water and use it for cooking)

1/2 cup Job's tears (washed, cleaned and drained)

1/2 cup brown sugar

2 gallons of water

Combine all the ingredients and bring the mixture to a boil. Lower the heat and simmer for 2 hours, stirring occasionally to prevent burning on the bottom of the pan. Serve hot with more sugar for those who prefer the dish a bit sweeter. Makes 20 servings.

The Chinese visiting scholars, the professors of Harvard Yenching Institute and the associated organizations on Divinity Avenue, Cambridge, Massachusetts, meet regularly on the last Friday of every month for a special lecture followed by discussions. This refreshing dish is served at the end of each meeting just as beer, cheeses and crackers are served by the New England Botanical Club at the end of their meetings.

Mung Beans (*Lü-dou* 綠豆)

In temperate China, mung bean is extensively cultivated as a summer crop for its seeds. There are several varieties or cultivars, some erect, others climbing. The escapees of the climbing variety produce small, black seeds with less mealy cotyledons. The flowering habit of the erect variety requires hand-picking to harvest because the fruits do not mature all together. Normally, after the first picking, the plants are left to grow in the field for several more weeks to wait for the green pods to turn black. In general, the seed coat of mung bean is green. Occasionally, there are plants bearing yellow seeds.

The earliest record of mung bean in Chinese literature appeared in an herbal completed in 975 A.D., four years after the First Emperor of Song Dynasty conquered Guangzhou (Map, Loc. 58). It has spread northward quickly, and in North China it has been used not only as an agricultural product, but also as a product of simple, light industry: the mung bean sprout industry, and the mung bean silk industry. Mung bean sprouts produced locally are common in American supermankets. Mung bean silk, however, is imported. In North China a dried, round, thin transparent sheet called *fen-pi* 粉皮 is common. Presently, both the thread-like *fen-si* 粉絲 (mung bean silk) and the thin sheet products are available in American Chinese stores.

These products are prepared by grinding the soaked mung bean, separating the starch grains from the tough, green cover through filtration and precipitation, and making a thin paste with the starch. In making the mung bean silk the cooked paste is poured into a bamboo sieve over a container of cold water, in which the hot puree solidifies into soft strings. Meter-long bamboo sticks are used to fish out and pick up the soft noodle-like material which is trimmed and hung with the sticks on drying frames. For export the mung bean silk is packed in 2 ounce or 4 ounce bundles, but in the marketplace in China it is always sold in bulk. It turns soft immediately in boiling water, and tolerates cooking without becoming mushy. Mung bean silk is used throughout China and numerous local recipes abound for cooking it alone, with vegetables and/or with meat (Figure 31).

Recently, the mung bean sprouts have become available in American supermarkets. They are used in salads in a similar manner as alfafa sprouts. In China there are two kinds of bean sprouts, the green bean (mung bean) sprouts (*lü-dou-ya* 綠豆芽) made of mung bean and the yellow bean (soybean) sprouts (*huang-dou-ya* 黃荳芽) prepared from soybean, used more in homes and schools than in restaurants. All bean sprouts are eaten cooked, but soybean sprouts must be cooked longer than mung bean sprouts.

Mung Bean Cold Drink (*Lü-dou tang* 綠豆湯)

Ingredients:

    1 1/2 cups mung beans (washed and drained)

    1 1/2 gallons water

    1/2 cup brown sugar

Combine the ingredients and bring to a boil on high heat for 15 minutes. Turn heat to low and simmer for 1 hour, or until the beans are very soft. Serve cold as a refreshing drink in hot seasons. This is a favorite summer treat throughout China, as illustrated by a personal experience which moved me deeply.

In August 1984, I was invited to give a ten-day short course to Chinese botanists working in various fields related to botany by the Institute of Botany, Academia Sinica, Wuhan, Hubei. Wuhan in August is said to be the hottest place in China. It was hot and humid, and I was scheduled for a three-hour lecture in the morning, plus a two-hour question and answer period in the afternoon. One day, as I finished my lecture, I was taken to the library where the librarian, wife of the Deputy Director of the Institute, gave me a large jug of cold mung bean drink, saying, "With your advanced age, and on such a hot day, you have talked straight for two and a half hours. Please drink this and be refreshed." At that time and place, it tasted better than ice cream soda!

The Chinese people learned to take such a drink believing it effective as a summer anti-stroke agent long before phytochemists ascertained the chemical composition of the mung bean. Now we know that 22% of the mung bean seed is protein, 0.8% fat, 59% of various carbohydrates, and rich in calcium, phosphor and iron compounds, carotene, thiamine, riboflavine and nicotinic acid. Chinese and Japanese scientists have isolated phosphatidylcholine, phosphatidylethanolamine, phosphatidylinositol, phosphatidylglycerol, phosphatidylserine, and phosphatic acid. A famous and successful Chinese practicianer, C.Y. Chen, prescribed two cups of mung bean drink daily for his diabetic patients before meals as a health tea and for a cure.

Mung Bean Crepe (*Lü-dou jian-bing* 綠豆煎餅)

Ingredients:

    1 cup mung beans (washed, soaked for 12 hours, and drained)

    1/2 cup all purpose flour

    2 eggs

    2 scallions (cleaned, thinly sliced across)

    2 Tbsp margarine

    1/2 cup water

    1/2 cup milk

    1/3 tsp salt

Combine the mung bean and water, and puree in a blender. Beat together the remainder of the ingredients with a hand beater until smooth. Grease a hot griddle. Pour the mixture by scant 1/4 cupful onto the griddle. Tilt it to let the batter run, and using a spatula, make a wholesome, thin pancake. Cook until the edges are dry. Turn and cook until golden and crisp. Serve hot. Makes 2–4 servings, depending on the appetite of the participants. In Wuhan a modification of this dish is prepared by cutting the folded crepes and cooking the strips with some meat, vegetables, and stock to make a soup.

Mung Bean Gruel (*Lü-dou-zhou* 綠豆粥)

Ingredients:

    1/2 cup mung beans (washed and drained)

1/2 cup short grain rice (washed and drained)

1/2 cup very coarse cracked wheat

10 cups water

Combine the ingredients and bring to a boil. Continue to cook for 10 minutes on high heat. Turn to medium-low heat and cook for 1 1/2 hours. By this time the mung beans should be split in the middle. Stir occasionally to avoid burning. The heat should be reduced as the mixture thickens.

This is a breakfast common in North China. Often sweet potato is cut into 1–2 cm cubes and added, and rice is replaced by millet. It is nutritious, tasty and chewy.

Stir-fried Mung Bean Sprouts (*Chao lü-dou-ya* 炒綠豆芽)

Ingredients:

1 lb fresh mung bean sprouts (washed and drained)

1/2 medium-sized onion (thinly sliced)

6 zanthoxylum (Sichuan peppercorns)

2 Tbsp vegetable oil

1/2 tsp salt

2 Tbsp water

Heat the oil and brown the onion. Add the zanthoxylum and stir in the sprouts. Add salt and water. Stir for 1 minute. Serve immediately. Makes 4 servings.

The secret to preparing this dish successfully is quick movement, slight cooking, and keeping the sprouts crisp. For personal use, some people prefer to add 1/2 teaspoons of vinegar before eating. For parties, one may like to add some color by putting in a few thin slices of green and/or red pepper to be cooked with the sprouts. Never cook mung bean sprouts with soy sauce, for it leaves an unpleasant stain to the dish.

Stir-fried Three Fresh (*Chao-san-xian* 炒三鮮)

Ingredients:

3/4 lb fresh mung bean sprouts (washed and drained)

1/3 lb Chinese leek (cleaned, washed, drained, and cut into 2 1/2 cm sections)

2 pork chops (cut the meat thinly to match the sizes of leek and bean sprouts, marinate the slices in soy sauce, shredded ginger, pepper, 1/2 tsp of salt, chopped onion, and all the cornstarch)

1 Tbsp soy sauce

1/2 cm piece fresh ginger (thinly sliced and shredded)

1/2 tsp salt

1 Tbsp cornstarch

1/2 cup chicken broth

1 tsp sesame oil

2 Tbsp vegetable oil

1 medium-sized onion (slice 3/4, chop 1/2)

A dash of black pepper

Slightly brown the onion slices. Add the marinated pork and stir. Add the chicken broth and continue to cook on medium heat for 5 minutes. Stir in the Chinese leek and bean sprouts. Add salt and sesame oil. Makes 4 servings.

The secret of preparing this dish is to have the three distinct colors: the red meat, green leek and white sprout, and have the flavor cooked into the meat, yet keeping the fresh colors and crisp texture of the vegetables.

Bean Silk Miniature Omelets (*Fen-si dan jiao* 粉絲蛋餃)

Ingredients:

2 oz mung bean silk 粉絲 (revived in boiling water and keep soaked before use)

1 large onion (sliced vertically, 1 cm across)

4 cups celery cabbage (cut into 4 by 2 cm pieces)

3 eggs

4 oz lean, ground beef

1/10 tsp ginger powder

1 tsp salt

2 scallions (finely chopped)

1 Tbsp soy sauce

6 zanthoxylum (Sichuan peppercorns)

3/4 cup water

4 Tbsp vegetable oil

1 tsp sesame oil

Combine the beef, ginger, scallions, soy sauce, 1/2 teaspoon of salt, and 1 tablespoon of vegetable oil. Mix well and put aside. Beat the eggs with a fork and 1/2 teaspoon of salt. Grease the skillet on medium heat. Pour 1 tablespoon of egg to make a 5 cm omelet, and immediately place 1 teaspoon of the beef mixture onto the center of the not yet fully cooked omelet and fold it once. Turn it over and move it to the side of the skillet. Repeat the procedure and make many miniature omelets. The early ones can be piled on all sides of the skillet with some turning to avoid burning, or have the cooked ones placed in a dish. When the preparing of the omelets is completed, take them out and clean the skillet. Pour in the remaining vegetable oil, and brown the onion slices and zanthoxylum. Stirred in the celery cabbage and the remaining salt. Cook for 2 minutes. Add water and the mung bean silk. Mix all ingredients well. Place the small omelets over the cabbage and the bean silk. Turn on high heat and cook covered for 3 minutes. Turn the heat to medium-low and cook for 15 minutes. Check occasionally and add some water if needed to avoid burning. Add sesame oil and serve hot. Makes 4 servings.

This is a flexible dish, which can be prepared for a party in advance and warmed up before serving. Ground pork may be used as a substitute for beef, or a mixture of two kinds of meat if preferred. More salt than suggested in the recipe may be added to taste.

Mung Bean Sprout Salad (*Liang-ban lü-dou-ya* 涼拌綠豆芽)

Ingredients:

>   1 lb fresh mung bean sprouts (remove the thin, basal, brown portion, wash, and drain)
>
>   2 oz fresh Chinese leek (cleaned and cut into 3 cm sections)
>
>   1 oz mung bean silk (revived in boiling water and keep soaked for an hour, cut into 3–5 cm sections)
>
>   2 cloves garlic (crushed and chopped)
>
>   1/2 cm fresh ginger (thinly shredded, or 1/10 tsp powder)
>
>   1 Tbsp soy sauce
>
>   1 tsp vinegar
>
>   8 cups water
>
>   1 1/2 tsp sesame oil
>
>   1/2 tsp salt
>
>   A dash of black pepper

Combine the garlic, ginger, soy sauce, vinegar, pepper and sesame oil to make a Chinese salad dressing. Mix these ingredients well and set the dressing aside. In a sauce pan boil the water. Add the prepared sprouts and leek sections and stir thoroughly. Cook for 1 minute. Drain and cool under running water.

Just before serving combine the sprouts, leek and bean silk with the dressing and salt. Mix well and serve cold. Some people like to add some yellow and red color to this dish. The yellow is achieved by beating an egg with a dash of salt to make a very thin sheet which is then shredded into 4 cm long, 5 mm wide strips. A few radishes or slides of red pepper placed along the side give the red color.

Red Beans (*Hong-dou* 紅豆)

The use of red beans can be traced to the first century of the Christian era, in *Shen-nong ben-cao jing* (Wu, 220–260). By the name and description, the species can be identified as the rice bean (= lesser red bean 赤小豆 *Delandia umbellata* [Thunb.] S. Y. Hu), a climber with slender oblong seeds, *ca.* 8–9 mm long, 3 mm across and conspicuously carunculate. Presently, this bean is used more in medicine than as a food. The Adsuki bean (= greater red bean 紅豆, *Azukia angularis* [Willd.] Ohwi), cultivated for food, is an erect herb with shorter and thicker seeds, *ca.* 8 mm long and 5–6 mm across, with very narrow and almost indistinct caruncle. The seeds of both species are shining, dark red. They are available in American Chinese stores, and the greater red bean is also sold in some natural food stores. Compared with the green bean, or mung bean (綠豆), the price of the red bean is about one-fifth higher, and its chemical composition contains slightly lower amounts of proteins, fats, carbohydrates, thiamine, and riboflavine, but higher amounts of calcium, phosphor, and iron compounds as well as nicotinic acid.

In Chinese cooking, the red bean paste is made of the greater red bean and is used in pasteries. Its Chinese name is *dou-sha* 豆沙 (bean sand), a short term and easy to pronounce. Canned *dou-sha* is available in Chinese stores. Many home-makers among Chinese Americans prefer to make their own *dou-sha*. The procedure is simple but time consuming. To make fresh *dou-sha*, boil 2 cups of the greater red bean in 10 cups of water for 2–3 hours, until the bean are broken. Pour the contents into a pan covered by a double layer of gauze. Carefully fold the edge of the gauze neatly, tie it securely, and suspend it over a pan to let the liquid and the fine particles of the broken bean drip into the pan. When the dripping stops, squeeze the gauze bag gently, open it and add a cup of boiling water. Then tie and hang it up again. Repeat this process five times. What is left in the bag can be relegated to the compost pile now. Set the pan aside overnight to allow the bean particles to settle. Pour off the clear portion and make a drink of it if desired. Transfer the bean to a bag made of strong, firm cloth and hang it up to let the liquid drip off. When the dripping stops, transfer the paste to a medium hot pan to warm it and to mix with 1/2 cup of sugar and 2 tablespoons of preserved rose petals. This is rose flavored *dou-sha* to be used for home-made pasteries and/or for desserts. The red bean called for in the following recipes refers to the greater red bean (Figure 29).

Oriental Red Pottage (*Dong-fang hong-dou-zhou* 東方紅豆粥)

Ingredients:

    1 1/2 cups red bean (cleaned, washed and drained)

    3/4 cup rice

    1/2 cup Job's tears (cleaned, washed and drained)

    1/3 cup raisins

1 gallon water

1/3 cup brown sugar

1/2 cup shelled chestnuts (fresh or dried and revived in water, broken into 3 or 4 pieces)

Combine the ingredients and bring to a boil. Reduce the heat to medium-low for another 1/2 hour, and then simmer for 2 more hours. Stir occasionally to avoid boiling over or burning at the bottom.

Remember that in the recipe for "Lotus Seed Refreshment" used in the Harvard Chinese scholars' informal gathering, red bean is one of the ingredients. In that recipe the lotus seed and the Chinese jujube are imported and naturally more expensive items. In the present recipe the ingredients are native to America (chestnuts cover the ground in some gardens). I have taken this refreshing dish to several potluck parties, and the pot has always returned empty.

Evidently, the ancient peoples in western and eastern Asia had learned the refreshing and revitalizing value of the "red pottage" about the same time. Jacob boiled lentils for a red pottage which he used to exchange for Esau's birthright (*Genesis* 25: 29–34). By the same era the Chinese used the greater red bean to prepare a "red pottage" for special treats. This tradition has been carried to the present. An elderly couple living in a home for senior citizens in Boston, prepare their staple food with equal amounts of rice, greater red bean, plus Job's tears in order to prevent their ankles from swelling, and for added strength and better memory. They usually cook enough for five days, and warm a portion for their main meals.

Steamed *Dou-sha* Wrappings (*Dou-sha-bao* 豆沙包)

Ingredients:

10 Ready-to-bake home-style biscuits (7.5 oz package)

1 cup home-made *dou-sha*

Dust a board with all purpose flour. Roll the *dou-sha* into a cylinder and cut it into 10 sections. Roll a biscuit flat, 6–8 cm in diameter. Place one portion of the *dou-sha* in the center and seal it. Cover the bottom of a steamer with a damp cloth and space the *dou-sha* wrappings on it. Cover and steam the contents on high heat for 20 minutes. Serve hot. Makes 5 or more servings. This is a simple and time-saving dish for the afternoon tea.

Soybeans (*Huang-dou* 黃豆)

Before the World War II, America was an importer of soybean. Now, she is an exporter.

Hong Kong, Taiwan, and even mainland China buy their soybean supply from USA. When I first visited Due West, South Carolina, in 1946, farms in the countryside were covered with cotton. Now, the same land is covered with soybean. Although soybean products have entered American supermarkets, several Chinese uses of soybean can introduce more variety and tastier food into many homes (Figure 35). Included as soybean products are: green immature soybean, soybean curd, soybean sheets, firm bean curd squares, soybean skin, and soybean sprouts. With the exception of soybean sheets, all the items are available in American Chinese stores. The Food Plants Research Institute, Amherst, Massachusetts, is in the process of introducing soybean sheets (千張皮) to American markets.

The green immature soybean used in China is the product of a special large seeded garden variety planted purposely for the green seeds. It appears in the market unshelled, and is called *mao-dou* 毛豆 (= *mao-tou*, hairy bean) because of the stiff hairs on the pods; or shelled, and is called *qing-dou* 青豆 (= *ch'ing-tou*, green bean) for the jade green color. The shelled green beans are generally sold to restaurants, while the less expensive unshelled beans are sold to the public. People buy them and shell them at home, or often cook them unshelled. Imported canned green soybean is available in American Chinese stores. Fresh green soybean is very tasty and nutritious. People with kitchen gardens can plant a couple of rows to enjoy soybean at this immature stage. All other items mentioned above are prepared from mature yellow soybeans, soaked and ground. Two types of special large-seeded black soybean are imported to meet the demand of Chinese Americans, and are used as special health food.

Boiled Fresh Soybean (*Zhu mao-dou* 煮毛豆)

Ingredients:

    4 lbs fresh, fully grown soybean pods

    2 Tbsp salt

    1 1/2 gallon water

Boil the water with salt. Add the soybean pods. Continue to cook for 1/2 hour on medium heat. Serve hot or cold. Makes 20 servings.

This is a very good dish for open house parties where fresh strawberries are served. A dish of unusual green tasty material provides in contrast with a bowl of red fruit provides a subject for conversation.

Green Soybean Pork (*Qing-dou chao-rou* 清豆炒肉)

Ingredients:

    1 1/2 cups shelled green soybeans

1 large red pepper (diced to match the soybean size)

1 large onion (cut into 1 cm square pieces)

3 pork chops (thinly slice the meat, marinate in soy sauce, ginger, garlic, salt, and cornstarch; boil the bones in 2 cups water and 1/2 tsp of salt to make 1 cup stock)

1 Tbsp soy sauce

2 cloves of garlic (crushed and diced)

1 Tbsp cornstarch

3 Tbsp vegetable oil

3/4 tsp salt

1 tsp sesame oil

A dash of ginger powder

Brown a teaspoon of onion in 2 tablespoons of oil. Stir-fry the marinated pork. Remove and clean the pan. Use the remaining oil to stir-fry the green soybean, onion and red pepper. Add the cooked pork and stock, and mix well. Cover and cook for 10 minutes. Add sesame oil. Serve hot. Makes 4 servings.

When fresh, green soybean first appears in the market, the above recipe makes an expensive dish served in fancy restaurants in large cities, such as Shanghai, Guangzhou, Beijing and Nanjing. Later in the season, when more green soybean is available in the market, the price comes down and the general public can enjoy it.

Bean Curd Fish (*Dou-fu yu* 豆腐魚)

Ingredients:

1 lb scrod (or haddock or fillet of sole)

1 box of bean curd

1 large onion (vertically sliced)

6–8 outer leaves of lettuce (washed and divided into 2 or 3 pieces each)

1 cm piece fresh ginger (thinly sliced and shredded)

4 cloves garlic (cut longitudinally once or twice)

6 zanthoxylum (Sichuan peppercorns)

3 Tbsp *dou-chi* 豆豉 (fermented black soybean)

4 Tbsp vegetable oil

2 Tbsp soy sauce

1/2 tsp salt

1 tsp vinegar

1/2 cup cooking sherry

1/2 cup water

Lightly brown the garlic and zanthoxylum. Stir in the onion, ginger, *dou-chi*, soysauce, and salt. Take the mixture out and turn off the heat. Place 2 sections of a broken bamboo chopstick parallel on the bottom of the pot to prevent sticking and burning. Lay 1 layer of the lettuce over the sticks and 1/2 of the fillet flat on top of the lettuce. Scatter 1/4 of the mixed spices (ginger, garlic, pepper, oil, salt, etc.) evenly over the fish. Take 1/2 of the bean curd in the left hand, and cut it down the middle and then crosswise into slices of 5 mm thick, letting them fall flat over the fish. Scatter 1/2 of the spice mixture over the bean curd and cover it with a layer of lettuce. Repeat the procedure by laying the remaining fillet over the lettuce, then 1/2 of the mixed seasonings, the remaining bean curd, and spices. Now, mix water, soysauce, sherry, and vinegar. Slowly pour the mixture over the material and along the wall of the cooker. Lay the remaining lettuce on top. Cover and cook on high heat for 5 minutes. Reduce the heat to medium for 15 minutes and then to medium-low for an additional 1/2 hour. Serve hot in the cooking vessel if it is shallow. Makes 4–5 servings.

This recipe is a modification of a dish called Earthenware Fish-head Bean Curd (*Sha-guo yu-tou dou-fu* 沙鍋魚頭豆腐), popular at sidewalk cafes and food stalls of Guangzhou and Hong Kong. Most Chinese recipes for cooking fish call for frying the fish first, which gives a fish smell throughout the house, more so in an apartment. This recipe eliminates the frying step, and gives the delicious result peculiar to the popular sidewalk dish. All my guests and members of the family like the dish tremendously, and some of them have tried the recipe with *yu-tou* 魚頭 (fish head) bought in Chinatown. Structurally, cooked with the fish head, the dish would be more like what one gets in the food stalls in Guangzhou. Bean curd is solidified protein with hardly any taste of its own. One of the important principles of cooking a good bean curd dish is to cook it for a long time with meat, fish or poultry. Recently, with the introduction of the Sichuan food, *Ma-po dou-fu* 麻婆豆腐 (pockmarked wife's bean curd), cooked with very hot pepper and powdered zanthoxylum (much of both), has become increasingly popular among graduate students, for example, Harvard Biology doctoral candidates. Not all people can tolerate hot dishes, however. Personally, I prefer Bean Curd Fish, for this dish is more tasty and nutritious than *Ma-po dou-fu.*

Soybean Sprout Soup (*Dou-ya-tang* 豆芽湯)

Ingredients:

   2 lbs fresh soybean sprouts

   1 ham bone from roast ham

   1 lb spareribs (cut into 3–4 cm pieces)

   2 large onion (longitudinally sliced)

6 zanthoxylum (Sichuan peppercorns)

2 gallons water

2 Tbsp vegetable oil

Lightly brown the onion and zanthoxylum. Add spareribs and stir-fry for 1 minute. Stir in the soybean sprouts, water and ham bone. Bring to a boil on high heat then reduce heat to medium-low and continue to cook for 1 hour. Then, simmer for another hour. Serve hot. Makes 8–10 servings.

Soybean sprouts and bean curd are the salvation of the Chinese people, as they are the most common and widespread foods for all, particularly for the farmers, working people, and young students of boarding schools. In a boarding school at Xuzhou (Map 16) in the 1920s, soybean sprouts alternating with bean curd were the daily main dish during the school year. For a couple of months in summer the yard-long beans, and in winter the celery cabbage and Chinese radish were also used. The institution food is prepared using the simplest recipes. The following recipe for soybean sprouts is worth trying. It may be prepared with or without pork or bacon, but is more tasty with one of them.

Soybean Sprouts with Bacon (*Dou-ya shao-rou* 豆芽燒肉)

Ingredients:

3 lbs bacon (cut into 4 cm long pieces)

5 lbs soybean sprouts

2 large onions (sliced coarsely)

10 zanthoxylum (Sichuan peppercorns)

2 Tbsp vegetable oil (use 5 Tbsp if without bacon)

1 cup water

1 Tbsp sesame oil

1 Tbsp salt

Fry the bacon on medium heat until the edges curl. Add onion slices, salt and stir-fry for 1 minute. Add the zanthoxylum, oil, and mix thoroughly. Stir in the soybean sprouts and add water. Bring to a boil. Turn heat to low and continue to cook for 40–60 minutes (the longer the better). Add sesame oil just before serving. Makes 10–12 servings.

Naturally, replacing 1 cup of water and salt with 1 cup of chicken broth would improve the resulting flavor. In China, chicken broth is a luxury, so it is never used for cooking in institutions.

Bean Curd *Hors d'oeuvre* (*Dou-fu-gan jiu-yao* 豆腐乾酒肴)

Ingredients:

10 pieces of *dou-fu-gan* 豆腐乾 (firm bean curd squares, available in Chinese stores)

1 cup soy sauce

1 Tbsp salt

1 cup water

1 tsp mixture of Five Spices (in the form of bottled powder, available in Chinese stores)

Combine all the ingredients. Boil on medium heat for 30 minutes, and then simmer for 1 hour, until the bean curd squares look brown. Remove the bean curd squares, cool and cut diagonally once, then slice each piece 4 mm thick. Serve with toothpicks at parties. Save the stock for future use. It can be used repeatedly.

Bean curd squares are a more refined product made in small molds and drained under heavy pressure, as compared with the bean curd in supermarkets which is prepared in large molds and drained under little pressure. The squares of the former are much firmer. This recipe is a modification of *hors d'oeuvres* served in Chinese bars or taverns. In taverns, large numbers of spiced bean curd squares are prepared ready for people with tight purse-strings, who may enjoy their drinks with *dou-fu-gan* (spicy bean curd) and peanuts while the opulent savor spicy pork liver and chicken or duck gizzards as *hors d'oeuvre*.

Bean Curd Sheet Pork (*Qian-zhang-pi chao-rou* 千張皮炒肉)

Ingredients:

3 pork chops (Slice the meat finely, marinate in soy sauce mixed with chopped ginger, garlic, salt, and cornstarch)

3 pieces of *qian-zhang-pi* (千張皮, thousand sheets skin = bean curd in form of sheets; cut into 5 cm strips and then slice into 3 mm shreds; or use one can of the material, available in American Chinese stores)

4 sticks of tender celery (clean, cut into 3 cm sections, then shred longitudinally to match the sliced meat and bean curd sheet sections)

1 large onion (finely sliced longitudinally)

1 cm ginger (chopped)

2 garlic cloves (chopped)

1/2 tsp salt

1 cup chicken broth

1 Tbsp sesame oil

2 Tbsp vegetable oil

1 Tbsp cornstarch

Saute the onion in vegetable oil. Stir-fry the marinated pork slices. Combine all the other ingredients except the sesame oil. Bring to a boil. Turn heat to medium-low and continue to cook for 20 minutes. Add the sesame oil and serve hot. Makes 6 servings. At present, only canned *qian-zhang-pi* 千張皮 is available in Chinese stores.

Black Soybean Ox Tail (*Hei-dou hui Niu-wei* 黑豆燴牛尾)

Ingredients:

    2 1/2 lbs ox tails (cut into 4–5 cm pieces)

    1 1/2 cup black soybeans (the larger the bean, the better)

    2 large onions (cut into 4 vertically sections)

    4 carrots (pared, cut into 4–5 cm sections)

    2 gallons water

    1 cm piece of ginger (shredded roughly)

    1 Tbsp salt

    (Optional) Sauce to dip the ox tail in:

    2 Tbsp soy sauce

    1 tsp vinegar

    1 scallion (chopped)

    2 cloves garlic (smashed and chopped)

    1 cm piece fresh ginger (finely chopped)

    1 tsp sesame oil

Combine the ingredients and bring to a boil. Turn the heat to medium-low and continue boiling for 3 hours. Combine the optional sauce ingredients for dipping the meat. This is a special Cantonese dish. Like yellow soybean, the black soybean is rich in protein, CHO, fats, carotene, Vitamins B12, B1, B2, flavones which become daizin and genisten on hydrolization, soyasapogenol A, B, C, D, E, choline, and organic acids. In addition, from the skin of the black soybean, chrysanthemin, Delphinidin-3-monoglucosides, pectin, and levulinic acid have also been isolated. The broth is especially good for senior citizens who suffer from dizziness and swollen ankles.

Watermelon Seeds (*Gua-zi* 瓜子)

In North China farmers plant two types of watermelon: *xi-gua* 西瓜 for the edible pulp, and *da-gua* 打瓜 for the seed. The cultivation of *xi-gua* (common watermelon) is a labor-consuming and intensive agricultural economy. Few farming families can afford to keep such a crop. For example, in the 1920s, in a village, say of fifty families, there might be only two families that had more labor than the land required, which might enable them to take care of a watermelon crop. When the young plants are 30 cm tall, each one must be hilled diagonally, guiding the vine to grow in one direction. The plants are not allowed to grow or to branch freely. They need constant care, fertilization and cultivation. Each vine is allowed to bear two fruits only. For pulp, the Chinese farmers have used all the cultivars common in America, and they have also developed many more types, including those with white pulp and black seeds, and those with red pulp and white seeds, narrowly, black along the margins. All the cultivars of *xi-gua* have oblong fruits, *ca.* 30–40 cm long, 20–30 cm in diameter, thin rind and small seeds, generally black, red, or white with black lines along the margin, 6–8 mm long, 5–6 mm wide.

The cultivation of *da-gua* (seed watermelon) in the agricultural economy of the 1920s was a sign of familial land surplus, i.e., a family that could allow a field to grow a cash crop rather than a staple food. Compared with watermelon cultivated for pulp, the *da-gua* type has smaller spherical fruits, *ca.* 20 cm in diameter, with thick rind, white and less sweet pulp, and large ovate seeds, 10–12 mm long, 8–10 mm wide, black with cream-yellow centers on both sides. The *da-gua* fruit is never sold. The harvest of *da-gua* is a community affair. Everyone in the village and any passing travelers are invited to eat as much watermelon as possible under the condition that they put the seeds into a clean half shell. A participant enters the field without any tools save his ability to detect the best or sweetest melon. He first taps the melons on the ground, chooses one, breaks it off the vine, holds it in one hand and smashes it with the other, then he splits the melon in two (the central portion is usually attached to the larger side). He frees the central portion and eats that first. He can walk in the field and find several good melons, bring them to one place, eat some and place the pulp of the melons in empty shells for taking home to the children and the elderly. The owner of the field collects the seeds, washes, drains and dries them on mats.

In Chinese culture, watermelon seeds are used just as almonds are used in the West, both as a nut to eat and as an ingredient in pastries. With the exception of the seeds of sliced watermelon sold on street corners, where the passersby buy and eat the slices and eject the seeds, all watermelon seeds are saved, washed, dried, and used for special occasions. In my village, watermelon and squash seeds, peanuts, and sunflower seeds are kept for the Chinese New Year, when people visit, chat, and crack seeds. During the Chinese New Year, tea is served, accompanied by a small, round tray with several compartments containing plain, roasted, red, and spicy, black watermelon seeds, sunflower seeds, various preserved fruits, candies, sweetened lotus seeds, and water-chestnuts.

In cities, converting the plain, raw watermelon seed into a commercial product sold as an important article for social gatherings belongs to the field of light food industry. In everyday life when friends visit, watermelon seeds are often served with hot tea. In the tea houses, most customers order tea and watermelon or sunflower seeds.

Special Subterranean Plant Food

The subterranean plant products of two American species, the sweet potato and the Irish potato have become the staple foods of certain groups of people in China. The sweet potato was introduced through the Spanish people and the Chinese immigrants of the Philippines. It has become the survival food of poverty-stricken people living in the less fertile areas of rural China. In urban areas the sweet potato is used as a treat. In northern China it is sold by itinerant venders who carry a charcoal-burning, earthenware urn to bake the potatoes. In southern China sweet potato cubes are boiled with brown sugar for refreshments. For the Chinese New Year, deep fried sweet potato chips are prepared by paring the potatoes, and cutting them diagonally to 3 mm thick slices, which are then steamed, dried, and cooked in vegetable oil.

In cities, the Irish potato is served as a vegetable by paring, slicing into 5 mm shreds, and quickly stir-frying it to produce a crisp dish. The Irish potato was first introduced by European and American missionaries, hence it is called *yang-yu* 洋芋 (foreigner's potato). In the homeland of the giant panda, it has become an important staple food to the ethnic groups living in the high Sino-Tibetan mountains, formerly called Eastern Tibet. There, since it is a staple food, the recipes for cooking Irish potato are simple and seldom vary among different people. A large quantity, say half a bushel, is cleaned, washed, and boiled whole with skin intact. Potato so prepared is served hot without salt or seasoning, and usually as the only item for the whole meal. The potatoes are fleshy farm products, hard to keep for long and too heavy to carry far. They are usually eaten by the people who cultivate them. In cities relatively little potato is consumed.

Carrot is called *hu-luo-bo* 胡蘿蔔 (radish of the Tatars), because it was introduced from Central Asia. Along with the sweet potato, it has become a survival food of poor people in rural areas. In cities it is sliced and cooked with celery cabbage or to add color in meat dishes. In urban areas, carrots are never eaten raw. Some rural people occasionally eat raw carrots.

Here, accounts are limited only to special underground plant products, including the enlarged winter buds of bamboo, the corms of sagittaria (Old World arrowheads), taro, water chestnut, the rhizomes of sacred lotus, the Chinese radish and the true yam.

Bamboo Shoots (*Sun* 筍)

People who patronize American Chinese restaurants know that bamboo shoots are a common ingredient in many dishes, for they are found in meat, fish, chicken and/or in vegetarian dishes. Millions of cans of bamboo shoots are imported annually from China via Hong Kong or from Taiwan for this purpose. Knowing the increasing market demand for the shoots and the economic importance of the culms in eastern Asia, the U.S. Department of Agriculture sponsored special expeditions to introduce bamboos from China and Japan. The reports and publications of McClure (1957), and of Young and Haun (1961) recorded that approximately 85% of the species with edible shoots were in cultivation in the United States. These include all the species that produce the canned shoots now imported by American Chinese stores. According to the above reports, the species of bamboos introduced between 1880 and 1925 should have produced enough shoots for the American market. The soybean was also introduced about the same time. Why is it that America is an exporting country of soybean but is still importing bamboo shoots? The answer lies in the cost of labor and lack of technique to free the tender, edible portion of the bamboo shoot from the many hard, leathery layers of the culm sheaths which are often covered with stinging hairs. Whereas the cultivation and harvest of soybean can be done by machines, bamboo shoots must be cut individually by hand, which is too costly to be profitable in America. However, this should not discourage growers from harvesting and eating their fresh bamboo shoots. Actually, in China the bamboo shoots are a delicacy of the season, and they are always expensive, particularly north of the 30th parallel where the climate and the shortage of farm land for staple food crops greatly limit the cultivation of bamboos.

In China the gourmets who can afford bamboo shoots never choose canned material. To them canned shoots are tasteless, for canning changes the consistency and destroys the aroma present in fresh shoots. In Beijing, Guangzhou, Chengdu, Nanjing, Shanghai, Wuhan and similar cities, the gourmets use fresh bamboo shoots in their dishes, and dried ones for soup. The dried shoots are generally prepared in the mountainous areas, for bamboo shoots appear so suddenly after a spring rain that they must be harvested and dried immediately while still young and tender. Bamboo shoot gatherers are young adults who take a couple of days' rations for a trip to the mountains where they stay in Taoist temples, collect the shoots in the natural bamboo forests, peel them to reduce the weight, and carry the tender portions back home to parboil and dry. In late April 1943, a young British geologist, Harry Whittington, the world's leading trilobite specialist, his wife, Dorothy, and I climbed the Zhaogong Shan (Map, Loc. 38). On top of the mountain we shared an isolated temple with bamboo shoot gatherers, and saw the loads of white, tender shoots gathered that day.

Chinese farmers have devised a crafty method of freeing the edible portion of the bamboo shoot, a method unknown to most of their countrymen including botanists! Professor F. A. McClure acquired his knowledge of bamboos through field work carried on among Chinese farmers and the bamboo shoot gatherers in the mountains. From them he learned how to free the tender portion from the leathery and often prickly outer layers. He taught me the trick in my office at the Harvard University Herbaria. This technique is recorded here for the first time in botanical literature.

In the early 1960s when preparing the manuscript on *THE BAMBOOS — A Fresh Perspective* (McClure 1966) for the Maria Moors Cabot Foundation, Harvard University, he often came to Cambridge, Massachusetts. One day he brought a bundle of fresh bamboo shoots to my office and said, "These are from my garden in Maryland." In great excitement, I took one and began to peel the culm sheaths, but with little success. With his characteristic amused smile and quiet voice he said, "Look!" as he took out his pocket knife. He took one shoot, and holding it on the desk with his thumb and index finger, he pierced the middle of the shoot to the septate center. He then moved the knife forward in line with the parallel veins of the culm sheath to the apex. Returning the knife to the original incision and moved it downward to open the basal portion in a similar manner. Like cracking a peanut to take out the seeds from its hard shell, the tender portion of the bamboo shoot was easily freed, and the culm sheaths fell all at once. Although the exposed portions of the culm sheaths are leathery and prickly, the inner basal tissue is tender, meristematic and breaks off easily. With deep gratitude to him for showing me the simple technique for preparing edible bamboo shoots, and an admiring delight of the illiterate but intelligent farmers who designed the method, I cleaned all the shoots in less than fifteen minutes, boiled them in the kitchenette of the Harvard University Herbaria and let everyone in the Herbarium try Dr. McClure's fresh bamboo shoots. Now it takes no more trouble to free the edible portion of a bamboo shoot than it does to peel a banana, and I can see no reason why fresh bamboo shoots will not appear on the vegetable stands in supermarkets throughout America in the future.

A bamboo shoot is comparable to an artichoke which is an unopened flower head with the tender edible portion covered by numerous thick and sharply pointed scales. Unlike the artichoke, a bamboo shoot is an unopened vegetative bud containing the primordia of a tall, often tree-like bamboo. In preparing an artichoke for food, one boils it with the scales, which have edible basal portions, in addition to the tender receptacle and the embryonic ovaries. In serving bamboo shoots, the scales (culm sheaths) are first removed, and the tender portion of some species is boiled to eliminate acridity.

The bamboo most widely cultivated for edible shoots is *mao-zhu* 毛竹 (*Phyllostachys pubescens* Mazel ex. de Lehaie). This species was introduced to Japan via the Ryukyu Islands in 1738. It was introduced to Europe from Japan in the 1880s. Ten years later, Rufus Fant obtained the species from Europe and planted it in his home at Anderson, South Carolina, USA. I visited this garden repeatedly between 1946 and 1980. Not using the buds to thin the growth, the bamboos have destroyed the garden and caused damage to its structures, because the species is characterized by its running rhizomes which can go through pavement, walls, etc. Bamboos belong to the family Gramineae. Like grasses, when let grow unchecked, they can be a menace. To harvest the shoots annually is to thin the growth and to keep it under control besides having crisp, tasty food.

In China the winter buds are ready for the market in Guangzhou and its vicinity during the Chinese New Year. The buds of large, mature bamboos are approximately 20 cm long and 7 cm in diameter. They are covered by 5 or 6 visible hairy scales (Figure 12b) and numerous concealed ones. The edible portion of a winter bud is a solid mass of meristematic cells, which are nutritious food. Along the Yangtze River Region in response to the spring rains, the meristematic cells begin to elongate, and numerous bamboo shoots suddenly emerge. They must be harvested immediately for the market. This crop is called *chun-sun* 春筍 (spring shoot) and portions of the edible tissue are septated. A bamboo grows only in height, not in width. The thickness of a bamboo stem is determined in the shoot.

The species of bamboo cultivated in America free from acridity are *Phyllostachys angusta* McClure, *P. dulcis* McClure, *P. viridis* (R. A. Young) McClure, and *P. vivax* McClure. The species that produce good food but with some acridity so that the shoots must be boiled first are *P. bambusoides* Sieb. et Zucc., *P. flexuosa* A. et C. Riviere, *P. nuda* McClure, and *P. pubescens* Mazel ex H. de Lehaie.

People who have bamboos on their property should make full use of their fresh shoots annually. Those who do not grow bamboos can still enjoy them. American Chinese stores carry some ready-made bamboo dishes in cans which can be bought and taken to pot-luck parties. A plain, cooked bamboo called *Yu-men-sun* 油燜筍, prepared from a bamboo with thin stems, makes a delightful addition to any party, and provides an interesting conversation piece. Another canned material containing shredded bamboo shoots and chopped Chinese mustard greens called *Dong-cai-sun-si* 冬菜筍絲 is a very good combination. A can of *Dong-cai-sun-si* stir-fried with an equal amount by volume of marinated pork slices makes a quick dinner for two or three people.

Bamboo shoots are a common ingredient for vegetarian dishes. Among the recipes already given, several call for bamboo shoots: Vegetarians' Delight, Chrysanthemum and Rice Crust, Buddha Disciples' Delight, and Red-in-snow and Bamboo Shoots. The Chinese temples of the Taoists and of the Buddhists are often situated in famous mountains where large bamboo gardens are located on the grounds. The monks grow bamboos for their own uses and as cash crops. In preparing fresh or dried bamboo shoots for the market, they save the tender, inner culm sheaths which are dried and used for food in winter. The dried culm sheaths are called *sun-yi* 筍衣 (garments of bamboo shoots), a delicacy of central China. This product is available in American Chinese stores now.

Bamboo Shoots and Pork (*Sun-si chao-rou* 筍絲炒肉)

Ingredients:

> 1 cup shredded bamboo shoots (fresh or canned)
>
> 1 cup shredded pork (marinated in ginger, soy sauce, garlic, salt and half of the onion)
>
> 1 medium-sized onion (thinly sliced)
>
> 1 garlic clove (mashed and chopped)
>
> 1 Tbsp soy sauce
>
> 1/2 cup chicken broth
>
> 5 mm slice of ginger (sliced and shredded)
>
> 2 Tbsp vegetable oil
>
> 1 tsp sesame oil

Brown half of the onion slices. Add the marinated pork. Cook for 2 minutes, stirring constantly. Add the bamboo shreds and the chicken broth. Stir, and cook covered for 10 minutes. Add sesame oil and serve 2. Chicken breast can be used in place of pork.

Bamboo Shoots and Virginia Ham (*Sun-pian huo-tui* 筍片火腿)

Ingredients:

> 1 cup thinly sliced bamboo shoots
>
> 1 cup thinly sliced Virginia Ham
>
> 1 cup broccoli (cut into bite sizes)
>
> 1/2 cup wood ear (revived in boiling water, cleaned and keep soaked)
>
> 3 Tbsp vegetable oil
>
> 1 medium-sized onion (coarsely sliced)
>
> 1/2 cup chicken broth
>
> 1/5 tsp salt

1 tsp sesame oil

Saute the onion. Add all the ingredients except the sesame oil. Stir and mix well. Cover and cook for 10 minutes. Add the sesame oil. Makes 3–4 servings.

Luo-bo (*Luo-bo* 蘿蔔, Chinese radish)

There is no evidence that wild radish is a native of China, but the cultivated radishes are indeed important elements in the Chinese culture. The name *luo-bo* 蘿蔔 appears in the *Er-ya* 爾雅, the earliest Chinese encyclopedia on natural and cultural objects. The authorship is traditionally attributed to Zhou-gong 周公 (Duke of Zhou), son of the first ruler of the Zhou Dynasty, who died in 1185 B.C.. Time and space have made no change to this name as they did with the names of many other food plants. China is well known for the diversity of her dialects, yet, *luo-bo* is a name used for radish throughout the country. Etymologically, *luo-bo* has become a basic term in Chinese language for fleshy roots, hence we have *hu-luo-bo* 胡蘿蔔 for carrot which is a later introduction through the Central Asian Silk Route, and *shan-luo-bo* 山蘿蔔 for pokeweed (*Phytolacca acinosa* Roxb.).

Horticulturally, numerous varieties and cultivars have been developed to adapt to the varied climatic and edaphic (soil) conditions. Although varieties with small red and/or white cylindrical roots are planted as spring crops, radishes are essentially a summer-autumn crop in temperate China and a winter crop in subtropical and tropical areas of China. In North China, the most extensively planted variety is the *hong-luo-bo* 紅蘿蔔 (red *luo-bo*, Figure 25e), which constitutes the major winter vegetable of the people. The *qing-luo-bo* 青蘿蔔 (green *luo-bo*), jade-green inside and out on one-third of the leaf-end and gradually changing to white inside and out towards the tail-end, is planted to a lesser extent. This variety commands a higher price for it is used as a fruit or as a salad for banquets.

As a fruit, the green *luo-bo* is pared, split vertically 3 to 5 times depending on the size of the radish, to 1/4 of its length, leaving the white portion intact to hold the sections. Then similar cuttings are made in a different direction so that the final vertical sections are about 8–12 cm long and 1 cm in diameter. After a big dinner, the host, holding the uncut, white portion of the green *luo-bo*, walks among his guests and passes it around as Americans do "after dinner mints". The slight pungent taste of the green *luo-bo* sections is refreshing and it helps digestion. In cities of North China, itinerant vendors carry the green *luo-bo*, prepared as described above, and Chinese olive in the evening to small streets, like the ice cream men in America. Hearing the voice of the vendor, people come out to buy the green *luo-bo* or both preparations for informal parties at home.

In southern and western China, people grow various cultivars of the white variety, which produces very large roots, 10–20 pounds in weight. I have seen large, juicy, watermelon-sized, white radishes which break into pieces upon being dropped to the ground. The large, elongated variety had been introduced to America from Guangzhou and is available in Chinese stores, and lately even in supermarkets. The green *luo-bo*, imported from China via Hong Kong (maybe from Beijing or Shanghai direct as direct flights now connect these cities with Boston, New York and San Francisco), is available in Chinese stores around the Chinese New Year time.

All portions of the *luo-bo* are used in China. In addition to using the roots as a vegetable, or an ingredient in pastries, the dried leaves, the old plants after flowering and fruiting, and the seeds are all used in medicine. Pharmacists and phytochemists in China worked out the chemical compositions of the roots and seeds of Chinese radishes and concluded that the roots are rich in many organic acids including: coumaric acid, caffeic acid, ferulic acid, phenylpyruvic acid, gentisic acid, hydroxy-benzolic acid, and various amino acids. Fresh roots also contain methyl mercaptan, Vitamin C and raphanusin. The seeds, in addition, contain erucic acid, linoleic acid, linolenic acid, glycerol sinapate, and also raphanin. In a report by the American Herbal Pharmacology Delegation to China (1975), on the evaluation of Chinese herbal remedies, the comment on the use of Chinese radishes is worthy of quoting here. "Experimentally, the antitumor activity has been confirmed, as well as the diuretic effects. Extracts of this plant have shown in vitro antibacterial, antifungal, and antiviral effects. The active antibacterial and antifungal principle has been identified as raphanin." (pp. 132–33). On discussing the issue of eating green *luo-bo* at parties in China, a colleague commented, "If this practice is introduced to America, the number of obese people would be reduced." This may prove to be an excellent idea. Whereas the Americans say, "An apple a day keeps the doctor away", the northern Chinese people say, "Eating pungent radish and drinking hot tea, let the starved doctors beg on their knees." (吃辣籮蔔喝熱茶，餓的大夫滿街爬。) The newly harvested red and green radishes can be quite pungent. Storage usually reduces the degree of pungency and renders the raw radish pleasant to taste (Figures 25d–i).

Northern Farmer's Salad (*Hua-bei luo-bo cai* 華北蘿蔔菜)

Ingredients:

    1 1/2 lb radish (usually red in northern China, the white ones available in America can be a good substitute; cut 3 mm slices, put 3–4 slices together and shred them along the long axes, so the finished pieces are about 3 mm wide on each side;

grated pieces are uneven in thickness and they do not make good salad; pieces
with skin taste better)

1 cup cross-sections of the inner leaves of celery cabbage (3mm wide)

1/2 medium-sized onion (thinly sliced vertically)

1/4 tsp salt

1 Tbsp light-colored soy sauce (生抽)

3–4 drops vinegar

1 Tbsp sesame oil

2 cloves garlic (crushed and chopped)

Combine all the ingredients and mix them well. More salt and/or some hot soybean
sauce may be added to taste. Makes 4 servings.

Central China Boiled Dinner (*Hua-zhong luo-bo-tang* 華中蘿蔔湯)

Ingredients:

3 lbs beef (cut into bite sizes)

6 cups Chinese radish (pared, cut irregularly to bite sizes, about 3 cm across)

4 cups celery (outer stalks and leafy portions of inner ones washed, and cut to 3 cm
pieces)

3 or 4 large potatoes (pared, washed, and cut irregularly into 3 or 4 pieces)

6 carrots (pared and cut to 3 cm sections)

3 medium-large onions (cut vertically to 4 pieces each)

4 large tomatoes (each cut into 4 or 5 pieces)

1 cm ginger (sliced thinly, then cut slices twice)

1 1/2 gallon water

1 Tbsp salt

(Optional) Two kinds of sauces as dips for meat in soup:

2 Tbsps fermented soybean sauce (either the mild type or the hot kind, served
with 1/2 tsp sesame oil)

2 Tbsps light-colored soy sauce (生抽; seasoned with 1/4 tsp lemon juice or vinegar,
1/2 tsp sesame oil, 1 crushed garlic clove, 1 tsp chopped ginger and chopped
onion or scallion)

Boil the water in a large container. Add all the ingredients except the optional
ones, and return to a boil. Turn the heat to low and continue to cook for three hours.
Makes 6–8 servings.

Material for one or both sauces should be ready and placed on the table. The sauces
may be put on the dishes of the participants for them to dip the meat, which becomes
tasteless after being boiled for 2–3 hours. This is a good dish for two working families,

for it can be cooked enough for two or three days, and warmed up just enough for each meal. Each time, some sauce for the meat would make the dinner more enjoyable.

North China Radish Soup (*Hua-bei luo-bo-tang* 華北蘿蔔湯)

Ingredients:

> 1 small radish or a section of radish about 10 cm long, 3–4 cm in diameter (sliced and then cut into strips as described for salad)
>
> 2 scallions (finely cut both white and green portions into 3–5 mm pieces)
>
> 1 slice ginger (5 mm thick, sliced and cut into slender pieces)
>
> 1 egg
>
> 1/2 cup all purpose flour
>
> 1 Tsp vegetable oil
>
> 2 1/2 cups water
>
> 1/2 tsp salt
>
> 1 tsp soy sauce
>
> 3–4 drops vinegar
>
> 1/2 tsp sesame oil

Slightly beat the egg and add flour. Use enough water to make a smooth dough. Dust a smooth surface with flour and roll the dough into a thin sheet about 30 cm in diameter. Dust some flour over the sheet and fold it once. Dust the exposed half and fold once again. Now, dust the surface facing you. Cut across the folded sheet with light touch, and unfold the sections. Freshly prepared soft noodles are lying on the working table. In northern China, a farmer's wife makes the family noodles daily (without eggs), the technique which a girl learns between 8 and 10 years old.

Brown the scallion and ginger with hot vegetable oil. Stir in the radish and salt, and cook for 2 minutes. Pour in the water and bring to a boil. Stir in the home-made noodles with the aid of a pair of chopsticks so that they will cook separately. On boiling, lower the heat and continue to cook, covered, for 2 minutes. Add soy sauce, vinegar, and sesame oil. Serves 1 person.

Cora D. Reeves was an ichthyologist, holding a doctorate degree from the University of Michigan. She went to China in the 1920s as a Methodist missionary and became Chairman of the Department of Biology, Ginling College, Nanjing, China. Upon retirement, she lived in Black Mountain, North Carolina, USA. There she told me one day that the best soup she ever had was the radish soup prepared by a colleague when she was sick in Nanjing. She recalled that the illness destroyed her appetite. The colleague who visited her brought a radish and some scallions, instead of flowers. Half an hour later a bowl of radish soup with homemade noodles was placed before her. The dish tasted so good that she finished it, and, as a result, her appetite was restored. Many people from China have similar experiences, for this dish is commonly served during convalescence.

Vegetarians' Three Treasures (*Su-san-xian* 素三鮮)

Ingredients:

    1 lb firm bean curd (1 box, available in supermarkets)

    1 lb Chinese celery cabbage (cleaned, washed, drained, and cut to bite-size)

    1 1/2 lb radish (sliced diagonally into 3 mm slices, then place 2–4 slices together and shred lengthwise)

    1 large onion (cut vertically into 5 mm slices)

    5 Tbsps vegetable oil

    1 tsp salt

    6 grains of zanthoxylum (Sichuan peppercorns)

    1 tsp sesame oil

    1/2 cup water

Use 2 tablespoonful of oil in a medium hot, deep frying pan. Holding the bean curd in the left hand, cut once lengthwise, and then slice across the two halves into 1 cm thick pieces. Let them fall into the pan, and brown one side, then the other. Remove the bean curd, clean the pan and pour in the remaining oil. Lightly brown the onion and zanthoxylum. Stir in the radish, cabbage, salt and water. Cook for 2 minutes. Cover the mixture with the browned bean curd and bring to a boil. Turn the heat to low and continue to cook for 10 minutes. Stir in the sesame oil. Makes 4 servings. Some people prefer to add more salt and/or some hot soybean sauce to taste.

Luo-bo, celery cabbage, and bean curd are the most common vegetables of the Chinese people. They have been praised by a leading Chinese Buddhist monk, Tai-xu Fa-shi 太虛法師, as the three treasures that keep the Chinese people alive.

Steamed Radish Cake (*Luo-bo-gao* 蘿蔔糕)

Ingredients:

    1 1/2 lb white radish (sliced into slivers, and boiled in 2 1/2 cups of water for 15 minutes)

    2 oz Chinese sausage (diced into 3 mm pieces)

    2 oz Virginia ham (diced to match the sausage in size)

    1/2 cup dried shrimp (softened in hot water, minced into 2–3 mm pieces; may be replaced by 4 oz fresh shrimp cut into small pieces)

1/2 oz black mushroom (*Xiang-gu* 香菇; revised and cut into 1 cm squares)

1 package of glutinous rice flour (available in American Chinese stores)

1 large onion (diced)

1 tsp salt

1/2 cup vegetable oil

1/2 cup roasted sesame seeds (crushed coarsely with a rolling pin)

1/2 cup chopped coriander leaves (about 3 plants, available in Chinese stores)

Mix the cooked radish including water, sausage, ham, shrimp, mushroom, onion, salt and oil. To the mixture stir in the glutinous rice flour and mix well. If too dry, add water. Pour the mixture into a generously greased cake pan and steam for 30 minutes. Cool slightly. Just before serving, sprinkle with sesame seeds and chopped coriander leaves. Serves 6–8.

This is a Cantonese dish prepared especially for the Chinese New Year Festival or special parties in winter.

Sagittaria Corms (*Ci-gu* 慈菇, Old World Arrowhead)

Species of arrowheads are predominantly aquatic perennials in America. Fernald (1950) recognized 16 species for northeastern North America. In China, however, one species (*Sagittaria sagittifolia* L.) is cultivated for its edible corms. (Figure 12a, 19, *Arnoldia* 30(1): 9–22. 1970). It is a minor cash crop, particularly in southern China, but is not commonly eaten anywhere in China. In fact, 99% of the population has never tasted it. The mealy fresh corms are used by Buddhists in their vegetarian dishes. In southern China the corms are gathered between late November and February for the New Year market. It can be eaten plain, as an ingredient in an entree, or as a dessert, and is available in Chinese stores in Boston around the Chinese New Year time.

Sagittaria Chips (*You-zha ci-gu-pian* 油炸慈菇片)

Ingredients:

1 lb sagittaria (pared and vertically sliced 3 mm thick)

6 cups vegetable oil

1 tsp salt

Heat the oil to medium hot and deep-fry the sagittaria slices to golden yellow. Remove and sprinkle with salt. Serve cool.

Sagittaria and Ham (*Ci-gu-pian Huo-tui* 慈菇片火腿)

Ingredients:

1 lb Virginia ham (thinly sliced to 3 x 2 cm x 3 mm thick pieces)

10 oz sagittaria from Chinese stores (pared and sliced to 4 mm thick pieces)

1 oz wood ear (*mu-er* 木耳; revived in boiling water, cleaned, and keep in water)

6 oz Chinese mustard, green fresh or pickled (cut to match the size of the pieces of sagittaria)

1 medium-sized onion (cut to match the other ingredients)

3 Tbsp vegetable oil

1/2 tsp salt

6 zanthoxylum (Sichuan peppercorns)

1/2 cup chicken broth

1/2 small can bamboo shoots (use half, cut thinly to match the other ingredients)

In a deep frying pan, brown the onion and zanthoxylum. Stir in all the other ingredients except the sesame oil. Mix well. Add the chicken broth, stir and bring to a boil. Turn the heat to medium, cover and cook for 5 minutes. Add sesame oil and serve hot. Makes 6 servings.

New Year Family Soup (*Xin-nian-geng* 新年羹)

Ingredients:

2 lb pork ribs (cut to 3–4 cm pieces)

1 lb sagittaria corms (pared and quartered)

2 large onions (cut into 4 pieces each)

1/2 lb bean curd bamboo (*fu-zhu* 腐竹; revived in cold water overnight)

1 section lotus rhizome (*ou* 藕, pared and cut crosswise into 5 mm thick slices)

1 oz wood ear (*mu-er* 木耳; revived, cleaned, and keep in water)

6 oz celery cabbage (cut into 3 cm pieces)

1 1/2 gallon water

1 tsp salt

Boil the water in a large pan. Add all the ingredients and bring to a boil again. Turn the heat to low and simmer for 3 hours. Makes 8 to 10 servings. This is a refreshing soup commonly prepared in Guangzhou and Hong Kong. It can be served informally, as members of the family gather for the New Year dinner. The early arrivals may be served a bowl of pre-dinner soup and play a game. Members of the party may stay after dinner to play and they may be served a bowl of New Year Family Soup or Oriental Red Pottage (see under "Red bean") for refreshment.

Sagittaria Pudding (*Ci-gu-bu-ding* 慈菇布丁)

Ingredients:

12 arrowhead corms (pared and ground)

2 Tbsp honey

2 Tbsp starch from lotus rhizome (*ou-fen* 藕粉; or substitute with tapioca, water chestnut starch or cornstarch)

1/2 cup water

Dissolve the starch in water. Add the ground sagittaria and honey. Steam for 30 minutes. Makes 2 servings. This recipe is adapted from a Chinese health food designed for a person suffering from a prolonged cough. In the original recipe, the sagittaria and honey are mixed in 2 cups of liquid prepared from boiling glutinous rice. The person is served one half of the preparation at a time.

Sacred lotus Rhizomes (*Ou* 藕)

The sacred lotus is cultivated in China as a multifarious crop. In the lakes, ponds, and the suburban lowlands of Beijing, Nanjing and other metropolises, it is planted mainly for ornamental purposes and secondarily for its edible portions. In rural areas of western and southern China the sacred lotus is cultivated side by side with rice as an important cash crop. The rootstocks can be eaten raw (rarely), cooked, fresh, dried, canned, or in the form of an extracted starch. The marketable material consists of two enlarged, terminal segments of a slender, creeping rhizome buried in the mud. It bears a strong apical bud (Figure 13a). Propagation is done vegetatively, by separating large, strong segments which are then buried in mud in April and May. The fleshy lotus rootstocks, dried sections in plastic bags, canned material, and starch extracts are all available in American Chinese stores, where the price of fresh lotus rhizome is comparable to that of choice steaks. The Cantonese people like to put fresh lotus rhizome in the New Year family soup (see the recipe for sagittaria). Along the Yangtze River people like to eat raw lotus root as a tid-bit. Sugar-preserved, dried slices are also available in Chinese stores at the Chinese New Year time.

The aerial portion of the sacred lotus is called *lian* 蓮 and its subterranean part is *ou* 藕. Due to its large, beautiful flowers, clean and unstained by the foul mud of its environment, *lian* has been a literary subject, an inspiration for poets, and a motif of artists in Chinese culture. *Ou*, on the other hand, has reached its full development in horticulture and culinary art in the hands of and by the ingenuity of the practical people. All parts of the plant are used in cooking, medicine and for animal feed. In addition to the information given on the uses of the flowers, seeds, and rhizomes in Chinese food, the leaves are also imported by Chinese tea houses owners and used to wrap chicken-rice or spare-ribs prior to steaming. The old receptacles of the fruits, the thin sections and the nodes of the rhizomes, the green embryos taken from the seeds, and the filaments of the stamens are all used in traditional Chinese medicine. Chinese phytochemists have ascertained that the rhizomes of lotus are rich in starch, protein, Vitamin C, several phenolic compounds (such as catechol, d-gallocatechol), tannins, and peroxidase. Over 20 alkaloids have been isolated from different organs of the species. Extracts from these materials have shown antitumor properties in animals.

On reviewing the Chinese use of the sacred lotus as a regular food in their *Edible Wild Plants of Eastern North America* in the 1940s, Fernald and Kinsey suggested the propagation of the American species, *Nelumbo lutea* (Willd.) Persoon, for similar purposes. Four decades slipped by and both the native and the introduced (Oriental) species are growing in America without being used. Now, with the findings of the phytochemists and the experimental scientists on the contents and merits of the seeds and the rhizomes of the lotus, and with the recent increasing demands for natural food and health food, lotus should be neglected no longer. Forty years ago, not much rice or soybean was cultivated in America. Now, America is self-sufficient in the production of rice and soybean. At present, the lotus seeds, rhizomes and dried leaves used in USA are imported from the Orient. We should look forward to seeing some local products of such material in American groceries (Figure 13a).

Fresh, Raw Lotus Rhizome (*Xian-ou-pian* 鮮藕片)

Ingredients:

2 lb fresh lotus rhizomes

Choose the largest segments. Cut off 1/10 from each end. Pare the choice sections and soak them in cold water for 10–15 minutes. Slice crosswise into 5 mm sections. Serve with toothpicks at parties. The end portions can be washed, pared and used for soup.

Stir-fried Lotus Slices (*Chao-ou-pian* 炒藕片)

Ingredients:

1 1/2 lb fresh lotus rhizome (remove the fibrous portions near the nodes, pare, and leave the pared segments in cold water; cut the pared rhizomes diagonally into 5 mm wide slices, then cut the slices 2 or 3 times along the long axis; keep the less desirable portions for soup)

1 onion (sliced vertically into 5 mm wide pieces)

6 zanthoxylum (Sichuan peppercorns)

4 Tbsps vegetable oil

1/2 cup water

1 tsp sesame oil

Lightly brown the onion. Add the zanthoxylum, partially breaking them between the fingers. Stir in the lotus slices. Add water and cook for 5 minutes stirring occasionally. Add sesame oil and serve hot. Makes 4 servings.

Lotus Starch Refreshments (*Ou-fen-yin* 藕粉飲)

Ingredients:

2 Tbsp lotus starch (mixed with 2 Tbsp water)

1 tsp honey (or sugar)

1 1/2 cup boiling water

Pour the boiling water into the container of lotus starch. Stir and mix well. Add the honey (or sugar, may be more to taste). Take as a late night snack. In Chinese families which include elderly people, an early morning drink is usually prepared so that they can have something hot an hour before getting up. Starch extracts of lotus, water chestnut, and the rootstocks of several species of *Cynanchum* cultivated in Shandong and Jiangsu are used in a similar fashion.

Taro (*Yu-tou* 芋頭)

Taro is a Polynesian name for an old world avoid. Its Chinese equivalent is *yu-tou* 芋頭, which means the head-like growth of *yu* 芋. Ethnobotanically, southern China must have been an early center of its cultivation because *yu* has become a basic term for the starchy underground growth of food plants introduced later. *Shan-yu* 山芋 (hillside *yu*) for sweet potato, and *yang-yu* 洋芋 (foreign *yu*) for Irish potato are good examples. Taro is known as "eddo" in the West Indies and "dasheen" in some part of the United States.

The edible portion of taro is the corm. The Chinese people call the central corm *yu-mu* (芋母, mother-of-yu), and the cormlets *yu-zi* (芋子, children-of-yu). In China, taro is a minor crop south of the Yangtze River. It is rarely planted in North China.

In South China, most cultivars of *yu-tou* prefer hot climate and moist soil. In Hong Kong, the escapees grow well in swampy areas and along streams. A crop matures in 7 to 8 months. However, in Guangzhou, a cold-tolerant cultivar grows between December and May, and the crop can be harvested in 180 days. It is locally called *zao-yu* (早芋, early taro). This cultivar bears corms about 6 to 7 cm long and 5–6 cm in diameter. Each corm bears 3 or 4 oblong cormlets 3.5 to 5 cm long and 2 to 4 cm in diameter with no aerial leaves. The cultivars planted north of the Yangtze River are very similar to the early taro of Guangzhou. Most taros in southern China produce large corms bearing 7 to 10 cormlets, some of which bear aerial leaves (Figure 12d).

Plain Boiled Taro (*Zhu yu-tou* 煮芋頭)

Ingredients:

2 lbs small corms or cormlets (2–4 cm in diameter, washed and drained)

1/2 gallon water

Sugar (placed in a container on the table)

Boil the taros until the larger ones are cooked. Serve hot or cold with the skin. Peel and dip in sugar. Serve at parties, as a refreshment or as a dessert.

Taro with Pork (*Yu-tou zhu-rou* 芋頭豬肉)

Ingredients:

1 1/2 lb pork (cut the meat into 2–4 cm irregular pieces)

1 large onion (cut irregularly to match the pork pieces)

2 lb taro (scrape off the skin, wash, and cut into 2–3 cm irregular pieces)

2 Tbsp soy sauce

1 Tbsp sugar

2 cm piece ginger (chopped)

1 tsp salt

1/2 cup water

2 Tbsps vegetable oil

Brown the onion and ginger. Stir in the meat. As it changes color, add soy sauce and sugar. Continue to cook on medium heat. Add the taro and water. Mix well. Cook for another 30 minutes, adjusting the heat to avoid burning. Makes 6 servings.

Stir-fried Taro (*Qing-chao yu-si* 清炒芋絲)

Ingredients:

1 1/2 lb taro (scraped, washed, and sliced into 1 cm pieces, shred the slices)

1 large onion (sliced vertically into 1 cm pieces)

1 cm piece of ginger (sliced and shredded)

1/2 cup vegetable oil

1 tsp salt

1/2 cup water

Brown the onion and ginger. Stir in the taro and salt. Add water and cook for 5 minutes. Serve hot. Makes 4 servings.

Water Chestnuts (*Bi-qi* 荸薺, *Ma-ti* 馬蹄)

Patrons of Chinese restaurants in America are familiar with water chestnuts for they are used in many dishes with meat, seafood, and poultry, often in combination with bamboo shoots and snow pea pods. A water chestnut (*bi-ji* 荸薺, *Eleocharis dulcis* [N. Burmam] Trin.) is the corm grown at the end of a long, slender, white rhizome of a leafless aquatic sedge. In northern Jiangsu where I was born, the species is wild in swampy areas near lakes. In famine years, the rural people used to go out in the spring to dig wild water chestnut (locally called *mao-di-li* 毛地栗, hairy ground chestnut) as famine food. When eaten raw, the corms are called *di-li-zi* 地梨子 (ground little pears).

In southern China, particularly in Guangzhou and its vicinity, water chestnut, fish, mulberry tree and silkworm culture form an important agricultural chain in the richest sector of the Pearl River Delta. This area has two seasons, the dry season beginning in November and the wet season starting the following March. Water chestnut and fish are raised simultaneously during the wet season, with the plants providing shelter and protection to young fish and utilizing the animal wastes as fertilizer. At the onset of the dry season, the fish crop is harvested first, the water drained, and then the water chestnut gathered, locally called *ma-ti* 馬蹄 (horse hoof). In winter the mud of the fish pond is moved over the bank where mulberry is cultivated. The annually enriched soil of the bank provides the best condition for mulberry growth, which is pruned to the ground in winter. The offshoots from the mulberry stumps produce the largest and thickest leaves possible for the silkworms, the pupae of which are used to feed the fish. Consequently, participants in such carefully planned and skillfully operated type of agriculture earn the highest rate of per capita income among farmers throughout China. Most of the fresh water chestnuts and the canned, candied, and/or dried, starchy products in American Chinese stores are produced from this area.

Fresh Peeled Water Chestnut (*Sheng bi-qi* 生荸薺)

Ingredients:
> 2 lb water chestnut (thoroughly washed, brush off the mud and as many of the scales as possible, drained)

Peel the individual corms. Wash and drain. Place the peeled corms in a dish. With some toothpicks, stick the centers of the top ones. Serve at parties.

This recipe is adapted from the practice of the itinerant Chinese vendors who peel the water chestnuts, string 5 or 10 on a thin, bamboo splint, lay the strings on a plate, and spread water occasionally to keep them fresh and crisp. In large cities of northern

China a vendor often carries peeled water chestnuts, green radish splints, and green canarium for people to buy as after dinner refreshments.

It is worthy of note that Chinese farmers of the water chestnut producing areas often eat the fresh products by rinsing off the mud and peeling the skin with their front teeth. From such an unhealthy practice, they are often infected by the liver fluke because the free swimming stage of the parasite may be attached to the scales of the corms. The precautions given in this recipe would eliminate the possible trouble caught by the farmers.

Plain Boiled Water Chestnuts (*Zhu bi-qi* 煮荸薺)

Ingredients:

> 3 lb water chestnuts (cleaned, with the broken or spoiled spots removed, washed, and drained)
>
> 8 cup water

Boil the water in a large container. Add the cleaned water chestnuts, bring to a boil and continue to boil for 10 minutes. Serve cold or hot at tea time or as a snack. This is by far the most common way of eating the water chestnut in China, for it keeps the crisp texture on boiling.

Jellied Water Chestnut (*Bi-qi-gao* 荸薺糕)

Ingredients:

> 1 lb fresh water chestnut (peeled and ground)
>
> 1 oz refined agar
>
> 1/2 cup sugar
>
> 1 cup water

Boil the water and dissolve the agar. Add the ground water chestnut and sugar. Bring to a boil and continue boiling for 5 minutes. Cool and place in the refrigerator to jell. Serve with or without cream. Makes 4 servings.

Water Chestnut Meatballs (*Bi-qi rou-yuan* 荸薺肉圓)

Ingredients:

> 1 1/2 lb ground meat (half beef, half pork)
>
> 1 small can water chestnuts (minced, or 1/2 lb fresh material if available, cleaned, peeled, washed, drained and minced)
>
> 2 lb celery cabbage (washed, drained, and cut crosswise into 3 cm long pieces)
>
> 1 Tbsps cornstarch

1/2 cup chicken broth

2 Tbsp soy sauce (light type, 生抽)

3 scallions (chopped fine)

1 cm piece ginger (minced)

3 Tbsp vegetable oil

1 tsp sesame oil

1/2 tsp salt

6 zanthoxylum (Sichuan peppercorns)

Combine the ground meat, water chestnuts, scallions, ginger, salt, soy sauce, and cornstarch. Mix well and make meatballs about 4 cm in diameter. Cook the meatballs in 2 tablespoonful of vegetable oil on medium heat until brown. Remove them and pour in the remaining oil, adding onion and Sichuan pepper. Stir in the celery cabbage. Add the chicken broth and bring to a boil, stirring occasionally. Place the meatballs on top of the cabbage, cover and simmer for 20 minutes. Serve hot. Makes 6 servings.

Yams (*Shan-yao* 山藥, *Shu-yu* 薯蕷)

The true yam is the fleshy root of the cultivated species in the genus *Dioscorea* L. (薯蕷屬). In American groceries, the term "yam" is applied to the sweet potato (甘薯, *Ipomoea batatas* [L.] Lam.) with red-orange flesh. Although the true yam and the sweet potato are both fleshy underground growths from herbaceous vines, the two species are very different. Yam (*shan-yao* 山葯, *Dioscorea opposita* Thunb.) is a monocot with square stems, opposite leaves, small unisexual greenish flowers and white, mucilaginous root-flesh, while the sweet potato is a dicot with round stems, alternate leaves, large, morning-glory-like bisexual flowers and red-yellow root-flesh without mucilage.

More than 30 species of wild yam are known in China. The fleshy, underground growth is gathered for food, for medicine, and/or for industrial uses. Three species are cultivated extensively in China. The root of *Dioscorea alata* L., cultivated in southern China, is sold fresh in the market. The root of *D. opposita* Thunb., cultivated in northern China, is processed and used in traditional Chinese medicine, more than as a fresh products for food. Both the fresh root of *D. alata* L. and the dried product of *D. opposita* Thunb. are available in American Chinese stores.

Yam has very important roles in Chinese culture. It appears in the earliest records on useful plants of China. It is treated not only as a starchy, edible product, but also as a special food for those physically feeble due to illness or debilitated due to age. Moreover, the Chinese yam is a major ingredient in all the prescriptions for diabetics. Recently phytochemists have identified dioscin, choline, free amino acids, d-abscisin II, polyphenol oxidase, Vitamin C and proteins in addition to 16% starch in the root.

Mannan and phytic acid were isolated from the mucilaginous material. Pharmacological studies of some species of *Dioscorea* L. indicate that *D. bulbifera* L. and *D. villosa* L. found anticancer activity in animals, and *D. opposita* Thunb. shows anti-inflammatory and antidiarrheal effects. Evidently, the Chinese empirical conclusions have certain scientific backing (Figures 12f and 20).

Yam Pork Broth (*Shan-yao rou-tang* 山藥肉湯)

Ingredients:

1 1/2 lb pork chops (cut into 4 x 3 cm pieces)

1 1/2 lb fresh white yam (scrape off the outer brown portion, wash, drain, and cut into slices to match the pork)

2 onions (cut vertically into 5–6 pieces, take 2 central portions and mince them)

1 1/2 cm fresh ginger (sliced thinly, cut 9/10 of the slides 3- or 4-times and shred the remaining)

1 gallon water

1 tsp salt

2 Tbsp soy sauce (light)

1 tsp sesame oil

1/2 tsp vinegar

2 cloves of garlic (mashed and chopped)

1/2 cup wood ear (木耳; revived in boiling water, cleaned, and keep in fresh water)

1/2 cup daylily bud (revived, with the hard, basal portion removed)

Boil the water in a large container. Add the pork, yam, onion and sliced ginger, salt, wood ear and daylily. Bring to a boil. Turn the heat to medium-low and continue to cook for 1 hour. Makes 4 to 6 servings.

Make a dip for pork with the soy sauce, vinegar, sesame oil, shredded ginger, and garlic. (Add hot soybean sauce if preferred.)

Yam Gruel for the Elderly

Ingredients:

1/2 cup dried Chinese yam slices

1/2 cup lotus seed

1/2 cup Job's tears

1/2 cup short grain rice

8 cups water

Wash and drain the yam slices, lotus seed, Job's tears (clean if needed), and rice.

Combine these with water and bring the mixture to a boil. Turn the heat to low and continue to boil for 2 hours. Serve one cup before bed-time and one cup an hour before breakfast. Keep refrigerated and warm enough for each serving. Sugar may be added to taste.

Candied Yam (*Ba-si-shan-yao* 拔絲山藥)

Ingredients:

    10 oz fresh yam (peeled and cut into angular bite-size pieces)

    6 Tbsps sugar

    1/2 cup water

    1 Tbsp sesame seeds

    4 cups vegetable oil

Use two pans. In one pan deep-fry the yam in the oil, stirring occasionally to keep the pieces from sticking. As the yam turns light brown, test one piece to ensure the center is cooked. Meanwhile, in the second pan make a syrup with the sugar and water. When the syrup turns yellow, test it by dropping a little into a cup of cold water. If it forms a ball, the syrup is ready. When the yam is properly cooked, remove it from the hot oil to prevent burning. To keep the outside crisp, though, try not to leave it out too long. Experience and planning are important in this recipe where timing is the secret to perfection. Transfer the yam into the boiling syrup, stirring occasionally so that each piece is thoroughly coated. Sprinkle sesame seeds over the coated yam. Serve hot, placing the yam dish in the center of the table and a bowl of cold water in front of each participant. Each person takes a piece of the coated yam, and dip it into the cold water to harden the syrup, making a crunchy coating over the softened yam. As the coated yam leaves the dish, the attached syrup forms a string (hence this dish is called *ba-si-shan-yao* 拔絲山藥, pulling string yam). At home in northern China, yam is served in a much simpler manner: boiled the yam with skin, then peeled and dipped in sugar as a refreshment or dessert. The candied dish prepared by the recipe described here is served at the close of banquets as a dessert.

SPICES AND FLAVORING MATERIALS

All good food depends upon the proper use of spices and flavoring materials. Chinese food is no exception. In connection with the use of spices in cooking Chinese food, I decide to record a personal experience as an introduction for explaining the fundamental Chinese idea of proper application of flavoring material. "Plants and Human Affairs" has always been a popular course in the Department of Biology at Harvard University. In the late 1940s, while Professor Paul C. Mangelsdorf was teaching the subject, a classmate asked for my help to complete her required term paper. Her project was on the Chinese invention of using pure plant protein precipitated from soybean. At that time relatively few American students knew the term "*dou-fu*" 豆腐 (bean curd). Apparently, her interest in the subject was a progressive one. For the project, we went together to Chinatown where I showed her the only production center of *dou-fu* available at the time. It was in a dim one-room area of a basement. There we watched the simple procedure of precipitating solid protein from soaked soybean. She recorded her observations and bought some samples, both *dou-fu* and *dou-fu-zha* 豆腐渣 (bean curd residue). Returning to the laboratory, I cooked a few dishes with the product and the by-product. She was delighted with the dishes and was confident that her report would be the most interesting. When she presented the report to the class, she also cooked a dish for the classmates to try. Unfortunately, she added a heaped teaspoonful of monosodium glutamate to flavor it. Naturally, the dish was spoiled and the image of *dou-fu* was tarnished among the Harvard students in economic botany for quite a time. Spices and flavoring materials are good only when they are properly applied. Too much would not only spoil the food, but might also be harmful to health.

Cultural and Historical Background

As an introduction it must be noted that a majority of the Chinese population, particularly the peasants, live on simple, plain food of plant origin. Little spice is used. The basic flavoring materials in the kitchen of a Chinese farmer are salt, sesame oil, and sometimes soy sauce and vinegar, or home-made, fermented grain or bean products. These people, particularly those living in northern China, may eat smelly, raw garlic and onion, but they seldom use these as spices for cooking. In cities, people use onion, ginger, garlic and/or *hua-jiao* 花椒 (zanthoxylum) in preparing some vegetable dishes and *da-liao* 大料 (major spices) for meat dishes. Presently the use of hot cayenne pepper is widespread, and certain geographical distinctions in its use have emerged. For example, in Sichuan people combine it with the numbing, pungent, native *hua-jiao* powder, in Shanxi it is mixed with vinegar to form a paste, and in Hunan people add chopped garlic and ginger to increase its stimulating potency. Such preparations were not known in China before the introduction of hot pepper in the seventeenth century, and now they are used more as relishes than as spices for cooking.

The use of spices as important items in Chinese culinary art has been developed by the relatively small number of people who could include animal products in their diets. In Chinese cooking, spices are used to disguise the rank taste of sheep and goat meat, to reduce the rich, oily texture of pork and some poultry, and to cover up the fishy smell of seafood. The earliest recorded native seasoning materials are *hua-jiao* and fermented grain and bean products from temperate China, and ginger and cassia from tropical China.

Hua-jiao first appeared in *Er-ya* 爾雅 (see under Chinese radish in previous chapter for explanation), and the name has been maintained in the continuously developing and changing Chinese culture. The species has become extensively cultivated throughout northern China and in the mountainous areas of western China. Since the plants are unisexual, certain trees growing in some farmers' kitchen gardens never bear fruit. The leaves of these trees are used by the owners in the preparation of fish, fermented grain and/or bean sauces.

The use of fermented grain and bean sauces as condiments was recorded in the annals of the Zhou Dynasty (1122–255 B.C.), when the duty of the Manager of the Royal Kitchen was to ready 1,200 earthenware jars of fermented products for palace consumption. From those early days to the present the simple words for flavoring materials, *jiao* 椒, *jiang* 醬, and *zao* 糟, have become generic terms in the Chinese language for the compound names of later inventions and introductions. Good examples are: *hu-jiao* 胡椒 for black pepper introduced from India over land, *hai-jiao* 海椒 for cayenne pepper brought to China over the ocean, *tian-mian-jiang* 甜麵醬 for a sauce prepared from wheat flour, *dou-ban-jiang* 豆瓣醬 for another sauce made from fermented, cooked soybean, *lao-zao* 醪糟 for a rice product fermented by yeast, and *hong-zao* 紅糟 for a fungal product of red-colored rice.

Ginger and cassia have been used by ethnic groups residing in South China since time immemorial, but their use was first recorded two or three hundred years later than that of *hua-jiao* for zanthoxylum (Figure 14a). The Chinese chronicles were first recorded in the Yellow River Region, from where the influence of historians moved southward to the Yangtze River and the Pearl River Regions. Ginger was recorded about 300 B.C. as *jiang* 薑, which has become a generic term for the pungent, fleshy, underground growth, such as *jiang-huang* 薑黃 for turmeric, and *shan-jiang* 山薑 for kaempferia. Cassia entered Chinese records about 200 A.D.. The name *gui* 桂 indicates that cassia was originally a product of Guangxi, where the more prominent city was Guilin 桂林 (cassia forest). Actually, all the cassia bark used commercially has been cultivated in the mountains of southern Guangxi from where it was transported to the coastal cities in China and Vietnam. Bark obtained from the various parts of the same tree is grouped, sorted and shipped under different commercial names, depending upon the age, position on the tree and the method of preparation. By the Chinese standard the best cassia, *rou-gui* 肉桂, is from the trunk of mature trees, and the less expensive material from the branches is called *gui-zhi* 桂枝. Actually both the Chinese cassia and the Saigon cinnamon are produced from plants cultivated by the people of the mountainous areas on the border of Guangxi and North Vietnam. The material shipped to Hong Kong and other cities in China becomes the Chinese cassia and that shipped to Saigon becomes the commercial Saigon cinnamon.

The use of spices to give food aroma and to make eating a pleasure is a function of cultural development which, in turn, depends much upon economic security, political stability, national communications and international exchanges. In Chinese history, such conditions were established in the Tang Dynasty (618–905 A.D.). An obvious increase in the number of kinds of spices was recorded in a publication dated 659 A.D.. The central government, with its capital in Chang-an 長安 (now called Xi-an 西安, Map, Loc. 19), encouraged business and diplomatic missions with the Middle East and religious pilgrimages to India. The spices introduced during this period moved slowly overland via the territory of the Tartars (Hu 胡 in Chinese). Consequently, these items all bear the name "*hu*", such as *hu-jiao* 胡椒 for black pepper, and *hu-lu-ba* 葫蘆巴 for fenugreek. The people of Chang-an, the national capital of Tang Dynasty, situated in the arid northwest and adjacent to the plateau region, had more meat in their diet than those of the farming lowlands. There, exotic spices were welcomed by governmental officers and the people alike. As the demand for the introduced spices increased, the ancient rulers requested and/or suggested to members of the diplomatic missions that spices be brought in exchange for Chinese gifts (silk, porcelain, jade, etc.). During the Sung Dynasty (960–1126 A.D.), the central government even established a Bureau in the Department of Foreign Affairs to oversee the care of the exotic, aromatic materials. Most of the spices used for flavoring food in the thirteenth century are still common in China today.

Common Spices Used in Chinese Food

The common spices available in the market or kitchen gardens are listed in the following outline. An asterisk (*) is placed after the extra-Chinese elements. The source species of these materials were originally introduced to China.

    A. Sauces and condiments from fermented grains and beans
        1. *Zao* 糟 — Fermented rice: red, white
        2. *Jiang* 醬 — Fermented beans and/or wheat
           Liquid: Soy sauce
           Paste: *Tian-mian-jiang* (wheat flour), *dou-ban-jiang* (soybean dominant with some wheat flour)
        3. *Dou-chi* 豆豉 — Fermented black soybeans
        4. *Liao-jiu* 料酒 — Cooking wine
        5. *Cu* 醋 — Vinegar
    B. Fresh, dried or pickled plant materials
        1. Root: Chinese angelica (dried)
        2. Rhizomes or bulbs: Garlic*, ginger, kaempferia, onion, turmeric*
        3. Leafy shoots: Basil*, coriander*, eryngo*, lotus leaves, perilla, rue*, scallion*, tea (red)
        4. Bark: Cassia
        5. Flowers and flower buds: Clove*, osmanthus, jasmine, rose petals
        6. Fruits: Black Chinese canarium, black pepper*, cayenne pepper*, fennel*, lesser galangal, *li-meng*, hot pepper*, long pepper*, *luo-han-guo*, star anise, tangerine peels (sometimes replaced by orange peels), *tsao-ko* (Amomum), zanthoxylum
        7. Seeds: Fenugreek*, mustard*, nutmeg*
    C. Sesame oil
    D. Sugar and other sweetening material used for flavoring
    E. Monosodium glutamate

To avoid any misleading impression on the use of spices in Chinese food, a few general observations must be made here. First, the above list, though quite comprehensive, is not exhaustive. Many items used locally as flavoring material at some times and as food material at other times, are not included here. For example, the small shepherd's purse plants are gathered by farming women before winter, stored with their harvest of such roots as Chinese radish and kept for flavoring the meat stuffings of *jiao-zi* 餃子 (Chinese dumplings), a special northern New Year dish. In the spring, as the shepherd's purse roots get tougher, the plants are gathered as potherbs, discarding the aromatic roots. Secondly, great differences in the use of spices occur among people of different regions, professions, and socio-economic status. In general, people in the cities, or in the arid northwest such as Lanzhou (Map, Loc. 20) and those in the hot, humid, tropical areas such as Guangzhou and Hong Kong (Map, Loc. 57 and 58) use more spices than farmers who usually eat plain and simple food. Thirdly, in China, many of the items used for flavoring have dual functions, both as spices and as medicines. The most obvious examples are the Chinese angelica, ginger, kaempferia, cassia, clove, and star anise. Moderate applications enhance aroma in food and give pleasure in eating. Over-indulgence can spoil the food and can even be toxic. Fourthly, an artificial flavoring material, monosodium glutamate (MSG), used to flavor dishes in all American Chinese restaurants and in many city-dwelling households, is required in most recipes contained in cookbooks currently published in China and/or in America. MSG was first introduced to China in the late 1920s. Its use should not be encouraged, for it has been reported that too much causes blindness. Sugar, vinegar, and cooking wine (actually, rice gin), are flavoring materials used in restaurants or some homes. The above industrial products are not included in the lists of plant materials.

Combined Spices and Spicy Liquids for Making Cold Cuts

In Chinese cooking, several spices are often used in combination. They may appear in the forms of powder, liquid, package, and ready-made sauces. The common ones are: *wu-xiang-fen* 五香粉 (Five Spices powder), *lu-shui* 滷水 (spicy liquid), *da-liao* 大料 (major spices), *yu-xiang* 魚香 (fish spices), and *sha-cha-jiang* 沙茶醬 (hot peanut sauce). The formula for each item may vary with people and places.

Five Spices powder: This combination consists of finely pulverized plant material sold in bottles. For 10 ounces of Five Spices powder, the approximate combinations are: 2.5 oz star anise, 2.5 oz fennel, 2 oz cassia, 1.5 oz clove, and 0.5 oz zanthoxylum. It is handy to have a supply in the pantry for preparing slow cooking meat for parties.

In the Chinese business area of large American cities, cooked duck and/or occasionally goose, barbecued spareribs, and oversized roast suckling pigs are sold in special stores. Such ready-made foods are prepared with pulverized Five Spices powder mixed with syrup or honey, rubbed over the surface an hour or so before cooking.

Spicy liquid: In Chinese cooking, *lu* 滷 means to boil in a spicy liquid which is used repeatedly. *Lu-shui* 滷水 is the preserved liquid. In cities, special shops prepare and sell different animal products as cold cut, cooked in spicy liquid. The spices used to prepare these products are cassia, clove, fennel, licorice, kaempferia, star anise, *tsao-ko* (Amomum), and zanthoxylum in South China; and black pepper, cassia, clove, fenugreek, fennel, galangal, ginger, and star anise in North China. In some recipes used in southern China, *luo-han-guo* 羅漢果 (Buddha's disciple fruit) and tangerine peels are also added. Normally these spices are placed in a tightly closed bag made of gauze and boiled with water, soy sauce, gin, rock sugar, onion and ginger for 30 minutes before adding the meat. The liquid is kept to the original amount by adding soy sauce and gin, and more spices are added once a week. Pork, beef, lamb, chicken, and duck can be prepared using the same liquid. Smaller articles such as heart, liver, stomach, firm bean curd, and wheat gluten, are cooked in a separate container with a small amount of the spicy liquid, which will not be returned to the original stock. All the material to be cooked in the stock must first be parboiled to 80% done and then transferred to the spicy liquid. Here is an example of *lu-shui* used in southern China: cassia, 2 oz; clove, 1/2 oz; kaempferia, 1/2 oz; licorice, 2 oz; *luo-han-guo*, a whole one; star anise, 1.5 oz; tangerine peels, 1/2 oz; and dried, red fermented rice, 3 oz; all safely enclosed in a gauze bag; ginger, 2 oz; and onion, 5 oz (both deep fried in 4 oz of vegetable oil); 10 lbs soy sauce; 5 lb gin; 6 lb sugar; and 7 lb water. Simmer the combined ingredients for 30 minutes before introducing any parboiled meat or poultry. Add salt and sugar to the *lu-shui* after using and increase the volume by adding soy sauce, gin and water in the above proportions. Bring to a boil, keep covered, and then refrigerate.

Major spices: *Da-liao* refers to a combination of several spices with star anise being the principal one. People buy the combination and make their spiced meat at home. They can also go to drug stores and ask for *da-liao*. The shop owners make the combination to which the local people are accustomed. *Da-liao* varies from place to place, except for the star anise which is a constant.

Fish spices: *Yu-xiang* is a set of spices prepared without *yu* (fish) and may not be used to cook fish dishes. Once, before knowing this term, I ordered the dish *Yu-xiang qie-zi* 魚香茄子 (Fish Spice Eggplant). After eating half of the dish, I asked the waiter, "Where is the fish?" From him I learned that *yu-xiang* is merely the name for a spicy sauce. A dish of eggplant cooked with this sauce contains no fish. The sauce consists of: salt (1/10 oz), soy sauce (1/2 oz), pickled cayenne red pepper (1 oz), chopped ginger

(1/2 oz), chopped garlic cloves (3/5 oz), chopped scallion (2/3 oz), vinegar (3/10 oz) and a tablespoon of gin. In Sichuan, hot, fermented broad bean sauce is sometimes used as a substitute for pickled, red cayenne pepper.

Hot peanut sauce: Lately, many bottled, ready-made sauces are available in American Chinese stores. A very popular one, Fujian *sha-cha-jiang*, is the principal spicy sauce of some Swatow dishes. It is extremely hot, containing pulverized and deep-fried peanuts, dried flounder, chopped dried shrimp, sesame paste (thinned in water), chopped garlic clove, minced scallion, cayenne pepper powder, mustard, and Five Spices powder, all mixed with a red, oily extract of hot pepper. This sauce may be used for cooking or as a condiment for dipping meat from a stew of a boiled dinner.

Natural Colors and Flavors

Three plant products are used in Chinese food more for coloring than for flavoring. These are the red fermented rice, the green leafy shoots of fenugreek and the yellow fruit of gardenia. The red fermented rice is used in the form of a powder or water extract.

Dried, leafy shoots of fenugreek are ground and available by the name of *xiang-cao* 香草 (fragrant herb) in market places in Lanzhou (Map, Loc. 20, Figure 14b and c). This powder is used to give the green color of a layered, steamed pastry.

The fruit of gardenia is soaked in water to produce a yellow dye. The liquid is used for dyeing bean curd in Guangzhou, and a concentrated form is used to color garden pea meal for a brilliant, yellow cake in Beijing.

Sugar and Other Sweetening Material Used as Spice

In many areas, sugar and other sweetening materials are used for flavoring various dishes.

Sugar (tang 糖): To most families in China, sugar is a luxury rather than a necessity. In the lower Yangtze River area, particularly in Shanghai, Suzhou and their vicinity, sugar is used to flavor meat. In Guangzhou and its vicinities, sugar is often added to a quick stir-fry dish of *jie-lan* 芥蘭 (*Brassica alboglabra* Bailey) in high quality restaurants. In making soup, a small piece of *luo-han-guo* 羅漢果 (*Siraitia grosvenorii* [Swingle] C. Jeffrey) is often added to enhance the flavor of the dish in the same area. Brown sugar (*zhe-tang* 蔗糖) is often partially charred to add color and flavor to meat dishes as well as to the puree in the preparation of different soy sauce for producing the *lao-tou-chou* 老頭抽.

Table Dips

On the dinner table there are small containers for various condiments. Some of the sauces may be mild, such as that prepared with chopped ginger, mashed garlic, soy sauce, vinegar, sesame oil, and minced scallion. Others, such as the freshly prepared Chinese mustard, may be very pungent. Still others can be extremely hot, such as the Sichuan red oil sauce prepared with the oily extract of cayenne pepper mixed with salt, zanthoxylum powder, mashed garlic, minced scallion, and sesame oil. It is advisable to take a bit and taste the sauce with the tip of a chopstick before dressing one's food with it.

Samples for the Application of Spices

Ten recipes for good home-made Chinese food are presented here to show the application of spices in preparing meat, seafood and poultry dishes.

Sherry Pork (*Zhu-rou* 豬肉)

Ingredients:
>1 1/2 lb pork roast (parboiled in water for 10 minutes, washed in cold water, and drained, keep the broth)
>5 oz garden pea shoots (washed, drained, and tendrils removed)
>1/2 cup cooking gin (or sherry)
>1 tsp salt (9/l0 tsp for cooking meat, 1/10 tsp for pea shoots)
>1 tsp sugar
>2 scallions (cut into 4 cm sections, use both the white and green portions)
>1 cm ginger (shredded)
>2/3 cup juice of red, fermented rice
>2 Tbsp vegetable oil
>3 cups water

Boil the pork in 3 cups of water for 10 minutes. Remove and score the surface in a 3 cm diamond pattern. Combine the meat, broth, onion, ginger, gin, salt and red fermented rice juice. Cook on medium heat for 30 minutes. Add sugar and simmer for 1 hour. Just before serving, stir-fry the pea shoots seasoned with 1/10 teaspoon salt. Transfer the meat onto a platter, pour the unthickened gravy over the meat and place the pea shoots alongside. Makes 4 servings.

Fish Fermented Rice (*Lao-zao yu-pian* 糟糟魚片)

Ingredients:

   1 1/2 lb fillet of sole (cut crosswise into 3 cm pieces, and marinated for 10 minutes
      in salt, ginger, beaten egg white and cornstarch)

   1 1/2 oz wood ear (revived in boiling water, cleaned, and keep in fresh water)

   2/3 cup chicken broth

   1 egg white

   3 cups vegetable oil (for frying fish)

   1 tsp salt

   1/2 tsp sugar

   1/2 onion (minced)

   5 mm ginger (minced)

   1/2 cup *lao-zao* (fermented rice and juice)

   1 tsp sesame oil

In a pan deep-fry the marinated fish to a golden brown. In another pan, saute the onion and add the fried fish, sugar, wood ear, and chicken broth. Mix well. Cover and cook for 3 minutes. Add the *lao-zao* and sesame oil. Makes 4 servings.

Chicken with Red Fermented Rice (*Hong-zao shao-ji* 紅糟燒雞)

Ingredients:

   3 lb chicken (cut into 4 cm pieces, and marinated in red fermented rice, sugar, soy
      sauce and sherry)

   1/2 tsp sugar

   1 Tbsp soy sauce (醬油, light)

   2 Tbsp dry sherry

   2 Tbsp vegetable oil

   1/2 tsp salt

   2 scallions (cut into 2.5 cm sections)

   1 cm ginger (shredded)

   1 cup bamboo shoots (sliced to 2.5 cm squares)

   1/2 cup chicken broth

   1 tsp sesame oil

Saute the scallion and ginger. Add the chicken with the marinated sauce and bamboo shoots. Stir in the chicken broth and bring to a boil. Simmer for 30 minutes. Add sesame oil and serve hot. Makes 5 or 6 servings.

Braised Duck in Red Fermented Rice (*Hong-shao ya* 紅燒鴨)

Ingredients:

    6 lbs duck (rubbed with Five Spices powder, cut into 6 pieces, then marinated in red fermented rice, soy sauce, sugar, salt and sherry)

    1 Tbsp Five Spices powder

    1 cup red fermented rice

    2 Tbsp sherry

    1 tsp sugar

    4 Tbsp vegetable oil

    6 scallions (cut into 3 cm sections)

    1 1/2 cm ginger (cut into 2 mm slices, then cut across twice)

    1 cup chicken broth

    3 Tbsp soy sauce

Brown the scallion and ginger in a Dutch oven or deep skillet. Stir in the duck with the marinate liquid and cook for 2 minutes. Pour in the chicken broth. Bring to a boil and cook for 5 minutes. Reduce the heat to medium-low and cook for 45 minutes. Makes 6 servings.

Spicy Fish (*Wu-xiang yu-pian* 五香魚片)

Ingredients:

    2 lb fillet of cod (cut into 6 x 3 cm pieces, then marinated in salt, pepper, Five Spices powder, minced ginger and onion, dark soy sauce, sugar, and sherry)

    2 Tbsp dark soy sauce (老抽, *lao-chou*)

    1 tsp salt

    1/10 tsp pepper

    1 tsp Five Spices powder

    1/2 medium-sized onion (minced)

    5 mm ginger (minced)

    1 tsp sugar

    3 cups vegetable oil

    1/2 cups dry sherry

    1 cup all purpose flour

Take the marinated fish out piece by piece and immediately coat it with flour and put it in the hot frying pan. Deep-fry the fish on high heat until each piece turns reddish-brown. Take out the cooked ones as needed. Serve at cocktail parties or take out for picnics.

Fish with Fermented Black Soybean

Ingredients:

1 1/2 lb fillet of sole (cut into 8–10 cm sections)

1/3 cup fermented black soybean (*dou-chi* 豆豉)

5 leaves of mint (chopped)

4 cloves garlic (minced)

5 mm ginger (minced)

1 Tsp dry sherry

1 tsp sesame oil

1/4 cup water

1 tsp oyster sauce

1/4 cup vegetable oil

Heat oil on high heat in a deep skillet. Add the fish sections and cook for 1 minute without stirring. Turn to the other side and cook likewise. Combine all the flavoring material except the sesame oil. Pour the mixture over the fish. Move the sections gently, trying not to break them into small fragments. Reduce heat and cook for 15 minutes on medium-low heat. Add sesame oil and serve. Makes 6 servings.

Chinese Angelica Chicken (*Dang-gui dun-ji* 當歸燉雞)

Ingredients:

4 lb chicken (washed, drained, and cut into as many pieces as desired, including neck, liver, and gizzard)

1 cm ginger (thinly sliced and shredded)

1 large onion (cut into 10 vertical sections)

1/2 cup Job's tears (cleaned, washed, and drained)

1/2 cup euryale seed (washed and cleaned)

1/2 cup jujube (washed)

1 Chinese angelica root (washed)

1 tsp salt

3 quarts water

1 Tbsp vegetable oil (or chicken fat if desired)

Saute the ginger and onion. Stir in the chicken and salt, and cook for 1 minute. Add the remaining ingredients and bring to a boil on high heat. Reduce the heat to medium-low. Cook for 1 hour. Now the angelica root is soft. Make 6 to 8 vertical cuts. Continue to cook for 1 hour. This is a popular dish for post-partum mothers. Many elderly Chinese also take this food at regular intervals. The broth is highly esteemed. The meat is often eaten with a soy sauce and sesame oil dip. Makes several single servings, or serves 4 to 8 persons.

Tea Eggs (*Cha-ye-dan* 茶葉蛋)

Ingredients:

    1 dozen eggs

    1 quart water (enough to cover the eggs)

    1 Tbsp salt

    3 sections star anise

    1 cm cassia bark

    8 zanthoxylum (optional)

    4 tea bags (or 3 Tbsp red tea)

Hard-boil the eggs. Crack the shells lightly all around and return them to the pot. Add more water to cover the eggs if needed, and add the spices and tea bags. Keep boiling for 5 minutes. Turn to low heat and simmer for 2 hours. Cool and keep in the liquid until time for serving.

This is a good dish for picnics and parties. The eggs can be served whole with the shell or in halves without the shell. Tea egg is a popular Chinese dish. The broken shell is taken off just before eating, leaving a smooth marbled egg, white with metted browning marking and delicious taste. Recipes for tea egg preparation differ with people and their geographic regions.

Chinese Tamales with Lotus Leaves (*He-ye-bao-ji* 荷葉包雞)

Ingredients:

    3 cup cooked rice

    3 fresh lotus leaves (dipped into boiling water to render the pieces flexible for wrapping and cut into 20 cm squares; may be replaced by revived dried leaves imported from China, which are available in Chinese stores)

    2 oz barbequed pork (diced into 5 mm cubes)

    2 oz peeled shrimp (diced)

    2 large, black mushrooms (revived in boiling water for 15 minutes; use the softened portion only, and diced)

    1 Tbsp soy sauce (light)

    1 Tbsp sesame oil

    1 Tbsp shortening

    1/2 tsp salt

Mix all the ingredients and wrap 4–5 tablespoons in each piece of lotus leaf. Steam for 15 minutes. Serve hot. Makes 4 servings.

Chicken Wrapped in Lotus Leaf (*He-ye zheng-ji* 荷葉蒸雞)

Ingredients:

    1 1/2 lb chicken breast (cut into 1 1/2 cm squares and marinate for 15 minutes in soy sauce, salt, ginger, onion, *Tian-mian-jiang*, sherry, sugar, and cornstarch)

    1/3 cup rice (washed and drained)

    6 fresh lotus leaves (parboiled to increase flexibility and cut into squares)

    1 small can of bamboo shoots (diced into 5 mm pieces)

    2 Tbsp diced pickled cucumber

    1 tsp salt

    1 Tbsp sherry

    1/2 tsp sugar

    1 Tbsp soy sauce (light)

    1 Tbsp *tian-mian-jiang*

    1/2 medium onion (minced)

    2 mm piece of ginger (minced)

    1 Tbsp cornstarch

Combine the marinated chicken, rice, bamboo shoots, and pickled cucumber. Wrap the mixture in lotus leaves. Steam for 40 minutes. This recipe makes 6 servings.

Wrapping food in lotus leaves in order to impart its pleasant flavor is a common practice in Chinese cooking. Spareribs, duck and other kinds of meat may be used instead of chicken.

HEALTH FOOD AND HERBAL TEA

For the conservation of health, the Chinese people use many unusual natural plant products in food and drink. These vegetal materials are not equivalent to the "health food" and "herbal tea" commonly consumed in Western countries. Two terms, *bupin* 補品 and *liangcha* 涼茶, are used in discussing the subject. *Bupins* are regular Chinese herbal medicines cooked in food and taken to cope with the stress and strain of life with environmental changes, and to replenish a person's natural immunity. Preparations made from these plant products are taken neither as food to assuage hunger, nor as medicine for curing specific ailments. They are often taken seasonally and occasionally for keeping the body in tone and for the prevention of diseases. *Liangchas* are decoctions prepared from 6 to 32 plant products, served as *liangcha* in simple shops or stands in large cities or marketplaces in large villages, or packages of plant material bought by people to decoct at home without any article of food.

After the normalization of the diplomatic relations between China and the USA in the early 1980s, large amounts of numerous kinds of *bupin* and *liangcha* were imported into America and sold in Chinese groceries, gift shops, and drug stores. Some of the products are beautifully packed in transparent, plastic boxes with colorful labels, explaining the contents, using with romanized words (in Latin or in Chinese pinyin) and giving the directions for cooking. Others items are imported in bulk and repacked locally into one-pound bags with the names and prices only, just as packages of peas, peanuts, and chestnuts in the same store.

Based on how these imported products are used, they can be categorized as *bupin* (補品, repairing material), or *liangcha* (涼茶, cooling tea). Most of the *bupin* products bought by consumers are boiled with chicken, duck, spareribs or ox-tail to make a tonifying broth for the family. Some packages contain many pre-weighed ingredients proportioned according to reputed ancient *liangcha* recipes. Recently, many requests have been received from economic botanists, officers of the US Customs, and the general public interested in Chinese food, particularly immunostimulating or immunomodulating food, for the identification of the *bupin* and *liangcha* ingredients. For the convenience of the professionals and the benefit of the users, the following accounts of *bupin* and *liangcha* have been prepared.

Bupin 補品 : Plant Esculents for the Conservation of Health

The Chinese ideograms representing *bupin* can be translated literally into "repair" (*bu* 補) and "substance" (*pin* 品). When used together, they refer to food taken to adjust or to tonify a person's physiological imbalance. The Chinese concept of physical well-being is to have a balanced *yin-yang* system. In Chinese terminology, *yin* 陰 refers to the blood (*xue* 血), and *yang* 陽 is expressed by the strength (*qi* 氣) of a person. A person in poor health and with low immunity is said to have deficient blood (*xue-kui* 血虧) and feeble vitality (*qi-xu* 氣虛). Food for improving the quality of the blood and for revitalizing the strength is taken as a *bupin*. It is considered a replenisher of one's natural power of immunity, or in modern terms, a booster of depleted electrolytes.

In some literature on Chinese medicine, *bupin* has been translated as "tonic". Generally, however, tonics are produced by pharmaceutical companies. The people who take *bupin*, consider it is a home-made food prepared with plant (rarely animal) products that may be used as ingredients in some tonic pills. *Bupin* is mild, and is prepared for helping normal persons to become healthier.

Characteristics of Chinese *Bupin*

Here *bupin* is interpreted as a home-made food, prepared with some medicinal herbs. It is used to replenish one's natural power of resistance to infectious diseases, to give strength, and to enrich the blood. Since environmental changes and aging affect the well being of a person, generally *bupin* is served at the time of seasonal or physiological changes for most users, and specially for persons recuperating from an illness and used continually for the elderly. For example, a chicken broth prepared with a ginseng root and some berries of matrimony vine is served to the entire family in November when the climate changes from warm to cold, and a different kind prepared with *Astragalus*, *Codonopsis* and *Lycium* (*huang-qi* 黃耆, *dang-shen* 黨參 and *gou-qi-zi* 枸杞子) is served to all members of the family in early summer. A broth made of silver ear

(Tremella) and chicken is often prepared for an elderly person in the family, taken before going to bed at night or waking up in early morning.

(1) *Limited Uses and Regional Differences*: The recent influx of *bupin* into Chinese stores in America is an unusual phenomenon that does not exist anywhere in China, nor was it so in America decades ago. Actually, most of the Chinese *bupin* are products of rare and endangered species, raised or gathered under high labor cost conditions. Here the laws of supply and demand apply: scarce commodity plus high demand equals high price. In China the use of *bupin* has been limited to those who can afford to buy small amounts from medicinal stores for special occasions. For example, outside China, ginseng is recognized as a Chinese *bupin*. However, in a book entitled *Common Herbal Medicine of Canton*, prepared by the Medical Division of the Regional Army, which contains 23 *bupins* among its 440 entries, ginseng is not included. Likewise, in the 1970 edition of the *Common Traditional Chinese Herbal Medicine of Shanghai*, ginseng is not included among its 480 entries. It is evident that in books on medicinal herbs for the common people, many *bupin* articles that are available to the Chinese consumers in America are absent. In China now, as it was 200 years ago, the use of *bupin* is limited to political leaders, professionals and rich merchants whose salaries are high enough to buy expensive, classical *bupin* or imported American ginseng.

Geographical variations have positive effects on the use of *bupin*. With their fresh food, good air, abundant sunshine, and more physically strenuous life and work, most rural people of China are in better health and they use no *bupin*. In general, the people in South China use more *bupin* than those in the North. Most of the shop owners and the *bupin* consumers among the overseas Chinese are people from southern China, and many of them emigrated via Hong Kong, where the use of *bupin* has become a popular custom. It is noteworthy that the current increase of *bupin* materials in American Chinese stores was due to importation via Hong Kong. After the adoption of an open door policy by the Chinese government, one of the dominant goals of business in China has been to increase foreign exchange. Rice fields have been converted to raise plants and animals which can be sold for higher prices on the foreign trade market. All factors combined, the availability of *bupin* is increasing and prices are decreasing in American Chinese stores.

(2) *Several Ingredients and Slow Cooking*: *Bupin* is prepared by slowly cooking several ingredients together. In China, the preparation of *bupin* is an art developed by affluent people. Before the invention of electric appliances, such as the slow cooking crock pots, a special device called *wu-geng-ji* (五更雞, early dawn cock) was in common use in Guangzhou. Families with elderly people who needed *bupin* first thing in the morning

prepared the ingredients before bedtime, and simmered them slowly throughout the night. The soft and soup-like nourishment was served hot the next morning at dawn when the cock crowed, hence the cooking device was called *wu-geng-ji.*

The object of preparing *bupin* is to provide people who are weak with some easily digested, nutritious food. Any of the following dried items can be added to a major ingredient such as rice, red beans, chicken or spareribs. Anyone who would like to try Chinese *bupin* should have in the pantry dried jujube, longan arils, Job's tears, lotus seed, dried lycium (berries of matrimony vine), lily bulb scales, and euryale seed.

(3) *Small Amounts and Occasional Usage*: Scarcity and high price have reduced the widespread use of *bupin* in China. Naturally, it also restricted the amount of *bupin* used. However, these external factors are not the only causes which help to eliminate the abuse of *bupin*. Actually, a large number of the imported products are prepared from poisonous species of plants. They are beneficial only when used in small quantities, and harmful otherwise. Since time immemorial, the Chinese people under daily stresses and strains, and with the threat of illness, have gradually learned through trial and error, which plants to use, the proper amounts to take and the correct methods to prepare them. The general practice is to take small amounts of several products and to prepare this mild *bupin* at times of seasonal and/or personal changes. For normal working people *bupin* should be used only occasionally. For the elderly and those in convalescence, mild bupin may be served daily.

Before the 1970s, the Chinese *bupin* was used only by the Chinese people and by some Asians who had adopted the Chinese culture. Even in regard to ginseng, the best known Chinese *bupin*, we find this statement: "The Chinese people constitute the world's primary ginseng consumer." (*Econ. Bot.* 30: 11. 1976). The International Ginseng Symposia sponsored by the Korean government and a pharmaceutical company, Pharmaton, Switzerland, and the advertising agents of these organizations have made ginseng popular in Europe and America. In the late 1970s, young people who tried to produce a "ginseng high" used large doses of ginseng powder concomitant with the intake of caffeine, several times a day. This led to a ginseng abuse syndrome described by Dr. Ronald K. Siegel, School of Medicine, University of California, Los Angeles, as "hypertension together with minor skin eruptions, restlessness, insomnia, and morning diarrhea" (Siegel, 1980). Siegel conducted an experiment with two groups of volunteers. The group of young people that took small amounts as a nutritional supplement to their diet did not have such adverse reactions.

The current influx of *bupin* products into the Chinese stores in America increases their availability to the public. The result of my personal investigation indicates that approximately three-fourths of the 72 common *bupin* products available in the American market (as listed below) are prepared from poisonous species. Many of them have gone through processes of detoxification before drying, and they are safe when used according to the conventional advice of small amount and occasional intake as food supplements. Nevertheless, the public should be warned against any *bupin* abuse. The Chinese *bupin* improves the health of a person at the cellular level, producing stimulatory activities on the granulocytes and macrophages at low concentration. None of the Chinese *bupin* products are hallucinogenic, nor do they produce "highs".

Common Chinese *Bupin* Available in America

The diversity of Chinese *bupin* (repair material) available in the American market is amazing, both in forms and in their distributions within the plant kingdom. In forms, they are: 4 kinds of fungal bodies, 32 roots, rhizomes, and/or other underground plant structures, 3 barks, 5 flowers, 13 fruits, and 9 seeds. Their distribution of the source species within the plant kingdom i.e.: 1 ascomycete, 3 basidiomycetes, 2 ferns, 55 dicots in 28 families, and 11 monocots in 6 families.

Over three-fourths of these *bupins* consist of fragments of vegetative organs of plants. The scientific identifications of these materials would be almost impossible with classical laboratory methods by any competent botanist, even with good facilities. Here a simple device is designed by which the users may be benefited from my sixty-years' experience in the study of Chinese *bupins*. The basic premises on which the device is established are: (1) The available *bupins* are classical Chinese herbs used in China since time immemorial. Their botanical identifications, and their chemical compositions, are well established in modern China. Unfortunately, this information is published in Chinese, and arranged by the number of strokes of the classical herbal names, followed by a Latinized botanical identification of the source species and the part or parts of the plant used. (2) The current dealers in the American market learned their trade through apprenticeship. They know their products by sight and their classical Chinese names. Actually, the management of a Chinese drug store and the arrangement of the crude drugs in it are very similar to that which one may find in Guangzhou or Hong Kong. These store owners can all read and write the Chinese names of their market products. Under this condition, an alphabetical list of the botanical names accompanied by the Chinese names in ideograms is given below. Botanists who need to identify the material, and readers who would like to buy and try some of the *bupins*, can simply show the Chinese names to a dealer and obtain the articles.

In American Chinese stores, errors in naming herbal items are rare. In my fifty-years' research on the *bupins* available in Boston, I have checked the identification of the articles sold in several herb shops and groceries. I found only one mistake where a *bupin* label was placed unintentionally on a package of clove. Apparently, this error was made by a clerk in a hurry, for when notified, the owner of the store was very appreciative and made the necessary correction immediately.

In the knowledge, marketing and using of Chinese *bupin*, a classical name of an item is an unifying means of communication between the Chinese scientists, practitioner/merchants and populace. Names of common *bupin* in Chinese ideograms are picturesque, and simple with 2 or 3 syllables. In the following list, one of these names follows the full botanical epithet of the source species with the Pinyin System of rhomanization and the family to which it belongs. During the past fifteen years, Steve Foster, author of *Herbal Bounty, Echinacea: Nature's Immune Enhancer* and *Herbs of Commerce* (Foster, 1984–92) and I co-operated in a research project on a detailed treatment of the source species of Chinese medicinal plants which are grown in American gardens. This co-operative research has enabled me to put the plus and minus signs, designating plants growing in (+) or absent from (–) America, before the market remark of the item, particularly the per unit retail price known in the winter of 1987, when the investigation was made. Evidently, there are more than three-fourths (78%) of the listed source species of Chinese *bupin* now grow in America. This information may be of special interest to growers and suppliers. Items in combined packages studied are not given a price. Different issues of each item in the list are separated by a semicolon (;).

A List of Common Chinese Bupins in the American Market

Abrus cantoniensis Hance; 雞骨草, Ji-gu-cao; Leguminosae; –, Canton Abrus, $0.99 per bag.

Achyranthes bidentata Blume; 牛膝, Niu-xi; Amaranthaceae

Aconitum carmichaeli Debx.; 白附子, Bai-fu-zi; Ranunculaceae; +, Aconite tuber

Adenophora tetraphylla (Thunb.) Fischer; 南沙參, Nan-sha-shen; Campanulaceae; +, Sha-shen

Angelica anomala Lallem. and **A. dahurica** (Fischer ex Hoffm.) Bentham et Hook. f.; 白芷, Bai-zhi; Umbelliferae; +, Fragrant angelica, $3.00 bag

A. sinensis (Oliver) Diels; 當歸, Dang-gui; Umbelliferae; +, Chinese angelica, $7.50–24.00

Areca catechu L.; 檳榔, Bin-lang; Palmae; –, Sliced betel-nut

Astragalus membranaceus (Fischer) Bunge; 黃耆, Huang-qi; Leguminosae; +, Astragalus, $15–35.00 lb

Atractylodes macrocephala Koidz.; 白朮, Bai-zhu; Compositae; –, Atractylodes root, $13.50 lb

Bupleurum chinense DC. and **B. scorzonerifolium** Willd.; 柴胡, Cai-hu; Umbelliferae; +, Hare's Ear, caudex and root, $16.00 lb

Carthamus tinctorius L.; 紅花, Hong-hua; Compositae; +, Safflower

Chaenomeles sinensis (Thouin) Koehne; 川木瓜, Chuan-mu-gua; Rosaceae; +, Chaenomeles, dried slices

Cibotium barometz (L.) J. Sm.; 金狗脊, Jin-gou-ji; Dicksoniaceae; –, Fern rhizome, sliced

Cinnamomum cassia Presl; 肉桂, Rou-gui; Lauraceae; –, Powdered, cassia bark

Cistanche salsa (C. A. Meyer) G. Beck; 肉蓯蓉, Rou-cong-rong; Orobanchaceae; –, dried broomrape, fleshy parasite

Citrus reticulata Blanco; 陳皮, Chen-pi; Rutaceae; +, Tangerine or mandarin orange peels, $5.00

Codonopsis pilosula (Franchet) Nannf.; 黨參, Dang-shen; Campanulaceae; +, Whole or sliced

C. tangshen Oliver; 防黨, Fang-dang; Campanulaceae; +, dried root, $5–25.00

Coix lachryma-jobi L.; 薏仁, Yi-ren; Gramineae; +, Job's tears, $1.50 per packet

Cordyceps sinensis Berk; 蟲草, Chong-cao; Ascomycetes; –, Dried sclerotia attached to host skin, the most expensive crude drug in Chinese stores

Crataegus pinnatifida Bunge; 山楂肉, Shan-zha-rou; Rosaceae; +, Sliced fruit, $11.00 lb

Cullen corylifolium (L.) Medicus; 補骨脂, Bu-gu-zhi; Leguminosae; +, Scurfy pea

Curculigo orchioides Gaertn.; 仙茅, Xian-mao; Amaryllidaceae; –, Dried whole rootstock

Delandia calcaratus (Roxb.) S. Y. Hu; 赤小豆, Chi-xiao-dou; Leguminosae; +, Rice Bean or Lesser Red Bean; $0.90 lb

Dendranthema vestitum (Hemsley) Ling; 菊花, Ju-hua; Compositae; +, Dried flower, $2.50 per bag

Dimocarpus longan Lour.; 龍眼肉, Long-yan-rou; Sapindaceae; +, Dried aril, $2.50 per 7 oz

Dioscorea opposita Thunb.; 山藥, Shan-yao, or 懷山, Huai-shan; Dioscoreaceae; +, Chinese Yam, dried slices of root, $15.00

Dipsacus asper Wall.; 川續斷, Chuan-xu-duan; Dipsacaceae; –, Sichuan teasel

Drynaria fortunei (Kunze) J. Sm.; 骨碎補, Gu-sui-bu; Polypodiaceae; +, Drynaria, rhizome

Eucommia ulmoides Oliver; 杜仲, Du-zhong; Eucommiaceae; +, Eucommia bark, $12.00 bag

Euryale ferox (Lour.) Steud.; 芡實, Qian-shi; Nymphaeaceae; +, Euryale, seed, shell removed, $2.50 per bag

Ganoderma lucidum (Leysser ex Fries) Karsten; 靈芝, Ling-zhi; Basidiomycetes; $20.00 per bag

Gastrodia elata Blume; 天麻, Tian-ma; Orchidaceae; –, Symbiotic, orchid and fungus **Armillaria mellea** (Vahl ex Fries) Quet, dried rhizomes

Glehnia littoralis F. Schmidt; 北沙參, Bei-sha-shen; Umbelliferae; –, Dried root, $5.00 per bag

Glycine max (L.) Merr.; 黑豆, Hei-dou; Leguminosae; +, Black soybean, $1.00 lb

Glycyrrhiza glabra L. and **G. uralensis** Fischer; 炙甘草, Zhi-gan-cao; Leguminosae; +, Chinese roasted licorice, $1.00 a box

Hordeum vulgare L.; 麥芽, Mai-ya; Gramineae; +, Sprouted barley, $3.95 per bag

Hylocereus undatus (Haw.) Britt. et Rose; 霸王花, Ba-wang-hua; Cactaceae; +, Night-blooming cereus, dried flower

Juglans regia L.; 核桃, He-tao; Juglandaceae; +, Broken walnut kernel

Ligusticum 'Chuanhsiung' Shan; 川芎, Chuan-xiong; Umbelliferae; –, Sichuan lovage, sliced caudex, $7.95 per box

Ligustrum lucidum Ait.; 女貞子, Nü-zhen-zi; Oleaceae; +, Chinese privet, Dried fruit

Lilium brownii F. E. Br. var. **colchesteri** Wilson; 百合, Bai-he; Liliaceae; +, Lily bulb, steamed and dried; $5.00 per bag

Liriope graminifolia (L.) Baker; 麥冬, Mai-dong; Liliaceae; +, Lily turf, dried root

Lonicera japonica Thunb.; 金銀花, Jin-yin-hua; Caprifoliaceae; +, Dried honey-suckle flowers and flower bud, $3.00 per bag

Lycium barbarum L. and **L. chinense** Miller; 枸杞子, Gou-qi-zi; Solanaceae; +, Lycium Berry, dried fruit of matrimony vine, $5.00-17.00 per lb

Magnolia officinalis Rehder et Wilson; 厚朴, Hou-po; Magnoliaceae; +, Dried bark, $18.00 lb

Microcos nervosa (Lour.) S. Y. Hu; 布渣葉, Bu-zha-ye; Tiliaceae; –, Dried leaves, $0.99 lb

Morinda officinalis How; 巴戟天, Ba-ji-tian; Rubiaceae; –, Morinda root, $19.00-30.00 per lb

Nelumbo nucifera Gaertn.; 蓮子, Lian-zi; Nymphaeceae; +, Sacred Lotus seed, with seed coat $3.95, without seed coat $3.50

Oryza sativa L.; 米芽, Mi-ya; Gramineae; + Sprouted rice, dried, $2–2.50 per lb

Paeonia lactiflora Pallas; 白芍, Bai-shao; Ranunculaceae; +, Root of white-flowered peony, outer rind removed

Panax ginseng C. A. Meyer; 人參, Ren-shen; Araliaceae; +, Chinese-Korean ginseng, $16.95 per box

P. quinquefolium L.; 西洋參, Xi-yang-shen; Araliaceae; +, American ginseng, cultivated $29.00–120.00 per lb, wild $165.00–675.00 per lb

P. wangianum Sun cv. 'Sanqi' Hort.; 三七, San-qi; Araliaceae; –, Sanqi ginseng

Platycodon grandiflorus (Jacq.) A. DC.; 桔梗, Jie-geng; Campanulaceae; +, Dried root, Balloon flower, $9.50–13.50 per lb

Polygonatum odoratum (Miller) Druce; 玉竹, Yu-zhu; Liliaceae; +, Jade Bamboo, dried rhizome, $3.75 per box

P. multiflorum (L.) All. (*P. cyrtonema* Hua) and **P. sibiricum** Redout (*P. chinense* Kunth); 黃精, Huang-jing; Liliaceae; +, Solomon's Seal, steamed and dried rhizome, $11.95 per box

Polygonum multiflorum Thunb.; 何首烏, He-shou-wu; Polygonaceae; +, Chinese Cornbind, Sliced dried rhizome, $6.95 per lb

Poria cocos (Schw.) Wolf; 茯苓, Fu-ling; Basidiomycete; –, China root, white slices, dried sclerotia, $13.50 per box

Potentilla anserina L.; 蕨麻, Jue-ma; Rosaceae; +, Silverweed, dried root tuber, local use in northwestern China, not available in the American market, used in Lanzhou, Gansu, Northwestern China, as a substitute for ginseng

Prinsepia uniflora Batal.; 奈仁肉, Nai-ren-rou; Rosaceae; +, Prinsepia seed, $9.95, per 7 oz

Prunella vulgaris L.; 夏枯草, Xia-ku-cao; Labiatae; +, Dried inflorescences of Selfheal, $1.50 per bag

Prunus armeniaca L.; 杏仁, Xing-ren; Rosaceae; +, Shelled apricot seed, $5.00 lb

Pseudostellaria heterophylla (Miq.) Pax ex Pax et Hoffm.; 太子參, Tai-zi-shen; Caryophyllaceae; +, Prince ginseng, dried whole root tuber, $3.90 lb

Rehmannia glutinosa (Gaertner) Libosch.; 地黃, Di-huang; Scrophulariaceae; +, Root, dried untreated, $3.00 per box, steamed and dried repeatedly $5.00 per box

Rosa laevigata Michx.; 金櫻子, Jin-ying-zi; Rosaceae; +, Cherokee rose, dried fruit, $10.50 per lb

Saussurea lappa (Decne.) Clarke; 廣木香; Guang-mu-xiang; Compositae; –, Himalayan Costus, sliced rootstock

Schisandra chinensis (Turcz.) Baill.; 五味子, Wu-wei-zi; Schisandraceae; +, Dried schisandra fruit

Siraitia grosvenorii (Swingle) C. Jeffrey; 羅漢果, Luo-han-guo; Cucurbitaceae; –, Dried whole fruit, $1.50 for two

Tremella fuciformis Berk.; 銀耳, Yin-er; Basidiomycete; –, Silver Ear, white mass of fruiting body, cultivated

Ziziphus jujuba Miller; 棗, zao; Rhamnaceae; +, Jujube, the market material all imported from China; 紅棗, Hong-zao, dried red jujube, $4.95 per lb; 蜜棗, Mi-zao, dried sugared jujube, $2.50 per lb; 黑棗, Hei-zao, dried black jujube, $9.50 per lb

The condition of marketing some of the *bupin* products in one-pound packages may lead some users to over-indulgence, as was the case of taking ginseng by some people in California (Siegal, 1979, 1080). The following precaution must be taken seriously for safeguard.

Amount of *Bupin* for Proper Application

The use of *bupin* is still in a transitional stage from ethnic practice to scientific understanding. A good example is established in the Chinese use of lotus seed as *bupin* since time immemorial. Various designs have been devised to cook lotus seed, some simple and other very fancy and time-consuming. Only recently, a plant physiologist, Jane Shen-Miller and her associates who germinated some radiocarbon-dated 1,288, 271-year-old lotus seed and raised the plants to flower in the University of California at Los Angeles, discovered a protein repair enzyme in the seed that keeps its longevity and viability. In a Chinese *bupin*, scientist has also discovered the protein repair enzyme.

Formerly, the practice of taking *bupin* was an art. The users depended upon the local apothecary for their supply. The proper amount of each ingredient for a formula was more or less a trade secret of the apothecary. The users learned of the names and efficacy of certain combinations from family practices and/or from friends and relatives. When they needed certain types of *bupin*, they simply went to the local store and told the clerk the name of the combination. The clerk carefully weighed the ingredients, wrapped them together, and sold the mixture for one serving. Under such conditions the users did not know the exact proportions of the ingredients in any *bupin* combination. Actually, the proportions varied with the recipes. Keeping trade secrets is a worldwide practice in pharmacology, so it was also in China.

However, in the latter half of the twentieth century, the situation in China had changed. Famous apothecaries holding family traditions were compelled to report their inherited trade secrets to the government, which in turn put the products onto the market. Recently available in American Chinese stores are single-serving *bupin* packets prepared from ancient formulas, containing the names of the ingredients in Chinese, sometimes with English translations and the appropriate weights. To get a general idea of the amounts of the various items in the single-serving packets, five combinations of *bupin* were purchased. Two of these have Chinese traditional apothecary's weight indications, which are converted to milligrams. The other three packages were opened one by one, and the contents carefully separated and weighed in US parmaceutical units (milligrams). In order to further help readers to appreciate the small amounts of *bupin* used in traditional Chinese practice, the conversion of milligrams to ounces is given here (1 ounce = 31,103 mg). One of these packages has the Chinese brand name translated into Cantonese System of Romanization, and the cooking directions and ingredients are given in English. In this case, the names unfamiliar to me are also placed in parentheses. The titles of the packages are my translation of the original names given in Chinese.

(1) *Four Ingredients Soup* (四物湯): Four ingredients are enclosed in this package. They are designed to be used for a delicious soup when cooked with chicken, duck, spareribs or fish:

Chinese angelica	5,600 mg
Sichuan lovage	4,700 mg
Steamed-dried Chinese foxglove root	10,800 mg
White peony root	1,200 mg

(2) *Cooling Bupin Mix* (清補涼): This mix is packed by Sun Wing Trading Company, Hong Kong. It gives cooking directions in English: "1. It makes a delicious soup with pork, best taken as a meal, and 2. Sweet soup can be made with sugar, best as a beverage." The names in parentheses after my identification are printed on the box. The term "Pearl Barley" is incorrect, and the name "Fox nut" is from an origin unknown to me.

Lily bulb	19,000 mg
Job's tears (Pearl Barley)	85,810 mg
Sliced yam (Dioscorea)	16,385 mg
Lotus seed	17,770 mg
Euryale seed (Fox nut)	10,385 mg
Jade bamboo (Polygonatum)	16,685 mg
Longan aril (Dried Longan)	2,325 mg

(3) *Complete Bupin Mix* (十全大補湯): This sample contains 10 ingredients beautifully packed in a gift box. The retail price is $4.95.

Whole codonopsis root	15,551 mg
Steamed *di-huang* (Rehmannia)	15,551 mg
Chinese angelica (Dang-gui)	15,551 mg
Roasted licorice	15,551 mg
Powdered cassia bark	15,551 mg
Astragalus root (Huang-qi)	15,551 mg
Sliced white peony root	15,551 mg
Sliced China root (Fu-ling)	15,551 mg
Sliced atractylodes root (Bai-zhu)	15,551 mg
Sichuan lovage (Chuan-xiong)	15,551 mg

(4) *Post-partum Beverage* (生化湯): This *bupin* mix is designed for enriching the blood and/or for dissolving any possible clots occurring at childbirth. It is a mild *bupin* which can be used repeatedly.

Chinese angelica	16,200 mg
Charred ginseng root	4,900 mg
Sichuan lovage	6,500 mg
Safflower	3,200 mg
Peach kernel (powdered)	2,500 mg
Licorice (roasted)	2,500 mg

(5) *Major Bupin Mix with Dried Gecko* (蛤蚧): This is one of the more expensive *bupin* mixes. The retail price is $11.50 for one package. In other *bupin* mixes, the consumers are given the opportunity to choose the animal ingredients they prefer. In this mix, the animal ingredient (gecko) is included.

Astragalus root	12,441 mg
Codonopsis root	12,441 mg
Chinese angelica	12,441 mg
Berry of matrimony vine	12,441 mg
Sliced China root	12,441 mg
Sliced Chinese yam	12,441 mg
Atractylodes root	12,441 mg
Steamed *di-huang* (Rehmannia)	12,441 mg
Roasted licorice	6,220 mg
Morinda root	9,331 mg
Eucommia bark	7,775 mg
Sichuan teasel	15,551 mg
Chaenomeles fruit (Mu-gua)	9,331 mg
Solomon's seal	15,551 mg
Lamb of Tartary rhizome	15,551 mg
Drynaria rhizome	15,551 mg
Shiny-leaved privet fruit	15,551 mg
Dried gecko (internal organs removed)	1 pair

The above analysis of five *bupin* mixes prepared in China and Hong Kong for the American market clearly indicates the small amounts of the 28 plant *bupin* products involved in these conventional recipes. A conversion of the pharmaceutical weighing system into one used in the kitchen would be more helpful to readers to visualize the applicable amounts than the milligrams given. One ounce equals

31.103 grams (= 31,103 milligrams). The majority of the above items (64.32%) use only between 2/5 and 1/2 ounce. The only product using slightly over 2 1/2 ounces is Job's tears in sample (2). In China Job's tears is an expensive cereal, often used to stuff a duck for special occasions. In the cases of Eucommia, Morinda, and Chaenomeles, the amounts used are 1/5 or 3/10 of an ounce. In the case of ginseng and safflower, only 1/10 of an ounce is used, whereas with peach kernel and longan aril, the amounts used are less than 1/10 of an ounce. The factors limiting the amounts of these products are: toxicity for the untreated peach kernel, and production costs for longan arils. In the humid tropics where longan trees grow, drying the arils in the rainy harvest season is a very tedious and difficult process, and keeping the dried material from molding and the sugar content from deteriorating are costly and painstaking procedures.

It is noteworthy that in the five *bupin* mixes, Chinese angelica is used four times, while Sichuan lovage and steamed foxglove root, and roasted licorice are used three times each. Peony root, Chinese yam, *dang-shen* (*Codonopsis tangshen*), *huang-qi* (*Astragalus membranaceus*), China root (*Poria cocos*), and *bai-zhu* (*Atractylodes macrocephala*) are used twice each. With the exception of the last two products, the source species of the frequently used *bupin* are grown in America, mostly for ornamental purposes.

Growers and users would be benefited by knowing the detoxification of several *bupin* products before they are put in the market. By nature, some of these species are poisonous plants. Their products are prepared with a certain detoxifying procedure. The Chinese foxglove root (*Rehmannia glutinosa* [Gaertn.] Libosch.), for example, has two forms in the American market: the raw, dried root and the steamed, dried root. In preparing the steamed product, dried root is mixed with brown rice gin (30% by weight). After the liquid is absorbed, the root is placed in a steamer, the wall of which is made with willow wood, and is steamed for eight hours until the material becomes uniformly black, inside and outside. Then it is taken out to dry in the sun. The dried root is steamed again for 4–8 hours, covered overnight to cool and is then taken out to dry in the sun again. When it is 80% dried, the root is sliced, thoroughly dried and preserved in a container for later use. Likewise, Solomon's seal (*Polygonatum sibiricum* Redout and *P. multiflorum* [L.] All.) is cooked with gin (100 rhizome: 50 gin by weight) in a double boiler. In the *bupin* mix analyzed above, these two items are black and relatively soft.

Home-made *Bupin* Dishes

In the Chinese groceries in America, both *bupin* mix and one-pound bags of individual ingredients are available. From the samples of *bupin* mixes discussed above, much variety is evident in the number of items used in preparing *bupin*. In general practice, three or four ingredients cooked with some chicken or pork bone is more common than using a large number of items. As one-pound bags reduce the costs of packaging and labor, it is more economical to have larger quantities of several common kinds of *bupin*, such as lily bulb, Job's tears, sliced yam, lotus seed, etc. in the pantry for frequent use (say twice a month). From my personal experience, I realize preparing *bupin* with non-vascular ingredients is quite different from preparing *bupin* with vascular plant products. Here they are discussed separately.

(1) *Preparation of Bupin with Fungal Products*: The history of using fungal *bupin* in China is as old as the Chinese culture. In the *Shen-nong Ben-cao Jing* 神農本草經 (*ca.* 200 A.D.), China root and two kinds of ganoderma were described, characterized, and accredited with the properties of maintaining youthfulness, giving strength and enhancing longevity. The natural growth of fungal species is associated with decaying wood in the forests. Due to the early and constant deforestation in China, the currently used products of Chinese fungal *bupin* are cultivated. The first scientific investigation into the cultivations of China root and silver ear, together with wood ear and black mushroom was made by Professor Y. Chen, Chairman of the Forestry Department of the Agriculture College, University of Nanking, after his return to China from the Arnold Arboretum for a year's sabbatical study. The result of his studies was published in his *Silviculture of Chinese Trees* (Chen, 1933). According to his findings, a package of China root in Boston means the cutting of pine trees in Anhui, China. Likewise, a box of silver ear or wood ear represents the falling of oak trees in Sichuan. A dish of black mushrooms indicates that in Fujian, Zhejiang, Jiangxi, or Anhui, the woody plants of an entire hillside have been completely fallen. After the removal of the coniferous and the lauraceous material from the deforested area, the hillside is sealed off from the public for 7 or 8 years. No unauthorized person is allowed to walk in the area until after the fungal crop has been harvested for the fourth or fifth time. The landowner, the management, and the hired workers protected the area and kept the trade secret of the cultivation and processing of the product. Such unwritten law was broken only during the Word War II, when eastern China was under Japanese occupation. Then the cultivation of fungal species, particularly the black mushroom, was taken to Japan, and methods were improved there. Now, Ganoderma is cultivated in sawdust (Figure 16a). At the present day, only the supply of winter-worm summer-herb (冬蟲夏草, *Cordyceps sinensis* Berk.) is a natural wild product.

China root (*fu-ling* 茯苓): The market product known as *fu-ling* consists of the sliced center portion of the sclerotium of *Poria cocos* (Schw.) Wolf (Basidiomycetes), a facultative parasite that grows on the roots or stumps of pines. The colors and sizes of the market slices vary according to their position on the sclerotium. The pure white, large, thin slices are from the center of the sclerotium, and the slightly pink or brown, very thick or rather thick slices are from the periphery. The white product commands a higher price.

The effective principles of China root which have been isolated and identified include: pachymic acid, tumulosic acid, β-pachyman, 3β-hydroxylanosta-7,9(11), 24-trien-21-oic acid, resin, chitin, proteins, fats, sterol, lecithin, glucose, adenine, histidine, choline, and various enzymes. Approximately 93% of the dry weight of *fu-ling* slices is polysaccharides, which are known for stimulating the production of macrophages, T-lymphocytes, and interferon (Figure 15a).

Fu-ling Congee

Ingredients:
  1/2 cup *fu-ling* slices (dried market material)
  1/2 cup sliced Chinese yam (dried market material)
  1/2 cup jujubes (dried red type, washed)
  1/2 cup rice bean (lesser red bean, cleaned and washed)
  1 cup ginkgo kernel (fresh material preferred)
  1/2 cup Job's tears (cleaned and washed)
  1 cup white rice
  1 1/2 gallon water
  2 Tbsp sugar

In a large pan, boil the water. Add all the above ingredients after cleaned and bring the mixture to a boil. Reduce the heat and continue to boil for 3 hours. Makes 6 to 8 servings.

This recipe is adapted from a *bupin* mix bought from a Chinese grocery exported by the China National Native Produce Corporation. The first six ingredients are packaged in a transparent plastic bag. On one side of the bag the names and properties of the ingredients are printed. The printed names are very different from those I have seen in botanical literature. My identifications of these materials are given in the parentheses after the printed names to show that popular names that go with Chinese products can be very misleading.

Cordyceps (winter-worm summer-herb): The full Chinese name of this product is *dong-chong xia-cao* 冬蟲夏草. In the market, it is generally shortened to *chong-cao* 蟲草. This product consists of the dried sclerotia of a parasitic fungus (*Cordyceps sinensis* Berk.)

which grows within the larva of a moth (*Hepialus armoricanus* Ober.) of the alpine meadows in western China. The fungal hypha enters the newly hatched larva through its breathing pores, and grows with the host, finally consuming the tissue of the larva and filling up its skin. The corpse remains in the ground, and, as the snow melts in the next spring, a sclerotium emerges from its head-end. The local people in western China climb the high mountains at day-break to search over the thin veneer of partially melting icy-snow for the slim, nail-like, brownish sclerotia attached to the skin of the dead worms. The collecting season is very short as the small worm-like sclerotium is quickly covered by the fast-growing vegetation of the alpine meadow. Furthermore, the market value of the product is reduced after asci (ascus) have developed in the sclerotia. Collectors and dealers are very strict concerning the developmental stage of the sclerotia, a factor that controls the price of the product.

The known chemical composition of air dried cordyceps includes: 25.32% protein, 28.90% carbohydrates (including various sugars and polysaccharides), 8.4% fats, (82.2% of which are unsaturated), 18.53% fiber, 10.84% water, 4.10% ashes, Vitamin B12, cordycepic acids, and cordycepin. The alleged properties of cordyceps are: warming, strengthening the vital organs and opening up the pathways for proper physiological functions (apertures of vitality in Chinese medical terms).

Cordyceps is the most expensive herbal remedy in Chinese medicine, more expensive than ginseng. In Chinese stores it appears in tiny bundles tied with a red thread. In China, the most popular dish of *bupin* is the duck-cordyceps broth. This dish is prepared with a freshly-killed and cleaned old duck. The head is split vertically and then resecured with six cordyceps sealed inside. The duck is then boiled in 1 1/2 gallons of water, some onion and ginger, and simmered for 3 hours. The broth and meat are served to the family at times of seasonal changes, in April and/or November.

Cordyceps Chicken Broth

Ingredients:

    12 cordyceps (rinsed slightly before cooking)

    1/2 chicken (about 2 lbs, cleaned, cut into 4 cm pieces)

    1 piece of ginger about 5 mm thick (shredded)

    1 medium-sized onion (cut into 8 vertical pieces)

    2 qts water

    1/2 tsp salt

To a crockpot of boiling water add all the ingredients. Bring to a boil and keep boiling for 2 hours. Remove the cordyceps prior to serving. Makes 4 servings. This dish is said to be especially good for people with anemia or impotence.

Ganoderma (*Ling-zhi* 靈芝): Ganoderma is a fragrant pore mushroom highly esteemed in Chinese culture. It is regarded as the elixir of life in Taoist practices, and the sign of auspice in art. It is one of the first mushrooms recorded in Chinese literature. Improved micrological techiques have made the cultivation of ganoderma a big success in China and Japan, for national uses and for exportation. Two types of whole ganoderma fruiting body and instant powder in tea-bags with measured daily dosage are available to elderly Chinese Americans. Many people bought it as an expensive Christmas or birthday present for retired parents. Users shared their experiences and eventually the Chinese people believed that ganoderma is a revitalization agent, effective to reduce rheumatic pains, to cure renal disorders and hypertension, to improve circulation, and to preserve a state of physiological equilibrium (homeostasis).

Phytochemists of China and Japan have isolated and identified the following effective principles from ganoderma: acid protease, alkaloids, amino acids, aminoglucose, coumarin, ergosterol, fatty acids, fungal lysozyme, lactone, mannitol, organic acids including fumaric acid and ricinoleic acid, polysaccharides, resin, and water soluble proteins.

Actually, the species of ganoderma (*Ganoderma lucidum* [Leysser ex Fries] Karsten, Figure 16a), cultivated and used in China, is widespread in the temperate zone of the Northern Hemisphere. In the Boston area, it appears on dead stumps of elm, maples and/or others deciduous trees between August and November. A student of a Chinese acupuncturist, Dr. James Su, collected the local specimens, which were distributed among Su's patients. The following recipe was given to me by Mrs. Su who assists her husband in his practice.

Lingzhi Tea

Ingredients:

    1/3 ounce air-dried native ganoderma (cut into pea-sized pieces)

    3 cups water

Boil the water in an enamel or Corningware pot. Add ganoderma and bring to a boil. Reduce heat and simmer until 1 cup of the decoction remains. Drink the liquid warm. To reuse the ganoderma residue, add 3 more cups of water and repeat the procedure. Save the liquid for the evening.

Here is some personal experience on using native and imported cultivated *lingzhi*. Between 1950 and1965 several large American elms were cut down, some along the Divinity Avenue in Cambridge and others in Babcock Street in Brookline, Massachusetts. Between 1970 and1980 there appeared some large fruiting bodies of ganoderma along the sidewalks and paths between buildings, some partially covered by glasses. At that time Dr. Shao-sung Chang Chao, Professor of Psychology (Emeratus), National Taiwan University, used powdered *lingzhi* bags imported from Taiwan regularly. I gave her my collections and she used them as a supplement of the imported material. I kept a few good specimens to be used as examples in my lectures on Chinese plant material used as immunostimulants. The deformed and small ones were cut into slices and cubes and put, a few pieces at a time, with the leaves of Eucommia, fruits of different hawthorns, flower- and fruit-spikes of common plantain to brew my morning tea.

Ganoderma is becoming popular among Americans who choose to take treatments of acupunctuirsts and/or from practitioners of tranditional Chinese medicine. One of these practitioners wrote, "I was using [ganoderma] with almost 40% of my patients." A thorough review of the subject appears in a book, *Reishi Mushroom* (Willard, 1990). In Chinese drug stores and groceries *Ganoderma japonicum* (Fries) Lloyd, a shiny black oblique mushroom on a stalk longer than the cap is sold for for $9.00–10.00 a one-pound bag and *G. lucidum* (Leysser ex Fries) Karsten, with a golden-red powdery cap on very short (hardly any) stalk costs $14.50–19.00 for the same amount. The easy availability and relatively low price may lead the user to over dosage. It should be remembered that ganoderma is toxic and abuse may lead to skin rash, loosened stool, dried mouth or slightly upset stomach.

Silver Ear (*Yin-er* 銀耳): Today, silver ear is the most popular *bupin* used in China, and by Chinese Americans. The Chinese name, *yin-er* means silver ear. It is the fruiting body of a club fungus (*Tremella fuciformis* Berkeley). When fresh, it is jelly-like and transparent. When dried it is ivory white and appears like crumpled paper. Formerly, it was the expensive, elegant *bupin* used by the royal family, government officers, rich merchants, and the elite. As the natural supply diminished gradually, people living in the mountains of northeastern Sichuan and northwestern Hubei began to cultivate it.

In the 1920s the local method of cultivating *yin-er* was to fall the broad-leaved trees that had trunks 5–8 cm in diameter in March. These trunks were sawn into 60 cm long sections that were piled up and covered with straw to "sweat". In June of the same year, the wood sections were laid flat over a bed. During the warmer season of the summer, on a clear day immediately after rain, the owner would go up the hill to pick up the fruiting bodies. The harvest was washed, cleaned, carefully strung on slender bamboo sticks and dried over charcoal fires. The smaller pieces were dried on a bamboo screen. In the past few decades, Chinese mycologists have helped to develop more effective techniques for cultivating *yin-er*. Today, in the *bupin* market of Hong Kong, and in Chinese stores of the world's metropolises such as Boston, New York, Washington, and San Francisco, artistically packed gift boxes of *yin-er* are plentiful.

In the market genuine *yin-er* should have the color of a moderately old ivory. The pieces are hemispherical, with the flat side 6–8 cm across, each with a brown central scar, indicating the point of attachment of the fungal fruiting body to the substratum (the rotten wood). This is an important character for recognizing genuine *yin-er*. In the market, there are artificial products which appear like *yin-er*, but lacks the brown central spot of attachment. Genuine *yin-er* can withstand cooking and simmering for hours and becomes soft with good texture and pleasant taste. Artificial *yin-er* cannot tolerate boiling. The material dissolves in the liquid and it tastes bitterish. I have had both the genuine and the artificial types of *yin-er* and studied the labels on the containers. The boxes packed in Hong Kong by dealers who received the material from China mainland where *yin-er* is cultivated on logs in deciduous forests, contain genuine *yin-er*, while the boxes marked "*yin-er* prepared in Taiwan" has the artificial product.

Yin-er (Tremella) is a mild *bupin* which can be used frequently. Its known chemical composition includes 7–10% protein, 0.6–1.28% fat, 60–65% carbohydrates (including various polysaccharides and mucilage), 4% inorganic salts (including those of sulphur, phosphorus, magnesium, calcium, potassium, iron, and sodium), 2.5% fiber and various vitamins.

Dried *yin-er* becomes soft immediately in boiling water. It is very light, and a little goes a long way. Normally, for 4 people, half of a large lump is enough for a dish. Before cooking, it must be revived by placing the material in a bowl and pouring boiling water over it. After reviving, the soft white "silver ears" must be separated individually from the central brown portion. Anything that does not become transparent after soaking in boiling water must be removed.

Yin-er Chicken

Ingredients:

> 1 piece of dried *yin-er* (8 cm across, revived in 5 cups of boiling water, and let stand until cold. Clean, remove impurities, and drain).
> 4 lbs chicken (washed, cut into pieces, and drained)
> 1 large onion (roughly sliced vertically)
> 1 1/2 cm piece of ginger (sliced thinly and shredded)
> 3/4 tsp salt
> 4 cups water

Boil the water. Add all the ingredients and bring to a boil. Stir and reduce heat to low for 2 hours. Makes 4–5 servings.

***Yin-er* Dessert**

Ingredients:

1 cup revived *yin-er*

1/4 cup sugar

1 Tbsp preserved *gui-hua* (osmanthus flowers, optional)

4 cups water

Combine all the ingredients (except the osmanthus flowers) and bring the mixture to a boil in a Corningware or enamel pot. Turn the heat to low and simmer for 2–3 hours. If the osmanthus flowers are available, stir in just before serving. Makes 4 servings.

As a dessert this dish can be served hot, warm or cold. It is a most common preparation for the elderly, served in the early morning before getting up or at bedtime.

(2) *Preparation of Bupin with Vascular Plant Products*: All living organisms share one essential quality, i.e. the desire to survive and to live long. To take nutrients and water for survival is necessary for all living organisms, but to select certain items and to develop techniques to render them more palatable is the unique characteristic of human beings. In Chinese culture, the desire for longevity has been a dominant force in the metaphysical (religious) and material development of the people. In these areas the Taoists have had a leading influence. From the very beginning, the Taoists built monasteries in the scenic mountains. They lived close to nature and used what they found in their surroundings. Consequently, they became the first to try many unusual animal and plant products. They discovered the beneficial effects of numerous natural products and they designed and/or invented many delicious foods and beverages. For example, in March 1938, long before the term "Kiwi fruit" was coined in New Zealand for the edible fruit of *Actinidia chinensis* Planchon, I was served a delicious wine made from it by the priests in a Taoist monastery of Qingcheng Shan 青城山, west of Guan-xian (Map, Loc. 38) in Sichuan Province . In its native home, the fruit is called *mao-li-zi* 毛李子 (hairy plum). Local people living in the mountains of western China ate the fruit and gathered it to sell to visitors. Missionaries in western China, in Chengdu (Map, Loc. 35) for example, used to make jam with it before the 1950s. Then, *mao-li-zi* wine was known only in one particular monastery. Now, the Guan-xian local government has patented and bottled *mao-li-zi* wine. Chinese Taoists believed in immortality and looked to nature to acquire longevity. Their influence in Chinese economic botany and culinary art is far-reaching. The first evidence of such influence is found in the *Shen-nong Ben-cao Jing* 神農本草經 (*The Classics of the Goddess of Agriculture*), compiled from traditional practices passed orally from father to son. This work was written down after the invention of the Chinese ideogram, copied as manuscripted, and printed repeatedly, with some alterations, up

to the present. In the volume I of the 1854 edition of this work, there are 125 entries: 18 are minerals including water; 14 are animal products; and 93 are plant materials. Eighty-six of the 93 plant products (92%) are characterized by the phrase "giving light weight and longevity when taken continuously". A majority of these products are still used for *bupin* now. Twenty of them are in the list of *bupin* common in the Chinese stores in Boston. Broomrape (*Cistanche salsa* [C. A. Meyer] G. Beck), eucommia, ginseng, jujube, Job's tears, lotus seeds, lycium berries and euryale seeds are some examples.

Approximately 94% of the common *bupin* products in the Boston area are vascular plants. Many of these products (20%) are cultivated as special crops, for example, the black soybean, Job's tears, Chinese yam, lily bulbs and lotus seed. Others are grown in orchards, such as the apricot seed, longan aril and lycium berries. In China these products are sold fresh for food. Another 20% of the *bupin* products from vascular plants contain a high concentration of aromatic substances so that they can only be used in small amounts as spices. These plants include Chinese angelica, ginseng, and other members of families Umbelliferae and Araliaceae. Two other fruits contain unusual material, which gives the strong sour taste in *mu-gua* 木瓜 (*Chaenomeles sinensis* [Dumonat de Courset] Schneider), and the extremely sweet taste in *luo-han-guo* 羅漢果 (*Siraitia grosvenorii* [Swingle] C. Jeffrey). Although there are many *bupin* products available in the markets, the majority of consumers use only a few which are usually the mild ones. For special needs, most people depend upon *bupin* mixes which contain the known ingredients and have cooking directions on the packages. The following accounts are recorded from personal experiences.

(a) *Selected Underground Bupin Products*: In the Boston area, the consumption of *bupin* comes largely from several root crops. The use of ginseng, Yunnan *san-qi*, astragalus (*huang-qi*), Chinese angelica (*dang-gui*), codonopsis (*dang-shen*), Chinese cornbind (*He-shou-wu*), Chinese foxglove (*di-huang*), jade bamboo (*yu-zhu*), northern *sha-shen*, peony root, detoxified Sichuan aconite, and Sichuan lovage are discussed here.

Ginseng (*ren-shen* 人參): The origin of the term ginseng is the French romanization of the Chinese vernacular name *ren-shen* 人參. In Chinese ethnomedicine, many fleshy underground growth (roots and stems) with tonifying effects are called "shen" 參, which was interpreted as the essence of earth good for the health of man (Hu, 1976). In ginseng, the first sound "ren" means "man". The original Chinese name for ginseng means the essense of earth dwelling in a root which have the form of a man. Believing in the Doctrine of Signature, from time immemorial, the Chinese people have regarded ginseng as the tonifying agent for man and it has been one of the most expensive herbal drugs in China before the 1980s.

Market products of ginseng are prepared from roots of two species in the genus *Panax* L.. The Chinese and Korean products are all prepared from the root of *Panax ginseng* C. A. Meyer, a tetraploid species native to eastern Asia. The American ginseng constitutes the root of *P. quinquefolium* L., another tetraploid species growing in eastern North America. Both species are perennial, aromatic herbs growing in the partially shaded broad-leaved deciduous forests, with fleshy white tap roots, short underground stems (rhizomes), one palmately compound aerial leaf (sometimes two leaves in cultivation), small greenish flowers in a terminal simple umbel and pea-sized red fruits. On account of continuous gathering of the roots for *bupin*, both species are on the endangered list of plants in their respective native areas. Actually, *P. ginseng* C. A. Meyer is hardly known in nature now. All the commercial products of Chinese and Korean ginseng are from cultivated (or partially cultivated) plants. In the cultivation of ginseng in America, fungicides and chemical fertilizers are used to insure crop success. The Chinese and Korean growers emphasize crop rotation. After a crop of ginseng is harvested, the field is not used again for growing ginseng for 10–15 years to get rid of the harmful soil fungi which cause root rots.

Before 1950, people of the international scientific community assumed that the use of ginseng and the medical credits given it were Chinese superstitions. However, after the World War II, improved research facilities have enabled economic botanists, phytochemists and medical scientists to isolate and identify the active components in ginseng, to observe the efficacy of its extracts for promoting cell division and prolonging life in laboratories, to test the capacity in reducing the effects of stress and strain in men as well as with experimentory animals and to make clinical observation on its effects on the central nervous system, on the cardiovascular and hematological systems, and on the endocrine system.

Between 1965 and 1985 several international ginseng symposia were held in Europe, Korea and Singapore. Botanists, biochemists, phytochemists, pharmacologists, physicians, plant physiologists and people working with tissue culture were invited to report the results of their studies of ginseng. Voluminous information was published after each symposium, ellucidating scientific truth backing what was before assumed to be myth created through Chinese experiences.

One of the interesting reports of the 1980 symposium held in Seoul, South Korea, was given by Dr. Ronald K. Siegel, Department of Psychiatry and Biobehavioral Sciences, School of Medicine, University of California, Los Angeles. He presented the result of a study of ginseng abuse syndrome among American young people who try to get "high" by taking large doses of ginseng powder. He observed that abusers of ginseng suffered from hypertension, nervousness, sleeplessness, skin eruptions, edema and morning diarrhea (Siegel, 1979 and 1980). Ginseng abuse syndrome is not known in the Orient. In China, the amount of ginseng used has been restricted by its price which has been prohibitively expensive, but by the American economic conditions and standard of living, the cost of ginseng fails to become a limiting factor. It must be remembered that ginseng is toxic, and too much ginseng is harmful to normal health.

International ginseng scientists discovered that the active principles in ginseng are different glycosides (ginsenosides). Like vitamins, there are many kinds of ginsenosides, each effects the users differently. Formerly, experienced Chinese ginseng users have recognized the different effects of Asian and American ginsengs and characterized the former as "warming" and the latter as "cooling". Today we know that the Chinese-Korean ginseng contains a stimulant ginsenoside while the American ginseng is rich in a tranquilizing ginsenoside. On account of the high price of ginseng, most of the Chinese consumers are people who have attained certain social and economic status that they can afford to buy the material. People of this age group prefer American ginseng because the Chinese and Korean ginseng cause insomnia.

The United States has exported American ginseng to China for 150 years, first wild ginseng and now largely cultivated material. Today, most American and Canadian ginseng growers send their dried roots to Hong Kong where they are sorted, trimmed, and graded by dealers prior to distribution for international trade as well as for importation to mainland China.

In the Boston area, the 1987 market prices for the top grade wild and cultivated American ginseng are $675 and $120 respectively, a difference of $500 per pound. For practical reasons users and dealers both want to know the distinguishing characters of the genuine wild ginseng and the cultivated materials. In 1980, I received a letter from a government drug inspector in Guangzhou asking for information for distinguishing wild and cultivated ginseng. In 1986 a similar request was received from an officer of the government of Singapore which received ginseng from China, Korea and USA via Hong Kong. Before the World War II, when the consumption of ginseng was limited to governmental officers, rich merchants and a few elites, business transactions of ginseng were in the hands of special dealers who formulated certain criteria for distinguishing the few kinds of product then available (Hu, 1976, 1979, 1980a and 1980b). Now, cultivated ginseng from East Asia and North American flood the market, in groceries as well as in herb shops. This condition has created problems for governmental agents in charge of trade regulations and the consumers who want to recognize the quality of the ginseng that they purchase in order to get their money's worth. For distinguishing the ginseng of Chinese and Korean origin from that of the American products, the size of the root and the presence of *lu-tou* 蘆頭 (persistent stem) are two good characters. In general, Asian ginseng farmers sow the seeds, transplant the seedlings when they are one-year old by hand, set them 8–12 cm apart and harvest them 4- to 6-years old, keeping the short rhizome (*lu-tou* 蘆頭) on the root. The American producers sow the seeds closely by machine, omit the transplanting, harvest the root 3- or 4-years old, with the rhizome removed with the aerial portion. Obviously, the market material with large roots each having a *lu-tou* is the Chinese or Korean ginseng and those with small roots without any *lu-tou* is the American material.

The distinguishing characters of ginseng are found both in the stem and the root. A genuine wild ginseng has a slender, zigzag stem bearing 10–30 prominent leaf-scars and a terminal bud. The root is relatively short, with stout branches marked with numerous horizontal lines (Figure 13f). In cultivation the ginseng field is carefully prepared with the application of fungicides and fertilizers (usually chemicals). The ginseng plants are made to grow faster and bigger than those in the wild population. They are harvested between 3 and 6 years. Under this condition, the stem of a cultivated ginseng is relatively short, stout, and with 3 to 6 leaf-scars terminated by 1 or 2 buds. The root of cultivated ginseng has relatively fewer horizontal lines (Figure 13g). The center of a horizontal line on the ginseng indicates the position of a small lateral root. The older the plant with the least disturbance, the more lateral roots and the more horizontal lines. The Chinese and Korean ginsengs are transplated when the roots are one-year old. They usually have long and stout roots white (due to bleaching) or brown (red ginseng, steamed or parboiled before drying). The American ginseng are short and small.

In nature, ginseng roots branch profusely and the tertiary and smaller branches tend to grow toward the surface of the soil for better aeration, drainage, and nutrient intake. During harvest, these tender rootlets are left in the soil. Due to the ancient Chinese belief in the Doctrine of Signatures and the Taoist teaching that a ginseng root represents the vital spirit of the earth and the crystallization of the essence of nature in the form of a root, which sometimes has the feature of a man, ginseng farmers and dealers have tried to capitalize on the "human-like" appearance of the roots. In transplanting the seedlings, ginseng farmers remove the lateral roots and place the year-old root at a 45° angle so that the growing ginseng will resume the figure of a bending man. More procedures are taken by dealers in trimming, tying, steaming, bleaching and even sugaring the roots before drying completely (Hu, 1976). Most American and Canadian growers send their dried ginseng roots to Hong Kong, where they are sorted, trimmed, and graded by dealers prior to distribution.

In Korea, fresh ginseng is available in the groceries. Some restaurants specialize in serving food cooked with ginseng. Taken as a food supplement, and with the amount eaten each time, fresh ginseng is not toxic. Actually, in the banquets of the International Ginseng Symposia at Seoul, sliced fresh ginseng roots were served just as carrots and celery served in America.

Ginseng Chicken Broth

Ingredients:

> 4 or 5 lb chicken (rubbed inside and out with salt, then washed out the salt and drain)
>
> 1 medium-sized or large ginseng (red or white, or 2 American ginseng, rinsed prior to cooking)
>
> 2 large onions (peeled and cut into halves)
>
> 1/2 cup lycium berries (rinsed and drained)
>
> 1/2 cup Job's tears (cleaned, washed, and drained)
>
> 1 cm ginger (sliced and shredded)
>
> 1 tsp salt
>
> 1 1/2 gallon water

Boil the water in a large container. Add the whole chicken and the other ingredients. Bring the mixture to a boil and reduce the heat. Simmer for 3 hours. Turn the chicken occasionally to cook evenly. Serve hot. Makes 8 servings.

Since much of the flavor is in the broth, make a mild sauce with 3 tablespoons of soy sauce, 1 teaspoon sesame oil, 1 finely chopped scallion, 1 crushed clove garlic, some finely shredded ginger and 1/2 teaspoon vinegar to be used as a dip to improve the taste of the meat.

Yunnan *san-qi* 雲南三七: The root of a special ginseng cultivated in Yunnan is known as *san-qi* 三七 (three-and-seven). It has become increasingly popular recently. In the market, it appears either whole as a small black top-shaped solid nut, 2–3 cm long, 1.5–2 cm across the top, or in 2 mm thick vertical slices. People in Yunnan have designed a special shiny maroon globular thick-walled crockery called *qi-guo* 氣鍋 (steaming crock) for cooking chicken with *san-qi*. This is a two layered structure like a Chinese *huo-guo* 火鍋, except its inner layer is an inverted funnel with a small opening 1.5 cm in diameter on top. A *huo-guo* is always made of metal and it cooks with burning charcoal while a *qi-guo* is made of pottery and cooks by steam. The material to be cooked is placed in the space between the layers of the crockery which is fitted on top of a boiler. Steam comes up the funnel, is forced down, and drops over the meat and cooks it. The finished dish is juicy, with the tender chicken and delicious soup flavored by *san-qi*.

Steamed *San-qi* Chicken

Ingredients:

4 chicken legs (cut into joints)

1 whole *san-qi* (or 6–8 slices of *san-qi*, available in Chinese stores)

1 medium-sized onion (cut into 8 vertical slices)

5 mm thick fresh ginger (thinly sliced)

1/2 tsp salt

1 tsp sesame oil

1 Tbsp soy sauce

Combine all the ingredients except the sesame oil and soy sauce. Place the mixture in the space between the two walls of the *qi-guo* and put on the cover. Fit the *qi-guo* to the mouth of a pot containing boiling water, leaving a space of 8–10 cm between the surface of the water and the bottom of the *qi-guo* to insure the movement of steam to the funnel of the crockery. Place an old dish towel between the boiler and the *qi-guo* to prevent the escape of steam. Cook on high heat for 15 minutes. Turn off the heat for 5 minutes. Take out the *san-qi* and slice the softened root. Return the slices to the *qi-guo*, cover and cook on medium-high for 30 minutes. Turn off the heat for 5 minutes before opening the *qi-guo* to avoid being hurt by the steam. Add sesame oil and soy sauce. Serve hot. Makes 4 servings.

Chinese angelica (*dang-gui* 當歸): *Dang-gui* is the root of *Angelica sinensis* (Oliver) Diels. It is the foremost herbal drug consumed in China. An analysis of a random selection of 100 recipes from the *Handbook of Chinese Pharmaceutics for Traditional Chinese Medicine* shows that both in frequency and in quantity, *dang-gui* is used more than any other medicinal product. *Dang-gui* is an aromtic, perennial herb which attains a height of 1 meter when in flower. The aerial portion dies back annually. The basal leaves are triangular-ovate in outline, ternately decompound, and the ultimate segments are ovate, lobate, and serrate. The best market products are cultivated in western Sichuan and the adjacent areas of Gansu and Shaanxi. In the Boston market, whole roots as well as trimmed and sliced products are available (Figure 15c).

In traditional Chinese medicine, *dang-gui* is said to have a warming property, used to enrich the blood and to improve circulation. Chinese and Japanese phytochemists have identified its contents as: volatile oils, butyl alcohol, sesquiterpenes, palmitic acid, stearic acid, unsaturated fatty acids, linoleic acid, β-sitosterol, sucrose, Vitamin B12, butylidene phthalide, n-Valerophenone-o-carboxylic acid, 2,4-Dihydrophthalic anhydride, folic acid, and nicotinic acid. Most of my friends take *dang-gui* occasionally as a food substitute. It can be cooked with chicken or spareribs.

Chinese Angelica Chicken Broth

Ingredients:

>   3–4 lb chicken (washed, drained, and cut into desired pieces)
>   1 cm piece ginger (thinly sliced)
>   1 large onion (cut into 8 vertical sections)
>   1/2 cup Job's tears (washed and drained)
>   1/2 cup lycium berries
>   1/2 cup jujube
>   1 *dang-gui* (whole, rinsed prior to using)
>   3 quarts water
>   1 Tbsp vegetable oil

Brown the ginger and onion slices. Stir in the chicken and cook for 1 minute. Add the remaining ingredients and bring to a boil on high heat. Reduce the heat to medium-low and cook for 1 hour. Now the angelica root is soft. Remove it and cut it vertically into 4–6 pieces. Return the pieces to the pot. Continue to cook for another hour. Makes 6 servings. This is a popular dish for postpartum mothers. Many elderly Chinese eat this dish at regular intervals, i.e. once a month. The meat is often dipped in a sauce prepared with soy sauce, sesame oil, chopped scallion and ginger.

Astragalus (*huang-qi* 黃耆): The taproots of several species of *Astragalus* L. are used as *huang-qi*; the common market product is from *A. membranaceus* (Fischer) Bunge. *Huang-qi* (*huangchy* in some older references) is mentioned in *Shen-nong Ben-cao Jing* which indicates that its use can be traced to prehistoric time, as can ginseng and *dang-gui*. The species is widespread in northern, northeastern and northwestern China. Unlike most Chinese species of Astragalus with purplish or bluish-pink flowers, *huang-qi* bears yellow flowers (huang = yellow). The market products are cut either longitudinally into 10–15 cm long sections or slantwise across the root. In recent years, Chinese scientists have identified choline, betaine, 2′,4′-dihydroxy-5,6-dimethoxyisoflavane, kumatakenin and many other chemical compounds in *huang-qi* and used various extracts on laboratory animals. They have found that the water extract of *huang-qi* has a positive diuretic effect on rats and dogs. Other experiments show its efficacy in lowering the blood pressure and the degree of blood sugar. Specifically, chemical components reported from *A. membranaceus* (Fischer) Bunge are coumarin, gallic acid, choline, betaine, several amino acids and different flavonoids. From *A. mongholicus* Bunge, kumatakenin, rhamnocitrin and sitosterol were isolated (Figure 15f).

Summer *Huang-qi* Soup

Ingredients:

 1/2 oz *huang-qi* (sliced)

 1/2 oz *dang-shen* (codonopsis, whole)

 1/4 cup lycium berries

 2 pork chops (washed and drained with bone)

 1 medium-sized onion (cut vertically twice)

 1/2 tsp salt

 5 mm ginger (shredded)

 8 cups water

Rinse the *huang-qi*, *dang-shen* and lycium berries. Combine all the ingredients and bring to a boil. Continue to boil on medium-low heat until 4 cups of soup remain in the pot. Remove the *huang-qi* and *dang-shen*, and eat the remaining ingredients with the soup. Makes 2 servings. This is a soup served frequently in early summer.

Codonopsis (*dang-shen* 黨參): The roots of several species of *Codonopsis* Wallich are sold as *dang-shen* in the Chinese herbal market. Dealers recognize the differences between the various products and label them *dong-dang* 東黨 for the material produced in northeastern China, *xi-dang* 西黨 for the material produced in Shaanxi and Gansu, *Lu-dang* 魯黨 for material produced in the Tai-hang Shan area of Shanxi, *Chuan-dang* 川黨 for products from Sichuan, and *Fang-dang* 防黨 for roots produced in Fang-xian, western Hubei. Consumers do not know of these differences because in the retail market, *dang-shen* is the only name used. The product found in Boston Chinese stores is largely *Fang-dang*. Sharing a common sound with ginseng (*ren-shen*), *dang-shen* is commonly regarded as the poor man's ginseng, and many doctors prescribe it as a substitute for ginseng. Consequently, the use of *dang-shen* reaches more people than that of ginsng both within China and among overseas Chinese *bupin* consumers.

The source species of *dang-shen* are perennial, climbing herbs with aerial stems dying back annually. The plant flowers in mid-August and September and by October the fruits are ripe. Seeds are gathered and sown in March. The seedlings are transplanted in September, October or the next spring. After the plants are old enough to produce flowers and seeds, the roots are harvested in autumn, at the end of the growing season. The harvested roots are dried partially in the sun, and then rubbed three or four times between the hands as they dry to ensure uniform consistency of the finished product. Today, all market products are from cultivated origin.

Historically, comparing with the herbal records of *dang-gui* (Chinese angelica), ginseng, ganoderma, and many other common *bupins*, *dang-shen* is relatively a newcomer in Chinese food and medicine. Its historical records can only be traced to the Qing Dynasty (1644–1908 A.D.). This condition is caused by ethnobotanical, cultural, political and ecological factors. Ethnobotanically, various ethnic groups gathered different species of plants for their own health care and named them independently. In my records there are over one hundred kind of thickened fleshy roots, rhizomes, and other underground growths of plant bearing the term "*shen*" (參). *Chuan-shen* 拳參 (fist *shen*) for *Polygonum bistorta* L., *tai-zi-shen* 太子參 (prince *shen*) for *Pseudostellaria heterophylla* (Miq.) Pax, *tu-ren-shen* 土人參 (local ginseng) for *Talinum paniculatum* (Jacq.) Gaertner, *dan-shen* 丹參 (scarlet *shen*) for *Salvia miltiorrhiza* Bunge, *sha-shen* 沙參 (sand *shen*) for *Adenophora stricta* Miq., and *pan-long-shen* 盤龍參 (coiling dragon *shen*) for *Spiranthes lancea* (Thunb.) Backer, are some examples. Among all the "*shens*" recorded in Chinese medicinal plants, only *dang-shen* is prescribed as a cheaper substitute for ginseng. This happened only after the seventeenth century when the ginseng producing area became the imperial property of the Qing (Ch'ing) Dynasty (1644–1908 A.D.), with the hillsides systematically guarded by the army which gave severe punishment by cutting off the hand or foot in case of any theft. Then, the annual harvesting was done and the distribution of the produce fell into government management. Ginseng became very rare and expensive. Practitioners began to prescribe *dang-shen* as a substitute for ginseng.

In the 1960s, Chinese pharmaceutical scientists began to investigate the chemical composition of *dang-shen* and the physiological effects of the drug on laboratory animals. Alkaloids, sucrose, glucose, insulin, starch, mucilage and resin were identified. From the *dang-shen* produced in Sichuan and Hubei, scutellarein glucoside and volatile oils have also been identified.

In traditional Chinese medicine, *dang-shen* is prescribed for tonifying the interior (*wu-zang* 五臟) and for improving vitality (*yi-qi* 益氣). Active research has been carried out in these areas. Alcohol and water extracts of market *dang-shen* were administered orally or by subcutaneous injection to normal rabbits. Increases in red blood cells and hemoglobin were evident. Similar experiments on rabbits whose spleens had been removed failed to show the same results. Investigators concluded that *dang-shen* improves the blood of animals through its effect on the spleen.

Alcohol and aqueous extracts used for intravenous and peritoneal injections of dogs under anesthesia decrease the blood pressure. Abdominal subcutaneous injections of these extracts increases the blood sugar in animals. When the extracts have been fermented and used for similar experiments, no effect on the blood sugar was evident. Extracts prepared from *dong-dang*, i.e. *dang-shen* produced from the Chinese Northeast, show no increase in blood sugar. The investigators thus concluded: (1) the sugar content of some *dang-shen* can increase blood sugar and give strength to the consumer, and (2) the *dang-shen* from northeastern China does not contain enough sugars to give similar benefit to the user. In the market, *dang-shen* from northeast China appears lighter in color (yellow) and in weight, while that produced in Gansu, Shaanxi, Shanxi, Hubei and Sichuan appears grayish yellow and heavier in weight. The Boston market product labeled *Fang-dang* represents products from Sichuan and Hubei, and has an aromatic smell, heavy weight, and good quality. It is the root of *Codonopsis tangshen* Oliver, a species known only in cultivation.

The consumers in the Boston area consider *dang-gui* (*Angelica sinensis* [Oliver] Diels) and *dang-shen* (*Codonopsis pilosula* {Franchet] Nannf., *C. tangshen* Oliver) of equal value for tonifying the blood and preserving the well-being of a person. In general practice *dang-gui* is prescribed for younger persons, particularly for women during menopause, and *dang-shen* for the elderly people. In preparing *bupin* with *dang-shen* the roots are never eaten. They are treated as soup bones, boiled in the stock and discarded before serving.

Codonopsis Broth

Ingredients:

    1/2 oz *dang-shen* (rinsed before cooking)

    1/2 cup Chinese yam (dried slices, available in Chinese stores)

    1/2 cup lycium berries (dried, color red, the old product with black color is not
      good)

    1/2 cup dried red jujube

    1/2 oz *huang-qi* (astragalus)

    1 1/2 lbs spareribs (cut into 3 cm pieces)

    4 quarts water

    1 tsp salt

Wash the dried material and drain. Combine the ingredients and bring to a boil. Simmer for 3 hours. Drink the broth and eat whatever can be chewed. Makes 4 servings. Many people take this broth regularly, once or twice a month.

Northern *Sha-shen* (*bei-sha-shen* 北沙參): The fleshy root of an umbelliferous seashore plant, *Glehnia littoralis* F. Schmidt ex Miquel, is used by its coexisting Chinese people living on the coast of the Yellow Sea as *sha-shen* 沙參 (sandy shore *shen*) since time immemorial. It has been cultivated to supplement the natural supply for local use. Recently, encouraged by the cooperative efforts of governmental institutes on seashore reclamation and on production of articles for exportation, the species is extensively cultivated on the shores of Hebei, Jiangsu, Liaoning and Shandong. When the crop is harvested the fresh root is washed, cleaned and dipped into boiling water. Then the outer skin is scraped off and the root partially dried, cut into 8–10 cm sections, and finally completely dried for the market.

Soon after the normalization of diplomatic relationship between China and USA, large amounts of *bei-sha-shen* 北沙參 (northern *sha-shen*) was imported into America from China. I was approached to give the botanical identification of the source species. Afterward, the specimens was kept for future reference. This material is very rich in starch, and after keeping for a while, the outer portions disintegrated into powder.

Northern *sha-shen* is a mild *bupin*. An alcohol extract given to rabbits lowers the body temperature with some analgesic effect. Water extracts increase the blood pressure in animals.

Cooling Summer Refreshment

Ingredients:
    1 cup rice
    1/2 cup glutinous rice
    1/2 cup lily bulbs
    1 cup dried red jujube
    3/4 cup mungbean
    1 cup northern *sha-shen* (dried Glehnia root)
    1/2 cup jade bamboo (*yu-zhu* 玉竹)
    1 1/2 gallons of water
    1/2 cup brown sugar

Wash and drain all the dried ingredients. Boil the water in a large kettle. Add the ingredients and bring to a boil. Lower the heat and simmer for 3 hours. Makes 10–12 servings. Have some extra brown sugar in a dish for those who like their refreshments sweet.

Chinese cornbind (*He-shou-wu* 何首烏): There are many Chinese plant names attached to legendary stories. *He-shou-wu* is one of these, and the story about it goes like this. Once there was a poor sick man known only by his family name, He (a century ago, poor insignificant people and women in rural China did not have personal names). In a famine, when the able-bodied of the area migrated to places where they could escape starvation through temporary jobs and/or by begging, he remained in the village, gathering wild plant material to assuage his hunger. One of the items he ate was a thick root, astringent and bitterish, growing on the hillsides, not yet having a name because it had not been used by the people. On eating this root, he did not only escape starvation, but also acquired improved health. By continuing to use it even when food became plentiful, his health was restored, his complexion brightened, and his hair gradually turned black (a sign of youthfulness). Eventually, he enjoyed a long life of strength and vitality, when neighbors knew what happened to him, they too tried the root, and called the plant *"He-shou-wu"* after him and his black hair (He 何 = the surname; *shou* 首 = head; *wu* 烏 = crow or black).

He-shou-wu (*Polygonum multiflorum* Thunberg) is a perennial twiner with very long reddish brown roots thickened and becoming fleshy in the middle. The plant can be recognized by its glabrous herbaceous stem, with alternate leaves on long petioles attached to membranous sheathy stipules (ocreae) 5–7 mm long. The laminas of the leaves are ovate-cordate, 7–8 cm long, 2–5 cm wide, entire and glabrous. The numerous flowers are small , 1 mm long, greenish-white and in terminal panicles. After blooming, the persistent sepals increase in size, becoming white and showy, with the outer ones winged. The fruits are shiny black, trigonous, and 2–3 mm long. It has been introduced into USA and it becomes showy in September in Boston area.

In China, before the 1930s, the species used to be common, growing on open hillside or climbing over thickets of villages, widespread along the Yangtze River region and thence southward to Guangdong and Guangxi and eastward to Japan. For medicinal purposes people merely gathered the wild roots in autumn, parboiled and dried them for the market.

Now, natural supply has dwindled and the market material is from cultivated sources. The species may be propagated vegetatively by small root tubers in late February and early March, or by seeds sowed between mid-March and early April. Seeds germinate 20 days after sowing. When the seedlings are 15 cm tall, they are transplanted. Seasonal care and adequate fertilizer are required for a good crop. The crop is harvested after 3 or 4 years , in autumn or early spring before the new shoots emerge. After digging, the roots are trimmed (with the root-lets and the thinner end-portions removed), washed, the larger ones cut into sections, and dried, parboiled or not. Fresh roots are 10–15 cm long, 3–12 cm in diameter, pinkish-brown or brownish-yellow on the outside, and whitish and mealy inside. Apparently some whimsical farmers made models and transplanted some young tubers into them in order to harvest some fanciful products assuming human forms. In the early 1980s a herbalist in the New Territories of Hong Kong bought two of such creations in Guangzhou and brought them to me in the Herbarium of the Chinese University of Hong Kong for identification. Such fancy forms have appeared also in some Chinese newspapers.

The market material contains spindle-shaped whole tubers or irregular pieces of the large roots, wrinkled, brown, bitterish and astringent to taste, and odorless. The *bupin* material in Boston stores are in slices, prepared by placing 12 kg of dried *He-shou-wu* in a mixture of 12 kg rice gin and 12 kg black soybean juice in a closed earthen vase and heated in a double boiler. The black soybean juice is prepared by boiling 10 kg black soybean in water for four hours, strained, and the residue is again boiled and strained to make up a total of 12 kg of juice. When the mixture is completely absorbed by the dried *He-shou-wu* the roots become soft and are sliced and dried for the market. Material so cured and sliced appears brownish black.

He-shou-wu was first recorded in Chinese herbals of the 10th century. Its bitterish tastes leaves a sweet aftertaste, a character which is interpreted in traditional Chinese medicine as a sign of warming. Small amount is nontoxic. It is used for premature graying of the hairs, for tonifying the imbalanced function of the liver, strengthening the kidneys, enriching the blood, and alleviating rheumatic pains. Between 1959 and 1977, Chinese medical scientists investigated the effect of *He-shou-wu* on cardiovascular diseases, reported clinical evidences of its efficacy for lowering blood pressure and for reducing the serum cholesterol level. Such news encouraged many communes to turn the rice fields into *He-shou-wu* plantations. Meanwhile, herbalists in North America publicized the material under the name *"fo-ti"* and middle aged people suffering from loosing their hair are buying *He-shou-wu* to make tea.

Having been interested in the economic botany of China, I naturally tried to use all available opportunities to investigate the production of *He-shou-wu*. In my lecture tour at Nanjing, September 1984, I heard about the cultivation of *He-shou-wu* in Central Jiangsu Province (Long. 120°E, Lat. 34°N). Professor F. X. Liu, a Botanist of the Jiangsu Institute of Botany, Dr. C. S. Zhong, Professor of Botany, Nanjing University, and two officers of the Provincial Government, accompanied me to visit a farm in Bin-hai District in Jiangsu. First, we were shown the construction of a building for making alchohol from the byproduct of *"He-shou-wu"* production. Then we were taken to the field. I was surprised to find that the crop locally called *"He-shou-wu"*, was not even a species of *Polygonum* L.. Actually, it was a species in family Asclepiadaceae, *Cynanchum wilfordii* (Maximowicz) Hemsley (S. Y. Hu 19782). This species is called *ge-shan-xiao* 隔山消 in the Chinese botanical literature *Iconographia Cormophytorum Sinicorum* (ICS 3: 484. Figure 4922. 1974). Actually, in northern China, particularly in the homeland of Confucius in Shandong Province, the root of *Cynanchum bungei* Decaisne, locally called *bai-shou-wu* 白首烏 (white-headed crow) and *di-hu-lu* 地葫蘆 (underground Lagenaria) has been recorded as *Tai-shan-he-shou-wu* 泰山何首烏, the *He-shou-wu* of the Sacred Mountain Tai-shan (ICS 3: 475. Figure 4903).

Obviously, in the Chinese market there are different plant products which carry the same name, *He-shou-wu*. The one with limited local use in Shandong and adjacent

Jiangsu is from the root of a different species of *Cynanchum* L.. The other one which has widespread use throughout China and abroad is the classic *He-shou-wu* prepared from the root of *Polygonum multiflorum* Thunb.. It is worthy of a note that the Cynanchum product prepared by the communes in Central Jiangsu and Shandong where the Jiangsu people obtained their stocks, is a white starch precipitated from the freshy root of *Cynanchum wilfordii* (Maximowicz) Hemsley. It is used as a *bupin* in a similar manner as are the starches obtained from the rhizomes of the sacred lotus and that of the corms of water chestnut. To avoid further market confusion, the Cynanchum product should carry the market name *"bai-shou-wu"* 白首鳥. I was informed that the residue from the starch extraction is used for making gin.

In l984, I learned another lesson about the production of *He-shou-wu* in China. For slightly over one thousand years since *He-shou-wu* was first recorded in a herbal of Sung Dynasty, the market material depended upon natural supply. Pharmacological discoveries of the active principles for lowering the serum cholesteral and increasing the coronary flow and pharmaceutical preparation of tablets and injections between the late 1960s and early 1970s suddenly exhausted the natural supply. Under government encouragement, a commune in Zhaoqing District of Guangdong Province (Long. 112°30'E, Lat. 23°N), which was famous for *He-shou-wu* product, converted the rice fields into *He-shou-wu* plantations. By early 1980s the harvest of *He-shou-wu* overflowed the market and the stock piled up in storage, which forced the farmers to stop *He-shou-wu* cultivation. This condition was unknown even to the people in Guang-zhou (Canton). In October 1984, when I lectured in South Chinese Agricultural University, the university administration tried to make arrangement for me to visit the famous area and to see *He-shou-wu* cultivation in Zhaoqing. At first, the university officer was told that on account of holding an American passport, I could not go to Zhaoqing since it was not yet open to international visitors. After much deliberation, he was told the truth: the communes in Zhaoqing had converted the *He-shou-wu* farms into rice fields again. Nevertheless, there is still sufficient *He-shou-wu* for consumption in China and for export, for the material in Chinese stores in Boston is the genuine preparation of the root of *Polygonum multiflorum* Thunb..

Chinese phytochemists have worked on the chemical contents of *He-shou-wu* and found 45.2% of the root is starch, 3.1% lipids, 3.7% lecithin. The acitve principles consist of anthraquinones, particularly chrysophanic acid, chrysophanol, emodin, rhein, and anthrone. Heat destroys rhein.

As a revitalizing agent, *He-shou-wu* may be used in combination with other *bupin* material, such as ginseng, *dang-gui*, *dang-shen*, etc. , or used alone. In a catalogue of 98 Chinese revitalizing agents and tonics, *He-shou-wu* appears in 8% of the pills prepared by multiple recipes each containing from 7 to 40 ingredients. One item called *shou-wu pill* 首烏丸 contains *He-shou-wu* alone. This pill is prepared for people who feel fatigue, lack of energy and memory, diziness, insomnia and tinnitus. A simple recipe is given below.

He-shou-wu Tea

Ingredients:

    1/2 cup *He-shou-wu* slices (available in Chinese stores, rinsed and drained)

    12 dried red jujube (rinsed and drained)

    2 Tbsp brown sugar

    10 cups water

Combine all the ingredients. Bring to a boil in a enamel or crock pot. Continue boiling on medium-low until 6 cups of liquid remains. Makes 6 servings. Take a cup daily before bed time if preferred. Eat 2 jujube with each cup of tea.

Chinese foxglove (*di-huang* 地黃): Those who are familiar with the western European common foxglove (*Digitalis purpurea* L.) can readily see the similarities between this species and its eastern Asian counterpart, *Rehmannia glutinosa* (Gaertner) Libosch.. Actually, the European foxglove has been introduced into China and it is called *mao-di-huang* 毛地黃 (hairy *di-huang*). Comparing the two species, the Chinese foxglove is smaller, with glandularly hairy leaves, and few purple flowers on a scape. The species is native to northern China, where it grows in waste areas, on city walls, and on hillsides along dusty highways, where the spring is dry and windy, and the summer is hot and rainy. It is a perennial herb, the aerial portions die back after flowering and fruiting, leaving the stout, ivory rootstock to over winter in the soil. Since time immemorial, this rootstock has been dug in late autumn or early spring and used as a hematinic agent called *di-huang* 地黃 (earth yellow). Children familiar with the nectar of the flowers usually pick them, remove the receptacles and suck the sweet juice at the base of the flower tubes. To them the plant is called *mi-guan-ke* 蜜罐棵 (honey-bottles herb).

Today, the supply of *di-huang* is from cultivated plants. The crop is propagated vegetatively. The rootstocks of the best plants in the field are dug up, cut into 3–6 cm sections, and planted immediately in a prepared bed with holes 6 cm deep, 10 cm apart, and in rows 17 cm parallel. In late October, the stock is dug out and burried in sand until next April or May when each piece is cut into 5–6 cm sections and planted in permanent fields in 40 cm rows, 30 cm hills and covered with earth 5–6 cm thick. The crop requires proper care in watering, weeding, hoeing the soil and seasonal application of fertilizer. Suckers and flower buds must be removed, leaving one main shoot to grow. The crop is harvested in early November, with the soil and aerial portion removed, and dried with or without washing. The best *di-huang* is produced in Henan Province.

There are two types of *di-huang* in the market, namely, the *sheng-di* 生地 (raw *di-huang*) and *shu-di* 熟地 (cured *di-huang*). *Sheng-di* consists of the dried root of *di-huang* prepared by farmers as described above. It appears grayish brown and is very hard. It is only used in complex formulas prescribed by doctors. *Shu-di*, the cured root of *di-huang* appears black and rather soft. This is the only kind of *di-huang* used as *bupin*, although it is also used in prescriptions. Both types of *di-huang* are available in American Chinese stores.

Detoxification of di-huang: The conversion of the dried hard *sheng-di* into the black soft *shu-di* requires special equipment. The process is performed in China by professional people. The raw material consists of dried *sheng-di* and rice gin in a proportion of 7:3 by weight. The equipment involves a double boiler prepared by fitting a large earthenware urn to a cooking pot. The dried root and gin are placed in the urn, the mouth loosely covered, and the cooking pot is heated for 8 hours. The urn is cooled overnight undisturbed. On the next day the ingredients are turned, cooked again for 4 to 8 hours until the root is uniformly black inside and out. Again the content is cooled overnight undisturbed. Then it is taken out of the urn and dried with frequent turning. When the material is 80% dry, it is cut into 5 cm x 3 mm pieces and then the drying process is completed.

In traditional Chinese medicine, *di-huang* is highly esteemed as an agent for enriching the blood, promoting circulation, giving strength, improving hearing and vision, and changing the color of hair from gray to black, the sign of rejuvenation. It is often used with Chinese angelica (*dang-gui*). By adding 1/2 oz of *shu-di* to the recipes already given for *dang-gui* would make a more wholesome *bupin*. However, in general practice, the consumption of *dang-gui* far exceeds that of *shu-di*. Personally *dang-gui* is always an item in my pantry, while I have never used *di-huang*.

Botanically, *di-huang* and *dang-gui* belong to different families. Chemically, their compositions are very different. *Dang-gui* is rich in volatile oils and unsaturated fatty acids. The primary contents of *di-huang* are: β-sitosterol and mannitol, stigmasterol, campesterol, rehmannin, catalpol, stachyose, arginine, glucose, sucrose, vitamin A, and aminobutyric acid. The following recipe is one easy to prepare for home use.

Homemade *Di-huang* Tincture

Ingredients:

    1 oz *shu-di*

1 oz lycium berries (good orange-red color)

1 sliver aloeswood (*chen-xiang* 沉香)

1 liter (fifth) whiskey

Add the first 3 ingredients to a bottle of whiskey. Wait for 10 days. Take 1 jigger (1 1/2 oz) straight or combined with ginger ale, once a day until finished. Add another bottle of whiskey to the same material and wait for 2 weeks before starting to use the tincture. This beverage is recommended for people suffering from general weakness, accompanied by pains of the side or abdomen, lack of appetite, and particularly for women with irregular menstruation.

Four-item Rejuvenating Tea

Ingredients:

3 oz *shu-di*

3 oz *dang-gui* (cleaned, whole Chinese angelica root)

3 oz *bai-shao* (sliced, white peony root)

3 oz Sichuan lovage (sliced rootstock)

Combine the ingredients and pulverize. Bring 1/10 of the powder to a boil in 2 cups of water. Turn the heat to medium low and continue to cook until 1 1/2 cups of the liquid remain. Take the tea once daily before breakfast.

In the analysis of samples of mixed *bupin* in the Boston market, the first recipe contains the same four ingredients as given in this tea. Evidently, this combination is commonly used for soup when it is cooked with meat or poultry, and for tea when the ingredients are decocted with water alone.

Jade Bamboo (*Yu-zhu* 玉竹): The market material of *yu-zhu* is the dried vertical slices of the slender rhizomes of a liliaceous plant (*Polygonatum odoratum* [Miller] Druce). In American Chinese stores, *yu-zhu* is sold in one-pound bags labeled "Polygonatum". Formerly the rhizomes of several species of *Polygonatum* Miller with annual aerial stems, alternate sessile entire leaves, axillary white belled-shaped hanging flowers and small purple-black berries were collected and used as *yu-zhu* in northern China. Now the market product is prepared from cultivated material grown in Henan, Jiangsu, Liaoning, Hunan and Zhejiang provinces. The crop is propagated vegetatively with fully grown rhizomes selected at the time of harvest in mid-August. For this purpose, the healthy rhizomes with copious roots and a large terminal bud are set aside and planted immediately in 3 cm deep grooves of a prepared bed. In November–December, a fertilizer consisting of human waste is applied in ditches dug between the rows. In the next growing season, similar application of fertilizer is given. As the aerial portions die in cold weather, the plants are covered with a layer of mixture of cow manure and clay 7 cm. In the early spring of the third year, as the new shoots emerge, it is fertilized again with human waste. In the autumn, the crop is ready for harvest.

The harvested rhizomes are dried in shade until a sticky material appears on the surface. They are then shaken lightly in a container to knock off the hair-like fibrous roots. The partially dried rhizomes are sorted by sizes, dried continuously in shade until they turn light yellow. Then they are kneaded between hands repeatedly so that they may dry uniformly smooth and shiny. Sliced jade bamboo is prepared by distributors who revitalize the dried rhizomes with water and cut them longitudinally by hand.

In traditional Chinese medicine, the properties ascribed to jade bamboo are sweet with slight bitter after taste, nontoxic, warming, nourishing the *yin* (vital essence), lubricating dryness, stopping thirst, enhancing wisdom, improving circulation, and producing good complexion, with the lungs, kidney, stomach and spleen being its target organs. Between 1954 and 1977, Chinese medical scientists have identified convallamarin, convallarin, kaempferol glycosides, quercitol, starches (25.6–30.6%), and mucilaginous substance from the rhizomes. In addition, Japanese phytochemists have isolated azetidine-2-carboxylic acid from the leaves. Clinical studies gave favorable reports of its efficacy for lowering the blood pressure by vasodilation and for hyperlipidemia. As jade bamboo is nontoxic, it can be used quite freely in making soup in combination with several *bupins* and spareribs, fish, or chicken. The recipe which I use often is given below.

Jade Bamboo Gruel for Vegetarians

Ingredients:

    1/2 cup jade bamboo (dried slices available in Chinese stores)

    1 cup rice (short grain preferred)

    1/2 cup glutinous rice

    1 cup dried jujube (red type)

    1/2 cup lotus seed (with the seed coat removed)

    1/2 cup lycium berries

    1 cup northern *sha-shen*

    1/2 cup Job's tears (cleaned, washed and drained)

    1 1/2 gallon water

    1/3 cup brown sugar (or more to taste, optional)

Combine the dry ingredients (except sugar), wash and drain. Pour the mixture into boiling water in a large container. Bring to a boil and then keep the material simmering for 3 hours. Add brown sugar (optional) and serve hot. Makes 6 to 10 servings. This is a nutritious gruel which may be prepared at weekends and warmed in small portions for weekdays.

Peony root (*bai-shao* 白芍): Dried slices of the root of the white variety of garden peony are sold in American Chinese stores both in *bupin* mixes and in one-pound bags. The peonies are essentially Old World species with the generic range extending from Algiers and Morocco northward through western Europe to Lapland, then eastward to Asia Minor, the Caucasus, the Himalayan region, China, Siberia, and Japan with two outlying species in western North America, from Washington southward to Oregon, Idaho, Nevada and California. In a revision of the genus *Paeonia* L., Stearn (1945) recognized 34 species, one-fourth of which occur in China. The source species for *bupin* is *Paeonia lactiflora* Pallas which "is a well-known inhabitant of most gardens ... there are many very attractive forms ... all hardy and will grow nearly anywhere" (Stearn, p. 144). In Chinese literature peony, was recorded in ancient Chinese folk songs, *The Classic of Songs* (*Shi-jing* 詩經), compiled before the time of Confucius (Hu, p. 1, 1971).

The present market supply is from cultivated material produced primarily in Zhejiang, Anhui, and Sichuan. The crop is propagated in early October by crown division and it requires much labor in transplanting, trimming the secondary and tertiary roots, applying proper fertilizers, and giving timely care. The crop can be harvested in summer or autumn when the plants are 3 or 4 years old. After the roots are cleaned and washed, the outer coarse skin of each root is scraped off with a bamboo knife. Then the roots are parboiled until flexible for improving the drying quality. After having been dried in the sun, the roots may appear brown, pink, or white, depending upon the depth of the skin removed in the scraping process. Slicing is done by distributors who first soak the dried roots in water until 80% of the outer tissues become hydrated, then leave the soaked roots in a container until every one is uniformly damp and flexible. Slicing is done by hand and the slices are dried in the sun. For prescriptions other than being used as *bupin*, peony root may be mixed with rice gin (in a proportion of 100 parts root to 20 parts gin by weight), steamed soft and then sliced, and roasted to brown or charred black, or stained with the fine powder of clay baked in an old fashioned stove.

In traditional Chinese medicinal practice, the target organs of peony root are the liver and spleen. It is supposed to soften the hardened liver tissue, to enrich the blood, to enhance the activities of the kidneys, and to benefit the urinary system. Modern Chinese medical scientists have made phytochemical analysis and clinical observations of the peony root. They have reported paeoniflorin, paeonol, paeonin, benzoic acid (*ca.* 1.07%), volatile oils, fatty oils, resins, tannins, various saccharides, starches, proteins, β-sitosterol, and triterpenes from peony root; and astragalin, kaempferol 3,7-diglucoside, gallotannin, 13-methyl tetradecanoic acid, and β-sitosterol in the flower; and tannin in the leaves. In general practice, peony root, Chinese angelica, Sichuan lovage, and Chinese foxglove root are esteemed as the "four precious" blood tonifying agents. Ready made *bupin* mixes with these ingredients are available in American Chinese stores.

Four Precious Tonifying Soup

Ingredients:

    1/10 oz sliced peony root (in ready mixed combination)

    1/5 oz Chinese angelica (a whole one or sliced)

    3/10 oz Sichuan lovage (thinly sliced)

    1/3 oz Chinese foxglove (black mass, already steamed and dried)

    6 oz spareribs

    1 medium-sized onion (sliced into 8 vertical sections)

    8 cups water

Combine all the ingredients and boil the mixture in a crockpot or any other non-metallic cooking ware. Cook slowly for 2 or 3 hours. Increase heat to boil off some liquid until 4 cups are left. Makes 2 servings. Use some soy sauce, ginger, garlic and sesame oil to make a dip for the spareribs.

Sichuan lovage (*Chuan-xiong* 川芎): Sichuan lovage is the rootstock of *Ligusticum wallichii* Franchet cv. 'Chuan-xiong' Shan. It is esteemed as one of the "Eight Precious Items" (八珍) in *bupin*. In Chinese literature, it is one of the first recorded aromatic herbs, called *mi-wu* (蘼蕪), gathered from the hillsides. It appeared in *Er-ya* 爾雅, the first Chinese encyclopedia on natural and cultural objects, and in ancient poems. From the folklore of northern China, we learn that it was used fresh for flavoring food and/or dried for making sachets to scent clothes. According to my investigation, the species is known only in cultivation. Personally, I have visited farms cultivating the crop in Guan-xian (Map, Loc. 38), as early as the spring of 1939, but I have never seen any wild plants in China, nor any herbarium specimens. It is not represented in the rich collections of Chinese specimens in the Harvard University Herbaria. Currently, all the market material is from cultivated sources, particularly from Sichuan, hence its market name *Chuan-xiong* (= Sichuan lovage).

Sichuan lovage is a strongly aromatic perennial herb with a short, stout, coarse, and knotty rootstock, carrot-like decompound leaves, all basal when the plant is young and small white flowers. As the object of the crop is to produce large rootstocks, plants are seldom allowed to flower. The crop is propagated vegetatively with healthy plants dug in February (about four months before harvesting). The stock is then taken up to the mountains, and planted immediately in a prepared flat bed (size 2.7 x 3 m) of clay soil under partial shade. The larger ones may be devided into halves before setting. The plants are thinned in March, leaving 8 to 10 leaders to each hill. In mid-July, on a cloudy day or in an early morning, the plants are dug out. After the leaves and rhizomes are removed, the rootstocks with the stems are tied into bundles, carried into a cool cave or a hut, laid on the straw and covered completely with straw until August, aerating once by moving the bottom bundles to the top. In mid-August, the seasoned stems are taken out, cut into 3 cm sections, each containing one good node with a strong bud. These sections are planted in properly prepared beds on land normally used for rice. The crop requires frequent weeding and heavy fertilization during the growing season, and as the cold weather sets in the dead leaves must be removed and the rootstocks protected by a mixture of fine soil and compost. The crop is harvested between late May and mid-June in Sichuan. The crop requires time, labor, good land and rich soil.

In American Chinese stores, Sichuan lovage is sold in various *bupin* mixes (see samples 1, 3, 4) and in one-pound bags. In traditional Chinese medicine, it is characterized as acrid, and believed to be warming, with the target organs being the liver and the bile. It is used for improving circulation, enhancing strength, and eliminating headaches and as a diuretic. Between 1956 and 1980, Chinese pharmacological and medical scientists made outstanding discoveries of the chemical compositions in *Chuanxiong* and their actions on the circulatory system. Phytochemists have isolated volatile oils, alkaloids, phenols, lactones and organic acids (ferulic acid, sedanoic acid) and a new compound (4-hydroxy-3-butylphthalide). Clinical observations proved the actions of Sichuan lovage on coronary circulation, on the peripheral blood vessels and blood pressure, and on platelet aggregation and thrombosis. They have reported its dilatory action on the coronary vessels with increased flow and decreased myocardial oxygen consumption; its effect on increasing the blood flow in the brain and the limbs, on producing a marked and prolonged hypotensive action; and its inhibitory action on platelet aggregation and the decreased size and the weight of the thrombin. Sichuan lovage stimulates blood circulation and removes stasis, exerting antithrombotic actions.

The following recipes are adopted from a *bupin* mix available in Chinese stores in Boston.

Four-precious Chicken Soup

Ingredients

    1 package of four ingredient *bupin* mix (containing of 3/10 oz Sichuan lovage, 1/3

oz cured *di-huang*, 1/5 oz *dang-gui*, and 1/10 oz *dang-shen*, wrapped in gause for
ready disposal before serving)

3 lbs chicken (washed and cut at the joints)

1 large onion (sliced into 8 vertical sections)

10 cups water

1 tsp salt

Boil the water in an enamel or a crock pot, add all the ingredients and bring to the
boil. Simmer for 1 1/2 hours. Serve hot. Makes 6 servings. Prepare a dip for the meat if
preferred.

Angina Soothing Tea

Ingredients:

1/2 oz safflower

1/2 oz Sichuan lovage

8 cups water

Combine the ingredients and bring to a boil. Lower the heat to medium low and
simmer until the liquid is reduced to 3 cups. Drink 1 cup before each meal, 3 times
daily.

(b) *Parasitic Whole Plant*: **Chinese broomrape** (*rou-cong-rong* 肉苁蓉, Figure 51):
The market material of *rou-cong-rong* is a dried product prepared from the fleshy,
parasitic plants growing on the roots of several desert shrubs, herbs or grasses in the
genera *Achnatherum* Beauv., *Kalidium* Moq., *Nitraria* L., *Reaumuria* L., *Caragana* Fabr.,
and *Haloxylon* Bunge. The parasitic plants gathered in April and early May are buried
in sand and dried there. Those collected in the autumn have higher water content.
They are thrown into the salt lake to soak for one or two years and then dried in the
sun. The material in the American market are black pieces prepared from the dried
parasitic plant treated with yellow gin. Before the application of gin, the material dried
in spring is soaked in water with one or two changes. The autumn harvest material
must be soaked in fresh water for several days, changing the water once or twice daily
until the center of each specimen is soft, and the salt washed out. The softened stems
are sliced, treated with gin (by the proportion in weight of 2 parts broomrape and 1
part yellow gin). The mixture is placed in a closed urn and then heated in boiling water
until the gin is completely absorbed. The treated broomrape is removed from the urn
and dried in the sun.

Broomrape was a food supplement for the ancient wandering nomads living the
deserts in northwestern China as ginseng was for the forest-dwellers of northern China.

Now, they are both precious *bupins* of the affluent city-dwellers. Broomrape was first recorded in the *Shen-nong Ben-cao Jing*, and it was characterized as warming, sweet, beneficial to the five viscera (heart, lungs, liver, kidneys, and stomach), tonifying the internal organs, regulating the *yin-yang* (electrolytes), and improving fertility. The market material is not toxic.

In a cookbook entitled *Therapeutic Recipes* by P. L. Wang and published in Taiwan (1968), broomrape is used to prepare a stock for a winter dish called *huo-guo* 火鍋 (= fire pot). The procedure and ingredients are given below.

Fire Pot with Broomrape Stock

To prepare the stock

Ingredients:

  1/2 oz broomrape (from Chinese herb shop)

  1 chicken (remove meat from the breast and legs)

  2 large onion (cut into 4 vertical pieces)

  2 oz dried laminaria (*hai-dai* 海帶, available in Chinese grocery; soaked in warm water and cleaned)

  3 quarts water

  1 1/2 tsp salt

  4 stalks celery (cut into 3 cm pieces)

Rinse the broomrape in cold water. Clean the chicken bones and skin and drain. Cut the softened laminaria into 2 cm square pieces. Boil the water and add all the ingredients. Bring the mixture to a boil and keep it simmering for 2 hours. The stock is ready for *huo-guo*.

To prepare the huo-guo

  Meat of the whole chicken (previously boned and skinned, cut into 4 x 1 cm pieces, score each piece thinly to increase the cooking surface, place the pieces in 4 plates)

  1 1/2 lb fillet of sole (cut across the fillet into 3 cm pieces, and place them next to the chicken meat)

  1 1/2 lb squid (wash, cut longitudinally into 2 pieces, rinse off sand and wash thoroughly, score the outer surface into 5 mm diamond pattern, then cut the scored half into 2 cm wide pieces, and place next to the fish)

  1 lb medium-sized shrimp (head removed, peeled, deveined, cut vertically 3/4 to near the tail, and place next to the squid)

  2 oz mung bean silk (*fen-si* 粉絲; revived in hot water, place in 4 separate dishes)

1 lb spaghettini (boiled 10 minutes, drained, mixed and coated with 1/2 cup corn oil, place next to the *fen-si*)

2 lb celery cabbage 白菜 (*bai-cai*; clean, drain, cut across the leaves into 4 cm pieces, and place in 4 other plates)

1 lb spinach (cleaned, drained, and placed next to the cabbage)

4 inner, tender stalks celery (cut slantwise into 2 cm wide pieces, and place next to the cabbage)

4 large onions (cut vertically into 6 pieces, and place the slices in another 4 dishes)

2 boxes bean curd (*dou-fu* 豆腐; cut lengthwise once, then slice into 1 cm thick pieces of *ca.* 3–5 cm square, and place them in the onion dishes)

To prepare the dipping sauce

1/3 cup light soy sauce (*sheng-chou* 生抽)

3 scallions (minced, then add to soy sauce)

1 1/2 cm piece fresh ginger (scrape off skin, mince finely, and add to sauce)

4 cloves garlic (press for juice, and dice the remaining solid portion, then add both the juice and pieces to sauce)

1 Tbsp sesame oil (add 3/4 to the above sauce)

1 tsp vinegar (add to sauce)

2 tsp fermented soybean hot sauce (*la-dou-ban-jiang*, optional, with 1/4 of the sesame oil placed in a separate dish, for people who like spicy taste).

Set the table with an electric skillet filled with a portion of the stock in the center. The dishes containing the various ingredients should be arranged in four sets, each between two participants, i.e., the meat/seafood dishes, the vegetable dishes, the noodle dishes, the bean curd/onion dishes will each be shared by two people. Keep the central skillet on high heat and have the stock boiling. Now, each participant take and place some dipping sauce in his/her small bowl. Starting with the chicken, each person takes a piece of the raw meat with the bamboo chopsticks, which do not conduct heat, holds the meat in the boiling stock until it is cooked. Then take it out and dip it in the sauce before placing it in the mouth. Place the onion in the stock to enhance the flavor. Cook the seafood, vegetable, bean curd, etc. in like manner. As the cooking and eating progress, the ingredients are consumed gradually, the amount of the stock in the skillet decreases and its flavor is enriched by the cooking of the meat, fish, and vegetables. Take some hot reserved stock from the kitchen pot and add it to the central skillet which should be kept nearly full all the time. Add the *fen-si* and spaghettini which will absorb the enriched flavor of the stock. Wait until the stock boils again, then take the soft *fen-si* and spaghettini into individual dipping bowls, add more sauce to taste and enjoy them. Finally, serve the stock from the skillet into each empty bowl and drunk it as a soup. Connoiseurs of Chinese food all agree that this delicious soup is the best part of the *huo-guo*.

Huo-guo is a winter dish for family and friends. There are many versions for preparing the stock and the ingredients. The above recipe is adapted from Wang's publication which calls for broomrape in the stock. Recently, I have used Chinese angelica in place of broomrape, and also 1 1/2 pounds of rump steak, with the trimmings added to the stock pot. The participants finished all the ingredients and enjoyed the soup tremendously. Actually, I have not met anyone who does not enjoyed this dish heartily, even a friend who had previously shown a disinclination toward Chinese food. After trying *huo-guo*, he commented that he had never had better soup at the end of a meal.

(c) *Barks Used for Bupin*: In the foregoing *bupin* list there are three kinds of bark from species of *Cinnamomum* Schaeffer, *Eucommia* Oliver and *Magnolia* L.. Here eucommia bark is selected for discussion, because of a large amount of information provided by recent investigations.

Eucommia (*du-zhong* 杜仲): Eucommia is a unique Chinese plant used for *bupin*. Morphologically, it is a deciduous tree bearing naked unisexual dioecious flowers. The male plants bear fascicles of anthers with the newly emerging leaves. The female ones have solitary naked ovaries each borne in the axil of a bract. The living tissues of the plant contain a hard rubber (gutta percha) in the bark, leaves, flowers and fruits. On account of these unique characteristics, the genus *Eucommia* Oliver, (typified by *E. ulmoides* Oliver) is a monotype in a family whose relationship with other dicots is obscure.

Eucommia is one of the ancient *bupin*, used by the Chinese people since time immemorial. In *Shen-nong Ben-cao Jing* (*ca*. 200 A.D.) it was characterized as having insipid taste, slightly acrid, growing in deep gorges, and good for backaches, revitalizing the internal organs, increasing prowess, strengthening the bones, muscles and tendons, fortifying determination, soothing the itching and damp condition of the urogenital organs, and stopping uncontrolled urinary drips. It also promises light weight and delayed aging when taken continuously. The market product is the bark of ten-year old or older trunks or branches. Local people also gather the newly emerging shoots for pot greens. Due to the uncontrolled, ruthless collection of the bark, natural stands of eucommia are rare. The current market supply is from cultivated trees.

Active scientific investigation into the phytochemistry and pharmacology of the commercial product of eucommia, laboratory experiments on the physiological activities of the extracts, and clinical observation of its efficacy for hypertension took place both within and outside China, particularly at The University of Hong Kong and the School of Pharmacy and the Department of Chemistry at the University of Wisconsin. In regard to the chemical composition, phytochemists have identified: gutta percha 6–12% (varying with age and location, more in the root than stem bark, the older stem contains more than the younger portion), pinoresinol glucosides 0.142%, alkaloids 0.066%, pectin 6.5%, resin 1.76%, fats 2.9%, organic acids 0.25%, ketose 2.15–3.5%, aldose, and chlorogenic acid. The seed contains fatty acids (67.38% linolenic acid, 9.97% linoleic acid, 15.81% oleic acid, 2.15% stearic acid, and 14.68% palmitic acid), and gutta percha (soluble in ethanol and acetone).

To ascertain the efficacy of eucommia, scientists in China used water and alcohol extracts for intravenous injections in anesthetized dogs and observed: (1) obvious lowering of blood pressure in the dogs; (2) water extracts perform better than alcohol extracts; (3) the roasted eucommia bark gives better results than the unroasted bark. Other scientists conducted experiments with eucommia on rabbits with cholesterol arteriosclerosis. In addition to confirming the antihypertensive efficacy of the extracts, they further explored the mechanics of hypotensive action by studying the blood vessels of the rabbits' ears. They concluded that the antihypertensive activity of eucommia lies in the peripheral vasodilation. In China, many more medical scientists have worked on eucommia, some on the diuretic actions and others on its toxicity. Eucommia is toxic and only a very weak solution is effective as an antihypertensive agent. Outside China, the new knowledge of the hypotensive nature of eucommia caught the pharmaceutical suppliers unawares. The shortage of eucommia caused the government to set regulations to prevent its exportation. Naturally, this condition first limited and finally stopped the eucommia supply to the people of Hong Kong and to all overseas Chinese.

In Hong Kong, a team led by Dr. Betty Chan of the University of Hong Kong used an ethanol extract of eucommia bark on rats and concluded that the extract causes arterial hypotension by direct action on the vascular smooth muscle. The results of the Hong Kong investigation triggered the enthusiasm of a research team in the School of Pharmacy and the Department of Chemistry at the University of Wisconsin, where the isolation and identification of the active antihypertensive principle in eucommia proved to be pinoresinol di-β-D-glucoside, which was synthesized by using coniferyl alcohol through the action of a microbe, *Caldariomyces fumago* Woronichin.

Eucommia was introduced to the United States 80 years ago and is widely cultivated as a rare plant in botanical gardens and arboreta. In 1979 I published an article on the morphology, ecology, cultivation, traditional Chinese uses, current research, and the history of the introduction of *Eucommia ulmoides* Oliver from China to Europe and

America. Many readers requested seed. Eucommia branches are very brittle, and in gathering the fruits it was inevitable that some small branches were broken. In collecting the fruits, all the twigs and leaves were saved, dried and kept in a closed jar. In 1980–81, I was asked to attend several international symposia and to deliver dozens of lectures. The pressure of preparing and delivering the lectures and the fatigue of thousands of miles of travel, not to mention the jet lags, caused me to be physically run down. After a complete medical examination, I was told that I had high blood pressure. Suddenly, the work of that review of eucommia came to my mind. Knowing that eucommia bark is used in China for hypertension, I made a tea with a handful of the dried leaves in an one-liter beaker in my laboratory, boiled it down to two-thirds of the original volume of water. That day I drank nothing but the eucommia tea. The next day I went back to the Harvard University Health Center and asked the same nurse to measure my blood pressure. She said, "Your blood pressure is down to normal. What have you done with yourself?" I told her about the eucommia-leaf tea and eventually I told my doctor about it. He had no objection of my experiment with myself, and henceforth, whenever I feel dizzy, I make some eucommia-leaf tea prepared as described above. I shared this experience with Professor K. C. Chang, Curator of East Asian Archaeology in the Peabody Museum of Archaeology and Ethnology, Harvard University. He has taken eucommia-leaf tea once a week for a year for the same purpose. In using the bark of a tree, man must kill it first. That is why eucommia has become an endangered species in China. In replacing eucommia bark with its leaves, man can have continuous supply of the hypotensive agent and also save the species.

Eucommia bark is available in many Chinese stores. The following recipe is given to me by a friend who uses the material frequently.

Eucommia Soup

Ingredients:

>2 pork chops
>1/2 oz eucommia bark (available in American Chinese herb shops)
>1/2 cup Job's tears (cleaned, washed and drained)
>1/2 cup euryale seeds
>1/2 cup red jujube
>8 cups water
>3 scallions

Wash and clean the ingredients. Boil the water and add all the ingredients. Bring to a boil and turn to low heat, simmering for 3 hours. Makes 2 servings.

This is a dish for a low salt diet. If shared with members of a family who can take salt, a dip made from soy sauce or oyster sauce can be served with the meat. For an older person with symptoms of high blood pressure, this dish may be served once a week.

A Rejuvenating Eucommia Tea

Ingredients:

> 1/3 oz eucommia bark (curred in gin; available in Chinese herb shops)
>
> 1/3 oz scurfy pea (curred by soaking in gin, roasted; available in Chinese drug stores)
>
> 20 walnuts (shelled)
>
> 2 cloves garlic (crushed)
>
> 1 tsp vinegar
>
> 3 cups water

Combine all the ingredients and bring to a boil in an enamel pot. Continue boiling on medium-low heat until 1 1/4 cup of the liquid remains. Drink warm in early morning. Use 2 more cups of water and cook in a similar manner once more. Drink warm at bedtime. This is a special recipe for a person with weak ankles, pain in the knees and/ or backache.

Eucommia with Lamb Kidney

Ingredients:

> 2 oz roasted eucommia (soak overnight in water, cut into 10 equal portions and wrap in gauze for easy removal)
>
> 2 lamb kidneys (cut lengthwise into 2 halves, carefelly remove the fibrous tissue attached to the inside, turn to the smooth surface, score along the vertical axis, and cut crosswise into 1 cm wide pieces)
>
> 7 scallions (white portion only)
>
> 1/10 oz *wu-wei-zi* 五味子 (dried schisandra fruit, available in Chinese herb stores; wash and drain)
>
> 3 cups water (used for soaking eucommia, saved for cooking)

Combine all the ingredients except the lamb kidneys. Bring the mixture to a boil. Lower the heat and continue to cook slowly until the mixture is boiled down to 1/3 of the original volume. Remove the eucommia. Drop the sliced lamb kidneys into the stock. Wait until it boils again. Turn off the heat and take the material warm before sunrise. Return to bed and rest for an hour before getting up. This recipe is designed for a weak person with backache and kidney trouble.

Eucommia Jujube Cookies

Ingredients:

8 oz eucommia bark (scrape off the outer, coarse portion and grate into sawdust, toast brown, pulverize and sieve)

2 cups jujube (soak for 2 hours, boil soft, save liquid for making a cold drink)

2 oz Sichuan teasel (soaked in 1/2 cup gin, dried, roasted, pulverized and sieved)

6 oz Chinese yams (pulverized, slowly cooked in 1 cup water)

Combine the ingredients which would be cemented by the boiled jujube flesh and the cooked yam. Mix well and cut into 50 pieces. Bake at 250° until done. Chew one piece before lunch. This recipe is adapted from two ancient formulae designed for preventing miscarriage in pregnant women.

(d) *Leafy Shoots and Leaves*: The imported articles used as *bupin* available in American Chinese stores are predominantly underground structures (50%) of plants: roots, bulbs, or rhizomes; fruits (15%); and seeds (16%). In China the situation is different. There, the rural people gather their own plant material to meet their health care needs, and the majority of the working population in cities gets their supplies from small herb shops dispensing fresh herbs brought by villagers, or from itinerant vendors. Professor F. A. McClure and Dr. T. M. Hwang (1934) first investigated the plants sold in such herb shops in Guangzhou (Canton). With the assistance of several students I studied the plants used for similar purposes in Chengdu (Map, Loc. 35) in the early 1940s. In the report of McClure and Hwang, 85.26% of the species used are aerial vegetative organs and 9.56% underground growths (roots and rhizomes). In our studies of the plants in Chengdu herb shops, we found that 60% of the items are young plants, leafy shoots, and leaves (Hu, 1945) and less than one-fifth (20%) are roots and rhizomes. Recently, three leafy shoots and/or leaves, formerly sold only in small local herb shops of southern China, have appeared on the shelves of Chinese groceries in Boston. Their names and botanical identifications are *ji-gu-cao* 雞骨草 (*Abrus cantoniensis* Hance), *bu-zha-ye* 布渣葉 (*Microcos nervosa* [Loureiro] S.Y. Hu), and *long-li-ye* 龍脷葉 (*Sauropus changianus* S. Y. Hu). All rural people in the vicinities of Guangzhou and Hong Kong recognize these species, and most urban children in the same areas are familiar with these local names because they have taken decoctions prepared from these herbs for health purposes since childhood.

Chicken bonewort (*ji-gu-cao* 雞骨草): Abrus is a perennial, leguminous vine with the annual wire-like leafy shoots (10–80 cm long) growing from a woody chestnut brown rootstock or rarely from the basal portion of a one-year old aerial stem. The long shoots are green before flowering, and they gradually turn chestnut brown. Short, simple branches bearing 2 to 7 leaves may appear on the lower portion of such annual stems. The leaves are alternate, even-pinnately compound, each with 7–16 pairs of small, oblong leaflets, 5–12 mm long, 3–4 mm wide, round and mucronate at the apex, oblique-truncate at the base. In August–September, short, hairy, raceme-like inflorescences appear, terminal to the leading shoots and/or to their short lateral branches. The flowers are small, 5–6 mm long, pink, blooming one at a time on a fleshy short lateral outgrowth of the rachis. The fruits are strongly compressed and sharply rostrate at the apex, containing 4–6 brown-gray maculate seeds. For extra cash, farmers or medicinal collectors go out to search for the plants in their natural habitats of the hillside, dig the woody rootstocks with their aerial shoots, remove the fruits if there are any, and sell the collection to dealers. On drying, the leaflets drop off, leaving the persistant rachises, subulate stipules and tiny hairy stipels. Together the twisted mass gives the fanciful appearance of small chicken bones, hence the local trivial name, *ji-gu-cao* 雞骨草 (chicken bonewort).

Before the 1950s, *ji-gu-cao* was only a folk remedy for jaundice in Guangdong and Guangxi. In the early 1950s Chinese medical scientists began to make phytochemical analysis and clinical observations on its efficacy for acute viral hepatitis. Simple preparations boiled with 60–85 g (2–3 oz) of *ji-gu-cao*, and 60 g (*ca.* 2 oz) of lean pork in 1,000 cc of water, and simmered down to one-third of the original volume, were administered to 44 patients, 3 times daily for 21 days. Over 95% of the patients were completely cured. Abrine, choline, sterol compounds, flavonoids, amino acids, and different sugars were reported from the dried material.

In American Chinese stores, *ji-gu-cao* appears in 600 g bundles tied in pink cellophane ribbons (Figure 15h). Every two bundles are sealed in a transparent plastic bag with the trade mark of the distributor, Overseas Trading Company of Hong Kong, printed in color. The cost of a 4-oz bag is US$0.99. In order to examine the contents more closely, I untied a bundle, placing the material in a pan and pouring boiling water over to soften it. The softened plants are spread on a tray and counted. The finding greatly distressed me, for 42 plants were uprooted and imported to America for only 50 cents retail price! What a brutal abuse of the natural resources this represents! The aerial growth of all the plants in the bundle is short and thin, more stunted than the herbarium specimens collected before the 1950s. This is a sign of the diminishing supply in nature for the larger plants were gathered earlier and only the smaller ones are available now. The species is endangered because of excessive uprooting. Protection of the natural population and cultivation of the species to meet China's growing internal need and for her exports abroad are urgent issues concerning *ji-gu-cao* (*Abrus cantoniensis* Hance). In 1960 Breteler published a *Revision of Abrus Adanson with Special Reference to*

Africa. Chinese workers have followed him and named *ji-gu-cao Abrus fruticulosus* Wallich ex Wight et Arnott. Botanists working on African species of *Abrus* Adanson have reached the conclusion that Breteler's specific limits are drawn far too widely and his *A. fruticulosus* sensu Breteler contains many distinct African species. In working with *ji-gu-cao* I have examined all the Indian and Chinese collections in the Harvard University Herbaria, including several isotypes and authentic specimens. I found that *Abrus cantoniensis* Hance from China differs from *A. fruticulosus* Wallich ex Wight et Arnott of Himalayan India in the apex of the leaflets and the color of the seeds. The Chinese species has leaflets round or retuse and apiculate at the apex and brown-gray maculate seeds while the Himalayan species has acute leaflets and shiny, black seeds. With these distinguishing morphological characters and for the correct identification of a medicinally useful species, it is better to use the scientific name, *A. cantoniensis* Hance, for *ji-gu-cao*.

Ji-gu-cao (Abrus) Jujube Beverage

Ingredients:

    2 oz dried *ji-gu-cao* (available in Chinese stores, washed and drained)

    12 jujube (available in Chinese stores, washed)

    10 cups water

    Honey or sugar (optional)

Combine the ingredients. Bring to a boil in a non-metal cookingware. Continue to cook in low heat until 6 cups of the liquid remain. Take warm. Makes 6 servings. Add sugar or honey to taste. Kept in refrigerator if one plans to use the preparation repeatedly. Jujube can be eaten, but *ji-gu-cao* must be discarded.

Rag leaf (*bu-zha-ye* 布渣葉): Two people taught me my first lessons of *bu-zha-ye*, a farm boy and a student of Chung Chi College, The Chinese University of Hong Kong. On March 8, 1968, I left Boston to begin field work towards the preparation of a flora of Hong Kong, and to teach botany in the Biology Department of Chung Chi College. At that time, the College was situated on a small farm with 300 acres of land granted by the Hong Kong Government, a much eroded grassy hillside for future extension. The government provided a train stop for the College between Sha Tin and Tai Po, two marketplaces with postal services. On May 4, I took a ten-minute ride from Chung Chi College to Sha Tin to explore the hillside above the marketplace. I climbed the steep hillside along a deep gully flanked on both sides by narrow terraces and irrigated by water guided from higher level by split bamboos for the cultivation of vegetables and cut flowers. Half way up on the dry hillside, I saw a shrub with mature fruits which I had not seen before. I tried to take a piece to use as a subject matter for conversation with the boy working in the garden. It was very hard to break off, because of its tough bast fibers. "May I please have the name of this plant ?" as I showed the specimen to the gardener. *"Bu-zha-ye,"* he kindly answered. *"Bu-zha-ye,"* I repeated after him and asked: "Is it good for anything?" He was apparently annoyed by such ignorant question, for to him, every adult knew about *bu-zha-ye*. He answered impatiently: *"Bu-zha-ye* is good for everything."

My second teacher was a student in the class on local flora of Hong Kong. In this class, the students were issued numbered lists on the trees, shrubs, vegetables, flowers, crop plants, fruits, important weeds, common grasses, and aquatic plants. Each plant was given a number, the scientific name, and the Chinese and English vernacular names. In the field studies, after the students gathered around a plant, they were first given a number in one of the lists so they could learn the names as they observed its characters. The day when we surrounded *Microcos nervosa* (Lour.) S. Y. Hu and after I gave the number, a girl named Li Feng-oi jumped with joy and said: "Oh, this is *bu-zha-ye!*" I asked her: "What do you know of *bu-zha-ye?*" She said: "When I was a child, whenever I had stomachache and indigestion, my mother always prepared me *bu-zha-ye* tea. I am so happy to know what it looks like and to see its flowers." *Bu-zha-ye* started Li Feng-oi's interest in medicinal plants and before she graduated from college, she wrote her senior thesis on the medicinal herbs of Hong Kong and prepared an exhibit in the Hong Kong City Hall on the subject.

Historically, the genus *Microcos* L. was first recorded in the writings of Dutch colonial administrators stationed in Malaysia and Sri Lanka in the late 17th and early 18th centuries. Johanne Burman (*Thesaurus Zeylanicus* 159. t. 74. 1737) described and illustrated a plant which the Dutch residents in Sri Lanka called "Kleine Cocos". Linnaeus in the *Species Plantarum* (1753) provided a generic name in Greek, *Microcos* L. and on the base of the inflorescence illustrated by Burman, named the species *Microcos paniculata* L.. Chinese botanists have applied this name in their account of the source species of *bu-zha-ye*. This creates a problem in the identification of a medicinal plant in China.

Careful examinations of all Asiatic specimens in the Harvard University Herbaria, including those from the woods of Sri Lanka (Ceylon), from the forests of Bombay Presidency, and from the open hillsides of Bangladesh (Khasia, Chittagong, Tenasserim), northern Burma, Thailand, Loas, and Vietnam, and southern China (Yunnan, Guangdong, Hainan Island and Guangxi), indicate that what were formerly named *Microcos paniculata* L. from these widespread regions with varied ecological conditions, involves several natural species. The specimen from Sri Lanka has ovate-lanceolate leaves graduately narrowed from the middle toward the apex, tappering into long acumens and uniformly stellate-tomentose beneath. In Guangzhou, the source species of *bu-zha-ye* has glabrous leaves. This species was named *Fallopia nervosa* Loureiro (*Flora Cochin-chinensis* 336. 1790). Another population in Bangladesh with glabrous leaves slightly broader above the middle in the apical portion than the basal half was observed and named *Microcos mala* Hamilton (Trans. Linn. Soc. London 13: 549 1822). Apparently, the range of this glabrous species with oblong leaves extends from Bangladesh eastward to northern Burma, northern Thailand, Yunnan, Guangxi, Guangdong, Laos and North Vietnam. Loureiro's specific epithet has the priority for the species and it was transferred into the genus *Microcos* L. and the combination, *M. nervosa* (Loureiro) S. Y. Hu (1987) has become the correct botanical name of *bu-zha-ye*. The material sold in the Chinese stores in Boston area belongs to this glabrous type. The medicinal use of *bu-zha-ye* for tea was first recorded by McClure and Hwang (1932) from Guangzhou where the people make an infusion with the leaves and take it internally to relieve indigestion and as a cooling drink. Nadkarni in *India Materia Medica* (1954) reported its use for indigestion, typhoid fever and syphilitic ulceration of the mouth and in small-pox, eczema and itches.

To my knowledge, the use of *bu-zha-ye* is only at the ethnomedicinal state and there is no record of its chemical compositions, laboratory experiments with animal assay, or clinical observations. Although *bu-zha-ye* has become an item of international trade, its use is still at an ethnobotanical level. The product in Chinese groceries in Boston area was packed in four-ounce plastic bags by Sun Wing Trading Company of Hong Kong, with printed labels marked "People's Republic of China origin". The contents are fully grown leaves, mostly broken. The amounts to be used for indigestion or colds vary between 15–30 g (1/2-1 oz). Like tea, the material is discarded after infusion.

Bu-zha-ye (Microcos) Pork Broth

Ingredients:

    1 oz dried *bu-zha-ye* (available in Chinese stores; place in a gauze bag)

    2 pork chops

    1/2 tsp salt

    1 medium sized onion (cut vertically into 4 pieces)

    1/2 cm sliced ginger (shredded)

    6 cups water

Combine the ingredients and bring the mixture to a boil in an non-metal pod. Keep simmer for 2 hours and then turn the heat to medium and reduce the liquid to 2 cups. Serve hot. Makes 2 servings. This is a dish suggested for persons suffering from poor digestion and with yellow complexion.

Dragon tongue (*long-li-ye* 龍脷葉): In Guangzhou and Hong Kong, the leaves of two species in the genus *Sauropus* Blume are used in food: *S. androgynus* L., cultivated and used fresh by owners of the plants, and *S. changianus* S. Y. Hu, locally called *long-li-ye*, available both fresh and dried in the maket places of Guangzhou (Canton) and Hong Kong.

Long-li-ye was recorded by McClure and Hwang (1934) as *Lung Lei Ip* (Cantonese romanization for "dragon tongue") without a botanical identification. They observed that the plants were cultivated in pots, the leaves were boiled with jujube and apricot seeds, and the liquid was taken internally for coughs. In the late 1940s, Prefessor Frederic C. Chang, Department of Chemistry, Lingnan University, Guangzhou, and his associates investigated the antibiotic properties of Chinese herbs and found the extract of *long-li-ye* inhibited the growth of the pathogenic species of *Streptococcus* Rosenback and *Straphylococcus* Rosenback. Specimens obtained from the Guangzhou market were brought to the Arnold Arboretum for identification. It proved to be an undescribed species and it was subsequestly described and named after Professor F. C. Chang (Hu, 1967).

Long-li-ye has had an international circulation since the 1960s. This issue is proved by a specimen received in 1964 for identification from the Inspector, Commonwealth Quarantine, Brisbane, Australia, who obtained the sample from a parcel sent from Kowloon, Hong Kong. The sample from Brisbane consisted of dried flowering leafy twigs. The commodity obtained from the Chinese groceries in Boston are leaves only, arranged in order and tied with a red cotton thread (Figure 15j) into bundles weighing between 32 and 53 grams (*ca.* 1–2 ounces). Two bundles are sealed in a plastic bag selling for US$1.25.

Long-li-ye is 20–40 cm high, evergreen and suffrutescent. The leaves are entire, oblong-oblanceolate, 5–8 cm long, 2.5–3.5 cm wide, rounded at the apex, and cuneate at the base, fancifully resembling the tongue of a small animal, hence the vernacular name *long-li-ye* (dragon tongue). The species is propagated by root division, often planted intermixed with vegetables in the New Territories of Hong Kong. It may be used fresh (100–300 g) or dried (60–120 g) (Figure 39).

Dragon Tongue Mixed Tea

Ingredients:

    1/2 bundle of *long-li-ye* (obtained from Chinese stores)

    6 dried red jujube (rinsed and drained)

    1/2 cup apricot seed (washed and drained)

    3 cups water

Combine the ingredients and bring the mixture to a boil in a non-metal cooking pot. Simmer for 2 hours. Turn the heat up to boil down the liquid to 2 cups if necessary. Take 1 cup before meal and another cup at bed time. Eat the jujube and apricot seeds. This is a soothing drink for people suffering from laryngitis or bronchitis.

(e) *Selected Flower Bupin*: During the hot humid middle summer days, people in Guangzhou and Hong Kong buy a ready-made mixture of flowers to make a decoction called *"wu-hua-cha"* 五花茶 (five flowers tea). More about this flower-drink will be discussed in *liangcha*. Here, the discussion is limited to the three flower products used as *bupin* available in American Chinese stores. These products are the dried flowers of honeysuckle (*Lonicera japonica* Thunb.), the night-blooming cereus (*Hylocereus undatus* [Haw.] Britt. et Rose), and the safflower (*Carthamus tinctorius* L.). Honeysuckle flower and safflower have widespread application throughout China. The flower of night-blooming cereus is limited to the tropical areas, especially in Guangzhou and its vicinities, including Hong Kong.

Honeysuckle flower (*jin-yin-hua* 金銀花): The earlist record of this honeysuckle (*Lonicera japonica* Thunb.) appears in an ancient Chinese herbal compiled in the late fourth century. The author, Tao Hong-jing 陶弘景, was a Taoist residing in the mountains of the lower Yangtze region where the species is semideciduous. As it retains some leaves in the winter, the local people call it *ren-dong-teng* 忍冬藤 (winter tolerant vine), which is the first recorded name. The current market name, *jin-yin-hua* 金銀花 (gold-silver flower), first appeared in a herbal of the Southern Song Dynasty (1127–1278 A. D.), illustrated with colored water-color paintings. By that time, *L. japonica* Thunb. was already a favorite in gardens of temples and of the gentry. Its very fragrant paired flowers open in the evening, appearing white the next morning, while the older flowers on the same branch have turned yellow. This later vernacular name, *jin-yin-hua* (gold-silver flower), refers to the yellow older flowers and the fresh white ones on the same plant, a condition which is often seen on garden favorites with fragrant white flower, such as the gardenia and the jasmine.

In American Chinese groceries honeysuckle flower appears in one-pound bags on the shelf of the tea section. The material consists of fully grown clavate flower buds, 2–3 cm long, with the enlarged apical portion 1.5–3 mm across, pubescent throughout with glandular hairs, and with the calyx attached to the basal end.

Honeysuckle flower is used as an antibiotic and a cooling agent in traditional Chinese medicine. It is one of the principal ingredients in *Yin-qiao je-du pian* 銀翹解毒片 (Honeysuckle-Forsythia Antibiotic Pill), a popular cold remedy in China and Hong Kong. Chinese phytochemists have isolated lonicerin, inositol and tannin from the market material and pharmacologists have performed experiments with laboratory animals to demonstrate its efficacy on inhibiting the growth of various bacteria.

Honeysuckle Flower Tea

Ingredients:

  2 Tbsp honeysuckle flower (available in Chinese stores)

  1 Tbsp green tea

  Sugar or honey (to taste)

  6 cups water

Place the flower and tea in a teapot. Boil the water and make tea. Many Chinese keep the tea in a thermos bottle for the day. In sumer season, I usually add some fresh mint and lemon to the above ingredients, cool the tea and place it in a bottle, and keep in the refrigerator for a very pleasant cool drink. Placed in a vacuum flask it can be taken out for a picnic.

Safflower (*hong-hua* 紅花): The petals of *Carthamus tinctorius* L. are collected daily, dried in the sun, and used as a *bupin* for tonifying the blood, especially for women. The plant is an erect and spinose herb with stiff stems branched before flowering. The leaves are alternate, sessile, lanceolate-ovate and with spiny teeth along the margin. The flowering heads are 2.5 cm in diameter with green spiny involucres, orange-red tubular petals deeply 5-lobed. Outside China, *Carthamus tinctorius* L. is cultivated for the red dye in the Near East and India and for the seed oil in California. In China safflower is cultivated for the dried petals used as a hematinic agent. Henan, Sichuan, and Jiangsu are the centers of production.

Safflower requires rich soil in relatively dry climate. It is propagated by seed, usually sowed in October, in rows 27–30 cm apart. The planting is thinned when the seedlings bear 3–5 leaves, finally leaving 30 cm between individuals for permanent growth. Young plants thinned off can be used as a potherb. During the growing season, the crop is fertilized 2 or 3 times. When the petals turn from yellow to red, they are gathered and dried in the sun or by heat.

Chinese phytochemists have identified the chemical composition of safflower and reported carthamin, carthamidin, safflon yellow, glycerides of arachidic acid, glyceryl oleate, glyceryl linolenate, glyceric acid, palmitin and stearin. In conventional practice safflower is used as an emmenagogue and a hematinic agent. Chinese medical scientists have reported their studies of the physiological activities. Aqueous extracts applied to the uterus of a guinea pig, rabbit, dog, and cat show postive excitement both in vivo and in vitro. Experiments with different animals under anesthesia show its efficacy in regulating the heart, increasing the blood flow of dog, constricting the blood vessels in rabbit's ear and inhibiting platelet aggregation in rats. Alcohol extract lowers the blood pressure of dog and cat, increases the tolerance to oxygen deficiency of mice and shows marked inhibitory action on thrombin and prothrombin. Safflower oil administered orally to rabbits with high cholesterol showed lowering of the total lipids, blood cholesterol, and glyceryl nitrate. Clinical trial of pills prepared from 1/2 oz safflower, 6/10 oz turmeric, 6/10 oz *dan-shen* (root of *Salvia miltiorrhiza* Bunge) and 1 oz *gua-lou* (fruit of *Trichosanthes kirilowii* Maxim.) were administered to 44 cardiovascular patients for four weeks. The therapeutic effect shows 99% improvement without any unfavorable side effects. Safflower is one of the more expensive *bupin*, but it has its merits.

Safflower Tincture

Ingredients:

   1 oz safflower petals

   1 cup gin

Place the safflower in the gin and heat the mixture in a water bath until half of the liquid has evaporated. Take half of the tincture. This recipe is designed for women with severe menstrual cramps. If the pain does not stop after the first dose, then take the remaining liquid.

Safflower Drinks

Ingredients:

   1 oz safflower petals (fine quality, finely shredded)

   1 oz Chinese angelica (*dang-gui*, finely chopped)

   1 oz sappan wood (ground)

   4 cups water

   1 cup gin

Boil the safflower and sappan wood in water until 2 cups of liquid remain. Add Chinese angelica and gin and continue cooking on medium-low heat until 1 full cup of liquid is left in the container. Take before meals. This drink is recommended especially to women suffering from irregular menstruation and general weakness.

   A few weeks ago a woman in her late twenties came to my office with a Chinese prescription, and asked me to translate the items listed in it. She told me that she felt the lack of energy and suffered from menoxenia. This homemade, nontoxic and non-addictive drink may be a helpful solution to many sufferers of menoxenia.

   A *bupin* mix called *sheng-hua-tang* 生化湯, sold in Chinese stores, is taken as a postpartum tonic by many Chinese mothers. It contains safflower, Chinese angelica, charred ginseng, Sichuan lovage, a small amount of peach kernel, and licorice.

Safflower is very light. It takes hundreds of flowers to make half ounce. Its production is labor-intensive, therefore, it is very expensive.

(f) *Selected Fruit Bupin*: Eight fruits are listed among the common Chinese *bupins* in the American market. Two of these species are cultivated as ornamentals (*Chaenomeles sinensis* [Thouin] Koehne and *Ligustrum lucidum* Aiton), and their fruits are gathered and used for medicinal purposes. The three species are cultivated in China primarily for their edible fresh or preserved fruits. These are *Crataegus pinnatifida* Bunge, *Dimocarpus longan* Lour., and *Ziziphus jujuba* Miller. The remaining three species, *luo-han-guo* (*Siraitia grosvenorii* [Swingle] C. Jeffery), lycium berries (*Lycium barbarum* L. and *L. chinense* Miller) and *wu-wei-zi* (*Schisandra chinensis* [Turcz.] Baillon) are cultivated specially for their dried fruits sold as *bupins*. The major portion of the following accounts of *luo-han-guo*, lycium berry and *wu-wei-zi* appears for the first time in the English language.

Buddha disciple fruit (*luo-han-guo* 羅漢果): *Luo-han-guo* is the extremely sweet fruit of a perennial, cucurbitaceous vine. It has been used as a common sweetening agent and home remedy in southern China since time immemorial. Its cultivation used to be the trade secret of a minority group now called *Zhuang-zu* 壯族 (= Strong Tribe), living in the mountains near Guilin in northeastern Guangxi (Map, Loc. 60). In ancient times, this ethnic group was attracted to *luo-han-guo* by its unusually sweet fruit which was gathered and used as a sweetening agent and/or for home remedies. As the knowledge of *luo-han-guo* spread, the demand increased, and the local people took some to the market to exchange for products or cash. The use of *luo-han-guo* gradually spread over the entire Cantonese speaking area. By the early 1900s the natural supplies had diminished to such a degree that the cultivation of local crops became profitable. People in Guangzhou and its vicinity all used the fruit, knew of its origin being Guangxi, yet none of them had ever seen the plant.

In 1932, an agriculturalist, Professor G. W. Groff, former Dean of the College of Agriculture, Lingnan University, Guangzhou, China, on a visit to Guilin by the invitation of Governor Li Zhong-ren 李宗仁, was given a few rootstocks of *luo-han-guo*. He brought the specimens to the Lingnan University campus which was located on the bank of the Pearl River in Guangzhou. This introduced specimen produced leafy branches but failed to flower in the humid lowland climate of Guangzhou. With the assistance of W. T. Swingle, an officer of the Division of Plant Exploration and Introduction, Bureau of Plant Industry, US Department of Agriculture, Washington, D.C., Groff was able to get a research grant from the National Geographic Society to study and collect the source species of *luo-han-guo* in Guangxi. He organized a *Luo-han-guo* Expedition Team, with Tam Ying-wah helping him to collect the herbarium and pickled specimens of the male and female flowers and the mature fruits for seed. On receiving the herbarium specimens, and with the advice of Professor E. D. Merrill, then Director of the Arnold Arboretum, Harvard University, the source species of *luo-han-guo* was described as *Momordica grosvenorii* Swingle in 1941, named in honor of Dr. Gilbert Grosvenor, President of the National Geographic Society. The seeds brought down to Guangzhou germinated and grew vegetatively at the Lingnan campus. During the dormant stage of the seedlings, half a dozen rootstalks were received by Swingle in Washington. So, *luo-han-guo* was introduced to the United States while China was at war with Japan.

Being a native of Guangxi, *luo-han-guo* has been the subject of research at the Guangxi Institute of Botany in the suburb of Guilin. A team of specialists at the Institute have made (1) extensive field investigations with the cooperation of the members of local communes, (2) intensive studies in the experimental gardens of the Institute aimed at understanding the biological characteristics and ecological requirements of the species, leading to the selection of superior cultivars for distribution to the farmers, and (3) phytochemical analysis of the fruit both in the laboratories of the Institute and in collaboration with the scientists elsewhere in China and abroad. Results of the studies were published in *Guihaia*, the journal of the Guangxi Botanical Club, between 1981 and 1984. Now, *luo-han-guo* is not only the best biologically and phytochemically known edible plant of China, but also a species with eight superior cultivars selected both from the wild types and from garden forms (Zhou, 1981).

The market material of *luo-han-gua* is a light, smooth, oblong, grayish-brown or olive-chestnut colored dried gourd, with the size and shape varying between that of a chicken or a goose egg. All parts of the fruit (the pulp, the seeds and the rind) are sweet. Being a cucurbitaceous fruit like the watermelon, *luo-han-guo* is a berry with a fleshy and juicy placenta, numerous seeds each covered by a rather hard outer seed coat, and a smooth outer rind. Unlike the watermelon, the sweetest portion of *luo-han-guo* is the outer rind.

Luo-han-guo grows best in rich loose soil on hillsides with good drainage in areas cool and damp in summer. It is propagated vegetatively by layering. For this purpose, in mid-August, strong 1–2-year old branches are selected from the pistillate and staminate plants in a ratio of 10 to 1, and laid. In November, these branches are separated from the mother plants and buried in sand to weather winter. At the next early spring they are planted 2.5 m apart, 2 in a hole 40 cm deep and 60 cm in diameter, filled with a mixture of compost and barn-yard manure. The hills are covered with rice straw.

Peas and beans are planted in the same field to prevent weeds. In the second year, a frame of 1.5–1.7 m high is needed. All suckers except one and the lateral shoots are removed, and only one vine of each plant is allowed to climb up the frame. Fertilization is applied before flowering and after fruiting and hand pollination carried out in early morning to enhance good bearing. The fruits are harvested in September–October, and placed on a wooden floor for 8–10 days to allow the color to change from green to yellow. The fruits are dried by warm air for 5–6 days, sorted, brushed, and wrapped by paper before packing in boxes for the market.

Although *luo-han-guo* is rich in sugars (25–38% of the flesh being sugars, especially frutose and glucose), the major sweetening element is a triterpenoid saponin, which is 300–500 times sweeter than cane sugar. Extracts of this compound are used for diabetic patients. Fresh *luo-han-guo* is very rich in Vitamin C. Phytochemists reported that from wild type, for every 100 g of fruit there is a maximum of 461.12 mg Vitamin C. With the fruits of two cultivars, the amounts vary between 339.68 and 389.32 mg. In drying, the fruit loses 9/10 of the Vitamin C content. The seed of *luo-han-guo* is rich in oil from various fatty acids including linoleic acid (52.3%), palmitic acid (14.7%), oleic acid (20.9%), stearic acid (7.1%), myristic acid (0.6%), capric acid (0.6%), and lauric acid (0.5%).

Formerly consumers of *luo-han-guo* believed it to be a soothing agent of the respiratory and digestive systems. It is taken in the form of tea or soup for colds, coughs, indigestion, dry stool, and dark urine. Recently, reports of clinical observations made in China and Japan indicate that *luo-han-guo* is used as an antihypertensive agent. In China it is also used by singers for the protection of their voices.

Luo-han-guo Tea

Ingredients:

    1/10 medium-sized, dried *luo-han-guo* (from Chinese store)

    1 Tbsp green tea (jasmine, *long-jing* (= dragon well) or any other brand)

    6 1/2 cups water

Boil the water. Use half of a cup to rinse the teapot to warm it. Empty the pot and place all the ingredients into it. Wait for 2 minutes and drink 1 or 2 cups. Keep the remaining tea in a thermos bottle for later use. If it is too sweet, dilute the tea with boiling water to taste.

Fujian Beverage for Cough

Ingredients:

    1/2 medium-sized *luo-han-guo*

    5 oz dried persimmon (available in Chinese stores)

3 cups water

Combine the ingredients and bring to a boil in a enamel or glass pot. Turn to low heat and simmer for 1 1/2 hours. Raise the heat and reduce the liquid to half the original volume. Take one-third, 3 times a day. This recipe is adopted from a folk remedy of the Fujian people.

Luo-han-guo Soup

Ingredients:

    1/2 medium-sized *luo-han-guo* (rind, pulp, and seed)

    1/2 oz wood ear (Auricularia; revived in boiling water and cleaned)

    1 large onion (cut vertically into 8 pieces)

    2 carrots (pared and cut into 2 cm sections)

    1 lb fresh lotus rhizome (available in Chinese stores, pared and cut slantwise into 4 mm thick sections, soak the sections in water for 1/2 hour and drain)

    6 oz sagittaria corms (fresh, ivory white, available in Chinese stores in winter)

    4 oz spareribs (cut into 4 cm sections)

    8 cups water

    1 tsp salt

Combine the ingredients and bring to a boil. Simmer for 2 hours. Makes 4 servings. This is a popular soup around Chinese New Year in the Boston area.

Eight Precious Beverage

Ingredients:

    1/2 *luo-han-guo* (dried, available in Chinese stores)

    1 oz longan aril (available in Chinese stores)

    1 oz dried dragon's tongue (*Sauropus changianus*, available in Chinese stores in Boston)

    1 oz dried jujube (rinsed and drained)

    2 oz northern *sha-shen* (dried, available in Chinese stores)

    1 medium-sized American ginseng (available in Chinese stores)

    2 oz white dried apricot seed (available in Chinese stores)

    12 cups water

Combine all the ingredients. Bring to a boil in a non-metal cooking ware. Simmer for 1 hour. By this time, the ginseng root should be soft. Use a plastic knife to cut the ginseng and return the slices to the pot. Continue to simmer for 2 more hours. Makes 10 servings. Serve hot or cold. Eat whichever ingredients can be chewed and swallowed.

This is an excellent *bupin* for humid hot summer days. It can be kept in the refrigerator for a week.

Lycium berries (*gou-qi-zi* 枸杞子): In the warm, humid areas of South China, *gou-qi-zi* (*Lycium chinense* Miller) is propagated vegetatively and cultivated in vegetable gardens as a winter crop for the edible, leafy shoots. In the dry, cold areas of North China, the same species is cultivated as a cash crop for its dried berries used as a *bupin* in traditional Chinese medicine. In the deserts of Ningxia (Map, Loc. 12), another species, *Lycium barbarum* L., is cultivated in orchards for the same purpose (Figure 46). By removing the suckers and careful pruning of one stem, each plant is trimmed to assume the habit of a small tree. These plantations are government enterprises and reportedly owned by local communes. Like coffee, the berries on a branch do not mature all at once. They must be individually hand-picked. The harvesting is done by temporary laborers who are paid five cents per pound. In Boston, the berries of both species are available, with the fruit of Ningxia *gou-qi-zi* (*L. barbarum* L.), being slightly larger and the price a little higher. Good market material contains golden-red dried berries. When the material turns chestnut-black, it is too old to be used.

In the orchards of Ningxia, selection for better cultivars was made and ten of the named forms are given below to show their morphological diversity (Figure 46a–j) and variation in chemical contents. They are all cultivars of *L. barbarum* L..

The accounts are translated from the original publication.

(a) cv. hemp leaf — Leaves green, lanceolate or linear-lanceolate, 5–12 cm long and 0.8–1.4 cm wide; fruits red, oblong-ellipsoid, 1.8–2.2 cm long and 0.6–1 cm in diameter; protein 20.8%, sugars 24.2%.

(b) cv. greater hemp leaf — Leaves dark green, lanceolate or linear-lanceolate, 6–12 cm long and 0.8–1.5 cm wide; fruits bright red, cylindrical, 2–2.6 cm long and 0.8–1.2 cm in diameter, the apex truncate; protein content 9.3%, sugars 36.6%.

(c) cv. white branch — Leaves dark green, on grayish-white branches, lanceolate, 2–5 cm long and 0.5–1 cm wide; fruits brilliant red, fusiform, 1.4–2 cm long and 0.6–1 cm in diameter; protein 15.4%; sugars 24.2%.

(d) cv. spherical berry — Leaves dark green, thick, lanceolate or linear-lanceolate, 5–8 cm long and 0.8–1.2 cm wide; fruits red, obovoid, 0.8–1.2 cm long and 0.6–1 cm in diameter; protein 17.5%; sugars 30.4%.

(e) cv. spherical-acute — Leaves green, lanceolate; fruits short, oblong, the apex acute, 0.8–1.4 cm long and 0.6–1 cm in diameter; rarely cultivated.

(f) cv. yellow-acute — Leaves yellowish-green, thin, lanceolate or linear-lanceolate,

5–8 cm long and 0.6–0.8 cm wide; fruits oblong-cylindrical, 1.8–2 cm long and 6–8 mm in diameter; protein 13.6%; sugars 42.8%.

(g) cv. yellow leaf — Leaves yellowish-green, thin, lanceolate; fruits narrow, oblong, round at the apex, 1.4-1.8 cm long and 0.4–0.8 cm in diameter; rarely cultivated.

(h) cv. round yellow leaf — Leaves yellowish-green, thin, lanceolate; fruits roundish, oblong, 1.5–1.8 cm long and 1.4–2 cm in diameter; rarely cultivated.

(i) cv. curly leaf — Leaves lanceolate, curly; fruits obovoid, 1.4–1.8 cm long and 0.6–0.9 cm in diameter, rarely cultivated.

(j) cv. little yellow — Leaves grayish-green, thick, lanceolate, 6–10 cm long and 0.7–1.2 cm wide; fruits orange, short, oblong, 1.2–1.5 cm long and 0.7–0.9 cm in diameter, fleshy, sweet; a good cultivar for the production of edible fresh fruits; protein 12.3%; sugars 52.3%.

Many of my friends have lycium berries in their pantries. An elderly professor, retired from the National University of Taiwan, soaks 20 berries in boiling water every morning. She drinks the liquid as tea during the day and eats the berries at night.

Gou-qi-zi (lycium berries) is a collective term for the berries of several species of *Lycium* L.. In ancient Chinese records it is affirmed to be an effective agent for tonifying the kidneys, strengthening the lungs, protecting the liver, and brightening eyesight. Asian phytochemists have identified carotene 3.39 mg, thiamine 0.23 mg, riboflavin 0.33 mg, nicotinic acid 1.7 mg, ascorbic acid, β-sitosterol, linoleic acid, zeaxanthin, betaine, and physalien from the fruits of various cultivars. Pharmacological experiments with carbon tetrachloride-induced lesion in mice proved its hepatoprotective action to the liver cells. Water extracts of lycium berries injected intravenously lower the blood pressure and excite the breathing of rabbits.

There are numerous recipes, both simple and complex, for using the lycium berry. Most of my friends use it in soup with Chinese angelica, ginseng, astragalus or with *yin-er* (Tremella).

Lycium Berries-Ginseng Broth

Ingredients:
    1/2 cup lycium berries (washed and drained)
    1 medium-sized ginseng root (red or white, American ginseng preferred for the
        elderly, break into pieces with a hammer)
    1/2 cup lotus seed
    1 medium-sized onion (cut into 6–8 vertical pieces)
    1/2 tsp salt
    8 cups water

3 or 4 lb chicken (washed and cut in pieces by the joints)

Combine the ingredients and bring to a boil. Simmer for 2 hours. Serve hot. A dip with soy sauce, sesame oil, chopped scallion and ginger may be prepared for the meat. Makes 4 to 6 servings.

Schisandra fruit (*wu-wei-zi* 五味子): In Boston Chinese stores *wu-wei-zi* is sold in thin, plastic, one-pound bags as are almonds, chestnuts, mung beans and peanuts. This product is the small, dried fruit of a deciduous, woody vine with light gray bark and alternate, papery, elliptical, and serrate leaves. Its flowers are dioecious, solitary on slender pedicels. After fertilization, the receptacle of a female flower elongates and projects the drupe-fruits developed from separate carpels in a spike-like cluster hanging on a delicate pedicel. The fruits ripen scarlet-red and are very ornamental. Birds like them. The species has been introduced into America and is hardy in New York City and New Jersey. It has been successfully cultivated in a farm in Amherst, Massachusetts.

Formerly, the *wu-wei-zi* consumed in China were gathered from wild plants, dried and distributed. Today, the market product is largely from cultivated material. As the species is dioecious, the surest way for a successful crop is through vegetative propagation, by cutting or layering. Cuttings of 12–15 cm long may be made from June to August. Layerings may be made in February or November, from two-year old twigs. Two- or three-year old young plants are transplanted to the permanent field with some shading, leaving 30 cm among hills, and 40 cm between rows. Proper weeding, watering and application of fertilizer are necessary. Mature red fruits are gathered in late October or early November, and dried in the sun for distribution.

The market material consisting of the dried drupe-like fruit of *Schisandra chinensis* (Turcz.) Baillon is produced in northern China. There is a species in central China, *S. sphenanthera* Rehder et Wilson (discovered by E. H. Wilson of the Arnold Arboretum), the fruit of which is also collected for distribution. Dealers in China called the fruit produced in northern China *bei-wu-wei-zi* 北五味子 (northern Schisandra fruit), and that of central China *nan-wu-wei-zi* 南五味子 (southern Schisandra fruit). However, consumers generally do not know the differences between the products.

The name *wu-wei-zi* means "five flavored drupelet". The plant as well as the product gets this name because the skin and flesh of the fruit taste sweet and sour, whereas the kernel is pungent and bitter, and when chewed, it leaves a salty after-taste. These are the five flavors recognized in Chinese cuisine.

In traditional Chinese health care, practitioners take the five flavors in plants as guidelines for prescribing and combining herbal drugs. They have developed a reticulum of principles which linked up the five flavors to five primary elements, to five internal organs and to the *yin-yang* theory. Regarding *wu-wei-zi*, its predominant taste, sour, is taken as an indication of its ability to penetrate the *yin* organs (the liver and the kidneys), and also of its astringent function, holding the body fluids from leakage. Since time immemorial, *wu-wei-zi* has been prescribed for people suffering from disorders of the liver and the kidney, and those with night-sweats and emissions. Trying to understand the nature and function of *wu-wei-zi*, scientists in China began to work with the market material in 1945. By 1974 they had identified the major chemical compositions of the substance and proved its efficacy by bioassay. According to their findings, the fruit of *wu-wei-zi* is rich in volatile oils, including sesquicarene, β2-bisabolene, β-chamigrene, α-ylangene. It also contains organic acids (particularly citric acid, malic acid, and tartaric acid), simple sugars and resin. The seed of *wu-wei-zi* contains 33% fats; schizandrin, r-schizandrin, e-schizandrin, p-schizandrin, pseudo-r-schizandrin, deoxyschizandrin, schizandrol, citral, sterol, Vitamins C and E, resin, tannins and saccharides. Pharmacologists have ascertained that proper amounts of *wu-wei-zi* extract significantly improved the mental activity and the working efficiency of the subjects tested. Physiologists have found that *wu-wei-zi* extracts improved the coordination and increased the physical prowess of the subjects, making them more alert and with better muscle coordination.

In this connection I should record a personal experience. In the 1960s, an employee of the Arthur D. Little Company often came to the Harvard University Herbaria to ask for my help to identify plant material of Asiatic origin. One day he brought a small vial containing several dried drupelets, and asked for its scientific name. As it was a *bupin* which I have seen in China, I wanted to know its use in America. I made a deal with him, i.e., I would give him the correct identification on one condition. He must first tell me honestly how the material is used in America. He told me that people in horse racing business use the material to feed their horses before the race. I gave him the name, *Schisandra chinensis* (Turcz.) Baillon immediately. The identification was made so quickly that it aroused his suspicion. To convince him that the scientific name was correct I took him to the herbarium and showed him several specimens with fruits which matched his material perfectly. We were both satisfied. Obviously, at that time when China was closed to USA, the Chinese *wu-wei-zi* had been imported via Japan by people in the horse racing business for improving the alertness and muscle coordination of their animals. It is only recently that the product appears in Chinese groceries as a *bupin*. Chinese scientists have also confirmed the fact that too much *wu-wei-zi* can inhibit mental activities of the animals under observation, cause excessive excitement, and can lead to insomnia and shortness of breath. Intravenous injections of water extract of *wu-wei-zi* increases the breathing rate of dogs and cats under anesthesia. An injection of weak alcoholic extract lowers the blood pressure significantly. In vitro experiments using *wu-wei-zi* preparations on cultured hepatocytes for assay of antihepatotoxic activity showed very strong protective effects in the carbon tetrachloride induced material. Observations of animals with liver damage suggest that *wu-wei-zi* protects liver cells, promotes cell regeneration, enhances detoxification of the liver by increasing hepatic glycogens and relieving fatty degeneration in the liver.

To my knowledge, *Schisandra chinensis* (Turcz.) Baillon flowers and fruits abundantly in New Jersey and Massachusetts. To gardeners who wish to use their fresh fruit, the following recipe is suggested.

Fresh Schisandra Beverage

Ingredients:

2 cups fresh, ripe, red fruit of *wu-wei-zi* from the garden

1/2 cup honey (more or less, depending on taste)

Steam the fruits so the skin breaks easily. Rub the fruits and remove the stones. Cook and add honey to taste. Simmer for 20 minutes on medium-low heat. Cool and bottle for future use. For 1 serving, use 2 tablespoons of the material and add either cold or hot water to make a refreshing drink. This recipe is especially good for those with respiratory disorders such as coughing with a cold.

Wu-wei-zi Tea

Ingredients:

1/2 oz ginseng (American ginseng for the elderly)

1/3 oz *wu-wei-zi*

1/3 oz asparagus root tubers (*mai-men-dong*, available in Chinese drug stores)

3 cups water

Combine the ingredients and bring to a boil in a enamel or glass pot. Turn the heat to medium-low and continue to cook until 1 1/2 cups of the liquid remain. Drink the liquid warm. Use some sweetening agent, such as honey, if desired. This recipe is suggested for the elderly or weak people who feel tired, unwilling to talk, or often have dizzy spells.

Wu-wei-zi Honey Mix

Ingredients:

1 lb dried *wu-wei-zi* (washed in 2 cups of water and drained, then covered in 2 cups of water overnight)

2 lb honey

6 cups water

Remove the pits from the fruit. Wash them with the remaining water and add the water to the flesh. Filter this mixture through cheese cloth. Combine the liquid and the honey and heat the mixture until 2 pounds and 4 ounces remain. Bottle and keep in a cold place. For each serving mix 1 tablespoon of the material with a cup of boiling water and take before meals. This is a recipe suggested for people suffering from emission.

(g) *Selected Seed Bupin*: Eight seeds are included in the list of common Chinese *bupin* in American stores. The betel nut (*Areca catechu* L.) is not native of China. It is well known as a masticatory in Southeast Asia and India, and it appears in many books on economic botany. Five other seeds, euryale, Job's tears, black soybean, lotus seed and the lesser red bean are discussed in the angiosperm seed section. It leaves seeds of apricot and of prinsepia for further treatment.

Apricot seed (*xing-ren* 杏仁): The seed of apricot (*Prunus armeniaca* L.) is the by-product of a tree crop cultivated for fruit. In China people who can afford to buy apricot as a fresh fruit generally throw the stones away. Servants and the less fortunate children gather the stones and sell them to collectors who use cheap labor to crack the shells (endocarp) and free the seeds (kernel) with the brown seed coat tightly covering the embryo (two cotyledons with the small radicle and plumule). In Chinese prescriptions, an apricot seed with the brown seed coat intact is called *bei-xing-ren* 北杏仁 (northern apricot seed). All Chinese apothecaries keep a supply of apricot seeds in this state.

Detoxificated apricot seed: The major portion of the annual production of apricot seed is detoxified, processed by being first subjected to boiling water to loosen the seed coat, which can then be rubbed off by hand, followed by soaking the white cotyledons in cold water with several changes to eliminate the bitter element and to detoxify them. Then the detoxified white cotyledons are dried for the market. In Chinese prescriptions, this decoated and detoxified material is called *nan-xing-ren* 南杏仁 (southern apricot seed) or *tian-xing-ren* 甜杏仁 (sweet apricot seed), which is also available in apothecaries. However, a greater portion of the detoxified material is used in pastry and for food. The bagged product in Chinese stores in Boston is this type.

In northern and eastern China where apricot trees are common, there is a cultivar that bears larger and sweeter fruits with seeds containing no bitter taste. As children, when we bought such fruits, we immediately cracked the shells after eating the fleshy outer portion and ate the cotyledons also. Locally this type of fruit was called *ba-dan-xing* 八旦杏. In some recent Chinese pharmacological publications there is a confusion regarding *ba-dan-xing*. Several authors have applied this term for the seeds of *Prunus dulcis* Miller, the sweet almond, which is known as *bian-tao* 扁桃 in Chinese botanical books (Chen, 1936). The sweet almond, though introduced into China in the second century as *"bian-tao"* (扁桃) or *"bian-he-tao"* (扁核桃), is very rare in China, cultivated only in a few modern botanical gardens and in Xinjiang (Chinese Turkestan in older literature). Seed of this species, the almond (*bian-tao-zi* 扁桃子), must not be confused with the apricot seed (*xing-ren*), as it was in *Chinese Herbal Remedies* (Leung, 1984, pp. 16–18). In Chinese literature, whenever *xing-ren* is mentioned, it means apricot seed, not the almond.

As a *bupin*, apricot seed is taken for tonifying the functions of the respiratory organs and for giving strength to the weak and elderly people. Apricot seed contains a cyanogenic glucoside, amygdalin (a poisonous element, *ca.* 3%), fatty oil (about 50%), protein and amino acids. On soaking and boiling, the toxic element, amygdalin, is hydrolyzed in two steps, first by the enzyme amygdalase and then by the enzyme prunase to benzaldehyde and hydrocyanic acid, the toxic element which is discarded with the water. The apricot seeds in Boston Chinese stores consist of ivory-white, heart-shaped cotyledons, 1–1.5 cm long and 0.7–1 cm wide, mostly separated, rarely in pairs. They are detoxified.

A Beverage for Soothing the Lungs

Ingredients:

> 1/2 cup white detoxified apricot seed (from the market)
> 1/2 cup sugared jujube (available in Chinese stores)
> 3 cups water

Combine the ingredients and bring to a boil in a non-metal pot. Continue to cook at medium-low heat to get 1 1/2 cups of the liquid, which is good for coughs.

Apricot Seed with Pig Lungs

Ingredients:

> 1/2 cup detoxified apricot seed (soaked in water for 2 hours and drained)
> 1 lb pig lungs
> 6 cups water
> 1/2 tsp salt

Combine the ingredients and bring to a boil in a crockpot. Simmer for 3 hours. Serve hot. Makes 4 servings.

Apricot Seed Gruel

Ingredients:

> 1/2 cup white detoxified apricot seed (soaked in water for 2 hours and drained)

1 cup rice (short grain preferred)

1/2 cup northern *sha-shen*

1/2 cup red jujube (rinsed and drained)

1/2 cup Job's tear (cleaned and washed)

6 cups water

Combine the ingredients and bring to a boil in a crockpot. Cook on low heat for 4 hours. Eat a bowl early in the morning and return to bed for an hour. This is a good *bupin* for an elderly or weak person suffering from swollen ankles and uncontrolled urination.

Prinsepia seed (*nai-ren* 柰仁, *nai-ren-rou* 柰仁肉): *Prinsepia* Endl. is a rosaceous genus endemic to northern and western China and the high mountains of Taiwan. Two species, *P. uniflora* Batal. and *P. sinensis* (Oliver) Oliver have been introduced into the United States. As their ornamental merits and economic importance are not outstanding, they have been kept as specimens in only a few botanical gardens and arboreta. Actually, prinsepia is poorly known in China. It does not have a stable name, and those names used in ancient records, local reports, market circulations and botanical references do not agree. The two Chinese names I put in parentheses are taken from the ancient classics and from a package bought in a Chinese store in Boston. The sudden appearance of this product in a Chinese supermarket in Boston amazes me.

The source species for the market product, *P. uniflora* Batal., is a low and spiny shrub, up to 1.5 m high, growing on the arid hillsides of Shaanxi, Gansu, Shanxi and Inner Mongolia. It has been collected from the barren hills of northern Jiangsu where it is called "*shan-tao*" (wild peach). The plant has two types of branches: a few elongated ones with alternate leaves, and numerous abbreviated branches with smaller leaves in fascicles. The leaves are linear-lanceolate, 3–6 cm long and 0.5–1.5 cm wide, shortly petiolate, subentire, or remotely serrate on the more vigorous suckers. The flowers are all solitary, axillary to the fasciculate leaves on short shoots. The fruits are drupaceous, small like those of the wild Nanking cherry, 1–1.5 cm in diameter, maturing in July and August when they are gathered by local residents. There is no record of its cultivation in China. The ripe fruits are eaten locally, and the seeds saved and sold to itinerant buyers, who eventually sell them to be processed for the market.

The market product consists of shelled seeds covered by a reddish-brown seed coat. In the supermarket it is sold in one-pound bags. The alleged properties are: sweet, cooling, and nontoxic, with the liver, heart and spleen being its functional targets. In traditional Chinese medicine, it is prescribed for expelling evil winds, dispersing inner heat, nourishing the liver, and brightening the eyes. It is used for tonifying the internal organs and for reinforcing will power. Its known chemical constituents are vegetable oils 32%, proteins 3.5%, fiber 55.11%, fats 7.5% and ashes 1.72%.

All the recipes for prinsepia seed published in Chinese give information on its external use for cataracts and other eye troubles. Fortunately, I met an elderly welfare recipient in Chinatown who kindly gave me a recipe on how people with poorly functioning livers use prinsepia seed. The Chinese people believe that there is a close association between eyesight and liver function. When a person can open the eyes only in dim light, he is considered to have his liver functioning poorly.

Bupin for Poor Eyesight

Ingredients:

  1/3 oz prinsepia seed (market material)
  1/2 oz Chinese angelica (*dang-gui*, available in Chinese stores)
  1/2 oz licorice
  4/5 oz *fang-feng* 防風 (root of the umbelliferous plant, *Saposhnikovia divaricata* [Turcz.] Schischk.)
  1 oz Chinese golden thread (*Coptis chinensis* Franchet)
  4 cups water

Combine the ingredients and bring to a boil in a non-metalic pot. Turn the heat to medium-low and continue to boil until 1 1/2 cups of the liquid remain. Strain and drink warm with or without honey. Add 3 cups of water to the residue and boil off the liquid to get 1 cup. Take warm at bedtime.

Apparently, prinsepia seed is very popular in Chinatown, for there is evidently insufficient supply in the stores. On March 18, 1987, when I tried to purchase some, the dealer told me that he had just sold the last bag and I had to wait for the next supply.

Chinese Herbal Teas: *Liangcha* and Parcelled Medicated Tea

The use of herbal tea is a popular health care system of the people throughout China, but the study of herbal tea is a little known area in Chinese economic botany. There are many types of Chinese herbal teas, and to my knowledge, there is no published literature on the subject. The entire subject awaits for much field investigation in various parts of China. In the tropical and subtropical areas along the Pearl River of southern China, the use of herbal tea has developed to a commercialized stage. Ready-made beverage called *liangcha* 涼茶 (cooling tea) is consumed in stores on side streets of large cities and the marketplace of large villages. Individual packages of material for preparing *liangcha* at home is available in America as well as in southern China. Recently, parcelled medicated herbal tea packed in colorful boxes and bags for international trade appear in Chinese stores in metropolises throughout the world. The result of on site investigations into *liangcha* practices and the source species in Hong Kong and Macao and identification of the materials used in the parcelled medicated herbal tea samples obtained in Chinese stores in Boston would provide a good foundation for future studies in the health care system of the common people in China. It is with this purpose in mind that a summary of our knowledge of the commercialised Chinese herbal tea is presented here in two parts: 1. *Liangcha*, which refers to the ready-made decoctions served in simple stores of cities or stands of villages in Hong Kong and its vicinity, and 2. parcelled medicated tea, which refers to the boxed or bagged material containing ten or a dozen parcels, bought by people to be used repeatedly at home.

Liangcha 涼茶

The Chinese ideograms of the name *liangcha* can be translated literally into "cooling" (*liang* = 涼) and "tea" (*cha* = 茶). The serving of tea at wayside stands is a seasonal summer practice throughout China. In most cases, the service can be traced to a kind-hearted lay Buddhist, who believes in "*ji-yin-de*" 積陰德 (stacking up charitable credits for the unseen world). Such drinkable materal is prepared from boiled water and tea. Any person can come to a cool tea stand and take a free drink to quench thirst. However, the custom of serving *liangcha* commercially is known only in the Pearl River region in South China, particularly in Guangzhou, Hong Kong and Macao, where the term *liangcha* refers to a drink prepared not with tea, but to a decoction infused from 18 or more kinds of wild plants collected locally. *Liangcha* in southern China is served on the side streets or small alleys in cities and shopping centers of large villages. Unlike *bupin*, which is prepared from using small amounts of expensive classical medicinal products cooked with meat, poultry or sea-food requiring special labor-consuming technique, *liangcha* is prepared from wild plants collected locally, even with portions of the plants discarded by farmers and/or by the affluent people. Good examples of the latter type are the vines of bitter melon from farmers after the harvest, and the stones of mango saved from the tables of rich people, gathered, dried and used for preparing *liangcha*. Normally, *liangcha* is served warm or cold at room temperature. For very little money, one can get a bowl or a glass of the ready-made *liangcha* to drink on the site where it is prepared. People drink *liangcha* more for health purpose than to quench thirst.

Our knowledge of *liangcha* is still at the ethnobotanical stage. Much of the materials given here are recorded as first-hand observations made in the vicinities of Guangzhou, Hong Kong and Macao between 1950 and 1994.

(1) *A Short History of Liangcha Research*: In the Pearl River region, the suppliers of the source material of *liangcha*, dealers of the decoction, and consumers of the beverage have formed an interdependent business relationship. This phenomenon has attracted the attentions of local physicians and economic botanists alike. The physicians would like to ascertain the causes of the continuous demand for *liangcha*, and the economic botanists wanted to know the species employed in the preparations. Here, a summary of the results of the on-site studies of a physician and my personal investigations is presented.

A former professor of the Medical School at Sun Yat Sen University, Guangzhou, China, Dr. Z. X. Zhuang (or S. C. Cheung in Cantonese spelling), was interested in the native medicinal plants in southern China. He moved to Kowloon in August 1938 to practice medicine and began to spent his spare time to investigate the medicinal plants of Hong Kong. In 1954, on the occasion of the Fourteenth Anniversay of the Association of Hong Kong-Kowloon Liangcha Dealers, he was asked to be the keynote speaker in the celebration. His subject was "The Scientific Explanation of the Efficacy of *Liangcha*". For this speech, he bought ten different *liangcha* mixes from various dealers and prepared a list of 47 species of plants with their vernacular names in Chinese. He mimeographed this list for distribution to members of the Association. On this list, he marked 22 items with numbers showing their presence in his samples. Names without numbers represented plants used in *liangcha* from his knowledge of the general practice.

From a physician's point of view, Dr. Zhuang reaffirmed that the obvious reason for people to take *linagcha* is economic. He maintained that *liangcha* is the most economical way for people to stay healthy and to prevent contagious diseases. He explained that when an epidemic strikes, the most susceptible people are the ones with poor digestion, irregular bowels, and insufficient urinary discharges. *Liangcha* is a shotgun practice for removing minor physical defects and for correcting physiological irregularities. He elucidated with examples that the multiple ingredients of *liangcha* include plants with laxative properties (such as *Ficus hispida* L., *Oroxylum indicum* [L.] Vent., *Helicteres angustifolia* L., and *Momordica charantia* L.), with stomachic actions (such as *Evodia lepta* [Spreng.] Merr., *Ilex asprella* HK. et Arn., *Cratoxylum ligustrinum* [Spach] Blume, *Microcos nervosa* [Lour.] S. Y. Hu, and *Cleistocalyx operculatus* [Roxb.] Merr. et Perry) and with diuretic and antipyretic effects (such as *Lophatherum gracile* Brongn., *Lygodium dichotomum* Sw., *Imperata cylindrica* [L.] P. Beauv., *Prunella vulgaris* L., and *Pandanus tectorius* Soland). He explained that the decoction prepared by boiling small amounts of many ingredients in large quantity of water for many hours, which creates certain chemical changes, is a safe and beneficial drink which keeps people healthy.

My own involvement with the study of *liangcha* began in 1957, the same year when I was asked to prepare a list of food plants of China to be distributed in Thailand. In November 1957, after the Ninth Pacific Science Congress which took place in Bangkok, I left for Hong Kong, spent one month with Mrs. A. T. Roy at Chung Chi College in the New Territories, and visited Macao for ten days. In Hong Kong and Macao, I was attracted by the *liangcha* establishments and began to study the subject. While in Macao, I bought a pacel of *liangcha* material from a local dealer. At Tai-po 大埔, a large village with a railway station in the New Territories of Hong Kong, I interviewed a very friendly *liangcha* dealer named Hu. Since we had the same surname, he treated me as a member of his family and let his children call me aunt. I made many visits to his shop and to his tiny farm reclaimed from the hillside by Tai-po-kao, where he raised a few chickens and several kinds of locally rare herbs. In my interview, I recorded his procedures documented with 22 samples, each with the Chinese names as he supplied. In reporting the scientific identifications of the source species of *Liangcha*, my collections are indicated by Macao and Tai-po.

In March 1968, I returned to Hong Kong and began field studies towards the preparation of a flora of Hong Kong. I collected 17,000 sets of specimens in eight years both in Hong Kong and Macao. These specimens were studied and identified at the Herbarium of the Arnold Arboretum (where the study set is deposited) and the Hong Kong Herbarium of the Agricultural and Fishery Department of the Hong Kong Government. A complete set was deposited in the Herbarium of the Department of Biology, The Chinese University of Hong Kong (CUHK), and the duplicates were distributed to the Royal Botanic Gardens at Kew, the National Herbarium at the Smithsonian Institution, and the Beijing Botanical Institute, Academia Sinica. In 1972, to meet the need for teaching field botany in a course called "Local Flora" to students of the Department of Biology, CUHK, I prepared "An Enumeration of the Vascular Plants of Hong Kong and the New Territories". Each scientific name is followed by a local Chinese name and an English name. All the species are placed in families which are arranged in a slightly modified Engler System. In addition to studying the diversity of Hong Kong plants, I also worked with the students of the Department of Biology, CUHK, helping them in their studies and identification of local medicinal plants used by herbalists. Subsequently, we built up a file of named collection of over 300 species of Hong Kong medicinal plants deposited in a special collection in the Herbarium of the Department of Biology, CUHK. Meanwhile, I became an Adviser to the Editorial Committee of the Chinese Medicinal Herbs of Hong Kong. We carried out field trips together and translated the Volume VI of their series of publications into English. All these experiences enabled me to acquire the Chinese names and local medicinal uses of the tropical Chinese species which eventually helped me to identify the source species of *liangcha*.

(2) *The Source Species of Liangcha used in South China*: A total of 67 species used for ready-made *liangcha* consumed in cities, towns and marketplaces of large villages in Hong Kong and its vicinity were identified. These species are listed alphabetically by their scientific names in bold typeface below. Each of the scientific names is followed by its local name in Chinese, its English name in parenthesis, the part of the plant employed for making *liangcha*, the number of species recognized by Zhuang from his ten samples, Hu's material gathered in Macao and Tai-po and two packets obtained in Chinese stores at Boston, which are indicated by Sun and Wong. The family to which each species belongs is added at the end of this information for those who are not familiar to tropical plants. Approximately 90% of the identified species are common plants of tropical and subtropical origin in southern China. The majority of the authors who first saw and described the species were botanists of the eighteenth century, such as Linnaeus, Loureiro, Osbeck, Roxburgh, Sprengel, and Thunberg. They went to collect in China, or received specimens from pioneer explorers in southern China (or Japan), where they were only allowed to gather common species near human dwellings and along paths of the open hillsides. Most of the listed species are still plentiful and available to local collectors who supply the market demand for *liangcha*. When I began to work on the vegetation of Hong Kong, the local *liangcha* suppliers used to take the first morning train with their collections from the villages of the New Territories to Kowloon and Hong Kong. Now, Hong Kong is so industrialized and commercialized that the villages by the railway have been changed into modern cities and those in isolated valleys are abandoned. The village people have moved into apartment buildings, supported by working income or by government resettlement grants. Now, the *liangcha* ingredients used in Hong Kong are all imported from the southern China mainland.

The *liangcha* dealers are friendly to the people they serve, and they are always willing to sell the material to people who want to prepare their decoction for their families at home. Recently, two famous brands of the *liangcha* packed in colorful plastic bags 18–20 cm square are available in American Chinese stores. Outside each bag are the important ingredients printed both in Chinese and their romanization, followed by the percentage in weight. The Wong Lo Kat (王老吉) brand is packed in Hong Kong and the 24-brand (*Er-shi-si-wei Liang-cha* 二十四味涼茶) is prepared in the hometown of Sun Yat Sen, 50 kilometers north of Macao. In the following identification list, the source species given in these two brands are indicated by Sun and Wong respectively.

Scientific Identification of *Liangcha* Source Species

Abrus precatorius L.; 相思子 (Love Pea); leafy stem; Zhuang 3. Leguminosae.

Achyranthes aspera L.; 土牛膝 (Achyranthes); whole plants; Tai-po, Macao, Sun. Amaranthaceae.

Acronychia pedunculata (L. f.) Miquel; 山油柑 (Acronychia); chopped wood from large branches; Tai-po, Sun. Rutaceae.

Adenosma glutinosum (L.) Druce; 毛麝香 (Musk Adenosma); whole plant; Zhuang record. Scrophulariaceae.

Alternanthera sessilis R. Br.; 白花子 (Alternanthera); whole plant; Zhuang record. Amaranthaceae.

Andrographis paniculata (Burm. f.) Nees; 穿心蓮, (Creat, Kiryat); whole plant; Sun. Acanthaceae.

Areca catechu L.; 檳榔 (Betel Nut); seeds; Zhuang record, imported. Palmae.

Artemisia apiacea Hance; 青蒿 (Celery Wormwood); whole plant; Zhuang 2. Compositae.

Berchemia lineata DC.; 老鼠耳 (Supple-jack); leafy shoots; Sun. Rhamnaceae.

Buchnera cruciata Buch.-Ham. ex D. Don; 白箭 (Buchnera); whole plant; Zhuang 4, rare now. Scrophulariaceae.

Centella asiatica (L.) Urban; 崩大碗 (Moneywort); whole plant; Zhuang record, much used. Umbelliferae.

Cleistocalyx operculatus (Roxb.) Merr. et Perry; 水榕花 (Water Banyan); inflorescence; Zhuang 5; Myrtaceae.

Colocasia esculenta (L.) Schott; 芋頭桿 (Taro Petioles); fleshy petioles of taro; Zhuang record, much used. Araceae.

Commelina nudiflora L.; 竹節草 (South China Dayflower); whole plant; Zhuang record. Commelinaceae.

Cratoxylum ligustrinum (Spach) Blume; 黃牛木葉 (Ox Yoke-wood); leafy branches; Zhuang 4, Sun, Wong. Guttiferae.

Desmodium styracifolium (Osbeck) Merr.; 金錢草 (Golden Coin Herb); Zhuang 5, Sun, Wong. Leguminosae.

D. triquetrum (L.) DC.; 葫蘆茶 (Bottle-gourd Tea); leafy shoots, with or without flowers and fruits; Sun. Leguminosae.

Elephantopus scaber L.; 苦地膽 (Elephants-foot); whole plant; Zhuang 2, Tai-po. Compositae.

Eleutherococcus trifoliatus (L.) S. Y. Hu; 苦竻蔥 (South China Eleuthero); leafy shoots; Zhuang 1. Araliaceae.

Evodia lepta (Sprengel) Merr.; 三椏苦 (Threefold Bitter); leafy branches; Zhuang record, Tai-po, Macao. Rutaceae. (= *Melicope pteleifolia* [Champ. ex Benth.] T. Hartley)

Ficus hispida L.; 牛奶仔 (Opposite-leaved Fig); fruits; Zhuang record. Moraceae.

F. microcarpa L. f.; 細葉榕 (Chinese Banyan Root); aerial roots, Zhuang record; leafy shoots, Wong. Moraceae.

Glossogyne tenuifolia Cass.; 金鎖匙 (Golden Key); whole plant; Zhuang report, very rare in the vegetation now, an endangered species. Compositae.

Helicteres angustifolia L.; 崗脂麻 (Hillside Sesame); flowering branches; Zhuang 3, Macao, Sun, Wong. Sterculiaceae.

Hypericum japonicum Thunb.; 田基黃 (Paddy St. John's wort); whole plant; Zhuang report. Guttiferae.

Ilex asprella (HK. et Arn.) Champ. ex Bentham; 崗梅根 (Plum-leaved Holly); roots, sliced; Wong (36% by weight of entire parcel). Aquifoliaceae.

I. pubescens HK. et Arn.; 毛冬青 (Downy Holly); leafy shoots; Sun. Aquifoliaceae.

I. rotunda Thunb.; 救必應 (Panacea Holly); Stems, sliced; Macao, Sun. Aquifoliaceae.

Imperata cylindrica (L.) P. Beauv.; 茅根 (Woolly Grass); rhizomes; Zhuang report, Tai-po, Macao. Gramineae.

Juncus alatus Franch. et Sav.; 燈心草 (Lamp-wick Rush); whole plant; Zhuang report. Juncaceae.

Laggera alata (D. Don) Sch.-Bip.; 六耳苓 (Six-ears Angled); whole plants; Sun. Compositae.

Lonicera confusa DC.; 山銀花 (Honeysuckle Vine); leavy shoots; Sun. Caprifoliaceae.

Lophatherum gracile Brongn.; 淡竹葉 (False Bamboo); whole plant; Zhuang 5, Tai-po, Macao, Wang. Gramineae.

Lygodium dichotomum Swartz; 海金沙 (Climbing Fern); leafy branches; Zhuang 7, Sun. Lygodiaceae.

Mangifera indica L.; 芒果核 (Mango Stone); stone of fruit, saved from table, cleaned, dried; Zhuang 2, Tai-po, Macao. Anacardiaceae.

Melastoma candidum D. Don; 野牡丹 (Bushy Melastoma); whole plant; Sun. Melastomaceae.

Microcos nervosa (Lour.) S. Y. Hu; 布渣葉 (Microcos); leafy shoots, also chopped wood; Zhuang 7, Tai-po, Macao, Wong. Tiliaceae.

Momordica charantia L.; 苦瓜莖 (Bitter-cucumber); old stems, used after harvest; Zhuang 3. Cucurbitaceae.

Morus alba L.; 冬桑葉 (Mulberry); chopped branches, and leaves gathered after frost; Zhuang, Tai-po, Macao, Wong. Moraceae.

Oroxylum indicum (L.) Vent.; 千層紙 (Thousand-fold Paper); white winged-seed; Zhuang 4, rare now. Bignoniaceae.

Osbeckia chinensis L.; 天香爐 (Chinese Osbeckia); whole plant; Zhuang 3. Melastomaceae.

Oxalis corniculata L.; 酸味草 (Creeping Oxalis); whole plant; Zhuang 3. Oxalidaceae.

Pandanus tectorius Soland.; 蘆刀根 (Pandanus Root); roots, chopped to pieces; Zhuang 2, Tai-po, Macao. Pandanaceae.

Phragmites karka (Retz.) Trin.; 水蘆強 (Reed); rhizomes, cut into pieces; Zhuang 2. Gramineae.

Phyllanthus cochinchinensis Spreng.; 鐵包金 (Iron Wrapping Gold); leavy shoots; Sun. Euphorbiaceae.

Plantago major L.; 車前草 (Common Plantain); whole plant, seasonal in Hong Kong, local material available only in spring, supply supplemented by material imported from China. Plantaginaceae.

Pogonatherum crinitum (Thunb.) Kunth; 金絲草 (Golden Hair Grass); whole plant; Zhuang 2, much used. Gramineae.

Polygonum chinense L.; 火炭母 (Trailing Smartweed); whole plant; Macao, Wong. Polygonaceae.

P. perfoliatum L.; 老虎刺 (Spiny Knotweed); leafy branches; Zhuang record. Polygonaceae.

Prunella vulgaris L.; 夏枯草 (Selfheal); whole plant; Zhuang record, not growing in Hong Kong; supply imported from China; Labiatae.

Psychotria rubra (Lour.) Poiret; 山大刀 (Red Psychotria); leafy shoots; Zhuang record. Rubiaceae.

Rosa laevigata Michx.; 金英強 (Cherokee Rose); flowering branches; Zhuang 4, Wong (18% of total weight). Rosaceae.

Rubus parvifolius L.; 茅�…(May Rasberry); leafy shoots with flowers and/or fruits; Tai-po, Macao. Rosaceae.

Sageretia thea (Osbeck) M. C. Johnston; 雀梅藤 (Hedge Sageretia); Tai-po, Macao. Rhamnaceae.

Sapindus mukorossi Gaertn.; 木患根 (Soap Berry Root); chopped wood; Zhuang 4, Tai-po, Macao. Sapindaceae.

Sapium discolor (Champ. ex Bentham) Muell.-Arg.; 山烏桕 (Hillside Tallow-tree); chopped branches; Tai-po. Euphorbiaceae.

Schefflera octophylla (Lour.) Harms; 鴨腳木皮 (Schefflera Bark); bark, chopped into 2 x 3 cm pieces; Zhuang 3, Tai-po, Macao. Araliaceae.

Scoparia dulcis L.; 山甘草 (Scoparia); whole plant; Zhuang record, Sun. Scrophulariaceae.

Solanum torvum Swartz; 水茄 (Golden Buttons); Tai-po. Solanaceae.

Triumfetta bartramia L.; 刺蒴麻 (Bur Bush); Tai-po. Tiliaceae.

Uraria crinita Desv.; 狐狸尾 (Fox's Tail Pea); whole plant; Zhuang record. Leguminosae.

Urena lobata L.; 藕頭婆 (Rose Mallow); leafy shoots; Tai-po. Malvaceae.

Vitex cannabifolia Sieb. et Zucc.; 牡荊葉 (Hemp-leaved Vitex); leafy shoots; Zhuang 5, Sun. Verbenaceae.

V. rotundifolia L.f.; 蔓荊子 (Creeping Vitex); fruits; Zhuang record. Verbenaceae.

Wedelia chinensis (Osbeck) Merr.; 蟛蜞菊 (Chinese Wedelia); whole plant; Zhuang record. Compositae.

Youngia japonica (L.) DC.; 還陽草 (Revitalizing Herb); whole plant with flowers; Zhuang record, Macao. Compositae.

Without an intimate knowledge of the native plants and their Chinese names, it would be impossible to identify most of the above source species of *liangcha*, because 26% of them are leafy shoots of local trees or shrubs, and 40% are herbaceous plants gathered before they flower. In addition to this predominantly leafy material, there are 7 items of chopped wood and/or sections of leafless branches, one bark of a tree, two rhizomes of grasses, one petioles of an aroid and three seeds. To botanists who work in herbaria or museums, all the leafy and woody material are sterile specimens, not excepted for identification or for keeping.

Various *liangchas* are prepared by complex traditional formulas, the number of ingredients varying in multiples of six or eight, for examples, 6, 8, 16, 24, etc. On the side streets of Guangzhou, Hong Kong and Macao, and in the marketplaces of smaller cities and towns, *liangcha* is prepared and served by professionals with special equipment to decoct and to sell the beverage in bowls or glasses. Before the introduction of soft drinks, people used to stop at the *liangcha* shops for a healthy refreshment, much as one would stop at an American drug store for a cup of coffee or coke. Now, people take *liangcha* more for health purposes than as a refreshment.

It should be pointed out that in the preparation of *liangcha*, some of the source species are more important than the others. As the use of *liangcha* is an essential healthcare issue of the populance of the tropical and subtropical China, the isolation of the important source species for the ready-made decoction sold in cities, towns and villages may lead to the discovery of new plant drugs for the control of epidemic diseases, such as the various types of Asia flus. Evidently, the more effective material would be used by more people in the *liangcha* business, as indicated by the large number of samples recorded by Zhuang in Hong Kong and Kowloon and Hu in Tai-po, the New Territories of Hong Kong, and by the remark on "much used". An analysis of the number of samples and the amount used indicates the following species are more important than the others: *Alternanthera sessilis* R. Brown, *Centella asiatica* (L.) Urban, *Cleistocalyx operculatus* (Roxb.) Merr. et Perry, *Desmodium styracifolium* (Osbeck) Merr., *Evodia lepta* (Sprengel) Merr., *Hypericum japonicum* Thunb., *Ilex asprella* (HK. et Arn.) Champ. ex Bentham, *Ilex rotunda* Thunb., *Lophatherum gracile* Brongn., *Lygodium dichotomum* Swartz, and *Microcos nervosa* (Lour.) S. Y. Hu.

Parcelled Herbal Tea in American Chinese Stores

In conjunction with my investigation of *bupin* and *liangcha* in Chinese stores at Boston, I saw some medicated tea in colorful boxes. I bought two samples, brought them to Harvard University Herbaria, and showed them to a visiting Chinese scholar, Professor D. Z. Chen, South China Institute of Botany, Academy of Science, Guangzhou (Canton), Guangdong, China. She was delighted to know that these products were available in Boston and wanted to take some immediately. She told me that in Guangzhou, she always keeps some handy at home and when she feels tired and lack of efficiency in her research, she simply makes a drink with the content of one parcel. The tea would restore her vitality and keep her from becoming sick. On hearing her explanation, I immediately opened a parcel, mixed it with one liter of boiling water, and simmered the mixture for an hour. The decoction appeared black and tasted bitterish. I added some sugar and divided it among the four of us. We drank with excitement.

On one side of the boxes, the major ingredients of the content were listed in Chinese, with their romanized name (the scientific epithet or the Cantonese transliteration of the local name), followed by the percentage of that item by weight. The Chinese names indicated that the ingredients were all classical crude drugs used in traditional Chinese medicine.

With the help of two licensed acupuncturists and herbologists, Dr. Kwok Lap Wong and Dr. Li Bang Zhao, who had clinics and/or herbal business in Boston's Chinatown, I obtained six more samples of the parcelled herbal tea which they recommended and recognized to be the common ones used by the populance in Guangzhou and its vicinity. They have both finished college and/or professional training in traditional Chinese medicine and practiced before coming to USA. They maintained that these medicated teas together with *liangcha* provided the fundamental health care of the general public in southern China, keeping them in tone between the change of weather, and saving them from going to the physicians with colds, headaches, pains, indigestion, sun strokes, and flus.

Eight samples had been carefully studied for the recognizing character from the external appearance and seven for the botanical identification of the source species in the internal contents. Each brand with identifiable content was assigned a number, from 1 to 7, for the purpose of reference in further discussions. Two of the samples contained a commercial tea (茶), which was the leaves of *Camellia sinensis* [L.] O. Ktze.; two others had fermented materials; one consisted of flowers only; one was designed for children, and one was a simple one made of three ingredients only. The last one was not identifiable for it gave no indication of the pulverized content. Descriptions of the samples were given by this loose grouping.

(1) *External Description of Parcelled Tea*: Five of the seven samples are packed in colorful rectangular boxes, varying between 12–15 x 7–10 x 4.5–6.5 cm in dimensions. The brand names used in the description are taken from the boxes directly or they are my translations of the Chinese names on the outside cover. In the parenthesis immediately after the accepted brand name are the Chinese name, its pinyin romanization and the number assigned to that brand to facilitate further discussion. Two of the samples are placed in paper bags or loose wrappings.

(a) *Boxes Containing Commercial Tea*: "Panacea Harmonizing Tea" (No. 1) (*Wan-yin Gan-he Cha* 萬應甘和茶). This brand name is my translation of the Chinese printed on both ends of the box. However, on the bottom of the box as well as its inside cover, the brand name given in Cantonese transliteration is "Yuen Kut Lam's Kam Wo Char" 源吉林甘和茶. The bottom of the box is densely covered with 31 Chinese names in the left column, 14 scientific names mixed with the Cantonese transliterations of 13 local names, and 4 English common names in the center column and the percentage by weight on the right column. The remaining space is covered by the indications, methods of preparation, the address of the manufacturer, the number of parcels, the weight of each unit and the net weight in total.

"Ho Yan Hor" (No. 2) (*He Ren Ke* 何人可): This brand name is taken from the box and it is printed on three sides. At one end of the box are 22 scientific names of the source species. On the other end is printed, "… a substitute for the ordinary tea or coffee … ideal for those who enjoy good foods and sumptuous meals … contains the extract of herbs …", and the directions.

(b) *Boxes Containing Fermented Material*: "Kanlu Tea" (No. 3) (*Gan Lu Cha* 甘露茶): This brand name is the Cantonese transliteration of the Chinese name printed on the top and on one end of the box. It is manufactured, packed and exported by the United Pharmaceutical Factory, Guangzhou, China. On the bottom of the box are the ingredients given in three columns, with the Chinese names on the left, pharmaceutical names in the middle and the percentage by weight on the right. At the bottom of the names "no poison" is written in capitals.

"Shen Chu Cha" (No. 4) (*Shen-qu Cha* 神麴茶): The brand name is printed on two sides of the box. On both ends of the box appears the trademark "Yang Cheng Brand" in Chinese and its transliteration, with a picture of a building in the center. At the bottom of the box, the ingredients are given in pharmaceutical terms without Chinese equivalents and the amounts used. This brand is manufactured by the United Herbal Tea Factory in Guangzhou and exported by the Guangzhou Medicines & Health Productcs I/E Corporation.

(c) *Herbal Tea in Loose Packages*: Two types of herbal tea are packed in local stores, each containing material for one serving. Customers can buy many packets for home use. These are the *"Wu Hua Cha"* 五花茶 (Five Flowers Tea) and *"Qi Xing Cha"* 七星茶 (Seven Stars Tea).

Five Flowers Tea (No. 5) (*Wu Hua Cha* 五花茶): This is the most prevailing home-made herbal tea used in Hong Kong during the hot summer seasons. In mid-June 1989, after a month's traveling to the Chinese natural reserve centers in Yunnan, Sichuan and Helongjiang, I became a guest of Dr. and Mrs. Paul P. H. But at the Department of Biology, The Chinese University of Hong Kong. The weather was unusually hot and humid. One afternoon, Mrs. Tammy But brought me a jug of black beverage saying, "Please drink this. It is good for your health in this weather. The children and I have already had some while you took a nap." Answering my inquiries about the drink, she told me that it is *"Wu Hua Cha"*, prepared from a package bought in the village market.

Seven Stars Tea (No. 6) (*Qi Xing Cha* 七星茶): This is a popular medicated tea used in southern China, designed for children who become lethargic, having a cold or indigestion. Packets prepared according to a Cantonese formula are wrapped loosely in brown paper, ready for customers in Boston' Chinatown.

(d) *A Simple Herbal Tea*: Canarium Onion Tea (No. 7) (*Lan Cong Cha* 欖葱茶): This brand name is a direct botanical translation of the Chinese appeared on the box. The original English name is not accepted purposely to avoid an early misnomer of the Chinese fruit, *gan-lan* 橄欖 (*Canarium album* [Lour.] Raeusch.). When the early European residents in Guangzhou (Canton) and Macao saw the Chinese canarium, which resembles the size and shape of a large olive of the Mediterranean area, they called it Chinese olive. Botanically, the Chinese canarium and the European olive belong to different orders in the plant kingdom. A canarium tree has compound alternate leaves and large panicles of tiny flowers with distinct petals, while the European olive has simple opposite leaves and small clusters of flowers with short corolla tubes. The original name printed on the box of this brand is "Olive Onion Herbs Tea". This name is misleading for there is no olive tree that yields edible fruit in tropical southern China and Southeast Asia. In *A Dictionary of the Economic Products of the Malay Peninsula*, Burkill (1935) used "Canarium" as an English name for the Malaysian species of *Canarium* L.. The correct translation of the local Chinese brand name should be "Canarium Onion Tea". It is better to discontinue the use of a former misnomer before it is widespread. This brand is manufactured and distributed by Chap Lan Tong in Hong Kong. On the bottom of the box, the nutrition facts are given in great details and the price is indicated by the barcode as required in any American food package.

(e) *Herbal Tea in Paper Bags*: Six Harmonizing Tea (*Liu-he Cha* 六和茶). This brand name is a direct translation of the Chinese name printed on the three sides of the sealed paper bag. From the Chinese pinyin of the ideograms, it is evident that this brand shares two words with the name of herbal tea no. 1: "Panacea Harmonizing Tea" (*Wan-yin Gan-he Cha* 萬應甘和茶), i.e., *"he cha"* 和茶 meaning "harmonizing tea". The bag paper is thin, brown, with red prints on the three sides, giving the Chinese brand name repeatedly, and the indications for its use as a cure for heat stroke, cold, gas pain and as a substitute for tea. This brand is manufactured by a private business enterprise in Tai Shan 台山, Guangdong, the homeland of the majority of overseas Chinese businessmen of the Chinatowns throughout the world. Only Chinese is used for giving information regarding the manufacturer, distributors, indication, and direction. The paper bag contains ten colorful paper boxes 8 x 5.5 x 2 cm in dimension. No ingredients are given anywhere and the content is so fragmentary that no identification of the source species is possible. For this reason, no number is assigned to this brand.

(2) *Internal Contents*: Every brand of the first seven parcelled Chinese herbal teas described above has its special ingredients and their contents appear different. The content of the eighth type is not identifiable. In conformity to the regulations of the American Maritime Customs, six brands print the major ingredients on the outside of the outer containers and/or the inner wrappings. Nevertheless, certain details are kept as trade secrets, and each brand has its specifications.

Panacea Harmonizing Tea (1): This brand contains ten small green boxes, 4.5 x 3 x 2 cm long, wide and high, with a red band covering three sides. On the red sides are printed the name and address of the original designer in a village of eastern Guangdong and the date (1968) of its packing in Hong Kong. Each small box is wrapped with very thin white Chinese paper 13–14 cm square. On the paper are printed green Chinese explanations of the nature of the tea, its indications for fever, headache, vomiting, pains in the bone, indigestion, and summer sun stroke with direction for serving. Inside the small box is black regular Chinese tea treated with an infusion of the

31 kinds of crude Chinese drugs, their identifications are marked "1" in the following list.

Ho Yan Hor (2): This brand contains ten white paper bags 11 x 9.5 cm long and wide, on one side of which are printed in red and green, the trade mark, the net weight of the tea, and the name and address of the producer in Perak, Malaysia. On the other side of the bag are given the indications, directions for preparation and pharmaceutical names of the 22 ingredients and their proportion by weight. The amount of commercial tea used is 63.5% by weight and that of licorice is 11.6%. The remaining 20 items used are in very small amounts (0.6–4.6%). All the ingredients are pulverized into fragments 1–5 mm long and sealed in a regular tea bag with a thread and a holder, on which is the trade mark.

Kanlu Tea (3): This brand contains ten parcels of white paper bags 11 x 8 cm long and wide. Printed green on one side are the brand name, indications for influenza, cold, headache, indigestion, gastrodistention after eating, and directions for preparing tea. Ginger may be added if preferred. On the other side of the bag, nine ingredients are listed in pharmaceutical terms. This is a tea made of both wild plant material and crude Chinese drugs. The major weight (50.47%) of the tea consists of the leaves of a local species of an araliaceous tree (*Schefflera octophylla* [Lour.] Harms.) The weights of the other 8 items vary between 7.48 and 3.73%. All the ingredients are broken into fragments, some coarse and others fine. A piece of medicated yeast cake, "Massa Medicata Fermentata", is broken into a white powder. This yeast cake is a compound material consisting of six known species of plants used in the fermentation process for its product. The identifications of these source species are credited to the brand as "3" in the list.

"Massa Medicata Fermentata", appeared in the list of pharmaceutical terms for *Shen-qu* 神麴 in *An Enumeration of Chinese Materia Medica* (Hu, 1980, using the Wade-Giles System of Romanization). The spelling in Pinyin System is *"shen-qu"*. This medicated yeast cake is the modern form of an ancient home remedy invented along the Yellow River region in northern China. It was first recorded in an early Chinese herbal compiled between 450–643 A.D.. The ideogram *"shen"* 神 in the name *"shen-qu"* 神麴 has many meanings in the Chinese language. In combination with *"qu"* 麴 (yeast) it can be translated as "divine yeast", "incredible yeast", or "yeast-of-gods". The market material of *shen-qu* consists of ochreous or white dry squares, 3 x 3 x 1 cm in dimension. It is used in many prescriptions designed for gas pain, indigestion and ailments of the digestive system. Medicated yeast cake is manufactured by special apothecaries and distributed as *"shen-qu"* 神麴.

Preparation of shen-qu: In the early Tang Dynasty when *shen-qu* was first recorded in an early herbal, the center of Chinese civilization was in the Yellow River region. The plant material used in the fermentation process for making *shen-qu* was native there then as it is now and the procedure has not altered very much either. The source species consist of three fresh wild plants and three cultivated ones. As recorded in *An Encyclopedia of Chinese Medicines* (Anonymous, 1977) the amounts used for the species are: 6 kg each for ground rice-bean (*Delandia umbellata* [Thunb.] S. Y. Hu) and apricot seed (*Prunus armeniaca* L., seed coat removed); 3 kg each for chopped fresh Chinese wormwood (*Artemisia apiacea* Hance), smartweed (*Polygonum hydropiper* L.), and cocklebur (*Xanthium sibiricum* Patrin ex Widder); 50 kg bran and 30 kg wheat flour (*Triticum aestivum* L.). The procedure is simple. Water is added to the mixture of the above ingredients to form a dough. A handful of the dough is kneaded to form a smooth ball which is pressed into a cake about 1 cm thick. The cakes are piled up and covered with straw to undergo fermentation by yeast cells attached to the fresh plants and/or by air-borne spores of micro-organisms in a warm room. When the cakes appear yellow they are cut into serving sizes, dried and stored for distribution.

Variations in shen-qu: There are variations in the amounts of the six ingredients used and in the methods of *shen-qu* preparation. One record goes like this. 2.2 kg each of apricot seed and rice bean are soaked and ground together. To the mixture, a mass of chopped smartweed (1.8 kg) and cocklebur (1.5 kg) are added and soaked together. Then 56 kg of wheat flour are added to make a dough for small cakes. Each cake is wrapped with hemp leaves (*Cannabis sativa* L.) and piled together for fermentation until they appear yellow. Procedures in drying, storage and distribution follow.

Shen Chu Cha (4): Two of the Chinese ideograms on the box of this brand are identical to those of Massa Medicata Fermentata (*shen-qu* 神麯). However, the contents of the materials are completely different. The medicated yeast cake described above for *shen-qu* is made of fresh and dried material of wild northern species. This branch labeled "Shen Chu Cha" 神麯茶 is manufactured in southern China and with the exception of wheat flour (38% by weight), all the other ingredients are crude drugs used in traditional Chinese medicine. The box contains 8 hard, olivaceous, rectangular blocks made in a mold, 5 x 4 x 2 cm in dimension. This mold has a design of two Chinese ideograms, " 羊城 " (*Yang-cheng* = Goat City) as the trade mark of the brand. Subsequently, the top side of the smooth olivaceous block is completely covered by the two elevated Chinese ideograms. Each block is wrapped in a piece of square white paper, 15 cm each way, on which are printed in Chinese the indications, direction for preparation and 14 ingredients with percentages by weight. It is advertized for improving digestion, strenthening the function of spleen, detoxicating alcoholic poison, and for cold with fever, gas pain, vomiting, dysentary and coughs.

Five Flowers Tea (5): The *Wu Hua Cha* (五花茶) purchased in Tai-po market by Tammy But contained dried flowering heads of a small-flowered white chrysanthemum, flowers of cream-colored honeysuckle, of purplish kudzu, and of black frangipani and tree cotton. The last two species are cultivated garden plants with fleshy corollas which drop to the ground after blooming and dried black. These are gathered, dried and sold to local herb dealers who sell *Wu Hua Cha* in parcels.

Five Flowers Tea is available in Boston's Chinatown, where the large purple flowers of kadzu are replaced by the yellowish dried flowers and buds of pogoda tree imported from China. Actually this species blooms abundantly in the Arnold Arboretum every year. Likewise, the honeysuckle vines on the fences in Cambridge, Massachusetts, flower and fruit annually. American people have not learned the use of these species other than horticulture. The botanical identifications of the flowers for preparing a healthy summer drink are indicated by the numeral 5.

Seven Stars Tea (6): This is a herbal tea designed for children and packed in loose parcels by Chinese herbologists in Chinatown. The ingredients are in large fragments 4–6 cm long, in loose bundles 4 x 2 cm in dimension, or in wholesome grains ready to be identified. The primary seven items used for this herbal tea are: (1) sections of the underground stem of licorice, (2) leaves of Lophatherum locally known as the false bamboo, (3) fruits of slightly germinated barley grains with husk, (4) fruits of sliced hawthorn, (5) polished seeds of Job's tears, (6) *shen-qu* and (7) *chan-hua* 蟬花 (Cicada flower), the branched sclerotium of an ascomycete attached to the nymphal skin of a cicada, the host of the fungus. Variations in adding nonpoisonous ingredients are common. The parcel I obtained from the Herbologist, Li Bang Zhao, contains pith of rice paper, stems and pith of lamp-wick rush, leaves of purple perilla, rice grain with hust, and popped Job's tears. He calls this formula the "Double-strength Seven Star Tea" (*Shuang-liao qi-xing cha*, 雙料七星茶). In the list of botanical identifications, numeral 6 is used to indicate all the source species.

Canarium Onion Tea (7): A box of this brand contains 10 paper bags 13 x 7.5 cm long and wide, one side light yellow and the other side white. On the yellow side, the information printed on the cover of the box is repeated. On the white side, direction for preparing the tea and the Hong Kong telephone and address of the manufacturer are printed. The contents of the bag consists of coarse fragments (1–2.5 cm sizes) of purple perilla leaves (40% by weight), of onion (30%) and of canarium fruits (30%). The botanical identifications of the source species are indicated by numeral 7 in the list.

Identification of Source Species of Parcelled Herbal Teas

A total of 71 species of botanical resource are identified from the seven samples of Chinese medicated tea described above. They are represented by 14 species in seven families of monocots, 55 species in 25 families of dicots and two parasitic fungi, a basidiomycete and an ascomycete. These species are arranged alphabetically by their scientific names in the following list. In each item, information about the Chinese name, English name (in parenthesis), and the part of the plant used in tea are given. A number or numbers representing the samples, and the family to which the species belong are supplied at the end.

Agastache rugosa (Fischer et Meyer) O. Ktze.; 藿香 (Agastache); leafy branches; 1, 2, 3. Labiatae.

Alisma plantago-aquatica L.; 澤瀉 (Water Plantain); 1. Alismataceae.

Allium fistulosum L.; 葱 (Welsh Onion); leafy plant; 7. Liliaceae.

Amomum tsaoko Crevost et Lemaire; 春砂 (Tsaoko); fruits; 4. Zingiberaceae.

A. villosum Lour.; 檳榔 (Grain-of-Paradise); 1. Zingiberacaeae.

Angelica anomala Lallem.; 白芷 (Sichuan Angelica); dried roots; 2. Umbelliferae.

A. pubescens Maxim.; 獨活 (Tuhuo); dried root; 4. Umbelliferae.

Areca catechu L.; 檳榔 (Betel Nut); seed; 1. Palmae.

Artemisia apiacea Hance; 青蒿 (Celery Wormwood); leafy shoots; 1, 3, 4. Compositae.

A. capillaris Thunb.; 棉茵陳 (Capillary Wormwood); very young leaves; 1. Compositae.

Atractylodes lancea (Thunb.) DC.; 蒼朮 (Cang Zhu); rootstocks; 2. Compositae.

Bombax ceiba L.; 攀枝花 (Tree Cotton Flowers); Dried flowers; 5. Bombacaceae.

Bupleurum chinense DC.; 北柴胡 (Chinese Thorowax); rootstocks; 2. Umbelliferae.

Camellia sinensis (L.) O. Ktze.; 茶葉 (Chinese Tea); Commercial tea; 1, 2. Theaceae.

Canarium album (Lour.) Raeusch.; 白欖 (Canarium); salt-preserved fruit, broken into pieces; 7. Burseraceae.

Cannabis sativa L.; 大麻葉 (Hemp Leaves); fresh leaves, used by some producers in the fermentation process in making medicated yeast (shen-qu); 3. Cannabaceae.

Chaenomeles sinensis (Thouin) Koehne; 川木 (Chinese Quince); sliced fruits; 1, 2, 4. Rosaceae.

Cimicifuga heracleifolia Komarov; 升麻 (Chinese Bugbane); dried rhizomes; 2. Ranunculaceae.

Citrus reticulata Blanco; 陳皮 (Mandarin Orange Peel); dried peels, saved after eating the fruit; 1. Rutaceae.

C. sinensis (L.) Osbeck; 青皮 (Orange Peel); dried immature fruit; 3, 4. Rutaceae.

Clematis chinensis Osbeck; 威靈仙 (Chinese Clematis); sections of vine; 1. Ranunculaceae.

Coix lachryma-jobi L.; 薏仁 (Job's Tears); polished seeds, both row and popped, used separatedly or together; 6. Gramineae.

Coptis chinensis Franchet; 黃連 (Chinese Golden Thread); rootstocks; 1. Ranunculaceae.

Cordyceps sobolifera (Hill.) Berk. et Br.; 蟬花 (Cicada Flower); branched sclerotium attached to the apical portion of the nymphal skin of the host, a cicada; 6. Ascomycete.

Crataegus pinnatifida Bunge; 山楂 (Hawthorn); sliced fruits; 1, 2. Rosaceae.

Delandia umbellata (Thunb.) S. Y. Hu; 赤小豆 (Rice Bean); Seeds, used in the preparation of medicated yeast; 1. Leguminosae.

Dendranthema x grandiflorum (Ramat.) Kitamura; 菊花 (Tea Chrysanthemum); dried small white flowers; 5. Compositae.

Dioscorea opposita Thunb.; 懷山 (Chinese Yam); sliced dried root; 4. Dioscoreaceae.

Elsholtzia splendens Nakai ex Maekawa; 香茹 (Aromatic madder); leafy shoots; 1, 3, 4. Labiatae.

Forsythia suspensa (Thunb.) Vahl; 連翹 (Forsythia); mature fruits; 1. Oleaceae.

Gardenia jasminoides Ellis; 梔子 (Gardenia); fruits; 1, 2. Rubiaceae.

Glycyrrhiza uralensis Fischer, **G. kansuensis** Chang et Peng, **G. inflata** Batalin, **G. glabra** L.; 甘草 (Chinese Licorice); roots or underground stems; 1, 2, 4. Leguminosae.

Hordeum vulgare L.; 大麥芽 (Germinated Barley); germinated seeds with husk; 2, 6. Gramineae.

Ilex latifolia Thunb.; 苦丁茶 (Broad-leaved Holly); leafy shoots; 1. Aquifoliaceae.

Juncus alatus Franch. et Sav.; 燈心草 (Lamp-wick Rush); Aerial portion of the leafless stems and pith, tied in bundles; 6. Juncaceae.

Ligusticum wallichii Franchet; 川芎 (Sichuan Ligusticum); sliced rootstock; 2. Umbelliferae.

Lindera aggregata (Sims) Kosterm.; 烏藥 (Chinese allspice); root; 1. Lauraceae.

Lonicera japonica Thunb.; 金銀花 (Honeysuckle Flower); Dried flowers and flower-buds; 5. Caprifoliaceae.

Lophatherum gracile Brongn.; 淡竹葉 (False Bamboo); whole plants, cut into sections 5–6 cm long; 1, 6. Gramineae.

Magnolia officinalis Rehder et Wilson; 厚朴 (Medicinal Magnolia Bark); dried pieces of bark; 2, 4. Magnoliaceae.

Mentha piperita L.; 薄荷 (Peppermint); leafy shoots; 2. Labiatae.

Microcos nervosa (Lour.) S. Y. Hu; 布渣葉 (Microcos); leaves; 1. Tiliaceae.

Morus alba L.; 冬葉 (Mulberry leaves); leaves gathered after frost; 1. Moraceae.

Nelumbo nucifera Gaertner; 蓮葉 (Sacred Lotus Leaves); dried leaves; 1. Nymphaeaceae.

Notopterygium franchetii Boiss.; 羌活 (Notopterygium); dried rootstock; 2, 4. Umbelliferae.

Oryza sativa L.; 穀芽 (Rice Grain); Rice with husk; 6. Gramineae.

Perilla frutescens (L.) Britt.; 蘇葉 (Purple Perilla Leaves); 1, 2, 3, 7, 40% of total weight. Labiatae.

Platycodon grandiflorus (Jacq.) A. DC.; 桔梗 (Balloon-flower); Root; 1, 2, 4. Campanulaceae.

Plumeria rubra L.; 紅雞蛋花 (Frangipani); dried flowers dropped after blooming; 5. Apocynaceae.

Pogostemon cablin (Blanco) Bentham; 藿香 (Patchouli); leafy shoots; 1, 2. Labiatae.

Polygonum hydropiper L.; 蓼 (Common Smartweed); fresh plants used in the fermentation process for making medicated yeast; 3. Polygonaceae.

Poncirus trifoliata (L.) Raf.; 枳殼 (Trifoliate Orange); fruits; 2. Rutaceae.

Poria cocos (Schw.) Wolf; 茯苓 (China Root); sclerotia; 4. Basidiomycetes.

Prunella vulgaris L.; 夏枯草 (Selfheal); Dried inflorescence; 2. Labiatae.

Prunus armeniaca L.; 北杏仁 (Apricot Seed); seed, soaked and with seed-coat removed; 1. (Apricot Seed); wholesome seed, with seed-coat; 2. Rosaceae.

Pueraria lobata (Willd.) Ohwi; 乾葛 (Wild Kudzu); flowers; 5. Leguminosae.

P. thomsonii Bentham; 甜葛 (Sweet Kudzu); root; 1. Leguminosae.

Rheum officinale Baill.; 大黃 (Medicinal Rhubarb); rootstock; 2. Polygonaceae.

Saposhnikovia divaricata (Turcz.) Schischk.; 防風 (Fang-feng); rootstock; 1, 2, 3. Umbelliferae.

Schefflera heptaphylla (L.) Frodin; 鴨腳木 (Schefflera); leaves; 3, 50% of total weight. Araliaceae.

Schizonepeta tenuifolia (Bentham) Briq. 荊芥 (Fine-leaved Schizonepeta); partially flowering and fruiting leafy plants; 2. Labiatae.

Scrophularia ningpoensis Hemsley; 玄參 (Ningpo Figwort); partially cured root; 2. Scrophulariaceae.

Scutellaria baicalensis Georgi; 黃芩 (Baical Skullcap); Rootstock; 1, 4. Labiatae.

Sophora japonica L.; 槐花 (Pagoda Tree); dried flowers and flower-buds; 5. Leguminosae

Tetrapanax papyriferus (HK.) K. Kock; 通草 (Rice Paper); pitch from the stem; 6. Araliaceae.

Tricosanthes kirilowii Maxim. 天花粉 (Tricosanthes Root); dried roottuber, poisonous; 1. Cucurbitaceae.

Triticum aestivum L.; 麥麩 (Wheat); bran 3; flour 3, 4, used in the fermentation process. Gramineae.

Xanthium sibiricum Patrim ex Widder; 蒼耳 (Cocklebur); fresh young plants, used in the preparation of medicated yeast; 3. Compositae.

Massa medicata fermentata; 神麴 (Medicated Yeast Cake); white powder and pieces of a special yeast; 3. Mixture of several species.

Conclusions

What herbal products do the Chinese people take to enhance immunity? Answers to this question vary with the cultural background and geographical distribution of the Chinese population, because their food habits are modified by climate, topography, and many other natural and cultural factors. We do know, however, that in their struggle for survival in a densely over-populated environment where unexpected natural calamities are frequent, famines often, wars common, and diseases widespread, the Chinese people have learned to take unusual herbs to enhance immunity and to prolong life. These valuable experiences have not only been handed down from one generation to the next, but also these precious heritages have migrated with the people. For example, the Chinese immigrants in the United States of America have established special stores in various metropolises to supply fresh, dried, and canned goods, spices and flavoring material to meet their cooking needs. Their habit of using immunostimulants has created the market demand for the importation of many Chinese herbs which are piled up on shelves of gift shops and groceries organized primarily for the Chinese people.

An investigation into the recently imported products sold as *bupin, liangcha* or parcelled medicated tea in the Chinese gift shops, groceries and special herbal stores in Boston has not only provided me information about the kind of herbs the Chinese people take to enhance immunity, but has also illustrated the availability of these items in America, and documented the botanical identifications and phytochemical findings basic to further research regarding plants for the conservation of health. The market prices and suggested recipes for preparing *bupin* dishes would be useful to those who like to try some of the plant products. The immunostimulant herbs used by the Chinese people in the Boston area can be grouped into three categories: the *bupin, liangcha* and parcelled medicated tea. Although the primary object of using these herbs is to conserve the health of normal people, the nature of the source supplies and the methods of their preparation and serving are different.

Bupins are plant products used in traditional Chinese medicine, and described in ancient herbals. Now, they are rare in nature, and are cultivated extensively in China. Sixty-six items of *bupin* from 73 source species are common in the market. Chinese pharmaceutical and medical scientists have carried out extensive research into the chemical composition of many of these items, studied the physiological properties of the active principles, and published numerous clinical observations. Their reports were published in Chinese. A selection of the 28 most commonly employed products, including 4 fungi, 11 roots, one parasitic flowering plant, one bark, 3 leaves and/or leafy shoots, 2 flowers, 3 fruits, and 2 seeds are discussed in detail and illustrated with 40 recipes showing how they can be prepared for personal and/or family use.

Bupins are prepared at home, with medicinal herbs in combination with food material, such as rice, meat, poultry, and seafood. They are used in small amounts as one would use spices in Western cuisine and the *bupin* dishes are often regarded as a special treat for the family. They are served seasonally or only occasionally. People deficient in vital energy, such as the elderly and those overburdened from the stresses and strains of everyday life, whose immune systems are weak, are given *bupin* more frequently to restore their vitality and raise their immunological parameters. The recipes have been tried by myself and a volunteer, Beverly Huckins, and the preparations have been shared by many colleagues at the Harvard University Herbaria. As a food, the preparations have agreeable tastes and excellent textures.

Liangcha and medicated tea are both vegetal preparations taken in the form of decoction. *Liangcha* is prepared with wild plants in complex formulas with 6 or 8 items, often in multiples of these numbers, up to 32 ingredients. In Hong Kong, Guangzhou (Canton) and their vicinity, *liangcha* is commercialized, sold in small stores on side streets or alleys in large cities and in marketplaces in large villages, with the boiling apparatus and several rice bowls filled with the liquid ready to be taken. People on the street stop by, pay a very small fee, take a bowl, drink the content while standing and walk away.

Medicated tea mixes are packets of commercial tea, impregnated with infusion of traditional Chinese medicine, or pulverized combination of from 3 to 22 herbs, bought by people for steeped at home. A total of 70 source species of medicinal material employed in medicated tea are identified. Six of these species are involved in the fermentation of *shen-qu* added to one formula. Medicated tea are prepared at home according to the instructions printed on the labels. The decoctions were tested, and their safety was proven. As for the taste, one must get used to it!

Unlike *bupin* products, only a few of the ingredients of *liangcha* and medicated tea have been studied by scientists. The research on *liangcha* and medicated tea is still at an ethnobotanical level. The identification of the source species provides a basis for future investigators who are qualified to separate the grain from the chaff. Moreover, 56% of the source species are already growing in American gardens as ornamentals. Recently, industrious farmers and gardeners have supplied Chinese groceries and American supermarkets with various cultivars of *Brassica juncea* and *B. rapa* while enterprising businessmen have built modern factories for producing nutritious products from soybean (*Glycine max*) and crisp sprouts from mungbean (*Phaseolus mungo*) for the consumption of the American public. Adventurous owners of herb farms and inventive herb dealers may find new sources of revenue from information provided here.

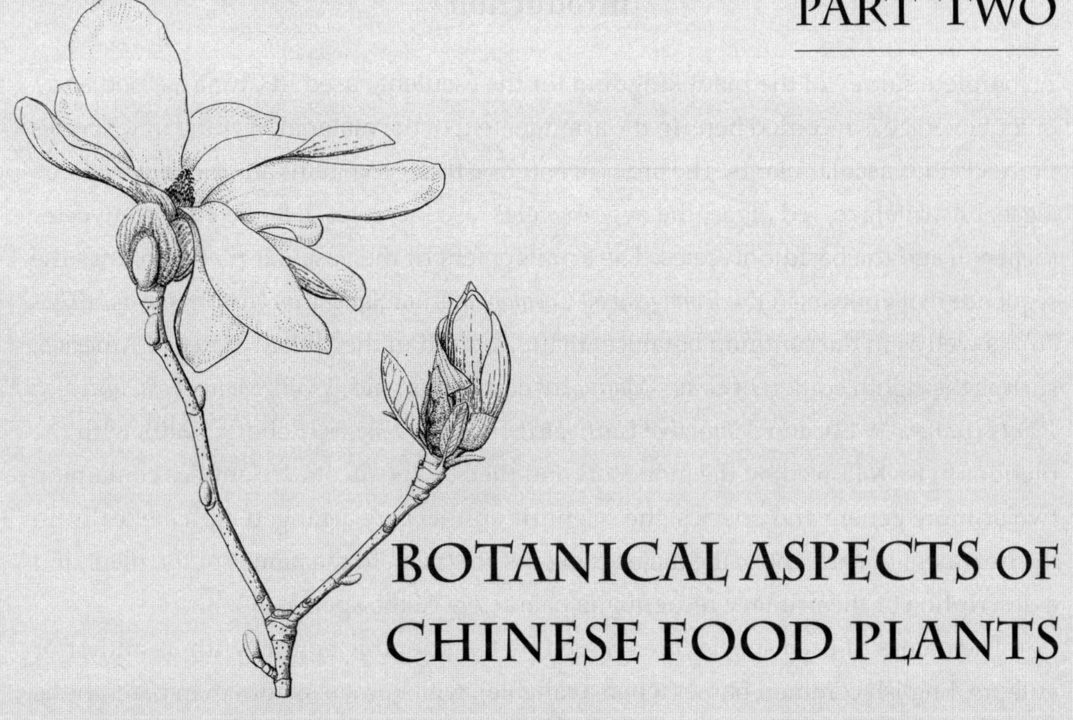

BOTANICAL ASPECTS OF CHINESE FOOD PLANTS

Introduction

A complete survey of the plant kingdom for the esculents used in China as food and/ or for beverage is recorded here. In the arrangement of the material, cryptogamic species proceeds the vascular plants. The order of entering the cryptogams are: monerans, green algae, brown algae, red algae, the zygomycetes, ascomycetes, lichenized ascomycetes (lichens), and the basidiomycetes. The arrangement of the vascular plants follows the sequence that appears in the *Iconographia Cormophytorum Sinicorum* (Anonymous, 1972–76), as well as in the common botanical manuals used in the United States of America, such as the eighth edition of *Gray's Manual of Botany* (Fernald, 1950), *Manual of Cultivated Plants* (Bailey, 1949), and *Manual of Cultivated Trees and Shrubs* (Rehder, 1940), with the pteridophytes followed by the monocots and then by the dicots. In families containing two or more genera and species, the scientific epithets are arranged alphabetically. In the treatment of each entry, the major concerns are given to the names of the plant and a description of the primary recognizing characters of the species.

The names of each species are entered by the scientific epithet with its author or authors, English common name (when available), well known synonym in parenthesis (if there is any), and one or several Chinese names commonly used by people in different areas within China. Each vernacular name in Chinese ideograms is proceeded by sound transliterations with both the *Pinyin* and the Wade systems of Romanization and immediately followed by a translation of their meaning in English (in most cases). After the names, a statement of the part of the plant used for specific purpose is added. These names represent the botanical identification of the food plants and the Chinese discovery of their use. An enumeration of these names was the actual contents of the *Food Plants of China* between 1957 and 1985. It served as a handy check list. Both sinologists and botanists interested in the studies of comparative enthnobotany would find a wealth of knowledge in the meaning of the names of edible plants originally given by many dozens of ethnic groups of the Chinese people throughout the country. The ethnic peoples, whether they belong to the major groups of Han (漢), Man (滿), Meng (蒙), Hui (回), and Zang (藏), Yao (瑤), Qiang (羌), Rong (戎), etc., have lived close to nature, and they have always given significantly meaningful names to economic plants. Examples are too numerous to deal with here, and a special article on this subject is in preparation for sinologists. However, it is worthy of note that formerly the rural people in China was largely illiterate and they could not write down the names they gave to plants. The recording of plant names were often performed by urban dwellers who had little or no knowledge of the plants. Sometimes such scholars caught the sound of a name and used some ideograms which completely missed the original. Names so recorded are hard to translate. Through my field studies with Chinese farmers in northern China and botanical explorations among many ethnic groups in the mountains of western and southern China, I have detected the original meaning and made corrections of some of such names. Good examples are the real meaning of *Shan-luo-bo* (山蘿蔔) for the corrupted ancient record of *Shang-lu* (商陸) in *Phytolacca esculenta* van Houtte, and the true significance of *Tao-jun-liang* (逃軍糧) for the corrupted name of *Tao-jin-niang* (桃金娘) in *Rhodomyrtus tomentosa* (Aiton) Hasskari. Yet, there are still some names of Chinese food plants with this type of origin. To such names, it is better to give the English name without translating, as Zedoary for *E-zhu* (莪朮) in *Curcuma zedoaria* (Bergius) Roscoe, and Erect Hedychium of *Li-ji* (立芨) in *Hedychium coronarium* J. Koenig. To those plants without an English common name, I merely give some incoherent English equivalents as in Sun Lotus for *Yang-he* (陽荷) in *Zingiber mioga* (Thunberg) Roscoe. To a very few others, only the phrase "a book name" is given, as for *Ji-cai* (蕺菜) in *Houttuynia cordata* Thunberg.

The description of the species was originally designed to cover a condensation of published information, ecological notes, or history of introduction. However, in the progress of the program, it became evident that information of many species recorded from areas of ethnic groups are lacking or insufficient, and the use of herbarium specimens was necessary. Eventually doubtful characters of native and introduced species were checked with collections in Harvard University Herbaria. Descriptions of little known species with insufficient recorded information are prepared by careful study of herbarium collections, sometimes supplemented by living specimens in the green house and experimental garden of the Department of Organismic and Evolutionary Biology. In such cases the descriptions provide botanical records for the species concerned, in addition to supplying information needed for completing the *Food Plants of China*. At the end of each description, additions of notes on geographical distribution, ecological background of introduction are given.

People interested in Chinese food plants should have little problem to acquire stocks. Approximately half of the species included in this work are available in the New World. Many of the species producing edible material in China have been introduced to America, some as ornamental plants. Nearly 19% of the food plants in China are extra Chinese elements introduced from the Old World tropics (India and Southeast Asia, 35%), Europe and the Mediterranean region (22%), tropical America (20%), Africa and South America (each 7%); North America, western and central Asia (each 4 to 5%), Pacific islands, and Australia (1 to 3 species). In China these introduced species belong to 54 families, with Fabaceae (Leguminosae) leading the list (20 species). Cucurbitaceae and Asteraceae (Compositae) next (14 species each), Myrtaceae and

Apiaceae (Umbelliferae) followed (11 species each), Solanaceae (9 species), Rosaceae (7 species) and Convolvulaceae (6 species). The other families have one or two species. Many of these species are from the New World.

NONVASCULAR PLANTS: MONERANS, ALGAE, FUNGI AND LICHENS

MONERANS

Cyanophytes: Blue-green Algae

Nostoc commune Vaucher — COMMON NOSTOC

Ge-xian-mi=Ko-hsien-mi (葛仙米, Grains of Immortal *Ge*); *Di-jiao-pi=Ti-chiao-p'i* (地角皮, Earth Corner Skin); *Di-er=Ti-erh* (地耳, Earth Ears), *Tian-xian-cai=T'ien-hsien-ts'ai* (天仙菜, Heaven Immortals Vegetable), *Di-ruan=Ti-juan* (地軟, Ground Soft). Whole thallus, gathered immediately after a summer storm, cleaned, put in soup, or seasoned for salad, especially served to people suffering from night blindness.

Thallus dark bluish green, 1–2 cm across, soft, slippery; appearing only on wet rocks after summer rain; dried material wrinkled, grayish brown, paper-like.

A widespread species growing only on boulders, appearing immediately after a summer rain, farmers gathering the material and eating it that day, seldom seen in the market.

Nostoc flagelliforme Bornet et Flahault — FILIFORM NOSTOC

Fa-cai=Fa-ts'ai (髮菜, Hair Vegetable). Whole mass of dry hair-like material, soaked, cleaned, cooked in soup, or used for a vegetarian dish called "Buddha Disciples' Delight"; available in Chinese stores in USA.

Thallus filiform, gray, in a tangled mass, becoming soft immediately when soaked in boiling water, appearing greenish black, soft when cooked.

Native to the arid hillsides of Lanzhou (Map, Loc. 20) and its vicinity, available only after the early summer rain; popular with Chinese Americans, used for many dishes around the Lunar New Year, for the sound of the vegetable, *fa-cai* 髮菜 is homophonic with the New Year greeting 發財 (=wishing you prosperous).

NONVASCULAR EUKARYOTES — ALGAE

Chlorophyceae: Green Algae

Enteromorpha compressa Greville

Gan-tai=Kan-t'ai (乾苔, Dry Moss); *Tai-cai=T'ai-ts'ai* (苔菜, Moss Vegetable); *Tai-tiao=T'ai-t'iao* (苔條, Moss Ribbon). Whole plants, slightly cooked.

A bright green, filamentous seaweed growing in shallow water in the quiet bay areas along the shores of Zhejiang and Fujian provinces; normally harvested in the dry winter season, and dried immediately on simple frames and lines; rain spoils the product, rendering it white; the market material having the appearance of crushed dark green grass from a lawn mower; very rich source of protein and minerals, and a cash supplement to the income of farmers who sell their surplus in large cities like Ningbo and Shanghai for 0.75–1.50 Chinese dollar per 375 g (*ca.* 1 lb).

Prasiola japonica Yatabe

Qi-cai=Ch'i-ts'ai (溪菜, Stream Vegetable). Entire plant, slightly cooked.

A fresh-water green alga growing in streams in the mountains of Fujian, the thalli is paper-like, irregular, wavy along the margin; local use only.

Ulva pertusa Kjellman, **U. lactuca** L. — SEA CABBAGE, SEA LETTUCE (Figure 16 c–d)

Hai-wo-ju=Hai-wo-chii (海萵苣, Sea Lettuce); *Shi-chun=Shih-ch'un* (石蒪, Stone Brasenia). Whole plant, slightly cooked.

Common green algae growing in shallow waters of the intertidal zone, on shells, rocks, twigs, or in rocky pools; thalli bright green, membranous, irregularly lobed, margin ruffled; *U. pertusa* grows better in northern China, called Sea Cabbage in Shandong; *U. lactuca* grows mostly in Fujian and Hong Kong, called Sea Lettuce locally; large amounts gathered in the dry season, dried on the shore, bagged for export.

Phaeophyceae: Brown Algae

Alaria crassifolia Kjellman

Xiao-kun-bu=Hsiao-k'un-p'u (小昆布, Lesser Kelp). Whole plant, cooked.

Purplish brown seaweed growing on subtidal rocks, the lower portion of the thalli pinnately deeply-lobed, the apical portion oblong, coriaceous, very thick along the midrib; not frequently used.

Chorda filum Lamouroux

Man-zao=Man-tsao (蔓藻, Creeping Seaweed). Whole plant.

A yellow-brown unbranched, rope-shaped seaweed 1–3 m long, 2–3 mm in diameter, sometimes spirally twisted; growing on intertidal rocks of North China sea coast.

Cladosiphon decipiens Okamura

Hai-yun=Hai-yün (海蘊, Refined Seaweed). Whole plant, cooked.

A filiform, slimy, light brown seaweed, 20–30 cm long, irregularly dichotomously branched; occurring along the North China coast.

Ecklonia bicyclis Kjellman

Hei-cai=Hei-ts'ai (黑菜, Black Vegetable). The blades, cooked.

A dark brown, coriaceous seaweed growing on intertidal and subtidal rocks, thalli 1 m or more long, pinnately and bilaterally lobed, the stipes terete, solid, the holdfast with dichotomously branched rhizoids.

Laminaria japonica Areschoug — Laminaria, Kelp

Hai-dai=Hai-tai (海帶, Sea-belt); *Dai-si=T'ai-ssu* (帶絲, Sliced Laminaria). The blades, regularly cut strips, slightly cooked and seasoned, or used for soups.

Large olive-brown, coriaceous seaweed, undivided, 2–6 m long, 20–40 cm wide, stipes terete; growing on subtidal rocks; commercially cultivated; both whole and shredded lamina available in American Chinese groceries.

Scytosiphon lomentarius J. Agardh

Xuan-zao=Hsüan-tsao (萱藻, Mother Seaweed). Blades, cooked.

Yellow-brown caespitose erect seaweed, simple branch cylindrical, deeply constricted at intervals; growing on middle intertidal rocks, holdfast discoid.

Sargassum enerve J. Agardh

Ma-wei-zao=Ma-wei-tsao (馬尾藻, Gulf Weed). Whole plant.

Yellow-brown seaweed, 40–80 cm long, main axes bearing slender, leaf-like retroflex lateral branches, 5–6 cm long, sharply dentate, the primary branches bearing globular vesicles 3–4 mm in diameter.

Turbinaria fusiforme Yendo (Syn. *Cystophyllum fusiforme* Harvey)

Yang-xi-cai=Yang-hsi-ts'ai (羊栖菜, Goat Floating Weed). Whole plant.

A large, dark brown seaweed, thalli 1 m long, with fleshy lateral clavate branches; growing on subtidal reefs.

Undaria pinnatifida Suringar — APRON-RIBBON VEGETABLE

Qun-dai-cai=Ch'ün-tai-ts'ai (裙帶菜, Skirt Belt Vegetable). Whole plant; cooked.

A larger yellow-brown seaweed, 1–1.5 m long, 50–100 cm wide, lanceolate, pinnately lobed, holdfast composed of dichotomously branched rhizoids, stipe winged.

Rhodophyceae: Red Algae

Campylaephora hypneoides J. Agardh

Niu-mao-shi-hua-cai=Niu-mao-shih-hua-ts'ai (牛毛石花菜, Ox-hair Gelidium). Whole plant.

Yellowish to dark red seaweed in tangled masses, 10–20 cm long, growing in intertidal zones; discoid holdfast consisting of rhizoidal cells, primary axis irregularly branched, the terminal segment parcepated or sickle-shaped; used for food or as a source for agar extraction.

Ceramium boydenii Gepp

Xian-cai=Hsien-ts'ai (仙菜, Fairy Vegetable); *Xian-cao=Hsien-ts'ao* (仙草, Fairy Herb). Whole plant, used raw.

Crimson-red to dark red seaweed, forming a mass of interwoven filaments 10 cm in diameter; epiphytic on other algae in intertidal zones, young plants dichotomously branched, branchlets 3–7, in whorls.

Chondrus elatus Holmes and C. ocellatus Holmes

Lu-jiao-cai=Lu-chiao-ts'ai (鹿角菜, Deer-horn Vegetable). Whole plants, eaten raw.

Purplish-red caespitose seaweed with subcylindric dichotomously branched thalli, 10–30 cm long, cartilaginous; growing on intertidal rocks by small disc-like holdfast; occurring in Taiwan and Fujian, pickled material available in American Chinese stores.

Eucheuma papulosa Cotton et Yendo

Ji-jiao-cai=Chi-chiao-ts'ai (雞腳菜, Chichen Feet Vegetable). Whole plant, Used raw.

Purplished-red membranous seaweed, 20–30 cm long, branches compressed, irregularly divided in broad branches, 2–3 cm wide, with the wider portion 6–7 cm; growing on subtidal dead coral.

Eucheuma spinosa J. Agardh

Qi-lin-zao=Ch'i-lin-tsao (麒麟藻, Unicorn Seaweed). Whole plant.

Purple-red cartilaginous seaweed, the axes cylindric, covered with conic tubercles, irregularly branched, the terminal segments acute, or obtuse; growing on subtidal rocks in Taiwan and Guangdong.

Gelidium amansii Lamouroux; **G. latifolium** Bornet; and **G. rigidum** (Vahl) Greville — ROCK-FLOWER VEGETABLE

Shi-hua-cai=Shih-hua-ts'ai (石花菜, Rock-Flower Vegetable); *Yang-cai=Yang-ts'ai* (洋菜, Sea Vegetable, Agar-agar). A product prepared from several species of *Gelidium, Eucheuma, Ceramium* and *Gracilaria.*

A purplish red, rigid, cartilaginous, subcylindrical or narrow-linear, caespitose seaweed, 10–20 cm long, forming dense tufts; holdfast filiform, repeatedly branched; abundant in the Yellow Sea.

Grateloupia flabellata Holm

Mi-hai-tai=Mi-hai-t'ai (米海苔, Rice Seaweed). Whole plant.

A small dark red fan-shaped seaweed, 5–6 cm long, the thalli compressed filiform, repeatedly dichotomously branched; growing on intertidal rocks.

Nemalion helminthoides (Velley) Batters var. **vermiculare** (Suringar) Tseng

Hai-suo-mian=Hai-so-mien (海索麵, Sea Vermicelli). Whole plant.

A multiaxial, deep brownish-purple seaweed, 6–20 cm long, 1.5–2.5 mm in diameter, thalli cylindrical, seldom branched; growing on surf-beaten rocks; common along the coast of China.

Porphyra tenera Kjellman; and **P. atropurpurea** (Olivi) De Toni; **P. yezoensis** Ueda

Zi-cai=Tzu-ts'ai (紫菜, Purple Vegetable). Whole plant, slightly cooked.

Membranous, purple or purple-red seaweed growing on intertidal rocks; thalli oblong, 20–30 cm long, 10–18 cm wide, margin undulate; commercially cultivated in the Yellow Sea.

NONVASCULAR EUKARYOTES — FUNGI

Zygomycetes: Bread Mold

Rhizopus stolonifer (Ehrenberg ex Fries) Lind (Syn. *Mucor stolonifer* Ehrenberg, *Rhizopus nigricans* Ehrenberg). (Growing on bean curd)
Dou-fu-ru=Tou-fu-ju (豆腐乳, Bean Curd Cheese). A fermented and seasoned product, with the bean-curd partially digested by the fungus; available in Boston in bottles and cans.

A white mold used in converting bean curd into the soft bean curd cheese with unusual aroma; in home-made bean curd cheese airborne spores of the fungus germinate and grow on the bean curd, completely covering it with the white mycelium; spices and salt added afterward.

Ascomycetes: Sac Fungi

Aspergillus oryzae (Ahlburg) Cohn and **A. soyae** Sakaguchi & Yamada et Yamada (Growing on cooked cracked wheat, rice, and soybean.)
Jiang-you=Chiang-yu (醬油, Soy Sauce); *Dou-chi=Tou-ch'ih* (豆豉, Spiced Soybean Preserve); *Dou-ban-jiang=Tou-pan-chiang* (豆瓣醬, Fermented Soybean). The entire plant body with soybean and spices, raw or cooked.

A blue mold used in the manufacture of soy sauce made with yellow soybean and cracked wheat, and in processing spiced soybean preserve with black soybean and various herbs.

Cordyceps sinensis (Berkeley) Saccardo — CORDYCEPS
Dong-chong-xia-cao=Tung-ch'ung-hsia-ts'ao (冬蟲夏草, Winter-worm Summer-herb); *Chong-cao=Ch'ung-ts'ao* (蟲草, Worm-herb). Dried sclerotia covered by dead worm skin; used to prepare chicken or duck broth, taken as a tonic.

The black sclerotia of a parasitic fungus emerged from the dead host, the larva of an alpine moth (*Hepialus armoricanus* Oberthur), gathered as the snow melts over the alpine meadows in western Sichuan, and adjacent Gansu, Qinghai and Yunnan; the market material neatly tied with red thread; the most expensive *bupin* sold in Chinese stores in Boston area.

Monascus purpureus Went

Hong-qu=Hung-ch'ü (紅麴, Red Brewer's Yeast); *Hong-zao=Hung-tsao* (紅糟, Red Fermented Grain). Mycelium and spores of fungus and fermented rice grain; used to cook chicken, to prepare brown-red bean-curd cheese, or to color pastry.

The market material consisting of red rice colored by the pigments, monascorubin and monascoflavin, secreted by the mycelia of the filamentous fungus (see Part I on red fermented rice for details).

Morchella esculenta (L. ex Fries) Persoon — MOREL

Yang-du-jun=Yang-tu-chün (羊肚菌, Goat Stomach Mushroom). Dried; for soup.

The fruiting body of a morel with wrinkled cap 4–8 cm long, 3–6 cm in diameter, white stipe 5.5–8 cm long, 2–4 cm in diameter; occurring in northern and western China.

Saccharomyces cerevisiae Moyen (growing on cooked rice)

Jiu-niang=Chiu-niang (酒釀, Mother-of-Wine); *Lao-zao=Lao-tsao* (糁糟, Fermented Grain); *Tian-jiu=T'ien-chiu* (甜酒, Sweet Liquor). A yeast culture made of cooked glutinous rice, eaten raw, or slightly cooked with sugar and sometimes with an egg; a very popular breakfast in West China.

Dry yeast balls imported from Taiwan or the mainland via Hong Kong, 2–2.5 cm in diameter, white; used for processing homemade fermented glutinous rice (See Part I for details).

Lichenized Ascomycetes: Lichens*

Cladonia alpestris (L.) Rabenhorst

Tai-bai-hua=T'ai-pai-hua (太白花, Tai-bai Peak Flower); *Que-shi-rui=Ch'üeh-shih-jui* 雀石蕊, Bird Cladonia). Aerial portion of the plant, cleaned, dried, pulverized, mixed with flour for making bread; small amount (600 g) washed, cleaned, boiled with a special strain of chicken with black skin and bone, or with pig feet, for a dish of special value to people with high blood pressure.

Thallus fruticose, grayish green to pale yellow, forming a compact clump 4–7 (–10) cm high, 2–3 cm in diameter, without conspicuous main branches, dichotomously to

---

* With the assistance of S. Hammer, a graduate student of Prof. D. Pfister, specialized in lichenology, Department of Evolutionary and Organismic Biology, Harvard University.

polychotomously branched, with small pores in the axils, branches cylindrical. hollow, firm, brittle, spongy when wet; apothecia dark brown, very rare, pycnidia flattened, red, terminal to the branches.

Native of northwestern and northeastern China and Inner Mongolia, growing on the floor of coniferous forest among some grasses and sedges, or in open pastures with mosses.

Cladonia macroptera Räsänen

Long-zhua-cai=Lung-chua-ts'ai (龍爪菜, Dragon's Claw Vegetable); *Da-chi-shi-rui=Ta-ch'ih-shih-jui* (大翅石蕊, Large Wing Rock Flower). Aerial portion of the plant, placed in boiling water to remove the bitter principle, washed, seasoned with salt, soysauce, a little garlic and ginger, and few drops of sesame oil, eaten as salad, information supplied by a Finnish Lichenologist, Professor T. Ahti, Botanical Museum, University of Helsinki, who was served the dish in Beijing.

Thallus fruticose, greenish-gray, up to 10 cm high, with horizontal portion at the ground level, sparsely branched, the branches deeply lacinate, 3–5 mm wide, with podetial squamules, stereome hard, horny, with some granules near the tips.

Common on Tai-bai Shan of Shaanxi, growing at the altitudes of 2,800 m, the species was first published from Honshu, Japan, and it has been observed on the Queen Charlotte Islands, British Columbia, Canada.

Lecanora affinis Eversmann — CHINESE MANNA

Gan-lu-yi=Kan-lu-i (甘露衣, Sweet Dew Lichen); *Feng-gun-yi=Feng-k'un-i* (風滾衣, Wind Rolled Lichen); *Shen-liang=Shen-liang* (神糧, God's Food). Whole thallus.

Thallus granulose, spherical, dull, pale greenish-yellow to white, closely appressed, 2–6 cm in diameter, slightly lobed, lobes flat, slightly fissured, exposing the white medulla; apothecia numerous, discoid, tan-brown.

A widespread species, growing on sandstone, granite, and calcareous rocks in open areas; individuals separated from the substrate, rolled by the wind, continuing to increase in size; rich in starch, used as an emergency food.

Lethariella cladonioides (Nylander) Krog

Hong-xue-cha=Hung-hsüeh-ch'a (紅雪茶, Red Snow Tea); *Jin-si-shua=Chin-ssu-shua* (金絲刷, Golden Wire Brush); *Lu-xin-cha=Lu-hsin-ch'a* (鹿心茶, Deer Heart Tea). Dry loose mass of dichotomously branched filaments, 1–3 g, used for tea, specially to people with fever and cough, or high blood pressure.

Almost globular loose masses of dichotomously branched filaments, 4–8 cm across,

4–5 cm thick; joints of the filament 1–4 mm long, center of the mass with a sign of attachment, occasionally showing a piece of the remain of woody tissue; colors of the mass generally grayish-white with reddish-orange terminal branches, the gray area sometimes appearing black. Endemic to the high mountains at 4,000 m on the border of Yunnan and Tibet; formerly used by local tribal people, now gathered and sold in the markets of Kunming.

Thamnolia vermicularis (Swartz) Acharius ex Schaerer

Di-cha=Ti-ch'a (地茶, Ground Tea), *Tai-bai-cha=T'ai-pai-ch'a* (太白茶, Tea of Tai-bai Shan); *Xue-cha=Hsiieh-ch'a* (雪茶, Snow Tea). Whole plant, used for tea, both fresh and dried.

Thallus fruticose, white, 5–6 cm long, 2–3 mm thick, terete, hollow, gradually attenuate at the distal end, occasionally branched, slightly astringent with a sweet after taste when chewed; fruiting material not yet known.

A widespread species, common in Shaanxi, Sichuan and Yunnan, growing with mosses on exposed soil of artic-alpine habitats.

Umbilicaria esculenta (Miyoshi) Minks

Shi-er=Shih-erh (石耳, Rock Ear); *Yan-rong=Yen-jung* (岩茸, Cliff Floss), *Yan-gu=Yen-ku* (岩菇, Cliff Mushroom). Whole thallus, highly esteemed for aroma and texture, being regarded as an article of food for the connoisseurs, steamed with lean meat and ginger, said to give strength to muscles and youthful complexion.

Thallus foliose, round when young, oblong or irregularly lobed later, 5–15 (–18) cm in diameter, attached to the substrate by a single cord, coriaceous, wavy or slightly torn along the margin, dull, grayish brown, rather smooth, slightly covered by white frost-like granules, cortex sometimes falling in patches, exposing the medulla, lower surface brown-black, with numerous verruculae; rhizomes branched, coral-like.

Native of eastern Asia, growing on exposed rocks and cliffs, with Lu Shan in Jiangxi and Huang Shan in Anhui being the famous areas of supply.

Usnea longissima Acharius

Song-luo=Sung-lo (松蘿, Pine Net); *Tian-peng-cao=Tien-p'eng-ts'ao* (天棚草, Sky Tent Herb); *Shan-gua-mian=Shan-kua-mien* (山掛麵, Mountain Vermicelli); *Xue-feng-teng=Hsiieh-feng-t'eng* (雪風籐, Snow Wind Vine); *Nü-luo=Nü-lo* (女蘿, Feminine Net). The hanging branches, cleaned, boiled in lye, washed, drained, broken into pieces, seasoned, used as salad.

Thallus string-like, pale greenish-yellow, soft, pendulous, 15–40 (–100) cm long, main branches short, 1–1.5 cm long, rugose, with annulus; cortex eroding away with age;

secondary branches with sparse soralia, lateral branchlets with unequal fibrils, smooth, isidia lacking, medulla wavy; apothecia discoid, 1–5 mm in diameter, ciliate, light brown, covered by grayish white powder; asci clavate; spores 8, oblong.

A widespread species in northern hemisphere, in China, growing on conifers of northeastern, and western provinces, Taiwan, and Inner Mongolia.

Basidiomycetes: Club Fungi

Agaricus bisporus (Lange) Singer — COMMON FIELD MUSHROOM (Syn. *A. campestris* of many authors)

Xi-yang-jun=Hsi-yang-chün (西洋蕈, Western Mushroom); *Bai-mo-gu=Pai-mo-ku* (白蘑菇, White Mushroom). Cultivated, used fresh and also canned for export.

Cultivated for the young white fruiting body, used fresh, canned for large cities and export.

Agaricus campestris L. ex Fries — MEADOW MUSHROOM, PINK BOTTON (Syn. *Psalliota campestris* [L.] Fries)

Mo-gu=Mo-ku (蘑菇, Mushroom). Both fresh and dried fruiting body used in northern China, the market material smaller, thinner than the Black Mushroom, with better flavor, highly esteemed by Chinese Americans.

White or light brown mushroom, caps hemispherical, 5–12 cm across, stipes 7–9 cm long, 1.5–2.5 cm across; dried material available in American Chinese stores.

Auricularia auricula (L.) Underwood — TREE EAR, WOOD EAR, AURICULARIA

Mu-er=Mu-erh (木耳, Wood Ear); *Wen-er=Wen-erh* (Cantonese pronunciation for *Yun-er=Yun-erh* [雲耳, Cloud Ear=Wood Ear of Yunnan]). Dried material, irregular foliaceous pieces, soaked in boiling water to soften, cooked with meat, or boiled in soup; imported material available in American Chinese stores.

Grayish black pieces 1–3 cm across, upper side smooth, opposite side velvety, recent medical researches approved its efficacy in preventing heart diseases.

Auricularia polytricha (Montagne) Saccardo — JUDA's EARS, AURICULARIA

Mao-mu-er=mao-mu-erh (毛木耳, Hairy Wood Ear); *Mu-er=Mu-erh* (木耳, Wood Ear=Cantonese name used in Chinatown of Boston). Large tough leathery dried material, requiring much boiling, used by Cantonese people for soup.

Irregularly foliaceous dried fruiting bodies, 3–5 cm across, grayish black, the opposite side brown; available in American Chinese stores.

Boletus edulis Bulliard ex Fries

Mei-wei-niu-gan=Mei-wei-niu-kan (美味牛肝, Delicious Calf Liver Mushroom); *Shan-wu-rong=Shan-wu-jung* (山烏茸, Black Luxuriant Growth-on-the Mountain). Used fresh locally.

A large brown or yellowish-brown, wild mushroom, caps subhemispherical, 7–15 cm in diameter; stipes 7–11 cm long, 2–3.5 cm across; occurring in central and western China.

Clavaria botrytis Persoon

Sao-zhou-gu=Sao-chou-ku (掃帚菇, Broom Mushroom); *Ji-guan-jun=Chi-kuan-chün* (雞冠蕈, Cock's Comb Mushroom). Fruiting body, cooked fresh, used locally.

A fleshy, stout fruiting body, subcylindrical on the basal half, branched above, the terminal segments red; wild, gathered locally.

Collybia albuminosa (Berkeley) Petch (Syn. *Termitomyces albuminosus* [Berkeley] Heim)

Ji-jun=Chi-chün (雞蕈, Chicken Mushroom); *Xia-zhi-jun=Hsia-chih-chün* (夏至蕈, Summer Solstice Mushroom); *San-da-jun=San-ta-chün* (三大蕈, Three Big Mushroom).

The three Chinese names for this mushroom describe its flavor, time for gathering the mushroom, and its appearance. It has a large cap 20–25 cm in diameter, smooth, white on the inside, and a stipe 20 cm long, enlarged at the middle. It has the texture and taste of a chicken. It appears in summer and is gathered wild. This is the favorite mushroom of food connoisseurs in western and southern China where it occurs in nature.

Dictyophora phalloidea Desvaux; **D. indusiata** (Ventenat et Persoon) Fischer — Netted Stinkhorn

Zhu-sun=Chu-sun (竹蓀, Bamboo Flower). Net of the fruiting body, gathered, dried, used in expensive vegetarian dinners.

A colorful fruiting body, 10.2 cm long, with a basal oval base, bell-shaped red cap on the lower margin hangs a white netted veil flaring at the bottom; the net gathered, dried, used as a delicacy.

Flammulina velutipes (M. A. Curtis ex Fries) Singer — Velvet Foot (Syn. *Collybia velutipes* [M. A. Curtis] Kummer)

Jin-gu=Chin-ku (金菇, Golden Mushroom). Cultivated and canned in Taiwan; available in Boston, 75 cents form a small can (155 grams).

Small, caespitose mushroom, occurring in cold weather; cap 1.5–6 cm across, nearly flat at advanced age, reddish-yellow or orange, stipe 2–6 cm long, 2–8 mm across; good taste, the stalk rather tough.

Ganoderma lucidum (Leysser ex Fries) Karsten and **G. japonicum** (Fries) Lloyd — LINGCHIH, GANODERMA (Figure 16a)

Ling-zhi=Ling-chih (靈芝, Divine Mushroom). Whole mushroom gathered, dried, chopped into pea-sized pieces, using 1/10 oz, boiled in water for tea; also pulverized; for mentally disturbed or the elderly persons.

A large mushroom appearing at the ground level, first white, changing to reddish brown with white margin; natural supply exhausted, now the imported material is black, with a flat cap obliquely attached to a stipe. The dried fruiting body is woody, used in small pieces for tea.

Hericium erinaceum (Bulliard ex Fries) Persoon — HEDGEHOGS

Hou-tou=Hou-t'ou (猴頭, Monkey Head).

A coarse, white mushroom covered with soft spine and strips hanging down in tufts, the entire fruiting body appears like a monkey head, hence the local name; used fresh locally as a delicacy.

Lentinus edodes (Berkeley) Singer — BLACK MUSHROOM (Syn. *Cortinellus edodes* [Berkeley] S. Ito et Imai)

Xiang-gu=Hsiang-ku (香菇, Fragrant Mushroom); *Dong-gu=Tung-ku* (冬菇, Winter Mushroom); *Song-gu=Sung-ku* (松菇, Pine Mushroom). Dried material available in Boston, much used in American Chinese dishes.

A cultivated mushroom, the cap purplish brown, turning black on drying, 4–5 cm across, the larger ones attaining 12 cm, stipes 3–10 cm; large amount imported from China and Japan; with pleasant aroma.

Pleurotus sajor-caju (Fries) Singer — PHOENIX MUSHROOM (Figure 16 b)

Feng-wei-gu=Feng-wei-ku (鳳尾菇, Phoenix Tail Mushoom). Recently cultivated in Hong Kong (Figure 16b).

Eccentric fruiting body, gray above, white on the gill side.

Poria cocos (Schweinitz) Wolf — FULING, CHINA ROOT (Syn. *Wolfiporia cocos* [Schweinitz] Ryvarden et Gilbertson)

Fu-ling=Fu-ling (茯苓, China Root). Sclerotia grown on pine wood; sliced interior white portion, available in Boston; cooked with lily bulbs, Job's Tear, Euryle seed, Chinese date, taken as a tonic by many Chinese in America.

Large oblong or fist-like sclerotia of a fungus growing in association with *Pinus massoniana* Lambert Formerly, when the supply depended upon natural material, people used to have the help of dogs to locate the sclerotia, which may have been buried 10–30 cm in the ground. Today some market material are from cultivated sources. The harvested sclerotia are taken to pharmaceutical houses where they are softened in water, trimmed and sliced into various sizes and thickness. The slices from the outer portion of a sclerotium are pink, and those from the center are snow white. The sclerotia growing around branches are called *Fu-shen* (茯神, Divine *Fu-ling*).

Tremella fuciformis Berkeley — SILVER EAR, TREMELLA
   Yin-er=Yin-erh (銀耳, Silver Ear). Cultivated on oak wood, very common in Boston's Chinatown, genuine material imported from the Chinese mainland makes clear soup after prolonged boiling (artificial product disintegrates on boiling); cooked with chicken for a broth, or taken as a sweet dessert.

   The market material of *Yin-er* consists of the dried hemispherical mass of the fruiting body, 5–8 cm across, with a dark brown center 1 cm in diameter, indicating the point of attachment to the substratum. Material without such a center is not genuine.

Tremella mesenterica Retzius ex Fries — GOLDON TREMELLA, YELLOW TREMELLA
   Jin-er=Chin-erh (金耳, Golden Tremella); *Huang-er=Huang-erh* (黃耳, Yellow Tremella). Material gathered from decayed tree trunks in northern and western China, available in Hong Kong; used by vegetarians to make immunomodulating cooling soup with lotus seed, lily bulbs, jujube, etc.

   The market material consisting of brain-like irregular dry mass, 4–5 cm in diameter, 2.5 cm thick at the center, flat on the contacting side, with the marking of attachment 1 cm long, 2 mm thick.

Ustilago esculenta P. Hennings — ZIZANIA SMUT
   Gu-hei-sui-jun=Ku-hei-sui-chun (菇黑穗菌, Zizania Smut)

   The market material consists of the enlarged young stems of a swamp grass, *Zizania caduciflora* Handel-Mazzetti. The morphogenetic region of the grass, infected by the fungus, becomes fleshy and is enlarged. The crop is harvested just before the sporification of the pathogene. A slight sign of black threads visible at the broken surface brings the price down. Before cooking, the shoot must be peeled. The taste and texture is reminiscent of mushroom and bamboo shoot.

Volvariella volvacea (Bulliard ex Fries) Singer — STRAW MUSHROOM

Cao-gu=Ts'ao-ku (草菇, Straw Mushroom). Raised on rice straw or partially decayed grasses; used fresh or canned, available in Chinese stores of Boston.

The cultivation of Straw Mushroom is a modern industry, even in Hong Kong, where it is used to supply local restaurants. Fresh material has gray caps 5–20 cm long, bell-shaped, some even covered by a volva.

CHAPTER VII

VASCULAR PLANTS: PTERIDOPHYTES AND SPERMATOPHYTES

PTERIDOPHYTES — FILICOPSIDS: TRUE FERNS

Helminthostachyaceae: Flowering Fern Family

Helminthostachys zeylanica (L.) Hooker

Qi-zhi-jue=Ch'i-chih-chüeh (七指蕨, Seven Finger Fern); *Ru-di-wu-gong=Ju-ti-wu-kung* (入地蜈蚣, Underground Centipede). Young fronds, used for vegetable.

Very interesting monotypic fern 30–50 cm high, rhizomes fleshy, creeping, 7 mm thick, bearing many fleshy roots 20–40 cm long, and 1 or 2 fronds at the apex; fronds on stipes 20–40 cm long, divided into the tripartite vegetative and the spicate fertile segments; the sterile segments ternately parted, 15–25 cm long and wide, pinnules subsessile, lanceolate, 10–18 cm long, 2–4 cm wide, subentire, acuminate, the fertile segment cylindrical, 6–8 cm long, with sessile sporangia in groups of 3–5. Growing in the forests of Old World Tropics, from India, Sri Lanka to Southeast Asia, and Australia, in China from Yunnan to Taiwan.

Blechnaceae: Blechnum Family

Blechnum orientale L.

Shan-jue-mao=Shan-chüeh-mao (山蕨貓, Mountain Fern Cat); *Wu-mao-jue=Wu-mao-chüeh* (烏毛蕨, Black Hair Fern). Young fronds used as food locally after cooking.

A large stiff fern up to 2 m tall, with short stout oblique rhizome, densely covered with linear-lanceolate dark brown scales, hence the Chinese name, Black Hair Fern; fronds pinnately compound, 50–120 cm long, 25–40 cm wide, pinnae linear-lanceolate, 10–20 cm long, 1.8 cm wide, bearing costal linear sori beneath.

Davalliceae: Davallia Family

Nephrolepis cordifolia (L.) Presl — Erect Sword Fern, Boston Fern

Qiu-jue=Ch'iu-chüeh (球蕨, Ball Fern); *Di-huang-pi=Ti-huang-p'i* (地黃皮, Ground Clausena). Tuber, used fresh or dried, for preparing a broth with chicken or pork.

A caespitose medium-sized fern with slender stolons, often terminated by a fleshy yellowish tuber; fronds oblong-lanceolate, 30–70 cm long, 3–5 cm wide, pinnately compound, the pinnae serrulate. Introduced from Guangzhou to Boston, now cultivated as an indoor plant in hotels throughout the world as Boston Fern.

Pteridiaceae: Bracken Family

Pteridium aquilinum (L.) Kuhn — Bracken, Brake, Eagle Fern

Jue=Chüeh (蕨, Fern); *Jue-qi-miao=Chüeh-ch'i-miao* (蕨其苗, Young Fern Frond); *Jue-tai-cai=Chüeh-t'ai-ts'ai* (蕨苔菜, Fern Frond Vegetable); *Jue-fen=Chüeh-fen* (蕨粉, Fern Starch). Young fronds, and starch from the rhizome.

A fire resistant fern growing on open hillsides; rhizome long creeping, hypogeal, 1 cm in diameter, containing starch grains; fronds deltoid-ovate, 30–40 cm long and across the base, tripinnately compound; stipes 20 cm long.

During famine, people dig and crush the rhizomes, mix the mass with water, remove the coarse material by sieving, precipitate the solid matter, and use it for food.

Ceratopteridaceae: Waterfern Family

Ceratopteris thalictroides (L.) Brongniart — Waterfern (Syn. *C. siliquosa* [L.] Copeland)

Shui-jue=Shui-chüeh (水蕨, Waterfern); *Qi=ch'i* (薺, Water Fiddlehead); *Long-xu-cai=Lung-hsü-ts'ai* (龍鬚菜, Dragon Whisker Vegetable)

Annual aquatic caespitose glabrous herbs 30–80 cm high, with light green herbage and short erect rhizomes bearing a rosette of delicate fronds and copious amount of roots anchoring the plant securely in mud, scales on the rhizome and young fronds cordate, translucent; stipes spongy, fleshy, subcylindrical, 10–40 cm long, laminas dimorphous, sterile fronds bipinnatifid, oblong-orbicular in outline, 10–45 cm long, 5–25 cm across, ultimate segments lanceolate, 6 mm wide; fertile fronds ovate-deltoid in outline, 15–40 cm long, 10–20 cm wide, deeply 2- or 3-parted, ultimate segments horn-like, 2 mm wide, the margin membranous, translucent, reflexed and meeting, covering the scattered brown sporangia. A monotypic pantropic species, in China growing along the margin of ricefields and irrigation ditches, with small buds in the axils of pinnae or pinnules, a means of vegetative reproduction.

Athyriaceae: Lady Fern Family

Anisogonium esculentum (Retzius) Presl — TENDER LADY-FERN (Syn. *Diplazium esculentum* [Retzius] Swartz; *Callipteris esculenta* [Retzius] J. Smith)

Cai-jue=Ts'ai-Chüeh (菜蕨, Vegetable Fern); *Shan-mang=Shan-mang* (山芒, Wild Fern); *Nen-ye-ti-gai-jue=Nun-yeh-t'i-kai-chüeh* (嫩葉蹄蓋蕨, Tender-leaved Athyrium). Young fronds, used as potgreen, southern China, the Philippines.

Large aquatic ferns 1–2 m high, rhizomes oblique, densely surrounded by enlarged persistent bases of stipes woven with copious amount of stiff black branched roots; fronds 10–15, bipinnate, deltoid-ovate in outline, 60–150 cm long, pinnae alternate, the middle ones 20–40 cm long, 10–12 cm wide, pinnules 10–12, lanceolate, 3–6 cm long, pinnately lobed and serrate, sparsely paleate; stipes 25–50 cm long, 4–6 mm thick, angular-furrowed, scales 1 cm long; sori linear, thinly indusiate on one or both sides of the vein. Widespread in the warm regions of China, from the Yangtze River area southward to Guangdong and Yunnan, growing along irrigation ditches or bank of streams near village; leaves succulent, cooked and eatern like spinach.

Athyrium multidentatum (Doellinger) Ching (Syn. *A. filix-foemina* [L.] Roth var. *multidentatum* Doellinger)

Hou-tui-ti-gai-jue=Hou-t'ui-t'i-kai-chüeh (猴腿蹄蓋蕨, Monkey-legged Athyrium); *Duo-chi-ti-gai-jue=Tuo-ch'ih-t'i-kai-chüeh* (多齒蹄蓋蕨, Many-toothed Athyrium). Young fronds, eaten as potherbs in northeastern China.

Plants 60–160 cm high, rhizomes short, suberect; laminas of fronds oblong-ovate in outline, 40–60 cm long, 20–30 cm across, tripinnate, ultimate segments 20–28 cm long, toothed, acuminate, base truncate; stipes and rachis straw-colored; sori 1–4 on each lobe. Growing in forest of Inner Mongolia and northeastern China at altitudes of 300–900 m.

Athyrium sinense Ruprecht — CHINESE LADY-FERN (Syn. *A. filix-femina* [L.] Roth var. *angustifrons* Kodama ex Mori).

Zhong-hua-ti-gai-jue=Chung-hua-t'i-kai-chüeh (中華蹄蓋蕨, Chinese Athyrium). Young fronts, gathered by rural people of Hebei for potherbs.

Plants 40–60 cm high, rhizomes short, suberect; laminas of fronds oblong-lanceolate in outline, 25–65 cm long, 12–25 cm wide, glabrous, bipinnate, ultimate segments sessile, serrate; stipes 20–25 cm long, straw-colored, base black, thickened; sori oblong, one on each lobe. Native of northern China, Korea and Japan, growing in forests at altitudes of 750–2100 m.

Aspleniaceae: Spleenwort Family

Neottopteris nidus (L.) J. Smith (Syn. *Asplenium nidus* L.) — BIRD-NEST FERN
Jue-cai=Chüeh-ts'ai (蕨菜, Fern Vegetable); *Wai-tou-cai=Wai-t'ou-ts'ai* (歪頭菜, Bending Head Greens); *Cao-jue=Ts'ao-chüeh* (巢蕨, Nest-fern).

Young fronds, cooked as emergency food in famines. Large epiphytic ferns growing under trees along streams; rhizomes short, erect, projecting the fronds in rosette; resembling large birds-nests, hence the common English name; fronds linear-lanceolate, entire, 95–115 cm long, 5–15 cm wide, with parallel linear sori, obliquely arranged near the midrib beneath. Planted in shade near building in Hong Kong.

Osmundaceae: Osmunda Family

Osmunda japonica Thunberg
Wei=Wei (薇, Osmunda). Young fronds, local uses by farmers only.

Large terrestrial ferns with dimorphic fronds; laminas bipinnate, pinnules of sterile fronds lanceolate, 4–6 cm long, 1.5–2 cm wide, serrulate; pinnules of fertile fronds linear, densely covered by brown sporangia.

Marsileaceae: Marsilea Family

Marsilea quadrifolia L.
Tian-zi-cao=T'ien-tzu-ts'ao (田字草, Square Herb); *Si-xian-ye=Ssu-hsien-yeh* (四仙葉, Four Sages Leaf). Young shoots eaten by farmers.

Small, delicate, aquatic ferns growing in rice fields or by the edge of ponds, rooting in mud; rhizomes wire-like, fronds 4-parted, floating on the surface of water in wet season, like a four-leaved clover; sori enclosed in small sporocarps emerging at the base of the stipes.

SPERMATOPHYTES — SEED-BEARING PLANTS

Gymnospermae: Naked-seed Plants

Cycadaceae: Cycas Family

Cycas revoluta Thunberg — SAGO PALM

Su-tie=Su-t'ieh (蘇鐵, Iron Revival); *Tie-shu=T'ieh-shu* (鐵樹, Iron Tree); *Feng-wei-jiao=Feng-wei-chiao* (鳳尾蕉, Phoenix Tail Cycas). Starch from the stem used in time of food shortage.

A palm-like evergreen low plant, with a solitary columnar trunk, 1–2 m tall, 12–20 cm in diameter, covered by leaf-bases, occasionally branched near the base; leaves pinnately compound, circinate in bud, linear-lanceolate-oblong, 0.5–2 m long, with hundreds of revolute leaflets, 6–18 cm long, 4–6 mm wide, the basal ones specialized into strong spines; each leaflet ending with a spine; flowers unisexual, dioecious.

Cycas siamensis Miquel — SIAMESE CYCAD

Shen-xian-mi=Shen-hsien-mi (神仙米, Immortal Rice); *Yun-nan-su-tie=Yun-nan-su-t'ieh* (雲南蘇鐵, Yunnan Cycad). Starch obtained from the trunk and root.

Palm-like plant with stout stem, 50–100 cm high; leaves pinnate, 1–2.5 m long, crowded at the end of the stem, leaflets 40–120 on each side of the stiff rachis, 2–18 cm long, 5–8 mm wide, acute and spinescent, shiny above, glabrescent beneath, slightly recurved along the margin, petioles spiny; flowers unisexual, dioecious, male cones ovate-oblong, 10–30 cm long, 4–8 cm across, microsporphylls ovate-rhomboid, 1–2.5 cm long, aciminate-caudate, yellow-lanate; megasporphylls 8–10 cm long, the enlarged apical portion rhomboid, 3–6 cm long and wide, reddish-brown, lanate, the distal half pectinate along the margin, ovules 2–4 on each side of the stalk, glabrous. Seed ovoid, golden-yellow, 2–3 cm long, 1.8–2.5 cm across. Native to the mountains of southern Guangxi and adjacent Thailand.

Ginkgoaceae: Ginkgo Family

Ginkgo biloba L. — GINKGO NUT, GINKGO, MAIDENHAIR TREE (Figure 17)

Bai-guo=Pai-kuo (白果, White Fruit); *Yin-xing=Yin-hsing* (銀杏, Silver Apricot); *Gong-sun-shu=Kung-sun-shu* (公孫樹, Grandfather-and-Grandson Tree). Ripe seed, roasted with hard cover; or only the soft portion cooked with meat or chicken, or with sugar for desserts.

A resinous deciduous tree up to 40 m high with a solitary trunk up to 1.5 m in diameter, producing sucker only when the crown is damaged; branchlets dimorphic, the elongated bearing alternate leaves, and the abbreviated terminated with a whorl of leaves surrounding a bud; leaves fan-shaped, shallowly bilobed at the truncate apex, 5–8 cm across the summit; flowers unisexual, dioecious, emerging with leaves on short shoots only; staminate ones pale yellow, consisting of numerous anthers in an oblong-cylindrical catkin; pistillate flowers inconspicuous, each with two naked ovules at the end of a slender stalk; seed drupe-like, yellow at maturity, the outer seed coat fleshy, acrid, ill-smelling, poisonous (may cause dermatitis), inner coat ivory-white, cartilaginous; kernel delicious cooked or roasted, light green when fresh; in Boston and New York the ripe seeds are ready to be gathered under the pistillate trees (see Part I for more details).

Podocarpaceae: Podocarpus Family

Podocarpus nagi Makino — BROADLEAF PODOCARPUS

Zhu-bai=Chu-pai (竹柏, Bamboo Juniper). oil extracted from seeds.

Evergreen tree up to 20 m high; branchlets slender, ascending; leaves opposite, ovate-lanceolate, 5–9 cm long, 1.5–2.5 cm wide, with no midrib; flowers unisexual, dioecious; staminate flowers consisting of single or branched cylindrical catkins axillary to normal leaves; pistillate ones solitary, consisting of one ovule at the apex of a bracteate stalk; seeds nut-like, globose, 1.2–1.5 cm in diameter, rich in oil.

Taxaceae: Yew Family

Torreya grandis Fortune — TORREYA NUT

Xiang-fei=Hsiang-fei (香榧, Aromatic Torreya Nut). Seeds with the fleshy outer coat removed, roasted, salted; available in Boston.

Large evergreen trees up to 25 m tall, with trunks up to 55 cm in diameter; leaves spirally arranged, appearing 2-ranked, linear, rigid, 1.2–2.5 cm long, 2–4 mm wide, abruptly spine-pointed; flowers unisexual, dioecious, axillary; staminate cones solitary, globose; pistillate cones in pairs; seed drupe-like, oblong to ovoid or obovoid, 2–4.5 cm long, 1.5–2.5 cm across, with a thin fleshy outer coat and a woody inner coat; endosperm slightly ruminate. Cultivated and selected for the edible endosperm since time immemorial; the market material consisting of roasted, salted nuts, with the outer coat removed; available in Boston; the known varieties and their Chinese names are given below:

var. **grandis**

Fei-zi=Fei-tzu (榧子, Torreya Nut). Nuts oblong.

var. **dielsii** H. H. Hu

Yuan-fei=Yuan-fei (圓榧, Round Torreya Nut). Nuts obovoid.

var. **merrillii** H. H. Hu

Yang-jiao-fei=Yang-chiao-fei (羊角榧, Goat-horn Torreya Nut). Nuts slender.

var. **non-apiculata** H. H. Hu

Dun-tou-fei=Tun-t'ou-fei (鈍頭榧, Round Head Torreya Nut). Nuts obtuse.

var. **sargentii** H. H. Hu

Cun-jin-fei=Ts'un-chin-fei (寸金榧, Inch-of-gold Torreya Nut). Nuts small, with the best aroma, highly esteemed.

Torreya nucifera Siebold et Zuccarini

You-fei=Yu-fei (油榧, Oil Torreya Nut). Seed roasted with salt.

A handsome evergreen tree up to 25 m tall, with spreading branches and reddish brown branchlets. A native of Japan, introduced into large Chinese cities for ornamental purposes; seeds used locally.

Pinaceae: Pine Family

Pinus armandii Franchet, and **P. koraiensis** Siebold et Zuccarini — Pine Nut, Piñon

Song-zi=Sung-tzu (松子, Pine Nut). Seeds collected from the forest.

Recently Chinese stores in the USA imported pine seeds from China. Chinese restaurants serve piñon with minced chicken or meat. The nuts are from two source species of 5-needle Chinese pines yielding wingless edible seeds, largely from *P. koraiensis.*

Pinus armandii Franchet — Chinese White Pine

Bai-song=Pai-sung (白松, White Pine); *Hua-shan-song=Hua-shan-sung* (華山松, Hua-shan Pine); *Wu-zi-song=Wu-tzu-sung* (五子松, Five Needle Pine). Native of north-western and western China.

Evergreen trees up to 20 m high, current years branchlets covered with red-brown hairs; leaves slender, 8–15 cm long, bright green; cones conic-oblong, 10–12 cm long, 5–8 cm across; seeds 1–1.5 cm long, 6–10 mm across, not dropping from mature cones.

A warm temperate and high altitude species growing from the middle Yellow River Region between Long. 110–112°E and Lat. 34–35°N, thence southward, crossing the

Tsinling Range to Sichuan, Yunnan and the adjacent areas in Gansu and southeastern Tibet in the west and to western Hubei and Guizhou in the east.

Pinus koraiensis Siebold et Zuccarini — HINGGAN RED PINE, KOREAN PINE

Hong-song=Hung-sung (紅松, Red Pine); *Hai-song=Hai-sung* (海松, Sea Pine). Native of the Hinggan Region, local people travelled on bicycles to the forest, gathered the seeds from the trees in August, sold to collectors of trade agents waiting in villages.

A pyramidal evergreen tree up to 30 m high; current year's branchlets covered with gray hairs; leaves stout-rigid, 6–12 cm long; cones conic-ovoid, 9–14 cm long, 6–8 cm across; seeds ovoid-trigonous, 1.2–1.6 cm long, 7–10 mm across, falling from mature cones.

A cold temperate species occurring in Tsinling and Heilongjiang.

Pinus bungeana Zuccarini ex Endlicher — LACE-BARK PINE

Bai-pi-song=Pai-p'i-sung (白皮松, White-bark Pine); *Hu-pi-song= Hu-p'i-sung* (虎皮松, Tiger Skin Pine). Seeds.

The only 3-needle pine in eastern Asia, endemic to the Tsinling Ranges thence northward to Beijing; distinguished by its white chalky white bark exfoliating in large scales, rigid leaves 5–10 cm long, conic-ovoid cones 5–7 cm long, 4–6 cm across, spiny umbo, and broad-ovoid winged seeds 1 cm long; introduced into American gardens.

Pinus densata Masters (Syn. *P. tabulaeformis* Carriere var. *densata* [Masters] Rehder; *P. wilsonii* Shaw)

Gao-shan-song=Kao-shan-sung (高山松, High Montane Pine); *Xi-kang-chi-song= Hsi-k'ang-ch'ih-sung* (西康赤松, Xikang Red Pine) Seeds, used locally in northern Yunnan.

Trees 30 m high, trunks 1.3 m in diameter, bark dark grayish-brown, fissured, current year's growth yellow-brown; leaves 2 or rarely 3 in a fascicle, stout, rigid, serrulate, 6–15 cm long, 1.2–1.5 mm across, slightly twisted, sheaths persistent cones oblique, ovoid, 5–6 cm long, 4 cm across, subsessile; seeds oblong-ovoid, 4–6 mm long, slightly compressed. Endemic to the mountains on the borders of Yunnan, Sichuan and eastern Tibet, known as 'Sikong Province'.

Ephedraceae: Ephedra Family

Ephedra intermedia Schrenk et C. A. Meyer — INTERMEDIATE EPHEDRA

Zhong-ma-huang=Chung-ma-huang (中麻黃, Intermediate Ephedra). Mature ovulate cones; picked from the plant and eaten as fresh fruit, not available in the market.

Semidesert dioecious shrubs 1 m or more high, with grayish-green photosynthetic stems, branches opposite or 3 in a whorl, internodes 3–6 cm long, 1–3 mm in diameter, finely striate; leaves membranous, 2–4 (–6) mm long, the basal 2/3 connate forming a sheath, the lobes 3 or 2, deltoid, scarious pollen-cones glomerate, rarely paired, ovoid or obovoid, consisting of 5 to 7 pairs or whorls of 3 scales, membranous along the margin, anthers 5–8, sessile, on slender column subtended by concealed tube, ovulate cones paired or 3 in a whorl, the uppermost 2 or 3 each bearing an ovule, integumental tubes twisted 4, mature cones red, the scales fleshy, 5–6 mm long, seeds 2 or 3, completely enclosed or slightly exserted.

Native of central Asia, the range extending eastward to Xinjiang, Qinghai-Tibetan plateau, Inner Mongolia, and northern and northeastern China, the stems gathered and sold as an inferior substitute of *ma-huang* used in traditional Chinese medicine.

Ephedra monosperma Gmelin ex C. A. Meyer — LITTLE EPHEDRA, MONGOLIA TEA

Xiao-ma-huang=Hsiao-ma-huang (小麻黃, Little Ephedra); *Dan-zi-ma-huang=Tan-tzu-ma-huang* (單子麻黃, One-seeded Ephedra). Dried mature seed-cones, mixed with tea or herbs for making Mongolia tea, taken with or without milk and cream.

Much branched, low, herb-like plants 5–15 cm high, the woody portion 1–5 cm long, internodes of branchlets 1–2 cm long, 1 mm in diameter; leaves opposite, scarious, the basal 1/3–1/2 connate; pollen cones solitary or paired, oblong-spherical, 3–4 mm long, 2–4 mm across, consistiong of 3 or 4 pairs of decussate scales with broad scarious margin, connate at the base, anthers 7 or 8, sessile, apical to the column, with the basal tube concealed, ovulate cones sessile, with 3 pairs of cone-scales, connate at the base, one rarely 2 bearing ovule, integumental tube curved; mature cone scales glaucous, red, 6–9 cm long, 5–8 mm across; seed solitary, exserted, trigonous-ovoid or oblong-ovoid, 5 mm long, 3 mm across base.

Ephedra sinica Stapf — CHINESE EPHEDRA, MA-HUANG

Ma-huang=Ma-huang (麻黃, Ma-huang), *Cao-ma-huang=Ts'ao-ma-huang* (草麻黃, Herb-like Ephedra). Dried mature seed-cones, mixed with various herbs for making tea in Mongolia.

Much branched small shrubs 20–40 cm high, branches green, internodes 2.5–4 (–5.5) cm long, 2 mm in diameter, striate; leaves opposite, brown, membranous, connate at the base forming a sheath, terminal lobes deltoid; pollen cones solitary, stalked, oblong, 14 mm long, consisting of decussate scales, anthers 7 or 8, shortly stipitate, terminal to the column with concealed basal tube; ovulate cones solitary, terminal to short lateral branchlets, ovoid, with 4 pairs of decussate scales, the basal 1/4–1/2 connate, terminal scales each bearing an ovule, integumental tube erose and villose at the tip; mature cones red, fleshy; seeds 2, trigonous-ovoid, 5–6 mm long, 2–3 m across, grayish brown, partially enclosed by the fleshy scales.

Growing in the semi-desert areas and rock-cliffs of the arid land in northwestern, western and northern China, thence extending westward to Xinjiang. The ethnobotanical history of this species runs parallel to the development of Chinese culture. From time immemorial, people gathered the "fruits" (mature ovulate cones) for tea. As the population increased and the supply becoming insufficient, people began to use the branches for tea. Gradually they discovered Ephedra Tea was good for fever, bronchial ailments, and they saved the material for illness. *Ma-huang* was listed in the earliest Chinese herbal, *Shennong-bencao-jing* (200 A.D.), and also was the first Chinese herbal drug studied chemically by early American-trained Chinese pharmacologists. Ephedrine has a worldwide medicinal application now.

Gnetaceae: Crettir Family

Gnetum indicum (Loureiro) Merrill

Ni-teng=Ni-t'eng (倪籐, Bago); *Shan-bai-guo=Shan-pai-kuo* (山白果, Mountain Ginkgo). Ripefruit, local use only.

Evergreen lianas common in tropical evergreen forests; stem with enlarged joints; leaves opposite, oblong elliptic, 10–20 cm long, 4–10 cm wide, coriaceous, entire; flowers unisexual; staminate ones in cylindrical catkin-like axillary clusters; pistillate ones in elongated branched paniculate spikes; seeds drupe-like, oblong-ellipsoid, 1.5–2 cm long, 1–1.2 cm across the middle, the fleshy outer coat ripe orange, smooth.

Angiospermae: Flowering Plants

MONOCOTYLEDONEAE

Typhaceae: Cat-tail Family

Typha latifolia L. — COMMON CAT-TAIL, BROAD-LEAVED CAT-TAIL

Pu-cai=P'u-ts'ai (蒲菜, Typha Vegetable); *Shui-la-zhu=Shui-la-chu* 水蠟燭, Water Candle); *Pu-cao=P'u-ts'ao* (蒲草, Pu Herb). Tender shoots, a delicacy.

A widespread, stout, upright, aquatic perennial herb, growing in marshes or along the edges of ponds, deeply rooted in mud; leaves sessile, linear, 1–2 m long, 5–

10 mm wide; flowers unisexual, monoecious, intermixed with long hairs, in 2 dense fuzzy cylindrical spikes at the distal end of the stipe, with the slender yellow staminate portion above the candle-like pistillate one; fruit a minute nutlet, inconspicuous, with copious hairs forming the candle-like brown complex, hence the vernacular name, Water Candle. The yellow tender center portion taken off from younger plants sold in large cities of northern Jiangsu for banquets, not a vegetable for the public.

Pandanaceae: Screw-pine Family

Pandanus forceps Martelli — HILLSIDE SCREW-PINE

Le-gu-zi=Le-ku-tzu (簕古子, Spiny Straps).

Large caespitose stiff perennial herb, growing in shade of monsoon mixed forests a mile up from the sea-shore; leaves linear, strap-shaped, 1–3 m long, 3–5 cm wide, with sharp saw-like margin armed with spines; flowers unisexual. Limited use, locally in Guangdong.

Pandanus utilis Bory, and **P. tectorius** Solander — COMMON SCREW-PINE

Lu-dou-shu=Lu-tou-shu (露兜樹, Open Sack Tree); *Lu-dao-shu=Lu-tao-shu* (蘆刀樹, Swamp Sword-Tree). Starch from fruit.

A shrub-like, much branched plant with numerous prop roots supporting the branched stem, growing along the seashore in sandy areas just above the tidal zone, among sedges and grasses; leaves crowded at the distal ends of the branches, sessile, sword-like, 1.5 m long, 3–5 cm wide, keeled, armed with sharp spines along the margin; flowers unisexual, the staminate ones in yellow branched spadices subtended by showy bracts, the pistillate flowers crowded in a solitary dense head; fruits syncarpous, oblong, 20 cm long, 15–20 cm in diameter, rough, consisting of 50–80 drupelets.

The book name, *Lu-dou-shu* (露兜樹), represents three meaningless ideograms carrying the sounds, in this case, missing the meaning of the original local name, *Lu-dao-shu* (蘆刀樹), meaning Swamp Sword Tree. I learned the name from an informant, a farming boy from Guangdong. When I asked him, "What does *Lu-dao-shu* mean?" After I was told the name, he held a leaf and put my hand on the margin saying, "This cuts like a knife." Subsequently, I understand that *Lu-dao-shu* means a tall monocot growing in marshy sea-shore, and bearing sword-like leaves which cut the hands of anybody that tries to gather them.

Potamogetonaceae: Pondweed Family

Potamogeton oxyphyllus Miquel

Ma-zao=Ma-tsao (馬藻, Horse Alga). Leaves and tender shoots, eaten locally.

A delicate, submersed, aquatic perennial plant, with slender and much branched stem; leaves with sheathing stipules, the laminas linear, 5–10 cm long, 2–3 mm wide, acuminate at the apex; flowers inconspicuous, in axillary spikes 5–15 mm long.

Potamogeton pectinatus L. — COMB PONDWEED

Shui-dou-er=Shui-tou-erh (水豆兒, Water Bean). Whole plant; used locally.

A delicate aquatic plant of a submersed, tangled mass; stem much branched, 1 mm in diameter; leaves sessile, with sheathing stipules, the laminas 3–10 cm long, 0.5–1 mm wide; flower small, inconspicuous, in moniliform spikes 1.5–3 cm long.

Aponogetonaceae: Aponogeton Family

Aponogeton natans (L.) Engler et Krause — APONOGETON (Figure 18)

Shui-yong=Shui-yung (水蕹, Water Block-up). Rhizome; limited local use.

A perennial plant growing in streams, ponds or rice fields; rhizome fleshy, tuber-like, covered by remanent petioles and roots; leaves long-petioled, floating or submerged, the laminas oblong, 5–11 cm long, 1.5–4 cm wide, obtuse at the apex, subcordate at the base, veins reticulate with 5 primary ones parallel; flowers small, in cylindrical spike 4–6 cm long; stamens 6, ovaries 3; fruits follicles.

Alismataceae: Water-plantain Family

Sagittaria sagittifolia L. — OLD WORLD ARROWHEAD, SWAMP POTATO, SAGITTARIA (Figure 19)

Ci-gu=Tz'u-ku (慈姑, 茨菰, Sagittaria); *Bai-di-li=Pai-ti-li* (白地栗, White Water Chestnut); *Jian-dao-cao=Chien-tao-ts'ao* (剪刀草, Scissors Herb). Corms, cooked.

An acaulescent aquatic perennial, cultivated in paddies as a vegetable crop for the edible corms; rootstock fleshy, bearing many axillary elongated stolons 50–70 cm long, each terminated by a bud which matures into the market *Ci-gu* (茨菰); leaves long-petioled, all basal, laminas sagittate, various in size and shape, up to 40 cm long, 15 cm wide; flowers white with green sepals, in whorl of threes on a terminal raceme, unisexual, the staminate ones above, the pistillate ones below; stamens numerous; pistils apocarpous; fruits strongly compressed achenes.

Hydrocharitaceae: Frog's-bit Family

Hydrocharis dubia (Blume) Baker — Asiatic Frog's-bit (Syn. *H. asiatica* Miquel)

Shui-bie=Shui-pieh (水鱉, Water Turtle); *Fu-cai=Fu-ts'ai* (苤菜, Frog's Bit). Leafy shoots, dried for winter uses locally.

Floating plants, stoloniferous; with numerous submerging roots 15–20 cm long, the branches binnate, filiform; leaves fasciculate, orbicular-cordate, 3–8 cm long and wide, green above, purple-red and with a central spongy patch beneath, entire, palmately 7–9 nerved; flowers white, unisexual, monoecious; staminate flowers 2 or 3 subtended by a funnelform spath 2-lobed at the distal end, pedicels 2 cm long, sepals 3, green, petals 3, suborbicular, clawed, 1–2 cm long, stamens 9, staminodes 3; pistillate flowers solitary, petals obovate, staminodes 6, ovary inferior, 6-locular, styles 3, 2-fid, hairy, stigmas 6, linear, yellow; fruits subglobose, 5–10 mm long, 4–8 mm across, fleshy, mucilaginous, seeds numerous, prickly. Widespread in eastern Asia, growing in ponds and paddy fields.

Ottelia acuminata (Gagnepain) Dandy — Yunnan Ottelia (Syn. *O. yunnanensis* [Gagnepain] Dandy)

Hai-cai-hua=Hai-ts'ai-hua (海菜花, Water Vegetable Flower); *Hai-cai =Hai-ts'ai* (海菜, Water Vegetable); *Hai-cao-tai=Hai-ts'ao-tai* (海草苔, Water Plant Scape). Young scapes, eaten as a delicacy in Yunnan.

Submerged acaulecent plants; leaves all basal, petioles varying with the depth of the water, laminas varying to sequence of their development and water, generally ovate to lanceolate, 5–11 cm long, 2.5–6 cm wide, acuminate, cuneate at base, primary nerves 5; flowers white, unisexual, blooming on the surface of water, scapes 0.3–4.5 m long, cylindrical, smooth, spathes 2- to 5-ridged, without wings; staminate flowers 40–50, fasciculate in a spathe, pedicels 4–10 cm long, sepals 3, petals 3, obcordate, 1–3.5 cm long, stamens 12, staminodes 3, pistillodes globose, white; pistillate flowers 2–7, rarely more in a spathe, styles 3, yellow, bifid, branches linear, 1.4 cm long, staminodes 3, ovary trigonous, unilocular, placentas 3, parietal; fruits fusiform-ellipsoid, fleshy, mucilaginous; seeds numerous, hairy at one end (comose). Common in the lakes and ponds at altitudes below 2,700 m where the water is less than 4 m deep, the petioles and scapes varying between 4 and 300 cm long; due to increased population and pollution, the species gradually disappeared from many lakes near Kunming.

Ottelia alismoides (L.) Persoon

Long-she-cao=Lung-she-ts'ao (龍舌草, Dragon's Tongue Herb), *Shui-bai-cai=Shui-pai-ts'ai*

(水白菜, Water Cabbage), *Shui-che-qian=Shui-ch'e-ch'ien* (水車前, Water Plantain). Scapes and leaves, eaten as potherbs.

Acaulescent perennial aquatic herbs; leaves all basal, petioles 30–40 cm long, laminas broad-ovate or oblong-cordate, 8–25 cm long, 3–18 cm wide; flowers white, tinged purple-green, sessile, perfect, solitary (rarely 2 or 3), in a spath with 5–10 vertical wings, sepals 3, green, petals 3, obovate, stamens 6, rarely 9, filaments glandularly hairy, ovary inferior, 6-locular, styles 6, bifid almost to the base; fruits fusiform-ellipticoid, striate. Widespread in eastern Asia.

Gramineae or Poaceae: Grass Family

Avena fatua L. var. **glabrata** Petermann — WILD OAT, TARTARY OAT
Qing-ke-mai=Ch'ing-k'o-mai (青稞麥, Green Wheat); *Ye-yan-mai=Yeh-yen-mai* (野燕麥, Wild Oat). Grains.

A caespitose annual grass, 80–120 cm high; leaves linear, 15–25 cm long, 4–12 mm wide; panicle pyramidal, 10–30 cm long, spikelets pedicellate, 18–25 mm long, 3-flowered, the florets readily falling; rachilla and lemma covered with long, stiff, brownish hairs, awns stout, geniculate, twisted below. Widespread in China, often growing in wheat fields.

Avena sativa L. — OATS
Yan-mai=Yen-mai (燕麥, Oat); *You-mai=Yu-mai* (油麥, Oil Wheat). Grains; cultivated in Inner Mongolia.

Very similar in appearance to the Wild Oat, differing in having larger leaves 30–50 cm long, 9–13 mm wide; 2-flowered spikelets, the florets not falling readily, lemma glabrous, and awn straight or absent; cultivated in northern China.

Bambusa beecheyana Munro — BAMBOO SHOOT
Beechey bamboo; introduced to USA, grown in California and Florida; important source of commercial bamboo shoots; several varieties and cultivars.

cv. 'Horse Tail'
Ma-wei=Ma-wei (馬尾, Horse Tail). Shoots of good quality.

Caespitose bamboo, culms up to 15 m tall, distal end upright; internodes 30–40 cm long, 5–7 cm in diameter; leaves lanceolate, 21–35 cm long, 3–4 cm wide; shoots conic, 30 cm long, sheath greenish yellow, sparsely hairy; good quality, no bitter taste.

var. **pubescens** (P. P. Li) Lin — Big Shoot

Da-tou-dian=Ta-t'ou-tien (大頭典, Big Head Shoot); *Da-sun-zhu=Ta-sun-chu* (大笋竹, Big Shoot Bamboo). Shoots used for canning.

Different from the above cultivar in having smaller stature, slightly bending crown, much more bristly pubescent; the culm 10 m high, with internodes 18–22 cm long; leaves broad-lanceolate, 18–23 cm long, 4–5 cm wide; winter shoots conic, 28 cm long, pubecent; meat bitter, requiring precooking, used for canning.

Bambusa gibboides W. T. Lin — TENDER SHOOT BAMBOO

Yu-du-nan=Yu-tü-nan (魚肚腩, Fish Belly). Shoot of good quality, having no bitter taste.

A caespitose bamboo 12 m high, middle internodes up to 50 cm long, 4–5 cm in diameter; leaves lanceolate, 18 cm long, 2–3 cm wide; shoots in the market 25 cm long, 6 cm across the base, sheaths green, covered with stiff brown hairs, meat of good taste, high quality.

Bambusa oldhamii Munro — OLDHAM BAMBOO

Lü-zhu=Lü-chu (綠竹, Green Bamboo). Shoots.

A caespitose bamboo 6–12 m tall, internodes 20–25 cm long, 3–12 cm in diameter; leaves oblong-lanceolate, 15–30 cm long, 3–6 cm wide, pubescent beneath. Widespread cultivation for culm; shoot used locally.

Bambusa stenostachya Hackel — THORNY BAMBOO (Syn. *B. spinosa* Roxburgh)

Le-zhu=Le-chu (簕竹, Spiny Bamboo); *Ci-zhu= Tz'u-chu* (刺竹, Thorny Bamboo). Winter buds, exposed above ground.

Caespitose spiny bamboo, cultivated as fences and for the strong thickwalled culms, 15–24 m high, the internodes 25–40 cm long, 8–15 cm in diameter; leaves linear-lanceolate, 10–25 cm long, 0.5–2 cm wide, spinulose-scabrous along the margin; shoots used only locally.

Bambusa vario-striatus (W. T. Lin) Chia et Fung — SWEET BAMBOO

Diao-si-zhu=Tiao-ssu-chu (吊絲竹, Hanging Red Silk); *Tian-sun-zhu=Tien-sun-chu* (甜筍竹, Sweet Shoot Bamboo). A Common bamboo in Canton, having no bitter taste; being the source of early shoot crop.

A caespitose bamboo 10 m high, internodes 30–25 cm long, 4–6 cm in diameter; leaves lanceolate, 20–25 cm long, 3 cm wide; the market shoots 20–28 cm long, greenish-yellow, densely hairy, high yield, meat tender, of good flavor, the favorite of Cantonese people.

Bambusa vulgaris Schrader ex J. C. Wendland — Common Bamboo

Qing-si-zhu=Ch'ing-ssu-chu (青絲竹, Green-silk Bamboo); *Long-tou-zhu=Lung-t'ou-chu* (龍頭竹, Dragon Head Bamboo); *Tai-shan-zhu=T'ai-shan-chu* (泰山竹, Tai-shan Bamboo). Winter buds exposed above ground; used locally.

Caespitose bamboo, culms 7–15 m high, 4–8 cm across, internodes uniformly green, 20–30 cm long, slightly glaucous when young; leaves lanceolate, 16–30 cm long.

var. **vittata** A. et C. Riviere — Stripe Bamboo

Huang-jin-jian-bi-yu=Huang-chin-chien-pi-yü (黃金間碧玉, Gold-and-green Jade). Winter buds; local use only.

An attractive caespitose bamboo much cultivated for landscape purposes, culms 10 m high, internodes 30–35 cm long, 4–6 cm in diameter; golden-yellow with green; leaves lanceolate, 20–28 cm long, 3 cm wide; shoots 20–28 cm long.

Cephalostachyum capitatum Munro

Kong-zhu=K'ung-chu (空竹, Hollow Bamboo); *Da-ba=Ta-pa* (打巴, the sound of the local Tibetan name). Young shoots, eaten locally.

Medium-sized bamboo, 13–15 m high, culms 1.5–4.5 cm in diameter, internodes 25–45 (–62) cm long, green, becoming yellow with age, culm-sheaths reddish-brown, 19–30 cm long, densely yellow-prickly, ligules black-brown, 1 cm high, sheath-blade 9–30 cm long, 5–11 mm wide, purple-green; branches numerous, filiform, uniform, fasciculate, without secondary branchlets; bearing 5 (–9) leaves, leaf-sheaths 6–8 cm long, laminas broad-lanceolate, 10–21 cm long, 2–4.3 cm wide, pilose. Native of the eastern Himalayan Region, growing spontaneously in the broad-leaved forests at the altitudes of 1,900–3,000 m in Guangxi.

Chimonobambusa metuoensis Hsueh et Yi — Tibetan Square Bamboo

Mo-tuo-fang-zhu=Mo-t'o-fang-chu (墨脱方竹, Square Bamboo in Mei-tuo); *Re-su-guo-ba=Je-su-kuo-pa* (熱宿戈巴, the Mei-tuo local name). Young shoots eaten by the local people.

Medium-sized bamboos 5–7 m high, culms 1–2.5 cm in diameter, consisting of 20–35 internodes, the lower ones tetragonous, the lower nodes surrounded by tubercular spines, lateral branches in threes, first appearing on the fifteenth node, the central one larger than the lateral two, generally the appearance of lateral branches numerous due to later branching, culm-sheaths purple-red, 8–13 cm long, bristly, sheath-blade minute; branchlets each bearing 2 or 3 leaves, leaf-sheaths 4–8 cm long, glabrous, dark purple, laminas lanceolate, 12–33 cm long, 1.5–4 cm wide, acute, glabrous. Endemic to southeastern Tibet, growing in broad-leaved forests at altitudes of 1,900–3,000 m.

Coix lachryma-jobi L. — JOB'S TEARS

Yi-yi=I-i, or *Yi-ren=I-jen* (薏苡, 薏仁, Job's Tears). Grains, pollished, raw or popped; both kinds available in Chinese stores in Boston; from summer crops cultivated in temperate China; plants growing wild in wet areas near villages in Hong Kong.

In tropical China such as undisturbed areas of Hong Kong, a tall, caespitose, perennial grass growing naturally along streams; the plant 1–1.5 m high, branches from the gound level; leaves linear-lanceolate, up to 30 cm long, 3 cm wide; flowers unisexual, monoecious, with the staminate cluster of spikelets and the solitary pistillate spikelets sharing a common peduncle and involucre; staminate spikelets 2-flowered, in twos or threes in a spike on slender peduncle 3–4 cm long, the base enclosed with the sessile pistillate spikelet in the green and coriaceous urceolate involucre becoming bony in fruit; pistillate spikelet 3-flowered, completely enclosed in the involucre with the two stylar branches exposed only; the market material representing the polished grain, ivory white, hemispherical, flat and deeply grooved on one side, 2–6 mm long, 4–6 mm wide, the brown, lateral canal representing the position of the sterile pistillate florets and the peduncle of the staminate spike (see Part I for more details).

Cymbopogon citratus (DC. ex Nees) Stapf — LEMON GRASS

Xiang-mao-cao=Hsiang-mao-ts'ao (香茅草, Aromatic Grass). Shoots before flowering, with the leaf-blades removed, used for seasoning soup; available in Boston.

A caespitose grass cultivated along edge of vegetable gardens, forming a dense clump 30–90 cm in diameter; leaves linear, 50–80 cm long, 1–1.5 cm wide, scabrous along the margin; seldom flowers; the market material consisting of fresh individual plant stalk with the leaf-tops removed; used to flavor soup and meat, too tough to be palatable.

Dendrocalamus asper (Schultes) Backer ex Heyne

Gao-she-zhu=Kao-she-chu (高舌竹, High Ligule Bamboo). Young shoots used for vegetable.

Caespitose bamboo forming dense clumps, culms 15–20 cm tall, internodes 30–45 cm long, 6–10 cm in diameter, covered by white waxy powder; leaves lanceolate, 20–30 cm long, 3–5 cm wide; shoots conical, covered with brown appressed acicular hairs. Cultivated around temples in Hong Kong, shoots used locally.

Dendrocalamus latiflorus Munro — Dried Bamboo Shoot (Syn. *Sinocalamus latiflorus* [Munro] McClure)

Da-ye-wu=Ta-yeh-wu (大葉烏, Big-leaved Black). Shoots of good quality, a late crop, used largely for dried products called *Sun-gan=Sun-kan* (笋乾, Dry Bamboo Shoot). Dried material available in Chinese stores in Boston.

Caespitose bamboo, 20–25 m tall, internodes 35–45 cm long, 5–7 cm in diameter; leaves large for bamboo, lanceolate, 30–35 cm long, 7–8 cm wide; shoots conical, yellow, the cone-scales ciliate; meat tender, good flavor.

Dendrocalamus tibeticus Hsueh et Yi

Xi-zang-mu-zhu=Hsi-tsang-mu-chu (西藏牡竹, Tibetan Male Bamboo), *Qian-lang-suo=Ch'ien-lang-so* (前郎索, the sound of the local Tibetan name); *A-wa=A-wa* (阿瓦, another Tibetan name). Young shoots used locally.

Tall caespitose bamboos 15–25 m high, calms 12–18 cm in diameter, consisting of 38–45 internodes, 40–45 (–60) cm long, green, each with a distal ring of gray hairs; branches appearing first on the fourteenth node, numerous, the central one the largest, 1.5 m long, 0.4–3 cm in diameter; culm-sheaths reddish-brown, 33–40 cm long, covered with black bristles, ligules 2–4 mm long, pilose, sheath-blade deltoid, 5–28 cm long, branchlets bearing 5–8 leaves, leaf-sheaths 6–11 cm long, glabrous, laminas broad-lanceolate, 10–32 cm long, 2.2–4.5 cm wide; inflorescences whip-like, with globose fascicles of 2–10 spikelets spaced 2–2.5 cm apart; spikelets purple, ovoid-oblong, 1–1.2 cm long, each consisting 3 or 4 florets, stamens 6, ovary ovoid, hairy, style 5–8 mm long, stigma 1, purple. Endemic to southeastern Tibet, growing in broad-leaved forests at altitudes of 1,200–1,500 m, cultivated locally.

Echinochloa crusgalli (L.) Beauvois — Barnyard Grass, Barnyard Millet

Two varieties yielding edible grain.

var. crusgalli

Bai-zi=Pai-tzu (稗子, Tare). Seed, gathered and used locally, not available in the market.

An annual weed, growing in open places, ditches, becoming troublesome in rice fields, often requiring hand picking; plant 50–130 cm tall, branched at the base; leaves linear, 5–40 cm long, 5–18 mm wide; flowers in purplish pyramidal panicles, 10–20 cm long, the branches 6–16 cm long; spikelets 3 mm long, tuberculate-hispid, awned.

var. frumentacea (Roxburgh) W. F. Wight — Japanese Barnyard Millet, Billion-dollar Grass

Hu-nan-ji-zi=Hu-nan-chi-tzu (湖南稷子, Hunan Millet). Grain.

Distinguished from the above variety by its broader leaves 2–2.5 cm wide, and more compact panicles 10 cm long, more turgid and awnless spikelets in shorter racemes 1.5 cm long; not as common as var. *crusgalli*.

Eleusine coracana (L.) Gaertner — FINGER MILLET, AFRICAN MILLET, RAGI
Ya-zhua-bai=Ya-chua-pai (鴨爪稗, Duck-feet Tare); *Ying-zhua-bai=Ying-chua-pai* (鷹爪稗, Fagle-claw Tare); *Long-zhua-su=Lung-chua-su* (龍爪粟, Dragon Claw Millet). Grain.

An annual grass 60–100 cm tall; leaves linear, 18–30 cm long, 5–10 cm wide; spikelets 5–6 flowered, sessile, on one side of a continuous rachis forming a spike 5–6 cm long, 1 cm across, 2–9 spikes forming a pseudoumbel at the end of a peduncle 30 cm long; occasionally cultivated in the warmer areas of China.

Fargesia melanostachys (Handel-Mazzetti) Keng f.
Hei-sui-jian-zhu=Hei-sui-chien-chu (黑穗箭竹, Black Spike Arrow Bamboo); *Niu-ma=Niu-ma* (牛麻, the sound of the local Tibetan name). Young shoots.

Caespitose bamboos 6 m high, rhizomes 0.9–1.5 cm in diameter, densely covered by deltoid scales, culms consisting 25 internodes, the middle ones 25–30 cm long, yellowish green, glabrous, slightly farinose below the nodes when young, branches 5–10, fasciculate, 30–50 cm long, 1–2 mm in diameter, uniform; culm-sheaths dark purple, 11–16 cm long, sparsely bristly, ciliate, ligules truncate, sheath-blade 1.2–8 cm long; branchlets bearing 1–3 leaves, leaf-sheaths purple, 2–4 cm long, glabrous; laminas lanceolate, 5–8 cm long, 4–6 mm wide; inflorescences in loose terminal panicles, spikelets purple, 6–25 mm long, stamens 3, styles 2, ovary oblong. Native to western Yunnan and southeastern Tibet, growing in pine and broad-leaved mixed forests at altitudes of 2,500–3,400 m.

Fargesia setosa Yi
Xi-zang-jian-zhu=Hsi-tsang-chien-chu (西藏箭竹, Tibetan Arrow Bamboo); *Niu-ma=Niu-ma* (牛麻, the sound of the local Tibetan name). Young shoots used locally.

Slender caespitose bamboos 1–7 m high, rhizomes 0.4–2.5 cm in diameter, densely covered by deltoid scales; culms consisting 16–30 internodes, the middle ones 18–28 cm long, green, subfarinose, the distal half setaceous when young; culm-sheaths red-purple or purple-green, densely covered by brown-yellow setae, the margin villose; branches 3–7, fasciculate, 60–80 cm long, 1–3 mm in diameter, farinose: branchlets bearing 3–5 leaves, leaf-sheaths 3–9 mm long, purple-green, hairy, laminas lanceolate, 4–17 cm long, 0.4–1.8 cm wide, pilose on both surfaces; flowers not known. Endemic to

Bomi in southeastern Tibet, growing in forests of *Keteleeria* Carriere at altitudes of 2,700–3,800 m.

Fargesia spathacea Franchet — GIANT PANDA BAMBOO

Guai-gun-zhu=Kuai-kun-chu (拐棍竹, Cane Bamboo); *Long-tou-zhu=Lung-t'ou-chu* (龍頭竹, Dragon-head Bamboo), *Shan-zhu=Shan-chu* (山竹, Hillside Bamboo). Seeds, gathered by the people on the hillside of central and western China, used for food.

Rhizomes creeping, enlarged at the end and curved up for the emerging shoots, culms 1–5 m high, internodes 10–22 cm long, 0.5–2 cm across, green or yellowish-green, often glaucous, with a conspicuously thickened ring below the nodes, shoots slender, sheaths purplish-red, veins and reticulations prominent, sheath-blades linear, dark purple; branches 7–13 on each node; leaves 1–3 on each branchlets, 6–10 cm long, 6–11 mm wide; flowering branchlets leafy, spikelets in short terminal panicles 2–4 cm long, each spikelet consisting of 2–4 florets , glumes apiculate or shortly awned, lemma 11–15 mm long, pilose, palea 11 mm long, bifid, stamens 3; seeds mature in late June. Endemic to central and western China, growing on hillsides at altitudes of 1,300–2,500 m; a common species for the food of the giant panda; culms strong, with curved-basal end, giving a fanciful resemblance of dragon's head, hence the trivial name '*Long-tou-zhu*', the same feature providing an good handle of a cane, and the culms are harvested for making stuffs, hence the local name "*Guai-gun-zhu*".

Hordeum vulgare L. — BARLEY

Da-mai=Ta-mai (大麥, Barley). In North China, polished grains used for preparing a summer drink, also ground with sorghum and soybean for a coarse, healthy, fiber bread flour.

Mai-ya=Mai-ya (麥芽, Malt). Sprouted grain showing tiny root only, dried; used in preparing *liangcha* as well as in prescriptions in traditional Chinese medicine, and in other beverages; imported material available in American Chinese groceries on the shelf for tea.

Annual or biennial plants 50–100 cm high, culms shiny, smooth; leaves linear, 15–25 cm long, 7–10 mm across middle; spikelets in compact cylindrical panicles, 3–8 cm long (excluding the awns), 1.5 cm across, 3 close at each node on an articulate rachis, glumes linear, hairy, shortly awned, lemma 5-nerved, awned or tricuspidate; grains variable according to the varieties; extensively cultivated, two varieties in northern China, and on the hillsides of western China.

var. **vulgare** — Barley

> *Da-mai=Ta-mai* (大麥, Big Grain). A staple food in northern China, ground with sorghum and soybean for a coarse, healthy, fiber-rich flour; pollished grains used for preparing summer drink; recently large amount used for beer industry.
>
> *Mai-ya=Mai-ya* (麥芽, Malt). Sprouted grain showing tiny root only, dried; an important ingredient in preparing *liangcha* and other beverages; imported material available in American Chinese stores.

A variety distinguished by the grains adnate to the lemma and palea, the lemma prolonged into coarse awns 8–15 cm long. In the lower Yellow River Region, especially in Shandong, Henan and northern Jiangsu, barley is sown in autumn, and harvested in June, with the young plants passing the winter under snow.

var. **nudum** J. D. Hooker

> *Yu-mai=Yü-mai* (玉麥, Jade Barley); *Luo-mai=Lo-mei* (裸麥, Naked Barley); *Qing-ke=Ch'ing-k'o* (青稞, Green Stem). The crop matures a week earlier then wheat and barley in northern China; partially matured ears gathered and roasted, ground into noodle-like material, eaten in famine years.

Plants similar in structure and appearance to the common barley, distinguished by the grains, being free from the lemma and palea. In northern Jiangsu, the crop matures 2 or 3 weeks earlier than barley; one of the important crops in the mountains of Sichuan, Gansu, and in the Qinghai-Tibetan plateau.

var. **trifurcatum** (Schlechter) Alefeld

> *Zang-qing-ke=Tsang-ch'ing-k'o* (藏青稞, Tibetan Barley); *San-cha-da-mai=San-ch'a-ta-mai* (三叉大麥, Trifurcate Barley). Grains; important food crop in Tibet.

Annual plants 100 cm high; leaves 18–23 cm long, 9–18 cm wide, panicles 5.5–8 cm long, 1.5 cm across, grains free from the lemma and palea, the lemma trifurcate, with the lateral lobes acuminate and recurves, shortly awned or not, the middle lobe clavate; cultivated at high altitude of western China.

Imperata cylindrica L. — Woolly Grass

> *Mao-gen=Mao-ken* (茅根, Imperata Root). Fresh rhizome; chewed raw by rural people for the sweet juice; *Mao-ti-gu=Mao-t'i-ku* (毛提穀, Pulling Pussy Grain). Very young flower bud, carefully drawn up from the ground and eaten by children in early spring; having a sweetish taste.

A very widespread and persistent weed in China, with fleshy, white, scaly subterranean rhizomes, sweet before the leaves unfold in the spring; aerial growth generally consisting of the linear leaves only, 30–50 cm long, (longer with plants growing crowded in fertile soil), 6–12 mm wide in the middle; in northern China plants flower in early spring before the leaves; in Hong Kong and Guangdong, flowers appear after the leaves are removed by fire or cutting; flowers small, in terminal cylindrical silky panicles 9–19 cm long, 1 cm across; spikelets all alike, awnless, in pairs, unequally pedicellate, on a slender continuous rachis, surrounded by long silky hairs, glumes membranous, lemma and palea thin and hyaline; seeds dispersed with the silky hairs; rhizomes used in tranditional Chinese medicine, available in American Chinese stores.

Oryza sativa L. — RICE

Dao=Tao (稻, Rice); *Gu-zi=Ku-tzu* (穀子, Field of Rice); *Mi=Mi* (米, Polished Rice); *Fan=Fan* (飯, Cooked Rice). Grain; the staple of the Chinese people south of the 35th parallel and in large cities north of it. It is an annual plant in most parts of China, may persist after harvesting in Guangdong and Hong Kong, where two or three crops are planted annually.

Flowering plants caespitose, 30–100 cm high; leaves linear, 8–25 cm long, 6–15 mm wide; flowers in terminal drooping panicle, 15–40 cm long; spikelets 1-flowered, laterally compressed, 7–10 mm long, 3–4 mm wide, lemma and palea papillose-roughened, clinging to the grain; the market rice mechanically polished, with the major portion of proteins, vitamins and minerals removed with the embryo and bran.

var. **glutinosa** Matsumura — GLUTINOUS RICE

Nuo-mi=No-mi (糯米, Glutinous Rice). Available in Boston.

Glutinous rice has short, chalky white grains which stick together when cooked. It differs from the common non-sticky rice chemically in the contents of the endosperm, in the maturation of which some carbohydrates turn to soluble starch and dextrin, having colloidal properties. Unlike wheat, the glutinous rice contains no true gluten, and it is fit for sweetmeats and fermentation (see Part I for details on Eight Precious Pudding and Fermented Rice).

Rice, like wheat and corn, is of hybrid origin, having a complex ancestry, and is polymorphic. It hybridizes readily with the wild species, *O. fatua* Koenig and *O. minuta* Presl of Southeast Asia, and produces many intermediate forms. The selections for desirable characters and the cultivation of special strains was directly related to the cultural development of the ancient people and to the scientific knowledge of recent agricultural scientists. In these areas the ancient farmers have made successful selections and cultivated strains for high yield, good taste, special aroma, and colorful grains, other than glutinous rice. Fifty years ago, the high-priced *xiang-dao-mi* (香稻米) with the aroma of "Basmati" of India and the short-grain rice known as Japanese rice in the current American market, were available in Shanghai. The red rice was cultivated in Guangdong and Yunnan. Likewise, the contemporary agricultural scientists have developed high yield and low stature strains which can grow in Shenyang (瀋陽, Long. 123°30′E, Lat. 43°40′N), extending the cultivation of rice eight degrees northward from where it used to be in the 1940s.

Panicum miliaceum L. — Broomcorn, Millet

The broomcorn is one of the oldest crop plants cultivated in China. In 1980, a botanist, Q. R. Wang. and an archaeologist, D. Y. Guo, reported the findings of a pile of *P. miliaceum* one meter square in storage at a Neolethic site in Gansu. Now, it is extensively cultivated in North China with two common varieties.

A caespitose annual grass, erect, much branched from the ground level , 60–120 cm tall; leaves linear, 20–35 cm long, 1–2 cm wide, with a herbaceous sheath covered by stiff long hairs (strigose-villose), flowers small, in loose panicles 15–30 cm long, nodding; spikelets 4.5–5 mm long, ovoid, acuminate; fruit shiny, stramineous, enclosed in chartaceous-indurate husk.

var. **effusum** Alefeld

Ji-zi=Chi-tzu (稷子, Millet). Widely planted; polished grain available in plastic bags in Boston.

A variety distinguished by its more compact panicles, non-sticky grain; extensively planted for its higher yield.

var. **glutinosa** Bretschneider

Shu-zi=Shu-tzu (黍子, Glutinous Millet). Less common; pollished grain used for making glutinous pastry.

A variety recognized by its more open and nodding panicles, sticky grain; only cultivated by people with more land who can afford to have the luxury for sweetmeats, and better brooms.

Phyllostachys angusta McClure

Shi-zhu=Shih-chu (石竹, Rock Bamboo). Shoots; free from acridity.

A medium-sized bamboo with creeping rhizome; culms 7.5 cm long, 2.5 cm in diameter, the internodes 15–19 cm long; shoots 20 cm long, 1.5 cm in diameter, pale yellow, strewn with small brown dots; good taste. This bamboo is called 'rock bamboo' for its hard texture. It is used in the manufacture of fine bamboo products in China, and has been introduced into USA from Zhejiang.

Phyllostachys bambusoides Siebold et Zuccarini — TIMBER BAMBOO (Syn. *P. reticulata* Koch)

Gang-zhu=Kang-chu (剛竹, Giant Timber Bamboo); *Ku-zhu=K'u-chu* (苦竹, Bitter Bamboo). Winter shoots; bitter, requiring boiling off the bitterness.

A tall and elegant bamboo with creeping rhizome; culms green, up to 24 m high; internodes 23–40 cm long, 7 cm in diameter; shoots 45–48 cm long, 2–3 cm in diameter, greenish to ruddy buff, densely spotted with dark brown throughout, bitter, edible after boiling. This bamboo was introduced into Japan in ancient time and is now the most useful species in China and Japan for its workability; most of the bamboo products in these countries are made of the culm of this species.

Phyllostachys congesta Rendle

Shui-zhu=Shui-chu (水竹, Water Bamboo). Shoots; slightly bitter.

A rather large bamboo with creeping rhizome; culms 8.5 m long, 5.5 cm in diameter; internodes 15–27 cm long, loosely farinose when young; shoots 25–30 cm long, 4.5 cm in diameter, edible, slightly bitter; introduced into USA from Zhejiang in 1907.

Phyllostachys decora McClure

Mei-zhu=Mei-chu (美竹, Beautiful Bamboo). Winter buds.

A slender bamboo with creeping rhizome; culms 8 m long, 3 cm in diameter, green, internodes 12–16 cm long, slightly ribbed-striate; shoots 20–30 cm long, 1.5 cm in diameter, dark green with pale green spots. Introduced into USA from Jiangxi by F. A. McClure in 1938.

Phyllostachys dulcis McClure — SWEET BAMBOO

Tian-zhu=Tien-chu (甜竹, Sweet Shoot Bamboo). Winter buds; used in central China.

A relatively large bamboo with creeping rhizome; calms 13 m high, 6.5 cm in diameter; internodes 20–33 cm long, dull green, often finely striped with cream or paler green, covered with white powder below the nodes; shoots 25–28 cm long, 1.5–2 cm in diameter, greenish cream color with brown spots, glabrous or with scattered stiff erect hairs, each culm sheath terminated by a strongly crinkled blade; available in early spring, with excellent cooking quality, a highly esteemed edible bamboo shoot in China.

Phyllostachys elegans McClure

Hua-zhu=Hua-chu (花竹, Flower Bamboo). Winter buds; shoots in the market.

An elegant bamboo with creeping rhizome; culms 8 m high, 6 cm in diameter; internodes

14–30 cm long, finely ribbed-striate, dull-green, speckled with small brown spots; shoots 32 cm long, 3.5 cm in diameter, olive-green, moderately maculate with small brown spots, each culm sheath terminated by a strongly crinkled blade. Introduced into USA by F. A. McClure from Guangdong in 1936 and again from Hainan Island in 1938.

Phyllostachys nidularia Munro

Sao-ba-zhu=Sao-pa-chu (掃把竹, Broom Bamboo). Supplying the largest shoot to the market.

A medium-sized bamboo with creeping rhizome; culms 10 m high, 3.5 cm in diameter, more or less arched under the weight of the foliage; internodes 17–38 cm long, green, retrose-setulose below the nodes; shoots 22–24 cm long, 2–4 cm in diameter, olive-green to pale green, covered with retrose-hirsute brown hairs near the base of each sheath, and the apex of the sheath terminated by a broad appressed inflated blade. Introduced repeatedly into USA by governmental plant explorers.

Phyllostachys nigra (Loddiges) Munro var. **henonis** (Mitford) Stapf ex Rendle

Jin-zhu=Chin-chu (金竹, Golden Bamboo); *Mao-jin-zhu=Mao-chin-chu* (毛金竹, Hairy Golden Bamboo). Shoots.

Bamboo 10–14 m high, middle internodes 34 cm long, green, glaucous and pilose, turning grayish-green with age; shoots harvested in May, sheaths pink-brown, dense-pubescent, auricles well developed, with purple-black curled hairs; leaves lanceolate, 4–10 cm long. Native to central China, cultivated on the hillsides below 1,400 m.

Phyllostachys pubescens Mazel ex H. de Lehaie (Syn. *P. edulis* H. de Lehaie)

Mao-zhu=Mao-chu (毛竹, Hairy Bamboo); Winter buds called *"sun"* (筍); *Dong-sun=Tung-sun* (冬筍, Winter Shoot); *Chun-sun=Ch'un-sun* (春筍, Spring Shoot). Bitter, boil off acridity before use.

A large bamboo with creeping rhizome; culms 13–20 m high, 5–7 cm in diameter, arched at the tip; internodes 10–21 cm long, pale green, densely velvety when young, gradually glabrescent; shoots cone-shaped, 20–30 cm long, 5–6.5 cm across the base, greenish smoky buff, densely maculated dark brown, covered by erect hairs, each sheath terminated by a triangular, strap-shaped stiff blade; the primary source species of the commercially produced canned bamboo shoots. This species is extensively cultivated in the warm temperate areas of China and Japan. The culm is used in heavy construction, well-drilling equipment, tubing for piping salt brine, water, gas, etc. The species was introduced into USA in the 1890s, and there are excellent groves in Anderson, S.C..

Saccharum officinarum L. — SUGAR CANE

Gan-zhe=Gan-che (甘蔗, Sugar Cane). Mature stem, chewed raw; source of sugar.

Tall, robust, perennial grass cultivated for the sugar industry; culms 3–5 m tall, 2–3 cm in diameter, solid, the pith containing sweet sugar juice, the outside straw-colored or purple; leaves linear, 100 cm or more long, 4–6 cm wide, midrib strongly elevated beneath, margin spinulose; flowers in large silky panicles 30–60 cm long, 10–20 cm across, consisting of thousands of small flowers; spikelets in pairs surrounded by tuft of soft hairs, awnless; of hybrid origin, chromosome 40.

Saccharum sinense Roxburgh — CHINESE SWEET CANE

Zhu-zhe=Chu-che (竹蔗, Bamboo Sugar Cane). Stem, freshly squeezed juice available on street corners and along sidewalks, sold as a refreshing drink, also used for tea in southern China. Dried bundles of the stem available in American Chinese stores.

Cultivated tough, tall grass, differing from the above species in being more hardy to insect attacks and more tolerant to drought, and in having taller and slimmer yellowish green stem, and long hairs on the axis of the panicle; of hybrid origin, chromosomes 45 or 46.

Secale cereale L. — COMMON RYE

Hei-mai=Hei-mei (黑麥, Rye). Cultivated; rare in China.

Caespitose annual grass, cultivated by people living in the northern mountains; culms 1 m or more high, densely villose; leaves linear, 15–25 cm long, 5–10 mm wide, pubescent; spikes 10–15 cm long, 1–1.5 cm in diameter, spikelets 2-flowered, 1.5 cm long, awned, arranged flatly against the rachis, glumes subulate, 1-nerved.

Setaria italica (L.) Beauvois — FOXTAIL MILLET

Xiao-mi=Hsiao-mi (小米, Small Grain, Millet); *Su=Su* (粟, Millet, a name used in ancient books).

A robust, caespitose, annual grass 1–1.5 m high; culms 8–10 mm thick; leaves linear-lanceolate, 15–35 cm long, 2–2.5 cm wide; panicles cylindrical, 10–22 cm long, 2–4 cm in diameter, nodding; spikelets subtended by 1–3 bristles; seed 2 mm across, clinging to the husk; cultivated in northern China, straw a good feed for horses.

Sorghum bicolor (L.) Moench — SORGHUM, KAOLIANG (Syn. *S. vulgare* Persoon),

Seeds; the tallest cereal crop plant known in China; many cultivars.

Robust annual, branched from the base, in cultivation only one stem allowed to grow; culms 2.5–3 m tall, 2.3 in diameter; leaves linear, 45–80 cm long, 4.5–6 cm wide, smooth, with a prominent midrib; panicles terminal, 20–45 cm long, 5–10 cm in diameter, various color and structure; spikelets paired, the fertile one 5–6 mm long; mature seed partially exposed. Used to be extensively cultivated in northern and northeastern China; today it is replaced by rice or maize in these areas. Two cultivars listed below.

cv. 'Caffrorum' (Syn. *S. vulgare* var. *caffrorum* [Retzius] F. T. Hubbard et Rehder)

Gao-liang=Kao-liang (高粱, Tall Cereal); *Shu-su=Shu-su* (蜀粟, Sorghum). Cultivated in North China for grains and straw, an important material for fuel and for building hunts.

cv. 'Saccharatum' (Syn. *S. saccharatum* Moench) — Sweet Sorghum

Tian-gao-liang=T'ien-kao-liang (甜高粱, Sweet Sorghum); *Tian-su-jie=T'ien-su-chieh* (甜粟楷, Sweet Tall Cereal). Stem; chewed raw as a substitute for sugar cane.

Thysanolaena maxima (Roxburgh) Kuntze

Zhong-zi-ye=Chung-tzu-yeh (粽子葉, Rice Tamale Leaf). Leaves; used for wrapping a mixture of soaked glutinous rice, ham, salted egg yoke, or other ingredients for pastry; the wrapped mixture boiled for hours, served hot or cold, particularly for the Dragon Boat Festival in June. The leaves hold the mixture while cooking, and also impart pleasant flavor.

A reed-like, caespitose grass growing in rocky disturbed areas with little soil, unfit for other species; culms 2–2.5 m tall, 8–10 mm thick; leaves bluish-green, crowded towards the shoot apex, 25–40 cm long, 5–8 cm wide; panicles purple, 40–70 cm long, open, bearing millions of minute flowers; spikelets 2-flowered, 0.5–1.5 mm long; a pioneer-species in denuded rocky areas in Hong Kong.

Triticum aestivum L. — WHEAT

Xiao-mai=Hsiao-mai (小麥, Wheat). Cultivated in North China, the staple food of the people in the Yellow River Region, sowed in September–October, harvested the next June; the first plant that entered my life, forming my first bite of food and my first lesson of the plant world. *Tian-mian-jiang=Tien-mien-chiang* (甜麵醬, Sweet Flour Sauce). A fermented product, the sauce served in North China with Beijing Duck). *Mian-jin=Mien-chin* (麵筋, Wheat Gluten). Gluten extracted from wheat flour; available in cans in American Chinese stores.

A caespitose annual, 1 m high at flowering and fruiting time; leaves linear, 25–35 cm long, 1–1.5 cm wide; spike compressed cylindrical, 7–11 cm long, 1–1.5 cm across;

spikelets 3–5-flowered, solitary, arranged flatly at each joint of a continuous rachis, awned or not; seed free from husk. Many strains cultivated in northern China where it was a staple food before 1940s. Other species of *Triticum* (*T. dicoccum* Schrank — Emmer, *T. durum* Desf. — Durum Wheat, *T. monococcum* L. — Einkorn, *T. turgidum* L. — Poulard Wheat) have been observed in the field or in herbaria containing Chinese collections.

Zea mays L. — Corn, Maize, Indian Corn

Yu-shu-shu=Yu-shu-shu (玉蜀黍, Corn, Jade Millet of Sichuan); *Bao-gu=Pao-ku* (包穀, Wrapped Grain); *Bang-zi=Pang-tzu* (棒子, Club Seed); *Bang-zi-mian=Pang-tzu-mian* (棒子麵, Corn Flour). *Bao-gu-fen=Pao-ku-fen* (包穀粉, Corn Flour). Maize is relatively a new addition to Chinese diet. In the 1940s, it was only the staple food of the ethnic groups living in the mountains of Sichuan and Yunnan. Today it has gradually replaced Sorghum in temperate zones of China. Before flowering, the plants of maize and sorghum have similar appearance.

Tall annual grasses 1–1.5 m high; stems 1–3 cm in diameter, the lower nodes with many poproots; leaves linear, 50–80 cm long, 3–7 cm wide; flowers unisexual, monoecious; the staminate ones in terminal open panicles 15–40 cm long, cream-yellow; pistillate spikelets sessile, densely arranged in many vertical rows on a fleshy cylindrical axis enclosed by numerous foliaceous bracts, axillary to normal leaves, with the silk-like styles protruding from the summit, such pistillate inflorescences maturing into the "ears of corn", with the central axes becoming woody. In China, maize is cultivated primarily for the grain; sweet-corn being rare; in Taiwan, a special strain cultivated for the canning industry, using of the tender young pistillate inflorescences; cans available in American Chinese stores.

Zizania caduciflora (Turcz ex Trin.) Handel-Mazzetti — Chinese Wild Rice, Zizania Shoot

Jiao-sun=Chiao-sun (茭筍, Zizania Shoot); *Jiao-bai=Chiao-pai* (茭白, Zizania White); *Jiao-gua=Chiao-kua* (茭瓜, Zizania Cucumber). Enlarged tender shoots infested by *Ustilago esculenta* P. Hennings; wild in Jiangsu, cultivated in paddies in Sichuan for the smut infested tender shoots, propagated by crown division. *Gu-mi=Ku-mi* (菰米, Zizania Seed), rare.

A robust, caespitose, aquatic perennial grass, growing in stagnant water along the edge of ponds or canals with stout rhizomes and roots buried in the mud; basal internodes 6–8 cm long, 1.5–2.5 cm in diameter; leaves crowded towards the base of the plant, sheaths stout, 30–35 cm long; laminas 70–130 cm long, 2–3 cm wide, the basal two-thirds gradually attenuate into a petiole-like portion joining the sheath, scaberulous;

flowers unisexual, monoecious, panicles 25–35 cm long; spikelets 1-flowered, falling readily. In cultivation, the plants seldom flower; the cucumber-like infested shoot harvested before any black vein appears, used as a delicacy.

Cyperaceae: Sedge Family

Carex kobomugi Ohwi — KOBOMUGI SEDGE

Sha-zan-tai-cao=Sha-tsan-t'ai-ts'ao (砂礦薹草, Sandbar Sedge). Starch from seed.

Stoloniferous grass-like herbs 10–20 cm high, culms trigonous, smooth, often bearing brown fibrous remnants of old leaf-sheaths; leaves coriaceous, 10–20 cm long, 4–6 mm wide, pungent, midrib prominent, papilliate-serrulate; flowers dioecious, spikelets numerous, in dense terminal oblong-ovate heads 4–6 cm long, 2–4 cm across, individual spikelets ovate, 1.5 cm long, consisting of 5 to 10 florets, bracts and scales herbaceous, serrate, staminate scales lanceolate, pistillate scales ovate-lanceolate, 1.5 cm long; perigynia coriaceous, shorter than the scales, 10–12 mm long; styles 3; achenes trigonous. Native to the seashore of eastern Asia, growing on sand-dunes; naturalized in eastern North America

Cyperus esculentus L. var. **sativus** Boeckeler — CHUFA, YELLOW NUT-GRASS

You-sha-cao=Yu-sha-ts'ao (油沙草, Oil Nut-grass). Tubers (corm), introduced from the Mediterranean Region to China in 1962, cultivated from Liaoning in the north to Guangxi and Yunnan down south.

Densely caespitose perennial herbs 30–60 cm high, with numerous tubers on short stolons, the tubers oblong, yellowish-brown, 1–1.8 cm long, 5–8 mm across, with obvious joints, aerial stems trigonous; leaves basal, flat, 30–60 cm long, 4–6 mm wide, basal sheaths brown; flowers small, in yellowish-brown loose terminal compound umbels subtended by 3–6 leaf-like involucral bracts; individual umbels consisting of 5–8 erect rays up to 12 cm long, terminated with 5–15 flattened spikelets; rachis wing-angled, spikelets 1–1.5 cm long, 1.6–1.8 mm wide, consisting of 10–20 florets, scales pale-yellow, membranous, ovate or ovate-oblong, 2.8 mm long, 5- or 7-nerved, obtuse; stamens 3; ovary oblong-trigonous, style 3-fid; nutlets 1.5 mm long. Native to Africa; cultivated in Europe and North America for the edible tubers, introduced to China to be cultivated in land areas too poor for wheat and beans.

Eleocharis dulcis (N. Burman) Trinius — WATER CHESTNUT (Syn. *E. tuberosa* [Roxburgh] Roemer et Schultes.

Bi-qi=Pi-ch'i (荸薺, 勃臍, Water Chestnut); *Di-li=Ti-li* (地栗, Ground Chestnut, a name used in northern China); *Ma-ti=Ma-t'i* (馬蹄, Horse-hoof, a name used in South China); *Tian-ci-gu=T'ien-tz'u-ku* (甜慈菇, Sweet Sagittaria, a name used in Chengdu). Cultivated extensively in fish ponds in southern China, gathered in low land wild populations in northern China; corms eaten raw and/or cooked, both fresh and canned material available in American Chinese stores. Starch obtained from corms dried, for export as well as for domestic uses, making a sweet early morning drink with sugar and boiling water for the elder and/or for pastry.

A perennial leafless sedge growing in marshes, with several wire-like rhizomes each terminated by a corm, 2.5 cm long, 3.5 cm across, white, fleshy and sweet inside, black with scales and a central conical bud outside; stems green, leaf-like, erects, cylindrical, (15–) 30–70 (100) cm long, 3–7 mm in diameter; leaves modified into 3 or 4 brown or black basal sheaths; flowers small, in a terminal cylindrical spike with imbricate scales, spikelets 1.5–4 cm long, 6–7 mm in diameter; fruit a nutlet. Imported fresh corms, preserved sugared dried slices and starch in packages, all available in American Chinese stores; sliced fresh corms adding crispiness to dishes, popular in Chinese restaurants in America; straw used for mats, baskets and even for hats.

Palmae or Arecaceae: Palm Family

Areca catechu L. — BETEL NUT, BETEL PALM, ARECA-NUT PALM
Bin-lang=Pin-lang (檳榔, Betel Nut). Seeds; imported. *Bin-lang-hua=Pin-lang-hua* (檳榔花, Betel Nut Flower); staminate flowers.

A tall tree with a slender trunk 10–30 m high, 15 cm in diameter; leaves crowded at the top of the trunk forming a crown 1.7–2 m across; flowers unisexual, monoecious, fragrant, with the staminate flowering branches above the pistillate ones; fruit ovoid, orange-red when ripe, 4–7 cm long, with the seed deeply enclosed in thick fibers. To free the seed, the fruit is opened and boiled. The betel nut was first recorded in ancient Chinese literature in the later half of the fourth century by Ji Han who was assassinated in 306 A.D.. Being the Prefect of an important center of ancient transportation, Xiang-yang, Ji described the tributary items brought to China from the tropical regions via Guangzhou (Canton). He gave a vivid account of the betel nut tree and the well established custom of chewing the nut wrapped in betel leaf and flavored with spicy powder. The species is extensively cultivated in southeastern Asia and Taiwan.

Arenga pinnata (Wurmb) Merrill — SUGAR PALM (Syn. *A. saccharifera* Labillardière)
Nan-ye-fen=Nan-yeh-fen (南椰粉, Southern Palm Starch); *Sha-mu-mian=Sha-mu-mien* (沙

木麵, Palm Flour); *Sha-tang-ye-zi=Sha-t'ang-yeh-tzu* (沙糖椰子, Sugar Coconut). Starch from stem.

A large palm with a single trunk 12 m high, 30–40 cm in diameter, covered by old leaf bases; leaves ascending, pinnate-compound, with up to 200 linear leaflets, 1–1.5 m long; flowers unisexual, monoecious, in separate clusters, the staminate branches appear after the pistillate ones.

The tree stores large amounts of starch in the stem, which converts into sugar before flowering. The staminate peduncle is tapped for the sweet juice, with a vessel suspended under the cutting; twice daily the juice is taken for condensation; each time a thin slice is taken off the peduncle to allow new flow. The sugar palm was recorded in the same work as was the betel palm. In that document, it described the starch obtained from the trunk, transported to northern China, and used for pastry.

Caryota mitis Loureiro — Low Fishtail Palm

Duan-sui-yu-wei-kui=Tuan-sui-yü-wei-k'uei (短穗魚尾葵, Short-spiked Fish-tail Palm); *Jiu-ye-zi=Chiu-yeh-tzu* (酒椰子, Wine Coconut). Sago obtained from the trunk; sap from the young inflorescence for sugar and wine.

Clustered palms 5–8 m high; leaves bipinnate, 1–3 m long, leaflets very numerous, obliquely widge-shaped, 10–20 cm long, without distinct midrib, truncate, irregularly dentate; petioles 30–60 cm long; inflorescences 30–40 cm long, pendulous, much branched, flowers monoecious, protandrous, staminate ones in triads, oblong, symmetrical, sepals 3, coriaceous, imbricate, petals 3, valvate, connate at base, stamens numerous, anthers oblong, apiculate at the apex, pistillate flowers globose, staminodes 0, ovary trilocular, 1 cell fertile; fruits globose, 1.2–1.5 cm in diameter, with stigmatic remains, the outer covering containing numerous stinging needle-like crystals, red at maturity, turning black. Native to tropical Asia, extending from India eastward to Malaysia, occurring in secondary forests of southern Guangxi.

Caryota ochlandra Hance — Fishtail Palm, Cluster Fishtail Palm

Guang-lang-mian=Kuang-lang-mien (桄榔麵, Fishtail Palm Flour). Starch from the stem.

A large palm, the stem 20 m high, 45 cm in diameter, smooth and ringed; leaves very large bipinnate, leaflets 15–20 cm long, rhomboid, fishtail-like, the terminal one fan-shaped, irregularly dentate; inflorescences 3 m long, hanging; flowers unisexual, in triads, two staminate ones with a pistillate one in the middle, stamens 110–155; fruit red, 2.5 cm in diameter. Cultivated for ornamental purposes; sago occasionally obtained from the stem.

Caryota urens L. — Toddy Palm, Wine Palm

Jiu-guang-lang=Chiu-kuang-lang (酒桄榔, Toddy Palm). Sap; fermented

A palm introduced from tropical Asia for ornamental purposes; stems up to 25 m high, 40 cm diameter, trunk smooth, ringed below; leaves bipinnate, 50 cm long, 30 cm wide, leaflets wedge-shaped, obliquely cut, toothed; inflorescences pendulous, flowers in triads, with two staminate ones on the sides, stamens 40–45; fruits spherical, reddish; seed sperical, 2 cm in diameter; used for beads. Sap tapped from young inflorescence used for making wine; tender young leaves edible; sago obtained from pith of stem used as flour.

Cocos nucifera L. — Coconut, Coconut Palm

Ye-zi=Yeh-tzu (椰子, Coconut). Fruit, rarely seen in the Chinese market in America.

A widespread tropical palm with rather slender trunk 20–30 m high; leaves 20–30 crowded at the end of the stem, 3–7 m long, 1–1.4 m wide, pinnately compound, the leaflets leathery, 30–80 cm long; inflorescences axillary, up to 2 m long, in simple-branched clusters among leaves, many staminate flowers on long branches, several triads near the base of each branch, with the central pistillate flower larger than the two lateral staminate ones; fruits ripe in autumn, a large drupe, smooth outside, with copious fibrous husk, and bony inner shell with three pores; embryo basal, attached with the hard endosperm and the liquid. In China coconut is rare, not available in most places; except sliced sugar preparations sold as candy.

Lodoicea maldivica (Gmelin) Persoon — Coco-de-Mer

Hai-di-ye=Hai-ti-yeh (海底椰, Coco-de-Mer). Native of the Seychelles Islands, cultivated in Thailand, thence the edible endosperm introduced into the Hong Kong market; used in soup, especially by the vegetarians of the area.

A large, robust, dioecious palm, trunk inconspicuously ringed; leaves costapalmate, sheath splitting opposite the petiole, costa reaching nearly to the margin, blade wedge-shaped at base, divided 1/4–1/3 into single-fold segments, each bifid and with drooping free ends; inflorescences interfoliar, massive; staminate one pendulous, rachillae massive, catkin-like, bracts forming pits each containing a staminate flower, stamens 17–22; pistillate one unbranched, peduncular bracts tubular, rachilla zigzag, bearing several empty sheathing bracts, the subsequent bracts subtending a sessile ovoid pistillate flower; fruits very large, ovoid, pointed, each bearing 1- (3)-pyrenes; seed large, 2-lobed, endosperm thick, hard, hollow, homogeneous.

Metroxylon rumphii Martius (Spiny-sheathed); **M. sagus** Rottb. (Smooth-sheathed)
— Sago Palm

Xi-gu-zi=Hsi-ku-tzu (西穀子, Sago-Grain). Starch obtained from stem felled before
flowering; the market material consisting of dried cooked globules (pearl sago)
prepared from the starch, imported; used for preparing sweet dishes.

Stout tree palms dying after fruiting, with stems emerging from underground stock
continuously, flowering stem 8–15 years old, trunks 10 m high, 30 cm in diameter;
leaves massive, 7–8 m long, pinnate, with numerous linear leaflets 50 cm long, 2.5–3
cm across middle; flowers polygamous, small, in large pyramidal panicles, the perfect
ones each with a scaly ovary; fruit like a small apple. Native to Malaysia and Indonesia,
growing in swampy area; cultivated for the starch from the stems and for roofing material
from leaves. Experienced Malaysian sago producers know the most profitable time to
fell the tree for the starch. Chinese immigrants to Malaysia have improved the production
of pearl sago by applying the methods of making *fen-si* (粉絲, Mungbean Silk) in the
industry (Burkill 1935, p. 1462). Sago was first recorded in Chinese food, as a gruel for
convalescence in A.D. 756–779.

Nypa fruticans Wurmb — Nypa Palm, Mangrove Palm

Shui-ye=Shui-yeh (水椰, Water Coconut). Young fruit, raw or preserved; sap used for
syrup.

A graceful, acaulescent palm growing in estuaries and brackish swamps in Hainan
and southern Guangdong, 4–7 m tall; leaves pinnate, 2.5–7 m long, on long petioles,
leaflets rigid, 50–80 cm long, 3–5 cm wide, with golden T-shaped scales on the midrib
beneath; flowers monoecious, panicles 1 m long, with the staminate catkin-like branches
lateral, and the pistillate one terminal, carpel 3; fruits drupaceous, spherical, 32–38
aggregated into a compact head. To people living near the mangroves, the nypa palm
is as useful as the coconut in the Pacific Islands, the young fruits for food, and the sap
tapped from the inflorescences for sugar, wine or vinegar.

Phoenix dactylifera L. — Date, Date Palm

Hai-zao=Hai-tsao (海棗, Ocean Jujube); *Fan-zao=Fan-tsao* (番棗, Foreign Jujube); *Yi-la-
ke-mi-zao=I-la-k'o-mi-tsao* (伊拉克蜜棗, Iraqi Honey Jujube). Imported from the Middle
East.

Tree palms 7–8 m high , the trunk covered with the persistent base of petioles; leaves
pinnate, the inner ones erect ascending and the outer ones arching; pinnae stiff, in pairs
or clusters, 40–45 cm long, glaucous; flowers small, dioecious, yellow, born in simple-
branched panicles; fruit oblong, 2.5–7 cm long, with thick sweet flesh and a slender

seed. The date was imported to southern China by ancient Arabian traders, hence it was first recorded as *Hai-zao* (jujube coming from the Ocean) or *Fan-zao* (foreign jujube). In the 1960–70s, much date was imported from Iraq into China in exchange for construction workers. Today, this date is known in China as Iraqi Jujube.

Phoenix loureiri Kunth (Syn. *P. hanceana* Naudin)

Kang-lang=K'ang-lang (康榔, Chaff Palm); *Ci-kui=Tz'u-k'uei* (刺葵, Spiny Palm). Mature fruits, eaten by children locally.

An indigenous Chinese palm, 1–3 m high, with unbelievable tolerance to ecological conditions; in Hong Kong, growing from the estuaries at sea level to inaccessible precipices on top of the hills at 800 m altitude; trunk 10–150 cm tall, 15–30 cm in diameter; leaves exceedingly spiny, up to 2 m long, arching, the leaflets in pairs or clusters, rarely solitary, 15–30 cm long, 10–15 mm wide, the basal ones specialized into very sharp spines; flowers dioecious, the panicles 60 cm long, subtended by a yellow waxy bract 15–20 cm long; fruit oblong, purple-black, 1.5 cm long, 6–8 mm in diameter, edible, a favorite of many birds.

Trachycarpus fortunei (J. D. Hooker) Wendland — Windmill Palm, Chinese Windmill Palm, Hemp Palm

Zong-lü=Tsung-lü (棕櫚, Chinese Palm); *Zong-shu=Tsung-shu* (棕樹, Palm Tree). Young inflorescence; seeds.

A medium-sized palm 1–15 m high, the trunk covered by fibrous stipules and stiff petioles of old leaves; leaves fan-shaped, 70 cm in diameter, the marginal half, or up to two-thirds divided, the petioles 1 m or more long, spiny along the sides, stipules fibrous, persistent; flowers pale yellow, small, dioecious; fruits spherical or kidney-shaped, 7–9 mm in diameter, black. A tea prepared from the seed, taken internally for keeping blood pressure normal.

Araceae: Arum Family

Amorphophallus rivieri Durieu — Devil's Tongue, Leopard Palm, Snake Palm

Mo-yu=Mo-Yü (魔芋, Devil's Tuber); *Mo-yu=Mo-yü* (墨芋, Ink Tuber); *Mo-dou-fu=Ma-tou-fu* (墨豆腐, Black Bean Curd); *Ju-ruo=Chü-jo* (蒟蒻, Amorphophallus); *Gou-zhua-yu=Kou-chua-yu* (狗爪芋, Dog Paw Taro). Tuber, ground and processed.

A very large and unusual aroid growing in forests, especially abundant in corn fields opened by the slash and burn system; the tuber irregularly globose, varying in size,

4–10 cm across, sending out a single ternately decompound leaf on a petiole varying between 20 to 80 cm long, green and maculated with irregular white-purple spots and patches, leaflets ovate, deeply lobed, 2–8 cm long; as the plant becoming mature enough to flower, the flowers appearing before the leaf, consisting of a scape 20–40 cm long, a large spathe green-purple outside, dark purple-black inside, funnel-shaped at the base, subtending an elongated-cylindrical fleshy spadix with green pistillate flowers below the yellow staminate ones on the sterile axis. After fruiting, the tuber disintegrates. All farmers and ethnic groups gather the tubers from the fields, grate or pound them, soak the material in water, remove the course portion, precipitate the starch and use it for making a jelly-like food. In Guangdong and Yunnan, various strains of *Amorphophallus* are cultivated. The processed product is used locally and also exported to Japan. Farmers named the plant *gou-zhua-yu* for its irregular lumpy tuber fancifully resembles the paw of a dog (*gou*=dog, *zhua*=paw, *yu*=taro, an underground tuber). The corms of a closely related species, *Amorphophallus mairei* Léveillé, used in Yunnan and adjacent areas of Laos for similar purpose.

Colocasia esculenta (L.) Schott — TARO (Syn. *C. antiquorum* Schott).

Yu=Yü (芋, Taro); *Yu-tou=Yu-t'ou* (芋頭, Taro Head); *Yu-zi=Yü-tzu* (芋子, Taro Son); *Yu-nai=Yü-nai* (芋奶, or 芋芿, the ancient and current trivial name of people in the maritime provinces); *Yu-tou-gan=Yü-t'ou-kan* (芋頭杆, Taro Petioles); *Yu-tou-hua=Yü-t'ou-hua* (芋頭花, Taro Scapes). Caudices (central axis bearing leaves and roots), corms (primary fleshy branches from the caudex), cormels (secondary fleshy branches from the corms), boiled, cooked with meat or seafood, and prepared for dessert; petioles, cooked at times of food scarcity.

Perennial acaulescent herbs 60–100 cm high, caudices corm-like, varying greatly in size and content, mealy after cooked, bearing corms and cormels, brown outside, covered by fibrous scales, white and slimy inside; leaves all basal, obliquely peltate, on fleshy erect stout petioles 15–90 cm long, sheathy at the base, laminas ovate, 20–50 cm long, 15–40 cm wide, acute or acuminate, base shallowly 2-lobed; flowers rarely produced in cultivation, unisexual, monoecious, numerous, crowded on a spadix subtended by an oblong, yellowish-green spathe 15–17 cm long, scapes shorter than the petioles, spadix 10 cm long, consisting of 4 portions, with the terminal 1 cm being the appendage, the yellow staminate portion below the appendage, the green pistillate portion at the base and a sterile section in between. Native of tropical Asia, extensively cultivated in China, especially in the southern provinces, propagated vegetatively by corms, each becoming the caudex bearing the leaves, seldom with a scape, and the corms with or without cormels. The following named cultivars are common in the vicinity of Guangzhou

(Canton), the ancient ones for better texture and good flavor, while the more recent ones selected for higher productivity.

cv. 'Betel Nut Taro' (*Bin-lang-yu=Pin-lang-yü* 檳榔芋, Betel Nut Taro); *Xiang-yu=Hsiang-yü* (香芋, Pleasant Flavor Taro).

Plants 1–1.5 m high, caudices large, oblong, dark brown outside, white with purple tissues inside, similar to the color of sliced betel nut (hence the name "*bin-lang-yu*"), corms few, slender on rhizome-like stalks; leaves comparatively small, 15–27 cm long, 15–22 cm across, purple-red along the margin and veins, apex slightly pointed, petioles stout, pale green. An excellent cultivar, with large central caudex, good texture, pleasant natural flavor, fine keeping quality, and high productivity, varying between 2.2 and 2.7 kg per plant, available in American Chinese stores.

cv. 'Buddha's Belly' (*Fo-du=Fo-tu* 佛肚, Buddha's Belly); *Du-miao=Tu-miao* (獨苗, Solitary Leaf-set); *Long-dong-zao-yu=Lung-tung-tsao-yü* (龍洞早芋, Early Taro of Dragon's Cave).

Plants 40–50 cm high; caudices oblong, yellowish-brown, inside white, lateral corms leafless, 4 or 5, oblong; leaves 35–50 cm long, 30–40 cm wide, petioles purple. An excellent cultivar, cold resistant, good texture and keeping quality, low productivity, average 450 g per plant.

cv. 'Dana Red Bud Taro' (*Da-na-hong-ya-yu=Ta-na-hung-ya-yü* 大嬭紅芽芋, Big Mother Red Bud Taro); *Da-na-yu=Ta-na-yü*, 大嬭芋, Big Mother Taro).

Plants 90–100 cm high, caudices large, subglobose, with thick dark brown skin and white inside, buds bright red, corms 7–10, comels 8–9, subglobose, leaves 40–50 cm long, rounded at the apex. A superior cultivar cultivated for hundreds of years, with good texture and fine keeping quality, medium high production, with an average of 1 to 2 kg per plant.

cv. 'Dog's Feet Taro' (*Gou-zhua-yu=Kou-chua-yü* 狗爪芋, Dog Feet Taro).

Plants 50–60 cm high, caespitose in appearance; caudices relatively small, almost indistinguishable from the large corms and forming a compact mass with them, corms 14–18, cormels 5–8, 9 or 10 of the corms bearing aerial herbage; leaves ovate-peltate, 38–42 cm long, 26–30 cm wide, petioles rather short and slender, 35–45 cm long, pale green. An ancient cultivar, requiring long growing season, resistant to unfavorable environmental factors, corms mealy but insipid, caudices usually dried in pieces for medicinal uses; productivity high, averaging 2.7 kg per plant.

cv. 'Big Lotus' (*Da-lian-hua=Ta-lien-hua* 大蓮花, Big Lotus Flower).

Plants 1.2 m high; caudices elongated-ovate-oblong, yellowish brown outside, white inside, lateral corms 8 or 9, several bearing aerial herbage, cormels many, winter buds white; leaves 40–43 cm long, 35–39 cm wide, petioles purple-green at the base. A more recent selection with 40 years' history of cultivation, cold resistant, corms less mealy, productivity average 900 g per plant.

cv. 'Hillside Taro' (*Shan-yu=shan-yü* 山芋, Hillside Taro).

Plants rather low, 40–50 cm high, caudices ovoid, yellowish-brown, inside white, corms 4 or 5, oblong, narrowed towards the base, each bearing 2 to 5 cormels, forming a loose mass with the cordex, buds white; leaves 25–35 cm long, 20 cm wide, petioles purple-red. A good cultivar, drought resistant, corms mealy, fine eating quality, low production, average 450 g per plant.

cv. 'Yellow Green' (*Qing-miao=Ch'ing-miao* 青苗, Green Herbage; *Qing-jia-yu=Ch'ing-chia-yü*, 青荚芋, Green Buds).

Plants 70 cm high, caudices cylindrical, corms 7–10, stalked, buds green; herbage yellow-green; leaves ovate, obtuse. An ancient cultivar, corms with good texture, sweet taste, fine keeping quality, rather high productivity, average 1.8–2.2 kg per plant.

cv. 'Variegated Loins' (*Hua-yao-hong-ya-yu=Hua-yao-hung-ya-yü* 花腰紅芽芋, Variegated Loins Red Buds Taro).

Plants 1–1.2 m high, caudices oblong, brown, inside white, lateral corms 5 or 6, stalked, half of them with 2 or 3 cormels each, forming a loose crown with the caudex; leaves 50–60 cm long, 40–50 cm wide, roundish at the apex, petioles pale green. An old cultivar bearing corms with medium eating quality but of high production, average 1.3–2.7 kg per plant.

Colocasia fallax Schott

Ye-yu-tou=Yeh-yü-t'ou (野芋頭, Wild Taro); *Jia-yu=Chia-yü* (假芋, False Taro). Petioles of leaves used for potherb in time of food shortage.

Perennial acaulescent herbs with creeping rhizomes and small corms 1–1.5 cm long; leaves all basal, on stout petioles 8–30 cm long, the basal portion 8–10 cm across, laminas ovate-suborbicular, peltate, 8–15 cm long, 5–12 cm across, abruptly acuminate, basal sinus 1–2 cm deep, lobes 2.5–6 cm long; flowering spathe yellow, spadix 7–8 cm long, terminal appendage 3 cm long, staminate portion 1.5–2 cm long, pistillate portion 1.5 cm long, middle sterile area 1 cm long, stamens truncate, ovary globose, ovules several.

Native to the eastern Himalayan Region, growing in swampy area under thickets, at altitudes of 8500–1400 m, occurring in southern Yunnan and adjacent Thailand.

Bromeliaceae: Pineapple or Bromelia Family

Ananas comosus (L.) Merrill — PINEAPPLE

Bo-luo=Po-lo (菠蘿, Pineapple); *Feng-li=Feng-li* (鳳梨, Phoenix Pear). Cultivated, for local uses and/or export in cans.

A stiff large perennial herb 1–1.5 m high; leaves sword-like, spiny, in a basal rosette, ascending, 70–90 cm long, 3–4 cm wide, prickly along the margin; flowers violet-reddish, densely aggregated into a globose head 5–8 cm in diameter, the distal end with sterile bracts; in fruiting the flowers forming a fleshy syncarp with fused inferior ovaries, and the apical portion developing into a leafy shoot.

Cultivated in tropical areas of China; Taiwan is one of the world's leading production center for pineapple.

Commelinaceae: Spiderwort Family

Commelina benghalensis L. — BENGAL DAY FLOWER

Bai-ri-sai-cao=Pai-jih-sai-ts'ao (百日晒草, Hundred Day Herb); *Fan-bao-cao=Fan-pao-ts'ao* (飯包草, Food Bag Herb). Young shoots, use for potherb.

An annual or perennial gregarious herb, growing in wayside irrigation ditches, 30–70 cm high, rooting at the lower nodes; leaves oblong-ovate, 3–7 cm long, 2–3 cm wide, base abruptly cuneate, apex acute or obtuse; flowers blue, bilaterally symmetrical, few, in a cymose cluster subtended by an oblique funnelform bract; stamens 6, 3 fertile; capsules oblong, 4–6 mm long.

Commelina communis L. — DAY FLOWER

Zhu-ye-cao=Chu-yeh-ts'ao (竹葉草, Bamboo Leaf Vegetable). Young shoots, gathered before flowering, used in soup or as potherb.

A much branched weed up to 1 m high; leaves lanceolate, 3–8 cm long; flowers blue, in a cymose cluster subtended by a folded ovate-cordate bract; stamens 6, 3 fertile; capsule oblong, 2–3 mm long.

Murdannia angustifolia (N. E. Brown) Hara (Syn. *Aneilema angustifolium* N. E. Brown) — WILD LEEK

Shan-jiu-cai=Shan-chiu-ts'ai (山韭菜, Bamboo Leaf Vegetable). Young shoots, gathered before flowering, used in soup or as potherb.

A small annual herb growing in lawns or gardens, 20–30 cm high at anthesis; basal leaves linear, 10 cm long, 5 mm wide, cauline leaves lanceolate, 1.5–4 cm long, 5–7 mm wide; flowers bright blue, in terminal branched cymose clusters; capsule subglobose, 3 mm in diameter.

Murdannia keisak (Hasskarl) Handel-Mazzetti

Liu-cao=Liu-ts'ao (疣草, Ulcer Herb). Young shoots.

A glabrous annual herb growing in damp places, rooting at the lower nodes; leaves lanceolate, 4–8 cm long, 5–10 mm wide, flowers blue-pink, 1–3 in axillary cymes; stamens 3, fertile, the filaments hairy; capsule ellipsoid, 10 mm long.

Murdannia nudiflora (L.) Brenan (Syn. *Aneilema bracteatum* C. B. Clarke) (Figure 8a)

Shen-xian-cao=Shen-hsien-ts'ao (神仙草, Herb-of-the-Goddess). Young shoots, used in soup; cultivated in some Chinese kitchen gardens in Boston.

A caespitose perennial herb, 30 cm high in flower; basal leaves 20–25 cm long, 1–1.5 cm wide; cauline leaves lanceolate, 4–7 cm long, 1–1.3 cm wide, acute; flowers blue, in terminal racemes, each one subtended by an ovate-orbicular membranous bract 5–8 mm long, falling at anthesis; fruits not observed.

Murdannia triquetra (Wallich) Brueckner

Shui-zhu-ye=Shui-chu-yeh (水竹葉, Water Bamboo Leaf); *Rou-cao=Jou-ts'ao* (肉草, Meat Herb). Young plants before flowering, used as potherb in Yunnan.

Perennial aquatic herbs, creeping, forming roots at the lower nodes, the herbage green with a purple-red tinge, internodes rather short, 1.5–2 cm long, 1,5 mm across; leaves linear-lanceolate, 1.5–3.5 cm long, 4–6 mm wide, acute, petioles sheathy, closed, 10–13 mm long, pubescent along the line of meeting, this line of hairs extending to the internode below; flowers pale-blue, 12 mm across at anthesis, solitary, terminal to the leafy shoot, sepals 3, linear-lanceolate, slightly herbaceous, the apex barbate, petals 3, obovate, 7 mm long, fertile stamens 3, anthers oblong, dark blue, versatile, staminodes hastate, ovary ovoid, 2 mm long, 3-locular, cells 2-ovulate; capsules trigonous-oblong, 5–7 mm long, 4 mm across. Widespread in the Himalayan Region, thence eastward in eastern Asia, growing in water along streams or in swamps.

Rhoeo spathacea (Swartz) Stearn (Syn. *R. discolor* [L'Heritier] Hance — Oyster Plant, Moses-in-a-Boat

Bang-hua=Pang-hua (蚌花, Oyster Plant). Leafy shoots and flowers. A native of Central America, extensively cultivated, sold in local markets fresh.

A purple-green succulent perennial herb; leaves ovate-lanceolate, 15–20 cm long, 2–2.5 cm wide; flowers white, concealed by a boat-shaped spathe; petals 3, stamens 6, filaments bearded; capsule 2-seeded.

Pontederiaceae: Pickerel-weed Family

Eichhornia crassipes (Martius) Solms-Laubach — Water Hyacinth

Shui-fou-lian=Shui-fou-lien (水浮蓮, Water Hyacinth). Young shoots.

A rhizomatous floating gregarious plant, semicultivated in protected areas, harvested, chopped, for feeding half-grown pigs; rooting at nodes, the roots feathery; leaves broad-ovate, 3–8 cm long, 2.5–7 cm wide, acute at the apex, petiole 30 cm long, strongly inflated; flowers violet, showy, in a large terminal raceme, the petals with a yellow-blue central patch; capsule ovoid, 8–10 mm long.

Monochoria hastata (L.) Solms-Laubach

Jian-ye-yu-jiu-hua=Chien-yeh-yu-chiu-hua (箭葉雨久花, Arrow-leaved Monochoria). Young shoots.

An erect caespitose aquatic herb growing in rice field or edge of a pond, rooting in mud, 40–80 cm tall; leaves ovate or triangular, 10–25 cm long, 5–10 cm wide, acuminate at the apex, petioles 25–50 cm long; flowers light blue, in terminal racemes 7–10 cm long; capsules oblong, 1 cm long.

Monochoria vaginalis (N. Burman) Presl ex Kunth

Ya-she-cao=Ya-she-ts'ao (鴨舌草, Duck Tongue Herb). Young rhizome.

A small aquatic herb growing in rice fields, 10–20 cm high; leaves variable in size and shape according to the depth of water in the fields, ovate-rhomboid to subcordate, 2–5.5 cm long, 0.5–5.5 cm wide, petioles up to 20 cm long, sheathing at the base; flowers light blue, petals oblong-ovate, 1–1.5 cm long; capsules ovoid, 1 cm long.

Liliaceae: Lily Family

Aletris spicata (Thunberg) Franchet — Chinese Stargrass

Fen-tiao-er-cai=Fen-t'iao-erh-ts'ai (粉條兒菜, Spaghetti Potherb); *Shi-zi-cao=Shih-tzu-ts'ao*

(獅子草, Lion Grass); *Xiao-fei-jin-cao=Hsiao-fei-chin-ts'ao* (小肺筋草, Lesser Lung-muscle Grass); *Fei-yong-cao=Fei-yung-ts'ao* (肺痈草, Lung-cancer Grass). Whole plants, boiled with sugar for a beverage for nourishing mothers to improve the quality of their milk.

Caespitose acaulescent perennial herbs; leaves linear, grass-like, 10–30 cm long, 2–5 mm wide; scape 30–60 cm tall, pilose, flowers yellowish-green, tinged pink, pilose, in a terminal raceme 6–30 cm long, basal bracts 2, perianth segments 6–7 mm long, connate at the base, stamens 6, inserted in the perianth lobes, ovary half inferior, ovoid; capsules ovoid, angular, pubescent. Native to temperate eastern Asia, growing on grassy hillsides at altitudes of 800–2,000 m, cultivated in Sichuan to meet the growing demand for the material in medicine.

Allium altaicum Pallas

Ye-shan-cong=Yeh-shan-ts'ung (野山葱, Wild Onion). Leaves.

A perennial herb growing in alpine meadows; bulbs ovoid, tunicate, 2–4 cm across; leaves 2–4, fistulose, 20–40 cm long; scape cylindrical, hollow, 40–100 cm long, 1–2 mm across the middle; flowers white, in a spherical umbell.

Allium ampeloprasum L. — LEEK, cultivars of the Porrum group (Syn. *Allium porrum* L.)

Xie-cong=Hsieh-ts'ung (薤葱, Leek). Newly introduced into cultivation, rare.

A large odorous herb with 6–9 flat keeled leaves, 1–1.5 cm wide, folded lengthwise, the lower portion together forming a pseudostem, white; flowers rose-pink, in a spherical umbel.

Allium ascalonicum L. — SHALLOT

Xiang-cong=Hsiang-ts'ung (香葱, Shallot). Whole plant, chopped, used as condiment, normally raw.

A relatively small fistulose onion cultivated for the leaves; bulbs small, aggregate, 2 cm long, 1.5 cm across the middle; leaves 15–30 cm long, 1 cm across; the market material representing the young plants of several cultivars.

Allium cepa L. — ONION

Yang-cong=Yang-ts'ung (洋葱, Foreign Onion); *Qiu-cong=Ch'iu-ts'ung* (球葱, Ball Onion). Cultivated.

An annual crop developed from a perennial species, cultivated for the spherical tunicate bulb varying in size, color and taste; leaves 4–6, fistulose, 30–80 cm long, 1–3 cm across the middle; scape up to 1 m tall, hollow; flowers greenish white, the stamens exserted; capsules trigonous; seeds black.

Allium chinense G. Don — CHINESE SHALLOT

Xie=Hsieh (薤, Shallot); *Xie-tou=Hsieh-t'ou* (薤頭, Shallot Bulb). Cultivated largely for the bulb, pickled material available in bottles in Boston; usually served as salt or sour preserves in banquet.

A caespitose onion cultivated for the young bulbs used for pickles, the bulb tunicate, 2 cm long, 1–1.5 cm across the middle; leaves 2–5, fistulose, 20–40 cm long, 1–3 mm in diameter, with vertical ridges; scape 25–35 cm long; flowers purple, the umbel subtended by 2 persistent bracts, stamens exserted. Cultivated in southern China.

Allium chrysanthum Regel — WILD ONION

Tian-cong=Tien-ts'ung (天蔥, Celestial Onion); *Huang-hua-jiu=Huang-hua-chiu* (黃花韭, Yellow-flowered Leek). Whole plant, gathered for vegetable by peasants in western Hubei.

Aromatic perennial herbs with oblong cylindrical or ovate-cylindrical bulbs covered by subcoriaceous brown outer scales; scapes 20–45 cm long, hollow, basal one-eighth covered by leafsheaths; leaves 3–6, tubular, hollow, shorter than the scape, flowers yellow, in spherical umbels, pedicels equal the length of the flowers, perianth campanulate, segments 6, ovate-oblong, the inner 3 broader, stamens 6, filaments subulate, deltoid at base, adnate to the base of and longer than the perianth, ovary globose, style filiform. Widespread to the hillsides of central and western China, extending westward to Tibet, growing in meadows at altitudes of 2,500–4,500 m.

Allium fistulosum L. — WELSH ONION, BUNCHING ONION, SPANISH ONION

Cong=Ts'ung (蔥, Scallion); *Da-cong=Ta-ts'ung* (大蔥, Big Scallion).

Extensively cultivated in China, especially in northern China, by a special technique of hilling up soil around the plant continuously, to encourage development of an elongated white portion up to 30 cm. long, 2 cm in diameter; eaten raw in winter; good keeping quality, available in Boston.

A caespitose perennial odorous herb, pseudostems consisting 5 or 6 leaves, fistulose, 30–50 cm long, 1–2 cm across the middle; flowers white, in a spherical umbels subtended by 2 white bracts. Propagated as annuals.

Allium hookeri Thwaites

Kuan-ye-jiu=Kuan-yeh-chiu (寬葉韭, Broad-leaved Leek). Leaves, fleshy roots eaten in Yunnan.

A caespitose perennial herb with numerous fleshy roots, 7–9 cm long, 4–5 mm thick; leaves 5, linear, 20–60 cm long, 5–10 cm wide, midrib prominent; flowers white, scapes 20–60 cm long, perianth segments 4–7.5 mm long, ovary obovoid; capsules 3-lobed, pericaps obcordate, 2.5 mm long, 3–4 mm across the top; seeds few, black.

Allium ledebourianum J. H. Schultes — TARTAR SCALLION

Xiao-cong=Hsiao-ts'ung (小蔥, Small Scallion); *Hu-cong=Hu-ts'ung* (胡蔥, Tartar Scallion); *Hui-hui-cong=Hui-hui-ts'ung* (茴茴蔥, Moslem Scallion); *Huo-cong=Huo-ts'ung* (火蔥, Fire Scallion); *Si-cong=Ssu-ts'ung* (絲蔥, Silk Scallion); *Mi-cong=Mi-ts'ung* (米蔥, Wheat Scallion). Whole plant.

A perennial herb occurring in alpine meadow; bulbs ovoid, two to several aggregate, slightly enlarged, 0.5–1 cm across; leaves 2, fistulose, 30–40 cm long, 2–7 mm across the middle; flowers purple, scapes 40–80 cm long, umbel subtended by a bract splitting into two at anthesis.

Allium lineare L.

Bei-jiu=Pei-chiu (北韭, Northern Leek); *Xian-cong=Hsian-ts'ung* (線蔥, Thread Onion). Young plants before flowering, used as potherbs or for seasoning meat dishes by cattleman and shepherd living in the arid plateau areas of northwestern China.

Perennial herbs 30–60 cm high, bulbs covered by brown fibers, cylindrical, 5 mm in diameter; leaves 3 or 4, laminas hollow, 10–20 cm long, 2–4 mm across; flowers pinkish-purple, numerous, in simple umbels terminating a scape 20–60 cm long, involucre 2-lobed, persistent, perianth segments 6, ovate-oblong, 3.5–5 mm long, stamens 6, exserted, the inner 3 filaments broadened and toothed at base, ovary obovoid, 3-locular, style columnar, stigma punctiform. Widespread in Eurasia, growing in the grassland of the arid region in China.

Allium macrostemon Bunge — CHINESE FIELD GARLIC (Syn. *A. nipponicum* auctt., non Franchet et Savatier).

Xiao-suan=Hsiao-suan (小蒜, Small Garlic); *Xie-bai=Hsieh-pai* (薤白, Shallot White); *Xiao-gen-suan=Hsiao-ken-suan* (小根蒜, Small Bulb Garlic), *Ye-suan=Yeh-suan* (野蒜, Wild Garlic). Young plants, gathered before flowering, mixed with flour, cooked in soup in northern Jiangsu.

Perennial herbs 30–70 cm high at anthesis, bulbs fleshy, subglobose-ovoid, covered by membranous outer tunic layers consisting of the lower portion of the leaf-sheaths; leaves 4 or 5, sheaths 10–20 cm long, tunicate, white, deteriorated after anthesis, laminas subcylindrical, trigonous, hollow, 15–30 cm long, 2–3 cm across; flowers purple-red, mixed with numerous sessile ovoid bulbils in a head 8–12 mm across, pedicels filiform, 1–1.2 cm long, perianth segment 6, 3–5 mm long, 1–2 mm wide, stamens 6, ovary trigonous; capsules seldom fully developed. Native of northern China, growing in fields, reaching banks of streams at altitudes of 3,200 m in eastern Tibet.

Allium mongolicum Regel — MONGOLIAN LEEK

She-cong=She-ts'ung (蛇蔥, Desert Onion); *Meng-gu-jiu=Meng-ku-chiu* (蒙古韮, Mongolian Leek). Young plants before flowering, an important vegetable and flavoring material for the meat diet in Inner Mongolia.

Caespitose perennial herbs 10–30 cm high, strongly pungent and smelly, bulbs covered by loose fibers; leaves all basal, narrow-linear, 10–15 cm long, 1–2 mm wide; flowers purple, in simple umbels terminal to scapes 10–30 cm long, involucre white, membranous, persistent, perianth segment 6, ovate-oblong, 6–9 mm long, each with a dark-purple mid-vein, stamens 6, exserted, filaments broadened and connate at base, adnate to the perianth, ovary subglobose-obovoid, style short. Widespread in central Asia, growing on the arid hillsides of Shaanxi, Inner Mongolia, and thence westward to Xinjiang.

Allium sativum L. — GARLIC

Da-suan=Ta-suan (大蒜, Garlic); *Hu-suan=Hu-suan* (胡蒜, Tartar Garlic); *Hu=Hu* (胡, the abbreviation of the foregoing name). Every portion of this species is eaten in China: the young plants called *Suan-miao=Suan-miao* (蒜苗, Garlic Seedling), being an important ingredient of genuine *Hui-guo-rou=Hui-kuo-jou* (回鍋肉, Return Cooking Pork); the scape called *Suan-tai=Suan-t'ai* (蒜苔, Garlic Scape), being a delicacy with pork in early spring; the bulb called *Suan-ban=Suan-pan* (蒜瓣, Garlic Clove), eaten raw or cooked daily by people in North China.

A strongly odorous herb extensively cultivated for the bulbs, young plants and the tender scapes; bulbs irregularly globose, usually divided into many cloves enclosed in thin tunicate scales; leaves flat, linear, attenuate towards the apex, 20–50 cm long, 1.5–2.5 cm wide, slightly folded along the midrib; scape wire-like, 50–60 cm long, the apical one-third being the whip-like bract; flowers few, purple-red, mixed with numerous bulblets (see Part I for more information).

Allium senescens L.

Shan-jiu=Shan-chiu (山韮, Mountain Leek); *Gu-cong=Ku-ts'ung* (古葱, Ancient Onion). Young bulbs and leaves, an important potherb to the cattleman of the arid region in China.

Perennial caespitose herbs 5–70 cm high, with horizontal rhizomes, bulbs 0.8–1.5 cm across, outer-coats membranous, black-brown; leaves all basal, narrow-linear, flattened, 10–35 cm long, 2–6 (–10) mm wide; flowers purplish-pink, in simple umbels terminal to scapes 20–65 cm long, involucre persistent, perianth segments 6, the inner 3 ovate, 4–6 mm long, stamens 6, exserted, the filaments connate at base and adnate to the perianth, ovary superior, 3-locular, style exserted. Widespread in Eurasia, growing on the hillside of northern China, Gansu and Xinjiang,

Allium tuberosum Rottler ex Sprengel — CHINESE LEEK

Many ways of cooking, extensively cultivated throughout China and in the kitchen gardens of many Chinese Americans; names varying with the parts employed: (1) Leafy shoots: *Jiu-cai=Chiu-ts'ai* (韭菜, Everlasting Vegetable); *Lan-ren-cai=Lan-jen-ts'ai* (懶人菜, Lazy Fellow's Vegetable). Fully grown leafy shoots cut repeatedly, cooked with egg or with bean sprouts, available in Boston. (2) Young scapes: *Jiu-cai-tai= Chiu-ts'ai-t'ai* (韭菜苔, Flowering Stalk). Gathered just before the scapes attain two-thirds the length of the leaves, highly esteemed by the Chinese people. This practice keeps the plants from seeding and overtaking the garden. (3) *Jiu-cai-hua=Chiu-ts'ai hua* (韭菜花, Flowering Umbels). The entire inflorescence preserved in salt, used throughout the year. (4) Tender decolorized plants. *Jiu-cai-huang=Chiu-ts'ai-huang* (韭菜黃, Yellow *Jiu-cai*). A delicacy produced for the Chinese New Year Festival by growing healthy rootstocks in dark hot houses in Beijing, by hilling up the growing plant gradually with sand in Chengdu, or by covering the plants with earthenware in Guangzhou (Canton).

A caespitose perennial herb; bulbs hardly enlarged, covered by fibrous reticulum representing the remains of leaf-sheaths; pseudostem cylindrical, 4–6 cm long, 4–6 mm thick, white and green above; leaves 4–5, flat, fleshy, without evident midrib, 30–50 cm long, 4–6 mm wide; flowers white, scapes 20–50 cm long, umbel subtended by a membranous bract splitting on one side. In Boston areas flowers appear in August; the crop can be harvested in June, and repeated monthly until November, the market material representing the leaf-blades attached to the pseudostem (a column formed by modified petioles.

Allium victorialis L. — WILD ONION

Ge-cong=Ko-ts'ung (各蔥, *Ge* Onion); *Shan-cong=Shan-ts'ung* (山蔥, Alpine Onion).

Tender unfolding leaves, gathered by the healthy young people of the Jia-rong (嘉戎) ethnic minority, dried in the sun, saved for special occasions such as entertaining guests or for festivals.

A perennial herb growing on the high mountains, especially in the deforested grassy slopes and alpine meadows; bulbs cylindrical, 10–15 cm long, densely covered by the reticulate fibers of former leaf-bases; leaves 2 or 3, oblong-elliptic, 8–20 cm long, 3–10 cm wide; scape 25–50 cm long, bracts splitting into 2, umbels spherical, flowers white; rather small, perianth segments elliptic, 4–5 mm long; capsules compressed-globose, 4 mm long, 5 mm across; seeds globose, shiny, one in each locule, pitted and reticulate.

Asparagus officinalis L. — Garden Asparagus

Shi-dao-bai=Shih-tao-pai (石刀柏, Asparagus); *Long-xu-cai=Lung-hsü-ts'ai* (龍鬚菜, Dragon Whisker Vegetable).

A hardy perennial herb cultivated for the edible spring shoots, with fusiform subterranean root tubers, stems 1–2 m high, much branched at flowering time; leaves scaly, subtending a fascicle of filiform modified stems (cladophylls), 1–1.5 cm long; flowers unisexual, yellow; fruit a red berry. Relatively recent introduction into China, very rare in the market.

Cardiocrinum giganteum (Wallich) Makino var. **yunnanense** Leichtlin ex Stearn — Yunnan Cardiocrinum

Bai-he-qi=Pai-ho-ch'i (百合七, Lily Hematinic); *Yun-nan-da-bai-he=Yun-nan-ta-pai-ho* (雲南大百合, Yunnan Cardiocrinum). Starch extracted from the bulb, used by the peasants living in the mountains of central and western China.

Bulbous perennial herbs 1–3 m high, the principal subterranean bulbs dying after flowering and fruiting, offset bulbs continuing the life of the individual, scales fleshy, erect stems simple, cylindrical, hollow, not rooting above the bulb; leaves in basal rosettes and cauline, ovate-cordate, 30 cm long, 20 cm wide, acute, entire, venation reticulate, petioles 25 cm long; flowers creamy-white with red-purple midveins, fragrant, in terminal racemes, bracts foliaceous, oblong, 7 cm long, 2–2.5 cm wide, perianth funnelform, segments 6, 15–17 cm long, stamens 6, 8 cm long, filaments unequal, anthers extrore, versatile, ovary oblong, 3 cm long, style 6 cm long; capsules oblong, 4 cm long, 3–3.5 cm across, loculicidal, valves toothed; seeds numerous, flat, half-moon-shaped, winged. Widespread in central and western China, growing in forests along streams at altitudes of 1,400–1,700 m.

Disporopsis aspersa (Hua) Engler ex Krause

Huang-jin-qi=Huang-chin-ch'i (黃金七, Golden Hematinic); *San-ban-jia-wan-shou-zhu=San-pan-chia-wan-shou-chu* (散斑假萬壽竹, Mottled False Disporum), *San-ban-zhu-gen-qi=San-pan-chu-ken-ch'i* (散斑竹根七, Mottled Bamboo-root Hematinic). Rhizomes, cooked with pork, used in central China.

Perennial herbs, the creeping subterranean rhizomes 3–10 mm thick, the terminal bud giving rise to a single stem; leaves alternate, subsessile, ovate-lanceolate or ovate-oblong, 3–8 cm long, 1–4 cm wide, acuminate, base subcordate-truncate, flowers yellowish-green, 1 or 2, axillary, perianth campanulate, 1–1.5 cm long, distal two-thirds lobed, lobes oblong, alternate with the inside bifid appendages, stamens 6, anthers subsessile, inserted to the sinuses of the appendages, ovary superior, oblong, 5 mm long; berries subglobose, purple-blue; seeds 2–4. Endemic to central and western China, growing in the shade of forests along valleys at altitudes of 1,300–1,700 m.

Disporopsis pernyi (Hua) Diels

Zhu-gen-qi=Chu-ken-ch'i (竹根七, Bamboo-root Hematinic); *Huang-jiao-ji=Huang-chiao-chi* (黃腳雞, Yellow-feet Chicken); *Zhu-gen-jia-wan-shou-zhu=Chu-ken-Chia-wan-shou-chu* (竹根假萬壽竹, Bamboo-root False Disporum). Rhizomes cooked with chicken for a special food given to postpartum mothers in Guizhou.

Perennial herbs 20–40 cm high, rhizomes cylindrical, 5–10 mm in diameter, stems mottled purple; leaves shortly petiolate, lanceolate, elliptic-lanceolate or ovate, 5–13 cm long, acuminate-caudate, base obtuse or rounded; flowers white, 1 or 2, rarely 3, in an axillary fascicle, perianth campanulate, 12–15 (–20) mm long, distal two-thirds 6-lobed, lobes oblong, inner appendages opposite the lobes, apex bifid, stamens 6, anthers subsessile, inserted to the apical sinuses of the appendages, ovary ovoid, 6–8 mm long; berries dark purple; seeds 1–3. Widespread in warmer regions of China, from Sichuan and Yunnan, extending eastward to Taiwan.

Fritillaria camschatcensis Ker-Gawler — Kᴀᴍᴄʜᴀᴛᴋᴀ Lɪʟʏ

Hei-bai-he=Hei-pai-ho (黑百合, Black Lily). Bulblets.

A large bulbous perennial herb 20–60 cm high; bulblets numerous, fleshy, tender, of the sizes of wheat grains and peas, attached to short thick axis, the larger ones above the smaller and with a hyaline membrane at the distal end, the smaller ones acute at the apex; leaves occurring on the distal half or one-third of the cylindrical stem, 3 or rarely 5 in whorls, those on the upper nodes opposite, and the uppermost one solitary, ovate-lanceolate, 2.5–8 cm long, 1–3 cm wide, subsessile, obtuse at the apex; flowers solitary, rarely two, campanulate, uniformly purple-black, perianth segments 6, oblong, 2.5 cm long, rounded at the apex; stamens 6, shorter than the perianth, styles 3, about same length as the perianth, clavate-falcate, reflexed, stigmas terminal, papillose; capsules oblong, 2.5 cm long, 2 cm across; seeds numerous.

Hemerocallis citrina Baroni, **H. lilioasphodelus** L., **H. minor** Miller, and **H. thunbergii** Baker — YELLOW DAYLILY. Several species cultivated as a cash crop in the lower Yellow River Region.

Jin-zhen-cai=Chin-chen-ts'ai (金針菜, Golden Needle Vegetable); *Huang-hua-cai=Huang-hua-ts'ai* (黃花菜, Yellow Flower Vegetable). Mature flower buds gathered just as they begin to open in early morning, scalded in boiling water, or steamed, sun-dried, shipped to Hong Kong, repacked in paper bags, available in all American Chinese stores.

Caespitose perennial herbs with some of the fibrous roots bearing fusiform tubers; leaves all basal, linear, 50–80 cm long, 1–2 cm wide; scapes branched at the distal end, 1–2 m (rarely more) long; flowers showy, yellow in all the species bearing edible mature flower-buds, opening one at a time, rarely two on the same scape, perianth trumpet-shape, lemon yellow to orange (see Part I for more information). All the ancestral stock of the daylilies grown in USA were from China, especially from central China. In USA, gardeners selected for color and shape of the perianth segments. In China, selections concentrated in the length of the mature flower-buds, the longer the better. Only yellow or orange flower-types are planted for food (Hu, 1968).

Hemerocallis fulva (L.) L. — RED DAYLILY

Xuan-cao=Hsuan-ts'ao (萱草, Mother Herb). Flower buds, rarely used; young leaves, used by some people living on hillsides.

Caespitose perennial herbs; leaves all basal, linear, 30–80 cm long, 8–12 mm wide; scapes 1 m high, branched at the distal one-fifth; flowers funnelform, perianth two-toned, orange with red pigment near center; a natural hybrid, triploid; very common on banks of expressways in eastern USA; not much used as food in China, where sentimental and literary significances are stressed.

Hosta plantaginea (Lamarck) Ascherson — PLANTAIN-LILY

Yu-zan=Yü-tsan (玉簪, Hosta); *Yu-zan-hua=Yü-tsan-hua* (玉簪花, Hosta Flower). Root, small amount, cooked with meat in western Hubei; leaves used in making pastry; flowers cooked as a delicacy, requiring parboiling to detoxify.

Perennial caespitose acaulescent herbs 25–60 m high; leaves ovate-cordate, laminas

15–25 cm long, 9–15.5 cm wide, on long and fleshy petioles; scapes terete, basal membranous bracts 4–6 cm long, 1.5–2 cm wide, flowers nocturnal, white, fragrant, showy, in terminal raceme, pedicels 1–2 cm long, subtended by 2 foliaceous bracts, perianth trumpet-shaped, tubes 5–6 cm long, limbs 6-lobed, the lobes oblong, 3–4 cm long, 12 cm wide, stamens 6, inserted on the perianth-tube, slightly exserted, ovary oblong, 1.2 cm long, style filiform, exserted; capsules cylindrical, 6 cm long, 1 cm across; seeds numerous, shiny, black, narrowly winged. Extensively cultivated in China for ornamental purposes.

Lilium brownii F. E. Brown, and **L. regale** Wilson

Bai-he=Pai-ho (白合, White Lily). Bulbs, used fresh or dried; starch obtained from the bulbs; both the dried scales and the starch are available in Boston.

Bulbous perennial herbs growing on grassy hillsides; bulbs consisting of imbricate fleshy ovate-elliptic scales 2–3 cm long, 2 cm wide; stem unbranched; leaves all cauline, lanceolate, 7–12 cm long, 2–3 cm wide; flowers white, showy, 1 to 4 terminal to the leafy stem, the perianth trumpet-shaped, stamens 6, anthers red-brown; capsules oblong, 5 cm long; seeds numerous, strongly compressed.

Lilium lancifolium Thunberg (Syn. *L. tigrinum* Ker-Gawler), and **L. speciosum** Thunberg — TIGER LILY

Juan-dan=Chüan-tan (捲丹, Recurved Red Lily); *Bai-he=Pai-ho* (百合, Hundred Union); *Suan-nao-shu=Suan-nao-shu* (蒜腦藷, Garlic Head Bulb). Fleshy scales of the bulb, eaten fresh or dried.

Bulbous perennial lily; bulbs ovate-spherical, 4–8 cm in diameter, scales white, broad ovate, 2 cm long and wide; stems 0.8–1.5 m high; leaves all cauline, linear-lanceolate, 3–8 cm long, 1.5–2 cm wide, bearing purplish-black bulbils in the axils; flowers showy, 3–6, subtended by leafy bracts, nodding, orange-red spotted purple-black, the perianth segments strongly reflexed.

Polygonatum odoratum (Miller) Druce (Syn. *P. officinale* Allioni); **P. inflatum** Komarov, **P. macropodum** Turczaninow, **P. involucratum** Maximowicz

Yu-zhu=Yü-chu (玉竹, Jade Bamboo). Rhizomes, gathered in autumn or early spring, cleaned, parboiled, drained, partially dried and rubbed between hands; the process repeated until the material is dried, flexible and slightly transparent; longitudinally sliced products available in Boston, labeled Rhizoma Polygonati Officinalis; often mixed in two-dollar packets with Chinese dates, Euryale seed, Job's tear, China-root and Chinese yam slices; used to make a health beverage, sometimes with pork chops added.

Perennial herbs 40–65 cm high, with creeping pale yellow rhizomes 0.5–1.3 cm in diameter, bearing numerous fibrous roots and one terminal winter bud from which emerges a solitary flowering stem; leaves alternate, subsessile, elliptic, rarely oblong, 6–12 cm long, 3–6 cm wide, acute or shortly acuminate, entire, glaucous beneath; flowers white, tinged green, 1–3 in axillary cymes on peduncles 1–1.4 cm long, perianth tubular, 1.4–1.8 cm long, distal end 6-lobed, lobes ovate, stamens 6, inserted to the middle of the perianth tube, ovary superior, style filiform, stigma capitate, berries globose, purple-black, 4–7 mm across. Native to temperate Eurasia, the market material largely from cultivated crops; people in various provinces gather rhizomes from botanically distinguishable local species: *P. inflatum* with 2–5 flowered cymes on peduncles 2–2.5 cm long, the perianth hairy inside, *P. involucratum* with 2-flowered cymes bearing 2 foliaceous bracts; and *P. macropodum* with 5 to 12-flowered cymes on peduncles 3–4 cm long.

Polygonatum sibiricum Delaroche ex Redouté

Huang-jing=Huang-ching (黃精, Yellow Elixir); *Ji-tou-shen=Chi-t'ou-shen* (雞頭參, Chicken Head *Shen*). Rhizomes containing starch and sugars, used for pastry; steamed and dried repeatedly, with sugar or honey added, eaten as candy; used in traditional Chinese medicine.

Perennial herbs 50–90 cm high, rhizomes pale yellow, the annual growth 4–10 cm long, the distal end thicker, fleshy, bearing one subterminal bud, resembling the head of a chicken, hence the common name *Ji-tou-shen*; leaves sessile, 4–6 whorled, linear-lanceolate, 8–15 cm long, 6–15 mm wide, slightly cirrhous, glabrous; flowers cream-white, turning pale yellow with age, 2–4 in axillary umbels, perianth tubular, hanging, 9–12 mm long, apex 6-lobed, lobes 4 mm long, stamens 6, inserted below the throat of the perianth, ovary superior, 3 mm long, style 5–7 mm long; berries black. Native to temperate eastern Asia, growing on the rich forest floor on the shady hillside; cultivated commercially in eastern and northern China.

Scilla scilloides (Lindley) Druce — CHINESE JACINTH (Syn. *S. japonica* Baker)

Jin-zao-er=Chin-tsao-erh (錦棗兒, Luxurious Jujube); *Di-zao-zi=Ti-tsao-tzu* (地棗子, Ground Jujube). Bulb, containing starch and sugars, soaked, cooked, eaten in famine years.

A perennial bulbous herb growing on open hillsides with poor soil and much stone; bulbs tunicate, ovoid, 2–3.5 cm long, rich in mucilage; leaves linear, 10–50 cm long, 0.3–1 cm wide; flowers small, rose-pink, in a dense raceme terminal to a scape 20–60 cm long. Capsules trigonous, 2–3 mm long; seeds black.

Tulipa edulis (Miquel) Baker — Edible Tulip

Shan-ci-gu=Shan-tz'u-ku (山慈菇, Mountain Sagittaria). Bulb.

Bulbous perennial herb growing in open hillsides; bulbs tunicate, ovoid, 1.5–2.5 cm across, covered by grayish brown papery scales, hairy on the inside; leaves opposite, cauline, the lower pair linear, 15–20 cm long, 3–13 mm wide, the upper pair (occasionally 3) smaller, 2–3 cm long; flower solitary, 1.8–2.5 cm long, white with purple lines; capsule spherical, 1 cm in diameter.

Taccaceae: Tacca Family

Tacca leontopetaloides (L.) O. Kuntze — Polynesian Arrowroot, East Indian Arrowroot, Salep (Syn. *T. pinnatifida* Forster et Forster)

Tian-dai-shu=T'ien-tai-shu (田代薯, Field Potato Substitute). Starch.

Tuberous herbs growing in tropical moist areas; leaves basal, large, oblong-elliptic, 10–25 cm long, 2–5 cm wide, cuneate at the base; flowers purplish, in a terminal umbel on a bending scape, subtended by filiform bracts. Tubers harvested soon after the leaves die down, and processed for starch.

Dioscoreaceae: Yam Family

Dioscorea alata L. — White Yam, Water Yam (Figures 12 f, 20c)

Shen-shu=Shen-shu (參薯, Ginseng Yam); *Da-shu=Ta-shu* (大薯, Big Yam). Cultivated; root, used for soup, occasionally available in Boston.

A twining vine of hybrid origin; tubers varying greatly in form and size, clavate, oblong-cylindrical, hand-like, generally weighing 2 kg (often more), brown outside, white, yellow or purple inside; stem 4-angled, winged, twining clockwise, bearing bulbils in the leaf-axils; leaves alternate below, opposite above, ovate-cordate; flowers unisexual, dioecious; staminate flowers in axillary panicles, pistillate flowers in axillary spikes, ovary inferior; capsules 3-winged, the wings 2–2.5 cm long, 1.5–2 cm wide; seed numerous, strongly compressed, winged. Tubers with many variations; Several cultivars selected in Canton:

1. cv. 'Early White Yam' (*Zao-bai-shu=Tsao-pai-shu* 早白薯).

Leaves sagittate; tubers fusiform, 20–25 cm long, 7–8 cm across the middle, exterior rough, grayish brown, white when young, with few fibrous roots, interior white, good quality; harvested in October.

2. cv. 'Late White Yam' (*Chi-bai-shu=Ch'ih-pai-shu* 遲白薯).

Leaves sagittate, tubers club-shaped, 40–48 cm long, 3–4 cm across the distal thick portion, exterior smooth, few fibrous roots, white when young, interior white, good quality; harvested in December.

3. cv. 'Greater White Yam' (*Da-bai-shu=Ta-pai-shu* 大白薯).

Leaves sagittate; tubers irregularly lumpy, oblong, 20–25 cm long, 15–18 cm across, exterior rough, grayish-brown, interior cream-white, rather coarse, harvested in November.

4. cv. 'Red Skin Yam' (*Hong-pi-shu=Hung-p'i-shu* 紅皮薯).

Leaves subcordate; tubers variable in shape and size, oblong or fusiform-obovoid, 25–28 cm long, 6–10 cm across, exterior with many fibrous roots, black-brown, cortex purple-red, interior white, rather fine; harvest in November.

5. cv. 'Broom Yam' (*Sao-ba-shu=Sao-pa-shu* 掃把薯).

Leaves cordate, petioles winged; tubers compressed-globose, the distal portion divided and lobed, 20–30 cm long, 20–35 cm across the lobed portion, light brown or red-black outside, cortex yellow or purple-red, interior tinged red-purple, inferior quality; harvested in December.

6. cv. 'Red Meat Yam' — *Hong-rou-shu=Hung-jou-shu* 紅肉薯).

Leaves cordate, tubers oblong or slightly lobate, 20–25 cm long, 5–10 cm across, exterior coarse, black-brown, fibrous roots numerous; cortex purple-red, meat light-red, fine; harvested in December.

Dioscorea bulbifera L. (Syn. *D. sativa* Thunberg, non L .) — AIR POTATO (Figure 20b)

Huang-du=Huang-tu (黃獨, Yellow Alone); *Huang-yao-zi=Huang-yao-tzu* (黃藥子, Yellow Drug). Tubers and bulbils, the inside yellow, bitter, edible only after cutting into pieces and soaking for several days with change of water, otherwise poisonous.

Perennial climbing herbs, stems twining counter-clockwise, bearing axillary angular purple-brown bulbils, root tubers spherical, 3–10 cm in diameter, black with numberous tough roots; leaves ovate-cordate, 7–22 cm long, 7–8 cm wide, acuminate-caudate, primary basal nerves 7–9; flowers small, yellowish green, both staminate and pistillate in fascicles of elongated spikes, pendulous; fruits oblong, 2.5 cm long, 1 cm across, reflexed; seeds brown, wings orbicular. Widespread in warmer areas of China, much used in traditional Chinese medicine; only used as an emergency food.

Dioscorea esculenta (Loureiro) Burkill (Syn. *D. sativa* L., p.p.)

Tian-shu=Tien-shu (甜薯, Sweet Yam); *Gan-shu=Kan-shu* (甘薯, Sweet Yam); *Mao-shu=Mao-shu* (毛薯, Hairy Yam); *Shan-shu=Shan-shu* (山薯, Hillside Yam); *Ma-wang=Ma-wang* (馬旺, the Thai ethnic name used in southern Yunnan). Tubers; a botanically little known species; planted by ethnic people living in the mountains of Hainan and Yunnan, with hardly any communication with the outside world.

A drought resistant perennial climber, known as domesticated plants of ethnic groups, tuberous roots yellowish brown, many to a stalk, each with a long, slender portion before the fleshy enlargement, stem cylindrical, spiny, the spines gradually reduced, after the middle only spiny near the nodes; leaves ovate-cordate, up to 15 cm long, 17 cm wide, acuminate and mucronate, palmately 9–13 nerved, petioles equal to or longer than the lamina, prickly; flowers unisexual, small, yellowish green, dioecious, spikes 40 cm long, perianth segments 1.75 mm long, stamens 6; capsules 2.7 cm long, 1.2 cm wide, retuse, base truncate; seeds winged. Widespread among the ethnic groups in Southeast Asia, unknown in nature; introduced from Hainan Island to the experimental garden of South China Agricultural University at Guangzhou (Canton) by Professor H. H. Hue; the largest tuber weighing 3 kg (11–12 lb) ; said to have very good texture and flavor.

Dioscorea hemsleyi Prain et Burkill

Nian-shan-yao=Nien-shan-yao (黏山藥, Glutinous Yam); *Bai-shan-yao=Pai-shan yao* (白山藥, White Yam); *Nian-shu=Nien-shu* (黏薯, Sticky Yam). A famine food of people living in the mountains, starch obtained from the root tubers used for pastries.

Deeply rooted pubescent perennial climbers, the stem turning counter-clockwise, each bearing 1–4 cylindrical or clavate root tubers, 20–40 cm long, 1–4 cm in diameter, yellowish-red outside, white and sticky inside; leaves ovate-cordate, glabrous above, hairy beneath; flowers purple-red, dioecious; stamens 6, shorter than the perianth, pistillate spikes 2–15 cm long, erect, solitary, staminodes 6, ovary inferior, hairy, style slender, stigmas 3, bifid; capsules oblong, 2–2.5 cm long, 1–1.5 cm across, pubescent, rounded; seeds oblong, including the wing 1.4 cm long, wings 4–9 mm long, 4–7 mm wide; endemic to Yunnan and adjacent areas in Guangxi and Sichuan.

Dioscorea japonica Thunberg — JAPANESE YAM (Figure 20a)

Shu-yu=Shu-yü (薯芋, Japanese Yam); *Ye-shan-yao=Yeh-shan-yao* (野山藥, Wild Dioscorea). Cultivated for tubers.

A perennial climber, the vine twining clockwise; tuber cylindrical, brown outside, white inside; stem bearing bulbils; leaves alternate below, opposite above the middle, elongated triangular, 5–10 cm long, 2–5 cm wide, hastate at the base, acuminate at the apex; flowers dioecious, staminate spikes fasciculate, erect, pistillate ones solitary, hanging, 8–12 cm long; capsules reniform, width of wing equal its length; seeds broad ovate, winged.

Dioscorea opposita Thunberg (Syn. *D. batata* Decaisne) — CHINESE YAM (Figure 20d)

Shan-yao=Shan-yao (山藥, Chinese Yam); *Jia-shan-yao=Chia-shan-yao* (家山藥,

Domesticated Dioscorea). Tuber, fresh or dried; sliced dried material available in Boston; a famous Beijing dessert in banquet called *Ba-si-shan-yao=Pa-ssu-shan-yao* (拔絲山藥, Pulling Yam Silk) is made of fresh Chinese Yam.

A perennial climber; tuber cylindrical, 50–90 cm long; stem twining clockwise, bearing bulbils in the leaf-axils; leaves opposite, rarely 3 in a whorl, broad ovate-triangular, shallowly lobed at the base; staminate spikes 2–4 verticillate, the rachis curly; capsule wings cuneate at the base.

Dioscorea pentaphylla L. — Palmate-leaved Yam

Wu-ye-shu-yu=Wu-yeh-shu-yü (五葉薯芋, Five-leaved Yam); *Mao-tuan-zi=Mao-t'uan-tzu* (毛團子, Hairy Ball). Starch extracted from the root tubers.

Pubescent perennial climbers with yellowish globose or hand-like root tubers; basal portion of the stem prickly, branchlets light brown, bearing axillary yellowish bulblets 1 cm in diameter; leaves palmately compound, leaflets (3-) 5, the middle ones obovate-lanceolate, 5–20 cm long, 2.5–8 cm wide, the lateral ones smaller, petioles 3.5–11.5 cm long, prickly; flowers yellowish-green, dioecious, staminate panicles 15–20 cm long, rachis prickly, branches racemose, 1–3 cm long, consisting of 20–50 flowers, perianth segments lanceolate, 1–1.5 mm long, fertile stamens 3, anthers longer than the filaments, staminodes 3, pistillate flowers in solitary or paired axillary racemes, pendulous, 10–20 cm long, perianth segments ovate-oblong, 2–3 mm long, staminodes 6, ovary inferior, stigmas 3, bifid; capsules oblong, 2 cm long, 1–1.3 cm wide, pubescent; seeds ovate, 3–4 mm long, wings membranous, 1 cm long. Widespread in eastern Himalayan Region, thence extending eastward to the Pacific islands; in China growing along hillside and steep valleys at altitudes of 500–1,500 m.

Musaceae: Banana Family

Musa acuminata Colla — Edible Banana (Syn. *M. cavendishii* Lambert ex Paxton, *M. chinensis* Sweet)

Xiang-jiao=Hsiang-chiao (香蕉, Banana). Fruit, fresh or dried,

A palm-like perennial herb, having a pseudostem formed by the leaf-sheaths, 1.5–2.5 m high; leaves erect or ascending, oblong, 1.5–2 m long, 50–80 cm wide, petioles stout, winged, 20–30 cm long; flowers yellow, in a large pendulous cluster, covered by numerous thick, imbricate ovate maroon bracts 20–30 cm long and wide, falling one after another at anthesis; fertile pistillate flowers at the basal portion, 10–12 in two rows; staminate ones at the distal portion, in two rows subtended by one bract; fruit a cylindrical-falcate berry, usually parthnocarpous.

Musa nana Loureiro — DWARF BANANA

Ya-jiao=Ya-chiao (芽蕉, Teeth Banana). Ripe fruits.

A cultivar of banana, bearing smaller fruit; horticulturally known as the Chinese banana or the Chinese dwarf banana.

Musa x **paradisiaca** L. — PLANTAIN (= *M. sapientum* L., *M. acuminata* Colla × *M. balbisiana* Colla)

Da-jiao=Ta-chiao (大蕉, Big Banana). Fruit, cooked.

A species of hybrid origin, bearing large seedless banana; edible after cooking; rarely cultivated in China.

Zingiberaceae: Ginger Family

Alpinia katsumadai Hayata

Cao-dou-kou=Ts'ao-tou-k'ou (草豆蔻, Round Cardamon). Fruit, spice.

A perennial aromatic herb 1–2 m high; leaves linear-lanceolate, 30–55 cm long, 2–9 cm wide, attenuate at both ends, hairy on both surfaces; flowers showy, white, in terminal racemes, lip tinged purple-pink; ovary covered by yellow hairs; capsule spherical, yellow, hairy.

Alpinia officinarum Hance — LESSER GALANGAL

Gao-liang-jiang=Kao-liang-chiang (高良薑, Lesser Galangal). Rhizome, in dried sections, used as spice.

A rhizomatous aromatic perennial herb 40–110 cm high; leaves cauline, linear-lanceolate, 20–30 cm long, 1.2–2.5 cm wide, caudate at the apex; flowers white with red lines, in terminal raceme, lip ovate, 2 cm long; capsule red, globose, 1 cm in diameter.

Alpinia oxyphylla Miquel

Yi-zhi-ren=I-chih-jen (益智仁, Enrich Wisdom Seed). Fruit and seed, used as spices; candied material sold in a chain-store in Hong Kong recently as raisins, with very pleasant flavor.

A perennial aromatic herb 1–3 m tall; leaves cauline, lanceolate, 25–35 cm long, 3–6 cm wide, caudate at the apex; flower white with red stripes, in terminal raceme, corolla lobes oblong, 1.8 cm long, hairy, lip 2 cm long, hairy.

Amomum cardamomum L. (Syn. *Elettaria cardamomum* [L.] Maton) — CARDAMON

Dou-kou=Tou-k'ou (豆蔻, Cardamon). Seed; spice in the Hong Kong market, used by the Indian and Portuguese residents especially. A species introduced from Southeast Asia, cultivated in Guangdong, Guangxi and Yunnan for the fruit, used more in medicine and less in food.

Perennial erect herbs, stem 2–3 m high; leaves lanceolate or oblanceolate, up to 23 cm long, and 7 cm wide, glabrous; scape short, stout, arising from the rhizome, bracts of spike 3 cm long, hairy, perianth yellow, the tube 2 cm long, the segments 1 cm long, lip yellow with red stripes, obovate, 1.5 cm long, hairy, ovary haiy; capsule compressed globose, 1.5 cm in diameter, gray.

Amomum hongtsaoko Liang et Fang — RED CARDAMON

Hong-cao-kou=Hung-ts'ao-k'ou (紅草蔻, Red Amomum). Fruits, used as spice.

Aromatic perennial herbs 1–2.5 m high, rhizome stout, 8 cm long, 3 cm in diameter, aerial stalk 2.5–3.5 cm in diameter, consisting of the leaf-sheaths of 11–14 leaves; laminas oblong-lanceolate, 10–80 cm long, 8–19 cm wide, acminate, glabrous, margin strigose, with a purple ligule at the juncture of the sheath, 1–2 cm long; flowers pale-orange, crowded into an oblong sessile head 13–28 cm long, borne on the rhizome near the aerial stalk, the basal portion covered with imbricate bracts 1–8 cm long, flowering bracts red, oblong, 3–4 cm long, pilose on the abaxial surface; flowers 5.5–7 cm long, sepals 2–3 cm long, 3-lobed, corolla pilose, lip with a red line on each side, stamens 2–2.5 globose. 1.5 cm long, style hairy; capsules subglobose, 3 cm long, 2.5 cm across, dark purple; Seeds numerous, light gray.

Amomum krervanh Pierre ex Gagnepain — SIAM CARDAMON

Bai-dou-kou=Pai-tou-k'ou (白豆蔻, White Cardamon). Fruit, spice.

An aromatic perennial herb 2–3 m high; leaves linear-lanceolate, 15–23 cm long, 5–7.5 cm wide, glabrous; flowers yellow, scape short, emerging from the rhizome, corolla tube 3 cm long, lip obovate, with red stripes; ovary hairy; capsules compressed globose, 1.5 cm long, light gray.

Amomum maximum Roxburgh

Jiu-chi-dou-kou=Chiu-ch'ih-tou-k'ou (九翅豆蔻, Nine-winged Cardamon). Fresh fruits eaten raw, or used as spice.

Caespitose perennial herbs 2–3 m high; leaves oblong-elliptic, 30–90 cm long, 10–20 cm wide, caudate, cuneate at base, glabrous above, pilose beneath, petioles 1–8 cm long, pilose, ligules 1–2 cm long, pilose, membranous along the margin; flowers white, on short scapes emerged from the rhizome and covered by imbricate scales, spikes globose, bracts brown, caducous, lip ovate-oblong, 3.5 cm long, entire, white, with red lines near the base; capsules ovoid, 2.5–3 cm long, 1.8–2.5 cm across, purple-green at maturity, with 9 prominent wings, hairy, calyx persistent.

Amomum subulatum Roxburgh — NEPAL CARDAMON

Xiang-dou-kou=Hsiang-tou-k'ou (香豆蔻, Fragrant Cardamon); *Ga-ge-la=Ka-ko-la* (嘎哥拉, the local name in eastern Tibet).

Perennial herbs 1–2 m high; leaves oblong-lanceolate, 27–60 cm long, 3.5–11 cm wide, caudate, base rounded or cuneate, glabrous on both surfaces, petioles 1–3 cm long, ligules 3–4 mm long, rounded; flowers yellow, in dense spike on short scapes 1–4.5 cm long, covered by imbricate purple-red scales subulate at the apex, lip oblong, 3 cm long, midrib yellow, villose; capsules globose, 2–2.5 cm in diameter, purple or red-brown, with 10 or more wavy wings, persistent calyx tubular and with subulate lobes. Native of eastern Himalayan Region, extending from Nepal to eastern Tibet, growing in broad-leaved forest at altitudes *ca.* 1,000 m.

Amomum tsao-ko Crevost et Lemaire — TSAOKO

Cao-guo=Ts'ao-kuo (草果, Yunnan Cardamon). Fruit, spice, available in Boston; people from Yunnan flavor chicken soup with whole fruits.

A perennial aromatic herb 2.5 m high; leaves oblong-lanceolate, 40–55 cm long, 15–20 cm wide; flowers in a crowded spike on a short scape arising from the rhizome; capsule red, oblong-ellipsoid, 2.5–4.5 cm long, 2 cm across the middle; the market material light brown.

Costus speciosus (Koenig) J. E. Smith — CRAPE GINGER

Zhang-liu-tou=Chang-liu-t'ou (樟柳頭, Camphor-willow Shoot). Young shoot; *Bi-xiao-jiang=Pi-hsiao-chiang* (閉鞘薑, Crape Ginger). Rhizome, spice.

An erect perennial herb 1–2 m high; leaves all cauline, oblong-lanceolate, 15–20 cm long, 6–7 cm wide, hairy beneath; flowers white, showy, the buds crowded in a head-like cluster 10–12 cm long, terminal to the leafy stem, opening one at a time; petals 5 cm long, lip obovate-orbicular, 6.5–9 cm long; capsule red, 1.3 cm long.

Curcuma aromatica Salisbury — TURMERIC

Yu-jin=Yü-chin (郁金, Turmeric). Rhizome, spice.

A perennial aromatic herb 1 m high; rhizome stout, yellow; leaves oblong-elliptic, 30–60 cm long, 10–20 cm wide, pubescent beneath; scape short, arising from the rhizome before or with the leafy stem; spike head-like, the buds concealed by numerous large, colorful bracts; perianth-tube funnelform, 2–2.5 cm long, lobes white-pink; lip yellow.

Curcuma domestica Valeton; and **C. longa** L.

Jiang-huang=Chiang-huang (薑黃, Turmeric). Rhizome, spice.

A perennial aromatic herb 60–80 cm high; rhizome yellow; leaves oblong-elliptic, 30–46 cm long, 15–18 cm wide, glabrous, petioles 45 cm long; scape apical to leafy shoot, spike cylindrical, 12–15 cm long, bracts colorful, imbricate, 3–5 cm long, green-white, the distal ones sterile, pink, petals 2 cm long, lip obovate, 1.2 cm long, white tinged yellow in the center.

Curcuma zedoaria (Bergman) Roscoe — ZEDOARY

E-zhu=O-chu (莪朮, Zedoary). Root tubers; rhizomes

The vegetative aspects similar to turmeric, differing in having slender stolons each terminated by a tuber; leaves strewn with purple spots along the midrib and near it; ellipsoid spike covered with bracts green at the basal portion and gradually changing to red toward the apex, the apical sterile bracts purple-pink; the perianth segments white, and the lip yellow. Propagated vegetatively by the rhizomes, the older ones also used as a source of commercial turmeric.

Hedychium coronarium J. Koenig — GINGER LILY, BUTTERFLY GINGER, WHITE GINGER

Li-ji=Li-chi (立芨, Erect Hedychium); *Jiang-hua=Chiang-hua* (薑花, Ginger Flower). Flower buds, used as vegetable.

A tall, robust perennial 1.5–2 m high, growing in damp areas along irrigation system; leaves elliptic-lanceolate, 20–40 cm long, 4–8 cm wide, pubescent beneath; spike sessile, terminal to the leafy stem, 10–20 cm long, imbricate bracts herbaceous, 4–5 cm long, 2.5–4 cm wide, each subtending a fascicle of 2 or 3 flowers with slender perianth-tube 8 cm long, lip large, obcordate, 6 cm long and wide, 2-lobed at the apex. Flower very fragrant, cultivated primarily for cutting flower; in Hong Kong country-side, farmers cut the shoots at day-break for the early morning market.

Kaempferia galanga L. — KAEMPFERIA

San-nai=San-nai or Shan-nai=Shan-nan (三奈, or 三余, Kaempferia); *Sha-jiang=Sha-chiang* (沙薑, Sand Ginger). Rhizome, dried sections, spice.

An acaulescent perennial aromatic herb 10–15 cm high; leaves 2–4, purple-green, oblong-suborbicular, 7–14 cm long, 4–9 cm wide, abruptly narrowed and acute at the apex; flowers purple, ephemeral, in a terminal fascicle, opening one at a time, with slender and long perianth tube 2.5 cm long, lobes 1.5 cm long, lip 2.5 cm long, deeply 2-lobed. Often cultivated for kitchen uses in pots by people living in apartments and boats in Hong Kong.

Zingiber mioga (Thunberg) Roscoe

Rang-he=Jang-ho (蘘荷, Wild Ginger); *Yang-he=Yang-ho* (陽荷, Sun-Lotus). Flower buds, young shoots; cooked.

A pungent perennial herb 0.6–1 m high; rhizome pale yellow; leaves linear-lanceolate, 20–30 cm long, 3–6 cm wide, attenuate at both ends; flowers white, scape arising from the rhizome, spike oblong, 6–7 cm long, imbricate bracts ovate-oblong, 4–5 cm long, perianth tube 4–5 cm long, the segment lanceolate, 2–3 cm long, lip pale yellow, obovate, 2–3 cm long.

Zingiber officinale Roscoe — GINGER, CANTON GINGER, TRUE GINGER

Jiang=Chiang (薑, Ginger); *Zi-jiang=Tzu-chiang* (子薑, 紫薑, Fresh Young Ginger, Purple Ginger); *Gan-jiang=Kan-chiang* (乾薑, Dry Ginger); *Mu-jiang=Mu-chiang* (母薑, Mother Ginger). Rhizome, very common spice, extensively cultivated in warmer regions of China.

A pungent perennial herb cultivated as an annual crop for the irregular lumpy fleshy rhizome; the aerial growth 40–80 cm high, the stem 1–1.5 cm in diameter; leaves cauline, lanceolate, 15–20 cm long, 2–3.5 cm wide; flowers seldom seen, greenish yellow, the scape arising from the rhizome, 12–13 cm high, spike ovoid, 4–5 cm long, bracts pale green, perianth tube 2–2.5 cm long, the lobes lanceolate, 2 cm long, lip yellow, striped purple and strewn with pale yellow spots. Propagated by rhizome, for local uses and exports.

Cannaceae: Canna Family

Canna edulis Ker-Gawler — EDIBLE CANNA, QUEENSLAND ARROWROOT

Jiao-yu=Chiao-yü (蕉芋, Canna Taro); *Ba-jiao-yu=Pa-chiao-yü* (芭蕉芋, Banana Taro). Starch from the rhizome.

A large erect perennial herb up to 2 m high, cultivated in damp areas for the stout rhizome containing edible starch; leaves oblong-elliptic, 30–60 cm long, entire, purplish beneath; flowers bright red, crowded in a terminal cluster, the 3 outer staminodes petaloid, uppermost at anthesis; ovary inferior; capsules spherical. A native of Central America, tuberous rhizome used by the people of Andean South America; cultivated in Queensland for the starch, hence the name Queensland Arrowroot.

Marantaceae: Arrowroot Family

Maranta arundinacea L. — ARROWROOT

Ge-yu-jin=Ko-yü-chin (葛郁金, Vine Turmeric); *Zhu-yu=Chu-yü* (竹芋, Bamboo Taro); *Fen-shu=Fen-shu* (粉薯, Starch Tuber). Starch from rhizome.

A caespitose perennial herb 40–90 cm high, the crown bearing many scaly oblong-ellipsoid tuberous rhizomes containing edible starch; leaves ovate-oblong, 10–20 cm long, 4–10 cm wide, on the slender petioles 5–10 cm long; flowers few, small, white, with 2 petaloid staminodes 8–10 mm long; capsules brown, 7 mm long. Used by the ancient people of Central America for the treatment of arrow poison, hence the common name, arrowroot.

Orchidaceae: Orchid Family

Gastrodia elata Blume — GASTRODIA

Tian-ma=T'ien-ma (天麻, Heaven Hemp); *Chi-jian=Ch'ih-chien* (赤箭, Red Arrow). Fleshy potato-like oblong rhizome; boiled in water for tea, or with chicken for soup.

A leafless saprophytic orchid growing on the floor of deciduous forests in high latitudes or altitudes, having a symbiotic existence with a fungus (*Armillaria mellea* [Vahl ex Fries] Quélet); rhizome potato-like, rather succulent, slightly enlarged at the bud-bearing end, oblong, 8–10 cm long, 2.5–3 cm in diameter, inconspicuously marked with 12–15 rings, with one stout red bud at the center of the thicker end; scape rose-pink, 50–100 cm tall, 5–10 mm in diameter, with a terminal simple spike 10–20 cm long; flowers small, numerous, perianth obliquely tubular, 1 cm long, lip 3-lobed, 5 mm long; ovary obovate-oblong, 5–6 mm long; capsules oblong, 1.5 cm long. Formerly a rare and expensive material for the rich and the sick people now cultivated on oak, willow or birch logs with fungal inoculum and covered by compost from forest.

Habenaria conopsea (Willdenow) Bentham (Syn. *Gymnadenia conopsea* [Willdenow] R. Brown)

Shou-shen=Shou-shen (手參, Hand Ginseng). Tuber, soup with meat.

A mucilaginous perennial terrestrial orchid, 30–80 cm high, with fleshy hand-like lobed tubers; leaves 4–7, all cauline, oblong-lanceolate, 6–15 cm long; flowers pink-purple, in a spike 6–15 cm long, lip 3-lobed, with a relatively long spur 1.5 cm long; capsule oblong, 1 cm long.

Habenaria delavayi Finet

Ji-zhua-shen=Chi-chua-shen (雞爪參, Chicken Foot *Shen*). Tubers.

A perennial terrestrial tuberous orchid growing along narrow paths in the mountains; tubers 1 or 2, obovate-oblong or clavate; leaves almost all radical, obovate-oblong; flowers white, medium-sized, the sepals pubescent on the outside, petals linear, twisting at the base, retrorse and falcate, lip deeply 3-parted, spur clavate, incurved. Used by the ethnic groups living in the mountains of Yunnan.

Nervilia fordii (Hance) Schlechter — Taro Orchid

Qing-tian-kui=Ch'ing-t'ien-k'uei (青天葵, Green Celestial Malva); *Zhen-zhu-cao= Chen-chu-ts'ao* (珍珠草, Pearl Herb), *Tian-kui=T'ien-k'uei* (天葵, Celestial Malva); *Yu-lan=Yü-lan* (芋蘭, Taro Orchid). Entire plant, used for tea or cooked with pork for broth.

Terrestrial perennial orchid, requiring annual dormacy, rhizome slender, 3–4 cm long, terminated by a tuber for starting a new individual, tubers white, fleshy, 5–20 mm in diameter, with thin scales; leaves solitary, borne at the end of a short erect subterranean stem, orbicular-cordate, 5–10 cm long, 8–12 cm wide, acute, rather succulent, dried membranous, palmately 8–12 nerved, inconspicuously pilose; flowers greenish, 2–4 (–10) on a leafless scape 15–30 cm high, with 2 or 3 purple-red bracts below the middle, sepals and petals similar in shape and color, lip undivided, embracing the clavate column, anther terminal. Endemic to southern China and adjacent Thailand; intensively collected for local uses and for export, a species on the verge of extinction.

Platanthera delavayi Schlechter

Ji-zhua-shen=Chi-chua-shen (雞爪參, Chicken Feet Tonic). Fleshy roots and the attached crowns, cooked.

A tuberous terrestrial perennial orchid 15–25 cm high, growing among the grasses of hillsides, the fleshy tubers fusiform; leaves one to each stem, sessile, lanceolate, 4–6 cm long, 1.2–1.4 cm wide, ascending, glabrous; flowers small, yellowish-green, the sepals 4–6 mm long, petals oblique ovate, lip linear-ligulate, 7 mm long, with a filiform spur 1.5 cm long, ovary glabrous, 1 cm long. A little known species, the tubers used only locally in Yunnan and Sichuan where it occurs.

Spiranthes lancea (Thunberg) Backer, R. C. Bakhuizen et Steenis (Syn. *S. australis* Lindley, *S. sinensis* [C. Persoon] Ames)

Pan-long-shen=P'an-lung-shen (盤龍參, Coiled Dragon Tonic). Fleshy root, cooked with chicken or pork; taken as a tonic.

A widespread, but very occasional, terestrial orchid 15–30 cm high; tuberous roots 3–6, cylindrical, 2–3 cm long, 4–5 mm thick, rather succulent; leaves 2–6, all basal, linear, occasionally oblanceolate, 3–15 cm long, 3–5 mm wide, tender green; flowers small, pink-white, spirally arranged in a terminal spike 5–10 cm long, perianth 3 mm long, ovary ellipsoid, hairy; capsules oblong, 6–7 mm long, 3 mm across the middle.

Vanilla planifolia Andrews (Syn. *V. fragrans* [Salisbury] Ames) — Vanilla

Xiang-guo-lan=Hsiang-kuo-lan (香果蘭, Sweet Fruit Orchid). Flavoring material obtained from the cured pods; used in ice cream and candies; the market material imported, mostly artificial.

A much branched scandent epiphytic orchid with wire-like aerial roots; leaves fleshy-succulent, oblong-elliptic, 15–20 cm long, 5–7 cm wide; flowers yellowish green, in axillary racemes, sepals and petals oblanceolate, 5 cm long, lip yellow with orange keel; capsules cylindrical, 12–15 cm long, fully grown but not ripe fruits picked, cured, the vanillin crystallizes on the outside. A native of Central America, with Mexico and Madagascar being the principal commerial producing areas; newly introduced to China.

DICOTYLEDONEAE

Saururaceae: Lizard's-tail Family

Houttuynia cordata Thunberg — Fish-smelling Herb

Yu-xing-cao=Yu-hsing-ts'ao (魚腥草, Fish-smell Herb); *Zhu-pi-gu=Chu-p'i-ku* (豬屁股, Pig Rump); *Ji-cai=Chi-ts'ai* (蕺菜, an ancient book name). Very young shoots including the white rhizome, gathered in early spring, quick-fried for vegetable; used extensively by various ethnic groups, with 33 different local names recorded in *An Encyclopedia of Chinese Medicines* (Anonymous, 1977); cultivated recently to meet the market demand; dried material available in American Chinese herb stores.

A rhizomatous, strongly odorous perennial herb, 5–20 cm high, rooting as the stem touching the ground; leaves alternate, ovate-cordate, 3–8 cm long, 3–6 cm wide, punctate, purple beneath, petioles 1–4 cm long, the lower portion bearing an elongated stipule on each side; flowers small, crowded in an oblong-cylindrical spike subtended by 4 white bracts 1.5 cm long; polygamous or unisexual, perianth wanting; anthers 3 or 4, ovary 3- or 4-carpellate; capsules cup-shaped, 3 mm across top, opening at the center, with horn-like persistent styles and stigmas on the edge; seeds globose, brown.

Piperaceae: Pepper Family

Piper betle L. — BETEL-PEPPER

Lü-ye=Lü-yeh (蔞葉, Betel Leaf). Fully grown leaves for chewing betel; *Qing-lü=Ch'ing-lü* (青蔞, Betel-leaf Green). Young shoots, for soup.

An aromatic climber growing in partial shade of trees near the homesteads of large families and in temples, the stem rooting when touching the ground; leaves alternate, dark green, subcordate with a broad sinus at the base, 5–15 cm long, 5–11 cm wide, acute or abruptly short-acuminate, palmately 5-nerved near the base; flowers minute, naked, dioecious, in nodding solitary spikes opposite the leaf, staminate spike 9 cm long, the pistillate ones 1.5–3.5 cm long, ovary adnate to the fleshy rachis with 4–5 stigmas shown outside the bracts; fruits embedded and coalescing to the axis. Native of Southeast Asia, the Chinese name similar to the Thai name, *Pelu*, showing an early introduction overland; used fresh for food, for chewing with betel-nut, after cured to render the leaves soft and uniformly yellow like the petals of sunflower, and to improve the aroma.

Piper longum L. — INDIAN LONG PEPPER

Bi-bo=Pi-po (畢拔, Long Pepper). Dried infloresence and young fruits, available in rural marketplaces; Chinese name "*Bi-bo*" originated from the translation of "pepper".

A perennial aromatic climber, rooting at the nodes; leaves entire, 4–9 cm long, 3–7 cm wide; flowers small, naked, dioecous, in slender spikes, concealed by peltate scales, staminate spikes 5–8 cm long, pistillate ones 2.5–5 cm long, smooth with the stigmas exposed only; fruits aggregate, the market material consisting of cylindrical multiple fruits at various stages of development, 2.55 cm long, 4–8 mm in diameter, brown, hard as a dried stick, very pungent; used as spice, pulverized, particularly in Gansu and Yunnan.

Piper nigrum L. — BLACK PEPPER, WHITE PEPPER

Hu-jiao=Hu-chiao (胡椒, Tartary Pepper); *Hu-jiao-mian=Hu-chiao-mien* (胡椒麵, Powdered Pepper). Fully grown fruits, dried, imported, common, available even in rural market places.

An aromatic climber, the nodes enlarged; leaves alternate, ovate-oblong or elliptic, 8–15 cm long, 5–9 cm wide, rounded at the base, palmately 5–7 nerved; flowers naked, dioecious, in pendulous cylindrical spikes 10–15 cm long, ovary globose, stigmas 3 or 4; fruits globose, 3–4 mm in diameter, red when ripe. The market material consisting of grayish black pepper, harvested when the fruits being fully grown but not yet red; white pepper with the outer pericarp removed.

Chloranthaceae: Chloranthus Family

Chloranthus japonicus Siebold et Zuccarini

Yin-xian-cao=Yin-hsien-ts'ao (銀線草, Silver Thread Herb); *Deng-long-hua=Teng-lung-hua* (燈籠花, Lantern Flower). Young shoots, used for potherb in northeastern China.

Perennial herbs 20–40 cm high, glabrous throughout, with creeping rhizomes bearing numerous fibrous roots, erect stems emerging from the terminal buds of the rhizomes, each bearing paired scales below the terminal whorl of four leaves and the inflorescence; leaves ovate-elliptic, 7–11 cm long, 4–7 cm wide, acute, base cuneate, apical 3/4 serrate, petioles 1–1.5 cm long; flowers white, sessile, apetalous, in a solitary spike terminal to the stem, stamens 3, filaments white, 4–5 mm long, connate at base and adnate to the abaxial basal side of the ovary, anthers 2, basal to the outer filaments, stigma truncate; fruits drupaceous, oblique, 2.5–3 mm long. Native to temperate eastern Asia, growing in shade of deciduous forests.

Chloranthus spicatus (Thunberg) Makino (Syn. *C. inconspicuus* Swartz)

Zhu-lan=Chu-lan (珠蘭, Pearl Orchid). Infloresences, added to tea to impart a sweet scent; *Zhu-lan-cha=Chu-lan-ch'a* (珠蘭茶, Chloranthus Tea), highly esteemed by the Chinese people.

A suffrutescent erect plant 30–60 cm high, generally planted in pots as house plants for the pleasant fragrance, particularly during the flowering season; nodes prominent, internodes 2–7 cm long; leaves deep green, opposite, elliptic, 4–12 cm long, 2–5 cm wide, sharply serrate, with a gland at the end of each tooth; flowers small, in terminal panicles of spikes, the spike 6–8 cm long, yellow at the beginning of anthesis, very fragrant, a staminate flower with 3 stamens, a pistillate one with green ovaries; fruit a pea-sized juicy red drupe, 4–5 mm in diameter. Native of subtropical China.

Salicaceae: Willow Family

Salix purpurea L. — Purple Osier

Po-qi-liu=Po-ch'i-liu (簸箕柳, Winnowing Fan Willow); *Hong-pi-liu=Hung-p'i-liu* (紅皮柳, Red Bark osier). Fresh leaves boiled in water for tea, used in northern China villages along the Grand Canal.

Shrubs (often trimmed into small trees) 3–4 m high, with slender tough branches purplish at first, becoming gray or olive-gray, glabrous; leaves lanceolate, 5–10 cm long, 8–15 mm wide, acuminate, base cuneate, serrulate, glabrescent, pale green and glaucous beneath, turning black on drying; flowers yellowish-red-black, woolly, appearing before leaves, dioecious, apetalous, in upright catkins 1.5–2.5 cm long, staminate flowers consisting of 2 stamens with connate filaments, red anthers and 1 gland, subtended by obovate bracts, purplish-black at the distal half, pistillate ones with subsessile tomentose ovary, stout style and bifid stigmas; capsules 2-valved, unilocular; seeds many, with long silky hairs. Widespread in temperate Eurasia, much cultivated in northern China, being an important tree crop, harvested annually at the ground level for the pliable twigs used for rough weaving with the bark, or for refined plaiting of winnowing fan, baskets and containers after the bark is removed.

Myricaceae: Wax-myrtle Family

Myrica adenophora Hance

Qing-mei=Ch'ing-mei (青梅, Green Myrica). Fruit, eaten raw.

An evergreen shrub 1–3 m high; branches covered by glandular hairs; leaves alternate, obovate-oblanceolate, 2–7 cm long, 5–10 mm wide, cuneate at the base, punctate, especially beneath; flowers dioecious, small, in globular heads; staminate head 1–2 cm long, pistillate ones 1 cm in diameter; fruit oblong-spherical, I cm long, white or red.

Myrica rubra Siebold et Zuccarini — YANGMEI, TREE STRAWBERRY

Yang-mei=Yang-mei (楊梅, Tree Myrica). Ripe red fruit, preserved in sugar, or used in preparing canned cold drinks; both products available in Boston.

An evergreen tree up to 15 m high, having a trunk 60 cm in diameter, branchlets glabrous and with glands; leaves obovate-oblanceolate, rarely elliptic, 6–16 cm long, 1–3.5 cm wide, glabrous, golden punctate beneath; flowers small, in axillary spikes, the staminate spike solitary or fasciculate, 1–3 cm long, the pistillate ones solitary, 0.5–1.5 cm long; fruits spherical, 1–1.5 (–3) cm in diameter, the excocarp fleshy, sour-sweet, red, white or in other colors according to the cultivar.

Note: When China was closed to the United States, it had continuous communication with England. One day in the early 1970s, Harold Hillier, a former member of the Council of the International Dendrology Society, sent me the paper cover of a Chinese can marked "Tree Strawberry". He asked for the identification of the contents and wrote: "How can a strawberry be associated with a tree?" My explanation to him was this: In the 1970s, English speaking had been a taboo in China for over 20 years. The people involved in foreign trade needed an English term for the tree crop called *"yang-mei"* (楊梅). Having neither sufficient knowledge of the plants involved, nor enough English language background, a person must have looked for the name of an introduced fleshy-rough, red fruit called *"cao-mei"* (草莓)*"* in a Chinese dictionary. It had seemed logical to him to translate the Chinese fleshy-rough red *yang-mei* as "Tree Strawberry."

Yang-mei has been a classical ancient Chinese fruit well known for its sour taste. The famous idiom, *"Wang-mei zhi-ke"* (望梅止渴, thirst stopped by looking at *mei*, the mouth waters when looking at the sour fruit).

Juglandaceae: Walnut Family

Carya cathayensis Sargent — CHINESE HICKORY

> *Shan-he-tao=Shan-ho=t'ao* (山核桃, Mountain Walnut). Fruit; gathered and used locally, not available in the market.

A deciduous tree up to 30 m high, trunk up to 70 cm in diameter, branchlets covered by brown glandular scars; leaves pinnate, the leaflets 5 or 7, oblong-lanceolate or obovate-lanceolate, 7–22 cm long, the apex acuminate, margin serrate, glandular-punctate beneath; flowers monoecious, staminate catkins 3 in a fascicle, 5–12 cm long, stamens 5–7, pistillate flowers 1–3, terminal to a leafy branch; fruits ovoid or spherical, 2.5–2.8 cm in diameter, densely covered with yellow glands, outer fleshy shell 4-valved; nut ovoid, 2–2.5 cm long, shell 1 mm in thickness, kernel 43–49% of the weight of the nut, 69–74% of the kernel being oil; good for cooking.

Carya illinoinensis (Wangenhaim) C. Koch — PECAN (Syn. *C. pecan* [Marshall] Engler et Graebner)

> *Mei-guo-shan-he-tao=Mei-kuo-shan-ho-t'ao* (美國山核桃, American Walnut) introduced in the early 1900s by the former Agriculture College, University of Nanking, rarely cultivated in the 1930s in Nanjing; imported nuts available in Hong Kong.

A deciduous tree 10–15 m high; leaves odd-pinnate, leaflets 11–17, oblong-lanceolate, falcate, 4.5–18 cm long, 2–4 cm wide, serrate or doubly serrate, hairy beneath; flowers; unisexual, monoecious, staminate catkins pendulous, the pistillate ones in terminal spikes; fruits oblong, 3–10 in a cluster, 4–6 cm long, the husk thin, splitting to the base; nuts 3.7–4.5 cm long, smooth, kernel sweet. Cultivated in some private gardens or botanical institutions in eastern China.

Juglans cathayensis Dode — Central China Wild Walnut

Ye-he-tao=Yeh-ho-t'ao (野核桃, Wild Walnut); *Shan-he-tao=Shan-ho-t'ao* (山核桃, Mountain Walnut). Ripe fruit, gathered, eaten locally, not available in the market.

Deciduous trees up to 25 m high, branchlets glandular-pubescent, pith laminate; leaves pinnate, 40–50 cm long, leaflets 9–17, sessile, ovate or ovate-oblong, 8–15 cm long, 3–7 cm wide, serrulate, sparsely stellate above, densely pubescent with pilose and stellate hairs beneath; flowers unisexual, monoecious, the staminate catkins 20–30 cm long, pedulous, the pistillate ones erect, densely glandular-pubescent, 3–6 cm long, bearing 3–10 inferior ovaries; fruits drupaceous, on pendulous clusters 10–20 cm long, bearing 3–8 ovoid hairy drupes 3–4.5 cm long, the stones ovoid-globose, pointed at the apex, with 6–8 broken spiny ridges. Widespread in northern, western and southwestern China; generally used as scions for grafting *J. regia* L..

Juglans mandshurica Maximowicz — Manchurian Walnut

Hu-tao-qiu=Hu-t'ao-ch'iu (胡桃楸, Walnut Catalpa); *Shan-he-tao=Shan-he-t'ao* (山核桃, Hillside Walnut). Kernel, rich in oil.

Deciduous trees 20 m high, branchlets glandular-pubescent, with lamellate pith; leaves pinnate, oblong in outline, up to 80 cm long, leaflets 9 to 17, ovate or ovate-lanceolate, 8–15 (–22) cm long, 3–7 cm wide, acuminate, base obliquely rounded or subcordate, serrulate, glabrescent above, pubescent beneath; flowers green, unisexual, monoecious, protandrous, staminate catkins appearing on last year's growth, 9–20 cm long, stamens 12 to each flower, pistillate flowers 4–10 in erect spike terminal to current year's growth; fruiting clusters 10–15 cm long, hanging, consisting of 5–7 drupes; fruits oblong-ovate, 3.5–7.5 cm long, 3–5 cm across, stones globose, ovoid or oblong, pointed, with 8 ridges and irregularly furrowed pits, shell thick. Native to northeastern China, growing in deciduous forests at altitudes of 500–1,000 m; introduced into western gardens in 1859.

Juglans regia L. — Walnut (English Walnut, Persian Walnut)

He-tao=Hai-t'ao (核桃, Walnut). Many varieties, extensively cultivated in the mountainous area of West China, some forms with very thin shells which can be broken by squeezing two nuts in one hand.

A deciduous tree 20–30 m high; branchlets glabrous; leaves 30–50 cm long, pinnately compound, leaflets ovate-oblong, the lateral ones obliquely oblong-lanceolate, variable in shapes and sizes, (4–) 6–13 (–18) cm long, obtuse or acute at the apex, entire with tuft of hairs in the nerve axils beneath; flowers monoecious, the staminate catkin (5–) 12–15

(–25) cm long, emerging before the leaves on last year's branchlets, green, pistillate spike 2 or 3 flowered, terminal to leafy branchlets, developed after the unfolding of the lower leaves on the twig, stigmas greenish yellow; fruits spherical-ovoid, hairy when young, glabrescent, 5 cm long, 4 cm across, often 2 in a cluster.

Note: General botanical references use "English Walnut" or "Persian Walnut". Recent floristic studies reveal that there is no indigenous species of *Juglans* L. in Europe and Iran (Flora iranica 121:5, 1976). The common walnut is of hybrid origin, and it hybridizes readily with the wild Chinese and American species (Rehder, 1940). Its first cultivation in Nanjing was between A.D. 222 and 277 (Hu, 1990), giving the origin as Xi-qiang (西羌). This area was called Eastern Tibet in the time of A. David and E. H. Wilson, as printed in their labels and publications. David discovered the Giant Panda in the area and Wilson introduced numerous plants from there into American gardens. I have botanized in the area, for four summers ascending the high mountains from various directions and am familiar with its vegetation and the people. In the isolated valleys of the Xi-qiang and the Gia-rong ethnic peoples, thin-shelled cultivars of *J. regia* L. and the wild population of *J. cathayensis* Dode grow side-by-side. The ancient Chinese records and current investigations both in China and in Iran indicate that *J. regia* L. is a selection of natural hybrids in western China.

Betulaceae (Corylaceae) : Birch Family (Hazel Family)

Corylus avellana L. — European Hazelnut
Xi-yang-zhen=Hsi-yang-chen (西洋榛, Filbert , Hazelnut). Rare in China.

A tall shrub 3–5 m high, branchlets glandularly pubescent; leaves suborbicular, 5–10 cm long and wide, abruptly acuminate, base cordate, pubescent beneath; flowers monoecious, staminate catkins naked during the winter, pendulous, 3–6 cm long, anthers pilose at the apex, pistillate flowers in head-like cluster enclosed in a scaly bract with the red styles protruding; fruit ovoid-subglobose, 1.5–2 cm long and across, surrounded by a leafy irregularly divided involucre toothed along the margin; shell brown. Rarely cultivated in China.

Corylus chinensis Franchet ex Trautvetter — Chinese Hazelnut
Zhen=Chen (榛, Hazel); *Hua-zhen=Hua-chen* (華榛, Chinese Hazelnut); *Shan-bai-guo=Shan-pai-kuo* (山白果, Hillside Ginkgo). Leaves used for tea in Inner Mongolia; fruits gathered and eaten locally.

Deciduous trees up to 40 m high, trunks 50 cm in diameter, branchlets pilose; leaves ovate-oblong, 10–18 cm long, 6–12 cm wide, base oblique cordate, apex acuminate, double-serrate, glabrous above, pubescent along the veins beneath; flowers unisexual, monoecious, staminate catkins 4–6 in a cluster, 1.5–2 cm long; fruits cluster subglobose 1.5 cm across, consisting of 4–6 nuts, each enclosed in a striate, tubular and contracted pubescent involucre; nuts subglobose, 1.5 cm in diameter. Native to Sichuan and the adjacent areas of Yunnan, Gansu and Hubei, thence northward to Inner Mongolia.

Corylus chinensis Franchet ex Trautvetter var. **fargesii** (Franchet) Wilson ex Hu
Yong-mao-zhen=Yung-mao-chen (絨毛榛, Tomentose Hazelnut). Fruits eaten in Hubei.

Deciduous trees 10 m high, branchlets pubescent; leaves ovate, 8–10 cm long, acuminate, base cordate; fruits 2–6 in a cluster, involucres much contracted above the nut into a short tube, divided at the distal half, villose or glabrescent; nuts subglobose, 1–2 cm in diameter. Native to central and western China, reaching northern Yunnan, growing in valleys of 900–2,200 m altitudes; introduced into western gardens in 1895.

Corylus heterophylla Fischer ex Trautvetter — NORTH CHINA HAZEL
Zhen=Chen (榛, Chinese Hazelnut). Fruit, gathered locally, rare.

A tall shrub or small tree 4–7 m high, branchlets and petioles glandularly hairy; leaves orbicular-ovate to obovate, 5–13 cm long and wide, abruptly acuminate, base subcordate or rounded, irregularly serrate, pubescent on the nerves beneath; fruits 1–4, involucre campanulate, longer than the nut, glandular-setose near base, lobes dentate; nuts subglobose, 1.5 cm across. Used locally in northern China where the species occurs in nature, no commercial cultivation.

Corylus heterophylla Fischer var. **sutchuenensis** Franchet
Zhen-zi=Chen-tzu (榛子, Hazelnut); *Chuan-zhen=Chuan-chen* (川榛, Sichuan Hazelnut). Fruits gathered and eaten locally.

Shrubs or small trees 7 m high, branchlets densely villose, glabrescent; leaves oblong or broad-ovate, 4–15 cm long, 5–8 cm wide, rounded or truncate and abruptly acute, base subcordate, irregularly double-serrate, often lobed above the middle, pubescent beneath, petioles 1–2 cm long; fruits 1–6 in a fascicle. Endemic to Sichuan and adjacent Hubei, growing in secondary forests of valleys at altitudes of 1,300–2,100 m; introduced to American gardens in 1909.

Corylus sieboldiana Blume var. **mandshurica** (Maximowicz et Ruprecht) C. K.
Schneider (Syn. *C. mandshurica* Maximowicz et Ruprecht) — JAPANESE HAZELNUT
Liao-zhen=Liao-chen (遼榛, Northeastern Hazelnut). Fruit, gathered and used by local people, not available in the market.

A deciduous shrub 3–5 m high, branchlets hairy; leaves elliptic or obovate-oblong, 5–10 cm long, rounded at base, doubly serrate, slightly lobed, pilose on the nerves beneath; nut 1–3, in tubular involucres 1.5–4 cm long, constricted above the nuts, lobes entire, densely covered with loosely decumbent bristles, the nut conical; limited to the northeastern provinces.

Corylus tibetica Batalin (Syn. *C. ferox* Wallich var. *thibetica* Franchet)

Hou-ban-li=Hou-pan-li (猴板栗, Monkey Chestnut); *Zang-ci-zhen=Tsang-tz'u-chen* (藏刺榛, Tibetan Spiny Hazelnut). Nuts, eaten in northwestern Hubei.

Deciduous trees 4–10 m high, branchlets sparsely pilose or glabrous; leaves broad-ovate, or ovate, rarely oblong, 4–12 cm long, 3–8 cm wide, abruptly acute, base rounded or subcordate, irregularly double-serrate, petioles 1.5–2.3 cm long; fruiting involucres armed with sharp branched spines, yellow-brown, nuts globose, 1.5 in diameter, locally available in September–October. Endemic to western China, growing in secondary forests of western Hubei at altitudes of 1,500–2,300; introduced into western gardens in 1879.

Fagaceae: Beech Family

Castanea henryi (Skan) Rehder et Wilson

Zhui-li=Chui-li (錐栗, Conic Chestnut); *Zhen-zhu-li=Chen-chu-li* (珍珠栗, Pearl Chestnut). Gathered locally in central China; fruit small, conic, of good flavor.

A deciduous tree 20–30 cm high, branchlets glabrous; leaves lanceolate or ovate-lanceolate, 12–17 cm long, 2–5 cm wide, apex long-acuminate, base rounded or cuneate, sharply serrate, lateral nerves 13–16 pairs, petioles 1–1.5 cm long; staminate spikes erect, axillary to the lower leaves, pistillate ones axillary to the upper leaves; involucres 3–3.5 cm across (including the spines); nut solitary, globose-ovoid, 1.5–2 cm in diameter. Occurring in central China.

Castanea mollissima Blume (Syn. *C. bungeana* Blume) — CHINESE CHESTNUT

Ban-li=Pan-li (板栗, Chestnut) ; *Li-zi=Li-tzu* (栗子, Chestnut). Extensively cultivated on the hillsides of the Yangtze River Region for the edible nut; roasted or used in cooking.

A deciduous tree 15–20 m high, branchlets pubescent; leaves elliptic-oblong, 9–18 cm long, 4–7 cm wide, base rounded or obtuse, apex short-acuminate, margin coarsely serrate, tomentose beneath, lateral nerves 10–18 pairs, petioles 1–1.5 cm long, pilose;

staminate catkins erect, 12–15 cm long, pistillate flowers 1 or 2, on the basal portion of the upper staminate catkin; involucres 4–6.5 cm across when mature, splitting irregularly, containing 2 or 3 nuts. Commercially cultivated on the hillsides for local uses and for export.

Castanea seguinii Dode

Mao-li=Mao-li (茅栗, Rough Chestnut). Gathered in warmer areas of China, consumed locally, rarely available in the market.

A small deciduous tree 6–15 m high, often having a shrubby habit due to cutting for cooking fuel; leaves elliptic-oblong, or lanceolate-oblong, 6–14 (–16) cm long, base rounded, subcordate or acute, coarsely serrate, lepidote-glandular and green beneath, lateral nerves 12–16 pairs; involucres 3–4 cm across, spines pilose, including 3–5 compressed nuts, 1–1.5 cm wide.

Castanopsis calathiformis (Skan) Rehder

Kao-si-li=K'ao-ssu-li (栲絲栗, Silk Castanopsis); *Bei-zhuang-kao=Pei-chuang-k'ao* (杯狀栲, Cup-shaped Castanopsis); *Da-ye-kao=Ta-yeh-k'ao* (大葉栲, Large-leaved Castanopsis). Starch obtained from the seed, used by ethnic groups in Yunnan.

Large trees up to 20 m high, branchlets glabrous; leaves oblong-elliptic, 15–25 cm long, 5–8 cm wide, obtuse, base cuneate, serrate or wavy; flowers small, staminate flowers in panicles, pistillate ones in spikes 10–16 cm long; fruiting involucres cupular, 9–13 mm across, scales deltoid, imbricate; nuts ovoid, 1.5 cm long, 8–10 mm across the middle. Endemic to southern Yunnan.

Castanopsis carlesii (Hemsley) Hayata

Xiao-hong-kao=Hsiao-hung-k'ao (小紅栲, Lesser Red Castanopsis); *Mi-zi-chai=Mi-tzu-ch'ai* (米子柴, Grain Cooking Fuel). Nuts; gathered and eaten locally in southeastern China.

Evergreen trees 8–15 m high, glabrous; leaves ovate or ovate-elliptic, 4–8 cm long, 1–3 cm wide, apex caudate or acuminate, base rounded or obtuse, margin entire or remotely serrulate above the middle, when young covered by rusty scales beneath, changing to gray with age, lateral nerves inconspicuous, petioles short, 5–9 mm long; staminate panicles crowded at the end of a branchlet, consisting of 3–6 spikes 4–8 cm long; pistillate spikes solitary, 7–9 cm long, involucres oblique-ovoid, 9–10 mm in diameter, pubescent, spines simple, short, stout and pubescent at base, more numerous toward the apical end; nut ovoid, 1 cm long, edible raw.

Castanopsis chinensis Hance — CHINESE CASTANOPSIS

Mi-zhui=Mi-chui (米錐, Grain Conic Chestnut); *Gui-lin-kao=Kuei-lin-k'ao* (桂林栲, Gui-lin Castanopsis); *Kao-li=K'ao-li* (栲栗, Castanopsis Nut). Nuts; edible raw, in western and southern China.

Evergreen trees 5–15 m high, branchlets glabrous; leaves ovate-lanceolate or elliptic, 7–12 cm long, 2.5–4 cm wide, acuminate at apex, rounded or oblique-obtuse at base, remotely sharp-serrate two-thirds of the apical portion, glabrous, lateral nerves 10–12 pairs, prominent, petioles 1.5–2 cm long; staminate panicles having 4–8 spikes 3–5 cm long, often appearing fasciculate, axis hairy; pistillate ones solitary, 6–13 cm long, involucres spherical, 2.8–3.5 cm in diameter, spines long, strong and branched; nut solitary, ovoid, 1–1.5 cm long, 8–12 mm in diameter; edible raw.

Castanopsis delavayi Franchet — DELAVAY'S CASTANOPSIS

Dian-zhui-li=Tien-chui-li (滇錐栗, Yunnan Conic Chestnut). Seeds; edible raw.

Evergreen trees up to 20 m high, branchlets glabrous or with white powder or scales; leaves obovate or ovate, 6–11 cm long, 3.5–7 cm wide, apex obtuse or shortly pointed, base cuneate or obtuse, margin remotely serrate above the middle, glabrous, with brown scales beneath, lateral nerves 6–10 pairs, petioles short, 1 cm long; staminate catkins erect, pistillate flowers loosely arrangeed in spikes, involucres ovoid or spherical, 1.5–2 cm across, containing a broad-ovoid or spherical nut 1–1.4 cm long, 0.8–1 cm in diameter.

Castanopsis eyrei (Champion ex Bentham) Tutcher — EYRE'S CASTANOPSIS

Tian-zhu-kao=T'ien-chu-k'ao (甜櫧栲, Sweet Castanopsis); *Si-li=Ssu-li* (絲栗, Silk Chestnut); *Shi-li-zi=Shih-li-tzu* (石栗子, Rock Chestnut). Ripe nut eaten locally in southern China.

Evergreen tree 20 m high, branchlets sparsely pubescent when young, glabrescent with age; leaves ovate-lanceolate, 5–10 (–13) cm long, 1.5–5.5 cm wide, apex caudate or acuminate, base rounded, rarely oblique-obtuse, entire, occasionally serrate above the middle, reddish brown scabrid beneath, petioles 1–1.3 cm long; staminate panicles with 4 to 9 spikes sometimes bearing leaves on the distal portion, the spikes 7–14 cm long, pistillate spikes solitary, axillary to mature leaves, 5–8 cm long; involucres oblong-ovoid, 2.5 cm long, 2 cm in diameter, the spines stout, hairy, rarely branched; nut solitary, ovoid-spherical, 1–1.4 cm in diameter, edible raw.

Castanopsis fargesii Franchet — CENTRAL CHINA CASTANOPSIS

Si-li-shu=Ssu-li-shu (絲栗樹, Silk Chestnut Tree); *Si-li-kao=Ssu-li-k'ao* (絲栗栲, Silk Chestnut Castanopsis). Nut, gathered locally in central China.

An evergreen tree up to 30 m high, with trunk 0.3–1 m in diameter; branchlets scabrid when young, glabrescent with age; leaves elliptic-lanceolate, 10–13 cm long, 2.5–3 (–5.5) cm wide, apex caudate or acuminate, base acute or obtuse, margin entire or with 3 to 5 small teeth at the apical one-third, rusty scabrid beneath, petioles 1–1.3 cm long; staminate panicles with 3 to 5 spikes, 5–8 mm long, often with leaves at the distal end, the axis sparsely scabrid, pistillate spikes solitary, 9–15 cm long, involucres subglobose, 1.5–2.5 cm in diameter, densely pubescent, spines stout, strong, branched, in 4–6 concentric rings; nut ovoid-globose, 1 cm long, 10–12 mm in diameter, the basal scar broad, edible raw.

Castanopsis fissa (Champion ex Bentham) Rehder & Wilson — HONG KONG CASTANOPSIS

Da-ye-li=Ta-yeh-li (大葉櫟, Big-leaved Oak); *Da-ye-zhui=Ta-yeh-chui* (大葉錐, Big-leaved Conic Chestnut) ; *Li-shuo=Li-shuo* (鱲蒴, Black Nut). Starch precipitated from the nut.

An evergreen tree 6–10 m high, young branchlets and leaves finely pubescent; leaves oblong-elliptic, 10–20 cm long, 4–7 (–9) cm wide, apex acute or short-acuminate, base acute or cuneate, margin remotely crenate, densely brown lepidote beneath, becoming less conspicuous with age, petioles 1.5–2.5 cm long; staminate spikes white, 10–16 cm long, axillary to bracts or leaves of the developing branches or in panicles, pistillate spike solitary, 10–15 cm long; involucres campanulate, 1.5–2 cm long, smooth, pilose, with small teeth in concentric rings, 1.5–2 cm long; nut ovoid, 1.4–1.6 cm long, 1 cm in diameter, smooth; starch used locally.

Castanopsis fordii Hance — HAIRY CHESTNUT

Mao-li-mu=Mao-li-mu (毛栗木, Hairy Chestnut Tree); *Shui-li=Shui-li* (水栗, Water Chestnut, corrupted to 水梨); *Nan-ling-kao=Nan-ling-k'ao* (南嶺栲, South China Castanopsis). Nuts, eaten locally.

A very distinct species recognized easily by its stout golden hairy branchlets, large leaves with very short petioles; an evergreen tree up to 30 m high; leaves oblong, 9–16 cm long, 2.5–5 cm wide, entire, short acuminate at apex, rounded at base, densely golden tomentose beneath, petioles 3 mm long and thick, hairy; staminate panicles sessile, 10–20 brush-like, golden hairy throughout, consisting of rusty erect spikes; pistillate spikes solitary, axillary to caducous membranous oblong bracts at the basal end of a developing branch growing simultaneously with the staminate panicles; involucre globose, 4–6 cm in diameter, generally only 3 or 4 near the base of the branch developed into mature fruits at the expense of the others on the same branch, spines numerous, sharp and long, completely concealing the involucre; nut compressed globose, 1.2 cm long, 1.5 cm in diameter, edible raw.

Castanopsis formosana (Skan) Hayata

Huang-mei=Huang-mei (黃楣, Yellow Lintel, the Hainan ethnic name); *Tai-wan-kao=T'ai-wan-k'ao* (台灣栲, Taiwan Castanopsis); *Huang-mei-kao=Huang-mei-k'ao* (黃楣栲, Yellow Lintel Castanopsis). Nuts, eaten in Hainan.

Evergreen trees 10–20 m high, trunks 50–70 cm in diameter, branchlets glabrous; leaves ovate-elliptic to ovate-lanceolate, 8–10 cm long, 3–4 cm wide, acuminate, the acumen often falcate, base acute or obtuse, glabrescent, silvery-gray beneath, coarsely, serrate; flowers small, 2 mm across, unisexual, staminate catkins in erect panicles, perianth segments 6, 1.5 mm long, 2.5 mm wide, hairy inside, stamens 8–12, pistillate flowers in erect spikes 4–10 cm long, perianth segments 6, ovary concealed by the involucre, with 3 stigmas exposed; fruits globose, 2.5–3 cm across, involucral spines 1 cm long, branched, hairy; nuts solitary, ovoid, enclosed, 1.5 cm long, 1.2 cm across, hairy. Native to the broad-leaved forests of Taiwan and Hainan, growing at altitudes of 200–700 m.

Castanopsis hystrix A. de Candolle

Kao-shu=K'ao-shu (栲樹, Castanopsis); *Xiao-hong-li=Hsiao-hung-li* (小紅栗, Lesser Red Chestnut). Nuts, used locally in Yunnan. .

An evergreen tree 25 m high, young branchlets sparsely covered with short rusty hairs and scales, glabrescent with age; leaves ovate-lanceolate, 7–12 cm long, 2–3.5 cm wide, entire, rarely with few teeth below the apex, densely brown lepidote beneath, petioles short, 5–8 mm long; staminate panicles with 3 or 4 spikes, more often axillary to bracts basal to leafy branchlets, spikes 6–8 (–15) cm long, pistillate spikes solitary, axillary to caducous bracts at the base of leafy branchlets, 4–12 cm long; involucres 3–4 cm in diameter, densely covered by strong spines coalescent at the base, mature involucre splitting into 4 valves; nuts ovoid, 0.8–1. 5 cm in diameter, basal scar very large; edible raw.

Castanopsis indica (Roxburgh) A. de Candolle — NORTHERN INDIA CASTANOPSIS

Hong-mei-zi=Hung-mei-tzu (紅楣子, Red Lintel); *Yin-du-kao=Yin-tu-k'ao* (印度栲, Indian Castanopsis). Nuts.

An evergreen tree 8–25 m high, young branchlets densely covered with golden hairs, changing to black-brown with age; leaves ovate-oblong, 9–15 cm long, 3–10 cm wide, remotely sharp serrate, base rounded, apex abruptly short acuminate, when young densely red-lepidote above, changing to gray-tomentose and lepidote beneath; spikes golden tomentose, the staminate ones axillary to bracts or developing leaves; panicles 6–17 cm long, pistillate spikes 15–22 cm long, often occurring on the same young shoot with staminate ones but below them, many flowers developing into fruits, involucres subglobose, 2–4 cm in diameter, densely covered by branched needle-like sharp and slender spines 10–12 mm long; nuts ovoid, 0.8–1.3 cm long, basal scar very large; edible raw; widespread in tropical China.

Castanopsis jucunda Hance

Wu-mei-kao=Wu-mei-k'ao (烏楣栲, Black Lintel Castanopsis). Fruits, sweet, used like chestnut in Hubei.

Trees 15 m high, branchlets glaucous and pubescent with brown hairs, winter buds velutinous; leaves elliptic, or oblong-lanceolate, 6–12 cm long, 3–5 cm wide, acuminate, base obtuse or rounded, coarsely crenate-serrate, with gray- or brown-waxy hairs. infructescences 11 cm long, involucres 1.5–2 cm across, densely prickly and villose with grayish-yellow hairs; nuts ovoid. Native to central and southern China.

Castanopsis kawakamii Hayata

Qing-gou-kao=Ch'ing-kou-k'ao (青鈎栲, Green Hook Castanopsis).

Seed gathered locally in warmer areas of China, used for food in Taiwan.

Evergreen tree 14–28 m high, branchlets glabrous; leaves ovate-oblong, 7–12 cm long, 2.5–4.5 cm wide, apex acuminate, base rounded or oblique-obtuse, entire or with 1 to 3 teeth below the apex, glabrous, petioles 1–1.5 cm long; staminate panicles with 4–6 spikes, 5–14 cm long, occasionally the shoot bearing leaves at the apical portion; pistillate spikes solitary, axillary, ovaries 3 in a bud, mature involucres globose, 3–4 cm in diameter, completely covered by branched spines; nuts compressed-globose, 1.2–1.5 cm long, 1.7–2.1 cm in diameter, basal scar very broad; edible raw.

Castanopsis lamontii Hance

Hong-gou-kao=Hung-kou-k'ao (紅鈎栲, Red Hook Castanopsis); *Shi-zhui-shu=Shi-chui-shu* (石錐樹, Stone Conical Chestnut Tree); *Bai-chuan=Pai-ch'uan* (白椽, White Beam). Nuts, used in South China locally.

A large tropical evergreen tree 8–15 m high, glabrous; leaves ovate-oblong, 8–14 cm long, 2.5–5 cm wide, abruptly acuminate, base oblique-obtuse or rounded, entire, lepidote beneath, staminate spikes solitary, axillary to bracts or leaves of developing branchlets, 5–9 cm long, pistillate ones solitary, involucres ovoid-conical, 3–4 cm long, 2–2.5 cm in diameter, apex abruptly rostrate, spines stout and short, in 4 interrupted concentric rings, inconspicuously pilose, densely hirsute on the inside. Named for Rev. A. Lamont of Hong Kong.

Castanopsis orthacantha Franchet (Syn. *C. concolor* Rehder et Wilson)

Mao-zhui-li=Mao-chui-li (毛錐栗, Hairy Conic Chestnuts); *Zhu-li= Chu-li* (豬栗, Hog Chestnut).

Very similar to Lamont's castanopsis, a large tree about 15 m high; leaves elliptic-lanceolate, 7–14 cm long, apex acuminate, the acumen often curved slightly; spikes stout, 15 cm long, involucres subglobose, 2.5–3.5 cm long, glabrous, spines shortly branched; nuts conical, 1–1.5 cm long, hairy, basal scar very large, edible raw.

Castanopsis platyacantha Rehder et Wilson

Hou-li=Hou-li (猴栗, Monkey Chestnut); *Bai-shi-li=Pai-shih-li* (白石栗, White Rock Chestnut); *Si-li=ssu-li* (絲栗, Silk Chestnut). Nuts; eaten in western China.

An evergreen tree up to 16 m high, trunk 0.30–1.20 m in diameter, glabrous; leaves elliptic-oblong, 8–16 cm long, 2.5–6 cm wide, subentire, remotely serrate half way above the middle, apex acuminate, base oblique-rounded, or obtuse, brown-lepidote when young, changing gray with age; staminate spikes solitary, in caducous bracts or developing leaves, 4–16 cm long; pistillate spikes solitary, axillary, 7–11 cm long, 2 to 4 ovaries developing to fruits, the remaining aborted; involucres oblique-ovoid or subglobose, spines stout, branched, base coalescent; nuts compressed globose, broad-ovoid or trigonous, basal scar very large, convex; edible raw.

Castanopsis sclerophylla (Lindley) Schottky

Ku-zhu=K'u-chu (苦櫧, Bitter Chestnut); *Ku-li=K'u-li* (苦栗, Bitter Chestnut); *Zhui-li=Chui-li* (錐栗, Conic Chestnut). Starch obtained from the nuts, used for preparing cold pudding.

An evergreen tree 5–10 m high, branchlets glabrous; leaves oblong-ovate, 7–14 cm long, 3–5.5 cm wide, acuminate, base oblique obtuse, serrate at the apical half, gray-lepidote when young; staminate spikes solitary, axillary to caducous scales or developing leaves, 7–13 cm long, rachis glabrous, pistillate spikes solitary, 5–13 cm long, involucres spherical-ovoid, 1.2 cm long, 1 cm in diameter, with no evident spines, scales in concentric rings; nuts subspherical, 1–1.4 cm in diameter; starch used for food.

Castanopsis tibetana Hance — MONKEY CHESTNUT

Da-ye-zhui-li=Ta-yeh-chui-li (大葉錐栗, Big-leaved Conic Chestnut); *Gou-kao=Kou-k'ao*

(鈎栲, Hook Castanopsis); *Hou-li=Hou-li* (猴栗, Monkey Chestnut); *Gou-li=kou-li* (鈎栗, Hook Chestnut). Nuts edible.

A gigantic evergreen tree 20–30 m high, branchlets glabrous; leaves oblong, (8–) 15–20 (–30) cm long, (4–) 5–8 (–10) cm wide, apical half coarsely sharp-serrate, base rounded, apex abruptly acuminate, lepidote beneath; staminate spikes axillary to bracts or lower leaves of developing branchlets, 10–15 cm long, glabrous; pistillate spikes solitary, 8–15 cm long, many ovaries developing to mature fruits; involucres densely covered with branched spines, slender, needle-like, 10–15 mm long; nuts conic, 1.5–1.8 cm long, 2–2.8 cm in diameter, brown, hairy; edible raw.

Castanopsis tribuloides (Lindley) A. de Candolle

Ji-li-kao=Chi-li-k'ao (蒺藜栲, Caltrop Castanopsis). Seeds, the content being 80% starch, 2.6% sugar, 3.09% polysaccharides; used both for food and making alchohol in Yunnan.

Large evergreen trees up to 25 m high, branchlets velutinous; leaves elliptic or ovate-elliptic, 10–18 cm long, 4–7 cm wide, acuminate, base oblique-rounded, entire, or the upper one-third remotely serrate, velutinous beneath; flowers small, pale yellowish-green, monoecious, rarely dioecious, developed from the upper axillary buds forming conspicuous panicles above the foliage, staminate spikes 5–15 cm long, erect, pistillate spikes 7–25 cm long, involucre 1-flowered, ovary 3-carpellate, styles 3; the fruiting involucres 2–3 cm in diameter (including the spines), spines sharp, connate at base; nuts globose, 0.7–1 cm in diameter. Widespread in eastern Himalayan Region, growing in forests at 900–1,800 m in Yunnan, thence eastward to Guangdong and southward to Vietnam.

Lithocarpus cleistocarpa (Seemen) Rehder et Wilson

Bao-li-shu=Pao-li-shu (包栗樹, Wrapped Oak). Seeds, roasted.

Evergreen tree 20 m high, branchlets stout, glabrous; leaves oblong-elliptic, 11–17 cm long, 3–6 cm wide, acuminate, base acute, entire, light gray, sparsely lepidote; fruiting branches 10–12 cm long, crowded with many fruits, involucres completely enclosing the nut, 1.5–2 cm long, scales imbricate; nuts compressed globose, 1.2–1.7 cm long, 1.4–1.9 cm in diameter, grayish hairy.

Lithocarpus dealbata (J. D. Hooker et Thomson) Rehder

Bai-li=Pai-li (白栗, Yunnan White Oak). Starch from fruits.

An evergreen tree up to 10 m high, branches densely covered with grayish-yellow hairs; leaves elliptic-oblong, 7–14 cm long, 2.5–5 cm wide, apex caudate or acuminate, base cuneate, entire, grayish-white beneath; staminate spikes 10–15 cm long, in loose panicles, the axes of which seldom bearing small leaves, pistillate spikes 10–15 cm long, 1 or 2 terminal to a branch, occasionally bearing some staminate flowers at the basal portion, the fruiting axes 10–20 cm long, crowded with mature fruits; involucres covering 2/3 or 3/4 of the nuts, densely hairly and with 6 concentric rings of acute scales; nuts broadly conic, 1–2 cm long, 1–1.6 cm in diameter across the base, scars broad, covering 1/4 of the nut. A little known species first reported from Nepal in 1865, current herbarium collections indicating the widespread occurrence from Nepal eastward to northern Burma, Yunnan-Thailand-Vietnam borders; nuts used as food by the ethnic groups locally, a good source species for food in the tropics.

Lithocarpus glaber (Thunberg) Nakai — Glabrous Tanoak (Syn. *Quercus glabra* Thunberg)

Diao=Tiao (椆, Lithocarpus); *Kao=K'ao* (栲, Lithocarpus); *Shi-tou-shu=Shih-t'ou-shu* (石頭樹, Rock Tree). Fruits, used in Hubei.

Evergreen tree 25 m high, branchlets velutinous-hairy; leaves oblong, elliptic-lanceolate, 5–12 cm long, 2–4 cm wide, acuminate-caudate, base acute, entire, rarely with few teeth at the distal end, coriaceous; infructescences 10–14 cm long, villose; involucres shallowly cupular, scales deltoid; nuts ovoid, 1.5–2 cm long, 1 cm across. Widespread in warmer regions of eastern Asia, growing on the sunny slopes at altitudes of 800 m.

Lithocarpus hancei (Bentham) Rehder

Ying-tou-shi-li=ying-t'ou-shih-li (硬頭石栗, Hard-cupped Lithocarpus). Seeds; used for food or for making alcohol in Yunnan.

Tall trees 18 m high, branchlets glabrous; leaves ovate-oblong, 7–14 cm long, 2.5–4.5 cm wide, acuminate-caudate, base cuneate; fruiting spikes 10 cm long, involucres saucer-shaped, scales deltoid; nuts ovoid, 1–2 cm long, 1.2–1.8 cm across. Widespread in subtropical areas of China.

Lithocarpus polystachyus (Wallich) Rehder

Tian-cha=T'ien-ch'a (甜茶, Sweet Tea); *Duo-sui-diao=To-sui-tiao* (多穗椆, Many-spiked Lithocarpus). Young shoots, dried for tea; starch from the nuts used for food.

Evergreen trees 7–15 m high, bark grayish brown, branchlets pubescent when young, glabrescent; leaves obovate-lanceolate or oblong, 8–17 (–20) cm long, 3–6 (–8) cm wide, acuminate-caudate, base cuneate and acute, entire, coriaceous, grayish-pilose beneath, petioles 1.5–2 cm long; flowers greenish-yellow, unisexual, monoecious, sessile, fasciculate in threes on slender spikes, staminate spikes often fasciculate, 7–9 cm long, erect, flowers 2–3 mm across, perianth segments pilose, stamens 8–10, on slender filaments, pistillode lanate, pistillate spikes 11–22 cm long, ovary subtended by scaly involucre, inferior, 3-locular; nuts numerous, cups shallow, scales deltoid, pubescent, glands ovoid, acorns shiny brown, 1.2–1.6 cm long, 1–1.5 cm in diameter. Common in southern China, a dominant species in forests of Guangdong and Guangxi.

Lithocarpus truncatus (King) Rehder et Wilson

Bao-tou-li=Pao-t'ou-li (包頭栗, Turban Oak). Starch obtained from the fruit, used in Yunnan.

Evergreen trees 6–20 m high, branchlets covered with white scales in the grooves; leaves ovate-elliptic 10–30 cm long, 4–8 cm wide, acuminate, base cuneate, entire, glabrous, veins red beneath; staminate panicles and pistillate spikes greenish-yellow, flowers 2–3 mm across; fruiting branches 10–12 cm long, densely covered with many fruits, involucres obconic, truncate at the apex, scales in 6–9 concentric rings; nuts compressed-ovoid, 1.3–2 cm long, 1.3–1.8 cm across the top, hairy. Growing on rock outcrops with pine and oak or on grassy savanna on the borderland of Yunnan-Thailand-Vietnam and northern India.

Quercus acuta Thunberg

Hong-cai-li=Hung-ts'ai-li (紅材栗, Red-wood Oak); *Xue-zhu=Hsüeh-chu* (血櫧, Blood Lithocarpus). Fruit, locally used in Taiwan.

Small deciduous trees 6–10 m high, branchlets densely covered with soft red-brown hairs; leaves glabrous, ovate-lanceolate, 8–20 cm long, 3–6 cm wide, abruptly caudate, entire, acute at the base, petioles and veins beneath red; staminate catkins very hairy, pendulous, axillary to caducous bracts basal to developing shoots, 2–16 cm long, the flowers yellow, several in a fascicle subtended by a membranous hairy, ovate-lanceolate bract 7–9 mm long, pistillate spike erect, short, axillary to normal leaves, the axis 1–2.6 cm long, densely golden lanate, glabrescent, bearing 2 to 5 flowers, involucres shallow, discoid, with 6 concentric rings; nuts oblong-ovoid, 1.5–2 cm long, 8–12 mm in diameter, the basal end with a corky protrusion, the apex with a prominent conic projection representing the persistent style. A native of Japan, introduced to Taiwan during the Japanese occupation, used by the ethnic people there in time of food shortage.

Quercus acutidentata (Maximowicz) Koidzumi

Bai-hua-li-shu=Pai-hua-li-shu (白花櫟樹, White-flowered Oak); *Rui-chi-hu-li=Jui-ch'ih-hu-li* (鋭齒槲櫟, Sharp-toothed Oak). Starch extracted from seed, used in western Hubei.

Deciduous trees 30 m high, branchlets glabrous; leaves oblong, rarely elliptic, 8–21 cm long, 5–9 cm wide, acuminate, base obtuse or rounded, grayish-stellate tomentose beneath; fruiting involucres cupular, 1–1.5 cm across, scales ovate-lanceolate; nuts oblong-ovoid, 1.5–2 cm long, 1–1.4 cm across base. Native to eastern Asia, growing in broad-leaved forests at altitudes of 500–2,100 m.

Quercus acutissima Carruthers

Ma-li=Ma-li (麻櫟, Hemp Oak); *Qing-gang=Ch'ing-kang* (青剛, Green-steel, the Sichuan name, referring to the enduring quality of the wood). Seeds, used as a coffee substitute in Yunnan.

A deciduous tree 15–20 m tall, branchlets hairy, glabrescent; leaves ovate-lanceolate 9–16 cm long, 3–4.5 cm wide, acuminate, base rounded or obtuse, sharply serrate, the teeth awned, lateral nerves 12–18 pairs; flowers monoecious, staminate catkins appearing with the unfolding of the winter bud, subtended by bracts, 5–6 cm long, pendulous, hairy, pistillate spikes consisting 2 or 3 flowers in small fascicles above the middle of the branchlet, in the axils of leaves, involucres cupular, covering half of the nut, 2–3 cm in diameter, bracts linear-lanceolate, 1–1.5 cm long, reflexed; nut oblong-ovate, 2 cm long, 1.5–2 cm in diameter, basal scar convex.

Quercus glauca Thunberg and **Q. schottkyana** Rehder et Wilson

Qing-gang-li=Ch'ing-kang-li (青剛櫟, Green Hard Oak); *Tie-diao=T'ieh-tiao* (鐵稠, Iron Oak); *Tie-li=T'ieh-li* (鐵櫟, Iron Oak). Fruits, gathered and used locally in Taiwan and Yunnan.

Decidious trees 15–20 m high, branchlets hairy, becoming glabrescent; leaves obovate-oblong, 6–13 cm long, 2–5.5 cm wide, abruptly acuminate at apex, base rounded or obtuse, serrate above the middle, glabrous above, covered by white hairs and scales beneath, lateral nerves 7–13 pairs; flowers monoecious, staminate catkins appearing with the unfolding of the winter buds, 4–6 cm long, pistillate spikes erect, stalked, peduncles 5–6 mm long, in the axils of apical leaves of the branchlets, with 2 or 3 flowers close to the apex of the peduncle, glabrous, involucres cupular, covering one-third to one-half of the nut, 9–12 mm in diameter, with 5–8 concentric rings; nuts ovoid-spherical, 1–1.6 cm long, 9–12 mm in diameter, glabrous, basal scar concave.

Quercus mongolica Fischer ex Turczaninow — MONGOLIAN OAK

Meng-li=Meng-li (蒙櫟, Mongolian Oak); *Zuo-li=Tso-li* (柞櫟, Shandong Silk Oak). Starch obtained from seed occasionally used, leaves used in Inner Mongolia for tea, boiled with the fruits of Siberian crabapple; in Shandong the leaves are used for feeding a special kind of silkworm, the source species of the yellowish-brown Shandong silk.

Deciduous trees up to 30 m high, branchlets glabrous, the second year's growth grayish-purple, with elevated lenticels; leaves crowded at the end of current year's growth, obovate or oblong-obovate, (6–) 10–14 (–20) cm long, 3–8.5 (–11.5) cm wide, obtuse or acute, narrowed toward the auriculate base, with 7–10 (–12) broad rounded or acute teeth on each side, pilose on the nerves, or glabrescent; flowers unisexual, the staminate flowers yellow, in pendulous catkins 5–8 cm long, pubescent, perianth segments 6 or 7, membranous, villose-ciliate, stamens 8, pistillate flowers 3–5, in pubescent erect spikes, 1–2 cm long, ovaries 3-celled, enclosed in globose involucres with imbricate ovate woolly scales, styles short, dilated at the distal end, the stigmatic surface on the inside; fruits subsessile, nuts oblong-ovoid, 2–3 cm long, 1–1.8 cm across, 1/3 enclosed by the thick cup, scales tuberculate, grayish-hairy, the upper 2–3 rows thinner, coriaceous, forming a short fringe. Native of northeastern Asia, growing on sunny slopes, introduced into American gardens in 1893.

Quercus phillyreoides A. Gray

Wu-gang-li=Wu-kang-li (烏崗櫟, Black Ridge Oak). Starch obtained from the nuts, eaten in Hubei.

Evergreen shrubs or small trees, 2–10 m high, branchlets stellate-pubescent, hairs yellow; leaves broad-elliptic to obovate-oblong, 2.5–7 cm long, 1.5–3 cm wide, obtuse, or acute, base rounded or subcordate, coriaceous, crenately serrate, often recurved, petioles short, 2–5 mm long, stellate-pubescent; infructescences short and stout, bearing 2 or 3 acorns, involucres hemispherical, 5–8 mm long, 1 cm across, tomentose inside as well as outside, scales deltoid; nuts ovoid, 1.2–2.2 cm long, 1.3–1.4 cm across, hairy at the distal end. Widespread in eastern Asia, growing on rocky hillsides at altitudes of 500–600 m, introduced into western gardens about 1862.

Quercus spinosa David ex Franchet

Ci-ye-li=Tz'u-yeh-li (刺葉櫟, Spine-leaved Oak), *Tie-xiang-shu=T'ieh-hsiang-shu* (鐵橡樹, Iron Oak).

Starch obtained from the nuts, used for food in Hubei.

Evergreen shrubs or small trees 15 m high, branchlets purple-brown, stellate-pubescent; leaves obovate or oblong, 2.5–5 cm long, 1–3 cm wide, apex rounded, base subcordate, coriaceous, entire or serrate, teeth spiny, petioles 2–3 mm long; infructescences 1.5–3 cm long, pubescent, bearing 1 or 2 acorns, involucres cupular, 5–8 mm long, 9–15 mm across, scales deltoid, pilose, hairs gray; nuts oblong, or ovoid, 1–1.5 cm long, 1 cm across. Native to western China, growing on rocky limestone hillsides at altitudes 1,000–2,500 m.

Quercus variabilis Blume

Shuan-pi-li=Shuan-p'i-li (栓皮櫟, Corky Bark Oak); *Bai-ma-li=Pai-ma-li* (白麻櫟, White Oak). Nuts ground, used for making soy sauce, and for preparing a cold pudding.

Deciduous trees 15–25 m high, bark black-brown, cork well developed up to 10 cm thick; branchlets pubescent; leaves white tomentose beneath, lanceolate-elliptic or obovate-oblong, 8–15 cm long, 2–6 cm wide, apex acute, obtuse or acuminate, base rounded, serrate, lateral nerves 14–18 pairs; staminate catkins appearing with the unfolding winter buds, pendulous, 5–8 cm long, golden hairy; pistillate flowers 2 or 3 on stalked axillary naked buds associated with leaves in the middle of a branchlet, involucres cupular, covering 2/3 of the nut, 1.5 cm long, 1.9–2.1 cm in diameter, bracts subulate, each with a median-longitudinal ridge; nuts ovoid-spherical, 1.6–1.9 cm long, 1.3–1.5 cm in diameter, basal scar broad, convex.

Ulmaceae: Elm Family

Aphananthe aspera (Thunberg) Planchon — Muku Tree

Cao-ye-shu=Ts'ao-yeh-shu (糙葉樹, Rough Leaf Tree). Fruits, gathered and used locally in Taiwan.

A deciduous tree, 20 m high, branchlets covered with asperous hairs throughout; leaves ovate or ovate-oblong, 5–13 cm long, 2.5–5.8 cm wide, acuminate, finely serrate, both surfaces rough; flowers monoecious, the staminate ones in dense corymbs basal to the young branchlet, the pistillate flowers in the axils of upper leaves, solitary, perianth segments linear, 5–6 mm long; fruit a small ovoid or subglobose drupe 8–10 mm in diameter, black, with some appressed hairs, the persistent style with 2 stigmas.

Celtis biondii Pampanini

Zi-dan-shu=Tzu-tan-shu (紫彈樹, Purple Bullet Tree) *Cu-ke-lang=Ts'u-k'o-lang* (粗殼榔, Coarse-shelled Celtis). Young shoots, gathered by people in Hubei for food; oil extracted from seed used for cooking.

Deciduous trees (or shrubs) up to 14 m high, branchlets rufous-tomentose; leaves ovate to narrow elliptic, 4–10 cm long, 2–5 cm wide, short acuminate, base obliquely obtuse or cuneate, the distal half crenate-serrate or entire, coarse-hairy, glabrescent; flowers polygamous; fruits 2 or 3, axillary, drupaceous, ovoid, 6 mm in diameter, on stalks 1–1.5 cm long. Native to central China, growing along hillsides at altitudes of 1,000 m; introduced into western gardens in 1894.

Celtis sinensis Persoon — CHINESE HACKBERRY (Syn. *C. japonica* Planchon)

Pu-shu=P'u-shu (朴樹, Chinese Hackberry). Ripe fruits, gathered by children and eaten locally.

Deciduous trees 4–10 m high, trunk light gray, smooth, branchlets hairy; leaves ovate, 3–10 cm long, apex acute, base oblique-obtuse, 3-nerved, serrate above the middle; flowers polygamous, solitary or in cymose axillary clusters, perianth segments 4, hairy, stamens 4, opposite the sepals, stigmas 2; fruit a subglobose drupe, 4–5 mm in diameter, reddish brown.

Hemiptelea davidii (Hance) Planchon

Ci-yu=Tz'u-yü (刺榆, Thorny Elm). Young leaves, gathered in North and East China, mixed with flour, steamed, seasoned with salt; sesame oil and ground garlic added.

Deciduous trees 5–8 m high, branchlets pilose, rigid, spinescent, spines 1.5–10 cm long; leaves elliptic or elliptic-oblong, 2–6 cm long, acute, subcordate at the base, penninerved, nerves 8–15 pairs, petioles very short, 2–5 mm long; flowers polygamous, 1–4 in a cyme, axillary to bracts or lower leaves of a developing branchlet, perianth segments 4 or 5, stamens 4 or 5; fruit a small oblique nutlet, with an oblique wing on the upper half, hence the generic name *Hemiptelea* (half-wing).

Ulmus davidiana Planchon — HAIRY ELM

Hei-yu=Hei-yü (黑榆, Black Elm); *Shan-mao-yu=Shan-mao-yü* (山毛榆, Hillside Hairy Elm). Fully grown but still tender fruits and leafy shoots, eaten by people living in the mountains of northern China.

Medium-sized deciduous trees 10 (–20) m high, current year's growth pubescent, older branchlets with 2 corky wings; leaves broad-obovate or elliptic-ovate, 4–10 cm long, 1.5–6 cm wide, short-acuminate, base rounded, margin double serrate, lateral nerves 12–20 on each side, hairy; flowers fasciculate on last year's growth; samaras obovate, 9–14 mm long, base cuneate; seed close to the apical notch. Native to northern China, growing on dry and exposed limestone hillside in Liaoning, Hebei, and Shanxi; introduced from Hebei to the Arnold Arboretum in 1910–1912.

Ulmus glaucescens Planchon — DROUGHT-RESISTANT ELM

Han-yu=Han-yü (旱榆, Drought-resistant Elm); *Hui-yu=Hui-yü* (灰榆, Gray Elm). Fully grown tender fruits and associated young leaves, eaten in northern China.

Small deciduous trees 5–18 m high, sometimes stunted into dwarf bushes, branchlets glabrous, second year's growth pale-yellow; leaves ovate to elliptic-lanceolate, 2.5–

5 cm long, 0.8–2.5 cm wide (on dwarf trees suborbicular, 8–15 mm long, 6–10 cm wide), acuminate or acute, base oblique-rounded, glabrous, dull bluish-green above, serrate, teeth simple, lateral nerves 7–8 on each side; flowers fasciculate, emerged from lateral winter buds with the terminal vegetative bud giving rise to a leafy branch; samaras elliptic-obovate, unusually large for elms of Chinese origin, 2–2.5 cm long, narrowed at base. Native to the dry hillsides of Inner Mongolia, Ningxia, Gansu, Shaanxi and Shandong; very close to *U. pumila* L., a species distinguished by its suborbicular samaras rounded at the base, only 1.5 cm long.

Ulmus parvifolia Jacquin — CHINESE ELM

Yue-yu=Yueh-Yü (躍榆, Surpass Elm); *Lang-yu=Lang-yü* (榔榆, Lofty Elm). Young leafy shoots, used in same areas and in like manner as the above.

A deciduous tree 15–25 m high, bark gray, smooth, branchlets pubescent; leaves with linear-lanceolate stipules, laminas elliptic, ovate, or obovate, 2–5 cm long, apex acute, base obliquely rounded, pubescent beneath when young, petioles short, 2–6 mm long; flowering in August or September; fruit a compressed nutlet, winged, elliptic-ovate, 1 cm long, glabrous, notched at the apex.

Ulmus propinqua Koidzumi

Chun-yu=Ch'un-yü (春榆, Spring Elm); *Bai-yu=Pai-yü* (白榆, White Elm). Young fruits that attain full size, steamed with flour, then seasoned with salt, sesame oil and galic.

A deciduous tree, 10–20 cm long, branchlets densely hairy; leaves obovate or elliptic, 3–9 cm long, double-serrate, lateral nerves 8–16 pairs, when young sparsely appressed pilose above, villose beneath; flowering before leaves, the flowers fasciculate; samaras 7–15 mm long, glabrous; seed close to the apical notch of the fruit.

Ulmus pumila L. — SIBERIAN ELM

Yu-qian-zi=Yü-ch'ien-tzu (榆錢子, Elm Coin). Young fruits attaining mature size, cleaned, mixed with flour, steamed, and then seasoned with salt, sesame oil, onion and garlic, eaten in northern China.

Yu-shu-pi=Yü-shu-p'i (榆樹皮, Elm Bark). Bark of an elm tree, stripped off in time of famines and eaten to assuage hunger and to prevent starvation.

Large deciduous trees 20–25 m high, bark of trunk rough, branchlets pubescent while young; leaves elliptic or elliptic-lanceolate, 2–7 cm long, acute or acuminate, oblique and obtuse at base, serrate; flowering before leaves, flowers small, inconspicuous, in fascicles; samaras suborbicular, 1–1.5 cm in diameter, with closed apical notch; seed slightly above the middle. The fully developed young fruit is a delight to people having no fresh vegetables for the winter season, eaten raw or cooked (See Part I for more information).

Moraceae: Mulberry Family

Artocarpus altilis (Parkinson) Fosberg — BREADFRUIT

Mian-bao-guo=Mien-pao-kuo (麵包果, Breadfruit). Ripe fruits; introduced, rare in China.

An evergreen tree 10–20 m high, trunk 60 cm in diameter; leaves ovate, 30–90 cm long, deeply pinnate-lobed; flowers monoecious, staminate ones crowded in a stiff clavate yellow spike 15–30 cm long, the pistillate flowers in oblong green heads; multiple fruits 12–20 cm long, weighing 5 kg, outside tough, warty, pulp mealy, sweet. Propagated by cuttings or root-sprouts; introduced in botanical gardens and agricultural institutions, no commercial plantings.

Artocarpus heterophyllus Lamarck (Syn. *A. integer* [Thunberg] Merrill — JACKFRUIT

Bo-luo-mi=Po-lo-mi (波羅蜜, Jackfruit). Ripe fruits.

An evergreen tree 5–15 m high, containing latex; leaves oblong or elliptic, 7–15 cm long, entire, glabrous; flowers monoecious, staminate spike axillary to normal leaves, 5–9 cm long, 2.5 cm in diameter, pistillate spike growing on the tree trunk or on primary branches, maturing into oblong multiple fruits 25–60 cm long, weighing 20 kg, green, covered with hexagonal knobs, with an unpleasant odor, pulp soft, white, sweet, edible raw; seeds edible roasted, resembling chestnut. Introduced into botanical and agricultural institutions, occasionally used in gardens of large hotels as an attraction to tourists.

Artocarpus hypargyreus Hance ex Bentham

Bai-gui-mu=Pai-kuei-mu (白桂木, White Laurel Tree). Ripe fruits, white, soft, very sweet, gathered locally, and eaten raw; farmers in the vicinity of Guangzhou (Canton) used to gather the fruits for sale in town for extra income.

An evergreen tree growing in undisturbed tropical forests, 10 m high, with latex, branchlets covered by rusty hairs; leaves oblong or obovate-oblong, 7–15 cm long, 5–8 cm wide, acuminate, rounded or obtuse at base, entire, white villose beneath; flowers monoecious, mixed with peltate bracts in a compact spherical or obovoid axillary spike on a peduncle 1–3 cm long, staminate heads 1.2–1.6 cm long, perianth segment 2 or 3, pistillate flowers smaller, perianth tubular; multiple fruits oblong-globose, 1.5 cm in diameter, snow white when ripe, delicious.

Artocarpus styracifolius Pierre (Syn. *A. bicolor* Merrill et Chun)

Er-se-bo-luo-mi=Erh-se-po-lo-mi (二色波羅蜜, Two-Colored Jackfruit). Fruits; sweet-sour, used for jam in Hainan Island; no local ethnic name being recorded. The Chinese name here used is a translation of the botanical name proposed by Merrill and Chun, made by W.Y. Chun or his students in the South China Institute of Botany.

Trees up to 20 m high, bark gray, rough, branchlets pubescent; leaves elliptic, oblong or obovate-elliptic, 4.5–7 cm long, 2–2.5 cm wide, acuminate, acumens 1.5–1.8 cm long, base cuneate, entire, those on off-shoots often pinnate-lobed, glaucous and hairy beneath, stipules subulate, 2 mm long; flowers unisexual, axillary, staminate inflorescences ellipsoid, ovoid or cylindrical, 6–12 mm long; multiple fruits globose, 4 cm in diameter. Native to Hainan Island and the mountains on the border of China and Vietnam.

Cudrania cochinchinensis (Loureiro) Kudo et Masamune (Syn. *Maclura cochinchinensis* [Loureiro] Corner)

Chuan-po-shi=Ch'uan-p'o-shih (穿破石, Penetrating Rock); *Gou-ji=Kou-chi* (枸棘, Hooked Spines); *Wei zhi=Wei chih* (萑芝, 偎枝, Cudrania); *Jia li zhi=Chia li chih* (假荔枝, False Litchi). Fruits, eaten raw or used for making wine.

Evergreen shrubs 2–4 m high, bark of root red-yellow, peeling frequently, branchlets grayish-brown, bearing stout recurved thorns 5–10 mm long; leaves oblong or obovate-oblong, 3–8 cm long, 2–3.5 cm wide, obtuse or abruptly short acuminate, base obtuse; flowers small, unisexual, dioecious, crowded into globose heads, solitary or in pairs, 6 mm in diameter, perianth segments 3–5, cuneate, unequal, hairy, pistillate head 1.8 cm in diameter, perianth segments 4, hairy; mature multiple fruits globose, 5 cm in diameter, hairy; achenes covered by fleshy persistent perianth. Common in southern China, thence the range extending southward to tropical Asia, eastern Africa and Australia.

Cudrania tricuspidata (Carrière) Bureau ex Lavallée (Syn. *Maclura tricuspidata* Carrière, *Cudrania triloba* Hance)

Zhe-shu=Che-shu (柘樹, Cudrania); *Zha-gu-ding=Cha-ku-ting* (榨古丁, Pressed Old Nail). Fruits; gathered locally, eaten in the field, not available in the market.

Lactiferous thorny deciduous shrubs or small trees 1–6 m high, bark brown, irregularly peeled, branchlets glabrescent, with hard, sharp, spines 0.5–3.5 cm long; leaves ovate to obovate, 3–8 cm long, 1.5–6 cm wide, acute or acuminate, base rounded or cuneate, entire or sometimes 3-lobed or tricuspidate; flowers small, dioecious, greenish-yellow, in globose heads, perianth of staminate flowers with 4 oblong segments and 2–3 bracts, stamens 4, filaments straight, ovary in pistillate flowers enclosed in the 4-parted perianth; syncarps subglobose, fleshy, 1.5–2.5 cm in diameter, yellow-red, fruits each with a crustaceous pericarp; seeds with thin endosperm, cotyledons folded. Widespread in eastern Asia, often cultivated for hedges; in northern China, farmers used to gather the young unfolding leaves to feed newly hatched silkworms in order to save the growing mulberry leaves; introduced into Western gardens in 1862.

Ficus auriculata Loureiro (Syn. *F. roxburghii* Wallich ex Miquel)

Da-guo-rong=Ta-kuo-jung (大果榕, Large Fruit Fig); *Mu-gua-guo=Mu-kua-kuo* (木瓜果, Tree Melon); *Yuan-ye-rong=Yuan-yeh-jung* (圓葉榕, Round-leaved Bayan); *Bao-zi-rong=Pao-tzu-jung* (包子榕, Wrapped-seed Bayan); *Da-ye-rong=Ta-yeh-jung* (大葉榕, Large-leaved Bayan). Mature fruits gathered, the seed processed and used for preparing jelly in Guangdong and Guangxi.

Trees 3–10 m high, branchlets stout, 1 cm in diameter, glabrescent; leaves ovate or suborbicular, coarsely serrate, 15–36 cm long, 15–27 cm wide, apex obtuse or rounded, base cordate or rounded; flowers (like all the figs) monoecious, enclosed in a hollow fleshy receptacle, depending upon small winged female wasps for fertilization, receptacle obconic, 1 cm long, 2 cm across the truncate end, basal bracts 3–7; mature syncarps obconic, truncate at the apex, 2.5–3 cm long, 3.5–4.5 cm across the top, puberulous, orifice concave.

Ficus awkeotsang Makino (Syn. *F. pumila* L. var. *awkeotsang* [Makino] Corner)

Ai-yu-zi=Ai-yu-tzu (愛玉子, Love Jade Seed). Gelatin purified from seed; recent immigrants from Taiwan introduced the material to Boston.

Creeping evergreen vines trailing on tree trunks or over rocks; branchlet reddish brown, pubescent; leaves oblong or broad-lanceolate, 6–15 cm long, 4–6 cm wide, acute at apex, obtuse at base, entire, hirsute beneath, trinervous; flowers mixed, staminate and gall or pistillate and gall; mature syncarps elliptic-obovoid or oblong, black, 6–10. 5 cm long, 3–5 cm in diameter, basal bracts 3, apical end conic-obtuse.

Ficus beecheyana Hooker et Arnott (Syn. *F. erecta* Thunberg var. *beecheyana* [Hooker et Arnott] King)

Mao-tian-xian-guo=Mao-t'ien-hsien-kuo (毛天仙果, Hairy Immortal Fruit).

Ripe fruit, gathered and eaten locally, not available in the market.

Deciduous trees or tall shrubs 1–8 m high, branchlets densely covered with stiff hairs; leaves ovate-oblong, 7–18 cm long, 3.5–9 cm wide, apex abruptly acuminate, base rounded, entire or inconspicuously remote-serrulate, strigose above, glabrous beneath, trinerved; mature syncarps solitary or paired, globose or pear-shaped, 1–1.8 cm long, basal bracts 3.

Ficus carica L. — COMMON FIG

Wu-hua-guo=Wu-hua-kuo (無花果, Flowerless Fruit). Ripe fruit.

Deciduous shrubs, 3–4 m high, branchlets stout, 4–6 mm in diameter, puberulous; leaves ovate-orbicular, 3- or 5-lobed, obtuse or rounded at apex, cordate, rounded or acute at base, inconspicuously crenate, scabrid above, pilose beneath; fleshy receptacle obovoid, 1.5–3 cm long and across the apical end, sessile or shortly stalked, basal bracts 3 or 4; mature syncarps green or purple, sweet and of good flavor. Rare in China; imported figs occasionally available in large cities; old leaves gathered just before falling, dried and used for tea.

Ficus chlorocarpa Bentham (Syn. *F. variegata* Blume var. *chlorocarpa* [Bentham] King)

Qing-guo-rong=Ch'ing-kuo-jung (青果榕, Green-fruited Fig). Ripe synconium, sweet, eaten in Hainan Island.

Trees 7–10 m high, with spreading crown, bark pale yellow, branchlets pubescent; leaves broad-ovate or oblong-ovate, 10–17 cm long, 6–8 cm wide, acuminate, base rounded, entire, palmately 5-nerved, pilose, glabrescent, petioles 5–6 cm long, stipules ovate, acuminate, 1–1.3 cm long, syconia cauliflorous, globose and constricted into a stalk at the base, 2.5 cm in diameter, apical end slightly compressed, red when ripe; achenes obovoid, verrucose. Native to southeastern China.

Ficus erecta Thunberg

Tian-xian-guo=T'ien-hsien-kuo (天仙果, Celestial Fairy Fruit). Fruit, gathered in the field and eaten locally.

Small deciduous trees or large shrubs up to 4 m high, branchlets glabrous, laticiferous; leaves obovate 10–20 cm long, 4–10 cm wide, apex acute, base obtuse or rounded, entire, glabrous; mature syncarps solitary, axillary, subglobose, 1–1.7 cm across, violet-black, basal bracts 3, stalk 1–2 cm long.

Ficus esquiroliana Léveillé — HONG KONG WILD FIG

Ma-an-shan-tian-xian-guo=Ma-an-shan-t'ien-hsien-kuo (馬鞍山天仙果, Saddle Mountain Fig). Fruit, ripens in March, covered by long yellow hairs, soft, very juicy, delicious.

Small trees 5–8 m high, branchlets densely yellow hispid, stout, 8–11 mm in diameter, hollow; leaves ovate, (10–) 15–23 cm long, (6–) 12–17 cm wide, rarely shallowly 3-lobed, abruptly short acuminate at apex, cordate or rounded at base, evenly dentate, petioles and the nerves beneath densely golden-hirsute, receptacles axillary, solitary or rarely two together, sessile, ovoid, 1.5–2 cm long, densely golden hirsute, basal bracts 3 to 5, foliaceous, acuminate; mature syncarps oblong, 3 cm long, 2 cm in diameter, soft, extremely juicy and sweet, the content can be sucked in, leaving the leathery skin empty. This little known species is an excellent source of fruit tree for planting in tropical hillsides (an original report of personal experience).

Ficus gibbosa Blume

Bai-rou-rong=Pai-jou-jung (白肉榕, White Flesh Banyan). Fruit; gathered locally and used raw.

Rather large tropical evergreen trees 5–20 m high, branchets glabrous; leaves oblong-rhomboid, (4–) 6–10 (–17) cm long, 3–4 (–7) cm wide, abruptly acute at the apex, oblique-obtuse at base, subentire, glabrous and grayish on both surfaces; fleshy synconia pea-sized, abruptly narrowed to a pedicel-like portion with basal bracts; mature syncarps yellow, globose, 1 cm in diameter.

Ficus henryi Warburg

Jian-ye-rong=Chien-yeh-jung (尖葉榕, Pointed-leaved Banyan). Fruits.

Rather small trees 4–12 m high; leaves elliptic-lanceolate, 7–10 (–20) cm long, 2–4 (–6) cm wide, thin papery, acuminate at apex, obtuse or cuneate at base, remotely serrate, glabrous above, densely glandular punctate beneath; fleshy synconia solitary or paired, axillary, ellipsoid, 6 mm long, 3 mm across the middle, sparsely pilose, stalked, bracts 3, triangular, apical to the stalk; mature syncarps globose, 2 cm in diameter. Known only from western Hubei and southern Sichuan.

Ficus oligodon Miquel

Ping-guo-rong=P'ing-kuo-jung (蘋果榕, Apple Fig). Mature fruits, eaten by the people in southeastern Tibet.

Small trees 5–10 m high, cauliflorous, trunks 10–15 cm in diameter, bark gray, smooth; leaves obovate-oblong, 10–25 cm long, 6–13 cm wide, glabrous above, rugose beneath, the distal two-thirds of the margin irregularly serrate, petioles 4–6 cm long; fruits fasciculate, pyriform, 2–3.5 cm across, dark red at maturity, distal end flat. Spontaneous in southwestern China, thence extending southward to Southeast Asia, growing in tropical forests at altitudes of 1,300–1,500 m.

Ficus pumila L. — Climbing Fig, Creeping Fig

Liang-fen-zi=Liang-fen-tzu (涼粉子, Cold Jelly Seed); *Bing-fen-zi=Ping-fen-tzu* (冰粉子, Ice Jelly Seed); *Gui-man-tou=Kuei-man-t'ou* (鬼饅頭, Devil's Steamed-bread); *Bi-li=Pi-li* (薜荔, Wall Litchi). Seed from ripe fruit; gathered, cleaned, soaked for a gelatin, with sugar added, to make a summer dessert.

A vigorously creeping evergreen fig growing attached to rocks, walls, tree trunks by means of exudation from the aerial roots, branches downy; leaves polymorphic, those on the vegetative creeping stem ovate-subcordate, 1–3 cm long, 0.8–2 cm wide, apex obtuse or acute, base oblique cordate, thinly papery, glabrous except on major veins beneath, leaves on erect branches bearing flowers oblong, ovate-oblong or obovate, 3–9 cm long, 1.5–4 cm wide, obtuse, rounded or rarely emarginate at apex, faveolate and pilose beneath; fleshy receptacles ellipsoid-subspherical, 1.5 cm long and across the middle, apex obtuse, densely covered by yellow hairs, pedunculated, bracts 3, apical to the stalk; mature syncarps turbinate, 4–8 cm long, 3–4 cm in diameter across the truncate end.

Ficus sarmentosa Buchanan-Hamilton ex J. E. Smith var. **henryi** (King) Corner (Syn. *F. foveolata* Wallich)

Bing-fen-shu=Ping-fen-shu (冰粉樹, Jelly Tree); *Ya-shi-liu=Ya-shih-liu* (崖石榴, Cliff Pomegranate). Fruits, eaten raw; seed for jelly.

A climbing fig, young branchlets pubescent, hairs brown, becoming grabrescent with age; leaves oblong-lanceolate, 6–21 cm long, 2–6 cm wide, entire, glabrous above, pilose beneath, the reticulum of veins and veinlets elevated beneath, the surface appearing faveolate; fleshy receptacles solitary or paired, axillary, subsessile, glabrescent; mature syncarps subglobose, 1.2–1.5 cm in diameter.

Ficus septica N. Burman

Da-wu-shu=Ta-wu-shu (大無樹, Big Nothing Tree); *Leng-guo-rong=Leng-kuo-jung* (棱果榕, Ridged-fruit Banyan). Fruit edible.

Evergreen trees 6–12 m high, branchlets stout, 4–6 mm in diameter, glabrescent; leaves oblong or obovate-oblong 15–25 cm long, 8–13 cm wide, apex abruptly acuminate, base obtuse or cuneate-acute, entire, punctate and papillose beneath; fleshy receptacles compressed globose, 10 mm long, 15 mm across, evenly white spotted, peduculate, stalks 3–4 mm long, bracts apical to the stalk; mature syncarps compressed-globose, 1.5 cm long, 2 cm across, with 8–11 vertical ridges, white spotted.

Ficus stenophylla Hemsley

Zhu-ye-rong=Chu-yeh-jung (竹葉榕, Bamboo-leaved Fig). Fruit.

Unusual evergreen shrubs 1–2 m high, branchlets pilose; leaves alternate, crowded at the end of the twigs, linear-lanceolate, 3.5–9 (–12) cm long, 0.5–1 (–2) cm wide, apex acuminate, base obtuse, entire, papillose and punctate beneath; fleshy receptacles obovoid, 5–7 mm long, 4–5 mm across, glabrous, apical scales protruding out, pedunculate, with the bracts apical to the stalk; mature syncarps ellipsoid, pointed at both ends, 10–12 mm long, 7–9 mm in diameter across the middle.

Ficus tikoua Bureau

Di-gua=Ti-kua (地瓜, Ground Melon). Ripe fruit; gathered locally and eaten raw.

A creeping fig growing along wayside, rooting at the nodes, branchlets pilose, 1–2 mm thick; leaves obovate-oblong, often panduriform, 2–8.5 cm long, 1.5–5 cm wide, apex acute, base obtuse or cuneate, subcordate, denticulate, sparsely scabrid above, short-hispid and scabrid beneath, petioles 2–5 cm long, often shorter than 1 cm long; fleshy receptacles paired, subspherical, 5–8 mm long, 4–5 mm across, pilose, pedunculate, bracts apical to the stalk; mature syncarps globose, 1.5–2 cm in diameter, red-black, very juicy and sweet.

Morus alba L. — WHITE MULBERRY, COMMON MULBERRY (Figure 21)

Sang-ye=Sang-yeh (桑葉, Mulberry Leaf); *Sang-shen-zi=Sang-shen-tzu* (桑椹子, Mulberry). Leaves; gathered after falling on the ground at the end of the growing season, dried in sun, put in storage, roasted before using, taken as tea by rural people in East and North China; fruit eaten raw.

A deciduous tree, much cultivated for silkworm industry, cut repeatedly for harvesting larger leaves from the suckers, young branchlets pilose, becoming glabrescent; leaves variable in size and texture, generally oblique-ovate, 5–10 (–20) cm long, 4–8 cm wide, apex acute or obtuse, base subcordate, glabrous above, sparsely pilose on the veins beneath; flowers dioecious, emerging before or with the leaves, staminate catkins in fascicles, cylindrical, yellow, 5–10 mm long or slightly longer, pistillate spikes axillary to bracts or leaves of a developing shoot, green, mature fruit an aggregate of numerous achenes each surrounded by the juicy perianth segments, changing from green to red and then to black, white-fruit trees rare in China.

Morus australis Poiret — SOUTHERN MULBERRY

Ji-sang=Chi-sang (雞桑, Chicken Mulberry); *Xiao-ye-sang=Hsiao-yeh-sang* (小葉

桑, Small-leaved Mulberry). Fruits, after turning black, eaten raw or used for making wine.

Deciduous shrubs or small trees 3–8 m high, bark grayish-brown; leaves ovate, 6–15 cm long, 4–12 cm wide, acuminate, base cordate, serrate, occasionally 3- to 6-lobed, scabrid above, pilose beneath, petioles 1.5–4 cm long, flowers small, yellowish green, unisexual, dioecious, staminate catkins 1.5–3 cm long, pendulous, perianth segments 4, stamens 4, pistillode globose, pistillate catkins oblong, erect, ovary ovoid, stigma 2-lobed; multiple fruits fleshy, oblong-ovoid, 1–2 cm long, turning from green to white (very rare) red and black at maturity. Common in southern China, southeastern Asia and India.

Morus nigra L. — BLACK MULBERRY

Hei-sang=Hei-sang (黑桑, Black Mulberry); *Yao-sang=Yao-sang* (藥桑, Medicine Mulberry). Ripe fruits, eaten in Xinjiang.

Deciduous trees 10 m high, bark dark gray, branchlets pubescent; leaves broad ovate, 6–12 (–20) cm long, 5–12 cm wide, acute or abruptly acuminate, base cordate, coarsely serrate, often 2 or 3 lobed, dark green and scabrid above, pubescent along the nerves beneath; flowers green, dioecious, the staminate catkins 2.5 cm long, pistillate ones half as long; syncarps oblong-ovoid, 2–2.5 cm long, dark purplish-red to black. Native to western Asia, cultivated in Xinjiang.

Cannabaceae: Hemp Family

Humulus lupulus L. — HOP

Pi-jiu-hua=P'i-chiu-hua (啤酒花, Beer Flower); *She-ma-cao=She-ma-ts'ao* (蛇麻草, Snake Hemp). Fully grown pistillate inflorescences used in brewing industry for flavoring beer and to give it a slight bitter taste; known to contain resin, aromatic and narcotic elements.

Perennial hispid discourse climbing herbs with several new shoots growing from the underground abstrct trimmed on frames; leaves opposite, ovate, 4–8 cm long and wide, often 3-lobed, serrate, punctate with yellow dots beneath; staminate flowers in panicles; pistillate ones subtended on large in adding-ovary cluster. Young infractescences gathered in the fall while the cup in full dried at 19–20°C, packed for distribution.

Humulus scandens (Loureiro) Merrill (Syn. *H. japonicus* Siebold et Zuccarini) —
JAPANESE HOP

La-la-yang=La-la-yang (拉拉秧, Tangled Vine); *Ju-ju-teng=Chü-chü-t'eng* (鋸鋸藤, Prickly
Vine); *Lu-cao=Lu-ts'ao* (葎草, a pharmaceutical name). Tender shoots for Potherb.

An annual scandent herb, the stem with recurved prickles; leaves alternate, long-
petiolate, petioles 6–14 cm long, prickles recurved, laminas suborbicular in outline,
palmately deep-lobed or 5-parted, 5–12 cm in diameter, segments elliptic, lanceolate or
obliquely elliptic-lanceolate, the central segment 5–9 cm long, 2–4 cm wide, the other
segments smaller; flowers dioecious, the staminate ones yellow, small, in large panicles
10–17 cm long, pistillate flowers in bracteated heads, the bracts green, lanceolate,
glandularly punctate; fruit a bilateral compressed nutlet 5 mm long and wide.

Urticaceae: Nettle Family

Boehmeria nivea (L.) Gaudichaud-Beauprè — RAMIE, CHINA GRASS
Zhu-ma=Chu-ma (苧麻, Ramie). Oil extracted from seed, used for cooking.

Perennial herbs 1–2 m high, branchlets densely villose; leaves ovate, 6–15 cm long,
5–13 cm wide, acuminate-caudate, base rounded or truncate, coarsely dentate-serrate,
lanate beneath, 3-nerved, petioles 2–8 cm long, villose; flowers yellow-green, in axillary
panicles, unisexual, monoecious, protandrous, staminate flowers pale yellow, perianth
segments 4, stamens 4, perianth of pistillate flowers tubular, hairy; achenes oblong, 1
mm long, densely pilose, stigma persistent. Native to China, extensively cultivated
south of the Yangtze River for the useful bast fiber.

Debregeasia edulis Weddell
Shui-ma=Shui-ma (水麻, Water Hemp). Fruit, eaten raw in Taiwan.

A deciduous shrub 2 m high, branchlets densely white pilose; leaves alternate, elliptic-
lanceolate, 4–16 cm long, 1–3 cm wide, apex acuminate, base obtuse, finely serrate,
rough and glandularly punctate, cystoliths in patches above, white tomentose and pilose
on the nerves beneath; flowers minute in globose heads of simple cymose cluster; mature
fruiting head yellow, juicy, 7 mm in diameter.

Debregeasia longifolia (Burman f.) Weddell
Chang-ye-shui-ma=Ch'ang-yeh-shui-ma (長葉水麻, Long-leaved Water Hemp). Fruits,
eaten in central China.

Deciduous shrubs 1–3 m high, branchlets dark brown, strigose; leaves lanceolate or oblong-lanceolate, 6–20 cm long, 0.5–3.5 cm wide, acuminate, base rounded, serrate, rough to touch; flowers unisexual, dioecious, crowded in sessile axillary heads, perianth segments of staminate flowers 4, ovate, 1. 5 mm long, stamens 4, pistillate flowers erect, 2–3 mm across; fruiting clusters fleshy, 4–5 mm across; achenes numerous. Native to tropical Asia, growing along streams in western and central China at altitudes of 1,400 m.

Elatostema edule C. B. Robison

Chi-che-shi-zhe=Ch'ih-ch'e-shih-che (赤車使者, Messenger-of-the-red-cart). Young shoots.

A perennial succulent herb 40–60 cm high, stems zigzag, slightly puberulous; leaves oblique ovate-lanceolate, 10–24 cm long, 5–9 cm wide, subsessile or shortly petiolate, apex caudate, base auriculate on the lower half, acute the upper half, serrate, cystoliths rod-shaped, conspicuous on the upper surface; flowers in compressed globose heads, staminate receptacles sessile or subsessile, 1.5 cm in diameter, pistillate receptacles flat; achenes broad-ellipsoid, 0.5 mm long.

Urtica angustifolia Fischer ex Hornemann (Syn. *U. dioica* L. var. *angustifolia* Ledebour)

Zhe-ma-zi=Che-ma-tzu (螫麻子, Stinging Hemp); *Xia-ye-xun-ma=Hsia-yeh-hsün-ma* (狹葉蕁麻, Marrow-leaved Nettle). Young shoots, boiled, drained, seasoned; eaten by rural people in northern China.

Perennial herbs 50–150 cm high, with stinging hairs, both simple and branched; leaves opposite, oblong-lanceolate or lanceolate, (5–) 8–12 cm long, 1.3–2.5 (–3.5) cm wide, acuminate, base-rounded, serrate, densely punctate with cystoliths, trinerved; flowers small, dioecious, in axillary fasciculate clusters, perianth segments of staminate flowers with 4 opposite stamens, pistillodes cupular, transparent, ovary in pistillate flowers oblong, stigmas penicillate; achenes oblong-ovoid, yellow, enclosed in two thickened persistent perianth segments. Native of cold temperate eastern Asia, occurring in northern China.

Urtica cannabina L.

Xie-zi-cao=Hsieh-tzu-ts'ao (蝎子草, Scorpion Plant); *Xin-ma=Hsin-ma* (焮麻, Burning Hemp); *Ma-ye-xun-ma=Ma-yeh-hsün-ma* (麻葉蕁麻, Hemp-leaved Nettle), *Ha-la-gai=Ha-la-kai* (哈拉蓋, Inner Mongolian ethnic name), Young shoots, used as potherbs in Inner Mongolia.

Caespitose perennial stinging herbs, 1–2 m high, stems ridged, bristly and pilose; leaves ovate in outline, 4–13 cm long and wide, palmatisect, the large segments pinnatifid, lanceolate, bristly on major veins, sparsely pilose on both surfaces; flowers greenish, unisexual, monoecious, in axillary spike-like panicles 12 cm long, perianth segments 4, 1.5 mm long, stamens 4, 2 mm long, pistillode cupular; achenes ovoid, 1.5–2 mm long, covered by bristly persistent perianth segments. A widespread weed, common in Eurasia.

Urtica laetevirens Maximowicz — Stinging Nettle

Zhe-ma=Che-ma (蜇〔螫〕麻, Sting Nettle); *Xun-ma=Hsün-ma* (蕁麻, Nettle); *Kuan-ye-xun-ma=K'uan-yeh-hsün-ma* (寬葉蕁麻, Broad-leaved Nettle); *Hu-ma-zi=Hu-ma-tzu* (虎麻子, Tiger Nettle); *Ha-la-hai=Ha-la-hai* (哈拉海, an ethnic name used in Tibet and Inner Mongolia). Young shoots, used for potherb in southeastern Tibet.

Perennial stinging herbs 0.3–1 m high, sparsely bristly and pilose, the nodes strigose; leaves cordate-ovate, 4–10 cm long, 2–6 cm wide, acuminate, base rounded or subcordate, coarsely dentate-serrate, sparsely bristly and pilose on both surfaces, trinerved, petioles 1.5–7 cm long, the lower ones the shorter; flowers small, greenish, monoecious, in axillary spike-like clusters 8 cm long, perianth segments 4, persistent and surrounding the achenes. Widespread in eastern Asia as a weed.

Urtica triangularis Handel-Mazzetti

San-jiao-ye-xun-ma=San-chiao-yeh-hsün-ma (三角葉蕁麻, Triangular-leaved Nettle); *Huo-ma=Huo-ma* (火麻, Burning Nettle). Young shoots, used for potherb in southeastern Tibet.

Stinging perennial herbs 0.6–1.5 m high, rhizomes stout, 1 cm in diameter, stems purplish, bristly, strigose; leaves deltoid or triangular-lanceolate, 2.5–11 cm long, 1–5 cm wide, acute, base truncate or subcordate, coarsely serrate, the basal teeth often double-serrate, densely bristly and pilose beneath, trinerved, petioles 1–5 cm long, densely bristly and strigose; flowers small, protandrous, monoecious, the staminate ones in axillary panicles, pistillate spike-like, axillary to upper leaves; achenes ovoid, slightly compressed, 2 mm long, brown, covered by bristly persistent perianth segments. Native of southwestern China, growing among grasses in partially shaded hillside at altitudes of 3,200–3,500 m.

Urtica urens L. — Burning Nettle

Ou-xun-ma=Ou-hsün-ma (歐蕁麻, European Nettle). Young shoots, used as potherbs in southeastern Tibet.

Soft-stemmed annual herbs 10–60 cm high, stems with abundant stinging bristles, middle internodes 3–6 cm long; leaves elliptical-oblong, 1.2–6 cm long, 1–3 cm wide, obtuse, base rounded, sharply serrate with 6–11 spreading teeth on each side, sparsely bristly on both surfaces, palmately 5-nerved; flowers small, unisexual, monoecious, in axillary spikes 0.5–2 cm long, perianth segments 4, strigose on the outside; achenes ovoid, 1.8 mm long, slightly compressed, covered by bristly persistent perianth segments. Widespread in Eurasia, growing along paths in forests and near houses at altitudes of 2,800–2,900 m in southeastern Tibet.

Proteaceae: Protea Family

Helicia nilagirica Beddome

Mu-zhu-guo=Mu-chu-kuo (母豬果, Hog Fruit); *Shan-hu-lu=Shan-hu-lu* (山葫蘆, Mountain Langenaria). Fruits, a favorite of the Kachin ethnic group in Yunnan.

Trees 5–10 m high, branchlets velutinous; leaves obovate-oblong, elliptic or oblong-lanceolate, (5–) 10–17 (–23) cm long, (2.8–) 4.5–9 cm wide, acuminate, base cuneate, entire or coarsely serrate; flowers yellow or white, paired and in velutinous racemes 12–16 cm long, bracts 1–2 cm long, perianth 4-lobed, 1.2–2 cm long, recurved, disk 4-lobed, stamens 4, ovary unilocular; fruits compressed globose, 25–35 cm long, 4 cm across, pericarp woody, rough, 4 mm thick. Native to eastern Himalayan Region, growing among trees on sunny hillside at altitudes 1,100–2,100 m.

Heliciopsis terminalis (Kurz) Sleumer (Syn. *H. henryi* [Diels] W. T. Wang)

Lao-shu-he-tao=Lao-shu-ho-t'ao (老鼠核桃, Field Rat Walnut); *Jia-shan-long-yan=Chia-shan-lung-yen* (假山龍眼, False Protea). Seeds, eaten as nuts by the people living in Hainan Island.

Trees 4–8 (–15) m high, trunk 1.5 m across the base; leaves lanceolate or oblong, 13–34 cm long, 4–9 cm wide, obtuse or rounded, base attenuate, entire, glabrous; flowers white, unisexual, dioecious, perianth segments 4, reflexed, stamens 4, ovary superior, style clavate, stigma punctiform; fruits subglobose, 3–4 cm long, 2.5 cm across, pericarp coriaceous, mesocarp spongy and fibrous, endocarp woody, rough; seed solitary, coat thin, cotyledons fleshy. Native to eastern Himalayan Region, growing in Hainan Island, Guangxi, southern Yunnan and the adjacent areas in Vietnam, Thailand and Burma.

Macadamia ternifolia F. Mueller — MACADAMIA NUT; QUEENSLAND NUT

Ma-ka-dan-ren=Ma-k'a-tan-jen (馬卡丹仁, Macadamia Kernal); *Ao-zhou-jian-guo=Ao-chou-chien-kuo* (澳洲堅果, Australian Nut). Kernel, eaten as nut.

Evergreen trees 10–20 m high, branchlets pilose; leaves subsessile, in whorls of 3 or 4, lanceolate or oblong-oblanceolate, 12–36 cm long, 2.5–5.5 cm wide, acute, base rounded, coarsely dentate-serrate, teeth prickly; flowers pink, densely covered by brown hairs, paired, 300–500 in axillary racemes longer than the subtending leaves, pedicels 5–6 mm long, mature flower buds clavate, 8 mm long, velutinous, anthesis beginning with the splitting of one side of the perianth, showing the middle of a bending style, perianth segments 4, divided to the base, recurved and rolling outward, exposing the 4 stamens, filaments lining the inside of the perianth segments, gibbous and verrucose throughout, ovary ellipsoid, densely velutinous, style velutinous below the middle, stigma clavate; fruits follicular, exocarp smooth and ligneous, mesocarp fibrous, endocarp papery; seed 1, spherical, 1.5–2.5 cm in diameter, nut-like, containing two large cotyledons, the commercial "Macademia Nut". Native to the lowlands of northeastern Australia; introduced to Hawaii in 1885, with the encouragement of the government, becoming an important orchard crop there; introduced into the botanical gardens in Guangdong, Hainan Island, and Yunnan as a drought resistant species for afforestation and for food in tropical China.

Loranthaceae: Mistletoe Family

Helixanthera parasitica Loureiro

Shu-shang-cha=Shu-shang-ch'a (樹上茶, Tea-on-Tree); *Wu-ban-ji-sheng=Wu-pan-chi-sheng* (五瓣寄生, Five-segmented Parasite). Leaves, used as a tea substitute in southern Yunnan.

Woody plants, parasitic on trees, branches up to 1.5 m long, loose, glabrous, turning black on drying; leaves opposite, papery, ovate, rarely ovate-lanceolate, 5–12 cm long, 3–4.5 cm wide, acute or obtuse, rarely short-lanceolate, base obtuse or rounded, petioles 5–12 mm long; flowers red-pink-yellow, subsessile, 40–60 in solitary or paired axillary spike-like racemes 5–10 cm long, the subtending leaves or bracts often absent, bracts ovate, receptacles obovoid, 1.5–2 mm long, sepals cupular, adnate to the inferior ovary, petals 5, 6–8 mm long, erect in bud, spreading and recurve at anthesis, stamens inserted to the middle of the petals, ovary unilocular, uniovulate, style 4–6 mm long; fruits red, oblong, 6 mm long, 4 mm across, densely papillose. Widespread in subtropical Asia, growing on oaks, chestnuts, camphors, tung-oil trees and various wild figs.

Santalaceae: Sandalwood Family

Buckleya graebneriana Diels

Mian-weng=Mien-weng (麵瓮, Flour Jar). Fruits, rich in starch, edible after cooking, also used for making wine.

A deciduous semiparasitic shrub 2 m high, branchlets pubescent; leaves opposite, oblong-obovate, 2–8 cm long, 1–3 cm wide, acute, pubescent on the nerves of both surfaces; flowers dioecious, the staminate flowers 4-merous, in terminal cymose clusters, pistillate flowers solitary at the shoot apex, ovary inferior, puberulous; fruit drupe-like, oblong, 1–5 cm long, yellow, persistent perianth segments 4, 1 cm long.

Buckleya henryi Diels

Mi-mian-weng=Mi-mien-weng (米麵瓮, Rice Flour Jar). Fruit.

A semiparasitic shrub 1 m high, glabrous; leaves opposite, papery, elliptic-lanceolate, 2–8 cm long, 1–3 cm wide, reddish-brown above; flowers dioecious, staminate cymes axillary, pistillate flowers solitary, ovary inferior, 1–1.5 cm long, persistent perianth segments 4, elliptic, 1 cm long.

Pyrularia edulis A. de Candolle

You-hu-lu=Yu-hu-lu (油葫蘆, Oil Bottle Gourd); *Tan-li=T'an-li* (檀梨, Sandalwood Pear). Fruit, used for extracting edible oil.

Deciduous shrubs or small trees 3–10 m high, young branchlets villose, becoming glabrescent; leaves alternate papery, ovate-oblong, 5–13 cm long, 3–6 cm wide, pubescent beneath; flowers polygamous, perianth segments 5 or 6, pubescent on the outside; ovary inferior; drupes pear-shaped, 4–5 cm long including the stalk, 2–5 cm in diameter.

Pyrularia sinensis Wu

Hua-tan-li=Hua-t'an-li (華檀梨, Chinese Pyrularia). Fruit used locally.

A deciduous shrub closely related to the above species, distinguished by having smaller leaves and oblong drupes; occurring in Sichuan.

Opiliaceae family: Opilia Family

Urobotrya latisquama (Gagnepain) Heipko

Shan-jie-lan=Shan-chieh-lan (山芥藍, Mountain Mustard Blue); *Lin-wei-mu=Lin-wei-mu* (鱗尾木, Scaly-tailed Shrub). Young inflorescences and flowers, gathered by people living in the mountains of southeastern Guangxi, and used as a delicacy.

Evergreen shrubs or small trees 4–7 m high, bark rough, yellowish-brown; leaves oblong-linear or linear-elliptic, 7–13 cm long, 2.5–5 cm wide, acuminate, base attenuate, acute, entire, coriaceous, glabrous, dried rugose, petioles 3–5 mm long; flowers yellow, tetramerous, in pseudoracemes 10–20 cm long, cauliflorous, before anthesis covered by imbricate bracts, every 2–4 flowers fasciculate and subtended by a caducous rhomboid, acute, ciliate bract 7–8 mm across, calyx obscure, perianth segments 4, oblong, distinct, reflexed, stamens 4, opposite perianth segments, filaments filiform, disk fleshy, ovary superior, cylindrical, unilocular, style 0, stigma truncate; fruits drupaceous, oblong, 2 cm long, 1 cm across. Native of the mountains on the Guangxi-Vietnam border, growing in shade along forest-margins in valleys.

Olacaceae: Olax Family

Erythropalum scandens Blume

Cai-teng=Ts'ai-t'eng (菜藤, Vegetable Vine); *Xing-teng=Hsing-t'eng* (腥藤, Fish-smelling Vine); *Niu-er-teng=Niu-erh-t'eng* (牛耳藤, Ox Ear Vine); *Chi-cang-teng=Ch'ih-tsang-t'eng* (赤蒼藤, Red-green Vine). Young shoots, used as potherbs in Guangxi.

Large lianas with fishy smell, trunk 9 cm across, climbing by tendrils; leaves ovate-deltoid, 6–20 cm long, 5–15 cm wide, acute, obtuse or rounded, base truncate or subcordate, palmately trinerved, petioles 3–6 cm long; flowers very small, cream-yellow, 3 mm across at anthesis, in axillary dichotomously branched cymes, calyx shallowly 5-toothed, corolla 5-parted to the base, the segments deltoid, 1 mm long and across the base, stamens same number and opposite the petals, anthers subsessile, ovary inferior, unilocular, style short, stigma 3-lobed; drupes oblong-ellipsoid, scarlet red, 1.5 cm long, 1 cm across, crowned with persistent calyx teeth and style. Native of southern China, extending eastward to the Philippines and southward to Southeast Asia, growing in shady areas of forests along ravines.

Balanophoraceae: Balanophora Family

Balanophora involucrata J. D. Hooker

Wen-wang-yi-zhi-bi=Wen-wang-i-chih-pi (文王一支筆, A Pen-of-King Wen); *Jie-mu-huai-tai=Chieh-mu-huai-t'ai* (借母懷胎, Fetus-in-Surrogate Mother); *Guan-yin-lian=Kuan-yin-lien* (觀音蓮, Goddess-of-Mercy Lotus); *Tong-qiao-she-gu=T'ung-ch'iao-she-ku* (筒鞘蛇菇, Sheathy Snake Mushroom); *Ji-sheng-huang=Chi-sheng-huang* (寄生黃, Yellow Parasite). Whole plants, gathered in autumn, dried, pulverized, taken for breakfast by sprinkling repeatedly on the top of a hard-boiled egg, a special food given to the aged persons with irregular heartbeats.

Yellowish-red fleshy parasites 5–15 cm high, glabrous throughout, having subterraneous tubers on rhizomes connected to the host by suckers; flowering stems springing up from oblong or globose tubers, each aerial stem bearing a median sheath, 5-lobed, lobes deltoid, 1–2 mm long; flowers numerous, small, unisexual, staminate flowers with 3–4 perianth-segments and 3 or 4 stamens, pistillate flowers naked, ovary oblong, unilocular, uniovulate, ovules without integument; fruits nut-like, seed with endosperm. Widespread in eastern Himalayan Region, growing in forests along streams in Yunnan, Sichuan, Guizhou, western Hubei and adjacent areas in Shaanxi; a species with interesting trivial names showing the deep knowledge of the mountainous people about the parasitic phenomenon of the plants, and comparing the sudden springing out of a flowering shoot from the tuber to the birth of a baby by surrogate mother.

Polygonaceae: Knotweed or Buckwheat Family

Coccoloba uvifera (L.) L. — Sea Grape

Hai-pu-tao=Hai-p'u-t'ao (海葡萄, Sea Grape). Fruit; newly introduced, rare.

A small tree 6–10 m high, branchlets glabrous; leaves coriaceous, orbicular or subobovate, 10–12 cm long, 16–18 cm wide, obtuse or retuse at the apex, cordate at base, margin undulate, veins red, ocreae funnelform, 1 cm long; flowers white, fragrant, 3–4 mm across, stamens 8, ovary trigonous-ovoid, styles 2; fruit enclosed in the fleshy persistent perianth, spherical-ovoid, 1–2 cm in diameter, red-purple. A native of tropical America, introduced to Taiwan, cultivated in gardens and parks.

Fagopyrum cymosum (Trevisan) Meissner (Syn. *Polygonum cymosum* Trevisan)

Jin-qiao-mai=Chin-ch'iao-mai (金蕎麥, Golden Buckwheat); *Qiao-mai-qi=Ch'iao-mai-ch'i* (蕎麥七, Buckwheat Hematinic), *Qiao-mai-dang-gui=Ch'iao-mai-tang-kui* (蕎麥當歸, Buckwheat Angelica). Whole plant, gathered in the fall in central China, dried and used for tea; highly esteemed locally.

Perennial herbs with stout woody, brown rootstocks yellow inside, stems 1 m high, much branched after anthesis, glabrous, hollow; leaves deltoid-cordate, 4–12 cm long and wide across the base, acuminate, base cordate or truncate, entire, pilose along the veins, ocreae membranous, tubular, 5–15 mm long, glabrous; flowers greenish-white, small, in terminal and axillary panicles, bracts triangular-lanceolate, 3 mm long, perianth-segments 5, oblong, stamens 8, ovary ovoid, styles 3; achenes trigonous-pyramidal, 5 mm long, shiny, black. Native to central and western China, growing along streams and on hillsides at low altitudes, about 500 m.

Fagopyrum esculentum Moench — Buckwheat

Qiao-mai=Ch'iao-mai (蕎麥, Buckwheat); *Tian-qiao=T'ien-ch'iao* (甜蕎, Sweet Buckwheat). Seed, cultivated in lowland.

An annual herb 40–80 cm high, cultivated for the grain, branching freely, the branches smooth, green or maroon; leaves triangular-ovate 5–10 cm long, 3–10 cm wide across the hastate base, entire, scabrid on the large nerves and along the margin, ocreae membranous, tubular, sheathing, truncate, 8–10 cm long; flowers pink-white, in axillary cymose clusters; stamens 8, styles 3; achenes trigonous-ovoid, 5 mm long, 3 mm across the base, pericarp smooth, yellowish-brown, the angles sharp. A minor crop in the plains of North China.

Fagopyrum tataricum (L.) Gaertner — TARTARY BUCKWHEAT

Ku-qiao=K'u-ch'iao (苦蕎, Bitter Buckwheat); *Ku-qiao-mai=K'u-ch'iao-mai* (苦蕎麥, Bitter Buckwheat). Seed, cultivated by people living in the high mountains of Sichuan-Tibetan borders, flour tastes slightly bitter.

Annual herbs 50–90 cm high, branchlets striate, purplish-green; leaves triangular-ovate, 2–7 cm long, 2,5–8 cm wide, abruptly acute, base cordate, entire, ocreae membranous, yellow-brown; flowers white-pink, in axillary clusters, perianth 1 mm long, stamens 8; achenes trigonous, elongate-ovoid, 6 mm long, 2 mm wide across the base, apex obtuse, pericarp rough, angles rounded, black; the starch having a bitterish taste, hence the local name "Bitter Buckwheat"; a crop grown by the ethnic groups living in the mountains in western China, westward to Nepal.

Oxyria digyna (L.) J. Hill — MOUNTAIN SORREL

Shen-ye-shan-liao=Shen-yeh-shan-liao (腎葉山蓼, Kidney-leaved Sorrel). Young plants, rich in Vitamin C, used in soup, with little cooking.

Very sour, perennial low herbs 10–20 cm high; leaves radical, long-petiolate, lamina reniform or suborbicular, 1.5–3 cm long, 2–4 cm wide, cordate, entire, glabrous; scape 15–20 cm long, flowers paniculate, perianth light green, stamens 6, styles 2; achenes compressed, winged.

Polygonum cuspidatum Siebold et Zuccarini

Hu-zhang=Hu-chang (虎杖, Tiger Stick); *Gang-ye-wu-zi=Kang-yeh-wu-tzu* (剛業霧子, Steel Leaves Fog Seed). Tender shoots, eaten as asparagus in western Hubei.

Perennial herbs with stout yellow rootstock, annual growth erect, 1–3 m high, spring sprouts fleshy, asparagus-like, irregularly mottled red, purple and green, much branched; leaves ovate-oblong, 5–12 cm long, 4–9 cm wide, abruptly acuminate-caudate, base rounded, entire; flowers white, unisexual, dioecious, in axillary fasciculate pseudoracemes or panicles, perianth-segments 5, outer 3 cucullate, the midrib developing into wings after anthesis, enveloping the fruit in pistillate flowers, stamens 9 in staminate flowers, exserted, staminodes in pistillate flowers 8 or 9, well developed, with sterile anthers included, ovary trigonous; achenes trigonous-pyramidal, shiny black. Widespread in eastern Asia, naturalized and becoming weedy in eastern North America.

Polygonum flaccidum Roxburgh

Shui-liao=Shui-liao (水蓼, Water Smartweed). Young shoots used as spice.

Annual herbs 80–100 cm high, branchlets pubescent, becoming glabrescent, nodes enlarged; leaves lanceolate, 3–8 cm long, 1–2 cm wide, the lower ones disintegrated at the flowering stage, green with a darker patch just below the middle, ocreae tubular, 1.2–1.5 cm long, pubescent, ciliate with bristles 1 mm long; flowers red, in 3–10 terminal loose racemes 1–10 cm long, achenes trigonous-ovoid, 2 mm long, pericarp reticulate.

Polygonum hydropiper L. — COMMON SMARTWEED

La-liao=La-liao (辣蓼, Pungent Smartweed); *Shui-liao=Shui-liao* (水蓼, Water Smartweed). Young plants, 15 cm high, with pungent taste like pepper, chopped and used as spice in northern China; one of the vernacular names reported same as that given to *P. flaccidum* Roxburgh of southern China; apparently young plants of both species are used by people in different localities for similar purposes.

Annual herbs 30–80 cm high, branchlets glabrous; leaves lanceolate, accuminate at both ends, punctate with dark glands on both surfaces, ocreae tubular, 1 cm long, pilose, ciliate with bristles 1–4 mm long; flowers greenish-pink, 3–5 in fascicles arranged in loose terminal or axillary pseudoracemes 6–8 cm long, perianth segments 4–5, punctate with purple-red glandular dots, stamens 6, ovary compressed ovoid, style 2- or 3-fid; achenes lenticular or trigonous, 2–3.5 cm long, dark brown. Widespread in Eurasia, growing in damp places, naturalized in North America.

Polygonum multiflorum Thunberg ex Murray

He-shou-wu=Ho-shou-wu (何首烏, Black Hair Ho). *Ye-jiao-teng=Yeh-chiao-t'eng* (夜焦籐, Night Shrivel Vine).

Perennial woody climber; stem vinish, 3–4 m long, branched, basal portion lignified. Leaves alternate, petiolate. Stipules tubular, surrounding the stem; laminas ovate, subcordate, 5–7 cm long, 3–5 cm wide, apex acuminate. Flowers white, small in terminal panicles; perianth 5-parted, enlarged in fruit; outer 3 segments thickened, winged at the back; stamens 2; styles 3. Achenes black, oblong-trigonous. Widely distributed from the Yellow River, to the Pearl River Region; the large underground rhizome used in Chinese medicine as tonics; the story goes as this, 'Poor man Ho was too sick to escape famine, dug wild root to stop starvation; the rhizome of this species saved his life; continuing to eat the rhizome, he was cured from illness and had black hair in old age; people called him and the vine *He-shou-wu*=Black Hair Ho; all tried to use the rhizome for the health purposes.

Polygonum viscosum W. Jameson

Xiang-liao=Hsiang-liao (香蓼, Fragrant Smartweed). Young plants, used as spice.

Annual herbs 50–100 cm high, branchlets covered by glandular hairs; leaves lanceolate, 5–13 cm long, 1.5–3.5 cm wide, apex acuminate, base cuneate, both surfaces strigose, ocreae tubular, membranous 7–15 mm long, hairy; flowers pink, in cylindrical terminal clusters, peduncles hairy; stamens 8, styles 1, achenes trigonous-ovoid, 2.5–3 mm long, black, shiny.

Rheum nanum Sievers — SMALL RHUBARB

Ai-da-huang=Ai-ta-huang (矮大黃, Short Rhubarb); *Ba-zhu-na=Pa-chu-na* (巴朱納, sound of an Inner Mongolian name). Starch obtained from the caudex and roots, used for food by the cattleman living on the steppes in northwestern China, Inner Mongolia and Xinjiang.

Perennial herbs 15–25 cm high, with stout straight tap root; radical leaves suborbicular, 8–10 cm in diameter, palmately nerved, stellate-pubescent, stipules foliaceous; flowers small, in terminal panicles, perianth segments in 2 whorls, the outer ones ovate, 4.5 mm long, 3 mm wide, the inner ones smaller; achenes black-brown, 10–12 mm long and wide, including the wing. Native of the Gobi Desert, growing on gravel ground.

Rheum officinale Baillon, and **R. palmatum** L. — CHINESE RHUBARB

Da-huang-gan-gan=Ta-huang-kan-kan (大黃竿竿, Rhubarb Stalk). Petioles, chewed by medicine collectors who stayed temporarily in the alpine meadows, leaf-blades poisonous.

Perennial herbs 1–1 5 cm high; rhizomes stout, yellow; radical leaves long petiolate, the petioles terete, fleshy, 15–30 cm long, 2–1 cm thick, laminas ovate-suborbicular, 30–60 cm long, and wide across the base, shallowly or deeply palmately lobed, the apex acute, base cordate, pilose beneath; ocreae tubular, membranous; flowers green-white, in terminal panicles, perianth segments 6, stamens 9, styles 3; achenes winged. Cultivated in the alpine meadows where wild plants grew, the rootstock harvested the fourth year or older, used in traditional Chinese medicine.

Rheum rhabarbarum L. (not *R. rhaponticum* L., which is often misapplied to this species) — RHUBARB

Da-huang-ye-bing=Ta-huang-yeh-ping (大黃葉柄, Rhubarb). Introduced into gardens of agricultural colleges and private citizens to meet special demands of foreign visitors, rare.

Perennial herbs, the flowering stems 1–1.5 m high, radical leaves in a rosette, petioles subterete, fleshy, 15–30 cm long, 3 cm thick, slightly flattened at the basal half; laminas ovate-suborbicular, 30–50 cm long and wide, shallowly lobed, base cordate, margin wavy. Native to the northeastern provinces of China, widely cultivated in temperate countries for the very sour edible petioles; propagated by rootstock division, flowering stems removed immediately when they appear in the field to ensure better harvest the next season.

Rheum undulatum L.

Bo-ye-da-huang=Po ych ta-huang (波葉大黃, Wavy-leaved Rhubarb). Leaves of cultivated plants used in northeastern China for vegetable.

Acaulescent perennial herbs 60–80 cm high at anthesis, rootstocks stout, 8–12 cm in diameter; radical leaves triangular-ovate, 10–16 cm long, 8–14 cm wide, obtuse, base subcordate, margin undulate, 5-nerved, petioles subcylindrical, 7–12 cm long, stipules sheathy, persistent, lower cauline leaves petiolate, upper ones sessile; flowers small, pedicillate, in terminal panicles, perianth segments 6, 2–3 mm long, outer 3 thick and short, reflexed after anthesis, stamens 9, ovary trigonous-ovoid, styles 3; achenes trigonous, winged. Native to northeastern China and adjacent Siberia.

Rumex acetosa L. — GARDEN SORREL

Suan-mo=Suan-mo (酸模, Sorrel); *Hong-suan-mo=Hung-suan-mo* (紅酸模, Red Sorrel); *Niu-er-da-huang=Niu-erh-ta-huang* (牛耳大黃, Ox-ear Rhubarb); *Jin-bu-huan=Chin-pu-huan* (金不換, No Exchange-for-Gold). Young leaves used as potherb.

Perennial sour herbs 50–100 cm high; leaves rather succulent, both basal and cauline, basal radical leaves sagittate, on slender elongated petioles 4–15 cm long, laminas oblong to broadly lanceolate, 5–9 cm long, 2–3.5 cm across, obtuse or acute, base sagittate, entire, cauline leaves subsessile, ocreae oblique-tubular, membranous; flowers small, unisexual, dioecious, green, tinged red, perianth-segments 6 in 2 whorls, oblong, the inner 3 larger, stamens 6 in staminate flowers, the inner segments of the pistillate flowers enlarged after anthesis, veiny, enclosing the fruit, ovary ovoid, styles 3; achenes trigonous, shiny black. Widespread in fields, meadows and hillsides in cold temperate northern hemisphere.

Rumex patientia L. — Monk's Rhubarb, Patience Dock

Ba-tian-suan-mo=Pa-t'ien-suan-mo (巴天酸模, Patience Dock). Young leaves and leafy shoots; reported being eaten only by people living in Hebei.

Perennial herbs 1–1.5 m high, branching after flowering, rootstocks stout, 1.7 cm across; radical and lower cauline leaves oblong or oblong-lanceolate, 15–30 cm long, 4–15 cm wide, obtuse or acute, petiolate, upper cauline leaves sessile, ocreae 2–5 cm long; flowers small, in fascicles forming large terminal panicles, perianth segments 6, the inner 3 enlarged, cordate after anthesis, 5–6.5 mm across, reticulate; achenes trigonous, brown, 3 mm long, enclosed in persistent perianth widespread in Eurasia, growing in northern China in waste places; the Chinese name being the translation of the common English "Patience Dock", apparently the ethnobotanical significance to Chinese people is slight, only meaningful to people who use English.

Chenopodiaceae: Goosefoot Family

Agriophyllum arenarium Beiberstein

Sha-peng-mi=Sha-p'eng-mi (沙蓬米, Sand Weed Grain); *Dong-qiang=Tung-ch'iang* (東廧, Eastern Wall); *Sha-peng=Sha-p'eng* (沙蓬, Sand Weed). Flour made of seed.

Annual herbs 15–60 cm high, branches covered by branched hairs; leaves sessile, lanceolate or linear, 1.3–8 cm long, 4–10 mm wide, with 7 strong vertical nerves running from the base upward, apex acuminate and strongly spinescent; flowers small, in axillary spiny and hairy fascicles, bracts each terminate with a strong spine, perianth membranous; fruits crowded in a head with the spiny persistent bracts, utricle compressed ovoid, narrowly winged; seed millet-like, compressed globose, smooth, 1–5 mm in diameter.

Atriplex hortensis L. — Orach, Garden Orach, Mountain Spinach

Shan-bo-cai=Shan-po-ts'ai (山波菜, Mountain Spinach). Young shoots, cooked as potherb.

Annual herbs 60–100 cm high, branchlets glabrous; leaves alternate, triangular-ovate, 5–10 cm long, 2–6 cm wide, dark green, mealy beneath, entire; flowers unisexual, staminate flowers yellow, in interrupted terminal panicle of spikes, perianth segments 5, pistillate flowers 2 or 3 in axillary fascicles, each flower with two free green bracts 6–7 mm long, 4–5 mm wide; fruit a utricle, enclosed in the persistent herbaceous bracts; seed compressed orbicular-discoid, embryo curved.

Atriplex rosea L. — SALTBUSH

Luo-la=Lo-la (落拉, a Xingjiang local name). Young shoots, cooked, and consumed by people in Xingjiang, not available in cities.

Annual herbs 30–50 cm tall, the entire plant gray and scurfy; leaves oblong or ovate, 2–4 cm long, 5–12 mm wide; flowers in lateral and terminal interrupted spikes, the staminate ones 8–15 cm long, the pistillate flowers 2 or 3 fasciculate, axillary to sessile foliaceous bracts, bracts subtending individual flowers connate at the base, the free portion dentate, 4 mm long, rhomboid; seeds compressed globose, enclosed in the persistent bracts.

Beta vulgaris L. — BEET, CHARD

Ciela Group: **B. vulgaris** var. **cicla** L. — CHARD, SWISS CHARD

Jun-da-cai=Chün-ta-ts'ai (君達菜, Chard); *Hou-pi-cai=Hou-p'i-ts'ai* (厚皮菜, Thick Skin Vegetable). Green leaves. Plants without swollen roots; cultivated for vegetable in southern China.

Hong-hou-pi-cai=Hung-hou-p'i-ts'ai (紅厚皮菜, Red Swiss Chard); *Huo-yan-cai=Hou-yan-ts'ai* (火焰菜, Fire Tongue Vegetable). Red leaves.

Crassa Group: **B. vulgaris** var. **rapa** Dumont — SUGAR BEET, GARDEN BEET

Tang-luo-bo=T'ang-lo-po (糖蘿蔔, Sugar Beet); *Tian-cai=T'ien-ts'ai* (甜菜, Sweet Vegetable). Roots yellowish-white, oblong or abconic; plants producing swollen fleshy roots; cultivated in northeastern China for sugar industry.

Tian-cai=T'ien-ts'ai (甜菜, Garden Beet). Roots purple-red, compressed globose, cultivated near large cities to supply foreign residents and used as a vegetable.

Annual herbs up to 1.4 m high at flowering time; leaves ovate or oblong-ovate, 30–35 cm long, 25 cm wide, ruffled, varying from light green to purple-red, the basal leaves forming a rosette, long petiolate, 15–20 cm long, 1–2 cm wide, the cauline ones progressively reduced in size upward; flowers small, green, in clusters of 2 to 8, hermaphrodite, perianth segments 5, stamens 5, ovary semi-inferior; fruits dry, irregular, developed from aggregates of several flowers cohering at base. Native to Europe and the Mediterranean Region, many cultivars; the following groups cultivated in China.

Chenopodium album L. — LAMB'S QUARTER, PIGWEED, GOOSEFOOT

Hui-hui-cai=Hui-hui-ts'ai (灰灰菜, Gray Vegetable); *Li=li* (藜, Pigweed). Young shoots gathered and washed, and after rubbing off the farinose powder, washed again, cooked as potherb by farmers in eastern China.

Very common annual weeds 40–80 cm high or higher, much branched; leaves glaucous, strongly farinose beneath, especially when young, rhombic or rhombic-ovate, 3–10 cm long, 2.5–5 cm wide, cuneate at base, irregularly remote-dentate; flowers perfect, perianth green, segments 5, stamens 5, ovary superior, stigmas 2; fruit a utricle; seed 1, lenticular, 1.2–1 5 mm in diameter, shiny black; embryo coiled around the mealy endosperm.

Chenopodium giganteum D. Don (Syn, *C. album* L. cv. 'Giganteum')

Zhang-li=Chang-li (杖藜, Cane Goosefoot). Young shoots used for potherb; Seeds used for food.

Stout plants up to 3 m high, basal portion of stem 5 cm across, ridged, green or purple-red, firm; leaves on primary stems rhomboid-ovate, 20 cm long, 16 cm wide, obtuse, base cuneate, irregularly wavy and serrate, leaves on branchlets smaller; flowers small, in terminal panicles, bending in fruit, sepals 5, green or purple-red; seeds 1.5 mm, black, surface reticulate. Native of northern India, widespread in gardens or growing as adventives in waste areas.

Chenopodium hybridum L. — MAPLE-LEAVED GOOSEFOOT

Da-ye-li=Ta-yeh-li (大葉藜, Big-leaved Chenopodium). Young shoots, nipped off repeatedly before flowering, used as spinach.

Upright glabrous herbs 40–150 cm high, much branched, green throughout, branchlets acutely angular; leaves broad-ovate or triangular-ovate, 6–17 cm long, 5–13 cm wide, with one of a few triangular-acuminate lobes, acute or acuminate, base rounded or truncate; flowers green, polygamous, several together forming loose panicles, perianth segments 5, ovate-lanceolate, obtuse, with a thick abaxial ridge; utricles lenticular, pericarp thin, easily rubbed off; seed solitary, black, 2–3 mm in diameter, with low ridges radiating from the center. Widespread in Eurasia, growing in rocky woods, thickets, clearings and waste places.

Corispermum puberulum Iljin

Sha-lin-cao=Sha-lin-ts'ao (砂林草, Gravel Forest Herb); *Lao-mu-ji-wo=Lao-mu-chi-wo* (老母雞窩, Fowl's Nest); *Ruan-mao-chong-shi=Juan-mao-ch'ung-shih* (軟毛蟲實, Soft-hairy Bugseed). Whole young plant gathered, chopped, dried for tea.

Stellate annual herbs 15–50 cm high; leaves linear, sessile, 5 cm long, 2–4 mm wide, acute, 1-nerved; flowers small, greenish, axillary to bracts, crowded, forming terminal or axillary spikes 5–6 cm long, 1 cm across, perianths 1–3, stamens 1–5; utricles flat, broad-elliptic, 3–4 mm long, 2–3.5 mm wide, distal end rounded, basal end cordate, stellate pubescent, rostrate. Native to temperate northeastern Asia, growing on the gravel of river beds.

Kochia scoparia (L.) Schrader — SUMMER CYPRESS, BELVEDERE

Sao-zhou-cai=Sao-chou-ts'ai (掃帚菜, Broom Vegetable); *Di-fu=Ti-fu* (地膚, Ground Cover). Young shoots, mixed with flour, steamed, seasoned, eaten in summer.

Annual herbs, 0.5–1.5 m high at flowering time, much branched: leaves alternate, lanceolate, 2–5 cm long, 3–7 mm wide, pilose; flowers perfect, in loose spikes, perianth segments 5, styles 2, free to the base; utricle compressed globose, 1 mm long, 1.3 mm in diameter; seed shiny black. Cultivated for ornamental purposes, young shoots nipped off to induce more branching and to improve the shape of the plant, also used as potgreen; old plants harvested in autumn for brooms.

Spinacia oleracea L. — SPINACH

Bo-cai=Po-ts'ai (菠菜, Spinach); *Chi-gen cai=Ch'ih-ken-ts'ai* (赤根菜, Red Root Vegetable). Young plants, cultivated, common.

Annual herbs 10–50 cm tall at the flowering and fruiting stages; radical leaves long petiolate, the petioles succulent, laminas ovate-sagittate, tender, dark green, 12–15 cm long, 6–10 cm wide, cauline leaves smaller; flowers dioecious, staminate flowers yellow, stamens 4, pistillate flowers green, fasciculate, ovary enclosed in fleshy horned bracts, after anthesis thickened, coalescent, spiny in fruit (smooth without spine in some cultivars).

Suaeda glauca Bunge — SEA BLITE

Jian-peng=Chien-p'eng (鹼蓬, Alkaline Weed). Young shoots, gathered locally, used as potherb in northern China.

Erect annual herbs 20–80 cm high, stems light green, striate, branches slender; leaves sessile, succulent linear-cylindrical, grayish-green, 1.5–5 cm long, 1–5 mm thick, gradually becoming smaller upward; flowers perfect, 2–5 in axillary fascicles solitary toward the shoot apex, perianth 5-parted, stamens 5, stigmas 2; fruit a compressed-globose utricle, 2 mm across the middle.

Suaeda salsa (L.) Pallas — SALT GOOSEFOOT, SEA BLITE (Syn. *S. maritima* [L.] Dumortier subsp. *salsa* [L.] Soo; *S. heterophylla* Kitagawa)

Huang-xu-cai=Huang-hsu-ts'ai (黃鬚菜, Yellow Whisker Vegetable); *Chi-jian-peng=Ch'ih-chien-p'eng* (翅鹼蓬, Winged Suaeda); *Xian-di-jian-peng=Hsien-ti-chien-p'eng* (鹽地鹼蓬, Saline Soil Suaeda). Young plants eaten in Hebei; farmers of Bohai Wan in Hebei harvesting the plants with mature seeds, breaking the vegetative growth into sections 1–5 mm long, exposing the shiny black smooth seeds, mixing this material with equal amount of cornmeal for making bread; the product is called *Huang-cai-pan-zi* (黃菜盤子) locally, information supplied by Professor Pong Dawen (彭大文) and material send from Beijing by Xu Gongping (徐公平).

Annual herbs 20–80 cm high, much branched, turning red in autumn; leaves linear, 0.8–3 cm long, 1–2 mm wide, obtuse, cross-section suborbicular, upper surface sulcate; flowers 3–5 in axillary fascicles, often interrupted, perianth segments 5, fleshy, 2.5 mm across, stamens 5, opposite the perianth segments, style 2-fid nearly to the base; seeds almost smooth. Widespread in Eurasia, growing by salt lakes and in saline soil in northern, northwestern and northeastern China.

Amaranthaceae: Amaranth Family

Achyranthes bidentata Blume

Tu-niu-xi=T'u-niu-hsi (土牛膝, Local Ox-knees); *Niu-xi=Niu-hsi* (牛膝, Ox-knees). Leafy shoots boiled with egg in western Hubei; roots cooked with meat for soup, imported dried material available in 1-pound packages in Chinese groceries in America.

Perennial herbs 60–80 cm high, with cylindrical fleshy roots, stems tetragonous, enlarged at the lower nodes; leaves opposite, oblong-elliptic or oblong-lanceolate, 4–15 cm long, 1.5–4.5 cm wide, acuminate, base acute, entire; flowers yellowish-green, small, in terminal spikes 10–15 cm long, turning 180° and pointing to the ground after anthesis, perianth herbaceous, ovate-lanceolate, awned; fruits enclosed in the persistent perianth and bracts. Widespread throughout China, growing in waste areas and along paths of hillside.

Alternanthera sessilis (L.) de Candolle — ALTERNANTHERA

Lian-zi-cao=Lien-tzu-ts'ao (蓮子草, Lotus Seed Herb). Whole plant, gathered in summer for the preparation of a cool drink; young shoots used as potherb.

Much branched weedy herbs rooting on nodes, growing in rice field or ditches, internodes 1–9 cm long, 2 mm in diameter, with two vertical rows of hairs; leaves opposite, elliptic-lanceolate, 2.5–6 cm long, 7–14 mm wide, apex acute or obtuse, base cuneate, both surfaces glabrous, often maroon-red beneath; flowers white, in short axillary spikes 5–8 mm long, perianth scarious, 2 mm long, the segments lanceolate, acute, filaments connate at the base; capsule compressed obcordate, 1.5 mm long and wide across the top, 1-seeded.

Amaranthus ascendens Loiseleur; **A. lividus** L. (Syn. *A. blitum* L.)

   Ao-tou-xian=Ao-t'ou-hsien (凹頭莧, Retuse-leaved Amaranth). Young shoots for potherb, gathered repeatedly during the growing season.

   Glabrous annual herbs 10–30 cm high, branchlets ascending; leaves pale green, ovate-rhomboid, 1.5–4.5 cm long, 1–3 cm wide, deeply retuse, base cuneate, entire, petioles 1–3.5 cm long; flowers polygamous, crowded in short axillary glomerules or terminal spikes or dense panicles 1–2 cm long, bracts and bracteoles membranous, rhomboid, perianth segments 3, membranous, rhomboid-lanceolate, stamens 3; utricles compressed-ovoid, 3 mm long, indehiscent, smooth, slightly wrinkled. Widespread weed, growing in waste areas, even in flower pods.

Amaranthus caudatus L. — Love-lies-bleeding, Tassel Flower

   Hong-xian-mi-cao=Hung-hsien-mi-ts'ao (紅莧米草, Red Amaranthus Weed). Young shoots, edible.

   Tall annual herbs up to 1.5 m highs with a central stem and numerous lateral branches at flowering and fruiting stages; leaves ovate, up to 12 cm long, 8 cm wide, gradually reduced in size upward; flowers polygamous, in terminal cylindrical, compact panicles, bracts and bractioles scarious, red, stamens 5, styles 3; utricles 1-seeded.

Amaranthus inamoenus Willldenow — Garden Amaranth

   Xian-cai=Hsien-ts'ai (莧菜, Amaranth). Young plants; a quick vegetable crop ready to harvest 25 days after sowing the seed, the market material consisting of young plants with or without the roots.

   Much branched annuals 25–30 cm high at flowering time; leaves ovate-suborbicular, up to 13 cm long, 11 cm wide, apex rounded, rarely emarginate, glabrous, petioles 2–6 cm long; flowers polygamous, in axillary glomerules, terminal to the stem, in interrupted compact spike-like panicle, bracts and bracteoles scarious, perianth 3, aciculate, stamens 3, ovary superior, styles 2 or 3; utricle 1-seeded, opening transversely; two forms cultivated throughout China.

Forma **ruber** Makino

Hong-xian-cai=Hung-hsien-ts'ai (紅莧菜, Red Amaranth). Leafy shoots, used as potherbs in summer; common in Chinese kitchen gardens in Boston area.

Leaves green with large central red patch, may be harvested in 25 to 40 days after sowing the seed.

Forma **viridis** Makino

He-bao-xian-cai=Ho-pao-hsien-ts'ai (荷苞莧菜, Purse Amaranthus). Extensively cultivated, leafy shoots and young plants available in summer, occasionally seen in Chinese groceries in Boston.

Leaves uniformly yellowish-green, may be harvested in 25 to 30 days after sowing (see Part I for more information).

Amaranthus paniculatus L. — FIELD AMARANTH

Xian-cai=Hsien-ts'ai (莧菜, Field Amaranth); *Ren-xian=Jen-hsien* (人莧, Man Amaranth); *Qian-sui-gu=Chien-sui-ku* (千穗穀, Thousand Ears Grain); *Ya-gu=Ya-ku* (鴉穀, Crow Grain); *Xian-ren-gu=Hsien-jen-ku* (仙人穀, Fairy's Grain). Young shoots and seed; young shoots used for potherb; seed popped and used for preparing candy during the Chinese New Year festival in villages of northern China.

A tall annual herb up to 2 m high with a central stem and numerous branches each terminating a flowering spike-like panicle; leaves ovate, the upper ones ovate-lanceolate, 4–13 cm long, 2–6 cm wide, apex obtuse, base obtuse or cuneate; flowers polygamous, in large panicle, branched at the base, cylindrical with numerous short spikes at the distal end, bracts and bracteoles subulate, apiculate, spinescent, green; utricle l-seeded. Cultivated for the young shoots; can be snipped off repeatedly; a few plants enough for a small family.

Amaranthus retroflexus L.

Fan-zhi-xian=Fan-chih-Hsien (反枝莧, Reverse Branch Amaranth). Tender shoots, used as potherb in Hubei.

Stout pubescent annual herbs 1–2 m high; leaves ovate or elliptic, 5–12 cm long, 2–6 cm wide, obtuse and mucronate, base acute, entire; flowers small, green, polygamous, in dense axillary and terminal panicles 1.5–6 cm long, bracts lanceolate, 4–6 mm long, longer than the perianth, awned, margin membranous, perianth segments 5, stamens 5, ovary superior, styles 3; utricles plano-concave, dehiscing transversely, seed 1, shiny black. Widespread in China as a weed.

Amaranthus tricolor L.

Yan-lai-hong=Yen-lai-hung (雁來紅, Wild Geese Return Red); *Lao-shao-nian=Lao-shao-nien* (老少年, Returning-to-Youth). Very young shoots; plants cultivated for ornamental purposes, rarely eaten.

Annual herbs 50–100 cm high, generally having a single central stem; leaves ovate-elliptic, 4–10 cm long, 2–7 cm wide, before flowering, the newly developed upper leaves gradually changing color from green to purple, then to brilliant red or yellow; flowers polygamous, in interrupted cylindrical panicles or axillary heads, bracts and bracteoles scarious, colorful; utricles 1-seeded. Cultivated for ornamental purposes, in thinning the young plants, the tender portion used for food.

Amaranthus viridis L.

Ye-xian-cai=Yeh-hsien-ts'ai (野莧菜, Wild Amaranth). Young shoots, used as potherb, occasionally available in the Hong Kong market.

Annual weed, stems 20–40 cm high, much branched, glabrous; leaves ovate-rhomboid, 2–9 cm long, 2–6 cm wide, apex triangular and retuse, petioles 3–6 cm long; flowers green, in axillary and terminal panicles, spike-like, interrupted, bracts and bracteoles green, scarious, stamens 3; utricles compressed globose, seldom open. Gathered from wild plants, not cultivated.

Celosia argentea L. — WILD COCK'S COMB

Ye-ji-guan-hua=Yeh-chi-kuan-hua (野雞冠花, Wild Cock's Comb); *Qing-xiang=Ch'ing-hsiang* (青葙, Green Celosia). Young plants.

An annual weed occurring in gardens, and particularly numerous in fields denuded by a flood, stems 30–90 m high, glabrous, branching after the formation of the first spike; flowers in terminal cylindrical spikes, 3–10 cm long, 1.5–2 cm in diameter, bracts and bracteoles scarious, white-pink, filaments of stamens and staminodes united at base forming a cup, ovary with several ovules; capsule dehiscent transversely; seeds reniform, black, shiny.

Phytolaccaceae: Pokeweed Family

Phytolacca acinosa Roxburgh (Syn. *P. esculenta* Van Houtte) — POKEWEED

Shan-luo-bo=Shan-lo-po (山蘿蔔, Mountain Radish); *Shang-lu=Shang-lu* (商陸, a corrupted name of the foregoing one). Young shoots, cooked twice, with water changes.

A large perennial glabrous herb with a radish-like fleshy root, stems 1–1.5 m high at flowering time; leaves elliptic-oblong, 12–25 cm long, 5–10 cm wide, apex abruptly acuminate, base acute; flowers white, turning pink, in terminal racemes up to 20 cm long, stamens 8, ovary with 8–10 carpels; fruit juicy, black, berry-like.

Nyctaginaceae: Four-O'clock Family

Boerhaavia diffusa L. (Syn. *B. repens* L.)

Zhu-er-yan=Chu-erh-yen (豬耳眼, Pig Ear-opening); *Huang-xi-xin=Huang-Hsi-hsin* (黃細心, Yellow Asarum); *Sha-shen=Sha-shen* (沙參, Sand Shen, the Hainan local name). Fleshy portion of thick root, locally called 'Sha-shen', roasted on open fire, eaten for its sweetish taste and nourishing contents.

Prostrating perennial herbs, branches loose, spreading, pilose; leaves opposite, estipulate, ovate or deltoid in outline, 1–4.5 cm long, and across the base, obtuse, or acute, base truncate-subrotundate, entire or wavy, pilose; flowers small, pink, 2–3 mm long, in axillary or terminal cymose clusters, glandular-pilose throughout, peduncles 1–5 cm long, hypanthia campanulate, 5-striate and glandularly pilose, perianth petaloid, divided, segments 5, orbicular, 1.5–2 mm in diameter, inserted on the rim of the hypathium, stamens 3, filaments filiform, ovary obovoid, style filiform, stigma discoid; fruit achene-like, oblong-obovoid, cuneate at the base, 3–3.5 mm long, 1 mm across the broad region, distinctly 5-ridged, the ridges-bordered with regular lines of transparent and globose glands; seed 1, coat membranous, brown, endosperm farinaceous. Pantropic, weedy, in northwestern Yunnan ascending to grassy slopes up to 3,400 m.

Aizoaceae: Mesembryanthemum Family

Tetragonia tetragonoides (Pallas) 0. Kuntze (Syn. *T. expansa* J. Murray) — NEW ZEALAND SPINACH (Figure 22)

Fan-xing=Fan-hsing (番杏, Foreign Apricot). Young shoots.

A vigorously spreading prostrate succulent annual herb attaining a spread of 1 m in diameter with little care; leaves alternate, deltoid-ovate, 1–7 cm long, 3–5 cm wide; flowers inconspicuous, solitary, axillary to leaves of growing shoot, flower-buds formed at the base of the receptacle, leading to a fascicle eventually, perianth greenish-yellow, 2–3 mm long, persistent; fruit angular, nut-like.

Portulaceae: Purslane Family

Portulaca oleracea L. — PURSLANE

Ma-chi-xian=Ma-ch'ih-hsien (馬齒莧, Horse-teeth Amaranth). Young shoots, used as potherb.

An aggressive succulent weed, growing vigorously, branching repeatedly, prostrate; leaves obovate-cuneate, 1–2.5 cm long, 1 cm across the truncate apex; flowers small, yellow, sessile, 1 at a time, 3–5 in a fascicle at the shoot apex, ephemeral; capsules globose, many-seeded, circumscissile dehiscing.

Basellaceae: Basella Family

Anredera baselloides (Humboldt, Bonpland et Kunth) Baillon — MADEIRA VINE

Luo-kui-shu=Lo-k'uei-shu (落葵薯, Basella Potato). Fresh tubers, rare.

Fast growing herbaceous vines, bearing root tubers; leaves alternate, fleshy, ovate, or lanceolate, 1.8–5.5 cm long, 1.5–4.5 cm wide, obtuse; flowers small, yellowish-white, 2–3.5 mm long, forming axillary or terminal spicate or racemose clusters 25 cm long, perianth segments 5, connate at base, stamens 5, ovary superior, subglobose-ovoid, styles 3; fruits globose, enclosed in persistent perianth. Native to South America, introduced to various cities, occasionally seen in Beijing.

Basella alba L. cv. 'Rubra' L. (Syn. *B. rubra* L.) — MALABAR SPINACH

Ruan-jiang=Juan-chiang (軟漿, Soft-juice); *Yan-zhi-cai=Yen-chih-ts'ai* (燕脂菜, Rouge Vegetable); *Luo-kui=Lo-k'uei* (絡葵, Vine Malva). Succulent leaves, used for soup in Guangzhou and Hong Kong.

A glabrous succulent climber, stems 4–5 m long; leaves alternate ovate-suborbicular, 3–12 cm long, 3–11 cm wide, acute or obtuse at apex, rounded at base, entire; flowers purplish-pink, in a spike 5–20 cm long, bracts 2, perianth 5-parted, stamens 5, styles 3 ; fruits ovoid, purple, fleshy, 5–6 mm long.

Caryophyllaceae: Pink Family

Gypsophila davurica Turczaninow ex Fenzl — NORTH CHINA BABY'S BREATH

Bei-si-shi-zhu=Pei-ssu-shih-chu (北絲石竹, Northern Gypsophila); *Cao-yuan-shi-tou-hua=Ts'ao-yuan-shih-t'ou-hua* (草原石頭花, Meadow Rock Flower). Young shoots gathered in spring, used for vegetable in northern China.

Caespitose perennial herbs 40–80 cm high, glaucous throughout; leaves opposite, sessile, linear-lanceolate, 2.5–5 cm long, 3–7 mm wide, acute, with 3–5 subparallel vertical nerves; flowers white, tinged pink, in loose terminal cymose panicles, calyx campanulate, 3–4 mm long, 5-nerved, 5-lobed, membranous between nerves, petals 5, obovate-oblanceolate, 6–7 mm long, retuse, stamens 10, filaments shorter than the petals, ovary oblong, styles 2; capsules globose, 4 mm in diameter; seeds reniform, tuberculate, the tubercles in rows. Common in meadows on the hillsides of cold temperate China and adjacent Siberia.

Gypsophila oldhamiana Miquel — Pink Baby's Breath
  Suan-ma-zha-cai=Suan-ma-cha-ts'ai (酸螞蚱菜, Sour Grasshopper Vegetable); *Shan-ma-sheng-cai=Shan-ma-sheng-ts'ai* (山馬生菜, Hillside Horse Salad); *Xia-cao=Hsia-ts'ao* (霞草, Rosy Herb); *Yin-chai-hu=Yin-ch'ai-hu* (銀柴胡, Silvery Hare's Ear). Young shoots; used for vegetable.

Perennial herbs 60–100 cm high, with stout rootstocks and grayish-blue herbage; leaves opposite, subsessile, oblong-lanceolate, 3–7 cm long, 4–15 mm wide, acute, base cuneate, entire, with 3–5 subparallel vertical nerves; flowers small, numerous, lilac-pink, in terminal corymbose panicles, calyx turbinate, 2–3 mm long, apical 1/3 lobed, petals 5, obovate-oblanceolate, 6 mm long, emarginate, stamens 10, slightly shorter than the petal, ovary ovoid, unilocular, styles 2; capsules globose-ovoid, slightly longer than the persistent calyx, 4-valved; seeds reniform, black, densely tuberculate, tubercles in rows. Native to northern and northeastern China and adjacent Korea.

Malachium aquaticum L. — Giant Chickweed (Syn. *Cerastium aquaticum* L.; *Myosoton aquaticum* [L.] Moench; *Stellaria aquatica* [L.] Scopoli)
  E-chang-cai=O-ch'ang-ts'ai (鵝腸菜, Goose Intestine Potherb). Young shoots.

Much branched biennial or perennial herbs, stems 20–80 cm long, branchlets pilose, glandular, matted; leaves ovate, 1–6 cm long, 1–3 cm wide, acute or acuminate, base rounded or subcordate, entire, glabrous; flowers white, solitary, rarely cymose, axillary, pedicels 5–15 mm long, glandular-hairy, sepals 5, ovate, petals 5, 4–5 mm long, apex bifid, stamens 10, anthers pink, ovary ovoid, styles 5; capsules ovoid, dehiscing into 5 valves, seeds many, compressed oblong, brown, verrucose. Widespread in the northern hemisphere, growing in meadows and thickets.

Pseudostellaria heterophylla (Miquel) Pax — Prince Ginseng
  Tai-zi-shen=T'ai-tzu-shen (太子參, Prince Shen). Root, used for soup, dried material available in Boston.

Erect perennial herb 5–20 cm high, with fusiform root tubers 1–4 cm long, 4–5 mm across the middle; leafy shoot emerging in May from the rootstock, the lower 3 or 4 nodes bearing opposite scales, the first 2 or 3 pairs of leaves oblanceolate, 2–6.5 cm long, 3–6 mm wide, acute at the apex, cuneate at base, the uppermost 2 pairs of leaves growing close together appearing verticillate, elliptic or ovate-lanceolate, 3–5.5 cm long, 1.5–2 cm wide, acute or acuminate at apex, acute at base; flowers of two types, the aerial flowers white, upright at the shoot apex, the cleistogamous flowers solitary or in simple fascicles in the axils of the bract-like leaves, aerial flowers with lanceolate sepals and obovate petals, 10 stamens, 3 styles; capsules ovoid, 6 mm long. An endangered species, now cultivated in Jiangsu and Shandong for domestic consumption and for export; used as a substitute of ginseng, called 'Prince Shen', both for the small size of the root, and for the mild natures; said to be good for children.

Silene conoidea L. — Catchfly

Mi-wa-guan=Mi-wa-kuan (米瓦罐, Rice Urn); *Mai-ping-cao=Mai-p'ing-ts'ao* (麥瓶草, Wheat Bottle Herb); *Deng-long-pao=Teng-lung-p'ao* (燈籠泡, Lantern Bubble); *Mei-hua-ping=Mei-hua-p'ing* (梅花瓶, Mei Flower Vase). Young shoots; used for potherbs.

Annual herbs 25–61 cm high, glandularly hairy throughout; leaves opposite, the basal ones spatulate, the cauline leaves lanceolate, 5–8 cm long, 5–10 mm wide, apex acute; flowers pink, few at the shoot apex, calyx tubes 2–3 cm long, the lower portion enlarged in fruit, persistent, with many prominent ridges, petals 5, each with 2 scales at the throat, stamens 10, styles 3; capsules ovoid.

Silene venosa (Gilibert) Ascherson — Bladder Campion

Gou-jin-mai-ping-cao=Kou-chin-mai-p'ing-ts'ao (狗筋麥瓶草, Dog-rib Wheat Bottle Herb). Young shoots, eaten in Helongjiang.

Perennial herbs 40–90 cm high; leaves opposite, subsessile, ovate-lanceolate, 5–8 cm long, 1–2.5 cm wide, acute or acuminate, base cuneate, entire and ciliate; flowers pink, in axillary sessile cymose fascicles forming terminal clusters, calyx broad-ovoid, inflated, membranous with 20 nerves and reticulate veinlets, 14–16 mm long, 7–10 mm across, apical lobes deltoid, ciliate, petals 5, 15–17 mm long, clawed, apex deeply bifid, without appendage, stamens 10, exserted, ovary ovoid, 3 mm long, styles 3, exserted; capsules subglobose, 8 mm across, dehiscing by 6 apical teeth; seeds reniform, 1.5 mm long, papillose. Widespread in Eurasia; growing in alpine meadows and by springs in northeastern China, extending southward through central Asian high mountains to northern India.

Stellaria media (L.) Cyrillo — CHICKWEED

E-er-chang=O-erh-ch'ang (鵝兒腸, Goose Intestine); *Fan-lu=Fan-lu* (繁縷, Tangled Herb). Young shoots, gathered and used locally as a potherb or cooked with rice.

An annual prostrating weed, stems 10–30 cm high, slender, much branched; leaves ovate, 0.5–2.5 cm long, 0.5–1.8 cm wide, entire, apex acute; flowers white, solitary, axillary to normal leaves, sepals 5, green, petals 5, deeply parted, almost to the base, stamens 10, styles 3 or 4; capsules ovoid-oblong, 3 mm long.

Nymphaeaceae: Water-lily Family

Brasenia schreberi J. F. Gmelin — WATER SHIELD

Chun-cai=Ch'un-ts'ai (蓴菜, or 蒓菜, Water Shield). Young unfolding leaves, gathered and used locally, especially famous in Hanzhou, used in soup for banquets.

Perennial floating herbs, the vines growing in ponds and small lakes, the rhizomes and roots buried in mud; leaves floating, laminas peltate, involute in bud, oblong-orbicular, 3.5–6 cm long, 5–10 cm wide, entire, petioles 25–40 cm long, hairy; flowers purple-red, solitary, on a slender scape, stamens 12–18, ovary superior, apocarpous.

Euryale ferox Salisbury — PRICKLY WATER-LILY, EURYALE (Figure 16 m)

Qian-shi=Ch'ien-shih (芡實, Euryale Seed); *Ji-tou-zi=Chi-t'ou-tzu* (雞頭子, Chicken Head Seed). Shelled seed, cooked with rice, lotus seed, lily bulb, and Job's Tear, for a delicacy, available in Boston.

A spiny stout perennial aquatic herb, growing in ponds and lakes, with sharp and stinging prickles throughout, rhizomes with copious roots buried in mud; leaves orbicular, unusually large, 1–1.3 m in diameter, purple beneath, petioles terete, 80–100 cm long, strigose and spinose; flowers solitary, apical to a petiole-like terete scape, never open, hidden in an ovoid receptacle 3 cm long, 4 cm in diameter, densely covered with recurved sharp spines, the outer segments olive-green, prickly, the inner segments gradually becoming perianth-like, purple-pink; stamens numerous, two-third of the distal hollow portion lined up from the center with radiating lines of stigmatic surfaces running upward and outward; ovaries apocarpous, embedded in the spongy tissue of the receptacle; mature nuts pea-sized, hard, spherical, the pericarp bony; embryo very small, situated in a hollow area, the edible portion chalk white, being the perisperm (see Part I for more information).

Nelumbo nucifera Gaertner — SACRED LOTUS

Lian=Lien (蓮, Lotus Plant); *Ou=Ou* (藕, Lotus Rhizome); *He-hua=Ho-hua* (荷花, Lotus

Flower); *Lian-zi=Lien-tzu* (蓮子, Lotus Seed); *He-ye=Ho-yeh* (荷葉, Lotus Leaf); *Lian-yong=Lien-yung* (蓮蓉, Mashed-cooked Lotus Seed); *Ou-fen=Ou-fen* (藕粉, Dried Rhizome Starch). Every portion of the plant used in Chinese food and medicine; fresh, dried and canned material of the above items all available in American Chinese stores.

A perennial rhizomatous aquatic herb with the rhizomes and roots buried in mud; leaves floating or emersed, involute in bud, unfolding in opposite directions, petioles terete, prickly, laminas peltate, 25–90 cm in diameter; flowers red or white, showy, solitary, on an erect terete scape, perianth segments spirally arranged, the inner ones becoming staminodes, stamens numerous, ovary apocarpous, embedded in an obconical spongy receptacle truncate at the top; fruit a nut, oblong, 1.5–2 cm long, the pericarp brown, bony; seed with reddish-brown seed coat, two white mealy cotyledons and a stalked green embryo. An important crop in the Lake Region of central China; every organ is used in food or medicine, even dried leaves imported from China via Hong Kong used as a wrapper for a special pastry, the Chinese tamale (see Part I for more information).

Nuphar pumilum (Hoffmann) de Candolle — YELLOW POND-LILY

Ping-peng-cao=P'ing-p'eng-ts'ao (萍蓬草, Floating Pond-Lily). Young tender rhizomes, used for potherbs in Yunnan and Hubei.

Perennial herbs with stout creeping cylindrical, scaly rhizomes 2–3 cm in diameter; leaves all radical, some submersed, others floating or aerial, the submersed leaves membranous, undulate along the margin, the floating and aerial laminas subcoriaceous, ovate-oblong, 6–17 cm long, 6–12 cm wide, obtuse or rounded, base subsagittate with broad and deep sinus, entire, glabrous above, pubescent beneath; flowers yellow, showy, 4–5 cm across, solitary, terminal to the scape, sepals 5 or 6, obovate, orbicular, concave, and coriaceous, petals numerous, obovate-cuneate, 8 mm long, stamens numerous, anthers linear, connectives truncate, ovary superior, oblong, stigmas sessile, discoid with 10 stigmatic rays, margin shallowly 10-lobed; capsules ovoid, 3 cm long, rather fleshy, crowned by the persistent sepals and stigmas, dehiscing irregularly; seeds oblong, 5 mm long, brown. Widespread in temperate eastern Asia.

Nymphaea tetragona Georgi — FOUR-ANGLED WATER-LILY

Shui-lian=Shui-lien (睡蓮, Sleeping Water-lily); *Zi-wu-lian=Tzu-wu-lien* (子午蓮, Midnight-Noon Water-lily). Rhizomes gathered by the people living in the arid region in northwestern China and used for food.

Perennial aquatic plants, rhizomes erect, stout, unbranched; leaves cordate-ovate to ovate-oblong, 5–12 cm long, 4–9 cm wide, mottled when young, with deep sinus; flowers white, opening in the afternoon, solitary, terminal to a scape, 3–5 cm across when fully open, sepals 4, ovate to broad-lanceolate, 1.5–4 cm long, persistent, petals 8–12, lanceolate, oblong or ovate, 2–2.5 cm long, stamens numerous, anthers introrse, ovary syncarpous, stigmas 5–8, radiating on a discoid style; fruits globose, berry-like, covered by persistent calyx; seeds globose, black, 2–3 mm in diameter. Widespread in temperate northern hemisphere, growing in lakes and pond from northeastern to southwestern China and Xinjiang.

Ranunculaceae: Buttercup Family

Aconitum carmichaeli Debeaux — Sichuan Aconite

Bai-fu-pian=Pai-fu-p'ien (白坿片, White Aconite Slice). Preparation made from young aconite tubers; the product appears white, almost translucent; used as a tonic in broth with chicken, especially for the elderly; available in American Chinese stores.

A perennial erect herb, stems 0.6–1.5 m high at flowering time, the annual aerial growth emerging from a subterminal bud of the subterranean tuber; leaves palmately parted and lobed, 6–11 cm long, 9–15 cm wide; flowers purplish-blue, zygomorphic, in a terminal raceme, upper sepal helmet-shaped, larger than the others, stamens numerous, carpels 3–5, free, hairy; follicle 1.5–1.8 cm long; seeds winged. An extremely poisonous species, containing an alkaloid, aconitine; the market material treated, detoxificated (see Part I for more information).

Clematis dioscoreifolia Léveillé et Vaniot (Syn. *C. paniculata* Thunberg, not of J. E. Gmelin).

Da-liao=Ta-liao (大蓼, Big Polygonum). Leaves, gathered and used locally in Taiwan.

A woody vine climbing over fences and thickets, branchlets pubescent, glabrescent with age; leaves opposite, pinnately compound, 10–15 cm long, the petioles and rachids twisting around supporting object for climbing, leaflets 5 or 7, ovate, 3.5–6.5 cm long, 2–4.5 cm wide, apex obtuse, base rounded; flowers white, in axillary and terminal panicles, perianth segments 4, petaloid, 1.2 cm long, stamens numerous, carpels many, apocarpous; achenes compressed ovoid, hairy, persistent style plumose, 3.6 cm long.

Clematis hexapetala Pallas — Mongolian Tea Clematis

Shan-mian-hua=Shan-mien-hua (山棉花, Hillside Cotton); *Shan-liao=Shan-liao* (山蓼, Hillside Polygonum); *Mian-tuan-tie-xian-lian=Mien-t'uan-t'ieh-hsian-lien* (棉團鐵線蓮,

Cotton-mass Clematis). Leaves and stems used in Inner Mongolia as a substitute for tea, or mixed with commercial tea; root gathered locally used as *Wei-ling-xian=Wei-ling-hsien* (威靈仙, an article of traditional Chinese medicine prepared from *C. chinensis* Osbeck of southern and central China).

Erect perennial herbs 40–100 cm high, caudex stout, clay-colored, bearing many fascicles of blackish-brown fibrous roots 8–16 cm long, 1–4 mm in diameter, branchlets cylindrical, striate, sparsely pilose; leaves opposite, ovate-deltoid in outline, pinnatisect or bipinnatisect, petioles 0.5–3.5 cm long, segments of lamina oblong-lanceolate or linear, 3–9 cm long, 4–10 mm wide, entire, acute or acuminate, prominently reticulate on both surfaces when dried; flowers white, 3–5 cm across, solitary or 3, terminal to the primary branch or the branchlets, the flower buds lanate, simulating a ball of cotton, perianth segments 6, rarely 4 or 8, obovate-oblong, 1.5–3 cm long, 5–8 mm wide, stamens numerous, filaments short and broad, anthers linear, equal to or longer than the filaments, pistils many, densely white woolly, especially so with the short styles, stigmas 1 mm long, recurved, pilose, with the stigmatic surface on the adaxial side; achenes glandular-rugose, compressed obovate, 4 mm long, 3 mm wide, hairy, margin thick and broad, persistent styles 2–2.5 cm long, slightly curved, white with villose hairs 2 mm long.

Native of northern China and adjacent Siberia, very variable in leaf-characters; browsed by cattle and camels, untouched by sheep and horses; this being the first record in English of its use as an article of tea.

Nigella damascena L. — Love-in-a-Mist; Fennel-flower

Hei-zhong-cao=Hei-chung-ts'ao (黑種草, Black Seed Herb). Leafy shoots used as potherb in Beijing and its vicinity.

Annual herbs 20–50 cm high, glabrous, much branched; leaves alternate, twice pinnatifid, ovate-oblong in outline, 8–15 cm long, 3–6 cm wide, ultimate segments linear-filiform, 0.6–2 cm long, up to 1.5 mm wide; flowers white, tinged blue, solitary, terminal to the leafy branches, 2.5–5 cm across, surrounded by finely divided involucres, sepals 5, petaloid, petals 5 to many in cultivation, oblong, clawed, the apex acute, bifid or toothed, stamens numerous, anthers oblong, ovary oblong, 5–6 mm long, styles 3–6, 5–6 mm long, recurved, stigmatic surface at the basal half; capsules oblong-subglobose, 2.5–3 cm long, with persistent involucre and styles, loculicidally dehiscing at the top. Native to southern Europe; introduced to Beijing for ornamental purpose.

A related species, *N. sativa* L., distinguished by its leaves with broader segments, flowers blue, 3.5 mm across, without involucres, capsules 3–7 carpels with spreading styles, and aromatic seeds, also cultivated in Beijing, seeds used for flavoring food.

Paeoniaceae: Peony Family

Paeonia lactiflora Pallas — Peony (Figure 23)

Bai-shao=Pai-shao (白芍, White Peony); *Shao-yao=Shao-yao* (芍藥, Peony). Root, with bark rubbed off, dried, used for preparing special health food, available in American Chinese stores.

Perennial herbs 60–80 cm high, the aerial growth glabrous, branched at flowering time; basal leaves twice ternately divided, leaflets oblong-elliptic or oblique ovate, 7.5–12 cm long, pilose beneath; flowers 1 to 3, terminal to the leafy stem, 5.5–10 cm across, stamens numerous, carpels 4 or 5, apocarpous, glabrous.

Lardizabalaceae: Lardizabala Family

Akebia quinata (Houttuyn) Decaisne — Five-leaved Akebia, Chocolate Vine

Mu-tong=Mu-t'ung (木通, Through Wood). Fruit, juicy and very sweet.

Deciduous woody vines; leaves palmately compound, leaflets 5, obovate, 2–8 cm long, 1–4.5 cm wide, apex retuse, base cuneate; flowers dark purple, in long-pedunculate corymbs emergying from abbreviate lateral shoots with the leaves, staminate flowers 3–6, situated at the distal portion of the rachis, stamens 6, pistillate flowers one to each cluster situated at the basal end of the rachis, perianth segments 3, carpels 3 or 6, only one rarely two developing into mature fruits, apocarpous; fruit a fleshy follicle, dehiscing on the adaxial suture, exposing a white jelly-like sweet juicy pulp embedding many black seeds, pericarp thick, purple and smooth outside, white and spongy inside; only the pulp edible.

Akebia trifoliata (Thunberg) Koidzumi var. **australis** (Diels) Rehder

Bai-mu-tong=Pai-mu-t'ung (白木通, White Akebia); *Ba-yue-gua=Pa-yueh-kua* (八月瓜, August Melon); *Di-hai-shen=Ti-hai-shen* (地海參, Ground Sea Cucumber). The juicy sweet pulp of the mature fruit.

Deciduous lianas, branchlets glabrous; leaves trifoliolate, leaflets ovate-oblong, 3–7 cm long, apex rounded and emarginate, base rounded or truncate, entire, glaucous beneath; flowers monoecious, purple-red, or light purple, in racemose clusters 15 cm long, growing in a fascicle of leaves on a spur, staminate flowers many, at the distal end, pistillate flowers 1–3 below, perianth segments 3, oblong, staminate flowers with 6 stamens and 3 pistillodes, pistillate flowers apocarpous, ovaries 3–6, stigmas sessile; fruits follicular, oblong, 8–13 cm long, 4 cm across the middle, grayish purple, pericarp fleshy, pulp juicy, embedding numerous seeds, dehiscent along the abaxial suture; seeds oblong, dark red. Native to eastern Asia, introduced into American gardens in 1907.

Decaisnea fargesii Franchet

Mao-er-shi=Mao-erh-shih (貓兒屎, Cat Feces); *Yang-jiao-zi=Yang-chiao-tzu* (羊角子, Goat Horns); *Gui-zhi-tou=Kui-chih-t'ou* (鬼指頭, Devil's Fingers). White pulp of the mature fruit, eaten in western China and Yunnan.

Deciduous shrubs 2–5 m high, glabrous throughout ; leaves pinnate, oblong in outline, 50–80 cm long, leaflets 13–25, ovate-elliptic, 5–14 cm long, 3.5–7 cm wide, acuminate or caudate, base obtuse or rounded, entire, glaucous beneath; flowers greenish, polygamous, in elongated racemes forming terminal panicles to lateral branches, panicles pendulous, 25–50 cm long, pedicels 1 cm long, perianth subcampanulate, 2.5–3 cm long, segments 6, petaloid, lanceolate, long-acuminate, nectaries 0, stamens 6 in the staminate flowers, monodelphous, staminodes in pistillate flowers distinct, pistils 3; fruits follicular, bluish-purple, glaucous, cylindrical, 5–10 cm long, 2 cm across, pericarp fleshy, dehiscent along the adaxial suture, pulp white; seeds numerous, embedded in the pulp in two rows, ovoid, flattened, 1 cm long, black. Endemic to western China, introduced to western gardens in 1893.

Holboellia coriacea Diels (Syn. *Stauntonia brevipes* Hemsley)

Ying-zhua-feng=Ying-chua-feng (鷹爪楓, Hawk's Talon Vine); *Po-gu-feng=P'o-ku-feng* (破骨風, Broken Bone Vine). Fruits, eaten in southwestern Hubei.

Evergreen lianas 2–5 m high; leaves alternate, trifoliolate, leaflets elliptic-oblong, 6–14 cm long, 2–4.5 cm wide, acuminate, base rounded or obtuse, entire, coriaceous, glabrous, petioles 2–6 cm long, petiolules 1–2 cm long, the middle ones the longer; flowers unisexual, 2 cm across, in loose axillary racemes, the staminate flowers white, stamens 6, distinct, the pistillate flowers purple; fruits fleshy, juicy, 4–6 cm long. Native to China, growing in the valleys south of the Yangtze River at altitudes of 600–1,700 m.

Holboellia fargesii Reaubourg — WILD MAN'S MELON

Wu-ye-gua-teng=Wu-yeh-kua-t'eng (五葉瓜藤, Five-leaved Gourd Vine); *Ye-ren-gua=Yeh-yen-kua* (野人瓜, Wild Man's Melon). Ripe fruit.

Glabrous woody vines 3–6 m long; leaves palmately compound, leaflets 5 or 7, oblong-obovate, 5–13 cm long, 2–6 cm wide, apex acute, base obtuse or cuneate, light gray beneath; flowers monoecious, white, fragrant, perianth campanulate, 2–3 cm long, segments 6, fleshy, petaloid, staminate flowers with 6 stamens and a pistillode, pistillate ones with 3 distinct carpels and small staminodes; fruits oblong, 8–12 cm long, 2 cm across, purple, edible. Native to central and western China; introduced into American gardens in 1907.

Holboellia latifolia Wallich

Wu-feng-teng=Wu-feng-t'eng (五風藤, Five Wind Vine); *San-ye-lian=San-yeh-lien* (三葉蓮, Three-leaved Lian). Mature fruits, eaten in Yunnan.

Evergreen lianas; leaves palmately compound, leaflets 3–7, ovate or obovate-oblong, 5–7 (–15) cm long, 2.5–4.5 cm wide, caudate or lanceolate, base rounded or obtuse, entire, green above, paler beneath; flowers appearing with or without leafy shoots from an axillary bud, fragrant, in fasciculate corymbs, unisexual, staminate flowers greenish white, perianth segments 6, fleshy, the outer ones narrowly oblong, 1.5–2 cm long, 1.2 cm wide, nectaries 6, stamens 6, pistillate flowers purple, perianth segments 2.2 cm long, 1.5 cm wide, staminodes 6, pistil apocarpous, ovaries 3; fruits purple, sausage-shaped, 5–7 cm long; seeds numerous. Native of the eastern Himalayan Region, growing in forests of Yunnan, Sichuan, Guizhou and the adjacent areas in Tibet at altitudes of 600–2000 m.

Sinofranchetia chinensis (Franchet) Hemsley

Chuan-guo-teng=Ch'uan-kuo-t'eng (串果藤, String-of-fruits Vine). Mature fruits, eaten like grapes in Yunnan.

Deciduous lianas 10 m high, branchlets glabrous, glaucous; leaves trifoliolate, petioles 6–20 cm long, terminal leaflets obovate-rhombic, 9–10 cm long, 7–10 cm wide, abruptly caudate, petiolules 3 cm long, base cuneate, lateral leaflets obliquely ovate, 7–11 cm long, petiolules 2–5 mm long; flowers small, white with purple stipes, unisexual, dioecious, in elongated racemes 6–20 cm long, peduncle 4–7 cm long, perianth segments 6, glands 6, stalked, in staminate flowers filaments connate at anthesis, separating afterward, anthers obcordate, extrorse, pistillodes 3, oblong, in pistillate flowers staminodes distinct, pistils apocarpous, ovary 3, oblong; fruits oblong, purple, berry-like, juicy, edible, insipid, 3 on a short stalk, 1.5 cm long, 1 cm across; seeds about 20, ovate-ellipsoid, flattened, 3 mm long, 2 mm wide, black. Endemic to central and western China, growing along margins of forests at altitudes of 1,350–1,900 m.

Stauntonia chinensis de Candolle — WILD WOOD-GOURD

Ye-mu-gua=Yeh-mu-kua (野木瓜, Wild Wood-gourd). Mature fruit, used as fresh fruit, or for making wine and/or jelly.

Evergreen lianas, glabrous; leaves palmately compound, leaflets generally 5, occasionally 3, sizes and shapes variable, oblong or obovate-oblong, 6–10 cm long, 3–5 cm wide, apex acuminate, base rounded or obtuse; flowers monoecious, greenish-puprle, with unpleasant odor, perianth segments 6 in 2 whorls, stamens 6, monodelphous, carpels in pistillate flowers 3, distinct; fruits oblong-subglobose, 6 cm long, 5 cm in diameter, juicy.

Stauntonia duclouxii Gagnepain

Yang-zhua-teng=Yang-chua-t'eng (羊爪藤, Goat-hoof Vine); *Yun-nan-ye-mu-gua=Yun-nan-yeh-mu-kua* (雲南野木瓜, Yunnan Wild Wood-melon). Pulp of fruit, sweet, of good flavor, eaten both by people in Yunnan and Hubei.

Evergreen lianas 2–5 m high, branchlets yellowish-brown; leaves palmately compound, leaflets 5–7, obovate or obovate-oblong, 6.5–9 cm long, 2.5–4.5 cm wide, shortly acuminate or acute, base obtuse, entire, glabrous, petioles 5–9 cm long, petiolules 1–3 cm long: flowers yellowish-green or white, fragrant, unisexual, monoecious, in loose racemes, perianth segments 6, the outer 3 larger, stamens 6, filaments connate into a column, anthers cuspidate, appendaged at the apex; fruits oblong-ovoid, 5–7 cm long, yellow. Native of the warmer areas south of the Yangtze River, from Zhejiang westward to Sichuan and Yunnan.

Stauntonia hexaphylla (Thunberg) Decaisne

Shi-yue=Shih-yüeh (石月, Moon-on-Rock); Fruit, eaten locally in Taiwan.

A woody climber; leaves palmately compound, leaflets 5 or 7, variable in size and shape, generally elliptic or obovate-oblong, 6–10 cm long, 3–4.5 cm wide, apex caudate, base rounded or obtuse; flowers in loose racemes, perianth segments 6, subequal, outer 3 oblong-lanceolate, inner ones linear-lanceolate, stamens 6, connate, carpels 3; fruit oblong-subglobose, 6 cm long, 4 cm in diameter.

Berberidaceae — Barberry Family

Berberis heteropoda Schrenk — Central Asian Barberry

Shan-li-zi=Shan-li-tzu (山李子, Hillside Plum); *Hei-guo-xiao-bo=Hei-kuo-hsiao-po* (黑果小檗, Black-fruited Barberry); *Ci-huang-bai=Tz'u-huang-pai* (刺黄柏, Spiny Phellodendron). Mature fruits, gathered, dried, used for preparing a beverage by boiling 31 g of the fruit in 4 cups of water, down to one cup of decoction, mixed with sugar, served especially to elderly persons with high blood pressure.

Deciduous shrubs 2 m high, branchlets terete, brown, armed by simple or branched spines 1–3 cm long; leaves broad-ovate, obovate or elliptic, 2–7.5 cm long, 2–4 cm wide, obtuse, base abruptly cuneate, petioles 0.6–3 cm long, entire or finely serrate, the teeth bristly, grayish-green above, glaucous beneath; flowers orange-yellow, fragrant, 1.2 cm across, in racemes 2–4 cm long, sepals 2, petals 6, obovate, nectariferous glands 6, stamens 6, shorter than the petals, ovary superior, 6-ovulate, stigma discoid; fruits oblong-globose, 12 mm long, purple-black, bloomy. Native to central Asia, growing on the hillside of Xinjiang. Introduced into western gardens in 1876.

Podophyllum emondii Wallich var. **chinense** Sprague — CHINESE MAY-APPLE

Gui-jiu=Kuei-chiu (鬼臼, Devil's Mortar); *Tao-er-qi=T'ao-erh-ch'i* (桃耳(兒)七, Peach Hematinic); *Qi-ye-lian=Ch'i-yeh-lien* (七葉蓮, Seven-leaved Lotus). Ripe fruits, sweet, eaten by people living in Hubei.

Perennial herbs, creeping, rhizomes cylindrical, 8–10 cm long, 5–8 mm in diameter, bearing numerous wire-like fleshy roots 10–15 cm long, 2–4 mm thick, and a terminal (rarely a subterminal) winter bud from which emerges the aerial growth, consisting of a fleshy cylindrical stem 20–30 cm long, 5–8 mm across, with 4–6 membranous basal prophylls 1–12 cm long, the innermost ones the longest, 2 leaves and a single flower at the apex; aerial leaves simple, suborbicular in outline, 6–15 cm long, 10–25 cm across the base, fundamentally tripartite, the segments irregularly trilobed, or undivided, the lobes deltoid or ovate-lanceolate, remotely serrate, villous beneath; flowers pink, emerging with the unfolding reflexed leaves, blooming before the expansion of the leaves, sepals 6, fugacious, petals 6, obovate, 3–4.5 cm long, 2.5–3 cm wide, stamens 6, filaments erect, white, anthers linear, ovary ellipsoid, ovules many, style obsolete, stigma peltate, with radiating stigmatic surfaces, lobed on drying; berries juicy, yellow-red, ellipsoid, 2–6 cm long, 2–2.5 cm across middle, rostrate with the persistent style-stigma; seeds oblong-ellipsoid, shiny chestnut brown, 5 mm long, 4 mm across, faveolate under lens. Native to central and western China, extending westward to Tibet and the Himayalan Region.

Note: Rhizomes used in ethnic medicine in Gansu, Shaanxi and Tibet; recent researches into the converted material from the very toxic extracts of Podophyllin and podophyllotoxin for treatments of cancerous growth, have created increasing demand and much uprooting which makes it now an endangered species; the name "Devil's Mortar" appeared in Chinese botanical literature referring to the deep scars on the rhizome, and the ethnic names "Peach Hematinic" to the attributed efficacy for enriching blood, and "Seven-leaved Lotus" to the lobed leaves and the solitary flower, all given by people in central and western China; these ethnobotanical names indicating the people's knowledge of the prominent morphological features of the plant, and their experience of the efficacy of the plant in solving their health problems.

Magnoliaceae: Magnolia Family

Magnolia denudata Desrousseaux (Syn. *M. heptapeta* [Buc'hoz] Dandy, *M. soulangeana* Soulange-Bodin (white-flowered form). (Figure 24).

Bai-yu-lan=Pai-yü-lan (白玉蘭, White Magnolia); *Yu-lan-hua-pian=Yü-lan-hua p'ien* (玉蘭花片, Magnolia Petals). Fresh petals of partially opened flowers, dipped in batter, deep fried, served while hot as a delicacy.

A deciduous tree 6–8 m high; branchlets stout, covered with gray hairs; leaves alternate, obovate-oblong, 10–18 cm long, 6–10 cm wide, apex acute, base cuneate, entire, puberulous beneath; flowers showy, fragrant, ivory-white, petals fleshy, stamens numerous, carpels spirally arranged, apocarpous; aggregate fruits 8–12 cm long, consisting of many 2-seeded follicles.

Illiciaceae: Illicium Family

Illicium verum J. D. Hooker — STAR ANISE

Ba-jiao-hui-xiang=P'a-chiao-hui-hsiang (八角茴香, Eight-horned Muslim Spice); *Ba-jiao=Pa-chiao* (八角, Eight Horns); *Da-hui-xiang=Ta-hui-hsiang* (大茴香, Big Muslim Spice). Ripe fruits, available in American Chinese stores.

An evergreen tropical tree up to 20 m high; leaves crowded at the shoot apex, obovate-oblanceolate, 5–11 cm long, 1.6–4 cm wide, apex acute or shortly acuminate, base cuneate, acute, entire; flowers deep red, solitary, axillary to normal leaves, perianth segments 7–12, stamens 11–20, carpels 8 or 9, in one whorl, apocarpous; fruits red-brown, 3.5 cm in diameter, consisting of 8 or 9 woody follicles; seeds brown, shiny.

Schisandraceae: Schisandra Family

Schisandra chinensis (Turczaninow) Baillon — SCHISANDRA, WU-WEI-ZI

Wu-wei-zi=Wu-wei-tzu (五味子, Five-flavored Drupelet). Fresh red fruit used for making soft drink in Beijing; dried drupelets sold in Chinese stores in Boston, used for preparing special health food dishes, or for medicated beverages; cultivated in Massachusetts; material imported by Japanese traders, sold to jockey clubs in the USA for feeding race horses.

A deciduous woody climber growing over trees, shrubs, fences; leaves alternate, papery, obovate or broad-oblong, 5–10 cm long, 2–5 cm wide, apex abruptly short acuminate, base acute, margin glandularly serrate, sparsely pilose on the nerves beneath;

flowers fragrant, dioecious, solitary or fasciculate, axillary, perianth white or pink, segment 6–9, staminate flowers with 5 stamens, pistillate flowers with numerous carpels, the axis elongating after anthesis, giving rise to a string of pea-sized drupelets, brilliant red at maturity, extremely sour. Native of northern China, the market material from cultivated sources.

Schisandra glaucescens Diels

Fan-ba-tuan=Fan-pa-t'uan (飯巴團, Lump-of-Cooked Rice). *Jin-shan-wu-wei-zi= Chin-shan-wu-wei-tzu* (金山五味子, Jinshan Schisandra). Ripe fruits.

Deciduous lianas, branchlets glabrous; leaves obovate or elliptic, 5–11 cm long, 2.5–5 cm wide, short-acuminate, base cuneate, or acute, papery, glaucous beneath, serrate; flowers pink, unisexual, dioecious, axillary, staminate flowers on pedicels 2–3.5 cm long, perianth segments 6–7, stamens 18–25, connate at the basal portion, pistillate flowers on pedicels 3–5 cm long, pistils 30–50, the receptacle elongating after anthesis, producing a spike-like cluster of fruits 4.5–7 cm long; drupelets subglobose, 9–12 mm long, 7–9 mm across; seeds compressed oblong. Native to central and western China, growing in forests at altitudes of 750–1,800 m; introduced into American gardens in 1907 by the Arnold Arboretum, Harvard University.

Schisandra incarnata Stapf

Xing-shan-wu-wei-zi=Hsing-shan-wu-wei-tzu (興山五味子, Xing-shan Schisandra). Ripe fruits.

Deciduous lianas; leaves elliptic, 6–10 cm long, 3–5 cm wide, acute, base cuneate, green beneath; flowers pink, unisexual, dioecious, solitary, axillary, staminate flowers on pedicels 1.6–3.5 cm long, perianth segments 7 or 8, elliptic-obovate, 1–1.7 cm long, 6–12 mm wide, stamens 29, the free portion of the filament equal the length of the anther, pistillate flower with 70 pistils, the elongated receptacles 2–5 cm long in fruit; drupelets scarlet, oblong, 1 cm long; seeds oblong, compressed. Endemic to western Hubei, growing in thickets at altitudes of 1,500–1,800 m.

Schisandra propinqua (Wallich) Baillon var. sinensis Oliver

Tie-gu-san=T'ieh-ku-san (鐵箍散, Iron Hoop Scattered); *Xiao-xue-teng=Hsiao-hsueh-t'eng* (小血籐, Lesser Blood Vine); *Cheng-chui-ye=Ch'eng-ch'ui-yeh* (秤錘葉, Steelyard Weight Leaf): *Zuan-yan-jin=Tsuan-yen-chin* (鑽岩筋, Boring-rock Muscle); *Zuan-gu-feng= Tsuan-ku-feng* (鑽骨風, Penetrating Bone Rheumatism). Ripe fruits, eaten in Hubei.

Deciduous glabrous lianas; leaves linear-lanceolate, 5–12 cm long, 1–3 cm wide, acuminate-caudate, base rounded or obtuse, remotely serrulate, pale green beneath; flowers yellow, solitary, axillary, unisexual, monoecious or dioecious, staminate flowers on pedicels 0.4–2.3 cm long, perianth segments 6–9, oblong-elliptic, 5–8 mm long, stamens 6–9, connate into a ball, pistillate flowers on pedicels 0.5–2.6 cm long, perianth segments 8–11, elliptic, 5–9 mm long, pistils 10–30, after anthesis the receptacle elongated to 3–7 cm long; drupelets red, sweet, edible. Native to central and western China, growing in forests at altitudes of 400–1,800 m; an endangered species, because of the increasing demand for the root and vines used as a remedy for rheumatism and muscular pains.

Annonaceae: Custard Apple Family

Annona cherimolia Miller — Cherimoya

Bi-lu-fan-li-zhi=Pi-lu-fan-li-chih (祕魯番荔枝, Custard Apple of Peru). Introduced to Taiwan in the 1920s.

A small semideciduous tree of the subtropics, 5–10 m high, growing best in arid semi-dessert condition, branchlets softly pubescent; leaves dark green, obovate, 10–15 (–25) cm long, apex obtuse, base rounded, velvety beneath; flowers greenish-white, fragrant, solitary or 2–3 in a nodding cluster, perianth segments 6, the three outer ones larger, stamens numerous, pistils apocarpous; fruits heart-shaped, 7.7–12 cm long, consisting of coherent carpels, covered with small knobs, the inside white, juicy, slightly acid; seeds many, brown, of the size of a bean. Propagated by budding or grafting; introduced to Taiwan in the 1920s, rare.

Annona muricata L. — Soursop

Niu-xin-guo=Niu-hsin-kuo (牛心果, Ox-heart Fruit). Newly introduced, rare; fruit sour, said to be unrivaled for preparing sherbet and refreshing drinks.

A tropical evergreen tree growing in the lowland, 8 m high; leaves obovate-elliptic, 8–15 cm long, acute at apex, glabrous beneath; flowers with fleshy perianth segments, the 3 exterior ones ovate-acute, the interior 3 smaller and thinner, rounded; fruits ovoid, heart-shaped or oblong, pulp white, cottony, juicy, highly aromatic; seeds bean-sized, brown.

Annona squamosa L. — Custard Apple, Sugar Apple, Sweet Sop

Fan-li-zhi=Fan-li-chih (番荔枝, Foreign Lychee). Imported ripe fruits, rather common in the Hong Kong market; small trees also observed in villages occasionally.

A small semideciduous tree 4–7 m high; branchlets puberulent; leaves oblong-lanceolate or lanceolate, 7–10 cm long, acute or shortly acuminate, glabrous; flowers 2–4 in small clusters borne upon the new branchlets, outer 3 perianth segments oblong, 2.5 cm long, apex rounded, the inner 3 ovate, minute, scale-like; fruits heart-shaped, round or ovoid, 5–7 cm long, yellowish green, tuberculate, covered with a whitish bloom, pulp white, custard-like, sweet, with delightful fragrance. Imported fruit highly esteemed.

Uvaria rufa Blume

Xiang-jiao-shu=Hsiang-chiao-shu (香膠樹, Sweet Gum Tree). Fruit.

A little evergreen scandent tropical shrub, covered with stiff brown stellate trichomes throughout; leaves alternate, leathery, ovate-oblong or oblong, 7–15 cm long, 3–6 cm wide, apex acuminate, base cordate, entire, rough to touch; flower-buds 2 or 3 in a bracteate cymose extra-axillary cluster half way on the side of an internode, only the terminal one developing into a flower, 2.5 cm across, perianth segments 6, the outer 3 green, ovate, 8 mm long and across, the inner 3 red, oblong, 6 mm long, 5 mm wide, stamens numerous, brown, pistils apocarpous, the stigmas white; fruit an aggregate of 1- or rarely 2-seeded indehiscent mature carpels on elongated stipes. First reported from Java.

Myristicaceae: Nutmeg Family

Myristica fragrans Houttuyn — Nutmeg

Rou-dou-kou=Jou-tou-k'ou (肉豆蔻, Meaty Cardamon). Seed used as spice.

A glabrous tropical evergreen tree commercially cultivated for the spice obtained from the fruit, 8–15 m high; leaves alternate, oblong-lanceolate, 8–12 cm long, glaucous beneath; flowers monoecious, yellow, 2–4 in axillary cymes, perianth urceolate, 3 cm long, 1.5 cm across, stamens monodelphous; fruit a drupe, yellow, the exocarp splitting into 2 halves at maturity; seed brown (the nutmeg) surrounded by red net-like fleshy aril (the mace).

Lauraceae: Laurel Family

Cinnamomum cassia (Nees) Nees ex Blume — Cassia, and **C. loureirii** Nees — Saigon Cinnamon

Rou-gui=Jou-kuei (肉桂, Cassia Bark); *Gui-zhi=Kuei-chih* (桂枝, Cassia branches); *Guan-gui=Kuan-kuei* (官桂, Official Cassia Bark); *Gui-xin=Kuei-hsin* (桂心, Cassia Heart); many other market names, all products from different parts of the same species, cultivated in the mountainous area bordering China and Vietnam; products shipped to Saigon carry the trade name "Saigon Cinnamon", those shipped overland to Guangzhou or by boat to Hong Kong have the trade name "Cassia".

An aromatic tropical evergreen tree 15–20 m high, branchlets softly pubescent; leaves alternate, oblong-elliptic, 8–12 (–20) cm long, 4–6 cm wide, shortly acuminate or obtuse at the apex, obtuse at base, entire, triplinerved; flowers small, greenish-yellow, in axillary cymose and terminal panicles, appearing with or slightly before the developing shoot; fruit a subglobose drupe, subtended by the persistent and slightly enlarged calyx, 9 mm in diameter, purplish black.

Note: A native of China, growing south of the Nanling Range. Since time immemorial the ethnic peoples living in southern China have used the bark as a spice and for medicinal purposes. They have learned to select various strains, and have developed certain techniques in preparing the products to meet the demands of different traders. For two centuries botanists in Europe and America have tried to work on the herbarium material to identify the species. In felling a tree, the producer sorts the products by the sizes of the trunk and branches and assigns the names. The strains and the age of the bark make the differences in the smells, or the chemical contents and the anatomical structures. The species as well as its products have spread with people throughout southeastern Asia. Blume's material was from a tree cultivated in Java. So are all other source species of cassia or Saigon cinnamon.

Here is a personal experience. One day in the 1960s, Professor Heber W. Youngken, Massachusetts College of Pharmacy, asked for my assistance to identify some commercial specimen from a large shipment of bark from Hong Kong, waiting to pass the inspection at the Maritime Customs. If we gave the identification as *C. cassia* (Nees) Nees ex Blume, the material must be returned to Hong Kong, and if we named it *C. loureirii* Nees, the shipment would be accepted. We had no characters to depend upon to be sure of the species, but we opted for *C. loureirii* Nees for humanitarian reasons. The fact is that all the commercial materials are from cultivated strains planted in limited areas. The French botanist maintained that the Saigon cinnimon was "either purchased from Chinese or Annamese merchants, and then brought into the port of Saigon." The mountains where cassia and Saigon cinnamon grow is known in Chinese geography as Shi-wan-da Shan (十萬大山, Hundred Thousand Big Mountains). Cassia is an ancient Phoenician name borrowed by the Arabian and Persian early traders between the East and West.

Cinnamomum chingii Metcalf

Xi-ye-xiang-gui=Hsi-yeh-hsiang-kuei (細葉香桂, Small-leaved Aromatic Cassia). Leaves used for flavoring canned food in Hubei.

Evergreen trees 20 m high, branchlets pilose; leaves elliptic or ovate-elliptic to lanceolate, 6–13 cm long, 2–3.5 cm wide, acuminate or acute, coriaceous, entire, trinerved, tomentose beneath; flowers greenish-yellow, in axillary cymose clusters, peduncles and pedicels pilose, perianth segments 6, 3 mm long, stamens 9, anthers 4-celled; drupes oblong, 1.5 cm long, 0.5–1 cm across, black, persistent receptacle cupular. Native to central and southeastern China, growing along roadsides at 450 m above sea level.

Cinnamomum wilsonii Gamble

Chuan-gui=Ch'uan-kuei (川桂, Sichuan Cinnamon); *Gui-pi-shu=Kuei-p'i-shu* (桂皮樹, Cassia Bark Tree). Branches and bark aromatic and pungent, used for condiment.

Medium-sized trees 2–16 m high, branchlets dark brown; leaves elliptic-oblong, 8–18 cm long, 3–5 cm wide, acuminate, base acute, entire, margin reflexed and cartilaginous, triplinerved, glabrescent and grayish beneath; flowers greenish-yellow, (1–) 3–5 in axillary cymes, peduncles slender, 1–6 cm long, pedicels filiform, pilose, 0.6–2 cm long, perianth segments 6, 4–5 mm long, pilose, stamens 9, filaments pilose, 2 glands near middle, ovary ovoid, style stout, stigma capitate; drupes oblong, 1.5 cm long, 8 mm across, black, on a truncate cupular receptacle. Native to central and southwestern China, growing in open country at altitudes of 400–1,000 m.

Litsea cubeba Persoon — FRAGRANT LITSEA

Mu-jiang-zi=Mu-chiang-tzu (木姜子, Tree Ginger Seed). Fruit.

A common deciduous small tree growing on the hillsides in tropical China, 4–15 m high, bark smooth, branchlets puberulous; leaves alternate, elliptic-lanceolate, 7–12 cm long, 1.4–3 cm wide, entire, both ends acute; flowers white fragrant, open in mid-winter, consisting of an pedunculate recemose clusters, each branch consisting of a pedunculate umbel with 5 flowers subtended by 2 pairs of rotundate bracts, each flower pedicellate, the one in the center of the umbel open first, perianth segments 6 in 2 whorls, the staminate flower with 9 (–11) fertile stamens, 6 outer ones with filaments devoid of glands, 3–5 inner ones each with a pair of submedian glands on the filament, and a pistillode in the center, pistillate flower with unequal perianth segments, the ones close to the bracts larger, the one near the center of the umbel smaller, 6 staminodes without glands on the filament, 3 inner staminodes each with a pair of glands on the filament, ovary globose, style obliquely oriented, stigma 2-lobed; fruits globose, drupaceous, 5 mm in diameter, after anthesis, the peduncles and pedicels elongated, 1–1.5 cm long with the fruiting peduncles 1–1.2 cm long and the fruiting pedicels 3–4 mm long. Widespread in southern China, with the pepper-like black fruits gathered, dried and used as a substitute for the imported cubeb, the unripe fruit of *Piper cubeba* L. f..

Litsea mollifolia Chun

Mao-ye-mu-xiang-zi=Mao-yeh-mu-hsiang-tzu (毛葉木香子, Hairy-leaved Fragrant Litsea). Young leaves, chopped for flavoring food.

Shrubs 2–3 m high, branchlets grayish-black, pilose; leaves oblong-lanceolate or oblong-elliptic, 10–17 cm long, 3–4 cm wide, both ends attenuate, sparsely pilose beneath; flowers white, unisexual, dioecious, in axillary umbels, peduncles short, pilose, perianth segments 6, pilose, punctate, fertile stamens 6, anthers 4, ovary in pistillate flowers ovoid, stigma subsessile, peltate; drupes red, subglobose, 5–6 mm long, 5 mm across, on hairy peduncles 8–10 mm long. Endemic to central China, growing by streams along margins of forests at altitudes of 300–1,800 m.

Persea americana Miller — AVOCADO, ALLIGATOR PEAR

Zhang-li=Chang-li (樟梨, Cinnamon Pear); *E-li=O-li* (鱷梨, Alligator Pear). Ripe fruit, introduced, very rare, imported product available in Hong Kong.

An aromatic evergreen tree of the tropics up to 10 m high, branchlets pubescent; leaves elliptic or oblong-elliptic 6–30 (–40) cm long, pubescent when young; flowers greenish-yellow, in terminal panicles, polygamous, small, perianth segments 6 in 2 whorls, fertile stamens 9, the filament of the outer 6 without glands, those of the inner 3 each with paired yellow glands, staminodes 3, pistil ovoid, style slender, stigma 1; fruits drupaceous, green, turning black, the skin leathery, the meat buttery when fully mature, shape and size various; very rich food, especially with high iodine content. A native of tropical America, introduced to Guangzhou in the late 1920s, as specimen trees in agricultural institutions and botanical gardens, fruit rarely used in China for food.

Papaveraceae: Poppy Family

Papaver somniferum L. — OPIUM POPPY

Ying-su=Ying-su (罌粟, Poppy Millet). Seed.

Annual erect herbs 60–150 cm high; the lower leaves petiolate, the upper cauline leaves sessile, ovate or oblong, 6–30 cm long, 2–12 cm wide, amplexicaul, irregularly lobed, the terminal segments dentate; flowers solitary, colorful and showy, scape terete, bending before anthesis, sepals green, 2, falling early, petals 4, 6–8 cm across at anthesis, white, pink, purple, some deep, others light, stamens numerous, filaments 10–15 mm long, anthers oblong, basifixed, ovary oblong-obovoid, 8–10 mm long, stipitate, glabrous, glaucous, truncate at the apex, stigmas 10, radiating from the center of a lobed disk, forming 10 stigmatic ridges 3–5 mm long; fruit a capsule, oblong, 3–5 cm long, 2.5–4 cm in diameter, truncate, stigmatic disk persistent, dehiscent by pores below the disk; seeds reniform, 1 mm long, reticulate and foveolate, gray. Widely cultivated in China in the 1890s, becoming illegal after the establishment of the Republic in 1911; few Chinese botanists have seen live specimens of the opium poppy.

Brassicaceae (Cruciferae): Mustard Family

The crucifers furnish important vegetables in China. Both in kinds of mustards, cabbages and radishes, and in acreage of water cress, China surpasses any others country of comparable size in the world. In the technique of crop management, the selection of cultivars, and in the culinary art of preparing the dishes, the Chinese farmers and housewives have attained the highest degree of success. In the following text, the common characters of the crucifers: herbaceous habit, four merous flowers, cruciform corolla and tetradynamous stamens are not repeated. Most of these Chinese vegetables are cultivars selected from open pollination hybrids and maintained by cultivation. Their taxonomic status are confusing. Botanists and horticulturalists have created many names. For the practical reason of convenience and simplicity, specific names are adopted and the decisions of recent Chinese horticultural botanists are honored about varieties.

Armoracia rusticana P. Gaertner, B. Meyer et Scherbius — HORSERADISH (Syn. *Cochlearia armoracia* L.)

Xi-yang-shan-yu-cai=Hsi-yang-shan-yü-ts'ai (西洋山俞菜, European Alliaria). Roots, rare; cultivated in botanical gardens or for supplying foreign residents.

A glabrous tall perennial herb 1 m or more high in flower; radical leaves 2 or 3, very large with long petiolate, laminas oblong, 20–30 cm long, 7–15 cm wide, petioles 30–35 cm long, terete, apex rounded, base oblique-cordate, evenly dentate, the teeth round, cauline leaves lanceolate, 14–20 cm long, 2.5–4 cm wide, dentate or pinnatifid, apex obtuse, base cuneate and becoming wings of the petiole; flowers white, numerous, small, 5–6 mm long and across, ovary oblong, style shorter than the capitate stigma; silicles small, oblong, 2–3 mm long, 1.5 mm across. Rare in China, cultivated for the stout rough pungent root, used as condiment, grated, personally I have not used it in China.

Brassica alboglabra Bailey (Syn. *B. oleracea* L. var. *alboglabra* [Bailey] Musil) — Chinese Kale, Chinese Broccoli (Figure 25c)

Jie-lan=Chieh-lan (芥蘭, Mustard Blue). Flowering shoots, developed in South China, bearing white flowers, available in Boston; annual crop cultivated in winter months in southern China.

Annual herbs, at flowering stage 35–60 cm high, glabrous and glaucous; leaves ovate-orbicular, rarely oblong-elliptic, 19–27 cm long, 16–27 cm wide, blue-green and glaucous, rarely gray-green, or grayish-green tinged yellow, smooth or wrinkled, some bullate, subentire, lobate and auriculate at the base, petioles 5–10 cm long, subterete; flowers white, in terminal racemes; siliques 3–9 cm long, beak 5–10 mm long; seeds rather large, 2 mm in diameter, grayish brown.

Jie-lan is the product of Cantonese horticulture and it followed the Cantonese emigrants to Southeast Asian and America. The crop is cultivated for the tender and crisp shoots. The plants bear no rossette of basal leaves. The tender shoots are harvested just as the first two flowers of the raceme open. Tender and crisp are the criteria for selection of good cultivars. At present 11 cultivars are recognized in Guangzhou and its vicinity.

1. cv. 'Lithocarpus Hill' (*Ke-zi-ling=K'o-tzu-ling* 柯子嶺).

Leaves ovate-orbicular, 19 cm long, 16 cm wide, gray-green, very glaucous, petioles 6 cm long, primary shoot 24 cm long at harvesting time, flowering cluster compact, 7 cm across; sown in June–July, harvested in September–February, portion cut 18 cm, weighing 90–130 g.

2. cv. 'Early Cockscomb' (*Zao-ji-guan=Tsao-chi-kuan* 早雞冠).

Leaves oblong, 28 cm long, 20 cm wide, wrinkled-bullate, very glaucous; petiole 9 cm long, primary shoot at harvesting time 35 cm long, weighing 135–180 g; sown in July–September, harvested in September–December, amount cut 30 cm.

3. cv. 'Greater Ball' (*Da-hua-qiu=Ta-hua-ch'iu* 大花球).

Leaves ovate-orbicular, 22 cm long, 21 cm wide, wrinkled, bullate, glaucous, petioles 6 cm long, primary shoot 26 cm long at harvesting time, flower-head 10 cm across; sowed in in August–September, harvest in October–December, amount cut 30 cm, weighing 90–180 g.

4. cv. 'Early Willow-leaf' (*Liu-ye-zao=Liu-yeh-tsao* 柳葉早).

Leaves oblong, 28 cm long, 17 cm wide, grayish green, smooth, glaucous, petioles 7 cm long, primary shoot at harvesting time 26 cm long; sown in August–September, harvested in November–January, portion cut 20 cm, weighing 90–120 g.

5. cv. 'Early Joint Star' (*Lian-xing-zao=Lien-hsing-tsao* 聯星早).

Leaves oblong, 20 cm long, 19 cm wide, rather thin, green, smooth, petioles 5 cm long, primary shoot 25 cm long, flower-cluster 6 cm across; sown August–September, harvested in November–January, amount cut 30 cm, weighing 90–135 g.

6. cv. 'Middle Season' (*Zhong-hua=Chung-hua* 中花).

Leaves oblong, 27 cm long, 19 cm wide, gray-green, hardly glaucous, petioles 8 cm long, primary shoot at harvesting time 32 cm long, flower-cluster 10 cm across; sown in September—November, harvested in November–February, portion cut 30 cm, weighing 135–180 g.

7. cv. 'Lotus Pond' (*He-tang=Ho-t'ang* 荷塘).

Leaves broad-elliptic, 22 cm long, 16 cm wide, green, smooth, barely gloucous, petioles 7 cm long, primary shoot 32 cm long at harvesting time; sown in September–November, harvested in November–February, portion cut 22 cm, weighing 110–125 g.

8. cv. 'Intermediate Late' (*Zhong-chi=Chung-ch'ih* 中遲).

Leaves large, ovate, 25 cm long, 18 cm wide, gray-green, hardly glaucous, tinged yellow, petioles 10 cm long, primary shoot 40 cm long at harvesting time, flower cluster 11 cm across; sown in September–November, harvested from December to the following March, portion cut 30 cm, weighing 120–150 g.

9. cv. 'Copper-shell Leaf' (*Tong-ke-ye=T'ung-k'o-yeh* 銅殼葉).

Leaves oblong-suborbicular, 19 cm long, 18 cm wide, light green, rather thin, hardly glaucous, petioles 10 cm long, primary shoot 28 cm long at harvesting time, flower cluster 9 cm across; sown in September–November, harvested in January–April, amount cut 20 cm, weighing 90–135 g.

10. cv. 'Late Flower' (*Chi-hua=Ch'ih=hua* 遲花).

Leaves suborbicular, 25 cm long, 23 cm wide, gray green, flat, smooth, hardly glaucous, petioles 11 cm long, primary shoot 35 cm long at harvesting time, flower cluster 9–10 cm across; sown in September–November, harvested in January–April, portion cut 35 cm long, weighing 135–180 g.

11. cv. 'Spring Pool' (*Quan-tang=Ch'üan-t'ang* 泉塘).

Leaves large and thick, orbicular, 22 cm long, 21 cm wide, gray-green, wrinkled, hardly glaucous, petioles 8 cm long, primary shoot at harvesting time 32 cm long, stout; sown in October–November, harvested in February–April, portion cut 25 cm long, weighing 220–215 g.

The sequence of the listing of the above cultivars is arranged by the time of sowing and harvesting. Like most cabbages and mustards, *B. alboglabra* grows well in a mild climate. In order to prolong the supplying season in tropical China, growers select for survival qualities in the seedlings that can grow in the hot humid summer between June and July and in the flowering plants with moderate shoot elongation in the rainy wet months of March and April. Their successes are witnessed by *Brassica* 'Lithocarpus Hill' and *Brassica* 'Early Cockscomb' which can be sown in June and July and consequently extend the harvesting season to September, and by *Brassica* 'Copper-shell Leaf', *Brassica* 'Late Flower' and *Brassica* 'Spring Pool', which extend the harvesting time to April. The last three named cultivars are selections made between 1950 and 1970. The first three cultivars have been in cultivation for many centuries.

The variations in the color and texture of the leaves clearly indicate the hybrid origin of *B. alboglabra* Bailey. The thick, firm, very glaucous gray-green or blue-green leaves of *Brassica* cv. 'Lithocarpus Hill' and cv. 'Early Cockscomb' show the relationship of the species to the European cabbage (*B. oleracea*), and the light green, hardly glaucous thin leaves of cv. 'Copper-shell Leaf' and the gray-green, barely glaucous leaves tinged yellow in cv. 'Intermediate Late' are characters of the South China mustard and cabbage (*B. chinensis* and *B. parachinensis*) which are always interplanted in the same bed with *B. alboglabra* in Guangzhou and its vicinities. Evidently, *B. alboglabra*, originated in similar manner as wheat and maize in ancient or medieval time when a form of European kale was brought to Guangzhou or Macao by the Arabian residents. The author of *Treatis of Canton Vegetables* (Anonymous, 1974) states that *jie-lan* has been cultivated since the eighteenth century, and *Brassica oleracea* var. *capitata* was introduced 200 years ago. Presently, the parental germplasm of European kale in *jie-lan* can be traced only to this point.

As a vetgetable in Chinese cuisine, the desirable qualities in *B. alboglabra* are tenderness, good flavor, crispness and firmness. To attain such qualities the tender shoots are cooked by stir-fry with little water added to ensure tenderness and good green color.

Brassica caulorapa Pasquale (Syn. *B. oleracea* L. var. *gongylodes* Kern) — KOHLRABI
    Pi-lan=P'i-lan (匹蘭, Piece Cabbage, or 擘蘭, Split Cabbage); *Qiu-jing-gan-lan=Ch'iu-ching-kan-lan* (球莖甘藍, Ball-stem Cabbage). Enlarged stem, common.

A biennial low and stout herb 40 cm high, the stem enlarging into a tuber 8–11 cm long, 10–12 cm in diameter just above the ground; leaves relatively few, oblong-elliptic, 20–28 cm long, 14–16 cm wide, dark green, glaucous, rounded at the apex, deeply parted at the base, petioles terete, 14–16 cm long, bearing small lobes at the apical end; flowers yellow; siliques 5–7.5 cm long, beaks short and thick. Origins unknown, introduced to China for several decades; eaten raw or cooked.

Brassica chinensis L. — PAKCHOI (Cantonese spelling for White Vegetable.) (Figure 25b)

Bai-cai=Pai-ts'ai (白菜, White Vegetable). Leaves with fleshy white petioles, dark green blades; a cultivar selected in South China.

An annual erect herbs cultivated as a leaf-crop in South China, 30–50 cm high at harvesting time; radical leaves forming a rosette with erect fleshy, white petioles, individual plants weighing 225–260 rarely up to 800 g, laminas varying with the cultivars, orbicular, oblong-ovate or obovate, 20–50 cm long, 15–20 cm wide, dark green generally, some cultivars having light green laminas, petioles 7–20 cm long, 2.5–5 cm wide across the base, 5–8 mm thick, cauline leaves clasping; flowers yellow, in elongated racemes; siliques cylindrical, 5–10 cm long, 7 mm across, beaks slender, 2 cm long; seeds brown or yellow.

Pakchoi was developed in tropical China about the seventh century, in the Tang Dynasty. It has followed the Cantonese emigrants to Southeast Asia and to the New World, and it is the common ingredient in American Chinese dishes. In Guangzhou 15 known cultivars have been recorded, 9 of which have very long history. Formerly, it was a crop for autumn and winter. Now, there are several cultivars that can be sown in January–March and are ready for the market in March–May and others sown in April–August for the market during June–October. Consequently, pakchoi in Guangzhou and Hong Kong is available throughout the year.

Brassica hirta Moench (Syn. *B. alba* [L.] Rabenhorst, *Sinapis alba* L.) — WHITE MUSTARD

Bai-jie=Pai-chieh (白芥, White Mustard). Common in restraurants and homes in North China.

An annual or biennial erect herb, cultivated for the seeds, up to 1 m high, branchlets hairy; lower leaves long petiolate, laminas obovate, 10–15 cm long, deeply parted or lobed; flowers yellow; fruits hairy, 2–3 cm long; seeds 1.5–2 mm in diameter, pale yellow or white. Cultivated in northern China, seed used both in food and medicine.

Brassica juncea (L.) Czernajew — MUSTARD GREENS, LEAF-MUSTARD, SWATOW MUSTARD, INDIAN MUSTARD

Annual or biennial herb, the most extensively cultivated species, with many cultivars; leaves uniformly green, the petioles and laminas of same color, harvested before the development of the flowering stem. Many garden forms, mostly grown from seeds, five distict types established in different regions of China are reported here.

1. var. **foliosa** L. H. Bailey (Syn. *B. juncea* var. *rugosa* [Roxburgh] Tsen et Lee; var. *integrifolia* [Ruprecht] Sinskaja) — SWATOW MUSTARD, BIG-STEM MUSTARD, WRAPPED HEART MUSTARD, BAMBOO MUSTARD (Figure 26c, d)
   Qing-cai=Ch'ing-ts'ai (青菜, Green Vegetable); *Da-jie-cai=Ta-chieh-ts'ai* (大芥菜, Big Mustard).

An annual herb extensively cultivated in southern and western China as a winter crop, the plants attaining 50–60 cm before developing flowering stems, and ready to be harvested; leaves oblong in outline, petioles very short, fleshy, broadened to fan-shaped with fleshy thick midrib, the laminas foliaceous, folded, lobed. This is the common mustard green used by the Chinese population in Boston area, cultivated in New Jersy and thence southward (available throughout the year) and harvested just before the development of the aerial stem and the cauline leaves. In the retail stores, the damaged portions of the outer leaves incurred in packing, shipping and storage are trimmed off (Figures 27d). A sauerkraut-like fermented sour and aromatic preparation of the halves and quarters of the plant are imported in large earthernware vessels from Taiwan or Hong Kong, called Sour Vegetable 酸菜(*Suan-cai=Suan-ts'ai*), and sold in American Chinese groceries. It is very popular among the Chinese Americans.

2. var. **gracilis** Tsen et Lee — HEMP MUSTARD
   Ma-cai=Ma-ts'ai (麻菜, Hemp Mustard); *Ku-cai=K'u-ts'ai* (苦菜, Bitter Mustard); *Xi-ye-jie-you-cai=Hsi-yeh-chieh-yu-ts'ai* (細葉芥油菜, Small-leaved Mustard).

An annual herb distinguished by the smaller leaves, distictly petiolate, the laminas with stiff hairs, and lobed serrate margin. Cultivated in Jiangsu.

3. var. **megarrhiza** Tsen et Lee (Syn. *B. napiformis* L. H. Bailey) — LARGE-ROOTED MUSTARD, BIG-HEAD CABBAGE
   Da-tou-cai=Ta-t'ou-ts'ai (大頭菜, Big Head Vegetable). Enlarged fleshy root, used for preserves, especially famous in Yunnan and Shandong.

A biennial pungent herb, distinguished by the large fleshy white obovoid-conical root, obovate or oblong dark green leaves, irregularly shallow-lobed and serrate, with stiff hairs on the nerves; cultivated in northern China and Yunnan.

4. var. **multisecta** L. H. Bailey (Syn. var. *multiceps* Tsen et Lee) — THOUSAND-LEAVES
   Xue-li-hong=Hsueh-li-hung (雪裡紅, Red-in-Snow); *Jiu-tou-niao=Chiu-t'ou-niao* (九頭鳥,

Nine-headed Bird); *Mei-gan-cai=Mei-kan-ts'ai* (霉乾菜, Mold Dried Vegetable); *La-cai=La-ts'ai* (辣菜, Pungent Vegetable); *Chun-bu-lao=Ch'un-pu-lao* (春不老, Spring Tender Green).

Leaves obovate-oblanceolate in outline, irregularly parted and lobed, petioles comparatively short, subterete, 1 cm in diameter, quite tender at harvesting time; flowers orange-yellow, 1.5 cm across. Seeds imported from Beijing and Lanzhou were sowed in Amherst and Cambridge, Massachusetts, with very satisfactory results, even over-wintered under some oak leaves gathered around the plants by wind. Under good cultivation the plant grows well, forming many-basal sessile shoots from the crown, thus the name "Nine-headed Bird".

5. var. **tumida** Tsen et Lee — SICHUAN PRESERVED VEGETABLE

Zha-cai=Cha-ts'ai (榨菜, Pressed Vegetable); *Si-chuan-zha-cai=Szu-chuan-chai-ts'ai* (四川榨菜, Sichuan Pressed Vegetable).

Young plants of this cultivar appearing similar in color and texture to the above variety, differing in the seedlings being smoother and more glabrous, and the stem of the rosette enlarged and the petioles lumpy before the initiation of the flower-bud. The enlarged stem, fleshy petiolar bases, and the central bud with primordia of stem and leaves are trimmed off the plant, washed, halved or quartered, withered slightly, and preserved with salt, powdered cayenne pepper and zanthoxylum pericarps under pressure. A town called Fuling in Sichuan is the center of production of this preserve which has a worldwide circulation, hence the name Sichuan Pressed Vegetable.

Brassica napobrassica Miller — RUTABAGA

Wu-qing-gan-lan=Wu-ch'ing-kan-lan (蕪菁甘藍, Rutabaga); *Yang-tou-cai=Yang-t'ou-ts'ai* (洋頭菜, Foreign Head Vegetable). Enlarged root; cultivated in large cities such as Shanghai and Guangzhou; also being introduced by early French missionaries to the homeland of the giant panda where the mountainous people lived in isolation have continued to plant it; rare elsewhere.

A glaucous biennial herb 50–60 cm high, with large solid, yellow-fleshy root; leaves dark green, the basal ones lyrate-pinnate, 45 cm long, 20 cm wide, the lobes decreasing in size downward, the terminal segment obtuse, irregularly dentate, petioles 14–20 cm long, slightly winged, cauline leaves clasping the stem; flowers light yellow; siliques 5–6 cm long, with very short beak. It is a winter crop in Guangzhou, sown in October and harvested in early March.

Brassica narinosa L. H. Bailey — BROAD-BEAKED MUSTARD

Ta-wo-cai=T'a-wo-ts'ai (搨窠菜, Spreading Vegetable); *Ta-di-cai=T'a-ti-ts'ai* (搨地菜, Flat-on-ground Vegetable); *Xue-li-qing =Hsueh-li-ch'ing* (雪裡青, Green-in-Snow). Leaves, especially good after frost, very hardy, able to overwinter in Boston.

A biennial herb 30–50 cm high at flowering and fruiting stage, glabrous throughout; basal leaves dark green with white petioles and major veins, forming a rosette spreading over the ground, laminas orbicular or obovate, 10–20 cm long and wide, entire, wavy, bullate, the petioles 8–10 cm long, cauline leaves obovate-orbicular, sessile, clasping the stem; flowers pale yellow, 10 mm across; silique short and stout, inflated, 2–4 cm long, with a short and stout beak 4–8 mm long; seed 1 mm in diameter, dark brown.

Brassica oleracea L. — WILD CABBAGE, CABBAGE, COLES, KALES

Gan-lan=Kan-lan (甘藍, Sweet Blue). Leaves, stem and flower-bud modification. Many varieties.

Annual or biennial herbs 1 m or more high at flowering and fruiting stage, glabrous throughout; leaves firm, dark green and bluish, glaucous, forms and sizes varying with the varieties; flowers pale yellow, racemose, 1.5–2.5 cm long, sepals erect; siliques cylindrical, 6 9 cm long, beaks short; seeds rather large, 1.5–2 mm in diameter, brown. Three distinct varieties common in China.

1. var. **botrytis** L. (Syn. *B. botrytis* (L.) Miller, *B. cauliflora* Garsault) — BROCCOLI, CAULIFLOWER (Figure 4b)

   Hua-cai=Hua-ts'ai (花菜, Cauliflower). Edible white flowering shoots, common in China; *Ye-hua-cai=Yeh-hua-ts'ai* (椰花菜, Broccoli). Edible green flowering shoots very rare in China.

Stem stout with oblong-elliptic leaves 20–30 cm long, 8–14 cm wide, obtuse at both ends, bearing a single terminal compact colorless or green head of flowering buds, formed by the condensed and consolidated fleshy flower-stems and malformed flower buds; the market material 15–20 cm across, with the subtending green leaves partially removed.

2. var. **capitata** L. — CABBAGE, HEAD CABBAGE

   Lian-hua-bai=Lien-hua-pai (蓮花白, Lotus Flower White); *Bao-xin-cai=Pao-hsin-ts'ai* (包心菜, Wrap-heart Vegetable); *Yang-bai-cai=Yang-pai-ts'ai* (洋白菜, Foreign Cabbage); *Gan-lan=Kan-lan* (甘藍, Sweet Blue). Extensively cultivated throughout China.

Stem short and stout, bearing a rosette of large spreading leaves; laminas obovate or orbicular, 15–40 cm long, the central ones forming a compact terminal ball of light green and gradually becoming white leaves; reportedly containing Vitamin U, good for healing stomach ulcers.

3. var. **gemmifera** Zenker — BRUSSELS SPROUTS

Qiu-ya-gan-lan=Ch'iu-ya-kan-lan (球芽甘藍, Ball-bud Cabbage); *Bao-zi-gan-lan=Pao-tzu-kan-lan* (抱子甘藍, Holding Baby Cabbage). Special buds, very rare.

Stem stout, rather tall, unbranched, the axillary buds of the middle section enlarged, remaining on the side of the stem after the falling of the leaves, becoming compact heads 3–5 cm in diameter, green.

Brassica parachinensis L. H. Bailey (Syn. *B. chinensis* var. *parachinensis* [Bailey] Tsen et Lee — MOCK PAKCHOI, FALSE PAKCHOI (Figure 25a)

Jian-gan-bai=Chien-kan-pai (箭杆白, Arrow-shaft White); *Su-zhou-qing=Su-chou-ch'ing* (蘇州青, Suzhou Green); *San-yue-bai=San-yüeh-pai* (三月白, Third Month White). Leaves.

An annual herb cultivated in eastern China for the radical leaves, sown in spring and summer; basal leaves well developed, closely congested, the laminas ovate or ovate-oblong, 15–30 cm long, apex rounded, base cuneate, strongly venose, petioles 6–15 cm long, greenish-white, the lower cauline leaves smaller, but with shape and petioles similar to the above, the upper cauline leaves ovate-lanceolate, not clasping; flowers yellow; seeds prominently pitted. Bailey's original material was from Nanjing and Shanghai; the three local names given above bespeak the significant characters of the species, namely the petioles long and white tinged green by "Arrow-shaft White", the area of cultivation by "Suzhou Green" (Suzhou is half way between Nanjing and Shanghai), and the time of harvesting the crop by "Third Month White". Since Chinese farmers and gardeners are accustomed to the lunar calender, the third month falls in early spring. Much used for preserves (see Part I for more information).

Brassica parachinensis Bailey cv. 'Caixin' (Figures 4b and 25a).

Cai-xin=Ts'ai-hsin (菜心, Vegetable Heart). Flowering shoots, a winter crop in South China.

A tropical annual crop cultivated in southern China for the flowering shoots; radical leaves poorly or not at all developed, never forming any rosette, laminas oblong, suborbicular or obovate, 16–30 cm long, petioles slender 9–15 cm long, green as the laminas or tinged white, the cauline ones always petiolate; 24 cultivars recorded, adopted for planting in spring, summer and autumn. Early cultivars sown in April–August, with the primary flower shoot appearing on the fourth or fifth node, continuing harvests lasting for 20–30 days; intermediate cultivars sowed in September–October, with the primary flowering shoot appearing on the seventh or the eighth node, continuing harvest lasting 20–30 days; late cultivars sowed in November–December, primary shoots appearing on the seventh or the tenth node, continuing harvests lasting for 10–15 days; two examples of each type are given below to show the variations in the leaf and the flower characters.

1. cv. 'Hurried Early Heart' (*Ji-zao-xin=Chi-tsao-hsin* 急早心)

Leaves oblong-elliptic or ovate, 16 cm long, 8 cm wide, yellowish-green, petioles 13 cm long, light green; primary shoots on the 4–5th nodes, 30 cm high; sown in May–August, harvest after 30 days, stem 1.5 cm across.

2. cv. 'April Nine Heart' (*Si-jiu-xin=Ssu-chiu-hsin* 四九心)

Leaves oblong-suborbicular, 22 cm long, yellowish-green, petioles 13 cm long, light green; primary shoot on the 4–5th nodes, 22 cm long, 1.5–2 cm across; sown any time between April and October, best season in May–September, harvest in 28–38 days, continuing harvesting for 10 days.

3. cv. 'October Heart' (*Shi-yue-xin=Shih-yueh-hsin* 十月心)

Leaves ovate, 22 cm long, 11 cm wide, yellowish-green, petioles 8 cm long, yellowish-green; primary shoot appearing on the 7–8th nodes, 32 cm long, 2 cm in diameter, yellow-green, first harvest after sowing 50–55 days, continuing harvests for 30 days, quality very good.

4. cv. 'Greater Ball Heart' (*Da-qiu-xin=Ta-ch'iu-hsin* 大球心)

Leaves ovate, 25 cm long, 13 cm wide, dark green, petioles 8 cm long, light green, primary shoot appearing on the 6–7th nodes, 32 cm high, 22 cm in diameter, flowering head 6–7 cm across, first harvest 50–60 days after sowing, continuing harvest for 30 days, medium quality.

5. cv. 'Yellow Tail Late' (*Huang-wei-chi=Huang-wei-ch'ih* 黃尾遲)

Leaves ovate or suborbicular, 20 cm long, 14 cm wide, yellowish-green, petioles 11 cm long, light green; primary shoot appearing above the tenth node, 27 cm high, stem 2 cm across, good quality.

6. cv. 'Third Month Green' (*San-yue-qing=San-yueh-ch'ing* 三月青)

Leaves broad ovate, 35 cm long, 20 cm wide, dark green, petioles 15 cm long, light green tinged white; primary shoot appearing on the 6–7th nodes, 30 cm high, 1.2 cm in diameter, first harvest 50–55 days after sowing, continuing harvest lasting 10–15 days.

Brassica pekinensis (Loureiro) Ruprecht — CELERY CABBAGE, CHINESE CABBAGE (Figure 26a and b)

Bai-cai=Pai-ts'ai (白菜, White Vegetable); *Da-bai-cai=Ta-Pai-ts'ai* (大白菜, Greater White Vegetable — for mature plant); *Xiao-bai-cai=Hsiao-pai-ts'ai* (小白菜, Lesser White Vegetable — for the seedlings and immature plants harvested in thinning); *Juan-xin-cai=Chüan-hsin-ts'ai* (捲心菜, Curly Heart Vegetable — for the head), all the above names used in northern China; *Shao-cai=Shao-ts'ai* (紹菜, Connected Vegetable); *Huang-ya-bai=Huang-ya-pai* (黃芽白, Yellow Bud White); the two last names used in Guangzhou; *Song=Sung* (菘, Loose-leaved Cabbage — ancient literature name).

A biennial vegetable crop extensively cultivated in northern China, the mature growth before flowering 50–60 cm high; radical leaves very well developed, forming a rosette 40–50 cm in diameter, the intermediate leaves erect, and the inner ones forming an oblong-cylindrical head changing from green to yellow-white, individual leaves 40–50 cm long, 35–50 cm wide, wrinkled and bullate, petioles broad, white, fleshy, the upper portion dividing into palmate nerves, cauline leaves sessile and clasping; flowers yellow, racemose; fruits 2.5–7 cm long, pericarp veiny, beak tapering; seeds dark brown, smooth. This is the most important vegetable for northern China, for the keeping quality of the tight head, every family gets a winter supply.

Brassica rapa L. — FIELD MUSTARD (Syn. *B. campestris* L.)

An erect, much branched and floriferous annual or biennial herb with yellow-green and slightly glaucous vegetative growth; radical leaves petiolate, lyrate-pinnatifid, 20–40 cm long, terminal segment ovate-subrotundate, entire or wavy, lateral segment 2 or 3 on each side, lower cauline leaves petiolate, upper cauline leaves sessile, auriculate or clasping, hispid; flowers bright yellow, racemose; siliques linear, 3–8 cm long, 2–3 mm across; seeds 1.5 mm in diameter, reddish brown or yellowish brown, finely alveolate. Three cultivars planted for vegetable and oil in western and northern China.

1. cv. 'Campestris' (*B. rapa* L. var. *campestris* [L.] Koch)

You-cai-tai=Yu-ts'ai-t'ai (油菜苔, Oil Rape Shoot). Flowering shoots, a spring crop of the Yangtze areas.

Vegetative growth yellowish-green, smooth; cultivated for the tender flowering shoots, a very popular vegetable in Chengdu and its vicinity.

2. cv. 'Oleifera' (*B. campestris* L. var. *oleifera* DC.)

You-cai=Yu-ts'ai (油菜, Oil Rape). Seed for oil, used for cooking and for light in Sichuan, Qinghai.

Vegetative growth similar to the above cultivar, cultivated in fields as a winter crop in Chengdu, flowered in early spring, harvested before getting the fields ready for planting rice. In the mountains of Qinghai, it is a summer crop flowering in August.

3. cv. 'Purpurea' (*B. campestris* L. var. *purpuraria* Bailey)

Hong-you-cai=Hung-yu-ts'ai (紅油菜, Red Oil Rape).

The vegetative growth brilliant purple-red; cultivated for the flowering shoot, sown and harvested at the same time as for cv. 'Compestris', the difference lies in the purple-red pigment in the cells.

Capsella bursa-pastoris L. — SHEPHERD'S PURSE

Ji-cai=Chi-ts'ai (薺菜, Shepherd's Purse); *Di-di-cai=Ti-ti-ts'ai* (地地菜, Ground Vegetable). Young plants with roots, a delicacy in northern and eastern China.

Pubescent annual or biennial herbs, 15–40 cm high at flowering and fruiting stage; radical leaves forming a rosette close to the ground, 6–15 cm across, petiolate, the laminas various in size and shape, spathulate or lyrate-pinnatifid, oblanceolate, occasionally entire, up to 10 cm long, 1.5 cm wide, cauline leaves progressively decrease in size, sessile, lanceolate; flowers small, white, in terminal racemes; fruit triangular, 4–7 mm across the truncate apex, with short persistent style, septum elliptic, 1 mm across the middle; seeds light brown, 1 mm long. A weed in lawns and gardens, survives the winter covered by snow, cultivated in Shanghai.

Cardamine leucantha (Tausch) O. E. Schulz

Bai-hua-shi-jie-cai=Pai-hua-shih-chieh-ts'ai (白花石芥菜, White-flowered Rock Mustard). Whole plant, used as a substitute for tea.

Pubescent perennial herbs 30–80 cm high in flower, growing in damp forested areas; radical leaves 2 or 3, petiolate, pinnately compound, leaflets 5 or 7, the terminal one elliptic-lanceolate, 4–6 cm long, 1–3 cm wide, serrate, petioles 2–6 cm long, cauline leaves similar but smaller; flowers white, in terminal racemes, 5–8 cm across; siliques hirsute, linear, 1.5–2.5 cm long, 1.5 mm in diameter, style persistent, 3–5 mm long; seeds brown, 2 mm long.

Cardamine lyrata Bunge

Shui-tian-sui-mi-ji=Shui-t'ien-sui-mi-chi (水田碎米薺, Paddy Water-cress). Young shoots, taken as potherb.

A glabrous perennial herb growing near water or in damp areas; with creeping stolens; leaves dimorphic, those on the creeping stem simple, orbicular-ovate, undulate along the margin, 2–3 cm in diameter, radical and cauline leaves pinnate-compound, or pinnatifid, 7 cm long, the terminal leaflets 6–25 mm long and wide, lateral segments 4–7 pairs, subsessile, the basal pair stipule-like; flowers white, small, 5–7 mm across; siliques linear, compressed, 3 cm long, 2 mm across, beak 4 mm long; seeds in one row, oblong, 2 mm long, winged.

Cardamine macrophylla Willdenow

Da-ye-sui-mi-ji=Ta-yeh-sui-mi-chi (大葉碎米薺, Large-leaved Bitter Cress). Young plants, used for potherbs in Hebei.

Perennial herbs 30–60 cm high, rhizomes creeping, erect flowering stems unbranched; leaves pinnate, 7–10 cm long, leaflets 9–11, elliptic or ovate-lanceolate, 3–8 cm long, 0.7–2 cm wide, obtuse or acuminate, serrate, pilose; flowers purple, in terminal racemes, sepals ovate-oblong, 5 mm long, green and tinged red, petals obovate, 7–8 mm long, clawed; siliques linear, 2–3.5 cm long, flattened; seeds oblong, 2 mm long, brown. Widespread in temperate China.

Cardamine regeliana Miquel; (Syn. *C. hirsuta* L.) — Bitter Cress

Sui-mi-ji=Sui-mi-chi (碎米薺, Broken Rice Capsella). Young shoots.

Sparsely hairy annual herbs growing along roadsides or on lawns; radical leaves pinnate compound, with hirsute petioles, the terminal leaflets suborbicular-ovate, lobate, 8–10 mm across, the lateral ones smaller; flowers white, small, in terminal racemes, corolla 2–3 mm long; siliques linear, 2–2.5 cm long, 1 mm across. Miquel's material was from Japan; with larger leaflets, the terminal ones 3–4 cm long and wide.

Cardamine urbaniana O. E. Schulz

Cai-zi-qi=Ts'ai-tzu-ch'i (菜子七, Herbal Hematinic); *Hua-zhong-sui-mi-ji=Hua-chung-sui-mi-ch'i* (華中碎米薺, Central China Bitter Cress). Young plants before flowering, used for vegetable.

Perennial herbs 30–100 cm high, with stout rootstock, branchlets pubescent; leaves pinnate, 7–20 cm long, leaflets 5–13, ovate-lanceolate or oblong, 3–6 cm long, 1–1.8 cm wide, acuminate, base acute, serrate, sparsely pilose; flowers pale purple, in terminal racemes, pedicels 3–10 mm long; siliques compressed-cylindrical, 5 cm long. Endemic to central China, growing in forests along streams at altitudes of 1,200–3,000 m.

Chorispora tenella (Pallas) de Candolle

Li-zi-cao=Li-tzu-ts'ao (離子草, Loose Seed Herb). Young shoots.

Glandular hairy annual herbs 8–30 cm high, much branched; radical and lower cauline leaves oblong-oblanceolate, 3–4 cm long, 3–5 mm wide, acute at apex, cuneate at base, pinnately shallow-lobed, upper cauline leaves sessile, dentate; flowers lavender coloured, small, 3 mm long; siliques cylindrical, 3–5 cm long, slightly curved, transversely septate, indehiscent, beaked; seed oblong, 1.5 mm long, light brown.

Eruca versicaria (L.) Cavanilles — ROCKET (Syn. *E. sativa* Miller).

Zhi-ma-cai=Chih-ma-ts'ai (芝麻菜, Sesame Vegetable). Young plant, used as potherb in northern China and Inner Mongolia.

Pubescent annual herbs 10–50 cm high; radical and lower cauline leaves obovate in outline, pinnatifid, 4–7 cm long, 2–3 cm wide, rounded at the apex, cuneate at base, petioles 2–4 cm long; flowers yellow with purple veins, 1–1.5 cm long; siliques cylindrical, 2–3 cm long, 5 mm across, beak short, broad; seeds globose-ovoid, 1.5–2 mm in diameter, light brown.

Erysimum cheiranthoides L. — WORMSEED MUSTARD

Xiao-hua-tang-jie=Hsiao-hua-t'ang-chieh (小花糖芥, Small-flowered Sugar Mustard). Young shoots, eaten in Hubei.

Pubescent annual herbs 10–50 cm high, pubescent with 2–4 branched hairs throughout; leaves subsessile, linear-lanceolate, 2–5 cm long, 2–5 mm wide, acute, base attenuate, entire or wavy-denticulate; flowers yellow, in terminal racemes, pedicels 2–5 mm long, sepals linear, 2–3 m long, erect, petals obovate, 3–5 mm long, unguiculate, stamens tetradynamous, ovary linear, pubescent, style short, stigma capitate; siliques linear, 2.5–4 cm long, pubescent. Widespread weed, growing in waste places and along paths in the northern hemisphere and northern Africa.

Eutrema wasabi Maximowicz — WASABI (Japanese name), ASIAN HORSERADISH

Shan-yu-cai=Shan-yu-ts'ai (山俞菜, Mountain Radish). Root.

Glabrous perennial herbs growing in damp forest floor, 30–80 cm high, stems stout, 4–6 mm in diameter; radical leaves 3–4, long petiolate, laminas reniform, 5–10 cm long, 6–11 cm across, coarsely dentate or undulate, petioles 20–30 cm long, cauline leaves ovate 4–5 cm long, 4 cm across the base, cordate; flowers white, racemose, 5–6 mm long; siliques poorly developed, 1 cm long, 2 mm across the middle, stipitate, beak 3 mm long.

Lepidium sativum L. — GARDEN CRESS

Du-xing-cai=Tu-hsing-ts'ai (獨行菜, Lonely Vegetable); *Mai-jie-cai=Mai-chieh-ts'ai* (麥楷菜, Wheat Straw Vegetable). Root, used as spice.

Glabrous, glaucous annual herbs 30–50 cm high; radical leaves pinnatifid, the segments toothed, cauline leaves becoming successively simple, linear, entire; flowers small, racemose; fruit oblong-ovate, winged, retuse.

Orychophragmus violaceus (L.) O. E. Schulz

Zhu-ge-cai=Chu-ke-ts'ai (諸葛菜, Premier Zhu-ge's Vegetable). Young plants; introduced from Beijing to Weston, Massachusetts, by Professor R. A. Howard in 1978, over-wintered outdoors, flowered, reseeded.

Glabrous, glaucous annual or biennial herbs 10–50 cm high; radical leaves petiolate, pinnatifid, 3–8 cm long, 1.5–3 cm wide, the terminal segment reniform or triangular, crenate, cauline leaves oblong, clasping; flowers purple, in terminal racemes, 2 cm across; silicles 7–10 cm long, 4-ridged, beak 1.5–2.5 cm long; seeds black-brown; native of northern China.

Pugionium cornutum (L.) Gaertner

Sha-jie=Sha-chieh (沙芥, Sand Mustard). Young leaves, cultivated.

Xerophytic annual or biennial herbs growing in the desert of northwestern China, 50–100 cm high; fleshy, basal leaves petiolate, pinnatifid, 10–20 cm long, 3–4 cm wide, the terminal segment ovate or oblong, entire or with 2 or 3 teeth, cauline leaves lanceolate-linear, entire; flowers yellow, in loose racemes, 1–1.5 cm across; silicles laterally compressed, 3–3.5 cm long, 3–5 mm wide, indehiscent, 1-seeded samara, each side with a lanceolate wing and 6–8 spines.

Raphanus sativus L. — RADISH, LUOBO (Figure 25d-i)

Luo-bo=Lo-po (蘿蔔, Chinese Radish). Fleshy roots, important vegetable throughout China; eaten raw, cooked or pickled; a cylindrical white type available in Boston Chinese stores throughout the year. *Lai-fu=Lai-fu* (萊服, Young Radish Plant). Whole plant, an ancient herbal name, now accepted as official in *An Encyclopedia Chinese Medicines* (Anonymous, 1977, p. 1800). *Luo-bo-ye=Lo-po-yeh* (蘿蔔葉, Radish Leaf). Cooked with meat or fish; partially dried and salted for preserves. *Luo-bo-zi=Lo-po-tzu* (蘿蔔子, Radish Seed). Seeds, roasted, pulverized, mixed with sugar for a sweet tea given to people (children specially) with gas pain.

Hispid or subglabrous annual or biennial herbs, 1 m high in flower, cultivated primarily as a summer crop in northern China, as a winter crop in southern China; radical leaves in a rosette, obovate-oblanceolate in outline, subentire and wavy, or pinnatilobed to pinnatisect, 20–50 cm long, 6–15 cm across, rounded at apex, cuneate or lobed at the base, petioles 2–7 cm long; flowers white, pink, or light lavender, racemose, 1–1.5 cm across; silicles fleshy, becoming spongy at maturity, cylindrical horn-like, 3–5 cm long, constricted, indehiscent; seeds globose-ovoid, 3 mm long, reddish-brown.

Raphanus sativus L. is a native of the Mediterranean Region. It is known in China only as cultigens. The exact time of this anthropogenous occurrence in China is not yet ascertained. The name *luo-bo* (蘿蔔) first appeared in a herbal prepared in the late seventh century (Mong, 684). Now, it is one of the mostly extensively cultivated vegetables in China, with the red types more prevailing in northern China, and the white ones in southern and western China. Current researches indicate that northern China, particularly in the vicinity of Beijing, has been the center of successful selections of unusual cultivars. The author of *An introduction of common vegetables of Beijing* (Anonymous, 1974) described 23 cultivars illustrated with two colored plates and 23 figures. Listed were criteria on selections for satisfactory eating and keeping qualities, high unit area productivity and good pricing to meet special market demands, such as early spring crops with short growing period and fancy types eaten as fresh fruits. Two samples of each type are given below to show the diversities in leaf-division, root-size, shape, weight and colors of outer skin and inner meat; and the time of sowing and harvesting.

I. Types for general supplies

A. Red root with white meat: Two from spring and two from summer crops,

1. cv. 'Four-leaved Radish' (*Si-ying=Ssu-ying* 四英).

Leaves 4–5, long petiolate, petioles purplish on the adaxial side, laminas subentire, oblong-obovate, wavy; roots obconical, 5 cm long, 2.5 cm across top, weighing 10 g; sown between mid-January and early February under cover, facing the sun, harvested between mid-March and mid-April.

2. cv. 'Five-leaved Radish' (*Wu-ying=Wu-ying* 五英).

Leaves 5–6, long petiolate, petioles purplish-red on the adaxial side; laminas subentire, wavy, obovate; fleshy roots obconical, 8 cm long, 3 cm across the top, weighing 18–14 g; sown in mid-March; harvested between late April and early May, if later becoming pithy.

3. cv. 'Cut-leaved Red' (*Hua-ye-hong-pao=Hua-yeh-hung-p'ao* 花葉紅袍).

Leaves 20, erect, laminas oblanceolate, pinnatisect nearly to the base, lobes 7–8 pairs, midrib and veins red; fleshy roots oblong, 9–14 cm long, 9–10 cm across, weighing 660 g; sown in late July, harvested in late October, per acre yield 4,000 to 5,000 kg; good for cooking.

4. cv. 'Rural Great' (*Nong-da-hong=Nung-ta-hung* 農大紅).

Leaves 18–20, suberect, petioles stout, red, laminas oblong-oblanceolate in outline, pinnatisect; fleshy roots light red outside, oblong-obovoid, 15–24 cm long, 14–15 cm diameter; weighing 950 g; sown in late July, harvested in middle or late October; per acre yield 2,300–2,600 kg; a natural hybrid selected in 1965 in Fengtai District (封台) of Beijing.

B. White root with white meat.

5. cv. 'Rock White' (*Shi-bai=Shih-pai* 石白).

Leaves 25, erect, petiolate, the petioles light green, laminas subentire or pinnatifid; fleshy roots subcylindrical, 25 cm long, 8 cm across, weighing 750 g; sown in late July, harvested in late October, per acre yield 1,980–2,300 kg; crisp, not pungent, good for cooking.

6. cv. 'Beauty Intense' (*Mei-nong=Mei-nung* 美濃).

Foliage of strong growth, leaves 40, light green, hispid throughout, the laminas pinnatisect, lobes 13 pairs; fleshy roots 55 cm long, 7 cm in diameter, cylindrical, the distal end gradually thinned, weighing 900–1,320 g; sown in late July–early August, harvested in late December; per acre yield 1,900–2,400 kg; good eating quality, not pungent.

C. Green root with green meat; good cooking and good carrying qualities, imported material available in American Chinese groceries around Chinese New Year time.

7. cv. 'Green Head Out' (*Lü-tou-qing=Lü-t'o-ch'ing* 綠頭青).

Leaves 25–30, erect, laminas pinnatisect, lobes 5–6 pairs; fleshy roots cylindrical, 25 cm long, 8–9 cm in diameter, much exposed; sown in late July, harvested in late October, per acre yield 1,980–2,400 kg; pungent, much used in radish-pickle industry.

8. cv. 'Wild Geese-egg' (*Tian-e-dan=T'ien-o-tan* 天鵝蛋).

Leaves 20, suberect-ascending, petioles obscure, laminas oblanceolate, pinnatisect, lobes 9–12 pairs; fleshy roots subspherical-oblong, 12 cm long, 11 cm in diameter, the meat changing from green at the apical portion gradually to white at the distal end, weighing 330–600 g, the larger ones up to 1.5 kg; sown in late July, harvested in late October, per acre yield 1,650–1,980 kg; fine cooking and keeping qualities.

II. Types for special supplies.

A. Fruit substitute: Attractive appearance and sweetish taste.

9. cv. 'Beautiful Heart' (*Xin-li-mei=Hsin-li-mei* 心裡美).

Rosette having 20 dark green suberect or spreading leaves, laminas oblanceolate, subentire or pinnatisect with 6–8 pairs of lobes; fleshy roots short cylindrical, 11 cm long, 11–12 cm in diameter, 10 cm in diameter, green with a white tail, inside brilliantly purple-red, straw-colored, or with mixed pattern of green-red-straw colors, weighing 350–450 g; sown in early August, harvested in late October, per acre yield 1,650 up to 2,310 kg; crisp, sweetish, good quality for eating raw, an important substitute for fresh fruits.

10. cv. 'Green-and-Green' (*Qing-pi-qing=Ch'ing-p'i-ch'ing* 青皮青).

Rosette leaves 12–13, suberect, petioles short, laminas subentire and wavy, obovate, 14–16 cm long, 8–12 cm wide, midribs green; fleshy roots oblong, 12–15 cm long, 11–12 cm in diameter, weighing 330–450 g; sown in eary August, harvested in late October, per acre yield 760–900 kg; crisp, juicy, slightly pungent at first, on keeping becoming sweetish, good eating quality fresh.

B. Pickle industry: Low water content and suitable size.

11. cv. 'Twine Leaves' (*Er-ying-zi=Erh-ying-tzu* 二英子).

Leaves 10, erect, distinctly petiolate, petioles green and tinged purple near the base, laminas subentire, wavy, oblanceolate; fleshy roots cylindrical, 13 cm long, 2–3 cm in diameter, white on outside and inside, rind thin, water content low, meat crisp, weighing 50–60 g; sown in late July–August, harvested after 45–60 days; per acre yields 725–990 kg; special selection to supply the radish-pickle industry.

12. cv. 'Purple Buds' (*Zi-ya-qing=Tzu-ya-ch'ing* 紫芽青).

Rosette of leaves spreading, distinctly petiolate, the petioles light green, gradually changing to purple, hence the trivial name "Heart-leaves Purple"; laminas pinnatisect, lobes 6–8 pairs; fleshy roots oblong-obconic, 14–19 cm long, 10 cm in diameter, green outside, light green inside, weighing 400–500 g; sown in early August, harvested in middle October, per acre yield 1,300–1,650 kg; crisp, less juicy, used primarily for salted preserves.

Note: Radish, an exotic element in Chinese diet, was introduced from the Mediterranean Region with ancient travellers, perhaps in seed form for gas pains, more than as a vegetable. Employing the germplasm, the people in northern China gradually developed different types of enlarged root which they called '*luo-bo*.' The Chinese *luobos* are very different from the radish in European and American gardens, in color, size and content.

A review of the illustrated manuals of vegetables of different cultural centers in China, such as Shanghai and Guangzhou, indicates that the diversity of forms reduced from north to south and the records on the fusion of types of cultivar has been from northern China, particularly Beijing and Shandong, to eastern, western and southern China, where some high yield types are cultivated for general supply for cooking and prickles. Evidently, people in eastern, western and southern China have not cultivated the technique or learned to appreciate the taste of eating raw *luobo* as fruits. In the mid-1920s and late 1930 English red radish and French black radish were introduced to eastern China, particularly to Shanghai and Nanjing. These were for supplying Western residents and/or for horticultural experiments. They were hardly known by the Chinese public.

Rorippa cantoniensis (Loureiro) Ohwi (Syn. *Nasturtium microspermum* de Candolle; *Rorippa microsperma* [de Candolle] Handel-Mazzetti)

Sha-di-cai=Sha-ti-ts'ai (沙地菜, Sandy Land Potherb). Young plants, gathered for potherb in central and eastern China.

Annual or biennial herbs, glabrous, 10–40 cm high; leaves basal and cauline, oblanceolate in outline, 2–6 cm long, 1–2 cm wide, pinnatipartite or pinnatilobate, the cauline leaves sessile, amplexicaul, auriculate; flowers yellow, small, subsessile, solitary, axillary to foliaceous bracts 0.5–2 cm long, sepals greenish, shorter than the oblanceolate petals, 2–3 mm long; siliques cylindrical, 6–8 mm long, 1–2 mm across, on pedicels 1–3 mm long; seeds many, ovoid, reddish-brown. Widespread in eastern Asia, growing in wetland, wayside, and margin of fields in central and southeastern China.

Rorippa globosa (Turczaninow) Thellung (Syn. *Nasturtium globosum* Turczaninow)

Shui-man-qing=Shui-man-ch'ing (水蔓青, Water Creeping Green); *Qiu-guo-han-cai=Chiu-kuo-han-ts'ai* (球果蔊菜, Ball-Fruit Wild Mustard). Young plants, gathered for potherb in Hubei.

Much branched annual herbs 50–100 cm high, basal portion of the stem lignified; leaves oblong-lanceolate, 5–10 cm long, 1–2.5 cm wide, acuminate, base cuneate, the cauline leaves amplexicaul, auriculate, coarsely crenate, strigose on both surfaces; flowers yellow, many in axillary and terminal racemes, pedicels 5 mm long, sepals ovate, petals obovate, unguiculate, equal the sepals in length; silicles globose, 2 mm in diameter, shortly rostrate; seeds many, minute, ovoid, pale brown. A widespread weed, growing in wetland and in fields of central and northern China.

Rorippa montana (Wallich) Small; **R. indica** (L.) Hiern

Han-cai=Han-ts'ai (蔊菜, Wild Mustard); *Tang-ge-cai=T'ang-ke-ts'ai* (塘葛菜, Bund Mustard). Young plants just beginning to flower, gathered and used for soup.

Annual weeds, 10–50 cm high, much branched, glabrous; basal leaves few, obovate, cuneate, irregularly lobed, the terminal segment 6–10 cm long; flowers small, greenish-yellow, petals 2 mm long; silicle linear, 2–2.5 cm long, 1–1.5 mm across.

Rorippa nasturtium-aquaticum (L.) Britton et Rendle (Syn. *Nasturtium officinale* R. Brown) — WATER CRESS

Xi-yang-cai=Hsi-yang-ts'ai (西洋菜, Western Vegetable); *Shui-tian-jie=Shui-t'ien-chieh* (水田芥, Paddy Mustard); *Dou-ban-cai=Tou-ban-ts'ai* (豆瓣菜, Split-bean Vegetable). Tender shoots, cultivated in stagnate water in the suburbs of Guangzhou, available in American Chinese groceries; used for vegetable, more common in beancurd soup; growing wild along small streams in Massachusetts.

A perennial creeping herb rooting in water, branches ascending, 20–40 cm high; leaves pinnatisect, the terminal segments 1–3 cm long, 1–2.5 cm wide; flowers white, in short terminal racemes, 3 mm across; silicles cylindrical, 1–2 cm long, 1.5–2 mm thick; seeded in early spring crop in Guangzhou.

Thlaspi arvense L. — PENNYCRESS

E-lan-cai=O-lan-ts'ai (遏藍菜, Stop Blue Herb); *Xi-ming=Hsi-ming* (菥蓂, Pennycress); *Da-ji=Ta-chi* (大薺, Greater Shepherd's Purse). Young shoots, a potherb.

An annual weed 20–60 cm high, much branched after flowering; radical and lower cauline leaves rather delicate, spathulate, subentire, 4–6 cm long, 0.5–1 cm wide, petiolate, cauline leaves much larger, thicker, oblong, 5–7 cm long, 1.5–2 cm wide, undulate or coarsely dentate, apex rounded, base sagittate or auriculate; flowers small, numerous, greenish white, 3 mm long; silicles strongly flattened, orbicular, oblong, or obovate, broadly winged, 1 cm in diameter, apex deeply emarginate.

Wasabia japonica (Miq.) Matsum. — JAPANESE HORSERADISH WASABI

Shan-yu-cai=Shan-yü-ts'ai (山俞菜, Wasabi); *Shan-kui=Shan-k'uei* (山葵, Hill Mallow). Large root, ground to make a sauce for table dip used in banquets; leaves cooked for vegetable.

Glabrous perennial herbs 40–50 cm high, with pungent fleshy roots; radical leaves long-petiolate, cordate-reniform, 10–15 cm long and wide, undulate-serrulate, roundish at the apex, cauline leavaes ovate-subcordate, 3–4 cm long; flowers white, in terminal simple racemes 3–5 cm long; sepals 4 mm, petals clawed; fruit a linear-oblong silique. Recently introduced from Japan and cultivated in the wetland along small streams near Dali, Yunnan.

Capparidaceae: Caper Family

Capparis spinosa L. (Syn. *C. cordifolia* Lamarck) — CAPER

Shan-gan-zi=Shan-kan-tzu (山柑子, Mountain Caper); *Ci-shan-gan=Tz'u-shan-kan* (刺山柑, Spiny Caper). Fruit.

Pubescent or glabrescent trailing shrubs 1 m high; leaves alternate obovate, broad elliptic or suborbicular, 2–4 cm long, 1.5–3 cm wide, entire, apex abruptly acute, base obtuse, stipules modified into two stout spines, one straight, the other recurved; flowers white, solitary, axillary, 3 cm across, on pedicels longer than the subtending leaves, sepals 4, ovate 1.5 cm long, petals 4, obovate, 1.5–2 cm long, stamens numerous, filaments 2.5 cm long, anthers dorsifixed; fruit a berry on a gynophore 3.5 cm long, longer than the diameter of the fruit.

This is the source species of commercial caper, a condiment, cultivated for the tender underdeveloped flower buds gathered in the morning, pickled in salt and vinegar. It has been reported growing spontaneously in Tibet, called *ci-shan-gan* (Wu, Fl. Xiz. 2:323. 1985).

Crataeva nurvala Buch.-Ham. (*C. religiosa* Forster) — SPIDER TREE

Yu-mu=Yu-mu (魚木, Fish Tree, named for the shape of the leaflets). Tender shoots, eaten by village people living in suburbs of South China where the species escaped from cultivation.

Semideciduous trees 10–15 m high; branchlets glabrous; leaves palmately trifoliolate, the middle one oblong-ovate, the lateral two obliquely ovate-elliptic, 7–12 cm long, 3–5 cm wide, apex acuminate, base acute, petioles 5–8 cm long; flowers showy, white in terminal racemes, pedicels 4 cm long; petals elliptic, 2.5 cm long, 1–1.2 cm wide, on slender claws, stamens 13–20, the filments 5–6 cm long, anthers basifixed, ovary oblong, on gynophore 5 cm long, stigma capitate, sessile or on very short style; fruit a large globose berry, pericarp brown, woody, 4–5 cm in diameter.

Note: In a class on local flora taught at the Chinese University of Hong Kong, responding to my question, "Why do people call this tree *'Yu-mu'*?" a student held out a branch and pointed to the leaflets saying, "Because of the resemblance of these leaflets to a fish." Native to Indo-Malaysian, cultivated in Hong Kong.

Stixis suaveolens (Roxburgh) Baillon

Ban-guo-teng=Pan-kuo-t'eng (斑果藤, Mottled-fruit Vine): *Liu-e-teng=Liu-o-t'eng* (六萼藤, Six Sepals Vine). Young leaves, used as a tea substitute in Yunnan; drupaceous fruits used locally.

Large lianas, branchlets stout, pilose; leaves oblong or oblong-lanceolate, (10–) 15–28 cm long, 4–10 cm wide, rounded or abruptly acuminate, base rounded, petioles 2–3 (–5) cm long; flowers pale yellow, 5–6 mm long, fragrant, 15–25 in erect pilose racemes or panicles, perianth segments 6, rarely 5, connate, the free portion erect, tomentose, stamens 40–80, anthers basifixed, ovary 4- to 10-ovulate, spherical, 3-locular, stipitated, style filiform; fruits drupaceous, covered with yellow hairs, oblong, 3–5 cm long, 2.5–4 cm across, pale yellow, mottled, verrucose. Native of eastern Himalayan Region, thence eastward to Hainan, Guangdong and adjacent areas in Vietnam.

Moringaceae: Moringa Family

Moringa oleifera Lamarck — HORSE-RADISH TREE

La-mu=La-mu (辣木, Pungent Tree). Root and leafy shoots used as condiment, cooked before application; half-ripe fruit, used for vegetable; seed fried, having the taste of peanuts.

Deciduous trees 10–30 m high, root pungent like mustard, branchlets pilose; leaves alternate, tripinnate, 25–50 cm long, leaflets on the ultimate pinnae opposite, 3–9, ovate or oblong, 1–2 cm long, 5–12 mm wide, entire; flowers yellowish-white, tinged pink, zygomorphic, in axillary cymose panicles, hypanthia cupular, sepals 5, reflexed, petals 5, unequal, one erect, 4 reflexed, fertile stamens 5, filaments distinct, anthers unilocular, staminodes 5, filiform, ovary stalked, 3-carpelled, unilocular, placentae parietal, ovules numerous, style slender, stigma punctiform; fruits pendulous, elongated, 20–50 cm long, rostrate, 3-valved, 9-ridged; seeds numerous, ovoid, 3-winged, 10 mm across. Native to tropical Asia, extensively cultivated in India, Malaysia, and Hainan Island.

Saxifragaceae: Saxifrage Family

Astilboides tabularis (Hemsley) Engler

Da-ye-zi=Ta-yeh-tzu (大葉子, Big Leaf); *Da-bo-gen-zi=Ta-po-keng-tzu* (大脖梗子, Thick Neck); *Shan-he-ye=Shan-ho-yeh* (山荷葉, Hillside Lotus-leaf). Young shoots and leaf-petioles, used as asparagus.

Very unusual perennial herbs 1–1.5 m high, with stout brown rhizomes 35 cm long,

2–3 cm in diameter, aerial growth hispid, consisting of one radical leaf and a smaller cauline one, terminated by the inflorescence; leaves peltate, orbicular, 18–40 (–100) cm in diameter, palmately 9 lobed, the lobes ovate, acute, dentate, petioles cylindrical, 30–60 cm long, the cauline leaf smaller; flowers white, small, in terminal panicles, calyx campanulate, 2.5 mm long, 4- or 5-lobed, lobes ovate, petals 4 or 5, obovate-oblong, stamens 8, 2 mm long, ovary 2 carpellate, half-inferior; follicles 2, 5 mm long; seeds winged, 1.5–2 mm long. Native to Liaoning, Jilin and the adjacent areas of Korea, growing in deciduous forests; an endangered species of a monotypic genus.

Penthorum chinense Pursh — CHINESE DITCH-STONECROP

Che-gen-cai=Ch'e-ken-ts'ai (扯根菜, Haul-the-Root Herb); *Shui-ze-lan=Shui-tse-lan* (水澤蘭 (湻藍), Water Eupatorium); *Shui-yang-liu=Shui-yang-liu* (水楊柳, Water Willow). Young shoots, used for potherb.

Perennial stoloniferous herbs 15–80 cm high, stolons wire-like, red, 6–10 cm long at flowering time, stems emerging from the terminal buds of stolons, decumbent at base, purple-red; leaves lanceolate, 4–11 cm long, 1 cm wide, acuminate, base attenuate, serrate; flowers small, yellowish-green, in terminal scorpioid panicles, primary and secondary axes and the pedicels all papillose, hypanthia subcampanulate, cupular at anthesis, perianth consisting of 5 deltoid segments inserted on the rim of the hypanthium, stamens 10, slightly exserted, anthers oblong, basifixed, pistils 5, adnate to the hypanthium, forming a star-like capsule with 5-horned stylar bases after anthesis; fruits red, 6 mm long, carpels dehiscing independently, each by an oblique circumcission, leaving the seeds attached to a solitary pendulous placenta hanging from the roof of the locule; seeds ellipsoid, minute, brown, glandularly papillose. Morphologically a very unique species; botanically very close to the American species first reported from Virginia, *P. sedoides* L.; historically recorded in Chinese herbals as a famine food plant in 1406, available evidences indicating an adventive taxon in China, further studies awaited to settle its taxonomic position.

Ribes alpestre Wallich ex Decaisne

Ci-li=Tz'u-li (刺李, Spiny Plum); *Da-ci-cha-biao=Ta-tz'u-ch'a-piao* (大刺茶藨, Big-spine Gooseberry). Fruits, eaten raw or used for wine in Hubei.

Deciduous upright shrubs 1–3 m high, branchlets with 3 stout spines on the nodes and prickly throughout; leaves cordate-suborbicular, 2–2.5 cm long, 2–3 cm wide, 3- or 5-lobed, the lobes incisely dentate, obtuse, petioles 2–3 cm long; flowers greenish or reddish, 1 or 2, axillary, hypanthia campanulate, glandular-bristly, petals white, upright, shorter than the sepals, stamens exceeding the petals, anthers with cup-shaped apical glands, ovary inferior; berries subglobose, 1.6 cm long, purple, glandular-bristly. Native to western China, thence westward to Tibet and the Himalayan Region, introduced into Western gardens in 1903.

Ribes burejense Fr. Schmidt — Gooseberry

Ci-li=Tz'u-li (刺梨, Prickley Pear). Ripe fruits gathered locally, not available in the market.

A spiny shrub 1 m high, branchlets yellow, densely covered by bristles and nodal prickles 0.5–1 cm long; leaves cordate or subcordate, 1.5–4 cm long, 1–5 cm wide, deeply 3- to 5-lobed, the lobes dentate, pubescent, glandular; flowers pale reddish-brown, 1 or 2 axillary, receptacle campanulate, sepals oblong, reflexed, petals oblong, 1/2 the length of the sepals; fruits prickly, green, juicy, 1 cm across.

Ribes emodense Rehder — Himalayan Currant

Tang-cha-biao=T'ang-cha-piao (糖茶藨, Sugar-tea Berry). Ripe fruits; eatan in western China.

A deciduous shrub 2 m high, young branchlets glabrous, red; leaves ovate-cordate, 3- or 5-lobed, the lobes acute, 15 cm long; flowers greenish-purple, racemose, the racemes 12 cm long, receptacles subcampanulate, pubescent, glabrescent, sepals broad-ovate, spreading; fruits red-black. Native of western China.

Ribes fasciculatum Siebold et Zuccarini var. **chinense** Maximowicz — Chinese Red Currant

Hua-cha-biao=Hua-cha-piao (華茶藨, Chinese Currant). Fruits, eaten raw or used for making wine.

Deciduous shrubs 1–1.5 cm high, branchlets gray, pilose, unarmed; leaves ovate or suborbicular, 4–11 cm long, 4–10 cm wide, 3- to 5-lobed, lobed deltoid-ovate, acute, dentate, base truncate or subcordate, pilose or glabrescent on both surfaces, petioles 1–3 cm long; flowers greenish-yellow, fragrant, unisexual, dioecious, staminate flowers 4–9, subumbellate, pistillate flowers 2 (–4); berries subglobose, scarlet, 6–9 mm in diameter, crowned with persistent sepals. Native to central, eastern and northern China; introduced into Western gardens in 1867.

Ribes grossularis L. — Gooseberry

E-mei=O-Mei (鵝莓, Gooseberry). Mature fruits; used in northern China.

Shrubs 1 m high, branchlets grayish-brown, prickly with simple prickles and armed with 3-cuspid spines; leaves ovate-cordate, 2–5 cm long, 3–5 cm wide, palmately lobed, crenate, pubescent on both surfaces; flowers pale yellow, 6 mm across, 1 or 2, axillary, pendulous, calyx campanulate, pilose, glandular, sepals oblong, 3 mm long, reflexed, petals ovate, 2 mm long, 2-fid at apex, stamens shorter than the sepals, ovary inferior, oblong, 2 mm long, villose and with long stout hairs terminated with black glands, style 6 mm long, hairy; berries yellow or red, hairy. Native to Europe, introduced to northern China; the Chinese name is a translation of the English common name.

Ribes longiracemosum Franchet — CHINESE BLACK CURRANT

Chang-chuan-cha-biao=Ch'ang-chuan-cha-piao (長串茶藨, Long-string Currant). Ripe fruits, eaten fresh or used for wine in Hubei.

Deciduous shrubs 3 m high, branchlets glabrous; leaves trilobate-cordate, 7–14 cm long and wide, acuminate, irregularly serrate, pilose along the nerves beneath, petioles 10 cm long; flowers greenish and red, perfect, in pendulous racemes 30 cm long, pedicels 5–7 mm long, hypanthia campanulate, sepals upright, ovate, shorter than the tube, stamens and style exserted; berries black. Native to western China, growing in thickets or forests at altitudes of 2,000–2,200 m, introduced into American gardens in 1908 by the Arnold Arboretum.

Ribes manshuricum (Maximowicz) Komarov

Gou-pu-tao=Kou-p'u-t'ao (狗葡萄, Dog Grape); *Shan-ma-zi=Shan-ma-tzu* (山麻子, Mountain Hemp). Fruit for jam.

Deciduous shrubs 1–2 m high, branchiets gray-brown, peeling; leaves aceriform, ovate-reniform in outline, 4–10 cm long, 4–9 cm wide, lobes 3, triangular, serrate, pubescent; flowers greenish-yellow, racemes 2.5–9 (–20) cm long; berries red, globose, 7–9 mm in diameter. Native of northeastern Asia, cultivated in USA in 1906.

Ribes pulchellum Turczaninow — BEAUTIFUL CURRANT

Xiao-ye-cha-biao=Hsiao-yeh-cha-piao (小葉茶藨, Small-leaved Currant). Fruit, eaten in northern China.

Deciduous shrubs 1–2 m high; branchlets brown, with paired prickles at the nodes; leaves ovate-cordate in outline, 1–3.5 cm long, deeply 3-lobed, truncate at base; flowers red, racemose, dioecious, staminate racemes 6 cm long, flowers saucer-shaped; berries red, globose, 5–6 mm in diameter, in racemose clusters 3 cm long. The closely related species, *R. diacanthum* Pallas, distinguished by leaves obovate, 1.5–3 cm long, 0.8–2 cm wide, cuneate at base, 3-lobed at apex, the lobes dentate. Native of northern China, cultivated in 1905.

Eucommiaceae: Eucommia Family

Eucommia ulmoides Oliver — EUCOMMIA, *DUZHONG* (Chinese market name)

Du-zhong=Tu-chung (杜仲, Eucommia). Young shoots, used as potherb; dried inner bark of 15-year or older trees, used for preparing a tonic broth with pork chops; fruits and old leaves used for making a health tea.

A monotypic species of unique deciduous trees 10–15 m high, every organ of the plant containing gutta-percha in form of white silky threads, branchlets smooth, with conspicuous punctiform lenticels; leaves papery, elliptic-ovate, 4–15 cm long, 2–6 cm wide, serrate, apex acuminate, base obtuse; flowers green, inconspicuous, appearing slightly before the leaves, dioecious, simple, a staminate flower consisting of a short androphore with 8–10 apiculate anthers at the summit, the pistillate flower consisting of a solitary stiped ovary bifid at the distal end with the stigmatic surface on the inside of the two lobes, each flower is subtended by a bract or a reduced leaf; fruit a strongly compressed oblong samara, 3.5–4 cm long, 1–1.5 cm across the middle, notched at the apex.

Morphologically, *E. ulmoides* is the most unique monotypic angiosperm as is *Ginkgo biloba*, a gymnosperm. Since time immemorial, the bark of eucommia has been used in traditional Chinese medicine as an agent for lowering blood pressure. It is taken as food, or in combination with other herbs. Subsequently, it has been known only in association with man. It is now widely cultivated in arboreta and botanical gardens as a rare plant outside China. An extensive review of the species was published in *America Journal of Chinese Medicine* (Hu, 1979), to which the readers can refer for more information. I and several of my friends are drinking eucommia leaf tea daily for hypotensive purposes.

Rosaceae: Rose Family

Amelanchier asiatica (Siebold et Zuccarini) Endlicher var. **sinica** Schneider — SHADBUSH, SUGARPLUM, AMELANCHIER

Hong-xun-zi=Hung-hsun-tzu (紅栒子, Red Shadbush); *Tang-di=T'ang-ti* (唐棣, Chinese Amelanchier). Fruit, gathered locally, eaten in Henan.

Small glabrous trees 3–5 (–15) m high, branchlets dark-purple brown, slender, spreading; leaves ovate to elliptic-oblong, 4–7 cm long, 2.5–3.5 cm wide, acute at apex, rounded or subcordate at base, finely serrate above the middle, tomentose and glabrescent; flowers white, racemes 4–5 cm long; fruit subglobose, 1 cm in diameter, bluish-black, 6–10 locular, 1 seed in each locule; occurring in northern and western China, intoduced into American gardens in the 1920s.

Chaenomeles sinensis (Dumont de Courset) Schneider (Syn. *Malus sinensis* Dumont de Courset, 1811; *Cydonia sinensis* Thouin, 1812; *C. sinensis* [Thouin] Koehne) — CHINESE QUINCE

Mu-gua=Mu-kua (木瓜, Wood Melon); *Tie-gan-hai-tang=T'ieh-kan-hai-t'ang* (帖梗海棠, Sessile-fruited Crab Apple). Fruits, used for tea given to people with gas pain; very sour, a little goes a long way.

Deciduous shrubs 5–10 cm high, branches spineless, branchlets pubescent, glabrescent, purple-brown; leaves elliptic-ovate, or oblong-elliptic, 5–8 cm long, 3.5–5.5 cm wide, both ends acute, sharply serrate, turning red in autumn, villose beneath while young, stipules glandular ciliate; flowers pink, solitary, 2.5–3 cm across, receptacle campanulate, glabrous; fruits oblong, 5–6 cm long, 4–5 cm in diameter, dark-yellow, woody, with sweet smell, turning black on drying.

Chrysobalanus icaco L. — ICACO, COCO PLUM

Jin-guo-mei=Chin-kuo-mei (金果梅, Golden Plum). Fruits, newly introduced, rare.

Large shrubs or small trees 7–10 m high; leaves obovate or obcordate, 3–7 cm long, 2.5–5 cm wide, glossy, leathery; flowers white, small, hairy, 4 mm across, in axillary cymes; fruits plum-like, pinkish-white, magenta-red or almost black, flesh white, cottony, having insipid taste, adhering closely to the large oblong seed. Native to Mexico, introduced into some tropical gardens and agricultural institutions.

Cotoneaster acutifolia Turczaninow — IRGAI (Mongolian name).

Hui-xun-zi=Hui-hsun-tzu (灰栒子, Gray Cotoneaster). Leaves used in Inner Mongolia for tea; leaves and fruits gathered in June–August in the Qinghai-Tibetan plateau, decocted into a strong tea for people suffering from malnutrition with bleeding gum.

Deciduous shrubs 4 m high, with slender spreading branches, all vegetative organs pubescent, glabrescent, two-year's growth reddish-brown; leaves elliptic-ovate or oblong-ovate, 1.5–5 cm long, 1.2–3.7 cm wide, acute, rarely acuminate, base broad cuneate or rounded, entire, dull green above, paler and pubescent on the nerves beneath; flowers pink, 2–5 in a terminal cyme on current year's growth, hypanthia obconical, pubescent, sepals deltoid, acute, pubescent and ciliate, petals erect, suborbicular, 3–4 mm long, 3–4.5 mm wide, stamens 18–20, equal or slightly shorter than the petals, ovary 2-carpellate, carpels 2-ovulate, hairy at the truncate apex, styles shorter than the stamens, stigmas punctiform; fruits pomaceous, obovoid-oblong or ellipsoid, 7–9 mm long, purple-black at maturity, the endocarp woody. Native of northern and northwestern China, common on the hillsides, introduced into the Arnold Arboretum from Beijing hillsides in 1883. Apparently, leaves of Mongolian species with pink flowers and black fruits, such as *C. melanocarpus* Loddiges, and those with white flowers and red fruits, such as *C. mongolicus* Pojarkova and *C. multiflorus* Bunge, are also gathered in Inner Mongolia and used for tea. The reporter employed a number of species without giving the specific epithets of the unidentified material.

Crataegus cuneata Siebold

Ye-shan-zha=Yeh-shan-cha (野山楂, Wild Hawthorn). Ripe fruit.

Shrub 1.5 m high, branches with spines 5–8 mm long, branchlets hairy; leaves broad-obovate, 2–6 cm long, 1–4.5 cm wide, apical half 3-lobed, the lobes incisely and irregularly serrate, base cuneate, becoming wings on the petiole, stipules ovate, toothed; flowers white, in terminal corymbs, 1.5 cm across; fruits compressed-globose, 1–1.2 cm long, red or yellow; children gather the fruits for making necklaces, wearing them first for fun and decoration, then keeping the fruits and eating them one after another.

Crataegus hupehensis Sargent

Hu-bei-shan-zha=Hu-pei-shan-cha (湖北山楂, Hubei Hawthorn). Fruits edible.

Small trees or shrubs 3–5 m high, branchlets thorny, glabrous; leaves ovate or ovate-oblong, 8–10 cm long, 4.5–8 cm wide, acute, base rounded, serrate or slightly lobate at the distal half, pilose above, glabrous beneath; flowers white, 1 cm across, in terminal corymbs on lateral shoots, pedicels 4–5 cm long, hypanthia bell-shaped, sepals deltoid, persistent, petals ovate, stamens 20; fruits subglobose, red, 2.5 cm in diameter. Native to central and western China, growing in thickets on the hillsides at altitudes of 140–800 m.

Crataegus maximowiczii Schneider

Mao-shan-zha=Mao-shan-cha (毛山楂, Hairy Hawthorn). Mature pea-sized fruits.

Deciduous shrubs or small trees up to 7 m high, pubescent throughout, thorns up to 3.5 cm long, branchlets purplish-brown; leaves oblong-ovate or rhomboid, 3.5–8 cm long, 3–6 cm wide, abruptly acute, rounded and abruptly acute at base, shallowly pinnate lobed, the lobes serrate, pilose above, villose beneath, petioles 6–10 mm long, stipules lanceolate, 6 mm long, dentate, each tooth terminated by a gland; flowers white, 1.2 cm across, in villose terminal corymbs 3–5 cm across; fruits oblong, becoming subglobose at maturity, 8–10 mm in diameter. Native to northern China, Inner Mongolia and adjacent areas in Siberia; introduced into Western gardens in 1904.

Crataegus pinnatifida Bunge — CHINESE HAWTHORN

Shan-zha=Shan-cha (山楂, Hawthorn); *Shan-li-hong=Shan-li-hung* (山裡紅, Red-in-the-Mountains). Fruit, commom in northern China, eaten raw, most popular as candied string of five on a bamboo stick called *Tang-hu-lu=T'ang-hu-lu* (糖葫蘆, Candied Hawthorn) in Beijing, and *Tang-qiu=T'ang-ch'iu* (糖球, Sugar Ball) in Xuzhou, northern Jiangsu; a jam called *Shan-zha-gao=Shan-cha-kao* (山楂糕, Hawthorn Cake) and dollar-sized thin sheets packed in a cylinder called *Shan-zha-pian=Shan-cha-p'ien* (山楂片, Hawthorn Flake) are both available in American Chinese stores.

Deciduous trees about 6 m high, branchlets purplish-brown, sometimes abbreviated into thorns, glabrous; leaves deltoid-ovate, 5–10 cm long, 4–7.5 cm wide, acute, base truncate or abrubly cuneate, pinnately lobed, the lobes double-serrate, pilose along the nerves beneath, petioles 2–6 cm long, stipules large, serrate; flowers white, 1.8 cm across, forming loose terminal corymbs 5–8 cm across, hypanthia obconical, stamens 20, anthers pink; fruits oblong-globose, 1–2.5 cm long, red, punctate, stones 3 or 4. Native to northern China, much cultivated for the edible fruit; introduced into Western gardens in 1880.

Crataegus sanguinea Pallas; et **C. dahurica** Koehne

Hong-guo-shan-zha=Hung-kuo-shan-cha (紅果山楂, Red Hawthorn); *Mian-guo-guo=Mien-kuo-kuo* (麵果果, Mealy Fruit). Mature fruits, mealy, lacking zest.

Shrubs or small trees 5–7 m high, branches spreading, unarmed or with short thorns, branchlets glabrescent; leaves rhomboid-ovate or broad-ovate, acute or shortly acuminate, pilose on both surfaces; flowers white, 0.9–1.5 cm across, 4 to 13 in terminal corymbs, bracts linear, glandularly serrate, caducous; sepals deltoid, petals orbicular, 5–6 mm long and wide, stamens 20, anthers pink or red, styles 5–6 mm long, ovary hairy; fruits subglobose, 1–1.5 cm in diameter, bright red or orange-red, persistent sepals reflexed. Native to northeastern Asia; the closely related species, *C. dahurica* Koehne, distinguished by glabrous leaves and fruits 6–8 mm in diameter, used in same area.

Cydonia oblonga (L.) Miller — COMMON QUINCE

Wen-po=Wen-p'o (榲桲, Quince); *Mu-li=Mu-li* (木梨, Wood Pear); *Tu-mu-gua=T'u-mu-kua* (土木瓜, Local Wood Melon). Fruits, yellow, with sweet smell, very sour, wrapped in dough and baked, eaten as dessert; sliced fruit dried, used for tea, for improving digestion.

Deciduous shrubs or small trees up to 5 m high, branchlets villose, glabrescent later, spineless, purplish brown; leaves entire, ovate-oblong, 5–10 cm long, 3–5 cm wide, acute or sometimes rounded, base subcordate, hairy beneath; flowers white tinged pink, solitary, terminal to a leafy shoot, receptacle campanulate-tubular, hairy, corolla 4–5 cm across, petals obovate, stamens *ca.* 20, styles 5, distinct; fruit pomaceous, yellow , 3–5 cm long, inconspicuously hairy, fragrant. Native of the Middle East and the Mediterranean Region, introduced overland via the silk-route in ancient times.

Eriobotrya cavaleriei (Léveillé) Rehder

Da-hua-pi-ba=Ta-hua-p'i-pa (大花枇杷, Large-flowered Loquat). Fruit.

Evergreen trees 9 m high, branchlets glabrous; leaves oblanceolate or lanceolate, 8–18 cm long, 3–7 cm wide, both ends acute, coriaceous, sharply but remotely serrate, yellowish-pilose beneath; flowers white, numerous, in terminal branched corymbs 9–12 cm across, pedicels 3–10 mm long, hypanthia obconical, 4 mm long, pilose, sepals deltoid-ovate, 2.5 mm long, obtuse, pilose and ciliate, petals obovate, 8–10 mm long, emarginate, stamens 20, 4–5 mm long, ovary glabrous, styles 2 or 3; fruits oblong-subglobose, 1–1.5 cm long, orange-red, verrucose. Native to central and southern China, growing in secondary forests on the hillsides.

Eriobotrya deflex (Hemsley) Nakai — TAIWAN LOQUAT

Tai-wan-pi-ba=Tai-wan-p'i-pa (台灣枇杷, Taiwan Loquat). Fruit.

Tropical evergreen trees 5–12 m high, branchlets stout, grayish-brown, new shoots and young leaves villose, becoming glabrescent; leaves coriaceous, crowded at apex of the shoot, long-elliptic, 10–19 cm long, 3–7 cm wide, irregularly serrate; flowers white, 10–12 in terminal panicles, villose and glabrescent, including the rachis, pedicels, receptacles and sepals, corolla 1.5–1.8 cm across, styles 3–5, united up to the middle; fruit globose, 1.5–2 cm in diameter, orange-red, glabrous.

Eriobotrya japonica (Thunberg) Lindley — LOQUAT, *PIBA* (Chinese market name)

Pi-ba=P'i-pa (枇杷, Pull-off-skin). Cultivated, the best crop from Suzhou, Jiangsu; canned material available in American Chinese stores.

Pi-ba-gen=P'i-pa-ken (枇杷根, *Piba* Root). Fresh root; cooked with pig's feet, and yellow-rice wine (黃米酒) or chicken, cooked slowly.

Small evergreen trees 4–10 m high, branchlets stout, dense-lanate, the hairs golden-brown; leaves large, dark green, obovate-oblong, or elliptic, 12–30 cm long, 3–9 cm wide, apex acute, base cuneate, remotely serrate above the middle, rusty-lanate beneath; flowers ivory white, fragrant, numerous, in large terminal panicles 10–20 cm long, lanate throughout, corolla 1.2–2 cm across, styles 5, distinct; fruits yellow, oblong-pyriform, 3–4 cm long, 2–5 cm across the middle, skin leathery, tough, flesh carnose, sweet-sour, juicy. Native of western China, wild specimens still occurring in the mountains of Sichuan and Yunnan, cultivated along the Yangtze River and thence southward, especially good in Suzhou, introduced to Japan by Buddhist monks in Tang Dynasty. Every part of the plant is used in traditional Chinese medicine.

Fragaria x **ananassa** Duchesne (Syn. *F. chiloensis* [L.] Duchesne x *F. virginiana* Duchesne) — STRAWBERRY, GARDEN STRAWBERRY; 2n=56

Cao-mei=Ts'ao-mei (草莓, Strawberry). Fruits; an octoploid from hybrid origin, introduced into China in the 1920s; not very common.

Acaulescent, stoloniferous, perennial herbs without upright stems, runners ready to form new plants; leaves pinnate-trifoliolate, leaflets cuneate-obovate, 3–8 cm long, 3–5 cm wide, dark green above, bluish-white beneath, coarsely dentate above the middle, petioles 8–15 cm long, stipules deltoid, adnate to the base of the petiole; flowers white, in branched cymose clusters on scapes 15–20 cm long, calyx 1.5–2 cm across, persistent, corolla 2.5 cm across, aggregate fruit with numerous achenes partially sunk in the fleshy red ovoid receptacle 2–4 cm long, 2.5–3 cm across the base.

Fragaria gracilis Losina-Losinskaja — WILD STRAWBERRY

Qian-xi-cao-mei=Chien-hsi-ts'ao-mei (缝細草莓, Slender Strawberry). Fruit, gathered by local people in northwestern China, not in cultivation.

Similar to the above species, except the hairs on the scapes and petioles appressed, persistent sepals reflexed in mature berry.

Fragaria moupinensis (Franchet) Cardot — RED BERRY (Ethnic name used in western China); 2n=28

Xi-nan-cao-mei=Hsi-nan-ts'ao-mei (西南草莓, Southwestern China Strawberry); *Hong-pao=Hung-p'ao* (紅泡, Red Berry). Fruit, used locally.

High mountain perennials, densely covered by golden villose hairs thoughout; leaves palmately trifoliolate, or 5-foliolate, leaflets rhombic-obovate, the middle one 1.5–2.5 cm long, rounded at the apex, cuneate at base, serrate-dentate above the middle, lateral leaflets oblique-oblong, 1–2 cm long, 6–12 mm wide, the lower margin evenly serrate-dentate, the upper margin toothed only above the middle, the extra pair (if present) very small, petioles 5–10 cm long, stipules membranous, golden-brown; flowers white, 1 or 2 in a simple cyme, scapes 6–12 cm long, pedicels 1–3 cm long, bracts tripartite, segments lanceolate, 5–11 mm long, prophylls 2, opposite, lanceolate, 4 mm long, corolla 1–1.8 cm across, calyx 1.5 cm across, deeply cut into 10 segments, 5-ovate-lanceolate, 6 mm long, 2 mm across base, 5-linear, 1 mm across the base, petals suborbicular,

6–8 mm in diameter, stamens 26, filaments shortly connate at the base, disk hirsute; receptacle with ovaries at anthesis 1.5–2 mm in diameter, styles fleshy, attached to the base of the ovary, receptacles with mature achenes globose, 5–10 mm in diameter, hirsute, enlarged and becoming oblong and juicy later. In the summer of 1939, I botanized in Moupin and saw Chinese wild strawberry for the first time. In commemoration of the occasion and based on specimens from the type locality, I made a detailed description of the very little known species.

Fragaria nilgerrensis Schlechter ex J. Gay; 2n=14

Huang-mao-cao-mei=Huang-mao-ts'ao-mei (黃毛草莓, Golden-hairy Strawberry). Fruit, gathered by local people. An East Asian species first discovered from the Himalayan Region, with a range extending from Nepal eastward to Yunnan, Sichuan, Hunan, Guizhou and the mountains in Taiwan and northern Vietnam.

Perennial creeping herbs, covered by silky yellow-brown hairs; leaves trifoliolate, leaflets firm, obovate-suborbicular, 1–4.5 cm long, 0.8–3 cm wide; flowers small, on scapes 5–8 cm high, corolla white, 1.3 cm across, calyx 10 mm across, sepals ovate-lanceolate; berries white or red, 1–1.5 cm across.

Fragaria nubicola (Hooker f.) Lindley — Tibetan Strawberry

Xi-zang-cao-mei=Hsi-tsang-ts'ao-mei (西藏草莓, Tibetan Strawberry). Fruit, gathered by the local people.

Perennial creeping herbs growing at high altitudes, the petioles and scapes covered with close-pressed silky hairs; leaves trifoliolate, the leaflets orbicular-ovate, 1–6 cm long, 0.5–3 cm wide; flowers 1–4 in a cymose cluster; fruits ovoid, calyx 12 mm across, sepals lanceolate, the outer 5 slightly smaller; growing along streams and meadows at 2,500–3,900 m.

Fragaria orientalis Losina-Losinskaja; 2n=28

Dong-fang-cao-mei=Tung-fang-ts'ao-mei (東方草莓, Oriental Strawberry). Fruits, gathered by the people living in the mountains of northern China.

Very similar to strawberry growing in some American gardens, distinguished by the small leaflets 1–5 cm long, 0.8–3.5 cm wide, hairy on both surfaces, peduncles 0.5–1.5 cm long, and smaller hemispherical berry 1.5 cm across the base.

Fragaria pentaphylla Losina-Losinskaja

Wu-ye-cao-mei=Wu-yeh-ts'ao-mei (五葉草莓, Five-leaved Strawberry). Fruit, eaten by local residents.

Villose perennial creeping plants 6–15 cm high; leaves 5-foliolate, leaflets firm, obovate-elliptic, 1–4 cm long, 0.6–3 cm wide, petioles 2–8 cm long, densely covered by spreading soft hairs; flowers (1–) 2–3 (–4), in a cymose cluster, sepals ovate-lanceolate, epicalyx lanceolate, the apex sometimes bifid; berry oblong, red, persistent calyx reflexed; occurring in northwestern and western China.

Fragaria vesca L. (Syn. *F. chinensis* Losina-Losinskaja); 2n = 14

Ye-cao-mei=Yeh-ts'ao-mei (野草莓, Wild Strawberry); *Ou-zhou-cao-mei=Ou-chou-ts'ao-mei* (歐洲草莓, European Strawberry); *Sen-lin-cao-mei=Sen-lin-ts'ao-mei* (森林草莓, Forest Strawberry). Fruit, used by residents of the forests, not yet in cultivation.

Perennial stoloniferous herbs, densely covered by spreading hairs; leaves trifoliolate, the leaflets obovate or rhombic, 1–5 cm long, 0.8–3.5 cm wide, coarsely dentate, hairy; flowers white, (1–) 2–6 (–6), scapes covered by spreading hairs, epicalyx occasionally bifid; berries hemispherical, purple-red, persistent calyx spreading, slightly reflexed; achenes with prominent veins; occurring in northern China, Korea, and adjacent areas.

Geum aleppicum Jacquin — Aleppo Avens

Shui-yang-mei=Shui-yang-mei (水楊梅, Water Myrica); *Wu-qi-zhao-yang-cao=Wu-ch'i-chao-yang-ts'ao* (五氣朝陽草, Five Vitalities Rejuvenating Herb); *Zhui-feng-qi=Chui-feng-ch'i* (追風七, Chase-out-Wind Hematinic). Young plants, gathered by the mountainous people in western Hubei, used for potherb, entire plant boiled with chicken in Yunnan, taken as a rejuvenator of vitality; plants in summer, roots in winter, used for tea in Shaanxi.

Perennial herbs 40–80 cm high at flowering and fruiting stage, hispid-villose throughout; basal leaves obovate-cuneate in outline, pinnately incised, the terminal segments ovate-rhomboid, or suborbicular, 5–10 cm long, 3–10 cm wide, 3-lobed, acute, base cuneate, coarsely serrate, lateral segments smaller, interrupted, cauline leaves all sessile, unequally incised, stipules foliaceous; flowers yellow, solitary, peduncles elongated and clavate, hypanthia subcampanulate, sepals 5, ovate-lanceolate, epicalyx 5, linear-lanceolate, inserted on the sinuses between sepals, stamens numerous, pistils apocarpous, on conical receptacles, ovary villose, style jointed and geniculate above the middle, the upper stigmatose section pilose, the lower persistent portion glabrous, uncinate and stiff; aggregate fruits subglobose, 1.5 cm in diameter; achenes strongly compressed, pilose, with hooked persistent style. Widespread in temperate Eurasia, growing in wetland meadows.

Malus asiatica Nakai — CHINESE APPLE

Sha-guo=Sha-kuo (沙果, Crab-apple); *Hua-hong=Hua-hung* (花紅, Flower Red); *Lin-qin=Lin-ch'in* (林檎, Forest Delight). Ripe fruit, available in the market.

Small deciduous trees 4–6 m high, branchlets stout, pubescent when young; leaves elliptic or ovate, 5–11 cm long, 4–5.5 cm wide, acuminate at apex, obtuse at base, finely serrate, densely pubescent beneath; flowers pink, 3–4 cm across, stamens 17–20, style 4, villose near base; fruit compressed-globose, 4–5 cm across, yellow with red tinge, base concave, apex with persistent calyx, good flavor. cultivated in northern China, many cultivars, cold resistant.

Malus baccata (L.) Borkhausen — SIBERIAN CRAB-APPLE

Shan-jing-zi=Shan-ching-tzu (山荊子, Hillside Thorn-fruit), *Shan-ding-zi=Shan-ting-tzu* (山釘子, Hillside Nail-fruit). Leaves used for tea; in Inner Mongolia, leaves and fruits gathered, dried, mixed with the leaves of local oaks, brewed in a cloth bag, taken with or without milk; fruits also used for making wine.

Deciduous trees 10–14 m high, branchlets glabrous, dark brown second year; leaves ovate-elliptic, 3–8 cm long, 2–3 cm wide, acuminate, serrate, the base obtuse or acuminate; flowers white, in umbel-like racemes, pedicels 3–4 cm long, calyx glabrous, corolla 3–3.5 cm across, stamens 15–20, styles 4–5, villose and connate at base; fruits yellow, globose, 5–10 rnm in diameter, without persistent calyx. Native of northern China, cultivated in New England, USA, for ornamental purposes.

Malus hupehensis (Pampanini) Rehder — TEA CRAB-APPLE

Ye-hai-tang=Yeh-hai-t'ang (野海棠, Wild Crab-apple); *Hua-hong-cha=Hua-hung-cha* (花紅茶, Flower Red Tea). Young leaves used for tea; ripe fruits for making wine.

Small trees 5–8 m high, the branches stiff, spreading, with many short flowering shoots, young branchlets pubescent, soon glabrous; leaves ovate-oblong, 5–10 cm long, 2.5–4 cm wide, serrulate, acuminate; flowers 3–7 in a terminal fascicle with leaves on short shoots, pink changing to white, fragrant, 3.5–4 cm across, pedicels slender, 3–4 cm long, sepals triangular-ovate, styles 3–4; fruits globose, 1 cm in diameter, greenish-yellow tinged red. Native of central China, common in American gardens and landscape.

Malus kansuensis (Batalin) Schneider

Ji-ling-zi=Chi-ling-tzu (雞 (鶺) 鴒子, Waterhen Egg); *Long-dong-hai-tang=Lung-tung-hai-t'ang* (隴東海棠, Gansu Crabapple). Ripe fruits; sour, eaten fresh or used for making jam or wine.

Shrubs or small trees 5–10 m high, branchlets pubescent, glabrescent, dark purple-brown; leaves ovate or broad ovate, 5–8 cm long, 4–6 cm wide, 3- or 5-lobed, the lobes deltoid, acute, base rounded, double-serrate, petioles 1.5–5 cm long; flowers white, 4–10 in corymbs, pedicels 2.5–3.5 cm long, hypanthia and sepals villose, corolla 1.5–2 cm across, stamens 20, ovary inferior, styles 3, glabrous; fruits ellipsoid or obovoid, 1.5 cm long, yellow or red. Native to the Tsinling Range, growing in forests along streams at altitudes of 1,500–3,000 m; introduced to American gardens in 1904.

Malus micromalus Makino (Syn. *M. baccata* x *M. spectabilis*)

    Xi-fu-hai-tang=Hsi-fu-hai-t'ang (西府海棠, West Court Crab-apple). Fruits, used fresh or preserved.

Small deciduous trees 3–5 m high; branchlets glabrescent; leaves elliptic-oblong, 5–10 cm long, 2–4 cm wide, acuminate, cuneate or rounded at base, serrulate; flowers pink, 4 cm across, receptacles villose, sepals ovate, pedicels 2–3 cm long; fruits subglobose, 1–1.5 cm across. Native of northern China, cultivated in New England.

Malus prunifolia (Willdenow) Borkhausen — PLUM-LEAVED APPLE, CHINESE APPLE

    Qiu-zi=Ch'iu-tzu (楸子, Catalpa Crab-apple); *Hai-hong=Hai-hung* (海紅, Ocean Red); *Hai-tang-guo=Hai-t'ang-kuo* (海棠果, Crab-apple). Fruits, used fresh or preserved.

Deciduous trees 15 m high, young branchlets tomentose; leaves elliptic, ovate-elliptic, 4.5–10 cm long, 3–3.5 cm wide, acute, crenate-serrate, pubescent young, glabrous later, petioles 1.5–3 cm long; flowers white suffused with pink, 3–4 cm across, receptacles tomentose, sepals ovate-lanceolate, corolla 3–4 cm across, stamens 20, styles 5; fruits pomiform, with persistent sepals, red, yellow or green. Native of Europe and western Asia, extending along the range to northern China, and cultivated there.

Malus pumila Miller (Syn. *M. communis* Poiret, *M. domestica* Borkhausen) — COMMON APPLE

    Ping-guo=P'ing-kuo (蘋果, Apple). Fruits, many cultivars introduced, extensively cultivated after 1950.

Deciduous trees up to 15 m high, usually kept 3–4 m high in cultivation, young branchlets tomentose; leaves elliptic or ovate-elliptic, 4–10 cm long, 3–3.5 cm wide, acute, crenate-serrate, pubescent young, glabrous later, petioles 1.5–3 cm long; flowers white, suffused with pink, 3–4 cm across, receptacles tomentose, sepals ovate-lanceolate, corolla 3–4 cm across, stamens 20, styles 5; fruits pomiform, with persistent sepals, red, yellow or green according to the cultivars. Native of Europe and western Asia.

Osteomeles schwerinae Schneider

Sha-tang-guo=Sha-t'ang-kuo (沙糖果, Sugar Fruit); *Lao-ya-guo=Lao-ya-kuo* (老鴉果, Crowberry). Fruit, gathered and consumed locally.

Graceful shrubs 1–3 m high, branchlets, petioles, flowering stalks and lower leaf-surfaces grayish pubescent; leaves odd-pinnate, oblong in outline, 5–7 cm long, leaflets 7–15 pairs, oblong or obovate, 5–10 mm long, 2–4 mm wide, acute, mucronulate, entire; flowers white, 3–5 in a terminal corymb, receptacles campanulate, sepals ovate-lanceolate, corolla 1 cm across, stamens 15–20, ovary 5-locular, styles 5, pubescent, ovules 1 in each cell; fruits globose-ovoid or subglobose, 6–8 mm in diameter, glabrous, bluish black. Native of western China, introduced into European and American gardens since 1908.

Photinia serrulata Lindley — PHOTINIA

Shi-nan=Shih-nan (石楠, Photinia); *Du-xing-qian-li=Tu-hsing-ch'ien-li* (獨行千里, Alone Traveling a Thousand Miles). Mature fruits, used for jam or wine.

Evergreen shrubs or small trees 3–6 m high, branchlets glabrous, with large terminal buds; leaves oblong-elliptic, 10–18 (–22) cm long, 3–6 cm wide, acuminate, base cuneate or rounded, glandularly serrate, pilose when young, glabrescent, petioles 2–4 cm long; flowers white, in large terminal corymbs 6–8 cm across, sepals 5, deltoid, petals 5, suborbicular, stamens ca 20, anthers purple, ovary half-inferior, styles 3, stigmas subglobose; fruits pomaceous, globose, 5–6 mm in diameter, red-purple. Common in central and southern China; introduced to Western gardens in 1804, cultivated in USA.

Potentilla anserina L.

Jue-ma=Chueh-ma (蕨麻, Fern-hemp); *Ren-shen-guo=jen-shen-kuo* (人參果, Ginseng Berry); *Chang-shou-guo=Zhang-shou-kuo* (長壽果, Longevity Berry). Root tubers, dried liver-color, available in the markets of Lanzhou, boiled with rice forming a tonic soup.

Stoloniferous perennial creepers, deeply rooted in the soil forming fleshy tubers at the distal ends; stolons wire-like, with 3–5 nodes, rooting and flowering readily on touching the ground; leaves all basal, oblong in outline, 4–10 cm long, pinnate, the rachis bearing small foliaceous fragments between the normal alternate leaflets, obovate-oblong, 0.5–2 cm long, progressively decreased in size downward, dark green and sparsely hairy above, silvery white densely hairy beneath, evenly sharp-incised-dentate, petioles short, stipules membranous; flowers yellow, 1.2–2 cm across, solitary, on hairy scapes 3–13 cm long, calyx 8 mm across, sepals ovate, 3 mm long, 2 mm wide, epicalyx smaller, 2.5 mm long, entire or tripartite, stamens 28–30, ovaries numerous, on a lanate globose receptacle; achenes obovoid, black-brown, smooth. A species with holarctic distribution in Asia, North America and Europe, extending southward to the Qinghai-Tibetan Plateau and the alpines of the Himalayan Region, including western Yunnan; tubers eaten by various ethnic groups in the Asian sector. Description prepared with many herbarium collections, including the specimens collected in Tibet by J. D. Hooker and T. Thomson. In December 1980, I ate the tuber for the first time in Lanzhou, Gansu, and in 1984, again in Qinghai. In commemoration of these occasions, this little known economic use of a widespread species is described in detail.

Potentilla chinensis Seringe — CHINESE SILVERWEED (Syn. *P. exaltata* Bunge)

Wei-ling-cai=Wei-ling-ts'ai (委陵菜, Potentilla); *Fan-bai-cai=Fan-pai-ts'ai* (翻白菜, White-beneath Vegetable); *Long-ya-cao=Lung-ya-ts'ao* (龍牙草, Dragon's Teeth Herb). Young plants, used for potherb in northeastern China.

Perennial herbs 15–30 cm high at anthesis, hirsute-villose throughout, caudex stout, continued to the tap root, lignified; radical and cauline leaves obovate in outline, pinnate or interruptely pinnate, leaflets 8–31, sessile, obovate-oblong, 1–5 cm long, 0.5–1.5 cm wide, pinnatifid, lobes deltoid-lanceolate, mucronate, green and glabrescent above, lanate beneath, petioles 3–5 cm long, the upper cauline ones sessile, stipules lanceolate, adnate to the petiole, often lobed; flowers yellow, 1 cm across, numerous, in terminal corymbose clusters, calyx lobes ovate, acute, epicalyx linear, petals obovate or obcordate, 5 mm long, stamens many, pistils numerous, receptacles villose; achenes ovoid, 1 mm long, persistent style subapical. Widespread in temperate China, especially along paths on dry hillsides.

Potentilla discolor Bunge

Fan-bai-cao=Fan-pai-ts'ao (翻白草, White-beneath Herb). Root, young shoots, gathered and consumed locally in northern China, not available in the market.

Hirsute perennial herbs 14–40 cm high at the flowering stage, with stout root system, some of the lateral ones becoming fleshy, fusiform, 8–10 mm in diameter at the middle; basal leaves pinnate, 5- to 9-foliolate, cauline leaves trifoliolate, leaflets linear-oblong, 1.5–5 cm long, 5–15 mm wide, lower leaflets smaller, ovate, incised-dentate, glabrescent above, villose beneath, stipules coriaceous, brown, strongly striate, those of the cauline leaves foliaceous; flowers yellow, 1–1.5 cm across, calyx lanate outside, lobes linear-lanceolate, epicalyx linear, petals obovate-suborbicular, receptacles villose, stamens 26–28, ovaries obovoid, styles stout, attached to the upper side; achenes glabrous.

Potentilla flagellaris Willdenow ex Schlechtendal — ASIAN CINQUEFOIL

Man-wei-ling-cai=Man-wei-ling-ts'ai (蔓委陵菜, Creeping Potentilla). Young shoots, eaten in northeastern China as potherb.

Perennial stoloniferous herbs 10–30 cm high at anthesis, coarsely villose throughout; leaves digitate, radial leaves and those on the stolons quinate, cauline leaves on erect flowering shoots ternate, leaflets obovate-cuneate, 2–3 cm long, 1–1.3 cm wide, rounded, apex rounded and deeply lobate-serrate, petioles 3–7 cm long, stipules brown, coriaceous, 3–5 dentate; flowers yellow, 1–2 cm across, solitary or paired on shoot apex, pedicels 2–4 cm long, calyx flat, 12–14 mm across, deeply 5-parted, segments lanceolate, 6–7 mm long, epicalyx triangular-subulate, 3–4 mm long, petals obovate-orbicular, (3–) 6–10 mm long, 5–7 mm wide, stamens many, pistils many, on villose receptacles, style lateral; achenes oblong-ovoid. Widespread in temperate eastern Asia and the alpine meadows of central Asia and northern India.

Potentilla fruticosa L. — SHRUBBY CINQUEFOIL, GOLD HARDBACK

Jin-lu-mei=Chin-lu-mei (金露梅, Golden Dew *Mei*). Young shoots, dried, used for tea in Hebei.

Deciduous shrubs 1–1.5 m high, branchlets red-brown, glabrescent; leaves pinnate, leaflets (3–) 5 (–7), oblong or ovate-lanceolate, 6–15 mm long, 3–6 mm wide, acute, base cuneate, entire, pilose; flowers yellow, solitary, 2–3 cm across, hypanthia villose, epicalyx lanceolate, sepals ovate-deltoid, petals obovate, stamens many, pistils numerous, densely villose, style enlarged upwardly, hidden in the villosity, attached below the middle of the carpel. A very variable species of cold temperate northern hemisphere and the high mountains of Asia and western North America; cultivated and much used in landscape gardening.

Prinsepia uniflora Batalin

Rui-he=Jui-he (蕤核, Prinsepia); *Nai-ren=Nai-jen* (奈仁, Prinsepia Seed). Fruit edible raw; seeds used for soup, cooked with spareribs for improving eyesight, available in Chinese groceries in Boston.

Low spiny deciduous shrubs 1–1.5 cm high, branchlets light gray, glabrous, with sharp thorns 6–10 mm long, the pith lamellate; leaves alternate on long shoots, fasciculate on short shoots, elliptic or oblong-linear, 2.5–6 cm long, 5–7 mm wide, acute or obtuse and mucronate, base acute, entire, or remotely serrulate; flowers white, 1–3 fasciculate with leaves on short shoots, pedicels 5–7 cm long, calyx cup-shaped, persistent, corolla 1.5 cm across, stamens 10, ovary glabrous, style lateral; drupe purple-red, 1–1.5 cm in diameter, stone ovoid, laterally compressed. A native of northern China, introduced into gardens in USA in 1911, now in cultivation as a rare plant.

Prunus spp. — The apricot, almond, cherry, *mei*, peach, plum and prune all belong to the genus **Prunus** L. They share these common characters: deciduous, alternate serrate leaves, stipulate, solitary or fasciculate flowers, cup-shaped receptacles, superior ovary developing into a drupaceous fruit with membranous excocarp, fleshy mesocarp (the edible portion), bony endocarp (the stone), containing one or rarely two seeds. Outside China, all but the almond are cultivated for the edible fruit; in China, many species are cultivated for the fruit, except a form of the apricot which has edible seeds.

Prunus armeniaca L. — APRICOT; 3 varieties and 1 cultivar described here.

1. var. **armeniaca**; Figure 27b.

Xing=Hsing (杏, Apricot). Extensively cultivated; fruits, eaten fresh or preserved; *Xing-ren=Hsing-jen* (杏仁, Apricot Seed). Seeds saved, with the stone cracked, the seedcoat removed, the bitter element soaked off, used in pastry, and in soup.

Trees up to 10 m high; leaves ovate-suborbicular, 5–9 cm long, 4–8 cm wide, abruptly acuminate, rarely acute, subcordate or rounded at base, glabrous, obtusely serrate; flowers solitary, buds red, subsessile, calyx red, reflexed at anthesis, corolla pink-white, 2.5 cm across, ovary softly hairy; fruit yellow, tinged red, globose, 2.5–3 cm in diameter, juicy, good flavor; stone smooth, dorsal suture rather straight, ventral suture roundish, sharply keeled, with a groove on each side at the base of the keel and a ridge along the groove. Cultivated in China over two thousand years. Taken to and cultivated by Chinese settled in Armenia after Zhang Qian's mission (張騫使西域) in Han Dynasty.

2. var. **ansu** Maximowicz (*P. ansu* Komarov; *Armeniaca ansu* [Komarov] Kostina)

Shan-xing=Shan-hsing (山杏, Wild Apricot). Fruits, rather dry, used for making preserves; seeds, soaked to remove the bitterness, dried, powdered, sugar added, used for refreshing drink, available in American Chinese stores.

3. var. **mandshurica** Maximowicz (*P. mandshurica* [Maximowicz] Koehne; *Armeniaca mandshurica* [Maximowicz] Skvortzov).

Dong-bei-xing=Tung-pei-hsing (東北杏, Apricot of the Northeast). Fruits, used for preserves; seeds used after detoxified by soaking, used in pastry or powdered for refreshing drinks.

Distinguished by having broad-oblong leaves, double serrate along the margin; flowers solitary, red; stones globose or oblong, 1.3–1.8 cm long and wide, slightly striate, the ventral suture flat and blunt with a dull ridge, dorsal suture obscure, rounded; flesh of poor quality, used for sugar preserves and jam.

4. cv. 'Badan'

Ba-dan-xing=Pa-tan-hsing (巴旦杏, Sweet-Kernel Apricot). Fruits globose, orange-yellow, 3–3.5 cm in diameter, juicy, sweet; stones smooth, slightly larger than those of var. *armeniaca*.

Rare in the market, consumers usually break the endocarp and eat the seed immediately after finishing the mesocarp.

Prunus avium L. (Syn. *Cerasus avium* [L.] Moench) — SWEET CHERRY

Yang-ying-tao=Yang-ying-t'ao (洋櫻桃, Foreign Cherry); *Tian-ying-tao=T'ien-ying-t'ao* (甜櫻桃, Sweet Cherry). Fruit, rare.

Trees introduced into cultivation, 10–20 m high, bark red-brown, birch-like, branchlets glabrous, red-brown; leaves ovate-orbicular, 6–15 cm long, 4–8 cm wide, abruptly acuminate at the apex, double-serrate, the teeth gladular; flowers white, 2.5–3.5 cm across, the calyx-lobes reflexed, glabrous; fruit globose, purple-black, 1–2.5 cm in diameter, smooth, sweet, on pedicels 2–6 cm long. Eurasian origin, introduced into the few port cities in northern China.

Prunus cerasoides D. Don

Yun-nan-ou-li=Yun-nan-ou-li (雲南歐李, Yunnan Plum). Fruit.

A glabrous trees 5–10 m high, branchlets stout, angular, grayish-red or purplish-brown; leaves ovate-elliptic or oblong-elliptic, 5–10 cm long, 2.5–4.5 cm wide, acuminate-caudate, rounded or cuneate at base, petioles 1–1.5 cm long, with 2 or 3 glands at the distal end; flowers pink-white, on a common peduncle 2–5 mm long, becoming longer in fruit, pedicels 2 cm long, calyx red, 15 mm long, corolla 2–2.5 cm across; drupe red, 1.2 cm in diameter; growing in woods along streams, a wild cherry, gathered by local people, not available in the market.

Prunus dulcis (Miller) D. A. Webb (Syn. *P. amygdalus* Batsch; *Amygdalus communis* L.) — SWEET ALMOND (Figure 27d)

Bian-tao=Pian-t'ao (扁桃, Flat Peach); *Bian-he-tao=Pian-ho-t'ao* (扁核桃, Compressed Stone Peach); *Ba-dan-xing=Pa-tan-hsing* (巴旦杏, a name confused with that of the sweet-kerneled cultivar of apricot in some books; not to be used). Seed; recorded in ancient Chinese literature, imported by Arabian traders to Guangzhou; now very rare in China.

Trees up to 8 m high, branchlets glabrous; leaves ovate-lanceolate, broadest slightly below the middle, 7–12 cm long, acuminate, base rounded, finely serrulate, petioles 2.5 cm long, glandular; flowers 1 or 2, pink-white, nearly sessile, sepals oblong, margin villose, corolla 2–3 cm across; drupes ovoid-ellipsoid, compressed, 3–6 cm long, velvety, dry, splitting open at maturity, stone smooth, slightly pitted, ivory-white (see Part I for more information).

Prunus domestica L. (Syn. *P. communis* Hudson) — COMMON PLUM, EUROPEAN PLUM
(Figure 27c, 3–4)
Yang-li=Yang-li (洋李, Foreign Plum); *Ou-zhou-li=Ou-chou-li* (歐洲李, European Plum). Fruit.

Trees 6–15 m high, branchlets pubescent, brownish-gray; leaves elliptic-obovate, 4–10 cm long, 2.5–5 cm wide, obtuse or acute at apex, rounded or cuneate at base, crenate, pubescent beneath; flowers white or greenish-white; 1–3 in a fascicle, calyx hairy, corolla 1.5–2.5 cm across, pedicels 1–1.2 cm long; drupes compressed-globose or ovoid, prominently sulcate, yellow, red or green, bloomy, the flesh adhering or free, stone compressed-globose, ventrally keeled, dorsally grooved, uneven, slightly foveolate.

Prunus hongpiensis C. L. Li
Xing-zi-shu=Hsing-tzu-shu (杏子樹, Apricot Tree); *Hong-ping-xing=Hung-p'ing-hsing* (紅坪杏, Hongping Apricot). Fruits.

Deciduous trees 10 m high, bark grayish-brown, irregularly fissured; leaves ovate-elliptic, 5–9 cm long, 2–4.5 cm wide, densely velutinous beneath, serrate; drupes globose, deeply furrowed along one side, 3.5 cm long, 3 cm across, yellow, densely velutinous. Endemic to Hongping in Sichuan; fruits mature in August, used locally; the discovery of this pubescent form with limited distribution in Hongping of Sichuan being a good evidence of the diversity of apricots in China and a support of the Chinese origin of cultivated apricot.

Prunus humilis Bunge — OULANA (Tartary name). PEKING CHERRY
Ou-li=Ou-li (歐李, Peking Cherry); *Suan-ding=Suan-ting* (酸丁, Sour Fellow). Fruits eaten locally, not available in the market; leaves gathered, dried and used by people in Inner Mongolia for tea called *"Wulagana chai"*.

Deciduous shrubs 0.3–1.5 m high, branchlets reddish-brown, current year's growth very slender, 1–2 mm in diameter, pilose; leaves elliptic, oblong, rarely obovate, 2.5–4 (–5) cm long, 7–15 (–20) mm wide, acute, rarely acuminate, base cuneate, serrate, the teeth glandular, petioles 2–3 mm long, stipules linear-filiform, 3–4 mm long, glandular along the margin; flowers white, tinged pink, 1–3 emerging from the lateral axillary buds with the leafy shoot from the central bud, pedicels 6–10 mm long, pilose, with green basal bracts and 0–3 prophylls, calyx campanulate at anthesis, the tube 3–4 mm long, pilose, 5-lobed, the lobes oblong-ovate, 2–3 mm long, glandular along the margin, slightly pilose, reflexed, corolla 1.5 cm across, stamens *ca.* 30, filaments unequal, pink at the base, anthers oblong-suborbicular, dorsifixed, ovary 1, rarely 2, ovoid, glabrous, style 5–6 mm long, stigma punctiform; fruits drupaceous, globose at maturity, 1.5 cm in diameter, bright red, juicy and sweet, without groove, stone smooth, elliptic-ovoid.

Native of northern China and Inner Mongolia, growing on exposed grassy slopes, common on the hillsides of Beijing, seeds gathered there, distributed by E. Bretschneider, flowered in France in 1873, was identified as *P. japonica* by Carrière (Rev. Hort. 1873: 457. fig. 41). Similar introduction flowered in Kew in 1886, and fruited abundantly in 1892. In 1894, J. D. Hooker commended, "It has been suggested that it is the parent of the cultivated *Prunus japonica*" (Bot. Mag, 120: t. 7335). Seeds of both taxa share the same name in traditional Chinese medicine as *Yu-li-ren* 郁李仁.

Prunus japonica Thunberg. (Syn. *Cerasus japonica* [Thunberg] Loiseleur) — FLOWERING
    ALMOND, JAPANESE BUSH CHERRY (Figure 27c, 1–2)
    Yu-li=Yü-li (郁李, Elegant Plum). Ripe fruit.

Shrubs 1.5 m high, branchlets slender, glabrous; leaves ovate or broad-lanceolate, 3–7 cm long, 2–3.5 cm wide, acuminate, glabrous or pubescent along the nerves beneath, glandularly serrate; flowers pink-white, 1.5 cm across, 2 or 3 in a fascicle, pedicels 5–12 mm long, stamens shorter than the petals, ovary glabrous; drupes dark red, glabrous, 1 cm in diameter, not sulcate; seeds used in traditional Chinese medicine.

Prunus kansuensis Rehder (Syn. *Amygdalus kansuensis* [Rehder] Skeels) — WILD PEACH
    Mao-tao=Mao-t'ao (毛桃, Hairy Peach); *Ye-tao=Yeh-t'ao* (野桃, Wild Peach). Ripe fruit,
    eaten locally only by children.

Low shrubs with spreading habit, growing in loess cliffs, in cultivation appearing as small trees 3–6 m high, with a smooth trunk covered by dark mahogany-brown bark; leaves lanceolate, 4–8 cm long, 12–20 mm wide, acuminate, acute at base, serrate, with 3–5 teeth to every cm of the margin; flowers white, the buds with a slight trace of pink, stamens 25–30, 1/3 the length of the petals, style villose, longer than the stamens; fruit subglobose, densely velutinous; stone shallowly grooved, not pitted, the grooves wide, regularly parallel in the basal half of the stone, Native of northwestern China, introduced into USA by Frank Meyer in 1914 in seed form.

Prunus mira Koehne

Guang-he-tao=Kuang-ho-t'ao (光核桃, Smooth Stone Peach); *Xi-zang-tao=Hsi-tsang-t'ao* (西藏桃, Tibetan Peach). Mature fruits.

Deciduous trees 3–10 m high, branchlets glabrous, green, turning brown with age; leaves lanceolate, 5–10 cm long, 1.2–4 cm wide, long-acuminate, base rounded, remotely crenate-serrate, villose along the midrib beneath, petioles 8–15 mm long, with 2–4 glands; flowers pinkish-white, 1 or 2 together, 2–2.5 cm across, calyx purple-red; drupes subglobose, 3–4 cm in diameter, densely tomentose, juicy, sweet; stone ovate-oblong, compressed, smooth. Native of western China, thence extending westward to Lhasa, growing in broad-leaved forests and along the margin of corn fields at altitudes 2,600–3,000 m.

Prunus mume Siebold et Zuccarini (Syn. *Armeniaca mume* Siebold, nom. nud.) — Mei-Hua, Mei, Japanese Apricot (Figure 27e)

Jin-mei=Chin-mei (浸梅, Liquid Preserved *Mei*-fruit). Ripe fruits washed and preserved in water with salt for annual supply in cooking (especially for duck) and as home remedy.

Chen-pi-mei=Chen-p'i-mei (陳皮梅, Orange Peel *Mei*). A soft dry black preserve without stone, prepared with sugar, orange peel and licorice, wrapped in paper with the brand name printed, popular tidbit among Chinese both in China and abroad; available in American Chinese stores.

Hua-mei=Hua-mei (話梅, Chatting *Mei*). A hard dry tidbit with stone of the fruit and covered by white powder of the dried preservatives, prepared with salt, a little sugar, and spices, served at informal gatherings and often carried when hiking, available in American Chinese groceries.

Small deciduous trees 2–5 m high, branchlets green, glabrous, some of which shortened and with a spine at the end; leaves broad-ovate or elliptic, 4–7 cm long, 3–5 cm wide, apex caudate, base rounded or obtuse, sharply serrate, petiole-glands sometimes wanting; flowers white, pink or red, fragrant, calyx-tube broadly campanulate, puberulous, the lobes round, corolla 2–2.5 cm across, ovary pubescent; drupes globose, 2–3 cm in diameter, yellow, very sour, the flesh adhering to the globose-ellipsoid pitted stone, strongly furrowed on the dorsal side. A native of China, cultivated for fruit and/or for ornamental purposes since time immemorial, selected cultivars introduced to Japan with Buddhism in the seventh century, thence to Western gardens in 1844, carrying the Japanese term "mume" for its original Chinese name *"mei"*. The fruit-bearing form is cultivated in Luogangdong 羅崗洞, a short distance from Guangzhou.

Prunus persica (L.) Batsch (Syn. *Amygdalus persica* L.) — PEACH (Figure 27a)

Tao=T'ao (桃, Peach). Fruit, fresh and/or preserved.

Trees 5–10 m high, branchlets glabrous, axillary buds 3, hairy, with the lateral ones being rounded, developing into flowers, the middle one lanceolate, developing into vegetative branchlet; leaves lanceolate, 8–15 cm long, 3–4 cm wide, broad above the middle, acuminate, base broad-cuneate, finely serrate with 5–10 teeth along every cm of the margin, glabrous, petioles 1–1.5 cm long, glandular; flowers pink, sepals pubescent outside, corolla 2.5–3.5 cm across; drupe subglobose, 5–7 cm across, tomentose, fleshy, juicy, the stone irregularly furrowed and deeply pitted. Many horticultural forms and varieties, a few outstanding ones bearing unusual fruits are listed below:

1. forma **aganopersica** (Reichenbach) Voss — FREESTONE PEACH

Li-he-tao=Li-ho-t'ao (離核桃, Freestone Peach). Stone separating from the flesh, the delicious *Xue-bu-tao* (血布桃) of Xuzhou, Jiangsu, belongs to this group.

2. forma **compressa** (Loudon) Rehder (Syn. *P. persica* var. *platycarpa* Bailey) — FLAT
   PEACH

Pan-tao=Pan-t'ao (蟠桃, Peach-for-Goddess). Fruit compressed, discoid, 2–3 cm between the base and apex, 4–5 cm across, extremely juicy, very sweet and delicious, the juice dropping with each bite; cultivation limited in area and amount.

3. var. **nectarina** (Aiton f.) Maximowicz (Syn. *P. persica* var. *nucipersica* Schneider) —
   NECTARINE

You-tao=Yu-t'ao (油桃, Oil Peach); *Hu-zi-lai=Hu-tzu-lai* (胡子來, Mongolian Peach). Fruit, rare in China.

Leaves more strongly serrate; fruit glabrous; individual trees occasionally originate from peach seed in northern China, the stone separating from the flesh.

4. forma **scleropersica** (Reichenbach) Voss — CLINGSTONE PEACH

Nian-he-tao=Nien-ho-t'ao (黏核桃, Clingstone Peach). The famous *Fei-cheng-tao* (肥城桃, Feicheng Peach) of Shandong belongs to this form.

Prunus pseudocerasus Lindley (Syn. *Cerasus pseudocerasus* [Lindley] G. Don) —
   CHINESE CHERRY; SOUR CHERRY

Ying-tao=Ying-t'ao (櫻桃, Cherry). Cultivated in eastern and northern China, fruit scarlet red, very juicy; eaten fresh.

Trees 6–8 m high, branchlets glabrous or pubescent; leaves ovate-elliptic, 6–15 cm long, 3–8 cm wide, double-serrate, teeth glandular, sparsely pubescent beneath, petioles 8 mm long, with 2 glands at the apical end, stipules 3- or 4-lobed; flowers white, 3–6 on a short peduncle, pedicels pilose, calyx broadly obconical, pubescent outside, the lobes oval or elongated-triangular, reflexed after anthesis, ovary and style glabrous; fruits globose, 1 cm in diameter, not sulcate. Cultivated in northern China since time immemorial; used in ancient religious ceremonies, recorded in the *Book of Rites* (*Li-ji*, 禮記), introduced to Western gardens since 1819.

Prunus salicina Lindley — CHINESE PLUM

Li=Li (李, Plum); *Jia-qing-zi=Chia-ch'ing-tzu* (嘉慶子, Excellent Rejoice Plum); *Yu-huang-li=Yü-huang-li* (玉皇李, Supreme Deity Plum). Fruit.

Trees 9–12 m high, branchlets glabrous, grayish-green, glossy; leaves oblong-oblanceolate or oblong-ovate, 6–8 (–12) cm long, 3–5 cm wide, acuminate or acute, crenate, glabrous, petioles 1–2 cm long, glands sometimes absent; flowers white, 2 or 3 in a fascicle, 1.5–2 cm across, pedicels 1–1.5 cm long, calyx campanulate, the lobes oblong, glabrous; fruit spherical or ovoid-cordate, 2–3.5 cm in diameter, yellow, red, sometimes green or purple, bloomy, sulcate; stones ovoid, adhering to the flesh, striate-sulcate.

Prunus simonii Carrière — APRICOT PLUM, SIMON'S PLUM

Hong-li=Hung-li (紅李, Red Plum); *Xing-li=Hsing-li* (杏李, Apricot Plum). Fruit; eaten in northern China.

A small tree 5–8 m high, branchlets glabrous; leaves oblong-lanceolate or oblong-obovate, 7–10 cm long, 3–5 cm wide, acuminate, cuneate at base, crenate, petioles with 2–4 glands; flowers pink, 2–3 in a fascicle, pedicels 2–4 mm long; fruits compressed-globose, 3–5 cm in diameter, red, the flesh yellow, adhering to the stone, good flavor, slightly astringent; stone small, globose, furrowed; native to northern China, introduced to France in 1627 and to USA in the 1880s, one of the earliest flowering tree in the Arnold Arboretum where the tree bears no fruit.

Prunus tomentosa Thunberg — HAIRY CHERRY

Mao-ying-tao=Mao-ying-t'ao (毛櫻桃, Hairy Cherry); *Shan-dou-zi=Shan-tou-tzu* (山豆子, Hillside Pea); *Shan-ying-tao=Shan-ying-t'ao* (山櫻桃, Wild Cherry). Mature fruits, eaten by local people.

Deciduous shrubs 1–3 m high, bark grayish-brown, exfoliate, branchlets tomentose, glabrescent; leaves obovate-elliptic, 5–7 cm long, 1–3 cm wide, abruptly acuminate, base acute, unequally serrate, pilose above, rugose, glandular along the margin; flowers white, slightly pinkish, before or with leaves, 1.5–2 cm across; fruits drupaceous, subglobose, 1 cm in diameter, scarlet. Native of eastern Asia, growing on exposed hillside; introduced into Western gardens, many varieties selected; much used as female parent in hybridization, many cultivars produced.

Prunus vaniotii Léveillé

Ku-tao=K'u-t'ao (苦桃, Bitter Peach); *Wu-mao-diao-li=Wu-mao-tiao-li* (無毛稠李, Glabrous Plum). Ripe fruits, eaten by people living in the mountains of western Hubei.

Deciduous trees 6–22 m high, branchlets pilose, glabrescent; leaves oblong-obovate, 4–12 cm long, 3–6 cm wide, acuminate, base rounded, serrate, petioles 1–2.5 cm long, with 1 or 2 glands on the adaxial surface, stipules linear, caducous; flowers white, 30–60 in terminal racemes 10–15 cm long, hypanthia cupular, glabrous, pilose inside, sepals deltoid, petals orbicular-obovate, 4–5 mm across, stamens 20–25, ovary glabrous, style shorter than the stamens; drupes globose, 5–6 mm in diameter, black, sweet, stones 4–5 mm long, slightly striate. Native to central and western China, growing in forests along streams at altitudes of 1,100–2,400 m.

Pyracantha fortuneana (Maximowicz) H. L. Li — CHINESE FIRE-THORN

Huo-ji=Huo-chi (火棘, Fire-thorn); *Huo-ba-guo=Huo-pa-kuo* (火把果, Torch Fruit); *Jiu-bing-liang=Chiu-ping-liang* (救兵糧, Saving Soldier Food). Fruit, gathered and used locally, not available in the market.

Evergreen shrubs 2–3 m high, branchlets dark brown, pubescent when young, the short shoots terminated into a thorn; leaves obovate-oblong, 1.5–6 cm long, 0.5–2 cm wide, apex rotundate, mucronate or emarginate, base cuneate, crenate; flowers white, in fasciculate cymes with leaves on spurs, corolla 1 cm across; fruits pomaceous, globose, 5 mm in diameter, with persistent calyx. Native to the Yangtze River Region and southward to Fujian and Guangxi, planted for hedges.

Pyrus betulaefolia Bunge — RUSSET PEAR; BIRTH-LEAVED PEAR

Du-li=Tu-li (杜梨, Russet Pear); *Tang-li=T'ang-li* (棠梨, Wild Pear). Fruit, strongly astringent, farm-children often collect and bury the fruits in wheat containers until the skin color changed to black, then eat them.

Trees 10 m high, young branchlets tomentose; leaves rhombic to oblong-ovate, 5–8 cm long, 2.5–3.5 cm wide, acuminate, base obtuse or acute, occasionally rounded, coarsely serrate, petioles 2–3 cm long; flowers white, 10–15 fasciculate with leaves terminal to spurs, pedicels 2–2.5 cm long; fruits subglobose, 0.5–1 cm in diameter, brown, dotted. Native to northern China, flowering in mid-April; used for stocks in grafting pears; 2n=32. Named cultivars developed from this species including:

1. cv. 'Red Sugar Bottle' (*Hong-tang-guan=Hung-t'ang-kuan* 紅糖罐)
Fruit compressed globose, 4–5 cm in diameter, crisp, sweet, late season; Hebei Province.

2. cv. 'Early Dotted Skin' (*Ma-pi-cao=Ma-p'i-ts'ao* 麻皮糙)
Fruit pomiform, russet with prominent white dots, 6–7 cm across, flesh white, crisp, juicy, sweet, early crop; widely cultivated in north Jiangsu Province, particularly in villages south of Xuzhou.

Pyrus bretschneideri Rehder (*P. ussuriensis* var. *chinensis* Kikuchi) — CRISP CHINESE PEAR
Bai-li=Pai-li (白梨, White Pear); *Guan-li=Kuan-li* (罐梨, Jug Pear). Fruit.

Small trees 5–8 m high, branchlets stout, hairy; leaves ovate or ovate-elliptic, 5–11 cm long, 3.5–6 cm wide, acuminate, base rounded or obtuse, sharply serrate, the setose teeth slightly appressed, floccose and glabrous, petioles 2.5–3 cm long; flowers white, floccose, glabrescent, pedicels 1.5–3 cm long, corolla 2.5–3 cm across; fruits globose-ovoid, 2.5–3 cm long, yellow, without persistent calyx-lobes, flesh crisp. Native of northern and northwestern China; local cultivars developed including:

1. cv. 'White Pear' (*Bai-li=Pai-li* 白梨)
Fruits obovoid, basal end convex, skin thin, yellow or brownish-yellow, flesh white, grits few, flavor excellent; Hebei.

2. cv. 'Honey Pear' (*Mi-li=Mi-li* 蜜梨)
Fruits ovoid or subglobose, basal end depressed, skin yellow, thin, dots small, flesh fine, juicy, sweet, flavor excellent; Shandong Province.

3. cv. 'Dang-shan Pear' (*Dang-shan-li=Tang-shan-li* 碭山梨)
Fruit obovoid, 8–9 cm long, 5.5–6 cm across the broad portion, flesh white, crisp, sweet, juicy, grits only near the center, medium-late season; Dangshan is the area on the borders of Anhui, Henan and Jiangsu provinces; popular specially in Xuzhou and Nanjing.

4. cv. 'Snow-flake Pear' (*Xue-hua-li=Hsüeh-hua-li* 雪花梨)
Fruit oblong-ovoid, skin rough, green, flesh white, large size, weighing up to 200 g; Hebei Province.

5. cv. 'Duck Pear' (*Ya-li=Ya-li* 鴨梨)

Fruit obovoid, the basal end oblique, yellowish-green, skin thin, flesh with fine texture, crisp, mid-season, good keeping quality, extensively cultivated in the plains of northern China with many local names, Duck Pear from Hebei especially popular.

6. cv. 'Yuan-wa Wood Pear' (*Yuan-wa-mu-li=Yuan-wa-mu-li* 袁窪木梨)

Fruit oval-oblong, unusually large for pears, 10–15 cm long, 9–11 cm across, often larger, skin green-yellow, dotted light brown, flesh white, juicy, crisp, sour-sweet, very good flavor; the largest pear I have ever seen was in Yuanwa Village, formerly in Jiangsu, now relocated to Anhui Province.

Pyrus communis L. (Syn. *P. sativa* DC.) — COMMON PEAR

Yang-li=Yang-li (洋梨, Foreign Pear). Newly introduced, fruit brought from the market needed to be ripened, flesh good when soft.

Trees 10–15 m high, the crown conical, branchlets sparsely hairy; leaves ovate-elliptic, 2–8 cm long, 3–5 cm wide, acute or short-acuminate, base rounded or subcordate, crenate, petioles 1.5–5 cm long; flowers white, 6–7 in a fascicle with leaves at the apex of a spur, petioles 1.5–3 cm long, receptacle villose, sepals deltoid-lanceolate, corolla 3 cm across; fruit turbinate or subglobose, the stalk 5 cm long, slender, after ripening fruit has soft juicy flesh; introduced from Europe and America; rarely cultivated.

Pyrus nivalis Jacquin — SNOW PEAR

Xue-li=Hsüeh-li (雪梨, Snow Pear). Fruit, very good quality.

Trees up to 16 m high, young branchlets tomentose; leaves elliptic or obovate, 5–8 cm long, 2–4 cm wide, acute, cuneate at base, entire, white-tomentose when young, petioles 1–3 cm long; flowers white, calyx tomentose, corolla 2–3 cm across; fruit subglobose, 5 cm across, yellowish-green, stalk 5 or 6 cm long. Native of southern Europe; rarely cultivated in China

Pyrus pashia Buchanan-Hamilton ex D. Don

Tang-li-zi=T'ang-li-tzu (棠梨子, Spiny Wild Pear); Chuan-li= Ch'uan-li (川梨, Sichuan Pear). Fruit.

Spiny trees up to 12 m high, young branchlets woolly, becoming glabrescent; leaves ovate or ovate-oblong, 4–7 (–12) cm long, 2–5 cm wide, acuminate, rounded at base, crenate or crenate-serrulate, those on suckers lobed, petioles 1.5–4 cm long; flowers white, 7–13 mixed with leaves at the apex of short shoots, pedicels 2–3 cm long, calyx with deltoid lobes (deciduous after anthesis), corolla 2–2.5 cm across, stamens 25–30, shorter than the petals, styles 3–5; fruit subglobose, 1–2 cm in diameter, brown dotted. Native of western China and the Himalayan Region, introduced to America in 1908; selected cultivars in western Yunnan called 'Black Pear' (*Wu-li=Wu-li*, 烏梨), used locally.

Pyrus pyrifolia (N. Burman) Nakai (Syn. *P. serotina* Rehder) — SAND PEAR
  Sha-li=Sha-li (沙梨, Sand Pear). Fruit with good flavor, containing too many grits to be a choice dessert fruit.

Trees 7–15 m high, with a compact rounded crown, branchlets dark brown, floccose when young, glabrescent later; leaves ovate or elliptic-ovate, 5–11 cm long, 4–6 cm wide, acuminate or acute, rounded or obtuse at base, slightly floccose, becoming glabrous, sharply setose-serrate with slightly appressed teeth, petioles 3–4.5 cm long; flowers white, floccose at first, glabrous later, 6–9 terminal to leafy short shoots, pedicels 3.5–5 cm long, corolla 2.5–3 cm across, styles 5, rarely 4, glabrous; fruits subglobose, 2–3 cm across, brown, dotted. Native of China, fruit of wild plants not palatable; many selected cultivars, including:

1. cv. 'Dotted Crisp' (*Ma-su-li=Ma-su-li* 麻酥梨)
  Fruits pomiform, russet, dotted, crisp; cultivated in Jiangxi Province.

2. cv. 'Cinnamon Yellow Pear' (*Huang-zhang-li=Huang-chang-li* 黃樟梨)
  Fruits globose, brown, sepals deciduous, flesh crisp, juicy; cultivated in Zhejiang Province.

Pyrus ussuriensis Maximowicz — CHINESE PEAR
  Qiu-zi-li=Chiu-tzu-li (秋子梨, Autumn Pear); *Shan-li=Shan-li* (山梨, Mountain Pear);
  Suan-li=Suan-li (酸梨, Sour Pear). Fruits and leaves of wild plants gathered, dried, used for tea in Inner Mongolia; many cultivars developed.

Trees up to 15 m high, branchlets glabrous, yellow-gray; leaves orbicular-ovate or ovate, 5–10 cm long, 4–6 cm wide, acuminate, rounded or subcordate at base, setosely serrate, petioles 2–5 cm long; flowers white, 5–7 mixed with leaves terminal to short shoot, glabrous, pedicels 2–5 cm long, sepals glabrous outside, villose inside, corolla 3–3.5 cm across, styles 5, distinct, the basal portion pilose; fruits subglobose, greenish-yellow, 2–6 cm across, hard, harsh, edible after ripening and turning black. Native of northern China; the known cultivars including:

1. cv. 'Anne's Pear' (*An-li=An-li* 安梨)
  Fruit compressed globose, greenish-yellow, sepals persistent; late season, good keeping quality. Hebei Province.

2. cv. 'Capital White' (*Jing-bai-li=Ching-pai-li* 京白梨)

Fruit globose, greenish-yellow, basal end slightly rounded, flesh soft, juicy, excellent flavor, good quality; cultivated in the mountainous area of northern China.

3. cv. 'Red Flower-Jug' (*Hong-hua-guan=Hung-hua-kuan* 紅花罐)

Fruit ovoid-subglobose, skin greenish-yellow, tinged red on the side facing the sun, sepals persistent, flesh juicy, sweet; low keeping quality, edible after ripening.

4. cv. 'Sandong Soft Pear' (*Shan-dong-mian-li=Shan-tung-mien-li* 山東麵梨)

Fruits ovoid or obovoid-globose, skin thin, yellow, flesh fine, grits few; for the mid-season market or late ripening, becoming soft, less juicy, low keeping quality.

5. cv. 'Wild Duck Pear' (*Ya-guang-li=Ya-kuang-li* 鴨廣梨)

Fruit irregular in shape, usually ovoid, sepals persistent, skin yellow-brown, rough, flesh soft, juicy, excellent flavor; late season; edible after ripening. Hebei Province.

Rhaphiolepis indica (L.) Lindley — HONG KONG HAWTHORN

Chun-hua=Ch'un-hua (春花, Spring Flower); *Che-lun-mei=Ch'e-lun-mei* (車輪梅, Wheel *Mei*); *Shi-ban-mu=Shih-pan-mu* (石斑木, Cliff Shrub). Fresh fruit, gathered by wood collectors in southern China.

Shrubs or small trees 1–4 m high; branchlets pubescent, glabrescent; leaves rather crowded near the shoot apex, elliptic, 4–8 cm long, 1.5–4 cm wide, obtuse or acute, cuneate at base, coriaceous, serrate; flowers pink becoming white, 6–25 in terminal corymbose panicle, corolla 1–1.3 cm across; fruit pea-sized, globose, fleshy, ripe black. Native of southern China, common on Hong Kong hillsides, the flowers opening at the Chinese New Year's time, for which it is called "Spring Flower", "Wheel *Mei*" for the corolla.

Rosa bella Rehder et Wilson — SOLITARY ROSE

Mei-qiang-wei=Mei-ch'iang-wei (美薔薇, Beautiful Rose). Fruit, used for making wine.

Shrubs 2–3 m high, branchlets covered by straight prickles, bristly near base; leaves pinnate, leaflets 7–9, elliptic or ovate, 1–2.5 cm long, acute or obtuse, serrate, glaucescent beneath, glabrous above with the midrib covered by stipitate glandular hairs, stipules glandularly ciliate; flowers bright, rose, 1–3 in a terminal cluster, 4–5 cm across, pedicels 5–10 mm long, with the receptacle stipitate-glandular, sepals leafy, caudate, entire; hips scarlet, ellipsoid, attenuate into a neck, 1.5–2 cm long. Native of northern China, introduced into American gardens in 1910.

Rosa davurica Pallas — Dahurian Rose

Shan-ci-mei=Shan-tz'u-mei (山刺玫, Mountain Thorny Rose). Fruit for jam or wine; leaves and fruits gathered for tea in Inner Mongolia.

Shrubs 1–2 m high, branchlets glabrous, the curved prickles paired; leaves pinnate, leaflets 5–7, oblong or elliptic, 1.5–3 cm long, 0.8–1.5 cm wide, acute or obtuse, serrate above the middle, glabrous above, glaucous and glandularly hairy beneath; flowers dark red, solitary, 4 cm across, pedicels with glandular hairs, hips red, globose or ovoid, 1–1.5 cm in diameter. Native of northern China, introduced into Western gardens in 1910.

Rosa laevigata Michaux — Cherokee Rose

Jin-ying-zi=Chin-ying-tzu (金櫻子, Golden Hips). Fruits and sliced root, used for health tea.

Evergreen spreading shrubs up to 5 m high, branchlets glabrous, armed with sharp curved prickles and bristles; leaves pinnate, usually trifoliolate, rarely with 5 leaflets, elliptic-ovate to ovate-lanceolate, 3–7 cm long, 1.5–4.5 cm wide, abruptly acute or short-acuminate, sharply and finely serrate; flowers white, showy, solitary, terminal to a shoot, 6–9 cm across, the pedicel, receptacle and sepals densely bristly; hips pyriform, orange, strongly bristly, 3.5–4 cm long, with persistent sepals. Native of warmer regions of China, spread first to the Middle East through the Arabian traders and then to the Mediterranean Region, introduced from Spain to Florida, and naturalized in Southeast USA, the land of the Cherokee Indians, hence the common name.

Rosa omeiensis Rolfe — Rose-of-Omei Mountain

E-mei-qiang-wei=O-mei-ch'iang-wei (峨嵋薔薇, Mt. Omei Rose). Fruit, used locally.

Upright shrubs 2–4 m high, branches reddish brown, prominently covered by flattened wide-based prickles, young shoots often bristly; leaves pinnate, leaflets rather small and many, 9–17, oblong or elliptic-oblong, 0.8–3 cm long, acute, cuneate, serrate, often pubescent along the midribs and rachis; flowers white, 2.5–3.5 cm across, pedicels and receptacles smooth; hips pear-shaped, bright red, 8–15 mm long, on a thickened yellow stalk. Native of western China, first recorded from Mt. Omei in Sichuan Province, introduced to European gardens in 1901, several selected horticultural forms cultivated in USA.

Rosa roxburghii Trattinnick — Roxburgh's Rose

Ci-li=Tz'u-li (刺梨, Prickly Pear); *Mu-li-zi=Mu-li-tzu* (木梨子, Wood Pear). Fruit used for herb tea, jam or wine; leaves used as a tea substitute.

Stout shrubs up to 2.5 m high, bark of older stem gray, exfoliate; branchlets sparsely armed with paired prickles; leaves pinnate, leaflets 9–15, oblong, 1–2 cm long, 0.5–1 cm wide, obtuse or acute, base cuneate, sharply serrate, glabrous beneath; flowers pale pink, 1–2 terminal to a shoot, 5–6 cm across, pedicels and receptacles prickly, sepals villose, persistent; hips globose, orange-red, 3–4 cm in diameter, densely prickly with very sharp spines. Very common in western China, introduced to American gardens in 1908.

Rosa rubus Léveillé et Vaniot — WHITE WILD ROSE

Tu-mi-zi=T'u-mi-tzu (荼藨子, Tumi, the local name for a white flowered plant).

Fruit used for jam and wine.

Sarmentose shrubs up to 6 m high, branchlets with small hooked prickles; leaves pinnate, leaflets 5–7, ovate-elliptic or obovate, 3–8 cm long, 1.5–4 cm wide, abruptly acuminate, sharply and coarsely serrate, pubescent beneath or nearly glabrous; flowers white, fragrant, 2.5–3 cm across, 3 or 4 in a dense corymb terminal to a short shoot, sepals ovate-lanceolate, deciduous; pedicels 1–2 cm long, tomentose, glandular; hips scarlet, subglobose, 8 mm across. Native to central and western China, introduced to America in 1907.

Rosa rugosa Thunberg — MEIKUEI, RUGOSE ROSE, MAGENTA ROSE, TURKESTAN ROSE

Mei-gui=Mei-kuei (玫瑰, Magenta-colored Rose). Fruit, fresh or dried; flower petals gathered early in the morning, mixed with sugar, used in pastry.

Upright shrubs 1–2 m high, branches stout, tomentose, densely bristly and prickly; leaves pinnate, leaflets 5–9, thick and firm, elliptic or elliptic-obovate, 2–5 cm long, 1–2 cm wide, acute or obtuse, serrate, lustrous dark green, rugose and glabrous above, pubescent and glaucescent beneath, petioles tomentose and bristly; flowers purple-red, solitary, terminal to a shoot, fragrant, pedicels bristly; hips depressed-globose, bright red, 2–2.5 cm across, with persistent sepals. Native of northern China, introduced into USA about 1845, now naturalized in the northeastern areas; a white flowered form often seen in gardens.

Rosa xanthina Lindley — CHINESE YELLOW ROSE, MANCHU ROSE

Huang-ci-mei=Huang-tz'u-mei (黃刺玫, Yellow Spiny Rose). Fruit.

Shrubs 1–3 m high, branchlets puberulous, brown, with straight stout prickles; leaves pinnate, leaflets 7–13, broad oval or suborbicular, 8–20 mm long, apex obtuse or rounded, serrate, glabrous above, slightly villose beneath; flowers yellow, solitary, semi-double, 4 cm across; hips subglobose, 1 cm in diameter, red, with persistent lanceolate sepals.

Rubus amabilis Focke — Pleasant Raspberry

Xiu-li-mei=Hsiu-li-mei (秀麗莓, Beautiful Berry). Fruit, gathered locally.

Shrubs 1–2 m high, branchlets glabrous, slightly prickly; leaves pinnate, petioles and rachis with subulate prickles, leaflets 7–11, ovate or ovate-lanceolate, 1–5.5 cm long, 0.7–2 cm wide, acute or acuminate, rounded at base, glabrous above, pubescent and often prickly along the veins beneath; flowers white, 4–5 cm across, nodding; aggregate fruits red, cylindrical, 1.5–2.5 cm long. Native of western China, introduced to American gardens in 1908.

Rubus biflorus Buchanan-Hamilton — Two-flowered Raspberry

Fen-zhi-mei=Fen-chih-mei (粉枝莓, Glaucous-branched Berry). Fresh fruit.

Upright shrubs 2–3 m high; stems and branchlets purple-brown, glaucous, white, covered with straight prickles; leaves pinnate, leaflets 3–5, ovate or elliptic, 2–4.5 (–10) cm long, 1–2.5 (–10) cm wide, the terminal ones broad-ovate or suborbicular, often shallowly 3-lobed, irregularly serrate, white or gray villose beneath, the petioles, rachis and midribs hooked-prickly; flowers white, 1–2 in axillary fascicles, 1.5–2 cm across, pedicels 2–3 cm long, prickly; aggregate fruits yellow, 1.52 cm across. Native of western China and the Himalayan Region, introduced to European gardens in 1818.

Rubus buergeri Miquel

Han-mei=Han-mei (寒莓, Cold Berry). Fruits gathered and used locally, not available in the market.

Evergreen climbers, stems rooting on touching the soil, branchlets densely covered by brown and white villose hairs with few prickles, up to 2 m long; leaves simple, suborbicular, 4–8 cm in diameter, base cordate, shallowly 5-lobed, irregularly serrate, villose beneath, stipules laciniate; flowers white, 1 cm across, 4–10 in crowded axillary racemes; aggregate fruits globose, 6–9 mm in diameter, red.

Rubus calycinoides Hayata — Mt. Morrison Raspberry

Yu-shan-xuan-gou-zi=Yü-shan-hsüan-kou-tzu (玉山懸鈎子, Yu Shan Raspberry). Fresh fruit, gathered and eaten locally.

Very unusual creeping raspberries of high mountain growing up to 15 cm high, the prostrating creeping stems wire-like, glabrescent, the upright annual growth 1–7 cm long, densely velutinous throughout and sparsely spinose with the prickles straight and hidden by the brown hairs; leaves simple, ovate, suborbicular or reniform, 1–2.5 cm across, 3-lobed, the lobes rounded, dentate, strongly rugose above, reticulate with raised veins, foveolate and densely woolly beneath; stipules foliaceous, laciniate; flowers white, 10 mm across, solitary, subtended by 2 or 3 bracts similar in shape and texture to the stipules, sepals golden woolly, anthers hairy; aggregate fruit orange, 15 mm across, stone of the drupelet reniform, 2 mm long, foveolate.

Rubus chroosepalus Focke

Mao-e-mei=Mao-o-mei (毛萼莓, Hairy-sepaled Berry). Fruits, eaten in Hubei.

Straggling shrubs, half evergreen, branchlets pubescent, armed with short and curved prickles; leaves suborbicular-cordate, 6–12 cm long, 6–8 cm wide, abruptly short acuminate, unequally serrate, sometimes shallowly lobulate, glabrous above, white-tomentose beneath, petioles 2–4 cm long, sparsely prickly; flowers apetalous, in terminal panicles 15–20 cm long, without prickles, sepals tomentose outside, purple inside, receptacles hairy, ovaries 12–15; aggregate fruits globose, black, 1 cm in diameter. Native to central and southern China, introduced to USA in 1900.

Rubus corchorifolius L. f. — JUTE-LEAVED RASPBERRY

Shan-mei=Shan-mei (山莓, Mountain Berry); *Shan-pao-zi=Shan-p'ao-tzu* (山泡子, Wild Berry). Fruit, eaten fresh or used for jam.

Upright shrubs 1–2 m high, suckering, stems prickly, branchlets red-brown, tomentulose and sparsely glandular when young, prickly; leaves simple, ovate or ovate-lanceolate, 3–8 cm long, 2–5 cm wide, irregularly double-serrate, occasionally lobed, puberulous above, villose beneath, petioles and nerves with hooked prickles; flowers white, solitary, 3 cm across, sepals villose; aggregate fruit red, globose. 10–12 mm across.

Rubus coreanus Miquel — KOREAN BLACKBERRY

Cha-tian-pao=Ch'a-t'ien-p'ao (插田泡, Rice-planting Blackberry). Fruit, eaten fresh or for making wine.

Arching shrubs up to 3 m high, branchlets reddish brown, glaucous, with stout hooked prickles; leaves pinnate, leaflets 5–7, ovate, or oblong-elliptic, 3–6 cm long, 1.5–4 cm wide, acute, broad-cuneate at base, unequally sharp-serrate, pubescent on the veins beneath, petioles and rachis prickly; flowers pink, small, 8–10 mm across, 3–7 in terminal corymbose clusters; aggregate fruit small, 5 mm in diameter, from red to black. Native to China, Korea and Japan, introduced into Western gardens in 1906.

Rubus ellipticus J. E. Smith — YELLOW RASPBERRY

Zai-yang-pao=Tsai-yang-p'ao (栽秧泡, Rice Planting Berry). Fresh fruit.

Upright shrubs 1–2 m high, branchlets stout, rough, densely covered with bristles mixed with some hooked prickles; leaves trifoliolate, petioles and rachis armed as the stems, leaflets oblong or suborbicular, the terminal ones 4–9 cm long, 3–8 cm wide, rounded or truncate at the apex, the lateral ones smaller, 1.5–7 cm long, 0.8–6 cm wide, glabrous above, gray tomentose beneath, sharply dentate; flowers white, 1 cm across, in axillary or terminal head-like panicles, calyx herbaceous; aggregate fruit yellow, 7–9 mm across.

Rubus flagelliflorus Focke — CENTRAL CHINA BLACKBERRY

Shao-hua-wu-pao=Shao-hua-wu-p'ao (少花烏泡, Few-flowered Black Berry). Fruits.

Climbing or prostrating and ascending shrubs, rooting as the branches touch the ground, branchlets glabrous, armed with curved prickles; leaves ovate-cordate or ovate-lanceolate, 9–12 cm long, 6–8 cm wide, acute, wavy and shallowly lobed, sparsely pilose above, tomentose beneath, stipules pinnatipartate, caducous, petioles 3–6 cm long, prickly; flowers white, small, in short axillary clusters, sepals ovate, reflexed, purple and glabrous inside, white along the margin, petals erect; aggregate fruits hemispherical, purple-black, ripening in June. Endemic to southwestern Hubei, introduced into American gardens in 1901.

Rubus ichangensis Hemsley et Kuntze — YICHANG RASPBERRY

Huang-pao-zi=Huang-p'ao-tzu (黃泡子, Yellow Raspberry). Fruit, used locally.

Scandent or prostrate shrubs, branchlets covered with stipitate glands and scattered curved prickles; leaves simple, ovate-lanceolate, 8–15 cm long, 4–7 cm wide, acuminate, base deeply cordate and slightly lobed, remotely mucronate-serrulate, glabrous, petioles 2–3 cm long, prickly like the midrib beneath; flowers white, 6–8 mm across, in an elongated cylindrical terminal panicle; aggregate fruit red, subglobose, 5–7 mm across, consisting of 10–12 drupelets, with good flavor.

Rubus idaeus L. — RED RASPBERRY; EUROPEAN RASPBERRY

Fu-pen-zi=Fu-p'en-tzu (覆盆子, Inverted Basin Berry). Fruit, fresh or preserved.

Deciduous shrubs up to 2 m high, branchlets finely tomentose when young, reddish-brown with few broad prickles at base; leaves pinnate, leaflets 3–5, broad ovate or oblong, 2–10 cm long, 1.5–4 cm wide, shortly acuminate, base rounded or subcordate, double-serrate, finely pubescent and glabrescent above, white tomentose beneath; flowers white, 1.5 cm across, in terminal glandless, tomentose and slightly prickly short racemes; aggregate fruit subglobose, 10–12 mm across, red.

Rubus idaeus L. var. **aculeatissimus** Regel et Tiling — Boruljian (Mongolian name), (Syn. *R. sachalinensis* A. Léveillé)

Sha-wo-wo=Sha-wo-wo (沙窩窩, Sand Raspberry). Ripe fruit; leaves and fruits gathered, dried in Inner Mongolia, used for tea.

Deciduous shrubs 40–100 cm high, branchlets prickly, tomentose when young; leaves broadly ovate in outline, trifoliolate, on offshoots often pentafoliolate, petioles 2–8 cm long, tomentose and prickly, with glandular strigose hairs, leaflets ovate or ovate-lanceolate, 3–10 cm long, 1.5–6 cm wide, acuminate, base round, subcordate, or cuneate and obtuse, coarsely double-serrate, white tomentose beneath, speculate on the nerves; flowers white, 2 cm across, several in terminal corymbs, pedicels 1–3 cm long, tomentose and glandular, prickly, hypanthia discoid, sepals triangular-lanceolate, caudate, 5–9 mm long, tomentose, aculeate, awned, lanate inside, petals oblanceolate, 8 mm long, stamens numerous, pistils apocarpous, lanate, styles subapical, stigmas punctiform; aggregate fruits red, drupelets tomentose. A taxon of circumpolar distribution, occurring on hillside of northern China and on altitudes of 1,500–3,000 m in Shanxi and Gansu.

Rubus lambertianus Seringe

Gao-liang-pao=Kao-liang-p'ao (高梁泡, High Ridge Berry). Fruits eaten in Hubei.

Straggling shrubs, semi-evergreen, branchlets pubescent, prickly; leaves ovate-cordate, 7–9 cm long, 7–8 cm wide, acute or short-acuminate, pilose above, strigose on major nerves and glabrescent beneath, petioles 3–5 cm long; flowers white, 8 mm across, in terminal panicles, peduncles 8–14 mm long, sepals deltoid-ovate, 8 mm long, acuminate, pilose, petals oblong-ovate, acute, base cuneate; aggregate fruits globose, mature in October–November. Endemic to central China, introduced into American gardens in 1907.

Rubus mesogaeus Focke

Lao-hu-pao=Lao-hu-p'ao (老虎泡, Tiger Berry); *Xi-yin-xuan-gou-zi=Hsi-yin-hsüan-kou-tzu* (喜陰懸鈎子, Shade-loving Blackberry). Fruits.

Deciduous shrubs with slender long terete branches, densely tomentose and with small prickles; leaves trifoliolate, terminal leaflets broad ovate or rhomboid, 6–7 cm long, 5–6 cm wide, shallowly lobed above the middle, acuminate, base rounded or subcordate, lateral leaflets smaller, obliquely subsessile, 4–5 cm long, 2.5–3 cm wide, glabrescent above, whitish, tomentose beneath, petioles 2–6 cm long, tomentose, prickly, stipules dissected; flowers white or pink, 1 cm across, in loose axillary corymbs, pubescent and prickly, sepals lanceolate, tomentose outside, sericeous inside, aggregate fruits black, 1 cm in diameter. Widespread in eastern Asia, growing in thickets at altitudes of 800–1,500 m; introduced into American gardens in 1907.

Rubus niveus Thunberg — HILL BLACKBERRY

Hong-pao-ci-teng=Hung-p'ao-tz'u-t'eng (紅泡刺藤, Prickly Red Berry Vine). Fruit; eaten fresh, used for jam or wine.

Shrubs 1–2 m high, branchlets tomentose when young, glabrescent and bloomy, with hooked prickles; leaves pinnate, leaflets 5–9, elliptic-ovate, 3–6 cm long, 0.5–3 cm wide, acute or acuminate, base cuneate or obtuse, coarsely serrate, glabrous above, white tomentose beneath, petioles and rachis with hooked prickles; flowers small, rose-purple, 5 mm across, in terminal panicles; aggregate fruit dark red becoming black, 1 cm in diameter.

Rubus palmatus Thunberg — THIMBLEBERRY

Qi-ye-mei=Ch'i-yeh-mei (槭葉莓, Maple-leaved Berry); *Xuan-gou-zi=Hsuan-kou-tzu* (懸鈎子, Nodding Hooked Berry). Fruit, eaten fresh or used for making wine.

Deciduous shrubs 1–2 m high, branchlets slender, glabrous, prickly; leaves simple, palmately 3–5-lobed, ovate-suborbicular, 5–10 cm long, 3–5 cm wide, the lobes acuminate, unequally coarse-dentate, softly pubescent beneath, prickly along the midrib, petioles 1.5–4 cm long, prickly; flowers white, solitary, 3 cm across; aggregate fruit yellow, hanging.

Rubus parvifolius L. (Syn. *R. triphyllus* Thunberg) — GRASSLAND RASPBERRY

Mao-mei=Mao-mei (茅莓, Grassland Berry). Fruit, eaten fresh or used for making wine.

Low arching prickly and hairy shrubs; leaves trifoliolate, rarely pinnate with 5 leaflets, the terminal one the largest, rhombic-suborbicular, the lateral ones smaller, obovate, 2–5 cm long, 1.5–5 cm wide, obtuse, base rounded or cuneate, unequally dentate, strigose-pilose above, glabrescent, white-tomentose beneath, petioles and rachis hairy and prickly; flowers purple-pink, small, 3–10 in terminal corymbs, prickly and pubescent; aggregate fruit red, with few mature drupelets.

Rubus peltatus Maximowicz — PELTATE-LEAVED RASPBERRY

Dun-ye-mei=Tun-yeh-mei (盾葉莓, Shield-leaved Berry). Fruit, gathered and consumed locally.

Erect shrubs 1–1.5 m high, stems red-brown, sparsely armed with prickles, branchlets green, glaucous; leaves simple, peltate, ovate-orbicular, 7.5–17 cm long, 6.5–15 cm wide, palmately 5- or 3-lobed, cordate at base, irregularly serrate, with strigose, rough and appressed hairs above, villose beneath, petioles 4.5–7.5 cm long, prickly, spines hooked; flowers white, solitary, 3–4 cm across, pedicels 2.5–4.5 cm long; aggregate fruit orange-red, oblong, 2–3 cm long, 1.5–2 cm across. Native of eastern Asia from Hubei to Japan; for the unusual large fruit, the germplasm may have added interest to the raspberry business.

Rubus phoenicolasius Maximowicz

Duo-xian-xuan-gou-zi=To-hsien-hsuan-kou-tzu (多腺懸鈎子, Glandular Raspberry). Fruits, eaten fresh in Hubei.

Upright shrubs up to 3 m high, branches, petioles and midribs densely covered with reddish-glandular bristles, branchlets sparsely prickly; leaves trifoliolate, rarely with 5 leaflets, ovate, 3.5–12 cm long, 2–6 cm wide, short-acuminate, base rounded, serrate, the teeth unevenly mucronate, pilose above, white-tomentose beneath; flowers pink, 5–8 mm across, 8 to 10 in densely pilose and glandular racemes, sepals lanceolate, pilose, petals obovate, shorter than the sepals; aggregate fruits subglobose, 1 cm in diameter, red. Widespread in eastern Asia, introduced into Western gardens in 1876.

Rubus pungens Cambessèdes — SPINY RASPBERRY

Ci-xuan-gou-zi=Tz'u-hsüan-kou-tzu (刺懸鈎子, Bristly Raspberry).

Fresh fruit, picked and eaten locally.

Deciduous creeping shrubs; branchlets purple-brown, glandularly hairy, prickly and bristly; leaves pinnate, leaflets 5–7, rarely 9, ovate or triangular, 2–3.5 cm long, 1–1.5 cm wide, the terminal one rhombic-ovate, 6 cm long, lobed and unequally double-serrate, sparsely villose and prickly beneath, petioles and rachis with hooked prickles; flowers pink, 2 cm across, 1–3, terminal to short shoots, sepals densely prickly, the lobes ovate-lanceolate, caudate, villose; aggregate fruit red, 1 cm in diameter. Native of central and western China, introduced to USA in 1907.

Rubus reflexus Ker-Gawler — VELUTINOUS BLACKBERRY

She-pao-le=She-p'ao-le (蛇泡簕, Snake Blackberry); *Xiu-mao-mei=Hsiu-mao-mei* (銹毛莓, Rust-hair Berry). Fruit, gathered and eaten locally.

Shrubby climbers, velutinous throughout, sparsely prickly, the spines straight or recurved; leaves simple, ovate in outline, 5–10 cm long, 4–11 cm wide, 3- or 5-lobed, the terminal lobe the larger, triangular-ovate, prominently reticulate; flowers white, 8–10 mm across, in axillary subsessile racemes; aggregate fruit subglobose, 1.5–2 cm across; purple-red, becoming black, sweet-sour. A native of Guangdong, never introduced into cultivation.

Rubus rosaefolius J. E. Smith — ROSE-LEAVED RASPBERRY

Kong-xin-pao=K'ung-hsin-p'ao (空心泡, Hollow Heart Berry). Fruit.

Shrubs 2–3 m high, branchlets hairy, prickly; leaves pinnate, leaflets 5–7, ovate-lanceolate, 3–5 cm long, 1–1.8 cm wide, acuminate, double-serrate, punctate on both surfaces, sparsely pubescent beneath; flowers white, 1 or 2, in leaf-axils; pedicels 1–2.5 cm long, bristly and prickly; aggregate fruit subglobose, 12–15 mm in diameter, bright red.

Rubus setchuenensis Bureau et Franchet — SICHUAN BLACKBERRY

Chuan-mei=Ch'uan-mei (川莓, Sichuan Berry). Fruit, eaten fresh.

Deciduous straggling shrubs 2–5 m high, branchlets terete, velutinous and stigose, unarmed; leaves simple, suborbicular, 6–17 cm long, 7–16 cm wide, 5- or 7-lobed, the lobes rounded, acute, irregularly dentate, rugose above, grayish-tomentose and reticulate beneath; flowers purple, in subsessile axillary or terminal panicles; aggregate fruit globose, rather small, 6–8 mm in diameter, black. Native of central and western China, introduced into Western gardens in 1898.

Rubus simplex Focke

Dan-sheng-mei=Tan-sheng-mei (單生莓, Solitary Berry). Ripe fruits, eaten in Hubei.

Perennials, with creeping rhizomes, aerial stem woody, erect, 25–30 cm high, pilose, prickly, without branches; leaves trifoliolate, the terminal leaflets ovate-lanceolate, 5–10 cm long, 0.6–4.5 cm wide, acuminate, strigose above, tomentose and prickly on the midrib beneath, serrulate, lateral leaflets smaller, petioles 6–10 cm long, armed with recurved prickles; flowers 2 or 3, axillary to the terminal or distal 2 or 3 lateral leaves, 2 cm across, sepals deltoid, acuminate-subulate, hairy and bristly, petals slightly longer than the sepals, stamens erect, ovaries many; aggregate fruits red. Native to western Hubei, Sichuan and Gansu, growing in forests at altitudes of 1,200–1,500 m.

Rubus suavissimus S. Lee — SWEET TEA

Tian-cha=T'ien-cha (甜茶, Sweet Tea). Leaves used for tea by the ethnic people living in the mountains in Guangxi since time immemorial, discovered by K. Lee, former director of Guangxi Botanical Institute, introduced into cultivation, commercial product of sweet tea is prepared from the species by that institute.

Subtropical shrubs up to 3 m high, young suckers purple red, prickly, branchlets glaucous, sparsely prickly; leaves very sweet, orbicular in outline, 5.2–11 (–16) cm long, 5–13 (–22) cm wide, palmately 5- or 7-parted, the segments lanceolate or elliptic, the median ones 4–9 (–19) cm long, 1.6–3.8 cm wide, acuminate, double-serrate, gray or brown hairy above, petioles 2–5 cm long, with 1 or 2 prickles, stipules 2–4 mm long, adnate to the petiole, persistent; flowers white, solitary, terminal to short leafy shoots, corolla 3–5 cm across, stamens numerous, ovaries densely gray hairy, styles 3, 4 mm long; aggregate fruit ovoid, red, edible.

The sweet element in the leaves is a glycoside, similar to stevioiside, 300 times the strength of cane sugar. The distribution of the species is now limited to the mountains at Long. 110°10′E, Lat. 23°N. It is propagated by the Guangxi Botanical Institute. Tea in one-ounce box is available in American Chinese groceries now.

Rubus sumatranus Miquel

Xian-mao-xuan-gou-zi=Hsien-mao-hsüan-kou-tzu (腺毛懸鈎子, Glandular Nodding Hooked Berry). Fruits eaten in Hainan Island.

Sprawling shrubs, branchlets and rachides armed with spreading glandular red hairs and hooked prickles 3–4 mm long; leaves pinnate, 5–12 cm long, leaflets (3–) 5–7, ovate-lanceolate, 2.5–5 cm long, 1.5–2.5 cm wide, acuminate, base obliquely rounded, irregularly serrate, villose along the nerves, the nerves armed with curved spines beneath, stipules linear-lanceolate; flowers white, solitary or several in a loose raceme, sepals ovate-lanceolate, 6–10 mm long, acuminate, glandularly hairy outside, villose inside, persistent and reflexed after anthesis, petals spatulate-oblong, stamens numerous, pistils many, apocarpous, oblong-ovoid, receptacles conical; aggregate fruits oblong-ovoid, 1–1.5 cm long, orange-red. Widespread in Southeastern Asia.

Rubus swinhoei Hance — CHINESE BLACKBERRY

Mu-mei=Mu-mei (木莓, Woody Berry); *Gao-jiao-lao-hu-niu=Kao-chiao-lao-hu-niu* (高腳老虎扭, High Bush Tiger Twist). Fresh fruit eaten locally.

Shrubs 2–3 m high, stems dark purple, young branchlets floccose-tomentose, with some curved bristles; leaves simple, broad-ovate, rarely ovate-lanceolate, 4–11 cm long, 2.5–3.5 cm wide, acuminate, base round, serrate, pubescent along the midribs above, densely villose and prickly along the nerves beneath; flowers white, small, 5 mm across, 3–9 in terminal racemes, pedicels and sepals villose and glandularly strigose; aggregate fruit purple-black, 1 cm in diameter. A widespread species in the warmer regions of China, occurring along the Yangtze River and thence southward to Guangxi and Yunnan.

Sanguisorba officinalis L. — BURNUT-BLOODWORT

Di-yu=Ti-yü (地榆, Ground Elm); *Shan-zao-shen=Shan-tsao-shen* (山棗參, Hillside Jujube Shen); *Meng-gu-zao=Meng-ku-tsao* (蒙古棗, Mongolia Jujube). Root and stem gathered, steamed, dried, mixed with brick-tea for making Mongolian tea.

Perennial glabrous herbs 30–80 cm high, with stout fusiform or cylindrical roots 0.5–2 cm in diameter; leaves pinnate, both radical and cauline, oblong in outline, 10–20 cm long, leaflets 9–15, ovate-oblong, 1–3 cm long, 0.7–2 cm wide, rounded and acute or obtuse, base cordate or subtruncate, dentate, the teeth apiculate, petiolules 2–10 mm long, some with stipels, cauline leaves smaller and with fewer leaflets, stipules foliaceous; flowers deep red or purple-red, small, numerous, in compact oblong-cylindrical spikes, 1–3.5 cm long, 6–12 mm in diameter, individual flowers subtended by colorful bract and 2 linear, hairy bracteoles, hypanthia turbinate, sepals 5, oblong-ovate, 1–1.5 mm long, stamens 4, filaments enclosed, anthers dark purple-black, ovary ovoid, hairy, style purple, 1 mm long, stigma tufted; achenes enclosed in the 4-angled hypanthia. Native of Eurasia hillside.

Note: A styptic to many ethnic groups, an important item in traditional Chinese medicine, presently, much of the market material from cultivated sources.

Sorbus alnifolia (Siebold et Zuccarini) K. Koch

Shui-yu-hua-qiu=Shui-yü-hua-ch'iu (水榆花秋, Water-elm Mountain Ash); *Qian-jin-shu=Ch'ien-chin-shu* (千觔樹, Thousand Ribs Tree). Fruit, eaten fresh or used for making wine.

Trees up to 20 m high, branchlets pubescent, red-brown, with conspicuous lenticels; leaves simple, ovate or oblong, 5–10 cm long, 3–6 cm wide, short-acuminate, base rounded, irregularly double-serrate; flowers white, 1–1.5 cm across, in terminal corymbs; fruits red, oblong or ovoid, 7–10 mm long. Central China; introduced into Western gardens in 1892.

Sorbus folgneri (Schneider) Rehder

Shi-hui-tiao=Shih-hui-t'iao (石灰條, Lime Twig); *Shi-hui-hua-qiu=Shih-hui-hua-ch'iu* (石灰花楸, Lime Mountain Ash). Ripe fruits, eaten fresh in western Hubei.

Trees 8–10 m high, branches spreading and arching, branchlets tomentose; leaves ovate or elliptic-ovate, 4.5–7 cm long, 1.7–4 cm wide acute or acuminate, base obtuse or rounded, finely serrate, white tomentose beneath, petioles 0.6–1.2 cm long; flowers white, in terminal branched corymbs 8–10 cm across the top, styles 3; fruits ellipsoid, 1.3 cm long, red. Native to central, western and southern China, growing in secondary forests on the sunny sides of hills at altitudes of 1,240–1,850 m; introduced into American gardens in 1901.

Sorbus pohuashanensis (Hance) Hedlund — Beijing Mountain Ash

Hua-qiu-shu=Hua-ch'iu-shu (花秋樹, Autumn Flower Tree). Fruit, used for making wine or jam.

Small trees up to 8 m high, branchlets grayish-brown, pubescent; leaves pinnate, leaflets 11–15, ovate-lanceolate, 3–5 cm long, 1.4–2 cm wide, acute, or acuminate, base obliquely rounded, serrate, rachis and midrib pubescent beneath, stipules herbaceous, coarsely dentate, persistent; flowers white, 6–8 mm across, in large terminal corymbose panicles; fruits red, 6–8 mm long. Originally recorded from the western hills of Beijing, introduced to USA in 1883.

Spiraea blumei G. Don

Zhen-zhu-mei=Chen-chu-mei (珍珠梅, Pearl *Mei*). Leaves used for tea in Guangxi.

Deciduous shrubs 1 m high, branchlets glabrous; leaves alternate, often fasciculate, obovate or ovate-rhomboid, 2–3.5 cm long, 1–2 cm wide, rounded or obtuse, base cuneate, shallowly 3- to 5-lobed above the middle, glabrous; flowers white, 3–8 mm across, corymbose, terminal to short shoots, sepals deltoid, petals obovate, stamens 18–20; follicles 5, subtended by persistent calyx. Native to eastern Asia, growing in limestone mountains in Guangxi; introduced into Western gardens in 1858.

Spiraea cantoniensis Loureiro (Syn. *S. reevesiana* Lindley) — REEVES SPIRAEA

Ma-ye-xiu-qiu=Ma-yeh-hsiu-ch'iu (麻葉繡球, Hemp-leaved Embroidered Ball); *Ma-ye-xiu-xian-ju=Ma-yeh-hsiu-hsien-chü* (麻葉繡線菊, Hemp-leaved Spiraea). Young shoots with flower-buds gathered, dried for tea.

Shrubs 1–2 m high, glabrous, branchlets slender, slightly arching, dark brown, very brittle; leaves rhombic-oblong to rhombic-lanceolate, obtuse or acute, base cuneate, incise-serrate; flowers white, 5–7 mm across, in a dense umbel-like terminal corymb on short branch, showy; aggregate fruit consisting of 5 follicles, in cultivation usually bearing no fruit. A native of China, common in New England gardens; introduced to Western gardens in 1824.

Spiraea trilobata L. — HARGAN CHAI, TEA SPIRAEA

Shi-bang-zi=Shih-pang-tzu (石棒子, Rock Stick); *San-lie-xiu-xian-ju=San-lieh-hsiu-hsien-chu* (三裂銹線菊, Trilobate Spiraea, a name found in botanical literature only). Fruits and seed, used in Mongolian tea.

Deciduous shrubs 1–2 m high, branches brown, vegetative organs glabrous throughout; leaves suborbicular to deltoid, 1.5–3 (–4) cm long and wide, varying greatly between those growing on new offshoots and those on old branches, incisely crenate-dentate, often 3-lobed, base cuneate, obtuse or subcordate, palmately nerved, bluish-green beneath, petioles 3–8 mm long; flowers white, 15–30 in umbelliform corymbs terminal to current year's growth, peduncles 5–10 mm long, pedicels 5–12 mm long, the lowest the longest, the lower ones with filiform prophylls, hypanthia obconical-discoid, 2 mm across the top, villose inside, glabrous outside, sepals 5, deltoid, acute, glandular-hairy inside, corolla 6–7 mm across, petals obcordate, 2 mm long, 3 mm wide, stamens 18–20, shorter than the petals, anthers discoid, dorsifixed, ovary glabrous, 5 carpels distinct, hairy along the adaxial suture, styles clavate, glabrous, stigmas punctiform; fruits forming a rosette of follicles, hairy along the dehiscing suture, seeds 4–6, fusiform, 1 mm long, brown. Native of northern China and adjacent Siberia and Turkestan, introduced into Western gardens in 1801.

Fabaceae or Leguminosae — Pea or Bean Family

Aeschynomene indica L. — Sensitive Joint Vetch

He-meng=Ho-meng (合萌, United Germinating). Young shoots gathered and used as potherb in Taiwan.

Annual herbs 30–100 cm high; leaves evenly pinnate, leaflets 20 or more pairs, sessile, oblong, 3–8 mm long, 1–3 mm wide, obtuse and mucronate, base rounded; flowers papilionaceous, 6 mm long, yellow with purple marking, 2 or 3 in a cymose cluster; legumes linear, compressed, 3–4 cm long, 3–4 mm wide, indehiscent, broken into 6 or 7 truncate 1-seeded sections, the segments ridged and rugose.

Apios fortunei Maximowicz — Potatobean, Groundnut

Tu-luan-er=T'u-luan-erh (土圞兒, Potato Bean); *Lao-jun-dan=Lao-chun-tan* (老君丹, Philosopher Laotzu's Elixir); *Tu-dan=T'u-tan* (土蛋, Earth Egg); *Luo-han-shen=Luo-han-shen* (羅漢參, Buddha Disciples' Shen). Thick tubers, gathered locally and used as emergency food.

Perennial climbers with tuberous root, branchlets sparsely pubescent; leaves pinnate, leaflets 3–7, ovate or lanceolate, 3–7 cm long, 1.54 cm wide, acute, or abruptly acuminate, base rounded; flowers greenish white, 6–8 mm long, in loose axillary pseudoracemes or terminal panicles 6–26 cm long; legumes linear, 7–8 cm long, 5–6 mm wide.

Arachis hypogaea L. — Peanut, Ground Nut (Figure 28)

Luo-hua-sheng=Lo-hua-sheng (落花生, Drop-Flower Borne); *Hua-sheng=Hua-sheng* (花生, Peanut). Seeds; roasted and eaten as a nut, cooked with meat, chicken, dry bean curd (豆腐乾) in regular dishes, or used as a source of vegetable oil.

Annual herbs 20–70 cm high, branchlets slightly pubescent; leaves even-pinnate, leaflets 4, oblong-obovate, 2.5–5 cm long, 1.5–2.5 cm wide, glabrous on both surfaces, ends rounded; flowers yellow, solitary, papilionaceous, the stalk bending after anthesis;

legumes indehiscent, developed underground, mature pericarp netted; seeds 1–3, with thin brown-pink seedcoat. A native of Brazil, extensively cultivated in China.

Astragalus membranaceus (Fischer) Bunge — HUANGCHY, ASTRAGALUS

Huang-qi=Huang-chi (黃芪, Huangchy); *Bei-qi=Pei-chi* (北芪, Northern Astragalus). Dried root, sliced. Used in combination with codonopsis (dang-shen) and lycium berries for making a tonifying soup with spareribs.

Graceful perennial herbs 60–150 cm high, branchlets villose; leaves pinnate, leaflets 21–31, ovate or oblong, 0.7–3 cm long, 0.4–1 cm wide, sparsely villose on both surfaces; flowers pale yellow or ivory white, papilionaceous, 10–12 in loose axillary racemes 7–8 cm long, on peduncles longer than the subtending leaf, calyx campanulate, 5–7 mm long, pilose with black or rarely white hairs, often glabrous, teeth deltoid, the lower one the longest, corolla 1.5–1.8 cm long, standard obovate, clawed, stamens 9 + 1, ovary stipitate, hairy; fruits inflated, hanging, 2–3 cm long, 0.8–1.2 cm wide, rostrate, the slender stipe long exceeding the persistent calyx, pericarp membranaceous. Native to northern China, much cultivated for the root, dried material available in American Chinese groceries; a form introduced by Steven Foster is grown in Massachusetts in late 1980s.

Azukia angularis (Willdenow) Ohwi — ADSUKI BEAN, GREATER RED BEAN (Syn. *Dolichos angularis* Willd., *Phaseolus angularis* [Willd.] W. F. Wight, *Vigna angularis* [Willd.] Ohwi et Ohashi) (Figure 29)

Hong-dou=Hung-tou (紅豆, Red Bean). Cultivated, Seed available in Boston, both in forms of dried grain and as canned paste; the grains cooked with rice to make a richer staple food; the paste used for pastries, especially for making Eight-precious Pudding.

Erect annual herbs 40–60 (–90) cm high, branchlets puberulent; leaves trifoliolate, the terminal one ovate, the lateral ones oblique ovate, 4–10 cm long, 2.5–5 cm wide, acute or acuminate, base rounded or cuneate, entire; flowers pale yellow, the wings lemon yellow, paired, in small axillary pseudoracemes on short peduncles, rachis short, 1–1.5 cm long, extrafloral glands on cushion between 2 flowers, bracts subulate, calyx campanulate, upper 2 lobes small, ciliate, wings slightly imbricate on the upper side, with the margin of the carculate keel petal below, keel curved almost 300 degrees, margin free except the apical half of the lower margin, forming a narrow tube enclosing the barbate half of the style, stigma lateral, appendage subulate; fruits cylindrical, 5–8 cm long, 8–10 mm across, glabrous; seeds 6–10, oblong, 4–7 mm long, 3-5 mm in diameter, dark maroon, ends truncate, hilum white, spongy, 3/4 the length of the long axis of the seed, caruncle very narrow, almost indistinct. Extensively cultivated as a minor crop in northern China as far north as Harbin in Helongjiang; very variable in seed color; hypogeal germination, erect when first flowering the green house, transplanted to the experimental garden, flowering again in June, in the hot weather of August suddenly flushing, becoming bushy and with climbing branches, bearing numerous flowering clusters; seeds planted in late June, flowering in August, becoming bushy as the transplanted specimen; introduced from northern China to USA in 1900.

Bauhinia hypehana Craib

Shen-zi-ye=Shen-tzu-yeh (腎子葉, Kidney-leaf); *Shuang-shen-teng=Shuang-shen-t'eng* (雙腎藤, Twin-Kidney Vine); *Bai-hua-long-xu-teng=pai-hua-lung-hsu-t'eng* (白花龍鬚藤, White-flowered Dragon Whisker); *Hu-bei-yang-ti-jia=Hu-pei-yang-t'i-chia* (湖北羊蹄甲, Hubei Bauhinia). Root and stem, stewed with pork kidneys, or cooked with pork, the preparations regarded as special health food in Hubei, people in Sichuan cook such material with pork intestine.

Very large and fast growing lianas, branchlets pubescent, tendrils opposite the leaves, pubescent; leaves unifoliolulate, the petioles articulate at the apex, laminas bilobed, 3–8 cm long, 4–9 cm wide, the lobes rounded, base cordate or subtruncate, pilose beneath; flowers white-pink, in terminal corymbs, calyx tubular, 1.3–1.7 cm long, 2-lobed, petals unguiculate, fertile stamens 3 or 4, ovary glabrous, stipitate; fruits flat, 14–30 cm long, 4–5 cm wide; seeds many. Native to central, western and southern China, climbing over tall trees and flowering above the tree top.

Bauhinia variegata L. — Camel's Foot; Hong Kong Orchid Tree

Yang-ti-jia=Yang-t'i-chia (羊蹄甲, Goat's Foot); *Gong-fen-yang-ti-jia=Kung-fen-yang-t'i-chia* (宮粉羊蹄甲, Pink Goat's Foot); *Yang-zi-jing=Yang-tzu-ching* (洋紫荊, Foreign Redbud). Young leaves, flower buds, flowers and young fruits used for vegetable in Hainan Island.

Small trees 3–5 m high, bark dark brown, almost smooth, branchlets pilose, glabrescent; leaves unifoliolate, laminas caudate-orbicular, 7–9 cm long, 9–11 cm wide, bilobed, the lobes entire, pilose beneath, with 9–13 primary palmate nerves; flowers showy, pink-purple with dark purple veins, 3–7 axillary and subterminal, in a dense pseudoracemose cluster, sepals spathiform, splitting along one side, softly pilose and glandularly punctate, corolla 10 cm across, petals obovate, oblong, 3.5–4 cm long, 1.5–2 cm wide, clawed at the base, fertile stamens 5, sterile ones 3–5, filaments 3–3.5 cm long, ovary stipitate, hairy along the sutures; fruits flat, 15–25 cm long, 1.5–2 cm wide, rostrate; seeds 3–15, discoid, 10–12 mm across, brown. Extensively cultivated in southeastern Asia, an important wayside tree in Hong Kong, much planted along the banks of highway, producing spectacular scenery in the countryside during March–April.

Cadelium radiatum (L.) S. Y. Hu — MUNGBEAN (Syn. *Phaseolus radiatus* L., *P. aureus* Roxburgh, *P. mungo* auctt., non L., *Vigna radiata* [L.] Wilczek) (Figures 30, 31)

Lü-dou=Lü-tou (綠豆 Mungbean, Green Bean). Seeds; preparations from the starch called *Fen-si=Fen-ssu* (粉絲, Mungbean Silk), and *Fen-pi=Fen-p'i* (粉皮, Mungbean Skin); the sprouts of the seed called *Lü-dou-ya=Lü-tou-ya* (綠豆芽, Mungbean sprout). The seed and its various products all available in Boston.

Erect or trailing annual herb 30–90 cm high, strigose throughout; leaves trifoliolate, the terminal leaflet broad-ovate, lateral ones strongly oblique-ovate, 5–12 cm long, 3–9 cm wide, acuminate, entire, stipules foliaceous, peltate, attached almost at the middle, both ends acute; flowers dull greenish-yellow, the apical curled portion of keel flushed purple, subsessile, paired at the apex of peduncle, rachis very short, with arrested growth at flowering and fruiting time, often resuming growth after the first fruits turning black, extrafloral glands on cushion between the paired flowers, bracteoles ovate-oblong, shorter than the calyx, lobes short, ciliate, the upper two almost united, standard twisted slightly to the left before anthesis, left wing strongly folded inward over the spur of the left keel petal, upper portion of the keel connate along the outer margin, the inner margin overlapped, forming a tube over the style, curved 300 degree, style twisted once at the base, then broadened, barbate, stigma positioned upward, with subulate apical appendage; fruits stiff, strigose, cylindrical-linear, 6–9 cm long, 5–6 mm across, ripened black, horizontal forming a 90° angle to the rachis and peduncle; seeds 10–14, green occasionally yellow or black, smooth or glandular-reticulate, 3–4.5 mm long, 3–4 mm across, the ends subtruncate, hilum acentric, white, covered by convex spongy cushion, 1.5–2 mm long, caruncle 0, micropyle evident, parahilum 1 mm long. Native to the mountains on the border of Yunnan, Burma and northeastern India, extensively cultivated as a summer crop in the lower Yellow River region. Two cultivars sowed in the field of winter wheat.

cv. 'Early Erect' — (綠豆, *Lü-dou=Lü-tou*, Green Bean). Erect herb, 60–80 cm high, flowering early and repeatedly, labor intensive crop, pods shatter readily, requiring hand pick twice before harvest; extensively cultivated.

cv. 'Decumbent Viny' — (拖秧綠豆, *Tuo-yang-lü-dou=T'o-yang-lü-tou*). Decumbent annual herbs, slightly viny at the tip of the stems, requiring longer growing season, fruits not shatter readily, requiring no hand picking, less productive, less acreage planted.

Seeds purchased from Chinese groceries in Boston germinate readily both in the green house and in the experimental garden. Germination is epigeal, and the primary leaves are sessile. Plants grown in June, began to flower at the third regular trifoliolate leaf. Flowers are self fertile. The stem continues to grow and with two or three leaves bearing axillary inflorescences. Finally only flowering clusters without any leaf. Rachis of lower infructescence resumes growth and bears paired flowers. Testa of seeds are green, smooth. A few seeds show glandular stripes or reticulations. Herbarium specimens from India with reticulate glandular texture are identified as *"Phaseolus mungo* L."* Further experimental studies of the taxonomic value of the color and texture of the testa in this group is awaited.

Cajanus cajan (L.) Millspaugh — Pegion Pea, Cajan, Red Gram, Dahl (Figure 32)
  Mu-dou=Mu-tou (木豆, Tree Pea). Seeds.

Upright shrubs 1–3 m high, branchlets pubescent; leaves pinnate-trifoliolate, leaflets lanceolate, 5–10 cm long, 1–3.5 cm wide, acuminate, both surfaces hairy, golden punctate beneath; flowers papilionaceous, yellow with red markings, 1.8 cm long, in an axillary cymose cluster, peduncles long; legumes linear, constricted between seeds, stiff hairy; seeds truncate-orbicular, carunculate. Native to tropical Asia, naturalized in Hong Kong.

Canavalia gladiata (Jacquin) de Candolle — Sword Bean
  Dao-dou=Tao-tou (刀豆, Sword Bean) Young fruits, pickled.

Annual climbers 2–3 m high, branchlets glabrous; leaves pinnate-trifoliolate, leaflets ovate, 8–20 cm long, 5–16 cm wide, acuminate, base rounded; flowers papilionaceous, purple-pink, 3–4 cm long, in axillary pseudoracemes; legumes sword-like, up to 38 cm long, strongly ribbed-ridged on the ventral side, the pericarp thick, rather woody; seeds reniform, red-brown. Native of Old World tropics, occasionally cultivated in tropical China. A closely related species, *C. ensiformis* (L.) DC. differs in having upright habit, 1–2 m high, smaller fruits containing white seeds, said to be poisonous, also cultivated in China for medicinal purposes.

Cassia nomame (Siebold) Kitagawa (Syn. *Cassia minosoides* L. subsp. *nomame* [Siebold] Ohashi)
  Dou-cha-jue-ming=Tou-ch'a-chüeh-ming (豆茶决明, Tea Cassia). Leaves used for tea.

Erect annual much branched herbs 10–60 cm high, branchlets pubescent; leaves even-pinnate, 4–8 cm long, leaflets small, 16–30 pairs, minute, linear-lanceolate, 5–7 mm long, 1.5–2 mm wide, acute and mucronate, base oblique; flowers yellow, 6 mm long, 1 or 2 in axillary clusters, stamens 4, rarely 5; legumes linear, strongly compressed, 3–8 cm long, 4–5 mm wide, consisting of 6–12 seeds.

Cassia occidentalis L. — COFFEE SENNA, STINKING WEED (Syn. *Senna occidentalis* [L.] Link)

Wang-jiang-nan=Wang-chiang-nan (望江南, Looking South beyond the Yangtze River); *Ye-bian-dou=Yeh-pian-tou* (野扁豆, Wild Flat Bean); *Li-cha=Li-ch'a* (黎茶, Li-Ethnic Tea). Seed, roasted brown, pulverized, using a small amount (3 g = one-tenth of an oz), make tea with brown sugar, used in Fujian as a tea substitute for people with high blood pressure.

Suffrutescent herbs 1–2 m high; leaves even-pinnate, leaflets 6–10 pairs, ovate, or ovate-lanceolate, 2–6 cm long, 1–2 cm wide, entire, pilose along the margin, petioles each furnished with a prominent gland above the base; flowers yellow, zygomorphic, few in axillary or terminal corymbs, sepals 5, petals 5, 10–12 mm long, stamens 10, upper 3 sterile, lowermost 3 larger, filaments curved upward, ovary linear, pilose, style filiform, curved upward, stigma truncate; legumes linear, strongly compressed and often slightly curved, septate, with broad margin, seeds 30–40, compressed-ovoid, verrucose except the depressed central portion. Widespread in the tropics.

Cassia sophora L. — (Syn. *Senna sophora* [L.] Roxburgh)

Jiang-mang=Chiang-mang (茳芒, Ancient name, meaning obscure); *Jiang-mang-jue-ming=Ching-mang-chüeh-ming* (茳芒决明, Jiang-mang Positive Bright). Young shoots and tender fruits, used for vegetable; seeds and roots used in Chinese medicine; first recorded in herbals of the fifth century.

Shrubs or subshrubs 1–2 m high, branchlets pilose; leaves even pinnate, leaflets 6–10, paired, ovate-lanceolate, 2.5–8 cm long, 1–3 cm wide, acute or acuminate, base rounded, strigose near the margin, petioles with a prominent gland near the base; flowers yellow, showy, in axillary corymbs, sepals 5, corolla 2 cm across, stamens 10, 7 fertile, 3 sterile, anthers of the lowermost two longer, ovary pilose; fruits compressed cylindrical, inflatted, 6–9 cm long; seeds ovoid, 2 mm long, brown, conspicuously reticulate, smooth after the membrane removed. Widespread in Asia, cultivated in northern China for the seed, naturalized in southern China.

Cassia tora L. — SICKLEPOD (Syn. *Senna tora* [L.] Roxburgh)

Jue-ming=Chüeh-ming (决明, Decisive Eye-sight Promoter); *Jia-lü-dou=Chia-lü-tou* (假

綠豆, False Mungbean); *Cao-jue-ming=Ts'ao-chüeh-ming* (草決明, Herb Cassia), *Jue-ming-zi=Chüeh-ming-tzu* (決明子, Cassia Seed); *Ye-hua-sheng=Yeh-hua-sheng* (野花生, Wild Peanut); *Ma-ti-jue-ming=Ma-t'i-chüeh-ming* (馬蹄決明, Horseshoe Cassia). Young plants, boiled with licorise to make a tea, used in Yunnan under the name of *Ye-hua-sheng*; seed, roasted, ground, used for beverage like coffee, said to improve eyesight; the first species of *Cassia* L. recorded in Chinese herbals.

Annual herbs 1–2 m high; leaves even-pinnate, petioles without a gland near base, leaflets (–4) 6, rachis with prominently stalked glands between opposite leaflets, laminas obovate or obovate-oblong, 1.5–6 cm long, 1–3 cm wide, rounded- or emarginated-mucronate, base rounded, pilose; flowers yellow, sepals 5, spreading, petals 5, obovate, 12 mm long, stamens 10, 7 well developed, ovary elongated, stipitate, pilose, style short and stout, stigma capitate; legumes subcylindrical, curved, 9–16 cm long; 4 mm across the middle; seeds numerous, oblong, the central thin portion narrow and long, the remaining part densely verrucose. Widespread in the tropics, common in areas south of the Yangtze River, cultivated for the seed in northern China; appearing in prescription of traditional Chinese medicine as *Jue-ming-zi* or *Ma-ti-jue-ming*.

Cicer arietinum L. — CHICK PEA

Ying-zhui-dou=Ying-ch'ui-tou (鷹嘴豆, Eagle-Beak Pea)

Erect annual herbs 20–60 cm high, densely glandular hairy throughout; leaves pinnate, leaflets 9–15, alternate, ovate-oblong, 8–17 mm long, 4–7 mm wide, acute, base rounded or obtuse, serrate, stipules foliaceous, dentate; flowers white or purplish, solitary, axillary on peduncles 1–2 cm long, calyx campanulate, 7 mm long, 5-toothed, petals slightly longer, stamens 9 and 1, ovary hairy; legumes inflated, oblong-subglobose, 2 cm long, hanging; seeds 1 or 2, ovoid-subglobose, 8 mm in diameter, white, (red or black in certain cultivars). Native to western Asia, cultivated in Europe, introduced into Hebei for food, rare in China.

Cullen corylifolium (L.) Medicus — SCURFY PEA (Syn. *Psoralea corylifolia* L.) (Figure 33)

Bu-gu-zhi=Pu-ku-chih (補骨脂, Repairing Bone Grease). Seed, used for preparing an aphrodisiac for man, by boiling equal amount (60 g = 2 oz) pulverized of roasted scurfy pea and nutmeg with 40 seedless dried Chinese jujube in 6 cups of water with 120 g (= 4 oz) ginger. The mixture is cooked to a paste; taking a tablespoonful after dinner.

An aromatic annual herb, up to 100 cm high, covered with short hairs and translucent dots throughout; leaves simple, ovate, 4.5–10 cm long, 3–6 cm wide, obtuse, base rounded or subcordate, irregularly coarse-serrate; flowers small, papilionaceous, white or purplish, 3–4 mm long, stamens 10, diadelphous; legumes small, subglobose, 5 mm long, indehiscent, black, pericarp thin, adhering to the seedcoat, punctate, foveolate, hilum deeply sunk in a depression, carunculate.

Delandia umbellata (Thunberg) S. Y. Hu — RICE BEAN, LESSER RED BEAN (Syn. *Dolichos umbellatus* Thunb., *Phaseolus calcaratus* Roxb., *Vigna calcarata* [Roxb.] Kurz, *Azukia umbellata* [Thunb.] Ohwi, *Vigna umbellata* [Thunb.] Ohwi et Ohashi)
Chi-xiao-dou=Ch'ih-hsiao-tou (赤小豆, Little Red Bean). Seed, used more for medicinal value than as an article of food; available in one pound bags in Chinese stores in large American cities; young fruits gathered from kitchen gardens, for vegetable.

Climbing herbs spreading to 3 m on open hillsides among grasses, branchlets hirsute, glabrescent with age; leaves trifoliolate, the terminal one ovate, the lateral two oblique-ovate, triangular-acuminate at the apex, 7–14 cm long, 3–8 cm wide, entire, finely puberulous on both surfaces; flowers light yellow, in axillary pseudoracemes up to 14 cm long on peduncles 6–14 cm long, secondary smaller cluster emerging latter and giving the appearing of fascicles of pseudoracemes, flowers paired with extrafloral glands on cushion between the pedicels, bracteoles linear, longer than calyx, persistent, calyx campanulate, lobes deltoid, ciliate, upper 2 almost united, standard 14 mm across, keel twisted, left one spurred, style barbate, stigma lateral, apical appendage subulate; fruits cylindrical, subglabrous, slightly curved at the distal end, 6–10 cm long, 5–6 mm across, ripened brown or brownish black; seeds 6–10, maroon or olive-ochreous, subcylindrical, 5–9 mm long, 3–4 mm across, subtruncate at the ends, hilum 2/3 the length of the long axis, conspicuously carunculate with 2 longitudinal ridges and a narrow slit between them.

Native of tropical southeastern Asia, growing over grassy slopes, flowering in May to August; seeds bought from Chinese stores in Boston planted in the Green House on the fifth floor of Biological Laboratories, Harvard University, in March 1991, transplanted into the Experimental Garden, near Harvard University Herbaria in late May, began to grow in June, producing massive vegetatively, failing to flower both in the Green House and in the Garden by the end of August; special selection for annual crop in the Yellow River Region in China being successful, hence seeds were introduced to the United States repeatedly in 1901–1906; germination hypogeal; seed color very variable, both maroon and straw-colored specimens gathered from one plant in Hong Kong; small plants, erect or with some branches climbing, bearing flowers and fruits in kitchen gardens (S. Y. Hu 10560).

Desmodium podocarpum de Candolle

Ye-huang-dou=Yeh-huang-tou (野黃豆, Wild Soybean); *Yuan-ling-ye-shan-ma-huang=Yuan-ling-yeh-shan-ma-huang* (圓菱葉山螞蝗, Rounded-rhomboid-leaved Tick-Clover). Seeds, used for food by people living in the mountains in western Hubei.

Shrubs 50–100 cm high, branchlets pilose; leaves trifoliolate, terminal leaflets suborbicular-rhomboid, 4–7 cm long, 3.5–6 cm wide, acute, base cuneate, pilose or glabrescent, petiolules 1–2 cm long, lateral leaflets-oblique-ovate, smaller, petiolules 1–3 mm long, stipules lanceolate, caducous; flowers small, purple-red, in terminal panicles 30 cm long, pedicels 2.5 mm long, calyx 5-toothed, corolla 4 mm long, stamens 9 + 1, ovary linear; legumes 1.6 cm long, 2-articulate, segments obovate-deltoid, pubescent. Native to the hillsides of western Hubei, growing at altitudes of 500–1,900 m above sea level.

Eriosma chinense Vogel — CHINESE ERIOSMA

Zhu-zi-li=Chu-tzu-li (豬子芷, Piglet Basket). Enlarged fleshy root, used to prepare a tonifying broth with pork.

Perennial herbs, velutinous throughout, with a deep long tap root, enlarged, becoming fleshy, fusiform at the basal end, 3–6 cm long, 1.5–2 cm across the top, black, sending out 1 or rarely 2 aerial shoots annually, 20–40 cm tall, the basal 1/3–1/2 leafless; leaves simple, lanceolate, 3–5 cm long, 0.5–1.5 cm wide, acute or obtuse, mucronate, obtuse or rounded at the base, entire; flowers small, papilionaceous, yellow, 7–8 mm long, 1–3 in axillary raceme; legumes oblong, 1 cm long, 4–5 mm across, pericarp black, strigose; seeds 2 or 3, oblong, 4 mm long, 2 mm across, shiny brown and maculate black, smooth, with prominent white caruncles. Growing on grassy slopes, dying back completely during the dry winter months, regenerating the aerial parts each spring; the eriosma in southern China is used in similar way as is ginseng in northern China.

Gleditsia sinensis Lamarck — CHINESE HONEY-LOCUST

Zao-jia-ren=Tsao-chia-jen (皂莢仁, Honey-locust Kernel). Hard, transparent material outside the cotyledons, cooked for dessert.

Large trees 15–20 m high, trunks smooth, gray, often armed with stout branched thorns up to 16–20 cm long; leaves pinnate, leaflets 6–14, obliquely ovate or oblong, 3–8 cm long, 1.5–3.5 cm wide, obtuse, serrate, glabrous; flowers small, green, polygamous, in axillary racemes on short shoots, stamens 6–8, ovary stipitate; legumes long and narrow, compressed, 10–30 cm long, 2–4 cm wide, brown, indehiscent, rich in saponin, extracted in water for washing silk. A native of China, introduced into Western gardens in 1774.

Glycine max (L.) Merrill — SOYBEAN (Figures 34 and 35)

Da-dou=Ta-tou (大豆, Giant Bean); *Huang-dou=Huang-tou* (黃豆, Yellow Bean). Seed, mixed with *gao-liang* and barley, ground together for flour, used to be the staple of the rural population in the Yellow River Region; soaked, cooked with salt, served as tidbits in northern China; used as source for all the following forms of Chinese food.

Huang-dou-ya=Huang-tou-ya (黃豆芽, Soybean Sprout). Soaked, sprouted under cover with daily wash, prepared at home in villages or as simple industry in cities; available in American Chinese groceries, used as vegetable or for making soup with bone and onion.

Dou-fu=Tou-fu (豆腐, Bean Curd); *Dou-jiang=Tou-chiang* (豆漿, Soybean Milk); *Dou-fu-nao=Tou-fu-nao* (豆腐腦, Soft Soybean Curd); *Dou-fu-gan=Tou-fu-kan* (豆腐乾, Dry Bean Curd, 5 cm firm squares); *Qian-zhang-pi=Ch'ien-chang-p'i* (千張皮, Thousand Sheet, thin sheets of firm bean curd). Different forms of food prepared from ground soaked soybean; available in American Chinese groceries.

Dou-fu-ru=Tou-fu-ju (豆腐乳, Bean Curd Cheese). Soft products of bean curd prepared after fermentation of *Rhizopus* (for the gray types) or *Monascus* (for the red types). *Jiang-you=Chiang-yu* (醬油, Soy Sauce). Various types of liquid product prepared after the fermentation of soybean with innoculation of *Apergillus*.

Dou-chi=Tou-ch'ih (豆豉, Fermented-and-seasoned Black Soybean). A dry and soft product used as spice.

Dou-you=Tou-yu (豆油, Soybean Oil). Cooking oil extract fom soybean. *Dou-ban-jiang=Tou-pan-chiang* (豆辦醬, Fermented Soybean Sauce). Paste, prepared from fermented cooked soybean and whole wheat flour, available in Boston, in cans only.

Erect strigose annual herbs 30–100 cm high, branches stout; leaves trifoliolate, leaflets rhombic-ovate, the lateral ones obliquely ovate, 7–14 cm long, 3–6 cm wide, acute or short acuminate, base rounded; flowers small, 3–4 mm long, papilionaceous, white or rarely purple, 2–5 in subsessile axillary raceme; legumes oblique-oblong, sharply pointed at both ends, 2- or 3-seeded, strigose; seeds oblong, varying in sizes and colors, generally pale yellow or black (see Part I for more information).

Glycine soja Siebold et Zuccarini

Lao-dou=Lao-tou (蟆豆, Wild Soybean); *Wu-dou=Wu-tou* (烏豆, Black Bean). Seed, gathered and used locally.

Annual climbers strigose throughout, branchlets slender; leaves trifoliolate, leaflets ovate-lanceolate, the lateral ones obliquely ovate, 1–5 cm long, 1–2.5 cm wide, acute, base obtuse, both surfaces villose; flowers purple-red, 4 mm long, in axillary fascicles; legumes oblong, 3 cm long, 4 mm wide, apex oblique and pointed; seeds 2–4, black.

Glycyrrhiza uralensis Fischer, **G. glabra** L., **G. inflata** Batalin — CHINESE LICORICE
Gan-cao=Kan-ts'ao (甘草, Sweet Herb). Underground growth, yellow, sweet; used for seasoning candies, and various dried fruits used as tidbits; ingredient of various bupin mixes prepared in packages, sold in Chinese groceries in Boston.

Suffrutescent perennial herbs 40–70 cm tall, rarely up to 1 m high, with creeping rootstocks red-brown on the outside, yellow and sweet inside; branchlets pubescent with straight and glandular hairs; leaves pinnate, leaflets 5–15, ovate or oval, 2–5 cm long, 1–3 cm wide, obtuse or acute, mucronate, base rounded, hairy and punctate, especially beneath; flowers papilionaceous, purplish, 10–15 mm long, in axillary racemes 5–12 cm long; legumes falcate or curved, 6–8 cm long, prickly and glandular; seeds 2–8, oblong-reniform, shiny, black.

Kummerowia striata (Thunberg) Schindler
Ji-yan-cao=Chi-yen-ts'ao (雞眼草, Chicken-eye Weed). Young plants, eaten in Shaanxi.

Annual (perennial in Hong Kong) herbs, prostrating or ascending, 5–30 cm high, branchlets pilose; leaves trifoliolulate, leaflets varying greatly in size and shape, oblong, obovate or elliptic, 5–15 mm long, 3–7 mm wide, pilose, obtuse and mucronate, lateral nerves 15–19 pairs, stipules membranous, strongly striate; flowers minute, white with pink standard, in axillary fascicles, opening on sunny mornings, calyx campanulate, 2 mm long, teeth ovate, subtended by 3 ovate-striate bracteoles, corolla 5 mm long, stamens diadelphous, falling with the petals after anthesis; fruits small, compressed-orbicular, 1-seeded, indehiscent, concealed by the persistent calyx; seed solitary, black, mottled brown, 2.5–3 mm long. A widespread weed in lawns and along waysides.

Lablab purpureus (L.) Sweet (Syn. *Dolichos lablab* L.) — HYACINTH BEAN, LABLAB BEAN
Mei-dou=Mei-tou (眉豆, Eyebrow Bean); *Bian-dou=Pien-tou* (扁豆, Flat Bean); *Que-dou=Ch'ueh-tou* (鵲豆, Magpie Bean). Young fruit, seed.

A much cultivated annual (perennial in Hong Kong) climber, strongly branched, the branchlets wire-like, glabrous; leaves trifoliolate, the terminal one triangularly ovate, the lateral ones strongly oblique-ovate, 5–10 cm long, 6–11 cm wide, entire; flowers white or red-purple, in axillary pseudoracemes on long or short peduncles; legumes oblong-falcate, 5–7 cm long, 1.5–2 cm wide, compressed; seeds 3–5, white or black, oblong-subglobose, 1–1.5 cm long, 0.8–1 cm across, with a long white spongy ridge along 3/4 of one side.

Lathyrus davidii Hance — DAVID'S VETCHLING

Shan-jiang-dou=Shan-chiang-tou (山豇豆, Wild Cowpea). Young shoot.

Large perennial climbers 80–100 cm high, the stem often reaching 3 m long; leaves pinnate, terminated by branched tendrils, stipules semisagittate with a prolonged basal lobe, leaflets 4–8, ovate or oblong, 3–10 cm long, 1.8–3 cm wide, acute, entire, glabrous; flowers yellow, 2 cm long, in axillary racemes 10–20 cm long; legumes cylindrical, 8–11 cm long; seeds oblong, brown. Wild in northern China.

Lathyrus japonicus Willdenow (Syn. *L. maritimus* Bigelow) — BEACH WILD PEA

Ye-wan-dou=Yeh-wan-tou (野豌豆, Wild Pea). Seed.

Perennial vines with fleshy creeping and forked rhizomes; stems stiffy, branching; leaves pinnate with terminal branched tendrils, stipules foliaceous, broad ovate, leaflets 4–10, elliptic or oblong-ovate, 1–5 cm long, 0.5–2.5 cm wide, entire; flowers purple-violet, 3–10 in axillary racemes, corolla 1.5–2.5 cm long; legumes firm, chartaceous, 3–7 cm long; seeds brown, subglobose, 4–5 mm in diameter.

Lens culinaris Medikus (Syn. *L. esculentus* Moench) — LENTIL

Bing-dou=Ping-tou (兵豆, Soldier Pea); *Xiao-bian-dou=Hsiao-pien-tou* (小扁豆, Small Flat Pea). Seed, cultivated with wheat in North China.

Pubescent annual or biennial climbers 10–40 cm high, branchlets slender; leaves pinnate, terminated by a single tendril, leaflets 8–14, obovate-oblong, 0.6–2 cm long, 2–5 mm wide, obtuse and mucronate; flowers small, papilionaceous, bluish-white, 1 or 2, at the apex of a slender axillary peduncle; legumes oblong, inflated, pale brown; seeds 1 or 2, lens-shaped. Introduced from the Mediterranean Region, rarely cultivated.

Lespedeza bicolor Turczaninow — TARTARY BUSH-CLOVER

Hu-zhi-zi=Hu-chih-tzu (胡枝子, Tartary Bush-Clover); *Sui-jun-cha=Sui-chün-Ch'a* (隨軍茶, Following Army Tea). Young shoots with leaves and flower-buds, gathered and used as tea substitute.

Deciduous shrubs 1–3 m high, branchlets sparsely pilose; leaves trifoliolate, the terminal leaflets larger, oval, broad-elliptic, or round-obovate, rarely oblong-lanceolate, (1-) 3–5 (–6) cm long, 1–4 cm wide, rounded or obtuse and apiculate, base rounded, sparsely pilose; flowers papilionaceous, purple, in axillary pseudoracemes or terminal panicles, calyx campanulate, unequally 5-lobed, lobes deltoid, the upper two almost connate to the apex, the lower one the longest, pubescent, standard 1.2 cm long, the wings and keel slightly shorter, clawed, stamens 9 and 1, ovary hairy; fruits oblique-ovate, 8–10 mm long, 4–5 mm wide, rostrate, prominently reticulate, slightly hairy; seeds olive-brown, maculate with darker brown, oblique-oblong, 4 mm long, 2.5 mm across, hilum on the oblique side, carunculate.

Widespread in temperate eastern Asia, growing on open grassy hillsides or margins of forests; introduced into Western gardens in 1856, cultivated in USA.

Lespedeza davurica (Laxmann) Schindler — Ox Bush-Clover, Mongolia Lespedeza Tea

Niu-zhi-zi=Niu-chih-tzu (牛枝子, Ox Bush-Clover), *Da-wu-li-hu-zhi-zi=Ta-wu-li-hu-chih-tzu* (達烏里胡枝子, Davuria Lespedeza, used in botanical literature, a translation of the specific epithet). Flowers and leaves, used in Inner Mongolia for tea.

Small deciduous shrubs 1 m high, branchlets pubescent, leaves trifoliolate, leaflets oblong, obovate, rarely lanceolate, 2–3 cm long, 7–10 mm wide, round, or obtuse and mucronate, base round, glabrous above, pubescent beneath, stipules setaceous, 3–4 mm long; flowers papillionaceous, greenish-yellow changing white, with purplish marking, in dense axillary pseudoracemes, sepals cupular, 5-lobed, the lobes lanceolate, 8–9 mm long, standard 10 mm long, wings and keel slightly shorter, stamens 9 and 1, ovary strigose; fruits obovate, 3 mm long, smooth and white strigose, 1-seeded; seed black, smooth, glossy, compressed oblong-rhomboid, 1.5 mm long, 1 mm across, hilum subterminal, rounded, carunculate.

Medicago hispida Gaertner — Bur Clover (Syn. *M. denticulata* Willdenow)

Mu-xu=Mu-hsü (苜蓿, Bur Clover); *Mu-ji-tou=Mu-chi-t'ou* (苜雞頭, Bur Clover Tops), *Nan-mu-xu=Nan-mu-hsü* (南苜蓿, Southern Clover). Young shoots, used as vegetable.

Suberect biennial herbs 20–40 cm high, branchlets sparsely pubescent; leaves trifoliolate, stipules foliaceous, ovate, prominently dentate, leaflets obovate, 1–1.5 cm long, 0.7–1 cm wide, truncate and retuse at the apex; flowers small, yellow, 2–6 in a head-like raceme terminal to a slender axillary peduncle; legumes spiral, the outer suture with hooked prickles; seeds 3–7, reniform, light brown.

Medicago sativa L. — Alfalfa

Zi-mu-xu=Tzu-mu-hsü (紫苜蓿, Purple Clover); *Xu-cao=Hsü-ts'ao* (蓿草, Clover Fodder); *Mu-xu=Mu-hsü* (苜蓿, Clover). Tender shoots, eaten in Shaanxi.

Perennial herbs 30–100 cm high, much branched and with deep root system, the taproot extending 2–3 m; leaves trifoliolate, leaflets obovate or oblanceolate, 1–2 cm long, 5 mm wide, obtuse and mucronate, distal end serrate, villose beneath, stipules lanceolate, 5 mm long, pilose; flowers purple, 8–25 in axillary heads, calyx campanulate, 5-toothed, corolla slightly longer than the sepals, standard reflexed, unguiculate; fruits curved 1–3 times, pilose, rostate; seeds 1–8, reniform.

Millettia speciosa Champion ex Bentham — SHOWY MILLETTIA

Shan-lian-ou=Shan-lian-ou (山蓮藕, Hillside Lotus Rhizome), *Niu-da-li=Niu-ta-li* 牛大力, Ox-strength); *Da-li-shu=Ta-li-shu* (大力薯, Great Strength Potato), *Ji-xue-teng=Chi-hsüeh-t'eng* (雞血藤, Chicken-blood Vine). Root, fresh or dried, boiled with pork to make a soup for strengthening the bones, used in southern China and Hong Kong.

Evergreen shrubs 1–3 m high, some branches vine-like, trailing, roots spreading, distal portions thickened, branchlets densely velutinous, all young growth densely strigose; leaves pinnate compound, ovate-oblong in outline, 10–30 cm long, leaflets 9–15, elliptic-oblong, the lower pairs ovate-oblong, 2–9 cm long, 1–3 cm wide, acute, or shortly acuminate, base obtuse or rounded, entire, petioles 1–8 cm long, thickened at base, velutinous, laminas glabrescent, shiny above; flowers white, showy, in terminal racemes or pseudopanicles, all parts densely pubescent, bracts ovate, pedicels 1 cm long at anthesis, terminated by 2 ovate prophylla, calyx campanulate, 1 cm long, shallowly 5-lobed, the lowest the longer, deltoid, acute, 3 mm long, the uppermost two the smaller, round; corolla papilionaceous, the standard 3 mm long, erect, inner surface with greenish guidelines and yellowish base, wings shorter than the keel, rounded at the apical end, stamens 9 and 1, pistil densely hairy, fruits sword-like, 7–21 cm long, 1.5–2.5 cm wide, densely velvelty, brown, acute at the apex, bearing 1 to 3 mature seeds and 6–8 aborted ovules; seeds compressed globose, shiny black, 1 cm in diameter, hilum linear, white, 2.5 mm long. Native of southern China, roots collected by local residents, in Hong Kong seen only in badland areas and isolated seashores.

Moghania philippinensis (Merrill et Rolfe) Li — SOUTHERN ASTRAGALUS (Syn. *Flemingia philippinensis* Merrill et Rolfe).

Qian-jin-ba=Ch'ien-chin-pa (千斤拔, Uprooting Thousand Pounds), *Tu-huang-qi=T'u-huang-ch'i* (土黃蓍, Local Astragalus Root). Root, sliced, cooked with pig-feet in a mixture of equal amount of water and cooking wine, a southern China recipe; the market material cut into 3–7 cm oblong slices.

Suffruticose, prostrating shrubs 1–2 m high, with straight stout rootstock up to 1 m long, branchlets slender, strigose-villose throughout, internodes 2–9 cm long, striate-sulcate, densely covered with brown hairs; leaves digitate-trifoliolate, leaflets ovate-oblong, 4–9 cm long, 1.5–4 cm wide, obtuse and mucronate, entire, glabrescent above, petioles 1.5–2.5 cm long, dense-velutinous, stipules subulate, 9–10 mm long, 1.5 mm across the base, strongly lineate with thick veins and strigose; flowers rose-pink, small, 6–8 mm long, strigose-villose and glandular throughout, crowded in terminal or axillary head-like clusters 2–3 cm long, bracts lanceolate, lineate as the stipules, hypanthia obconical, 2 mm long and across the top, calyx deeply lobed, segments unequal, the lowest the longer, 6 mm long, lanceolate-caudate, strigose, lateral ones 4 mm long, uppermost segment deeply bifid, corolla papilionaceous, slender, 6 m long; fruits turgid-subglobose, 8–10 mm long, 5–6 mm across, strigose and glandular; seeds 2, spherical, 2 mm in diameter, black, caruncle white. Native of southern China, thence extending eastward to the Philippine Islands.

Mucuna sempervirens Hemsley

Guo-shan-long=Kuo-shan-lung (過山龍, Dragon Crossing the Mountain); *Niu-ma-teng=Niu-ma-t'eng* (牛麻藤, Ox-hemp Vine); *Chang-chun-you-ma-teng=Ch'ang-ch'un-yu-ma-t'eng* (常春油麻藤, Evergreen Oil-hemp Vine). Seeds.

Vigorous lianas, cauliflorous, branchlets pubescent; leaves trifoliolate, the terminal one ovate-oblong, the lateral ones strongly oblique, ovate-oblong, 7–12 cm long, 4–5.5 cm wide, acuminate, entire, glabrous; flowers rose-purple, showy, 6.5 cm long, numerous, in large hanging pseudoracemes; legumes compressed linear-oblong, 7–12 cm long, 1–1.2 cm wide, ochreous; seeds 8–12, compressed oblong-discoid, 10–12 mm long, 8–10 mm wide, the ends rounded or truncate, hilum short, surrounded by a ridge, caruncle white.

Phaseolus coccineus L. — SCARLET RUNNER-BEAN (Syn. *P. multiflorus* Lamarck)

Hua-dou=Hua-tou (花豆, Flower Bean); *Hong-hua-cai-dou=Hung-hua-ts'ai-tou* (紅花菜豆, Red Flowered Bean). Seeds, rare.

Perennial climbers, the vines up to 7 m long; leaves trifoliolate, the terminal one ovate, lateral ones oblique ovate, 5–10 cm long, 4–6 cm wide, acute, base rounded, entire, glabrous; flowers bright scarlet, showy, 1.8–2.5 cm long, in axillary racemes 10–12 cm long, on peduncles 10 cm long, the keel curved spirally with the style, positioning the stigma to the visiting insect; legumes flat, long and falcate, containing 4–6 large oblong seeds 2 cm long, white or black and mottled red.

Phaseolus limensis Macfadyen — LIMA BEAN

Li-ma-dou=Li-ma-tou (利馬豆, Lima Bean). Seed, introduced, rare.

Robust twining or erect annual herbs; leaves trifoliolate, leaflets broadly ovate, up to 12 cm long, acute or lanceolate; flowers white, in cylindrical pseudoracemes 10–15 cm long, on short peduncles 2–3 cm long; legumes 7–10 cm long, 2 cm wide, sharply beaked, prominently ridged along the ventral suture; seeds compressed-oblong, 2 cm long, 1.2 cm wide, white. Native of tropical America, introduced by the former Lingnan University to Guangzhou and its vicinity in the 1920s.

Phaseolus lunatus L. — BUTTER BEAN, SIEVA BEAN, SMALL LIMA BEAN

Jin-jia-dou=Chin-chia-tou (金莢豆, Golden Pod Bean). Seeds.

Slender and slightly pubescent annual herbs, twining or erect; leaves trifoliolate, thin, leaflets broad-ovate, the lateral two oblique, 5–8 cm long, entire, acute; flowers white, rather small, 0.8–1 cm long, in paniculate racemes of various length; legumes 5–8 cm long, 1–1.5 cm wide, slightly pubescent, sharply beaked, dehiscing into twisting valves; seeds rhomboidal in outline, white. Native of tropical America, introduced by agricultural institutions, rare.

Phaseolus vulgaris L. — STRING BEAN, WAX BEAN, SNAP BEAN

Si-ji-dou=Ssu-chi-tou (四季豆, Four-season Bean); *Cai-dou=Ts'ai-tou* (菜豆, Vegetable Bean); *Yun-dou=Yun-tou* (雲豆, Cloud Bean). Tender fruit, very common; seeds introduced into Hong Kong, for food and medicine.

Erect or scandent annual herbs, branchlets finely pubescent; leaves trifoliolate, terminal one ovate, broadly ovate or rhombic, lateral two obliquely ovate, 4–16 cm long, 3–11 cm wide, acute or acuminate; flowers white, purplish, in short axillary pseudoracemes; legumes variable in sizes and shape, the common ones in China subcylindrical, slightly curved and beaked, hanging, 10–12 cm long, 1 cm across; seeds slightly compressed oblong, 10–14 mm long, 7–8 mm across, smooth, white, light-brown, often mottled, hilum small, 2 mm long, on the slightly concave side of the long axis, caruncle white. Native of tropical America, introduced, selected to fitting varied climatic conditions of China, becoming one of the commonest vegetable throughout the country; the cultivar with long flat bean such as cv. "Kentucky Wonder" is known only in agricultural institutions.

Pisum sativum L. — PEA (Figure 36)

Much branched glabrous biennial or annual herbs, scandent; leaves pinnate, terminated with branched tendrils, stipules foliaceous, the basal lobes dentate, leaflets 2–6, oblong, entire, 2–5 cm long, 1–2.5 cm wide, obtuse and mucronate; flowers white or pinkish with purple-red wings, solitary or 2 terminal to the peduncle, the standard erect; legumes 4–10 cm long, 1–1.3 cm wide, compressed oblong, shortly beaked; seeds 2–7, spherical, 5–8 mm in diameter, white or pale brown and mottled olive-green. Native of Eurasia, extensively cultivated throughout China as a field crop mixed with wheat for mature seed, and as garden crops for young shoots, immature pods, or fully grown green seed; several popular varieties:

1. var. **sativum** — Garden Pea, Green Pea, Common Pea, English Pea

Wan-dou=Wan-tou (豌豆, Garden Pea). Young seed; *Wan-dou-tou=Wan-tou-t'ou* (豌豆頭, Pea Shoot). Leaf-buds, fully grown tender pods, young and mature seed (see Part I for more information).

2. var. **arvense** (L.) Poiret — Field Pea

Wan-dou=Wan-tou (豌豆, Field Pea). Mature seed; planted with winter wheat; flowers with pinkish standard, and purple wings; fruits usually with 2 or 3 mottled round seeds.

3. var. **macrocarpa** Seringe — Snow Pea

Xue-dou=Hsüeh-tou (雪豆, Snow Pea); *He-lan-dou=Ho-lan-tou* (荷蘭豆, Dutch Pea). Young pods, used more in American Chinese restaurants than in China; flowers white, pods large, 5–7 cm long, harvested early, before the full development of the seed; mature seed ivory-white, wrinkled.

Psophocarpus tetragonolobus (L.) de Candolle — Wing Bean

Si-leng-dou=Ssu-leng-tou (四楞豆, Four-angled Bean). Young fruit, seed, tuberous root.

Much branched, glabrous twining perennial herbs with edible tuberous roots; leaves trifoliolate, leaflets ovate, the lateral ones slightly oblique-ovate, 5–14 cm long, 3–9 cm wide, acuminate, entire or occasionally lobed; flowers yellow-pink or violet, rather large, 2–5 cm long, the standard upright, 2.5 cm across, several forming a crowded raceme at the apex of peduncles 2–10 cm long, occasionally solitary; legumes rather large, 10–20 cm long, with 2 wavy wings along each suture, the wing 5 mm wide. A native of tropical Africa, introduced to India and southeastern Asia, rare in China.

Pueraria edulis Pampanini — Edible Kudzu

Ge-gen-fen=Ke-ken-fen (葛根粉, Pueraria Root Starch). Starch from the enlarged fleshy root; noodles made from the starch, transparent.

Perennial lianas with fleshy tuberous roots rich in starch, branchlets pubescent; leaves trifoliolate, the terminal one rhombic, three lobes, the lateral two strongly oblique, 8–15 cm long, 6–10 cm wide, all the lobes acuminate, stipules sagittate, 1 cm long; flowers purple-red, 1.5 cm long, in axillary pseudoracemes, the calyx glabrous, corolla 1.5 cm long; legumes linear, 6 cm long, sparsely pubescent.

Pueraria lobata (Willdenow) Ohwi (Syn. *P. thunbergiana* [Siebold et Zuccarini] Bentham) — KUDZU, PUERARIA

Ge-gen=Ke-ken (葛根, Pueraria Root); *Ye-ge=Yeh-ke* (野葛, Wild Kudzu). Root, cooked with Chinese dates, and sliced yams for a soup.

Ge-hua=Ke-hua (葛花, Pueraria Flower). Flowers and flower buds, gathered, dried, sold for an ingredient of *Wu-hua-cha* (五花茶, Five Flowers Tea) in Guangzhou and Hong Kong; packed mixture available in American Chinese stores ($1.80 per packet, August 1991 price).

Perennial lianas velutinously strigose throughout; leaves trifoliolate, leaflets ovate-orbicular, shallowly lobed, the lobes rounded, 5–19 cm long, 14–18 cm wide, acuminate, stipules peltate; flowers purple-red, in axillary pseudoracemes 10–20 cm long; legumes compressed-linear, 5–8 cm long, densely strigose, the hairs golden.

Pueraria montana (Loureiro) Merrill

Ge-ma-mu=Ke-ma-mu (葛麻母, Mother-of-Kudzu). Starch from root.

Perennial lianas with fleshy tuberous root and velutinously strigose throughout; leaves trifoliolate, middle leaflets broadly ovate or rhombic, the lateral ones oblique-ovate, unlobed, acuminate, stipules lanceolate; flowers purple-red with prominent basal yellow patch inside the standard, in axillary panicles; legumes flat, linear, 4–9 cm long, rusty strigose.

Pueraria thomsonii Bentham — SWEET KUDZU, CULTIVATED PUERARIA

Gan-ge-teng=Kan-ke-t'eng (甘葛藤, Sweet Kudzu); *Fen-ge=Fen-ke* (粉葛, Mealy Pueraria). Root, cultivated, potato-like, pared, sliced, used for soup with pork chops; imported material available in American Chinese groceries.

Much branched lianas, densely strigose throughout; leaves trifoliolate, leaflets ovate, the lateral ones oblique-ovate, 10–21 cm long, 9–18 cm wide, acuminate, irregularly lobed, stipules aciculate-oblong; flowers purple, 1.3 cm long, in axillary pseudoracemes; legumes oblong flat, 15 cm long, velutinous strigose.

Rhynchosia volubilis Loureiro — Rat-eyes

Lu-huo=Lu-huo (鹿藿, Deer Vine); *Lao-shu-yan=Lao-shu-yen* (老鼠眼, Rat-eyes). Seed, used locally.

Perennial scandent herbs, covered with a mixture of velutinous villose, white strigose and short glandular trichomes throughout, branchlets wire-like, continuing growth after flowering; leaves trifoliolate, the terminal ones rhombic-suborbicular, the lateral ones strongly oblique-ovate-oblong, 2–6 cm long, 2–5.5 cm wide, apex obtuse, or rounded, mucronate, base cuneate or round, entire; flowers small, yellow, in axillary pseudoracemes, solitary at the initial stage of development, the foliaceous bracts deciduous; legumes 2-seeded, oblong, 1–1.5 cm long, 7–8 mm across, apiculate, redbrown at maturity, dehiscent and exposing 2 shiny seeds on the venture suture for a long time, similar to the eyes of a rat (hence the vernacular Chinese name "Rat-eyes"); seeds subglobose-reniform, 3 mm long, slightly less than 3 mm across, shiny black, funiculus stout, compressed, carcuncle white, hilum concave.

Robinia pseudoacacia L. — Black Locust, Yellow Locust

Yang-huai=Yang-huai (洋槐, Foreign Pogoda Tree). Flower, washed mixed with flour, steamed, seasoned, eaten in North China.

Large trees 10–25 m high, often appearing shrubby in China due to cuttings for fuel, trunks 50–70 cm in diameter, bark dark gray, deeply and irregularly furrowed, branchlets prickly with persistent stout specialized stipular spines; leaves pinnate, oblong in outline, leaflets 10–25, oblong, 2–6 cm long, 1–2 cm wide, retuse and mucronate, base rounded; flowers white, fragrant, in hanging racemes 10–20 cm long; legumes flat, oblong, 4–10 cm long, 1.5 cm across, brown, with a prominent ridge along the ventral suture, shortly beaked; seeds many, compressed oblong-reniform, 6 mm long, 3 mm across, smooth, brown-mottled, hilum on the upper 4/5 of one side of the long axis, funiculus long, caruncle white, scanty.

Sesbania grandiflora (L.) Poiret — Scarlet Wisteria Tree, Vegetable Humming-bird

Mu-tian-qing=Mu-t'ien-ch'ing (木田青, Tree River Hemp). Flowers and young fruit.

A very interesting species of short-lived, soft-woody tree 4–10 m high, branchlets puberulent; leaves even pinnate, linear-oblong in outline, 20–40 cm long, glabrous, leaflets 16–60, oblong, 1–2.5 cm long, 8–10 mm wide, obtuse and mucronate, base rounded, petioles 1.59 cm long, terete; flowers white-pink or rose, 2–5 in loose axillary racemes or terminal panicles, 7–12 cm long, calyx tubular-campanulate, the partially developed flower-buds curved, sickle-like; legumes compressed linear, 20–60 cm long, tetragonous due to the ridges along the sutures, hanging, pericarp reticulate, ochreous; seeds numerous, oblong-reniform, chestnut-black, 6–7 mm long, 3–4 mm wide, hilum orbicular, in a depression on one side of the long axis. Extensively cultivated in southeastern Asia, naturalized in the East Indies, perhaps native of northeastern Africa, rare in cultivation in Yunnan and Guangdong.

Sophora japonica L. — PAGODA TREE, CHINESE SCHOLAR TREE

Huai-shu=Huai-shu (槐樹, Pagoda Tree). Branchlets; twigs boiled in water, into which poaching an egg, drinking the liquid and eating the egg as a home remedy for stopping bleeding.

Huai-hua-mo=Huai-hua-mo (槐花末, Sophora Flower-bud). Flowers and buds gathered just as the few flowers of the panicles starting to open, dried, used as an important ingredient of the Five Flowers Tea (五花茶, *Wu-hua-cha*), a special drink for the hot summer season in Hong Kong.

Huai-dou-zi=Huai-tou-tzu (槐豆子, Sophora Seed). Endosperm, cooked with sugar, eaten as dessert in North China.

Deciduous trees 15–20 m high, trunk with smooth gray bark, branchlets green, puberulous with soft white trichome; leaves pinnate, oblong in outline, 15–25 cm long, leaflets 9–15, oblong-ovate, 2.5–7.5 cm long, 1.5–3 cm wide, acute and mucronate, base obtuse or rounded, entire; flowers yellowish-white, in terminal panicles 15–30 cm long, showy and lasting for many weeks; fruits cylindrical, 5–8 cm long, stipitate, constrict between seeds, fleshy and indehiscent; seeds oblong, the cotyledons green, covered by transparent endosperm. A native of China, introduced to Japan with Buddhism as many other cultivated plants.

Sophora viciifolia Hance

Lang-ya-ci=Lang-ya-tz'u (狼牙刺, Wolf-teeth Spines); *Bai-ci-hua=Pai-tz'u-hua* (白刺花, White Spine Flower); *Ma-ti-zhen=Ma-t'i-chen* (馬蹄針, Horse-hoof Needles). Flowers.

Deciduous shrubs 1–2.5 m high, branches stiff, erect, tuberculate, bark reddish-brown, branchlets short, spiny, glabrescent; leaves pinnate, oblong-linear in outline, leaflets 11–21, oblong or elliptic, 5–12 mm long, 4–5 mm wide, rounded, emarginate and mucronate, sparsely pilose beneath, stipules aciculate; flowers white, often tinged blue, calyx campanulate, 3–4 mm long, purple-blue, dense pilose, corolla 15 mm long, standard spatulate-obovate, reflexed; fruits 2.5–6 cm long, constricted between seeds, beaked, densely pilose: seeds 1–7. Native to the provinces along the Yellow and Yangtze

Rivers, growing on sandy sunny hillsides at altitudes of 1,000–1,500 m; introduced into Western gardens in 1897.

Stizolobium capitatum (Sweet) 0. Kuntze (Syn. *Mucuna capitata* [Sweet] Wight et Arnott) *Li-dou=Li-tou* (貍豆 or 藜豆, Wild Cat Bean); *Gou-zhua-dou=Kou-chua-tou* (狗爪豆, Dog's Paw Bean). Young pods as vegetable, seed used for making bean curd; very rare in China.

Annual climbers, softly pubescent throughout, the vines twining counter-clockwise; leaves trifoliolate, the terminal one rhombic-ovate, the lateral ones oblique ovate, 6–9 cm long, 4.5–7 cm wide, obtuse, mucronate; flowers purple-rose, in subsessile head-like axillary racemes; legumes subcylindrical, slightly curved at the distal end, 10–15 cm long, 1.5 cm across; seeds 5 or 6, grayish-white, 1.5 cm long, hilum with white caruncles. Introduced from southeastern Asia, rarely cultivated.

Tamarindus indicus L. — TAMARIND (Figure 37)

Luo-wang-zi=Lo-wang-tzu (羅望子, Net Look Seed, a book name); *Suan-jiao=Suan-chiao* (酸角, Sour Bean). Fruit, very sour.

Evergreen trees 6–20 m high, branchlets glabrescent, brown-black; leaves even pinnate, leaflets 14–40, oblong or elliptic, 1–2.5 cm long, 4–9 mm wide, rounded or retuse, base oblique rotundate, glabrous; flowers in axillary or terminal racemes, with 4 reflexed white sepals, 3 petals, yellow with red veins and greenish margin, 3 fertile stamens with the filaments connate below the middle; fruits subcylindrical, fleshy, indehiscent, 3–15 cm long, 2–3 cm across, brown, the pulp rich in tartaric acid, very sour. Native of Africa, common in India, cultivated in Yunnan; one tree has grown in the garden of the Department of Biology, The Chinese University of Hong Kong for 20 years but never flowered.

Trigonella foenum-graecum L. — FENUGREEK

Hu-lu-ba=Hu-lu-pa (葫蘆巴, Fenugreek); *Xiang-cao=Hsiang-ts'ao* (香草, Spice Herb); *Xiang-dou=Hsiang-tou* (香豆, Aromatic Pea); *Ku-dou=K'u-tou* (苦豆, Bitter Pea). Ground seed used as spice, young shoots dried, ground, the powder used in layers of pastry in northwestern China; available in American Chinese stores.

Pleasantly scented annual herbs, 30–80 cm high, sparsely pubescent; leaves trifoliolate, leaflets subequal, obovate-cuneate, 1–3.5 cm long, 0.5–1.5 cm wide, apex rounded and dentate; flowers white, 1 or 2 in subsessile axillary fascicle, papilionaceous, 1–1.5 cm long; legumes cylindrical, cinnamon-brown, 5–11 cm long, 5 mm across, attenuated at the distal end; seeds oblique-rectangular, truncate at both ends, 3–4 mm long, 2–3 mm wide, 2 mm thick, brown, smooth, finely glandularly punctate, each side with an oblique furrow, dented at the juncture of the meeting furrows, hilum punctiform, situated in the dent, cotylendons wrapped in transparent endosperm.

Vicia bungei Ohwi (Syn. *V. tridentata* Bunge)

San-chi-ye-wan-dou=San-ch'ih yeh-wan-tou (三齒野豌豆, Three-toothed Wild Pea). Young plants used for potherb in Shaanxi.

Biennial climbers 20–40 cm high, stems slender, tetragonous, much branched, glabrous or pilose; leaves pinnate, terminated by tendrils, leaflets 6–10, oblong or obovate-oblong, 1–2.5 cm long, 3–8 mm wide, retuse or truncate, mucronate, base rounded, pilose beneath, stipules sagittate, 2–3 mm long; flowers purple-blue, in axillary racemes 2–2.5 cm long, calyx campanulate, 7–8 mm long, pilose, teeth lanceolate, corolla 2–3 cm long, standard retuse, ovary-stipitate, golden hairy, style villose; legumes flat, oblong, 2.5–3.5 cm long, 6–8 mm wide; seeds 3–8, subglobose, 3 mm across. Widespread in northern China.

Vicia cracca L. — Vetch, Bird Vetch, Cow Vetch

Shao cai=Shao-ts'ai (紹菜, Nest Vegetable); *Cao-teng=Ts'ao-t'eng* (巢藤, Vetch). Young shoot including small flowering buds, cooked fresh or dried for winter use.

Annual climbers puberulous with appressed hairs; leaves pinnate, with terminal tendrils, leaflets 8–24, elliptic, 1–3 cm long, 2–8 mm wide, acute-obtuse and mucronate; flowers pinkish-blue, 1 cm long, 7–15 in axillary racemes, papilionaceous; legumes flat, oblong, pointed obliquely at the distal end, 1.5–2.5 cm long; seeds 2–5, black, spherical, smooth, 2 mm in diameter, hilum 2 mm long, curved in conformity with the shape of the seed, apparently without caruncle.

Vicia faba L. — Broad Bean, Horse Bean, Italian Bean

Can-dou=Ts'an-tou (蠶豆, Silkworm Bean, so named because the beans are ready for the market as silkworms begin to form cocoon); *Hu-dou=Hu-tou* (胡豆, Tartatian Bean). Seeds; many ways of cooking; fried cotyledons imported from China available in American Chinese groceries.

Hu-dou-ban-jiang=Hu-tou-pan-chiang (胡豆瓣醬, Broad Bean Sauce). A very hot sauce made in Pixian of Sichuan from fermented broad bean with unusual flavor; occasionally available in American Chinese stores as *Pi-xian-dou-ban-jiang=Pi-xian-tou-pan-chiang* (郫縣豆瓣醬, Pixian Bean Sauce), used in Sichuan.

Erect, biennial herbs, generally unbranched or branching only at the ground level, 50–100 cm high; leaves pinnate, with a short filiform vestige of a tendril at the apex of the rachis, leaflets 2–6, almost alternate, oblong-elliptic, 4–6 cm long, 2.5–4 cm wide, apex oblique or roundish, apiculate, glaucous; flowers white with purple-violet blotch, 3 cm long, 0.8–1 cm across, 1–4 in subsessile axillary fascicles; legumes oblong-cylindrical, slightly pointed at both ends, 5–10 cm long, 1.5–2 cm across; seeds generally 2 or 3, up to 6, compressed rectagular, 2 cm long, 1.2–1.5 cm wide, 5–7 mm thick, the hilum-end truncate and grooved, the opposite end rounded, hilum black, 8–10 mm long, in the groove of the trucate end.

Vicia hirsuta (L.) S. F. Gray

Xiao-chao-cai=Hsiao-ch'ao-ts'ai (小巢菜, Lesser Vetch); *Qiao-yao=Ch'iao-yao* (翹搖, Rising Shaking); *Bai-hua-shao-cai=Pai-hua-shao-ts'ai* (白花苕菜, White-flowered Vetch). Young shoots used for potherb.

Annual climbers 10–30 cm high, much branched; leaves pinnate, terminated by tendrils, leaflets 8–16, oblong-obovate, 0.5–1.5 cm long, 1–4 mm wide, truncate, retuse and mucronate, base cuneate, glabrous; flowers white, 2–5 in axillary racemes, rachis and pedicels pubescent, calyx campanulate, 5-toothed, pilose, petals 5 mm long, ovary hirsute, style hairy; legumes flat, 7–10 mm long, yellow, hairy; seeds 1–2, brown. Widespread weed in temperate northern hemisphere.

Vicia pseudo-orobus Fischer et Meyer

Jia-xiang-wan-dou=Chia-hsiang-wan-tou (假香豌豆, False Sweet Pea). Tender shoots, used as potherb in Hubei.

Perennial climbers 50–200 cm high; leaves pinnate, terminated by tendrils, leaflets 4–10, ovate or ovate-oblong, 3–6 cm long, 1–2.5 cm wide, acute, base rounded, pilose beneath, stipules hastate; flowers purple, numerous, in axillary racemes 5–9 cm long, calyx tubular, membranous, 5-toothed, pilose, petals 10–12 mm long, ovary stipitate, glabrous, style glandular; legumes flat, elliptic, 3 cm long; seeds 1 or 2. Widespread in temperate eastern Asia.

Vicia sativa L. — Spring Vetch, Tare

Da-chao-cai=Ta-ch'ao-ts'ai (大巢菜, Greater Vetch); *Jiu-huang-ye-wan-dou=Chiu-huang-yeh-wan-tou* (救荒野豌豆, Emergency Wild Pea); *Wei=Wei* (薇, Vetch, the classical name in ancient Chinese herbals). Young shoots.

Annual climbers 25–50 cm high; leaves pinnate, terminated with tendrils, leaflets 8–16, obovate-oblong, 0.8–2 cm long, 3–7 mm wide, apex truncate and emarginate, base cuneate, sparsely pubescent, hairs yellow; flowers purple-red, 1 or 2, axillary to upper leaves; legumes linear, compressed, 2.5–4.5 cm long; seeds 8–10, brown, globose.

Vicia unijuga A. Broun

Wai-tou-cai=Wai-t'ou-ts'ai (歪頭菜, Slanting-head Vegetable). Young shoots, gathered and used locally.

Perennial herbs 1 m high, branchlets velutinous; leaves pinnate, leaflets 2, ovate-elliptic, 3–10 cm long, 1.5–4 cm wide, apex acuminate or acute, base obtuse, stipules foliaceous, sagittate; flowers purple or purple-red, 1.5 cm long, showy, 6–12 in axillary racemes, peduncles 20–25 cm long; legumes oblong, 3–4 cm long, brown; seeds compressed globose, brown.

Vigna sinensis (L.) Savi ex Hasskarl — YARD-LONG BEAN, ASPARAGUS BEAN (Syn. *Dolichos sinensis* L.) (Figures 11a and 38)

Dou-jiao=Tou-chiao (豆角, Bean Horn); *Jiang-dou=Chiang-tou* (豇豆, Vigna). Tender pods used as vegetable, extensively cultivated throughout China, occasionally available in American Chinese groceries.

Glabrous annual climbers; leaves trifoliolate, ovate-rhombic, lateral ones strongly oblique, 15–12 cm long, 4–7 cm wide, acute, rounded or cuneate; flowers light purple, rarely yellowish-lavender, paired at the apex of slender axillary peduncle; legumes hanging, cylindrical, light green, 20–30 cm long, 8–10 mm across, torulose (swollen at intervals) at maturity, indehiscent; seeds oblong-reniform, reddish-brown, some cultivar with white marking near the hilum, 15–20 in a pod. Garden origin in China; growing well in many kitchen gardens in Boston.

var. **sesquipedalis** (L.) Ascherson et Schweinfurth — YARD-LONG BEAN (Syn. *Vigna sesquipedalis* (L.) Fruwirth 1898, *Vigna unguiculata* [L.] Walpers subsp. *sesquipedalis* [L.] Verdcourt 1970)

Chang-jiang-dou=Ch'ang-chiang-tou (長豇豆, Long Vigna); *Dou-jiao=Tou-chiao* (豆角, Bean Horn). Tender fully grown fruit, used for vegetable, fresh or pickled, cultivated mostly in southern China, common in American Chinese groceries.

A variety with longer and thinner fruits, 45–70 (–90) long, 7–8 mm across, generally dark green, firm and crisp, the seeds of some cultivar light brown, one end larger than the other.

Vigna unguiculata (L.) Walpers — COWPEA, BLACK EYE PEA, CATJANG, CHINA PEA (Syn. *Dolichos unguiculatus* L. 1753, *Phaseolus cylindricus* L. 1754, *Dolichos catjang* Burman

f. 1768, *Vigna cylindrica* [L.] Skeels 1913, *V. unguiculata* [L.] Walpers subsp. *cylindrica* [L.] van Eseltine 1913)

Fan-dou=Fan-tou (飯豆, Food Bean); *Mei-dou=Mei-tou* (眉豆, Eyebrow Bean); *Jiang-dou=Chiang-tou* (豇豆, Vigna); *Bai-dou=Pai-tou* (白豆, White Bean), *Hua-fan-dou=Hua-fan-tou* (花飯豆, Pinto Bean). Seed.

Rambling and trailing annual herbs; leaves pinnately trifoliolate, terminal leaflets ovate, the lateral ones obliquely broad-ovate, 5–8 cm long, 5 cm wide, acute, base obtuse; flowers yellowish-white-purple, 2, rarely 3, crowded at the distal end of a long peduncle; legumes firm, cylindrical, 8–13 cm long, 6–7 mm across, containing 6–8 seeds, dehiscent; seeds oblong-reniform, 8–12 mm long, 7–9 mm across, white, often with black eye, sometimes mottled.

There is much confusion in botanical and agricultural literature about this species. Linnaeus did not have a clear idea about it. He doubtfully cited *"Cacara nigra* Rumph. amb. V. 381. t. 138 f ?"* after the description. Rumphius's plate apparently illustrates a velvet bean. It does not fit the description of a plant with pods in heads. This character is shown in Walpers's plate 23 when he made the combination, *Vigna unguiculata* (L.) Walpers. His material has short pods with less than 10 seed, a characteristic of cowpea.

Vigna vexillata (L.) A. Richard

Ye-jiang-dou=Yeh-chiang-tou (野豇豆, Wild Vigna). Rootstock, cooked with pork bone for a bupin (an immunostimulating broth).

Perennial climbing or trailing herbs velutinous-strigose throughout, rootstock woody, branchlets wire-like; leaves trifoliolate, leaflets ovate or rhombic-ovate, 4–8 cm long, 2.5–4.5 cm wide, acuminate, base cuneate or rounded; flowers pink-purple, 2–4 crowded at the distal end of peduncles 9–12 cm long; legumes held erect, cylindrical, 9–11 cm long, 5 mm across; seeds oblong, shiny-black.

Wisteria sinensis (Sims) Sweet — WISTERIA (Syn. *Rehsonia sinensis* [Sims] Stritch)

Zi-teng-hua=Tzu-t'eng-hua (紫藤花, Wisteria). Flower and flower buds, washed, drained, mixed with flour, steamed, seasoned; eaten in northern China only (see Part I for more information).

Woody lianas, branchlets puberulous; leaves pinnate, leaflets 7–13, ovate-elliptic, 4.5–11 cm long, 2–5 cm wide, acuminate, base obtuse or rounded, pubescent with white hairs when young, glabrescent; flowers showy, light purple, in large hanging terminal pseudoracemes; legumes clavate, slightly compressed, 10–15 cm long, hairy, late-dehiscent; seeds discoid when fully developed, axis parallel to the hilum 12 mm long, the one crossing the hilum 11–15 mm long, 2 mm thick, under-developed seeds oblong, axis parrallel to hilum 8–10 mm long, the other one 5–7 mm long, brown-black.

Oxalidaceae: Oxalis Family

Averrhoa bilimbi L. — BILIMBI, TREE SORREL

San-ren=San-jen (三稔, Three Prosperity). Fruit.

Large cauliflorous evergreen trees up to 14 m high, branchlets velutinous-hispid throughout; leaves pinnate, 30–40 cm long, leaflets 23–45, lanceolate, 5–8 cm long, 1.3–2.5 cm wide, acuminate-caudate, base obliquely truncate; flowers red-purple, 12–14 mm long, in racemose panicle 12–14 cm long, sepals velutinous; fruits greenish-yellow, 8–10 cm long, with 5 obscure angles.

Averrhoa carambola L. — CAROMBOLA, BILIMBI

Yang-tao=Yang-t'ao (洋桃, Foreign Peach). Ripe fruit, available in American Chinese groceries.

Large evergreen trees with spreading low branches; leaves pinnate, 10–15 cm long, leaflets 5–11, ovate or oblong, 3–6.5 cm long, 2–3 cm wide, abruptly short acuminate; flowers small, red-purple, 6 8 mm long, in small axillary or terminal corymbose panicle, 2–6 cm long, sepals glabrescent; fruits oblong, 5–8 cm long, 5–6 cm across the middle, yellow or brownish, deeply 5-furrowed and 5-ridged, star-like in cross-section, with pleasant apple smell, sour; seeds embedded in jelly mass of aril, obliquely ovate in outline, compressed, 9–12 mm long, 6–7 mm wide, 2 mm thick, shiny, smooth, raphe prominent along one side.

Linaceae: Flax Family

Linum usitatissimum L. — FLAX

Ya-ma-you=Ya-ma-yu (亞麻油, Linseed Oil); *Zhi-ma=Chih-ma* (脂麻, Oil Hemp). Seed.

Graceful erect annual herbs, 30–80 cm high; leaves small, alternate, sessile, linear-lanceolate, 2–3 cm long, acute, entire; flowers blue, solitary, 7–10 mm across; capsules globose, 6–8 mm in diameter; seeds 10, flat, oblong, reddish-brown.

Zygophyllaceae: Caltrop Family

Nitraria sibirica Pallas — SIBERIAN NITRARIA

Bai-ci=Pai-tz'u (白刺, White Thorn). Fruit.

Low, spiny, gray-white bushes, branchlets sparsely short pilose, some hairs glandular; leaves simple, 2–3 fasciculate, fleshy, oblanceolate-spatulate, 2–3 cm long, 3–6 mm wide, apex round, base cuneate, entire; flowers small, sessile, greenish-yellow, 8 mm across, in terminal panicles with scorpioid branches, stamens 10–15, ovary villose, 3-locular; fruits drupaceous, pea-sized, red turning black, sour, gathered locally, not available in the market.

Tribulus terrestris L. — BURNUT, CALTROP, PUNCTURE VINE

Ji-li=Chi-li (蒺藜, Caltrop). Seed, gathered and consumed locally.

Prostrating annual herbs, branches 30–100 cm long, branchlets hispid with long and short trichomes throughout; leaves even-pinnate, 3–5 cm long, leaflets 6–14, oblong, 6–15 mm long, 2–5 mm wide, acute or obtuse, oblique-rotundate at base, entire; flowers yellow, solitary, 1–1.5 cm across, stamens 10, ovary strigose; fruits strongly spiny, strigose and verrucose, rugose, 1 cm across, dehiscent septicidally into 5 segments, each segment bearing 2–4 strong spines and bristles, enclosing the seeds of two locules.

Rutaceae: Rue Family (Orange Family)

Acronychia oligophlebia Merrill

Gong-jia=Kung-chia (貢甲, an aboriginal name); *Shan-bai-gan=Shan-pai-kan* (山白柑, White Acronychia). Ripe fruit, eaten fresh, including the rind and the seeds.

Evergreen trees 16 m high, bark yellowish-gray, shallowly furrowed; leaves unifoliolate, laminas obovate-oblong, rarely oblong-oblanceolate, 7–18 cm long, 3.5–7 cm wide, obtuse or rounded, base cuneate, entire, punctate, petioles 1–2 cm long, gradually enlarged toward the distal end, articulate; flowers fragrant, small, greenish-white, in axillary panicles, sepals 4, minute, petals 4, ovate-elliptic, 2–3.5 mm long, stamens 8, filaments glabrous, ovary superior, 4-locular, style columnar; fruits yellow, smooth, subglobose, 8 mm in diameter. Endemic to the mountains bordering Guangdong and Vietnam, extending eastward to Hainan Island.

Acronychia pedunculata (L.) Miquel

Shan-you-gan=Shan-yu-kan (山油柑, Hillside Oil Tangerine). Fruit, eaten raw, consumed locally, not available in the market.

Small trees often having shrubby appearance, up to 10 m high, branchlets glabrous; leaves unifoliolate, opposite, oblong-elliptic, 6–15 cm long, 2.5–6 cm wide, obtuse, base acute, densely punctate beneath, petioles 1–2 cm long, articulate at the distal end; flowers greenish-yellow changing to white, 5–8 mm across, in axillary corymbose panicles, sepals

4, petals 5, linear, stamens 8, ovary hairy, 4-locular; fruits subglobose-pyriform, 8–10 mm long, 6–8 mm across, greenish-yellow, succulent, excellent flavor.

Citrus aurantium L. — SOUR ORANGE, BITTER ORANGE, SEVILLE ORANGE
Suan-cheng=Suan-ch'eng (酸橙, Sour Orange). Fruit.

Small evergreen trees, 3–4 m high, branchlets green, trigonous, thorny; leaves unifoliolate, coriaceous, elliptic, oblong or obovate-oblong, 5–10 cm long, 2.5–5 cm wide, glabrous, punctate, petioles broadly winged, articulate at the distal end; flowers fragrant, white, solitary or a few clustered at the end of a branchlet, stamens 25, the filaments slightly connate at base; fruits globose, 7–8 cm in diameter, orange-red, rind thick, rough, segments 10, hard to separate, pulp sour, bitter, poor eating quality.

var. **amara** Engler
Dai-dai-hua=Tai-tai-hua (代代花, Generation Flower). Flowers gathered, dried, used for flavoring tea.

Fruits orange-red, depressed-globose, 4.5 cm long, 5 cm across, segments 10, pulp sour.

Citrus limon (L.) N. Burman — LEMON
Ning-meng=Ning-meng (檸檬, Lemon). Fruit.

Tall thorny evergreen shrubs or small trees, 2–3 m high, branchlets greenish-gray, smooth, glabrous; leaves unifoliolate, elliptic, oblong, 5.5–9 cm long, 4–5 cm wide, obtuse or acute, punctate, crenate, petioles articulate, wings obscure; flowers purple-pink outside, white, in axillary cluster, fragrant; fruits ellipsoid, 7–11 cm long, yellow, pulp very sour; cultivated in agricultural institutions; not popular with the Chinese people.

Citrus limonia Osbeck — MANDARIN LIME, BANGAPUR LIME, LEMANDARIN
Li-meng=Li-meng (黎檬, Limeng); *Yi-mu-zi=I-mu-tzu* (宜母子, Fitting-for-mother). Fruits, very sour, every portion used; juice used in the preparation of ginger, a delightful tidbit for pregnant mothers.

Li-meng-bing=Li-meng-ping (黎檬餅, Lemandarin Cake). Fruits, after the juice pressed out and seeds removed, steamed, then preserved with salt and used for cooking chicken, a famous Cantonese dish, called *"Limeng* Chicken".

Medium-sized thorny shrubs, seldom left to grow into small trees; leaves unifoliolate, elliptic, both ends rounded, petioles disarticulate, wings obscure; flowers light purple outside, white inside, stamens 20 or more; fruits depressed, 3.5–4 cm across, smooth, rind thin, readily separated, segments 8–10, juice orange, very sour; seeds few, ovoid, cotylendons green.

Note: The specific epithet of *C. limonia* Osbeck was derived from a Cantonese local name "*Yi-mu-zi*" as it sounded to a Swede, Peter Osbeck, who was in Huangpu (黃埔) collecting botanical specimens in the early 1750s, when he learned the name of the plant from his informant. The Chinese people generally give names to plant with some specific significance. In this case, the sour fruit of the plant, especially a preserve, *Limeng* ginger (黎檬薑), prepared from it, has been a favorite tidbit of pregnant mothers. In Guangzhou and its vicinity, the juice of *C. limonia* is pressed out and used in the manufacture of a ginger preserve called "*Mei-jiang*" (梅薑, *Mei*-flavored ginger), a treat which pregnant mothers like especially. Early writers who understood the ethnobotanical origin of the name, recorded the fruit as "*Yi-mu-zi*" (宜母子, Mother's Delight), while others who did not know the significance merely used ideograms that sounded close to the name given and wrote "*Li-meng-zi*" (黎檬子, the name used in most current botanical literature). Two cultivars were identified in the 1930s.

1. cv. 'Myerslemon' Wong

Xiang-li-meng=Hsiang-li-meng (香黎檬, Aromatic Limeng). Fruit oblong, 7–8 cm long, 7 cm across, rind reddish-yellow; Native of China, introduced to America in 1908.

2. cv. 'Hunglemon' Wong

Hong-li-meng=Hung-li-meng (紅黎檬, Red Limeng). Fruit round, 4.5 cm in diameter, dull red, picked before fully ripe, preserved in salt and used for culinary purposes.

Citrus maxima (N. Burman) Merrill — POMELO, POMMELO (Syn. *C. grandis* [L.] Osbeck)
You-zi=Yu-tzu (柚子, Pomelo). Fruit.

Large evergreen trees 10–15 m high, branchlets angular, puberulous, occasionally weakly spiny; leaves unifoliolate, elliptic or ovate-elliptic, 8–20 cm long, finely serrulate, petioles disarticulate, broadly winged; flowers solitary or in small clusters 2–2.5 cm long, petals reflexed, ovary globose; fruits yellow, very large, subglobose or pyriform, 10–25 cm long, rind thick, the inside cottony, 1–1.5 cm thick, pulp rather dry, sweet, of good flavor. Several cultivars reported in the 1930s.

1. cv. 'Big Red'

Da-hong-pao=Ta-hung-p'ao (大紅抛). Fruits spherical or oblong-globose, rind smooth, 10 mm thick, hard to remove, segments 16, reniform, pulp pink; seeds numerous, cuneate, creamy white. Cultivated at Beigan in Pingyang of Zhejiang Province.

2. cv. 'Four Seasons'

Si-ji-pao=Ssu-chi-p'ao (四季抛, Four Times Flowering). Trees flowering four times, beginning in early April, flowering every 20 days, the best fruits from the early blossoms; fruits obovoid, 12 cm long, 9 cm across the broad portion, yellow, rind smooth, 1 cm thick, segments 12, reniform, pulp juicy, good flavor; seeds few, small, cuneate, yellow. Cultivated in Pumen, Zhejiang Province.

3. cv. 'Shangyuan' Hort ex Y. Chen

Xiang-yuan=Hsiang-yuan (香圓, Fragrant Round). Large tree 10 m high, trunk 30 cm in diameter, branchlets spiny; leaves elliptic, 6 cm long, 4 cm wide, petiolar wings 2 cm wide; fruit oblong, 6 cm long, 4.5 cm in diameter, rind rough, segments 10, pulp white-gray, poor quality, not much eaten as a fruit. Cultivated in the Lower Yangtze provinces, used more as an ornament placed before the ancestor shrines.

4. cv. 'Pinshanyou' Hort ex Y. Chen

Pin-shan-you=P'in-shan-yu (坪山柚, Pinshan Pomelo). Trees with round crown, with spreading branches, trigonous branchlets; leaves 7–8 cm long, 4–5 cm wide, wings of petioles 10–12 mm wide; fruits yellow, obovoid, 14 cm long, 12 cm across the broad portion, rind coarse, 1 cm thick, segments 15, 8 cm long, 4.5–5 cm thick, central axis hollow, pulp of good quality; cultivated in southern Fujian Province.

5. cv. 'Shatinyou' Hort ex Y. Chen

Sha-tian-you=Sha-t'ien-yu (沙田柚, Shatian Pomelo). Fruit pyriform, 12 cm long, 9 cm across the broad end, the distal end round, yellow, rind rough, 1 cm thick, hard to remove, segments 12, reniform; seeds many, large, cuneate, cotyledons white. Cultivated in Shatian, Yong District, Guangxi Province. Estimated to be the best pomelo in the Chinese market.

6. cv. 'Wentanyou'

Wen-dan-you=Wen-tan-yu (文旦柚, Actor Wen's Pomelo). Trees with rounded crown, branchlets puberulous, spiny; leaves broad ovate, 7–12 cm long, petiolar wings foliaceous; fruits subglobose, 8–9 cm in diameter, rind smooth, 12 mm thick, segments 18, reniform, with tough parchment-like cover, pulp pale yellow. First known from Changtai, Fujian Province, now widely cultivated in central China and Taiwan.

Citrus medica L. — Citron, Turunj (Sanskrit)

Ju-yuan=Chü-yüan (枸櫞, the translation of a Sanskrit name). Fruit, rarely eaten.

Evergreen small trees with spreading crown, branchlets spiny; leaves unifoliolate, oblong or ovate-oblong, 8–15 cm long, 3.5–6.5 cm wide, obtuse or mucronate, crenulate, petioles short, inconspicuously articulate, wings obscure; flowers solitary or in axillary clusters, purple-red outside, white inside, stamens 30 or more; fruits ovoid-oblong or subglobose, 10–20 cm long, lemon yellow, rind rough, aromatic.

cv. 'Sarcodactylis' (Syn. *C. medica* var. *sarcodactylis* Swingle)

Fo-shou-gan=Fo-shou-kan (佛手柑, Buddha Hand). Fruit, sliced 15 g, cooked with 30 cm fresh pig intestine (cleaned, sectioned).

Differing from the citron in its retuse leaves and fruits with finger-like segment at the distal half.

Citrus mitis Blanco — CALAMONDIN (Syn. *C. microcarpa* Bunge)

Si-ji-ju=Ssu-chi-chu (四季橘, Four-season Orange); *Jin-ju=Chin-chü* (金橘, Golden Orange). Fruit. Generally cultivated for ornamental purposes (sold around the Chinese New Year season), except in Ningpo, Zhejiang Province, where it is cultivated for the fruit, used in the preparation of a sugar preserve called *"Jin-ju-bing"* (金橘餅, Golden Orange Cake), consumed in China and for export.

Small compact evergreen shrubs 1–2 m high, branchlets slender; leaves unifoliolate, elliptic, 3–6 cm long, 1.5–2 cm wide, emarginate, wavy or crenulate, petiolar wings very narrow; flowers solitary or in pairs, white, stamens 20; fruits depressed subglobose, 1.5–2 cm across, slightly concaved at both ends; rind golden, smooth, thin, loose, segments 15, reniform, central core hollow, pulp juicy, sour; seeds 1 or 2 in each segment, cotyledons green.

Citrus x paradisii Macfadyen — GRAPEFRUIT (See *C. maxima* [N. Burman] Merrill; *C. sinensis* [L.] Osbeck).

Pu-tao-you=P'u-t'ao-yu (葡萄柚, Grape Fruit). Fruit; developed from mutants, first reported from the West Indies; rarely cultivated in China, introduced into some agricultural institutions.

Evergreen trees 10–15 m high; leaves unifoliolate, ovate, 7–15 cm long, 3.5–6 cm wide, crenulate, glabrous, petiolar wings broad; flowers white, 2–20 in loose clusters; fruits borne in clusters (hence the name "grape fruit"), globose or depressed globose, 10–14 cm in diameter, rind smooth, yellow, relatively thin, adhering, segments 11–14, pulp very juicy, white or pink, central core hollow. Said to have been developed from a seedling sport in West Indies, much cultivated in Florida and Texas, USA. A tree bearing fruit similar to the New World grape fruit used to grow in my yard at Chengdu, Sichuan, where I lived between 1938 and 1946. The fruits had thin rind, very juicy and sour pulp, and often two in clusters. Chinese growers did not propagate the tree because people dislike the sour taste.

Citrus reticulata Blanco — MANDARIN, TANGERINE, SATSUMA ORANGE (Syn. *C. nobilis* Loureiro, *C. deliciosa* Tenore)

Gan=Kan (柑, Mandarin Orange); *Gan-ju=Kan-chü* (柑橘, Orange). Fruits.

Small compact evergreen trees 3 m high, branchlets slender, weakly spiny; leaves unifoliolate, lanceolate or ovate-lanceolate, 5–8 cm long, 3–4 cm wide, subentire or crenulate, petioles articulate, wings obsolete; flowers small, creamy-white, solitary or fasciculate, stamens 18–25, 3–5 filaments connate at the base, ovary 9- to 15-locular; fruits depressed globose or subpyriform, 5–8 cm in diameter, deep orange or orange-red, rind loose, segments separate easily, pulp juicy, with good flavor; treated as cultivars here.

Numerous somatic mutants have been observed and cultivated by gardeners and farmers of Senzhou in Zhejiang Province, Zhangzhou in Fujian Province, and Chaozhou in Guangdong Province. Tanaka and Hayata treated these taxa as species, C. C. Hu changed them to varieties, and Hiroe listed them as subforms. These taxa are propagated vegetatively. According to current interpretation, they are clones which are treated as cultivars.

1. cv. "Erythrosa'
    Zhu-ju=Chu-chü (朱橘, Scarlet Mandarin) (Syn. *C. erythrosa* Tanaka, *C. reticulata* var. *erythrosa* [Tanaka] C. C. Hu).

Small evergreen trees, 50-year-old plant 6 m high, branchlets occasionally spiny; leaves elliptic, 4–6 cm long, 2–2.5 cm wide, subentire or wavy-crenulate, petiolar wings very narrow; fruits depressed globose, 3 cm long, 4 cm across, distal end slightly concave, rind rough, red, loose, segments 7, reniform, skin very tender, central core hollow, pulp red-orange, juicy; seeds few, ovoid, cotyledons green. Cultivated in Ningpo, thence spread to the Lower Yantze Region.

2. cv. 'Kinokuni'
    Ru-ju=Ju-chü (乳橘, Milk Orange) (Syn. *C. kinokuni* Tanaka, *C. reticulata* var. *kinokuni* C. C. Hu).

Compact evergreen shrubs or small trees; leaves ovate, 4 cm long, 2 cm wide, petiolar wings obscure; fruits small, depressed globose, rind rough, thin, loose. segments 11, reniform, pulp orange, very sweet, hence sold in Shanghai as *Mi-ju* (蜜橘, Honey Mandarin Orange).

3. cv. 'Ponki'
    Tian-ju=T'ien-chü (甜橘, Sweet Mandarin Orange) (Syn. *C. nobilis* var. *ponki* Hayata, *C. ponki* Tanaka).

Small spineless tree, 6-year-old plant 4 m high; fruits depressed-globose, 2.5 cm long, 3 cm across, rind yellow, smooth, segments 10, reniform, pulp pale orange, very juicy and sweet; poor keeping quality.

4. cv. 'Suavissima'

Ou-gan= Ou-kan (甌柑, Ningpo Tangerine) (Syn. *C. suavissima* Tanaka).

Spineless shrubs or small trees; fruits oblong or subglobose, 4.5 cm long, 5 cm across, distal end slightly concave, basal end truncate, rind orange, rough, thin, easily removed, segments 10, reniform, central core small, pulp orange, very juicy, good flavor, excellent keeping quality; seeds few, each with several embryos, cotylendons white or light green.

5. cv. 'Subcompressa'

Zao-ju=Tsao-chü (早橘, Early Mandarin Orange) (Syn. *C. nobilis* var. *subcompressa* Tanaka).

Small tree with upright branches; leaves elliptic, acute at both ends; fruits depressed-globose, 3.5 cm long, 4 cm across, distal end slightly concave, basal end rounded, rind orange, segments crescent-shaped, pulp orange, juicy, sweet but slightly sour; seeds many, ovoid, polyembryonic; harvested in October; good for catching early market; poor keeping quality.

The *Huang-yan-mi-ju=Huang-yen-mi-chü* (黃嚴蜜橘, Huangyan Honey Mandarin Orange) is a good example of this cultivar.

6. cv. 'Succosa' (Tanaka)

Tian-tai-mi-ju=T'ien-tai-mi-chü (天台蜜橘, Honey Mandarin Orange of Tiantai) (Syn. *C. succosa* Tanaka).

Shrubs or small trees of compact habit; leaves elliptic, obtuse at both ends; fruits small, depressed-globose, 2.5 cm long, 3 cm across, orange color, rough, rind very thin, segments 9, reniform, skin thin, pulp deep orange, very juicy and sweet, with good flavor.

7. cv. 'Suhoiensis'

Si-hui-gan=Ssu-hui-kan (四會柑, Sihui Mandarin Orange) (Syn. *C. suhoiensis* Tanaka).

Evergreen shrubs, 13-year-old plant 1.8 m high, branchlets slender; fruits depressed globose, 3 cm long, 4 cm across, rind thin, smooth, yellow, segments 12, kidney-shaped, pulp pale orange, juicy, very sweet; seeds many, polyembryonic, cotylendons green.

8. cv. 'Tangerina')

Hong-ju=Hung-chü (紅橘, Red Tangerine) (Syn. *C. tangerina* Tanaka).

Small trees; fruits depressed globose, 3.5 cm long, 4.5 cm across, distal end concave, basal end narrowed, rind smooth, red-orange, loose, segments 9–11, reniform, skin thick, central core hollow, pulp orange; seeds many, small, ovoid, polyembryonic, cotyledons green. This cultivar was developed in Fujian, and now it is extensively cultivated in Zhejianag Province where it is locally called "*Fu-ju=Fu-chü*" (福橘, Fujian Tagerine). It is also cultivated in other provinces of the Yangtze River Region.

9. cv. 'Tankan (Hayata)'

Jiao-gan=Chiao-kan (蕉柑, Banana Mandarin Orange) (Syn. *C. tankan* Hayata).

Spineless small trees; fruits subglobose, 3.5 cm long, 4.5 cm across, both ends rounded, rind smooth, deep orange, loose, segments 10, kidney-shaped, skin very thin, orange, central core small; seeds few or none, small, polyembryonic, embryos green. This cultivar was first cultivated in Chaozhou (潮州) in eastern Guangdong Province, and the adjacent Quanzhou of Fujian Province. In the late 17th and the 18th centuries, people of this general area migrated to Taiwan, and the cultivar was carried along. As the area of cultivation extended and the volume of production increased, various local names were given to it. In Shanghai market, the fruit is called "*Xian-luo-mi-ju*" (=*Hsian-lo-mi-chü* 暹羅蜜橘, Siamese Honey Mandarin Orange), and in Taiwan it is known as *Tong-gan=T'ung-kan* (桶柑, Bucket Mandarin Orange).

Citrus sinensis (L.) Osbeck — SWEET ORANGE (Syn. *C. aurantium* L. var. *sinensis* L.)
   Cheng=Ch'eng (橙, Orange); *Tian-cheng=T'ien-ch'eng* (甜橙, Sweet Orange); *Guang-gan=Kuang-kan* (廣柑, Guangdong Mandarin Orange, the name used in Chengdu, Sichuan Province); *Guang-ju=Kuang-chü* (廣橘, Cantonese Tangerine, the name used in Nanjing).

Glabrous compact evergreen trees 3–5 m high, branchlets green, angular, generally spineless; foliage dark green; leaves unifoliolate, elliptic, 4–10 cm long, 2–5 cm wide, shortly acuminate, base obtuse, entire, punctate, petioles articulate, wings very narrow, often obscure; flowers white, axillary, solitary or 2–6 fasciculate, stamens ca. 20, the filaments connate beyond the middle into bundles; fruits globose, 5–6 cm in diameter, rind smooth, yellow, adhering, hard to be removed, smooth, yellow or dull yellow, segments 9–14, central core solid, pulp yellow, juicy, sweet; seeds many to none, cotyledons white.

Xinhui (新會) and its vicinities was the center of cultivation of the sweet orange.

During the Qing Dynasty, when people moved from Guangdong to Sichuan, orange plants were brought along. Now, Sichuan Province is self-sufficient in its citrus fruit production. Many horticultural forms are propagated by budding or grafting. They can be classified into two groups:

Round Orange Group: Inside the distal end of the fruit without evident underdeveloped extra carpels giving the shape of a navel.

1. cv. 'Liucheng'

Liu-cheng=Liu-ch'eng (柳橙, Willow Orange).

Fruits globose, 5–6 cm in diameter, ring rough, with some longitudinal grooves, segments 10, central core large, pulp deep yellow; seeds many, harvest in November and December, cultivated in Guangzhou and its vicinities.

2. cv. 'Sekkan'

Xue-gan=Hsüeh-kan (雪柑, Snow Orange) (Syn. *C. sinensis* f. *sekkan* Hayata).

Fruit oblong-globose, 7 cm long, 6 cm across, rind smooth, orange, pulp juicy, good taste and fine flavor; seeds none. Cultivated in Zhaozhou in eastern Guangdong Province, harvest in January and February.

3. cv. 'Sunwuitincheng'

Xin-hui-tian-cheng=Hsin-hui-t'ien-ch'eng (新會甜橙, Xinhui Sweet Orange).

Fruits subglobose, 3.5 cm long, 5 cm across, rind smooth, dull golden, hard to be removed, segments 10, kidney-shaped, skin thin, colorless, pulp golden, very juicy and sweet; seeds few, polyembryonic. Xinhui, a district on the Pearl River Delta, southwest of Guangzhou, is honored as the capital of Chinese citrus fruit production. I was there in 1935. In order to demonstrate the superior quality of sweetness of the product, an owner cut an orange across the middle, placed a section facing a piece of paper saying, "The sugar in the juice of this orange holds water so that the paper is not wet." Evidently, the Xinhui stock of sweet orange has been introduced to Taiwan which sends large supplies of oranges to Hong Kong around the Chinese New Year's time. In Hong Kong market, the fruit is known as Taiwan Orange.

4. cv. 'Tatincheng'

Da-tian-cheng=Ta-t'ien-ch'eng (大甜橙, Big Sweet Orange) (Syn. *C. sinensis* var. *tatincheng* Wong).

Fruits compressed globose, 9 cm long, 9.5 cm across, rind coarse and thick, grayish yellow, segments 10; seeds numerous, cotyledons white; harvest in December.

Navel Orange Group: Inside the distal end of the fruit has additional underdeveloped carpels, and the apex of the fruit with a navel.

5. cv. 'Hushuae Orange'

Xin-hui-qi-cheng=Hsin-hui-ch'i-ch'eng (新會臍橙, Xinhui Navel Orange).

Small trees with low spreading branches, 2.5 m high; leaves elliptic, 7.5 cm long, 3 cm wide, acuminate; fruits globose, 7 cm long, 7.5 cm across, underdeveloped carpels forming an evident navel at the distal end.

Note: First discovered near Xinhui, south of Guangzhou, in 1933. In an expedition of agricultural products, Benemerato (1936) found it quite common in various orchards in Guangdong Province. He published an illustrated report about it in 1936.

6. cv. 'Washington Navel'

Mei-guo-qi-cheng=Mei-kuo-ch'i-ch'eng (美國臍橙, Washington Navel Orange) (Syn. *C. sinensis* var. *brazziliensis* Tanaka).

Fruit globose, 5 cm in diameter, rind thick, segments 9, skin thick, central core small but solid, ending with prominent navel consisting of underdeveloped additional carpels of various sizes, pulp orange, juice orange; seedless. Originally cultivated in Brazil, now grown principally in California; introduced to China via Japan in 1921, first cultivated in Zhejiang Province; market material imported from California, especially the brand named SUNKIST. In Nanjing, the market name of the fruit is *"Mei-ju=Mei-chü"* (美橘, American Orange) while in Shanghai, it is called *"Hua-qi-mi-ju"* (=*Hua-ch'i-mi-chü* 花旗蜜橘, Stars and Stripes Flag Honey Orange).

Clausena anisum-olens (Blanco) Merrill

Ji-pi-guo=Chi-p'i-kuo (雞皮果, Chicken Skin Fruit). Fruits, used fresh or for making jam in Guangxi Province.

Large evergreen shrubs, branchlets pubescent; leaves pinnate, leaflets 7–13, obliquely oblong, 5–12 cm long, 2–3 cm wide, acuminate, base cuneate, glabrescent; flowers yellowish-white, in terminal panicles 16 cm long; fruits globose, pale yellow, 1–1.4 cm in diameter, containing 1–4 seeds.

Clausena lansium (Loureiro) Skeels — WAMPI (Syn. *C. wampi* [Blanco] D. Oliver)

Huang-pi=Huang-p'i (黃皮, Yellow Skin). Ripe fruit, eaten fresh or preserved.

Aromatic evergreen trees 4–10 m high, glandularly pubescent throughout, branchlets verrucose; leaves pinnate, leaflets 5–13, alternate, obliquely oblong-elliptic, 6–13 cm long, 2.5–6 cm wide, acuminate, base oblique, the upper portion rounded, the lower one cuneate, evenly punctate beneath, entire, wavy, or remotely crenulate; flowers small, 5 mm across, greenish-yellow, fragrant, in terminal panicles 20–30 cm long, 10–20 cm across, stamens 8–10, ovary villose and glandular; fruit subglobose, 1.2–3 cm in diameter, dull yellow.

Fortunella hindsii (Champion ex Bentham) Swingle — MOTHER-OF-KUMQUAT

Jin-dou=Chin-tou (金豆, Golden Pea). Ripe fruit, gathered and eaten on hillside, not available in the market.

Small evergreen shrubs 1–4 m high, growing in windy grassy slopes, branchlets angular, armed with strong spines; leaves alternate, shiny green, coriaceous, punctate, oblong, elliptic or obovate, 3–7 cm long, 1–3 cm wide, obtuse and emarginate or rounded, base cuneate or obtuse; flowers small, white, 8 mm across, 1 or 2 axillary; fruits globose, 8–11 mm across, rind smooth, golden-orange, firm, pulp none; seeds usually 1 or 2 fully developed, large, 6–10 mm across. "Golden Pea" is a fitting name for the plant, for the small fruits are eaten like peas, skin and all, as did our neolithic ancesters.

Fortunella margarita (Loureiro) Swingle — KUMQUAT, NAGAMI KUMQUAT, OVAL KUMQUAT

Jin-ju=Chin-chü (金橘, Gold Orange). Fruit, the skin is the best part of a fresh Kumquat, both fresh and preserved material available in American Chinese stores.

Evergreen shrubs 2–3 m high in cultivation, branchlets smooth, spineless; leaves unifoliolate, oblong-lanceolate, 5–9 cm long, 2–3 cm wide, acute, entire or remotely crenulate, dark green above, punctate, petioles articulate, narrowly winged; flowers small, 7–8 mm long, white, fragrant, solitary or in small axillary fascicles, stamens 20–25, filaments unequal, connate into several bundles; fruits oblong, 2.5–3.5 cm long, rind smooth, orange, loose, segments 4 or 5, pulp none; seeds large.

Glycosmis citrifolia Lindley

Shan-xiao-ji=Shan-hsiao-chi (山小桔, Small Wild Calamondin). Pink juicy ripe fruit, picked up by children and eaten, not available in the market.

Glabrous spineless, evergreen shrubs 1–3 m high, growing in exposed grassy slopes; leaves unifoliolate and trifoliolate, leaflets elliptic, 4.5–13 cm long, 2–5 cm wide, acuminate, base acute, entire, hardly punctate beneath; flowers white, small, 4–5 mm across, in axillary cymose panicles or fascicles, ovary globose, punctate; fruits pink, juicy, pea-sized, 8–10 mm long, 10–12 mm across, very sweet. Native of Hong Kong hillsides, the branchlets bear both flowers and mature fruits between November and February. This is the first record of its fruits being eaten by man as food.

Glycosmis hainanensis Huang

Hai-nan-shan-xiao-ji=Hai-nan-shan-hsiao-chi (海南山小桔, Hainan Glycosmis). Fruit, eaten fresh locally.

Shrubs 2–3 m high; leaves unifoliate, laminas oblanceolate, rarely oblong, 7–15 cm long, 2.5–6 cm wide, abruptly narrowed and caudate, base cuneate, entire, petioles 1–1.5 cm long, articulate; flowers white, subsessile, in short axillary cymose clusters, peduncles and rachis velutinous, petals elliptic, 4 mm long, stamens 10, connectives punctate at the tip; fruits red, globose, 7 mm in diameter. Endemic to Hainan Island, growing in dense forested area along streams.

Murraya koenigii (L.) Sprengel — CURRY-LEAF TREE

Ga-li-ye=Ka-li-yeh (咖喱葉, Curry leaf). Fresh leaves minced and used in preparing curry dishes by Portuguese residents of Hong Kong and Macao.

Aromatic evergreen shrubs or small trees 3–5 m high, glandularly punctate throughout; leaves pinnate, leaflets 17–31, chartaceous, lanceolate or ovate-lanceolate, 2–5 cm long, 0.5–2 cm wide, acuminate, base obtuse or rounded, finely serrulate; flowers small, 8 mm long, in terminal panicles, stamens 10; fruits oblong-subglobose, 1 cm long; seeds 1 or 2. Native to tropical Asia, introduced into Hong Kong by Portuguese residents.

Orixa japonica Thunberg

Chou-chang-shan=Ch'ou-ch'ang-shan (臭常山, Stinking Common-in-Mountain); *Chou-shan-yang=Ch'ou-shan-yang* (臭山羊, Stinking Goat); *Ri-ben-chang-shan=Jih-pen-ch'ang-shan* (日本常山, Japanese Changshan); *Bai-hu-jiao=Pai-hu-chiao* (白胡椒, White Pepper). Fruits, used for condiment by the people living in the mountains of western Hubei Province.

Deciduous shrubs 1–3 m high, branchlets pilose when young; leaves obovate, or elliptic-oblong, 3–17 cm long, 2–9 cm wide, short-acuminate or obtuse, base cuneate and obtuse, entire rarely crenulate, punctate; flowers small, yellowish-green, 5–6 mm across, 4-merous, unisexual, dioecious, emerging from lateral winter flowering buds, staminate flowers 3–12 in sessile pseudoracemose clusters 1.5–4 cm long, rachis villose, bracts membranous, obovate-oblong, punctate, usually cymbiform, pedicels 4–5 mm long, bractioles 2, supermedian, sepals lanceolate, petals obovate-oblong, stamens exserted, pistillode discoid, pistillate flowers solitary, very rarely 2, staminodes linear, yellow, ovary superior, apocarpous, thick-discoid, styles adaxial, stigmas capitate; fruits follicular, aggregates 1.5 cm in diameter, follicles compressed-hemispherical, adaxial face concave, 1 cm long, 7 mm across, conspicuously striate, the lower striae reticulate, pericarps in two distinct layers, the outer layer herbaceous, olivaceous, the inner layer cartilaginous, cream-white, separated readily at maturity of the fruit. Native of eastern Asia, growing in shade of broad-leaved deciduous forest at altitudes of 750–1,300 m; introduced into Western garden in 1870.

Ruta graveolens L. — COMMON RUE

Chou-cao=Ch'ou-ts'ao (臭草, Ill-smell Herb); *Yun-xiang=Yun-hsiang* (芸香, Rue). Young shoots, used in Cantonese cooking, to add flavor to soup.

Strongly aromatic perennial herbs 50–80 cm high, punctate throughout; leaves decompound, 6–12 cm long, ultimate segments obovate, 1–2 cm long, entire or crenulate; flowers greenish-yellow, 2 cm across; petals 4 or 5, stamens 8–10, ovary 4-carpellate, subglobose, glabrous, punctate, the apex 4-lobed; capsules subglobose, with 4 conical apical projections, 6–8 mm across the middle, loculicidally dehiscent from the inside of the projections; seeds black, shiny.

Zanthoxylum ailanthoides Siebold et Zuccarini (Syn. *Fagara ailanthoides* [Siebold et Zuccarini] Engler)

Yue-jiao=Yueh-chiao (越椒, Vietnam Pepper). Young leaves, used in Taiwan.

Deciduous trees 15 m high, branchlets often hollow, sparsely spinose; leaves pinnate, 25–60 cm long, leaflets 11–27, papery, ovate-elliptic, 7–13 cm long, acuminate-caudate, base rounded, shallowly serrate, the teeth glandular; flowers unisexual, small, greenish-white, in terminal coumpound corymbose panicles; folicles red; seeds black, globose, shiny. Native to the forested area of southeastern China.

Zanthoxylum armatum de Candolle (Syn. *Z. planispinum* Siebold et Zuccarini; *Z. alatum* sensu Forbes et Hemsley, Rehder et Wilson, non Roxburgh; *Z. alatum* var. *planispinum* Rehder et Wilson)

Zhu-ye-jiao=Chu-yeh-chiao (竹葉椒, Bamboo-leaved Zanthoxylum); *Zhu-ye-hua-jiao=Chu-yeh-hua-chiao* (竹葉花椒, Bamboo-leaved Sichuan Peppercorn); *Gou-jiao=Kou-chiao* (狗椒, Dog Zanthoxylum). Pericarps, used for condiment; seeds for oil.

Aromatic deciduous shrubs or small trees 2–3 m high; branchlets armed with curved prickles broadened at the base, the prickles on the stem becoming woody; leaves pinnate, rachis winged and spiny; leaflets 3, rarely 5, or up to 9, lanceolate or oblong-lanceolate, the terminal 3 sessile, giving the appearance of bamboo leaves, hence the vernacular name "*Zhu-ye-jiao*" (竹葉椒); flowers unisexual, dioecious, small, inconspicuous, in axillary cymose panicles 2–6 m long, perianth segments 6–8, stamens 6–8, ovary of the pistillate flowers 2- to 4-carpellate, generally only 1 or 2 fertile; fruits with 1 or 2 follicles, pericarp verrucose, with large glands, red; seeds ovoid, black, shiny; growing in western and southern China; pericarps used locally for spice.

Zanthoxylum bungeanum Maximowicz — ZANTHOXYLUM, NORTHERN CHINA PEPPERCORN

Hua-jiao=Hua-chiao (花椒, Peppercorn). Dried pericarp, spice. (See Part I for more information).

Hua-jiao-ye=Hua-chiao-yeh (花椒葉, Zanthoxylum Leaf). Fresh leaves, gathered and used as spice in villages of northern China, for cooking fish and also in preparing home-made sauce from fermented soybean and wheat flour.

Deciduous shrubs or small trees 3–7 m high, trunks and branchlets armed with strong prickles, broad at the base; leaves pinnate, leaflets 5–11, ovate, or ovate-oblong, 1.5–7 cm long, 1–3 cm wide, crenate, teeth glandular, velutinous at the nerve angles beneath; flowers unisexual, small, yellowish-green, in axillary or terminal panicles, perianth segment 4–8, ovary sessile; follicles globose, red-purple, glandular, verrucose; seeds black, attached to the edge of the pericarp. Native of northern China, widely cultivated in kitchen gardens, leaves and dried pericarps used in cooking; considered to be the first spice used in Chinese culinary culture.

Zanthoxylum schinifolium Siebold et Zuccarini — WILD ZANTHOXYLUM

Xiang-jiao-zi=Hsiang-chiao-tzu (香椒子, Aromatic Zanthoxylum); *Ya-jiao=Ya-chiao* (崖椒, Cliff Zanthoxylum); *Qing-hua-jiao=Ch'ing-hua-chiao* (青花椒, Green Zanthoxylum). Pericarp, used for spice in Hebei.

Deciduous shrubs 1–3 m high, armed with prickles, branchlets purple, glabrous; leaves pinnate, leaflets 11–21, lanceolate or elliptic-lanceolate, 1.5–5 cm long, 0.7–1.5 cm wide, acuminate or acute, occasionally retuse, base acute or obtuse, serrate, punctate between teeth; flowers small, yellowish-green, unisexual, dioecious, in terminal or axillary corymbs 3–8 cm across, sepals 5, 0.5 mm long, petals 5, oblong, 1–1.5 mm long, stamens 5 in staminate flowers, ovary 3-carpellate, style obscure, stigmas 3; follicles purple-red, shortly rostrate; seeds bluish-black, shiny. Widespread in eastern Asia, growing in forests.

Zanthoxylum simulans Hance — SICHUAN ZANTHOXYLUM, SICHUAN PEPPERCORN (Figure 14a)

Chuan-jiao=Ch'uan-chiao (川椒, Sichuan Pepper). Dried pericarp, spice, available in American Chinese stores.

Aromatic deciduous shrubs 1–3 m high, growing on the semidesert arid hillsides of western Sichuan, branchlets prickly; leaves pinnate, rachis with curved spines, leaflets 5–9, ovate-orbicular, oblong-ovate or rhombic-ovate, 2.5–6 cm long, 1.8–3.5 cm wide, crenulate, punctate, strigose above; flowers unisexual, dioecious, small, inconspicuous, in axillary or terminal corymbose panicles 1–5 cm long, perianth segments 5–8, stamens 5–7; fruits red-purple, follicles 1 or 2, the pericarp verrucose with large glands outside, white and smooth inside; seeds globose, black, shiny.

Between 1950 and 1980, when important ports like Shanghai in eastern China and Tsingtao in northern China were kept inactive in international trade, some Sichuan businessmen established distribution centers in Hong Kong for the export of special Sichuan products. *Hua-jiao* (花椒, Zanthoxylum) and *Zha-cai* (榨菜, preserved spicy tender shoots of *Brassica juncea* (L.) Czernajew var. *tumida* Tsen) thus could be sent continuously to overseas Chinese dealers. Consequently, the Sichuan Peppercorn became the only brand used in Chinese cookbooks published in America.

Burseraceae: Torchwood Family

Canarium album (Loureiro) Raeuschel — CHINESE CANARIUM

Gan-lan=Kan-lan (橄欖, Canarium); *Bai-lan=Pai-lan* (白欖, White Canarium); *Qing-guo=Ch'ing-kuo* (青果, Green Fruit). Fruit; eaten raw, used for tea and preserves; the preserved material is available in American Chinese stores.

Large resiniferous evergreen trees 10–20 m high, branchlets glabrous; leaves pinnate, 10–30 cm long, leaflets 9–15, petiolulate, ovate-oblong, 6–18 cm long, 3–8 cm wide, shortly acuminate, base obliquely rotundate, entire; flowers small, creamy-white, in axillary cymose panicles 10–15 cm, long; fruits drupaceous, ellipsoid, yellowish-green, 3 cm long, 1.5–2 cm across. Native of tropical China, extensively cultivated in southern China.

Canarium bengalense Roxburgh — BENGAL CANARIUM

Fang-lan=Fang-lan (方欖, Square Canarium). Fruit, and oil from seed.

Trees 15–25 m high, trunks 75–120 cm across; branchlets stout, 1–1.5 cm in diameter, with conspicuous lenticels, hairy; leaves pinnate, leaflets 11–13 (–21), oblong or obovate-lanceolate, 10–20 cm long, 4.5–6 cm wide, acuminate, sparsely pilose, hairs yellow; flowers small, dioecious, staminate ones in narrow corymbose-racemes, 30–40 cm long, segments 3–4 cm long, 7-flowered, stamens 6, filaments connate to the middle, disk cupular, strigose, ciliate, pistillate flowers in clusters 5–8 cm long; drupes green, ellipsoid, 1.8–2 cm long, 1 cm across. Native of the eastern Himalayan Region, occurring in Yunnan and Guangxi Provinces.

Canarium pimela Koenig — BLACK CHINESE CANARIUM (Syn. *C. nigrum* Engler)

Wu-lan=Wu-lan (烏欖, Black Canarium); *Lan-chi=Lan-ch'ih* (欖豉, Preserved Black Canarium). Fruit; harvested and with the stone removed, pickled; the material appearing black, soft, used for cooking.

Lan-ren=Lan-jen (欖仁, Canarium Kernel). Seed; used in southern China as walnut is used in northern China dishes, lightly fried and then mixed with white chicken meat, formerly a popular ingredient of Cantonese dishes at banquets, now has been replaced by the imported cashew nut.

Large resiniferous evergreen trees 10–16 m high; leaves pinnate, 30–60 cm long, leaflets 15–21, oblong or ovate-elliptic, 5–15 cm long, 3.5–7 cm wide, abruptly short-acuminate or obtuse, base shortly oblique obtuse, entire; flowers small, in axillary panicles; drupes oblong, 4–5 cm long, 2.5–3 cm across, purple-black.

Canarium strictum Roxburgh

Yang-rui=Yang-jui (漾蕊, the Bourong ethnic name); *Yang-duan=Yang-tuan* (漾短, the Thai ethnic name); *Tian-lan=T'ien-lan* (滇欖, Yunnan Canarium).

Fruits, eaten by the Bourong and the Thai ethnic groups living in Yunnan Province.
Large evergreen trees 50 m high, bark light gray, branchlets velutinous, glabrescent; leaves pinnate, leaflets 11–13, ovate-lanceolate or elliptic, 10–20 cm long, 4–6.5 cm wide, acuminate, acute or obtuse, base oblique obtuse; flowers small, unisexual, dioecious, staminate flowers in subterminal panicles 15–40 cm long, stamens 6, filaments connate at the basal 1/2–3/4, pistillate flowers in racemes 7–20 cm long, slightly elongated after anthesis, 10–20 cm long, bearing 1 or 3 fruits, ovary 3-locular; drupes black at maturity, obovoid or ellipsoid, 3.5–4.5 cm long, 1.7–2.3 cm across, stone smooth. Native of eastern Himalayan Region, growing in mixed tropical forest at altitudes of 400–1,300 m.

Canarium tonkinense Engler — TONKIN CANARIUM

Yue-lan=Yueh-lan (越欖, Vietnam Canarium). Pickled fruit, for culinary use.

Trees 15 cm high, trunks 70 cm across, branchlets pubescent; leaves pinnate, leaflets 11–13 (–15), ovate or oblong, 13–20 cm long, 6–8 cm wide, acuminate, acumens 1–1.5 cm long, base oblique-rounded; flowers small, greenish, dioecious, in large panicles 20–30 cm long, densely pubescent, perianth segments 5–6 mm long, stamens 6, filaments connate, disk cupular, fleshy; drupes oblong, 3.2 cm long, 2 cm across middle, purple-brown; seed usually 1, edible.

Meliaceae: Mohagny or Chinaberry Family

Aglaia odorata Loureiro — Mock Lime

Mi-zai-lan=Mi-tsai-lan (米仔蘭, Millet Lan); *Yu-zi-lan=Yu-tzu-lan* (魚子蘭, Fish-egg Lan); *Sui-mi-lan=Sui-mi-lan* (碎米蘭, Broken Rice Lan); *Shan-hu-jiao=Shan-hu-chiao* (山胡椒, Hillside Pepper). Much cultivated shrubs, the flowers used in Hebei for flavoring tea.

Much branched evergreen shrubs, 1–7 m high, young branchlets pilose, hairs brown and stellate; leaves pinnate, ovate in outline, 5–12 cm long, leaflets 3–5, obovate-oblong, 2–7 cm long, 1–3 cm wide, acute, the rachis winged; flowers small, unisexual or polygamous, golden yellow, extremely fragrant, staminal inflorescences large, in axillary panicles, never fully open, bud-like, 2 mm in diameter, falling with the 3 mm long pedicels, calyx discoid, 4-lobed, petals fleshy, obovate-orbicular, imbricate, slightly concave, staminal tube similar to the petals in color and texture, campanulate-funnelform, basal half 4-ridged-furrowed, spongy, distal half collar-like, petaloid, anthers 4, sessile, inserted on the inside of the collar, pistillate inflorescences spike-like, axillary, 2–3 cm long, flowers sessile or subsessile, ovary superior, glandular; fruits oblong, 12 mm long, 10 mm across, berry-like, punctate, rostrate, apical portion stellate hairy; seed hemispherical, 6 mm across, arillodial, sarcotesta fleshy. Native to the mountains on the border of China, Thailand, Vietnam and eastward to Hainan Island; a common garden plant, cultivated outdoor in Hong Kong, flowering from January to August, generally only staminate plant.

Lansium parasiticum (Osbeck) Sahni et Bennet — Langsat, Duku, Dookoo (Syn. *L. domesticum* Coorrea)

Lan-sa-guo=Lan-sa-kuo (蘭撒果, Langsat, a direct translation of the common English name [adopted from Indonesian] by sound). Fruits, preserved in syrup after skin removed.

Lactiferous, cauliflorous, pubescent, evergreen trees 12–20 m high; leaves alternate, pinnate, 30–50 cm long, leaflets 5–7, oblong-elliptic, 13–21 (40) cm long, 6–8 (–18) cm wide, entire, acuminate; flowers small, 4–5 mm across, pale yellow, in hanging solitary or fasciculate racemes 10–30 cm long; fruits oblong-ovoid or subglobose, 2–4 cm long, 2.5–3 cm across, pale yellow, the rind thin, lactiferous, aril fleshy, white, sweet or sourish; seeds 1 or 2, bitter. Native of Malaysia and Indonesia, introduced to Taiwan before 1950.

Sandoricum koetjape Merrill — Sentol, Santol, Kechapi (Syn. *S. indicum* autt, non Cav.)

Suan-ming-guo=Suan-ming-kuo (酸名果, Santol). Fruit, introduced to Taiwan before he
    1950s.

Large, semideciduous, lactiferous tree 15–30 m high, trunks columnar, smooth,
branchlets and all new growth velutinous, pubescent throughout; leaves trifoliolate,
leaflets petiolulate, oblong-elliptic or ovate-oblong, 6–25 cm long, 3–15 cm wide, equal-
sided, wine-red when young, lateral nerves 10–14 pairs; flowers pale yellow, small, 5-
merous, in axillary panicles 10–20 cm long, calyx cupular, shortly 5-lobed, petals 5,
free, reflexed, staminal tube 5-toothed at the apex, teeth bifid, anthers 10, included,
alternate with the teeth, ovary 5-locular, style terete, stigma 5-lobed; fruits drupaceous,
yellow, velutinous, depressed-globose, 5–6 cm in diameter, exocarp thin, mesocarp
fleshy, white, juicy, sour-sweet, stone tough, laterally compressed; seeds usually 1,
conformable to the stone, cotyledons red. Native of Indonesia, extensively cultivated
in southeastern Asia, planted in Taiwan before 1950.

Toona sinensis (Jussieu) M. J. Roemer — CEDRELA (Syn. *Cedrela sinensis* Jussieu)
    Xiang-chun=Hsiang-chun (香椿, Fragrant Ailanthus). Tender leafy shoots, only small
    amount used, chopped for making omelets; an early spring favorite in northern
    China.

Deciduous, strongly aromatic trees 10–15 m high, spreading by suckers emerging
from the root, branchlets puberulous; leaves pinnate, 30–90 cm long, leaflets 11–32,
oblong, 8–15 cm long, 3–5 cm wide, abruptly acuminate, base obtuse, petiolulate; flowers
small, white, fragrant, in terminal panicles; capsules oblong-ellipsoid, 1.5–2.5 cm long;
seeds numerous, oblong, winged at one end. Native of China, extensively cultivated
for the edible young shoot, introduced to USA in 1862, growing well in Boston area,
increasing 2–3 m annuallly, propagated by root suckers.

Walsura robusta Roxburgh
    Ge-she-shu=Ko-she-shu (割舌樹, Cut-the-tongue Tree). Fruit, eaten in southern Yunnan.

Trees 4–25 m high, branchlets glabrous; leaves pinnate, ovate in outline, 15–30 cm
long, leaflets 3–5, oblong-elliptic, terminal one 7–16 cm long, 3–7 cm wide, lateral ones
smaller, acuminate or caudate, base obtuse; flowers white, small, 4–6 mm long, in axillary
and subterminal panicles 5–15 cm long, sepals 5, petals 5, stamens 10, filaments connate
below the middle, anthers introrse, disk cupular, ovary 2- or 3-locular, hairy; fruits
globose or ovoid, 1–2 cm in diameter, pericarps fleshy, velutinous; seeds 1 or 2, arillate.
Widespread in Indo-Malaysia, thence northward to southern China; growing in the
tropical forests of southern Yunnan, Guangxi, Guangdong and Hainan Island.

Euphorbiaceae: Spurge Family

Aleurites moluccana (L.) Willdenow — CANDLENUT, CANDLEBERRY TREE, INDIAN WALNUT, COUNTRY WALNUT

Shi-li=Shih-li (石栗, Rock Chestnut). Seed, used in making curry.

Large evergreen trees 10–15 m high, stellate tomentose throughout the annual growth; leaves ovate, 10–20 cm long, 5–17 cm wide, entire, occasionally lobed, petioles 5–12 cm long, with 2 glands at the distal end; flowers unisexual, monoecious, small, white, in terminal panicles 10–15 cm long, stamens 15–20, ovary 1-locular; drupes ovoid or subglobose, 5 cm in diameter.

Native of the dry limestone cliffs of Malaysia, used locally for lighting by pounding the seeds with cotton into a mass, attached to bamboo-splints, lit at night, hence the name "candleberry". People in Java use the seeds as food roasted or after fermentation; the Indian and Portuguese population in Hong Kong use the product in their curry.

Antidesma bunius (L.) Sprengel — CHINESE LAUREL, BUNI, WUNI (Indonesian names). *Wu-yue-cha=Wu-jüeh-ch'a* (五月茶, May Tea). Leafy shoots, used for tea.

Graceful evergreen trees, 8–10 m high, bark grayish, smooth, branchlets grabresent; leaves oblong or obovate-oblong, 6–16 cm long, 2–6 cm wide, both ends obtuse; flowers small, unisexual, dioecious, in simple or branched spikes, stamens 3, ovary glabrous, styles 3; fruits drupaceous, subglobose, green, red, black, all in the same cluster, 5–6 mm in diameter. Growing from the Himalayan Region eastward to Indonesia; appearing to be a natural element in the vegetation of Hong Kong and adjacent areas of southern China; cultivated in Java for the pleasant-tasting small dark-red to black ovoid fruits.

Baccaurea ramiflora Loureiro

Huo-guo=Huo-kuo (火果, Fire Fruit). Fruits, tasting sour-sweet, available in the market of Longzhou, Guangxi Province.

Cauliflorous trees or shrubs 4–10 m high, bark dark gray, branchlets glabrous; leaves often crowded at the shoot apex, oblong-oblanceolate, 10–20 cm long, 4–8 cm wide, acuminate, base cuneate, entire, chartaceous, petioles 1–4 cm long; flowers small, unisexual, dioecious, in axillary racemes, pubescent, perianth segments 4–6, ovary ovoid, hairy, stigma bifid; fruits ovoid or spherical, in clusters 30 cm long, yellow-red, juicy. Native to the mountains of southern China and adjacent Vietnam.

Bischofia javanica Blume — TOOG

Chong-yang-mu=Ch'ung-yang-mu (重陽木, Double-nine Tree). *Qie-dong= Ch'ieh-tung*

(茄苓, a name used in Fujian and Taiwan). Fruit eaten in Shaanxi and Taiwan, not available in the market.

Deciduous trees 10–35 m high, bark smooth, grayish brown, branchlets glabrous; leaves trifoliolate, leaflets ovate, oblong or ovate-elliptic, 7–15 cm long, 4–8 cm wide, acuminate, base acute or obtuse, serrate; flowers small, unisexual, dioecious, in axillary panicles 15–20 cm long, stamens 5, rudimentary ovary peltate; ovary 3- rarely 4-locular; fruits subglobose, 10–13 mm in diameter, brown.

Cleidiocarpon cavaleriei (Léveillé) Airy-Shaw

Hu-die-guo=Hu-tieh-kuo (蝴蝶果, Butterfly Fruit); *Shan-ban-li=Shan-pan-li* (山板栗, Mountain Chestnut); *Zhu-you-guo=Chu-yu-kuo* (豬油果, Lard Nut); *Hou-guo=Hou-kuo* (猴果, Monkey Nut). Seeds, boiled; oil extracted from seed.

Large evergreen trees up to 30 m high, trunks 50–100 cm in diameter; leaves elliptic-oblong, 11–16 cm long, 2–6 cm wide, acuminate, base cuneate, entire, stellate-pubescent beneath when young, glabrescent, petioles 2–5 cm long, with 2 black glands; flowers yellow-white, unisexual, monoecious, paniculate, the pistillate flowers at the basal end, ovary 2-locular, uniovulate, stigmas 3; fruits oblique, 3–3.8 cm long, 2–2.7 cm across, with persistent calyx. Native to the mountains of southwestern China and adjacent northern Burma and Vietnam.

Euphorbia hirta L. — Garden Euphorbia (*Chamaesyce hirta* [L.] Millspaugh)

Guo-lu-wu-gong=Kuo-lu-wu-kung (過路蜈蚣, Creeping Centipede); *Fei-yang-cao=Fei-Yang-ts'ao* (飛揚草, Flying-spreading Herb); *Da-fei-yang=Ta-fei-yang* (大飛揚, Greater Flying-spreading). Young shoots, used as food in Taiwan.

Erect or ascending annual lactiferous herbs 15–40 cm high, stems terete, the lower internodes 4–6 cm long, hirsute or glabrescent; foliage with a reddish hue, leaves opposite, oblong-lanceolate, 2–4 cm long, 1–1.5 cm wide, acute, base oblique, crenulate or subentire, glaucous and sparsely hirsute beneath; flowers very small, monoecious, cyathia crowded in axillary cymose heads 5–8 mm in diameter, sessile or pedunculate; capsules trigonous, 1.5 mm across, sparsely pilose, readily popped open, scattering the seeds, hence the vernacular name *"Fei-yang-cao"* (Flying-spreading herb). A common weed in gardens and waste places near villages, used for medicinal purposes in Hong Kong; only recorded as food in Taiwan.

Euphorbia humifusa Willldenow (cf. *E. tashiroi* Hayata)

Xiao-chong-er-wo-dan=Hsiao-ch'ung-erh-wo-tan (小蟲兒臥單, Small Insects' Bedding-sheet); *Di-jin=Ti-chin* (地錦, Ground Brocade). Young shoots, eaten in Taiwan.

Low diffused herbs spreading over the ground to 15–20 cm in diameter, branchlets slender, reddish, glabrous; leaves opposite, small, sessile, oblong, 5–10 mm long, 4–6 mn wide, apex rounded, base obliquely obtuse, serrulate, green, tinged purplish-red; flowers small, cyathia hirsute, axillary, involucre obconic, 5 mm long, red, 4-lobed, ovary 3-locular, styles 3, bifid; capsules subglobose; seeds trigonous-ovoid, 1 mm long, brown-black.

Mallotus furetianus (Baillon) J. Mueller

Shan-ku-cha=Shan-k'u-ch'a (山苦茶, Hillside Bitter Tea); *Mao-cha=Mao-ch'a* (毛茶, Hairy Tea). Leaves used for tea in Hainan Island; said to have good flavor.

Shrubs or small trees 10 m high, branchlets pilose; leaves alternate, the upper ones almost opposite, oblong-obovate or obovate-lanceolate, 5–15 cm long, 2–6 cm wide, acute or caudate, base cuneate and subcordate, entire or wavy and denticulate, barbate in nerve angles and glandular punctate beneath, petioles 1–3 cm long; flowers small, apetalous, unisexual, dioecious, staminate flowers 1–5, fasciculate, in a pseudoraceme 4–10 cm long, peduncles densely velutinous-stellate, pedicels 2 mm long, perianth tunicate in bud, splitting into 3 or 4 segments at anthesis, 1.5 mm long, stamens 25–45, filaments filiform, unequal, thickened and glandular at the apex, bearing a theca on each side, pistillate flowers in terminal racemes 7–10 cm long, perianth splitting open along one side, 4.5 mm long, ovary densely covered by echinate processes, style columnar, stigmas 3, recurved; capsules globose, 1.4 cm in diameter, bristly and densely punctate with orange glands; seeds globose, coat crustaceous, spotted-patched. Widespread in southeastern Asia, growing in forests along streams, reported from Hong Kong and Hainan Island.

Manihot esculenta Crantz — Cassava, Manico, Tapioca Plant (Syn. *M. utilissima* Pohl, *Jatropha manihot* L.)

Mu-shu=Mu-shu (木薯, Tree Potato); *Xi-gu-mi=Hsi-ku-mi*, (西穀米, Western Rice); *Zhen-zhu-mi=Chen-chu-mi* (珍珠米, Pearl Grain). Enlarged root; and products prepared from its starch.

Erect suffrutex 1.5–3 m high, bearing fleshy tuberous roots; leaves alternate, suborbicular in outline, 10–20 cm in diameter, palmately 3- to 7-parted, segments oblong-lanceolate, acuminate, entire, petioles 30 cm long; flowers yellowish-white, tinged purple, unisexual, monoecious, apetalous, borne in axillary racemes or panicles, stamens 10, ovary 3-locular, styles 3, connate at the base; capsules oblong, 1.5 cm long, 6-ridged. Native of Brazil, introduced and cultivated in Guangzhou and its vicinity.

Phyllanthus emblica L. — EMBLICA, MYROBALAN

Yu-gan-zi=Yü-kan-tzu (餘甘子, After-taste Sweet); *You-gan-zi=Yu-kan-tzu* (油甘子, Oil Sweet Seed). Mature fruit, the fleshy outer cover, eaten fresh or preserved in liquid.

Deciduous shrubs or small trees 1–4 m high, trunk gray, smooth, branchlets slender, falling like a compound leaf; leaves alternate, 2-rank, oblong, 1–2 cm long, 3–5 mm wide, glabrous; flowers unisexual, monoecious, yellowish-green, 3–6 in axillary fascicles, perianth segments 6, stamens 3; capsules subglobose, fleshy outside, bony inside, pale yellow when mature, pleasant taste, on drying dehiscent forcefully, throwing the seeds 30–50 cm away.

Ricinus communis L. — CASTOR BEAN

Bi-ma=Pi-ma (蓖麻, Castor Bean). Seed; roasted, oil extracted, occasionally used for cooking in villages of northern China. The raw seed is extremely poisonous.

Tall annual herbs 2–3 m high, branchlets glaucous; leaves alternate, peltate, orbicular in outline, 15–30 cm across, palmately 5- to 11-lobed, the lobes ovate-oblong, acuminate, serrate, petioles terete, 10–30 cm long; flowers unisexual, monoecious, apetalous, pseudoracemose, stamens numerous, ovary 3-locular, locules uniovulate, styles 3, red, bifid; capsules subglobose, 1–2 cm across, smooth or covered with soft spine-like processes, dehiscent loculicidally; seeds oblong, 8–10 mm long, gray, mottled, caruncle white, on one end of the long axis. Native of Africa, extensively cultivated in China, sometimes appearing weedy.

Sauropus androgynus (L.) Merrill — JAVA SAUROPUS

Shu-zi-cai=Shu-tzu-ts'ai (樹子菜, Bush Vegetable); *Shou-gong-mu=Shou-kung-mu* (守宮木, Palace-security Shrub). Planted in kitchen gardens; leaves and young shoots used in Guangzhou and Hong Kong for soup, used especially in summer.

Glabrous shrubs 1–2 m high, branchlets green; leaves alternate, ovate-elliptic, rarely lanceolate, 3–7 (–10) cm long, 1.5–2.5 (–3.5) cm wide, acute, petioles 2–4 mm long, stipules inconspicuous, triangular, 2 mm long and across the caudate base; flowers small, unisexual, monoecious, on current year's growth, the staminate clusters axillary to lower leaves, green, pistillate ones maroon, axillary to middle leaves, upper leaves of the shoots subtending no flowering clusters, staminal column very short, ovary 3-locular, styles 3, bifid; capsules subglobose, 1.2 cm in diameter; seeds oblong-trigonous, 7 mm long. This is the first record of the species being cultivated as a food plant in China, hence a voucher specimen is cited (S. Y. Hu 8751, Nov. 12, 1969).

Sauropus changianus S. Y. Hu — Dragon's Tongue, Chinese Sauropus (Syn. *S. rostratus* auctt., non Miquel) (Figure 39)

Long-li-ye=Lung-li-yeh (龍脷葉, Dragon's Tongue, so named for the oblong-obvate emarginate leaves). Leaves and young shoots used for preparing pork broth in Guangzhou and Hongkong; dried material available on shelves for tea in American Chinese stores.

Cauliflorous low suffrutex 15–30 (–40) cm high, branchlets softly glandularly pillose, annual growth 5–6 cm long, 3 mm across, bearing 5–7 leaves, internodes 3–15 mm long; leaves oblong-oblanceolate, 7–17 cm long, 2–5 cm wide, rounded and retuse-mucronate, base cuneate, entire, glabrous, petioles 3–4 mm long, stipules conspicuous, triangular-cordate, 6 mm long, 4 mm across the cordate base, scarious, persistent, awned; flowers small, monoecious, fasciculate above the leafscars of 3- or 4-year old growth, subtained by persistent bracts, staminate flower green, 2 mm across, on filiform pedicels 4 mm long, pistillate flowers 5 mm across, maroon, pedicels 2 mm long, styles 3, sessile, bifid, the branches recurved; capsules pea-sized, concealed in the persistent sepals (S. Y. Hu 5258).

Callitrichaceae: Water Starwort Family

Callitriche stagnalis Scopoli — Water Starwort, Water Chickweed

Shui-ma-chi=Shui-ma-ch'ih (水馬齒, Water Portulaca). Whole plant, eaten in Taiwan.

Small aquatic plant, branchlets delicate, 10–20 cm long; leaves dimorphic, the submersed leaves linear-lanceolate, 1 cm long, 1–2 mm across, 1-nerved, the floating ones obovate-spathulate, 8 mm long, 5 mm wide, 3-nerved; flowers small, naked, unisexual, monoecious, staminate flower consisting of 1 stamen, pistillate flower a single 4-locular ovary, bearing 2 distinct filiform styles; fruits nut-like, at maturity separating into 4 winged 1-seeded segments. Widespread in northern hemisphere.

Empetraceae: Crowberry Family

Empetrum nigrum L. — Black Crowberry

Yan-gao-lan=Yen-kao-lan (岩高蘭, High Rock Lan). Ripe fruit, eaten in northeastern China.

Small creeping evergreen shrubs 20–50 cm high, branchlets hairy; leaves small, alternate, sessile, linear-oblong, 5 mm long, 1–1.5 mm wide; flowers unisexual, dioecious, 1–3, axillary, apetalous, perianth segments 3, dull-red; berries subglobose, 5–7 mm in diameter, purple-black, sweet; seeds 7–11, each enclosed by an endocarp, like the seed of hollies.

Anacardiaceae: Cashew Family

Allospondias lakonensis (Pierre) Stapf — FALSE SOAP-BERRY (Syn. *Poupartia chinensis* Merrill)

Jia-mu-huan=Chia-mu-huan (假木槵, False Soap-berry); *Yang-shi-shu=Yang-shih-shu* (羊屎樹, Goat Droppings Tree); *Jia-suan-zao=Chia-suan-tzao* (假酸棗, False Sour Jujube). Fruit, red, with wine flavor, eaten fresh.

Deciduous trees 7–10 m high, branchlets glabrous; leaves pinnate, 20–45 cm long, leaflets 11–25, alternate, oblong-lanceolate, 6–10 cm long, 1.5–3 cm wide, acuminate, base obliquely rounded, entire; flowers creamy-white, small, polygamous, in terminal panicles 15–30 cm long, the terminal one of the cymules fruit-bearing, stamens 8–10, ovary 4- or 5-locular, style thick, decurrent on the back of the carpel, connivent above; fruits drupaceous, in loose terminal clusters, oblong-subglobose, 8–12 mm long, 8–9 mm across, both ends truncate, red, with 4 large cavities in the mesocarps at the distal end giving 4 depressions alternate with 4 prominent dome-like structure on drying, endocarp bony, star-like; seed one in each locule, cotyledons oily.

Anacardium occidentale L. — CASHEW, CASHEW NUT

Yao-guo=Yao-kuo (腰果, Kidney Fruit); *Gang-ru-shu=Kang-ju-shu* (槓如樹, the Taiwan translation of Cashew Tree). Imported nuts, gradually replacing Canarium Nut in Cantonese dishes; raw nuts poisonous.

Lactiferous evergreen trees 10–12 m high, containing an irritant skin-poison, especially in raw seeds, a substance destroyed on heating; leaves simple, alternate, oblong-ovate or obovate, 10–20 cm long, 8–15 cm wide, glabrous, rounded or slightly emarginate, base cuneate or obtuse, coriaceous; flowers small, fragrant, polygamous, greenish-white, becoming rose-pink, in terminal panicles 20–25 cm long, stamens 10, in staminate flowers 9 short, 1 long with red anther projecting above the corolla; ovary 1-locular, uniovulate, style simple, exserted; fruits kidney-shaped, indehiscent, grayish-brown, terminal to a bright red or yellow fleshy stalk (cushion or cashew apple), 6–20 cm long, 4–8 cm across, pulpy, fragrant, edible fresh; seed kidney-shaped, testa reddish brown, enclosing 2 large white cotyledons 2.5–3 cm long. Native to tropical America, introduced to tropical Asia, now India being a producing center; cultivated in Fujian, Hainan Island, Guangdong, Guangxi and Yunnan of China.

Bouea macrophylla Griffith — GANDARIA (SUNDANESE), KUNDANG (JAVA NAME)

Gan-da-li-guo=Kan-ta-li-kuo (干達利果, Gandaria). Fruit, introduced to Taiwan before 1950

Resiniferous evergreen trees up to 20 m high, bark grayish-brown, finely fissured, giving out gummy drops when cut; leaves opposite, coriaceous, ovate-oblong or elliptic-lanceolate, 11–30 cm long, 4–8 cm wide, acute or acuminate, base acute, glabrous; flowers very small, pale yellow, polygamous, in axillary panicles, sepals and petals 3–4, stamens 3–5, ovary unilocular, 1-ovulate; fruits drupaceous, subglobose, 3–5 cm long, 3–4 cm across, endocarp leathery; seed with blue-violet cotyledons. Native of Malaysia, where it is cultivated as an estimable village fruit tree.

Buchanania latifolia Roxburgh

Dou-fu-guo=Tou-fu-kuo (豆腐果, Beancurd Fruit). Seed, ground, used for making beancurd in Yunnan.

Deciduous trees 13–15 m high, branchlets stout, slightly pubescent; leaves broad oblong, 12–24 cm long, 6–10 cm wide, rounded or emarginate, base rounded or obtuse, entire, glabrescent; flowers white, small, in terminal panicles 20 cm long, velutinous, sepals 5, petals 5, 2.5 mm long, stamens 10, fertile carpels 1, hairy, pistillodes 4 or 5; drupes compressed ellipsoid, 9 mm long, 6 mm across, black. Native to southeastern Asia, growing in deciduous forests in valleys at altitudes of 750–900 meters in Yunnan, Hainan Island and Guangdong.

Choerospondias fordii (Hemsley) S. Y. Hu — SOUR JUJUBE (Syn. *Poupartia fordii* Hemsley)

Nan-suan-zao=Nan-suan-tsao (南酸棗, Southern Sour Jujube) Ripe fruit, used locally by rural people, not available in the market.

Glabrous deciduous trees 8–10 m high, branchlets stout, 5 mm across, annual growth 3–5 cm long, bearing 5 to 7 pinnate leaves; petioles terete, 4–7.5 cm long, leaflets 9–11 (–13), opposite, petiolulate, terminal ones elliptic, the lateral ones obliquely ovate-lanceolate, 4–10.5 cm long, 2–5 cm wide, acuminate, the acumen 1–1.5 cm long, base acute or rounded, shiny-green above, glaucous beneath, entire, occasionally remotely crenate; flowers purple-maroon, small, dioecious, axillary to basal bracts or lower leaves of current year's growth, the staminate flowers in delicate panicales equal or shorter than the petioles, calyx cupular, the lobes rounded, ciliate, petals 5, reflexed, stamens 8–10, pistillate flowers solitary, on very short pedicels axillary to lower leaves, staminodes much shorter than the petals, pistil subglobose, styles 5, distinct, inserted on the shoulder of the ovary; fruits drupaceous, subglobose-oblong, 2.2–2.5 cm long,

1.8–2 cm across, ripe yellow, exocarp thin, mesocarp slimy and very sour, endocarp with 5 germination pores.

Choerospondias pubinervis (Rehder et Wilson) S. Y. Hu — Xi-Suan-Zao (Syn. *Spondias axillaris* Roxburgh var. *pubinervis* Rehder et Wilson)
  Xi-suan-zao=Hsi-suan-tzao (西酸棗, Western Sour Jujube). Ripe fruit.

Large deciduous trees 15–25 m high, trunks 1 m in diameter, bark gray, deeply fissured and persistent, branchlets softly pubescent throughout; leaves pinnate, leaflets 9–11, the terminal ones elliptic or oblong, the lateral ones obliquely ovate-oblong, 3.5–12.5 cm long, 1.5–4.5 cm wide, pubescent, particularly the midribs and lateral nerves hispid beneath, petioles rachis, petiolules pubescent, accuminate-caudate, the acumen 1–2 cm long, base acute or obliquely rounded, entire or sometimes serrate; flowers dioecious, red-brown, the staminate ones in axillary panicles 6–7 cm long, subtended by bracts or lower leaves, calyx cupular, pubescent, stamens 8–10, pistillode obscure, pistillate flowers in short racemes, 4- or 5-flowered, occasionally 1-flowered by abortion, then the peduncles 1–2 cm long; fruits oblong, 2.3–2.7 cm long, 1.3–1.5 cm across.

Cotinus coggygria Scopoli var. **glaucophylla** C. Y. Wu — Smoke Tree
  Fen-bei-huang-lu=Feng-pei-huang-lu (粉背黃櫨, Glaucous Smoke Tree). Young leafy shoots used for potgreen in Yunnan.

Deciduous shrubs 2–7 m high, branchlets glabrous; leaves ovate-oblong, 3.5–10 cm long, 2.5–7.5 cm wide, emarginate, base rounded or subcordate, entire, glabrous, glaucous beneath, petioles 1.5–3.3 cm long; flowers greenish-yellow, inconspicuous, polygamous, in loose terminal panicles, sepals 5, 1 mm long, petals 5, 1.6 mm long, stamens 5, disk yellow, ovary globose, styles 3; fruits small, oblique, drupaceous, 4–4.5 mm across, the fruiting panicles showy, with the pedicels of the sterile flowers lengthened and plumose. The species of Eurasian distribution, this variety widespread in northern, western and southwestern China, growing along streams at altitudes of 1,300–2,400 m in Shanxi, Shaanxi, Gansu, Sichuan and northern Yunnan.

Dracontomelon duperreanum Pierre — Asian Pheasant Tree (Syn. *D. dao* sensu auctt, non *D. dao* [Blanco] Merrill et Rolfe).
  Ren-mian=Jen-mien (人面, Man Face, named for the markings on the stone giving a fanciful resemblance of a man's face). Fruit, preserved in sugar, available in American Chinese stores.

Evergreen trees 10–20 m high, with dense rounded crown and rather smooth bark; branchlets softly pubescent, bearing no latex, with slightly gummy drops; leaves large, pinnate, 30–45 cm long, leaflets 11–17, alternate, oblong-lanceolate, 6–12 cm long, 2.5–4 cm wide, acuminate, entire; flowers small, greenish white, 5-merous, pentacylic, in terminal panicles with the lower branches subtended by normal leaves; stamens 10, ovary globose-discoid, pilose, styles 5, attached separately to the shoulder of the ovary, with the distal portions of all 5 connate, giving rise to a column and the capitate stigma, ovules 2 in each locule, hanging from the inner suture on the roof of the carpel; fruits compressed globose, drupaceous, 2 cm across, brownish yellow, pulp fibrous, taste sweet-sour, good flavor, stones flattened, subtriangular-discoid, the top with 5 unequal germination poles radiating slightly off the center, the two larger ones each occupying one-third of the top portion of the disk, the 3 smaller ones evenly distributed below, producing a fanciful resemblance of a human face, hence the Chinese vernacular name *"Ren-mian"* (Man Face).

Mangifera indica L. — MANGO

Mang-guo=Mong-kuo (芒果, Mango). Fruit; the stone saved after eating the flesh, dried and used in the preparation of *liangcha* (涼茶, cooling tea).

Large evergreen trees 10–15 m high, bark grayish brown, scaly, branchlets stout, the current year's growth 5–8 cm across, pilose, terminal buds stout, developing into flowering panicles, another subterminal bud leading to a vegetative shoot after blooming; leaves simple, large, crowded at the distal end of a branchlet, coriaceous, elliptic-oblong, 10–40 cm long, 3–6 cm wide, acuminate, base obtuse, entire, with 21–23 lateral nerves, the reticulation of veinlets clear, petioles 3–6 cm long; flowers rather small, 5–6 mm across, polygamous, fragrant, sepals and petals 4–5, stamens 1 to 5, all fertile or 1 long and fertile, ovary ovoid, style slender; fruits various in sizes and shapes, drupaceous, oblong or ovoid-ellipsoid, often oblique at one end, 5–12 cm long, the exocarp leathery, the mesocarp juicy and good flavor when ripe, with latex when green, stone compressed oblong, 4–6 cm long, 3–4 cm wide, 1–2 cm thick, endocarp fibrous, woody, very tough; seed 1.

Mangifera persiciforma Wu et Ming

Suan-guo=Suan-kuo (酸果, Sour Fruit); *Tian-tao-mu=T'ien-t'ao-mu* (天桃木, Celestial Peach Tree); *Bian-tao=Pien-t'ao* (扁桃, Compressed Peach) . *Mai-ga=Mai-ka* (嘜咖, the Guangxi ethnic name). Fruit, eaten by various ethnic groups living in Yunnan and Guangxi Provinces.

Evergreen trees 10–19 m high, branchlets glabrous, leaves lanceolate or linear-lanceolate, 10–20 cm long, 2–3.8 cm wide, acute or acuminate, base acute, margin wavy; flowers small, greenish-yellow, in terminal panicles 12–19 cm long, sepals 4 or 5, ovate,

2 mm long, petals 4 or 5, lanceolate, 4 mm long, disks pulvinate, 4- or 5-lobed, fertile stamens 1, staminodes 2–3, ovary globose, glabrous; fruits peach-like, slightly compressed, 5 cm long, 4 cm across, stone large. Endemic to the mountains on the border of Yunnan-Guangxi-Guizhou, growing near villages at attitudes of 600–800 m.

Mangifera sylvatica Roxburgh

Dong-mang-guo=Tung-mang-kuo (冬芒果, Winter Mango); *Tian-tao=Tien-t'ao* (天 桃, Celestial Peach). Fruit, eaten in Guangxi.

Evergreen trees up to 20 m high, branchlets glabrous; leaves oblong or oblong-lanceolate, 15–20 (–25) cm long, 3–6 cm wide, obtuse or shortly acuminate, base obtuse, entire, coriaceous, petioles 3–7 cm long; flowers small, yellowish-white, polygamous, in terminal panicles, sepals 5, petals 5, fertile stamens 1; fruits drupaceous, kidney-shaped, 6–7 cm long, 4 cm across. Native to eastern Himalayan Region, growing in forests along streams in Guangxi, with fruits hanging on the tree in winter, hence the local name "*Dong-mang-guo*" (冬芒果, Winter Mango).

Pistacia chinensis Bunge — CHINESE PISTACHIO

Huang-lian-ya=Huang-lien-ya (黃蓮芽, Chinese Pistacia Shoot); *Huang-er-cha=Huang-erh-cha* (黃兒茶, Yellow Tea). Young shoots used as potherb, leaves for tea, in northern China.

Tall deciduous trees 10–30 cm high, with beautiful autumn red foliage, branchlets pubescent, winter buds red, the terminal one vegetative, the subterminal ones floral; leaves pinnate, leaflets 10 to 12, lanceolate, 5–8 cm long, 1.5–2 cm wide, acuminate, base cuneate, entire; flowers small, unisexual, dioecious, apetalous, blooming before leaves, in subsessile panicles; drupes small, obovoid-subglobose, 6 mm in diameter, green, red, turning purple-blue.

Pistacia vera L. — PISTACHIO, PISTACHIO NUT

Wu-ming-zi=Wu-ming-tzu (無名子, No-name Seed); *A-yue-hun-zi= A-yueh-hun-tzu* (阿 月渾子, Arabian Obscure Nut); *Kai-xin-guo=K'ai-hsin-kuo* (開心果, Open-heart Nut, Chinese ideograms meaning happiness).

Spreading deciduous trees 7–9 m high, branchlets stout, anuual growth 1.5–15 cm long, 3–6 mm across, glabrescent, bearing terminal vegetative buds and subterminal and axillary floral buds; leaves pinnate, leaflets 3, rarely 1 or 5, oblong, ovate-orbicular, 4–8 cm long, 3.5–6.5 cm wide, apex acute, rounded, or sometimes obcordate or truncate, the terminal one petiolulate, the lateral ones sessile; flowers apetalous, dioecious, in lateral sessile panicles 5–8 cm long; fruits reddish, drupaceous, obliquely ovoid,

2–2.5 cm long, 1.5–2 cm across, apex conic-acute; seeds compressed ovoid, containing green cotyledons.

Native of western Asia, Iran being the center of production; introduced in ancient times to southern China through the Arabian traders, hence the name "Arabian Obscure Nut". Either due to the fact of keeping trade secret by the traders, or of the difficulty of translating "pistachio" into Chinese language, the earliest record of the material was "No-name Seed". Recently in Hong Kong and in American Chinese groceries, pistachio is sold with a Chinese name "*Kai-xin-guo*" (開心果, Open-heart Nut), for the bony shell of the commercial nut is partially split.

Spondias pinnata (L. f.) Kurz — Hog-Plum

Bin-lang-qing=Pin-lang-ch'ing (檳榔青, Betel Green); *Huang-jin-guo=Huang-ching-kuo* (黃獠果, Yellow Deer Fruit); *Ji-wan-zi=Chi-wan-tzu* (雞杬子, Chicken Canarium). Fruits.

Deciduous, resiniferous and glabrous trees 5–8 m high, bark smooth, light gray, when bruised all parts have turpentine smell; leaves pinnate, 30–40 cm long, leaflets 5–11, opposite, ovate-oblong, 7–12 cm long, 3–5 cm wide, acuminate, base of the terminal ones obtuse or rarely cuneate, that of the lateral ones rounded unequally, entire, lateral nerves 20–22 pairs, reticulation of the veinlets prominent; flowers greenish-white, blooming before leaves, small, 4–5 mm across, polygamous, in large subsessile panicles up to 45 cm long; fruits oblong-ellipsoid, 3.5–5 cm long, 2.8–3.5 cm across, smooth, yellowish-brown, with the smell of rotting apples, edible but tough, of poor quality, exocarp lactiferous, mesocarp fibrous, endocarp woody, thick. Common in the tropical forests of southern Yunnan and Hainan Island.

Aquifoliaceae: Holly Family

Ilex latifolia Thunberg cv. 'Kudingcha' S. Y. Hu

Ku-ding-cha=K'u-ting-ch'a (苦丁茶, Bitter Fellow Tea). Fully-grown fresh leaves, collected from small trees cultivated in plantations converted from former rice fields, Ta-po District, North East Guangdong province; parent plants grown at a farm situated at end of stream, trees 10 m high, trunk 60 cm in diameter, the female one bearing fruits 7 mm in diameter.

Evergreen trees up to 20 m high; trunks 60 cm in diameter. Leaves elliptic-oblong, 9–25 cm long, 3.5–8 cm wide, base obtuse, apex short acuminate, rarely acute, finely serrate. Flowers fasciculate, axillary to leaves on 2- or 3-year old branches; perianth connate at base, sepals 4, petals 4, stamens 4, ovary superior, without style. Fruits ripe red, 6–7 mm in diameter. Pyrenes 4, oblong-elliptic, irregularly wrinkled and pitted. Known only in cultivation, now planted in plantations as Kudingcha.

Ilex yunnanensis Franchet — BABY TEA, YUNNAN HOLLY

Wa-er-cha=Wa-erh-ch'a (娃兒茶, Baby Tea). Young shoots gathered locally in West China, used as a substitute for tea.

Evergreen shrubs 1–3 m high, branchlets ferruginous-villose; leaves alternate, ovate or ovate-lanceolate, 2–3.5 cm long, 1–2 cm wide, acute, base rounded, aristate-crenulate-serrate; flowers white, small, 4-merous, unisexual, dioecious, staminate flowers in 1- or 3-flowered cymes, pistillate ones solitary, very rarely in 2- or 3-flowered cymes; fruits drupaceous, red, solitary, globose, 5–6 mm in diameter, persistent calyx ciliate, endocarps 4, smooth, leathery, each containing 1 seed with tiny embryo and oily endosperm.

Celastraceae: Bitter-sweet Family

Celastrus kusanoi Hayata

Nan-she-teng=Nan-sheh-t'eng (南蛇藤, Southern Snake Vine). Fruit; eaten in Taiwan.

Deciduous scandent shrubs 3–4 m high, branchlets pubescent; leaves alternate, oblong-elliptic or suborbicular, 7–10 cm long, 5–11 cm wide, rounded and abruptly short-accuminate, cuspidate, base rounded, finely serrate, glandularly pubescent along the midrib and lateral veins beneath; flowers dioecious, yellowish-green, 3–9 in axillary cymes on pubescent peduncles shorter than the petioles of the subtending leaves; capsules small, globose, 5–10 mm in diameter, with prominent persistent style.

Staphyleaceae: Bladdernut Family

Euscaphis japonica (Thunberg) Kanitz — EUSCAPHIS

Ye-ya-chun=Yeh-ya-chun (野鴉椿, Wild Crow Ailanthus). Young leaves; used as potherb in Taiwan.

Deciduous trees 3–8 m high, bark gray with vertical furrows, branchlets glabrous, crushed portion yielding a foul smell; leaves opposite, pinnate, 13–32 cm long, leaflets 5–9 (–11), ovate, 5–11 cm long, 1–5 cm wide, acuminate, base rounded, sharply serrate, pubescent along the midrib beneath; flowers white, small, 4 mm across, 5-merous, in large terminal panicles 7–10 cm long and across, polygamous, only the terminal ones of the upper cymules fruit-bearing, perianth segments in 2 series, all persistent in fruit, ovary 3-carpellate, apocarpous, styles free, connate only at the stigma-level; fruits and the infrutescent red, conspicuous in the landscape, pericarps coriaceous, the outer surface striate and loosely reticulate; seeds subglobose, 1 to each follicle, shiny black, 4 mm in diameter, faveolate under lens.

Staphylea bumalda de Candolle

Sheng-gu-you=Sheng-ku-yiu (省沽油, Staphyllea Oil); *Zhen-zhu-hua=Chen-chu-hua* (珍珠花, Pearl Flower); *Shuang-hu-die=Shuang-hu-tieh* (雙蝴蝶, Paired Butterflies). Oil extracted from seed used for cooking.

Deciduous shrubs or small trees 2–10 m high, branches spreading; leaves trifoliolate, opposite, leaflets elliptic or elliptic-ovate, 3–7 cm long, 1.5–4 cm wide, short-acuminate, base cuneate, serrulate, teeth mucronate, pilose above, strigose on the midrib beneath, petioles 3–7 cm long; flowers white, 5-merous, in loose terminal panicles 5–7 cm long, sepals oblong, petals obovate, 8 mm long, stamens 5, filaments hairy; capsules inflated, 2–3 cm long, 2-lobed, seeds yellow, globose, 5 mm in diameter. Native to temperate eastern Asia; introduced into Western gardens in 1812.

Icacinaceae: Icacina Family

Mappianthus iodoides Handel-Mazzetti

Ding-xin-teng=Ting-hsin-t'eng (定心藤, Setting-heart Vine); *Tian-guo-teng=T'ien-kuo-t'eng* (甜果藤, Sweet-fruit Vine). Fruit; eaten fresh in Yunnan.

Lianas, climbing by stout tendrils, branchlets pubescent; leaves oblong-elliptic, 8–17 cm long, 3–7 cm wide, acuminate-caudate, base rounded; flowers yellow, hairy, unisexual, dioecious, in axillary cymes, staminate flowers with cupular and 5-toothed calyx, petals funnelform, fleshy, hairy, stamens 5, anthers ovoid, introrse, pistillodes villose, pistillate flowers not yet known; drupes oblong, 2–3 cm long, 1–1.7 cm across, with appressed stiff hairs, orange-red, sweet; seed 1. Native to southern China and adjacent Vietnam; growing in steep valleys at altitudes of 800–1,800 m.

Pittosporopsis kerrii Craib

Jia-hai-tong=Chia-hai-tung (假海桐, False Pittosporum). Seeds eaten by the people living in the mountains of southern Yunnan.

Shrubs or small trees 4–7 (–17) m high, bark red-brown, branchlets glabrous; leaves alternate, oblong, 12–22 cm long, 4–8.5 cm wide, acuminate, base obtuse, petioles 1.5–2.5 cm long; flowers greenish-white or pale yellow, hairy, fragrant, in axillary cymes, sepals 5-lobed, persistent and enlarged after anthesis, petals 5 , spatulate, stamens 5, anthers basifixed, ovary oblong, unilocular, 2-ovulate, style geniculate, persistent; fruits drupaceous, globose, stone oblong. A monotypic genus endemic to the mountains along the borders of Burma, Thailand, Yunnan, Laos and Vietnam; growing along streams at altitudes of 350–1,600 M.

Aceraceae: Maple Family

Acer mono Maximowicz (Syn. *A. pictum* auctt., non Thunberg)

Di-jin-qi=Ti-chin-ch'i (地錦槭, Ground Embroidery Maple). Young leaves used for potherb; old leaves gathered and used as tea substitute.

Deciduous trees up to 20 m high, branchlets glabrous, becoming yellowish-gray and slightly fissured the second year; leaves alternate, suborbicular in outline, 5- or 7-lobed, 8–15 cm across, cordate or subcordate, the lobes ovate-triangular, acuminate, entire, with axillary tufts of hairs beneath, palmately 5-nerved; flowers small, greenish-yellow, blooming before the expansion of the leaves, in terminal corymbs 4–6 cm across, polygamous, most of the flowers staminate, only the terminal one of the cymules with fertile ovary, 8–10 mm across, petals 4, stamens 8, inserted on the annulus, pistillate flowers smaller, ovary compressed, triangular, stigma filiform, fruits two samara, wing-tips 4.5 cm apart, wings twice the length of the nuts.

Acer saccharum Marshall — Sugar Maple

Qi-shu-tang-jiang=Ch'i-shu-t'ang-chiang (槭樹糖漿, Maple Syrup). Liquid sweet material for market, condensed from juice of certain maple tree in North-East United States. *Qi-shu-tang=Ch'i-shu-t'ang* (槭樹糖, Maple Sugar). Solid sweetening material produced from watery liquid gathered from trunk of maple trees, four feet above ground, in March and April, both products available in Hong Kong.

Large tree up to 40 m high, with spreading branches, bark gray, furrowed. Leaves opposite, 8–14 cm across, cordate, palmately 3–5-lobed, lobes acuminate, coarsely dentate. Flowers small, on pendulous pedicels, in terminal subsessile corymbs, sepals connate; petals wanting; stamens exserted in staminate flowers; ovary 2-carpellate, styles 2. Fruit a 2-winged compressed samaras.

Acer truncatum Bunge — Shantung Maple

Yuan-bao-shu=Yuan-pao-shu (元寶樹, Silver-ingot Tree); *Gua-zi-cha=Kua-tsu-cha* (瓜子茶, Melon Seed Tea). Young leaves; used for tea.

Deciduous trees 8–10 m high, branchlets glabrous or glabrescent; leaves opposite, semiorbicular in outline, 5-lobed, 5–10 cm long, 8–12 cm wide, base truncate-subcordate, the lobes triangular, acuminate; flowers polygamous, greenish-yellow, in terminal corymbose panicles, only the terminal ones of the cymules with fertile ovary, stamens

8, ovary 2-carpellate; samaras with broad wings, the large nuts about as long as the wings, 5 cm between the wing-tips.

Sapindaceae: Soapberry Family

Arytera littoralis Blume

Bian-guo-mu=Pien-kuo-mu (扁果木, Flat-fruited Tree); *Bin-mu-huan=Pin-mu-huan* (濱木患, Seashore Soapberry). Tender leafy shoots, used for potherb by the Thai ethnic people in southern Yunnan.

Trees 8–20 m high, branchlets pilose; leaves pinnate, oblong in outline, 12–35 cm long, leaflets 8–12, oblong-lanceolate, 8–18 cm long, 2.5–7.5 cm wide, acuminate, base obtuse or rounded, entire, glabrous, glandular at vein-axils; flowers yellowish-white, fragrant, small, in axillary or terminal panicles 6–14 cm long, calyx cupular, 5-toothed, hairy, ciliate, petals 5, disk glabrous, stamens 8, filaments pilose, ovary obovoid, 1.5 mm long, hairy, 2-locular; fruits consisting of 2 separate carpels, or with only one developed, oblong, 1–1.5 cm long, 5–10 mm across, orange-red, turning to black with age; seed 1, oblong, arillate, coat-red. Widespread in eastern Himalayan Region, extending eastward to the Philippines, growing in hillside-thickets at altitudes of 540–1,180 m in southern Yunnan, Guangxi, Guangdong and Hainan Island.

Cardiospermum halicacabum L. — Balloon-vine

Feng-chuan-ge=Feng-ch'uan-ko (風船葛, Wind-sail Vine); *Dao-di-ling=Tao-ti-ling* (倒地鈴, Reverse Bell). Tender shoots, eaten in northern China.

Annual vines 1–3 cm long, branching extensively, climbing by branched tendrils; leaves biternately compound, deltoid in outline, 5–12 cm long and across the base, ultimate segments ovate or rhomboid, 4–8 cm long, 1.5–2.5 cm wide, coarsely serrate; flowers small, white, polygamous, sepals 4, petals 4, stamens 8, ovary superior, 3-locular, cells uniovulate, styles short, 3-cleft; capsules obovoid, inflated, veiny, loculicidally dehiscent, segments boat-like; seeds shiny black, with a white heart-shaped spot. A pantropic weed, cultivated in northern China primarily for ornamental purpose.

Dimocarpus longan Loureiro — Longan (Syn. *Euphoria longan* [Loureiro] Steudel, *E. longana* Lamarck)

Long-yan=Lung-yen (龍眼, Dragon Eye) ; *Gui-yuan=Kui-yuan* (桂圓, Cinnamon Round). Aril of seed, transparent, white when fresh, very juicy and sweet, becoming brown and black on drying; different prepared forms of the fruit (dried fruit, dried aril, shelled and canned aril) available in American Chinese stores; the dried round fruits, thrown over the bride and groom in weddings, picked up by young people as a good omen; leaves used for tea.

Evergreen trees 8–10 m high with compact dark round crown, branchlets pubescent; leaves alternate, pinnate, 15–30 cm long, leaflets 4–12, alternate, oblong-elliptic, oblique, 6–20 cm long, 2.5–5 cm wide, wavy, entire; flowers small, polygamous, greenish-yellow, 4–5 mm across, in large terminal velutinously stellate-tomentose panicles, stamens 8, the filaments pubescent, ovary short, cordate, 3-lobed, warty and pilose, stigmas 2, due to abortion, one carpel developed; fruits globose, 12–18 mm in diameter, having the structure of an one-seeded indehiscent achene, pericarp brown-cinnamon, crustaceous, rather smooth at maturity, the inside fleshy, juicy, sweet white transparent aril being the edible portion; seed 1, large, subglobose, or oblong, 10–14 mm in diameter, shiny-black, hilum white, oblong, 8–10 mm long, looking at it from the hilum-side, the seed giving a fanciful resemblance to the white and black structure of an eyeball, hence the Chinese vernacular name, *"Long-yan"* (Dragon Eye).

Handeliodendron bodinieri (Léveillé) Rehder

Ya-jiao-ban=Ya-chiao-pan (鴨腳板, Duck-feet Tree); *Zhang-ye-mu=Chang-yeh-mu* (掌葉 木, Palmate-leaved Tree). Oil from seed; used for cooking in Guangxi.

Deciduous shrubs or trees 4–12 m high, bark gray; leaves opposite, palmately compound, leaflets 4 or 5, elliptic, 3–12 cm long, 1.5–6 cm wide, the middle one the largest, acuminate, base attenuate, entire, glabrous, punctate with dark-red or black dots; flowers white, in terminal panicles 10 cm long, sepals 5, petals 4 or 5, pubescent outside, stamens 7–8, unequal, ovary 3-locular, cells 2-ovulate; capsules red, 2–3.5 cm long, 1–1.2 cm across; seeds black, partially covered with white aril. Endemic to the limestone areas of Guangxi and adjacent Guizhou, growing in mixed forests at altitudes of 400–800 m.

Lepisanthes rubiginosa (Roxburgh) Leenhouts (Syn. *Erioglossum edule* Blume; *E. rubiginosum* [Roxburgh] Blume)

Chi-cai=Ch'ih-ts'ai (赤材, Red Timber). Fruit; sweet, eaten in southern Yunnan and Hainan Island.

Trees or shrubs 2–7 m high, densely velutinous; leaves pinnate, oblong in outline, 20–40 cm long, leaflets 5–17, oblong or ovate-oblong, 8–20 cm long, 2.5–6 cm wide, obtuse, base oblique rounded, entire; flowers white, fragrant, in terminal panicles, sepals 5, petals 4–5 mm long, clawed, villose outside, stamens 8, filaments pilose, ovary hairy,

3-locular, hairy; fruits drupaceous, solitary or 3, separated, 1.2–1.4 cm long, changing from red to black; seeds oblong, without aril. Widespread in the Pacific Islands.

Litchi chinensis Sonnerat — LITCHI, LEECHEE, LYCHEE

Li-zhi=Li-chi (荔枝, Litchi). White juicy aril of the seed; both fresh and dried fruit available in Boston; fresh material from Florida, sells at $4.50 a pound in Chinatown, June, 1985, rated as the queen of fruits in South China. Introduced to Florida by G. W. Groff of former Lingnan University, Guangzhou, China.

Medium-sized evergeen trees 8–15 m high, branchlets puberulous; leaves pinnate, 10–25 cm long, leaflets alternate, elliptic or oblong-lanceolate, 6–15 cm long, 2–4 cm wide, acuminate, base acute, entire, shiny above, dull gray and glabrous beneath; flowers apetalous, small, greenish white, 2–3 mm across, in large terminal pubescent panicles 15–30 cm long and wide across the base, calyx cupular, truncate, petals oblong, stamens 6–10, usually 8, the filaments hairy, ovary 2- or 3-lobed, hairy, styles 2-lobed, through abortion 1 achene-like fruit with fleshy aril developed; fruits ovoid, 2.5–3.5 cm long and across the base, the pericarp papery, the inside smooth, outside with hexagonal-conic shields each with a warty protuberance in the center, brittle, reddish-purple at maturity, the aril white, transparent, very juicy and sweet, with very poor keeping quality, highly esteemed as the queen of Chinese fruits. Native of China, many selected cultivars in Fujian and Guangdong Provinces, natural progenitor not yet known, introduced to USA repeatedly, successful only when some soil with certain associated mycorrhiza came together.

Nephelium lappaceum L. — RAMBUTAN

Shao-zi=Shao-tzu (韶子, Rambutan); *Mao-long-yan=Mao-lung-yen* (毛龍眼, Furry Longan); *Hong-mao-dan=Hung-mao-tan* (紅毛丹, Rambutan). Fruit, the edible portion being the aril.

Evergreen trees 10–20 m high, branchlets velutinous; leaves alternate, pinnate, 15–45 cm long, leaflets generally 5–7, oblong-elliptic, 6–18 cm long, 2.5–7.5 cm wide, acute or shortly acuminate, base obtuse or rounded, entire, puberulous beneath; flowers white, apetalous, small, polygamous, in large panicles 10–30 cm long, rusty hairy, perianth segment 4–6, stamens 5–8, inserted onto a disk, filaments woolly, ovary 2- or 3-lobed, villose, style 1, inserted between the lobes, deeply bifid or trifid, through abortion only one carpel developed into an achene-like indehiscent arillate fruit; fruits ovoid, red, 3.5–8 cm long, 2–5 cm in diameter, beset with tubercles each ending in a soft uncinate spine; seed 1, 2.5–3.5 cm long, 1–1.5 cm in diameter, aril white, transparent, juicy, much esteemed in southeastern Asia; inferior to litchi (lychee) in sweetness and flavor.

Xanthoceras sorbifolia Bunge

Wen-guan-guo=Wen-kuan-kuo (文冠果, Scholar Hat Fruit); *Ya-mu-gua=Yai-mu-kua* (崖木瓜, Cliff Wood Melon). Seed.

Deciduous shrubs or small trees, 3–8 m high, branchlets glabrescent, red-brown, with conspicuous lenticels and fissures; leaves pinnate, 15–20 cm long, leaflets 9–19, sessile, elliptic-lanceolate, 2–6 cm long, 1–2 cm wide, tomentose with stellate hairs beneath, sharply aristate-serrate, acute; flowers white, polygamous, blooming slightly in advance of the leaves, in showy pseudoracemes, rarely with simple basal branches, rachis 5–10 cm long, being the continuation of the central portion developed from a mixed terminal bud, with the vegetative branches initiated below, pedicels each subtended by 2–4 involucre-like foliaceous bracts, sepals oblong, yellow, tomentose, petals obovate, clawed, pink or yellow at the base, disk 5-lobed, each with an orange corneous appendage, stamens 8, ovary globose, densely yellow-pubescent, style filiform, stigma capitate; fruits globose, 3.5–6 cm in diameter, pericarp thick, the outer surface rugose and pubescent, the inner portion ligneous, dehiscent loculicidally; seeds 2 in each locule, black, angular, 9–12 mm across. A native of northern and northwestern China, introduced into European gardens in 1866. Herbarium specimens indicate its cultivation in USA as early as 1914.

Sabiaceae: Sabia Family

Meliosma rhoifolia Maximowicz

Shan-zhu-rou=Shan-chu-ju (山豬肉, Mountain Pork). Young shoot, eaten in Taiwan.

Medium-sized trees 10 m high, branchlets glabrous, axillary buds naked, velutinous, wider than long; leaves pinnate, 20–35 cm long, leaflets 7–17, oblong-lanceolate, the basal one ovate, 3–15 cm long, 2–3.5 cm wide, acuminate-caudate, base cuneate or subrotundate, aristately serrate, glabrous, lateral nerves 8–13 pairs; flowers small, white, in very large terminal panicles 17–25 cm long and across the middle, velutinous, perianth segments imbricate, stamens 5, 2 fertile, ovary sessile; fruits drupaceous, globose, 4–5 mm in diameter.

Sabia schumanniana Diels

Hei-dou-ban-cai=Hei-tou-pan-ts'ai (黑豆瓣菜, Black Split-bean Vegetable); *Nü-er-teng=Nü-erh-t'eng* (女兒藤, Maiden Vine); *Si-chuan-qing-feng-teng=Si-chuan-ch'ing-feng-t'eng* (四川清風藤, Sichuan Sabia). Young shoots, used for potherb.

Evergreen lianas 2–3 m high, roots fleshy, pale yellow, branchlets yellowish-green, almost glabrous; leaves lanceolate or oblong-lanceolate, 5–12 cm long, 1.53 cm wide, acuminate, base obtuse or rounded, glabrous, membranous and white along the margin; flowers greenish-purple, campanulate, solitary or 3–6 in axillary cymes, calyx 5-parted, petals 5, obovate-oblong, stamens 5, opposite the petals; fruits globose or reniform, 6–7 mm across, blue, prominently reticulate. Native to central and western China, growing over thickets along paths or on trees by streams at altitudes of 1,000–1,400 m.

Balsaminaceae: Touch-me-not Family

Impatiens noli-tangere L. — SNAPWEED

Ye-zhi-ma=Yeh-chih-ma (野芝麻, Wild Sesame); *Shui-jin-feng=Shui-chin-feng* (水金鳳, Water Golden Phenix), *Hui-cai-hua=Hui-ts'ai-hua* (輝菜花, Glorious Herb Flower). Seed; used as sesame, hence the local name "*Ye-zhi-ma*", given by the people living in the mountains of western Hubei.

Succulent annual herbs 40–100 cm high; leaves alternate, ovate-elliptic, 5–10 cm long, 2.5 cm wide, obtuse or acuminate, base cuneate, acute, crenate, petioles 2–3 cm long; flowers showy, yellow- and red-spotted at the throat, zygomorphic, 2 or 3 in cymose clusters on slender peduncles, calyx green in bud, at anthesis the lateral 2 small, ovate, acute, green, the superior one orbicular, petaloid, keeled, notched and apiculate, the inferior one saccate and prolonged backward into a long spur, petals 2, lateral, lobed, stamens 5, filaments fleshy, connate at the distal half, the inner surface extending into a tent, covering the ovary and separating it from the ring of anthers, the entire androecium falling in one body while the flower in full bloom, ovary superior, ovules many, style obscure; capsules linear-oblong, dehiscing explosively and coiling elastically; seeds numerous, brown. Widespread in Eurasia, growing in shady and damp areas.

Rhamnaceae: Buck-thorn Family

Berchemia flavescens (Wallich) Brongniart (Syn. *B. hypochrysa* Schneider)

Niu-er-teng=Niu-erh-t'eng (牛兒藤, Cattle Vine). Ripe fruit; eaten in Shaanxi.

Deciduous twining shrubs, branchlets glabrous; leaves ovate, or oblong, 6–13 cm long, 4.5–7.5 cm wide, rounded and mucronate or short-acuminate, base rounded, entire, glabrous, dried golden-yellow beneath; flowers greenish-yellow, 2 mm across, in terminal panicles 5–10 cm long, sepals deltoid, petals narrow, erect, incurved, stamens 5, anthers exserted, ovary concealed by the fleshy disk; drupes oblong, 8–10 mm long, black-red, sweet, edible. Native to western China and the Himalayan Region; introduced into American gardens in 1920.

Berchemia floribunda (Wallich) Brongniart (Syn. *B. giraldiana* Schneider)

Niu-bi-quan=Niu-pi-ch'uan (牛鼻圈, Cattle Nose-ring); *Gou-er-cha=Kou-erh-ch'a* (鈎兒茶, Hook Tea). Root; cooked with pig's feet or egg in Shaanxi, a special preparation served to people with rheumatic pains.

Deciduous twining shrubs; leaves ovate or ovate-oblong, 4–8 cm long, 2–3 cm wide, obtuse or rounded, mucronate, base rounded, entire, green, papillose above, grayish-white and glaucous beneath; flowers greenish, 5 mm across, in terminal panicles; fruits oblong, 7–10 mm long, 5 mm across, red and turning black. Native to southern and western China and the Himalayan Region, growing in thickets at altitudes of 500–1,600 m; introduced into American garden as *B. giraldiana* Schneider in 1911; being confused by some authors with *B. racemosa* auctt, non Siebold et Zuccarini.

Berchemia lineata de Candolle

Gou-er-cha=Kou-erh-ch'a (鈎兒茶, Hook Tea); *Lao-shu-shi=Lao-shu-shih* (老鼠屎, Rat Droppings). Fruit, eaten in Taiwan, recorded before 1950.

Arching, low shrubs 1–2 m high, with trailing branches, branchlets slender, brownish, pubescent; leaves small, oblong rarely ovate, 0.5–1.5 cm long, 3–10 mm wide, obtuse and mucronate, base rounded, upper surface glandularly rugose-maculate between the lateral nerves; flowers white, small, 4 mm long, 2–4 in axillary fascicles near the shoot apex, with the subtending leaves reducing in sizes progressively, sepals linear, acute, petals oblong, stamens 5, disk cupular, ovary globose, glabrous; fruits drupaceous, oblong, 4–5 mm long, 2–3 mm across, black at maturity, having a fanciful resemblance of the droppings of rat, hence the Cantonese vernacular name "Rat Dropping". Native of southeast China; young shoots used in folk medicine locally called *Lao-shu-er* (老鼠耳, Rat ear).

Berchemia racemosa Siebold et Zuccarini — SMOOTH SUPPLE-JACK

Huang-shan-teng=Huang-shan-t'eng (黃鱔藤, Yellow Eel Vine). Fruit; eaten in Taiwan, recorded before 1950.

Glabrous deciduous scandent shrubs 2–4 m high, branchlets terete; leaves ovate, 2.5–8 cm long, 1.2–5 cm wide, acute, base rounded, entire, glaucous beneath, lateral nerves 7–9 pairs, petioles 1–2.5 cm long; flowers small, 4 mm across, flower buds conical and acute, greenish white, 5-merous, in large terminal panicles 5–15 cm long with racemose lower branches, sepals ovate-lanceolate, caudate, 4 mm long, petals oblong, shorter than the sepals; stamens hardly exserted, ovary 2-locular, style 2-parted; drupes ovoid or oblong, 5–6 mm long, changing from green to red and finally black. Native of Japan and Taiwan, introduced to USA in the 1880s.

Berchemia yunnanensis Franchet — Yunnan Supple-Jack

Ya-gong-teng=Ya-kung-t'eng (牙公藤, Teeth-man Vine); *Jin-gang-teng=Chin-Kang-t'eng* (金剛藤, Diamond Vine); *Yun-nan-gou-er-teng=Yun-nan-kou-erh-t'eng* (雲南鈎兒藤, Hook-Vine of Yunnan). Ripe fruit.

Completely glabrous large liana 3–5 m high, branchlets yellowish-brown; leaves oblong-ovate, 2–7 cm long, 1–2.5 (–3.5) cm wide, obtuse and awned or rounded and mucronate, rarely acuminate, base rounded or subcordate, entire, papillose beneath, lateral nerves 8–10 pairs, petioles 2–3 cm long; flowers pale yellow, fragrant, 6 mm across, 2–4 in fascicles or in cymes forming a compact terminal subcylindrical panicle 3–4 cm long, 2 cm across, flower buds pentagonous-ovoid, acute, sepal rotate, the lobes triangular-caudate, 2.5 mm long, petals erect, obovate, 1.5 mm long, involute, covering the filament, anthers ovate, exerted; drupes cylindrical-oblong, 7–10 mm long, 3–4 mm across the middle, black when mature.

Hovenia dulcis Thunberg — Japanese Raisin Tree

Guai-zao=Kuai-tsao (拐棗, Zigzag Jujube); *Ji-ju-zi=Chi-chu-tzu* (雞距子, Cock's Spur); *Jin-gou-li=Chin-kou-li* (金鈎梨, Golden Hook Pear); *Zhi-ju=Tsu-chu* (枳椇, a book name). Fresh zigzag peduncle of the fruit cluster, very sweet and good flavor; occasionally available in village markets in western China.

Large deciduous trees 15 m high, branchlets reddish brown, puberulous; leaves alternate, ovate, 8–16 cm long, 6–11 cm wide, acuminate, base rounded, serrate, trinerved, pilose along the nerves; flowers small, greenish yellow, in axillary cymose panicles, the pedicels and secondary floral axis becoming fleshy, zigzag, reddish-brown, edible, as the fruits mature; fruits nut-like, apical to the fleshy pedicels in similar manner as the cashew nut; seeds flattened, one to each nut, testa shiny dark brown, cotyledons flat, broad.

Rhamnus erythroxylon Pallas

Liu-ye-shu-li=Liu-yeh-shu-li (柳葉鼠李, Willow-leaved Mice Plum); *Hei-ge-lan=Hei-ke-lan* (黑格蘭 [隔攔], Black Separated Block). Leaves; gathered locally and used for tea in Shaanxi.

Deciduous shrubs 2 m high, branchlets glabrous, each ending with a sharp spine; leaves linear, 3–10 cm long, 2–10 mm wide, acuminate, base cuneate, remotely serrulate, the teeth mucronate; flowers greenish-yellow, 10–20 in fascicles, sepals 5, petals 5; drupes globose, 5–6 mm in diameter, black; pyrenes 2 or 3, seeds obovoid. Native to northwestern China and Inner Mongolia, thence extending northward to Siberia; introduced into Western gardens in 1823. The original ideograms given in the source book " 隔蘭 ", meaning "resist and orchid", makes no sense; the ethnobotanical meaningful homonymous ideograms as seen in the square bracket, refer to the black fruits and the thorny branches which block up the approaching predators.

Rhamnus utilis Decaisne, and **R. davurica** Pallas — CHINESE BUCKTHORN

Dong-lü=Tung-lü (凍綠, Frozen Green); *Da-lü=Ta-lü* (大綠, Big Green); *Shu-li=Shu-li* (鼠李, Mice Plum). Young shoots, gathered for tea.

Shrubs or small trees 4–10 m high, some branches abbreviated and modified into thorns, branchlets glabrous; leaves various in shapes and sizes, elliptic-oblong, elliptic-obovate or ovate-oblong, 3–12 cm long, 2–5 cm wide, acuminate, base rounded or cuneate, remotely crenulate-serrulate; flowers greenish white or pale yellow, 2 to 5 in axillary fascicles at the basal portion of short shoots, 4- or 5-merous; fruits drupaceous, subglobose, 6–7 mm across, black at maturity, containing 2–4 pyrenes with the endocarps distinct, each enclosing a single seed. Both species occurring in northern China with the range of *R. utilis* extending to the Yangtze River Region and that of *R. davurica* to Inner Mongolia, northeastern China and adjacent Siberia. The technical characters which botanists use for distinguishing the species are that *R. utilis* have elliptic-oblong leaves with shorter petioles 5–12 mm long, and larger fruits 6–7 mm across, while *R. davurica* has elliptic-obovate to oblong leaves with longer petioles 6–25 mm long and fruits 6 mm across. Local people gather the young leaves for same purposes, small amount of the tender leaves for tea and large amount of older leaves for green dye, hence the vernacular names "Frozen Green, Big Green".

Sageretia thea (Osbeck) M. C. Johnston — HEDGE SAGERETIA (Syn. *S. theezans* [L.] Brongniart)

Que-mei-teng=Chueh-mei-t'eng (雀梅藤, Bird Mei Bush); *Suan-wei=Suan-wei* (酸味, Sour Taste); *Mi-zai-le=Mi-tsai-le* (米仔簕, Millet Thorn); *Dui-jie-ci=Tui-chieh-tz'u* (對節刺, Opposite Thorn), Young shoots for tea; fresh fruits picked and eaten locally, not available in the market.

Arching shrubs 1–2 m high, branchlets densely velutinously pubescent, the strong suckers bearing many abbreviate side shoots becoming thorns; leaves alternate, occasionally nearly opposite, ovate, oblong, 1–4 cm long, 1–2 cm wide, acute, obtuse or rounded, base round or subcordate, finely serrulate, petioles short, pilose; flowers very small, 3 mm across, greenish-white, sessile, the flower-buds globose, pilose, millet-like, in axillary spikes 2–5 cm long or terminal spicate panicles, sepals rotate, 4- or 5-lobed, the lobes triangular, acute, 1.5 mm long, petals half as long, erect, incurved covering the short filament, anthers cordate, exerted, ovary globose, glabrous, surrounded by cupular disk, style short and stout, stigmas 3, subcapitate; fruits berry-like, black, globose, 5 mm in diameter, stigma persistent; seeds strongly compressed, flat, obcordate, 4 mm long and wide, glossy-brown. Introduced to western gardens in 1908.

The local names of the species show the observations of the people in regard to the habit (vine-like with thorns), taste of the fruit (sour), and the millet-like small flower-buds.

Ziziphus jujuba Miller — Jujube, Chinese Date (Syn. *Z. vulgaris* Lamarck).

Zao=Tzao (棗, Jujube); *Hong-zao=Hung-tzao* (紅棗, Red Jujube), *Hei-zao=Hei-tsao* (黑棗, Black Jujube); *Bei-jing-mi-zao=Pei-ching-mi-tzao* (北京蜜棗, Peking Honey Jujube); *Wu-he-hong-zao=Wu-ho-hung-tzao* (無核紅棗, Pitted Sugared Jujube).

Spinescent trees up to 10 m high, root suckers often giving young plants a shrubby appearance, branchlets glabrous, glaucous, of two types: spiny vegetative long shoots woody, 4–5 mm in diameter, often zigzag, with paired spines, one straight and stout, pointing forward and the other shorter, curved backward, the slender spineless flowering shoots usually emerged from spurs, 10–12 cm long, 1 mm in diameter, deciduous; leaves ovate; 1.5–5 (–7) cm long, 1–2.5 (3.5) cm wide, obtuse, base slightly unequal-rounded, trinervous, crenulate, stipules on long vegetative shoots modified into spines, those on the short deciduous shoots callose or subulate, 1.5 mm long, petioles 0.3–2 cm long, the longest near the basal portion of the shoot, progressively reduced in length toward the distal end; flowers pale yellow, small, 2–5 in axillary cymes on peduncles shorter than the petiole of the subtending leaf, flower-buds subglobose, slightly depressed, pentagonous, 2 mm across, calyx rotate, 5-lobed, the lobes triangular, 1.5 mm long, acute, petals shorter, obovate, clawed, folding around the filament, disk annular, anthers cordate, exerted, ovary very small, globose, styles 2, stigma on the inner surface; fruits drupaceous, oblong, truncate-rounded at both ends, sizes variable, 1.5–3 (–5) cm long, at maturity the excocarp changing from green to creamy yellow, the mesocarp fleshy and slightly sweet, after ripening and drying changing the skin maroon; stone ellipsoid or obovoid, sharply pointed at both ends, the more so at the distal end, protruding into a stiff spine 3–6 mm long, the sizes variable according to the sizes of the fruit, 1.2–2.5 cm long, 5–12 mm across the middle, rugose, irregularly ridged, pitted, and tuberculate, without evident germination pores; seeds usually 1, obliquely compressed ovoid, 9–12 mm long, 5–6 mm wide, 4 mm thick, testa brown or chestnut, smooth, cotylendons completely surrounded by waxy and oily endosperm.

A native of northern China where it is extensively cultivated by the house.

The fruit ripens in the fall and is consumed fresh locally. For cash, rural families sold their products to collectors for preserves ciruclated nationally and internationally. In American Chinese stores, four types of preserved Chinese jujube are available: two shriveled with intack skin, a dull maroon *hong-zao* (Red Jujube) and a black *hei-zao* (Black Jujube), and two scored skin sugar preserves, the *mi-zao* (Honey Jujube) and the seedless *Wu-he-zao* (Pitted Jujube). From the materials bought in Boston stores, it is evident that the producers of these products have selected the larger drupes for the sugared materials with the average size being 3.2 cm long, 2.2 cm in diameter, containing stones 2.2–2.5 cm long, 7–12 cm across, each bearing on seed. The smaller fruits are sun dried for *hong-zao* or parboiled and then dried for *hei-zao*. These products are much smaller than the sugared material, with an average dimension of 2.4 cm long, 1.6 cm across, and stones 10–16 mm long, 4–6 mm across, containing no seed. In preparing the sugared products, the excoarp of the drupes are shallowly scored 40–50 times by hand holding small bundle of needles set 1–2 mm apart, before the application of sugar in solution. (See Part I for more information).

Ziziphus mauritiana Lamarck — INDIAN JUJUBE (*Z. jujuba* [L.] Lamarck, non Miller). *Tian-ci-zao=T'ien-tz'u tsao* (滇刺棗, Yunnan Prickly Jujube). Fruit.

Evergreen small trees 3–6 m high, branchlets yellow-villose, 3–4 mm in diameter; with stout paired spines 3–5 mm long, the longer one straight and pointing forward, the short one recurved; leaves ovate-suborbicular, 3–8 cm long, 2–4 cm wide, obtuse or rounded and erose, base unequally rotundate, trinerved, crenulate-denticulate, glabrous above, densely golden-tomentose beneath; flowers greenish-yellow, 7–35 in an axillary cymose clusters, peduncles, pedicels, and flower buds yellow-tomentose, sepals rotate, 5-lobed, the lobes deltoid, 1.8 mm long, petals obovate, incurved and enclosing the filaments, anthers exserted, disk annular and lobed, ovary glabrous, style 2-cleft; drupes subglobose-oblong, pale yellow at maturity, fleshy, crisp, not sweet, wild forms 1 cm in diameter, the Hong Kong market material imported from Thailand 3–4.5 cm long, 2–3 cm across the middle.

Widely distributed in Southeast Asia and India; occurring in Hainan Island, Guangdong and Yunnan; fruit used locally.

Vitaceae: Grape Family

Parthenocissus tricuspidata (Siebold et Zuccarini) Planchon — BOSTON IVY, JAPANESE IVY

Di-jin=Ti-chin (地錦, Ground Tapestry); *Pa-qiang-hu=P'a-ch'iang-hu* (爬牆虎, Climbing Wall Tiger). Stem; after leaves fall, chew as sugarcane; used in Taiwan, reported before 1950.

Large lianas climbing over cliffs or walls in cultivation, branchlets rough with short internodes and branched tendrils ending with discoid adhesive tips; leaves variable, cordate in outline, 10–20 cm long and wide, trilobed with acuminate and coarsely serrate lobes, or on young plants and basal shoots 3-foliolate with petiolulate leaflets, the middle one elliptic or obovate elliptic, the lateral two obliquely oblong with unequal sides, or on much branched growth produced in late season, ovate and coarsely serrate, glabrous, with few strigose hairs on the veins beneath; flowers green, perfect, in compound cymose clusters opposite the leaves, calyx minute, collar-like, petals 5, 3–5 mm long, spreading at first, then recurved, the apical three-fifths incurved along the margin and hooded at the apex, stamens 5, filaments suberect, anthers dorsifixed, ovary conical, the basal portion yellow and 5-lobed, 2-locular, locules 2-ovulate, style short and stout, stigma punctiform; berries bluish-black, bloomy, 6–8 mm across.

Native of central China with the range extending eastward to Japan; introduced to Western gardens in 1862, very common and naturalized in New England; fruits eaten by birds, scattering the species by buildings.

Vitis amurensis Ruprecht

Shan-pu-tao=Shan-p'u-t'ao (山葡萄, Mountain Grape). Fruit.

Deciduous lianas climbing up to 15 m high, branchlets glabrescent; leaves broadly ovate-orbicular in outline, 4–17 cm long, 3.5–18 cm wide, 3- or 5-lobed with rounded sinuses, the middle lobe broad-ovate, shortly acuminate, apiculate, dentate, green and pubescent on the veins beneath; flowers small, dioecious, in panicles 8–13 cm long, opposite the leaves, rachis and secondary axis sericeous, staminate flowers with calyx 1 mm across, petals spreading, 5 mm across, filaments filiform, anthers basifixed, pistillode discoid, lobed, pistillate flower buds oblong, 2 mm long, the petals falling off together, staminodes 5, ovary globose, 1 mm in diameter, style 1 mm long, stigma punctiform; fruits subglobose, 8 mm in diameter, black; seeds 2 or 3, glossy-brown, the two-seeded ones hemispherical and pointed, the plane side with two parallel depressions, the rounded side with a central disk leading by a vein three-fourth of the vertical surface to the hilium at the pointed side; occurring in northern China, introduced into Western gardens about 1854.

Vitis balanseana Planchon

Xiao-guo-ye-pu-tao=Hsiao-kuo-yeh-p'u-t'ao (小果野葡萄, Small-fruited Wild Grape). Fruit.

Scandent shrubs climbing over thickets, young growth arachnoid, covered by cobweb-like whitish hairs, soon becoming glabrous, branchlets striate-sulcate, 3–4 mm in diameter; leaves cordate-suborbicular in outline, 3.5–9 cm long, 3–9 cm wide, remotely crenulate, the teeth callose, acute, base broadly sinuate-cordate, pentanerved; flowers dioecious, very small, in panicles 5.5–13 cm long, calyx discoid, petals 1 mm long, falling off together, stamens of staminate flowers 5, filaments filiform, pistillodes globose, 0.5 mm in diameter, ovary of pistillate flowers very small, ovoid, 1 mm long including the short style, stigma discoid; berries pea-sized, 4–5 mm in diameter; seeds 2 or 3, one side rounded, the other side in conformity with the number of seeds, plane or angular, with two parallel depressions, the opposite side rounded, with an oblong center in a vertical groove, hilum punctiform, testa shiny-castaneous; occurring in tropical southern China, Hainan Island, Guangdong and Guangxi.

Vitis betulifolia Diels et Gilg — Sweet Wild Grape

 Hua-ye-pu-tao=Hua-yeh-p'u-t'ao (樺葉萄萄, Birch-leaved Grape). Fresh fruit; collected and eaten in the field.

Deciduous lianas, branchlets arachnoid, glabrescent; leaves oblique-ovate or broadly ovate, 5–10 cm long, 4–9 cm across, acuminate, base shallowly cordate or subtruncate-cordate, dentate-serrulate, subglabrous above, velutinous beneath, petioles 3–5 cm long; flowers small, yellowish-green, in arachnoid panicles 5.5–7 cm long; fruits black, glaucous, globose, 7–12 mm long, sweet. Wild in deciduous forests, growing in altitudes of 1,500–2,700 m of central and western China and Yunnan.

Vitis davidii (Romanet) Foëx

 Ci-pu-tao=T'zu-p'u-t'ao (刺葡萄, Spiny Grape). Fruit.

Deciduous lianas, branchlets prickly, the prickles straight or recurved, 2–4 mm long, tendrils branched; leaves broadly ovate or suborbicular, 5–15 cm long, 6.5–14 cm wide, occasionally inconspicuously 3-lobed, apex acuminate, dentate, pilose along the nerves beneath; flowers small, in panicles 5–15 cm long, petals connate at apex, caducous; berries bluish-purple, 11.5 mm in diameter; a species widely distributed in central China, along the Yangtze River and hence southward to Fujian.

Vitis ficifolia Bunge

 Sang-ye-pu-tao=Sang-eh-p'u-t'ao (桑葉葡萄, Mulberry-leaf Grape). Fruit.

Deciduous lianas, branchlets arachnoid, glabrescent with age, tendrils 20 cm long, branched; leaves ovate, 11–25 cm long, 7–13 cm wide, 3-lobed, acute, base cordate, coarsely serrate, tomentose beneath, petioles 4.5–12 cm long; flowers small, in axillary panicles 10–15 cm long, calyx discoid, petals 2 mm long; berries globose, 6–9 mm in diameter, blue-black, edible. Native to central, eastern and northern China, growing in thickets at altitudes of 1,200–1,700 m.

Vitis flexuosa Thunberg

Ge-liu=Ko-liu (葛藟, White-leaved Bramble). Fruit.

A deciduous liana, branchlets slender, villose when young; leaves deltoid-ovate, 3.5–11 cm long, 2.5–9.5 cm wide, acute or shortly acuminate, base subcordate or truncate, unequally dentate, pilose beneath; flowers small, 2 mm across, panicles 6–12 cm long, axis sericeous; berries 6–8 mm across, purple-black. Native to the warm areas of China, from the Yangtze River provinces hence southward to Guangdong and Yunnan; introduced to Western gardens in 1900.

Vitis labrusca L. — FOX GRAPE, CONCORD GRAPE

Hu-pu-tao=Hu-p'u-t'ao (狐葡萄, Fox Grape, a translation of the English common name). Fruits, sweet and with good flavor, cultivated in northern China.

High-climbing, deciduous, coarse lianas, young branches and leaves densely white-rusty-tomentose, branchlets bearing tendrils at each node; leaves thick, quadrate-orbicular or deltoid-ovate, 7–16 cm long, 5–15 cm wide, occasionally 3-lobed, coarsely dentate, teeth mucronate, rusty-tomentose beneath; flowers small, greenish-yellow, in axillary panicles 5–10 cm long, calyx discoid, petals 5, valvate, stamens 5, opposite the petals, anthers ellipsoid, ovary superior; berries globose, 1.5–2 cm in diameter, purple-black, or amber-white, juicy and with very pleasant flavor; seeds 5–8 mm long, 4–6 mm across. Native to eastern North America, introduced to northern China, rarely cultivated.

Vitis piasezkii Maximowicz

Fu-ye-pu-tao=Fu-yeh-p'u-t'ao (複葉葡萄, Compound-leaved Grape). Fruit; gathered and consumed locally.

Deciduous lianas, branchlets and petioles rufous-pubescent when young, glandularly bristly; leaves variable even on the same branch, ovate-cordate in outline, simple, slightly or deeply lobed, or palmately compound with 3 or 5 leaflets, 4–9 cm long, 3–8 cm across, when compound the middle leaflets petiolulate, rhombic-ovate, 5–12 cm long, acute or short-acuminate, cuneate, the lateral ones sessile, strongly oblique, smaller, coarsely and angularly dentate, floccose beneath, glabrous above; flowers small, in panicles 10–15 cm long and across; berries 1 cm across, black, bloomy. Native of northwestern China, drought resistant, introduced to Western gardens in 1807.

Vitis quinquangularis Rehder

Mao-pu-tao=Mao-p'u-t'ao (毛葡萄, Hairy Grape). Fruit.

Deciduous lianas up to 8 m high, new growth white-maroon arachoid; leaves ovate, 10–15 cm long, 6–8 cm wide, occasionally 3- or 5-lobed, apex acute, base subcordate, or truncate, serrate, densely velutinous beneath; flowers greenish yellow, in panicles 8–11 cm long; berries 6–8 mm in diameter, purple-black. Native of and widely distributed in China, from the Yangtze River provinces southwestward to Guangxi and Yunnan in the south and Sichuan and Gansu in the west; often climbing clifts.

Vitis romanetii Romanet

He-pu-tao=He-p'u-t'ao (黑葡萄, Black Grape). Fruit.

Vigorously growing lianas climbing over thickets in low altitudes, young branchlets floccose-pubescent and glandularly bristly, purplish when young; leaves ovate, shallowly 3- or 5-lobed, 9–20 cm long, 8–14 cm wide, dentate, base cordate; flowers greenish-yellow, in panicles up to 25 cm long; berries globose, 1 cm across, black-purple. A native of the Tsinling Range, which extends from the borders of Gansu, and Shaanxi, Sichuan eastward to northern Hubei, western Henan and Jiangsu; introduced into Western gardens in 1881.

Vitis vinifera L. — EUROPEAN GRAPE, WINE GRAPE

Pu-tao=P'u-t'ao (葡萄, Grape). Fruit; fresh, dried, or converted into wine; extensively cultivated now.

Deciduous lianas, introduced and cultivated in China; branchlets glabrous or floccose when young; leaves suborbicular, 7–15 cm across, cordate, 3- or 5-lobed with rounded sinuses, the lobes coarsely dentate, glabrous or floccose beneath; flowers greenish-yellow, in large panicles; berries usually ellipsoid, or subglobose, black and bloomy to red and greenish. Probably native of the Caucasian Region, cultivated in Europe and western Asia, introduced to China from Central Asia about A.D. 200, production was limited because formerly China did not use grape for wine (rice and various other grains were the source of alcoholic fermentation).

Elaeocarpaceae: Elaeocarpus Family

Elaeocarpus hainanensis Oliver

Shui-shi-rong=Shui-shih-jung (水石榕, Water Rock Banyan); *Shui-liu-shu=Shui-liu-shu* (水柳樹, Water Willow). Fruits of the size of Canarium, eaten in Hainan Island.

Small trees, branchlets glabrous; leaves crowded at the distal end of the shoots, linear-lanceolate, 7–15 cm long, 1.4–2.8 cm wide, acute, base attenuate, serrulate, petioles 1–2 cm long, flowers white, 2–6 in axillary racemes, bracts persistent, ovate, 8–14 mm long, pilose, sepals 4 or 5, lanceolate, 1.8–2.4 mm long, pilose, petals obovate, base acute, deeply divided, the segments filiform, disk annular, hairy, stamens numerous, anthers awned, ovary glabrous, 2-locular, cells 2-ovulate; drupes ellipsoid, 2–3 cm long, 8–12 mm across, glabrous. Endemic to Hainan Island, common in forests of steep valleys; cultivated in Hong Kong, attractive in the landscape of the Chinese University of Hong Kong.

Elaeocarpus serratus L. — Ceylon Olive

Xi-lan-gan-lan=Hsi-lan-kan-lan (錫蘭橄欖, Ceylon Olive). fruit; pickled, used as a substitute of olive.

Trees 11–14 m high, branches spreading with upright suckers, branchlets glabrescent, with prominent lenticels and vertical wrinkles; leaves simple, obovate-elliptic, 6–14 cm long, 2.5–5 cm wide, apex obtuse, abruptly short-acuminate, or rounded, base cuneate and obtuse, crenulate-serrate, glabrous above, with large glands at the vein-angles beneath; flowers white, 1.2–1.4 cm across, in axillary racemes 5–9 cm long, sepals green, lanceolate, 5 mm long, petals obovate, lobed and fringed, stamens many, anthers opening by apical pores, the abaxial lobe of the apex white-setose, ovary ovoid, style columnar, pubescent at the lower half, stigma punctiform; drupes oblong-ellipsoid, 2–3 cm long, 1.5–2 cm across, stone ellipsoid, 1.5–2.5 cm long, 1–1.2 cm across, bony, extremely rugose and warty, base rounded, apex obtuse and shallowly 3-clift; seed 1. A native of Sri Lanka, extensively cultivated there as a garden plant; introduced to Taiwan before 1950.

Sloanea hemsleyana (Ito) Rehder et Wilson

Fang-li=Fang-li (仿栗, Like-Chestnut). Oil; extracted from the seed, used for cooking in western Hubei.

Evergreen trees 15 m high, branchlets glabrous; leaves ovate-oblong, 9–22 cm long, 3–7.4 cm wide, acuminate, base obtuse, serrate, petioles 1.5–3 cm long; flowers greenish-white, 1 cm across, 5–13 emerged from terminal buds in mid-summer, densely velutinous throughout, pedicels 3–8.5 cm long, sepals 4, ovate, 6 mm long, distal one-fifth of the petals shallowly and irregularly divided, stamens numerous, hairy, anthers oblong, introrse, the connective long-produced at the apex, ovary ovoid, tetragonous, densely velutinous, 3 mm long, style columnar, 4-ridged, pilose, stigma punctiform; fruit woody, globose, 2.5–3 cm long, prickly, dehiscing by 4 or 5 valves, the bristles 5–

12 mm long; seeds 2–4, oblong, arillate. Native to central and western China, growing in forests in deep valleys at altitudes 500–1,000 m.

Tiliaceae: Linden Family

Microcos nervosa (Loureiro) S. Y. Hu (Syn. *M. paniculata* auctt., non L.)

Bu-zha-ye=Pu-cha-yeh (布渣葉, Cloth-refuse Leaf); *Po-bu-ye=P'o-pu-yeh* (破布葉, Rag Leaf). Leafy branchlets; used for tea in southern China; dried leaves tied in bundles available in the tea section of Chinese groceries in America.

Small evergreen trees 4–10 m high, often shrubby due to repeated cuttings, branchlets and leaves pubescent with stellate hairs, leaves rough to touch, ovate-oblong, 10–15 cm long, 4–8 cm wide, acuminate, base rounded and trinerved, serrate, stipules lanceolate; flowers greenish-yellow, small, 2–3 subtended by large bracts in terminal panicles, stellate-hairy, sepals 5, oblong, 5 mm long, petals 5 or wanting, oblong, 2–3 mm long, with a gland at the base, stamens numerous, on a short receptacle, ovary 3-carpellate, cells 4-or 6-ovulate, style conical; fruit a nut, obovate, 7 mm long, the pericarp fibrous. Native to the hillsides of southern China.

Malvaceae: Mallow Family

Abelmoschus esculentus (L.) Moench — OKRA, GUMBO, GOMBO (Syn. *Hibiscus esculentus* L.)

Ka-fei-kui=Ka-fei-kuei (咖啡葵, Coffee Mallow). Young fruits; mature seeds, roasted and used as a coffee substitute.

Annual hispid herbs 1–2 m high, branchlets hispid and sparsely bristly; leaves cordate in outline, 15–30 cm in diameter, 3- to 7-lobed, the lobes serrate; flowers yellow with purple centers, showy, solitary, axillary to normal leaves, sepals, petals, and staminal column united at base; capsules cylindrical, 10–20 cm long, 1–2 cm across, ridged and furrowed, constricted above the base, apex attenuated into a beak; seeds numerous, globose, 3–4 mm in diameter, white young.

Abelmoschus manihot (L.) Medikus. (Syn. *Hibiscus manihot* L.; for more synonyms, see Hu, 1955)

Qiu-kui=Ch'iu-kuei (秋葵, Autumn Mallow); *Huang-shu-kui= Huang-shu-kuei* (黃蜀葵, Yellow Sichuan Mallow). Flowers used in soup, put into the boiling soup just before serving; root boiled with pork to make a broth.

Perennial herbs 1–2 m high, sparsely hispid; leaves palmately 5- to 9- parted, the segments oblong-lanceolate, 8–18 cm long, 1–6 cm wide, hispid; flowers showy, yellow with purple eye, calyx spathaceous, 5-lobed, deciduous; capsules oblong-conical, 4–5 cm long, 2 cm across the base, the persistent calyx lanceolate, 5 mm across the base; seeds numerous, reniform, 3 mm long, 2.5 mm across the broad end, smooth, under lens with minute vertical rows of glandular tubercles.

Gossypium herbaceum L. — Levant Cotton

Mian-you=Mien-yiu (棉油, Cotton-seed Oil); *Cao-mian=Ts'ao-mien* (草棉, Herbaceous Cotton). Seeds for extrating cooking oil; roasted, pulverized, oil extracted after applying hot water. (I have observed the procedure and eaten bread prepared with cotton oil; perhaps this is the first report of farmers eating cotton oil as a substitute of sesame oil in the cotton-producing areas of northern China.

Low annual herbs 0.5–1 m high, tomentulose throughout; leaves subcordate, 5-lobed, divided above the middle, the lobes ovate, acute, stellate-hirsute above, tomentulose beneath, sparsely villose on the nerves, petioles hirsute-villose, stipules linear, tardily caducous; flowers solitary, pedicels 1.5–2 cm long, hirsute-villose, involucral bract ovate, united at base, subentire or 3-toothed, 1.5–2 cm long and wide, calyx truncate, petals twice as long as the bracts; capsules ovoid, beaked; seeds oblique-ovoid, fuzzy, lint white. Not known in a wild state, probably indigenous to northern Arabia or northern Africa, widely cultivated in northern China.

Note: Growing up in a cotton growing village in northern Jiangsu, I have observed that farmers roasted and pulverized cotton seed, and applied hot water to extract the oil. The top grade oil was used as a substitute of sesame oil in cooking, and the remaining portion for lubrication of farming implements, especially the ox-cart. The amount used was small, and any toxic effect of which was never heard.

Gossypium hirsutum L. — Upland Cotton

Mei-mian=Mei-mian (美棉, American Cotton); *Da-lu-mian=Ta-lu-mien* (大陸棉, Continental Cotton). Seeds for oil.

Annual herbs up to 1.5 m high, branchlets villose, the hairs tufted; leaves broad-cordate in outline, half-way 3-lobed, the lobes broad-triangularly ovate, abruptly acuminate, glabrescent, punctate and villose along the nerves above, sparsely villose beneath, the hairs tufted, petioles hirsute, stipules ovate-falcate; flowers solitary, pedicels 1.5–2 cm long, hirsute, involucral bracts free at base, 7- to 9-toothed, 4 cm long, 2.5 cm wide, calyx 5-toothed, the teeth deltoid, acute or acuminate, ciliate, petals twice as long as the bracts; capsules ovoid, beaked, 4- or 3-locular; seeds fuzzy, the lint white, firmly adherent to the seed. Native of central America, introduced to China in the 1920s, much cultivated now.

Hibiscus sabdariffa L. — ROSELLE, RED SORREL, JAMAICA SORREL

Mei-gui-qie=Mei-kuei-ch'ieh (玫瑰茄, Rose Eggplant), *Shan-qie=Shan-chieh* (山茄, Hillside Eggplant). Immature fruits; picked before any portion becoming woody, with the fleshy red calyx, used for making jams, sauces, and acid drinks; dried material in packages (370 g) importd from Taiwan, available in American Chinese groceries.

Purplish-maroon suffrutescent shrubs 1–2 m high, with arching branches, branchlets subterete, glabrous; leaves alternate, variable in shape and size, the lower ones ovate, undivided, gradually changing in upper part of the stem to palmately 3- or 5-lobed or parted, 6–16 cm long, 3–15 cm wide, segments lanceolate or elliptic, 1.5–12 cm long, 0.5–5 cm wide, acute or acuminate, serrete, the teeth callose, often red, petioles 3–14.5 cm long, strigose-pilose along a narrow groove above, stipules linear-subulate, 8 mm long; flowers yellow with dark crimson eyes, solitary, axillary to normal or reduced leaves, sometimes giving the appearance of a raceme, flower buds maroon-crimson, strigose, connate at the base, pedicels 4–5 mm long, tomentose, epicalyx 8, lanceolate, 10–12 mm long, 0.5–2 mm wide, strigose-ciliate, sepals 5, ovate-lanceolate, brilliant purple-red, with 3 dark veins, tomentose on the inside, thickened and enlarged immediately after anthesis, 3.5 cm long, petals 5 cm long, stamens monadelphous, pistil 5-carpellate, pilose; capsules subglobose-ovoid, 2 cm long, 1.5 cm across, strigose, loculicidally dehiscent; seeds many, grayish-brown, trigonous-reniform, 5 mm long, 4.5 mm across, rugose, tuberculate, scabrid-hairy. A native of tropical Africa, extensively cultivated as an annual crop in the tropics, and naturalized in many warm countries.

Hibiscus schizopetalus (M. T. Master) J. D. Hooker — LANTERN FLOWER, CORAL HIBISCUS

Diao-deng-hua=Tiao-teng-hua (吊燈花, Hanging Lantern Flower); *Deng-long-hua=Teng-lung-hua* (燈籠花, Lantern Flower); *Lie-ban-zhu-jin=Lieh-pan-chu-chin* (裂瓣朱槿, Schizopetalous Red Hibiscus). Flowers, used in Taiwan in food, reported before 1950.

Glabrous evergreen shrubs 3 m high, branchlets slender; leaves elliptic or oblong, 4–7 cm long, 1.5–4 cm wide, acute or shortly acuminate, base obtuse, basal half entire, distal half serrate; flowers scarlet, solitary, pendulous, pedicels slender, 10–14 cm long, articulate at the middle, epicalyx 8, very small, 1 mm long, ciliate; calyx spathulate, tubular, 1.5 cm long, shallowly 5-toothed, petals 4–5 cm long, reflexed, laciniate, staminal tube 9–10 cm long; capsules oblong-cylindrical, 4 cm long, 1 cm in diameter; seeds smooth; native of east tropical Africa, cultivated in southern China, seldom fruit.

Hibiscus syriacus L. — Rose-of-Sharon

Mu-jin=Mu-chin (木槿, Tree Hibiscus). Flowers; used for soup.

Deciduous shrubs 2–4 m high, branchlets light gray; leaves ovate-rhombic, 5 cm long, 2–4 cm wide, often 3-lobed, base cuneate, irregularly dentate; flowers white, lilac, purple-red, solitary, axillary, involucre bracts 6 or 7, linear, 7–9 mm long, very sparsely stellate-pilose, calyx campanulate, 1.4–1.7 cm long, densely short stellate-tomentose, 5-toothed, the lobes triangular, 6–8 mm long, corolla campanulate, petals obovate, 3.5–4.5 cm long, ciliate and stellately villose outside, staminal tube 2.5–3 cm long, antheriferous to the base; capsules oblong-ellipsoid, densely golden stellate-tomentose, 1.5–2 cm long, 1.3 cm in diameter, rostrate; seeds glabrous, with a row of white hairs. Said to be native of China; however, Chinese specimens are all from cultivated sources.

Hibiscus tiliaceus L. — Beach Hibiscus, Cuban Bast

Huang-jin=Huang-chin (黃槿, Yellow Hibiscus); *Huang-mu-jin=Huang-mu-chin* (黃木槿, Yellow Rose-of-Sharon); *Tong-ma=T'ung-ma* (桐麻, Copper Hemp); *Gang-ma=Kang-ma* (港麻, Harbor Hemp); *Hai-ma=Hai-ma* (海麻, Seaside Hemp). Roots and tender shoots, eaten by the fishing population living along the shores of the South China Sea.

Small trees 10 m high, often shrub-like due to repeated cutting, stellate-pubescent throughout; leaves cordate or suborbicular, 7–15 cm long and across, acute, rarely rounded, entire, grayish-white beneath, stipules oblong, 2–3 cm long, 1.2–1.5 cm across the base, falling long before the leaves; flowers yellow with a bright purple-red center, showy, 6–8 cm across, subtended by two stipule-like basal bracts, pedicels, hypanthia and calyx grayish-stellate-tomentose, sepals 5, lanceolate, 10–14 mm long, epicalyx 10, deltoid, 3–4 mm long, corolla funnelform, petals 5 cm long, stamens monadelphous, anthers unilocular, basifixed, dehiscing horizontally, ovary hairy, style red, glandular; capsules hairy, loculicidal-dehiscent; seeds numerous, dark brown, reniform, 4 mm long, smooth, with glandular lines.

Hibiscus trionum L. — Flower-of-an-Hour

Ye-xi-gua-miao=Yeh-hsi-kua-miao (野西瓜苗, Wild Watermelon Seedling); *Xiang-ling-cao=Hsiang-ling-ts'ao* (香鈴草, Aromatic Bell Herb). Seed; gathered by farmers working in the field and eaten immediately for the sesame-flavor; not available in the market.

Erect annual herbs 20–60 cm high, verrucosely pubescent with stellate hairs through-out; leaves ovate in outline, digitately parted, segments obovate-oblong, pinnatisect, with broad sinuses, punctate above; flowers yellow with dark purple eye, solitary, on pedicels 2.5 cm long, calyx campanulate, 1 cm long, green, with 10 purple veins, inflated and membranaceous in fruit, epicalyx linear, 1 cm long, petals funnelform, 2 cm long, stamens numerous, monadelphous, anthers reniform, ovary hairy; capsules oblong-subglobose, enclosed in the inflated, strigose-hispid persistent calyx; seeds reniform, black, glandular-papillose, rugose, having the taste of sesame when chewed raw. Widespread in temperate China, especially in cotton or soybean fields; the Chinese vernacular names referring to the similarity to the aromatic taste of the seed, and of the leaves to those of young watermelon, and the bell-shaped persistent calyx.

Malva neglecta Wallroth — COMMON MALLOW, CHEESES

Ye-kui-cai=Yeh-kuei-ts'ai (野葵菜, Wild Mallow Vegetable). Leaves and young shoots, gathered and used locally, not available in the market; *Mu-mo-zi=Mu-mo-tzu* (木磨子, Little Millstone). Young fruit; children gather the young fruits and eat for fun.

Small perennial herbs sparsely stellate-pubescent, with strong and deeply penetrating tap root reaching 20 cm deep in the soil, the basal portion of stem woody, branchlets decumbent; leaves suborbicular-reniform, 1–3 cm long, 1–4 cm wide, crenate-denticulate, shallowly 5- or 7-lobed, petioles long, 3–12 cm long; flowers small, white with 3 or 5 purple-pink veins, 1.5 cm across, calyx campanulate, persistent, petals 5–6 mm long, staminal tube pubescent, anthers white, styles 12–15, purple-pink; fruits discoid, 5–8 mm across, at maturity separating into 12–15 round-reniform indehiscent schizocarps.

Malva verticillata L. — WINTER MALLOW (Syn. *M. crispa* auctt., non L.)

Dong-xian-cai=Tung-hsien-ts'ai (冬苋菜, Winter Amaranth); *Dong-han-cai=Tung-han-ts'ai* (冬寒菜, Winter Cold Vegetable); *Kui-cai=Kuei-ts'ai* (葵菜, Mallow Green). Leaves; cultivated for the leaves, harvested continuously in winter, the market material consisting of bundles of leaves, only the leafblades used in cooking, usually with bean curd in soup or by quick stir-fry, a popular winter potherb in Chengdu, West China.

Annual or biennial herbs, 0.5–1 m high at anthesis, branchlets stellate-villose; basal leaves reniform, cauline leaves 5-lobed, the lobes of the lower leaves rounded, those of the upper ones triangular, 5–11 cm wide, crenulate-dentate, petioles 5–8 cm long, glabrescent; flowers subsessile, in compact fascicles, epicalyx, linear-lanceolate, 5–6 mm long, calyx slightly inflated, sparsely stellate-hirsute, staminal tube setose above; fruits discoid, 5–7 mm in diameter, schizocarps 10–11, back smooth, rugose along the rounded angles, palmately striate on the sides. A native of China, extensively cultivated, particularly in Sichuan.

Note: *Malva crispa* L. has been used by authors of Chinese vegetables. This species is a native of the Scandinavian Peninsula. It has crisped ovate-orbicular leaves finely double- or triple-serrulate, with every 2 mm of the margin consisting of 3–5 teeth. Its carpels are reticulate on the back. Although some Chinese specimens have crisped leaves, they are reniform with crenate-dentate margin, every 9–13 mm consisting of 2–3 teeth. Their carpels are smooth on the back. There is no material evidence that genuine *M. crispa* L. exsists in China.

Thespesia lampas (Cavanilles) Dalzell ex Dalzell et Gibson — HAIRY PORTIA TREE
   Xiao-jin=Hsiao-chin (肖槿, Mimetic Mallow); *Bai-jiao-tong=Pai-chiao-t'ung* (白腳桐, White-foot Tong). Young shoots and flowers.

Shrubs or small trees 1–3 m high, branchlets ferruginously stellate-velutinous; leaves ovate, 8–13 cm long, 6–13 cm wide across the base, tricuspidate, rounded, subcordate and 5- or 7-nerved at base; flowers yellow with red center, in axillary cymes or solitary, epicalyx bracts 5, subulate, calyx cyathiform, truncate, 5-toothed, corolla campanulate, petals 6.5 cm long, connate at base; fruits ovoid, 2.5 cm long, slightly pentagonous; seeds ovoid, black, glabrous, papillose, with a halo of brown hairs around the hilum. Native of Old World tropics, extending from the Philippines to East Africa, in China occurring in Hainan Island and the muddy seashores of Hong Kong.

Bombaceae: Bombax Family

Adansonia digitata L. — AFRICAN BAOBAB
   Hou-mian-bao-shu=Hou-mien-pao-shu (猴麵包樹, Monkey Bread Tree). Leaves used for vegetable as spinach; fruit pulp chewed, sucked, or for making a lemonade-like drink; seed cooked, tastes like almond.

Deciduous trees 30 m high, with short and stout trunks and many branches; leaves palmately compound, crowded at the end of branches, leaflets 5–7, oblong elliptic, or oblong-obovate, 5–16 cm long, 4–6 cm wide, acuminate or acute, entire, sparsely stellate-pubescent, petioles 10–20 cm long; flowers white, opening before leaves, solitary, axillary, pendulous on long peduncles, calyx 5-lobed, petals 5, obovate, 10 cm long, spreading, reflexed, stamens numerous, the basal half of the filaments connate, ovary superior; fruits oblong-cylindrical, pendulous, indehiscent, 10–30 cm long, with a hard woody shell and mucilaginous yellowish-white pulp; seeds many, embedded in rows in the pulp, attached to fibrous strands. Native to tropical Africa, introduced to the Tropical Botanical Garden in Yunnan.

Bombax ceiba L. — RED SILK-COTTON TREE; TREE COTTON (Syn. *Gossampinus malabarica* [de Candolle] Merrill; *Bombax malabaricum* de Candolle)

Mu-mian=Mu-mien (木棉, Tree Cotton); *Pan-zhi-hua=P'an-chih-hua* (攀枝花, Hold-on-branch Flower). Flower; after anthesis, flowers dropping on the ground and turning black, picked up, dried in the sun, sold to dealers and used as one of the five ingredients of *Wu-hua-cha=Wu-hua-ch'a* (五花茶, Five Flowers Tea); much used in hot and humid summer season in Guangzhou and Hong Kong; mixed material imported from Hong Kong available in American Chinese groceries.

Tall deciduous trees, trunks with pyramidal spines, branchlets stout, 8–10 mm across; leaves palmately compound, leaflets 5–7, petiolulate, elliptic or oblong, 10–20 cm long, 5–7 cm wide, acuminate, base obtuse, entire; flowers large, brilliant red, showy, blooming before the leaves, calyx cupular, 3.5–4.5 cm long, sericeous inside, petals fleshy, oblong, 8–10 cm long, stellate, stamens numerous, several seriate, the innermost ones 5 branched, each branch bearing 1 anther, intermediate 10 shorter, outermost series in 5 bundles, ovary 5-carpellate; capsules oblong-ellipsoid, 10–15 cm long, 4.5–5 cm wide, pericarp woody, loculicidally dehiscent, filled with brownish cotton; seeds numerous, covered with cotton. Native of southeastern Asia, introduced to southern China for landscape purposes; flowers picked up locally, dried, available in the market.

Durio zibethinus J. Murray — DURIAN

Liu-lian=Liu-lien (榴槤, Durian); *Ci-shao-zi=Tz'u-shao-tzu* (刺韶子, Prickly Elegant Fruit). Fruit; imported into the Hong Kong market and to Taiwan.

Medium-sized ramiflorous trees 20–40 m high, trunks columnar, bark brown, rough, flanky, branchlets densely covered with reddish-yellow scales; leaves simple, oblong, 5–25 cm long, 2.5–9 cm wide, caudate-acuminate, base rounded, densely covered with golden scales beneath; flowers rather large, showy, yellowish-white, 3–30 in pendulous, densely scaly, cymose clusters, epicalyx splitting irregularly into 2–4 segments and falling off, calyx campanulate, ventricose and saccate at base, 2–3 cm long, deciduous, petals 4–5, oblong-spathulate, with claw, 1–2 cm long, 3–5 cm long, 1.5–4 cm wide, stamens numerous, pentadelphous, anthers reniform, ovary ovoid-oblong, hairy and scaly, 4- or 5-locular, style 3–4.5 cm long, stigma capitate; fruits globose, ovoid-oblong, 15–30 cm long, 13–15 cm in diameter, both ends rounded, 5-valved, fleshy-fibrous, spiny, the spines coarse, pyramidal, 5- or 7-gonous, sharp, densely scaly; seeds 2–6 in each locule, ovoid-oblong, arillate, the aril complete, fleshy, pulpy, white or pale yellow, strongly smelling. One has to learn to like it, once one becomes used to it, one loves it;

inferior fruits preserved with sugar or cooked with salt; fruit common in Hong Kong market.

Pachira aquatica Aublet — GUIANA CHESTNUT, WILD COCOA TREE (Syn. *P. macrocarpa* [Chamisso et Schlechtendal] Walpers).

Gua-li=Gua-li (瓜栗, Guiana Chestnut). Seed; boiled or roasted.

Trees 5–20 m high, with loose crown and glabrous, brown branchlets; leaves palmately compound, leaflets 5–11, ovate-elliptic or elliptic-oblong, 13–28 cm long, 4.5–8 cm wide, petioles 11–15 cm long; flowers solitary, large, showy, axillary to leaves near the shoot apex, on stalks 2 cm long, stellate pubescent and glabrescent, calyx campanulate-tubular, subtruncate, persistent, with 3–5 basal glands, petals 5, 15–20 cm long, reflexed, outside greenish-yellow, sparsely velvety, inside white, smooth, stamens 200–250, staminal tube rather short, divided into many parcels, each containing 8–10 filaments, yellow below, changing to deep red, anthers oblong-linear, versatile, ovary cylindrical, style longer than the stamens, stigma small, 5-lobed; capsules pyriform or subglobose-ellipsoid, 12–14 cm long, 4–6 cm in diameter, heavy, shell thick, woody, pulp fleshy; seeds numerous, irregular, 2–2.5 cm long, 1–1.5 cm across, dark brown, with white spiral markings. Native of tropical America, introduced to the Tropical Botanical Garden in Yunnan.

Sterculiaceae: Sterculia Family

Firmiana simplex (L.) W. F. Wight — CHINESE PARASOL TREE

Wu-tong=Wu-t'ung (梧桐, Firmiana); *Qing-tong,=Ch'ing-t'ung* (青桐, Green-barked Tong). Seed; roasted and eaten as nuts.

Deciduous trees 10–15 m high, trunk 30–40 cm in diameter, bark smooth, greenish-gray, branchlets stout, close at the end of older branches; leaves ovate-orbicular in outline, 20–30 cm across, the apical half 3- or 5-lobed, base broadly cordate, palmately 7-nerved, stellate pubescent beneath; flowers greenish-yellow, monoecious, apetalous, perianth segments 5, linear, 10 mm long, reflexed, villose, anthers of staminate flowers 10–15, crowded at the apex of a columnar androphore, ovary of the pistillate flower with 5 separate carpels; fruits follicular, the 5 follicles opening long before the seeds reaching maturity, pericarps papery, 7–10 cm long, with 2 or 3 pea-sized seeds on the margin. Native of China, much cultivated, often appearing in Chinese paintings of artists who tried to depict *Paulownia*, the tree on which the male phoenix rests in Chinese mythology.

Sterculia foetida L. — GREAT STERCULIA, KELUPANG, INDIAN ALMOND

Xiang-ping-po=Hsiang-p'ing-p'o (香蘋婆, Aromatic Pingpo). Seed; eaten in Hainan Island, after roasting, having the taste of chestnut.

Trees 30 m high, trunk gray, smooth, peeling off in small pieces, branchlets stout, 1 cm across, crowded at the shoot apex; leaves palmately compound, petioles unequal, the lower ones longer, giving young trees the appearance of an umbrella, leaflets 7–11, elliptic, (9–) 17–28 cm long, (3.5–) 5–8.5 cm wide, acuminate-caudate, the acumen up to 3 cm long, base obtuse; flowers flesh-colored, 3 cm across, appearing with the young leaves, in erect racemose panicles, 8–10 cm long, perianth four-fifth divided, the lobes linear, pubescent, staminal column curved, hairy, anthers 12–15, crowded at the head-like terminal group 4 mm across, ovary pubescent, 5-carpellate; fruits follicular, oblong, oblique, 4–6.5 cm long, 2.5–3.5 cm across, pericarps pubescent, woody-fibrous, striate-reticulate; seeds black, shiny. Widespread in the Old World tropics, occurring in southern Yunnan and Hainan Island of China.

Sterculia lanceolata Cavanilles — SCARLET STERCULIA (Figure 40)

Jia-ping-po=Chia-p'ing-p'o (假蘋婆, False Sterculia); *Qi-jie-guo=Ch'i-chieh-kuo* (七姐果, Seven Sisters Fruit). Seed; after roasting.

Evergreen trees 5–15 m high, branchlets glabrous, 2–4 at the apex of a branch; leaves, unifoliolate, the petiole articulate at the distal end, leaflets elliptic, 7–15 cm long, 2.5–7 cm wide, acuminate, base obtuse, entire; flowers small, apetalous, greenish-pink, polygamous, in slender panicles axillary on leafy shoots, or fasciculate on spurs, staminate flowers blooming first, sepals 5, lanceolate, 6 mm long, reflexed, staminal column filiform, curved at the apex, anthers 10 crowed at the apex of the staminal column forming a ball, pistillate flowers few, terminal to the panicle, ovary stipitate, 4–5-carpellate, ovules 2–5; fruits follicular, the pericarps opening into star-like hanging follicles, brightly yellow or red, iridescent hairy; seeds black, 1–5 to each follicle. Native of southern China, common in Hong Kong, growing in forests along streams; very ornamental with its beautiful hanging fruits.

Sterculia nobilis J. E. Smith — NOBLE BOTTLE TREE, PINGPO

Ping-po=P'ing-p'o (蘋婆, Pingpo); *Feng-yan-guo=Feng-yen-kuo* (鳳眼果, Phoenix Eye Fruit). Seeds, boiled or roasted.

Ping-po-ye=P'ing-p'o-yeh (蘋婆葉, Pingpo Leaf). Leaves; used fresh for wrapping rice mixed with marinated meat to make a steamed dish (the Chinese tamale), very tasty, a popular homemade food of Guangzhou.

Trees 10–30 m high, branchlets sparsely stellate pubescent when young, glabrescent; leaves unifoliolate, closely arranged at the apex of a shoot, petioles slightly thickened at both ends, laminas oblong or obovate-oblong, 8–25 cm long, 5–15 cm wide, abruptly short acuminate or rounded due to mechanical injury, base rounded, entire; flowers pink, apetalous, polygamous, in short panicles, perianth 1 cm across, 5-lobed to the middle, the lobes lanceolate, softly pubescent, staminal column curved, ovary hairy, stipitate; fruits follicular, paricarps scarlet red, inside brownish-red, dried brown, parchment-like, 4–8 cm long, 3–5 cm across, hairy; seeds black, 1–5, subglobose, oblong, 2–3 cm long, 1.5–2 cm across, with pleasant taste when boiled. A native of the tropical forests in Guangdong and southern Yunnan where its local name is *"ping-po"*. The Chinese ideograms represent the sound of the ethnic name. After the pattern of adopting Cantonese local names for English names in *kumquat* for *Fortunella margarita* (Lour.) Swingle, *loquat* for *Eriobotrya japonica* (Thunberg) Lindley, and *longan* for *Dimocarpus longan* Loureiro, *pingpo* is taken for the edible seed of *Sterculia nobilis* J. E. Smith. In southern China, the pericarp of the species is used for a medicated tea, prepared by boiling 3 pingpo husks, 6 sugared jujubes, 1 dried tangerine peel in 3 cups of water, decocted to 1 cup of juice, given to a person suffering from dysentery with blood in the stool.

Sterculia pexa Pierre — WOOLLY PINGPO

Jia-ma-shu=Chia-ma-shu (家麻樹, Homestead Hemp Tree); *Jiu-ceng-pi=Chiu-ts'eng-p'i* (九層皮, Nine-layered Bark); *Mian-mao-ping-po=Mien-mao-p'ing-p'o* (棉毛蘋婆, Hairy Pingpo). Seed; eaten locally.

Trees 5–10 m high, branchlets chalky-white; leaves palmately compound, leaflets 7 or 9, sessile or shortly peticulate, oblanceolate or obovate-oblong, 9–23 cm long, 4–8 cm wide, acuminate, pilose or glabrescent beneath, lateral nerves 22–32 pairs; flowers greenish-white, apetalous, polygamous, in panicles 16 cm long, perianth subcampanulate, 5-lobed to the middle, the lobes 6 mm long, stellate outside, staminal column curved, anthers ca. 20 in a globose head, ovary globose, 5-carpellate, densely villose, cells 20-ovulate; fruits follicular, follicles 4–9 cm long, 2.5–3.5 cm wide, the pericarp woody and fibrous, hispid and hirsute; seeds oblong, 1.5 cm long, 1 cm across, rounded at both ends, shiny black. A native of the tropical forests on the mountains bordering Guangxi, southern Yunnan, northern Vietnam and Thailand.

Theobroma cacao L. — CACAO (tree), COCOA (defatted product), CHOCOLATE (sugared product).

Ke-ke=K'o-k'o (可可, Cocoa) ; *Qiao-ke-li=Ch'iao-k'e-li* (巧克力, Chocolate). Products prepared from seeds, imported material very expensive, rarely used; successful introduction of the species has been made in Hainan and Yunnan.

Small trees 4–10 m high, bark smooth, branchlets pubescent; leaves simple, elliptic, oblanceolate or obovate-oblong, 20–30 cm long, 7–10 cm wide, caudate-acuminate, base obtuse, sparsely stellate-tomentose on the lower surface, entire; flowers pale yellow, appearing on the trunk of the tree or of large branches, 1.5 cm across, pedicels 2 cm long, calyx pinkish, star-like, sepals 5, lanceolate, acute, petals smaller, consisting of 2 portions, the basal portion obovate, 3–4 mm long, white with 2 purple guide lines and apical concave pouch, the distal portion spatulate, yellow, 2–3 mm long, delicately attached to the pouch, reflexed, staminodes 5, opposite the sepals and surrounding the style, fertile stamens 5, the filaments bending outward, projecting the anthers concealed in the corresponding petals, ovary 5-carpellate, 5-celled, ovules many in each cell; fruits large, indehiscent, red-orange, elliptic-ovoid, 15–20 cm long, 6–8 cm across, with 10 broad ridges and 10 shallow furrows, pericarp fleshy; seeds numerous, ovoid, 2–4 cm long, 1–2 cm across, covered by whitish, sugary, mucilaginous pulp developed from the testa, removed through fermentation and dried for the market. Native of tropical America, now cultivated in Asia, Africa, and the Pacific islands.

Dilleniaceae: Dillenia Family

Dillenia indica L. — Simpoh (Malay name)

> *Wu-ya-guo=Wu-ya-kuo* (五椏果, Five-branched Fruit, the Chinese aboriginal name for Dillenia). Sour fruit, having the taste of unripe apple, eaten by people living in the mountains of southern Yunnan and adjacent Guangxi.

Evergreen trees 30 m high, branchlets stout, sericeous and glabrescent; leaves oblong, 15–30 cm long, 6–12 cm wide, acute or obtuse, base rounded or cuneate, dentate, strigose along the midrib above and on the major nerves beneath, petioles 3–7 cm long; flowers white, solitary, terminal, 15–20 cm across, pedicels 4–8 cm long, pilose, sepals 5, yellowish-green, ovate, fleshy, 1 cm thick at the base, petals obovate, 7–9 cm long, stamens numerous, in 2 series, the inner ones longer, anthers dehiscing by apical pores, carpels 8–10, glabrous, distinct; fruits globose, 8–10 cm across, enclosed in the fleshy persistent calyx. Widespread in the Indo-Malaysian region, growing in the tropical rainforest of Yunnan and Guangxi.

Dillenia pentagyna Roxburgh

> *Niu-pang=Niu-p'ang* (牛旁, Niupang, the sound of the Hainan ethnic name); *Xiao-hua-*

wu-ya-guo=Hsiao-hua- wu-ya-kuo (小花五椏果, Lesser-flowered Dillenia); *Wu-shi-di-lun-tao=Wu-shih-ti-lun-t'ao* (五室第倫桃, a translation of the specific epithet). Fruit; eaten fresh or in the form of jam in Hainan Island.

Deciduous trees 6–15 (–25) m high, trunks up to 1 m in diameter, gray, smooth, exfoliate, branchlets glabrous; leaves oblong, rarely obovate-oblong, 20–50 cm long, 10–20 (–30) cm wide, those on young shoots attaining 120 cm long, 40 cm wide, rounded, obtuse or acute, base cuneate or decurrent, glabrous above, appressed pilose on the nerves beneath, shallowly wavy-serrulate, lateral nerves 30–50 on each side, extending slightly beyond the margin, petioles 1–5 cm long; flowers yellow, 2–7 fasciculate on abbreviate shoots, pedicels 6 cm long, glabrous, sepals green, tinged purple, elliptic, 8–12 mm long, 5–8 mm wide, ciliate, corolla 2–3 cm across, petals 5, obovate-elliptic, 1.5–2 cm long, 5–10 mm wide, stamens numerous, in 3 series, the outer 20 sterile, the middle series 60–90 , 2.5–4 mm long, recurved, the inner 10, 6–9 mm long, anthers extrorse, dehiscing vertically; fruits compressed-globose, indehiscent, seeds 1 or 2, black, ovoid, 5 mm long, 3.5 mm across, with aril. Widespread in Southeast Asia, growing in forests and open hillsides of Yunnan and Hainan Island.

Dillenia turbinata Finet et Gagnepain

Shan-niu-pang=Shan-niu-p'ang (山牛旁, Hillside Niupang); *Da-hua-wu-ya-guo=Ta-hua-wu-ya-kuo* (大花五椏果, Large-flowered Five-branched Fruit); *Da-hua-di-lun-tao=Ta-hua-ti-lun-t'ao* (大花第倫桃, Large-flowered Dillenia). Fruit; gathered by people living near the forest where the trees grow naturally.

Evergreen trees 30 m high, velutinous-strigose throughout, branchlets stout; leaves alternate, obovate-oblong, 15–30 (–40) cm long, 8–12 (–15) cm wide, obtuse or rounded, rarely acute, base obtuse or cuneate, remotely serrate, lateral nerves 15–22 (–40) on each side, strigose on the nerves above, sparsely strigose beneath, petioles 2–4 cm long; flowers yellow, solitary, showy, 10–13 cm across, sepals 5, velutinous, persistent, fleshy, petals 5, obovate, 5–7 cm long, 3.5–4.5 cm wide, stamens numerous, filaments red, shorter than the anthers, in 2 series, the outer series 310–325, the inner series 25, carpels 8 or 9, distinct, each containing 40–45 ovules; fruits berry-like, subglobose, 4–5 cm in diameter, dark red; seeds 1 or more, obovoid, 6 mm long, dark brown, glabrous, without aril. Endemic to the mountains bordering Yunnan, Guangxi, Guangdong and Vietnam, very common in the forest at lower elevation on Hainan Island.

Actinidiaceae: Actinidia Family

Actinidia arguta (Siebold et Zuccarini) Planchon — BOWER ACTINIDIA, BOWER VINE

Ruan-zao-zi=Juan-tsao-tzu (軟棗子, Soft Jujube); *Ruan-zao-mi-hou-tao=Juan-tsao-mi-hou-t'ao* (軟棗獼猴桃, Soft Jujube Monkey Peach); *Teng-gua=T'eng-kua* (藤瓜, Vine Gourd). Ripe fruit; very sweet and juicy; not available in the market.

Large lianas climbing over trees up to 30 m high, branchlets villose, with brown laminate pith; leaves membranous, ovate-suborbicular or oblong, 6–13 cm long, 5–9 cm wide, abruptly short acuminate or retuse due to mechanical injuries, base rounded or subcordate, rarely cuneate, serrate, glabrous with tufts of brown or white hairs in the nerve-angles beneath; flowers white, 3–6 in cymose axillary clusters, 5-merous, stamens numerous, styles many, filiform; berries oblong, 1.5–2 cm long, glabrous, soft, juicy, sweet, with good flavor. Native of northern and northeastern China and adjacent areas in Korea; introduced into Western gardens in 1874; grows well in Boston area where fruits become soft on the vine as the leaves turn yellow, so soft and juicy that one can suck the entire content off, leaving the skin empty.

Actinidia callosa Lindley var. **henryi** Maximowicz

Xiao-yang-tao=Hsiao-yang-t'ao (小羊桃, Small Goat Peach). Fruit; eaten fresh or used for wine by people living in the mountains of western Hubei.

Deciduous lianas climbing up to 7 m high, branchlets purple-red, with yellow hairs and lamellate pith; leaves elliptic-ovate or oblong-ovate, 5–10 cm long, 3–6.5 cm wide, acuminate, base rounded or obtuse, setosely serrulate, setose along the nerves and barbate in nerve-angles beneath, petioles 3–5 cm long; flowers yellow or white, fragrant, 1.5–2 cm across, on slender glabrous pedicels shorter than the petioles; fruits oblong, greenish-brown, spotted. Native to central and western China, growing on hillsides at altitudes of 600–1,400 m; introduced into American gardens in 1907.

Actinidia chinensis Planchon — CHINESE GOOSEBERRY, MAOLIZI (=Hairy Plum, the ethnic name of people living in western China), KIWI FRUIT (New Zealand commercial name).

Mao-li-zi=Mao-li-tzu (毛李子, Hairy Plum); *Mi-hou-tao=Mi-hou-t'ao* (獼猴桃, Monkey Peach); *Yang-tao=Yang-t'ao* (羊桃, Goat's Peach). Fresh fruit. *Mao-li-zi-jiu=Mao-li-tzu-chiu* (毛李子酒, Actinidia Wine).

A wine formerly developed and used by the Taoist monks of Qingcheng Mountain, a famous Chinese Taoist center in Sichuan, now patterned in Guan Xian and available to tourists in Sichuan. Personally, I was first treated to the wine at the temple in March 1938 and again by friends of Sichuan University in the 1980s.

Large deciduous lianas climbing over trees up to 20 m high, branchlets densely velutinously villose, pith white, laminate; leaves papery, suborbicular or obovate-orbicular, 5–17 cm long, acute or retuse, base subcordate, bristly serrate, with brown-stellate hairs beneath; flowers unisexual, dioecious, or polygamous, white, changing to pale yellow, calyx and pedicels brown-villose, stamens numerous, ovary superior, golden-strigose, consisting of 30–40 carpels, styles free, 1 cm long, base golden-hairy, bilaterally compressed, the distal half gradually broaden, with stigmatic surface on the adaxial side; fruit a large berry, oblong, the larger ones 7 cm long, 5.5 cm across, central axis white and spongy, containing 35–40 upright vascular bundles on the periphery, each bundle with 12–15 pairs of seed-bearing vascular strains branching off, mesocarps light-green and juicy, excocarp parchment-like and hairy on the outside; seeds numerous (ca. 840, 20–30 in each carpel), paired, oblique-obovoid, 2 mm long, 1–1.4 mm across middle, reticulate-faveolate, hilum punctiform with the raphe running down along a ridge to the middle. Native of China, introduced from Ichang, Hubei, to New Zealand in 1911. Now, the New Zealand product called kiwi fruit is sold in USA for an average of 50 cents a piece. In 1986, C. F. Liang and A. R. Ferguson, in the *New Zealand Journal of Botany*, recognized *Actinidia deliciosa* (Chevalier) Liang et Ferguson as the botanical name for the kiwi fruit on the strength of its hispid trichomes on the excarp, peduncle and petiole. They also proposed *Actinidia chinensis* var. *rufopulpa* (Liang et Huang) Liang and Ferguson for a taxon native of Hubei, with reddish pulp, locally called *Hong-rou-yang-tao=Hung-jou-yang-t'ao* (紅肉羊桃, Red Meat Goat's Peach).

Actinidia coriacea Finet et Gagnepain

Ge-ye-mi-hou-tao=Ko-yeh-mi-hou-t'ao (革葉獼猴桃, Leather-leaved Actinidia). Fruit; eaten fresh, or used for jam or wine.

Deciduous lianas, branchlets glabrous, pith white or pale yellow, leaves ovate-oblong or elliptic-oblong, 4–13 cm long, 2–4 cm wide, acute or acuminate, base obtuse, glabrous, serrate, with red-glandular teeth above the middle; flowers red, unisexual and dioecious, staminate flowers 3–5 in short racemes axillary to small bracts basal to new growth, sepals oblong, petioles 5, suborbicular, stamens numerous, pistillate flowers solitary, axillary to small bracts or leaves, ovary oblong, hairy, styles 12–16, distinct, filiform, stigma oblique-clavate; fruits oblong-globose, 2 cm long, brown, juicy and sweet when fully ripe. Native to western Sichuan, hence extending southward to Yunnan, growing in thickets and along margin of forests at altitudes of 350–1,500 m; introduced to Western gardens in 1908.

Actinidia kolomikta (Maximowicz et Ruprecht) Maximowicz

Shen-shan-mu-tian-liao=Shen-shan-mu-t'ien-liao (深山木天蓼, Deep Valleys Woody Sky

Smartweed); *Gou-zao-mi-hou-tao=Kou-tsao-mi- hou-t'ao* (狗棗獼猴桃, Dog-jujube Actinidia); *Mao-ren-shen=Mao-jen-shen* (貓人參, Cat Ginseng). Fruit; gathered and eaten locally.

Deciduous lianas, climbing up to 15 m high, branchlets with lamellate pith; leaves ovate-oblong, 6–12 cm long, 5–7 (–12.5) cm wide, caudate, base truncate, rarely rounded, setose, serrate, grayish and pubescent on the nerves beneath; flowers white to pink blotched, changing to purple-red, particularly in the staminate plants, polygamous, fragrant, 3–5 in axillary cymes, sepals ovate, acute, glabrous, petals oblong, rounded, stamens numerous, staminodes in the pistillate flowers with sterile anthers, ovary oblong-ovoid, glabrous, styles 3–5 mm long, stigmas 8–12 (–15); fruits oblong-ovoid, 2–2.5 cm long, greenish-yellow, smooth. Native to northwestern China, extending northeastward to the Amur River region, from where the species was first reported by Maximowicx and Ruprecht in 1855.

Actinidia latifolia (Gardner et Champion) Merrill — BROAD-LEAVED ACTINIDIA

Kuo-ye-mi-hou-tao=K'uo-yeh-mi-hou-t'ao (闊葉獼猴桃, Broad-leaved Actinidia); *Duo-hua-mi-hou-tao=Tuo-hua-mi-hou-t'ao* (多花獼猴桃, Many-flowered Actinidia); *Teng-li=T'eng-li* (藤李, Vine Plum). Fruit; eaten fresh, or used for jam or wine.

Deciduous lianas or sprawling shrubs, young branchlets pilose, pith white, older branchlets hollow; leaves ovate-suborbicular or oblong-ovate, 8–14 cm long, 4.5–10 cm wide, acuminate, base rounded or subcordate, subentire or remotely serrulate, stellate-villose beneath, petioles 2–8 cm long, flowers pale-yellow, dioecious, in compound axillary cymose clusters, peduncles 3–5 cm long, pedicels 2 mm long, sepals 5, oblong, persistent, petals 5, ovate-orbicular, 7–9 mm across, stamens numerous in staminate flowers, pistillode oblong, densely villose, pistil in pistillate flower oblong, pilose, style 18–20, connate at the base into a short column, stigma oblique discoid; fruits globose, 2 cm in diameter, brown, glabrescent. Native to southeastern China, Taiwan and Hainan Island, growing on sunny hillside.

Actinidia polygama (Siebold et Zuccarini) Maximowicz — SILVER-VINE

Mu-tian-liao=Mu-t'ien-liao (木天蓼, Woody Sky Smartweed); *Tian-liao=Tien-liao* (天蓼, Sky Smartweed), *Ge-zao=Ke-tzao* (葛棗, Vine Jujube); *Ge-zao-mi-hou-tao=Ke-tzao-mi-hou-t'ao* (葛棗獼猴桃, Vine-jujube Actinidia). Shoot, cooked for vegetable; fruit, gathered and eaten fresh.

Deciduous lianas, climbing up to 5 m, branchlets pilose, pith white, solid; leaves broad-ovate to ovate-oblong, 8–14 cm long, 4–8.5 cm wide, acuminate, base rounded or subcordate, appressed serrate, the distal half or the entire leaves silver-white or pale yellow, sometimes green mottled white or greenish-yellow, setose on the veins beneath, petioles bristly; flowers white, fragrant, 1.5 cm across, 1–3 in axillary cymes, sepals 5, pilose, petals 5–6, oblong-ovate, 1.2 cm long, stamens numerous, anthers sagittate, dorsifixed, ovary glabrous, bottle-shaped, style filiform, stigmas 18–20; fruits oblong-ovoid, smooth, without spot, yellow, turning cherry-red, 1 cm in diameter, rostrate, with persistent sepals. Widespread in temperate eastern Asia, introduced to Western gardens in 1861; the Chinese name first appeared in the *New Tang Pharmacopoeia* (A.D. 659) as *mu-tian-liao* (Woody Sky Smartweed), because of the sharp stinging taste of the unripe fruits.

Actinidia suberifolia C. Y. Wu

Shuan-ye-mi-hou-tao=Shuan-yeh-mi-hou-t'ao (栓葉獼猴桃, Suberose-leaved Actinidia). Fruit; eaten fresh in Yunnan.

Deciduous lianas, branchlets velutinous, pith brown; leaves oblong, 9–13.5 cm long, 4–6.5 cm wide, acute or acuminate, base rounded, margin remotely serrulate, rugose and pilose along the nerves above, densely cinnamic-tomentose along the nerves and elsewhere stellate beneath; flower orange, dioecious, in velutinous cymose corymbs, sepals 5, ovate, petals 5, obovate, 7 mm long, 4 mm wide, stamens numerous, pistilode globose, villose; fruits globose, 1.5 cm in diameter, densely rusty-hairy. Endemic to southern Yunnan, growing among thickets at altitudes 900–1,200 m.

Actinidia tetramera Maximowicz

Shan-yang-tao=Shan-yang-t'ao (山洋桃, Hillside Actinidia), *Si-rui-mi-hou-tao=Szu-jui-mi-hou-t'ao* (四蕊獼猴桃, Four-stamened Actinidia). Fruits.

Deciduous lianas, climbing to 17 m high, branchlets glabrous, pith brown, lamellate; leaves often crowded at the shoot apex, ovate-oblong, 5–10 cm long, 2.5–4 cm wide, acuminate, base obtuse or rounded, finely serrate, setose along the veins and barbate at nerve-angles beneath, petioles 1.5–3 cm long, flowers white or red, fragrant, polygamous, 2 or 3 fasciculate, sepals 4 or 5, green, ovate, ciliate, persistent, petals 4 or 5, obovate, 6 mm long, glabrous; fruits oblong-globose, 1.5–2 cm long, golden yellow, turning brown, smooth, hanging on slender stalks. Native to central, and northwestern China, growing on hillsides at altitudes of 1,400–1,500 m, introduced into American gardens in 1927.

Saurauia tristyla de Candolle (Syn. *S. oldhamii* Hemsley)

Shui-dong-ge=Shui-tung-ko (水東哥, Water East Brother). Fruit, gathered and used in Taiwan.

Shrubs or small trees growing in partial shade along streams of tropical forest, young growth ferruginous, covered by glandular stellate indumentum and squamulose-acute scales, branchlets glabrescent, sparsely scabrid, broken surface showing rubber-like material similar to that of *Eucommia*; leaves obovate-elliptic, 11–17 cm long, 4–9 cm wide, acuminate, base rounded or obtuse, subaculeate-serrate, squamulose on the veins, especially beneath; flowers pink changing white, in axillary cymes, perianth fleshy, containing mucilage, sepals unequal, glabrescent, oblong-orbicular, base coherent, stamens numerous, the filaments connate at the base and adnate to the corolla, ovary glabrous, styles 3; fruits white, sweet, juicy, globose-ovoid, 6 mm across, with persistent calyx.

Theaceae: Tea Family

Camellia gigantocarpa H. H. Hu et T. C. Huang

Bo-bai-da-guo-you-cha=Po-pai-ta-kuo-yu-ch'a (博白大果油茶, Large-fruited Bobai Oil Tea). Oil; extracted from seeds.

Trees 5–10 m high, branchlets hairy; leaves thinly coriaceous, oblong or obovate-oblong, 7–13.5 cm long, 3–9.5 cm wide, shortly acuminate, base obtuse, glabrous, the primary nerves impressed; flowers white, solitary, subterminal, 12 cm across, bracts and outer sepals yellow-sericeous, petals 6 or 7, emarginate, stamens numerous, ovary 3-locular, styles 3, hairy; capsules reddish-brown, subglobose, 7–17 cm in diameter, pericarp woody, up to 2 cm thick, dark brown. Endemic to Bobei area of Guangxi Province; cultivated for edible oil from seed.

Camellia latilimba H. H. Hu

Li-cha=Li-ch'a (梨茶, Pear Tea); *Da-you-cha=Ta-yu-ch'a* (大油茶, Greater Oil Tea). Seed; for edible oil.

Glabrous evergreen trees 1–6 m high , branchlets yellowish-brown; leaves coriaceous, oblong, 8–13 cm long, 3–3.5 cm wide, acuminate, base obtuse or cuneath; flowers white, solitary, terminal, 3 cm across, sepals coriaceous, petals thick and tough, connate at base, filaments of stamens united at base, ovary 4- to 6-locular; capsules globose, 5–6 cm in diameter, yellowish-brown; seeds ca. 10. Endemic to southern Fujian, cultivated for the edible oil.

Camellia oleifera Abel

You-cha=Yu-ch'a (油茶, Oil Tea). Seed; containing 30% oil, edible or used in industry.

Shrubs on exposed grassy slopes, small tree in forests, 3–4 cm high, branchlets pubescent, new growth appearing in March, brilliant yellow-green; leaves coriaceous, elliptic-oblong, 3–5 (–9) cm long, 2–3 (–4) cm wide, obtuse or acute, base obtuse or cuneate, serrulate; flowers pink in buds, white when fully open, sessile, 1 or rarely 2, terminal, axillary to apical leaves, calyx ciliate, petals 5 to 7, 4–5 cm across when fully open, stamens numerous, filaments yellow, connate, ovary densely woolly, style columnar, 1/3–1/2 divided; capsules globose, 1.8–2.2 cm in diameter, loculicidally dehiscent, the sections often falling with the seeds. Native of South China, extensively cultivated south of the Yangtze River for the edible oil; in Hong Kong, the blooming season being October–December, when last year's fruits mature.

Camellia polyodonta Hou

Wan-tian-hong-hua-you-cha=Wan-t'ien-hung-hua-yu-ch'a (宛田紅花油茶, Red-flowered Wan-tian Oil Tea). Seed; for edible oil.

Small trees 8 m high, branchlets glabrous; leaves coriaceous, oblong, 8–14 cm long, 3–6 cm wide, acuminate-caudate, base obtuse or rounded, finely serrate, reticulum of the veins impressed above, sparsely villose beneath; flowers solitary, rose-pink, 5–10 cm across, sessile, sepals 15, pubescent, petals 5 to 7, obcordate, villose outside, filaments hairy, ovary villose, 3-locular, style 3-cleft; capsules globose, 4.5–10 cm in diameter, loculicidally dehiscent.

Camellia semiserrata Chi

Guang-ning-you-cha=Kuang-ning-yu-ch'a (廣寧油茶, Oil Tea of Guangning). Oil extracted from seed.

Trees, up to 12 m high, glabrous throughout except parts of the flowers; leaves elliptic-oblong, coriaceous, 9–15 cm long, 3–6 cm wide, acuminate, base obtuse, serrate above the middle; flowers white, solitary, 6.5 cm across, terminal, bracts and sepals deciduous, petals 6 or 7, obovate, entire, stamens numerous, filaments connate, the tubes 3.5–4.6 cm long, ovary velutinously pubescent, style 3- to 5-cleft; capsules ovoid, 4.5 cm across, dehiscing into 3 to 5 valves; seeds 2.5 cm long. Native of Guangdong and Guangxi.

Camellia sinensis (L.) 0. Kuntze — TEA (Syn. *Thea sinensis* L.)

Cha=Ch'a (茶, Tea); *Hong-cha=Hung-ch'a* (紅茶, Red Tea); *Lü-cha=Lü-ch'a* (綠茶, Green Tea); *Xiang-pian=Hsiang-p'ian* (香片, Fragrant Tea).

Young leaves of tea plants; extensively cultivated. Green Tea is prepared from fresh tender unfolding leaves; Red Tea is prepared from slightly fermented leaves; Jasmine

Tea (i.e. Fragrant Tea) is Green Tea scented with Jasmine flowers, the better grade retains the aroma with the flowers removed after one or two treatments; the ordinary kind retains the flowers. Many kinds of tea from China are available in American Chinese stores.

Shrubs or small trees 1–5 m high, branchlets glabrous; leaves thinly coriaceous, variable in shapes and sizes, 3–12 cm long, 1–5.5 cm wide, acuminate or rounded, base acute, cuneate or rounded, serrate; flowers white, 2.5–3 cm across, 1–4, axillary, pedicellate, calyx glabrous or sparsely soft pubescent, ciliate, imbricate, petals 7 or 8, stamens numerous, filaments partially connate, ovary 3-locular, style columnar, 3-cleft at the apex; capsules globose, trilobed when fully developed, 1.5 cm long, 1.5–3 cm across, chestnut brown, loculicidally dehiscent, pericarp rather thin, 1 mm thick; seeds 1 or 2 in each locule, subglobose or hemispherical, smooth, bright brown, persistent sepals 5–8 mm across.

This species is a native of the Nanling Range in southern China, where it is an evergreen tree if allowed to grow naturally. In Hong Kong, the plant commences to send out new shoots in March. Vegetative shoots develop best in summer, and flowers from these shoots bloom in October–November. It has been cultivated since time immemorial, always on slopes too steep for cereal and vegetable crops. Being planted near the wild types and/or closely related species, natural hybridization happened. Subsequently, the species is polymorphic. Selections for the size and texture of the leaves which effect productivity of the crop, and for color and aroma which lead to better quality and price of the product have led to the following varieties and cultivars. Observations of the living specimens made at the Hong Kong Government Experiment Farm at Tai Mo Shan are supplemented by herbarium collections of the Arnold Arboretum, Harvard University.

1. var. **assamica** (J. W. Master) Kitamura — Assam Tea

Pu-er-cha =P'u-erh-ch'a (普洱茶, Assam Tea). Leaves elliptic, 8–16 cm long, 2.5–6 cm wide, acuminate, sharply serrate; flowers 1–4, pedicels long, sepals glabrous inside, petals 7–9, style cleft at the distal end; fruits globose or trilobed, 1.2–3 cm across, persistent calyx 1 cm across.

2. var. **bohea** (L.) Pierre — Fujian Tea

Dun-ye-cha=tun-yeh-ch'a (鈍葉茶, Obtuse-leaved Tea). Leaves oblong-elliptic, 3–8 cm long, 2–4 cm wide, obtuse, dark green; flowers solitary, sepals hairy or glabrous, petals 5 or 6, style columnar; capsules subglobose or trilobed, pericarp thin, crustaceous. Native of the Wuyi Mountain (武夷山), which is the water-divide betweem Fujian and Jiangxi Provinces, and famous for its 36 unique peaks, 72 scenic areas, and tea cultivation. Herbarium specimens from Japan generally belong to this variety. "Bohea" is the early English spelling of the Fujian dialect of the famous tea-producing mountain, Wuyi.

3. var. **cantoniensis** (Loureiro) Pierre — Cantonese Tea

Mao-e-cha=Mao-eh-ch'a (毛萼茶, Hairy Sepal Tea). Leaves elliptic-oblong, 5–11 cm long, 2–5 cm wide; flowers 1 or 2, axillary or terminal, sepals villose inside, pillose outside, petals 7–9, style columnar, the distal 1/4 divided. Native of Guangdong and Guangxi.

4. var. **viridis** (L.) Pierre — Pointed Leaf Tea

Jian-ye-cha=Chien-yeh-ch'a (尖葉茶, Pointed Leaf Tea). Low shrubs 1 m high; leaves elliptic-lanceolate, 4–10 cm long, 1–3 cm wide, acuminate, base cuneate, serrate, light green; flowers 1–4, sepals hairy.

5. cv. 'Japonica' — Japanese Tea

Ri-ben-cha=Jih-pen-ch'a (日本茶, Japanese Tea). Shrubs 1 m high; leaves oblong, 1.5–5 cm long, 1–2.5 cm wide, obtuse, base acute or obtuse, sharply serrate; flowers solitary; fruits globose or 3-lobed, 1.5–2.5 cm across, pericarp rough. Stock from Japan (S. Y. Hu 5527, 5528).

6. cv. 'South China Small' — Small-leaved Tea

Xiao-ye-cha=Hsiao-yeh-ch'a (小葉茶, Small-leaved Tea). Low shrubs 1 m high; leaves small, elliptic, 1.5–3.5 cm long, 0.8–1.2 cm wide, acute, obtuse or shortly acuminate, serrate; flowers solitary; fruits globose or trilobed, 1–2 cm across, persistent calyx glabrous, fruiting pedicels 5–8 mm long. Stock from local farmers (S. Y. Hu 5530).

Clusiaceae: Garcinia Family

Cratoxylum ligustrinum (Spach) Blume

Huang-niu-mu=Huang-niu-mu (黃牛木, Ox Yokewood); *Huang-niu-cha=Huang-niu-ch'a* (黃牛茶, Yellow Ox Tea). Leafy shoots, used for herb tea.

Small trees, or shrubs (due to the cutting made by drug or fuel collectors), up to 10 m high, growing on sea-side cliffs and along streams inland, bark light brown, smooth, exfoliate, branchlets slender, bearing green foliage and persistent pericarps of old fruits; leaves opposite, elliptic, 5–10 cm long, 2–3 cm wide, acute, base obtuse or acute, entire; flowers coral-red, 1–7 in axillary cymes, 1 cm across, sepals 5, green, imbricate, obtuse, petals 5, obovate; stamens numerous, yellow, the filaments connate into 3 bundles, alternating with 3 yellow glands, ovary ovoid, 3-locular, styles 3, ovules numerous in each cell; capsules ovoid, 8–12 mm long, two-thirds covered by persistent calyx, loculicidally dehiscent; seeds winged, 6 mm long.

A native of southern China, wood hard, with fine grain, used for making ox-yoke, leafy shoots gathered by herb collectors for a cooling tea, hence the vernacular names, "Ox Yokewood" and "Yellow Ox Tea".

Cratoxylum prunifolium Dyer

Ku-ding-cha=K'u-ting-ch'a (苦丁茶, Bitter Tea). Leaves and young shoots, used as a tea substitute.

Pubescent trees 5–12 m high, velutinous throughout, branchlets fissured; leaves subcoriaceous, opposite, oblong-elliptic 3.5–9 cm long, 2–3.5 cm wide, acute or obtuse, base rounded or obtuse, entire, villose beneath; flowers orange-red, 4–6 in racemose clusters emerging from lateral buds on last year's growth, or solitary to lower leaves of current year's growth, densely villose, sepals 5, hairy, petals 5, obovate, 1 cm long, stamens numerous, filaments connate at the basal half, forming a linear band, glands small, ovoid, ovary glabrous, styles 3; capsules ovoid-ellipsoid, 8–14 cm long, 6–8 mm across, one-fourth covered by the persistent calyx.

Native of the Indo-Malayan Region, only known from southern Yunnan in China.

Garcinia cowa Roxburgh

Yun-nan-shan-zhu-zi=Yun-nan-shan-chu-tzu (雲南山竹子, Yunnan Mangosteen). Fruit; large, edible.

Evergreen tree up to 20 m high, rich in sticky yellow latex, branchlets glabrous, gray; leaves elliptic-lanceolate, 7–12 cm long, 3–5 cm wide, acuminate-caudate, entire; flowers yellow, 1 cm across, polygamous, staminate flowers 4–12, mostly in axillary fascicles, sepals ovate, fleshy, yellow, petals oblong, yellow with reddish inner face, stamens many, filaments connate, forming a column, anthers sessile, pistillate flowers 2–8 in a fascicle or solitaty and axillary to leaves, ovary subglobose, stigmas radiate, glands papillose; fruits rounded or ovoid, pointed at the distal end, 3–4 cm in diameter, 5–8 grooved at the upper half, orange, pulp around the seeds orange, sour.

Native of southeastern Himalayan Region, occurring in Burma, Bengal, Assam and northern Thailand; in China known only from southern Yunnan.

Garcinia mangostana L. — MANGOSTEEN, MANGGIS, MESETOR

Yang-luo-han-guo=Yang-lo-han-kuo (洋羅漢果, Imported Buddha Disciple Fruit); *Feng-*

guo=Feng-kuo (鳳果, Phoenix Fruit); *Mang-ji-shi=Mang-chi-shih* (莽桔柿, Mangosteen). Fruit; imported into Hong Kong, rare, one of the best tropical fruits.

Slow growing evergreen trees up to 10 m high, beginning to bear flowers when 15-year old, known in cultivation only; leaves coriaceous, elliptic-oblong, 15–20 cm long; flowers solitary, rose-pink; fruits globose, 5–8 cm in diameter, reddish-purple, rind smooth, leathery, pulp snow white, enclosing 5–7 seeds.

Native of Indonesia, cultivated in southeastern Asia, fruit imported to Hong Kong; sliced husks exported from Singapore to China for medicinal uses.

Garcinia multiflora Champion ex Bentham — GUANGDONG MANGOSTEEN

Shan-ji-zi=shan-chi-tzu (山桔子, Wild Mangosteen). Fruit; gathered by fuel gatherers working on hillsides; very delicious.

Evergreen trees 5–17 m high, branchlets rich in white latex (turning yellow on exposure); leaves opposite, coriaceous, obovate-oblong, 7–15 cm long, 2–5 cm wide, acuminate or acute, base obtuse, entire; flowers yellow, 4–6 in terminal cymose cluster, unisexual, dioecious, 4-merous, sepals roundish, tinged red, petals obovate-spatulate, 1.5 cm long, concave, pistilate flower with 4 sets of staminodes, ovary tetragonal, 5–6 mm across, terminated by a large discoid yellow stigma; fruits subglobose, 3–4 cm in diameter, greenish-yellow.

The species was first described from Hong Kong, but it is now extremely rare there. A former Head of the Police Department, the New Territories, Hong Kong, is very knowledgeable of the flora in the area. His residence was located in a protected area on the seashore near the Plover Cove Reservoir. When he first moved to the property, he discovered a tree bearing some fruit which he did not recognize. He tried the fruit and it was delightful. I was invited to tea and asked for the identification of the species. The appearance of the fruit resembles a lime, but it is smaller and perfectly spherical. It tastes sweet and with a delightful flavor. Personally, I think that it tastes better then the imported mangosteen bought in the Hong Kong market. In connection with the Flora of Hong Kong Project, I have collected seventeen thousand sets of botanical specimens from Hong Kong and the New Territories; and this was the only plant of the species I have observed.

Garcinia oblongifolia Champion ex Bentham — SOUTH CHINA MANGOSTEEN

Huang-ya-guo=Huang-ya-kuo (黃牙果, Yellow Teeth Fruit); *Ling-nan-shan-ji-zi=Ling-nan-shan-chi-tzu* (嶺南山桔子, Lingnan Wild Mangosteen). Fruit; round, ripe yellow, good flavor, sweet; gathered by residents of the hillside and eaten immediately, not available in the market; the latex from the rind stains the teeth of the eater yellow, hence the trivial name, "Yellow Teeth Fruit".

Evergreen trees 8–10 m high, trunk 15–20 cm in diameter, bark dark gray, smooth, branchlets glabrous, with yellow resinous juice; leaves coriaceous, opposite, oblong, 5–10 cm long, 2–2.5 cm wide, acute or obtuse, base acute, entire; flowers small, yellow, unisexual, dioecious, 3 in terminal cymose clusters or fasciculate on older branchlets, 4-merous, sepals 4, oblong, 3–4 mm long; petals 4, obovate, 7–8 mm long, staminate flowers pedicellate, stamens yellow, numerous, sessile, crowded on a square fleshy receptacle; pistillate flowers subsessile, pedicels 2–3 mm long, staminodes fasciculate, ovary globose, 6-carpellate, ovule 1 in each cell, stigma recurved, crested; fruits globose, 2–3 cm in diameter, seeds arillate, the pulp very sweet and with very pleasant taste. The yellow resin stains the hands, lips and teeth, hence the vernacular name "Yellow Teeth Fruit", for those who are not careful in eating the aril, their teeth would be stained yellow (Figure 41).

Garcinia paucinervis Chun et How

Jin-si-li=Chin-ssu-li (金絲李, Golden Thread Plum). Ripe fruits, eaten in Guangxi by people living in the hillside, not available in the markets.

Evergreen trees 30 m high, with yellow latex throughout, trunks 80 cm in diameter, branchlets tetragonous, purple-red when young; leaves opposite, oblong or obovate-oblong, 8–15 cm long, 3–6.5 cm wide, acuminate or acute, base obtuse, entire, coriaceous; flowers unisexual, small, in axillary or terminal cymose clusters, staminate flowers 4–10, pedicels 3–5 mm long, bracts 2 or 3, sepals 4, petals 4, stamens many, filaments connate; fruits oblong, 3–4 cm long, 1–2 cm across, yellow, tinged red.

Native to the mountains on the borders of Yunnan, Guangxi and Vietnam, flowering and fruiting twice annually, the first blooming season in May–June with the fruits maturing in October–November, fruits of autumn flowers mature in March–April.

Garcinia xanthochymus J. D. Hooker ex T. Anderson — SOUR MANGOSTEEN

Dan-shu=Tan-shu (蛋樹, Egg Tree). Fruit; introduced to Taiwan before 1950.

Middle-sized evergreen resiniferous trees, trunks straight, branchlets drooping, drying sharply 6–8 costate; leaves opposite, coriaceous, oblong-linear, 15–35 cm long, 5–10 cm wide, acuminate or acute, base rounded or obtuse, petioles 2.5 cm long, 6–9 mm in diameter, rugose, lateral nerves 35–45, obvious; flowers white, 1–1.5 cm across, 2–8 in fascicles from the axils of fallen leaves, pedicels 2.5–4 cm long, sepals unequal, orbicular, concave, fleshy, petals 7–8 mm long, suborbicular, spreading, stamens of staminate flowers numerous, in 5 bundles, alternate with 5 fleshy glands, ovary of pistillate flowers ovoid, 5-locular, stigmatic lobes 5, spreading, entire; fruits subglobose, dark yellow, apple-sized; seeds oblong 1–4, arillate, the aril (pulp) pleasantly sour, good for making drinks and sherbets. Native of southeastern Asia, cultivated in Taiwan before 1950.

Mammea americana L. — MAMEY, MAMMEE, MAMMEE APPLE
  Man-mei-guo=Man-mei-kuo (滿美果, Mamey Fruit). Fruit; introduced to Taiwan before 1950.

Polygamous trees up to 20 m high, branchlets stout, 4–6 mm thick; leaves thick coriaceous, oblong-obovate, 10–20 cm long, 6–9 cm wide, rounded or retuse, base obtuse, entire, lateral nerves numerous (over 60 on each side), reticulations conspicuous; flowers white, fragrant, solitary or 2–4, axillary on young shoots, sepals 2, petals 4–6, stamens numerous, ovary of pistillate flowers ovoid, 4-locular, stigma peltate; fruits subglobose, 10–15 cm in diameter, rind 2 mm thick, outside rough, reddish brown, the pulp firm, juicy, subacid and pleasant, better stewed; seeds toxic to fish and chicken. Native of the West Indies, cultivated as a yard tree in gardens of tropical America, introduced to Taiwan before 1950.

Hypericaceae: Hypericum Family

Hypericum ascyron L. — ST. JOHN'S-WORT
  Niu-xin-cha=Niu-hsin-ch'a (牛心茶, Ox-heart Tea); *Huang-hai-tang=Huang-hai-t'ang* (黃海棠, Yellow Crab Apple). Leafy shoots, gathered by the people living in northeastern China, dried and used for tea.

Perennial erect herbs 50–100 cm high, branchlets tetragonous; leaves opposite, sessile, oblong-lanceolate, 5–10 cm long, 1–3 cm wide, obtuse or acute, base slightly amplexicaul, entire; flowers yellow, in terminal cymose clusters, sepals 5, ovate, corolla 3–8 cm across, petals 5, stamens numerous, the filaments connate at base into 5 fascicles, ovary ovoid, styles 5-fid above the middle; capsules ovoid, 1.5–2 cm long, 8–10 mm across the base, septicidally dehiscing. Widespread in temperate eastern Asia and Eurasia.

Violaceae: Violet Family

Viola acuminata Ledebour — CHICKEN LEG VIOLET (Syn. *V. micrantha* Turczaninow [1832], non Presl [1822])
  Ji-tui-cai=Chi-t'ui-ts'ai (雞腿菜, Chicken Leg Vegetable); *Ji-tui-jin-cai=Chi-t'ui-chin-ts'ai*

(雞腿菫菜, Chicken Leg Violet). Young shoots, eaten by people living in northeastern China.

Erect perennial herbs 15–40 cm high, rootstocks stout, erect or oblique, erect stems 2–6, emerging in early spring from the ground; leaves ovate-cordate, 3.5–5.5 cm long, 3.2–4 cm wide, acuminate, margin crenate, petioles of caulines equal or shorter than the laminas, stipules oblong, 1.5–2 cm long, 3–10 mm wide, pinnatisect; flowers white or light purple, solitary, on slender axillary peduncles 4–6 cm long, bracts 2, 3–5 cm from the base, sepals linear-lanceolate, base truncate, petals 11–15 mm long (including the spur, 2–3 mm long); capsules glabrous.

Viola chinensis G. Don — Chinese Violet

Jin-jin-cai=Chin-chin-ts'ai (菫菫菜, Violet Green). Young plants, gathered and used locally, not available in the market.

Perennial acaulescent herbs 6–10 cm high, glabrescent, rootstocks buried deeply in the soil, sending out leaves and flowers in early spring; leaves all radical, ovate-triangular or subcordate-lanceolate, quite variable in shapes and sizes, obtuse, base subcordate, crenate; flowers purple, zygomorphic, solitary, on slender scapes, sepals 5, ovate-lanceolate, petals 5, spur 4 mm long; capsules oblong-ovoid, loculicidally dehiscent; seeds light brown.

Widespread in northern China, where people gather the young plants for food, particularly in times of famine.

Viola collina Besser — Hairy Fruit

Qiu-guo-jin-cai=Ch'iu-kuo-chin-ts'ai (球果菫菜, Ball-fruited Violet); *Mao-guo-jin-cai=Mao-kuo-chin-ts'ai* (毛果菫菜, Hairy-fruited Violet). Young plants; gathered in early spring for potherb in northeastern China.

Acaulescent perennial herbs 3–8 (–20) cm high, rootstocks stout, erect or oblique, 2–7 cm long, 0.8–3 cm across; leaves all radical, suborbicular or broad ovate, 1–3.5 cm long, 0.8–3 cm wide, acute or obtuse, base cordate, crenulate, both surfaces villose, petioles 1.5–5 cm long at anthesis, 4–15 cm long in fruit, winged, hairy, stipules lanceolate, 1–1.4 (–2) cm long, basal portion adnate to the petiole, margin serrulate; flowers small, violet or white, on short peduncles 2.5–3 cm long, bending downward after anthesis, bracts 2, subulate, above-medium, sepals hairy, oblong, obtuse or rounded, petals 4–5 mm long including the spur, ovary hairy; capsules globose, densely pubescent, close to the soil on bending peduncles. Widespread in temperate eastern Asia and Eurasia.

Viola verecunda A. Gray — NORTH CHINA VIOLET

Jin-cai=Chin-ts'ai (菫菜, Violet Vegetable). Young shoots; eaten in northeastern China, gathered in early spring, soon as the leaves appear.

Glabrous perennial herbs, aerial stems 8–15 (–20) cm long, rootstock erect or oblique, up to 2 cm long, bearing numerous fibrous roots; leaves rosulate and cauline, ovate-cordate, some basal leaves reniform, 1–3.6 cm long, 1.5–3.8 cm wide, obtuse or subacute, basal sinuses broad, crenate, petioles 1–1.8 cm long, stipules lanceolate, lower half adnate to the petiole of basal leaves, free in cauline leaves; flowers white or lilac, solitary on erect stems, on peduncles 6–7 cm long, bracts 2, two-thirds above base, sepals 5, ovate-lanceolate, basal appendage minute, saccate, petals 6–8 mm long, hairy, stamens 5, ovary glabrous; capsules oblong, 7–8 mm long. Widespread in temperate eastern Asia.

Viola yedoensis Makino — PURPLE GROUND-NAIL

Zi-hua-di-ding=Tzu-hua-ti-ting (紫花地丁, Violet-flowered Earth Nail); *Guang-ban-jin-cai=Kuang-pan-chin-ts'ai* (光瓣菫菜, Glabrous Petal Violet). Young plants, eaten in northeastern China.

Acaulescent perennial herbs with stout tap root; leaves deltoid to linear-lanceolate, 2–6 cm long, 5–10 mm wide, obtuse or rounded, base truncate, obtuse or cuneate, pilose, shallowly dentate, petioles 2–10 cm long, stipules membranous, free portion deltoid; flowers violet, solitary, on peduncles longer than the petioles, bracts 2, medium, sepals 5, petals 5, unequal; capsules oblong, 6.5–10 mm long. Widespread in temperate eastern Asia.

Flacourtiaceae: Flacourtia Family

Dovyalis hebecarpa (G. Gardner) Warburg — CEYLON GOOSEBERRY (Syn. *Aberia gardneri* Clos)

Xi-lun-cu-li=Hsi-lu-ts'u-li (錫倫醋栗, the translation of Ceylon Gooseberry). Fruit, introduced to Taiwan.

Deciduous, spinose shrub 4–5 m high, branchlets pubescent when young, glabrescent and grayish-white on aging, suckers with strong spines 5–6 cm long; leaves alternate, ovate, 3–8 cm long, 2–4 cm wide, entire, acute, base rounded, densely villose beneath; flowers small, inconspicuous, apetalous, unisexual, dioecious, stamens of staminate flowers 10 to many, 5 mm across, ovary of pistillate flowers subglobose, styles distinct; fruits globose, 2–3 cm in diameter, maroon-purple, the rind thin, velvety, pulp purplish, edible (too acid to eat raw), excellent for jams and jelly.

Native of Sri Lanka, cultivated in the moderately dry areas of the tropics, does not grow well in humid region.

Flacourtia indica (N. Burman) Merrill — GOVERNOR'S PLUM, MADAGASCAR PLUM
Ci-li-mu=Tz'u-li-mu (刺籬木, Spiny Hedge Plant). Fruit; edible, not available in the market.

Deciduous shrubs or small trees 2–10 m high, with straight or branched thorns on the trunk and branches; leaves obovate or oblong-obovate, 2–5 cm long, 1.5–3 cm wide, rounded, base cuneate, crenate-serrate; flowers small, 5 mm across, apetalous, dioecious, in short racemose clusters, sepals 4–6, stamens numerous, ovary in pistillate flowers ovoid, styles 5–6, bifid at the tip; fruit drupaceous, dull red, oblong-globose, 8–12 mm in diameter, with 6–8 furrows; seeds 5–8. Widespread, growing near villages in tropical southeastern Asia; in China, reported from Guangdong and Guangxi.

Flacourtia inermis Roxburgh — LOBI-LOBI (the Indonesian name), BATOKO-PLUM
Luo-bi-mei=Lo-pi-mei (羅比梅, Lobi Plum, the name formed by partial translation of the Indonean name). Fruit, introduced to Taiwan before 1950.

Low spineless trees 5–15 m high, branchlets shortly velutinous-pubescent; leaves ovate-elliptic, 10–20 cm long, 5–9 cm wide, acuminate, base rounded or obtuse, coarsely serrulate, the teeth obtuse and glandular at tip, both surfaces glabrous; flowers small, bisexual, in short axillary racemes, perianth segments 3–5, subglabrous outside, densely hairy inside, stamens 15–20, 3–4 mm long, ovary 4- or 5-locular, styles distinct, 4 or 5; fruits cherry red, globose, 2 cm across, too sour to be eaten raw, good for jam and jelly. Cultivated in Indonesia, introduced to Taiwan before 1950.

Flacourtia rukam Zollinger et Moritzi — SPINY HEDGE PLUM
Da-ye-ci-li-mu=Ta-yeh-ts'u-li-mu (大葉刺籬木, Large-leaved Spiny Hedge).

Fruit, used to make jam.

Trees 5–10 m high, branchlets pilose; leaves ovate-oblong or elliptic-lanceolate, 6–12 cm long, 4–6 cm wide, acuminate, base obtuse or rounded, crenuate, lateral nerves 5–11 on each side, petioles 6–8 mm long; flowers unisexual, apetalous, in short axillary racemes, pilose, sepals 4–5, ovate, stamens in staminate flowers 25, filaments filiform, disk fleshy, 8-lobed, ovary in pistillate flowers globose, partially 6-celled, ovules 2 on each parietal placenta, styles distinct; fruits fleshy, globose, 1.4 cm in diameter; seeds 12. Widspread in Southeast Asia and India, a common and useful plant in Hainan Island.

Xylosma congestum (Loureiro) Merrill — PIERCING TREE OIL

Zha-mu=Cha-mu (扎〔柞〕木, Piercing Tree). Oil, extracted from the seed, used by people living in the limestone region of Guangxi.

Evergreen trees 10 m high, trunk armed with piercing branched spines, branchlets glabrous or pilose; leaves ovate, 3–8 cm long, 2–5 cm wide, acuminate, base obtuse or rounded, serrate; flowers small, greenish-yellow, unixexual, dioecious, in axillary racemes; fruits berry-like, black, globose, 3–4 mm in diameter; seeds 2–3.

Common in southern China, growing on sunny hillsides, thickets, or banks between fields. The correct Chinese ideogram for the name of this species is *"Zha"*(扎, for piercing). The ethnobotanical origin of the Chinese name *"Zha-mu"*(扎木, Piercing Tree) refers to the branched spines on the trunk and their effect on the approaching animals. The printed ideogram (柞) in botanical books is homonymous to *zuo-shu* (柞樹), an oak tree in northern China, the leaves of which are used for feeding silkworm in the production of Shandong silk (see *Xinhua Zidian*, 1987, p. 569, 613). For avoiding future confusion, the ideogram " 扎 " should be applied in the Chinese name of this species.

Passifloraceae: Passionflower Family

Passiflora edulis Sims — PURPLE GRANADILLA, PASSION FRUIT (Figure 42)

Xi-fan-lian-guo=Hsi-fan-lien-kuo (西番蓮果, Passion Fruit). Fruit, newly introduced, cultivated in gardens and in small yards forming an arbor to give shade; fruits used for preparing cool drink.

Perennial liannas climbing by axillary tendrils, branchlets angular, glabrous; leaves alternate, suborbicular in outline, 2–11 cm long, 4–10 cm across, 3-lobed, subcoriaceous, base rounded, broadly serrate, petioles with glands, stipules linear, subulate, 1 cm long; flowers white, solitary, actinomorphic, 7 cm across, sepals 3 cm long, basally connate, forming a receptacle, petals 5, 2.5 cm long, distinct, corona fleshy, filaments 4 or 5 series, the outer two series filiform, 1.5–2.5 cm long, the others progressively shorter, operculum concave, cup-shaped, stamens 5, opposite petals, anthers and pistil on a androgynophore, ovary unilocular, with 3 carpels, parietal placentation, ovules numerous, style distinct, stigma subcapitate; fruits purple, ovoid, 5 cm long; seeds numerous. Native of Brazil, cultivated in Taiwan, Hong Kong and Guangzhou.

Passiflora foetida L. — RUNNING POP, WILD WATERLEMON, LOVE-IN-A-MIST

Long-zhu-guo=Lung-chu-kuo (龍珠果, Dragon Pearl Fruit). Ripe fruit; with good flavor, gathered and consumed locally, not available in market.

Herbaceous vines, glabrous throughout, branchlets and foliage with ill-odor; leaves membranous, suborbicular in outline, 3- or 5-lobed, 4–15 cm long, 4–12 cm wide, acuminate, base subcordate with broad sinus; flowers solitary or 2–4, fasciculate, axillary, 3–5 cm across, peduncles 4–6 cm long, bracts involucrate, 2–4-pinnate, the ultimate segments filiform, sepals ovate-oblong, petals slightly shorter, white-pink or lilac-purple, corona filaments white, tinged purple at the lower portion, operculum short, white, punctate purple, androgynophore greenish-yellow, ovary oblong, glabrous, styles white, clavate, stigma subglobose, green, nutant; fruits globose, yellow to red, 3 cm long, 2 cm across, sparsely hirsute. Native of tropical America, naturalized in Hong Kong, Hainan Island, and Guangdong.

Begoniaceae: Begonia Family

Begonia fimbristipula Hance — Purple Back Tea Begonia

> *Zi-bei-tian-kui=Tsu-pei-t'ien-kuei* (紫背天葵, Purple-backed Begonia). Whole plant; dried, packed and sold in monasteries of Guangdong and Hong Kong.

Delicate small, tuberous, acaulescent perennial, growing in partial shade on rocks and crevices of tropical foggy peaks; tubers subglobose, 5 mm across; leaves 1 or 2, ovate-cordate, 3–7 cm long, 2–6 cm wide, acuminate, irregularly serrate, with appressed stout hairs on both surfaces, purple beneath, petioles slender, hispid; flowers pink, monoecious, 2–4 terminal to a slender scape 10–15 cm long, perianth segments 4 in the staminate flowers, 3 in the pistillate ones; capsules trigonous, unequally winged. Native to the high mountains of southern China, gathered for tea, used locally and for export, an endangered species.

Begonia hayatae Gagnepain

> *Yuan-guo-qiu-hai-tang=Yuan-kuo-ch'iu-hai-t'ang* (圓果秋海棠, Round-fruited Begonia). Stem; gathered and used in Taiwan.

Succulent rhizomatous glabrous perennial herbs 50–70 cm high, occasionally up to 1 m high, branchlets stout; leaves obliquely oblong, 15–20 cm long, 5–7 cm wide, acuminate, base unequally cordate and auriculate, palmately 6- or 7-nerved, remotely denticulate, petioles 3–4 cm long, stipules lanceolate, acute, 1 cm long, caducous; flowers white, monoecious, 4 or 5 in axillary cymes, perianth segments 4 in the staminate flowers, 6 in the pistillate ones, the outer pairs larger, 1.2 cm long, 8 mm wide, stamens numerous, ovary 3-locular, styles 3, bifid at the apex; capsules wingless, berry-like, depressed globose, 12 mm long, 14 mm across. Endemic to Taiwan.

Cactaceae: Cactus Family

Epiphyllum oxypetalum (de Candolle) Haworth — DUTCHMAN'S PIPE CACTUS

Tan-hua=T'an-hua (曇花, Ephemeral Flower). Flower; gathered the day after blooming, dried, used in making soup.

Spineless succulent cactus 1–3 m high, old stems subcylindrical, woody, flowering branches articulate, flat, leaf-like, 15–40 cm long, 5–6 cm wide, undulate along the margins, dark green, with slightly thickened median vertical axis; flowers nocturnal, white, fragrant, solitary, sessile, pendent from subapical marginal areoles, 20–30 cm long, 10–20 cm across, the perianth tube 12–15 cm long, with scattered scales, bent forward 90°, segments numerous, outer ones linear, increasing in width progressively to the inside, the inner ones petaloid, oblong-elliptic, 6–8 cm long, 2–3 cm wide, stamens numerous, inserted on the throat, ovary inferior, unilocular, ovules on parietal placentae; generally dropped after anthesis, fruits not seen in China. Native of tropical America, cultivated in gardens or as potted plants in southern China, and in green houses elsewhere in China.

Hylocereus undatus (Haworth) Britton et Rose — NIGHT-BLOOMING CEREUS, QUEEN-OF-THE-NIGHT (Figure 43)

Ba-wang-hua=Pa-wang-hua (霸王花, Usurper's Flower). Flower; an escaped species growing over walls and cliffs near Guangzhou; flowers gathered soon after blooming, dried, used in making broth with bone and meat; factory workers returning to the mainland for vacations often bring back dried flowers of the species as well as life chicken as present for the family.

Perennial succulent climbers growing on tree trunks, clifts or walls by aerial roots; stems green, stout, up to 7 m long, 10–12 cm across, winged, the wing and angle 1–2 cm wide, undulate, areoles 4–5 cm apart, with brown spines 3–4 mm long; flowers white, nocturnal, fragrant, 30 cm long, perianth tubes 15 cm long, the segments numerous, outer ones greenish tinged purple, the inner ones petaloid, white, stamens numerous, slightly shorter than the style, ovary inferior, style white, stout; fruits oblong, 10–12 cm across, red, edible. Native of tropical America, escaped and naturalized in tropical China.

Opuntia dillenii (Ker-Gawler) Haworth — PRICKLY PEAR (Syn. *O. stricta* Haworth var. *dillenii* [Ker-Gawler] L. Benson)

Xian-ren-zhang=Hsian-jen-chang (仙人掌, Fairy's Hand); *Xian-tao=Hsian-t'ao* (仙桃, Fairy Peach). Ripe fruit; gathered locally and eaten, not available in the market.

Strongly spiny succulent perennial herbs, 0.5–2.5 m high, old stems woody, cylindrical, fleshy segments flat, ovate-oblong, 15–20 (–40) cm long, 1.5–2 cm wide, green when young, becoming grayish-green, areoles 2–6 cm apart, slightly elevated, villose when young, spines brown, stout, 2–6 cm long; flowers bright yellow, solitary, 2–6 cm in diameter, the perianth segment obovate, stamens numerous, filaments greenish-yellow, ovary obovoid, style erect, white; fruits obovoid, 5–8 cm long, smooth, purplish-red, fleshy, the pulp dark red, juicy; seeds numerous.

Opuntia ficus-indica (L.) Miller — INDIAN FIG (Syn. *O. megacantha* Salm-Dyck).
Feng-long-guo=Feng-lung-kuo (鳳龍果, Phoenix Dragon Fruit); *Huo-long-guo=Huo-lung-kuo* (火龍果, Fire Dragon Fruit). Ripe fruit; available in Kunming, Taiwan and Hong Kong.

Largely fleshy erect cactus up to 5 m high, with a trunk 80 cm in diameter, stem segments obovate-oblong, 20–50 cm long, 10–30 cm across, bluish-green; spines variable, 1–2 cm long, white or off-white, sometimes lacking; flowers yellow, 6–7 cm long, 5–7 cm across; perianth segments several series, stamens numerous, ovary inferior; fruits pink-purple, oblong, 10–12 cm long, 8–10 cm across the middle, smooth, with flexible triangular bracts, 2 cm long, 3.5 cm across the base; flesh white, with numerous small black seeds; available in Hong Kong markets; said to have been introduced from Thailand.

Elaeagnaceae: Oleaster Family

Elaeagnus angustifolia L. — OLEASTER
Sha-zao=sha-tsao (砂棗, Sand Jujube). Fruit; used fresh, in Lanzhou, dried material common in the market; an important source of Vitamin C for the people living in the arid region of northwestern China.

Deciduous small trees or tall shrubs, 5–10 m high, some branchlets shortened into stiff thorns, current year's growth silvery white, second year's growth and branches rusty black; leaves elliptic-lanceolate, rarely oblong, 4–8 cm long, 0.8–2.5 cm wide, acute, obtuse or rounded, base obtuse, entire; flowers silvery outside, orange inside, 2–3 rarely 1 in an axillary fascicle, perianth campanulate, 6–7 mm long, 5 mm across the middle, apex 4-lobed, lobes deltoid, stamens 4, ovary inferior; fruits drupaceous, oblong, 10–15 mm long, 8–9 mm across. An important woody species of the arid region of northern and northwestern China, with an extensive range westward to the Mediterranean Region. A garden cultivar is called Russian Olive in American gardens.

Elaeagnus bockii Diels

Chang-ye-hu-tui-zi=Ch'ang-yeh-hu-t'ui-tzu (長葉胡頹子, Long-leaved Elaeagnus). Fruit, not available in market.

Evergreen shrubs 2 m high, branchlets armed with short stout thorns, current year's growth densely covered with brown peltate scales throughout; leaves elliptic, 5–11 cm long, 0.8–2.8 cm wide, acute, base obtuse, entire, the margin recurved when dried; flowers in axillary clusters, perianth funnelform, 10 mm long, 3 mm across the middle, apex 4-lobed, stamens 4, enclosed, ovary inferior, style stellate pilose; fruits oblong, 9–10 mm long, red, pendulous. Endemic to central China.

Elaeagnus courtoisii Belval — HAIRY OLEASTER

Mao-mu-ban-xia=Mao-mu-pan-hsia (毛木半夏, Hairy Oleaster). Fruit; eaten fresh or used for wine.

Deciduous shrubs 2 m high, densely covered with yellow stellate-villose hairs throughout; leaves elliptic-oblong, 4–9 cm long, 1–4 cm wide, abruptly acuminate or obtuse, base obtuse or rounded; flowers small, pale yellow, solitary, axillary, perianth tubular, 4-lobed, the tube 5 mm long, lobes deltoid, 3 mm long, hairy inside, stamens 4, ovary inferior, style glabrous; drupes red, oblong, 1 cm long, densely silvery-scaly. Native of central and eastern China.

Elaeagnus glabra Thunberg

Yang-nai-zi=Yang-nai-tzu (羊奶子, Goat Nipple); *Gui-xiang-liu=Kui-Hxiang-liu* (桂香柳, Cassia Fragrant Willow). Ripe fruit; eaten locally, not available in the market.

Evergreen subscandent shrubs up to 6 m high, branchlets lustrous brown, rugose due to the persistent brown peltate scales, occasionally with thorns; leaves oblong-elliptic, 4–11 cm long, 1–4 cm wide, abruptly acuminate, base rounded, upper surface lustrous, sparsely covered with brown scales, lower surface copper-colored, densely covered with brown scales, some large and others small; flowers 1–7 in axillary clusters, densely covered by glandular brown scales, perianth narrow-funnelform, 7–9 mm long, 2–3 mm across the middle, tube narrowed towards the base; fruits ellipsoid, 1.5–2 cm long, 6–8 mm across, longitudinally ridged and furrowed, densely covered by glandular brown scales. Native of eastern Asia, being reported from eastern China, Japan, Korea. Distinguished by very long and slender fruit-stalk, 2–3 times longer than the length of the drupe.

Elaeagnus henryi Warburg

Yi-chang-hu-tui-zi=I-chang-hu-t'ui-tzu (宜昌胡頹子, Ichang Elaeagnus). Fruit; gathered and eaten, or used for wine.

Evergreen scandent shrubs 3–5 m high, branchlets angular, rugose, densely covered with brown scales; leaves oblong, elliptic or obovate, very variable in size, 2–12 cm long, 1.5–7 cm wide, abruptly acuminate, base rounded, obtuse, or acute, entire, upper surface almost glabrous, scales very sparse, lower surface silvery, evenly covered by pale brown scales; flowers fasciculate, perianth tubular, 10 mm long, 3 mm across the middle, apex 4-lobed, rugosely covered by brown scales; fruits ellipsoid, 2 cm long, red when mature, covered by light brown scales. Endemic to central and western China.

Elaeagnus lanceolata Warburg

Pi-zhen-ye-hu-tui-zi=P'i-chen-yeh-hu-t'ui-tzu (披針葉胡頹子, Lanceolate-leaved Elaeagnus). Fruit; used in central China for food.

Evergreen shrubs 4 m high , branchlets grayish-rugose; leaves subcoriaceous, lanceolate, 3–9 cm long, 1–2.5 cm wide, acute, base obtuse or rounded, shiny, subglabrous above, densely covered by brown scales beneath; flowers solitary or fasciculate, perianth tubular, 7–10 mm long, 2–3 mm wide, apex 4-lobed; fruits oblong, 1–1.5 cm long, 6–7 mm across, red, covered by brown scales. Endemic to central and western China.

Elaeagnus loureiri Champion ex Bentham — SOUTH CHINA OLEASTER

Deng-diao-zi=Teng-tiao-tzu (燈吊子, Hanging Lantern Berry); *Diao-zhong-teng=Tiao-chung-t'eng* (吊鐘藤, Hanging Bell Vine). Fresh fruit; gathered by people living on the hillsides, not seen in the market.

Evergreen scandent shrubs 2–3 m high, branchlets rugose, covered by persistent brown scales, some shortened into stout thorns, a means for anchorage and climbing more than for protection; leaves elliptic or oblong, 4–9 cm long, 2–4 cm wide, acuminate, base obtuse or rounded, subglabrous above, evenly but incompletely covered by peltate scales, variable in sizes and color; flowers solitary, perianth broad campanulate, 12 mm long, 6–8 mm across; fruits ellipsoid, 2 cm long, 8 mm across, red at maturity. Endemic to southern China, common in the dry season, deciduous forest of Hong Kong.

Elaeagnus magna (Servettaz) Rehder

Yin-guo-hu-tui-zi=Yin-kuo-hu-t'ui-tzu (銀果胡頹子, Silvery Fruit Elaeagnus). Mature fruit; eaten locally by people living on hillsides, not available in the market.

Deciduous shrubs 1–3 m high, thorny, branchlets silvery; leaves membranous, obovate-elliptic, 4–10 cm long, 1–4.5 cm wide, obtuse, base cuneate, with silvery scales above, some persistent, silvery-white beneath; flowers white, 1–3 on new shoots, pedicels 2–3 mm long, perianth tubular, 8–10 mm long, yellow inside, stamens 4, style stellate hairy; fruits oblong, 12–16 mm long, with silvery scales, finally pink. Endemc to central China, Guangdong and Guangxi.

Elaeagnus multiflora Thunberg

Mu-ban-xia=Mu-pan-hsia (木半夏, Oleaster); *Si-yue-zi=Ssu-yüeh-tzu* (四月子, Fourth Month Fruit). Fresh fruit; gathered and eaten locally, not available in the market.

Deciduous shrubs 2–3 m high, branchlets rugose, densely covered by brown scales, some shortened into thorns; leaves membranous, oblong or elliptic, 3–7 cm long, 0.8–4 cm wide, shortly acuminate to obtuse, base obtuse, rounded, or cuneate, stellate-pubescent above, glabrescent, silvery beneath with scattered large brown scales; flowers yellowish-white, covered by silvery and brown scales outside, perianth tube campanulate, 10–12 mm long, 3 mm across the middle, distinctly constricted above the ovary; fruits oblong, 1.5 cm long, pendulous, the stalk 1–3 times longer than the length of the drupe.

Elaeagnus oldhamii Maximowicz

Shi-hu=Shih-hu (柿糊, Little Persimmon); *Yi-wu=Yi-wu* (宜吾, Good-for-me). Fruit; gathered by people living on hillsides and eaten fresh.

Small evergreen tree 3 m high, branchlets angular, rugose, densely covered with brown scales, thorns stout, 1–2.5 cm long, bearing 1 or 2 leaves; leaves very distinct for *Elaeagnus*, obovate, rarely emarginate or obcordate, 2.5–5 cm long, 1.5–2 cm wide, rounded or emarginate, base cuneate, glabrescent above, covered with silvery and scattered brown scales beneath, entire; flowers silvery, small, 1 or 2, axillary, perianth campanulate, 5–6 mm long, 3 mm across, apical half divided into 4 lobes, lobes deltoid; fruits globose, silvery, 6–8 mm long. Endemic to Taiwan.

Elaeagnus umbellata Thunberg

Niu-nai-zi=Niu-nai-tzu (牛奶子, Cow Nipple); *Tian-zao=T'ien-tsao* (甜棗, Sweet Jujube). Fruit; eaten locally in eastern and northern China, not available in the market.

Spreading deciduous shrubs up to 4 m high, branchlets angular, silvery, becoming gray with age; leaves variable, elliptic or oblong, 3–7 cm long, 1–3 cm wide, acute, obtuse or shortly acuminate, base obtuse or rounded, with silvery scales above when young, glabrescent, silvery beneath, with some brown scales also; flowers yellowish-white, very fragrant, with silvery scales outside, perianth funnelform, 1.2 cm long, 2 mm across the middle, gradually narrowed toward the base, style scaly; fruits subglobose, 6–8 mm across, scaly, finally red.

Hippophaë rhamnoides L. — Sea-Buckthorn, Shaji

Sha-ji=Sha-chi (沙棘, Sand Thorny Bush); *Cu-liu=Tz'u-liu* (醋柳, Vinegar Willow); *Suan-ci=suan-tz'u* (酸刺, Sour Thorn). Leaves for tea, ripe fruits for refreshing drink, seeds for oil.

Deciduous small trees up to 10 m high, branchlets angular and ridged, young growth covered with silvery scales and woolly stellate hairs, thorns sharp, 2.5–6 cm long, often with short secondary thorns 5–10 mm long; leaves linear-lanceolate, 2–6.5 cm long, 2–11 mm wide, obtuse or acute at both ends, entire, both surfaces covered with silvery scales, glabrescent above; flowers dioecious, blooming before the leaves, staminate flowers sessile, perianth reduced to 2 valvate lobes, stamens 4, filament short, pistillate flowers pedicellate, perianth with 2 minute lobes, style filiform, stigma cylindrical; fruit drupaceous, golden, subglobose, 6–8 mm across; seeds shiny chestnut-colored, oblique ovoid, slightly bilaterally compressed, 3–4 mm long, 2–3 mm wide, smooth and densely punctate, each side with a vertical pole-to-pole groove. Native of the desert areas of northern, and western China.

Note: In connection with conservation and protection of the Yellow River Region, and with the support of the World Bank, *shaji* has been capitalized in China. The National Shaji Commission and Yellow River Regional Shaji Committee were organized in Xian for operations of extensive planting of *shaji* in the desert and for the preparation of carbonized refreshing drinks, tea bags and oil from the seed.

Lythraceae: Loosestrife Family

Lythrum salicaria L. — Spiked Loosestrife

Qian-qu-cai=Ch'ien-ch'u-ts'ai (千屈菜, Thousand Bending Vegetable); *Shui-liu=Shui-liu* (水柳, Water Willow). Young shoots; cleaned, boiled, washed again, seasoned and eaten; recorded in the *Famine Herbal* (Zhu, 1407), recent record for food not yet available; tea for dysentery patients.

Perennial erect herbs 1 m high, pubescent throughout; leaves sessile, opposite, or 3 in whorls, lanceolate, 3.5–6.5 cm long, 1–1.1 cm wide, acuminate or acute, base cordate or amplexicaul; flowers magenta, in axillary cymose clusters forming interrupted terminal spikes; hypanthia cylindrical, tubular, the tube 5 mm long, striate, apical end with 6 subulate appendages on the striae and 6 deltoid sepals alternate with the appendages, petals 6, obovate-elliptic, 6–8 mm long, inserted below and opposite the appendages; stamens 12, 6 long, 6 short, inserted near the base of the hypanthium, the short ones opposite the sepals (included), the long ones opposite the petals (exserted); ovary conical-ovoid, on a stout gynophore, style filiform, stigma capitate; capsules ovoid, pericarp membranous; seeds numerous, brown, obovoid, angular, 1.5 mm long. Widespread in Eurasia, common in damp areas, wet meadows; naturalized in eastern North America.

Rotala indica (Willdenow) Koehne — Tooth-cup

Jie-jie-cai=Chieh-chieh-ts'ai (節節菜, Joint Potherb). Young shoots; eaten locally by people living near rice fields.

Annual weeds in paddies, 10–15 cm high, branchlets tetragonal, glabrous, much branched; leaves opposite, obovate-oblong, 6–10 mm long, 3–5 mm wide, subsessile; flowers small, 1.5–2 mm long, in axillary spikes 6–12 mm long, bracts obovate, 3–5 mm long, bracteoles 2, calyx campanulate, scarious, 4-toothed, corolla pink, very small, stamens 4, ovary superior, 1 mm long; capsules oblong, 1.5 mm long, striate; seeds wingless. Widespread in central and southern China, also occurring in Japan and the Philippines.

Punicaceae: Pomegranate Family

Punica granatum L. — Pomegranate

Shi-liu=Shih-liu (石榴, Pomegranate). Juicy seed coat; extensively cultivated, many cultivars, some for the ornamental flowers, mostly for the edible juicy seeds.

Shrubs 2–5 m high, branchlets often ending in a sharp thorn, spurs many; leaves opposite or fasciculate, oblong-obovate, 2–8 cm long, 1–2 cm wide, obtuse or retuse, base cuneate, entire; flowers orange-red, solitary or 2–3 in a fascicle, polygamous, hypanthium fleshy, the upper portion 5-lobed, valvate, persistent, petals membranous, crumpled in bud, inserted on the throat of the hypanthium, staminate flowers blooming slightly earlier, the base of the hypanthium obconical, dropping after anthesis, the pistillate flowers appearing 1–2 weeks later, hypanthium urceolate, the base rounded, stamens numerous, ovary 1, inferior, placentae two types, the lower ones free-central, and the upper ones pariental; fruit a large berry-like capsule, subglobose, pericarp leathery, dehiscing irregularly; seeds numerous, testa fleshy, juicy, often red, the inner seed coat woody.

Lecythidaceae: Lecythis Family

Bertholletia excelsa Humboldt et Bonpland — BRAZIL NUT; CREAM NUT, NIGGERTOES

Ba-xi-li=Pa-hsi-li (巴西栗, Brazil Chestnut). Imported, rare.

Very large trees of the Amazon forests up to 40 m high, branchlets stout, finely puberulous; leaves simple, alternate, oblong, 15–40 cm long, 10–15 cm wide, abruptly acuminate or rounded, base rounded, entire, with numerous (over 30) lateral nerves, the reticulation prominent on both surfaces, glabrous except along the midrib above; flowers creamy-white, 5 cm across, in terminal panicles 15–20 cm long, bracts 2, pilose, calyx cupular, deeply clifted into 2 sepals, petals 6, unequal, stamens numerous, united to a thick flap on an androphore at the lower side of the flower, ovary inferior, subglobose, style subulate, stigma minute; fruits globose, woody, 12–15 cm in diameter, weighing 900–1,800 g, with an apical plug (operculum) developed from the hardened calyx; seeds 12–24, 3-angled, brown, testa bony, 3–4 cm long, 2–2.5 cm across the broad back, rich in oil and protein. Introduced to some tropical gardens in China.

Rhizophoraceae: Mangrove Family

Carallia brachiata (Loureiro) Merrill

Zhu-jie-shu=Chu-chieh-shu (竹節樹, Bamboo-joint Tree); *Yan-zhi-guo=Yen-chih-kuo* (胭脂果, Rouge Fruit); *E-shen-mu=O-shen-mu* (鵝腎木, Goose-kidney Tree). Fruit; eaten in Yunnan.

Evergreen trees, 10–13 m high; leaves leathery, ovate, ovate-oblong, or elliptic, 6–15 (–18) cm long, 3–9 (–10) cm wide, acuminate, base acute, entire, petioles 6–10 mm long; flowers white, in axillary cymose corymbs, hypanthia funnelform-campanulate, 3–4 mm long, sepals 6–8, petals suborbicular, 2-lobed and toothed, stamens twice as many as the petals, ovary inferior, 4-locular, stigma capitate; fruits subglobose or ellipsoid, 1 cm in diameter; seed 1, reniform. Widespread from Madagascar via India, Burma, Malaysia, eastward to Australia, in China growing in tropical forests of Yunnan, Guangxi, Guangdong, Hong Kong and Hainan Island, at altitudes of 500–900 m.

Kandelia candel (L.) Druce — CANDEL; FIVE-PETALED MANGROVE

Shui-bi-zai=Shui-pi-tsai (水筆仔, Water Pen); *Jia-teng=Chia-t'eng* (加藤, Candel). Young root; recorded as food in Taiwan before 1950.

Evergreen shrubs 1–3 m high, growing in shallow waters at the estuaries in Hong Kong, branchlets glabrous, with distinct annular stipule-scars; leaves opposite, oblong or obovate-oblong, 4–11 cm long, 2–4.5 cm wide, rounded, base obtuse, stipules 1–2 cm long, covering the young buds, caducous; flowers white, in axillary dichotomous cymes, hypanthia hemispherical, 5 mm across, sepals 5, valvate in bud, fleshy, reflexed, petals reflexed, deeply 2-lobed, each lobe divided into 6–8 filiform segments, stamens numerous, ovary inferior, 2–4 locular, ovules in pairs, style simple; fruits fleshy, ovoid, 1 cm long, with 1 viable seed germinating on the plant, breaking the apical portion of the pericarp and sending out the fleshy clavate hypocotyl 10–15 cm long, hanging on the plant for months. An important component of the mangroves in Guangdong and Hong Kong.

Nyssaceae: Sour Gum Family

Nyssa javanica (Blume) Wangerin — SOUTH CHINA SOUR GUM
Sai-bao=Sai-pao (洒堡, the Taron ethnic name); *Yun-nan-zi-shu=Yun-nan-tzu-shu* (雲南紫樹, Yunnan Nyssa). Fruit, eaten by some of the ethnic groups of Yunnan.

Deciduous trees 20 m high, branchlets purple-brown, glaucous; leaves ovate-oblong, or ovate-lanceolate, 10–15 cm long, 3.5–5 cm wide, entire, acuminate, base obtuse, densely tomentose, glabrescent, densely papillose, petioles 1.5–3.5 cm long, furrowed and papillose above; flowers greenish-white, unisexual or polygamous, staminate flowers sessile, numerous, forming an axillary head 0.8–1.2 cm long, on a peduncle 1–4 cm long, sepals 5, villose, petals 5, obovate, pilose, stamens 10 in 2 series, the inner ones shorter, disk pulvinate, pistillate flowers few, on peduncles 2 cm long, ovary inferior, hairy; fruits drupaceous, globose or obovoid, 1.5–2.5 cm long, 2 cm across, crowned with persistent calyx, purple-red, turning black. Wide-spread in eastern Himalayan region, eastward to Indonesia; growing in damp areas of forests at altitudes of 800–2,100 m in Yunnan, Guangdong and Guangxi.

Combretaceae: Combretum Family

Terminalia bellirica (Gaertner) Roxburgh — BELLERIC MYROBALAN
Bi-li-le=Pi-li-leh (毗黎勒, Belleric). Kernel, only a few used, too many being toxic.

Large trees 18–35 m high, branchlets velutinous; leaves crowded at the end of shoots, ovate or obovate, 18–26 cm long, 6–12 cm wide, obtuse or shortly acuminate, base rounded or cuneate, glabrous, punctate above, petioles 3–9 cm long, with 2 median glands; flowers pale yellow, small, polygamo-dioecious, in axillary spikes 5–12 cm long, pubescent, hairs reddish, staminate flowers above the bisexual ones, perianth 6.5 mm long, cupular below, 5-lobed, stamens 10, exserted, disk 10-lobed, hairy, ovary inferior, unilocular, style stout at basal half, hairy; drupes ovoid, 2–3 cm long, 2–2.5 cm across, ridged, hairy; seed 1. Widespread, from India to Malaysia, growing in forests of shady valleys at altitudes of 540–1,350 m in southern Yunnan.

The Chinese name is a direct translation of the ethnic term shared by all the people living in the mountains of eastern Himalayan Region.

Terminalia catappa L. — INDIAN ALMOND, SINGAPORE ALMOND
Lan-ren-shu=Lan-jen-shu (欖仁樹, Canarium Kernel Tree). Seeds, eaten in Hainan Island.

Deciduous trees 20 m high, bark brown, branches spreading; leaves obovate, 12–25 (–30) cm long, 10–15 cm wide, obtuse, rounded or acute, base cuneate and subauriculate, pilose when young, glabrescent, petioles with discoid glands, one on each side; flowers small, greenish-white, polygamous, in axillary spikes 8–15 cm long, staminate flowers on the distal end, perianth 5-lobed, velutinous inside, stamens 10, exserted, 2.5–3 mm long, fertile flowers few, on the basal portion of the rachid, ovary inferior, 2-ovulate, style short, columnar, stigma punctiform, disk hairy; fruits drupaceous, compressed-ellipsoid, 2.5–5 cm long, ridged or narrowly-winged along two sides, olivaceous-black, outer portion of the pericarp fibrous, stone lignified; seeds oblong, matured in winter.

Myrtaceae: Myrtle Family

Eugenia uniflora L. — PITANGA, SURINAM CHERRY
Fan-ying-tao=Fan-ying-t'ao (番櫻桃, Foreign Cherry); *Bi-dang-qie=Pi-tang-ch'ieh* (畢當茄, Pitanga). Fruit.

Large, compact, glabrous shrubs or small trees up to 5 m high, branchlets slender, 1–1.5 mm in diameter, light gray; leaves opposite, glandular, punctate, ovate or ovate-lanceolate, 2–5 cm long, 1–2 cm wide, acuminate or obtuse, base acute, entire, dark green shining above, paler beneath, petioles 2–5 mm long; flowers white, rather small, 8–10 mm across, slightly fragrant, 3 to 5 appearing fasciculate as the axillary buds unfolding, before the development of the vegetative shoot, pedicels very tender, 1–2.2 cm long, hypanthia subglobose, calyx patelliform, shallowly 4-lobed, the lobes deltoid, 2 mm across, slightly enlarged and persistent in fruit, petals white, tinged pink on the outside, obovate-oblong, clawed, 5 mm long, ciliate, stamens ca. 40, filaments free, ovary inferior, 2-locular, ovules 10–14 in each cell, berries red, obovate, 1–2.5 cm in diameter, 8-ribbed; seeds 1 or 2 to each fruit, large, endosperm mealy, embryo straight. Native of Brazil; introduced to tropical Chinese gardens by different individuals; in Hong Kong, the Chinese name representing a straight translation of the English common name, Pitanga; in mainland tropical gardens its name is descriptive of its cherry-like red fruit and its foreign origin.

Feijoa sellowiana Berg — FEIJUA, PINEAPPLE GUAVA

Fei-hou=Fei-hou (妃后, Feijua, a translation of original name in sound). Fruit.

Shrubs or small tree 3 m high, with white-gray foliage, branchlets pubescent, current years' growth short, 1–2 mm in diameter; leaves opposite, obovate-oblong, 2–5 cm long, 1–2 cm wide, entire, obtuse, rounded or retuse, base acute, tomentose and punctate above, white lanate beneath; petioles 3–4 mm long, lanate, with 4 glandular setae on the adaxial side; flowers purplish-red, showy, 3–4 cm across, solitary, developed at the base of leafy shoots, on pedicels 1.5–2 cm long, terminated by 2 deltoid prophylls, hypanthia campanulate, 8 mm long, calyx lanate, 4-lobed, lobes orbicular, 4–5 mm long, persistent, petals 4, oblong, 15–17 mm long, punctate, stamens numerous, red, long exserted, filaments free, 16–20 mm long, anthers oblong, versatile, ovary 4-locular, ovules several in each cell, style equal filament in length, red; berries oblong or spherical, 3–8 cm long, dull green or tinged red, glaucous, pulp white, jelly-like around the seeds. Native of South America, introduced to the tropical botanical garden in Yunnan after 1950.

Myrtus communis L. — COMMON MYRTLE

Xiang-tao-mu=Hsiang-t'ao-mu (香桃木, Scented Peach Tree). Fruit.

Strong-scented evergreen shrubs 1–3 m high, branchlets slender, 1–1.5 mm in diameter; leaves opposite, glabrous, punctate, ovate-elliptic, rarely ovate-lanceolate, 2–5 cm long, 1–2 cm wide, acute, base obtuse or rounded, entire, petioles short, 1–2 mm long, with 2 or 3 glandular setae on each adaxial side of the base; flowers white, tinged red, 2.5 cm across, solitary, axillary to normal leaves, pedicels slender, 1.5–2 cm long, terminated by 2 oblong prophylls, hypanthia turbinate, densely punctate, sepals 5, deltoid, persistent and enlarged in fruit, petals orbicular, 8–10 mm across, stamens numerous, longer than the petals, anthers versatile, each with a prominent gland at the apex, ovary inferior, ovules numerous, style columnar, longer than the stamens; berries bluish-black, globose-oblong, 1 cm in diameter. Native of the Mediterranean Region and western Asia; introduced into tropical Chinese gardens after 1950.

Psidium guajava L. — GUAVA

Fan-shi-liu=Fan-shih-liu (番石榴, Foreign Pomegranate). Fruit; cultivated and naturalized in South China, fruit eaten raw; available in the market.

Small evergreen trees 4–5 m high, the trunk exfoliate, light brown, branchlets angular, pubescent; leaves opposite, light green, oblong or ovate-oblong, 7–12 cm long, obtuse or shortly acuminate, base rotundate, pilose beneath; flowers white, 2.5 cm across, solitary or 2–3 in axillary cymes, hypanthia pyriform, glabrous, 1 cm long, sepals 4 or 5, rotundate, persistent, petals 4 or 5, oblong, 1.5–2 cm long, 7–15 mm wide, stamens numerous, inserted on the disk, ovary inferior, 4- or 5-locular, ovules numerous; fruits fleshy, pyriform, 2.5–6 cm long, yellowish-green; seeds numerous, woody, buried in white flesh. Native to Central America, cultivated and naturalized in Hong Kong and Guangzhou.

Psidium littorale Raddi — PURPLE STRAWBERRY GUAVA (Syn. *P. lucidum* [Degener] Fosberg)

Cao-mei-fan-shi-liu=Ts'ao-mei-fan-shih-liu (草莓番石榴, Strawberry Guava). Fruits.

Small trees or shrubs up to 7 m high, bark smooth, grayish-brown, branchlets terete, glabrous; leaves elliptic or obovate, 5–10 cm long, 2–4 cm wide, acute, base obtuse, thickly coriaceous, glabrous; flowers white, solitary, 2.5 cm across, stamens numerous, ovary inferior; fruits ovoid or globose, purple-red, 2.5–4 cm long, fleshy, white or red, sweet. Native to Brazil; introduced into Hainan Island and Yunnan.

Rhodomyrtus tomentosa (Aiton) Hasskarl — DOWNY MYRTLE, HILL GOOSEBERRY (Figure 44)

Tao-jun-liang=T'ao-chun-liang (逃軍糧, Escaped Solders' Food); *Tao-jin-niang=T'ao-chin-niang* (桃金娘, a corrupt term close to the sound of the first name). Ripe fruit; gathered on the hillside and eaten raw, very seedy, not available in the market; formerly, British residents in Hong Kong picked the berries for home-made jam.

Evergreen shrubs 1 m high, growing on grassy hillsides, up to 3 m high when growing along the magin of woods, branchlets tomentose; leaves opposite, subcoriaceous, oblong or obovate-oblong, 3–6 cm long, 1.5–3.5 cm wide, obtuse, rounded or retose, base obtuse, prominently trinerved, tomentose beneath; flowers 1–3 in axillary cymes, purplish-red, 2 cm across, hypanthia turbinate, 6 mm long, with 2 ovate bracts at base, calyx 4 or 5 lobed, the lobes rotundate, 4–5 mm long, tomentose; petals 4 or 5, rounded, tomentose outside, stamens numerous, ovary inferior, 3-locular, cell with 2 rows of ovules, style filiform, stigma capitate; fruits subglobose, purple-violet, 1.4 cm in diameter, juicy; seeds numerous, testa lignified.

Syzygium aqueum (N. Buman) Alston — WATER ROSE-APPLE

Fan-gui-pu-tao=Fan-kuei-p'o-t'ao (番鬼蒲桃, Foreign-devil's Rose-apple). Fruit; introduced in Hong Kong and southern China, rare.

Large spreading trees up to 10 m high, bark smooth, gray, branchlets smooth; leaves almost sessile, opposite, ovate-oblong or elliptic-oblong, 6–10 (–20) cm long, 2.5–5.5 (10) cm wide, obtuse, base rounded and subcordate; flowers white, in axillary or terminal racemes 1–5 cm long, consisting of 4 or 5 flowers, hypanthia hemispherical, abruptly contracted into a pseudostalk, calyx 4-lobed, the lobes rounded, erect at anthesis, petals 4, subrotundate, unguiculate at base, 5 mm across, stamens numerous, ovary inferior, 2-locular, multiovulate, style 1 cm long; fruits white or pink, turbinate, 4 cm across the flattened top, crisp, juicy, persistent calyx lobes fleshy, incurved; seeds several or none. Native of Indo-Malaysia, extensively cultivated in southeastern Asia, rare in China.

Syzygium aromaticum (L.) Merrill et Perry — CLOVE (Syn. *Eugenia caryophylla* Thunberg)

Ding-xiang=Ting-hsiang (釘香, Nail Spice); *Ji-she-xiang=Chih-she-hsiang* (雞舌香, Chicken-tongue Spice). Dried flower buds; used in cities, rarely by rural people.

Mu-ding-xiang=Mu-ting-hxiang (母釘香, Mother of Clove). Fruit, available in cities, especially in medicinal shops.

Trees 6–20 m high, branchlets greenish-white; leaves coriaceous, elliptic, obovate-elliptic, or lanceolate, 7–12 cm long, 2.5–4.5 cm wide, acuminate, base cuneate and acute, entire; flowers red, in short terminal trichotomous cymose panicles, hypanthia clove-shaped, fleshy, 1–1.5 cm long, calyx 4-lobed, the lobes erect, acute, petals 4, greenish, falling early in the form of a hemispherical calyptra 6 mm across, stamens numerous, appearing in 4 yellow masses, filaments white, ovary inferior, 2-locular, cells multiovulate, style stout, thickened at the base, 4–4.5 mm long; fruits oblong-ellipsoid, 2.5–3 cm long, 1.3–1.5 cm across, persistent calyx lobes fleshy, pericarp thin, pulpy, fleshy; seeds solitary, pinkish-purple, oblong. Native of Moluccas, cultivated in Southeast Asia, imported to China in Sung Dynasty to the imperial court as tributes for exchange of silk and other precious Chinese goods, being under governmental monopoly, used particularly for the preparation of emergency remedies (Hu, 1990, pp. 495 and 513).

Syzygium buxifolium Hooker et Arnott

Chi-nan=Chih-nan (赤楠, Red Phoebe); *Yu-lin-mu=Yu-lin-mu* (魚鱗木, Fish-scale Bush). Ripe fruit, juicy, black, edible, also used for making wine.

Evergreen shrubs 1–3 m high, with compact habit and dark green foliage, branchlets 4-angled, glabrous; leaves opposite, coriaceous, obovate or broadly oblong, 2–4 cm long, 1.5–2 cm wide, obtuse or retuse, base cuneate; flowers white, small, in axillary and terminal cymes, 1–2 cm long, sessile, hypanthia obconic, 2 mm long, truncate, calyx lobes inconspicuous, petals round, 1 mm across, falling off early, stamens many, ovary inferior, style short; fruits globose, ripened shiny black, 8–12 mm in diameter; seed 1, large, cotyledon fleshy. Native of Hong Kong.

Syzygium cumini (L.) Skeels — JAVA PLUM, JAMBOLAN

Hai-nan-pu-tao=Hai-nan-p'u-t'ao (海南蒲桃, Hainan Rose-apple). Fruit, locally eaten in South China.

Evergreen trees 10–12 m high, branchlets grayish white, terete; leaves oblong-elliptic, 5–12 cm long, 2.5–5.5 cm wide, acuminate, base acute, entire; flowers white, in terminal and axillary panicles 3–8 cm long, hypanthia cupular, abruptly contracting at base, 2 mm long and across the truncate top, perianth segments falling together, stamens numerous, ovary inferior, 2-locular, cells multiovulate, style subulate; fruits oblong, 8 mm long, 6 mm across, persistent calyx collar-like; seeds several. Native of the Indo-Malaysian Region, from India to China and the Philippines.

Syzygium jambos (L.) Alston — ROSE-APPLE, MALABAR PLUM

Pu-tao=P'u-t'ao (蒲桃, Rose-apple). Fruits, common in villages, occasionally available in the market.

Evergreen trees 10–12 m high with spreading crown, dark green foliage, large white attractive flowers, growing by streams or in places of sufficient moisture; leaves opposite, coriaceous, oblong-elliptic, 10–20 cm long, 2.5–5 cm wide, long-acuminate, base cuneate and obtuse, lateral nerves conspicuous beneath, forming a marginal vein at their union; flowers white, 1–3 in terminal cymes, showy, 4–5 cm across, hypanthia obconic, 14–15 mm long, calyx lobes 4, 7–8 mm wide, rounded; petals 4, suborbicular, 1.2–1.5 cm long, 1–1.2 cm wide, stamens numerous, the filaments 2–3.5 cm long, ovary inferior, style slender; fruits subglobose or ovoid, 2.5–4 cm in diameter, greenish- yellow, rather hollow when ripe, making a noise on shaking; seeds 1 or 2, cotyledons large, fleshy; the edible portion with good flavor, but very little meat.

Syzygium malaccense (L.) Merrill et Perry (Syn. *Eugenia malaccensis* L.)

Nan-yang-pu-tao=Nen-yang-p'u-t'ao (南洋蒲桃, Malaysian Rose-apple). Fruit.

Medium-sized trees 5–20 m high, branchlets glabrous, terete, yellowish-green turning reddish-brown with age; leaves elliptic-oblong or obovate-oblong, abruptly short acuminate, base obtuse, 12–23 (–50) cm long, 5.5–8 (–20) cm wide, entire, glabrous; flowers red, in dense clusters in the axils of fallen leaves, hypanthium obconical, 6 mm across the top, contracted and extending into a pseudostalk 6–8 mm long, calyx lobes 4, broadly ovate, green tinged red, petals 4, red, rounded and unguiculate at base, stamens numerous, filament red, ovary inferior, 2-locular, cells multiovulate, style straight, wine-colored, gradually thickened toward the base; fruits red, rarely yellowish-white, ellipsoid-globose, 5–8 cm long, flattened at the apical end, persistent calyx fleshy, incurved, seed 1, globose, 2.5–3.5 cm across, brown, the edible portion white, juicy. Native to the Indo-Malayan Region, introduced to Hong Kong, and some tropical gardens in China.

Syzygium samarangense (Blume) Merrill et Perry — JAVA APPLE, SAMARANG ROSE APPLE (Syn. *S. javanicum* Hort.; *Eugenia javanica* Lamarck)

Yang-pu-tao=Yang-p'u-t'ao (洋蒲桃, Foreign Rose-apple); *Jin-shan-pu-tao=Chin-shan-p'u-t'ao* (金山蒲桃, San Francisco Rose-apple, a name used by Cantonese growers); *Lian-wu=Lien-wu* (蓮霧, Water Apple, the name used in Taiwan). Fresh fruit.

Trees 5–15 m high, trunk 50 cm in diameter near the base, branchlets terete; leaves opposite, shortly petiolate, oblong, 10–25 cm long, 5–12 cm wide, acuminate or obtuse, base rounded, entire, petioles 2–3 mm long; flowers white, fragrant, 3–4 cm across, in loose terminal racemes, hypanthia turbinate-campanulate, abruptly contracted at the base and extending into a pseudostalk 2–4 mm long, calyx lobes rounded, 4 mm long, 8 mm broad, petals and stamens falling early, filaments yellow, 1 cm long, ovary inferior, style greenish-yellow, 2.5 mm long; fruits pink, depressed pyriform or turbinate, 3–5 cm long, 4.5–5.5 cm across the top, often seedless, the edible portion milkywhite, juicy.

Melastomataceae: Melastome Family

Medinilla radiciflora C. Y. Wu

Suan-jiao-gan=Suan-chiao-kan (酸腳杆, Sour Ankle). Fruit; eaten by people living in Yunnan.

Shrubs or small trees 2–5 m high, branchlets tetragonous, terete with age, bark splitting vertically; leaves opposite, ovate-lanceolate, 15–24 cm long, 3–5.5 cm wide, caudate, base rounded or obtuse, 3- or 5-nerved, entire, slightly pilose and scabrid beneath, petioles 5–10 mm long; flowers pink, in cymose panicles emerging from the caudex, pedicels 4 mm long, pilose, calyx cupular, pilose, petals 4, obovate, 4.5 mm long, stamens 8, ovary inferior; fruits urceolate, 8 mm long, 7 mm across, pilose and verrucose; seeds numerous, cuneate. Native of southern Yunnan and Hainan Island, growing in forests along valleys at altitudes of 420–950 m.

Melastoma dodecandrum Loureiro

Di-nian=Ti-nien (地稔, Ground Berry). Ripe fruit, not available in the market, gathered and eaten locally in South China.

Perennial creeping herbs lying on dry slopes and often over bare rocks, basal portion of the stem woody, the young portion hairy, internodes slender, 3–5 cm long; leaves opposite, ovate or obovate-oblong, 1–4.5 cm long, 0.8–2.4 cm wide, obtuse, base acute, trinerved, sparsely strigose along the margin above and on the nerves beneath; flowers purple-pink, very pretty, 2.5–3 cm across, blooming only in sunshine, 1–3 in terminal cymes, hypanthia subglobose, 5 mm in diameter, strigose, sepals 5, lanceolate, 3 mm long, petals suborbicular-ovate, 10 mm in diameter, stamens 10 in 2 sets, 5 purple, sterile, the connective prolonged at the base into a bow, saccate at the end, 5 yellow, fertile, the connective auriculate at base, ovary inferior, 5-locular, multiovulate, style slender; fruits berry-like, ripe purple, globose, 7–10 mm in diameter, sweet, juicy.

Melastoma normale D. Don

Zhu-gu-nian=Chu-ku-nien (豬姑稔, Young Pig Berry); *Mao-nian=Mao-nien* (毛稔, Hairy Berry); *Zhai-yao-pao=Chai-yao-p'ao* (窄腰泡, Narrow Waist Berry); *Zhan-mao-ye-mu-dan=Chan-mao-yeh-mu-tan* (展毛野牡丹, Spread-hairy Melastoma). Ripe fruits, splitting, exposing the red content; recorded edible, but few people eat the fruit now.

Suffrutescent herbs 2–3 m high, densely covered with appressed rough hairs throughout; leaves opposite, ovate-elliptic, acuminate, 4–10.5 cm long, 1.4–3.5 (–5) cm wide, entire, palmately 5-nerved, petioles 5–10 cm long; flowers purple-red, showy, 3–7 (–10) in terminal corymbs, hypanthia urceolate, 1–1.6 cm long, sepals ovate-lanceolate, 1.5 cm long, margin fimbriate, petals obovate, 7 mm long, stamens 10 in 2 series, ovary inferior; fruit a rough berry, subglobose, 6–8 cm in diameter, irregularly splitting open, exposing the minute seeds on fleshy placenta. Widespread in southern China and Indo-Malayan Peninsula, growing in open hillsides at altitudes of 500–2,800 m.

Melastoma polyanthum Blume

Jiu-ping-guo=Chiu-p'ing-kuo (酒瓶果, Wine-bottle Berry); *Cui-sheng-yao=Ts'ui-sheng-yao* (催生藥, Abortifacient Herb); *Duo-hua-ye-mu-dan=Tuo-hua-yeh-mu-tan* (多花野牡丹, Many-flowered Melastoma). Fruits.

Shrubs 1 m high, densely covered with coarse fimbriate scales and hairs throughout, branchlets tetragonous; leaves opposite, lanceolate, 5.4–13 cm long, 1.6–4.5 cm wide, entire, pentanerved, petioles 5–10 cm long; flowers pink to purple-red, showy, 10 or more in terminal corymbs, hypanthia campanulate, 1.6 cm long, sepals lanceolate, 1.5 cm long, very hairy on both surfaces, caducous, petals obovate, 2 cm long, ciliate, stamens 10 in 2 sets, the longer ones purple, the shorter ones yellow, ovary inferior; capsules urceolate, truncate, 6–8 mm long, 5–7 mm across, placenta red, fleshy; seeds numerous. Widespread in southern Asia, extending eastward to Australia; growing on exposed hillsides at altitudes of 300–1,830 m; rural people in Yunnan boil the roots with black pepper, drink the tea for abortion, hence the local name *"Cui-sheng-yao"*.

Melastoma sanguineum Sims

Lang-gou-li=Lang-kou-li (狼狗脷, Police Dog Tongue); *Mao-nian=Mao-nien* (毛稔, Hispid Berry). Ripe fruits; eaten by people living in Hainan Island; a very common species on Hong Kong hillsides, where the people name it "Police Dog Tongue" for the leaf-shape.

Shrubs 1–2 m high, branchlets loose, densely yellow-strigose; leaves opposite, ovate-lanceolate, 10–14 cm long, 2.5–5 cm across the base, acuminate, base obtuse or rounded, with 5 prominent subpalmate nerves, the upper surface glabrous but uniformly and sparsely bristled with hair tissue at the base, lower surface bristle along the nerves; flowers purple-red, large, showy, 1–3 terminal to lateral shoots, hypanthia subglobose, 1–1.5 cm long and across, densely bristled, sepals 5–7, deltoid-lanceolate, shorter than the hypanthia, petals 5–7, obovate-spatulate, 3–5 cm long, stamens 10 in 2 series, ovary inferior, 5- to 7-locular, multiovulate; fruits hemispherical, 1–1.6 cm long, 1–1.5 cm across, covered with curved bristles, carnose, irregularly split on one side, exposing the scarlet red placentae and numerous seeds.

Trapaceae: Trapa Family

Trapa bicornis Osbeck — TRAPA, LING (Syn. *T. bispinosa* Roxburgh)

Ling=Ling (菱, Ling); *Wu-ling=Wu-ling* (烏菱, Black Ling). Fruit; gathered in September, boiled and eaten; common in South China, grown in partial protection.

Floating annual herbs, stems wire-like; leaves of two types, the submerged pinnatifid, the segments filiform, the floating leaves rosulate, rhombic-deltoid, 2–8 cm long, 3–10 cm wide, obtuse or acute, base broadly cuneate and rounded, irregularly coarse-dentate, hairy on the nerves beneath, glossy above, petioles 5–14 cm long, inflated and fusiform at the middle; flowers white, solitary on an axillary scape 3 cm long, hypanthia cupular, calyx deeply 4-lobed, the lobes ovate, hairy, petals 4, oblong, stamens 4, ovary half-inferior, 2-locular, cell uniovulate, disk cristate; fruits drupaceous, the fleshy pericarp early deteriorating, the nut 2-horned, horns representing the positions of 2 sepals, triangular in outline, purple-red or greenish, 4–6 cm between the apice of the horns, the middle slightly elevated; seeds meally, 2-cotyledons unequal, 1 minute and scaly. Cultivated in ponds and canals of the rice area, from the Yangtze River southward in China.

Trapa natans L. — Waternut, Water Caltrop (Syn. *T. incisa* Siebold et Zuccarini var. *quadricaudata* Glück)

Ling-jiao=Ling-chiao (菱角, Ling Horn); *Ye-ling=Yeh-ling* (野菱, Wild Trapa). Fruits, common in the market, particularly during the period of the Mid-Autumn Moon Festival (the full moon day in September or October).

Very aggressive aquatic annual herbs growing in quiet streams and ponds throughout the country; leaves of 2 types, the submersed remote, with capillary segments, the floating ones rosulate, deltoid, 1.5–5 cm long and wide, inflated portion of the petiole 4 cm long; flowers white, axillary to floating leaves, sepals 4, hairy, petals 4, stamens 4, glands nearly entire; fruits with 4 horns, 2 long and with sharp spines, 2 short ones recurved. One of the first recorded plants in ancient Chinese classics, every portion of the plant used by the Chinese people, as food, for medicine, and in religious ceremonies.

Various forms are reported in several provincial and local floras as species, *T. bispinosa* Roxburgh, *T. japonica* Flerov from Inner Mongolia, *T. maximowiczii* Korshinsky and *T. quadrispinosa* Roxburgh are some examples.

Cynomoriaceae: Cynomorium Family

Cynomorium songaricum Ruprecht

Suo-yang=So-yang (鎖陽, Lock-up Yang); *Xiu-tie-bang=Hsiu-t'ieh-pang* (銹鐵棒, Fusty Iron-club). Starch extracted from the stem, used for pastries.

Fleshy parasitic plants growing in association with desert shrubs (*Nitraria sibirica* Pallas), young subterranean plants subglobose or oblong, 6–15 mm in diameter, sparsely covered by minute triangular scales, the aerial portion of flowering plants 15–40 (–100) cm high, 3–6 cm in diameter, with small scales spirally arranged, the scales 0.5–1.5 cm long and wide, the distal clavate portion bearing small polygamous dark purple flowers, fragrant, subtended by scales; staminate flowers each consisting of 1 stamen and an irregularly 4-lobed perianth, the pistillate flowers each consisting of the pistil with or without a staminode, perianth 6-lobed, the lobes lanceolate, ovary inferior or half-inferior, uniovulate, style erect, stigma punctiform; achenes numerous, white, with persistent style, 1 mm across; seed 1, red, testa firm. Native of the desert area in Inner Mongolia and northwestern China, thence extending westward to central and western Asia.

Araliaceae: Aralia or Ginseng Family

Aralia chinensis L. — CHINESE ANGELICA TREE

Ci-long-bao=Tz'u-lung-pao (刺龍包, Prickly Dragon); *Cun-mu=Tz'un-mu* (楤木, Aralia Tree). Young leaves from the unfolding winter buds, used as potherb.

Coarse prickly shrubs or small trees 3–5 m high, branchlets yellow-villose, spinescent; leaves bi- or tri-pinnate, ultimate pinnae with 5–11 leaflets, and two single leaflets at the base, leaflets ovate, 5–12 cm long, 3–8 cm wide, acute or shortly acuminate, serrate, strigose above, villose beneath; flowers white, small, numerous in terminal compound panicles of umbels 30–60 cm long, rachis villose, pedicels 4–6 cm long, sepals 5, petals 5, stamens 5, ovary inferior, 5-locular, styles 5; fruits black, globose, 3 mm across, pentagonous. Widespread in China, particularly in the warmer regions.

Aralia cordata Thunberg

Tu-dang-gui=T'u-tang-kuei (土當歸, Local Angelica). Leaves used as potherb; roots eaten as parsnips.

Perennial herbs up to 2 m high, with white fleshy roots like parsnip; leaves bipinnate, ultimate pinnae 3- or 5-foliolulate, leaflets broad-ovate or oblong-ovate, 4–10 cm long, 3–10 cm wide, acute, base obliquely cordate, finely serrate, veins pubescent on both surfaces; flowers white, in compound terminal and axillary panicles of umbels; fruits black, fleshy, 5-gonous, 3 mm across.

Eleutherococcus senticosus (Ruprecht et Maximowicz) Maximowicz — ELEUTHERO

(Syn. *Acanthopanax senticosus* [Ruprecht et Maximowicz] Harms)
Ci-wu-jia=Tz'u-wu-chia (刺五加, Eleuthero). Stem and root, used for tea.

Upright prickly shrubs 2–3 m high, the stem of suckers densely yellow-bristly and prickly, glabrescent and grayish-white with age; leaves palmate, leaflets 5, elliptic or ovate-lanceolate, 8–12 cm long, 3–5 cm wide, acuminate, doubly serrate, pubescent beneath; flowers greenish-yellow, in globose umbels 3–4 cm across, peduncles 5–7 cm long, pedicels 1–2 cm long, sepals 5, minute, petals 5, ovate, stamens 5, ovary inferior, 5-locular, style columnar; fruits ovoid-globose, 8 mm across, 5-gonous. Native of northeastern China and adjacent Siberia; introduced to USA by the Arnold Arboretum in 1872, the original plant still living in good condition.

Eleutherococcus trifoliatus (L.) S. Y. Hu. (Syn. *Zanthoxylum trifoliatum* L.; *Acanthopanax trifoliatus* [L.] Voss)

San-ye-ci-wu-jia=San-yeh-tz'u-wu-chia (三葉刺五加, Three-leaved Eleuthero); *Bai-le=Pai-le* (白簕, White Prickly Stem). Young tender shoots, eaten by people in rural Guangdong in times of food shortage; in Hong Kong, root used for home-made tincture.

Arching and climbing shrubs 1–7 m high, branchlets sparsely covered with stout recurved prickles; leaves palmately trifoliolate, leaflets petiolulate, broad ovate, 4–10 cm long, 3–6 cm wide, acute, base obtuse, finely serrulate; flowers greenish-yellow, umbels 3–10 in terminal panicles, calyx 5-toothed, petals 5, stamens 5, ovary inferior, 2-locular, styles 2, united to the middle; fruits compressed globose, 5 mm across, mature black. Native of tropical China, introduced to Europe in 1773.

Panax ginseng C. A. Meyer — Chinese Ginseng, Korean Ginseng (Syn. *P. schin-seng* Nees)

Ren-shen=Jen-shen (人參, Ginseng). Root, eaten fresh as carrot in small amount, or boiled with chicken for a broth, use a large one or two small ones for the whole family.

Perennial herbs 30–80 cm high, producing 1 or 2 aerial leaves annually; root fleshy, fusiform, much branched, with horizontal lines indicating the positions of rootlets; stems short, stout with the cultivated products, slender with wild stock, normally with 1 (rarely 2 in cultivation) terminal winter bud; leaves twice-palmate, primary segments 2 to 4 depending upon the age and the growth condition of the plants, ultimate leaflets elliptic, the central 3 larger, 8–12 cm long, 3–5 cm wide, the lateral two broad-elliptic, 4–5 cm long, 2–3 cm wide; flowers yellowish-green, in simple umbel on a stipe with the lower portion united with the petiole of the compound leaf, polygamous, sepals 5, minute, petals 5, ovate, small, stamens 5, ovary inferior, 2-locular, styles 2; fruits drupaceous, scarlet red; pyrenes 1 or 2. Native of northern China and adjacent Korea, growing in partial shade of deciduous forests, associated with species of *Tilia* L.; now wild population very rare.

Panax quinquefolium L. — American Ginseng, Sang

Xi-yang-shen=Hsi-yang-shen (西洋參, Western Ginseng); *Hua-qi-shen=Hua-chi-shen* (花旗參, Star-Stripes Flag Sang). Root, imported annually or cultivated in China, used in similar manner as the Chinese Ginseng; seeds introduced by various growers from USA to China recently.

Perennial herbs, at flowering stage 25–45 cm tall, fleshy roots white, carrot-like, often forked due to excessive development of a lateral root; under natural growing condition, that of a 10-year old plant 5–10 cm long, 1–1.5 cm across the thickest portion, with numerous corky horizontal lines indicating the numbers and positions of lateral roots; stems subterranean, zigzag, rough, 2–3 (–5) cm long, 3–6 mm across, with distinct subdeltoid concaved alternate leaf-scars and 1 (very rarely 2) terminal bud developed in the axil of the solitary palmately compound leaf; features of ginseng leaves varying greatly with the age of the plants and their ecological conditions, leaves of seedlings bi- or trifoliolulate, those of 2-year old plants palmately compound, with 2 petiolulate tri- or pentafoliolulate primary units (palmae), leaflets obovate, the central ones the largest, 8–12 cm long, 4.5–7 cm wide, the lateral ones progressively smaller, acuminate, base obtuse or rounded, serrate; flowers greenish-yellow, beginning to appear in some 3-year old plants, polygamous, the first year's flowers all staminate, hypanthia obconical, hypathia of pistillate flowers subglobose, longer than the broad ovate sepals and petals; fruits red, lobatesubglobose; pyrenes (seeds) 1–3, white, with short viability.

Native of eastern North America, from eastern Canada southwestward to the Ozark Mountains in Missouri and Arkansas. American ginseng was first recorded in Chinese herbals in 1757 (Hu, 1990). For more than 200 years the importation has continued and the quantity has increased steadily, as shown by the data released by the Horticultural and Tropical Product Division Commodity Programs, FAS, USDA. Hong Kong is only an exchange port, eventually the American ginseng is distributed to Chinese consumers within China and abroad.

The earliest port of entry of American ginseng to China was Guangzhou. Later, after Shanghai became an international business center, it shared with Guangzhou the volume of importation. After 1937, Hong Kong has been the only port from which American ginseng enters China. At present, most American growers send their annual product to Hong Kong, for sorting and redistribution (see Part I for more information).

Panax wangianum Sun cv. 'Sanchi' — SANCHI (Syn. *P. pseudo-ginseng* Wallich var. *wangianus* [Sun] Hoo et Tseng; *P. notoginseng* [Burkill] F. H. Chen).
San-qi=San-ch'i (三七, Three-and-seven); *Tian-qi=Tien-ch'i* (滇七, or 田七, Yunnan Sanchi); *Ren-shen-san-qi=Jen-shen-san-ch'i* (人參三七, Ginseng Sanchi). Whole or sliced root, boiled with chicken for a broth, or steamed in a special cooker designed for chicken-sanchi dish; various imported sanchi products available in Chinese stores in America.

Perennial herbs 30–70 cm high, with fleshy short obconical or fusiform tap-root 2–3 (–5) cm long, 1–2 (–3) cm across the broad end, and numerous lateral roots and rootlets;

stem subterranean, short and stout, bearing 1 terminal winter bud capable of developing into a twice-palmately compound leaf when young, and with the primary petiole combined with the lower portion of the scape forming a common stalk at the flowering phase, 10–30 cm long, 3–12 mm across; leaves palmately compound, consisting of 1, 2, 3 or 4 palmae at different stages of development, petiolules terete, 10–15 cm long, 2–3 mm across, leaflets 3–7, elliptic-oblong, 5–14 cm long, 2–6 cm wide, abruptly acuminate, base rounded or oblique, serrate, the teeth bristly, sparsely strigose along the veins on both surfaces, or glabrescent, scapes solitary, terete, 20–30 cm long above the palmae; flowers yellowish-green, numerous (100 or more in cultivation) in simple umbel, polygamous, with the staminate flowers situated on the periphery of the umbel and the fertile ones in the center, stamens 5, ovary 2-carpellate, inferior; fruits drupaceous, red, 6–9 mm long, strongly compressed and appearing 2-lobed, pyrenes 1, 2 or 3, trigonous-globose, endocarp thick-coriaceous, ivory-white.

Native of western and southwestern China, growing in rich soil, good drainage, and partial shade of the hillside deciduous forests, with some hemlocks; in nature, at flowering stage, the plants usually bear 3 palmae, each with 7 leaflets, hence the vernacular name, *Sanchi* (*san* (three) for the 3 petiolules, *chi* (seven) for the 7 leaflets on each stalk, see Hu (1978, 1980) for details about the nomenclatural problems, ecological conditions and morphological structures).

Pentapanax henryi Harms

Xiu-mao-wu-ye-shen=Hsiu-mao-wu-yeh-shen (銹毛五葉參, Rusty-hairy Five-leaved Shen); *Ma-chang-zi-shu=Ma-ch'ang-tzu-shu* (馬腸子樹, Horse-intestine Tree). Young shoots used for potherb in Yunnan.

Small trees, or tall shrubs 2–8 m high; leaves pinnate, leaflets 3–5, ovate or ovate-oblong, 6–14 cm long, 3–8 cm wide, acute or shortly acuminate, base rounded, obtuse or subcordate, serrate, rusty-barbate in the nerve-angles beneath, petioles 5–20 cm long, petiolules of the terminal leaflets 1.5–3 cm long; flowers white, small, in large terminal panicles of umbels 2–30 cm long, rusty hairy, involucral bracts ovate, 0.5–1 cm long, calyx 5-toothed, petals 5, deltoid, 1.5 mm long, stamens 5, ovary inferior, syncarpous, 5-locular, styles 2–5; drupes ovoid, purple-black, 6–7 mm in diameter. Native of Hubei, thence westward to southern Yunnan.

Apiaceae (or Umbelliferae): Parsley (or Carrot) Family

Aegopodium alpestre Ledebour — GOUTWEED

Xiao-ye-qin=Hsiao-yeh-chin (小葉芹, Small-leaved Celery); *Shan-qin-cai=Shan-chin-ts'ai*

(山芹菜, Hillside Celery); *Dong-bei-yang-jiao-qin=Tung-pei-yang-chiao-chin* (東北羊角芹, Northeastern Goat-horn Celery). Young plants, gathered extensively in Liaoning hilly areas for food, cooked with flour, in soup or for pickles.

Perennial herbs 20–60 cm high, with creeping rhizomes thickened at the nodes, from which emerge the roots and erect stems, the stems hollow, glabrous; leaves both basal and cauline, deltoid in outline, 5–10 cm long, on petioles up to 30 cm long, ternately pinnate, ultimate segments ovate-lanceolate, 1–4 cm long, 0.5–2 cm wide, acuminate, the distal ones pinnately lobed, serrate, glabrous, sometimes scabrid along the veins beneath, upper cauline leaves sessile, sheathy; flowers white, tinged pink, polygamous, in terminal and axillary compound umbels 4–6 cm across, in fruit up to 12 cm across, without involucre, secondary umbels 10–16, umbellets 1 cm across, flowers 15–20, sepals obscure, petals obovate, apical end incurved, stamens 5, ovary inferior; fruits ovate-oblong, meriocarps laterally compressed, glabrous, ribs thin, vittae inconspicuous. Widespread in cold temperate eastern Asia.

Anethum graveolens L. — DILL

Shi-luo=Shih-luo (蒔蘿, Dill); *Xiao-hui-xiang=Hsiao-hui-hsian* (小茴香, Lesser Muslim Spice). Young leaves, used as spice; fruit, used in traditional Chinese medicine, roasted, pulverized, soaked in gin, taken for improving digestion.

Annual or biennial aromatic herbs, glabrous, 1–1.5 m high, stem 1.5 cm in diameter, green with silvery vertical stripes; leaves ternately decompound, ovate-oblong in outline, 10–26 cm long and wide, petioles 5–7 cm long, sheathy at the base and with membranous stipules, 2.5 cm long; flowers small, yellow, 5-merous, in terminal compound umbels 6–15 cm across, consisting 40–46 umbellets, devoid of involucre and involucel, ovary inferior, 2-carpellate, ovules solitary; schizocarps flattened, the lateral ridges winged, mericarps oblong, 4–5 mm long, 3–4 mm wide (including the wings).

Native of southwestern Asia, introduced by Arabian traders; first mentioned in Chinese herbal of Tang Dynasty (756 A.D.); Jiangsu and Anhui are the present producing centers for market demand.

Angelica omeiensis Shan et Yuan — ASPARAGUS ANGELICA

Du-huo-gan-gan=Tu-huo-kan-kan (獨活桿桿, Angelica Shoot). Young shoots; before the appearance of any leaves, eaten as cucumbers by hunters and medicinal herb-collectors living in the spruce and fir forests and alpine meadows of western Sichuan during the warm season.

Perennial herbs 1.5 m high, with one stout stem 2.5–3.5 cm in diameter, hollow, purplish outside, glabrous; leaves ternately 2- or 3-pinnate, the petioles with foliaceous white stipules 8–15 cm long, 4–5 cm wide on each side, laminas broadly triangular-ovate in outline, ultimate segments pinnatisect, 10–12 cm long, 4–5 cm wide, the lobes ovate-elliptic, 1.5–4 cm long, 8–15 mm wide, acuminate, base oblique, sharply serrate, pilose on major veins on both surfaces; flowers lavender, in large compound umbels terminal to the primary and lateral shoots, consisting of 22–40 umbellets, involucral bracts linear, 3 cm long, 1 mm wide, pilose, peduncles of umbellets 6–12 cm long, pilose, bracteoles linear-filiform, 1–2 cm long, longer than the pedicels, flowers polygamous, 20–80 to each umbellet, the outer ones perfect, pedicels 5–10 mm long, pilose, hypanthia glabrous, calyx teeth inconspicuous, petals 1 mm long, unguiculate, fully grown but immature schizocarps obovate, 9 mm long, 6 mm wide, with well-developed wings, 5 prominent ridges and 7 resinous grooves. Native of western Sichuan, common in the alpine meadows at altitudes of 3,500–4,000 m, the emerging stems being an important source of vitamins and minerals of the small number of temporary summer residents of the alpine meadow; the first report of this wild food was based upon S. Y. Hu 1422 and 2684, determined by F. T. Pu as *A. omeiensis* Shan et Yuan.

Angelica polymorpha Maximowicz

Guai-zi-qin=Kuai-tzu-ch'in (拐子芹, Staff Celery). Young plant; gathered from the hillside in northeastern China, considered to be an important source of green vegetable in Liaoning Province.

Perennial herbs up to 1 m high, stems solitary, cylindrical, hollow, ridged, purple, glabrous except the portion below the inflorescence; leaves both radical and cauline, on petioles up to 15 cm long, with broad sheath at the base, up to 47 cm long including the petiole, laminas deltoid in outline, 2- or 3-times pinnate, the primary and secondary pinnae petiolate, the petioles curved upward forming arches, ultimate segments ovate-cordate, acute, irregularly double-serrate, teeth apiculate, scabrid along the margin, elsewhere glabrous, upper cauline leaves reduced progressively upward, the petioles sheathy completely, purple; flowers white, in terminal axillary compound umbels 4–10 cm across, bracts 0 to 3, linear, secondary umbels 8–22, scabrid, umbellets 1 cm across, containing 20 flowers, involucral bracts 7–10, filiform, scabrid, sepals obscure or very small, subulate, caducous, petals obovate, claw obscure, apex incurved, stylopodium thick-discoid; fruits dorso-ventrally flattened, rectangular-oblong, 6–7 mm long, 4–5 mm wide, cordate at the base, ribs of mericarps close to each other, lateral ribs broadly winged, membranous, 1.5–2 mm wide, wider than the seed, furrows very narrow, each with 1 vitta, commissural vittae 2. Native to cold temperate eastern Asia.

Angelica sinensis (Oliver) Diels — Chinese Angelica, Dang-gui (Syn. *A. polymorpha* Maximowicz var. *sinensis* Oliver)

Dang-gui=Tang-kuei (當歸, Proper Restoration). Enlarged root; cultivated in the alpine meadows of Yunnan, Sichuan and Gansu, third year old roots harvested in autumn, cleaned, dried in shade for a week, sorted by sizes, and dried over low heat; whole roots, elegantly prepared thin slices, and bundled rootlets, all available in American Chinese stores; used for the restoration of strength and for the improvement of circulation, in the form of broth prepared with Chinese jujube, lycium-berries, laminaria and chicken or pork, boiled first and then cooked with low heat for 3–4 hours.

Aromatic perennial herbs 40–150 cm high, aerial stem purplish-red; basal leaves ternately decompound, ovate in outline, 15–20 cm long and wide, ultimate leaflets ovate or oblong-lanceolate, 1–2 cm long, 5–15 mm wide, shallowly 3-lobed, serrate; flowers white, small, in axillary and terminal compound umbels, involucre 0–2, primary division of the inflorescence consisting of 9–13 secondary umbellets on unequal stalks subtended by linear involucels, pedicels 12–26, pilose; mericarps suborbicular, 4–6 mm long, 3–4 mm wide, lateral ridges winged. Native of the Tsinling Range, hence extending southwestward to the mountains of Gansu, Hubei, Sichuan, Guizhou and Yunnan. Natural population is rare, market products are all from cultivated crops. It has been used in China since the time immemorial. It is recorded in the first Chinese dictionary of terms, *Erh-ya* 爾雅 (ca. 3 B.C.) as Mountain Celery (*shan-qin*, 山芹). In the first Chinese herbal, the *Herbal Classics of the Divine Plowman* (*Shen-nong ben-cao-jing*, 神農本草經), the name *dang-gui* was used, and under this name, the product has spread worldwide in many different forms. For many years I have received annual long distant calls from the American public interested in Chinese plant products asking for the meaning of *dang-gui*. *Dang*=proper or should, and *gui*=return or restore. Thus *dang-gui* is a herb used for helping the circulation, to make the blood return to the proper channel, and to restore the correct order in the circulation system (see Part I for more information).

Anthriscus cerefolium (L.) Hoffmann — Chervil

Xue-wei-cai=Hsüeh-wei-ts'ai (雪維菜, Chervil). Leaves; used for condiment by foreign residents in large cities, and by restaurants serving western food.

Erect annual or biennial aromatic herbs, 14–80 cm high, the stem much branched; leaves ternately decompound, ovate-deltoid in outline, 8–10 cm long and wide, the ultimate segments triangular-ovate, pinnately lobed, the lobes 3 mm long, 1 mm wide, obtuse; flowers small, white, in terminal compound umbels, ultimate umbellets 3–8 flowered, ovary linear; schizocarps needle-like, 7–8 mm long, 1 mm across, beaked at the apex. Native of Europe and western Asia, introduced by European residents of modern cities in China.

Anthriscus nemorosa (M. Bieberstein) Sprengel — WOODLAND CHERVIL (Syn. *A. aemula* [Woronow] Schischkin; *A. sylvestris* [L.] Hoffmann var. *aemula* Woronow).

Lin-di-e-shen=Lin-ti-o-shen (林地峨參, Woodland High Shen); *E-shen=O-shen* (峨參, High Shen); *Shui-hu-luo-bo=Shui-hu-lo-po* (水胡蘿蔔, Water Carrot). Young shoots, an important wild vegetable in Liaoning and Jilin Provinces; root used for pickles by "Chinese Koreans" (the people living in eastern Jilin with Korean language and culture).

Perennial herbs 50–120 cm high, stems stout, much branched; leaves deltoid in outline, 7–12 cm long and across the base, ternately decompound, the ultimate segments 3–7 mm long, 2–4 mm across, serrate; flowers polygamous, small, white, in terminal compound umbels, peduncles 4–7 cm long, with or without bracts, secondary umbels 7–11, involucral bracts 5, ovate-lanceolate, ciliate, umbellets 1–1.5 cm across, consisting of 10–15 flowers, 4–11 fertile, sepals obscure, petals unequal, the outer ones of the outside flowers larger, 3.5 mm across, stamens 5, ovary inferior, pilose; fruits oblong, 7–8 mm long, 2–3 mm across, spiculate and muricate, mericarps laterally flattened, ribs inconspicuous, rostrate, 1-vitta in each furrow, commissural vittae 2. Widespread in eastern Asia, growing in meadows and in rich soil of forests along streams; roots used in Sichuan under the name "*E-shen*".

Apium graveolens L. var. **dulce** (Miller) Persoon — CELERY

Qin-cai=Ch'in-tsai (芹菜, Celery). Leaves, particularly the petioles, extensively cultivated throughout the country; Chinese criteria for the selection of cultivar stress strong flavor and slender petioles used in quick fry-cooking.

Annual or biennial herbs 20–150 cm high, glabrous throughout; basal leaves deitoid oblong in outline, 70–80 cm long, bipinnate, the ultimate leaflets ovate, 2–4.5 cm long, deeply trilobate, lobes rhombic, dentate, petioles 10–30 cm long, subcylindrical, broadly grooved on the adaxial side, slightly sheathy at the base, cauline leaves triparted, obovate-cuneate, shallowly 3-lobed and dentate; flowers greenish-white, small, in axillary and terminal compound umbels devoid of involucre and involucel, ultimate umbellets very small, 5 mm across, consisting of 10–12 minute flowers 1 mm across, on pedicels 2–5 mm long; schizocarps very small, subglobose, 1–1.5 mm long and across, slightly compressed. Cultivated throughout the country, the edible petiole thin, used only for cooking.

Carlesia sinensis Dunn

Shan-hui-xiang=Shan-hui-hsiang (山茴香, Hillside Fennel); *Jia-fang-feng=Chia-fang-feng* (假防風, False Fang-feng). Young plants, used for flavoring vegetable and meat dishes in Shandong.

Perennial aromatic herbs 10–30 cm high, with stout carrot-like tap root, rootstocks covered by fibrous remains of disintegrated herbage; leaves rosulate, on slender petioles 4–6 cm long, laminas oblong-ovate in outline, 3–9 cm long, ternately pinnate, ultimate segments (3–) 5–8 (–15) mm long, 1 mm wide, acute, glabrous; flowers white, in terminal or axillary compound umbels 4–6 cm across, involucral bracts 5–8, linear, 5–10 mm long, secondary umbels 10–20, peduncles 1.5–3 cm long, bracts many, linear, 3–5 mm long, pedicels 2 mm long, sepals ovate-linear, petals obovate, apex acuminate, incurved, stamens 5, ovary inferior, styles filiform, divergent, stylopodium conical; fruits oblong, 5–6 mm long, strigose, mericarps laterally flattened, ribs slightly raised, vittae 1 opposite ribs and 3 between ribs, 4 on the commissural side. A monotypic genus, poorly represented in herbaria, an endangered species.

Carum carvi L. — CARAWAY

Mian-xiang-cai=Mien-hsiang-ts'ai (麵香菜, Caraway); *Zang-hui-xiang=Tsang-hui-hsiang* (藏茴香, Tibetan Fennel). Young plants, chopped and used for flavoring; Fruit, rarely used in Chinese cooking.

Erect, glabrous, slender, biennial or perennial herbs 20–80 cm high, with fusiform fleshy roots, stem much branched; leaves pinnately decompound, oblong in outline, 8–14 cm long, 3.5 cm wide, ultimate segments linear, 2–3 mm long, 1–2 mm wide, petioles 5–8 cm long, with broad basal sheaths; flowers very small, pinkish-white, in large and small compound umbels, primary divisions 10–14, involucre 2 or 3, linear, 4–5 mm long, ultimate umbellets 15- to 18-flowered, pedicels 2–4 mm long; schizocarps oblong, 5 mm long, 2 mm across, with strong ribs. Widespread in the northern hemisphere, occurring along paths in hillsides and in the meadows of northern China and the Qinghai-Tibetan Plateaus.

Centella asiatica (L.) Urban — MONEYWORT

Beng-da-wan=Peng-ta-wan (崩大碗, Break Large Bowl); *Ji-xue-cao=Chi-hsüeh-ts'ao* (積雪草, Centella). Whole plant; used for the preparation of *liang-cha* (涼茶) with many other herbs to make a cool drink, common on streets or in market places of South China.

Stoloniferous creeping perennial herbs, puberulous; leaves simple, alternate or subfasciculate, reniform, 1–5 cm across, base cordate, dentate, petioles 2–5 (–20) cm long, stipules membranous, adnate to the base of the petiole; flowers small, hidden by the leaves, inconspicuous, violet-maroon (almost black), sessile, 3, crowded to the apex of fasciculate peduncles 3–4 mm long, involucral bracts green, with red margin, ovate, hypanthia subglobose, sparsely villose, petals 5, red, stamens 5, anthers black, ovary inferior, 2-carpellate, the carpels strongly compressed, schizocarps suborbicular, 3 mm in diameter, reticulate. A very common species occurring along waysides, in gardens, on steep hillsides, in wet banks of rice fields and in cracks at the foot of rocks in tropical China.

Coriandrum sativum L. — Coriander, Chinese Parsley, Cilantro (Spanish)

Yuan-sui=Yuan-sui (芫荽, Coriander); *Xiang-cai=Hsiang-ts'ai* (香菜, Aromatic Vegetable). Young or fully grown plants; harvested before flowering, in villages of northern China leaves are washed, chopped small, mixed with cooked and salted soya bean and sprinkled over cooled fish or meat; seed used as a spice.

Strongly aromatic annual or biennial herbs 30–100 cm high, stems branched, striate; basal and lower cauline leaves pinnate or bipinnate decompound, ultimate segments ovate or rhombic, 1–2 cm long, 1–2 cm long, deeply lobed, cuneate, upper cauline leaves subsessile, deeply triparted, the segments linear, petioles slender, 3–15 cm long; flowers pink-white, small, in axillary and terminal compound umbels, the petals unequal, those on the outside of an umbellet much larger than those on the inside, stamens 5, ovary inferior, globose, 2-carpellate; schizocarps globose, 2 mm in diameter. Native of southern Europe, extensively cultivated in China, much used in Chinese cooking, general eaten raw.

Cryptotaenia japonica Hasskarl

Ya-er-cai=Ya-erh-ts'ai (鴨兒菜, Duck Vegetable); *Han-qin=Han-chin* (旱芹, Highland Celery). Young shoots, gathered and consumed, not available in the market. Erect glabrous perennial herbs 30–60 cm high; basal and lower cauline leaves ternately parted, triangular in outline, the terminal segments rhombic-ovate, the lateral ones slightly oblique-ovate, 3–10 cm across the broad portions, petioles 5–17 cm long, sheathy at the base; flowers small, white, in loose irregular subumbelliform panicles, ultimate umbellets 2- to 4-flowered, involucre linear-subulate, 1.5 mm long, sepals inconspicuous, petals rounded, caducous, 1 mm across, inferior ovary obconic, 1 mm long, glossy; schizocarps oblong-ellipsoid, 4–6 mm long, 2 mm across the middle, with strong ribs. Widespread in Eurasia and North America.

Cuminum cyminum L. — CUMIN, CUMMIN, ZIRAN (Persian), JIRA (Sanskrit).

Zi-ran-qin=Tzu-jan-ch'in (孜然芹, Ziran Celery). Fruit; used in the form of powder as a flavoring material in Xinjiang.

Annual or biennial herbs 15–30 cm high, glabrous throughout except the fruit, stems branched at the ground level; leaves ternately divided, deltoid in outline, the ultimate segments linear or filiform, 1–5 cm long, 0.2–0.5 mm across, petioles 1–2 cm long, sheathy at the basal half; flowers small, white, turning lilac or rose, in subsessile compound umbels at anthesis, 2–3 cm across, ultimate umbellets 3–5 flowered, involucral bracts 3–6, linear, 1–5 cm long, apiculate, involucels 3–5, pedicels unequal, sepals subulate, petals oblong, retuse and incurved, stamens shorter than the petals, ovary oblong, glandular, ridged; schizocarps oblong-eilipsoid, 6–7 mm long, densely setaceous and bristly. Native to Egypt, cultivated throughout the Mediterranean Region, Iran, and India, introduced to Xinjiang and cultivated for local use.

Czernaevia laevigata Turczaninow (Syn. *Angelica laevigata* Franchet, non Fischer)

Liu-ye-qin=Liu-yeh-ch'in (柳葉芹, Willow-leaved Celery); *Ji-zhua-qin=Chi-chua-ch'in* (雞爪芹, Chicken-feet Celery). Young plants, large amount gathered annually in Liaoning for potherb between April and June.

Biennial erect herbs 90–120 cm high, stems hollow, ribbed, branched; leaves both radical and cauline, on long petioles, laminas ovate-deltoid in outline, 30–50 cm long, 25–40 cm wide, pinnate, lateral pinnae petiolate, ovate, 4–10 cm long, 1–4 cm wide, pinnatisect, segments lanceolate, acuminate, base of terminal segments decurrent, serrate, strigose along the nerves above, glabrous beneath, cauline leaves gradually reduced in size above the middle of the stem, petioles sheathy; flowers white, in terminal compound umbels 10–30 cm across, primary umbels 20–40 (–70), pilose, bracts 0 or 1, sheath-like, umbellets 1–2.5 cm across, involucral bracts 10, lanceolate, sepals obscure, petals obovate, apex strongly incurved, those of the marginal flowers larger, stylopodium discoid; fruits dorso-ventrally flattened, mericarps oblong-suborbicular, 3–4 mm long and wide, lateral ribs strong winged, vittae 3–5 in furrow, (7–) 0–14 in commissural side. Native to northeastern China and the adjacent land in Siberia; leaves very variable, two varieties recognized by local botanists; both used for food:

1. forma **latipinna** Chu — *Kuan-ye-liu-ye-qin=K'uan-ye-liu-yeh-ch'in* (寬葉柳葉芹, Broad-leaved Willow Celery). Ultimate leaf-segments ovate, 2–4 cm wide, incised-lobate.

2. forma **exalatocarpa** Chu — *Wu-yi-liu-ye-qin=Wu-yi-liu-yeh-ch'in* (無翼柳葉芹, Wingless Willow Celery). Lateral ribs of the mericarps almost wingless.

Daucus carota L. — CARROT

Hu-luo-bo=Hu-lo-po (胡蘿蔔, Carrot); *Jin-sun=Chin-sun* (金筍, Golden Shoot, a Cantonese name). Root, extensively cultivated, regarded as the life-saver of many people in times of food shortage in the 1960s.

Biennial herbs, the aerial portion with strong smell, roots fleshy, cylindrical and tapping gradually toward the distal end, yellow-orange, stems much branched, appearing before flowering; leaves basal and cauline, basal leaves rosulate, the outer ones the larger, triangular-ovate in outline, first ternately and then pinnately decompound, up to 22 cm long, 20 cm wide, the cauline leaves 26–30 cm long, 14–20 cm wide, becoming progressively smaller upward, the ultimate segments deeply lobed, cuneate, the lobes linear, 1–2 mm long, acute, petioles various, those of the basal leaves longer than the laminas, the cauline petioles becoming shorter progressively upward, sheathy at the base, membranous along the sheeths; flowers white, small, in terminal and axillary compound umbels 10–15 cm across, primary segments 100–150, peduncles becoming progressively shorter to the inside, forming a lace-like mosaic, umbellets consisting of 40–50 flowers, 1–1.5 mm across, sepals obscure, petals suborbicular-obovate, unequal, the outer ones of the outermost umbellets larger, bilobed, incurved and cristate, stamens 5, filaments twice as long as the petals, ovary inferior, subglobose-oblong, bicarpellate, carpel with 5 hispid primary ribs and 4 smooth furrows, involucres and involucels triangular, pinnate, segments linear-lanceolate, aciculate, strigose, margin membranous; schizocarps oblong-ellipsoid, 5–6 mm long, 3 mm across, each side with 4 curved ribs developed from the smooth furrows, armed with 10–12 hooked or barbed bristles, the mericarps separated along the adaxial face (commissure), leaving the axis intact, not separating into two carpophores; peduncles of the primary umbellets turning upward after anthesis and curved inward as the fruits mature, forming an interlocked ball of prickly fruits. Wild populations of the species widely distributed in America, Africa and Eurasia; recorded in Chinese herbals in late Yuan Dynasty (after 1330 A.D.).

Eryngium foetidum L. — ERYNGO

Yang-yuan-sui=Yang-yuan-sui (洋芫荽, Foreign Coriander); *Yang-xiang-si=Yang-hsiang-ssu* (洋香絲, Foreign Silk-spice). Young shoots, chopped into very thin slices, used as spice.

Perennial or biennial, strongly aromatic, glabrous herbs, 15–20 cm high; leaves all basal, coriaceous, oblanceolate, 10–18 cm long, 2–3 cm wide, acute, cuneate, prickly serrate; flowers small, green and whitish, subsessile, in bracteated heads situated at the juncture of the branching of the scape and its secondary axes, involucre foliaceous, consisting of 4–6 lanceolate and spinose bracts, bracteoles longer than the flowers, ovate-lanceolate, crowded with the flowers in an ovate-oblong head 4–8 mm long, 5 mm across, sepals ovate-oblong, 0.5 mm long, aciculate at the apex, persistent, petals slightly shorter, largely concealed by the sepals, stamens shorter than the petals, anthers on very short filaments, concealed by the sepals and petals, ovary inferior, subglobose, densely verrucose, styles 2, 1 mm long, exserted, after anthesis the axis of the inflorescence slightly elongated into a fruiting column 1.5 cm long, 5 mm across; schizocharp subglobose, 1.5 mm long and across, densely covered by hyaline white tubercles, mericarps separated at the commissure, showing no evident carpophores. Native of tropical America, cultivated in Guangdong, Hainan Island, Guangxi and Yunnan, rarely seen in the market.

Ferula tunshanica Su — TONG SHAN ASAFETIDA

Tong-shan-a-wei=Tung-shan-a-wei (銅山阿魏, Tong Shan Asafetida). Young plants, gathered in early spring, dipped in boiling water, washed in cold water, seasoned for salad; fleshy root, gathered in autumn, cleaned, partially dried, pickled for winter use, eaten as a vegetable.

Glabrous and glaucous perennial herbs, 1.5 m high at flowering stage, stems branched 50 cm above the ground; radical and lower leaves 3- or 4-times divided, ultimate segments ovate, 1.5 cm long, 1 cm wide, pinnately lobed, petioles 10–16 cm long; flowers small, in compound paniculate umbels, involucral bracts 5, ovate or linear-lanceolate, caducous, umbellets 5–10 on stalks 0.5–6 cm long, involucels lanceolate, florets 3–12, sepals subulate, petals rhomboid, incurved; schizocarps oblong, 1 cm long, dark brown, ribbed, lateral ribs slightly thickened. Native to eastern China, particularly in Jiangsu, Anhui and Shandong, growing in rock crevices.

Foeniculum vulgare Miller — FENNEL

Hui-xiang=Hui-hsiang (茴香, Muslim Spice); *Huai-xiang=Huai-hsiang* (懷香, Muslim Spice, another form of writing the sounds of the first name); *Yu-xiang=Yu-hsiang* (魚香, Fish Spice, named by people who use the leaves in cooking fish); *Xiang-si-cai=Hsiang-ssu-ts'ai* (香絲菜, Spice Silk Vegetable, named for the filiform leaf-segments, the smell, the green color and the uses in cooking). Leaves and seed; introduced and common in kitchen gardens.

Perennial, aromatic, glabrous herbs up to 2 m high at the flowering stage, much branched; basal and cauline leaves triangular-ovate in outline, decompound, 30–50 cm across, the ultimate segments filiform, 2–10 cm long, petioles sheathy at the basal portion; flowers pale yellow, very small, 2 mm across, in compound umbels 4–10 cm across at anthesis, peduncles terete, 10–15 cm long, striate, primary divisions 30–40 on stalks 1–10 cm long, becoming longer after anthesis, sepals inconspicuous, petals ovate, incurved at the apex, stamens exserted, ovary obconic, 1 mm long, stigmas on very short styles or sessile on the conic stylopodia; schizocarps oblong-ellipsoid, 5 mm long, 2 mm across, straw-colored, with prominent lighter ribs.

Glehnia littoralis (L.) F. Schmidt (Syn. *Phellopterus littoralis* L.)

Shan-hu-cai=Shan-hu-ts'ai (珊瑚菜, Coral Vegetable); *Long-xu-cai=Lung-hsu-ts'ai* (龍鬚菜, Dragon Whisker Vegetable); *Liu-jiao-cai=Liu-chiao-ts'ai* (六角菜, Six Horn Vegetable). Leafy shoots; people living in different areas learnt to eat the plant independently and gave their own names.

Bei-sha-shen=Pei-sha-shen (北沙參, Northern Sand Shen). Root; produced in seashore farms of Shandong, used for preparing a broth with meat or chicken, available in American Chinese stores.

Very mild aromatic perennial herbs 10–20 cm high at flowering time, covered with grayish villose trichomes throughout, with very long cylindrical tap root 1–1.5 cm in diameter; leaves basal and cauline, ternately divided, triangular-ovate in outline, the terminal segments rhombic-orbicular or obovate, the lateral segments oblong-ovate or trilobate, 6–10 cm long, 2–4 cm wide; flowers white, small, polygamous, in axillary and irregularly branched terminal compound umbels, each umbel consisting of 10–20 umbellets on peduncles 5–30 mm long, involucres linear-lanceolate, membranous, involucels filiform, umbellets subglobose, consisting of 20–30 sexually differentiated flowers; sepals lanceolate, 1 mm long, petals obovate-oblong, incurved at the apex, 1 mm long, staminate flowers with obconical villose hypanthia, 5 exerted fertile stamens, discoid stylopodia, pistillate flowers shortly pedicellate or sessile, ovary oblong-globose, densely villose, 1 mm long, styles 2, stylopodia thick-discoid; schizocarps subglobose-obovoid, 8–14 mm long, 10 mm across, mericarps unequally developed, some with 5-winged ribs, villose along the margin, corky, others with 4-winged ribs.

Native of the coastal beach of eastern Asia, from northern Japan to Hainan Island, growing on sand dune, cultivated on the seashores of Liaoning, Hebei, Shandong and Jiangsu; description made from specimens grown on undisturbed sandy beaches of Hong Kong.

Heracleum moellendorffii Hance — Cow Parsnip

Dong-bei-niu-fang-feng=Tung-pei-niu-fang-feng (東北牛防風, Northeastern Cow Parsnip); *Da-ye-qin=Ta-yeh-ch'in* (大葉芹, Large-leaved Celery); *Lao-shan-qin=Lao-shan-ch'in* (老

山芹, Old Mountain Celery). Young plants, gathered in Liaoning annually for potherbs.

Perennial herbs 1 m or more high at anthesis, with slight aromatic taste, roots conical, grayish-yellow, stems cylindrical, hollow, striate; leaves both radical and cauline, on long petioles, sheathy at the base, laminas ovate in outline, 16–19 cm long, 15–16 cm across the base, ternate or pinnate, lateral leaflets pedicellate, ovate, 8–9 cm long, 5–6 cm wide, pinnatilobed, acuminate or acute, base cordate, terminal leaflets palmately parted, 12–13 cm long and wide, segments pinnately lobed, sharply serrate; flowers white, in terminal and lateral compound umbels 10 cm across, involucral bracts 1 or 2, lanceolate, primary umbels 10–20, bracts 3–5, linear-subulate, umbellets containing 10–20 flowers, sepals subulate, petals unequal, stylopodia conical; fruits oblong-suborbicular, 7–8 mm long, 7 mm across, meriocarps oblong-suborbicular, 7–8 mm long, 7 mm wide, 1 vitta in each furrow, 2 vittae in the commissural side. Native to northern and northeastern China and adjacent areas in Korea.

Hydrocotyle benguetensis Elmer (Syn. *H. ranunculifolia* Ohwi).

Zhi-xue-cao=Tzu-hsueh-ts'ao (止血草, Stop Bleeding Herb). Leaves; gathered and used in Taiwan.

Creeping perennial herbs with slender much branched stems, rooting at the nodes; leaves alternate or fasciculate, orbicular-reniform, 1–3 cm across, deeply 3-lobed, the lobes obovate, crenate, hairy, palmately 5- or 7-nerved, petioles 2–10 cm long, stipules membranous, ovate, cordate; flowers white, small, subsessile, in simple, solitary axillary umbels, peduncles 2–5 cm long, calyx lobes inconspicuous, petals 5, ovate, acute, stamens 5, ovary inferior, styles 2, stylopodia discoid; schizocarps laterally compressed, suborbicular, 1 mm long, 1.3 mm across, carpophores wanting.

Ligusticum sinense Oliver — CHINESE LOVAGE

Hao-ben=Hao-pen (藁本, Chinese Lovage); *Shan-yuan-sui=Shan-yuan-sui* (山芫荽, Mountain Coriander); *Wei-xiang=Wei-hsiang* (蔚香, Great Aromatic Herb). Young plants before the emergency of the flowering stalk, gathered in early spring and used as vegetable; root for tea.

Perennial aromatic erect herbs 1–1.5 m high, with rough subterranean cylindrical caudex 3–8 cm long, 0.7–2 cm across, internodes of stems 5–23 cm long, second joint being the longest; leaves ternately pinnate compound, the radical ones rosulate, cauline leaves alternate, deltoid in outline, 8–15 cm long, bipinnate, the ultimate segments 1.5–4 cm long, 1.2–2 cm wide, sparsely serrate, the apex acuminate, petioles 9–20 cm long, sheathy; flowers white, small, in terminal and axillary compound umbels 4–12 cm across, consisting of 20–40 umbellets, peduncles of umbellets 4–10 cm long, involucral bracts 8–10, linear, 1 cm long, 1 mm or less wide, umbellets consisting of 35–60 flowers, greenish-yellow, finally white; schizocarps broad oval, 2 mm long, 1 mm wide. Native of temperate northern and western China, has been used by the Chinese people since time immemorial.

Note: Specimens collected in western Sichuan with the aid of local medicinal herb-collectors (Figure 1), grow in the shade of mixed forest at the altitude of 3,500 m, with strong tap-root, 30 cm long, 2–4 cm in diameter. Such roots are highly esteemed as herb tea. A botanist from China, specialized in family Umbelliferae, Professor F. T. Fu, Jiangsu Institute of Botany, told me that in his field work, whenever specimens were plentiful, that night he would eat Chinese lovage as vegetable.

Ligusticum wallichii Franchet

Chuan-xiong=Chuan-hsiung (川芎, Sichuan Lovage). Rootstocks; dried thin slices, used in *bupin* mixes, available in one pound bags in American Chinese stores.

Strongly aromatic perennial herbs 30–60 cm high, rootstocks yellowish-brown, with several winter buds, each developing into an aerial shoot with prominent nodes, basal and cauline leaves; the leaves ternately decompound, triangular-ovate in outline, 12–15 cm long, 10–15 cm across the base, pinnae 4–5 pairs, ovate-lanceolate, 6–7 cm long, 5–6 cm across the base, ultimate segments linear-lanceolate, 2–5 cm long, 1–2 mm wide; flowers white, small, in compound umbels, polygamous, staminate flowers alone, or in larger plants, the pistillate flowers in the terminal umbel, and the staminate ones in the axillary umbels, involucres with 3–6 linear bracts 0.5–2.5 cm long, umbellets 7–20 on peduncles 2–4 cm long, sepals obscure, petals obovate-obcordate, 1.5–2 mm long, incurved at the apex, staminate flowers with 5 fertile stamens, disks glandular, hypanthium of the pistillate flower oblong, equal the petals in length, styles 2, 2–3 mm long, recurved; schizocarps ellipsoid, 3 mm long, ribs slightly winged. Native of China, now known only in cultivation.

Oenanthe javanica (Blume) de Candolle (Syn. *O. stolonifera* [Roxburgh] de Candolle)

Shui-qin=Shui-ch'in (水芹, Water Celery). Young shoots; eaten locally, not available in the market.

Rhizomatous perennial herbs 15–40 cm high, mild-aromatic; basal and caulines leaves triangular-ovate in outline, pinnately compound, ultimate segments ovate-lanceolate, 2–5 cm long, 1–2 cm wide, unequally crenate-dentate, petioles 7–15 cm long; flowers small, white, in terminal compound umbels with 7–12 umbellets, each umbellet containing 10–12 staminate and pistillate flowers, sepals triangular-lanceolate, 0.5 mm long, petals obovate, 1 mm long, clawed, the apex caudate and incurved, the first umbel terminate the primary stem, bearing pistillate flowers with urceolate hypanthium developing into fruits, the umbels developed from the axils of cauline leaves bearing staminate flowers with conic hypanthium and no style; schizocarps oblong, 3 mm long, 2 mm across, with persistent styles. Common in China, often near water, also occurring in southeastern Asia and India.

Ostericum grosseserratum (Maximowicz) Kitagawa (Syn. *Angelica grosseserrata* Maximowicz)

Sui-ye-qin=Sui-yeh-ch'in (碎葉芹, Cut-leaved Celery); *Xiao-ye-qin=Hsiao-yeh-ch'in* (小葉芹, Small-leaved Celery); *Shan-qin-cai=Shan-ch'in-ts'ai* (山芹菜, Hillside Celery); *Da-chi-dang-gui=Ta-ch'ih-dang-kui* (大齒當歸, Large-toothed Dang-gui). Young plants; gathered in Liaoning and Jilin for potherbs; roots gathered in Fujian (福建) and are used in northern China as ginseng, hence the names *"Fu-shen"* (福參) and *"Jian-shen"* (建參).

Perennial erect herbs 70–100 cm high at anthesis with stout tap-root, stems terete, striate, hairy below the nodes, branched only on flowering; leaves both radical and cauline, ovate-deltoid in outline, petioles 6–8 cm long, sheathy at the base, laminas 10–16 cm long, 7–10 cm wide, 3-times ternately compound, the terminal pinnae pinnate, ultimate segments ovate, 1–4.5 cm long, 1.5–4 cm across, deeply parted or pinnately lobed, acute, with large-dentate and coarse hairs along the nerves; flowers white, in terminal and lateral branched compound umbels, bracts 4–8, lanceolate, secondary umbels 8–17, scabrid, peduncles unequal, umbellets 10–20, bracts 6–10, linear, sepals deltoid, petals obovate, long-clawed, incurved, stylopodia discoid, styles filiform; mericarps flat, oblong-orbicular, 6 mm long, 4–4.5 mm wide; plants with citrus smell when young and anise smell at maturity. Widespread in temperate eastern Asia.

Ostericum sieboldii (Miquel) Nakai (Syn. *Angelica miqueliana* Maximowicz; *Ligusticum nipponicum* Wolff)

Shan-qin=Shan-ch'in (山芹, Hillside Celery); *Shan-qin-cai=Shan-ch'in-ts'ai* (山芹菜, Hillside Celery Vegetable). Young plants; gathered by people living in Liaoning for vegetable.

Perennial erect herbs 1 m or more high at anthesis, stems hollow, branched at the distal portion only, ridged; leaves both radical and cauline, 2- or 3-times pinnate, on slender petioles sheathy at the base, pinnae and pinnules petiolulate, ultimate segments ovate, 3–12 cm long, 2–6 cm wide, acute or acuminate, base cordate, or truncate, ultimate segments cuneate, serrate, glabrous; flowers white, in compound umbels 4–8 cm across, peduncles 7–13 cm long, scabrid, bracts 1 or 2, lanceolate, umbellets 1 cm across, containing 20 flowers, bracts 6–10, lanceolate, sepals ovate, petals broad obovate, claws short, incurved at the apex, stamens 5, ovary inferior; fruits compressed, ovate-oblong, 4–5.5 mm long, 3.5–4 mm wide, lateral ribs strongly winged. Widespread in temperate eastern Asia.

Ostericum viridiflorum (Turczaninow) Kitagawa (Syn. *Angelica viridiflora* [Turczaninow] Bentham)

Er-jiao-qin=Erh-chiao-ch'in (二角芹, Two-horned Celery); *Lü-hua-shan-qin=Lü-hua-shan-ch'in* (綠花山芹, Green-flowered Hillside Celery). Young plants, said to be a delicious wild potherb in northeastern China.

Biennial or perennial erect herbs 50–100 cm high, stems purple-red as emerging from the ground, becoming green, sharply ridged, branched from the middle; leaves both radical and cauline, on stout petioles 16 cm long, laminas deltoid-ovate in outline, 15 cm long, 16 cm across base, once or twice ternately compound, pinnae pinnate, stalked, the ultimate segments sessile, ovate, 3–7 cm long, 1.5–6 cm wide, acuminate, base rounded, sharply serrate, stigose along the nerves, upper cauline leaves with sheathy petioles; flowers greenish-white, in terminal loose clusters of compound umbels 4–9 cm across, the central (terminal) umbel subsessile, the lateral ones branched, bracts 2 or 3, umbellets 1 cm across, containing 10–20 flowers, involucral bracts 5–9, linear, sepals ovate, petals obovate, white, claws long, incurved at the apex, stylopodia conical; fruits oblong, 5–6 mm long, 3.5–4 mm wide, ribs sharp, slightly winged, lateral ribs strongly winged. Native to northeastern China and adjacent Siberia.

Pastinaca sativa L. — PARSNIP

Mei-guo-fang-feng=Mei-kuo-fang-feng (美國防風, American Fangfeng). Root; introduced, rare.

Robust, sparsely pubescent biennial herbs 1.5 m high, cultivated as a root vegetable like carrot, before harvest leaves all basal; stem sharply ridges, sparsely hispid, 1.5 cm in diameter; basal and lower cauline leaves pinnate, oblong-ovate in outline, 20–35 cm long, 12–19 cm wide, leaflets sessile, 4 or 5 pairs, ovate-oblong, lobed, 5–11 cm long, 3.5–7 cm wide, obtuse, base obliquely subcordate, the terminal segment cuneate, petioles subterete, shorter than the laminas, basal sheath 3–4 cm long; flowers greenish-yellow, in compound umbels 4–9 cm across, umbellets 15–30, consisting of 5–40 flowers, the first compound umbel bearing fertile flowers with oblong hypanthia equal the length of the petals, the umbels developed after the terminal one bearing polygamous flowers; schizocarps glabrous, dorsally flattened, the lateral ribs broadly winged, strongly ribbed, oil-tubes solitary, 2–4 on the commissural side.

Native of Eurasia, grown in cool temperate areas, when left in ground to over winter, the roots becoming sweeter in the spring; roots white, 20–30 cm long, 4–10 cm across apical end, a very good substitute of ginseng and *dang-gui* (Chinese angelica); rare in China.

Petroselinum crispum (Miller) Nyman ex A. W. Hill — Parsley (Syn. *P. sativum* Hoffmann)

Yang-yuan-sui=Yan-yuan-sui (洋芫荽, Foreign Coriander). Leaves, introduced, cultivated for hotels serving western dishes, used in salad or as garnish.

Glabrous biennial herbs 40–90 cm high; basal and lower cauline leaves ternately decompound, broadly triangular in outline, 10–13 cm long, ultimate segments flat or crisped and curled, cuneate-ovate, 3–5 cm long and wide, stalked, deeply trilobed, the lobes dentate, petioles terete, very shortly sheathy at the base; flowers very small, greenish-yellow, 1 mm across, in compound umbels 2.5–4 cm across, consisting of 5–10 umbellets, sepals inconspicuous, petals ovate, apex narrowed and incurved, the first umbel bearing pistillate flowers with oblong hypanthia, the axillary umbels polygamous, with the outer flowers having oblong hypanthia, the staminate flowers with obconic hypanthia; schizocarps obovate, laterally compressed, 2–3 mm long, 1.5 mm across, mericarps curving, contiguous only at the base and apex. Grown in Northern and central Europe, rarely cultivated in China.

Pimpinella arguta Diels — Central China Anise

Jian-chi-hui-qin=Chien-ch'ih-hui-ch'in (尖齒回芹, Sharp-toothed Anise). Young shoots used as vegetables by the people living in the mountains of the homeland of the Giant Panda.

Perennial herbs 1 m high, rootstock short, producing one or two winter buds giving rise to 1 or 2 raments, each having 2 basal leaves and the flowering stem, internodes 5–15 cm long, 4–5 mm in diameter, striate-sulcate; basal leaves and the lower cauline leaves twice ternately divided, the ultimate segments ovate-rhomboid, 1.5–3 cm long, acuminate-caudate, cuneate at base, sharply serrate, the teeth apiculate, pilose on major nerves beneath, petioles 2–10 cm long; flowers small, greenish-white, in compound umbels 5–7 cm long, 6–8 cm across the top, consisting of 7–9 umbellets, bracts 4, linear-lanceolate, falcate, peduncles of the umbellets 3.5–6 cm long, umbellets consisting of both flowers and 2–5 tertiary umbellets, pedicels 2–4 mm long at anthesis, pilose, hypanthia compressed, ovate-oblong, 1 mm long, sepals deltoid and acuminate, corolla 2 mm across, petals white, incurved.; schizocarps oblong, 3 mm long, 1.8 mm across, glabrous, smooth, without evident ridges and grooves.

A native of central and western China. The leaves appear similar to that of *Ligusticum sinense* Oliver, inflorescences different. The above description is drawn from an isotype of the species, with the Chinese name taken from botanical literature published recently.

Pleurospermum camtschaticum Hoffmann (Syn. *P. uralense* auctt., non Hoffmann)

Leng-zi-qin=Leng-tzu-ch'in (棱子芹, Ridged-fruited Celery). Young plants, gathered in spring, washed, drained, chopped with onion, seasoned with soybean sauce for salad in Helongjiang and Jilin.

Perennial herbs 1–2 m high at anthesis, stems 2–3 cm in diameter, strigose and glabrescent, ridged; leaves both radical and cauline, ovate-deltoid, 15–30 cm long and across the base, ternately decompound, ultimate segments ovate or lanceolate, 2–6 cm long, 0.5–2.5 cm wide, lobate and dentate, strigose on the nerves, petioles of basal and lower cauline leaves 15–30 cm long; flowers polygamous, in terminal compound umbels 10–20 cm across, involucral bracts numerous, linear-lanceolate, 2–8 cm long, pinnate or entire, reflexed, deciduous, umbellets 20–60, 4–7 cm across, pedicels 10–12 mm long, strigose, sepals 5, ovate, petals oblong-ovate, 2.5–3 obtuse; fruits ovoid, 7–10 mm long, 4–6 mm wide, ridges narrowly winged, tuberculate. Widespread in the shade of forests in temperate eastern Asia.

Sanicula chinensis Bunge — CHINESE SANICLE

Bian-dou-cai=Pian-dou-ts'ai (變豆菜, Variant Bean Vegetable); *Shan-qin-cai=Shan-ch'in-ts'ai* (山芹菜, Wild Celery); *Ya-zhang-qin=Ya-chang-ch'in* (鴨掌芹, Duck-feet Celery); *Zi-hua-qin=Tzu-hua-ch'in* (紫花芹, Purple-flowered Celery). Young plants, gathered before flowering, dipped in boiling water, washed, seasoned for vegetable; extra collection pickled with salt for future uses.

Glabrous perennial herbs 30–100 cm high; leaves both radical and cauline, reniform or cordate in outline, ternately digitate, segments obovate or rhomboid, 3–10 cm long, 4–13 cm wide, lateral segments deeply lobed, acute, base cuneate, the lateral ones oblique, sharply serrate, petioles 7–30 cm long, cauline leaves shortly petiolate or subsessile; flowers small, polygamous, white or greenish-yellow, in simple, 2- or 3-chotomous umbels, involucres foliaceous, 8 mm long, umbellets 2 or 3, involucral bracts 8–10, ovate or linear, florets 6–8, subsessile, sepals lanceolate, petals retuse and inflexed, stamens 5, ovary inferior, prickly, hooked; fruits densely prickly-hooked. Widespread in temperate eastern Asia.

Sanicula rubriflora F. Schmidt — Purple-Flowered Sanicle

Zi-hua-bian-dou-cai=Tzu-hua-pian-dou-ts'ai (紫花變豆菜, Purple-flowered Sanicle); *Da-ye-qin=Ta-yeh-ch'in* (大葉芹, Large-leaved Celery); *Zi-hua-qin=Tzu-hua-ch'in* (紫花芹, Purple-flowered Celery); *Ya-zhang-qin=Ya-chang-ch'in* (鴨掌芹, Duck-feet Celery). Young plants; apparently in many places of Liaoning and Jilin, people gave the same names to the young plants of *S. chinensis* and *S. rubriflora*, and prepare them for food by similar methods.

Perennial herbs 20–50 cm high, rootstocks oblique, with numerous fibrous roots, the only terminal winter bud giving rise to 5–6 leaves and 1 or 2 scapes; leaves all radical, the outer 2–4 smaller, on petioles 4–7 cm long, the inner 3–4 much larger, on petioles 15–40 cm long, laminas reniform, 4–15 cm wide, ternately digitate, segments unequal, obovate, 2–9 cm long, 1.5–7 cm wide, lobate, the lobes sharply serrate, the teeth aciculate, base cuneate; scapes terete, terminated by 2 opposite, foliaceous, palmatisected bracts and 3 umbels, peduncles 5–10 cm long, involucral bracts 6–8, linear, 1–2 cm long; flowers purple-red, subsessile, polygamous, sepals linear, prickly at the apex, petals exceeding the sepals, obovate, incurved at the apex, ovary of the pistillate flowers prickly-hooked; fruits oblong, 4–6 mm long, mericarps hooked prickly, usually not separated.

Native to cold temperate eastern Asia, growing in rich forest soil and wet meadows of Liaoning, Jilin, Helongjiang and adjacent areas in Siberia and Korea, thence extending eastward to Japan; the ternately divided young leaves bearing some resemblance to some beans, for the very different flowers and fruits developed later, and the observant users of the sanicles called them "Variant Bean Vegetable" (變豆菜) as recorded in the *Famine Herbal* (Zhu, 1407).

Saposhnikovia divaricata (Turczaninow) Schischkin — Fangfeng (Syn. *Siler divaricatum* [Turczaninow] Bentham et Hooker f.; *Ledebouriella seseloides* [Hoffmann] Wolff).

Fang-feng=Fang-feng (防風, Preventing Infection); *Shan-qin-cai=Shan-ch'in-ts'ai* (山芹菜, Hill Celery). Leaves, flowers; gathered locally and at different seasons, boiled for a tea, given to persons with a cold and in pain.

Robust perennial herbs 30–80 cm high, rootstocks bearing one terminal winter bud covered by brown fibrous remains of decayed petioles, stems solitary; basal and lower cauline leaves ternately decompound, ovate-deltoid, 7–19 cm long and across the base, ultimate segments ovate-lanceolate, 1.5–3 cm long, 2–7 mm wide, petioles 2–6.5 cm long, upper cauline leaves gradually decreasing in sizes and divisions; flowers yellow, small, in terminal compound umbels, each with 5–9 umbellets, without involucres, involucels linear-lanceolate, pedicels 4–9, sepals deltoid, obvious, petals obovate, incurved, apex retuse, ovary inferior, styles 2; schizocarps flat, oblong-ovate, 3–5 mm long, 2–2.5 mm wide, lateral ribs winged, stylopodium conical. Native of northern China; market supply cultivated in northeastern China and Inner Mongolia. *Fang-feng* from Sichuan (= *Chuan-fang-feng* 川防風, *Ligusticum brachylobum* Franchet) used locally.

Sium sisarum L. — SKIRRET

Ze-qin=Tse-ch'in (澤芹, Swamp Celery). Root; gathered and used locally, not available in the market.

Rhizomatous, glabrous, aromatic perennials 30–80 cm high, roots tuberous, clustered, stems much branched; lower cauline leaves odd-pinnate, leaflets 3–11, various in shape and size, ovate to lanceolate, 2.5–8 cm long, 3–10 mm wide, upper cauline leaves trifoliolate, leaflets linear or lanceolate; flowers small, 2 mm across, in compound terminal umbels consisting of 5 or 6 umbellets on stalks 10–15 mm long, each with 10–12 flowers, involucel-bracts lanceolate, 2.5 mm long; schizocarps 3 mm long, ribbed, mericarps curved, 3 ribs on the back, 1 on each side. Said to be native of eastern Asia, cultivated for the edible roots; for the same Chinese name, *S. suave* Walter is used in *Flora Reipublicae Popularis Sinicae* (55[2]: 155. 1985.)

Spuriopimpinella brachycarpa (Komarov) Kitagawa (Syn. *Pimpinella brachycarpa* [Komarov] Nakai).

Zhi-zhu-xiang=Chih-chu-hsiang (蜘蛛香, Spider Spice); *Ming-ye-qin=Ming-yeh-ch'in* (明葉芹, Shiny-leaved Celery); *Duan-guo-hui-qin=Tuan-kuo-hui-ch'in* (短果回芹, Short-fruited Anise). Young plants; gathered between April and June, blanched in boiling water, cut into bite-size, pickled in brine, said to be one of the best wild edible plants in northeastern China.

Erect perennial herbs 60–120 cm high, rootstocks short, stout, bearing numerous fibrous roots and one terminal winter bud, upright stems striate, pubescent at the nodes; leaves both radical and cauline, ovate-deltoid, up to 16 cm long and wide across the base, petioles up to 15 cm long, laminas ternately twice divided, the ultimate segments ovate-rhomboid, 3–8 cm long, acuminate, base cuneate, the lateral ones oblique at the base, double-serrate, glabrous or strigose on the nerves, upper cauline leaves trifoliolate, shortly petiolulate or sessile; flowers small, white, polygamous, in terminal compound umbels, secondary umbels 9–13, peduncles 2–4 cm long, involucral bracts filiform, florets 9–25, sepals lanceolate, petals obovate, retuse and incurved at the apex, clawed, stamens

5, ovary inferior, styles filiform, stylopodia conical; fruits subglobose, 3 mm in diameter, mericarps with inconspicuous ribs. Common in the forests of Liaoning and Jilin, extending thence northward to Siberia.

Torilis japonica (Houttuyn) de Candolle — HEDGE-PARSLEY

Qie-yi=Ch'ieh-i (竊衣, Stealthy-on-cloth); *Hua-nan-he-shi=Hua-nan-ho-shih* (華南鶴蝨, South China Crane-lice). Young plants, gathered before flowering, used for potherb in Liaoning.

Slender erect annual herbs 20–80 cm high, stems much branched, strigose throughout; leaves both radical and cauline, ovate in outline, the laminas 2–6 cm long and wide across the base, 2- or 3-pinnate, the ultimate segments oblong-lanceolate, corsely serrate; flowers white or tinged pink, in loose terminal compound umbels 1–4 cm across, bracts 4–7, 0.5–2 cm long, primary umbels 4–11, peduncles unequal, umbellets consisting of 7–12 flowers, involucral bracts 7–8, sepals deltoid-lanceolate, petals unequal, the outer ones of the marginal flowers larger, recurved at the apex, stamens 5, ovary bristly, styles short, stylopodia conical; fruits oblong, 3–4 mm long, bristly with hooked prickles and warty, ribbed, the secondary ribs more prominent. Widespread weed in China and eastern Asia.

Cornaceae: Dogwood Family

Cornus capitata Wallich (Syn. *Dendrobenthamia capitata* [Wallich] Hutchinson)

Shan-li-zhi=Shan-li-chih (山荔枝, Mountain Lychee). Fruit; gathered and eaten by people living on the hillside.

Small evergreen trees 6 m high, branchlets white villose; leaves opposite, thinly coriaceous, elliptic-lanceolate, 5.5–10 cm long, 2–3.4 cm wide, acute or shortly acuminate, entire, both surfaces pubescent, the hairs appressed; flowers yellow, in compact heads subtended by 4 white petaloid obovate and cuneate bracts 3–4 cm long, 2–3 cm wide, calyx tubular, 4-lobed, the lobes obtuse; aggregate fruits compressed-globose, dark red, on stout stalks 4–7 cm long.

Cornus hongkongensis Hemsley (Syn. *Dendrobenthamia hongkongensis* [Hemsley] Hutchinson)

Xiang-gang-si-zhao-hua=Hsiang-kang-ssu-chao-hua (香港四照花, Hong Kong Dogwood). Fruit edible; the plant is extremely rare in Hong Kong; an endangered species there.

Small shrub-like evergreen tree, 3 m high, branchlets covered with short brown hairs, glabrescent; leaves opposite, thick-coriaceous, elliptic or ovate-elliptic, 6–12 cm long, 3–6 cm wide, acuminate, entire; flowers yellow, in crowded heads subtended by 4 white petaloid bracts, broad-oblong, acute, calyx tubular, truncate at the apex; aggregate fruits globose, yellow or red, on peduncles 4–10 cm long.

Cornus kousa Buerger ex Miquel — Kousa (Syn. *Dendrobenthamia japonica* [de Candolle] Fang)

  Si-zhao-hua=Ssu-chao-hua (四照花, Four-shining Flower); *Shan-li-zhi=Shan-li-chih* (山荔枝, Mountain Lychee). Ripe fruits; used locally, not available in the market.

Tall shrubs or small trees 3–5 m high; leaves opposite, oblong-elliptic or ovate-elliptic, abruptly long-acuminate; flowers small, yellowish-green, in dense globose head subtended by 4 large petaloid white bracts, terminal to a leafy branchlet, sepals 4, petals 4, stamens 4, ovary inferior; fruits drupaceous, connate into a globose fleshy head 1.5–2.5 cm in diameter, red and sweet. Native of eastern Asia, two varieties in China.

1. var. **angustata** Chun — Narrow-leaved Chinese Dogwood (Syn. *Dendrobenthamia angustata* [Chun] Fang)

  Ye-li-zhi=Yeh-li-chi (野荔枝, Wild Lychee). Mature fruit, rare.

Evergreen (?) shrubs, branchlets pubescent, hairs short, yellow; leaves oblong-elliptic, 7–12 cm long, 2.5–4.5 cm wide, pilose above, glabrescent, pilose with appressed white hairs beneath; flowers pilose, disk cupular. Native of Zhejiang, Jiangxi and northern Guangdong.

2. var. **chinensis** Osborn — Chinese Dogwood (Syn. *Dendrobenthamia japonica* [de Candolle] Fang var. *chinensis* [Osborn] Fang)

  Qing-pi-shu=Ch'ing-p'i-shu (青皮樹, Green Bark Tree); *Liang-zi=Liang-tzu* (涼子, Cool Fruit). Ripe fruit; gathered and eaten locally.

Beautiful deciduous trees, branchlets pubescent, hairs white; leaves ovate-elliptic, 5–12 cm long, 3–7 cm wide, pilose or glabrescent; fruits 2–2.5 cm in diameter, scarlet red, hanging on slender stalks 7–8 cm long.

Native of the Qinling Range separating Shaanxi and Sichuan; introduced to USA in 1907, common in gardens, school yards and parks in New England, flowering in June, fruits mature in October, devoured by many birds.

Cornus officinalis Siebold et Zuccarini — Chinese Cornel, Japanese Cornelian Cherry (Syn. *Macrocarpium officinale* [Siebold et Zuccarina] Nakai)

Shan-zhu-yu=Shan-chu-yu (山茱萸, Wild Cornel); *Yu-rou=Yu-jou* (萸肉, Cornel Meat). Fruit, eaten raw or used for jam; now extensively cultivated in China for domestic use as a stomachic, and for export to Japan and other eastern Asian countries, to overseas Chinese; available in American Chinese stores.

Deciduous shrubs or small trees 3–4 m high, bark light brown, flaky, branchlets dark brown; leaves opposite, ovate-elliptic or ovate-lanceolate, 5–12 cm long, 2.5–6 cm wide, acuminate-caudate, appressed hairy on both surfaces; flowers yellow, small, 5 mm across, appearing before leaves on last year's twig, 25–45 in a sessile head 2 cm across, subtended by 4 ovate-orbicular hairy brownish-yellow bracts 5–6 mm long, 4–5 mm wide, pedicels 6–7 mm long, pubescent, sepals 4, inconspicuous, petals lanceolate, 2 mm long, stamens 4, glands thickly discoid, ovary inferior, hairy, style slender, stigma punctiform; fruits drupaceous, oblong, 1.5 cm long, 1 cm across, scarlet-red, hanging on slender stalks 1 cm long, very sour.

Native of eastern Asia, introduced into Western gardens in 1877, extensively distributed in central, eastern, western and northwestern China; an important material used for elderly persons unable to control the urine, taken in form of tea with ginseng, *yi-zhi-ren* (益智仁, *Alpinia oxyphylla* Miquel), and *bai-zhu* (白朮, *Atractylodes macrocephala* Koidzumi); Zhejiang (the seat of the ancient Kingdom of Wu) being the primary center of production, hence the name "*Wu-zhu-yu*" (吳茱萸) in some prescriptions.

Cornus walteri Wangerin

Mao-lai=Mao-lai (毛萊, Hairy Dogwood). Oil, extracted from seed.

Deciduous trees 6–14 m high, bark dark gray, splitting into longitudinal strips; leaves opposite, elliptic or oblong, 4–10 cm long, 2.7–4.5 cm wide, acuminate, softly pubescent on both surfaces; flowers white, 1.2 cm across, in terminal corymbose panicles, petals lanceolate, ovary hairy; drupes black, 6 mm in diameter. Native of China, widespread in central, western and northwestern regions, introduced to USA in 1907.

Cornus wilsoniana Wangerin

Guang-pi-shu=Kuang-p'i-shu (光皮樹, Smooth Bark Tree). Oil, extracted from the seed.

Deciduous shrubs or small trees 5–18 m high, bark smooth; leaves opposite, elliptic, 3–9 cm long, 1.5–5.8 cm wide, acuminate, finely pubescent and papillose; flowers white, 9 mm across, in terminal corymbose panicles, petals lanceolate, 5 mm long, ovary villose; drupes bluish-black, globose, 6 mm in diameter. Native of central China, thence southward to Guangdong.

Helwingia japonica (Thunberg) F. Dietrich — HELWINGIA

Qing-jia-ye=Ch'ing-chia-yeh (青荚葉, Green Pod-on-leaf). Young leafy shoots, gathered and eaten locally, not available in the market.

Deciduous shrubs growing on grassy hillsides, 1–3 m high, branchlets purplish-green; leaves alternate, ovate-elliptic, rarely lanceolate, 3–13 cm long, 1.5–9 cm wide, acuminate, serrate; flowers small, green, dioecious, in small fascicles on the midrib of normal leaves, staminate flowers 5–12, sepals obscure, petals 3 or 4, ovate-deltoid, 2 mm long and wide, stamens 3 or 4, exserted, pistillode globose, distal portion 3–4 parted, pistillate flowers usually solitary, pedicels 1.5 mm long, petals 4, ovate, 2 mm long, ovary inferior, subglobose, 2.5 mm long and wide, without staminodes, style conical, the distal portion 4- rarely 5-parted with the stigmatic surface lining the adaxial side; fruits drupaceous, similar in structure to the fruits of holly, with fleshy mesocarp and lignified separate endocarps each enclosing a seed, forming the pyrenes, drupes oblong, 8 mm long, 5 mm in diameter; pyrenes 1–4, trigonous-oblong, 5–6 mm long, 2 mm across, rugose, reticularly striate, sulcate and pitted.

Ericaceae: Heath or Rhododendron Family

Arctostaphylos rubra (Rehder et Wilson) Fernald (Syn. *Arctous ruber* [Rehder et Wilson] Nakai)

Hong-bei-ji-guo=Hung-pei-chi-kuo (紅北極果, Red North Pole Berry). Fruit, gathered and consumed locally, not available in the market.

Prostrating shrubs 4–20 cm high, the principle stems woody, wirelike, 3–5 mm in diameter, much branched, forming a matlike colony, bark reddish-brown, flaky, erect, current year's growth 1–7 cm long, bearing 5–7 leaves terminated by a large ellipsoid winter bud covered by thickly coriaceous glabrous scales; leaves alternate, obovate-lanceolate, 1–4.5 cm long, 5–15 mm wide, acute, base cuneate, serrate and sparsely ciliate, glabrous, petioles 5–12 mm long, broad as if winged by the decurrent leaf-base; flowers white, in terminal 3- to 6-flowered racemes or panicles, developed from the same terminal bud and at the same time as the vegetative shoot, calyx 4- or 5-lobed, persistent, corolla urceolate, stamens 10, included, anthers awned, opening by terminal pores; fruits bright red, drupes berrylike, smooth, 9–12 mm across; pyrenes 3–5, 3–4 mm long, endocarp thick-coriaceous. Amphi-northern-Pacific, a very interesting pattern of distribution, occurring in northwestern North America and northeastern Asia, thence extending southward on high mountains to Gansu, western Sichuan, and northwestern Yunnan, at altitudes of 3,300–4,000 m.

Gaultheria veitchiana Craib — CHINESE TEABERRY

Hong-tang-cha=Hung-t'ang-cha (紅糖茶, Brown Sugar Tea). Leafy stems; shoots of the low shrub, gathered by the handful as the lumber men return to camp from work, the fresh material placed over the charcoal fire in the lumber camp with a popping noise from the bursting surfaces of the leathery green leaves; the partially burnt freshly roasted material is thrown into a pot of boiling water, a tea with good natural flavor and a sweetness of brown sugar is made.

Evergreen hispid low shrubs 30–50 cm high, with creeping subterranean stolons; branchlets verrucose; leaves oblong-elliptic, 4.5–10 cm long, 2–4.5 cm wide, acuminate or acute, base obtuse or rounded, serrate, coriaceous, shiny glabrous above, evenly hispid with dark brown hairs beneath, petioles 3–4 mm long; flowers white, in axillary racemes 2–5 cm long, consisting of 7–20 bracteate flowers, bracts ovate, 5 mm long, acute, bracteoles 2, hypanthia hemispherical, calyx lanceolate, corolla subglobose-urceolate, 4 mm long, stamens 10, anthers awned, opening by apical pores; ovary half-superior; capsules completely covered by fleshy bright blue enclosure developed from the hypanthium and calyx. Native of western China, growing on the edge of spruce forests at altitudes of 2,600–3,300 m; a common tea of the lumberjacks.

Rhododendron oldhamii Maximowicz — OLDHAM'S AZALEA

Jin-mao-du-juan-hua=Chin-mao-tu-chuan-hua (金毛杜鵑花, Golden Hair Rhododendron). Petals, gathered and used in Taiwan, recorded before 1950.

Partially evergreen hispid shrubs 1–3 m high, covered by red-brown hairs throughout; leaves elliptic-lanceolate or ovate-oblong, 3–8 cm long, 1.2–4 cm wide, acute and mucronate, base cuneate or obtuse; flowers dark red, 1–3 on pedicels covered with hispid and glandular red hairs, corolla funnelform, 4 cm long, stamens 10, the filaments glandular papillose at the basal portion, ovary superior, style papillose near base; capsules ovoid, 8 mm long, hairy. Endemic to Taiwan.

Vaccinium bracteatum Thunberg

Wu-fan-shu=Wu-fan-shu (烏飯樹, Black Rice Tree); *Qing-jing-fan=ch'ing-ching-fan* (青精飯, Youthful Strength Rice). Leafy shoots, boiled with rice to dye it black, believed to enhance revitalizing property; dried black rice available in monasteries of Hong Kong; a common practice of the rural people in South China.

Evergreen, tropical shrubs or small trees, 1.5–3 m high, branchlets pubescent; leaves elliptic or ovate-oblong, 2.5–6 cm long, 1–1.5 cm wide, acute, base obtuse, serrate, pubescent along the midrib; flowers white, hanging, in axillary racemes 2–6 cm long, bracts foliaceous, hypanthia hemispherical, hairy, sepals deltoid, acute, corolla cylindrical-urceolate, 5–7 mm long, stamens 10, unawned, opening by terminal pores, ovary inferior, hairy, style columnar, stigma punctiform; berries globose, 4–6 mm in diameter, black, full of seeds, unfit for eating fresh. Native of East Asia.

Vaccinium myrtillus L. — BILBERRY, WHORTLEBERRY

Hei-guo-yue-ju=Hei-kuo-yueh-chü (黑果越橘, Black-fruited Bilberry). Fruit, gathered and eaten locally in the mountains of Xinjiang, and Mongolia.

Very low deciduous shrubs 15–30 cm high, branchlets glabrous, ridged; leaves ovate-oblong, 1–2.8 cm long, 6–8 mm wide, obtuse or acute, rounded at base, serrulate, the teeth glandular; flowers greenish-pink, solitary, axillary to normal leaves, hanging, calyx entire, collarlike, corolla campanulate-urceolate, 3–4.5 mm long, stamens 10, filaments very short, anthers comparatively long, base saccate, benting forward at the middle and attenuate, awned, ovary inferior, glabrous and glaucous, 5-locular; fruits black, bluish-glaucous, juicy, the juice red. A holarctic species, in China occurring only on the high mountains of Inner Mongolia and Xinjiang, berries eaten locally by the people there.

Vaccinium oxycoccos L. — SMALL CRANBERRY (Syn. *Oxycoccus microcarpus* Turczaninow)

Mao-hao-dou=Mao-hao-tou (毛蒿豆, Hairy Wormwood Pea); *Xiao-guo-hong-mei-tai-zi=Hsiao-kuo-hung-mei-t'ai-tzu* (小果紅莓笞子, Lesser Red Berry). Fruit, used locally where it occurs, not available in the market.

Small, delicate, trailing, evergreen plants with flexible slender stems, fibrous branched roots, branchlets pubescent; leaves coriaceous, ovate-oblong, 3–8 mm long, 1–3 mm wide, acute or obtuse, base rounded, entire, margin slightly recurved, glaucous beneath; flowers roseate, 1 or 2 racemose, bracteate, terminal to leafy branchlets, pedicels filiform, 3–3.5 cm long, with 2 median bracteoles, nodding, hypanthia hemispherical, 2 mm across, sepals rounded, glandularly ciliate, corolla deeply 4-parted, segments oblong, 5–6 mm long, reflexed, stamens exserted, awnless, filaments stout, flattened, 1.5 mm long, hispid, anthers rugose, 1.5 mm long, with smooth and straight apical tubes 2 mm long, ovary inferior, style columnar, stigma punctiform; berries red, globose, 5–6 mm in diameter, glaucous. Widely distributed in Eurasia and northern North America; in China occurring in high mountains in Xinjiang and Mongolia, growing on boggy and peaty soil.

Vaccinium vitis-idaea L. — MOUNTAIN CRANBERRY, COWBERRY, LINGBERRY

Yue-ju=Yueh-chu (越橘, Cowberry). Fruit, eaten locally where available.

Evergreen shrubs 10–30 cm high, with creeping subterranean rhizomes, branchlets pilose; leaves coriaceous, obovate, 1–2 cm long, rounded or emarginate, base acute, obtuse, rounded or cuneate, dark green and glossy above, dotted with glandular hairs beneath, crenulate, teeth slightly recurved; flowers white-pink, 2–5 in subterminal bracteated racemes, pedicels short, bracteoles submedian, hypanthia subglobose-obconical, glabrous, sepals ovate, glandularly ciliate, acute, corolla campanulate, 5–6 mm long, the distal one-third 4-lobed, the lobes ovate, stamens 8, awnless, filaments hispid, thecae short, 1/2 as long as the terminal tubes, ovary inferior, style exerted, stigma punctiform; fruits globose, 7 mm in diameter, red. Native of the northern hemisphere, with holarctic distribution; in China occurring on the high mountains of Inner Mongolia, Xinjiang, and western Sichuan.

Myrsinaceae: Myrsine Family

Ardisia cornudentata Mez — JADE MOUNTAIN ARDISIA (Syn. *A. morrisonensis* Hayata)
Yu-shan-zi-jin-niu=Yu-shan-ts'u-chin-niu (玉山紫金牛, Mt. Morrison Ardisia). Fruit, gathered and eaten in Taiwan.

Small, evergreen trees or shrubs 3–6 m high, branchlets densely ferrugineously glandular-hairy throughout when young, glabrescent with age; leaves alternate, subcoriaceous, elliptic or lanceolate, rarely oblanceolate, 7–18 cm long, 2.5–6 cm wide, acuminate or acute, base cuneate or acute, sharply serrate or remotely dentate, glabrous above, evenly dotted with glandular hairs beneath, petioles 5–15 mm long, glandularly hairy; flowers pink-white, in subumbelliform short, simple racemes (rarely branched) terminal to specialized axillary shoots with 1 or 2 leaves at the distal ends, racemes unbracteolate, with 8–14 flowers, the rachis and pedicels densely glandularly hairy, sepals imbricate, rounded, glandularly ciliate, corolla conical in bud, starlike when fully open, 10 mm across, punctate, stamens 5, filaments short, anthers sagittate, acute, base saccate, connectives punctate, ovary superior, style filiform, stigma punctiform; fruits black, globose, 5 mm in diameter, persistent calyx 6 mm across, glabrous. Endemic to Taiwan.

Ardisia quinquegona Blume — ASIATIC ARDISIA
Luo-san-shu=Lo-san-shu (羅傘樹, Official Silk Umbrella Tree); *Shan-tao-ye=Shan-t'ao-yeh* (山桃葉, Hillside Peach-leaf); *Ji-yan-qing=Chi-yen-ch'ing* (雞眼青, Chicken-eyes Green). Young leaves, used for tea.

Undershrubs or small trees 2–3 m high, glandularly lepidote throughout; leaves elliptic-lanceolate, 7–10 cm long, 2–3 cm wide, attenuate at both ends, entire; flowers pink, rarely tinged white, in subumbelliform axillary clusters, peduncles 3–4 cm long, calyx cupular, 5-lobed, the lobes triangular, acute, persistent, corolla subrotate, 4–5 mm across, lobes ovate, acuminate, stamens 5, inserted on the corolla tube and opposite the lobes, ovary hemispherical, glabrous, style filiform, 4 mm long, stigma punctiform; fruits compressed-globose, 5-angled, the angles rounded, changing from green to yellow, red and finally black. Common in the forests of southern China, Taiwan and Hainan Island.

Ardisia sieboldii Miquel

Shu-qi=Shu-chih (樹杞, Tree Lycium). Fruit. eaten in Taiwan before 1950.

Evergreen trees or shrubs 3–10 m high, branchlets covered by brown scales; leaves obovate-elliptic or oblanceolate, 7–13 cm long, 3–5 cm wide, shortly acuminate, obtuse or rarely rounded, base cuneate and acute, glabrous above, brown lepidote beneath, entire; flowers white, protogynous, with the style exposed from the flower-buds, small, numerous, in axillary corymbose panicles 3–7 cm across, peduncles and pedicels lepidote, sepals ovate, acute, punctate, corolla rotate, 7 mm across, petals oblong, 3–4 mm long, stamens broadly sagittate, the filament short, ovary superior, style slender, stigma punctiform; fruit globose, 7 mm in diameter, red-black. Native of East Asia, from Japan to Taiwan, occurring in Zhejiang and Fujian of mainland China.

Ardisia solanacea Roxburgh (Syn. *A. humulis* auctt, non Vahl)

Suan-tai-cai=Suan-t'ai-ts'ai (酸苔菜, Sour-shoots Vegetable); *Pa-lei=P'a-lei* (帕累, the Thai ethnic name). Tender leafy shoots; parboiled, washed, drained, and seasoned for salad, a favorite of the Thai living in southern Yunnan.

Shrubs or small trees, 6 m high, branchlets glabrous; leaves elliptic-lanceolate, or oblanceolate, 12–20 cm long, 4–7 cm wide, acute, obtuse or rounded, base acute or decurrent, entire, punctate; flowers pink, in axillary branched cymes, peduncles 5–10 cm long, pedicels 1–3 cm long, calyx slightly connate at the base, lobes ovate-reniform, 3 mm long, punctate, petals ovate, 9 mm long, acute, punctate, stamens 5, inserted to the base of and equal to the length of the corolla, ovary globose; fruits globose, 7–9 mm in diameter, dark red tinged black, punctate. Native to tropical Asia, extending from India, Singapore to southern Yunnan.

Embelia laeta (L.) Mez — SOUR EMBELIA

Suan-teng-guo=Suan-t'eng-kuo (酸藤果, Sour Vine Berry). Mature fruit; picked up and eaten in the field, not available in the market; *Ru-di-long=Ju-ti-lung* (入地龍, Underground Dragon), Root; *Suan-teng-tou=Suan-t'eng-t'ou* (酸藤頭, Sour Vine Shoot).

Young leafy shoots. Both the roots and young leafy stems used for a tea, taken for improving digestion.

Scandent evergreen shrubs 1–3 m high, branchlets glabrous, reddish brown; leaves obovate-oblong, 2–6 cm long, 1–2 cm wide, rounded or obtuse, base cuneate, obtuse, entire, taste sour, petioles 5–7 mm long; flowers white, small, dioecious, 4-merous, 3–8 in axillary short racemes or fascicles 5–10 mm long, calyx 1 mm across, punctate, corolla rotate, 4 mm across, petals oblong, punctate outside, densely papillose inside, stamens 4, exserted in the staminate flower, ovary superior, ovoid in the pistillate flower, style filiform, stigma capitate; fruits pea-sized, with persistent style, very pretty on the plant, changing from green to red and then purple-black, edible, very sour. Native in East Asia; in China occurring in the Yangtze River Region, thence southward to Guangdong and Yunnan.

Embelia longifolia (Bentham) Hemsley

Chang-ye-suan-teng-guo=Ch'ang-yeh-suan-t'eng-kuo (長葉酸藤果, Long-leaved Embelia); *Mo-gui-xi=Mo-kui-hsi* (沒歸息), and *Mu-gui-shi=Mu-kui-shih* (木桂拾), both are the aboriginal names of the ethnic groups living in southern Yunnan. Fruits; picked from the plant and eaten locally.

Sprawling glabrous shrubs 3 m high; leaves oblanceolate, 6–12 cm long, 2–4 cm wide, acute, obtuse, or abruptly acuminate, entire, sparsely punctate; flowers greenish-pink, dioecious in short axillary racemes 1 cm long, calyx connate at the basal half, lobes acute, petals ovate-elliptic, obtuse, 2 mm long, stamens in the staminate flowers twice the length of the petals, ovary in the pistillate flowers glabrous; fruits red, subglobose, 1 cm in diameter, on stalks 1 cm long. Native to southern and southwestern China, growing along the margin of forests at altitudes of 1,200–1,300 m.

Embelia ribes Burman f.

Bai-hua-suan-teng-zi=Pai-hua-suan-t'eng-tzu (白花酸藤子, White-flowered Embelia); *Suan-teng=Suan-t'eng* (酸藤, Sour Vine); *Hei-tou-guo=Hei-t'ou-kuo* (黑頭果, Black Head Fruit); *Qiang-zi-guo=Ch'iang-tzu-kuo* (槍子果, Bullet Fruit). Fresh fuits; picked from plant and eaten fresh; young shoots, eaten as salad or cooked for potherb by the Thai people living in southern Yunnan.

Straggling shrubs 3–6 m high, branchlets glabrous; leaves ovate-elliptic or oblong, 5–8 (–10) cm long, 3.5 cm wide, acuminate, base rounded, entire, glabrous, petioles 5–10 mm long; flowers small, greenish-white, in terminal panicles 15–30 cm long, primary axis pilose, calyx tube 1 mm long, lobes deltoid, acute, hairy and papillose, staminate flowers with 5 stamens inserted median to the petals, ovary of the pistillate flowers ovoid, glabrous, style slightly shorter than the petals, stigma capitate; fruits subglobose, 3–4 mm in diameter, red changing purple-black. Widespread in Southeast Asia, from India to Indonesia, thence northward to southern China, very common in Hong Kong, growing on open hillsides at altitudes 100–2,000 m.

Embelia sessiliflora Kurz

Suan-ji-teng=Suan-chi-t'eng (酸雞藤, Sour Chicken Vine); *Ye-mao-suan=Yeh-mao-suan* (野貓酸, Wild Cat Sour); *Duan-geng-suan-teng-guo=Tuan-keng-suan-t'eng-kuo* (短梗酸藤果, Short-stalked Sour Vine Fruit). Fruit, eaten fresh, sweet-sour; tender shoots used for salad or cooked for potherb.

Sprawling vine 3–5 m high, branchlets pubescent; leaves ovate-elliptic or oblong, 6–11 cm long, 2.5–5 cm wide, obtuse or acuminate, base rounded, entire, glabrous; flowers small, sessile, greenish-white, in terminal panicles 10–15 cm long, calyx connate at base, lobes deltoid, pilose, punctate, persistent, petals elliptic, 1.5–2.5 mm long, pilose and punctate outside, papillose inside, stamens inserted to the middle of the corolla, ovary glabrous, stigma bifid; fruits globose, 3 mm in diameter. Widespread in eastern Himalayan Region, growing along paths and margin of forests at altitudes of 1,400–2,800 m in Yunnan and Guizhou.

Embelia subcoriacea (C. B. Clarke) Mez

Da-ye-suan-teng-zi=Ta-yeh-suan-t'eng-tzu (大葉酸藤子, Large-leaved Embelia); *A-lin-xi=A-li-hsi* (阿林稀, the aboriginal name used in western Yunnan). Fruits; eaten locally.

Straggling shrubs or small trees 3–5 m high; leaves obovate or obovate-elliptic, 8–15 cm long, 3.5–6.5 cm wide, acute or abruptly acuminate, base cuneate, entire, punctate; flowers unisexual, small, 3 mm long, in axillary racemes 3–5 cm long, pilose, sepals acute, punctate, petals ovate or oblong, hairy inside, stamens in staminate flowers longer than the petals, pistillate flowers not yet known; fruits red, compressed globose, 0.8–1 cm in diameter. Native to eastern Himalayan Region, common in forests at altitudes of 1,400–2,300 m in southwestern Yunnan, Guangxi and Guizhou.

Embelia vestita Roxburgh

Mi-tang-guo=Mi-t'ang-kuo (米湯果, Rice Soup Fruit); *Bai-la-shu=Pai-la-shu* (白蠟樹, White Wax Tree); *Mi-hua-suan-teng-zi=Mi-hua-suan-t'eng-tzu* (密花酸藤子, Dense-flowered Embelia). Fruits, sweet-sour, eaten fresh with sugar added.

Straggling shrubs 5 m high, branchlets pilose; leaves ovate-oblong, rarely elliptic-lanceolate, 5–10 cm long, 2–3.5 cm wide, acute, obtuse or acuminate, base obtuse or rounded, serrulate, punctate beneath, petioles 4–8 mm long; flowers unisexual, very small, 2 mm long, in axillary racemes 2–6 cm long, pilose, sepals connate at the base, petals oblong-elliptic, apex rounded or emarginate, stamens 5, adnate to the petals, ovary in pistillate flowers ovoid, stigma subcapitate; fruits subglobose, 5 mm in diameter, red, punctate. Native of eastern Himalayan Region, growing in rocky hillsides at altitudes of 1,000–1,700 m in western Yunnan.

Maesa argentea (Wallich) A. de Candolle

Yin-ye-du-jin-shan=Yin-yeh-tu-chin-shan (銀葉杜莖山, Silvery-leaved Maesa). Fruit, slightly sweet.

Shrubs 1–2 m high, or small trees 5 m high, branchlets pilose; leaves ovate-elliptic, 12–17 (–22) cm long, 5–9 (–11) cm wide, acuminate, base obtuse and decurrent, coarsely serrate, pilose along the major nerves above, villose and pilose beneath; flowers white, small, 2–3 mm long, in axillary panicles 1–5 cm long, hairy, stamens in staminate flowers inserted in the middle of the corolla and opposite the petal, ovary inferior, style included, stigma 3-fid; fruits globose or ovoid, 3–4 mm in diameter, fleshy, white, pink or red; seeds many, black. Native to eastern Himalayan Region, extending to central India, eastward to the dense forests of Yunnan and Sichuan.

Maesa japonica (Thunberg) Moritzi ex Zollinger (Syn. *M. doraena* auctt., non Blume)

Bai-hua-cha=Pai-hua-ch'a (白花茶, White-flowered Tea); *Shan-gui-hua=Shan-kui-hua* (山桂花, Hillside Osmanthus); *Du-jing-shan=To-chin-shan* (杜荊山, a classical name associated to this species). Fruits, eaten in Taiwan; leaves used as a common tea substitute in Guangdong.

Small evergreen scandent shrub 1–3 m high, branchlets glandular papillose when young, glabrous later; leaves elliptic, lanceolate or oblong, 5–16 cm long, 2–5.5 cm wide, acuminate, base acute, coriaceous, glabrous above, sparsely glandular beneath, the distal portion remotely coarsely serrate, basal portion entire; flowers white, in axillary racemose panicles 1–3 cm long, rachis and pedicels papillose, pedicels 2 mm long, with 2 apical bracteoles, sepals suborbicular, corolla tubular, 4 mm long, lobed, the lobes rounded, stamens included, ovary half inferior, protogynous, with the style exposed from the ovoid bud; fruits globose, turning from yellow to white, 4–5 mm in diameter, juicy, sweet, persistent sepals and style evident.

Maesa montana A. de Candolle

Jin-zhu-liu=Chin-ch'u-liu (金珠柳, Golden Pearl Willow); *Shan-di-du-jin-shan=Shan-ti-tu-chin-shan* (山地杜莖山, Montane Maesa). Young leaves; used for tea by people of southeastern Tibet.

Shrubs 1–3 m high, branchlets pubescent; leaves elliptic, oblong-lanceolate, rarely ovate, 7–14 (–23) cm long, 3–7 (–9) cm wide, acute, or acuminate, base obtuse or rounded, serrate or remotely crenate-serrate, teeth glandular; flowers white, small, numerous, in axillary racemose panicles 7–10 cm long, hairy, pedicels 1–3 mm long, corolla subcampanulate, 4 mm long, glandularly striate, stamens included, ovary inferior; fruits globose, 4 mm in diameter, yellow turning white, juicy. Common in ravines of western and southeastern China, growing along margin of forests at altitudes of 650–2,800 m.

Maesa perlaria (Loureiro) Merrill (Syn. *M. sinensis* A. de Candolle)

Ji-yu-dan=Chi-yu-tan (鯽魚膽, Carp's Egg Mass). Young leafy shoots; used as a tea substitute in Hainan.

Shrubs 1–3 m high, branchlets pubescent; leaves elliptic-ovate, 6–9 cm long, 3–5 cm wide, acuminate or acute, base obtuse, distal half coarsely serrate, both surfaces pilose and glabrescent; flowers small, white, unisexual, dioecious, in sessile axillary panicles, rachis, pedicels, hypanthia and sepals pilose, hypanthia obconical, sepals deltoid, acute, corolla subrotate, 5-lobed, the lobes rounded, staminate flowers bearing 5 stamens inserted to the corolla tube and opposite to the lobes, pistillode pulvinate, style more or less developed, not branched, pistillate flowers bearing staminodes with sterile anthers, ovary half inferior, style 3-branched at the distal end, stigmas subcapitate; fruits berrylike, juicy, oblong-globose, with the persistent sepals attached 1/3 below the rostrate apex, white and slightly transparent, the clusters giving the impression of the egg-mass of carps, hence the vernicular name *"Ji-yu-dan"*; seeds numerous, angular, glandularly brown and densely papillose dried. Common in the subtropical forests of the maritime provinces in southeastern China and Hainan Island.

Primulaceae: Primrose Family

Lysimachia fortunei Maximowicz — SWAMP-LOOSESTRIFE

Zhen-zhu-cai=Chen-chu-ts'ai (珍珠菜, Pearl Vegetable); *Xing-xu-cai=Hsing-hsu-ts'ai* (星宿菜, Star Vegetable, referring to shape of the flower); *Hong-gen-cao=Hung-ken-ts'ao* (紅根草, Red Root Herb); *Ni-qiu-cai=Ni-ch'iu-ts'ai* (泥鰍菜, Eel Vegetable, referring to the habitat where eels grow).

Perennial erect herbs 30–70 cm high, with creeping rhizomes, growing by ponds and along the banks of rice-field; leaves subsessile, alternate, lanceolate or oblanceolate, 5–11 cm long, 2–3 cm wide, acuminate at both ends; flowers white, small, in crowded terminal racemes 10–20 cm long, rachis and pedicels glandularly puberulous, bracts lanceolate, acute, calyx deeply 5-parted, the lobes oblong, membranous and glandularly ciliate along the margin, corolla starlike, 7–8 mm across, 3/4 lobed, lobes oblong, stamens included, ovary superior, style short; capsules globose, 2–2.5 mm in diameter, persistent calyx and style prominent. Native of East Asia, widely distributed in China, apparently in association with rice culture; many local names and herbal uses for improving circulation, reducing pain and as a diuretics.

Sapotaceae: Sapodilla Family

Butyrospermum parkii Kotschy — SHEA TREE

Niu-you-shu=Niu-yu-shu (牛油樹, Butter Tree). Seed; used for making Shea butter.

Lactiferous trees 10 m high, ferruginous-tomentose throughout, trunk 40–100 cm in diameter, bark dark grayish-black, rough, inner bark red, branchlets stout, rugose; leaves crowded at the end of shoots, stipulate, oblong, 15–22 cm long, 4.5–6 cm wide, rounded or retuse, base obliquely rounded or obtuse, coriaceous, with 25–29 prominent lateral nerves on each side, entire, petioles 6–7 cm long, terete; flowers yellowish, small, appearing before the leaves, in crowded fascicles developed from lateral buds below the terminal vegetative bud, pedicels 1–2.5 cm long, calyx campanulate, deeply 8-lobed, the lobes ovate-lanceolate, 8 mm long, densely ferruginous-lanate, persistent, petals not much longer, imbricate, stamens exserted, opposite the petals, anthers elliptic, versatile, staminodes fimbriate, alternate with the fertile stamens, ovary superior, 8- to 10-locular, style subulate, stigma punctiform; fruits ellipsoid, 3–5 cm long, 2.5–4 cm across; seed solitary by abortion, filled with thick cotyledons, exalbuminous.

Native to xeric tropical Africa, introduced into Yunnan, cultivated in the tropical desert valley in Yuanjiang area; the seed containing 42% shea butter, the melting point being 35.36°C, with the fatty acids being 49.9% oleic acid, 40.4% stearic acid, 5.2% linoleic acid, 3.2% palmitic acid, 0.9% arachidic acid, and 0.4% myristic acid (Liao, et al., 1990).

Chrysophyllum cainito L. — STAR APPLE

Xing-ping-guo= hsing-p'ing-kuo (星蘋果, Star Apple, a translation of the English common name). Introduced to Taiwan before 1950.

Beautiful tropical evergreen lactiferous trees 10–20 m high; new branchlets including the leaves and flowering fascicles densely covered with appressed brown silky hairs throughout, older branchlets light gray, glabrescent; leaves oblong-elliptic, 1–15 cm long, 2–6 cm wide, acuminate-caudate, base obtuse or acute, entire, coriaceous, glabrescent above, pubescent beneath; flowers small, purplish-white, 3–35 in axillary fascicles, pedicels 8–12 mm long, calyx cupuliform, 3 mm across, 5-lobed, the lobes imbricate, pubescent, corolla subrotate, 5-lobed, the lobes ovate, 2.5 mm long, obtuse, stamens 5, subsessile, inserted at the median-base of the coralla lobes, anthers ovoid, ovary superior, globose, style very short, stigma discoid, 5-parted, two lobes each; fruits fleshy, globose, 6–10 cm in diameter, light green-purple, smooth, pulp white, sweet, translucent; seeds 3–8, testa brown, glossy. Native of the West Indies and Central America, introduced into botanical gardens in China, rare; cross-section of fruit star-shaped like that of an apple, hence the common name "Star Apple".

Eberhardtia aurata (Pierre ex Dubard) Lecomte

Xiu-mao-suo-zi-guo=Hsiu-mao-suo-tzu-kuo (銹毛梭子果, Rusty-hairy Shuttle Fruit); *Shui-mu-ji-guo=Shui-mu-ch'i-kuo* (水母雞果, Water Hen Fruit); *Xue-jiao-shu=Hsueh-chiao-shu* (血膠樹, Blood Gum Tree); *Shan-pi-ba=Shan-p'i-pa* (山枇杷, Mountain Loquat). Seed; for the extraction of cooking oil.

Trees 7–15 m high, branchlets velutinous; leaves oblong or obovate-oblong, 12–24 cm long, 4.5–9.5 cm wide, acuminate, base obtuse or rounded, glabrous above, densely velutinous beneath, lateral nerves 16–23 on each side, petioles 2–3.5 cm long, velutinous; flowers small, creamy-white, fragrant, fasciculate, axillary, sepals 2–4, bifid, 5–7 mm long, hairy, petals 5, filiform, apex incurved, 2–3 mm long, petaloid appendage 4–5 mm long, stamens 5, filaments deltoid, anthers ovate, staminodes 5, filaments subulate, sterile anthers sagittate, ovary hairy; fruits drupaceous, globose, rusty hairy, with persistent style; seeds 3–5, compressed, chestnut-brown, 2–2.3 cm long, 1–1.5 cm wide, hilum oblong, endosperm thin, cotyledons oblong. Endemic to the mountains on the borders of Guangdong-Guangxi-Vietnam; growing in mixed forests at altitudes of 1,190–1,350 m.

Madhuca pasquieri (Dubard) Lam

Mu-hua-sheng=Mu-hua-sheng (木花生, Tree Peanut); *Chu-nai-mu=Ch'u-nai-mu* (出奶木, Milky Tree). Seed; for the extraction of cooking oil.

Lactiferous large trees 30 m high, branchlets velutinous, glabrescent; leaves coriaceous, obovate or obovate-oblong, 6–16 cm long, 2–6 cm wide, obtuse or acuminate, base broad-acuminate or obtuse; flowers yellowish-white, 2 or 3 in axillary fascicles, pedicels 1.3–3.5 cm long, hairy, sepals 4, ovate, 3–5 mm long, hairy inside and outside, petals 6–11, oblong, corolla tube 1.5 mm long, stamens 12–22, filaments subulate, 1 mm long, anthers 1.5–2.5 mm long, ovary ovoid, hairy, 6- to 8-locular, styles 8–10 mm long; fruits oblong, 2.5–3 cm long, style persistent, pericarp fleshy; seeds 1–5, oblong, without endosperm, cotyledons flat, oily. Endemic to the mountains on the border of Yunnan, Guangxi and northern Vietnam.

Manilkara hexandra (Roxburgh) Dubard (Syn. *Mimusops hexandra* Roxburgh).

Yuan-xi-guo=Yuan-hsi-kuo (猿喜果, Monkey's Delight). Fruits, used in Taiwan; *Tie-xian-zi=T'ieh-hsian-tz'u* (鐵線子, Iron-thread Seed). Seed, used for extracting edible oil in Hainan Island.

Lactiferous evergreen tree 6–20 m high, branchlets gray, glabrous; leaves obovate or oblong, 5–12 cm long, 2.5–5 cm wide, obcordate or emarginate, rarely rounded, base cuneate or rounded, coriaceous, entire, often recurved; flowers pale yellow or white, 2–5 in axillary fascicles, pedicels 7–10 mm long, calyx deeply lobed, the lobes ovate, reflexed, in 2 series of 3 or 4, 3 mm long, the outer lobes glandularly rugose, the inner ones tomentose, corolla rotate, 8–10 mm across, tubes 1.5 mm long, lobes rather complex, by the vascular supply pattern the corolla lobes being 12 or 16 in 2 series, lobes of the outer series reflexed, 3 mm long, deeply 3-parted, the middle segment obovate-oblong and unguiculate, the lateral segments lanceolate and acuminate, lobes of the inner series oblong, 1.5 mm long, 2- or 3-cuspidate, erect, bending forward slightly and forming a crown (staminodes), stamens 6 or 8, inserted at the base of the middle segment of the outer lobes, ovary superior, subglobose, 8- or 9-locular, locule 1-ovulate, style cylindrical, 1 cm long, stigma punctiform; fruits yellow-red, sweet, oblong, 1.4 cm long, 8 mm across, persistent calyx and style conspicuous. Native of Southeastern Asia, from India and Sri Lanka eastward to Thailand, Vietnam, Hainan and Taiwan.

Manilkara zapota (L.) P. van Royen — Sapodilla, Naseberry, Chico (Syn. *Achras sapota* L., *Manilkara zapotilla* [Jacquin] Gilly, *Manilkara achras* [Miller] Fosberg)

Ren-xin-guo=Jen-hsin-kuo (人心果, Man-heart Fruit). Fruit; imported edible fruits common in Hong Kong market; small trees occasionally observed in villages of the New Territories of Hong Kong and in botanical gardens of South China.

Kou-xiang-tang=K'ou-hsiang-t'ang (口香糖, Chewing Gum). Imported product prepared from chicle, a gum obtained from the latex of the trunk.

Laticiferous evergreen trees 10–25 m high, young branchlets including the leaves and flowers covered with floccose brown hairs, older growth gray, stout; leaves clustered at the end of twigs, coriaceous, elliptic-oblong, 10–15 cm long, 2.5–6 cm wide, entire, acute or roundish; flowers solitary, axillary, with the pedicels and calyx ferruginously woolly, pedicels 10–25 mm long, calyx subcampanulate-oblong, 8–9 mm long, 7–8 mm across, divided almost to the base, the segments ovate, fleshy, in 2 series of 3 each, corolla white, tubular, the distal 1/3 divided into 12 lobes in 2 series, the inner lobes ovate, bifid or acute at the apex, the outer ones oblong, rounded and irregularly dented at the apex, stamens 6, inserted median on the corolla tube, ovary superior, pubescent, style exerted, stigma discoid, inconspicuously lobed; fruits ovoid, 6–8 cm long, 5–7 cm wide, rusty-brown, mottled, the flesh yellow-brown, very good flavor when fully ripe; seeds 0–12, hard, laterally compressed, glossy, black-brown, with a linear hilum along the inner angle. Native of the Yucatan Peninsula; extensively cultivated for the edible fruit, taken by the Spanish people to the Philippines, thence spread in southeastern Asia; the gum (chicle) extracted from the tree forms the basis of chewing-gum industry, produced in Mexico, British Honduras and Guatemala, used chiefly in the United States; introduced by American soldiers to China during World War II, becoming popular now.

Pouteria annamensis (Pierre) Baehni

Tao-lan=T'ao-lan (桃欖, Peach Canarium); *Da-he-guo-shu=Ta-ho-kuo-shu* (大核果樹, Large Stone Fruit). Fruit; with sweet taste and good flavor, eaten by people in Hainan Island.

Trees 10–20 m high, branchlets densely pubescent; leaves oblong or obovate-oblong, 7–17 cm long, 3–5 cm wide, acute or rounded, rarely emarginate, base cuneate, glabrous, petioles 1.5–4.5 cm long; flowers white, in axillary fascicles, velutinous, calyx 3 mm long, corolla 4 mm long, 5-lobed, the lobes 1.5 mm long, fertile stamens 5, ovary hairy, 5-locular, style 2.5 mm long, fruits pomiform, 2.5–4 cm long; seeds 4 or 5, shiny brown, compressed, 12 mm long. Endemic to the mountains on the border of Vietnam, Guangdong and Yunnan, extending eastward to Hainan Island.

Pouteria grandifolia (Wallich) Pierre

Long-guo=Lung-kuo (龍果, Dragon Fruit); *Tao-lan=T'ao-lan* (桃欖, Peach Canarium); *Ma-ji-kang=Ma-ch'i-k'an* (麻雞康, the Thai ethnic name). Fruit, eaten by the Thai ethnic group living in southern Yunnan.

Trees 40 m high, bark pale gray, leaves alternate, obovate-oblong, (7–) 10–30 cm long, 6–10 cm wide, acute or acuminate, base obtuse, petioles 1.5–4 cm long; flowers white, 3–10 in axillary fascicles, pedicels 2–3 mm long, hairy, calyx velutinous, 5-lobed, the lobes ovate or orbicular, 2.5–3 mm long, inner ones fimbriate, corolla 3–4.5 cm long, fertile stamens 5, inclosed, inserted to the throat of the corolla, anthers cordate, basifixed, staminodes 5, filiform, ovary conical, hairy, style cylindrical, stigma punctiform; fruits globose, 4–5 cm in diameter, yellow, pericarp fleshy; seeds 2–5, brown. Native to the eastern Himalayan Region, growing in the rain forests at altitudes of 500–1,180 m in southern Yunnan, thence eastward to Guangdong and the adjacent areas in Thailand and Burma.

Xantolis stenosepala (Hu) P. Royen

Ji-xin-guo=Chi-hsin-kuo (雞心果, Chicken Heart Fruit); *Tian-ci-lan=Tien-tz'u-lan*
(滇刺欖, Yunnan Prickly Canarium); *Xia-ye-he-bao-guo=Hsia-yeh-ho-pao-kuo* (狹葉荷
包果, Narrow-leaved Pouch Fruit); *Ma-bang=Ma-pang* (罵榜, the Thai ethnic name
used in southern Yunnan). Ripe fruits; eatern locally in Yunnan.

Evergreen trees 6–15 m high, trunk 10–20 cm in diameter, bark brown, branchlets
pilose; leaves oblong-lanceolate (5–) 7–15 (–18) cm long, 2.5–6 cm wide, acuminate,
base acute; flowers small, 2–5 in axillary fascicles, 5-merous, calyx 5-lobed, hairy inside
and outside, petals 6.5 mm long, stamens inserted to the base of petals; fertile anthers
sagittate, sterile ones subulate, ovary globose-ovoid, velutinous, 5-locular, style 12 mm
long, hairy at the base; fruits drupaceous, oblong-ovoid, 3 cm long, 2–3 cm across,
hairy, pericarp hard; seeds 1 or 2, brown, 2.5 cm long, 1.3 cm across. Endemic to southern
Yunnan, growing in tropical forests at altitudes of 1,150–1,700 m.

Ebenaceae: Ebony Family

Diospyros kaki L. f. — Persimmon, Kaki

Shi-zi=Shih-tzu (柿子, Persimmon). Fresh fruit; two types available in northern China.
Hong-shi=Hung-shih (蔑柿, Dead Ripe Persimmon), soft persimmon, ripened
naturally, juicy, very sweet; *Lan-shi=Lan-shih* (濷柿, Treated Persimmon), hard
persimmon, crisp, sweet, prepared by soaking the harvested mature fruits in
lukewarm lime water for several days; the market product appearing greenish-
yellow, with the texture of an apple but having a sweet persimmon taste.
Shi-bing=Shih-ping (柿餅, Dried Persimmon Cake). Compressed, brownish-black, dried
persimmon cake, covered with white powder, sweet; available in American Chinese
groceries; much used in traditional Chinese medicine.

Large deciduous trees 5–14 m high, current-year's branchlets grayish-green,
pubescent; leaves obovate-elliptic, 6–12 cm long, 3.5–5.5 cm wide, pubescent above,
grayish-white beneath; flowers rather inconspicuous because of the large foliaceous
calyx hiding the white corolla, unisexual, dioecious, 4-merous, staminate flowers 8 mm
long, in axillary cymes, calyx 4-fid, pubescent on both surfaces, corolla tubular-
campanulate, silky, stamens 14–24, inserted on the base of corolla, filaments hairy,
anthers hastate, pistillode globose, pistillate flowers solitary, pendulous, foliaceous calyx
tomentose, lobes various in shapes and sizes, ovate-acute or suborbicular, 1.5–2.5 cm
long, persistent, corolla broadly campanulate, pubescent, 1 cm long, shallowly 4-lobed,
staminodes 8, hairy, ovary depressed tetragonous-subglobose, 8-locular, style subulate,

4-parted halfway, stigmas erect, 2-parted; fruits a large juicy berry, golden-yellow and firm, strongly astringent, after ripening soft and reddish golden, sweet, edible, sizes and shape various, generally depressed ovoid or quadrangular-globose. Native of central and western China, recorded in the ancient *Book of Rites* (《禮記》) as a fruit used in various religious and governmental ceremonies; numerous cultivars, the following being the most outstanding:

1. cv. 'Constrieta' Tsen (*Pan-shi=Pan-shih* 盤柿, Disk Persimmon). Fruits large, compressed; the finest product in the market.

2. cv. 'Costata' André (*Si-leng-shi=Ssu-leng-shih* 四稜柿, Tetragonal Persimmon). Fruits with four ridges.

3. cv. 'Mazelii' Mouillefert (*Ba-leng-shi=Pa-leng-shih* 八稜柿, Octagonal Persimmon). Fruits with eight ridges.

4. cv. 'Silvestris' Makino (*You-shi=Yu-shih* 油柿, Oil Persimmon). Fruit with many ridges.

Note: In Chinese culture, every part of a persimmon plant is useful, especially in traditional medicinal practice. The root (*Shi-gen=Shih-ken* 柿根) is used as a cardiotonic, especially when cooked with *Euphorbia humifusa* Wallich and pork. The wood has manifold uses, being the best kind for furniture. Bark of the trunk when roasted, pulverized, and mixed with oil is used for dressing burns. The flowers, dried and pulverized, is spread on broken small-pox. The windfall green fruits are used by farmers to dye and tannin cloth for summer wear. A product called *"Shi-qi"* (*=Shih-ch'i* 柿漆, Persimmon Tannin), a gummy material, prepared from an aqueous extract of the very astringent young fruits is used as a hypotensor. The persistent calyx discarded from the table is gathered, cleaned, trimmed and sold to herb stores as *Shi-di* (*=Shih-ti* 柿蒂, Persimmon Calyx) to be used in specific prescriptions for gas pains. Chinese and Japanese phytochemists have isolated betulinic acid, hydroxytriterpenic acid, oleanolic acid, and ursolic acid from persimmon calyx used in medicine.

Diospyros lotus L. — Date Plum, Wild Persimmon
  Ruan-zao=Juan-ts'ao (軟棗, Soft Date); *Jun-qian-zi=Chün-chien-tzu* (君遷子, Wild
    Persimmon). Ripe fruit; small black fruits dropped on the ground with the falling
    leaves, eaten by children.

  Large deciduous tree 20–25 m high, trunks 1–1.3 m in diameter, bark gray, deeply
  fissured and broken into irregular pieces, branchlets glabrous; leaves elliptic-oblong,
  6–10 cm long, 3–5.5 cm wide, acuminate, base obtuse, glabrous above, pubescent

beneath; flowers white, unisexual, dioecious, staminate flowers 5 or 6, in axillary cymes, calyx patelliform, 5 mm across, deeply 4-lobed, corolla urceolate, 5–6 mm long, 3 mm across, 4-lobed, lobes rounded, stamens 16, anthers pubescent, awned at the apex, pistillode globose, pistillate flowers solitary, calyx foliaceous, 2 cm across, deeply 4-lobed, corolla broadly tubular, deeply 4-lobed, lobes orbicular, rolling back, staminodes 8, glabrous, ovary globose, 6 mm in diameter, style short, pilose; fruits globose, 1–1.5 cm in diameter, changing from green to yellow, and purplish-black, edible only after turning black; dried fruit available in markets of villages and small towns. Native of China, used as stock for grafting persimmon, introduced to Japan for the same purposes, and into USA in late 19th century.

Diospyros oldhamii Maximowicz — Taiwan Wild Persimmon (*D. oldhamii* var. *chartacea* Hayata; *D. taitoensis* Odashima)

Tai-wan-shi=Tai-wan-shih (台灣柿, Taiwan Persimmon); *Tai-dong-dou-shi=Tai-tung-tou-shih* (台東豆柿, Taidong Pea Persimmon). Fruit, used locally in Taiwan.

Small deciduous trees, branchlets glabrous; leaves chartaceous, elliptic, 3–8 cm long, 2–3 cm wide, acuminate, base rounded, acute, entire, glabrous above, yellow-hirsute at the nerve-angles and occasionally sparsely pilose beneath; flowers white, glabrous, unisexual, dioecious, staminate flowers 5 or 6 in axillary cymes, calyx patelliform, 3–4 mm across, deeply 4-lobed, corolla tubular-urceolate, 5–6 mm long, pistillate flowers solitary; fruits depressed-globose or ellipsoid, 2–3.5 cm long, 2–2.8 cm across, calyx persistent.

Symplocaceae: Sweetleaf Family

Symplocos paniculata (Thunberg) Miquel — Asiatic Sweetleaf

Ji-gu-tou=Chi-ku-t'ou (雞骨頭, Chicken Bone); *Bai-tan=Pai-t'an* (白檀, White Tan). Seed; oil extracted from seed used for cooking.

Deciduous shrubs 4 m high, bark pale brown, exfoliate, branchlets pilose; leaves elliptic or obovate-oblong, 3–11 cm long, 2–4 cm wide, acuminate, base cuneate, serrulate, sparsely pubescent beneath; flowers white, fragrant, 8–10 mm across, in terminal panicles 4–8 cm long, sepals 5, 2 mm long, ciliate, corolla rotate, 4–5 mm long, 5-lobed, stamens 30, adnate to the corolla, filaments connate into 5 fascicles, ovary inferior, 2-locular; drupes ovoid, bright blue, 5–8 mm long, crowned with persistent calyx; stone 1-seeded. native to eastern Asia, introduced into Western gardens in 1875.

Oleaceae: Olive Family

Chionanthus retusus Lindley — FRINGE TREE

Cha-ye-shu=Ch'a-yeh-shu (茶葉樹, Tea-leaf Tree); *Liu-su-shu=Liu-su-shu* (流蘇樹, Tassel Tree). Leaves, gathered in northern China, used for tea.

Small deciduous tree 2–6 m high, branchlets glabrescent; leaves opposite, oblong, obovate or ovate-elliptic, 3–10 cm long, 2–4.5 cm wide, rounded, often retuse, base cuneate or rounded, entire; flowers greenish-white, fragrant, polygamodioecious, in terminal compound cymose panicles 3–8 cm across the top, peduncles filiform, pedicels 8 mm long, calyx patelliform, deeply 4-lobed, corolla 1.5–2 cm long, divided almost to the base, lobes linear, slightly broader toward the distal half, 12–18 mm long, 1–2 mm wide, apex obtuse, stamens 2, inserted to the corolla tube, anthers ovate, subsessile, mucronate, ovary superior, ovoid, style short, stigma capitate; fruits drupaceous, oblong-ellipsoid, 8–14 mm long, 7–9 mm across, dark blue and glaucous; seed solitary. Native of China, Korea and Japan, introduced to western gardens in 1845.

Jasminum sambac (L.) Aiton — JASMINE; MOLI (an ethnic name of the Yunnan people).

Mo-li=Mo-li (茉莉, Jasmine). Flowers; gathered in early morning, used for flavoring tea; the best grade of tea treated one or two times with fresh flowers which are removed; the lower grade retains the flowers.

Evergreen vines 1–3 m high, in cultivation trimmed to shrub-like habit, branchlets pubescent, glabrescent; leaves simple, opposite, ovate-oblong, 3–9 cm long, obtuse at both ends; flowers white, fragrant, 3–5 in terminal cymose clusters, sepals 8, filiform, corolla shortly salverform, the tube shorter than the suborbicular lobes, often double; fruits berrylike, seldom seen. Native of Yunnan and adjacent mountains of Guizhou and Guangxi; first known as cultivated plants, introduced by ancient Arabian traders to Arabia, hence to European gardens.

Ligustrum japonicum Thunberg — WAX-LEAF PRIVET, JAPANESE PRIVET

Nü-zhen=Nü-cheng (女貞, The Virgin). Tender shoots, eaten locally in Taiwan before 1950.

Shrubs or small trees 3–6 m high, branchlets puberulous when young, glabrescent, lenticellate; leaves broad-ovate or ovate-oblong, 4–10 cm long, 1.5–4 cm wide, obtusely short-acuminate or acute, obtuse or rounded at base, entire, coriaceous; flowers white, small, in large terminal panicles of current year's growth, 6–15 cm long, pyramidal, pedicels 1–2 mm long, calyx campanulate, 1.5 mm long and across the subtruncate apex, corolla salverform, tube 3 mm long, limb 4-lobed, 5–6 mm across, stamens 2, exserted, ovary superior, globose, glabrous, style cylindrical, undivided, stigmas linear, lateral; fruits black, oblong, 10–12 mm long, 6–8 mm across, persistent calyx prominent. Native of Japan, introduced into Western gardens for ornamnetal purposes in 1860.

Olea europaea L. — Olive

Qi-dun-guo=Ch'i-tun-kuo (齊墩果, Olive); *Yang-gan-lan=Yang-kan-lan* (洋橄欖, Foreign Canarium). Fruit; introduced and cultivated for edible oil, rare.

Evergreen trees 6–7 m high, branchlets trigonous, silvery scurfy; leaves subcoriaceous, opposite, lanceolate or oblanceolate, 1.5–5 cm long, 8–12 mm wide, acute and mucronate, base cuneate, entire, densely covered by silvery scales beneath; flowers white, fragrant, small, subsessile, in axillary cymose panicles 2–6 cm long, calyx cupular-obconical, 1.5 mm long, 3 mm across, subtruncate, corolla spreading, tube very short and completely concealed by the calyx, lobes oblong, 4 mm long, 2 mm wide, valvate in bud, stamens 2, exserted, ovary superior, style very short, stigma subcapitate; fruits drupaceous, oblong, 2–2.5 cm long, shiny black and dropping on the ground at maturity. Native of the Mediterranean Region, introduced to China repeatedly, for the edible fruits and as a source of olive oil.

Osmanthus fragrans (Thunberg) Loureiro — Sweet Olive, Osmanthus, Guihua

Gui-hua=Kuei-hua (桂花, Osmanthus). Flowers; gathered for seasoning tea; mixed with sugar or honey, bottled and kept for making pastries, regarded as a delicacy; added to rice gin for preparing *guihua* liquor, available in American Chinese stores.

Evergreen trees 10–15 m high, with a spreading compact crown and a trunk up to 1 m in diameter (in cultivation often shrubby, 1–2 m high, kept in pots in northern China), branchlets angular, ochreous, lenticulate; leaves coriaceous, opposite, elliptic, 4–12 cm long, 2–4 cm wide, abruptly acuminate, base obtuse or acute, subentire or remotely serrate above the middle; flowers small, very fragrant, pale yellow, gold-orange or reddish-orange, in axillary fascicles on second year's growth, calyx discoid, 1 mm across, inconspicuously 1- to 4-toothed, corolla spreading, 3–4 mm across, deeply 4-lobed, lobes oblong, 2 mm long, valvate in bud, stamens 2, poorly developed (in cultivation), generally hidden by the short corolla tube, ovary superior, ovoid, style short, stigma capitate (in many flowers style pointed and without stigma); fruits drupaceous, often abscent, when present, oblong, 1–1.5 cm long, purple-black. Native of the Nanling Range, with Guilin (桂林, Osmanthus Forest) being the center of osmanthus cultivation.

In China, the *guihua* culture is associated with the Mid-autumn Festival (中秋節), the Chinese Thanksgiving, a time of family reunion.

Gentianaceae: Gentian Family

Nymphoides cristata (Roxburgh) 0. Kuntze — LESSER WATER SNOWFLAKE (Syn. *Limnanthemum cristatum* [Roxburgh] Grisebach)
Yin-lian-hua=Yin-lien-hua (銀蓮花, Silver Lotus); *Shui-pi-lian=Shui-p'i-lien* (水皮蓮, Water Surface Lotus). Young shoots, eaten in Taiwan.

Floating, perennial herbs with submersed rhizomes, stems slender, each with one long internode sending the flowers to the water surface and many short ones floating; basal leaves rosulate, on long petioles, cauline leaves on short petioles, laminas suborbicular-ovate, 2.4–8 cm long, 4–7 cm across the middle, subcordate and deeply bilobed at the base, apex obtuse or rounded, green above, purple-red and roughened with convex dots (hydathodes) beneath; flowers small, white, several in fascicles axillary to cauline leaves, pedicels filiform, 2–3 cm long, calyx 4 mm long, lobed to the base, petals equal calyx in length, entire, stamens 5, inserted to the short corolla tube, anthers sagittate, ovary superior, unilocular, ovules many; capsules subglobose, 6 mm in diameter; seeds several, scabrous, 1.5 mm across. Native of southeastern Asia, occurring in Guangdong, Taiwan and Hainan; in appearance very close to *N. hydrophylla* (Loureiro) O. Kuntze and *N. indica* (L.) O. Kuntze, distinguished by its small flowers (with petals equal sepals in length) and rough seeds.

Nymphoides peltatum (S. G. Gmelin) 0. Kuntze — YELLOW FLOATING HEART, WATER POPPY (Syn. *Limnanthemum nymphoides* Hoffmansegg et Link)
Xing-si-cai=Hsing-ssu-ts'ai (荇絲菜, Floating Silk Herb); *Xing-cai=Hsing-ts'ai* (莕菜, Floating Heart Vegetable). Young leaves; gathered before fully unfolded, used in soup; cultivated in a special area in the West Lake of Hanzhou, young leaves used for serving summer visitors of this famous city, the slippery texture and unusual form, being a good subject for table talk, no particular flavor.

Floating perennial herbs rooting in mud, stems wirelike, branched repeatedly, internodes 5–15 cm long; leaves oblong, reniform or suborbicular, 2–8 cm; long, 2–7 cm across the middle, base cordate, margin undulating, green above, purple-red and rough beneath; petioles sheathing at the base; flowers yellow, ephemeral, opening early on sunny mornings, just above the floating foliage, 2–5 fasciculate in the axils of cauline leaves, with 1 blooming at a time, 3 cm across, sepals lanceolate, 10–12 mm long, ciliate, corolla 5-lobed, hairy at the throat, lobes fimbriate along the margin, stamens 5, exerted, the anthers hornlike, base sagittate, ovary superior, with 5 nectars at the base, style short, stigma funnelform, lobed; capsules oblong, or ellipsoid, 1.5–2 cm long, 6–8 mm across, calyx and style persistent; seeds discoid, hairy.

Widespread in ponds and slow moving streams, particularly from the Yangtze River Region, thence southward to Yunnan. Its ancient record can be traced in one of the songs in the *Book of Odes* (《詩經》); presently, young unfolding leaves served to international guests and tourists in hotels of Hanzhou (Map, Loc. 28).

Apocynaceae: Dogbane Family

Amalocalyx yunnanensis Tsiang

Mao-che-teng=Mao-ch'e-t'eng (毛車藤, Hairy Cart Vine). Young fruits; gathered and eaten by the ethnic people living in the mountains of southern Yunnan and adjacent areas of Laos and Burma.

Evergreen, lactiferous lianas, tomentose throughout; leaves obovate or elliptic-oblong, 5–15 cm long, 2–10.5 cm wide, acute and mucronate, entire, base subcordate-auriculate, petioles 1–1.5 cm long; flowers light purple-red, 15–20 in corymbose panicles, peduncles 6–12 cm long, pedicles 5–15 mm long, calyx campanulate, 5–7 mm long, glandular, sepals oblong, 4–5 mm long, 3 mm wide, corolla funnelform, fully grown buds 3 cm long, clavate, tubes 2 cm long, limb 3–4 cm across, lobes ovate, stamens 5, inserted to the middle of the corolla tube, filaments short, barbate, anthers sagittate, disk annular, 5-lobed, ovary 2-carpellate, multiovulate, style short, stigmatic clavuncles oblong; follicles oblong-cornute, 11 cm long, 1 cm across the middle; seeds orbicular, comose, hairs yellowish-brown. Endemic to the southernmost area of Yunnan, and the adjacent land in Laos, Thailand and Burma; growing in forests at altitudes between 800–1,000 m.

Apocynum venetum L. — DOGBANE

Luo-bu-ma=Lo-pu-ma (羅布麻, Netting Hemp); *Ze-qi-ma=Che-ch'i-ma* (澤漆麻, Marsh-varnish Hemp); *Ye-cha=Yeh-ch'a* (野茶, Wild Tea). Leafy shoots; gathered, steamed, and dried, used for tea in Inner Mongolia.

Erect, glabrous, richly lactiferous perennial herbs 0.5–2 m high, with pink subterranean rhizomes, branches purple-red; leaves opposite, elliptic-lanceolate, or ovate-oblong, 1–5 cm long, 0.5–1.5 cm wide, obtuse and apiculate, base roundish, petioles 3–6 mm long, with subulate glands at the base; flowers pink or purplish-pink, in terminal cymose clusters, calyx deeply 5-lobed, corolla subcampanulate, distal two-thirds densely papillose, deeply 5-lobed, 6–8 mm long, bearing 5-triangular appendages below the throat opposite the lobes, stamens 5, filamens short and barbate on the inside, anthers sagittate, thecae below the acute apex, annulus fleshy, lobed, ovary 2-carpellate, clavuncle ellipsoid, adherent to the lower portion of the anther; fruits of 2 long follicles 8–15 cm long, 2–3 mm in diameter; seeds numerous, ovate-oblong, 2–3 mm long, comose, the hairs white, soft, 1.5–2 cm long. Native of temperate Eurasia, common in lowland and near lakes, a very important hypotensive agent, applied for its cardiotonic and diuretic activities, now extensively used throughout China.

Carissa carandas L. — KARANDA

Ci-huang-guo=Tz'u-huang-kuo (刺黃果, Spiny Yellow Fruit). Fruit; introduced and cultivated, rare.

Evergreen shrub or climber, lactiferous, thorny, stems irregularly bent, branchlets glabrous, spines 3–5 cm long, bifurcate; leaves opposite, broadly ovate or oblong, occasionally suborbicular, 3–4 (–7) cm long, 2–3.5 cm wide, obtuse, rounded, abruptly acute, base rounded or cuneate, coriaceous; flowers white, salverform, fragrant, calyx 2–3 mm across, 5-lobed, lobes acute, corolla 1.5–2 cm across, tube 1.5–2 cm long, lobes contort, stamens 5, inserted on the corolla tube, ovary superior, 2-locular, 4-ovulate, style 8–10 mm long; fruits baccate, ellipsoid, 1–2 cm long, red turning to black; seeds 4, dorsal-ventrally compressed, oblong, 5–6 mm long, 2–3 mm wide. Native of India, cultivated in villages in tropical China as hedges, fruits used locally, not available in the market.

Carissa macrocarpa (Ecklon) A. de Candolle — NATAL PLUM (Syn. *C. grandiflora* [E. H. Meyer] A. de Candolle) (Figure 45)

Da-hua-jia-hu-ci=Ta-hua-chia-hu-tz'u (大花假虎刺, Large-flowered Crown-of-thorns). Introduced, fruit used for jam.

Evergreen shrub 1 m high, much branched, spreading, 3 m across, branches with dense foliage and forked spines, stout, 3–4 cm long; leaves dark green, ovate, 3–7 cm long, 2.5–5 cm wide, abruptly acute, base rounded; flowers white, 1–5 in terminal subsessile cymose clusters, bracts small, ovate, acute, pedicels 2–3 mm long, calyx obconic, 5-lobed, lobes obovate-oblong, 6 mm long, 2 mm wide, corolla salverform, tubes 1.5 cm long, hairy inside, limbs 3–5 cm across, lobed to the throat, lobes obovate, rounded at apex, cuneate, stamens 5, inserted at or below the middle of the corolla tube, anthers lanceolate, glabrous, ovary ovoid, 2-locular, ovules 8–10 in each locule, style, 4 mm long, upper half clavate, stigmatic surface on ridges below the acute apex furnished with penicillate white hairs; fruit scarlet, ovoid-ellipsoid, 4–5 cm long, sour, persistent sepals conspicuous; seeds 6–16, orbicular. Native of South Africa, cultivated in Guangdong, Hong Kong, and Taiwan, for ornamental purposes; fruits used locally, not in the market.

Ecdysanthera rosea Hooker et Arnott — SOUR CREEPER

Suan-teng=Suan-t'eng (酸藤, Sour Vine). Leaves; used in Taiwan.

Lactiferous evergreen climber growing over thickets by villages, branchlets wirelike, purplish-black, current year's growth glabrous, leafy and with a terminal loose corymbose panicle of very small pink-purple flowers; leaves elliptic or obovate-oblong, 3–7 cm long, 1–4 cm wide, abruptly acuminate, base acute, cuneate, rarely obtuse, entire, glaucous beneath; flowers small, panicles 7–10 cm across, bracts and bracteoles deltoid, acute, calyx patelliform, 2 mm cross, deeply lobed, lobes triangular, acute, corolla bell-shaped, 4–5 mm long, 5-lobed to the middle, stamens inserted near the base of the corolla tube, anthers long sagittate, the connective broadened and produced a truncate apex, included, ovary hairy, surrounded by a cupular gland, 2-carpellate, carpels distinct, style short and stout, distal end globose, adherent to the anthers below the thecae, apex ovate-triangular, bifid; follicles divergent, beanlike, 10–17 cm long, lenticulate, attenuate toward the apex; seeds numerous, compressed, oblong, 12–14 mm long, 3 mm wide, comose, the hairs silvery-white, 5 cm long, testa coriaceous, velutinously lanate. Native of southeastern Asia, used in folk medicine for indigestion and rheumatic pains; only report for food from Taiwan before 1950.

Hunteria zeylanica (Retzius) Gardner et Thwaites (Syn. *H. corymbosa* Roxburgh)

Zi-lan-shu=Tzu-lan-shu (子欖樹, Baby Canarium); *Huang-yang=Huang-yang* (黃羊, Yellow Goat). Fruits, eaten in Hainan Island.

Lactiferous trees 8–20 m high, trunks 10–25 cm in diameter, branchlets glabrous; leaves opposite, coriaceous, oblong-elliptic, or ovate-oblong, 9–16 cm long, 2.5–5 cm wide, acuminate-caudate, base rounded or obtuse, lateral nerves 30 pairs, parallel; flowers greenish-white, fragrant, 10–15 in terminal corymbose-panicles 2–4 cm across, calyx obconical, 5-lobed, lobes ovate, 1.5 mm long, corolla salverform, tubes 7–10 mm long, the distal 1/3 slightly enlarged, hairy inside, limbs 5–6 mm across, contorting clockwise in bud, stamens 5, inserted to the distal enlarged portion of the corolla tube, stamens ovate-lanceolate, base rounded, ovary ovoid, 2-carpellate, carpels distinct, style filiform, enlarged at the distal end, acute at the apex, not adhering to the anthers; berries in pairs, obovoid, shortly stipitate, or ellipsoid, 1.5–2.5 cm long, 1.2–1.8 cm in diameter, olive-green turning to orange-red; seeds 1 or 2. Native of Southeast Asia, from eastern India to Hainan Island.

Melodinus henryi Craib

Si-mao-shan-cheng=Ssu-mao-shan-ch'eng (思茅山橙, Si-mao Melodinus). Mature fruits; eaten fresh locally.

Robust evergreen lianas, branchlets pubescent; leaves oblong-elliptic or lanceolate, 6–19 cm long, 2.6–6.5 cm wide, acute or acuminate, base obtuse, petioles 6–10 cm long; flowers white, in terminal and axillary cymose corymbs 4–5.5 cm long, calyx obconical, sepals orbicular, 2 mm long, ciliate, corolla contorted in bud, salverform, 12 mm long, tubes 8 mm long, velvety, limbs 9–10 mm across, corolla lobes suborbicular, 4–5 mm across, coronal scales inserted at the throat, stamens 5, median on the tube, ovary superior, clavuncle ovoid, acute; fruits berrylike, ellipsoid, 8.5 cm long, 4 cm across middle, orange-red. Endemic to southern Yunnan and adjacent areas in northern Burma and Thailand; growing over hillsides at altitudes of 760–2,800 m.

Melodinus khasianus Hooker f.

> *Jing-shan-shan-cheng=Ching-shan-shan-ch'eng* (景山山橙, Jing-shan Melodinus). Ripe fruits, eaten locally in central and western Yunnan.

Glabrous lactiferous evergreen lianas; leaves elliptic-oblong, 6–11.5 cm long, 1.5–4 cm wide, shortly accuminate, base acute; flowers white, in terminal corymbs 2.5–6.5 cm long, sepals orbicular, 3 mm across, ciliate, corolla 8 mm long, limbs 1.2 cm across, lobes ovate, coronal scales rectangular, erose; fruits subglobose, 5.5 cm long, 4 cm in diameter, red-orange. Widespread in eastern Himalayan Region, growing in margins of forest at altitudes of 1,600–1,900 m in central and western Yunnan.

Melodinus suaveolens Champion ex Bentham — Mountain Orange, Monkey Creeper

> *Shan-cheng=Shan-ch'eng* (山橙, Mountain Orange); *Ma-liu-teng=Ma-liu-t'eng* (馬騮藤, Monkey Creeper); *Hou-zi-guo=Hou-tsu-kuo* (猴子果, Monkey Fruit). Ripe fruit; collected from wild plants of the size of a large orange, with sweet, red pulp; available in the fruit market of Hong Kong, boiled with pork for soup.

Lactiferous evergreen glabrous lianas; leaves opposite, coriaceous, ovate-oblong or elliptic, 5–8 cm long, 2–4 cm wide, attenuate at both ends; flowers white, fragrant, in terminal cymose panicles, pedicels short, pilose, calyx subglobose-cupular, scabrid, 3 mm long, 4 mm across, lobes rounded, margin membranous, corolla salverform, tubes 1–1.2 cm long, puberulous, limbs 1.5–2 cm across, 5-lobed, lobes oblique, falcate, with 2 irregular teeth, throat with scale-like yellow crown, cupular, stamens 5, inserted to the middle of the corolla lobe, anthers sagittate, hairy, glutinated to the rounded clavuncle, ovary 2-carpellate, 2-locular, multiovulate; fruit fleshy, globose, 5–8 cm in diameter, orange-red at maturity, rind thick, pulp red, sweet, edible; seeds numerous, compressed oblong, 12 mm long, 8 mm wide, 3 mm thick, brown, irregularly striate-fissured, testa thick, firm, rather spongy, breakable. Native of southern China, first reported from

Hong Kong where the species is very common, growing over rocks and trees in the hillsides, along streams; not in cultivation yet.

Asclepiadaceae: Milkweed Family

Cynanchum auriculatum Royle ex Wight (Syn. *C. caudatum* sensu auctt., non [Miquel] Maximowicz)

Niu-pi-xiao=Niu-p'i-hsiao (牛皮消, Ox-hide Melt); *Lao-niu-piao=Lao-niu-p'iao* (老牛瓢, Cow Gourd-ladle). Young shoots; gathered, boiled, washed, and eaten as potherb in Taiwan before 1950; young fruits, children opening the fully grown young fruit called "*Lao-niu-piao*", and eating the tender seeds in northern Jiangsu.

Perennial lactiferous climber, branchlets pubescent, roots fleshy, enlarged, irregularly fusiform; leaves opposite, cordate, 4–12 cm long, 3–10 cm wide, pubescent beneath; flowers white, small, 20–30 in an axillary corymbose cluster, sepals ovate-oblong, corolla rotate, the lobes reflexed, hairy inside. corona cupular, lobed, lobes oblong, fleshy, obtuse, with a triangular-ligular scale inside each lobe, pollinia pendulous, stigma conical, 2-cleft; fruits follicular, follicles lanceolate, 8 cm long, 1 cm wide; seeds ovate-oblong, coma white.

Cynanchum bungei Decaisne

Bai-shou-wu=Pai-shou-wu (白首烏, White-Hair Black); *Tai-shan-he-shou-wu=Tai-shan-he-shou-wu* (泰山何首烏, *He-shou-wu* of Tai Mountain); *Di-hu-lu=Ti-hu-lu* (地葫蘆, Ground Bottle-gourd). Root; starch extracted from the enlarged fleshy root, sold in the market as *He-shou-wu-fen*, used with sugar as a health-food, for the aged persons especially.

Perennial climbing herbs with stout tuberous roots 5–15 cm long, 1.5–3.5 cm across the middle; leaves opposite, lanceolate-hastate, 3–8 cm long, 1–5 cm wide across the base, acuminate, base cordate, hirsute on both surfaces; flowers white, in axillary corymbs, sepals lanceolate, corolla rotate, lobes reflexed, softly hairy inside, corona 5-lobed, the lobes lanceolate, longer than the column, each lobe with a ligulate scale inside, pollinia pendulous, base of stigma 5-angled, top entire; follicles 2 or 1, lanceolate, glabrous, attenuate at the distal end, 9 cm long, 1 cm across middle; seeds ovate, 1 cm long, 5 mm wide, coma white, 4 cm long.

Native of northern China, growing between latitudes of 34–42°N. The species was first cultivated in Shandong Province for the production of starch which is called "*He-shou-wu*" (何首烏). It was introduced and cultivated by a commune in central Jiangsu. However, this name has been applied to *Polygonum multiflorum* Thunberg in ancient

Chinese herbals since the tenth century and in current botanical literature. To avoid confusion, current local practice must give way to historical application of *He-shou-wu* to *P. multiflorum* Thunberg.

Cynanchum forrestii Schlechtendal

Bai-long-xu=Pai-lung-hsu (白龍鬚, White Dragon Whisker). Rootstocks, including the crown and several wirelike roots; gathered from the hillsides, dried, sold in the local market in Yunnan, used in soup, making a very tasty beverage.

Perennial erect herbs, branches rarely ramified, densely pubescent at the distal portion; leaves opposite, ovate-oblong, 4–5 cm long, 1.5–4 cm wide, abruptly short acuminate, base subcordate or rounded; flowers yellow, small, in axillary corymbs, sepals lanceolate, corolla 3 mm across, rotate, lobes ovate, ciliate, corona fleshy, triangular; follicles seldom paired; seeds compressed, coma 2 cm long. Endemic to Yunnan.

Cynanchum otophyllum Schneider

Qing-yang-shen=Ch'ing-yang-shen (青羊參, Taoist Goat Shen). Root, boiled with pork or spare-ribs for a special dish.

Lactiferous perennial herbaceous vines with tufts of cylindrical roots, grayish black on the outside, 8 mm in diameter; leaves membranous when dried, ovate-lanceolate, 7–10 cm long, 4–8 cm wide, acuminate, base cordate, tomentose on both surfaces, petioles equal the laminas in length; flowers white, in axillary cymose clusters, peduncles 3–10 cm long, those of the early flowers longer, pedicels 5–8 mm long, calyx with 5 glands on the inside, corolla rotate, 3 mm across, lobes oblong, slightly pilose, pollinia ovoid, ovary glabrous, styles 2, stigma globose; follicles lanceolate, 8 cm long, 1 cm across the middle; seeds ovate, 6 mm long, 3 mm wide, comose, hairs 3 mm long. Widespread in southwestern China, growing on grassy hillsides at altitudes of 1,400–2,800 m.

Cynanchum wilfordii (Maximowicz) Hemsley

Ge-shan-xiao=Ke-shan-hsiao (隔山消, Melt Across-the-mountain); *Ge-shan-jiao=Ke-shan-ch'iao* (隔山撬, Cross-the-mountain). Fleshy root, used for extracting starch.

Perennial climber with fleshy fusiform and branched root 10 cm long, 2 cm in diameter, branchlets pubescent; leaves opposite, papery, ovate, 5–6 cm long, 2–4 cm wide, acuminate, base cordate, both surfaces pubescent, becoming black when dried, basal nerves 3–4; flowers pale yellow, 15–20, in short panicles of fascicles, sepals pubescent, oblong, corolla rotate, lobes rounded, obtuse, glabrous outside, hairy inside, corona shorter than the column, the lobes quadrangular, apex truncate, narrowed at base, pollinia oblong, pendulous, style slender, stigma slightly ridged; fruits follicular, generally unpaired, gradually attenuate at the distal end, base narrowed at base, 12 cm long, 1 cm across middle; seeds dark brown, ovate, 7 mm long, comae silky white, 2 cm long.

Telosma cordata (Burm. f.) Merrill

Ye-lai-xiang=Yeh-lai-hsiang (夜來香, Night Fragrant); *Ye-xiang-hua=Yeh-hsiang-hua* (夜香花, Night Fragrant Flower). Entire flower cluster with some in full bloom and others in buds, soaked in water to enable visiting insects to crawl out, then dropped in boiling soup just before serving.

Deciduous tropical perennial climbers, branchlets yellowish-green, puberulous; leaves membranous, ovate-oblong or ovate, 6.5–9.5 cm long, 4–8 cm wide, shortly acuminate, base cordate, slightly pubescent on the nerves, petioles 1.5–5 cm long, distal ends with 3–5 small glands; flowers greenish-yellow, fragrant, especially so at night, calyx hairy, sepals ovate-lanceolate, with 5 glands at the base, corolla salverform, hairy at the throat, lobes oblong, 6 mm long, 3 mm wide, ciliate, corona membranous, adhering to the anthers, pollinia oblong, erect, ovary glabrous, carpels distinct, styles short, stigma capitate, 5-angled; follicles lanceolate, 7–10 cm long, glabrous; seeds ovate, 8 mm long, coma silky. Native of southern China, cultivated.

Convolvulaceae: Morning-glory Family

Calonyction aculeatum (L.) House — Prickly Ipomoea (Syn. *Ipomoea alba* L.)

Yue-guang-hua=Yueh-kuang-hua (月光花, Moonlight Flower). Leafy shoots and fleshy sepals eaten as potherbs; dried flowers used for soup and also in pastries in Yunnan.

Lactiferous annual twining vines, stems prickly; leaves ovate-cordate, 6–10 (–20) cm long, 6–10 (–18) cm wide, caudate-subhastate, entire or slightly lobate; flowers white, showy, opening at night, solitary or 2–3 in axillary cymes, sepals 2 cm long, ovate-oblong, base fleshy, awned, corolla salverform, tubes 10–12 cm long, limbs 8–11 cm across, shallowly lobed, stamens 5, slightly exserted, disks cupular, ovary 2-locular, cells 2-ovulate, style filiform; capsules ovoid, 3 cm long, 2.5 cm across, rostrate, fruiting stalks thickened at the distal end. Native to tropical America, widespread in the tropics presently.

Calonyction muricatum (L.) G. Don (Syn. *Ipomoea muricata* [L.] Jacquin)

Tian-qie=T'ien-ch'ieh (天茄, Celestial Eggplant); *Ding-xiang-qie=Ting-hsiang-ch'ieh* (丁香茄, Clove Eggplant). Leafy shoots; used for potherb; young fruit for jam.

Robust annual twining vines, stems muricate, green when young, brownish-red with age; leaves cordate, 5–7 cm long and wide, caudate, pilose above; flowers purple, showy, solitary or paired, on thickened-clavate pedicels, sepals oblong-lanceolate, 1.2 cm long, caudate, corolla funnelform, tube 4.5 cm long, limbs 4 cm across, lobed, stamens 5, included, disk cupular, ovary glabrous, 2-locular; capsules globose, 1.2 cm long, 1.4 cm across, rostrate; seeds 4, smooth, black. Native to tropical America, introduced to the Philippines by the Spanish people and thence to tropical China by overseas Chinese residing in Manila; recorded in the famous *Famine Herbal* (救荒本草, *Jiu-huang-ben-cao*), published in 1407.

Convolvulus chinensis Ker-Gawler — CHINESE BINDWEED (Syn. *C. arvensis* sensu; auctt., non L.)

Fu-fu-miao=Fu-fu-miao (伏伏苗, Creeping Potherb); *Tian-xuan-hua=Tien-hsuan-hua* (田旋花, Field Morning Glory). Rhizomes with a tender shoot bearing few young leaves; gathered from sorghums fields in early spring to be used in limited amount only, as too much will cause diarrhoea; mixed with cracked wheat and ground beans to make a thin gruel to assuage hunger; used annually in early spring by rural people in North China.

Perennial trailing weeds growing in sorghum or cotton fields, with white subterranean rather tender rhizomes furnished remotely with small scales; leaves alternate, ovate-oblong, 2.5–5 cm long, 1–3.5 cm wide, entire; flowers pink, solitary, calyx green, 5-parted, corolla funnelform, 2 cm long, stamens 5, ovary superior, stigma 2-lobed; capsules conical; seeds 4. Common in lowland-fields of northern China; first reported here as a food, and I myself used to eat it in a northern Jiangsu village.

Ipomoea aquatica Forsskal — AQUATIC MORNING-GLORY, WATER SPINACH (Syn. *I. repens* [Vahl] Poiret)

Weng-cai=Wung-ts'ai (蕹菜, Water Spinach); *Teng-teng-cai=t'eng-t'eng-ts'ai* (藤藤菜, Vine Herb); *Wu-xin-cai=Wu-hsin-ts'ai* (無心菜, Hollow Potherb); *Tong-cai=T'ung-ts'ai* (通菜, Macaron Potherb). Tender leafy shoots, obtained from plants cultivated in aquatic gardens in southern and western China; available in American Chinese Stores, occasionally observed in kitchen gardens of Chinese people living in Boston area. (Figure 6a).

Annual herbs in general cultivation, becoming perennial in swampy areas or stagnant water, forming large floating colonies, stems hollow, 5–12 mm across the middle of internodes, glabrous, rooting at the nodes; leaves various in sizes and shapes, ovate, triangular-subsagittate, or linear, 6–20 cm long, 1–8 cm wide, acute, obtuse or acuminate, base cordate, hastate or subsagittate, petioles 5–17 cm long, glabrous; flowers white, solitary or 3–5 in axillary cymes, peduncles 1.5–9 cm long, pubescent at the base, bracts scale-like, pedicels 1.5–5 cm long, glabrous, sepals ovate, 7–8 mm long, obtuse, mucronate, corolla funnelform, 3.5–5 cm long, stamens unequal, filaments hairy, ovary conic, glabrous; capsules ovoid or depressed-globose, 8 mm long, 10 mm across; seeds 1 or 2 in each cell, obliquely oblong, 6 mm long, 4 mm across, smooth, very evenly puberulous. A native of tropical Asia, extensively cultivated in warmer regions of China, being the commonest summer vegetable which every family can afford.

Note: The annual yield of the crop is high, with an average of 5–7.5 metric ton (mt) for one-fifteenth of a hectare (=1 Chinese *mu*). After the first harvest, the plants keep on branching and growing, the supply continues for 3–8 months. It takes 60 days from seeds or cuttings to produce the first crop. Once the plants are established in the field, marketable tender shoots can be harvested every 2 or 3 weeks (even 5–7 days with some cultivars), each time 1 mt per *mu*, continuing for many months. Many cultivars have been established. The criteria for selection of good cultivars are: tender crunchy stem, high yield, contrast of pleasant colors, and compact tissue structure for chewing. Seven cultivars from Canton were published and are given here, the first five are selected for high yield, crunchy stem and contrast of colors, and the last two are selected for compact chewy stems, with low yield.

1. cv. 'Big Green Stem' (*Da-gu-qing=Ta-Ku-ch'ing* 大骨青).

Plant 55 cm high, spreading 20 cm, stem greenish-yellow, internodes 5.5 cm long, 1 cm in diameter; leaves ovate, 10–15 cm long, 7–8 cm wide, dark green, petioles 8.5 cm long, harvesting time March–August, once every 2–3 weeks, each time 1 mt, total annual output 5–7.5 mt per *mu*.

2. cv. 'Big White Stem' (*Da-ji-bai=Ta-chi-pai* 大雞白).

Plant 55 cm high, spreading 35 cm, stem greenish-yellow, internodes 5.5 cm long, 1.5 cm in diameter; leaves ovate-hastate, 10–15 cm long, 5–7 cm wide, acute, petioles 15 cm long, harvesting time April–September, once every 15–20 days, each time 1 mt, total annual yield 7 mt per *mu*.

3. cv. 'Big Yellow Stem' (*Da-ji-huang=Ta-chi-huang* 大雞黃)

Plant 45 cm high, spreading 30 cm, stem yellowish-white, internodes 5 cm long, 1.5 cm in diameter; leaves ovate-cordate, 10–17 cm long, 4–6.5 cm wide, yellowish-green, petioles 14 cm long, harvesting May–September, annual yield 6 mt per *mu*.

4. cv. 'Thin Shell' (*Bo-ke=Po-k'o* 薄殼)

Plant 60 cm high, spreading 25 cm, internodes 9.9 cm long, 1.7 cm in diameter;

leaves ovate-cordate, 18–20 cm long, 8–10 cm wide, green, petioles 17 cm long, harvesting time May–September, once every 15–20 days, each time slightly over 1 mt, annual yield 7 mt per *mu*.

5. cv. 'Sword-leaved Hollow Vegetable' (*Jian-ye-tong-cai=Chien-yeh-t'ung-ts'ai* 劍葉通菜)

Plant 40 cm high, spreading 28 cm, internodes 2–3 cm long, 0.8 cm in diameter, rather stiff; leaves linear-lanceolate, 6–17 cm long, 1–3 cm wide, dark green, petioles 10 cm long; harvesting time late April to October, once every 7–10 days, each time 1/4–1/3 mt, total annual yield 6 mt per *mu*.

6. cv. 'Iron-wire Stem' (*Tie-xian-geng=T'ieh-hsian-keng* 鐵線梗)

Plant 20 cm high, spreading 30 cm, stem slender, stiff, light green, internode 3 cm long, 0.5 cm in diameter; leaves lanceolate, 13 cm long, 3 cm wide, dark green, petioles 5 mm long, light green; propagated by cuttings from former crops, 50–60 days ready for first harvest, from May to October, once every 5 to 7 days, each time 1/10–1/3 mt, annual yield 2.5–3 mt per *mu*.

7. cv. 'Slender Hollow Stem' (*Xi-tong=Hsi-t'ung* 細通)

Plant 16 cm high, spreading 35 cm, stem slender, firm, purplish-red, internodes 2.8 cm long, 0.6 cm in diameter; leaves rather small, triangular-lanceolate, 11 cm long, 3.5 cm wide, dark green, petioles 6.5 cm long, purple-red; harvesting time May to December, once 5–7 davs, annual yield 2.5 mt per *mu*.

Ipomoea batatas (L.) Lamarck — SWEET POTATO (Syn. *Convolvulus batatas* L.)

Fan-shu=Fan-shu (番薯, Foreign Yam); *Hong-yu=Hung-yu* (紅芋, Red Taro); *Bai-yu=Pai-yu* (白芋, White Taro); *Gan-shu=Kan-shu* (甘薯, Sweet Yam) *Shan-yu=Shan-yu* (山芋, Hillside Taro). Enlarged fleshy root; extensively cultivated, has become the poor man's staple food in villages of northern China, and in time of food shortages in town also.

Lactiferous trailing perennial herbs in the tropics, cultivated as an annual crop by cuttings, branchlets rooting at the nodes; enlarged roots fleshy, fusiform or oblong, with the flesh white, orange or purple depending on the type of cultivars, rich in starch (changing into sugar on storage); leaves cordate or deitoid-ovate, 5–12 cm long and across the base, entire or shallowly lobed; flowers pink-lavender inside, white outside, 2–5 in axiilary cymes on peduncles 8–10 cm long, bracts lanceolate, 2–5 cm long, deciduous, pedicels 5–7 cm long, calyx shallowly campanulate, 7–8 cm long, 8–11 mm across the top, lobed almost to the base, lobes ovate, acute, corolla funnelform, 3.5–5 cm long, 5–6 cm across the wavy margin, stamens 5, inserted 2 mm above the base of the corolla tube, filaments unequal, 8–12 mm long, the basal 1/3 hirsute, anthers oblong, 3 mm long, sagittate at the base, ovoid, glabrous, 2 mm across base, style filiform, 16 mm long, glabrous, stigma globose, weakly 2-lobed; capsules depressed globose; seeds 2–4, glabrous, in cultivation often bearing no fruits and seeds after flowering. Native of tropical America, introduced to southeastern China via the Philippines during the Spanish rule of that country; flowers seldom seen in northern China; in Hong Kong, the species flowers freely in October.

Ipomoea cairica (L.) Sweet — CAIRO MORNING-GLORY

Wu-zhao-jin-long=Wu-chao-chin-lung (五爪金龍, Five-clawed Golden Dragon); *Wu-zhao-long=Wu-chao-lung* (五爪龍, Five-clawed Dragon). Starch extracted from the root; eaten by people in southern Yunnan.

Evergreen glabrous perennial twining vines climbing over bamboo thickets and trees, older plants bearing farinaceous root tubers; leaves palmately 5-sect, segments ovate-lanceolate or ovate-oblong, the middle one the largest, 4–5 cm long, 2–2.5 cm wide, acuminate and mucronate, entire or wavy, petioles 2–8 cm long; flowers purple, showy, sepals unequal, imbricate, 5–9 mm long, the outer two smaller, corolla funnelform, plaited and contorted in bud, 5–6 cm long, stamens unequal, filaments hairy, ovary glabrous, style filiform, stigma capitate, 2-lobed; capsules globose, 1 cm in diameter; seeds black, 5 mm long.

A native of tropical and subtropical Africa, weedy, widespread in the disturbed areas of the tropics, common in Hong Kong, Yunnan, Guangxi, Guangdong and Taiwan; in Hong Kong climbing fences and covering small trees along the newly constructed highways, seldom bearing fruits; a white-flowered form occasionaly observed.

Ipomoea mauritiana Jacquin (Syn. *I. digitata* sensu auctt., non L.)

Qi-zhao-long=Ch'i-chao-lung (七爪龍, Seven-clawed Dragon). Starch in root; extracted mechanically and used locally, not available in the market.

Robust lianas with stout roots, branchlets wirelike, glabrescent; leaves suborbicular-cordate in outline, 6–10 cm long, 10–15 cm across the base, deeply 5–7-parted, segments oblong or ovate, the middle lobes 5–7.5 cm long, 2–3 cm wide, acuminate or obtuse, the lateral ones smaller gradually as they become the side lobes, green above, glaucous beneath, sparsely pilose along the midribs above; flowers pink-violet, solitary or 3–7 in axillary cymose clusters on peduncles 5–6 cm long, bracts caducous, pedicels 10–14 mm long, puberulous, calyx subglobose, 10–15 rmn across, deeply lobed, the lobes ovate-oblong, glabrous, corolla funnelform, 5–6 cm long, 8 cm across the spreading subentire limb, stamens with hairy filaments, ovary subglobose, style filiform; capsules ovoid, 10–12 mm long, 8–10 mm across; seeds oblong-hemispherical, brown-black, with tufted cotton-like long white hairs several times longer than the length of the seed, attached to a curved ridge opposite the hilum. Native to tropical America, occurring in Taiwan, Hainan, Hong Kong, and along the seashores of Guangdong.

Ipomoea staphylina Roemer et Schultes

Hai-nan-shu=Hai-nan-shu (海南薯, Hainan Yam). Fleshy tubers eaten by the people living in Hainan Island; starch content very low.

Large lianas bearing fleshy root tubers, some branches lignified, branchlets glabrous, rarely verrucose; leaves broad-ovate or oblong, 8–15 cm long, 5–11 cm wide, acute, base truncate or subcordate, entire, pilose, glabrescent; flowers white or pink, in axillary corymbs on peduncles 6 cm long, pedicels 4–6 mm long, bracts caducous, sepals 6 mm long, corolla funnelform, 2 cm long; capsules ovoid, 1 cm long, pericarp coriaceous, dehiscing into 4 valves; seeds barbate on one end. Widespread in Indo-Malaysian Region; growing on open hillsides in southern Yunnan, Guangxi, Guangdong and Hainan.

Merremia hungaiensis (Lingelsheim et Borza) R. C. Fang (Syn. *Ipomoea hungaiensis* Lingelsheim et Borza, *I. wilsonii* Gagnepain)

Huang-hua-tu-gua=Huang-hua-t'u-kua (黃花土瓜, Yellow-flowered Ground Gourd); *Shan-tu-gua=Shan-t'u-kua* (山土瓜, Mountain Ground Gourd). Enlarged root; used in Yunnan.

Perennial lianas with 2–3 reddish-brown tuberous fleshy roots, ovoid or globose, white inside, rich in starch and latex, branchlets wirelike, glabrous; leaves alternate, elliptic, oblong or ovate, 3–11 cm long, 1–5 cm wide, entire, apex acute, base rounded or subcordate, petioles 1–3.5 cm long, puberulous; flowers yellow, solitary or 2–3 in axillary cymes on peduncles 2–6 cm long, sepals oblong, 7–14 mm long, the outer two shorter than the inner three, obtuse, margin membranous, corolla funnelform, 3.6–6 cm long, stamens 5, filaments subequal, broadened and hairy toward the base, ovary conical, 2-locular, style filiform, stigma capitate, 2-lobed; capsules compressed-globose, 1–1.3 cm long; seeds 1–4, 5–7 mm long, covered by dark brown hairs. Native to southwestern China, starch extracted and used locally, not available in the market.

Boraginaceae: Borage Family

Antiotrema dunnianum (Diels) Handel-Mazzetti

Gou-she-cao=Kou-she-ts'ao (狗舌草, Dog's Tongue Herb). Root and leaves; eaten by the people living in the mountains of Yunnan.

Perennial herbs 9–30 cm high, pubescent throughout, stem branching before flowering; radical leaves spathulate-elliptic, 3.5–18 cm long, 1–5 cm wide, obtuse, base cuneate, with appressed hispid hairs on both surfaces, cauline leaves elliptic, sessile; flowers blue, small, in terminal scorpioid panicles, the branches racemelike, pedicels 2–3 mm long, calyx 3 mm long, deeply 5-lobed, the lobes linear-lanceolate, corolla 5-lobed, the lobes 2 mm long, tubes 4.5 mm long, with 5 scalelike appendages inside, at the middle, stamens 5, exserted, ovary 4-lobed; fruits each consisting of 4 nutlets, reniform, 2 mm long, densely verrucose, concave on the adaxial side. Native to southwestern China, occurring in grassy slopes of the hillsides at altitudes of 1,950–2,500 m.

Brachybotrys paridiformis Maximowicz

Shan-qie-zi=Shan-ch'ieh-tzu (山茄子, Hillside Eggplant). Young shoots, used for potherb in northeastern China like *Asparagus* in America.

Perennial herbs 30–40 cm high, with creeping rhizomes 3 mm in diameter, early growth asparagus-like, the lower leaves scalelike, brown, median leaves spatulate, 9 cm long, majority of the normal leaves crowded at the shoot apex, 5 or 6, obovate-elliptic, 7–17 cm long, 2.5–8 cm wide, acute or acuminate, base cuneate, sparsely hispid above, pilose beneath; flowers purple, 1 cm across, 3–6 in a solitary cyme, peduncles 10 cm long, pedicels 4–6 cm long, sepals 5, subulate-lanceolate, 1.5 cm long, 3 mm wide, densely pilose, petals funnelform, tube 4 mm long, limb 1 cm across, with 5 appendages at the throat, stamens 5, exserted, ovary 4-cleft, style slender; nutlets 4, black, shiny, 3 mm long, tetragonous, hairy. Native to northeastern China and adjacent Siberia.

Cordia dichotoma J. G. Forster

Po-bu-zi=P'o-pu-tzu (破布子, Rag Seed); *Po-bu-mu=P'o-pu-mu* (破布木, Rag Tree). Seed; gathered by people living in the mountainous areas of Taiwan and Tibet, oil (51% reported) extracted and used for cooking, local use only, not available in the market; fruits eaten by some ethnic groups of these areas also.

Deciduous trees up to 10 m high, branchlets ochreously tomentose when young, glabrescent with age; leaves oblong, ovate, or obovate, 8–12 cm long, 4–10 cm wide, obtuse, base rounded; flowers small, white changing yellow, subsessile, many in loose terminal scorpioid panicles, appearing dichotomous in fruit, calyx campanulate, 6–8 mm long, glabrous at anthesis, shallowly 4- or 5-lobed, the tubes fleshy, lobes deltoid, 1–2 mm long, margin membranous, corolla rotate, the tubes 4 mm long, lobes obovate, reflexed, 4–5 mm long, rounded, base narrowed, stamens 2–5, inserted at the throat between corolla lobes, filaments subulate, glabrous, anthers oblong, ovary obovoid,

4 mm long, beaked, bilocular, styles 2, 1–5 mm long, bifid, stylar branches 4 mm long, winged; fruits drupaceous, oblong or ovoid, 10–15 mm long, yellow, orange or red, persistent calyx cupular, in 8–12 mm across, margin erose, mesocarp containing a gelatinous substance, water with a jelly-like material accumulated along the cross-section, endocarp bony, brown, stone laterally compressed and with a ridge, trigonous, 9 mm long, 5 mm across the middle; seeds usually one fully developed, inside chalk-white. Native to the Old World tropics, common in southern China.

Ehretia thyrsiflora (Siebold et Zuccarini) Nakai (Syn. *E. acuminata* R. Brown var. *obovata* [Lindley] Johnston)

Da-gang-cha=Ta-kang-ch'a (大崗茶, Large Slope Tea). Young shoots, used as potherb in Hainan Island.

Trees 10 m high, bark dark gray, irregularly fissured, branchlets pilose; leaves oblong-elliptic or obovate-oblong, 6–10 cm long, 4–6 cm wide, acute, base obtuse or rounded, serrulate, glabrous or sparsely pilose above, petioles 2.5–3.5 cm long; flowers white, small, fragrant, in terminal cymose panicles 10–20 cm long, calyx obconic, 5-lobed, lobes rounded, 1 mm long, ciliate, corolla 3–4 mm long, 5-lobed, stamens exserted, inserted to the base of the corolla tube, ovary 2-locular, cells 2-ovulate, style 1.5 mm long, bifid, stigma capitate; fruits drupaceous, pea-sized, orange-red, including 2 stones, alveolate; seeds 2 in each stone.

Verbenaceae: Verbena Family

Avicennia marina (Forsskal) Vierhapper — BLACK MANGROVE

Hai-lan-zi=Hai-lan-tzu (海欖子, Sea Canarium); *Hai-dou=Hai-tou* (海豆, Sea Pea). Seed; gathered, soaked, used by people in Taiwan, Fujian, and Guangdong.

Evergreen shrubs or small tree 1.5–6 m high, branchlets angular, glabrous, with prominent nodes; leaves ovate-oblong or obovate, 2–7 cm long, 1–3.5 cm wide, rounded or obtuse, base cuneate, obtuse, entire, glabrous above, pubescent beneath; flowers small, 3-bracteate, sessile, in terminal head-like clusters, outer bracts pubescent, inner ones glabrous, long ciliate, calyx 5-lobed, lobes ovate, 3 mm long, hairy, corolla brownish-yellow, 4-lobed, tube 2 mm long, lobes 2 mm long, stamens 4, inserted on the throat of the corolla tube and alternate with the lobes, filaments very short, anthers oblong, ovary ovoid, pubescent, imperfectly 4-locular, ovules 4; fruits ovoid, 12 mm long, 2-valved, loculicidally dehiscent. Native of southeastern Asia and the Pacific area, growing in seashores and tidal streams, with numerous pneumatophores of uniform thickness.

Callicarpa japonica Thunberg

Nü-er-cha=Nü-erh-ch'a (女兒茶, Maiden Tea); *Zi-zhu=Tzu-chu* (紫珠, Purple Pearl). Young leaves; used as a substitute for tea.

Deciduous shrubs 1–2 m high, branchlets densely covered by yellow stellate hairs when young, glabrescent; leaves ovate or ovate-elliptic, 7–15 cm long, 3–5.5 cm wide, abruptly long-acuminate, base acute, serrate, thin chartaceous, punctate beneath, stellate pubescent along the midrib above; flowers small, purple-pink or white, many in axillary corymbose cymes, the peduncles and secondary axes stellate pubescent, pedicels and flower-buds scabrid-glandular, calyx cupular, with 4 minute teeth, coralla tube 2 mm long, limb 4-lobed, 5 mm across, stamens 4, exerted, anthers oblong, opening by apical pores, ovary globose, style filiform, longer than the stamens, stigma capitate; fruits purple, berries globose, 3 mm in diameter, persistent calyx discoid; pyrenes 4, plano-convex, 2 mm long, 1 mm across the middle, endocarp smooth, coriaceous. Native of eastern Asia, common on the hillsides of China; introduced into European gardens in 1845 for the purple berries, several horticultural forms developed in America.

Callicarpa kochiana Makino (Syn. *C. mollis* sensu Liu, non Siebold et Zuccarini)

Bai-tang-zi-shu=Pai-t'ang-tzu-shu (白棠子樹, White Wild Pear). Ripe fruit; gathered and eaten in Taiwan.

Deciduous shrubs, stellate-hairy throughout; leaves narrowly oblong or broadly lanceolate, 15–25 cm long, 5–8 cm wide, acuminate, base acute, glandularly serrulate, densely stellate-tomentose beneath; flowers lavender, small, in large axillary dichotomously branched cymose panicles, calyx villose, 4- or 5-parted, the segments linear, corolla 2 mm long, stamens exserted, 3 times longer than the corolla, ovary glandular; fruits white, 2 mm across. Endemic to Taiwan.

Clerodendron fragrans Ventenat — FRAGRANT GLORYBOWER

Chou-mo-li=Ch'ou-mo-li (臭茉莉, Stinking Jasmine); *Chou-mu-dan=Ch'ou-mu-tan* (臭牡丹, Stinking Peony). Root, gathered, dried, cooked with pork for the elderly persons, to give strength and to remove pain and stiffness of the muscles and joints.

Shrubs (1–) 2–3 (–4) m high, any broken vegetative portion giving out stinking smell, roots rather thick and fleshy, branchlets stout, trigonous, young growth densely brownish-tomemotose; leaves rather ovate-deitoid, 7–10 (–22) cm long, 5–11 (–21) cm wide, acute or shortly acuminate, base subtruncate or subcordate, remotely and irregularly coarse dentate, both surfaces sparsely hirsute, hairs mixed with glands, petioles 2–10 cm long, subterete, pubescent; flowers light purple, or white and tinged lavender on the outside, in compact terminal head-like cymose panicles 6–9 cm across, calyx purple-red, foliaceous, corolla variable, in the simple-flowered variety the corolla tube slender, 2–3.5 cm long, the limbs 8–10 mm across, deeply lobed to the throat, in the double-flower variety, the tubes short, and the limbs globose, 1–1.5 cm across, stamens 4, long exserted, anthers sagittate, ovary superior, style exserted; fruits globose, 8–10 mm in diameter, subtended by colorful foliaceous persistent calyx; pyrenes 4.

A native of the warm areas of China, from the Yangtze River southward to Yunnan, Guangxi, Guangdong and Hong Kong; individual flowers having the form of wild and cultivated jasmimes, and with a foul smell, hence the Chinese vernicular name "Stinking Jasmine".

Gmelina arborea Roxburgh

> *Tian-shi-zi=Tien-shih-tzu* (滇石梓, Yunnan Gmelina); *Lao-ke-sao=Lao-k'e-sao* (老可嫂, the Thai ethnic name). Flower; the colorful and fragrant flowers, gathered locally and used by the Thai of southern Yunnan for flavoring and coloring pastries.

Deciduous trees 15 m high, trunk 30–50 cm across, bark gray, smooth, branchlets velutinous; leaves broad-ovate, 2–22 cm long, 10–18 cm wide, caudate-acuminate, base shallowly cordate, rarely cuneate and decurrent, with 2–4 prominent glands on the upper surface at the base of major veins, petioles 5–14 cm long; flowers showy, yellow, velutinous outside, purple inside, in terminal panicles, peduncles 15–30 cm long, calyx campanulate, 3–7 mm long, 5-toothed, corolla bilabiate, 4 cm long, upper lip entire or 2-fid, lower lip 3-lobed, the middle lobe large, stamens exserted, ovary glabrous, 4-locular, style exserted; drupes ellipsoid or ovoid-ellipsoid, 1.6–2 cm long, yellow, smooth; seeds usually one. Widespread in the eastern Himalayan Region; occurring in deciduous forests in southern Yunnan, Guangdong and Hainan Island at altitudes of 760–1,300 m.

Vitex negundo L. var. **cannabifolia** (Siebold et Zuccarini) Handel-Mazzetti (Syn. *Vitex cannabifolia* Siebold et Zuccarini)

> *Mu-jing=Mu-ch'ing* (牡荊, Vitex Bush). Seed; gathered and used for food in Taiwan

Deciduous shrubs 1–2.5 m high, branchlets tetragonous; leaves opposite, digitately compound, 5-foliolate, the upper leaves subtending flowering branches often 3-foliolate, leaflets lanceolate, or ovate-elliptic, 3–15 cm long, 1–3 cm wide, acuminate, base acute or obtuse, entire or coarsely serrate, green and finely pilose above, densely grayish-white-tomentose beneath; flowers small, numerous, blue-lavender, in terminal panicles 10–30 cm long, 8–22 cm across, calyx campanulate, minutely 5-toothed, corolla zygomorphic, tube 2.5 mm long, limb 2-lipped, the upper lip 2-lobed, the lower lip 3-lobed, stamens 4, didynamous, anthers attached at tip, thecae divaricate, ovary superior, style filiform, stigmas 2; fruits ovoid, 3 mm long, 1.8 mm across the rounded apex, 8/10 covered by the persistent calyx, black; pyrenes 4. Native of eastern Asia, common throughout China; introduced into Western gardens in middle 1750.

Vitex vestita Wallich ex Schauer

Chao-dou-shu=Ch'ao-tou-shu (炒豆樹, Roasted Bean Tree); *Huang-mao-jing=Huang-mao-ch'ing* (黃毛荊, Rusty-hairy Vitex). Fruits; roasted, having the flavor of roasted broadbean (*Vicia faba* L.), hence the vernicular name "*Chao-dou-shu*".

Shrubs or small trees 1–3 m high, branchlets tetragonous, velutinous; leaves ternately compound, leaflets elliptic or oblong, 3–10 cm long, 1.5–5 cm wide, acute or shortly acuminate, base obtuse or rounded, sparsely strigose above, tomentose and punctate beneath, entire, petioles 2–6 cm long, strigose, petiolules 1–10 mm long; flowers creamy-white, in axillary cymose corymbs, calyx truncate, tomentose and punctate, enlarged and discoid after anthesis, corolla cylindrical, 1–1.2 cm long, pilose and punctate, stamens enclosed, filaments hairy, ovary punctate; fruits oblong, 6–9 mm long, 4–8 mm across. Widespread in southeastern Asia, growing in thickets of dry hillside in southern Yunnan at altitudes of 580–1,400 m.

Lamiaceae or Labiatae: Mint Family

Elsholtzia ciliata (Thunberg) Hylander — AROMATIC MADDER (Syn. *E. patrini* [Lepechin] Garcke; *E. cristata* Willdenow)

Xiang-ru=Hsiang-ju (香薷, Aromatic Madder); *Yu-xiang=Yu-hsiang* (魚香, Fish Spice); *Chou-jing-jie=Ch'ou-ching-chieh* (臭荊芥, Stinking Catnip). Leafy shoots; used for spice, especially for cooking fish dishes.

Annual aromatic herbs up to 1.5 m high, growing close to abandoned villages or in gardens, branchlets 4-angled; leaves opposite, ovate or lanceolate, 1.5–6 cm long, 0.5–2.5 cm wide, acuminate, base cuneate and decurrent, prominently dentate, pilose on veins and densely punctate beneath; flowers lavender, in compact terminal, one-sided spike-like panicles 3–8 cm long, bracts suborbicular, 6 mm in diameter, villose, ciliate, the midrib extending into an awn-like projection 2–3 mm long, verticils many-flowered, shortly pedunculate, calyx tubular, 2 mm long, densely villose and with yellow glands, 5-toothed, the teeth setose, corolla tubular, 5 mm long, villose, faintly bilabiate, upper lip 4-lobed, the lobes subequal, the middle 2 lobes connate at base. the lower lip consisting of 1 large lobe, inferior in bud, stamens 4, didynamous, exserted, filaments 4–6 mm long, glabrous, anthers 2-locular, confluent, ovary superior, on a pulvinate disk with a front gland taller than the carpels, style gynobasic, stigma bifid; nutlets 4, oblong, 1 mm long, glandular, persistent calyx 4 mm long, hyaline at base.

Elsholtzia densa Bentham

Chou-xiang-ru=Ch'ou-hsiang-ju (臭香薷, Stinking Aromatic Madder); *Mi-hua-xiang-ru=Mi-hua-hsiang-ju* (密花香薷, Dense-flowered Aromatic Madder). Leafy shoots; used as spice by people in Tibet for cooking lamb.

Annual herbs 20–60 cm high, pubescent with multicellular straight hairs throughout; leaves elliptic-lanceolate, 2–10 cm long, 1–2.5 cm wide, acuminate, base cuneate, coarsely serrate, punctate beneath; flowers lavender, verticils in dense terminal cylindrical pseudoracemes 1–7 cm long, calyx campanulate, shallowly 5-toothed, inflated and persistent in fruit, corolla bilabiate, upper lip erect, notched, lower lip 3-lobed, stamens 4, didymous, anthers subglobose, style glabrous, stigma bifid, disk 4-lobed; nutlets generally 2, tuberculate. Widespread in northeastern and western China, thence westward to central Asia and northern India.

Elsholtzia penduliflora W. W. Smith

Da-huang-yao=Ta-huang-yao (大黃藥, Greater Yellow Herb); *Huang-yao=Huang-yao* (黃藥, Yellow Herb). Seed; mature seed gathered, roasted, oil extracted and used for cooking.

Subshrubs 1–2 m high, branchlets tetragonous, sparsely tomentose and glandular; leaves membranous, lanceolate or ovate-lanceolate, 6–15 cm long, 2.5–3.8 cm wide, acuminate, base attenuate or subcordate, serrulate, densely punctate beneath; flowers white, small, in terminal racemes 5–15 cm long, verticils 6- to 12-flowered, calyx 3 mm long, glandular, 10-nerved, teeth deltoid, corolla 5.5 mm long, bilabiate, upper lip retuse, lower one 3-lobed, the middle lobe suborbicular, stamens 4, the anterior pair longer, exserted, ovary glabrous, style gynobasic, apical end 2-lobed, hypogynous nectiferous disk lobed, the anterior lobe enlarged; nutlets oblong, yellowish-brown. Endemic to southwestern Yunnan.

Eriophyton wallichii Bentham ex Wallich

Mian-shen=Mian-shen (綿參, Soft Shen). Root; used in food by people living in north-western Yunnan.

Perennial herbs 10–20 cm high, root fleshy, cylindrical, stems tetragonous, tomentose when young; leaves subsessile, variable, the lower ones small, scale-like, glabrous, the middle ones opposite, rhombic-oblong, 3–4 cm long and wide, acute, base cuneate, serrate above middle, tomentose on both surfaces; flowers light-purple to pink, 6 in a sessile verticil, calyx campanulate, membranous, 8 mm long, villose, 5-toothed, lobes equal, 7 mm long, 10-nerved, corolla 2.5–3 cm long, bilabiate, the upper lip galeate, the lower one 3-lobed, stamens 4, exserted, anthers 2-celled; nutlets yellow, 3 mm long. Native to the alpine meadows of the Himalayan Region, growing among broken rocks in the high mountains of Yunnan, Sichuan, Tibet at altitudes of 3,400–4,700 m.

Lamium album L. — WHITE-FLOWERED DEADNETTLE

Duan-bing-ye-zhi-ma=T'uan-ping-yeh-chih-ma (短柄野芝麻, Short-stalked Wild Sesame). Young shoots, eaten as vegetable in Tibet.

Perennial herbs 30–60 cm high, stems setaceous, hollow; leaves ovate-suborbicular or oblong-lanceolate, 2.5–6 cm long, 1.5–4 cm wide, acute or acuminate, base cordate, dentate-serrate, setaceous on both surfaces, petioles 1–6 cm long; flowers yellowish-white, 8–9 in dense axillary verticils, calyx subcampanulate, 1–1.3 cm long, setaceous, lobes lanceolate, corolla 2–2.5 cm long, bilabiate, the upper lip obovate, the lower one 3-lobed, mid-lobe reniform, retuse; nutlets 3–3.5 mm long, warty. Widespread in Central Asia, growing along margin of forests in northwestern China and Inner Mongolia.

Leonurus artemisia (Loureiro) S. Y. Hu — SOUTH CHINA MOTHERWORT (Syn. *L. heterophyllus* Sweet)

Hong-hua-ai=Hung-hua-ai (紅花艾, Red-flowered Artemisia); *Yi-mu-cao=I-mu-ts'ao* (益母草, Benefit Mother Herb). Young shoots, cooked with some rice to make a pottage, with some brown sugar added, given to children.

Annual herbs 40–120 cm high, pilose throughout, growing in waste places near villages, stems 4-angled, 4–6 mm in diameter; leaves before flowering all basal, in a rosette, cauline leaves opposite, ovate-orbicular in outline, 3–12 cm long and across, triparted, the segments of the lower leaves ovate, lobate, those of the upper ones lanceolate, irregularly dentate, glandularly punctate beneath; flowers red, in dense axillary fascicles forming sessile verticils, bracts and bracteoles bristly, calyx campanulate, 5-lobed, the lobes bristly, corolla bilobate, 12 mm long, the upper lip arcuate, rotundate and slightly emarginate, the lower lip 3-lobed, stamens 4, in 2 pairs, exserted, anthers 2-locular, thecae parallel, ovary superior, on a disk slightly thicker in front but without elevated gland, style filiform, gynobasic, stigma bifid, pointed; nutlets 4, trigonous. A native of southern China, the plants flower in April–May and complete their life-cycle before the hot season. The northern China Motherwort flowers in late summer and autumn.

Note: Authors of older Chinese botanical works have applied the name of a very different Siberian species, *L. sibiricus* L., to both this spring flowered species of southern

China and to the autumn flowered species widespread in northern China which Linnaeus named as *L. tataricus* L. These classical mistakes should be corrected. Linnaeus's material from Siberia, *L. sibiricus* L., is very different from these two Chinese species (Hu, 1974).

Lycopus lucidus Turczaninow

Di-gua-er-miao=Ti-kua-erh-miao (地瓜兒苗, Ground Melon Top); *Di-shen=Ti-shen* (地參, Essence-of-the-Earth); *Di-sun=Ti-sun* (地筍, Earth Bamboo Shoot). Thickened underground rhizomes; gathered locally and used in northern China and in Yunnan.

Mild aromatic perennial herbs 50–100 cm high, with short knotty rhizomes each ending with a cylindrical or fusiform scaly tuber 5–8 cm long, 10–12 mm across the middle, stems tetragonous, rooting on the nodes on touching the ground; leaves lanceolate or elliptic, 4–12 cm long, 1–4 cm wide, acuminate at both ends, coarsely sharp-serrate, the teeth 5 mm long, awned, glandularly punctate beneath; flowers white, small, in axillary prickly verticils, bracts and calyx lobes sharply aciculate, calyx broadly oblique-campanulate, 5-lobed, lobes ovate-lanceolate, awned at the apex, corolla 3 mm long, glabrous, barbate on the inside of the throat, faintly bilabiate, glabrous and shiny punctate, the upper lip 2-lobed, the lower lip 3-lobed, stamens 2 fertile, anthers with parallel thecae, staminodes clavate, ovary superior, style filiform, exserted, stigma-bifid; nutlets small, smooth, obovoid-trigonous, truncate-rounded at the apex. Native of northern China, thence extending eastward to Japan; growing in wet areas.

Mentha arvensis L. — FIELD MINT

Bo-he=Po-ho (薄荷, Mint). Young shoots used for tea, cultivated in kitchen gardens; wild population widespread, polymorphic.

Perennial aromatic herbs with white creeping rhizomes, the stem rooting when touch the ground, tetragonous; leaves opposite, ovate-elliptic, 2–5 cm long, 1–3 cm wide, acute, base obtuse, narrowed into the wing at the upper portion of the hairy petiole, margin serrate at the upper two-thirds, hirsute along the veins on both surfaces, the hairs multicellular, sparcely pilose above; flowers in subsessile verticils, the bracts lanceolate or linear, villose, pedicels 3–4 mm long, villose or subglabrous, calyx campanulate, 3 mm long, the lobes deltoid, suddenly narrowed, pilose, flowers pistillate or hermaphrodite, corolla lilac, white or pink, weakly 2-lipped, with 4-subequal lobes, the upper lobe wider and emarginate, stamens 4, exserted, style longer than the filaments; nutlets 4, light-brown. Widespread in the northern temperate Eurasia, growing on banks of streams and irrigation ditches.

Note: In recent Chinese botanical work, *M. haplocalyx* Briquet is applied to this variable octoploid species growing in China. Briquet's material was from Sri Lanka (*Thwaites* 2077). It has lanceolate leaves attenuate at both ends, evenly hispid with short unicellular hairs, hispid along the veins. Its hispid tubular calyx has long subulate villose lobes three times longer than the tube. After comparing the Chinese collections with those from Europe and Sri Lanka, it appears more reasonable to apply *M. arvensis* L. to plants growing in China, particularly to those growing in high latitudes and altitudes.

Mentha arvensis L. forma **piperascens** Holmes
Jia-bo-he=Chia-po-ho (家薄荷, Cultivated Mint). Leafy shoots; dried, salted, sold in packets, used as tea, available in American Chinese stores.

Perennial aromatic herb 60–80 cm high, planted throughout the country, by seed or cuttings of the rhizomes or lower stems, harvested twice annually, in mid-July and mid-October; source of commercial menthol and oil.

Mentha x **piperita** L. (*M. aquatica* L. x *M. spicata* L.) — PEPPERMINT
La-bo-he=La-po-ho (辣薄荷, Pungent Mint). Young shoots; used as spice or for tea.

Aromatic perennial herbs 60–80 cm high, glabrescent, often purplish-tinged, herbage with pungent odor; leaves ovate-lanceolate, 2–4 cm long, 0.5–2 cm wide, acute, base rounded, serrate, glabrous above, sparsely villose on the major nerves beneath, punctate on both surfaces, petioles 5–7 mm long; flowers lilac, pink or white, in shortly pedunculate glabrous verticils crowded at the shoot apex into spike-like panicles. Sterile. Native of Europe, cultivated in Beijing and Nanjing.

Mentha X **piperita** L. var. **citrata** Ehrhart — LEMON MINT
Ning-meng-liu-lan-xiang=Ning-meng-liu-lan-hsiang (檸檬留蘭香, Lemon Mint). Leafy shoots; used as spice in cooking or for tea.

Leaves mostly entire, herbage with a lemon odor when crushed. Introduced from Europe, cultivated in Beijing, Hanzhou and Nanjing.

Mentha spicata L. — SPEARMINT
Liu-lan-xiang=Liu-lan-hsiang (留蘭香, Spearmint). Leafy shoots; for tea.

Glabrous aromatic herbs 30 cm high, with strong sweet scent; leaves subsessile, broadly ovate-oblong, 2.3 cm long, 1–1.5 cm wide, acute, base cordate, glabrous above, rarely sparsely hairy on the principle nerves beneath, punctate; flowers lilac, pink or white, the verticils glabrous, sessile, in terminal spicate panicles 4–6 cm long, 1–1.5 cm across, calyx campanulate, 2.5 mm long, tube and lobes about equal in length, the lobes deltoid, aciculate at the apex, glabrous or with 1 or 2 hairs on the margin, corolla 4 mm long, 4-lobed, stamens 4, long exserted, style longer than the filaments; nutlets trigonous-ovoid, light brown, smooth, shiny. Native of Europe, extensively cultivated in China, the escaped plants naturalized occasionally.

Mesona chinensis Bentham — Chinese Jelly Herb

Liang-fen-cao=Liang-fen-ts'ao (涼粉草, Jelly Herb). Whole plant; soaked in water and jelled; both dried plants and canned product available in American Chinese stores.

Perennial herbs 30–100 cm high, the flowering branches bearing several lateral branchlets, internodes 4–7 cm long, the lower ones subterete, glabrescent, the upper ones tetragonous, villose, the hairs multicellar; leaves ovate, 1–5 cm long, 0.8–2.5 cm wide, acute, base rounded and abruptly acute, serrate, sparsely pubescent above, more so beneath, especially on the nerves; flowers white on purple calyx, in terminal cylindrical racemose panicles, verticils 6- to 10-flowered, subtended by two broad-ovate bracts 4–5 mm long, cordiform and caudate at apex, pedicels filiform, calyx 2-lipped, villose, the upper lip 3-fid, the lower lip entire, ovate, acute, corolla 3 mm long, bilabiate, the tube broadly campanulate, the upper lip shallowly 3-fid, the lower lip boat-shaped (navicular), entire, stamens 4, didynamous, exserted, the filament turning upward, the base of the upper appendiculate with a spur, anthers 2-locular, thecae confluent, ovary 4-parted, on pulvinate disk with prominent front lobe, style gynobasic, filiform, stigma bifid, the upper lobe shorter; nutlets 4, hidden within the urceolate reticulate persistent calyx.

Ocimum basilicum L. — Sweet Basil, Karalaka (Sanskrit).

Luo-le=Luo-lo (羅勒, Basil); *Lan-xiang=Lan-hsiang* (蘭香, Orchid Spice); *Zhai-xiang-cai=Chai-hsiang-ts'ai* (寨香菜, Fortress Sweet Herb); *Jiu-ceng-ta=Chiu-tseng-t'a* (九層塔, Nine-story Pagoda); *Ji-ji-cao=Chi-chi-ts'ao* (吉吉草, Fortune Herb). Young shoots; used for seasoning food.

Annual (perennial in Hong Kong and Guangdong) aromatic herb 20–80 cm high, the stem of perennial plants woody at the base, annual branchlets herbaceous, subglabrous; leaves opposite, ovate-elliptic, 2–5 cm long, 1–2.5 cm wide, irregularly dentate or subentire, punctate, petioles hirsute; flowers white with lilac lower lip, verticils 6-flowered, in terminal hirsute racemose panicles 5–18 cm long, 1–2 cm between verticils, bracts ovate, calyx ovoid-campanulate, 5-lobed, the upper lobe ovate, margin winged, ciliate, deflexed after anthesis; corolla tube shorter than the calyx, limbs bilabiate, the upper lip 4-fid, the lower lip slightly longer, entire, stamens 4, didymous, declinate, the filament of the lower pair longer, those of the upper pair geniculate, appendiculate with a spur, anthers ovoid-reniform, thecae confluent, ovary deeply 4-parted, disk pulvinate, style filiform, gynobasic, shortly bifid at the apex, stigmas marginal; nutlets 4, oblong-ovoid, 2 mm long, smooth. Native of tropical Asia, introduced into China before the third century.

Note: The Chinese name appearing to be a translation of the Sanskrit karalaka or an abbreviation of it into *"luo-le"*. The name was changed into *"Lan-xiang"* (蘭香) later, because the sound *"le"* tabooed the name of a Tartarian warrior, *Shi-le* (石勒), who conquered northern China and became the first King of Late Chao Dynasty (A.D. 273–333).

Perilla acuta (Thunberg) S. Y. Hu — WILD PERILLA (Syn. *Ocimum acutum* Thunberg; *Perilla frutescens* Britton var. *acuta* [Thunberg] Kudo; *Perilla frutescens* Britton var. *arguta* [Bentham] Handel-Mazzetti)

Zi-su=Tsu-su (紫蘇, Purple Perilla); *Ye-zi-su=Yeh-tsu-su* (野紫蘇, Wild Perilla); *Xiang-si-cai=Hsiang-ssu-ts'ai* (香絲菜, Aromatic Sliver Vegetable). Young shoots, sliced into slivers for seasoning meat or fish dishes; leaves used for tea.

Aromatic annual herbs 30–100 cm high, with purplish-green or purple herbage, tetragonous stems villose when young, glabrescent later; leaves ovate or ovate-orbicular, 4–12 cm long, 2.5–10 cm wide, acute or acuminate, base rounded and acute, margin serrate below the upper 2/3, green or purplish-green above, purple beneath, punctate, sparsely pilose or glabrescent; flowers pink or white, in loose terminal or axillary pseudoracemes with two flowers at each verticil, calyx shallowly campanulate or subglobose, villose, 5-lobed, corolla tube exserted; nutlets 4, light brown, reticulate, enclosed in persistent calyx 4–5 mm long, gibbous at the base. Native of central and southern China, growing along paths of hillsides, naturalized in eastern North America, extending from Massachusetts southward to Florida, Texas, Missouri, Arkansas and Kansas.

Perilla crispa (Thunberg) S. Y. Hu — COCK'S COMB PERILLA (Syn. *Ocimum crispum* Thunberg; *Perilla frutescens* Britton var. *crispa* [Thunberg] Handel-Mazzetti; *Perilla nankinensis* [Loureiro] Decaisne)

Xiang-si=Hsiang-ssu (香絲, Aromatic Sliver); *Hong-zi-su=Hung-tzu-su* (紅紫蘇, Red Perilla); *Hui-hui-su=Hui-hui-su* (茴茴蘇, Moslem Perilla); *Zheng-xiang-si=Ch'eng-hsiang-ssu* (正香絲, True Aromatic Sliver); *Ji-guan-zi-su=Chi-kuan-tzu-su* (雞冠紫蘇, Cock's Comb Perilla). Young shoots, sliced into slivers for seasoning food, or used for tea.

Annual herbs 50–100 cm high, with maroon-red herbage, young growth softly villose, the indumentum purplish and multicellular, stems tetragonous, the older portion subterete and glabrescent; leaves deltoid-orbicular, crisped, 3–11 cm long, 3–8 cm wide, acuminate-caudate, the acumen 1–2 cm long, margin sharply dentate or pectinate, the teeth 5–9 mm long, hirsute along the major nerves on both surfaces, glabrous or sparsely punctate beneath; flowers purple-pink, in slender terminal or axillary pseudoracemes, 6–14 cm long, verticils 2-flowered, each subtended by a glabrous and ciliate ovate-acuminate bract, calyx 5-toothed, corolla 1/2 exposed; nutlets 4, reticulate, light brown, enclosed in persistent calyx 5–8 mm long.

Perilla maxima (Arduino) S. Y. Hu — Field Perilla (Syn. *Melissa maxima* Arduino [1763]; *Ocimum frutescens* L. 1762, non L. 1753; *Perilla ocymoides* L. [1764]).

Bai-su=Pai-su (白蘇, White Perilla); *Bai-si-cai=Pai-ssu-ts'ai* (白絲菜, White Sliver Vegetable); *Ren=Jen* (荏, Field Perilla); *Xiang-su=Hsiang-su* (香蘇, Aromatic Perilla); *Jia-su=Chia-su* (家蘇, Domesticated Perilla); *Qing-su=Ch'ing-su* (青蘇, Green Perilla). Leaves used to wrap a mixture of seasoned rice and meat for a dish similar to the dolma of the Middle East, cooked with grape leaves. *Su-zi=Su-tzu* (蘇子, Perilla Nutlets). Seed; source of oil.

Large, robust, green annual herbs 100–150 cm high, stem tetragonous, much branched; leaves ovate-elliptic or triangular, 5–15 cm long, 3–10 cm wide, crenulate-serrate, acuminate, base acute, hirsute along nerves, sparsely pubescent, punctate beneath; flowers small, white, in dense, terminal or axillary pseudoracemes 2–8 cm long, calyx villose, corolla very short, equal the calyx lobes in length; nutlets 4, oblique-globose, brown, reticulate, enclosed in persistent tubular calyx 10–12 mm long. Cultivated as a summer crop for the seed as a source of oil; all parts of the plant used in herbal medicine; being one of the plants recorded in ancient Chinese classics.

Plectranthus amboinicus (Loureiro) Sprengel — Sour Mint, Country Borage (Syn. *Coleus amboinicus* Loureiro)

Fan-gui-ning-meng=Fan-kuei-ning-meng (番鬼檸檬, Foreigner's Lemon). Young shoots, taken from kitchen garden and used fresh for adding flavor to soup.

Much branched aromatic perennial herbs up to 1 m high at flowering time, hairy throughout; leaves ovate to ovate-deltoid, 4–5 cm long, 3.5–4 cm across, petiolate, petioles 5–12 mm long, densely pubescent and glandular on both surfaces; flowers lilac-mauve with white corolla tube, calyx 4–6 mm long, corolla 8–9 mm long, the upper lip 2 mm long, the lower one 4 mm long; stamens 4, inserted on the corolla throat. Native of tropical Africa, widely cultivated in tropical Asia; the Hong Kong boat people planted it in pots near their cooking areas.

Prunella vulgaris L. — HEAL-ALL, CARPENTER-WEED

Xia-ku-cao=Hsia-k'u-ts'ao (夏枯草, Summer Withered Herb). Fruiting clusters and leafy shoots; used for tea, dried bundles available in American Chinese groceries.

Perennial herbs 15–40 cm high, rooting and forming new ramets at the lower nodes; leaves ovate-elliptic, 1.5–5 cm long, 1–2.5 cm wide, acute, base rounded and acute, entire or remote-serrate, leaves of new ramets all basal, ovate, obtuse, base truncate, crenulate, petioles equal or up to twice the length of the laminas; flowers purple, in compact cylindrical terminal pseudoracemes 2–6 cm long, 1–1.5 cm across, bracts reniform-cordate, each subtending 3 subsessile flowers, verticils 6-flowered, calyx tubular-campanulate, unequally bilabiate, the upper lip shallowly 3-lobed, the lower lip 2-toothed, corolla ascending, 12 mm long, glabrous, dilated below the throat, the limbs bilabiate, upper lip erect, arched, ruffled, the lower lip reflexed-spreading, 3-cleft, stamens 4, ascending under the galeate upper lip of the corolla, the upper pair shorter, filament bifid at the apex, the short arm bearing the anther, thecae 2, divergent, disk cupular, ovary 4-cleft, style filiform, gynobasic, apex bifid; nutlets subtrigonous obovoid, glossy brown, decorated with 4 dark double and parallel lines from the acute base to the rounded apex, the areola attached to a white conical caruncle. A species widespread in the northern Hemisphere, in China growing spontaneously along paths of grassy hillsides from the Yellow River Region southward to Yunnan, Sichuan, and eastward to Taiwan.

Scutellaria amoena C. H. Wright

Tian-huang-qin=T'ien-huang-chin (滇黃芩, Yunnan Skullcap). Leafy shoots; used in Shaanxi for tea.

Perennial herbs 12–45 cm high, pubescent throughout; leaves opposite, oblong-ovate, 1.4–4 cm long, 0.5–1.5 cm wide, acute or obtuse, base obtuse, rounded or cuneate, crenate or subentire, pubescent and punctate beneath; flowers blue or purple, hairy and glandular, in terminal racemes, verticils 2-flowered, bracts foliaceous, 5–10 mm long, calyx campanulate, 3 mm long, bilabiate, with a helmet-like projection on the upper side, corolla 2–3.5 cm long, the tube curved and dilated at the throat, bilabiate, the upper lip entire, arched, the lower one 3-lobed, stamens 4, filaments hairy, ovary glabrous, style pilose, stigma bifid, disk fleshy; nutlets subglobose, black, papillose. Native to western China, growing in thickets of hillside at altitudes of 1,000–1,300 m.

Scutellaria baicalensis Georgi — BAICAL SKULLCAP

Huang-jin-cha=Huang-chin-ch'a (黃金茶, Golden Tea); *Huang-qin=Huang-ch'in* (黃芩, Baical Skullcap). Leafy branches gathered, steamed, dried in shade, mixed with dried fragments of *Ma-huang* (*Ephedra sinica* Stapf or *E. monosperma* C. A. Meyer), steeped in boiling water for tea in Inner Mongolia; the dried root being the important article in traditional Chinese medicine called *"Huang-qin"*, used as a vasoactive agent for improving blood circulation.

Perennial herbs 20–60 cm high, with stout caudex connected to the tap-root, brown on the outside, golden yellow inside, very bitter, branchlets pilose; leaves opposite, subsessile, lanceolate or linear-lanceolate, 1.5–4.5 cm long, 3–12 mm wide, acute or obtuse, base rounded, entire, ciliate, punctate beneath; flowers purple, reddish-purple or purple-blue, solitary in the axils of gradually reduced leaves, giving the appearance of terminal raceme, pedicels bending at the distal end, the flowers appearing one-sided on the rachis, calyx campanulate, shallowly 2-lobed, coriaceous, with a horizontal protuberance across the middle of the tube, increasing in size after anthesis and giving a helmet-like appearance to the persistent calyx in fruit, corolla bilabiate, densely glandularly hairy, 2–2.5 cm long, the tube curved, gibbous at the base, and dilated at the throat, the upper lip galeate, slightly retuse, the lower lip 3-lobed, the middle lobe rounded, stamens 4, the upper pair with shorter filaments and 2-thecous anthers, the lower pair with longer filaments and 1-thecous anthers, ovary deeply 4-parted, glabrous, style filiform, stigma 2-fid; nutlets 4, brown, tuberculate.

Native of northern East Asia, growing on open hillsides, common in Inner Mongolia, northern and western China; introduced into European gardens for ornamental purpose, appeared in the 1922 August flower show at the British Royal Society at London, a favorite for rock gardens in America also.

Note: The Chinese names of the species quoted above illustrate the problem of folk etymology of food and medicinal plants in China. The ancient people living among the plants of the hillside discovered the economical uses of Baical Skullcap and employed it for a health tea, calling it *"Huang-jin-cha"* (Golden Tea) for the species has golden yellow roots. Eventually this use spread to population centers and the root became a medical commodity which needed a written name for perscription. The learned man in town having neither knowledge of the plant nor the hillside practices, gave an ideogram for the product which captured the sound but missed the meaning of the folk name. This name has been accepted by early Chinese botanists as the base for the generic name of *Scutellaria* L. and the subfamily *Scuatellarioideae* Briquet.

Scutellaria barbata D. Don

Ban-zhi-lian=Pan-chih-lien (半枝蓮, Half-branch Flower); *Gan-lu-zi=Kan-lu-tzu* (乾露子, Dry Exposing Seed). Young shoots; gathered in early spring, pickled in salt, rubbed between the hand, dried in the sun and stored for winter use.

Erect perennial herbs 15–35 cm high, with creeping subterranean rhizomes, stems tetragonous, glabrous; leaves triangular-ovate or lanceolate, 1–3.5 cm long, 5–15 mm across the base, obtuse, base truncate, glabrous, remotely and inconspicuously crenate; flowers rather showy, purplish-blue, in terminal and axillary one-sided spike-like pseudoracemes 3–5 cm long, verticils 2-flowered, one on each side of a node, each subtended by a small leaf or a bract, calyx campanulate, bilabiate, the lips entire, the upper one galeate, corolla 10–12 mm long, the tube slender, saccate at base, gradually dialate upward, 3–5 mm across the throat, limb bilabiate, upper lip entire, rounded, galeate, lower lip 3-lobed; stamens 4, ascending under the upper lip, thecae of the lower pair 1 locular, those of the upper pair 2-locular, cordate, subdivaricate; nutlets 4, compressed-subglobose, tuberculate, 1 mm in diameter. Native of China, growing in lowland and on banks of springs and canals throughout the country.

Scutellaria scordifolia Fischer ex Schrank

Bing-tou-huang-qin=Ping-t'ou-huang-ch'in (併頭黃芩, Paired Skullcap). Leafy shoots, used for tea in Shanxi.

Perennial herbs 15–50 cm high, glandular and hairy throughout, with creeping subterranean rhizome bearing white opposite scales and buds at the nodes, aerial stems simple or branched; leaves opposite, on short petioles, laminas variable in size and shape, oblong, linear-oblong or lanceolate, occasionally deltoid, 1.5–4 (–5.5) cm long, 2–16 (–20) mm wide, obtuse, base rounded or truncate, remotely or inconspicuously crenate, glabrous above, pilose beneath, punctate; flowers purplish-blue, solitary, axillary to normal or reduced leaves, often appearing on one side of the tetragonous stems, pedicels 2.5 mm long, calyx campanulate, 3.5 mm long, 2-lipped, upper lip with a crest, corolla 1.7–2.5 cm long, densely glandular-villose, the tube slender, geniculate at the base, throat 6–7 mm across, upper lip galeate, lower lip 3-lobed, stamens 4, ovary 4-cleft, style filiform; nutlets oblong, 1–1.5 mm long, tuberculate. Widespread in cold temperate eastern Asia.

Stachys adulterina Hemsley — HUBEI ARTICHOKE

Di-can-zi=Ti-ts'an-tsu (地蠶子, Ground Silkworm). Tubers; used for vegetable, cooked or pickled, by the same recipe as used for Jerusalem Artichoke.

Perennial herb 60–100 cm high, with rhizomes thickened at the distal end into fusiform tubers 3–4 cm long, 1–1.5 cm in diameter, smooth, white like a silkworm (hence the vernicular name), stem tetragonous, villose on the nodes; leaves variable, the lower 2 or 3 pairs oblong-elliptic, 5–7 cm long, 2–3 cm wide, the middle ones ovate-lanceolate,

7–9 cm long, 1.5–3 cm wide, acute or acuminate, base rounded or subcordate, serrate; flowers red, in terminal pseudopanicles consisting of 5–7 verticils, the lowest verticil subtended by small lanceolate leaves 1.5–2.5 cm long, the upper bracts becoming smaller progressively, rachis and the short pedicels pubescent, calyx turbinate, 5-toothed, the teeth subequal, deltoid, acute, ciliate, corolla 15 mm long, the tube 8 mm long, not dilated below the throat, the limb bilabiate, upper lip erect, galeate, hairy, lower lip 3-cleft, reflexed, stamens 4, anthers examined all sterile. Endemic to western Hubei and adjacent area in Sichuan.

Stachys affinis Bunge (1833, March) — CHINESE ARTICHOKE (Syn. *S. sieboldii* Miquel)

Cao-shi-can=Ts'ao-shih-ts'an (草石蠶, Herbaceous Rock Silkworm); *Gan-lu-zi=Kan-lu-tzu* (甘露子, Sweet-as-dew); *Di-gu-niu=Ti-ku-niu* (地牯牛, Ground Slug); *Bao-ta-cai=Pao-t'a-ts'ai* (寶塔菜, Precious Pogoda Vegetable). Tubers; fusiform, slightly constricted repeatedly, white when fresh; pickled material black (called "*Bao-ta-cai*"), bottled material available in American Chinese groceries.

Perennial herbs 30–120 cm high, hirsute throughout, stems tetragonous, the base furnished with short rhizomes thickened at the end into fleshy fusiform tubers 3–5 cm long, 10–12 mm across the middle, slightly constricted at the nodes; leaves varying in shape and size, ovate-cordate or ovate-oblong, 2.5–9.5 cm long, 1.5–3.5 cm wide, acute or shortly acuminate, base cordate, serrate, petioles of leaves on the middle of the stem 1–2 cm long, becoming progressively shorter upward; flowers rose-purple, in interrupted terminal spicate-panicles 10–15 cm long, verticils 6-flowered, subsessile, the lowest verticil subtended by lanceolate sessile leaves 4 cm long, 1 cm wide, calyx turbinate, 6–7 mm long, 5-toothed, teeth subequal, ovate, acute and aciculate, corolla 12–13 mm long, the tube 9 mm long, not dilated below the throat, limb bilabiate, upper lip erect, arched, hairy, lower lip 3-lobed, reflexed, stamens 4, ascending under the upper lip, anthers ovoid, 2-locular, thecae divergent; nutlets 4, obovoid, dark brown, 1.5 mm across the broad end, faintly tuberculate. Native of northern China, cultivated for the edible tubers.

Note: In many Chinese botanical publications, the authors applied *S. sieboldii* Miquel (1865) to this species, because of *S. affinis* Fresenius (1833 from Egypt). However, recent bibliographical research has ascertained the fact that *S. affinis* Bunge from China was published in March 1833 while that of Fresenius was in December 1833. Bunge's binomial has the priority.

Thymus mongolicus (Ronniger) Ronniger (1934) — PLATEAU THYME (Syn. *T. serpyllum* L. ssp. *mongolicus* Ronniger, 1930)

Di-jiao-ye=Ti-chiao-yeh (地椒葉, Ground Spice Leaf); *Di-jiao=Ti-chiao* (地椒, Ground Spice); *Shan-jiao=Shan-chiao* (山椒, Spice-on-the mountain, or Wild Spice). Leafy shoots; used for tea or as a spice in northwestern China.

Perennial prostrating herb, rooting at the nodes, erect stems 2–7 (–14) cm high, filiform, 0.5–1 mm in diameter, uniformly white pilose, hairs recurved; leaves ovate-elliptic or oblong, 4–10 mm long, 2–5 mm wide, glabrous, punctate, obtuse, base cuneate into a short and ciliate petiole 1–2 mm long, 3-nerved beneath; flowers rose-purple, in terminal capitate clusters, verticils 6- to 10-flowered, pedicels 2 mm long, densely hispid, calyx green, campanulate, bilabiate, tube 1.5 mm long, nerves pilose, throat barbate, lobes 2.5 mm long, upper lip foliaceous, glabrous, unequally 3-toothed, lower lip 2-toothed, uniformly setose along the margin, corolla 8 mm long, slightly dilated at the throat, pilose, tube 5 mm long, limb bilabiate, upper lip glabrous, punctate with shiny brown glands, lower lip 3-lobed, stamens 4, didymous, exserted, anthers red or yellow; nutlets oblong, chestnut-colored, smooth, locked in the brown persistent calyx by the stiff barbate hairs at the throat. Endemic to the grassland of the arid plateau of northwestern China, the type material of the species was collected from southern Gansu where the ethnic people gather the shoots for tea, used especially for colds.

Thymus serpyllum L. — Lemon Thyme, Wild Thyme

Ye-bai-li-xiang=Yeh-pai-li-hsiang (野百里香, Wild Hundred-mile Fragrant); *Qian-li-xiang=Ch'ien-li-hsiang* (千里香, Thousand-mile Fragrant). Leafy shoots; used fresh or dried locally as spices.

Mat-forming perennial aromatic herbs with long slender creeping non-flowering branches, woody at the base, rooting at the nodes, red-brown, with recurved hairs only along the ridges, upright stems 3–15 cm long; leaves oblong, ovate or obovate, 7–15 mm long, 1–7 mm wide, obtuse or acute, the base cuneate and hairy, merging into a short petiole; flowers rose-purple, crowded in terminal capitate cluster consisting of verticils subtended by normal leaves, the lowest verticil consisting of 1 flower on each side, the upper ones of 3- or 5-flowers cymes, pedicels 4 mm long, pilose, bracteoles linear-subulate, 3–4 mm long, hairy, calyx green, 6–7 mm long, tubular-campanulate, bilabiate, throat barbate, nerves setose, upper lip 3-toothed, teeth lanceolate, setose along the margin, two teeth of lower lip parallel, curved upward slightly, subulate, strongly setose, corolla bilabiate, upper lip obcordate, pilose outside, lower lip 3-lobed, stamens 4, exserted, anthers sterile. Eurasian in distribution, Chinese specimens from northern and northeastern China having much larger leaves than those from northern Europe and Japan.

Thymus vulgaris L. — COMMON THYME, GARDEN THYME

She-xiang-cao=She-hsiang-ts'ao (麝香草, Musk Herb); *Bai-li-xiang=Pai-li-hsiang* (百里香, Hundred-mile Fragrant). Leaves; dried and bottled, mostly used by foreign residents in large cities.

Low shrubs, grayish throughout, 18–30 cm high, densely white tomentose, branchlets filiform, annual growth 5–10 cm long, 0.5–1.5 mm thick, basal woody portion cylindrical, 3–4 mm in diameter; leaves subsessile, opposite, often appearing fasciculate due to short axillary branches, linear or ovate-lanceolate, 5–10 mm long, 0.5–2.5 mm wide, acute, entire, margin recurved, veins obscure, both surfaces densely pilose and punctate; flowers whitish-lilac, gynodioecious, verticils 2-, 6- or 10-flowered, crowded into capitate clusters, calyx campanulate, green, 3–4 mm long, villose and punctate, bilobate, throat barbate, upper lip 3-toothed, teeth deltoid, acute, ciliate, lower lip 2-toothed, teeth subulate-aciculate, setose, corolla equal or slightly longer than the calyx, hairy, 4–5 mm long, bilabiate, upper lip erect, glabrous, punctate, stamens 4, exserted, anthers pink, thecae parallel, style red; nutlets chestnut brown, smooth, persistent calyx straw-colored, the lower portion slightly saccate. Native of the Mediterranean Region and the Middle East; introduced in the Nanjing Botanical Garden in the 1930s.

Solanaceae: Nightshade Family

Brugmansia arborea (L.) Lagerheim — ANGEL'S TRUMPET

Tian-shi-hao=T'ien-shih-hao (天使號, Angel's Trumpet). Plants occasionally observed in villages, imported fruits available in Hong Kong.

Shrubs or large herbs, 2 m high, minutely pubescent; leaves ovate-elliptic, 20–30 cm long, 10–15 cm wide, entire; flowers white, hanging, calyx sheathy, 15–17 cm long, 7 cm across the middle, obliquely and unequally opened into 3–4 apical lobes; corolla trumpet-shaped, 30 cm long, 16–20 cm across the apex; fruits fusiform, orange-red. Native of Brazil.

Capsicum annuum L. — GREEN PEPPER, RED PEPPER, CAYENNE PEPPER, CHILI

La-jiao=La-chiao (辣椒, Hot Pepper); *Hai-jiao=Hai-chiao* (海椒, Pepper-across-the-ocean); *Qing-jiao=Ch'ing-chiao* (青椒, Green Pepper); *Jiao-zi=Chiao-tzu* (椒子, Pepper). Fruit; extensively cultivated, an important summer vegetable in northern, western and central China; used green or red, fresh, dried and/or pickled, many unusual dishes and recipes developed in special areas, such as Sichuan, Shanxi and Hunan.

Annual herbs 40–100 cm high, branchlets glabrous; leaves alternate, ovate-oblong,

3–10 cm long, 2–6 cm wide, acuminate, base altenuate, margin wavy; flowers creamy white, solitary, axillary, usually nodding, calyx campanulate, 5-toothed, green, persistent in fruit, corolla rotate, 10–15 mm in diameter, 5-lobed; stamens 5, inserted near the base of the short corolla tube, anthers bluish, partially exposed, ovary superior, 2- or 3-locular, ovules numerous, style short, stigma capitate; fruits berry-like, nodding or erect, hollow, the pericarp fleshy and firm, ripe red, sometimes pungent; seeds white, compressed-orbicular.

Native of Central America, introduced to the Philippines by the Spaniards and then into China by overseas Chinese residing there in the seventeenth century. First mentioned in Chinese botanical records written in A.D. 1621; now, a major summer vegetable crop. Two varieties with few cultivars selected in China are given below:

1. var. **grossum** Bailey — Bell Pepper, Sweet Pepper, Pimento

Shi-tou-fan-jiao=Shih-t'ou-fan-chiao (獅頭番椒, Lion-head Foreign Pepper); *Tian-fan-jiao=T'ien-fan-chiao* (甜番椒, Sweet Foreign Pepper); *Deng-long-jiao=Teng-lung-chiao* (燈籠椒, Lantern Pepper). This group of cultivars is less popular in China, available only in large cities.

Flowers solitary; fruits oblong, pomiform or compressed, 5–8 cm long, 4–10 cm across, base concave, pungent or slightly so. Introduced repeatedly since the 1930s; cultivated primarily in the vicinity of large metropolises to supply hotels established for tourism; used in salad dishes. Two special cultivars selected in northern China for low stature, spring planting and early harvest in cold areas such as Beijing and thence northward.

a. cv. 'Little Dwarf' (*Xiao-ai-yang=Hsiao-ai-yang* 小矮秧).

Plant up to 47 cm high; transplanted into protected fields in late April; fruits pomiform, 7 cm long, 6.5 cm across basal portion; available for market between middle or late May and July.

b. cv. 'Dumpling Pepper' (*Bao-zi-jiao=Pao-tzu-chiao* 包子椒).

Plant 40–50 cm high, with few branches; transplanted into open fields in early May; fruits strongly compressed, 5–7 cm high, 6–10 cm across basal portion; available between middle June and July.

2. var. **longum** L. H. Bailey. — Hot Pepper; Chili Pepper

La-jiao=La-chiao (辣椒, Hot Pepper); *Jian-jiao=Chien-chiao* (尖椒, Pointed Pepper); *Niu-jiao-jiao=Niu-chiao-chiao* (牛角椒, Ox-horn Pepper); *Yang-jiao-jiao=Yang-chiao-chiao* (羊角椒, Goat-horn Pepper).

Flowers solitary; fruits cylindrical-cornute, often slightly curved at the distal end,

10–13 cm long, 1–2.5 cm across the basal portion, base truncate-rounded, apex pointed, very pungent; the most popular variety in China, extensively cultivated; used green, red, fresh or preserved, dried whole or pulverized, and pickled whole, or made into sauces.

Capsicum frutescens L. — Shrubby Hot Pepper (Syn. *C. frutescens* L. var. *conoides* Bailey)

Zhi-tian-jiao=Chih-t'ien-chiao (指天椒, Skyward-pointing Pepper); *Xiao-mi-la=Hsiao-mi-la* (小米辣, Millet Hot Pepper); *Ye-la-zi=Yeh-la-tzu* (野辣子, Wild Hot Pepper); *Ye-la-shu=Yeh-la-shu* (野辣樹, Wild Hot Pepper Tree). Fruit; fresh or dried, used for condiment in tropical areas of China; cultivation relatively rare.

Perennial herbs 20–150 cm high, lower portion of the stems woody; leaves ovate or ovate-lanceolate, 1.5–12 cm long, 1–2.5 cm wide, acute or acuminate, base acute, more or less decurrent to the petiole, glabrous; flowers white, paired or fasciculate, axillary, calyx cupular, truncate, corolla 0.8–1 cm across; fruits erect, 1.5–3.5 long, red, very hot. Native of Central America, introduced and planted primarily for ornnmental purposes, requiring little care; presently widely distributed in tropical China, often appearing wild in Hainan and Yunnan; two cultivars reported in Treatise of Canton Vegetable (Anonymons, 1974).

l. cv. 'Green Skyward' (*Qing-la-zhi-tian-jiao=Ch'ing-la-chih-t'ien-chiao* 青蜡指天椒)
   Plants 80 cm high, much branched; leaves 5.5 cm long, 2.5 cm wide; fruits ovate-oblong, 2.5–3.5 cm long, 7–8 mm across, green, becoming red at maturity; available between April and October; appearing in kitchen gardens.

2. cv. 'White Skyward' (*Bai-la-zhi-tian-jiao=Pai-la-chih-t'ien-chiao* 白蜡指天椒)
   Plants 115 cm high; leaves 5.5 cm long, 2.5 cm wide; fruits oblong-ovoid, 2.3 cm long, 8 mm across, white, maturing yellowish-white; harvested between April and October.

Cyphomandra betacea (Cavanilles) Sendtner — Tree Tomato, Tamarillos, Palo de Tomate

Shu-fan-qie=Shu-fan-ch'ieh (樹番茄, Tree Tomato). Ripe fruits.

Tree-like shrubs 3–5 m high, densely tomentose throughout, trunks 4–10 cm in diameter, branchlets herbaceous; leaves alternate, large, ovate-cordate, 8–20 cm long, 7–15 cm wide, acuminate, base cordate, entire; flowers pinkish-white, 8–12 in loose axillary cymose panicles, peduncles 4–5 cm long, pedicels filiform, 1.5–3 cm long, calyx campanulate, 4 mm long, shallowly 5-lobed, the lobes broader than long, corolla subrotate, 1.5–2 cm across, tubes 2 mm long, lobes lanceolate, 7–8 mm long, tomentose along the margin, stamens 5, anthers oblong, 5 mm long, opening by apical pores, ovary ovoid, 2-locular, ovules numerous, style 6 mm long, stigma punctiform; fruits egg-sized, orange-red, ovoid or ovate-ellipsoid, 7–8 cm long, 5 cm across, exocarp tough, mesocarp meaty, yellow; seeds numerous, compressed, 3 mm long, 2.5 mm across, yellow, enclosed in jelly-like purple-violet pulp, under lens reticulate.

Native of Peru, grows rapidly in tropical and subtropical mountains, the fruit highly esteemed by the people of the inter-Andean regions; cultivated in New Zealand and Sri Lanka, fruit occasionally sold in Hong Kong, tasting better when stewed (cooked with skin and seed removed); plants begin to flower 1.5–2 years old, bearing period 5–6 years, a low-labor cost vegetable crop for mountainous areas.

Lycium barbarum L. — NINGXIA LYCIUM-BERRY (Syn. *L. halimifolium* Miller) (Figure 46)

Ning-xia-gou-qi=Ning-hsia-kou-ch'i (寧夏枸杞, Ningxia Matrimony Vine). Fruit; produced in modernized governmental farms in which the plants are trimmed into small trees for the convenience of cultivation and new harvesting. Fruits dried in one day, by a special construction. The products now constitutes the major revenue for the regional government, oriented towards export; the fruit is available in American Chinese stores.

Gou-qi-cha=Kou-ch'i-ch'a (枸杞茶, Lycium Tea). Persistent calyx; and peduncle of fruit; a by-product of the governmental lycium farm, in form of tea bags, sold with the fruit nationwide and worldwide.

Stout deciduous arching shrubs 2–3 m high, branchlets thorny; leaves alternate or in fascicles, elliptic-lanceolate or ovate-rhombic 2–3 cm long, 2–6 mm wide, acute, base cuneate, entire; flowers pink or rose-purple, fasciculate with leaves at the end of spurs, pedicels 5–15 mm long, calyx cupular, 4–5 mm long, 2–5-lobed, corolla funnelform, the tube longer than the lobes, narrowed below the middle, 1–1.5 cm long, limb 5-lobed, ciliate, stamens 5, filaments villose; berries red, rarely yellow, 10–20 mm long, 5–10 mm across. A species of Eurasia distribution, growing in the arid region of northwestern China, the area closely associated to the development of ancient Chinese civilization, it has been recorded in the earlist classics, poems, and herbals. It has long been a species in kitchen gardens of the people. After 1950, new management techniques were introduced into governmental institutions for the development of orchards for the production of lycium-berry. Ten cultivars were established in Ningxia, two of the following selections have yellow leaves, and two others bear yellow berries.

1. cv. 'Hemp Leaf' (*Ma-ye-gou-qi=Ma-yeh-kou-ch'i* 麻葉枸杞, Hemp-leaved Lycium-berry)

Leaves green, lanceolate or linear-lanceolate, 5–12 cm long, 8–14 mm wide; fruits red, oblong-ellipsoid, 1.8–2.2 cm long, 6–10 mm in diameter; known dry-weight content: protein 20.8%, sugars 24.2%.

2. cv. 'Greater Hemp Leaf' (*Da-ma-ye-gou-qi=Ta-ma-yeh-gou-ch'i* 大麻葉枸杞, Large Hemp-leaved Lycium-berry)

Leaves dark green, lanceolate or linear-lanceolate, 6–12 cm long, 8–15 mm wide; fruits bright red, oblong-cylindrical, 2–2.6 cm long, 8–12 mm in diameter, the apex truncate; known content: protein 9.3%, sugars 36.6%.

3. cv. 'White Branch' (*Bai-tiao-gou-qi=Pai-t'iao-kou-ch'i* 白條枸杞, White Branch Lycium-berry)

Leaves dark green, on grayish-white branches, lanceolate, 2–5 cm long, 5–10 mm wide; fruits brilliant red, fusiform, 1.4–2 cm long, 6–10 mm in diameter; known content: protein 15.4%, sugars 24.2%.

4. cv. 'Spherical Berry' (*Yuan-guo-gou-qi=Yuan-kuo-kou-ch'i* 圓果枸杞, Round Fruit Lycium-berry)

Leaves dark green, thick, lanceolate or linear-lanceolate, 5–8 cm long, 8–12 mm wide; fruits red, obovoid, 8–12 mm long, 6–10 mm in diameter; known content: protein 17.5%, sugars 30.4%.

5. cv. 'Spherical Acute-head' (*Jian-tou-yuan-guo-gou-qi=Chien-t'ou-yuan-kuo-kou-ch'i* 尖頭圓果枸杞, Pointed-head Round Fruit Lycium-berry)

Leaves green, lanceolate; fruits short, oblong, the apex acute, 8–14 mm long, 6–10 mm in diameter, rarely cultivated.

6. cv. 'Yellow Acute-head' (*Jian-tou-huang-ye-gou-qi=Chien-t'ou-huang-yeh-gou-ch'i* 尖頭黃葉枸杞, Pointed-head Yellow-leaved Lycium-berry).

Leaves yellowish-green, thin, lanceolate, or linear-lanceolate, 5–8 cm long, 6–8 mm wide; fruits oblong-cylindrical, 18–20 mm long, 6–8 mm in diameter, known content: protein 13.6%, sugars 42.8%.

7. cv. 'Yellow Round-head' (*Yuan-tou-huang-ye-gou-qi=Yuan-t'ou-huang-yeh-kou-ch'i* 圓頭黃葉枸杞, Round-headed Yellow-leaved Lycium-berry).

Leaves yellowish-green, thin, lanceolate; fruits narrow, oblong, rounded at the apex, 1.4–1.8 cm long, 4–8 mm in diameter, rarely cultivated.

8. cv. 'Yellow Berry' (*Huang-guo-gou-qi=Huang-kuo-kou-ch'i* 黃果枸杞, Yellow-fruit Lycium-berry).

Leaves yellowish-green, thin, lanceolate; fruits rounded, oblong, 1.5–1.9 cm long, 1.4–2 cm in diameter; rarely cultivated.

9. cv. 'Curly Leaf' (*Juan-ye-gou-qi=Chuan-yeh-kou-ch'i* 捲葉枸杞, Curly-leaved Lycium-berry).

Leaves lanceolate, curly; fruits obovoid, 1.4–1.8 cm long, 6–9 mm in diameter, rarely cultivated.

10. cv. 'Little Yellow' (*Xiao-huang-guo-gou-qi=Hsiao-huang-kuo-kou-ch'i* 小黃果枸杞, Little Yellow-fruit Lycium-berry).

Leaves grayish-green, thick, lanceolate, 6–10 cm long, 7–12 mm wide; fruits orange, short, oblong, 1.2–1.5 cm long, 7–9 mm in diameter, fleshy, sweet; a good cultivar for the production of berries to be used as fresh fruit; known content: protein 12.3%, sugars 52.3%.

Lycium chinense Miller — Chinese Matrimony-vine

Gou-qi=Kou-ch'i (枸杞, Matrimony-vine). *Gou-qi-zi=Kou-ch'i-tzu* (枸杞子, Lycium-berry). Dried fruit; market product consisting of smaller fruits than those from Ningxia (Map, Location12–13), available in American Chinese stores.

Gou-qi-tou=kou-ch'i-t'ou (枸杞頭, Lycium Green-shoot). Tender portion of the stem; specially cultivated as an annual crop in tropical or warm temperate areas of China for winter or early spring market, only the tender growing area and the leaves on the woody portion are used in soup with very little cooking, rarely used as potherb (see Part I for more information).

Thorny deciduous shrubs 1 m high, with slender arching branchlets; leaves alternate, often appearing in fascicles because of the axillary short branches, ovate or ovate-rhombic, rarely lanceolate at the end of long-shoots, 1.5–5 cm long, 5–17 mm wide, entire, obtuse; flowers 1–4, in leaf-fascicles on short shoots, calyx campanulate, 3–4 mm long, 3- to 5-lobed, corolla purple, rotate, tube 9–12 mm long, the lobes ciliate, stamens 5, filaments barbate at base; berries ovoid or ovate-ellipsoid, 5–15 mm long, 8–9 mm across. Native of eastern Asia, growing in thickets, graveyards, and by fences. Extensively cultivated in Guangdong and Hong Kong by cuttings, stored in shade of irrigation ditches in summer and planted in late autumn; harvested for the Chinese New Year Festival, when the stems are 25–30 cm high, cut and tied in one-pound bundles for the market.

Lycopersicon lycopersicum (L.) Karsten ex Farwell — Tomato, Love Apple (Syn. *L. esculentum* Miller) (Figure 49)

Fan-qie=Fan-ch'ieh (番茄, Foreign Eggplant); *Yang-shi-zi=Yang-shih-tzu* (洋柿子, Foreign Persimmon); *Xi-hong-shi=Hsi-hung-shih* (西紅柿, Western Red Persimmon). Fruit; extensively cultivated, becoming a common summer vegetable, eaten cooked, rarely raw.

Glandular, hairy herbs 1–2 m high, stems clambering, requiring support; leaves ovate-oblong in outline, 7–12 cm long, 5–8 cm wide, irregularly pinnate or pinnatifid, segments unequal, coarsely dentate; flowers cream-white, in racemelike cymes above the leaf-axils, calyx green, corolla rotate, anthers forming a central cone, produced into sterile apical tips; fruits juicy, red or yellow, size and shape varying with the cultivars. Native of the Pacific South America, introduced into China in the early 1920s, now extensively cultivated as a summer vegetable crop; a most popular dish being tomato-fried egg.

Nicandra physaloides (L.) Gaertner — Apple-of-Peru

Bing-fen=Ping-fen (冰粉, Ice Jelly); *Tian-zhu=Tien-chu* (田珠, Field Pearl); *Jia-suan-jiang=Chia-suan-chiang* (假酸漿, False Physalis). Seed; used for preparing a transparent cooling refreshment.

Coarse erect annual herb 0.4–1.5 m high; leaves ovate or elliptic, 4–12 cm long, 2–8 cm wide, acute or acuminate, base obtuse, sinuate-toothed, or angled, sparsely pubescent; flowers on axillary and terminal panicles, pale-blue, calyx of distinct sepals, 5-angled, sagittate, enlarged and inflated after anthesis, persistent, acute, base cordate-auriculate, petals campanulate, 4 cm long, shallowly 5-lobed; berries 1.5–2 cm in diameter, yellow, rather dry; seeds reniform, pale brown, 1 mm across. Native of South America, cultivated in Yunnan.

Physalis alkekengi L. var. **francheti** (Master) Makino — Chinese Lantern

Suan-jiang=Suan-chiang (酸漿, Sour Sap); *Hong-gu-niang=Hung-ku-niang* (紅姑娘, Red Maiden); Berry; very rare, observed only in mountainous area of western Sichuan.

Tall anuual herbs 30–100 cm high, branchlets glabrous; leaves alternate or paired on the same side of the stem, ovate, ovate-rhombic or suborbicular, 5–15 cm long, 2–8 cm wide, obtuse, base rounded and abruptly cuneate, subentire, or undulate and coarsely and remotely dentate, ciliate; flowers cream-yellow, solitary, axillary, nodding, calyx campanulate, 6 mm long, 5-lobed to the middle, lobes lanceolate-acuminate, densely ciliate, corolla rotate, 1.5–2 cm across, shallowly 5-lobed, ciliate; fruits included in inflated persistent calyx, bladder-like, bright orange or vermillion-red, 3–4 cm long, 2.5–3.5 cm across base, prominently 10-nerved, veinlets reticulate, berries globose, 1–1.5 cm in diameter.

Physalis angulata L. — Husk Tomato (Syn. *P. ciliata* Siebold et Zuccarini) (Figure 47)
Duan-dou=Tuan-tou (端豆, Pea-berry); *Tian-pao-cao=Tien-p'ao-ts'ao* (天泡草, Celestial Bladder Herb); *Deng-long-cao=Teng-lung-ts'ao* (燈籠草, Lantern Herb); *Da-tou-pao=Ta-t'ou-p'ao* (打頭泡, Strike Forehead Bubble). Ripe berry; plants grown in soybean field, berry enclosed in brown inflated persistent calyx, falling under the spreading branchlets during harvest time, picked up by farmers and eaten in the field, not available in the market; fully grown fruits with green bladder-like calyx, taken from the branch by children, blown and made a quick strike of the forehead creating a loud exciting noise, hence the vernacular name "Strike Forehead Bubble".

Annual herb 30–45 cm high, sparsely pubescent or glabrous, branchlets spreading, mature fruits lying on the ground; leaves broad-ovate, 3–7 cm long, 2–5 cm wide, acuminate or acute, base rounded and abruptly acute, petioles 2–5 cm long; flowers pale yellow, solitary, axillary, hanging, calyx broadly campanulate, 4–5 mm long, the distal portion 5-lobed, the lobes deltoid, acuminate, ciliate, corolla funnelform, 11–12 mm long, shallowly lobed, margin softly hairy, anthers yellow, oblong, 2 mm long, style columnar, stigma punctiform; fruits globose, 1–1.5 cm across, green, turning purple at maturity, sour, persistent inflated calyx straw-colored, subglobose, 2.5 cm across, prominently reticulate. Native of tropical American, growing in cotton or soybean fields as a weed.

Physalis peruviana L. — Cape Gooseberry, Barbados Ground Cherry
Xiang-duan-dou=hsiang-tuan-tou (香端豆, Sweet Pea-berry); *Deng-long-guo=Teng-lung-kuo* (燈籠果, Lantern Fruit). Fruit; cultivated, rare.

Much branched annual or perennial herbs with 1 m spread, densely glandular-pillose throughout; leaves triangularly ovate, 6–10 cm long, 4–7 cm wide, subentire, or coarsely remote-dentate, acuminate, base rounded or subcordate; flowers pale yellow, with purple markings, calyx broad-campanulate, 15 mm long, slightly over half 5-lobed, the lobes deltoid-caudate, 9 mm long, corolla campanulate, 15–20 mm long, 10–15 mm across, subentire, truncate, ciliate, stamens 5, purple-blue, anthers 3.5–4 mm long; fruits globose, 15–20 mm in diameter; berries sweet-scented, yellow, persistent calyx bladder-like, 2.5 cm long and across, straw-colored, reticulate. Native of South America, introduced into botanical gardens and agricultural institutions in the 1930s; not common.

Physalis pubescens L. — GROUND CHERRY, DOWNY GROUND CHERRY

Mao-suan-jiang=Mao-suan-chiang (毛酸漿, Hairy Sour Sap); *Jia-deng-long=Chia-teng-lung* (假燈籠, False Lantern); *Lü-deng=Lü-teng* (綠燈, Green Lantern). Fruits, eaten by children.

Decumbent annual herbs, softly villose or glandular-viscid throughout, 30–60 cm high; leaves ovate-cordate, 3–8 cm long, 2–6 cm wide, acuminate, base rounded, irregularly course dentate; flowers pale yellow, solitary, axillary, on pedicels 5–6 mm long, calyx densely villose, campanulate, 5-mm long, 5-lobed, tube 2 mm long, lobes lanceolate-subulate, 3 mm long, corolla folded, funnelform, 8 mm long, truncate, shallowly lobed, villose, stamens 5, anthers purple, style terete, stigma subcapitate; berries globose, 1 cm in diameter, enclosed in inflated persistent calyx, 2–3 cm long, 2–2.5 cm across the base, villose, prominently reticulate. Native of the New World, naturalized throughout China, fruit eatern locally, not available in the market; entire plant used in folk medicine.

Solanum japonense Nakai

Ye-hai-qie=Yeh-hai-ch'ieh (野海茄, Wild Oceanic Nightshade); Leafy shoots, used for potgreen in northern Yunnan.

Perennial herbaceous vines 1–1.5 m high, branchlets slightly pilose; leaves deltoid-lanceolate, or ovate, 3–8.6 cm long, 2–5 cm wide, caudate-acuminate, base rounded and acute, margin wavy or occasionally 3- or 5-lobed; flowers purple, in axillary cymose corymbs, calyx cupular, 5-toothed, corolla rotate, 1 cm across, 5-lobed, lobes 4 mm long, stamens 5, anthers oblong, 2.5–3 mm long, dehiscing by apical pores, ovary ovoid, styles 5 mm long, stigma capitate; berries globose, 1 cm in diameter, bright red; seeds reniform, 2 mm across. Native to eastern Asia, widespread in temperate China, occurring along hedges, hillsides and waste areas, at altitudes of 2,400–2,900 m.

Solanum melongena L. var. **esculentum** Nees — EGGPLANT, BRINJAL (Malaysian) (Figure 48)

Qie-zi=Chieh-tsu (茄子, Eggplant); *Ai-gua=Ai-kua* (矮瓜, Short Gourd). Fruit; extensively cultivated, a major summer vegetable in China, many forms, some cultivars with white skin.

Annual herbs 50–150 cm tall, densely covered by purplish or gray stellate hairs throughout; leaves alternate, ovate-oblong, 10–20 cm long, 5–10 cm wide, irregularly lobed or angular, obtuse, base oblique-subrotundate, or subtruncate, hairs purple on the upper surface, gray beneath; flowers purple, partially nodding, solitary, ultra axillary, calyx campanulate, purplish-green, 1 cm long, irregularly 5-lobed, corolla rotate,

2–2.5 cm across, stamens 5, inserted at the base of corolla, anthers yellow, dehiscent by apical pores, ovary superior, ovoid, villose, style columnar, basal portion stellate, stigma subcapitate; fruits baccate, firm, size, shape and color varying with the cuitivars. Native to tropical Asia, a wild prickly variety growing in Bangladash.

Note: The introduction of the cultivated eggplants into China was through ethnobotanical channels. Ancient Chinese records show that it came both overland by the silk route and in an unrecorded manner of cultural diffusion among the Asian peoples. It was first recorded in A.D. 609 in the *Miscellaneous Records of Daye Reign* (A.D. 605–617) of Sui Dynasty as Gourd-of-the-Kunlun Mountain (崑崙瓜, *Kun-lun-gua=K'un-lun-kua*), a name which indicates the exact direction of its movement overland from West to East. In Tang Dynasty, Mong in *Health Food and Curative Herbs* (A.D. 684) recorded it as "*Luo-su*" (落蘇), a name without an obvious Chinese meaning. Evidently, it represents the sound of its original name of the subtropical Asian mountainous tribes, accepted by their Chinese neighbors. Mong was a resident of a small town in the mountainous area of northwestern Henan. Again in Tang Dynasty, Chen (A.D. 739), in *Unrecorded Materia Medica* adopted the present name "*Qie-zi*" which is used throughout China and has been the basionym of many cultivars developed in the country. Chen was a resident of the coastal area, Wenzhou in Zhejiang. In the eighth century, people from the coastal China, particularly from Quanzhou (Amoy) in Fujian down south to Swatow in Guangdong and up north to Ningpo (Wenzhou) in Zhejiang had established connections with people in southeastern Asia where the brinjal (eggplant) was a common vegetable. The etymological origin of *qie=ch'ieh* (茄) for eggplant is likely its Malaysian name "*brinjal*" (Burkill, 1935).

In southern China such as Guangzhou (Canton), cultivars have been selected for firm texture, elongated and slender form, and green and white color. By arranging the early or late planting schedules, the market supply can be prolonged from April to December. In northern China such as Beijing, small dwarf cultivars for short growing season were selected so that the young plants flower between the fourth and the sixth leaves. Nine of the following ten cultivars are from Guangzhou, three for early harvests, three for summer supplies and two for autumn crops, and the last two are from Beijing, a dwarf early bearing cultivar and another one with the longest known slender eggplant.

1. cv. 'Early Red' (*Zao-hong-qie=Tsao-hung-ch'ieh* 早紅茄, Early Red Eggplant) (Figure 48a).

Plant 70 cm high, stem purple-red; leaves of flowering section oblong-ovate, 20 cm long, 13 cm wide, dark green with purple tinge, petioles and veins purple-red; fruits cylindrical, 27–31 cm long, 3–4 cm in diameter, slightly curved at the distal end, purple-red; sowed in September–October, harvested between April and June.

2. cv. 'Early Green' (*Zao-qing-qie=Tsao-ch'ing-ch'ieh* 早青茄, Early Olive-green Eggplant) (Figure 48b).

Plants 70 cm high, stems olive-green; leaves elongate-ovate, 28 cm long, 18 cm wide, green, wavy, veins green; fruits clavate and slightly curved, 28–35 cm long, 4 cm in diameter, pointed at the apex, olive-green.

3. cv. 'Light Bulb' (*Dian-deng-dan=Tien-teng-tan* 電燈膽, Electric Light-bulb) (Figure 48c).

Plants 80 cm high, stem purple-red; leaves ovate-suborbicular, 25 cm long, 18 cm wide, dark green tinged purple, shallowly wavy, petioles and nerves purple; fruits short and plump, smooth like an electric light-bulb, 15 cm long, 7 cm across the broad area, purple-red.

4. cv. 'Extended Vision' (*Yuan-jing-qie=Yuan-ching-ch'ieh* 遠景茄, Far-view Eggplant) (Figure 48e).

Plants 80 cm high, stems purple-black; leaves ovate-suborbicular, 25–30 cm long, 11–17 cm wide, green tinged purple, petioles and veins purple-red; fruits purple, clavate, 20–30 cm long, 4–6 cm across the broad portion, dark purple-red with the persistent calyx olive-green and tinged purple.

5. cv. 'Hanging Purse' (*He-bao-qie=Ho-pao-ch'ieh* 荷包茄, Hanging Purse Eggplant) (Figure 48d).

Plants 70 cm high, stems purple; leaves ovate, 22 cm long, 15 cm wide, olive-green tinged purple, petioles and nerves dark purple; fruits cylindrical, distal portion rounded and slightly thickened, 20–25 cm long, 4–5 cm across the thick portion; dark purple-red.

6. cv. 'White Dragon' (*Sha-long-bai-qie=Sha-lung-pai-ch'ieh* 沙龍白茄, Sand Dragon White Eggplant) (Figure 48f).

Plants robust, 90 cm high, stems olive-green; leaves oblong-ovate, 24 cm long, 16 cm wide, olive-green, margin lobed, petioles and veins olive-green; fruits slender and curved, 30–36 cm long, 3 cm in diameter, pointed at the distal end, white, skin thin, persistent calyx tubular, light green, a superior cultivar.

7. cv. 'Goddess-of-Mercy Fingers' (*Guan-yin-shou-zhi=Kuan-yin-shou-chih* 觀音手指, Goddess-of-Mercy Fingers) (Figure 48h).

Plants 70–100 cm high, stems with short internodes, olive-green; leaves ovate, 24 cm long, 14 cm wide, green, margin deeply lobed, petioles and veins green; fruits very long, cylindrical, 30–37 cm long, 2 cm across, light green, skin uneven, seeds few, a good cultivar.

8. cv. 'Crane Cave Fall' (*He-dong-qiu-qie=Ho-dung-ch'iu-ch'ieh* 鶴洞秋茄, Crane Cave Autumn Eggplant) (Figure 48i).

Plants tall, 100 cm high, stem green; leaves ovate, 27 cm long, 15 cm wide, olive-green, rather deeply lobed along the margin, petioles and veins green; fruits slender, distal end pointed, 28–33 cm long, 3–4 cm across the middle; gray-green, good flavor, high production, a superior cultivar.

9. cv. 'Pomegranate-on-fire' (*Huo-shi-liu=Huo-shih-liu* 火石榴, Fire Pomegranate).

Plants rather dwarf, 50–60 cm high; leaves ovate, 20–24 cm long, 12–18 cm across the basal portion, margin shallowly 2-lobed; fruits appearing between the fourth and sixth leaves, subspherical, 8–9 cm long, 10–11 cm across the middle, shiny dark purple, with purple persistent calyx and petiole; selected for greenhouse planting in the vicinity of Beijing, sown in August, transplanted in September, fruit ready for the market in December.

10. cv. 'Wire Eggplant' (*Xian-qie=Hsien-ch'ieh* 線茄. Wire eggplant or Thread Eggplant).

Plants 70–80 cm high, stems purple; leaves elongated-ovate, 20–22 cm long, 12–13 cm wide, shallowly lobed or subentire along one margin; fruits slender, cylindrical-curved, 38–44 cm long, 2.5–3 cm across, dark purple-black, inside white, tender, good quality; a common summer crop, sown in late April, transplanted in June, harvested from late July to November in Beijing area; similar cultivar bearing slightly shorter and thinner fruits (36 cm long, 2.2 cm across) in Shanghai and Ningpo called *"Tiao-qie"* (=*T'iao-ch'ieh* 條茄, Ribbon Eggplant), appearing in mid-June to November.

Solanum nigrum L. — BLACK NIGHTSHADE

Long-kui=Lung-kuei (龍葵, Nightshade); *Hei-duan-dou=Hei-tuan-tou* (黑端豆, Black Pea-berry). Young shoot; cooked, washed and seasoned; ripe fruit, eaten by children in northern Jiangsu.

Widespread annual weedy herbs 50–150 cm high, in Hong Kong appearing as perennials with fresh branchlets growing off old plants and beginning to flower and fruit in winter, primary stems 1–2 cm in diameter, sparsely prickly, branchlets glabrous, green; leaves alternate, ovate or ovate-oblong, 4–8 cm long, 2.5–5 cm wide, both ends acuminate, margin wavy or coarsely dentate; flowers white, in ultra axillary cymes, peduncles 1–2.5 cm long, calyx green, 5-lobed, the lobes ovate, corolla rotate, 5–6 mm in diameter, stamens 5, filaments short, anthers oblong, 2 mm long, dehiscing by oblique terminal slits, ovary globose, style dilated, stigma discoid; berry globose, shiny black at maturity, very juicy, edible; seeds many, discoid, 1 mm in diameter. Chromosome count of plants from southern China being diploid and that of northern China hexaploid;

some authors applied *S. nigrum* L. var. *pauciflorum* Liou to plants growing in Guangzhou and Hong Kong.

Solanum photeinocarpum Nakamura et Odashima

Bai-hua-cai=Pai-hua-ts'ai (白花菜, White-flowered Vegetable); *Kou-zi-cao=K'ou-tzu-ts'ao* (扣子草, Button Herb); *Gu-niu-cai=Ku-niu-ts'ai* (古鈕菜, Ancient Button Vegetable); *Shao-hua-long-kui=Shao-hua-lung-k'ui* (少花龍葵, Few-flowered Nighshade). Leafy shoot; used as potherbs in Yunnan.

Slender herbs 1 m high; leaves ovate or oblong, 4–8 cm long, 2–4 cm wide, acuminate, base acute and decurrent becoming wings on the petiole, entire, wavy, or irregularly dentate, sparsely pilose on both surfaces; flowers white, 1–6 in subumbelliform axillary cymes, calyx 2 mm long, 5-toothed, corolla rotate, 7 mm across, 5-lobed, lobes ovate, 2.5 mm long, stamens 5, dehiscing by apical pores, ovary globose, style hairy, stigma capitate; berries black, spherical, 5 mm in diameter. Native to southeastern Asia, growing along streams and damp areas in the margin of forests at altitudes of 100–1,420 m.

Solanum tuberosum L. — Potato, Irish Potato, White Potato

Yang-yu=Yang-yu (洋芋, Foreign Taro); *Shan-yao-dou=Shan-yao-tou* (山藥豆, Yam Pea); *Ma-ling-shu=Ma-ling-shu* (馬鈴薯, Horse-bell Potato). Tubers; extensively cultivated, a staple food for people living in the mountains.

Annual herbs 1 m high, stems rather tender, bending and rooting near base, glabrous or hairly, bearing irregular tubers at the basal end; leaves decompound, ovate in outline, 10–25 cm long, leaflets 3 or 4 pairs, mixed with small irregular lobes on the rachis, ovate or oblong, 1–6 cm long, acute, base rounded, both surfaces pilose, more so beneath; flowers white or bluish, 10–14 in terminal cymose panicles, pedicels 2–2.5 cm long, pilose, calyx green, hairy, tube campanulate, 4 mm long, lobes linear-lanceolate, 6–10 mm long, corolla rotate, 2.5–3 cm across, hairy outside, shallowly lobed, stamens 5, filaments 1 mm long, glabrous, anthers yellow, oblong, 6 mm long, dehiscing by oblique terminal pores; berries globose, yellowish-green, 2 cm in diameter (seldom seen). Native of the Andes, extensively cultivated, propagated by tubers.

Scrophulariaceae: Figwort Family

Mazus japonicus (Thunberg) 0. Kuntze (Syn. *M. rugosus* Loureiro)

Lan-qi-cai=Lan-chi-ts'ao (爛薺菜, Rotten Shepherd's Purse); *Tong-quan-cao=T'ung-ch'uan-ts'ao* (通泉草, Announcing Fountain Herb). Whole plant; gathered and consumed in early spring as a famine food in northern China.

Annual or biennial herbs 6–20 cm high at flowering and fruiting stage; leaves basal, forming a rosette, spatulate, 2–6 cm long, base cuneate, irregularly coarse-dentate, cauline leaves smaller, the lower ones obovate and cuneate, the upper ones ovate, sessile; flowers small, ephemeral, purple-blue, calyx green, tube campanulate, 2 mm long, lobes lanceolate, 4 mm long, corolla 10 mm long, bilabiate, upper lip lavender, short and narrow, lower lip expanded, 3-lobed, 5–6 mm across, white with yellow and white capitate hairs at the center, stamens didymous, all fertile, anthers parallel, thecae divergent, ovary globose, 2-locular, ovules numerous, style filiform, stigma 2-lipped, lamellate, the lower lip longer and recurved; capsules subglobose, 4 mm in diameter, included in foliaceous persistent calyx, dehiscing loculicidal; seeds minute, cylindrical, testa papillose.

Melampyrum roseum Maximowicz — ASIAN COW-WHEAT

Shan-luo-hua=Shan-lo-hua (山蘿花, Hillside Net Flower). Root; used for tea in northwestern China.

Densely branched annual herb 30–60 cm high, scabrid throughout, dried black; leaves ovate-elliptic, lanceolate or linear-lanceolate, 2–8 cm long, 0.2–3 cm wide, acuminate, base obtuse, subentire; flowers purple-red, in terminal loose racemes 2–10 cm long, bracts foliaceous, green or purple-red, decreasing in size upward progressively, entire or fringed at the base, calyx campanulate, 5 mm long, with 5 subulate teeth, corolla 1.5–2 cm long, bilabiate, the upper lip galeate, 2-lobed, the lower one 3-lobed, biconvex, stamens 4, anthers oblong, ovary 2-locular, style filiform; capsules ovoid, 10–13 mm long; seeds 2–4. Native to temperate eastern Asia; very closely related to *M. lineare* Desrousseaux of eastern North America.

Mimulus tenellus Bouge — EASTERN ASIAN MONKEY FLOWER

Gou-suan-jiang=Kou-suan-chiang (溝酸漿, Ditch Sour Sap). Leafy shoots; used for potherbs in northwestern China.

Perennial glabrous creeping and ascending herbs 40 cm high, branchlets 4-angled; leaves opposite, ovate-deltoid, 1–2 cm long, 8–15 mm wide, acute, base rounded, remotely serrate, petioles 1–2 cm long; flowers yellow, solitary, axillary, calyx tubular, 4–7 mm long, minutely 5-toothed, persistent and inflated in fruit, corolla 10–15 mm long, almost bilabiate, lobes rounded, stamens 4, included; capsules oblong, 3–4 mm long; seeds small, smooth. Growing along ditches in temperate eastern Asia at altitudes of 570–1,200 m.

Rehmannia glutinosa (Gaertner) Liboschitz ex Fisher & C. A. Meyer — CHINESE FOXGLOVE

Di-huang=Ti-huang (地黃, Chinese Foxglove); *Mi-guan-ke=Mi-kuan-k'o* (蜜罐稞, Honey Bottle Plant); *Gan-di-huang=Kan-ti-huang* (乾〔干〕地黃, Dry Di-huang). Dry uncured root in market. *Sheng-di-huang=Sheng-ti-huang* (生地黃, Partially Cured Di-huang). Material partially cured in air dry container. *Shu-di-huang=Shu-ti-huang* (熟地黃, Thoroughly Cured Di-huang). Black and soft, used both in prescription and for *Bupin*.

Perennial caespitose herbs 15–30 cm high at flowering time, glandularly pubescent throughout, root fleshy. Leaves obovate-oblanceolate, petiolate; petioles 1–2 cm long; lamina 3–10 cm long, apex round, base cuneate, margin sublobate, serrate. Flowers pink-purple, in terminal short raceme on a scape slightly longer than a larger leaf; sepals tubular-urceolate, 5-lobed, the lower lobe larger; corolla tubular, 4 cm long, unequally 5-lobed, the upper 2 lobes reflexed, lower 3 lobes straight; ovary superior, 2-locular, capsule unilocular, ovoid. Widely distributed in Northern China; extensively cultivated along the Yellow River Region for the medicinal root; the black cured kind used for *Bupin*.

Striga asiatica (L.) 0. Kuntze — Witchweed (Figure 50)

Du-jiao-jin=Tu-chiao-chin (獨腳金, One-foot Gold). Whole plant gathered from the grassy hillside, used for tea or cooked with pork chop for soup in southern China, especially esteemed in Hong Kong.

Small inconspicuous erect parasitic herbs 10–15 cm high, strigose throughout, seldom with branches; leaves small, linear or scale-like, 1 cm long, 1–2 mm wide; flowers golden, red or white, varying between different populations, solitary, axillary, sessile, 2-bracteate, sepals tubular, with 10 ridges, 5-toothed, corolla tubular, the tubes 8 mm long, curved and gibbous below the throat, limbs bilabiate, 2–3 mm across, upper lip 2-lobed, superior in bud, stamens 4, enclosed, anther with 1 theca, ovary superior, style clavate, stigma punctiform; capsules oblong-ellipsoid, loculicidally dehiscent; seeds numerous, minute. Widely distributed in the Old World tropics, growing on exposed glassy slopes in southern China; has been recorded as a parasite on corns in North Carolina (R. F. Britt 3001, 1963).

Bignoniaceae: Bignonia Family

Catalpa bungei C. A. Meyer — North-China Catalpa

Qiu-shu=Ch'iu-shu (楸樹, Catalpa); *Jin-si-qiu=Chin-ssu-ch'iu* (金絲楸, Golden Thread Catalpa). Unfolding buds, pickled; flowers, stir-fried.

Deciduous trees 15 m high, bark dark gray, broken into scales, branchlets brown;

leaves opposite or in threes, whorled, deltoid-ovate, 6–15 cm long, 6–12 cm wide, acuminate, base truncate or broad-cuneate, entire or with few teeth near the base, glabrous, petioles 2–8 cm long; flowers white, tinged pink and with purple spots inside, 3–12 in terminal corymbose racemes, sepals 2-lobed, corolla 3–3.5 cm long, campanulate, 2-lipped, the upper lip 2-lobed, the lower one 3-lobed, fertile stamens 2, included, ovary superior; capsules cylindrical, 25–50 cm long, 5.5 mm across, dividing into 2 valves after maturity; seeds numerous, oblong, compressed, winged, bearing a tuft of white hair at each end. Native of northern China, introduced into Western gardens in 1877.

Catalpa ovata G. Don — CATALPA

Zi=Tzu (梓, Catalpa, an ancient literary name); *Huang-hua-qiu=Huang-hua-ch'iu* (黃花楸, Yellow-flowered Catalpa); *He-qiu=Ho-ch'iu* (河楸, River Catalpa); *Shui-tong=Shui-t'ung* (水桐, Water Paulownia). Young shoots, used for potherb in Yunnan and Jiangsu.

Deciduous trees 10–13 m high; leaves broad ovate or suborbicular, 10–13 cm long, 7–25 cm wide, 3- to 5-lobed, lobes acuminate, glabrescent, palmately 5- to 7-nerved; flowers pale yellow, striped yellow and spotted dark violet inside, showy, in terminal panicles, calyx 2-lipped, corolla campanulate, 2.5 cm long, the limbs 5-lobed, the upper two smaller, fertile stamens 2, included, ovary superior, 2-locular; capsules cylindrical, 20–30 cm long, 5–9 mm across, villose when young, seeds numerous, minute, with horizontal wings giving an oblong appearance, terminated by silky white hairs. A native of central China, introduced into Japan before the sixth century, cultivated near Buddhist temples, spreading from Japan to Western gardens.

Orobanchaceae: Broom-rape Family

Cistanche salsa (C. A. Meyer) G. Beck; **C. deserticola** Ma — CISTACHE (Figure 51)

Rou-cong-rong=Jou-ts'ung-jung (肉蓯蓉, Fleshy Cistache); *Tian-da-yun=T'ien-ta-yun* (甜大芸, Sweet Big Cistache); *Xian-da-yun=Hsien-ta-yun* (鹹大芸, Salty Big Cistache). Slices of dried fleshy stem of the parasitic species; cooked with meat and vegetables to prepare a bupin, taken as a special treat.

A stout rough parasitic plant 10–30 cm high, growing on the roots of various shrubs and grasses in the deserts of northwestern China, stems fleshy, cylindrical, yellow, without branch, 5–20 mm in diameter, covered with ovate to lanceolate imbricate bracts; flowers yellowish-purple, crowded in a terminal spike 5–20 cm long, 5–7 cm in diameter, bracts ovate, acuminate, villose or glabrescent, 2–3.5 cm long, bracteoles lanceolate, hairy, calyx campanulate, yellow or white, 10–12 mm long, 5-lobed, lobes deltoid, ciliate, vilose, corolla campanulate, 2.5–3 cm long, shallowly 5-lobed, tube creamy-white, lobes light purple, 5 mm long and wide, stamens 4, paired, enclosed, filaments villose at the base, thecae parallel, ovary unilocular, with 4 parietal placenta; capsules ovoid, enclosed in persistent calyx, 2-valved; seeds minute, numerous.

Orobanche coerulescens Stephan — Blue-flowered Orobanche

Lie-dang=Lieh-tang (列當, Orobanche); *Du-gen-cao=Tu-ken-ts'ao* (獨根草, Single-rooted Herb); *Tu-er-tui=T'u-erh-t'ui* (兔兒腿, Rabbit Leg). Young plant; gathered, steamed, dried, used for tea, or cooked with lamb for soup.

Parasitic herbs 15–40 cm high, fleshy, appearing only at the flowering stage, arachnoid-lanate throughout, with subterranean stout caudex and solitary cylindrical, yellow-brown scape, 5–10 mm in diameter, covered by ovate-lanceolate scales 0.8–2 cm long, 2–6 mm across the base; flowers numerous, bluish-purple, lavender or rarely pale yellow, in terminal spikes 5–10 cm long, individual flowers subtended by an ovate-lanceolate bract and 2 bracteoles, sepals subcampanulate, bifid nearly to the base, each lobe 2-toothed, corolla shallowly bilabiate, 2 cm long, the tube slightly curved and enlarged at the throat, the upper lip round, the lower lip 3-lobed, stamens 4, enclosed, inserted halfway on the corolla tube, anthers glabrous, obovoid, rounded at one end, cuspidate at the other end, ovary oblong, style cylindrical, placenta 4, stigma capitate; capsules oblong, 1 cm long, seeds numerous, dust-like. Native of the semi-desert areas of temperate eastern Asia, growing on sandy or exposed hillside, parasitic on different species of *Artemisia* L.

Pedaliaceae: Pedalium Family

Sesamum orientale L. — Sesame (Syn. *S. indicum* L.)

Zhi-ma=Chih-ma (枳麻, Thorny Hemp); *Zhi-ma=Chih-ma* (芝麻, Fragrant Hemp); *Hu-ma=Hu-ma* (胡麻, Tartary Hemp). Seed; used in various pastries.

Zhi-ma-jiang=Chih-ma-chiang (芝麻醬, Sesame Butter). Seed; roasted, ground like peanut butter, with good aroma; bottled material available in American Chinese groceries.

Zhi-ma-you=Chih-ma-yiu (芝麻油, Sesame Oil); seed roasted, ground into a paste, then thoroughly mixed with boiling water in a container; with gentle, continued shaking, the aromatic oil skimmed off used in salad and cooking; bottled product imported from China available in American Chinese stores.

Robust annual herb 1 m or more high, stiffly short pubescent and glandular throughout, stem 8–12 mm across, tetragonous, simple or rarely with few branches; leaves varying with their position, the lower ones ovate-oblong, 10–20 cm long, 7–10 cm wide, occasionally deeply lobed or trifoliolate, remotely dentate, acuminate, base rounded, those on the middle and above becoming smaller gradually, ovate-oblong or lanceolate, subentire, petioles stout, hairy; flowers white or purplish-white, solitary, axillary to normal or smaller leaves, pedicels 2.5–3 mm long, with a linear hairy basal bract subtending a globose glandular hollow structure, calyx consisting of 5 lanceolate villose sepals 3–4 mm long, persistent, corolla villose, 3 cm long, the basal 3–4 mm of the tube narrow, the remaining 2 cm subcampanulate, 8–15 mm across the throat, limb bilabiate, the upper lobes roundish, the lower lip longer, stamens 4, didymous, the fifth one represented by a staminode, ovary superior, conical, 2-carpellate, style filiform, stigma 2-fid, ovules numerous, placenta axile; capsules tetragonous-oblong, erect, 2.5–3 cm long, 1 cm across, scabrid-velutinous, apex rostrate, dehiscing loculicidal; seeds compressed ovoid, released from the terminal pores of the capsule, testa white or black according to the cultivars. Native of the Old World tropics, recorded in the first known Chinese herbal, extensively cultivated throughout the country, for food, oil and medicine; straw used for fuel.

Acanthaceae: Acanthus Family

Dicliptera chinensis (Vahl) Nees

Gou-gan-cai=Kou-kan-ts'ai (狗肝菜, Dog Liver Herb). Fresh leaves; used in soup; young shoots used as potherb.

Perennial herbs 30–50 cm high, erect or ascending, pilose throughout, branchlets articulate, internodes 3–7 cm long; leaves opposite, unequal, one larger than the other of the same node, ovate or elliptic, 2.5–5 cm long, 1–3 cm wide, shortly acuminate, base obtuse or acute, entire; flowers pink-purple, small, sessile, in axillary peduculate cymes subtended by 2 subulate stiff common bracts 5–6 mm long, bracts 2 to each cyme, foliaceous, opposite, unequal, obovate or suborbicular, 10–12 mm long, rotundate and apiculate at the apex, hairy, cymes each 2-flowered, bracteoles oblong, 4 mm long, calyx 5-lobed almost to the base, sepals lanceolate, 4 mm long, corolla bilabiate, 12–14 mm long, hairy, tube slender, lips reflexed, the upper one larger, shallowly notched, lower lip 3-lobed, stamens 2, exserted, filaments hairy, inserted to the throat of the corolla, anthers 2-locular, one theca below the other; disk cupular, ovary superior, 2-locular, cell 2-ovulate, style filiform, exerted, stigma 2-lobed, the lower lobe recurved; capsules clavate, stipitate, 5 mm long, hairy, placenta loosened from the base on dehiscence; seeds discoid, 1.5 mm in diameter, brown, papillose. Native of tropical Asia, common near villages; cultivated in Hong Kong for meeting local market needs.

Hygrophila lancea (Thunberg) Miquel (Syn. *Justicia lancea* Thunberg).

Shui-suo-yi=Shui-suo-yi (水蓑衣, Water Straw-cloak). Young shoots; gathered and used in Taiwan before 1950.

Hygrophilous perennial herbs 60 cm high, branchlets glabrous, nodes with few setae, subquadrangular; leaves lanceolate, 5–10 cm long, 5–15 mm wide, obtuse, base attenuate, entire, glabrous, occasionally sparsely pilose on the major veins beneath; flowers pale purple, sessile, fasciculate, axillary, bracts ovate-lanceolate, 1 cm long, apex subulate, bracteoles linear, 5 mm long, calyx consisting of lanceolate sepals, 15 mm long, subulate at the apex, corolla 10–12 mm long, puberulous without, ventricose at the apex, limb bilabiate, upper lip erect, 2-toothed, lower lip 3-lobed, rugose, stamens 4, didymous, exserted, anthers 2-celled, disk obscure, ovary superior, oblong, ovules many, style filiform, pilose near the base; capsules oblong-cylindrical, 12–14 mm long, light brown, retinacular curved, acute; seeds many, discoid, 1 mm in diameter, brown, glandular, margin white, mucilaginous.

Peristrophe roxburghiana (Schultes) Bremekamp — RED SILK-THREAD

Hong-si-xian=Hung-ssu-hsian (紅絲線, Red Silk-thread); *Guan-yin-cao=Kuan-yin-ts'ao* (觀音草, Goddess-of-Mercy Herb). Leafy shoot; with or without flowers, used for making soup or for tea, available in village markets in Hong Kong.

Perennial herbs 20–40 cm high, pilose throughout; leaves opposite, ovate, 4–9 cm long, 2–4 cm wide, subentire, acuminate, base cuneate; flowers rose-purple, bracts foliaceous, ovate, 2 subtending a small fascicle, 2 cm long, 10–12 cm wide, base subcordate, apex acute, villose, peduncle 3–4 mm long, bracteoles 2, lanceolate, 6 mm long, 1.5 mm wide across the base, calyx 5-parted to the base, sepals lanceolate, 4–6 mm long, pilose, corolla bilabiate, tube slender, 1.8 cm long, limb 2-lipped, upper lip rounded, 1 cm in diameter, lower lip narrower, the apex shallowly 3-lobed, pilose without, stamens 2, exserted, filaments subglabrous, anthers 2-celled, connective narrow, thecae in a straight line, 1 high, the other low, 3 mm long, ovary ovoid, style filiform, stigma 2-lobed, the lobes recurved; capsules clavate, stipitate, 1 cm long, pilose, acute; seeds discoid, tuberculate, almost black.

Native of southern China and adjacent Himalayan region, growing in damp shady ravines, now cultivated in wet vegetable beds after anthesis, the corolla hanging on the filiform red style, hence the vernacular name "Red Silk-thread".

Plantaginaceae: Plantain Family

Plantago major L. — Common Plantain

Che-qian-cao=Ch'e-chien-ts'ao (車前草, Cart Trail Herb). Young plants and/or flowering and fruiting spikes, gathered for tea, used fresh or dried, for elderly persons especially, a mild diuretic agent, enhancing the elimination of the water of senior citizen with high blood pressure. Young plants chopped with pork and used as filling of *jiao-zi* (餃子) in northeastern China.

Perennial acaulescent herbs 15–40 cm high, with short button-like erect caudex and strong fibrous roots; leaves all basal, forming a rosette 15–40 cm in diameter, upright in crowded grassy area, laminas ovate-oblong or suborbicular, 5–15 cm long, 3–12 cm wide, obtuse or rounded, base subrotundate, suddenly cuneate and decurrent into the petioles 5–15 cm long, with 5–7 primary palmate nerves, margin wavy and remotely crenate, stiples membranous, hidden at the ground level; flowers small, arranged in spicate cluster terminal to an axillary scape 8–25 cm long, protogynous, bracts ovate, acute, green, rachis 5–16 cm long, sepals 4, persistent, corolla tubular, membranous, 4-lobed, the lobes reflexed, ovate, stamens 4, epipetalous alternate with the corolla lobes, anthers oblong, mucronate, ovary unilocular, ovules numerous, style slender, stigmatic surface linear, mucilaginous-hairy, covering eight-tenths of the style; capsules ellipsoid, 3 mm long, mucronate with the conical base of the persistent style, dehiscing circumscissilely; seeds many, black, the testa forming a mucilaginous layer in moisture.

Cosmopolitan, weedy in gardens and along paths, hence the Chinese name "Cart Trail Herb". A companion species *P. asiatica* L. growing in damp areas, especially along streams and paddies, recognised by the elliptic leaves wavy and erose along the margin, pubescent spikes, and sepals purplish-green, carinate in the middle, membranous along the margin, occasionally mixed with the common species in market products.

Rubiaceae: Madder Family

Canthium horridum Blume — Thorny Bramble

Zhu-du-mu=Chu-tu-mu (豬肚木, Hog Stomach Bush). Fruit; used in South China.

Thorny shrubs 2–3 m high, branchlets slender, scabrid, some specialized into thorns 0.4–3 cm long; leaves opposite, ovate-elliptic, 2–3 (–5) cm long, 1–3 cm wide, acuminate, base rounded and abruptly acute, entire, sparsely scabrid on major nerves, petioles short, 2–3 mm long, hairy, stipules deltoid, yellow-strigose; flowers small, white, solitary or 2 in an axillary fascicle, pedicels 2–3 mm long, glabrous, bearing subterminal spathaceous and ciliate prophylls, hypanthia obconical or subcampanulate, 2 mm long, calyx collar-like, 5-toothed, ciliate, corolla 6 mm long, tube 3 mm long, limb 5-lobed, lobes lanceolate-caudate, reflexed, white and turning yellow with age, stamens 5, inserted on the throat, filaments broadened at the base, anthers sagittate, dorsifixed, ovary inferior, styles filiform, stigma capitate, globose; fruits drupaceous, 1.5–2.5 cm in diameter, ripened yellow. Native of southeastern Asia, occurring in Yunnan, Guangxi, Guangdong and Hainan in China; fruits used by various ethnic groups, not available in the market, becoming popular in cities.

Coffea arabica L. — Coffee, Arabian Coffee

Ka-fei=K′a-fei (咖啡, Coffee); *Xiao-li-ka-fei=Hsiao-li-K′a-fei* (小粒咖啡, Small-grain Coffee). Seed; roasted, ground, brewed for beverage; introduced to Guangzhou (Canton) in the 1910s, fruiting herbarium specimens dated 1917, after 1950 governmental institutions established for greater production; imported product used in hotels catering for international tourists. Entering the twenty-first century, many people in China use coffee.

Glabrous evergreen shrubs or small trees up to 5 m high when unpruned, flowering branchlets green, internodes 3–4 cm long, 3–4 mm in diameter; leaves shiny green, subcoriaceous, elliptic (5–) 10–15 (–20) cm long, (1.5–) 4–6 (–7.5) cm wide, acuminate, base acute, entire, undulate, stipules interpetiolar, deltoid, acuminate; flowers white, fragrant, 2–4 in a subsessile cymose axillary cluster, pedicels short, up to 4 mm long in fruit, hypanthia subglobose, calyx represented by a narrow collar with 5 small teeth, corolla salverform, tube 6–10 mm long, limbs 1.5–2 cm across, deeply 5-lobed, the lobes oblong, 1 cm long, 2.5–3 mm across, acute, stamens 5, white, completely exserted, inserted at the throat between corolla lobes, anthers linear-sagittate, 8 mm long, dorsifixed, ovary inferior, 1 mm across, 2-locular, cell 1-ovulate, style filiform, bifid, stigmas 2, lining the adaxial surfaces of the stylar lobes; fruits drupaceous, oblong, 10–15 mm long, 8–10 mm across, yellow changing to crimson at maturity, crowned with the persistent calyx, mesocarp (pulp) yellow, fleshy, endocarp fibrous (parchment); seeds oblong-ellipsoid, 8–12 mm long, endosperm copious, corneous, embryo small; approximately 1,000 seeds in 450 g (1 lb), the best aroma among commercial coffee products.

Native of Ethiopia, an upland species growing in the forests at altitudes of 1,500–2,000 m. The indigenous African people gathered the mature fruits, dried and chewed them since time immemorial. History relates that in Arabia, a beverage was prepared with fermented sweet pulp. Coffee prepared from dried, roasted and ground seeds for a stimulating or refreshing beverage was recorded in the middle of the fifteenth century in Arabia. Coffee reached Cairo in 1510, Venice in 1616, and to England in 1650. In eastern Asia, coffee was first planted by the Dutch in Java in 1690, with seeds from Yemen. In the New World, it reached Surinam in 1718, Cayenne Island on the coast of French Guina in 1722, to Jamaica via Kew Gardens in 1730. Coffee was introduced to China in the early 1900s. Herbarium specimens indicate that in 1917, coffee plants have flowered and fruited in the campus of Lingnan University at Canton (Guangzhou), for educational purposes. It was only after the 1950s that coffee plantations were established by governmental institutions in Yunnan and Hainan Island.

This species is the major source of commercial coffee bean. Two varieties are recognized from the forests of Ethiopia.

1. var. **arabica** (Syn. *C. arabica* var. *typica* Cramer)

Vigorous growing plants with sturdy habit, fruiting branchlets horizontal or drooping; leaves narrow-elliptic, brown-tipped when young. Many cultivars have been selected, with best examples:

cv. 'Blue Mountain' — developed in Jamaica and much grown in the West Indies and in Kenya.

cv. 'Nyasa' — common in Nyasaland and Uganda.

cv. 'Kent's' — originated in Mysore and widely planted in India.

cv. 'Mocha Extra' — originated in Arabia.

2. var. **bourbon** Choussy

High-yielding plants with slender habit, fruiting branchlets borne stiffly at an acute angle, often weighed down by fruits; leaves broad elliptic, green-tipped when young. Many cultivars selected for local plantings, the outstanding ones:

cv. 'Mundo Nove' — in Brazil.

cv. 'Cuturra' — in Colombia.

cv. 'Ground Bourbon' in Tonkin, North Vietnam.

Coffea canephora Pierre ex Froehner — ROBUSTA COFFEE (Syn. *C. robusta* Linden)
Zhong-li-ka-fei=Chung-li-k'a-fei (中粒咖啡, Intermediate Grain Coffee). Seed; for beverage, inferior in aroma, high productivity, used in coffee blends.

Robust glabrous shrubs or small trees 4–10 m high, fruiting branchlets bending; leaves broad-elliptic or obovate, 9–17 (–30) cm long, 5–7 (–15) cm wide, corrugated or undulating, shortly acuminate, base rounded or broadly cuneate, petioles stout, 1–2 cm long, stipules interpetiolar, triangular, acute, 5 mm long; flowers white, fragrant, subsessile, 3–6 in axillary cymose clusters, pedicels short, prophylla sub-basal, hypanthia subglobose, calyx very short, corolla tube 1 cm long, limbs 2.5–3 cm across, 6- or 7-lobed, lobes 1–1.5 cm long, 5–7 cm wide, stamens 6, anthers 1 cm long, sagittate, apiculate, basifixed, style filiform, bifid at the distal end, stylar arms 5 mm long, ovary inferior, 2-locular; fruits drupaceous, subglobose, 1–1.5 cm long, 1.2 cm across, crimson when ripe, drying black, held on the tree until harvested, in heavy clusters of 20 or more drupes to one node; seeds ellipsoid, 7–9 mm long, pressed together by grooved flattened surface, approximately 1,500 seeds in 450 g (1 lb) of coffee bean.

Native of tropical Africa. People of Uganda grew this species for chewing long before its discovery by European explorers. The fresh drupes were parboiled and dried for keeping. The twin seeds of a drupe signifying brotherhood are used in their religious ceremonies. Cultivars selected from this species are important in the coffee plantation of tropical Africa and Asia, for they are vigorous and nematode resistant. They are suited for low altitude planting. Plants were first introduced from Belgium Congo to Brussels in the late 1890s, and were distributed to Java and Singapore between 1898 and 1900. Flowering and fruiting specimens were collected from private gardens in Nodoa, Hainan Island in April 1929. It is an important stock for the manufacture of instant coffee.

Coffea liberica Bull ex Hiern — Liberian Coffee (Syn. *C. excelsa* A. Chevalier)
Da-li-ka-fei=Ta-li-k'a-fei (大粒咖啡, Large Grain Coffee). Seed; for beverage, with a bitter flavor, generally used as a filler with other coffee.

Vigorously growing evergreen shrubs or tree, 5–17 m high, fruiting branchlets bearing the largest flowers and fruits of the cultivated coffees; leaves broadly oblong-elliptic or obovate-elliptic, 15–30 cm long, 5–15 cm wide, acuminate, base cuneate, petioles 1–2 cm long; flowers white, fragrant, star-like, in axillary fascicles of 1–3 cymes, each consisting of 1–4 flowers, opening at irregular intervals, self-sterile, hypanthia subglobose, calyx small, corolla salverform, tubes 1.5–2 cm long, lobes spreading, 1.5–2 cm long, stamens 6–9, anthers 1 cm long, dorsifixed, the connectives dark brown, style 2.5 cm long, bifid, stigma 5 mm long; fruits drupaceous, maturing one year after flower, subglobose-oblong, 19–21 mm long, 15–17 mm across, streaked red at maturity, drying black, approximately 800 seeds in 450 g (1 lb) of coffee beans.

Native of the hot humid lowland forests of Liberia, introduced to England in the 1870s, hence it was introduced to southeastern Asia, to Sri Lanka in 1873 and Malaya in 1875. It flowered and fruited in Hainan Island in April 1929. It is susceptible to fungal and insect attacks, and it often produces spongy seeds. It is the least important of the cultivated species of coffee, supplying 1 percent of the world's total coffee production.

Gardenia angusta (L.) Merrill — Common Gardenia (Syn. *G. thunbergii* L.) (Figure 52)
Shui-heng-zhi=Shui-heng chih (水橫枝, Over-water Branches); *Zhi-zi=Chih-tzu* (枝子, Gardenia Fruit); *Huang-zhi=Huang-chih* (黃枝, Yellow Gardenia). Fruit; from which

an orange dye is obtained for coloring bean curd in Guangzhou; animal assay indicating that the dyed yellow bean curd enhance the life span of mice; tea for people suffering from acute hepatitis.

Completely glabrous evergreen shrubs 1–3 m high; leaves opposite, elliptic or obovate-elliptic, 3–10 cm long, 1–3 cm wide, acuminate or obtuse, cuneate at base, coriaceous, petioles 2–5 mm long, stipules tubular, obliquely opened on one side, green, turning brown with age; flowers very fragrant, white, turning yellow, solitary, terminal to short branchlets bearing 2 pairs of leaves, pedicels 3–6 mm long, covered by tubular stipule-like bracts, hypanthia turbinate, 6-ridged, 10–12 mm long, 4 mm across the top, sepals 6, 10–17 mm long, 1 mm wide, obtuse, persistent, corolla salverform, tubes 3 cm long, limbs 3–6 cm across, deeply 6-lobed, lobes contorted in bud, widespread at anthesis, obovate-cuneate, 2–3 cm long, 8–12 mm wide, stamens 6, exserted, sessile, dorsifixed, inserted at the throat between corolla lobes, anthers linear, 12–14 mm long, the base hidden in the corolla tube, ovary inferior, disk annular, style slender, stigma clavate, 4–5 mm across, striate, with 2 or 3 stigmatic lines running down the sides; fruits berry-like, 10–20 mm long, 10–15 mm across the middle, ellipsoid or subglobose, orange when ripe, placenta juicy; seeds discoid, 5 mm long, 2 mm wide, testa tuberculate.

Native of southern China, common along the streams of Hong Kong, flowering in early summer, giving refreshing fragrance in the air, hence the ancient fishing community named their village "*Hong* (=fragrant) *Kong* (=harbor)"; the ethnic name, "*Shui-heng-zhi*", descibes the growing condition, with the branches arching over water. Three garden natural forms are observed in Hong Kong, all with double flowers and bearing no fertile stamens.

cv. 'Jasminoides' (Syn. *Gardenia jasminoides* Ellis)
  Bai-chan=Bai-ch'an (白蟬, White Cicada); *Zhi-zi-hua=Chih-tzu-hua* (梔子花, Garden Gardenia).

  Leaves elliptic, oblong-elliptic, rarely obovate-elliptic, 5–12 cm long, 2.5–4.5 cm wide, acuminate, the acumens 5–10 mm long; flowers doubled, corolla 6.5–8 cm across.

cv. 'Rotundifolia'
  Yuan-ye-zhi-zi=Yuan-yeh-chih-tzu (圓葉梔子, Round-leaved Gardenia).

  Leaves obovate-rotundate, or ovate, 3–6.3 cm long, 2–3.5 cm wide, obtuse or rounded at the apex, the base acute; flowers doubled, corolla 4.5–5 cm across.

cv. 'Silver Variegated'
  Bian-ye-xiao-zhi-zi=Pian-yeh-hsiao-chih-tzu (變葉小梔子, Small Variegated Gardenia)

Leaves variegated, linear-lanceolate, 1.5–4 cm long, 3–10 cm wide, acute, base acuminate; flowers doubled, 6.5 cm across.

Ixora chinensis Lamarck — CHINESE IXORA

Long-chuan-hua=Lung-chuan-hua (龍船花, Dragon Boat Flower); *Shan-dan=Shan-tan* (山丹, Mountain Cinnabar); *Wu-yue-hua=Wu-yue-hua* (五月花, May Flower). Flowering clusters; fresh flowers gathered from the hillsides in early summer, boiled with pork chops for soup, a seasonal dish of health food in rural Guangdong.

Evergreen shrubs 0.5–1.5 m high, branchlets glabrous; leaves opposite, oblong or obovate-elliptic, 6–13 cm long, 1.5–3.5 cm wide, obtuse, base cuneate, occasionally subcordate-rounded, petioles 2–5 mm long, stipules interpetiolar, collar-like; flowers cinnabar-red, 20–50 in crowded terminal cymose, corymbs 3–8 cm across, pedicels 1–2 mm long, glabrous, hypanthia turbinate, calyx minutely 4-lobed, corolla salverform, tubes slender, 3–3.5 cm long, limbs 12–15 mm across, deeply 4-lobed, lobes obovate-orbicular, 5–6 mm long, stamens 4, subsessile, inserted on the throat between corolla lobes, anthers linear-sagittate, 2.5 mm long, basifixed, ovary inferior, style filiform, exserted, apex bifid, stigmatic surface marginal; fruits berry-like, subglobose, 7–8 mm in diameter, red turning black.

Native of southern China, common on the hillsides, beginning to bloom in the fifth month of the lunar calender, with the traditional Dragon Boat Festival falling on the fifth day, hence the vernacular name "Dragon Boat Flower". Much cultivated for landscape purposes in Hong Kong.

Morinda citrifolia L. — INDIAN MULBERRY, AWL TREE, MORINDA

Hai-ba-ji=Hai-pa-chih (海巴戟, Seashore Morinda). Young leaves; used as vegetable; ripe fruits, eaten in Hainan Island.

Stout evergreen shrubs or small trees 2–3 m high, branchlets tetragonous, with prominent joints; leaves opposite, oblong or broadly elliptic, occasionally ovate, 10–30 cm long, 5–14 cm wide, glabrous, acute or shortly acuminate, base cuneate, or rounded, petioles 5–12 mm long, stipules interpetiolar, broadly deltoid, 5–12 mm long, free, obtuse and apiculate, membranous; flowers white, sessile, heterostylous, 20–50 united into a tight terminal capitulum, 15–25 mm across, on stalks 1–2 cm long, 8–14 mm thick, due to simultaneous development of an axillary bud the final appearance as opposite to a leaf on the same node, hypanthia jointed at the base, basal bracts 0–2, foliaceous, 6–10 mm long, adnate to hypanthia, calyx truncate, corolla salverform, tube 15–17 mm long, limbs 2 cm across, 6-lobed, lobes imbricate in bud, lanceolate, 1–2 cm long, throat pilose, stamens 5, inserted to the middle of the corolla tube, anthers oblong, dorsifixed, partially exserted, disk annular, ovary 4-locular, cell 1-ovulate, style filiform, bifid at the apex, stylar branches oblong, stigmas lining the adaxial surface; fruits syncarpous, subglobose, 2.5–4 cm across, pulp juicy, each segment consisting of 4 pyrenes, endocarp cartilaginous; pyrenes obovoid, pointed to the distal end, the broad top portion occupied by a large air chamber, a special adaptation for seed dispersal by floating in sea water, the lower portion compressed-attenuate, 7 mm long, 5 mm across; seeds discoid, winged at the end. Spontaneous on the seashores in tropical southeastern Asia, the western Pacific islands and Australia, occurring in Hong Kong, Hainan and Taiwan; in Taiwan used for extracting a yellow color from the root before 1950.

Morinda officinalis How — CHINESE HERBAL MORINDA

Ji-yan-teng=Chi-yen-t'eng (雞眼藤, Chicken-eye Vine); *Ba-ji-tian=Pa-chi-t'ien* (巴戟天, Chinese Morinda); *Ji-chang-feng=Chi-ch'ang-feng* (雞腸楓, Chicken Intestine Feng). Root with intermittent enlarged sections of the cordex; fresh or dried, cooked with pork for a broth taken as a health food; dried material in one-pound package available in American Chinese stores.

Evergreen lianas with the cortex of root enlarged intermittently into flesh sections 1–2 cm in diameter, grayish-yellow outside, roughened with vertical striae and transverse cracks, branchlets pilose; leaves opposite, oblong, 6–10 (–14) cm long, 2.5–4.5 (–6) cm wide, acute or acuminate, base obtuse or rounded, rarely cuneate, sparsely pilose or glabrescent above, strigose on the nerves beneath, petioles 4–8 mm long, stipules tubular, truncate; flowers white, sessile, 2–10 jointed together forming a subglobose head (capitulum) 5–9 mm across, enlarged after anthesis, peduncles 3–10 mm long, hypanthia hemispherical, coalescent at the base at anthesis, calyx collar-like, truncate, corolla funnelform, tubes short, limbs deeply 4-lobed, densely villose-lanate inside, shoulder cornute and changing to glandular cristate at anthesis, stamens partially exserted, disk annular, ovary 4-locular, cell 1-ovulate, style very short, apex bifid, stigma obscure; fruits syncarpous, yellow-red, subglobose, 6–11 mm across, fleshy, containing many pyrenes, endocarp woody, rugose, irregularly pitted; pyrenes compressed-ovoid; seeds oblong, wingless.

Native to the Nanling Range in southern China, cultivated by the ethnic people living in the mountains, by-cuttings or division of rootstocks. The roots are harvested 4–5 years after planting, washed, spread in the sun to dry, individually beat with a wooden club when 60–70 percent dry, returned to the sun until completely dried. The final product appears grayish-yellow, rough, with vertical furrows and transverse grooves, often broken into 1–3 cm sections holding together by the central xylum cord, having the appearance of the intestine of a chicken, hence the vernacular name *"Ji-chang-feng"* (雞腸楓, Chicken Intestine Feng).

Mussaenda pubescens Aiton f.

Yu-ye-jin-hua=Yu-yeh-chin-hua, (玉葉金花, Jade-leaf Gold-flower); *Bai-zhi-shan=Pai-chih-shan* (白紙扇, White-paper Fan); *Bai-hua-cha=Pai-hua-ch'a* (百花茶, Hundred-flowered Tea). Leaves; gathered, dried, used fresh in summer for preparing a cooling tea; saved and used as a tea substitute by the ethnic groups living in the mountains of Guangdong and Hainan Island.

Evergreen shrubs with some sprawling twining branchlets, densely pilose throughout; leaves opposite, occasionally whorled, ovate-elliptic, 5–8 cm long, 1–3 cm wide, acuminate, base acute or obtuse, entire, petioles 3–8 mm long; flowers golden-yellow, subsessile, in dense terminal corymbose clusters, hypanthia obconic, 2–3 mm long, sepals 5, lanceolate-subulate, 4 mm long, an early flowering unit often with a specialized sepal, enlarged, becoming petioloid and foliaceous, white, ovate-elliptic, 2.5–4.5 cm long, 1.5–3 cm wide, acute or obtuse, palmately 3- to 5-nerved, corolla tubular-salverform, tubes 1.5–2 cm long, uniformly barbate inside, limbs 5–6 mm across, 5-lobed, lobes ovate-deltoid, glandularly papillose above, stamens 5, included, ovary inferior, style glabrous, deeply 2-fid, stigma punctiform; fruits oblong-globose, 8–10 mm long, 6–8 mm across, fleshy, black, with a discoid apical scar; seeds numerous, angular, foveolate.

Paederia scandens (Loureiro) Merrill — PAEDERIA (Syn. *P. foetida* Thunberg, non L., 1767)

Ji-shi-teng=Chi-shih-t'eng (雞屎藤, Chicken Droppings Creeper); *Ban-jiu-fan=Pan-chiu-fan* (斑鳩飯, Dove's Food); *Chou-teng=Ch'ou-t'eng* (臭藤, Stinking Creeper). Very tender young leafy shoots; gathered and used for pastry, tea or soup, agreeable flavor developed by cooking; a favorite wild plant food in Hong Kong.

Large perennial twiners climbing over thickets or fences up to 5 m high, much branched, the branchlets flexuous, any tender part foetid when bruised; leaves opposite or occasionally 3, ovate, ovate-oblong, and lanceolate toward the end of the branchlets, 3–10 (–15) cm long, 2–5 (–7) cm wide, acute or acuminate, base rounded or acute, petioles 1–3 (–7) cm long, stipules intrapetiolar, triangular, deciduous; flowers small, sessile, polygamous, in axillary cymose clusters as well as in terminal cymose panicles, hypanthia obconical, calyx 5-lobed, lobes deltoid, acute, persistent, corolla tubular-campanulate, 12 mm long, densely silvery, velvety-tomentose, purple-red inside, densely villose, limbs spreading, 5-lobed, stamens 5, filaments short, anthers oblong, basifixed, disk hemispherical, ovary 2- (often 1-) locular, cell 1-ovulate, style branched, stigmas linear, exserted; fruits pea-sized, spherical, shiny, ripe light brown, the outer cover thin and brittle, free from the remaining enclosed portion of the fruit, pyrenes 2 (often cotyledons), hypocotyl well developed.

Widespread in eastern and southeastern Asia; in warm regions of China, extending from the Yangtze River southward to Guangdong, Hong Kong, Taiwan and Hainan, often weedy, climbing over thickets and fences, even over mangroves on Hong Kong seashore.

Paederia stenobotrya Merrill — WHITE PAEDERIA (Syn. *P. scandens* [Loureiro] Merrill var. *tomentosa* auctt., non Handel-Mazzetti)

Bai-ji-shi-teng=Pai-chi-shih-t'eng (白雞屎藤, White Chicken Droppings Creeper). Root; ground with soaked soybean for a milk, cooked and given especially to people with jaundice; leafy shoots boiled with sugar for a tea.

Scandent perennials densely white hirsute-villose throughout, twining over tall grasses or shrubs; leaves opposite, ovate-oblong, 6–13 cm long, 3–7.5 cm wide, acuminate, acute, base cordate, hispid above, white lanate beneath, margin erose, petioles 1–5 cm long, hirsute, stipules ovate-deltoid, acute and apiculate; flowers light lavender, in interrupted cymose panicles, sessile, hypanthia obconical, calyx 5-lobed, lobes triangular-ovate, acute, 1.5 mm long, corolla campanulate, tube 5–6 mm long, 2.5–4 mm across, abruptly narrowed, silvery-velvety without, sparsely pilose within, limbs 5-lobed, stamens 5, anthers oblong, introrse, dorsifixed, ovary inferior, style branches filiform, stigmas linear; fruits globose, pubescent, straw-colored, 4–6 mm in diameter. Native of the warm areas in China, growing in secondary vegetation on the hillside, the dense tomentum gives a grayish-white appearance to the plant, hence the vernacular name *"Bai-ji-shi-teng"*.

Vangueria edulis Vahl — VOA-VANGA (the ethnic name of Madagascar); TAMARIND OF THE INDIES (Syn. *V. madagascariensis* F. Gmelin).

Wan-jia-guo=Wan-chia-kuo (萬加果, Voa-vanga). Fruit, introduced into few botanical gardens, little known in China.

Deciduous shrubs 2–3 m high, branchlets glabrous, 3–4 mm in diameter; leaves opposite, ovate or broadly elliptic, 8–20 cm long, 5–10 cm wide, obtuse or shortly acuminate, base rounded or acute, petioles 1–2 cm long, stipules interpetiolar, tubular; flowers greenish-white, in loose pubescent axillary scorpioid clusters 4–5 cm across, 5-merous, pedicels filiform, 1–2 mm long, hypanthia obconic, calyx truncate, 5-toothed, corolla 7 mm long, tube campanulate, 4 mm long, densely villose, limbs 5-lobed, stamens exserted, filaments buried in dense hairs, anthers oblong, 1 mm long, apiculate, ovary

5-locular, cell 1-ovulate, style columnar, stigma capitate; fruits drupaceous, consisting of 5 pyrenes, compressed globose, 4 cm across; pyrenes oblong, 15 mm long, 9 mm across. Native of Madagascar, introduced to Europe in 1809, thence to tropical botanical gardens; very rare in China.

Caprifoliaceae: Honeysuckle Family

Abelia engleriana (Graebner) Rehder

Shen-xian-cai=Shen-hsian-ts'ai (神仙菜, Vegetable-of-the-Immortal); *Shen-xian-ye-zi=Shen-hsian-yeh-tzu* (神仙葉子, Immortal's Leaf); *Duan-zhi-liu-dao-mu=Tuan-chih-liu-tao-mu* (短枝六道木, Short-branched Abelia). Leaves and leafy shoots; used for food by people in Shaanxi.

Deciduous shrubs 1–2.5 m high, with spreading slender branches, branchlets puberulous; leaves ovate, elliptic-ovate, or elliptic-lanceolate, 1.5–4 cm long, 0.5–1.5 cm wide, acuminate or acute, base cuneate, serrate, both surfaces pilose; flowers rosy-pink, one or several terminal to short branchlets, sepals oblong-elliptic, 8 mm long, persistent, corolla tubular below the middle, campanulate above, base gibbous, stamens 4, inserted to the middle of the corolla tube, anthers versatile, ovary inferior, 3-locular, 1-cell fertile, 1-ovulate, fruits drupaceous, achenelike, crowned by the persistent sepals. Native to central and western China; introduced into American gardens in 1908 by the Arnold Arboretum of Harvard University.

Lonicera henryi Hemsley — Central China Honeysuckle

Xi-ye-ren-dong=Hsi-yeh-jen-tung (細葉忍冬, Small-leaved Honeysuckle). Flowering shoots and flowers; used for tea.

Partially evergreen twining or prostrating shrubs, branchlets densely strigose; leaves oblong-lanceolate or lanceolate, 4–8 cm long, 1.5–4.5 cm wide, acute or acuminate, base truncate, rounded or subcordate, glabrous and pubescent on the midrib of both surfaces, ciliate; flowers yellow and purplish-red outside, sessile, in axillary pairs and also crowded at the shoot apex in racemose clusters, bracts linear-subulate, 2–3 mm long, ciliate, hypanthia glabrous, glaucous, calyx 5-lobed, lobes ovate-triangular, 1 mm long, corolla bilabiate, tube ventricose, 1.5 cm long, glabrous outside, pilose inside, limb bilabiate, upper lip 1-lobed, linear, 5 mm long, lower lip 4-lobed connate, 5 mm wide, stamens 5, inserted at the throat, anthers oblong, exserted, ovary inferior, style filiform, hirsute below the middle, stigma capitate, exserted; fruit pea-sized, bluish-black. Native to the mountains of central and western China, introduced to American gardens in 1908, moderately hardy.

Lonicera japonica Thunberg — JAPANESE HONEYSUCKLE (Figure 53)

Jin-yin-hua=Chin-yin-hua (金銀花, Gold-silver Flower). Leafy shoots and flower buds; gathered just before anthesis, dried material imported from China available in American Chinese groceries, used as a tea substitute.

Half evergreen lianas much cultivated throughout China, branchlets hispid; leaves ovate-oblong, 3–8 cm long, 1–3.5 cm wide, acute or short-acuminate, base rounded, or subcordate, pubescent on both surfaces, ciliate, petioles 3–6 mm long; flowers sessile, in axillary pairs, white, turning yellow, protogynous, opening slightly and exposing the stigma in late afternoon, very fragrant, apparently pollinated by moth, kept wide open and bilabiate in the morning, becoming yellow in the afternoon, bracts foliaceous, bracteoles orbicular, 1 mm across, ciliate, hypanthia 1 mm long, calyx 5-lobed, lobes 1 mm long, corolla tube slender, 1.5–2 cm long, hairy inside and outside, limbs 5-lobed, bilabiate, upper lip 1-lobed, linear, 1.5.–2 cm long, lower lip 4-lobed, connate to the middle, 6–7 mm across, stamens 5, inserted on the throat, anthers oblong, exserted, style filiform, glabrous; fruits pea-sized, shiny black, juicy, staying on the plant over winter in Boston. Native to eastern Asia, introduced into Western gardens in 1806, escaped and became weedy in eastern North America.

Sambucus formosana Nakai — TAIWAN ELDER (Syn. *Ebulum formosana* [Nakai] Nakai)

Mo-gu-xiao=Mo-ku-hsiao (冇骨消, No Bone Melt). Fruit; recorded to be edible in Taiwan before 1950.

Suffrutescent herbs 2–3 m high, branchlets glabrous, 4–6 mm across; leaves opposite, pinnately compound, 12–30 cm long, leaflets alternate, lanceolate, 10–15 cm long, 2–5 cm wide, acuminate, base oblique, rounded serrate, papery, glabrous; flowers white, turning yellowish, in large terminal compound cymose corymbs 15–20 cm across, with prominent scarlet red sessile or stipitate cupular or campanulate glands at the juncture of secondary or tertiary branching, hypanthia obconical, 0.8 mm long, calyx 5-lobed, lobes deltoid, corolla rotate, 2–3 mm across, limbs 5-lobed, stamens 5, alternate with the corolla lobes, inserted anthers subglobose, ovary inferior, style obscure, stigma capitate; fruits berry-like, orange, ovoid, 2.5 mm across, juicy; pyrenes 3, ovoid-subtrigonous. Endemic to Taiwan, common in thickets.

Sambucus williamsii Hance — CHINESE BLACK ELDER

Jie-gu-mu=Chieh-ku-mu (接骨木, Bone-healing Wood); *Bao-gun=Pao-kun* (寶棍, Precious Stick). Young leaves; gathered and boiled, washed, seasoned, eaten as vegetable in Inner Mongolia.

Glabrous deciduous shrubs 3–6 m high, branchlets striate-sulcate, with copious yellow pith; leaves opposite, pinnate compound, leaflets (3–) 5–7 (–11), broad-elliptic or elliptic-oblong, 5–12 (–16) cm long, 2–4 cm wide, acuminate-caudate, base rounded, acute, or oblique-acute, coarsely serrate, flowers white, turning pale-yellow, numerous, in terminal subglobose panicles 7 cm across, hypanthia obconical, calyx 5-lobed, lobes deltoid-lanceolate, corolla rotate, 4 mm across, 5-lobed, falling together, stamens 5, anthers oblong, ovary inferior, style very short, stigmas capitate, 2 or 3; fruits berry-like, blue-purple, juicy, 4–5 mm in diameter, crowned with persistent calyx and stigmas; pyrenes rugose. Native to northern China and adjacent areas of Korea, introduced to American gardens in 1938.

Viburnum cordifolium Wallich ex de Candolle — CHINESE HOBBLE-BUSH

Xin-ye-jia-mi=Hsin-yeh-chia-mi (心葉莢蒾, Cordate-leaved Viburnum); *Jia-mi=Chia-mi* (莢蒾, Viburnum). Fruits, gathered and eaten locally, not available in the market.

Deciduous shrubs or small trees 2–5 m high, branchlets stout, densely scurfy-tomentose when young; leaves opposite, cordate or ovate, 7–12 (–16) cm long, 5–9 (–16) cm wide, acuminate, base cordate or rounded, crenate-serrate, pubescent on major nerves with branched hairs of both surfaces; flowers white, small, in terminal flat corymbs, without sterile marginal flowers, peduncles, primary, secondary and tertiary axes, pedicels, and sepals all covered with branched hairs and yellow glands, hypanthia obovate-oblong, 2 mm long, glabrous, calyx 5-toothed, 3 mm across, lobes triangular, acute, corolla rotate, 7–9 m across, 5-lobed, lobes oblong, apex rounded, disk conical, style 0, stigmas capitate, 2- or 3-lobed; fruits scarlet red, ovate-oblong, 6–8 mm long, 5–6 mm across, crowned with persistent calyx and stigmas. Native to the eastern Himalayan Region, spontaneous in the broad-leaved evergreen and deciduous mixed forests in Yunnan and Sichuan, extending eastward to Guangxi and westward to Bhutan and Nepal.

Viburnum dilatatum Thunberg

Tu-lan-tiao=T'u-lan-'t'iao (土蘭條, Native Hobble Bush); *Jia-mi=Chia-mi* (莢蒾, Viburnum). Fruits, eaten by people of Shaanxi and Gansu.

Deciduous shrubs 2–5 m high, branchlets densely pubescent; leaves suborbicular, obovate or ovate, 3–10 cm long, 2–8 cm wide, acute, base rounded or obtuse, dentate, teeth mucronate, stellate pubescent beneath, glandular near the base, petioles 1–1.5 cm long; flowers white, in terminal umbelliform corymbs 4–12 cm across, pilose, hypanthia 1 mm long, hispid and glandular, corolla rotate, hairy, stamens 5, exserted, anthers white, ovary inferior; fruits oblong-ovoid, red, 6-8 mm long; pyrenes ovate, 5–7 mm long, 4–6 mm wide, striate-sulcate. Widespread in eastern Asia, introduced into Western gardens in 1845.

Viburnum ichangense (Hemsley) Rehder — ICHANG VIBURNUM

Yi-chang-jia-mi=I-ch'ang-chia-mi (宜昌荚蒾, Ichang Viburnum); *Dui-ye-san-hua=Tui-yeh-san-hua* (對葉散花, Opposite-leaved Loose Flower). Fruit; eaten fresh by people living in the mountains of western Hubei.

Deciduous shrubs 2–3 m high, branchlets pubescent with stellate hairs; leaves opposite, ovate-elliptic to ovate-lanceolate, 3–7 cm long, 1.5–4 cm wide, acuminate, base rounded or broadly cuneate, dentate; flowers white, small, 2–4 mm across, in terminal corymbs, calyx 1.5 mm long, 5-toothed, stellate pubescent, corolla rotate, 5-lobed, stamens 5, shorter than the corolla, ovary inferior; fruits globose-ovoid, red, turning black, 7 mm long, stones compressed. Native to central and western China, introduced into Western gardens in 1901.

Viburnum mongolicum (Pallas) Rehder — MONGOLIAN WHITEROD, MONGOLIAN VIBURNUM

Bai-nuan-tiao=Pai-nuan-t'iao (白暖條, White Warm Rod); *Meng-gu-jia-mi=Meng-ku-chia-mi* (蒙古荚蒾, Mongolian Viburnum). Mature fruit; gathered and eaten as grape; the leaves and fruits spread in the sun, stacked up at night, repeated for 2 days, roasted, placed in a willow wickerwork, stirring every 2–3 hours until the material becoming red, mixed with commercial brick tea before brewing the beverage Mongolian style.

Shrubs 1–2 m high, the vegetative growth stellate-pubescent, older branches grayish-white; leaves opposite, ovate-oblong, 2–6 cm long, 1-2.5 cm wide, acute or obtuse, base rounded, dentate, the teeth mucronate, sparsely pubescent on both surfaces, petioles 3–8 mm long; flowers yellowish-green, sessile, 8–12 in terminal cymose clusters, hypanthia cylindrical, glabrous, calyx discoid, 5-toothed, corolla cylindrical-campanulate, 6 mm long, the tube 5 mm long, lobes semiorbicular, stamens 5, filaments short, anthers oblong, ovary inferior, unilocular; fruits blue-black, compressed, oblong-ellipsoid, 10 mm long, 8 mm in diameter; pyrenes light brown, compressed-elliptic; seed black, flattened-ovoid, alveolate; rich in oily endosperm. Native of the arid region of northern China and Inner Mongolia, growing on the hillsides between 1,800–3,200 m altitude, specimens from altitudes of 2,600–3,200 m with much smaller leaves; introduced into western gardens in 1785.

Valerianaceae: Valerian Family

Valerianella olitoria (L.) Pollich — Lambs-lettuce, Corn-salad
Ye-ju=Yeh-chu (野苣, Wild Lettuce); *Fa-guo-ma-lan-tou=Fa-kuo-ma-lan-t'ou* (法國馬蘭頭, French Horse Aster). Tender shoots; used for salad, or for quick-fry potherb; cultivated in Shanghai.

Delicate annual or short-lived biennial herbs 10–30 cm high, with slender, forked, angular stems, pilose along the angles, internodes 6–10 cm long, 2 mm in diameter; leaves basal and rosulate, cauline and opposite, in 3–4 pairs, sessile, exstipulate, glabrous, spatulate, or oblong-oblanceolate, 10 cm long, 2 cm wide, entire, ciliate near the base; flowers purplish-blue, in cymose terminal clusters with foliaceous lower bracts, calyx obsolete, corolla funnelform, 2 mm long and across the 5-lobed limb, tube gibbose, stamens 3, barely exserted, ovary 3-locular, one fertile and 1-ovulate, 2 abortive and empty; fruits oblique-rhomboidal, 2–4 mm across, the fertile side larger and acute at the apex, with turgid corky back; seed solitary, oily. Native to Eurasia with holarctic distribution; cultivated in Europe for salad, introduced to the vicinity of Shanghai since the 1880s (Anonymous, 1959).

Dipsacaceae: Teasel Family

Scabiosa tschiliensis Grüning — North China Scabiosa (Syn. *S. comosa* auct., non Fischer ex Roemer et Schultes, *S. fischeri* auct., non de Candolle)
Shan-luo-bo=Shan-lo-po (山蘿蔔, Mountain Radish); *Hua-bei-lan-pen-hua=Hua-pei-lan-pen-hua* (華北藍盆花, North China Blue Pot Flower). Young shoot; eaten by the local people in northern China.

Perennial herbs with woody rootstock bearing 1, 2, or rarely more buds capable of developing into vegetative shoots with radical leaves, or flowering shoots with opposite cauline leaves, 30–60 cm high; radical leaves 8–11, with petioles 1–8 cm long, laminas ovate-elliptic, 3–9 cm long, 0.8–3 cm wide, acute, base cuneate, coarsely dentate-crenate or irregularly lobed, cauline leaves 4–8 cm long, 1.5–3.5 cm wide, acute, the lower ones cuneate and petiolate, upper ones sessile, irregularly lobed or parted, the ultimate segments 1.5–4 mm wide; flowers blue-purple, in dense terminal capitula 2–5 cm across, peduncles 8–30 cm long, involucral bracts green, linear-oblong, densely pubescent, outer ones 1–2 cm long, inner ones ovate, acute, receptacles conical, 1–1.5 cm long, florets zygomorphic, the marginal ones much larger, hypanthia ellipsoid, apex strongly constricted, calyx discoid, 5-lobed, corolla pubescent, tubes 4–8 mm long, limbs unequally 5-lobed, the larger lobes of the marginal flowers 10–12 mm long, spathulate, lobes of the inner small flowers ovate or oblong, 1–2 mm long, stamens 4, exserted, ovary inferior, style filiform, stigma capitate-discoid, exserted; achenes ellipsoid, beaked. Native of northern China, first recorded from Xiao-wu-tai Shan in Hebei. The species is now recorded from Shandong, northeastern China and Inner Mongolia.

Cucurbitaceae: Gourd Family

Benincasa hispida (Thunberg) Cogniaux — Winter Melon, White Gourd (Syn. *B. cerifera* Savi) (Figures 54–55)

Dong-gua=Tung-kua (冬瓜, Winter Melon); *Bai-gua=Pai-kua* (白瓜, White Gourd). *Mao-gua=Mao-kua* (毛瓜, Hairy Gourd). Fruit; young and hairy, cooked for vegetable; large and mature, up to 95 cm long, 30 cm in diameter, the rind covered with white powder, often sold in pieces; both types available in American Chinese groceries.

Annual climbers, densely hispid throughout, tendrils with 2 or 3 branches; leaves reniform, 10–30 cm wide, shallowly palmate lobed, apex acute-acuminate, base subcordate, with broad sinus, margin lobate, the lobes broadly triangular, acute, dentate; flowers orange-yellow, monoecious, solitary, staminate flowers opening three weeks before the pistillate ones, during the middle of the season, a pistillate flower occurring between 2 or 3 staminate ones, approaching the closing time of the growth season, all the flowers being pistillate, calyx green, hairy both inside and out, 5-lobed, the segments lobate, reflexed, corolla rotate, 5–8 cm across, deeply 5-lobed, the lobes obovate, hairy outside, stamens 3, anthers sigmoid, disk hairy, pistillode obscure, ovary of pistillate flower inferior, oblong, densely hispid, calyx and corolla similar to that of the staminate flower, disk hairy, style short, slightly branched, stigmas 3, fleshy; fruits varying in shape and size, the largest ones 75–95 cm long, 28–30 cm in diameter, weighing 13–18 kg (30–40 lbs), rind green with waxy cover, flesh white, pulp white, spongy; seeds numerous, white. Native land uncertain, extensively cultivated in China and southeastern Asia, available in American Chinese groceries, during the Chinese New Year time, price becoming as high as that of the best steak; two varieties recognized by botanists in China.

1. var. **hispida** — Dong Gua (冬瓜) (Figure 55 a–d)

The largest known fruit-bearing vegetable, with some cultivars producing market material up to 95 cm long, weighing up to 18 kg. Four cultivars from Guangzhou (Canton) illustrating the variations of form and size are given below.

a. cv. 'Big Green Rind' (*Da-qing-pi=Ta-Ch'ing-p'i* 大青皮, Great Green Skin)

Vines extending 5 m long, internodes average 20 cm long; leaves 20 cm long, 30 cm wide, petioles 14 cm long; first pistillate flower appearing between the 18th and 22nd nodes; fruits cylindrical, 95 cm long, 28 cm wide, wall 6 cm thick, meat white, skin olive-green, total weight 13–18 kg. An ancient cultivar selected by farmers; seeds sowed in January–March (mostly in February), harvest between 175–185 days, one of the best producer, much cultivated.

b. cv. 'Gray Bushel' (*Hui-dou=Hui-tou* 灰斗, Gray Bushel)

Vines extending 4.5 m, internodes 18 cm long; leaves 20 cm long, 32 cm wide, petioles 14 cm long; first pistillate flower appearing between the 16th and the 19th nodes; fruits subglobose-oblong, 40 cm long, 33 cm in diameter, wall 5 cm thick, rind gray-green with white wax powder, total weight 11–15 kg. A cultivar with uncertain origin, widely cultivated; seeds sowed in February–March, harvest after 150–160 days, with good keeping quality.

c. cv. 'Ox Spleen' (*Niu-pi=Niu-p'i* 牛脾, Ox Spleen)

Vines shorter than the two cultivars above, 4 m long, internodes 18 cm long, many secondary and tertiary branches; leaves 24 cm long, 27 cm wide, petioles 14 cm long; first pistillate flowers between the 13th and 16th node; fruit clavate, 70 cm long, 14 cm in diameter, rind dark green, wall 5 cm thick, total weight 5–6 kg. A cultivar known for 100 years; seeds sowed in February–March, harvest after 120–130 days, a good cultivar for supplying the summer market.

d. cv. 'Yellow Climber' (*Huang-deng=Huang-teng* 黃登, Yellow Climbing-up)

Vines 3.5 m long, internodes 14 cm long; first pistillate flower appearing on the 8th–11th nodes; fruits oblong, 26 cm long, 17 cm in diameter, olive-green, waxy, total weight varying between 3.5–6.5 kg. A cultivar with over 100 years' history, with smaller stature, compact high quality fruit, good keeping quality.

2. var. **chiehqua** How — FESTIVAL GOURD (Figure 55 e-j)

Jie-gua=Chieh-kua (節瓜, Festival Gourd). *Mao-gua=Mao-kua* (毛瓜, Hairy Gourd). Young fruit; harvested before fully grown, a major summer vegetable in Hong Kong; available in American Chinese groceries almost throughout the year; market during the Dragon Boat Festival, hence the local name "Festival Gourd"; in American Chinese stores sold as *Mao-gua* for the dense hairy skin.

Annual climber, tendrils 3-branched; leaves 5–7 lobes, dentate, hispid throughout; flowers yellow, similar in size and structure as those of the species; fruits cylindrical or clavate, 15–23 cm long, 4–8 (–10) cm in diameter. The variety has a history of 300 years in the vicinity of Guangzhou; seed sowed in December–February, the first crop ready for April–May, and throughout the year; six cultivars are given below (Figure 55 e–j).

a. cv. 'Big Vines' (*Da-teng=Ta-teng* 大藤, Large Vine)

Vines 5.5 m long, much branched; leaves 20 cm long, 27 cm wide, petioles 14 cm long; first pistillate flower appearing on the 7th–14th nodes; fruits cylindrical, 21 cm long, 6 cm in diameter, uniformly green, with darker stripes toward the distal end, average weight of market material 580 g (1.3 lb). A very productive summer crop, seeds sowed in February–August, harvest after 50–60 days, continuing to bear for two months.

b. cv. 'Pineapple' (*Bo-luo-zhong=Po-luo-chung* 菠蘿種, Pineapple Type)

Vines 4.5 m long, beginning to branch from the first or second node; leaves 21 cm long, 20 cm wide, petioles 14 cm long; first pistillate flower appearing on the fifth or sixth node; fruits cylindrical, 23 cm long, 11 cm in diameter, yellow-green, hispid, average weight 450 g, mature fruit for seed covered with waxy powder. A cultivar with long history, seeds sowed between December and late February, in 80–100 days ready for the market.

c. cv. 'Twin Carps' (*Zi-li-yu=Tzu-li-yu* 孖鯉魚, Paired Carps)

Vines 3.5 m long, internodes short, branching system strong, new branches emerging even in the axil of cotyledon; leaves 20 cm long, 20 cm wide, 5–7-lobed, petioles 10–13 cm long; first pistillate flower on the third or fourth node; fruits cylindrical, 21 cm long, 6 cm in diameter, green, slightly pointing at the distal end, hispid, mature fruit without waxy powder, marketing material weighing 450 g. A more recent selection, known for several decades, seeds sowed in January–August, harvesting after 40–75 days, a good cultivar for long season of production, heavy bearing, and good quality.

d. cv. 'Patched Rind' (*Hei-pi-qing=Hei-p'i-ch'ing* 黑皮青, Black Rind Green)

Vines 4.5 m long, internodes short, branches numerous; leaves 21 cm long, 21 cm wide, petioles 14 cm long; first pistillate flower on the fourth-eighth nodes; fruits variable in size, 18–21 cm long, 6–7 cm in diameter, dark green with vertical stripes and patches, rounded at the apical end, fully mature fruits without waxy powder. A cultivar with long history, extensively cultivated, seeds sown in December–February, harvesting in April–June, continuing to bear for 40 days.

e. cv. 'Seven Stars' (*Qi-xing-zai=Ch'i-hsing-tsai* 七星仔, Seven Little Stars)

Vines 3.5 m long, internodes short, branching system strong; leaves 18 cm long, 18 cm wide; first pistillate flower on the third-seventh nodes; fruits cylindrical, 21 cm long, 6 cm across, green with greenish-white spots and patches, marketing material 315 g. A favorite of farmers for its smaller stature, tolerance of hot summer and late season planting, seeds sowed in March–August, harvesting after 35–80 days, continuing to bear for 35–40 days.

f. cv. 'Spotted Pineapple' (*Qi-xing-bo-luo-zai=Ch'i-hsing-po-luo-tsai* 七星菠蘿仔, Seven-star Little Pineapple)

Vines 4 m long, power of branching medium; leaves 22 cm long, 22 cm wide, petioles 12 cm long; first pistillate flower on the fifth-seventh node; fruits oblong or short-cylindrical, 15 cm long, 7 cm in diameter, green with paler spots, hispid, the hairs white, when fully grown covered with waxy powder, marketing material 490 g. A cultivar known for four decades, seeds sown in February, harvested after 75–85 days, continued bearing period 50–70 days; its superior qualities being resistant to cold weather, early crop, available for the market in April–May, thick rind, good for handling.

Citrullus battich Forskå, (1775) — WATERMELON (Syn. *Momordica lanata* Thunberg, 1794; *Citrullus vulgaris* Schrader ex Ecklon et Zeyher, 1836; *Citrullus lanatus* [Thunberg] Matsumura et Nakai). (Figure 56)

Xi-gua=Hsi-kua (西瓜, Melon-from-the-west); *Da-gua=Ta-kua* (打瓜, Smashed Melon). Mature fruit; eaten as fresh summer fruit. *Gua-zi=Kua-tzu* (瓜籽, Melon Seed). Seed; Roast or treated with spices, very popular throughout China, especially during the Chinese New Year season; two types (a large black seed with white center and a small red seed), available in American Chinese stores.

Annual herbaceous climbers, trailing on the ground in cultivation, lanate throughout, branchlets subterete, tendrils 2- or 4-branched; leaves triangular-ovate, 8–20 cm long, 5–15 cm wide, deeply 3-parted, the segments pinnate, irregularly lobed with deep-round sinus, dentate; flowers pale yellow, monoecious, staminate flowers opening weeks before the pistillate ones, solitary, pedicels 2–3 cm long, calyx 5-lobed, the lobes linear, corolla rotate, tube short, lobes obovate, stamens of staminate flowers 3, distinct, the anthers sigmoid, pistillode subglobose, staminodes of pistillate flowers well developed, the sigmoid anthers sometimes bearing pollen grains, ovary inferior, oblong, hairy, style short, stigmas 3, fleshy; fruits baccate, varying greatly in shape, size, color, pulp, and seed. Native of Africa, extensively cultivated in China, for juicy pulp and/or for edible seed. The Yellow River Region, where wheat being the major crop, has been the primary production area, with Dezhou (德州, in Shandong) having the best watermelon, and Xuzhou (徐州, in Jiangsu) producing the largest amount of seeds. In Guangxi a special red-seeded cultivar is cultivated for local consumption of the pulp and for export of the seed to Guangzhou and Hong Kong, and thence abroad. Three cultivars are given below.

1. cv. 'Pulp Xigua' (*Pu-tong-xi-gua=P'u-t'ung-hsi-kua* 普通西瓜, Common Watermelon) Fruit; cultivated for the edible pulp.

Fruits oblong, rarely spherical, 30–40 cm long, green with darker vertical markings; pulp red or rarely pale yellow, very sweet; seeds black, rarely marbled, 6–8 mm long, 5–6 mm wide. Extensively cultivated for summer consumption as a fresh fruit, often sold in slices along the street; seeds often gathered by children or the less prosperous people, eaten raw or roasted.

2. cv. 'Red-seed Xigua' (*Hong-zi-xi-gua=Hung-tsu-hsi-kua* 紅籽西瓜, Red-seeded Watermelon)

Fruit relatively small, generally oblong, 18–20 cm long, sometimes almost round, 16 cm in diameter, skin yellowish-green with darker vertical markings, rind rather thin, pulp red, very juicy but not as sweet as the black-seeded ones; seeds red, 6–7 mm long, 5 mm wide. Cultivated in Guangxi, pulp consumed locally, seed exported to Guangzhou and Hong Kong, thence exported to Chinese stores abroad for the Chinese New Year season, very popular among Chinese Americans from southern China.

3. cv. 'Dagua' (*Da-gua=Ta-kua* 打瓜, Strike Melon).

Fruits spherical, skin green with vertical bands, rind thick, pulp white, firm, not very sweet; seeds black with ovate white center, 10–12 mm long, 8–10 mm wide. Limited cultivation as a field crop for the seed, during harvest time, early in the morning, the field is opened for the public to enjoy the cool melon with one condition, leaving the seeds gathered into containers made of the empty melon shells for the owner to collect them, a cooperative way for harvesting the crop in rural China.

Cucumis melo L. — MELON

Gua=Kua (瓜, Gourd). Ripe fruit; eaten fresh.

Annual climber, hispid throughout, stems striate-sulcate, tendrils simple (unbranched); leaves orbicular, ovate or reniform in outline, 8–12 cm long, 3–7 cm wide, palmately nerved and shallowly lobed, the lobes rounded; flowers yellow, 2 cm across, monoecious, the staminate flowers fasciculate, pedicels 0.5–2 cm long, hypanthia turbinate, 7–8 mm long, calyx lobes subulate, corolla rotate, the lobes roundish, stamens in the staminate flowers 3, distinct, anthers sigmoid, one bearing 1-theca, the other two each with 2-thecae, connective projected at the apex; ovary of the pistillate flower inferior, oblong, staminodes well developed, style terete, short, stigma 3-lobed, fleshy; fruits varying in shape, size, and texture according to the varieties and cultivars. Assumed to be native of Africa and southwestern Asia, introduced into China by intermediate ancient itinerant travels.

Note: Before 1930, the cultivation of melon in China was in the lower Yellow River area where wheat, soybean and cotton constituted the primary crops. There, different cultivars of a special type with thin-skin and sweet meat called *"Tian-gua"* (甜瓜, Sweet Gourd) were selected. The entire fruit, except seeds, is eaten. The meat of most cultivars of this type is crisp and sweet. That of a large melon with yellow skin called *"Mian-gua"* (麵瓜, Mealy Gourd) is soft and less sweet, adapted for senior citizens. The earliest record of melon in Chinese herbals can be traced back to late fifth century with the name *"Yue-gua"*, (越瓜, Gourd from South China), indicating its introduction from overseas by Arabian traders. However, in the soybean and cotton fields of the lower Yellow River Region, there existed a wild type with small hairy globose fruits, 3 cm in diameter, when young uniformly green and hairy, intensively bitter, at maturity changing to yellow, smooth, with a pleasent smell, numerous seeds, thin skin, little flesh, and finally becoming soft. It was a weed in soybean or cotton fields. Farmers pull the plants off at any stage of their development. Any one plant left in the field could produce dozens of yellow ripe fruits and numerous green ones. As children playing with soybean harvesters, we used to gather the yellow fruits, knead them between palms to make the inside mushy and throw them over playmates, calling them *Ma-pao* (馬炮, Horse Bullet). Chinese botanists named and described the taxon *Cucumis bisexualis* Lu et Wang (1948). It seems very possible that this spontaneous species has contributed the unique features of thin rind, pleasant musky smell, and bitter taste to many Chinese cultivars of Sweet Melon. In the middle seventeenth century a cultivar related to *Yue-gua* was recorded as *Sheng-gua* (生瓜, Raw Gourd). Apparently all the Chinese cultivars were developed in China.

After 1930, cultivars of two varieties of *C. melo* L. have been introduced from America and Iran, but these are rarely seen in the market. In the early 1940s, Wendell Willkie, Vice-President of USA, made a famous around the world trip in seven weeks in the middle of World War II. He entered China at Urumqi in Xinhing and visited Lanzhou next. The seed of Honeydew he brought to China was used to start an important crop in Lanzhou. This is the only American commercial melon available in some large cities of China, sold as *Bai-lan-gua* (白蘭瓜, White Lanzhou Melon).

Persian melon has been cultivated in Xinjiang without the exact record of its introduction. Early Chinese officials and travellers called it *"Ha-mi-gua"* (哈密瓜, Hami Melon). After the establishment of airplane transportation, a small amount of this melon was brought to Beijing and appeared in the market as *Ha-mi-gua*. The above review indicates that cultivars of five distinct varieties of edible melons are available in China.

1. var. **chinensis** Pangalo — CHINESE SWEET MELON

Tian-gua=Tien-kua (甜瓜, Sweet Melon); *Xiang-gua=Hsiang-kua* (香瓜, Fragrant Melon). Ripe gourd; many forms with various colors and shapes, mostly crisp, sweet, with thin edible skin, eaten as fresh fruit; cultivated in northern China.

Fruits medium-sized, oblong or obovoid, 8–16 cm long, 6–12 cm across the middle, rind smooth, thin, palatable, flesh sweet, crisp and with pleasing flavor. The best market products are cultivated in the lower Yellow River Region, with many horticultural forms. The large ones are of various shades of green and tinged yellow, the choicest type has greenish flesh and bright red seeds. Another smaller melon with uniform canary yellow rind, crisp, sweet, white flesh is occasionally seen in Nanjing and Hangzhou. The hairy bitter young fruits and the thin rind of various cultivars in this variety are features indicating its relationship with the spontaneous taxon, *C. bisexualis* Lu et Wang.

2. var. **conomon** (Thunberg) Nakai — Oriental Pickling Melon (Syn. *C. conomon* Thunberg) (Figure 57).

Cai-gua=Ts'ai-kua (菜瓜, Vegetable Gourd); *Yue-gua=Yueh-kua* (越瓜, Cantonese Gourd); *Bai-gua=Pai-kua* (白瓜, White Gourd); *Shao-gua=Shao-kua* (稍瓜, Pseudo-melon). Young fruit; used for pickles; large fully grown fruits 30 cm long, 5–7 cm across, whitish green used by farmers for salad or soup.

Fruits cylindrical-oblong or clavate, rind thin, flesh firm, neither bitter when young, nor sweet at maturity; with two distinctive cultivars.

a. cv. 'Cai-gua' (菜瓜, Vegetable Gourd). Fruits oblong, 16–20 cm long, 10–12 cm across, smooth, green, crisp, edible raw like cucumber.

b. cv. 'Shao-gua' (稍瓜, Pseudo-melon). Fruits cylindrical or clavate, 25–35 cm long, 6–8 cm in diameter, pale green, flesh firm, used for cooking or pickles.

3. var. **flexuosus** (L.) Naudin — SNAKE MELON (Syn. *C. flexuosus* L.)

Sheng-gua=Sheng-kua (生瓜, Raw Gourd); *Yang-jiao-gua=Yang-chiao-kua* (羊角瓜, Goat-horn Gourd). Fruit; used for pickles, recorded in the herbals of early seventeenth century, rarely cultivated now.

Leaves suborbicular or cordate, 8–15 cm in diameter, irregularly wavy and serrate, petioles prickly; pistillate flowers on long pedicels; fruits clavate, 45–60 cm long, 4 cm in diameter, rind tough, dark green, with 10 pale stripes, flesh firm, not palatable raw.

4. var. **inodorus** Naudin — Honeydew

Bai-lan-gua=Pai-lan-kua (白蘭瓜, White Lanzhou Melon). Fruit; rare and very expensive; imported American material available in Hong Kong fruit market.

Vines strong; leaves large, palmately lobed; fruits large, subglobose, smooth, 15–20 cm in diameter, rind thick, not palatable, flesh greenish-white, very sweet when fully ripe, outer portion firm, inner portion soft, hollow; seeds ivory; introduced from USA in the early 1940s.

5. var. **reticulatus** Seringe — Persian Melon

Ha-mi-gua=Ha-mi-kua (哈密瓜, Hami Melon). Fruit; rare and expensive.

Vines strong; leaves large; fruits ellipsoid, yellowish-green with dark green vertical patches and bands, 18–25 cm long, 10–14 cm across the middle, crisp, mildly sweet, rind smooth or slightly netted, not palatable; seeds ivory. Introduced from central Asia, cultivated in Xinjiang, Hami being one of its large cities on the Silk Route, west of Lanzhou.

Note: In addition to the above five varieties with edible fruits, two varieties have been introduced and cultivated occasionally for ornamental purposes or as articles of virtu. These are: var. *agrestis* Naudin (*C. acidus* Jacquin, *C. callosus* [Rottler] Cogniaux et Harms) recorded as *San-leng-gua= San-leng-kua* (三棱瓜, Trigonous Gourd), or *Ma-pao-gua=Ma-pao-kua* (馬泡瓜, Horse Bubble Gourd), with small oblong or globose fruits, devoid of fragrance and not palatable; var. *dudaim* (L.) Naudin (*C. dudaim* L.) recorded as *Xiang-gua=Hsiang-kua* (香瓜, Sweet Smelling Gourd), with small fruit of the size of an orange, flattened at the ends, marbled; kept as curio.

Cucumis sativus L. — Cucumber (Figure 58)

Huang-gua=Huang-kua (黃瓜, Yellow Gourd); *Qing-gua=Ch'ing-kua* (青瓜, Green Gourd, the Cantonese name); *Hu-gua=Hu-kua* (胡瓜, Tartary Gourd). Fruit; fully grown yet young and tender cucumber used for salad; introduced into China repeatedly. Cucumbers was first recorded in China in early seventh century as *Hu-gua* (胡瓜) which indicated its introduction overland by the Central Asian Silk Route. Later, on observing the yellow color of the mature seed-fruits, its name was changed into *Huang-gua* (黃瓜), which was recorded in early eighth century.

Annual climber, strigose and prickly throughout, tendrils simple; leaves ovate-cordate, 7–15 cm long and across the base, shallowly palmate-lobed, margin sharply serrate and ciliate; flowers yellow, fasciculate, monoecious, the staminate flowers appearing two weeks in advance, hypanthia funnelform, sepals 5, linear, stamens 3 in staminate flowers, filaments short, hairy, anthers sigmoid, two 2-thecous, the third one

1-thecous, pistillode discoid, ovary in pistillate flowers inferior, oblong-fusiform, spinescent, staminodes well developed; fruits cylindrical, curved, or clavate, (7–) 30–40 (–60) cm long, rough, often prickly, dark green when young, becoming yellow and smooth at maturity; seeds numerous, white.

Native of tropical Asia, extensively cultivated throughout China, forming an important summer vegetable, available even in village market between May and July; used raw or cooked, fresh or pickled; fresh juice or decoction obtained from shoots, leaves and/or roots used locally for dysentery and boils; many Chinese cultivars selected for the best result of local environmental conditions; more cultivars in northern China (such as Beijing) with narrowly cylindrical fruits (20–) 30–50 (–66) cm long, 3–4 cm across, rough and prickly at the distal four-fifths and suddenly narrowed into a smooth handle at the basal one-sixth; on the contrary, cultivars in southern China (such as Hong Kong) are shorter, uniformly cylindrical throughout, with both ends rounded, and few scattered prickles, (13–) 19–20 (–28) cm long, 4 cm across; a few examples of Chinese cultivars are given below, with the fruit characters referring to the fully grown ones ready for the market.

1. cv. 'Peking Prickle' (*Bei-jing-ci-gua=Pei-ching-tz'u-kua* 北京刺瓜, Beijing Prickly Cucumber)

   Fruits narrowly clavate-cylindrical, with 10 rows of prickly tubercles, prickles white, wall thick 11 mm across, crisp, good flavor, seeds few, cold-resistent quality good, may be used for winter crop in greenhouse, two types selected: the larger type 33–45 cm long, 3.5–4 cm across, ready for the market in late April or early May, unit yield high; the smaller type 25–33 cm long, 3–3.5 cm across, with fewer prickly tubercles and less evident ridges.

2. cv. 'Shandong Prickle' (*Ning-yang-ci-gua=Ning-yang-tz'u-kua* 寧陽刺瓜, Prickle Cucumber from Ningyang of Shandong)

   Fruits clavate-cylindrical, 35–60 cm long, 3.4–4 cm across, dark green, tubercle small and numerous, wall thick, pulp small, excellent eating quality, strong cold and disease resistance.

3. cv. 'Truncate Head' (*Jie-tou-gua=Chieh-t'ou-kua* 截頭瓜, Cut Head Cucumber)

   Fruits clavate-cylindrical, slightly narrowed in the middle, (30–) 35–40 (–50) cm long, 3–4 cm across, smooth, white prickles not on tubercles, wall relatively thin and pulp large, harvest in June, continuing for 2 months.

4. cv. 'Peking Whip' (*Bian-gua=Pien-kua* 鞭瓜, Whip Cucumber)

Fruits elongated clavate-cylindrical, 33–60 (–66) cm long, the longest known cucumber, light green, yellowish at the distal end, smooth, without tubercle and prickle, wall thin, pulp large, eating quality poor, high resistance to heat and diseases.

5. cv. 'Fangshan Field Cucumber' (*Fang-shan-di-huang-gua=Fang-shan-ti-huang-kua* 房山地黃瓜, Field Cucumber-of-Fangshan)

Fruits short-cylindrical, 12–15 cm long, 5 cm across, tubercles small, prickles black, wall greenish-white, pulp large, harvest in late June, used for salad or pickles, tolerent to heat and drought, resistant to insects and disease.

6. cv. 'Canton Green' (*Da-qing=Ta-ching* 大青, Big Green)

Fruits short-cylindrical, 19–21 cm long, 4 cm across, green, smooth, prickles black, scattered in the middle portion, available in two seasons, spring crop sown in February and harvest in April–May, autumn crop sown in September and harvest in October–November.

7. cv. 'Shanghai Small' (*Xiao-huang-gua=Shiao-huang-kua* 小黃瓜, Small Cucumber)

Fruits oblong, 7 cm long, 4 cm across, green with 10 light stripes, cross-section triangular, prickled tubercles numerous, harvest in late May to July, poor eating quality, used for pickles; cultivated over 30 years in the vicinity of Shanghai, said to have been introduced from Harbin.

Cucurbita foetidissima Humboldt, Bonpland et Kunth — Buffalo-Gourd

Chou-gua=Ch'ou-kua (臭瓜, Fetid Gourd); *Han-sheng-you-gua=Han-sheng-yu-kua* (旱生油瓜, Xerophytic Oil-gourd). Starch and oil; starch obtained from root, oil extracted from seed; used for cooking in Shaanxi.

Perennial climbers, roots stout, fusiform, stems robust, striate, rough, with foul smell, tendrils 4- or 8-branched; leaves ovate-cordate, 20–25 cm long, 10–16 cm wide, triangular-caudate, serrate, strigose above, hispid beneath, palmately 5-nerved, petioles 5–7 cm long; flowers yellow, solitary, axillary, unisexual, monoecious, stamens 3 in staminate flower, anthers united into a column 2 cm long, filaments short, pistillate flowers 6 cm long, sepals lanceolate, strigose, corolla funnelform, lobed above the middle, style columnar, distal end 3-fid, stigmas 3, 2-lobed, ovary hairy; fruits oblong-globose, 6–10 cm long, 6.5 cm across middle, green-yellow-white striped and patched, stalks 3 cm long; seeds flat, oblong-ovate, 6–12 mm long, 4–7 mm wide, rich in protein (30–35%) and oil (34%).

Native to arid North America (Mexico and adjacent USA), drought tolerant, introduced and cultivated in the barren land of northwestern China, propagated by nodal rooting.

Cucurbita moschata (Duchesne) Poiret — PUMPKIN, CROOKNECK SQUASH

Nan-gua=Nan-kua (南瓜, Gourd-from-South); *Fan-gua=Fan-kua* (番瓜, Foreign Gourd); *Fan-gua=Fan-kua* (飯瓜, Meal Squash). Mature or young fruit; mature fruit cooked with rice or millet for meals, young ones cooked and seasoned as vegetable.

Annual trailing vines, stems strong, 2–5 m long, rooting on nodes, covered with rather soft hairs throughout, tendrils 3- to 5-branched; leaves broad-ovate or ovate-orbicular, 20–40 cm long and wide across the base, shallowly 5-lobed, the middle lobes triangular, serrate, palmately nerved, petioles hollow, strigose; flowers solitary, yellow, large, showy, monoecious, protandrous, hypanthia funnelform, sepals 5, linear and enlarged at the apical portion, often becoming leafy, corolla campanulate-funnelform, 10 cm or more across, lobes crinkly, stamens 3 in the staminate flowers, filaments distinct at the distal end, broadened and partially connate at the base, anthers linear, connate, two 2-thecous, one 1-thecous, ovary of pistillate flowers inferior, oblong or compressed-globose, style stout, stigmas 3, fleshy; fruits pyriform, oblong or crooknecked, the basal portion solid, the apical portion hollow, rind smooth, yellowish and marbled green, often glaucous at maturity, flesh mealy, golden-yellow, weighing 3.5–31 kg, seeds numerous, edible roasted.

Native of Central America, introduced into China in Ming Dynasty; now extensively cultivated throughout China. Three selected cultivars are given below.

1. cv. 'Pillow Squash' (*Zhen-tou-gua=Ch'en-t'ou-kua* 枕頭瓜, Chinese Pillow Squash)

Fruits subcylindrical-oblong, 25 cm long, 16 cm in diameter, slightly constricted in the middle, truncate and concaved at both ends, dark green when young, dull-yellow and marbled-netted at maturity, flesh yellow-orange, mealy and sweet, individual squash weighing 3.5 kg, good keeping quality.

2. cv. 'Ox-Leg Squash' (*Niu-tui-gua=Niu-t'ui-kua* 牛腿瓜, Ox-leg Squash)

Fruits clavate, with crookneck, 50 cm long, basal two-thirds cylindrical, 15 cm in diameter, solid, the distal one-third enlarged, hollow, 30 cm across, rounded, slightly concave at the apex, dark green when young, yellowish-brown and marbled-netted green at maturity, flesh mealy and sweet; individual squash weighing 27–31 kg, with very good keeping quality.

3. cv. 'Round Box Squash' (*He-gua=Ho-kua* 盒瓜, Box Squash)

Fruits compressed-globose, irregularly lobed and grooved, 18 cm long, 24 cm in diameter, both ends concaved, green when young, dull-yellow and netted-patched at maturity; individual squash weighing 3–13 kg, slightly mealy and sweet, good quality for storage and transport.

Cucurbita maxima Duchesne ex Lamarck — AUTUMN SQUASH, PUMPKIN, BUTTERCUP
Squash

Sun-gua=Sun-kua (筍瓜, Bamboo-shoot Squash); *Bai-yu-gua=Pai-yu-kua* (白玉瓜, White
Jade Pumpkin); *Da-gua=Ta-kua* (大瓜, Big Pumpkin); *Yin-du-nan-gua=Yin-tu-nan-kua*
(印度南瓜, Indian Pumpkin). Fruit; mature fruits, cooked, texture rather plain, rarely
used in Chinese dishes, cultivated, recorded from Beijing and Shanghai, cultivated
to meet demands of foreign residents.

Strong annual trailing vines, slightly prickly, tendrils 3- or 5-branched; leaves large,
orbicular or reniform, 15–28 cm long, 17–32 cm wide, almost entire, serrate, obtuse,
petioles stout, 15–20 cm long, hollow, densely strigose; flowers yellow, solitary,
monoecious, staminate flowers on peduncles 10–20 cm long, hypanthia campanulate,
sepals lanceolate, 1.8–2 cm long, setaceous, corolla funnelform, lobes ovate, obtuse,
crinkly, reflexed, stamens 3, filaments short, anthers forming a column, two 2-thecous,
one 1-thecous, ovary of pistillate flowers inferior, oblong or globose, style short, stigmas
2 or 3; fruits spherical to oblong, 18–30 cm long, 12–20 cm across middle, orange or
pale-yellow, or compressed subglobose, green and marbled-netted with paler color,
flesh moderately dry, not mealy and sweet; seeds white, smooth, margin narrow.
Probably native of South America, introduced to China from India.

Cucurbita pepo L. — SUMMER SQUASH, VEGETABLE MARROW, SUCCHINI SQUASH

Xi-hu-lu=Hsi-hu-lu (西葫蘆, Western Lagenaria); *Bei-gua=Pei-kua* (北瓜, Northern
Squash); *Jiao-gua=Chiao-kua* (絞瓜, Turning Gourd). Fruit; young ones cooked for
vegetable; full-sized inmature fruits sliced into long strings, dried on ropes or fences,
kept for winter uses in northeastern China; mature fruits cooked whole, cutting one
end for making an opening to remove the seeds, then turning a pair of chopsticks to
loosen the content into noodle-like strips, seasoned for vegetable, hence the vernicular
name *Jiao-gua* (Turning Gourd).

Annual trailing or erect plants, with stout prickles throughout the vegetative
organs, tendrils with several branches; leaves ovate-triangular in outline, 15–30 cm
long, 12–25 cm wide across the basal portion, palmately 5- or 7-lobed, the lobes ovate
and acuminate at the apex, sharply serrate; flowers yellow, solitary, monoecious,
protandrous, staminate flowers on stout peduncles 3–6 cm long, strigose and prickly,
hypanthia subcampanulate, sepals linear, corolla funnelform, 5-lobed, lobes involute
and crinkly, apex acuminate, ovary of pistillate flowers inferior, stigmas 3; fruits oblong
or slightly curved cylindrical-clavate, rind uniformly dark green, stalk strongly ridged-
grooved, expanded at the distal end, flesh white turning pale-yellow, coarse, moist;

seeds white, margin broad and thickened. New world origin, introduced and much cultivated in northern and northeastern China.

Gynostemma pentaphyllum (Thunberg) Makino — GYNOSTEMMA

Jiao-gu-lan=Chiao-ku-lan (絞股藍, Twisting Vine); *Qi-ye-dan=Ch'i-yeh-tan* (七葉膽, Seven-leaved Bile, the Guangxi ethnic name, indicating bitter taste). Leafy shoots used for tea.

Herbaceous perennial climbers, branchlets tender, striate-sulcate, slightly hairy, tendrils 2-branched; leaves palmately compound, leaflets variable in number, (3–) 5–7 (–9), elliptic-lanceolate, the middle ones 7–12 cm long, 3–4 cm wide, the lateral ones smaller progressively, acuminate, serrate, petioles 3–7 cm long; flowers small, dioecious, greenish-yellow, in axillary panicles 10–15 (–30) cm long, sepals 5, deltoid, corolla rotate, 5–6 mm across, the lobes ovate, lanceolate, stamens 5 in staminate flowers, subsessile, ovary of pistillate flowers inferior, styles 3, bifid; fruits globose, 5–8 mm in diameter, fleshy, ripe black; seeds 2, compressed-oblong, 4 mm long, rugose.

A troublesome weed in fields of northern China, thence southward to southeastern Asia. Leaves used for tea in Taiwan; it was recorded for tea before 1950; now, tea bags prepared by the Zhuang (壯) Ethnic Institute of Traditional Chinese Medicine, containing young leafy shoots with tendrils, broken into 4–10 (–20) mm pieces, dried and attractively packed in boxes, each containing 40 bags, with instructions in three languages (Chinese, Japanese and English), are available in American Chinese stores as *Jiao-gu-lan Tea*, it is exported by the Yao (瑤) Ethnic Institute of Food Manufacture, advertised as "Ginseng of the Tropics".

Hodgsonia heteroclita (Roxburgh) J. D. Hooker et Thomson — HODGSONIA

You-zha-guo=Yu-cha-kuo (油渣果, Fat-cracklings Fruit). Seed; used for oil, or roasted and eaten as a nut; raw nut toxic.

Large climbers, vines up to 20 m long, internodes 10–20 cm long, 8–12 cm in diameter, tendrils 2- to 5-branched; leaves ovate-cordate in outline, palmately 5-parted, 15–20 cm in diameter, the lobes ovate, acuminate, entire, glabrous, petioles 6–10 cm long, twisted or looped; flowers yellowish-white, showy, dioecious, the staminate flowers 5–7 in axillary racemes, peduncles 12 cm long, pedicels 6–10 mm long, pistillate flowers solitary, pedicels 2–2.5 cm long, hypanthia tubular, sepals deltoid, corolla rotate, 7.5–15 cm across, deeply 5-lobed, apex truncate and fringed, the fringes 15–20 cm long, twisted and hanging, stamens 3, filaments short, anthers connate, thecae linear, conduplicate and contorted, ovary inferior, globose, 2.5 cm in diameter, velvety, and covered with glandular warts, 3-carpellate, unilocular, with three parietal-placentas, each bearing 2 rows of ovules, style columnar-clavate, stigmas 3, bifid; fruits pomiform, 12–18 cm, 20–26 cm across, pale rose-brown, rind thin, firm, pulp spongy and juicy, holding 6 seed-complex, 7–10 cm long, 5–7.5 cm wide, 1–3 cm across the back, covered by fibrous coat and a woody wrapping with 1–3 seeds; seeds compressed, winged, with white seed coat, 2 large, flat, oily cotyledons.

Native of the eastern Himalayan Region, from Sikkim eastward to Yunnan and northern Vietnam, cultivated in the Tropical Botanical Garden at Xi-shang-ban-na (西 雙版納) in Yunnan.

Lagenaria siceraria (Molina) Standley — WHITE-FLOWERED GOURD, BOTTLE GOURD, HULU, CALABASH GOURD (Syn. *L. vulgaris* Seringe)

Hu-lu=Hu-lu (葫蘆, or 壺盧, used in medicinal practice); *Bao=Pao* (匏) or *Bao-gua=Pao-kua* (匏瓜, Lagenaria, the classical name); *Pu=P'u* (蒲) or *Pu-gua=P'u-kua* (蒲瓜, Lagenaria, the ethnic name of southern China); *Ku-hu-lu=K'u-hu-lu* (苦壺盧, Bitter Lagenaria). Young fruits, used as vegetable; mature fruits of the bitter variety, boiled with jujube for tea, given to people with swollen ankles.

Much branched annual climbers, glandularly hairy throughout, branchlets glabrescent, tendrils 2-branched; leaves cordate-reniform, 10–35 cm long and across the base, shallowly lobed, sharply serrate, apex acute, or acuminate, petioles 16–20 cm long, with 2 glands at the distal end; flowers white, solitary, axillary, monoecious, on peduncles longer than the petioles, hypanthia funnelform, 2 cm long, 1 cm across the distal end, sepals lanceolate, 5–8 mm long, corolla rotate, limbs 6–7 cm across, deeply 5-lobed, lobes suborbicular-obovate, wavy, apex acute, base cuneate, staminate flowers with stamens 3, filaments distinct, anthers coherent, twisted and forming an oblong column, two 2-thecous, one 1-thecous, pistillate flowers with inferior ovary, style columnar, stigmas 3, bifid; fruits lageniform, pyriform, cylindrical, clavate or compressed-globose, varying in size and shape according to varieties and cultivars, at maturity ochreous, hollow, rind firm and hard; seeds numerous, white, obovate, margin thickened, ends truncate.

Native of Old World Tropics, appearing in the earliest Chinese botanical records, extensively cultivated for vegetable, utensils, and for ornamental and medicinal purposes. Market material of varieties bearing edible fruits are harvested green, while the rind is tender, and the flesh white and crisp. Every part of the plant is used in traditional Chinese medicine; four Chinese varieties with some cultivars are given below.

1. var. **siceraria** — BOTTLE GOURD, HULU, PU (Figure 59)

Hu-lu=Hu-lu (葫蘆, Hulu, the name used in northern China); *Pu=P'u* (蒲, the name used in eastern Guangdong); *Bian-pu=Pian-p'u* (扁蒲, the name used in Guangzhou). Tender young fruits, used as vegetable.

Fruits in the market pale green, or maculate with white patches, pyriform, suddenly narrowed above the middle, distal end rounded or truncate, 20–30 cm long, 12–16 cm across the apical end, weighing 0.9–3 kg, available between June and September in Guangzhou, in July–August in northern China; mature fruits used for containers, dippers, and musical intrument; being a sign for the practice of tranditional Chinese medicine.

2. var. **hispida** (Thunberg) Hara — HERCULES CLUB, HUZI, CYLINDER GOURD

Hu-zi=Hu-tzu (瓠子, Huzi, the northern Chinese name); *Pu-gua=P'u-kua* (蒲瓜, Pugua, the Cantonese name). Young fruits, used for vegetable.

Fruits of market material pale green, cylindrical or clavate, 55–60 cm long, 12–13 cm across the middle, flesh white, weighing 700–900 g, some cultivars with crookneck. Ripe fruits ochreous, rind firm, woody, used for utensils.

3. var. **cougourda** (Seringe) Hara — LONG-NECKED HULU (Syn. *L. vulgaris* f. *cougourda* Seringe)

Chang-jing-hu-lu=Chang-ching-hu-lu (長頸葫蘆, Long-necked Hulu). Young fruit; cooked for vegetable.

A variety with strong vines up to 3 m long; leaves suborbicular, 15 cm long, 24 cm wide; fruits clavate, 45–48 cm long, 12 cm across the large apical portion, 1 cm across the slender handle, available for the market in June–September, individual gourd weighing 675–700 g; having a long history of cultivation in Guangzhou.

4. var. **microcarpa** (Naudin) Hara

Xiao-hu-lu=Hsiao-hu-lu (小葫蘆, Little Hulu). Ripe fruits, used for ornamental.

Fruits small, lageniform (strongly constricted at the middle) or subglobose-depressed at both ends, 8–10 cm long or across the middle; two cultivars are common in northern China.

a. cv. 'Yahulu' (*Ya-hu-lu=Ya-hu-lu* 亞壺盧, Lageniform Hulu)

Fruits lageniform, 8–10 cm long, 3 cm across the distal portion, ripe yellow, rind hard; used as toy or for decoration.

b. cv. 'Youhulu' (*You-hu-lu=Yu-hu-lu* 油壺盧, Katydid Cage Hulu)

Fruits compressed-globose, 7–9 cm high, 10–11 cm across, smooth, shiny light brown, rind hard, firm, 4–5 mm thick; used as a decorative cage for a special green, singing, short-winged grasshopper (Katydid) locally called *You-zi* (油子), hence the gourd is called *You-hu-lu*. In general, a *You-hu-lu* is very artistically carved, with patterns of animals and plants for good ventilation. The shortwing Chinese katydids are day-time singers, and raising them being the hobby of men-of-leisure living in large cities. Chinese *you-hu-lu* occasionally seen in American antique shops, sold as "Cricket Cage".

Luffa acutangula (L.) Roxburgh — Ridgy Luffa

Leng-si-gua=Leng-ssu-kua (棱絲瓜, Ridgy Silk Gourd); *Yue-si-gua=Yueh-ssu-gua* (粵絲瓜, Cantonese Luffa); *Si-gua=Ssu-kua* (絲瓜, Silk Gourd, the name used in American Chinese groceries). Young fruit; with the ridges and the associated skin pared level, cut into angular bite-size pieces, cooked slightly with marinated meat or chicken, or in soup, tender and crispy, available in most American Chinese groceries throughout the year.

Very coarse, much branched annual climbers, stems 3–6 m long, tendrils 3-branched; leaves cordate-orbicular, 15–20 cm long and across the base, shallowly 5- or 7-lobed, the lobes broad-triangular, wider than long, acute, serrate, both surfaces strigose; flowers yellow, monoecious, staminate flowers many, in cymose clusters on peduncles 10–15 cm long, stamens 3, free, two 2-thecous, one 1-thecous, anthers much convolute, pistillate flowers solitary, on very short pedicels, corolla rotate, 5 cm across, ovary inferior, ellipsoid, densely villose-hirsute and glandular, style columnar, stigmas 3, 2-fid, fleshy; young fruits cylindrical-clavate, with 8–12 strong and rough ridges and deep furrows, fully grown fruits 35–75 cm long, dark green, at maturity turning light brown; seeds numerous, black, ovate-oblong, reticulate, margin not winged, poisonous, 15–20 seeds causing vomiting and diarrhea.

Native of Old World Tropics, all parts of the species used for medicine by the ethnic peoples living in the borders of China, Vietnam, Burma, Yunnan; recorded in Chinese herbal in the middle the 12th century; extensively cultivated in Guangdong and Guangxi, for local uses as a summer vegetable, and for export. As vegetable crops, several cultivars have been selected in Guangzhou; for local consumption, selections were made for slender fruit with soft rind and crisp white flesh; for export, selections were made for rough hard rind, smaller size and short growing time leading to quick harvest and early market.

1. cv. 'Soft Rind' (*Yuan-pi=Yuan-p'i* 軟皮, Soft Skin); *Qing-pi-si-gua=Ch'ing-p'i-ssu-kua* (青皮絲瓜, Green Skin Luffa)

    Young fruits in the market cylindrical, 40–50 cm long, 4.5–5.5 cm in diameter, olive-green or yellow-green, 10-ridged, occasionally with 11 or 12 ridges, skin soft and thin, flesh thick and crisp. An excellent cultivar, can be cultivated for spring, summer and autumn seasons, individual fruit weighing 200–500 g.

2. cv. 'Long Luffa' (*Chang-si-gua=Ch'ang-ssu-kua* 長絲瓜, Long Luffa)

Market fruits cylindrical, 75 cm long, 4–4.2 cm in diameter, rind wrinkled, soft, green, with 10–12 strong ridges, dark green, flesh white, individual fruit weighing 495 g. An excellent cultivar planted for early market, harvesting 42–45 days after sowing the seed in March–April.

3. cv. 'Dark Rind' (*Wu-pi-si-gua=Wu-p'i-ssu-kua* 烏皮絲瓜, Black Skin Luffa)

Market fruits cylindrical-clavate, 40 cm long, 4–4.2 cm in diameter, rind dark green, with 10 ridges, individual fruits weighing 200–220 g; an excellent cultivar for early spring and late fall plantings.

4. cv. 'Hard Rind' (*Ying-pi-si-gua=Ying-p'i-ssu-kua* 硬皮絲瓜, Hard Rind Luffa)

Market material short-clavate, 30 cm long, 5.8 cm across the thicker portion, rind grayish-green, rough to touch and hard, with 8–10 ridges, flesh thick and white, individual fruit weighing 90–225 g; a favorite cultivar of farmers, planted for exporting the products.

Luffa aegyptiaca Miller — SPONGE GOURD, VEGETABLE SPONGE (Syn. *L. cylindrica* of most authors, not of Linnaeus)

Si-gua=Ssu-kua (絲瓜, Silk Gourd); *Shui-gua=Shui-kua* (水瓜, Water Gourd). Young fruit; partially peeled and cooked.

Much branched annual climbers, finely hairy throughout; leaves ovate in outline, deeply 5-lobed, the terminal lobe elongated-triangular, 2- or 3-times longer than wide, 8–12 cm long, acuminate, serrate, warty above, strigose beneath; flowers yellow, monoecious, staminate flowers 15–20, the buds crowded in the form of a head, opening one after another, leaving a long rachis resembling that of a raceme, hypanthia short campanulate, 5–9 mm long, hairy, 5-lobed, the lobes ovate-lanceolate, 0.8–1.3 cm long, 4–7 mm across the base, corolla rotate, 5–9 cm across, lobes suborbicular, 3–4 cm long, hairy, stamens 5, filaments 6–8 mm long, hairy at the base, anthers much convolute, pistillate flowers solitary, pedicels 2–10 cm long, ovary oblong-cylindrical, densely hairy, stigmas fleshy; fruits cylindrical, 15–55 cm long, 5–8 cm in diameter, fleshy before maturity, changing thin, yellow-brown, containing straw-like reticulate fibrous mass, dehiscing by apical operculum; seeds black, numerous, oblong, flat, 10–12 long, 8 mm wide, with thin marginal wing.

Native of Old World Tropics, cultivated near dwellings, on fences, over frames close to doors forming cooling shades in tropical villages of China. There, the use of the young fruits as vegetable is secondary, seldom seen as a vegetable crop. Two cultivars have been observed in Guangzhou.

1. cv. 'Long Water Gourd' (*Chang-shui-gua=Ch'ang-shui-kua* 長水瓜, Long Water Gourd).

Fruits at the edible stage 55 cm long, 5 cm in diameter, green, with dark green vertical stripes, flesh 1 cm in thickness, white; individual fruit weighing 675 g; seeds sowed in April–May, with the continuous branching, it takes 80–90 days before any fruit appears; bearing period continues for 5 months; poor quality for keeping and transportation, generally consumed by growers.

2. cv. 'Short Water Gourd' (*Duan-shui-gua=Tuan-shui-kua* 短水瓜, Short Water Gourd).

Fruits at edible stage oblong-cylindrical, 25 cm long, 5.5 cm across, both ends rounded, rind light green, with deep green vertical bands, flesh 1.4 cm thick, white; individual fruit weighing 450 g, harvesting period lasting for 150 days from August to November in Guangdong; poor quality for keeping and transportation, consumed by growers.

Momordica charantia L. — BITTER CUCUMBER, BITTER MELON, BALSAM PEAR (Figure 60)
Ku-gua=K'u-kua (苦瓜, Bitter Melon); *Lai-pu-tao=Lai-p'u-t'ao* (癩葡萄, Scabby Grape); *Hong-gu-niang=Hung-ku-niang* (紅姑娘, Red Maiden); *Jin-li-zhi=Chin-li-chih* (錦荔枝, Elegant Lychee). Fruit; fully grown green fruit, cooked with meat or fish making a delightful dish, tasting slightly bitter, available in American Chinese groceries; the juicy red pulp of mature fruit, sweet, eaten by growers in northern China.

Annual climbers, sparsely covered by multicellular straight hairs throughout, stems slender, striate-sulcate, tendrils unbranched; leaves reniform, orbicular in outline, 3–12 cm long and across, 5- or 7-parted, the segments ovate-oblong, pinnate-lobed, remotely serrate; flowers pale-yellow, monoecious, solitary, on slender elongated peduncles bearing a foliaceous, reniform bract slightly below the middle, hypanthia shallowly funnelform, 5-lobed, lobes ovate-lanceolate, 4–6 mm long, corolla rotate, 1.5 cm across, deeply lobed, the lobes obovate, hairy, stamens 3 in staminate flowers, filaments free, anthers sigmoid, two 2-thecous, one 1-thecous, ovary inferior, fusiform, warty, style columnar, stigmas 3, fleshy, bifid; fruits obovoid, oblong, or subcylindrical, 10–20 cm long, 5–11 cm across the basal end, tubercled and deeply furrowed, young pericarp bitter, mature pulp orange, juicy, sweet; seeds numerous, rectangular, 1.5–2 cm long, 1–1.5 cm across, with a groove along the margin and 2 teeth at each end, pale brown.

Native of tropical Asia, recorded as an edible plant in a Chinese herbal of the middle fifteenth century, cultivated in southern China as a vegetable crop, 3 cultivars known in Guangzhou area.

1. cv. 'Broad Shoulder' (*Da-ding-ku-gua=Ta-ting-k'u-kua* 大頂苦瓜, Big Shoulder Bitter Melon)

Market material obconical, 20 cm long, 11 cm across the basal end, green, tubercles long, prominent, flesh 1.3 cm thick, individual fruits weighing 225–540 g. A favorite cultivar selected for a wide range of seasonal adoptability, can be planted as spring crop in February–March, summer crop in April and autumn crop in July–August, and for the slightly bitter taste of the flesh.

2. cv. 'Smooth Body' (*Hua-shen=Hua-shen* 滑身, Slippery Body)

Market material subcylindrical-obconical, 25 cm long, 6 cm across the basal end, distal end pointed, smooth, deep green, shiny, tubercles few, flesh 8–9 mm thick, individual fruit weighing 225 g. Good quality for keeping and transport, strongly bitter, available in the market in May–July.

3. cv. 'Long Body' (*Chang-shen=Ch'ang-shen* 長身, Long Body).

Fruits in the market elongated-cylindrical, 30 cm long, 5 cm across, slightly pointed at the distal end, green, tubercles and ridges prominent, flesh 8 mm thick, compact, individual fruits weighing 225–450 g, good keeping and transporting quality, flesh bitter with sweet after taste; available in May–July.

Sechium edule (Jacquin) Swartz — CHAYOTE, CHRISTOPHINE (Figure 61)

Zhun-ren-gua=Chun-jen-kua (隼人瓜, Chayote); *Fo-zhang-gua=Fo-chang-kua* (佛掌瓜, Buddha's Palm Gourd); *Fo-shou-gua=Fo-shou-kua* (佛手瓜, Buddha's Hand Gourd). Young fruit; fully grown light green fruits, cooked for vegetable.

Much branched perennial herbaceous climbers with tuberous roots, tendrils 3- to 5-branched; leaves reniform-cordate, 10–20 cm long and wide, shallowly 3- or 5-lobed, scabrid above, sparsely hairy beneath; flowers small, greenish-white, monoecious, floccose in buds, staminate flowers 10–30, in spike-like pseudoracemose panicles, buds globose, hypanthia cupular, 5-lobed, lobes lanceolate, stamens 5, filaments united into a column, 3 mm long, papillose, divided at the apex, anthers 5, sigmoid, two 2-thecous, one 1-thecous, pistillate flowers solitary, sharing the same leaf-axil but developed later than the staminate pseudopanicle, ovary inferior, obovoid-ellipsoid, floccose, uniovulate; fruits compressed-obovoid, 5–10 cm long, 6–8 across the broad axis of the distal end, irregularly lobed, pale green, fleshy; seed 1, white, smooth, 4 cm long, 2 cm across. Native of tropical America, earliest specimen gathered in Yunnan in 1936, cultivated in Hong Kong, fruits occasionally available in the market.

Siraitia grosvenorii (Swingle) C. Jeffrey — LUOHANGUO (Syn. *Momordica grosvenorii* Swingle) (Figure 62)

Luo-han-guo=Lo-han-kuo (羅漢果, Buddha's Disciple Fruit). Ripe fruit; portions used in making soup; a special tea for soothing the throat of singers, very sweet; used as a substitute for sugar, both the dried fruit and the preparation available in American Chinese stores.

Robust perennial climbers with fleshy oblong or ovoid rootstocks 7–23 cm long, 6–12 cm across, branches emerging from the caudex annually, the young growth covered with yellowish-brown hairs mixed with black glands throughout, tendrils 2-branched; leaves triangular-ovate, 12–25 cm long, 5–17 cm across the base, acuminate-caudate, broadly cordate, wavy or remotely inconspicuously dentate, petioles (3–) 5–7 (–9) cm long; flowers yellow, 2.5–3.5 cm across at full anthesis, dioecious, staminate flowers 6–10 in axillary racemose cluster, peduncles 7–13 cm long, pedicels 5–15 mm long, hypanthia cupular, 4–5 mm long, 8 mm across, 5-lobed, lobes deltoid-lanceolate, 3 mm across the base, apex caudate-lanceolate, corolla 5-parted, segments oblong-ovate, 1–1.5 cm long, 7–8 mm across, acute, annula scaly, stamens 5, filaments distinct, 4 in two pairs, 1 odd, the filaments thickened and pilose at the base, anthers sigmoid, two 2-thecous, one 1-thecous, pistillate flowers solitary or 2–5 in cymose clusters, ovary inferior, sepals linear, corolla deeply 5-parted, segments acuminate and subulate at the apex, staminodes 5, well developed, 2–2.5 mm long, occasionally bearing pollen grains, ovary oblong-ovoid, 10–12 cm long, 5–6 mm across, basal end rounded, distal end slightly narrowed, densely pubescent and glandular, style 2.5 mm long, stigmas fleshy, bifid; fruits oblong, spherical or pyriform, 5–8 cm long, 4–6.5 cm across, pilose, villose or velutinous, smooth or inconspicuously striate, or with 6–11 ridges faintly radiating from the stalk, at maturity, smooth, drying light brown, with thin and brittle rind 1 mm in thickness, intensively sweet; seeds oblong, ovate or suborbicular, margin thickened, rough, deeply grooved.

Native to the mountains of northeastern Guangxi, extensively cultivated by the Zhong ethnic people and used as a sweetening agent since time immemorial; four superior cultivars reported in 1979 are given below, all requiring artificial pollination for fruit production.

1. cv. 'Changtan Fruit' (*Chang-tan-guo=Ch'ang-t'an-kuo* 長灘果, Luo-han-guo of Long Sandy Bank).

Leaves 17–22 cm long, 12–19 cm wide, petioles 5.5–8 cm long; flowers pale yellow; fruits oblong or ovoid-oblong, with 9–11 faint vertical lines, 7.5 cm long, 5.3 cm across; average weight of fresh fruits 68.3 g.

2. cv. 'Lajiang Pear' (*La-jiang-guo=La-chiang-kuo* 拉江果, Luo-han-guo of Lajiang).

Leaves 16–18 cm long, 8–13 cm across; fruits oblong-pyriform, densely velutinous, 6.5 cm long, 5.3 cm across, average weight of fresh fruits 77.37 g; an excellent cultivar with a large range of ecological adoptability, can be planted in low hillsides.

3. cv. 'King Luohanguo' (*Dong-gua-han=Tung-kua-han* 冬瓜漢, Winter Melon Fellow).

Leaves 14–25.5 cm long, 9–16.5 cm wide, petioles 5–6.5 cm long; fruits cylindrical-oblong, densely pilose, 6.6 cm long, 5.4 cm across, with 6 ridges radiating from the stalk at the basal end; average weight of fresh fruits 85.20 g, a good cultivar, with large fruits and high productivity.

4. cv. 'Green Ball' (*Qing-pi-guo=Ch'ing-p'i-kuo* 青皮果, Green Rind Luo-han-guo).

Leaves cordate, acute, petioles 3.5–6 cm long; flowers yellow, with red glandular hairs; fruits globose, reticulate, pilose, 5.38 cm long, 5.6 cm across, average weight of fresh fruits 74.05 g; a cultivar with a wide range of ecological adaptability, good for plantings on hillsides and plains, currently the most widely cultivated type, covering 75% of the known acreage.

Solena amplexicaulis (Lamarck) Gandhi ex Saldanha et Nicholson — SHED GOURD
(Syn. *Melothria heterophylla* [Loureiro] Cogniaux, based on *Solena heterophylla* Loureiro).
Mao-gua=Mao-kua (茅瓜, Shed Gourd); *Pu-gua=P'u-kua* (埔瓜, Plain Melon). Ripe fruit; gathered and consumed locally in Taiwan before 1950.

Small perennial climbers with fusiform fleshy roots, 10–25 cm long, 1–3 cm in diameter, branchlets slender, subglabrous tendrils simple; leaves ovate-triangular, hastate, cordate-quinquangular, or rarely subovate, 5–13 cm long, 2–7 cm across the base, 3- or 5-lobed, rarely entire, remotely denticulate, petioles 5–10 mm long; flowers pale yellow, monoecious, protandrous, staminate flowers 5–15 in corymbose cymes, hypanthia campanulate, calyx lobes minute, deltoid, corolla lobes deltoid, reflexed, stamens 3, filaments slender, anthers subglobose, hairy, pistillate flowers solitary, axillary, calyx lobes deltoid, corolla lobes reflexed, disk annular, papillose and pilose, ovary inferior, ellipsoid, 3–4 mm long, 4 mm across the middle, densely yellow floccose-strigose and glandular-punctate, style columnar, stigmas 3, fleshy; fruits orange-red, oblong-ellipsoid, 2.5–6 cm long, 1.5–4.5 cm across the middle; seeds 10–20, oblong-ellipsoid or obovoid, smooth. Widespread in southeastern Asia, common in southern China; roots used in traditional Chinese medicine.

Trichosanthes anguina L. — SNAKE GOURD

She-gua=She-kua (蛇瓜, Snake Gourd); *Mao-wu-gua=Mao-wu-kua* (毛烏瓜, Hairy Black Gourd). Young fruit; fully grown slender twisted gourd 50–100 cm long, used as vegetable, cultivated in Wuhan, Hubei Province.

Annual climbers, pubescent with mixed trichomes, pilose, strigose and glandular throughout, tendrils 2- or 3-branched, pilose; leaves ovate-reniform, 8–16 cm long, 3–12 cm across the base, subentire or shallowly 3- to 7-lobed, remotely denticulate, apiculate, petioles 3.5–8 cm long, pilose; flowers white, monoecious, staminate flowers 8–10 in racemose loose clusters on peduncles 5–12 cm long, hypanthia tubular, 2.5–3 cm long, 4–5 mm across, calyx lobes deltoid, 2 mm long, corolla rotate, 4 cm across, deeply 5-parted, the segments oblong-ovate, long-fringed along the margin, the fringes filiform, stamens 3, filaments short, free, anthers sigmoid, coherent, two 2-thecous, one 1-thecous, pistillate flowers solitary, sharing the same leaf-axil with a staminate inflorescence, ovary inferior, 3 cm long, 1 mm across, densely glandular-pilose, the trichomes reddish-brown, increasing in length fast, becoming curved and twisted, when ripe 1 m long, 3–4 cm in diameter; seeds 10–12, oblong, 11–17 mm long, 8–11 mm across, grayish-brown, thickened and angular along the margin. Native of tropical Asia, cultivated in China, tender fruits used as vegetable, root purgative.

Zehneria indica (Loureiro) Keraudren (Syn. *Melothria indica* Loureiro)

Mao-bao-er=Ma-pao-erh (馬㼲兒, an ancient aboriginal name, meaning not clear); *Lao-shu-la-dong-gua=Lao-shu-la-tung-kua* (老鼠拉冬瓜, Mice Pulling Winter Melon); *Lao-shu-gua=Lao-shu-kua* (老鼠瓜, Mice Melon); *Tu-bai-lian=T'u-pai-lien* (土白蘞, Local Bai-lian, the name for the fleshy root). Fruit; eaten locally in Hainan Island; root boiled with chicken for postpartum mothers to replenish blood.

Delicate perennial trailing or climbing herb with fleshy thickened roots, much branched, stems 1–2 m long, tendrils unbranched; leaves ovate-deltoid, 3–8 cm long, acute or acuminate, base variable, subtruncate or hastate-sagittate, entire or irregularly crenate, rough to touch; flowers small, white, unisexual and monoecious, 5 mm across, solitary or fasciculate, corolla campanulate, staminate flowers each bearing 3 stamens, distinct, 2 anthers 2-thecous, 1 anther 1-thecous, ovary in pistillate flowers inferior; fruits ovoid or oblong, 1.2–2 cm long; seeds smooth. Widespread in eastern and southeastern Asia, extending to Polynesia.

Campanulaceae: Bluebell Family

Adenophora tetraphylla (Thunberg) Fischer — ADENOPHORA (Syn. *A. verticillata* Fischer)
Nan-sha-shen=Nan-sha-shen (南沙參, Southern Shashen); *Sha-shen=Sha-shen* (沙參,
Shashen); *Pao-shen=P'ao-shen* (泡參, Froth Shen). Dried tap root; fusiform or
cylindrical, 10–15 cm long, often with outer skin rubbed off, pale brown, or whitish-
brown, light, breaking readily, taste bitter-sweet, with vertical and horizontal
wrinkles, sold in one-pound bags in American Chinese groceries; used with jujube,
lotus rhizome, and pork chops with bone for soup at the time of change of seasons
from cold to hot.

Perennial erect herbs with fleshy roots and short caudex from which emerging the
annual growth, stems 50–150 cm high, unbranched before flowering, glabrous; leaves
various, radical leaves of vegetative ramets reniform on elongated petioles, cauline
leaves 3–6 verticillate, subsessile, ovate-elliptic below, gradually changing to lanceolate,
2–14 cm long, 1.5–3 cm wide, acute, serrate; flowers blue or purplish-blue, in terminal
pseudopanicles, pedicels filiform, equal or slightly longer than the corolla, hypanthia
obconical, calyx lobes 5, persistent, corolla campanulate, hanging, 8–10 mm long,
stamens 5, filaments 5–6 mm long, the basal half thickened and hairy, anthers oblong,
disk tubular, ovary inferior, style clavate, shallowly divided; fruits obovoid, 8 mm long,
5 m across the truncate apex, dehiscing by 3 basal valves; seeds many, oblong, angular,
smooth.

A widespread species native of eastern Asia, growing on the sunny slopes among
grasses, extensively cultivated in China for the roots, with the best products coming
from Anhui, Jiangsu and Zhejiang. The market material consisting of dried tap-root,
fusiform or cylindrical, 15 cm long, 1 cm across, pale yellow or brown, with numerous
horizontal wrinkles, light, loose in structure, breaking readily, and taste bitter-sweet.
Actually, the market material consists of roots of more than one species of the genus,
gathered by people in various areas of China. The best known ones are *A. capillaris*
Hemsley from central and western China, *A. khasiana* Hooker f. et Thomson (*A. axilliflora*
Borbas) from Yunnan, *A. pereskiifolia* (Fisher ex Roemer et Schultes) G. Don from
northern China, *A. polyantha* Nakai from northern China and Inner Mongolia, *A. potaninii*
Korshinsky and *A. stricta* Miquel from eastern China. The market material of the
roots of these species, especially with the outer skin removed, are very hard to
distinguish.

Codonopsis lanceolata (Siebold et Zuccarini) Bentham et J. D. Hooker
Shan-hai-luo=Shan-hai-luo (山海螺, Mountain Periwinkle); *Nai-shen=Nai-shen* (奶參, Milk

Shen); *Ru-shu=Ru-shu* (乳薯, Milk Potato); *Tu-dang-shen=T'u-tang-shen* (土黨參, Wild Tangshen). Root; dried or fresh, cooked with pig-feet to prepare a dish for nursing mothers.

Lactiferous perennial climbing herbs, lateral branches short, slender, purplish, glabrous, branchlets 2 to 4; leaves crowded at the end of small lateral shoots, ovate-oblong or lanceolate, 3–10 cm long, 1.5–4 cm wide, acute at both ends, entire; flowers purple, creamy-white on the outside, 1 or rarely 2, terminal to the leafy branchlets, calyx consisting of large leafy sepals, ovate-lanceolate, 3 cm long, 1.5 cm wide, acute, corolla broadly campanulate, 4 cm long, 3 cm across, shallowly 5-lobed, lobes deltoid, reflexed, margin scabrid, stamens 5, inserted on the hypanthia opposite the sepals, filaments broadened at the base, anthers lanceolate-oblong, 6 mm long, thecae auriculate at the base, ovary inferior, subglobose, 13 mm across, style short, stigma 3-fid; capsules subglobose, 2 cm in diameter, persistent calyx foliaceous.

Widespread in eastern Asia, growing on hillsides and margins of forest; cultivated in Helongjiang, Zhejiang, Jiangxi, and Guangxi.

Codonopsis pilosula (Franchet) Nannfeldt — TANGSHEN (Syn. *Campanumoea pilosula* Franchet)

Dang-shen=Tang-shen (黨參, Ginseng of Shangdang, abbreviated, from the next name); *Shang-dang-ren-shen=Shang-tang-jen-shen* (上黨人參, Ginseng of Shangdang); *Liao-shen=Liao-shen* (遼參, Tangshen from Liaoning); *Shi-tou-shen=Shih-t'ou-shen* (獅頭參, Lion-head Tangshen). Dried root; market material represent dried root with the caudex; used to prepare soup with spareribs, astragalus roots, sliced Chinese yam, dried jujube, and lycium berries as a health food.

Lactiferous perennial herbs with fleshy tap roots 1–1.7 cm across, terminated by rough tuberculate caudex, annual growth of stem 1–2 m long, twining, much branched, hirsute, leaves on the prime shoots and secondary branchlets alternate, those on smaller branchlets appearing opposite, ovate, 1–7 cm long, 0.8–5.5 cm wide, acute or obtuse, base subtruncate-cordate, entire or wavy, strigose above, villose beneath, petioles 0.5–4 cm long, pilose-strigose; flowers yellowish-green, solitary, terminal to branchlets, hypanthia cupular, green, calyx 5-lobed, lobes oblong-lanceolate, obtuse, corolla campanulate, 1.8–2.3 cm long, 1.8–2.5 cm across, yellowish-green, with purple-violet spots on the inside, shallowly 5-lobed, stamens 5, filaments broadened at the base, 5 mm long, anthers oblong, 5–6 mm long, ovary inferior, 3-locular, ovules numerous, style short, stigmas 3; capsules hemispherical, conical at the distal end; seeds numerous, ovoid, smooth, brown.

Native of northern and western China, extensively cultivated, the root has been used as a substitute of ginseng, since the middle 17th century. The name, *Shang-dang-ren-shen* (上黨人參, Ginseng of Shangdang), was first recorded in 1695. By abbreviation, it became *Dang-shen* (黨參). Actually, roots of different species of *Codonopsis* Wallich ex Roxburgh have been gathered from wild populations or cultivated sources and sold as tangshen. The best known ones are *C. nervosa* (Chipp) Nannfeldt from Sichuan, and southern Gansu, *C. tsinlingensis* Pax et Hoffmann from the Tsinling mountains in Gansu and Shaanxi thence southward to western Sichuan, *C. tubulosa* Komarov from western Guizhou, Sichuan and Yunnan, and *C. viridiflora* Maximowicz from Qinghai, Shaanxi, and Gansu. Experienced dealers are able to distinguish the products partly from the areas of their production and partly by their sizes, colors, textures and tastes. Consumers do not know the difference.

Codonopsis tangshen Oliver — Sᴜᴄʜᴜᴀɴ Tᴀɴɢsʜᴇɴ

Chuan-dang-shen=Chuan-tang-shen (川黨參, Tangshen of Sichuan); *Chuan-dang=Chuan-tang* (川黨, Tangshen of Sichuan, abbreviated from the preceeding name); *Fang-dang=Fang-tang* (防黨, Tangshen from Fangxian in Western Hubei). Dried root; used in the same way as with *dang-shen*; the market material in American Chinese stores consists of this type mostly.

Perennial climbers with fusiform-cylindrical fleshy roots 15–30 cm long, 1–1.5 cm across, grayish-yellow, smooth, only the upper 1 or 2 cm with few lines, annual growth emerging from the rough caudex, 1–3 m long, 2–3 mm thick, lateral branches 15–50 cm long, tertiary branchlets 1–5 cm long, glabrous; leaves alternate, ovate or ovate-lanceolate, 2–8 cm long, 0.8–3.5 cm wide, obtuse or acute, base cuneate or obtuse, rarely cordate, wavy and faintly serrate, petioles 0.5–2 cm long; flowers solitary, pale-green or yellowish, purple at the basal portion, hypanthia cupular, calyx deeply parted, segment oblong-lanceolate, 1.4–1.7 cm long, 5–7 mm wide, acuminate, corolla campanulate, 1.5–2 cm long, 2.5–3 cm across, with purple spotted inside, shallowly 5-lobed, lobes deltoid, stamens 5, filaments 7–8 mm long, anthers 4–5 mm long, ovary inferior, style columnar, 7–9 mm long, enlarged at the apex, stigma 3-lobed; capsules subglobose, conical at the distal end, 2–2.5 cm in diameter; seeds numerous, oblong, yellowish-brown, shiny.

A native of the mountains on the borders of Sichuan, Hubei, Hunan and Guizhou. It was already in cultivation in the late nineteenth century, with the nomenclatural type being gathered from a farm in western Hubei and the specific epithet taken from the name given by a farmer.

Platycodon grandiflorus (Jacquin) A. de Candolle — BALLOON FLOWER

Jie-geng=Chieh-keng (桔梗, Balloon Flower); *Ling-dang-hua=Ling-tang-hua* (鈴鐺花, Hand-bell Flower); *Bao-fu-hua=Pao-fu-hua* (包袱花, Wrapper Flower); *Si-ye-cai=Ssu-yeh-ts'ai* (四葉菜, Four-leaved Vegetable); *Ji-ni=Chi-ni* (薺苨, an ancient classical name, original meaning obscure). Young shoots collected in March–April, used as potherb; roots gathered after the leaves withered in the fall, soaked, sliced, partially dried, pickled in salt, garlic, red pepper, and zanthoxylum pericarp, a special preparation in northern China; dried root, use for soup or tea, available in one-pound packages in American Chinese stores.

Lactiferous perennial, glabrous-glaucous, erect herbs 40–100 cm high, with fleshy carrot-like roots up to 20 cm long; leaves opposite or in whorls of 3, ovate-lanceolate, 2–7 cm long, 1–3 cm wide, acute, base obtuse, serrate, glaucous beneath; flowers blue or white, one or several, terminal to the aerial branch, showy, hypanthia obconical or subglobose, calyx 5-lobed, lobes 5–9 mm long, acute or caudate, corolla broadly funnelform-campanulate, 1.5–4 cm long, 4–6 cm across, 5-lobed, 1–1.5 cm long, 9–12 mm across the base, stamens 5, filaments filiform, 3–4 mm long, base densely barbate, anthers linear, 7–8 mm long, ovary inferior, 5-locular, multi-ovulate, style filiform, gradually thickened, 5-lobed, stigmas spathulate; capsules ellipsoid or ovoid-conical, 1.5–2 cm long, 1.3 cm across the middle; seeds compressed-oblong, 1.5 mm long, 0.1 mm wide, shiny, brown.

Native of termperate eastern Asia, widespread in China, occurring even on the foggy hill-tops in Hong Kong, much cultivated for the roots, used in traditional Chinese medicine; frequently seen in American gardens for ornamental purposes.

Wahlenbergia marginata (Thunberg) A. de Candolle — BLUE FLOWER SHEN (Syn. *W. gracilis* Auctt., non [G. Forster] A. de Candolle, for the species of Southern Hemisphere.)

Lan-hua-shen=Lan-hua-shen (藍〔or 蘭〕花參, Blue Flower Shen); *Wa-er-cai=Wa-erh-ts'ai* (娃兒菜, Baby Vegetable); *Lan-hua-cao=Lan-hua-ts'ao* (藍花草, Blue Flower Herb); *Hu-lu-cao=Hu-lu-ts'ao* (葫蘆草, Lagenaria Herb); *Jin-xian-diao-hu-lu=Chin-hsien-tiao-hu-lu* (金線吊葫蘆, Golden-thread Holding Hulu); *Niu-nai-cao=Niu-nai-ts'ao* (牛奶草, Milk Herb). Entire plants including the fleshy root and caudex (called "*Shen*"); used to cook with chicken or pork for postpartum mothers; young plants for vegetable.

Small caespitose lactiferous perennial herbs, erect, 10–50 cm high, with white carrot-like root 10 cm long, 4 mm across, stems slender, glabrous; leaves variable, subsessile, the lower ones crowded, spatulate or oblanceolate, 1–4 cm long, acute, undulate, median ones linear, the uppermost ones bract-like; flowers small, blue turning purplish, solitary, protandrous, on pedicels 1–5 cm long, hypanthia obconical, 2.5 mm long, glabrous, calyx 5-lobed, corolla subfunnelform, 5–8 mm long, apical 2/3 divided, stamens 5, anthers oblong, ovary inferior, 3-locular, cells multiovulate, style clavate, 3-fid at the apex, stigmas 3, lanceolate; capsules turbinate, 6–8 mm long, truncate at the apex, ridged and reticulate, the areoles of the pericarp hyaline-membranous.

Widespread in the warm temperate and the subtropical areas of eastern Asia, occurring along roadside, by paths on the hillsides, in sandy areas, on rock crevices where some soil is collected, and even among grasses in less disturbed areas of lawns in college campus. It seems that the tiny dustlike seeds carried with the debris to the upper atmosphere by storms, mushroomed and settled down, germinated and established wherever ecological conditions permitted. The first record of the species in Chinese herbals appeared in the middle fifteenth century. At present over 20 vernacular names have been recorded, some as potherbs while others are used for soup, the roots boiled with chicken or spareribs. The great variety of local names seems to indicate that ethnic groups discovered the uses of the plants independently.

Asteraceae (or Compositae): Aster (or Sunflower) Family

Anaphalis margaritacea (L.) Bentham et J. D. Hooker — Pearly Everlasting
Mao-nü-er-cai=Mao-nü-erh-ts'ai (毛女兒菜, Hairy Girl Vegetable); *Lai-xiao=Lai-hsiao* (蘋蕭, Anaphalis); *Zhu-guang-xia-qing=Chu-kuang-hsiang-ch'ing* (珠光香青, Pearly Anaphalis); *Da-ye-bai-tou-weng=Ta-yeh-pai-t'ou-weng* (大葉白頭翁, Big-leaved White-haired Gentleman); *Da-huo-cao=Ta-huo-ts'ao* (大火草, Big Fire-herb); *Yi-mian-qing=I-mien-Ch'ing* (一面青, One-side Green). Whole plant; cooked with pork or green-shelled duck eggs, used by ethnic groups living on the mountains in Taiwan, and those in mountains of Sichuan and Guizhou on the mainland.

Caespitose, erect, aromatic, perennial herbs 10–100 cm high, young growth woolly throughout, new plants emerging from the base of upright stems, rhizomes 10–15 cm long, the distal end of each rhizome turning upwards and continuing the growth into the leafy and flowering aerial shoots; leaves alternate, sessile, linear-lanceolate, 5–10 cm long, 0.3–1.5 cm wide, acute, entire with grayish- or brownish-white persistent woolly hairs beneath, arachnoid and becoming glabrascent above; flowers very small, monoecious, all tubular, heads forming corymbose clusters terminal to the stems, 4–8 mm in diameter, phyllaries snow-white, ovate-oblong, scarious, receptacles smooth, florets pistillate, the central 5 or 6 staminate corolla filiform, 4 mm long, ovary inferior, style very slender, stigma bifid; achenes very small, oblong-terete, 0.3 mm long, papillose, brown; pappus capillary, bristly.

Widespread in the cold temperate Northern Hemisphere; in Asia, the species occurs in the alpine slopes and ridges in western China and in Taiwan.

Arctium lappa L. — GREAT BURDOCK, EDIBLE BURDOCK (Syn. *Lappa edulis* Hort.)
Niu-bang=Niu-pàng (牛蒡, Ox Flank); *Bang-weng-cai=Pang-wung-ts'ai* (蒡翁菜, Burdock); *Xiang-er-cao=Hsiang-erh-ts'ao* (象耳草, Elephant-ear Herb); *Niu-bang-geng=Niu-pang-keng* (牛蒡根, Burdock Root). Root and leaves; used for vegetable.

Biennial herbs 1–1.5 m high with white fleshy tap-root, before flowering all leaves radical, in rosette, stems 1–2 cm in diameter; radical leaves ovate-cordate, 15–32 cm long, 30–40 cm wide, rounded-obtuse and apiculate, base sinuate-cordate, cauline leaves alternate, ovate, becoming progressively smaller, green and slightly scarid above, arachnoid and glandular beneath, wavy and inconspicuously serrulate, petioles stout, shorter than the laminas; flowers purple, all tubular, perfect, in subglobose heads forming corymbose terminal clusters, heads 1.5–3 cm in diameter, involucre green, arachnoid when young, phyllaries numerous, lanceolate, aciculate-hooked at the apex, receptacles scales subulate, corolla subequally cleft, lobes elongated; achenes smooth, oblong, flattened, narrowed toward the basal end, 6–7 mm long, 3 mm across the broad end, gray, with horizontal black patches; pappus short, golden, bristly, falling individually. Native of temperate Eurasia, wild on the hillsides of western China, cultivated in northern China for the seed used in traditional Chinese medicine.

Artemisia brachyloba Franchet — MONGOLIA SAGEBRUSH
Yan-hao=Yen-hao (岩蒿, Cliff Artemisia); *Shan-hao=Shan-hao* (山蒿, Mountain Artemisia). Roots and leafy shoots, used in Inner Mongolia for tea.

Suffrutescent plants 20–50 cm high, basal stems 1–1.5 cm in diameter, lignified, the branchlets slender, villose; leaves deltoid or ovate in outline, 2–4 cm long, 1.5–2 cm across the base, the lower ones bipinnatisect, the upper ones subtending the short flowering branches becoming progressively decreased in size and divisions, ultimate segments linear, acute or obtuse, margin recurved, glabrous above, white tomentose beneath; flowers heterogamous, small, in capitula 2.5–4.5 mm long, 2–3.5 mm across, in loose spikelike axillary or terminal clusters, involucres ovoid, up to 4 mm long, 3.5 mm across, phyllaries 3- or 4-seriate, ovate, obtuse, tomentose, margin scarious, florets yellow, the outer 8–12 pistillate, with tubular-filiform corolla, the inner 18–22 bisexual, corolla tubular-campanulate, 5-lobed; achenes brown, 1.5 mm long, glabrous; pappus inconspicuous. Common on grassy hillsides or on rocky cliffs, in arid and semidesert areas.

Artemisia lactiflora Wallich ex de Candolle — WHITE MUGWORT

Jiao-cai=Chueh-ts'ai (角菜, Horn Vegetable); *Ya-jiao-cai=Ya-chaio-ts'ai* (鴨腳菜, Duck-feet Vegetable); *Tian-cai-zi=T'ien-ts'ai-zu* (甜菜子, Sweet Potherb); *Si-ji-cai=Su-chi-ts'ai* (四季菜, Four-season Vegetable); *Bai-bao-hao=Pai-pao-hao* (白苞蒿, White-bracted Artemisia); *Ya-jiao-ai=Ya-chiao-ai* (鴨腳艾, Duck-feet Artemisia). Young shoots; freshly pinched tender portions from plants in kitchen garden, used year round in Hong Kong, available in special restaurants as a delicacy served with Peking Duck (personal experience).

Caespitose perennial herbs 60–150 cm high in flower, strongly aromatic, young growth sparsely pilose, glabrescent, stems 3–5 mm in diameter, striate; leaves ovate in outline, 10–18 cm long, 6–12 cm wide, ternately parted, the terminal segments 3-lobed, fancifully resembling duck's feet (hence the local name "Duck-feet Vegetable"), lobes serrate, apex rounded or acute; flowers small, all tubular, in small globose heads on spikelike branches 2–12 cm long, often interrupted, together having the appearance of a loose panicle, heads 3–4 mm across, consisting of 10–12 florets, involucre scarious, persistent, receptacle flat, outer 3 or 4 florets pistillate, without perianth, the remaining 7–8 florets pale yellow, corolla 2 mm long, the distal half enlarged, shallowly lobed; achenes obovate, 1 mm long, without pappus.

Native of southern China, and eastern Himalayan region, often cultivated in kitchen gardens, and/or in pots, propagated by stock division, young shoots growing out from old stocks continuously, introduced to America by Chinese immigrants for the same purpose.

Artemisia selengensis Turczaninow ex Besser. — ARTEMISIA SHOOT

Lou-hao=Lou-hao (蔞蒿, Slender Artemisia); *Hao-cai=Hao-ts'ai* (蒿菜, Artemisia Vegetable); *Lü-hao=Lü-hao* (驢蒿, Ass Artemisia); *Shui-hao=Shui-hao* (水蒿, Water Artemisia). Tender leafless shoots; gathered as they emerge from the soil, eaten as a delicacy in northern China.

Aromatic perennial herbs 60–150 cm high, with short creeping rhizome, young growth arachnoid throughout, the lower portion of the primary stem and the upper leaf-surface becoming glabrescent with age, stems 5–9 mm in diameter, striate, with pith; leaves numerous, ternately parted, segments of the lower leaves pinnately lobed, herbaceous, deteriorated long before anthesis, the middle cauline leaves with subcoriaceous entire segments 3–15 cm long, 2–6 mm wide, green above, gray tomentose beneath, accuminate and acute, petioles winged; occasionally some samples with remotely serrate segments; flowers small, all tubular, in oblong heads on axillary and terminal racemes together forming a compact ellipsoid brush-like panicles 30–45 cm long, 3–7 cm across the middle, terminal to the annual leafy stems; capitula oblong, 4 mm long, arachnoid, phyllaries few, imbricate, ovate, florets about 15, the 5 outer ones pistillate, 10 perfect, corolla 4 mm long, the distal half broadened, subcampanulate, anthers yellow, apical appendages triangular-subulate, white; seeds ellipsoid, brown, rostrate at the apex.

Endemic to northern China, growing along banks of streams and undisturbed waste areas such as unattended graves, leafy shoots having a strong scent, asses graze on the new growth (hence the name "*Lü-hao*").

Aster ageratoides Turczaninow — White Aster (Syn. *A. trinervius* sensu Forbes et Hemsley, non D. Don)

Ye-fen-tuan-er=Yeh-fen-t'uan-erh (野粉團兒, Wild Piece-of-dough); *Shan-bai-ju=Shan-po-chu* (山白菊, Hillside White Chrysanthemum). Young leaves; used locally, not available in the market.

Caespitose perennial herbs 40–100 cm high, with glabrous short off-shoots developed from the caudex before anthesis, stems pilose, 2 mm in diameter, branched only before flowering; basal leaves disintegrated early, cauline leaves elliptic-lanceolate, 8–14 cm long, 2–3 cm wide, acuminate at both ends, triplinervious, remotely serrate or serrulate, sparsely pilose with appressed hairs on both surfaces and with some glands beneath; flowers white, in heterogamous heads arranged in loose corymbs, capitula radiate, 1.5–2 cm across, involucres obconical-hemispherical, 7–10 mm across, phyllaries imbricate, in 3 series, outer ones ovate, inner ones oblong, coriaceous, ciliate, ray florets 10–14, ligules white, tinged purple or pink, 10–11 mm long, 2–2.5 mm wide, discoid florets numerous, yellow, corolla tubular, 4.5–5.5 mm long, stamens 5, ovary inferior, style slender, branches flattened, with triangular papillose apical appendages; achenes ovate-oblong, 2–2.5 mm long, grayish-brown, pappus shiny-brown.

Native of eastern Asia, extensively distributed in China, growing on hillsides, very variable; many varieties recognized.

Aster scaber Thunberg — Rough Aster (Syn. *Doellingeria scabra* [Thunberg] Nees)

Dong-feng-cai=Tung-feng-ts'ai (東風菜, East Wind Potherb); *Shan-ge-lu=Shan-ke-lu* (山蛤蘆, Lizard Reed); *Shan-bai-cai=Shan-pai-ts'ai* (山白菜, Hillside Cabbage); *Xian-bai-cai=Hsien-pai-ts'ai* (仙白菜, Immortal Cabbage). Young shoots; used for potherb.

Robust perennial herbs 1–1.5 m high, stems glabrescent, 3–7 mm in diameter, developed from the apical buds of short rhizomes; leaves scabrid, dentate, the radical ones triangular-ovate, 5–12 cm long, 2.5–12 cm across the base, on petioles longer than the laminas, disintegrated before anthesis, cauline leaves ovate-cordate, middle ones 8–14 cm long, 6–14 cm wide, on petioles shorter than the laminas, upper leaves progressively smaller, changing from ovate to oblong-elliptic on lateral branches, base truncate or with broad sinus, often decurrent forming wings on the upper half of the petiole; flowers in heterogamous heads forming loose terminal corymbose clusters, capitula 18–24 mm in diameter, involucres hemispherical, 4–5 mm across, phyllaries imbricate, in 3 layers, ovate-oblong, margin membranous, obtuse or rounded at apex, ligulate florets 10, tube 3–3.5 mm long, limbs 11–15 mm long, corolla of tubular florets 5.5 mm long, campanulate, the lower portion suddenly narrowed, stamens 5, anthers oblong, base obtuse, ovary inferior, style slender, stigmas 2, with deltoid hairy appendages at the apex; achenes oblong-obovoid, with 6 vertical ridges; pappus yellowish-brown, 3.5–4 mm long, scabrous.

Native of estern Asia, extensively distributed in China, especially in the north, growing on hillsides; fleshy roots and leaves used locally as a home remedy for snake bites.

Carthamus tinctorius L. — Safflower, False Saffron (Figure 63)

Hong-hua=Hung-hua (紅花, Red Flower); *Hong-hua-cai=Hung-hua-ts'ai* (紅花菜, Red Flower Potherb); *Hong-lan-hua=Hung-lan-hua* (紅藍花, Red-flowered Indigo); *Ci-hong-hua=Tz'u-hung-hua* (刺紅花, Prickly Red Flower). Young plants; used for potherbs by ethnic Rong (戎) in western Sichuan; dried corolla used for tea, especially for mothers during the postpartum period, material imported from China and distributed in mixed-tea packages avaible in American Chinese stores.

Stiff, spinose erect annual or biennial herbs, 60–90 cm high, stems striate, basal portion lignified, branching above the middle, all branches terminated by a flowering head; leaves varying by position, radical leaves in rosette, oblanceolate, pinnatifid, petiolate, herbaceous and non-spinose, lower cauline leaves elliptic or lanceolate, 4–12 cm long, 1–3 cm wide, sessile, acute, spinose at the apex, upper cauline leaves ovate-oblong, amplexicaul, sinuate and spinose or prickly; flowers orange-red, all tubular, in large terminal solitary ovoid capitula 2–3 cm across the base, narrowed into a bottleneck from which emerging the corolla, phyllaries prickly, the outer ones thickly coriaceous, enlarged and becoming foliaceous, elliptic or obovate and rounded at the base, varying by the cultivars, inner ones lanceolate, acute, receptacles flat, paleate, paleae filiform, hyaline and snow-white, connate at the base, 8–11 mm long, curved inward following the vertical contour of the involucre, corolla 3.5 cm long, the basal 2.5 cm filiform, then suddenly enlarged and immediately divided into 5 unequal linear lobes, 5–7 mm long,

0.5–1 mm wide, exposed from the aperture of the involucre, stamens 5, subsessile, anthers forming a slender tube 4 mm long, apical appendages ovate, 1.5 mm long, slightly incurved and forming a hollow dome collecting pollen grains, style 3 cm long, stigmatic lobes 1 mm long, slightly exposed from the anther-tube, apical appendages 2 mm long, connate and forming a papillose column, completely exposed; achenes obovoid, 8 mm long, 5 mm across the broader portion, inconspicuous 4-striate, basal end oblique, white, without pappus. Native of the Mediterranean Region, from western Asia (East Mediterranean) to the Persian Gulf, recorded in Chinese herbals of the early 11th century, now commercial cultivation centered in Sichuan, Henan and Zhejiang, for the dried corollas which are harvested by hand daily in the early morning.

Chrysanthemum coronarium L. var. **spatiosum** Bailey — EDIBLE CROWN DAISY, EDIBLE GARLAND CHRYSANTHEMUM (Syn. *Chrysanthemum spatiosum* Bailey).

Tong-hao=T'ung-hao (茼蒿, Crown Daisy); *Peng-hao=P'eng-hao* (蓬蒿, Horseweed Mugwort); *Hao-cai=Hao-ts'ai* (蒿菜, Crown Daisy Potherb). Young plants; tender stem with radical leaves only, used for quick cooked potherbs or put in boiling soup just before serving; cultivated in kitchen gardens of Chinese American in Boston area; with strong odor, but once a person learns to eat it would love the vegetable.

Glabrous aromatic annual herbs 60–90 cm high, young leafy plants tender like spinach, stems terete, 3–9 mm in diameter, striate, rather succulent; leaves herbaceous, radical ones spatulate, wavy, pinnately lobed or coarse serrate, cuneate with winged petioles or subsessile, middle cauline leaves obovate-oblong, 8–12 cm long, 1.5–4.5 cm across, pinnate partitate, segments pinnately lobed, sessile and subamplexicaul or auricular; flowers pale yellow, both radiate and discoid forming large head terminal to leafy branchlets, capitula 4–6 cm in diameter, involucres hemispherical, 1.8 cm across, phyllaries ovate-oblong, with brown scarious margin, ligulate florets 1.5–2 cm long, limbs obovate-oblong, 1–1.5 cm long, 5 mm wide, toothed at the apex, tubular florets 1 cm long, all perfect, anther-tube included, stigmatic branches recurved, exposed, apical appendages short, truncate, papillose; seeds oblong-obovoid, angular and ridged, narrowly winged along one ridge, apex truncate, base cuneate, light brown. The species, *C. coronarium* L. is a native of the Mediterranean Region, cultivated for ornamental purposes in large cities throughout China. Records in ancient Chinese herbals can be traced to the early seventh century when it was used as food. Apparently, the variety with less cut and rather succulent leaves, *C. coronarium* var. *spatiosum* Bailey was developed in China, even with selected polyploid forms; two cultivars common in the vicinity of Guangzhou forming the important winter crops of the area.

1. cv. 'Greater Tonghao' (*Da-ye-tong-hao=Ta-yeh-t'ung-hao* 大葉茼蒿, Large-leaved Crown Daisy).

Market material 21 cm high, spread 28 cm across; leaves 18 cm long, 16 cm wide, thick with wrinkles, margin shallowly lobed, petioles 1.4 cm long; sowed in September–January, harvested in November–March; flowers appearing in mid-February, seeds mature next April; excellent eating quality.

2. cv. 'Lesser Tonghao' (*Zhai-ye-tong-hao=Chai-yeh-t'ung-hao* 窄葉茼蒿, Narrow-leaved Crown Daisy).

Market material 20 cm high, spread 18 cm across; leaves 12 cm long, 5 cm across, think and smooth, margin deeply lobed; sowed in September–December, flowers appearing in early February, seeds mature in March, good eating quality.

Chrysanthemum X morifolium Ramatuelle — FLORIST'S CHRYSANTHEMUM (Syn. *C. sinense* Sabine; *Dendranthema morifolium* [Ramatuelle] Tzvelev).

Ju-hua=Chu-hua (菊花, Chrysanthemum); *Ju-hua-miao=Chu-hua-miao* (菊花苗, Chrysanthemum Shoot); *Hang-ju=Hang-chu* (杭菊, Hangzhou Chrysanthemum); *Bai-cha-ju=Pai-ch'a-chu* (白茶菊, White Tea Chrysanthemum). Young shoots; chopped, cooked with rice, jujube and lycium-berry for a gruel; fresh white ligulate florets from large heads, freed from the receptacle, washed and soaked in water to get rid of possibly harbored insects, used for special dishes in banquet as a delicacy; dried commercial products from Hangzhou, used with green tea or as a substitute for tea, available in one-pound packages in American Chinese groceries.

Strongly aromatic caespitose perennial herbs 60–100 cm or more high, pilose throughout, stems terete, striate-sulcate, branched at tne distal portion after anthesis, forming new shoots after the removal of the old flowering trunks; leaves all cauline, the lower ones ovate, 5–10 cm long, 3–7 cm wide across the basal portion, 3- or 5-lobed with rounded sinus, lobes coursely dentate or serrate, acute, base subtruncate and shortly decurrent, petioles, stipules present or not, foliaceous and often divided; flowers both heterogamous, shape, size and color varying greatly according to the cultivars, capitula solitary or in loose terminal corymbose clusters, involucres cupular, phyllaries imbricate, oblong, with scarious margin, receptacles flat, without paleae, ligulate flowers sterile, the limbs oblong, filiform, straight or curved, succulent, yellow, white, red, maroon, uniform or bicolored, tubular flowers yellow, the numbers varying from numerous to none. Like wheat and corn, chrysanthemum is a species with obscure origin. It is a hexaploid, with *Chrysanthemum indicum* L., *C. lavandulaefolium* (Fischer) Makino, and *C. boreale* Makino being some of its progenitors. The name can be traced to the beginning of the Christian Era in Chinese herbals. At present, the cultivation of chrysanthemum in China can be divided into two areas, as an art by the affluent people of leisure, for the unique forms, large or unusual sizes, and beautiful colors, and as a crop by rural people for the supply of dried flowers for tea and also for herbal medicine, and the dried roots and leafy shoots for medicinal purposes. As a crop, Zhejiang, Anhui, Henan and Sichuan are important areas of production.

Chrysanthemum nankingense Handel-Mazzetti — NANKING CHRYSANTHEMUM SHOOT
   (Syn. *C. boreale* Auctt., non Makino)
   Ju-hua-lao=Chu-hua-lao (菊花荖, Chrysanthemum Shoot); *Ju-hua-nao=Chu-hua-nao* (菊花腦, Chrysanthemum Brain). Tender shoots; young leafy plants, nipped off top portion with 4–5 leaves, used for potherb, cooked by quick fry.

Perennial caespitose herbs up to 1 m high; leaves ovate, 5–7 cm long, 4–6 cm across the basal portion, pinnatipartite, segments lobed and dentate, dark green and sparsely pilose on both surfaces, obtuse or acute, base subcordate or truncate, suddenly narrowed into petiolar wings, petioles 1–2 cm long, stipules evident, green; flowers yellow, radiate and discoid, capitula 1–1.5 cm across, in loose corymbose clusters, involucre hemispherical, phyllaries 4-seriate, imbricate, herbaceous, margin membranous, brown, outer ones pilose, receptacles conical, without paleae, ligulate florets 5 mm long, pistillate, tubular florets 3 mm long, with the exception of the central 4 or 5 all staminate, corollas papillose, the distal half enlarged and campanulate, staminal tubes included, apical appendages triangular-ovate, stigmas of the central perfect flowers exposed, recurved, apical appendage small, truncate, papillose; achenes black, oblong, 1.4 mm long, truncate, basal end oblique, striate and tuberculate. Growing in Nanjing in a semiwild condition by walls or along hedges of families with large yards, gathered in spring as new shoots emerging, introduced from Nanjing by Lucy Hsie, wife of Professor C. S. Sie, former Dean, Agricultural College, Nanking University, before 1950, to her kitchen garden in Hightstown, New Jersey; propagules from this garden transplanted to the experimental garden of the Department of Organismic Evolutionary Biology, Harvard University, Cambridge; young shoots harvested from June to September, then left to flower; limited cultivation as an annual crop from seed in Shanghai since the early 1920s (Anonymous, 1959).

Cichorium endivia L. — ENDIVE
   Ju-ju=Chu-chu (菊苣, Chrysanthemum Lettuce); *Ku-ju=K'u-chu* (苦苣, Bitter Lettuce). Leaves; used in salad, newly introduced, cultivated for large hotels with foreign visitors, rare in China.

Lactiferous subglabrous annual or biennial herbs, at anthesis 30–100 cm high; basal leaves in a rosette, oblong, lobed, curled, dentate to deeply pinnatifid, in cultivation forming a pale yellowish-green ellipsoid head 10–12 cm long, 2.5–4 cm across the middle, cauline leaves successively reduced in size; flowers violet-blue, all ligulate, bisexual, peduncles of the terminal capitula strongly thickened, involucres 1–1.4 cm long, 4–10 mm across, phyllaries 2-seriate, lanceolate, the inner ones twice as long as the outer, receptacles flat, without paleae; achenes obovoid, striate; Pappus of scuffy scales. Native of southern Europe, rarely cultivated in China.

Cichorium intybus L. — CHICORY (Figure 64)

Ye-ku-ju=Yeh-k'u-chu (野苦苣, Wild Endive). Leaves; cooked as potherbs; roots, making a substitute of coffee, gathered from roadsides and used in Xinjiang and northerneast China locally.

Lactiferous, hairy, rather stiff perennial herbs 20–120 cm high; basal leaves in rosette, lanceolate, 7–30 cm long, 1–12 cm wide, runcinate, pinnatifid, irregularly lobed, or dentate, acute at apex, cuneate at base, shortly petiolate, cauline leaves gradually decreasing in size, sessile, amplexicaul; flowers azure-blue, occasionally pink or white, all ligulate, perfect, capitula numerous, in axillary fascicles, opening one at a time, 2–3 cm across, involucres cylindrical, phyllaries in 2 series, lanceolate, outer ca 8, inner ca 5, twice as long as the outer ones, receptacles flat, without paleae; achenes obovate, 2–3 mm long, angular, pale brown, apex truncate. Native of the Mediterranean Region, naturalized in Xinjiang, used locally for a tea served especially to enhance the normal function of the liver.

Cirsium japonicum de Candolle — JAPANESE THISTLE

Da-ji=Ta-chi (大薊, Greater Thistle); *Ci-ji-cai=Tz'u-chi-ts'ai* (刺薊菜, Prickly Thistle Vegetable); *Mao-ji=Mao-chi* (貓薊, Cat Thistle). Young leafy shoots; used for potherbs.

Robust spiny herbs, roots fusiform, branchlets hairy; leaves sessile, the basal ones oblanceolate in outline, the cauline ones ovate-oblong, amplexicaul, 15–20 cm long, 5–10 cm wide, pinnately deep-lobed with broad sinuses, the lobes dentate, teeth prickly, sparsely sericeous on both surfaces; flowers purple, all tubular, in large solitary subglobose capitula terminal to the central or lateral branches, 2–5 (–8) cm in diameter, involucres campanulate, phyllaries imbricate, linear, the outer ones green and spiny, the inner ones membranous, purple-tinged, receptacles flat, bristly, florets bisexual, petals 1.5–2 cm long, 5-lobed, lobes linear, acute, stamens 5, filaments hairy, anthers sagittate at the base, apical appendages lanceolate, ovary glabrous, stylar branches cohering; achenes oblong, 4 mm long, 5-ridged; pappus white, silky, plumose,

1.5–2 cm long. Widespread in southeastern China, Japan, southward to southeastern Asia, sometimes occurring on sandy shores.

Cirsium segetum Bunge — LESSER THISTLE (Syn. *Cephalonoplos segetum* [Bunge] Kitamura, *Breea segeta* [Bunge] Kitamura)

Ci-er-cai=Tz'u-erh-ts'ai (刺兒菜, Prickly Potherb); *Xiao-ji=Hsiao-chi* (小薊, Lesser Thistle); *Ci-ci-ya=Ts'u-ts'u-ya* (刺刺芽, Spiny Tender Shoot); *Ci-ji-ya=Tz'u-chi-ya* (刺薊芽, Thistle Shoot). Young plants; tender shoots with rosulate leaves, boiled, washed, drained, seasoned for salads; or cooked with cracked cereals and taken as a famine food in northern China.

Spiny perennial herbs 10–60 cm high, with wirelike subterranean white rhizomes, the terminal buds developing into solitary aerial shoots, new growth arachnoid; leaves of young plants rosulate, carrying upward with the elongating stems becoming alternate, oblong or oblong-lanceolate, 2–10 cm long, 1–2.5 cm wide, sessile, subentire to coarsely serrate, spinulose along the margin and spiny on the teeth, sharply spiny on the apex, subglabrous above, arachnoid beneath; flowers purple-red, all discoid, tubular, capitula subglobose, 1–4 (–6) on each stem, becoming bowl-shaped after anthesis, phyllaries imbricate, the outer ones ovate, firm, with a longtitudinal rib (vitta) terminated by a spine, the inner ones becoming progressively longer, lanceolate-linear, patent, obtuse and erose, receptacles flat, densely setaceous, florets hermaphrodite, sometimes unisexual, corolla 2.5–2.8 cm long, basal 1.5–2 cm filiform, distal portion slightly broadened, divided, lobes 5–6 mm long, anthers of staminate and fertile florets 3 mm long, caudate, apical appendages oblong-ovate, acute; achenes oblong, compressed, pappus sordid-white, several series, plumose, 2 cm long at anthesis, 3.5 cm long and falling in masses. Native to eastern Asia, a common weed in neglected areas and in fields of grain sorghum and cotton in northern China; entire plants gathered by drug collectors, used as diuretics, lactagogue, or a cooling agent.

Crassocephalum crepidioides (Bentham) S. Moore

Ge-ming-cai=Ko-ming-ts'ai (革命菜, Revolution Potherb); *Jia-tong-hao=Chia-t'ung-hao* (假茼蒿, False Crown Daisy); *Liang-zi-cai=Liang-tzu-ts'ai* (涼子菜, Cooling Vegetable); *An-nan-cao=An-nan-ts'ao* (安南草, Vietnam Weed). Young plants, used for potherbs.

Annual herbs 30–90 cm high, the young growth tender and smooth, with immature capitula on bending shoot apices, primary stem 3–10 mm in diameter, dried with ribs and furrows, sparsely pilose in the furrows; leaves petiolate, elliptic, 9–15 cm long, 3.5–8 cm wide, sharply serrate, the teeth unequal, or irregularly pinnatifid at the base, the lobes ovate, serrate, sinuses broad, sparsely pilose or glabrescent above, hairs on the nerves beneath multicellular, acuminate on both ends; flowers purplish-pink, 2 to 4 young capitula on glandular tomentose common stalk, peduncles very slender, 2–3.5 cm long, subtended by linear-filiform bracts, involucres herbaceous, pilose, campanulate, 10–12 mm long, calyculate at base with filiform-subulate bracteoles 2–3 mm long, phyllaries in 1 series, persistent and reflexed on denuded receptacle, linear, 10–12 mm long, 1 mm across base, strongly veined, scarious along the margin, acute or obtuse, glandular and pilose-papillose at the apex, receptacles flat at anthesis, becoming hemispherical denuded, subalveolate with white network defining the pitted areolae, florets all tubular, corolla filiform, longer than the phyllaries at anthesis, base hardly enlarged, distal end slightly broadened and 5-cleft, stamens completely concealed, style of some capitula all exserted, long branched, with subulate apical appendages; achenes oblong-cylindrical, 2 mm long, the full ones reddish-brown, with 10–12 blunt ribs and white hairs in the furrows, annulus white, well developed, empty achenes of the same head light brown, striate, the striae pilose, evidently the ribs of the mature achenes arising from the spaces between the striae; pappus copious, soft, white, capillary, longer than the corolla, denuded receptacle 6 mm across, with prominent interalveolar white lines. The species was first recorded from tropical Africa in 1849. In China, it has been recorded from Hainan Island, Guangdong, Yunnan and Guangxi. In Guangzhou, it is used as a potherb, and in Guangxi for medicinal purposes. All records were made no earlier than the 1930s.

Cynara scolymus L. — GLOBE ARTICHOKE; BUR ARTICHOKE

> *Ju-ji=Chu-chi* (菊薊, Chrysanthemum Thistle). Young capitula; basal fleshy portion of phyllaries (involucral bracts) and the young receptacle; introduced to agricultural and botanical institutions, rare in China.

Erect spiny herbs, 1–2 m high at anthesis, young growth arachnoid-tomentose, stems 1–2 cm across, striate-ribbed; leaves rosulate on young plants, altermate on stems, ovate-lanceolate in outline, 30–50 cm long, 10–20 cm wide, pinnatifid or parted, soft hairy, glabrescent above, grayish-tomemtose beneath, occasionally spiny, segments spiny, segment lanceolate, caudate, base sessile, often deccurrent into wings on the stem; flowers purple-blue, all discoid, in loose large capitula terminal to the stem, ovoid, 6–10 cm across the base, phyllaries ovate, spiny at the apex, base fleshy, 1–5 cm long, 1–2 cm across the base, receptacles flat, setaceous, 2–2.5 cm in diameter, florets numerous, corolla 4.5–5 cm long, the basal 3 cm filiform, the distal 1.5–2 cm slightly thickened, divided, membraneous, anther-tubes exposed, apical appendages rounded, style branches coherent, papillose, exposed; achenes oblong, 4-angled, glabrous, pappus

2–2.5 cm long, plumose, white, several series, falling with the achene. Native of the Mediterranean Region, extensively cultivated in southern Europe and California; propagated largely by suckers removed from the mother plant.

Echinops latifolius Tausch — Chinese Globe Thistle (Syn. *E. dahuricus* Fischer)
Lou-lu=Lou-lu (漏蘆, Leaking Hut); *Lan-ci-tou=Lan-tzu-t'ou* (藍刺頭, Blue Globae Thistle). Young Plants; used as emergency food.

Spiny, arachnoid perennial herbs 50–100 cm high, with stout persistent root-stock, stems 4–8 cm thick, striate-ribbed; leaves of young plants rosulate, partially becoming cauline with the elongating solitary stem, obovate-oblanceolate in outline, pinnatifid, 10–20 cm long, cauline leaves ovate or oblong, 5–15 cm long, 3–11 cm wide, pinnatifid, segments oblong-ovate, pinnatisect, lobes triangular, strongly spiny, apex spiny, grayish-green above, arachnoid, lanate beneath, sessile, auriculate, the auricles spiny; flowers purplish-blue, capitula single-flowered, sessile, with numerous setaceous short bracts connate at the base, numerous sessile capitula, aggregated in dense globular capitate inflorescences 4–5 cm in diameter, involucre of individual capitulum cylindrical, phyllaries imbricate, the outer series short, setaceous, connate at the base, intermediate series linear, parted into filiform segments 5–9 mm long, inner series consisting of 7–9 oblanceolate bracts 9–10 mm long, lacinate along the margin, corolla tubular, 12–14 mm long, basal 5 mm filiform, the upper portion enlarged, 7–9 mm long, divided, segments linear, obtuse, tinged blue, stamens 5, caudate, basal appendage setaceous, anther-tube 6 mm long, apical appendage ovate, acute, style thickened before divided, branches 2, recurved, exposed; achenes cylindrical and somewhat tetragonal, small; pappus coroniform, 2 mm long, connate at the base. Native of temperate eastern Asia, growing on dry grassy hillsides.

Elephantopus mollis Humbold, Bonpland et Kunth — White-flowered Elephant's Foot
(Syn. *E. tomentosus* Auctt., non L.)
Bai-hua-di-dan-cao=Pai-hua-ti-tan-ts'ao (白花地膽草, White-flowered Elephant's Foot Weed); *Bai-hua-di-dan-tou=Pai-hua-ti-tan-t'ou* (白花地膽頭, White-flowered Elephant's Foot Shoot). Whole plant; used in herbal tea.

Erect perennial herbs 15–60 cm high, hirsute-hispid throughout, stems initiated from basal portion of stumps, 4–8 cm across, striate-sulcate; leaves all cauline or basal with short internodes, ovate-oblanceolate, 5–20 cm long, 3–6 cm wide, acute or shortly acuminate, base attenuate, forming narrow wings on the petioles and broadened, clasping the stem, crenate-serrate, pilose above, densely pilose-villose and resin-dotted beneath; flowers white, all tubular, 4 in a capitulum, 18–30 capitula aggregated in pedunculate leafy-bracteate glomerules 1–2.5 cm across, these glomerules arranged in loose determinate paniculate-inflorescences, foliaceous bracts of glomerules cordate, 0.7–1 cm wide, pilose and resin-dotted, capitula 8 mm long, sessile, phyllaries 7 or 8, imbricate, lanceolate, outer ones 4 mm long, acute, pilose, inner ones 6 mm long, aciculate, glabrous, florets 4, 8.5 mm long, basal 4 mm of corolla filiform, distal 2 mm unequally 5-cleft, stamens 5, sagittate, anther-tube 1 mm long, style branches filiform, recurved; achenes obovate-oblong, 2.5 mm long, ribbed, pilose; pappus chaffy and abruptly awned, 4 mm long, 5 in one series. Native of tropical Americans.

Note: The type collection of this species was collected from Caracas of Venezuela. It is a new adventive species in southern China, occurring in Taiwan, Hainan, Hong Kong, Guangdong and Yunnan, growing along paths of moist woodlands. It has been reported in Chinese botanical works as *Elephantopus tomentosus* by various authors. All Chinese collections so identified have been annotated by monographers of the genus *Elephantopus* L. as *E. mollis* HBK. The genuine *E. tomentosus* L. is a species of eastern North America, with basal leaves in rosettes. I have seen no material evidence that this species has been introduced to China yet.

Emilia sonchifolia (L.) de Candolle — Tassel Flower

Yi-dian-hong=I-tien-hung (一點紅, Red Drop); *Yang-ti-cao=Yang-t'i-ts'ao* (羊蹄草, Goat-feet Herb); *Ye-xia-hong=Yeh-hsia-hung* (葉下紅, Leaves Red Beneath). Young plant gathered in Taiwan, eaten as potherbs locally.

Annual and/or perennial lactiferous herbs 10–50 cm high, young vegetative growth pubescent with long moniliform hairs, glabrescent and glaucous with age, much branched, branchlets slender; leaves variable in shape and size according to their positions, basal leaves spatulate with suborbicular laminas 1–3 cm long, 1.5–4 cm wide, wavy and remotely dentate, base rounded and suddenly cuneate into wings to the distal portion of petioles 3–4.5 cm long, lower cauline leaves lyrate, sessile, 3.5–10 cm long, with the terminal segments deltoid-subcordate, 2–4.5 cm long, 2.5–5 cm wide, wavy and remotely dentate, the teeth ended with aggregates of bubble-like cells, median cauline leaves pinnatifid, base sagittate or auricular, upper cauline leaves ovate-lanceolate, sessile, subentire or remotely dentate; flowers purple-red, all discoid, capitula solitary, terminating the developing shoot, the lower and later flowering ones on the same peduncle one (rarely two) on slightly longer stalks, together these capitula giving the appearance of loose corymbs of heads to a plant, involucres subcampanulate, 12–14 cm long, phyllaries 8–12, in one series, hardly overlapping along the margin, connate at the base, segments linear, acute, membranous along the margin, scarious at the apex, persistent and reflex, receptacle flat, glabrous, naked, 3–5 mm across, florets slightly longer than the involucral bracts, polygamous, corolla 7–8 mm long, filiform and white below, distal 2 mm 5-cleft, lobes obtuse, scabrious, anthers of most flowers included, base obtuse, apical appendages deltoid, acute, styles of few central florets equal the petals in length, branches short, recurved, appendages obscure; achenes cylindrical, truncate at both ends, 3 mm long, 5-ribbed, ribs pilose; pappus copious, white, 5–6 mm long, silklike. A pantropic weed, perhaps native of Africa, young shoots used as salad throughout southeastern Asia, said to have slightly acid taste with a touch of bitterness, very delicate; a tea is prepared for persons with dropsy.

Erechtites valerianaefolia (Wolf) de Candolle — Fireweed

> *Shan-bo-qin=Shan p'o-chin* (山坡芹, Hillside Celery); *Shan-ai-cai= Shan-ai-ts'ao* (山艾菜, Wild Artemisia Vegetable). Young plants; gathered before flowering, used for potherb locally.

Annual herbs 1–1.5 m high, very tender when young, primary stems 0–8 mm in diameter, striate-sulcate, glabrescent; leaves ovate or elliptic in outline, 5–12 cm long, 2.5–5 cm wide, acute at both ends, the lowers ones irregularly coarsely dentate, those above the middle of the stem pinnatifid, rarely lyrate, segments lanceolate, dentate irregularly, senuses rounded, sparsely pubescent on the nerves beneath; flowers small, yellow, many in cylindrical capitula on filiform peduncles 1–2 cm long, subtended by linear bracts 3–15 mm long, forming loose terminal or lateral cymose or corymbose clusters, involucres subcylindrical, calyculate with linear bracts, phyllaries in 1 series, linear, 6–8 mm long, acute, herbaceous, striate, membranous along the margin, persistent and reflexed after anthesis, receptacles naked, 2.5–3 mm across, denuded ones rugose, florets tubular, all fertile, marginal ones pistillate, inner ones perfect, corolla filiform, not thickened at the base, slightly enlarged towards the apex, 5-cleft, lobes obtuse and papillose, apical anther-appendages of perfect flowers exposed, triangular, acute, stylar arms linear, truncate and papillose at the apex; achenes ellipsoid, 2.5 mm long, tapering at the end, 10-ribbed, light brown, 10 furrows dark brown and pilose. Native of tropical America; new adventives in Taiwan, Hainan, and Guangdong.

Farfugium japonicum (L.) Kitamura var. **formosanum** (Hayata) Kitamura (Syn. *Ligularia tussilaginea* [Burman f.] Makino).

> *Gao-wu=Kao-wu* (橐吾, Taiwan Farfugium). Young plant and roots, eaten locally in Taiwan before 1950.

Evergreen, glabrescent perennial herbs 50–65 cm high at anthesis; leaves all radical, long-petiolate, laminas involute when young, reniform, 4–13 cm long, 4.5–14 cm across the base, shallowly 3- or 5-lobed, repand-angulate, remotely dentate, petioles slightly broadened but not sheathing at the base, 7–18 cm long, striate-sulcate; flowers light yellow, both ligular and tubular, 2–7 heads loosely arranged in cymose corymbs terminal to the scapes 30–40 cm long, with two linear bracts, the lower one 12–15 cm above the base, capitula 2 cm long, 3–4 cm across, involucres rugose, calyculate with 4 or 5 lanceolate bracts, phyllaries 1-seriate, linear-lanceolate, herbaceous, margin scarious, ligular florets 5 to 7, tubular corolla 8–9 mm long, base not evidently broadened, distal half campanulate, shallowly 5-lobed, the lobes triangular, 1.5 mm long, anthers sagittate-caudate, the tube 3 mm long, completely exserted, apical appendage deltoid, acute, stylar arms recurved and coiled, apex truncate; achenes cylindrical, 4 mm long, strigose; pappus copious, 8–9 mm long, rough, with some short setae at the base. Endemic to Taiwan. The species, *Farfugium japonicum* (L.) Kitamura, a native of Japan and eastern China, has larger and firmer leaves, higher scapes, larger capitula, and shorter brown pappus.

Gnaphalium affine D. Don — Chinese Cudweed (Syn. *G. multiceps* Wallich)

Ai-mo-mo=Ai-mo-mo (艾饃饃, Cudweed Pastry); *Ye-er-ba=Yeh-erh-pa* (葉兒粑, Leaf Wrapper); *Shu-qu-cao=Shu-ch'u-ts'ao* (鼠麯草, Mice Yeast Herb). Young shoots gathered before flowering, ground, mixed with rice flour, seasoned, for a steamed pastry.

Annual or biennial herbs 4–30 cm high, woolly throughout; young plants with rosette of radical leaves, deteriorated at anthesis, cauline leaves alternate, sessile, spatulate or oblanceolate, 2.5–4 (–6) cm long, 4–8 (–12) mm wide; flowers yellow, capitula small, subglobose, 2 mm in diameter, numerous, glomerate, involucres subcampanulate, phyllaries bright golden, scarious, 3-seriate, imbricate, receptacles flat, 1 mm across, denuded ones roughened with elevated disks, naked, florets mostly pistillate, with filiform corolla, often unbranched style, truncate at the apex, 2–4 central florets perfect, with tubular-cylindrical corolla, 5-lobed at the apex, anthers sagittate, caudate at the base, apical appendage sometimes exserted, stylar arms 2, truncate; achenes oblong, 0. 7 mm long, glossy brown, papillose; pappus capillary, white, bristly, 1-seriate, shorter than the corolla. Widespread weed, common in warm temperate and subtropical areas of China, extending westward to Nepal and eastward to Japan; earliest record in Chinese herbals can be traced to the fifth century.

Gynura aurantiaca (Blume) de Candolle — Velvet Plant; Purple Passion

Zi-rong-teng=Tzu-jung-t'eng (紫絨藤, Purple Velvet Vine). Tender shoots without flowers; used in soups, available in American Chinese groceries.

Perennial herbaceous vines, densely covered by velvety-purple hairs throughout, young stems erect, clambering with age, up to 2.5 m high, lateral branchlets 30–50 cm long; leaves ovate-elliptic, 5–20 cm long, 2–8 cm wide, acuminate, base of lower cauline leaves rounded or auriculate and suddenly narrowed to a petiole, coarsely dentate, upper cauline leaves sessile, auriculate; flowers orange-yeilow, aging purplish, all discoid, with unpleasant odor, attracting house flies; capitula solitary or 3–5 forming an open cymose-corymbose cluster, involucres cylindrical-campanulate, 10–12 mm long, densely purple-velvety, phyllaries uniseriate, coherent at base, calyculate with linear bracts 5–8 mm long, herbaceous with scarious margin, acute, receptacles flat, 2 mm across, supporting 20 florets, corolla filiform, 11–12 mm long, prominently broadened at the base, distal 2–2.5 mm campanulate, 5-lobed, lobes 1 mm long, deltoid, acuminate, anthers caudate at base, apical appendages partially exposed, deltoid, acute, stylar arms flat, apical appendage subulate, 1 mm long, papillose, acute; achenes linear, 10-ribbed, 3 mm long; pappus white, longer than the corolla, capillary and bristly.

Native of Indonesia, widely cultivated and naturalized in southeastern Asia.

Gynura bicolor de Candolle — Red Groundsel

Di-huang-cai=Ti-huang-ts'ai (地黃菜, Ground Yellow Vegetable); *Hong-cai=Hung-ts'ai* (紅菜, Red Vegetable); *Hong-feng-cai=Hung-feng-ts'ai* (紅鳳菜, Red Phoenix Vegetable); *Shui-qian-cai=Shui-ch'ien-ts'ai* (水前菜, Water Front Vegetable); *Zi-bei-tian-kui=Tsu-pei-t'ien-kui* (紫背天葵, Purple Celestial Mallow); *Guan-yin-xian=Kuan-yin-hsien* (觀音莧, Goddess-of-Mercy's Amaranth); *Dong-feng-cai=Tung-feng-ts'ai* (東風菜, East-wind Potherb). Roots; cooked with sliced pork; young shoots used as a potherb.

Perennial herbs 1–1.5 m high, stems 1–2 cm in diameter, fleshy when young, glabrous or sparsely pubescent; leaves ovate-elliptic or oblanceolate, 5–30 cm long, 3–7 cm wide, the lower ones irregularly dentate, the upper cauline leaves occasionally lobed and the lobes dendate, sessile, lyrate-sagittate; flowers bright orange, the first appearing capitulum terminating the flowering branch, solitary or two, the capitula developed later in the leaf-axiles forming a loose corymbose panicle, involucres campanulate, 12–14 mm long, sparsely pubescent with multicellular trichomes, calyculate with linear bracts 3–4 mm long, phyllaries uniseriate, connate at base, linear, acuminate or acute, glandular-pilose at the apex, herbaceous with scarious margin, receptacles flat, 5–6 mm across, denuded ones alveolate, corolla 14–15 mm long, filiform and suddenly broadened at both ends, the distal enlarged portion campanulate, 5-lobed, 3 mm long, lobes deltoid, acute, anthers caudate at base, the tube exserted, apical appendage prominent, triangular, stylar arms exceeding the anther-tube, apical appendage subulate, 2 mm long, acute, papillose; achenes glossy brown, cylincrical, slightly narrowed at both ends, glabrous, 10-ribbed, under high power lens appearing papillose; pappus white, copious, longer than the phyllaries. From the collections in Harvard University Herbaria, the species is native to the mountains of southern China, northern Burma, and Vietnam.

Note: When Roxburgh received a specimen in 1790, he was misinformed of its native home as being Moluccas, which he recorded. His statement was accepted by many subsequent authors blindly. Actually, the species is extensively distributed in Yunnan, Sichuan, Guangxi, Guangdong, and the mountains on the borders of Burma, Thailand, and Vietnam. The various ethnic names here recorded indicate widespread uses discovered by many tribes living in the mountains.

Gynura divaricata (L.) de Candolle — Chinese Groundsel

Bai-zi-cai=Pai-tzu-ts'ai (白子菜, White-seeded Vegetable); *Ji-cai=Chi-ts'ai* (雞菜, Chicken Vegetable); *Di-bai-zi-cai=Ti-pai-tzu-ts'ai* (地白子菜, Ground White-seeded Vegetable); *Bai-bei-san-qi=Pai-pei-san-chi* (白背三七, White-back Hematonics); *Da-fei-niu=Ta-fei-niu* (大肥牛, Bit Fat Cattle). Leafy shoots; boiled with brown sugar, in the boiling liquid, poach eggs to make a delicious hematonic dish.

Succulent perennial herbs with robust rootstocks sending primary stems 0.5–1.5 m long, 5–10 mm in diameter, erect at first, clambering with age, terminated by a cluster of capitula, secondary branches emerged in the next growing season shorter and thinner, with the leafy portions being 5–30 cm long, 3–5 mm in diameter, having a scape-like portion terminated by a few capitula; leaves alternate, varying in shape and size and density of the indumentum according to ecological conditions and their position on the plant, the lower leaves oblong, or ovate-orbicular, 5–10 cm long, 2–6 cm wide, obtuse, the base roundish and abruptly cuneate into the petiole, densely tomentose and appearing whitish beneath, less pubescent above, the upper leaves elliptic or oblanceolate, 5–9 cm long, 1–2 cm wide, acute, base attenuate, less hairy beneath, both types petiolate, the uppermost leaves sessile, spatulate and often auriculate, all leaves remotely dentate, and with dense reticulum of glandular superficial lines on both surfaces; flowers bright golden-yellow, conspicuous in the landscape, all discoid, capitula 1–6 on slender-peduncles, with 1, 2, or without linear bracts, forming a loose terminal cymose cluster, involucres campanulate, 8–12 mm long, calyculate with triangular or linear bracts, phyllaries uniseriate, linear-lanceolate, connate at base, herbaceous, hairy, rugose, striate, with vertical glandular lines, margin scarious, acute or acuminate, apex papillose, persistent and reflexed after anthesis, receptacles 4–5.5 mm across, naked, alveolae inconspicuous, center with a stiff projection, corolla filiform, 9–11 mm long, basal 1 mm thickened, distal 3 mm campanulate, 5-lobed, lobes triangular, acuminate, with 1 or 2 glandular lines, anthers sagittate, apical appendages deltoid, equal to the corolla lobes or exceeding the latter, stylar arms plane-concave, apical appendages subulate, 1.5 mm long, papillose; achenes shiny chestnut-brown, cylindrical, 4.5 mm long, 10-ribbed, the ribs rugose, the furrows glandular and setaceous; pappus capillary, copious, longer than phyllaries. A native of Guangdong, the holotype was collected from Huangpu by Peter Osbeck, where many recent specimen have been gathered. The species is frequent on the open hillsides in Guangdong and Hong Kong, often growing in rocky areas. In Hong Kong, it also occurs close to the seashore, on rock cliffs subjected to ocean spread. Having rootstocks buried in the soil, the species is fire-resistant, often seen with perennial grasses after the hillside fires. It has been observed in kitchen gardens of security guards of Kowloon railway station, of fishing community in Ping Chau and in Wuzhou in Guangxi. Its use in food is at the ethnobotanical level. It has not been observed in groceries or in village market places.

Helianthus annuus L. — COMMON SUNFLOWER (Figure 65)

> *Xiang-ri-kui=Hsiang-jih-kuei* (向日葵, Sun-facing Mallow); *Zhuan-lian=Chuan-lien* (轉蓮, Turning Lotus). Seeds; roasted, cracked like unsalted peanuts, at home and outside in parks and scenic areas, in temples and theaters; after harvest, root and leaves prepared for tea.

Tall robust annual herbs 1–3 m high, hispid throughout, stems tough, 2–4 cm in diameter, with copious pith; leaves alternate, ovate-cordate, 10–30 cm long and wide across the base, acute, coarsely serrate; flowers yellow, both ligulate and tubular, capitula few, in cultivation often solitary, on stout and long peduncles, heterogamous, involucres saucer-shaped, phyllaries nearly equal, imbricate, outer ones oblong-ovate, inner ones oblong-lanceolate, long attenuate and caudate, herbaceous, scabrid, ciliate, receptacles flat, chaffy, bracts subtending all discoid florets oblong, tricuspidate, the middle lobe longer, hispid at tip, folded and winged on the ridge towards the base, ray florets pistillate or sterile, ligules lanceolate, 2–4 cm long, acute, the basal tubular portion 4 m long, weakly pilose, disk florets over one thousand, perfect, protandrous, those situated on the middle of the radius of the receptacle blooming first, hence advancing outward and inward, with the central ones opening last, corolla goblet-shaped, 7 mm long, the basal 2 mm strongly narrowed, densely strigose, the remaining 5 mm subcylindrical, slightly thinner at the middle, glabrous, 5-lobed at the apex, stamens 5, inserted to the base of the cylinder, filaments thickened at the apex, anthers completely exserted, the thecae black, connectives and the apical appendages ochraceous, stylar arms flat, appendage prominent, attenuate, papillose; achenes compressed obovoid, black or gray with white and black stripes, rarely white, 1–1.5 cm long; pappus represented by two scale-like deciduous structure.

A species native of North America, was first recorded in a Chinese herbal compiled in the early seventeenth century, *Qun-fang-pu* (Wang, 1630), with these names: *Zhang-ju* (丈菊, Ten-feet Chrysanthemum), *Xi-fan-ju* (西番菊, Western Foreign Chrysanthemum), and *Ying-yang-hua* (迎陽花, Welcome-the-sun Flower). Today, it is extensively cultivated throughout China, in small scale along margins of gardens or fields, banks of canals, or less fertile newly opened odd lots, for the production of "seeds" used as a special snack food. Parched sunflower seed is consumed in parties, tea houses, scenic areas, parks, and in special festivities in China as are popcorn, peanut, and candy bar in the ball-fields and theaters in USA. In the rural areas, various parts of the plant have been utilized as home remedies or for tea. After the terminal large head is harvested, the root and the leaves are gathered for tea, the flowers of smaller immature heads are dried and used as home remedy for toothache and for improving eye-sight, the pith from the stems are used for diabetic patients, and even the receptacle and seed hulls are used for people suffering from constipation.

Helianthus tuberosus L. — JERUSALEM ARTICHOKE, GIROSOLE

Yang-jiang=Yang-chiang (洋姜, Foreign Ginger); *Ju-yu=Chu-yu* (菊芋, Chrysanthemum Taro). Tubers; boiled and used as vegetable, fresh material, washed, left in air for a day, pickled.

Erect perennial herbs 2–3 m high, readily becoming an aggressive weed, hispid throughout, with tuberous fleshy rhizomes; leaves opposite at first, becoming alternative, ovate-elliptic, 10–15 cm long, 3–9 cm wide, acuminate, roundish and attenuate at base, strigose with hairs thickened at base above, pilose beneath, coarsely serrate; flowers yellow, both ligulate and tubular, capitula several in corymbose cluster terminating the stem, with the first one being the oldest, 5–6 cm across, involucres campanulate, 1 cm long, 1.5 cm across, phyllaries 2- or 3-seriate, lanceolate, 10–12 mm long, acuminate, strigose, ciliate, ligulate florets 2–3 cm long, 3–5 mm across the middle, acute, pilose on both surfaces, pistillate or sterile, tubular florets bracteate, bracts (chaff) scarious, glabrous below, herbaceous, pilose and ciliate at the distal half, corolla goblet-shaped, 5 mm long, basal 1 mm narrowed, glabrous, the remaining portion cylindrical, pilose at the base, glabrous and 5-lobed at the apex, lobes deltoid, anthers 5, exposed, thecae black, appendages deltoid, ovary compressed-oblong, 4 mm long, pilose, stylar arms flat on the adaxial surface, pilose on the abaxial surfaces, reflexed and coiled, the appendages linear-subulate, pilose; achenes brown, compressed-obovoid, 4 mm long, 2.5 mm across the pilose top, the basal end obliquely obtuse; pappus 2 or 3, lanceolate-subulate, rounded and lobed at the base, ciliate. A polyploid species originated in temperate North America, introduced to Europe in the early seventeenth century, becoming popular in England in the 1020s, the introduction into China is recent, with flowering specimens collected in Guangzhou in 1917, in Zhejiang in 1924 and in Sichuan in 1938. Most early Chinese collections were made around temples. Evidently the species is widespread, but only in small lots. The Chinese people like meally tubers such as potato and taro. The tubers of this species contain insulin rather than starch grains, and few farmers cultivate it for food.

Hemistepta lyrata (Bunge) Bunge (Syn. *H. carthamodes* [Buchanan-Hamilton] O. Kuntze, *Saussurea carthamoides* Buchanan-Hamilton ex de Candolle, *S. affinis* Sprengel)
Ni-hu-cai=Ni-hu-ts'ai (泥胡菜, Earth Tartary Potherb); *Tu-ni-tiao=T'u-ni-t'iao* (禿妮條, Bald-headed Maid Shoot); *Ye-ku-ma=Yeh-k'u-ma* (野苦麻, Wild Bitter Hemp). Young plants; an emergency food, with the white tomentum rubbed and washed off, boiled with cracked cereal.

Biennial herbs 30–80 cm high at anthesis, arachnoid and resinously dotted throughout; leaves all sessile and alternate, those of young plants basal, rosulate, oblanceolate, 7–14 (–16) cm long, 3–4 (–10) cm wide, sinuate-pinnatifid, the sinuses deep and round, apical segments deltoid-ovate, dentate, the lower ones decreasing in size gradually, the proximal ones linear, the lower cauline leaves similar to the basal ones, the upper ones becoming smaller, with fewer lateral lobes, and lanceolate terminal segments, all leaves densely tomentose beneath, arachnoid, glabrescent and with resinous reticulations above; flowers purplish-pink, all discoid and alike, involucres subglobose, 8–12 mm long, 10 mm across, phyllaries 8-seriate, the outer ones ovate-deltoid, acute, each furnished with a membranous glandular crest, outermost ones arachnoid, 2 mm long, gradually increasing in length, the innermost phyllaries lanceolate-caudate, 12 mm long, glabrous, ecristate, the distal end purple-pink, receptacles discoid, flat, 6–8 mm across, setaceous, setae white, 6–7 mm long; florets filiform, corolla 12–14 mm long, the base suddenly enlarged and cupular, the distal 3 mm petaloid, deeply 5-cleft, segments subequal, linear-filiform, 2.5 mm long, obtuse at the apex, stamens 5, anthers caudate, the tubes 2 mm long, exserted, straw-color, apical appendage acute, style longer than the staminal tube, the distal end thickened, stylar arms oblong, erect, smooth, without apical appendage; achenes glossy chestnut-brown, oblong, 4 mm long, truncate, slightly narrowed and with oblique scar at the base, 12-ribbed; pappus copious, snow-white, plumose, 9 mm long, connate at base. Native of eastern Asia, weedy, occurring in fields near roads and vacant lots, widespread in China, eaten by people living in villages along the lower Yellow River in famines.

Ixeris chinensis (Thunberg) Nakai

Shan-ku-mai=Shan-k'u-mai (山苦蕒, Hillside Bitter Potherb); *Nai-jiang-cai=Nai-chiang-ts'ai* (奶漿菜, Milk Vegetable); *Xiao-ku-mai=Hsiao-k'u-mai* (小苦蕒, Lesser Ixeris). Root and young plant; eaten locally in northwestern China.

Perennial lactiferous herbs 10–35 cm high, branched at ground level at anthesis; leaves mostly basal, few cauline, linear-oblanceolate, 3–20 cm long, 1–2 cm wide, acute or rounded, base cuneate, entire or irregularly pinnatifid, cauline leaves smaller, amplexicaul, acuminate; flowers white, capitula in loose corymbs, involucres cylindrical, 7–9 mm long, 3–4 mm across, phyllaries in 2 series, outer 4 ovate, 1 mm long, inner 8 linear, 6–8 mm long, obtuse, corolla all ligulate, 8 mm long, 2 mm across, tubes 4 mm long, stamens 5, anthers dark green-black, style 8 mm long, branches recurved; achenes fusiform, slightly curved, 3.5 mm long, red-brown, beak 3 mm long; pappus white, setaceous, 6 mm long. Widespread and weedy in temperate eastern Asia.

Ixeris denticulata (Houttuyn) Stebbins (Syn. *Youngia denticulata* [Houttuyn] Kitamura)

Ku-mai-cai=K'u-mai-ts'ai (苦蕒菜, Bitter Potherb); *Qiu-ku-mai-cai=Ch'iu-k'u-mai-ts'ai* (秋苦蕒菜, Fall Bitter Potherb); *Pan-er-cao=P'an-erh-ts'ao* (盤兒草, Dish Herb). Young plant; used for potherb.

Erect lactiferous perennial herbs, glabrous throughout, 20–80 cm high at anthesis, stems 2–4 mm across, terete, the lower portion lignified; leaves sessile, varying in size and shape according to their position, radical leaves rosulate, oblong, spatulate or lyrate in outline, 5–16 cm long, 2–4 cm wide, obtuse or rounded at the apex, margin shallowly lobed and denticulate, base attenuate gradually into the petiole-like portion, broadened and often auriculate at the insertion to the stem, cauline leaves oblong, pinnatifid or slightly lobed, irregularly pinnatifid and sharply denticulate or serrate, auriculate-amplexicaul, the upper cauline leaves decreased in size progressively, oblong or lanceolate, often ovate-caudate and amplexicaul; flowers yellow, all ligulate and perfect, capitula few flowered, 2–4 on slender foliaceous-bracteated stalks forming a cymose cluster, several of such clusters sometimes giving the appearance of a corymb, involucres oblong-cylindrical, 6–7 mm long, 2 mm across, calyculate, phyllaries 7–9, 1-seriate, the outer ones crested, the inner ones obtuse, herbaceous with scarious-white margin, 6–7 mm long, persistent, starlike, usually not reflexed, receptacles 1–1.5 mm across, denuded concaved; florets 6 to 8, corolla 1 cm long, tube 3–4 mm long, glabrous, ligules 6–7 mm long, toothed at the apex, stamens 5, anther-tubes exposed, apical appendages glossy, style slender, distal portion exserted, pilose, stylar arms long, straight, slender, pilose on the abaxial surface, without apical appendage; achenes dark brown or black, lanceolate, 10-ribbed, ribs shortly strigose, apical beaks 0.5–1 mm long, discoid at the distal end, apiculate at the center, pappus white, unequal, 3–5 mm long, 1-seriate. Native of eastern Asia, very common on exposed hillsides, flowering September–November, occasional in June in Hong Kong; used in local herbal medicine, eaten as a vegetable in Jiangsu Province.

Kalimeris indica (L.) Schultz-Bipontinus — (Syn. *Boltonia indica* [L.] Bentham, *Aster indicus* L.)

Ma-lan-tou=Ma-lan-t'ou (馬蘭頭, Horse Aster Green); *Ma-lan=Ma-lan* (馬蘭, Horse Aster); *Ji-er-chang=Chi-erh-chang* (雞兒腸, Chicken Intestine); *Hong-geng-cai=Hung-keng-ts'ai* (紅梗菜, Red Stalk Vegetable). Young shoots; used for potherb, a favorite spring wild plant dish in Nanjing and Shanghai, now cultivated in the vicinity of Shanghai to meet local market demand; whole plants, fresh or dried, used in herb tea.

Perennial herbs 15–60 cm high, with new shoots emerging from the rootstock in the form of glabrous tender-rhizomes bearing 4 or 5 deltoid ciliate scales, the terminal growth continuing into aerial stem, the mature stem wirelike, 1.5–2 mm in diameter, purplish, sparsely pilose; leaves elliptic or lanceolate, rarely obovate, 3–10 cm long, 0.5–5 cm wide, acute or obtuse, mucronate, base subsessile, cuneate, margin with 2 or 3 remote teeth, sparsely pilose or glabrescent, upper cauline leaves smaller, lanceolate, entire; flowers both ligulate and discoid, the ray flowers purple-violet, tubular florets yellow, capitula solitary, rarely two, on foliate stalks, involucres cupular, 4–9 mm across, phyllaries 3-seriate, oblong or obovate-oblong, 4–5 mm long, herbaceous with scarious often pink margin, ciliate at the apex, receptacles conic, denuded ones areolate, areolae apiculate at the center, ray florets pistillate, ligules oblong, 10 mm long, 3 mm across the middle, cuneate into a short tube 1 mm long, pilose, disk flowers goblet-shaped, 3 mm long, basal one-third narrowed, apex 5-lobed, lobes recurved, acute, stamens 5, anthers obtuse at the base, the tubes exserted, yellow, apical appendages deltoid, smooth, ovary pilose, with 1 series of setae at the summit, style stout, stylar branches short, flattened, slightly exserted, apical appendage short, strongly papillose-hairy; achenes obovoid-trigonous, 1 mm long, 0.3 mm across the shoulder, dark blackish-brown, with 3 light brown ribs, the ribs setaceous, rugose on the shoulder; pappus setaceous, coronate to the ovary, deciduous in fruit. Native of eastern Asia, common in the Lower Yangtze region, growing along roadsides, river banks and in lowland meadows.

Lactuca formosana Maximowicz

Hua-ku-cai=Hua-k'u-ts'ai (花苦菜, Flowering Bitter Vegetable). Young plant; gathered and eaten locally in Taiwan before 1950.

Lactiferous biennial herbs 45–100 m high, stems 6–8 mm in diameter, branched at the distal half; leaves sessile, the lower cauline ones obovate-oblong in outline, 12–14 cm long, 5–8 cm wide, shortly acuminate, cuneate and narrowed into winged petioles slightly clasping the stem at the insertion, laminas lyrate-pinnatifid, the terminal segments deltoid, the lateral ones in 2 to 4 pairs, obovate, irregularly mucronate-dentate, glabrous and scabrid along the margin above, glaucous and strigose on the veins beneath, median cauline leaves oblong-elliptic, 7–11 cm long, 5–8 cm wide, lyrate-pinnatifid, auriculate-amplexicaul, the terminal segments reniform, the lateral ones obovate, 2- or 3-paired, irregularly incised-dentate, the upper cauline leaves ovate-oblanceolate; flowers pale-yellow, becoming purplish with age, all ligulate and perfect, capitula 5 to 7 in subumbellate clusters, on bracteate stalks 1.5–3 cm long, involucres cylindrical, 12–15 mm long, 5 mm across, phyllaries herbaceous, glaucous, obtuse, imbricate, 3-seriate, outer ones ovate, 5 mm long, the inner ones linear, 14 mm long, receptacles flat, glabrous, florets 14–16 mm long, the tube 4 mm long, dilated and woolly at the distal end, ligules 10–12 mm long, 0.8 mm wide, 5-toothed at the apex, anthers sagittate, the tube 2 mm long, apical appendages obtuse, glabrous, style slender, slightly exposed, pilose, the branches short, erect, abaxial surface pilose, no appendages; achenes shiny-black, strongly flattened and winged, elliptic, 4–5 mm long, 2 mm across the middle, each side with a median longitudinal rib, apical beak 2.5–3 mm long, pappus multi-seriate, white, 7 mm long, bristles simple, falling separately. Endemic to Taiwan; formerly several mainland collections have been incorrectly identified to be this species on the basis of superficial resemblance of the leaves. The mainland collections lack the flattened achenes, winged and ribbed, as do those from Taiwan.

Lactuca sativa L. — LETTUCE, GARDEN LETTUCE, ASPARAGUS LETTUCE

Wo-ju=Wo-chu (萵苣, Lettuce); *Sheng-cai=Sheng-ts'ai* (生菜, Salad). Leaves or the crisp portion of stems; quick fry potgreen, or cooked with meat.

Glabrous annual herbs 100 cm high or higher at anthesis, basal portion of stem 2–6 cm in diameter, firm with fleshy pith, elongating before flowering; leaves various, the basal ones rosulate, heading or not, orbicular or lanceolate, 18–40 cm long, 6–20 cm wide, entire, crinkled or runcinate-pinnatifid, crisp, median cauline leaves orbicular, oblong or ovate, auriculate-amplexicaul, subentire, remotely apiculate-dentate, 8–10 cm long, 3–4 cm wide, obtuse, upper leaves reduced in size, changing into numerous suborbicular or ovate bracts 5–10 mm in diameter or across the base; flowers pale yellow, all ligulate, in numerous capitula on bracteated stalks giving the appearance of large panicles, involucres cylindrical at anthesis, becoming ovoid in young fruits, 5–6 mm across the middle, phyllaries herbaceous, imbricate, 4-seriate, the outer ones ovate, obtuse, glaucous, the inner ones ovate-lanceolate, 8 mm long, acuminate, receptacles smooth, florets 1 cm long, corolla 8 mm long, tube filiform, 4 mm long, distal end broadened and villose, ligules obtuse, glandular at the apex, anthers caudate at base, exserted, apical appendage obscure, ovary 2 mm long, style slender, scarid at the distal end, slightly thickened, unbranched or with very short deltoid arms; achenes compressed obovate-elliptic, 3.5 mm long, gray or black, 10-ribbed, long beaked, the beak filiform, discoid with umbrella-like white bristle-pappus. Native to the Mediterranean Region, used for salad; recorded in Chinese herbals of the seventh century, Chinese selection emphasized the improvement of the enlarged stems called *"Wo-sun"* (萵筍), cultivated as vegetable crops throughout the country; cultivars of the leafy types representing recent introductions made independently through the agricultual colleges located in Guangzhou and Nanjing, and by foreign residents to Shandong.

1. var. **angustana** Bailey — Asparagus Lettuce (Syn. *L. sativa* L. var. *angustata* Irisch ex Bremer)

   Wo-sun=Wo-sun (萵筍, Lettuce Shoot); *Wo-zhu-sun=Wo-chu-sun* (萵竹筍, Asparagus Lettuce). Fleshy pith of the enlarged stem; cooked, pickled, rarely eaten raw, leaves for feeding chicken and geese.

Market material clavate, 20–30 cm long, 5–6 cm in diameter, with leaves removed except tender ones at the shoot apex; leaves firm, oblong-lanceolate, 20–30 cm long, 4–6 cm wide, never forming head, irregularly sinuate-dentate, teeth apiculate, upper leaves gradually changing into bracts, auriculate-amplexicaul, lanceolate-attenuate; large acreage, good selections and wide variety of uses developed in the Yangtze River Region, nine cultivars recorded from Shanghai (Anonymous, 1959); introduced from China to the experimental gardens of Cornell University before 1920, extensively cultivated throughout China, one of the important early summer vegetable; several forms identified by variations in color pattens as *White Wo-sun* (白萵筍) for the pale green leaves, and *Purple-leaved Wo-sun* (紫葉萵筍) for the purplish leaves, and by the shape of the leaf-apices as *Pointed-leaved Wo-sun* (尖葉萵筍) for the attenuate-accuminate leaves; seeds harvested, cleaned, kept in well-ventilated place, used in Chinese herbal medicine, prescribed for improving the quality of milk for mothers, stimulating hair growth and enhancing recovery from kidney ailments.

2. var. **capitata** L. — Common Lettuce

   Jie-qiu-wo-ju=Chieh-ch'iu-wo-chu (結球萵苣, Head Lettuce); *Qing-sheng-cai=Ch'ing-sheng-ts'ai* (青生菜, Green Salad). Leaves; cooked in soup or quick fried, rarely used as salad by people in China.

Market material a globose leafy head 15–18 cm in diameter; before harvest the plants 23–27 cm high, with the spread 24–29 cm across; leaves orbicular, rather thick, green, 20 cm in diameter, slightly crinkly, the inner ones forming a tight head, weighing 180–360 g each; cultivated mostly for supplying foreign residents in large cities.

3. var. **crispa** L.

Zhou-ye-wo-ju=Chou-yeh-wo-chu (皺葉萵苣 Crinkly-leaved Lettuce); *Ruan-wei-sheng-cai=Juan-wei-sheng-ts'ai* (軟尾生菜, Soft-tailed Lettuce). Leaves; cooked or used as salad, rather rare.

Plants at harvesting time 25 cm high, spreading 27 cm across; leaves suborbicular, thin, yellow green, 18 cm long, 17 cm wide, very crinkly, folding loosely, inner ones forming a loose head, market material weighing 120–180 g each, introduced from abroad recently.

4. var. **oblongifolia** Lamarck — Romaine Lettuce

Chang-ye-wo-ju=Chang-yeh-wo-chu (長葉萵苣, Long-Leaved Lettuce); *Niu-li-sheng-cai=Niu-li-sheng-ts'ai* (牛脷生菜, Ox-tongue Lettuce). Leaves and fleshy portion of the stem, cooked or eating raw.

Plants erect, 40 cm high at harvest time; with spread of 50 cm across; leaves obovate, 28 cm long, 16 cm wide, green, slightly crinkly, never heading; market material weighing 180–210 g each, cultivated for a long time, having a more bitter taste than other garden forms.

Petasites japonicus (Siebold et Zuccarini) Maximowicz — Sweet Coltsfoot, Butterbur
Feng-dou-cai=Feng-tou-ts'ai (蜂斗菜, Bee Bushel Vegetable). *She-tou-cai =She-t'ou-ts'ao* (蛇頭菜, Snake-head Herb). Petioles of young leaves; cultivated in Nanjing and Shanghai to meet the market demand of the Japanese residents; fresh juice extracted from leaves and root used as an emergency rememdy for snake bites in villages.

Dioecious arachnoid perennial herbs 15–70 cm high, rhizomes slender; leaves radical, appearing after anthesis, laminas cordate-reniform, 25–80 cm across the base, irregularly mucronate-dentate, petioles terete, fleshy, 20–60 cm long, 1.5 cm in diameter, succulent, hairy; flowers on scapiform stem covered with many foliaceous bracts with parallel-nerves, capitula numerous, in racemose or thyrse-like terminal cluster, involucres campanulate, calyculate, 5–6 mm long, phyllaries 1-seriate, herbaceous, slightly thickened after anthesis, persistent, reflexed, 10 mm long, 2 mm wide, receptacles flat, smooth, 5 mm in diameter, staminate florets tubular, base slightly enlarged, tube 8 mm long, the distal 2 mm broadened, 5-toothed, anther-tube included, style specialized for pollen dispersal, thickened and strigose at the distal end, divided into 2 deltoid-acute and hairy lobed, bearing no seed, aborted ovary with white pappus, pistillate heads bearing florets with filiform corolla, unequally 4- or 5-lobed, segments filiform, style filiform, equal or longer than the corolla, shortly 2-fid at the apex; achenes glabrous, 3.5 mm long, glandular-dotted, no hairs, no ribs, shiny brown; pappus snow white, 12 mm long, capillary. Native to eastern Asia, occurring in all the prvinces of the Yangtze River Region, and in Korea and Japan; introduced into Western gardens, growing extremely well in the damp, shaded area of the Arnold Arboretum, Harvard University.

Picris hieracioides L. — BITTERWEED, OX TONGUE (Syn. *P. hieracioides* L. var. *japonica* Regel ex Herder; *P. japonica* Auctt.)

Mao-lian-cai=Mao-lien-ts'ai (毛蓮菜, Hairy Lotus Vegetable). Young shoots; ethnic food, reported to have been eaten by people in Taiwan and Jiangsu, gathered and used locally, not available in the market.

Biennial or perennial herbs with rosette of radical leaves and coarse rough-bristly leafy stems 15–100 (–140) cm high, terminated by the first cluster of flowers, 4–10 mm in diameter, hairs 2–3 mm long, thickened at the base, branched and hooked at the apex; radical leaves few to eight, rarely over 10, rosulate, spathulate, 7–12 cm long, 1–2 cm wide, rounded or obtuse, base cuneate, gradually narrowed into the wings of the petiole, broadened before the insertion, sparsely strigose on both surfaces, the hairs stout but short, hairs on the petioles and midribs beneath longer and branched-hooked at the summit, cauline leaves all sessile, the lower and middle ones larger and broader than the radical and the upper cauline leaves, oblanceolate, oblong, elliptic 4–16 cm long, 1–4 cm wide, obtuse, base auricular, or amplexicaul, sinuate-dentate, teeth broadly deltoid and apiculate; flowers yellow, all ligulate, 16–30 to a capitulum, 3 cm across at anthesis, the first capitulum terminated the growth of the primary shoot, 1 to 4 capitula forming a corymbose cluster, peduncles of individual capitula 1.5–4 cm long, densely pubescent with bristles and soft white multicellular hairs, the bristles often 2- to 4-branched and hooked, secondary flowering branches developing from the axils of cauline leaves proceeding from the upper ones downward, stopped to the middle of the primary stem, rarely beyond, tertiary flowering branches sometimes developed, involucres campanulate, 10–15 mm long, 10–14 mm across, phyllaries imbricate, 3-seriate, the outer ones lanceolate, herbaceous, the middle ones linear-lanceolate, herbaceous, the inner-most ones linear, herbaceous with scarious margin, all herbaceous portion pubescent with both soft white hairs and stout bristles often branched and hooked at the apex, receptacles smooth, glabrous, center of areolae of denuded receptacle often apiculate, florets 18–20 mm long, corolla-tubes 2 mm long, the upper portion broadened and tomentose, ligules linear, 2.5 mm across the middle, truncate at the apex, 3-lobed, lobes 2 mm long, glandular at the apex, the lateral lobes each cleft into short segments, stamens 5, inserted at the hairy portion of the corolla, filaments short, anthers caudate, tubes 3–4 mm long, apical appendages deltoid, style slender, stylar arms slender, hairy on the abaxial surface, twisted, apical appendages absent; achenes ellipsoid, reddish-brown, 3 mm long, 14-ribbed, transversely muricate over and between the ribs; pappus in two rows, the outer bristly, the inner plumose, white. An Eurasian species, adventive to eastern North America; specimens from these regions examined with a good binocular and excellent light showing constant morphological characters of flowers and achenes, variations of leaves within any area appearing greater than those from widely separated regions; apparently a weedy species dispersed unintentionally by man; the flowering heads gathered in Tibet and used as a remedy for coughs; the entire plants collected in Inner Mongolia, used as a cooling agent for influenza and as an analgesics for breast tumor; a very poor food plant, the hooked hairs hurt the throat for a long time after eating.

Scorzonera glabra Ruprecht — Serpent Root (Syn. *S. ruprechtiana* Lipschitz et Krascheninnikov; *S. austriaca* Willdenow var. *glabra* Ruprecht)
Ya-cong=Ya-ts'un (鴉蔥, Crow Onion); *Lao-wu-zui=Lao-wu-chui* (老鳥嘴, Crow's Beak). Roots and young plants; eaten as an emergency food in northern China.

Lactiferous perennial herbs 15–30 cm high, with stout vertical rootstocks densely crowned by fabrous remains of former leaves, stems scapose, solitary, rarely 2 or 3, emerging from different winter buds on the rootstock, bearing several foliaceous bracts; leaves all radical, lanceolate or linear-lanceolate, 12–30 cm long, 6–20 mm wide, acuminate, base cuneate and attenuate into the broadened sheath-like base, glaucous beneath, entire or wavy, on each side of the midrib with two vertical nerves running from the base to the apex, reticulation of veinlets clear beneath; flowers pale yellow, all ligulate, capitula solitary, 3.5–4 cm long at anthesis, involucres oblong-cylindrical, 2.5 cm long, 1 cm across before anthesis, phyllaries 5- or 6-seriate, the outer ones ovate-triangular, 8 mm long, 6 mm wide, inner ones varying from oblong-lanceolate to linear, 1–2.5 cm long, 4 mm wide, margin membranous, apex obtuse, glabrous or hairy, receptacles smooth, 5–6 mm across, alveolate, areolae apiculate at the center, florets 30–40, 3–3.5 cm long, tubular portion of corolla 15 mm long, hairy at the distal end, ligules 14–15 mm long, 4–5 mm wide, 5-lobed at the apex, lobes with 2 prominent veins and glandular, stamens 5, filaments hairy, anther-tubes 5 mm long, apical appendage oblong, obtuse, style concealed by the anther-tube, stylar branches slender, 14 mm long, pilose on the abaxial side, stigmatic surface to the end on the adaxial surface, achenes cylindrical, ochreous, 1 cm long, slightly narrowed at both ends; pappus plumose, white, unequal, 10–13 mm long.

Serratula chinensis S. Moore — CHINESE SAWWORT

Ya-ma-cai=Ya-ma-ts'ai (鴨麻菜, Duck-hemp Vegetable); *Cai-ji-miao=Ts'ai-chi-miao* (菜雞苗, Vegetable-chicken Shoot); *Ma-hua-tou=Ma-hua-t'ou* (麻花頭, Hemp-flower Shoot); *Guang-dong-sheng-ma=Guang-tung-sheng-ma* (廣東升麻, Cantonese Bugbane). Young shoots; eaten by the people living in the mountains of Lechang (樂昌) in northern Guangdong.

Perennial stout herbs 70–100 (–150) cm high, with stout short rootstock bearing fleshy fusiform roots and a rosette of radical leaves, disintegrated at anthesis, cauline leaves oblong-elliptic, 9–15 (–22) cm long, 3.5–7 (–8) cm wide, acuminate and awned at tip, base cuneate and attenuate into wings of petiole 1.5–2.5 (–4.5) cm long, upper cauline leaves smaller, those below the capitula 2–3 cm long, serrate, teeth apiculate, scabrid above, arachnoid and glabrescent, with multicellular hairs and numerous resinous dots beneath; flowers purple-rose, all tubular and perfect, in large solitary capitula on peduncles 6–8 cm long, with foliaceous bracts, involucres ovoid, 3 cm long, 3–4 cm across at anthesis, phyllaries 7-seriate, leathery, the outer series ovate, middle ones ovate-oblong, glabrous with scarious margin, inner series linear, rounded at the apex, florets 3 cm long, glabrous, corolla-tube 17–18 mm long, the upper 3–4 mm broadened, the lobes linear-filiform, 7–9 mm long, stamens 5, inserted at the base of the broadened portion of the corolla-tube, anthers shortly caudate and glandular at base, anther-tube exerted, 8 mm long, the apical appendage oblong and rounded at the tip, style slender, long-exserted, papillose-hairy, undivided or shortly cleft at the truncate apex; achenes compressed-obovate-oblong, 6–7 mm long, 2 mm wide, grayish-ochreous, glabrous, inconspicuously striate, base oblique, apex truncate, dentate, pappus 4-seriate, unequal, varying from 3 to 16 mm long, often the base of 2 or 3 connate, bristle serrulate. Endemic to the warm temperate mountains of China, uprooted by the people for the fleshy root and gathered by others for the edible shoots, poorly represented in herbarium collections, a good example of a species endangered before it is fully understood scientifically.

Serratula coronata L. — CROWN-SAWWORT (Syn. *S. coronaria* Pallas).

Wei-ni-hu-cai=Wei-ni-hu-ts'ai (偽泥胡菜, False Hemistepta); *Huang-cao=Huang-ts'ao* (黃草, Yellow Herb). Young plants; boiled, soaking off the bitter taste, drained, seasoned and eaten cold, recorded in Flora of Jiangsu; rootstocks used locally for dying yellow, hence the name "Yellow herb".

Robust perennials, 50–140 cm high at anthesis, primary stem 5–8 mm in diameter, with the growth terminated by the first capitulum, branches appearing in the upper-most leaf-axil and proceeding thence downward for more flowers; leaves both radical and cauline, with the former disintegrated at anthesis, the cauline leaves all sessile, oblong-elliptic, 10–20 (–30) cm long, 7–11 (–18) cm wide, pinnatisect, segments alternate, 6 or 7 on each side, ovate-elliptic or lanceolate, 3–9 cm long, 1–3.5 cm wide, the terminal ones larger and wider, the basal ones auriculate or linear and stipule-like, serrate and setaceous along the margin, sparsely white pubescent on both surfaces; flowers rose-purple, heterogamous, the outer row sterile and filiform, capitula solitary, 4–5 cm across, involucres campanulate, 2–3 cm long, 2.5 cm across the top, phyllaries 12-seriate, imbricate, the outer ones broad-ovate, middle ones oblong-lanceolate, both apiculate with a spine, the inner ones linear-lanceolate, scarious, apex not spiny, receptacles discoid, densely setose, fertile florets 2.8 cm long, corolla tubular, the basal filiform portion 9 mm long, the next portion suddenly broadened, 7 mm long, the apical portion 7 cm long, 5-cleft, segments linear-filiform, stamens 5, inserted to the base of the broadened portion of the corolla tube, filaments 6–7 mm long, anthers caudate, the tube 5 mm long, exserted, pale yellow, apical appendage oblong, style slender, slightly thickened at the summit, hairy, stylar branches long and slender, pilose on the abaxial side, stigmatic surface wavy on the adaxial ridge to the end without appendage; achenes compressed-oblong, 6 mm long, 1.5 mm wide, slightly narrow and with an oblique scar at the basal end, smooth, ochreous, truncate at the distal end; pappus several series, finely serrulate and plumose, 11–14 mm long. Native of temperate eastern Asia and northern Europe, the appearing of the first capitulum limiting the growth of the primary stem, flowering heads from the axillary branches often producing the corymbose appearance for the plant.

Silybum marianum (L.) Gaertner — MILK THISTLE, HOLY THISTLE

Shui-fei-ji=Shui-fei-chi (水飛薊, Water Fly Thistle). Oil extracted from achenes; used for cooking.

Fast growing annual or biennial herbs 1–2 m high, stems columnar, arachnoid, prickly; leaves mostly radical, few cauline, basal leaves rosulate, petiolate, oblong, 30–50 cm long, 12–25 cm wide, acute, base cuneate, pinnatisect, lobes spiny, green, maculate white, tomentose beneath, cauline leaves smaller, sessile; flowers purple-pink or white, numerous in large terminal solitary capitula, involucres subglobose, 3–5 cm across, phyllaries imbricate, bristly at apex, florets perfect, corolla tubular, lobes linear, 5 mm long, stamens 5, anthers sagittate at base, style pilose; achenes ovate, 5–6 mm long, flat; pappus setaceous, white, in several series, connate at the base forming a ring. Introduced from Europe recently, cultivated in northern China.

Sonchus arvensis L. — FIELD SOW-THISTLE (Figure 66)

Shan-Ku-mai=Shan K'u-mai (山苦麥, Hillside Bitter Wheat); *Ku-mai-cai=K'u-mai-ts'ai* (苦麥菜, Bitter Wheat Herb); *Ye-ku-mai=Yeh-k'u-mai* (野苦賈, Wild Bitter-lettuce). Young plants; used by people living in the mountains of the Nanling Range in southern China, as are the dandelions by the European and American peoples.

Perennial herbs 15–150 cm high, rhizomatous, and with stout rootstock, lower portion of stem 5–10 mm in diameter, glabrous; leaves all sessile, the radical ones oblanceolate in outline, 15–34 cm long, 2–6 cm wide, subentire and dentate to pinnatipartite, the terminal lobe semiorbicular, rounded at the apex or obtuse, the other lobes when present deltoid, cauline leaves oblong or pinnatifid, apex acute, base auriculate, dentate, all teeth apiculate; flowers yellow, the first capitulum terminating the elongation of the primary stem, all lateral branches flower-bearing, exceeding the first flowering branch in length, at the initial stage of the inflorescence, the buds and bracts white tomentose-floccose, glabrescent, leaving numerous long gladular processes discoid at the end, capitula 2 to many, pedunculate, subumbelliform-corymbose, involucres cylindrical-campanulate 10–14 mm long, phyllaries 38–50, 4-seriate, the outer ones herbaceous, ovate-lanceolate, the apex obtuse, the base thickened with age and spongy, covered with a dense line of glandular processes along the midrib, the innermost ones linear, acuminate, 10 mm long, with glandular processes at the distal half, corollas all ligulate, 16–17 mm long, the tube filiform, distal half hairy, ligules linear, 5-toothed at the apex, hairy at the basal half, anthers 3.5 mm long, glandular at the base, apical appendages deltoid, obtuse, brown, style branches exserted, subterete, 4 mm long, hairy except the stigmatic surface on the adaxial side, without apical appendage; achenes oblong, 2.5–3.5 mm long, brown, 12-ribbed, the ribs rugose; pappus snow-white, persistent, denuded receptacles 4–5 mm across, smooth or rugose. A weed on highway banks and rock cracks, occasionally occurring in gardens as small young plants in southern China; a polypioid, 2n = 36 and 54 reported.

Sonchus brachyotus de Candolle — BITTER SOW-THISTLE

Ku-mai-cai=K'u-mai-ts'ai (苦麥菜, Bitter Wheat-field Herb); *Ku-ma-cai=K'u-ma-ts'ai* (苦媽菜, Bitter Mother Herb). Young rosettes, washed, drained, mixed with wheat flour, steamed, cooled, seasoned with mashed garlic, chopped onion, salt, vinegar and soysauce, an early summer favorite of farmers in northern Jiangsu.

Perennial herbs 30–60 cm high at anthesis, with subterraneous rhizomes 3–5 cm long, bearing a terminal bud, developing into a rosette, stem erect, terete, the lower portion glabrous, the upper portion arachnoid; leaves firm, light green often with purplish spots, the radical leaves oblong-lanceolate, 10–20 cm long, 2–6 cm wide, glabrous, spinulose-dentate, cauline leaves auriculate or cordate-amplexicaul sinuate-subruncinate or entire; flowers yellow, all ligulate, in capitula 3–5 cm across, on bracteate peduncles forming an umbelliform corymb, arachnoid, involucres campanulate, 1.5–2 cm long, phyllaries 4-seriate, outer ones oblong-ovate, 6 mm long, arachnoid, inner ones linear, 14 mm long, pubescent, florets 2.3 cm long, the tube 1.3 cm long, distal pilose, ligules narrow, 1 mm wide, 5-lobed at apex, stamens short, anther tube 2 mm long, base caudate, apical appendages short, round, style concealed, stylar arms slender, gray, pilose, without apical appendage; achenes compressed-oblong or subtetragonous, ribbed and transversely muriculate; pappus snow-white, capillary, in 2 series, 1.2 cm long.

Native of temperate eastern Asia, common in poor alkaline soil, dry slopes and grave-yards; the light color and firm texture of leaves very similar to those of *S. arvensis* L., here treated as a distinct species for the lacking of glandular processes, the thin phyllaries and the geographical separation.

Sonchus oleraceus L. — Common Sow-Thistle

Ku-ju-cai=K'u-chu-ts'ai (苦苣菜, Bitter Lettuce Vegetable), *Ku-cai=K'u-ts'ai* (苦菜, Bitter Vegetable). Young plants; gathered locally by rural people, not available in the market.

Annual or biennial herbs 10–140 cm high, the stems simple or branches, the lower 4–12 mm in diameter; leaves all sessile, the radical ones rosulate, oblanceolate in outline, attenuate gradually into the wings of petiole-like base, undivided or lyrate, the cauline leaves, 10–20 cm long, the apical portion semiorbicular or deltoid, 6–7 cm wide, cauline leaves amplexicaul, with foliaceous dentate long basal lobes, ovate, rhomboid or oblanceolate in outline, runcinate, pinnatifid to pinnatisect, 10–16 cm long, 4–7 cm wide, the terminal lobes long acuminate or cordate, lateral lobes ovate or lanceolate, all lobes irregularly dentate, the teeth with broad base, prolonged and sharply apiculate; flowers pale yellow, all ligulate, the first capitulum terminating the elongation of the primary stem, on short and bracteate peduncle, solitary or paired, rarely in three and umbelliform-corymbose, all lateral branches flower-bearing, longer than the primary shoot, often bearing tertiary flowering branches, together giving the impression of a large compound panicle, at the initial stages of the development of the inflorescences, the young bud with its bracts and capitula arachnoid, the falling hairs leaving long glandular processes discoid at the apex, densely covering the peduncle, few on the phyllaries, involucres campanulate, 11–12 mm long, phyllaries 4-seriate, the outer ones herbaceous, ovate or lanceolate, 2–5 mm long, connate, thickened, forming a spongy disk, the innermost ones linear, acuminate, 9–10 mm long, scarious toward the base, persistent and reflexed on the denuded rugose receptacle, the center of each areole conical, corolla unequal in length, 10–13 mm long, the outermost ones the longest, the tube 6 mm long, the upper half long villose, the hairs tangled and after anthesis, the entire corolla-mass falling together, ligules 5–7 mm long, 1 mm wide, 5-toothed at the apex, stamens 5, exserted, anther-tube 2 mm long, apical appendages deltoid, black, stylar branches setaceous, black on the outside, stigmas flat, without appendage; achenes chestnut-brown, compressed-obovoid-oblong, 2.5–3 mm long, 0.75–1 mm wide, 12-ribbed, ribs rugose; pappus snow-white, 5–8 mm oblong, persistent, 2n = 32.

A weedy species, occurring in gardens, edge of village woods and wastes areas with rich soil; in southern and western China, gathered for food by peoples of varied ethnic background. The glandular processes on the inflorescences and the thickened-spongy phyllaries, very similar to those of *S. arvensis* L., sharing the same geographical range, but growing in very different ecological conditions and having distinct leaf-colors and textures.

Stevia rebaudiana (Bertoni) Hemsley — CAA-EHE, KAA-HEE (Syn. *Eupatorium rebaudianum* Bertoni)

Tian-ye-ju=T'ien-yeh-chu (甜葉菊, Sweet-leaved Chrysanthemum). Extracts; steviocide extracted from the leaves, used for sweetening food and beverage; recent introduction.

Much branched perennial herbs 1 m high, pubescent throughout, lower portion of stem lignified; leaves obovate-spatulate, 2–11 cm long, 1.5–4 cm wide, rounded, base cuneate, crenate above the middle; flowers white, capitula in simple cymose clusters, involucres consisting of 5–6 scarious lanceolate phyllaries, 5 mm long, hispid and glandular, florets tubular, bisexual, 4–6 in each capitulum, receptacle smooth, corolla 4 mm long, glandular, 5-lobed, stamens 5, anthers with oblong apical appendages, stylar branches hairy, exserted; achenes small, fusiform, black, ribbed, glandular; pappus setaceous, 4 mm long. Native to the highlands of Paraguay, introduced and cultivated by clonal propagation through axillary shoot proliferation, spreading to 27 provinces in China in a few years; the steviocide is said to be 300 times as sweet as conventional sugar.

Taraxacum officinale Weber ex Wiggers (s. l., including *T. mongolicum* Handel-Mazzetti and *T. sinicum* Kitagawa) — COMMON DANDELION (Figure 67)

Pu-gong-ying=P'u-kung-ying (蒲公英, Dandelion); *Po-po-ding=P'o-p'o-ting* (婆婆釘, Grandmother's Nail); *Di-ding=Ti-ting* (地釘, Ground Nail); *Huang-hua-cao=Huang-hua-ts'ao* (黃花草, Yellow-flower Herbs); *Jin-zan-cao=Chin-tsan-ts'ao* (金簪草, Golden Hairpin Herbs). Young plants; gathered before signs of flowering visible, cleaned, cooked by stirred-fried, rarely available, for the plants are collected for medicinal uses.

Acaulescent perennial herbs with stout and deep tap-root, lactiferous, rootstock black and rough outside, rather fleshy inside; leaves oblanceolate in outline, sessile, forming a spreading rosette or quite erect depending on the growing conditions, extremely variable in size, shape and margin, the outer leaves of 2-year-old plants spatulate or oblanceolate, the inner leaves with a large terminal lobe and sharp sinuses, 8–13 cm long, 1–2 cm wide, plants 3–5 years old with runcinate leaves, those on denuded sites spreading almost flat on the ground, those growing among grasses suberect-spreading, 12–30 cm long, 2–5 cm wide, with many deltoid lobes, with broad sinuses, lobes varying in width, most plants with lobes 1.5–2 cm wide at the base, few plants with lobes 1–4 mm wide at the base, midribs of the spreading rosettes reddish-green, those of the erect plants light-green; flowers yellow, numerous, in solitary capitula terminal to scapes 8–30 cm long at anthesis, scapes white woolly in bud, sparsely hairy with long multicellular hairs especially at the distal portion, involucres oblong-cylindrical, 2 cm long, 8–10 mm across, phyllaries 4-seriate, the outer 3 series ovate-lanceolate, 6–12 mm long, 2–3 mm wide, herbaceous, green with very narrow membranous margin, glabrous, the innermost series linear, 18–20 mm long, 1.5–2 mm wide, dark green, scabrid along the margin, erose and glandularly pappilose at and below the apex, receptacles discoid, 8–10 mm across, strongly rugose, areolae each with a semicircular abaxial wall enclosing a hemiorbicular shiny depression with spongy hollow elevation black at the center, florets all ligulate, unequal in length, the outermost ones the longest, 16–28 mm long, the inner ones reduced in length gradually, corolla tubes 5–6 mm long, the distal half sparsely villose, outer ligules 10–12 mm long, 1.5 mm wide, each with a longitudinal median band on the outside, apex truncate, 5-lobed and glandular, stamens 5, anthers caudate, tubes 3 mm long, constricted at the apex, appendages deltoid, glandular, ovary oblong, 1.5 mm long, the apical one-third suddenly narrowed and becoming discoid at the end, the basal 1 mm subtetragonous, with 4 strong ribs and two minor ones in between, all 12 ribs transversely rugose, style slender, the exposed portion setaceous, branches long and recurved-rolling, abaxial surfaces setaceous, stigmatic surface yellow, without appendages; achenes compressed ellipsoid, 6–7 mm long, 1.5–2 mm across the thick portion, rugose on the ribs and spinulose on the broad distal portion, beaked and papillose, the beak suddenly extending to filiform and glabrous portion 8–9 mm long, discoid at the end; pappus white, persistent, capillary, unequal, 7–9 mm long.

A common weed of holarctic distribution, affected largely by wind and attachment to animals, particularly by man.

Note: The species is a polypoid, reproduced often by apomixis, very variable, the shape, margin, color and length of scape all varying with the age of the individual and its ecological condition. Numerous species have been described on such variations, and the twentieth century has been an era of such nomenclatural explosion. For *Taraxacum* Weber, over 2,400 epithets at the species and subspecies levels have been recorded in *Index Kewensis*, described by unstable characters which other botanists cannot check. Such increase in the number of names creates a condition that professional botanists with a dandelion in hand cannot identify it to species. For example, Richards and Sell (1976) could only list the 289 species credited to Europe (Fl. Eur. 4: 332–343), including *T. officinale* Weber ex Wiggers, the generic type. Surrounding me, there grow thousands of common dandelions for me to study the specific characters and to monitor the phase changes of individuals. In the fenced experimental ground of the Department of Organismic and Evolutionary Biology, Harvard University, I studied the phenotypic characters of a population before the plants flower. My findings are: (1) All one-year-old plants bear spatulate or oblanceolate leaves remotely dentate. (2) Three-fourths of the 2-year-old or older plants branch deep in the soil (Figure 67 b), having 2 to 4 raments. (3) The sizes of the leaves and the incision of the laminas are strongly effected by the age of the plant and the environment, the older the plant, the deeper the marginal cuts, and the narrower the segments. On plants with 2 or more raments, the outer leaves of the smaller rament are often subentire, oblanceolate or spathulate. (4) The color of tne midribs, "petioles" and scape is strongly effected by their exposures to light and temperature. The tender basal portion of leaves and scapes are all pink, the midribs and scapes of plants partially covered by leaves are green, and the same parts of plants exposed on bare ground are rose-purple. Observations of the flowering and fruiting characters of a population on a lawn in an area of 5 by 2 meters, on the south side of the red brick building of the Department of Biochemistry and Molecular Biology were made in mid-April. (1) The number of capitula of a flowering plant varies from 4 to 35. At this time of the year, only plants 2-year-old or older bloom. (2) At anthesis the scapes are about the same length of the subtending leaves, slightly bending below the middle; during the development of the achenes, the length of the scape do not increase much; after the maturity of the achenes, the fruiting scape elongates, doubling the length of the flowering stage. (3) After anthesis, the ligules shrivel and the inner series of phyllaries folding slightly, the ovaries developing to full achene-sizes in 2–3 days with the apical beak elongating. (4) The elongated beaks of the achenes pushing the withered corolla off in a whole mass; color of the achenes turning partially blown; the spinules on the achenes forming early, before the full elongation of the beak. (5) Herbarium specimens uprooted at anthesis have short scapes, and 2 days after anthesis have achenes with short beaks. With these facts about *T. officinale* Weber ex Wiggers, the Chinese plants described and illustrated as *T. mongolicum* Hand.-Mazz. and *T. sinicum* Kitag. in *Icanographia Cormophytorum Sinicorum* (4: 679–680. 1975) and reported in various floras in China as edible plants growing in gardens and wayside can be matched by many dandelions on the lawns and along the fence of the experimental ground in Harvard University. They are here recognized as members of the polymorphic *T. officinale* Weber ex Wiggers.

Tragopogon porrifolius L. — SALSIFY, VEGETABLE OYSTER, OYSTER PLANT

Bo-luo-men-shen=Po-lo-men-shen (婆羅門參, Brahman's Vital Root); *Xi-yang-bai-niu-bang=Hsi-yang-po-niu-pang* (西洋白牛蒡, European White Burdock); *Suan-ye-bo-luo-men-shen=Suan-yeh-po-lo-men-shen* (蒜葉婆羅門參, Garlic-leaved Brahman's Vital Root). Roots; used locally by the people in Xinjiang, wild or cultivated; cultivated in gardens near Shanghai to meet the market demand of Western residents.

Grass-like biennials, 30–80 (–125) cm high at anthesis, the stems with prominent nodes, the internodes 5–12 mm across, the underground portion consisting of strong tap-root and rough rootstock; leaves radical and cauline, all sessile, linear-lanceolate, 15–30 cm long, 5–12 mm across the middle, gradually attenuate at the distal end, widened and sheathing at the base; flowers lilac to red-purple, in solitary capitula terminal to the primary stem or lateral shoots which becoming inflatted peduncles 19–24 cm long, 5–8 mm in diameter, slightly below the flowering head, involucres broadly campanulate, glaucous, 3.5–6 cm long, 1.5–2 cm across at anthesis, phyllaries 8–10, uniseriate, lanceolate, 3–5.5 cm long, 7–8 mm across the base, acuminate-caudate, tomentose along the distal margin and the inside, receptacles flat, 1.5–1.8 cm across, the denuded ones strongly rugose and tuberculate, outer florets equal the phyllaries in length, the tubes white, 5 mm long, softly hairy at the distal end, ligules linear, 8 mm long, colorful, apical lobes oblong, obtuse, glandularly pilose, anthers caudate, tubes 6 mm long, of darker color than the ligule, apical appendages short, glandular, ovary ovoid, striate-sulcate, distal half scabrid on the striae and gradually attenuate into the glabrous beak, style short, hidden in the anther-tube, stylar branches filiform, recurved and coiled, pilose and purple, without apical appendate; achenes angular-fusiform, 1.4 cm long, squamose-muricate on the ribs, the distal end gradually passing into the cylindrical peak 15–16 mm long, floccose-lanate at the end, the base strongly oblique, forming a semiorbicular projection, pappus plumose, brownish-white, unequal, the 5 long and stout ones 22–26 mm long, the apical 1/4 covered by purplish spicules, 5–7 short ones in between 16–18 mm long, plumose throughout. An Eurasian species spontaneous in the Mediterranean Region eastward to Xinjiang (formerly known as Chinese Turkestan), widely cultivated for its edible root.

Youngia japonica (L.) de Candolle

Huang-an-cai=Huang-an-ts'ai (黃鵪菜, Yellow Quail Vegetable); *Huang-hua-cai=Huang-hua-ts'ai* (黃花菜, Yellow Flower Vegetable); *Huan-yang-cao=Huan-yang-ts'ao* (還陽草, Revitalizing Herb); *Ye-jie-cai=Yeh-chieh-ts'ai* (野芥菜, Wild Mustard). Young plants gathered before flowering, and used as potherb in northwestern China; fresh material available in Hong Kong and Macao market-places under the name *Huan-yang-cao* (還陽草, Revitalizing Herb), used as an ingredient for cooling tea.

Annual or perennial herbs 20–60 cm high, branchlets pilose; leaves all radical and rosulate before flowering, oblanceolate or lyrate, 8.5–13 cm long, 0.5–2 cm wide, pinnatifid, the terminal lobe larger, the lateral ones reduced in size progressiviely downward, acute or rounded, base cuneate; flowers yellow, capitula many in loose terminal corymbose panicles, involucres oblong-cylindrical, green, phyllaries in 2 series, the outer ones 5, ovate-deltoid, the inner ones 8, lanceolate, receptacles flat, florets 14–17 in each capitulum, corolla ligulate, 5–7 mm long, truncate, 5-foothed, the short tube pilose, stamens 5, anthers green, style and stigma yellow; achenes fusiform, with 11–13 ridges, 2 to 4 more stout and spiny; pappus white, 2–3 mm long. Widespread and weedy in eastern Asia.

In the following citations of literature, the references without English titles are my translations, each followed by its original ideographs with the sounds given in the Pinyin System. For a period of thirty years, especially between 1960 and 1980, authorships were not credited to individual scientists participating in the research activities. Instead, their institutions appeared in the authors' position on the title pages. Publications of this nature are arranged by the dates of the issues under anonymous, each with the institution given in parenthesis. The Chinese names are not repeated in publications with titles in English or Latin. In the publications using "Tomus" for volume, and II or 2 for number of issue, the original printings are taken as they appeared on the title page, without change for uniformity.

Aaronson, S. 1986. A role for algae as human food in antiquity. *Food and Foodways* **1**: 311–315.

———. Fungal parasites of grasses and cereals: their role as food or medicine, now in the past. *Antiquity* **63**: 247–257.

Anderson, E. N. 1988. *The Food of China*, pp. xvi + 1–311. New Haven: Yale University Press.

Anonymous. 1949. *Manual of Cutivated Plants.*

——— (Shanghai Institute of Agricultural Science). 1959. *Species and Cultivars of Shanghai Vegetables* (上海蔬菜品種志, *Shanghai shucai pingzhong zhi*), pp. 1 + 9 + 374, figs. [1–].

——— (Hebei Agricultural University). 1961. *Vegetable Gardening* (蔬菜栽培學, *Shucai zhaipi xue*), pp. 1–2 + 1–234, figs. 110. Beijing: Agricultural Press.

——— (Institute of Botany, Academia Sinica). 1972–1983. *Iconographia Cormophytorum Sinicorum*, Vol. 1 (1972), pp. i–vi + 1–1157, figs. 1-1730; Vol. 2 (1973), pp. iv + 1–1321, figs. 1731–3954; Vol. 3 (1974), pp. i–iii + 1–1038, figs. 3955–5414; Vol. 4 (1975), pp. [iii] + 1–932, figs. 5415–6380; Vol. 5 (1976), pp. [iii] + 1–1146, figs. 6831–8374; Supplement I (1982), pp. viii + 1–806, figs. 8375–8758; Supplement II (1983), pp. xii + 1–879, figs. 8765, 8773–9082. Beijing: Science Press.

——— (Yunnan Institute of Botany). 1972. *Economic Plants of Yunnan* (雲南經濟植物,

Yunnan jingji zhiwu) (1981 ed.), pp. 1–574, figs. 1–328. Kunming: Yunnan People's Press.

—— (Beijing Institute of Agricultural Research). 1973. *An Introduction of Common Vegetable of Beijing* (北京主要蔬菜品種介紹, *Beijing zhuyao shucai pinzhong jieshao*), pp. 1–6 + 1–239, figs. [1–225], color plates 4. Beijing: Agricultural Press.

—— (Canton Vegetables Editorial Committee). 1974. *Treatise of Canton Vegetables* (廣州蔬菜品種誌, *Guangzhou shucai pinzhong zhi*), pp. 1–7 + 1–454, figs. [350]. Shanghai: Shanghai People's Press.

—— (Guangdong Institute of Botany = South China Institute of Botany, Academia Sinica). 1974–1977. *Flora Hainanica*, Tomus III (1974), pp. [i] + 1 + 1–629, figs. 527–942.; Tomus IV (1977), pp. [i] + I + 1–644, figs. 944–1272. Beijing: Science Press. (For Vols. 1 and 2, see W. Y. Chun, et al.)

—— (Instituto Botanico Boreali-occidentali, Academiae Sinicae). 1974–1981. *Flora Tsinlingensis*, Tomus I (pars 2) Spermatophyta (1974), pp. 1–647; Tomus II, Pteridophyta (1974), pp. iii + 1–246; Tomus I (pars 3) Spermatophyta (1981), pp. [ii] + 1–500. Beijing: Science Press.

—— (Institutum Sylviculturae et Pedologiae, Academiae Scientiarum Sinicae). 1975–1977. *Flora plantarum herbaceanun Chinae boreali-orientalis*, Tomus 3 (1975), pp. ii + 242, figs. 100; Tomus 5 (1976), pp. ii + 1–186, figs. [wthout numbers]; Tomus 6 (1977), pp. ii + 1–308, figs. 1–120; Tomus 11 (1981), pp. ii + 1–220, figs. 91. Beijing: Science Press. (For Vols. 1, 2, 7, see T. N. Liou).

—— (Research Institute for Culinary Technology). 1975. *Cantonese Cooking* (粵菜烹飪, *Yuecai pongren*), pp. 1–270. Reprinted in Hong Kong.

—— (Hubei Institute of Botany). 1976–1979. *Flora Hupehensis*, Tomus I (1976), pp. 2 + 1–508; Tumus 2 (Fi Shuxia, et al., 1979), pp. 2 + 1–522. Wuhan: Hebei People's Press.

—— (Chinese Cookbooks Editorial Committee). 1976–1981. *Chinese Recipes: From Guangdong* (中國菜譜：廣東, *Zhongguo caipu: Guangdong*) (1976), pp. 1–355, color plates [1–12]; *Chinese Recipes: From Jiangsu* (中國菜譜：江蘇, *Zhongguo caipu: Jiangsu*) (1979), pp. 1–302, color plates [1–12]; *Chinese Recipes: Shaanxi* (中國菜譜：陝西, *Zhongguo caipu: Shaanxi*) (1981), pp. 1–310. Beijing: Economic Press of the Treasury Department.

—— (Jiangsu New Medical College). 1977–1978. *An Encyclopedia of Chinese Medicines* (中藥大詞典, *Zhongyao da cidian*), Vol. I (1977), pp. vi + 15 + 1–1489; Vol. II (1977), pp. 13 + 1491–2754. Shanghai: Shanghai People's Press. Supplement (1978), pp. 1–764. Shanghai: Shanghai Science Technology Press.

—— (Yunnan Institute of Botany, Academia Sinica). 1977–1979. *Flora Yunnanica*, Tomus 1 (1977), pp. [7] + 1–870; Tomus 2 (1979), pp. [2] + 1–889. Beijing: Science Press.

—— (Wuhan Institute of Botany, Academia Sinica). 1980. *Shennongjia zhiwu*, pp. 1–467, photos 63, illustrated. Wuhan: Hubei People's Press.

——. 1981. *Map of the People's Republic of China* (Book No. 12014–1126). Beijing: Cartographic Publishing House.

——. 1984. *Selected Royal Dishes*, pp. 1–160, illustrated. Hong Kong: Tai Dao Publishing Ltd.

——. 1987. *Xinghua Zidian* (新華字典, New China Dictionary), pp. 1–88 + 1–640. Hong Kong: Commercial Press.

Bailey, L. H. 1920. A collection of plants from China. *Gentes Herb.* **1**: 1–49, figs. 1–17.

Baranov, A. L. 1965. Preparation of dried vegetable marrows for winter use in north Manchuria. *Econ. Bot.* **19**: 68–69, figs. 1–6.

——. 1967. Wild vegetables of the Chinese in Manchuria. *Econ. Bot.* **21**: 140–155, figs. 1–6.

Bates, D. M., R. N. Robinson and C. Jeffrey. 1990. *Biology and Utilization of the Cucurbitaceae*, pp. xiv + 1–485.

Burkill, I. H. 1935. *A Dictionary of the Economic Products of the Malay Peninsula*, Vol. I (A–H), pp. xi + 1–1220; Vol. II (I–Z), pp. 1221–2402. London: Crown Agents.

Camp, W. H., W. R. Boswell and J. R. Magness. 1957. *The World in Your Garden*, pp. 1–231. Washington, D.C.: National Geographic Society.

Chang, S. T. and W. A. Hayes (eds.). 1978. *The Biology and Cultivation of Edible Mushroom*. New York: Academic Press.

Chen, C. Q. (739 A.D.). *Unrecorded Materia Medica* (本草拾遺, Bencao shiyi). Original copies lost, material recored in S. Z. Li (1693).

Chen, C. Y. 1975. *A Manual of Chinese Dietary Healing* (養生食經, Yangsheng shijing), pp. 1–20 + 1–1129. Hong Kong: Shanghai Bookstore.

Chen, S. L. (934 A.D.). *A Herbal on Edible Material* (食性本草, Shixing bencao), fascicles 10.

Chen, Y. 1933. *Silviculture of Chinese Trees*, pp. 60–65 (on the cultivation of China root).

——. 1937. *Illustrated Manual of Chinese Trees and Shrubs*, pp. 2 + 2 + 6 + 14 + I-XXXXIV + 1–106 + 1–13 + 1–1191 + Index of Chinese names (1–64) + Index of scientific names (1–67) + Index of English names (1–6) + Appendix [I]: Alphabetical list of morphological terms with Chinese equivalents (1–18) + Appendix [II]: Abbreviations of author's names (1–13). Nanking: Agricultural Association of China.

——. 1959. *Illustrated Manual of Chinese Trees and Shrubs*, reprint with Supplements, pp. [I] + 11 + 1–60. Shanghai: Shanghai Technology Press.

Cheung, S. C. 1954. *Efficacy of Liangcha* (涼茶治效從科學的解説, Liangcha zhixiao cong *kexue di jieshuo*), Special Publication of the Fourteenth Anniversary of Hong Kong Crude Drugs and Liangcha Dealers Association, pp. [1–3].

Cheng, W. C. 1983–1986. *Woody Plants of China* (中國樹木誌, *Zhongguo shumu zhi*), Vol. 1 (1983), pp. 6 + 1–20 + 44, photos 1–40, figs 1–400; Vol. 2 (1985), 9 + 931–9368 + 1–103, figs, 401–1215. Beijing: China Forestry Press.

Chun, W. Y., et al. 1964–1965. *Flora Hainanica*. Tomus I (1964), pp. x + 1–517 + 1–26, figs. 1–285; Tomus II (1965), pp. vii + 1–470 + 1–26, figs. 286–526. Beijing: Science Press.

Church, M. B. 1920. Laboratory experiments in the manufacture of Chinese ang-khak in the United States. *Journ. Ind. Eng. Chem.* **12**: 45–46.

Davis, E. W. 1983. Notes on the ethnomycology of Boston's Chinatown. *Bot. Mus. Leaflets.* **29(1)**: 59–67, plates 7–8.

Delfs, R. A. 1974. *The Good Food of Szechuan*, pp. 1–124. N.Y.: Kodansha Intern.

Fernald, M. L. and A. C. Kinsey. 1943. *Edible Wild Plants of Eastern North America*, pp. i–xii + 1–425, plates I–XXV, figs. 1–129; revised by R. C. Rollins (1958). N.Y.: Harper & Row.

Foster, S. 1984. *Herbal Bounty*, pp. 1–192. Salt Lake City: Gribbs Smith.

———. 1991. *Echinacea: Nature's Immune Enhancer*, pp. I-x. 1–150. Rochester, Vermont: Healing Arts Press.

———. 1992. *Herbs of Commerce*, pp. 1–78. Austin, Texas: Am. Herbal Products Ass.

——— and W. D. Gray. 1970. *The Use of Fungi as Food and in Food Processing*, pp. i–v + 1–113. Cleveland: CRC Press and Chemical Rubber Co.

Greenhalgh, P. 1980. Production, trade and markets for culinary herbs. *Trop. Sci.* **22(2)**: 159–188.

Hamilton, R. A. and W. B. Storey. 1956. Macadamia nut production in the Hawaiian Islands. *Econ. Bot.* **10**: 92–100.

He, S. Y. 1988. *Flora Hebeiensis* **2**:1–676.

Heiser, C. B. 1976. *The Sunflower*, pp. xxvi + 1–198. Norman, OK: University of Oklahoma Press.

Hill, A. F. 1952. *Economic Botany — A Textbook of Useful Plants and Plant Products*, pp. i–xii + 1–560, figs. 1–250. N.Y.: McGraw-Hill.

Hill, L. D. (ed.) 1979. *World Soybean Research*, pp. i–xvii + 1–1073. Danville, Illinois: Interstate Printers.

Hou, K. Z. 1954. Comments on the correct botanical names of several vegetables long cultivated in South China. *Acta Phytotax. Sin.* **3(1)**: 75–91.

———. 1956. *Flora of Canton* (廣州植物誌, *Guangzhou zhiwu zhi*), pp. v + 1–958, figs. 415. Beijing: Science Press.

Hu, S. Y. 1937. The cultivation and uses of water chestnut (*Eleocharis dulcis*) in Canton and vicinity. *Lingnan Biol. Seminar Rep.* **1**: 1–5.

—. 1937. Plant esculents used for the conservation of health (中國補品之研究). M. Sc. thesis, Lingnan University, Canton, China.

—. 1940. China-root — Fu-ling (茯苓) or T'u-fu-ling (土茯苓). *Journ. West China Border Res. Soc.* **12** (B): 80–86.

—. 1942a. *An Annotated List of Vascular Plants of Chengtu*, pp. 1–50. Sichuan: Canadian Miss. Press.

—. 1942b. *Ethnobotany of the Gia-rong People*, pp. 1–25. Unpublished manuscript.

—. 1945. Medicinal plants of Chengtu herb shops. *Journ. West China Border Res. Soc.* **15(B)**: 95-176, map.

—. *Ilex omeiensis* and related species. *Ic. Pl. Omei.* **2**: t.157–173.

—. 1948. Shien — Some noteworthy edible herbs of China. *Herbalist* **14**: 30–36, plates I-II.

—. 1949–1950. The genus *Ilex* in China. *Journ. Arn. Arb.* **30** (1949): 233–287; **31** (1950): 39–80, 214–236.

—. 1954–1956. A monograph of the genus *Philadelphus. Journ. Arn. Arb.* **35** (1954): 275–333; **36** (1955): 25–109, 325–368; **37** (1956): 15–90.

—. 1955. *Malvaceae, Flora of China, Family 153*, pp. 1–80, plates 1–24, index.

—. 1956. Climbing the trails of the giant panda. *Appalachia* **1956**:164–172.

—. 1956. Malva — A herb of high nutritive value. *Herbalist* **22**: 22–30.

—. 1957. *An Enumeration of the Food Plants of China*, pp. 1–33, mimeographed; distributed in the Ninth Pacific Science Congress; introduction published in the *Proc. Ninth Pac. Sci. Congr.* **4**: 289–290.

—. 1957. Oriental hollies. *Nat. Hort. Mag.* (Holly Handbook) **36**: 31–64.

—. 1967. The economic botany of dragon-tongue [*Sauropus changianus* S. Y. Hu-Euphorbiaceae]. *Econ. Bot.* **21**: 288–292.

—. 1968. Uses of daylily as food and in medicine (Daylily Handbook). *Am. Hort. Mag.* **47(2)**: 214–218.

—. 1974. A contribution to our knowledge of *Leonurus* L., I-mu-ts'ao, the Chinese motherwort. *Journal of the Chinese University Hong Kong* **2(2)**: 365–378, fig., map.

—. 1975. The tour of a botanist in China. *Arnoldia* **35**: 264–295.

—. 1976. The genus *Panax* (ginseng) in Chinese Medicine. *Econ. Bot.* **30**: 11–28, figs. 1–7.

—. 1977. Knowledge of Ginseng from Chinese records. *Journal of the Chinese University Hong Kong* **4**: 281–305.

—. 1978. The ecology, phytogeography and ethnobotany of ginseng. *Proc . Second Intern. Gins. Symp.*, pp. 149–157.

—. 1979. A contribution to our knowledge of Tu-chung — *Eucommia ulmoides*. *Am. Journ. Chin. Med.* **7**: 5–37.

———. 1980a. The metasequoia flora and its phytogeographic significance. *Journ. Arn. Arb.* **61**: 41–94.

———. 1980b. Biological and cytological foundation for better ginseng to more people. *Proc. Third Intern. Ginseng Symp.*, pp. 171–179.

———. 1987. *Chinese Food Plants used to Reinforce the Immune System*, pp. 1–19, distributed by the Arnold Arboretum in an exhibit on food plants of China, February 1987.

———. 1988. Nomenclatural changes for an economically important plant from China (re: *Microcos nervosa* [Lour.] S. Y. Hu). *Journ. Am. Arb.* **69**: 77–80.

———. 1990. History of the introduction of exotic elements into traditional Chinese medicine. *Journ. Am. Arb.* **71**: 487–525.

———. L. Rüdenberg and P. Del. Fredici. 1980. Studies of American ginseng. *Rhodora* **82**: 672–686.

Hua, C. 1958. *A Complete Cookbook for Homemade Food (Cantonese, Jiangsu, Beijing and Sichuan styles) for Four Seasons* (粵蘇京川四季烹飪 = 家庭食譜全書, *Yue Su Jing Chuan siji pengren = Jiating shipu quanshu*), pp. 1–102 + 1–121 + 1–105 + 1–106. Hong Kong: Shanghai Bookstore.

Huang, P. S. [1984]. *The Secret of the Efficacy of the Incredible Ling-zhi* (*Ganoderma*) (in Chinese without date, translated into English by S. Y. Hu in 1984), pp. 1–4.

Hymowitz, T. 1970. On the domestication of the soybean. *Econ. Bot.* **24**: 408–421.

Jeffrey, C. 1980. *The Cucurbitaceae of Eastern Asia*, pp. 1–60. London: Roy. Bot. Gard. Kew.

Kay, E. D. 1979. *Food Legumes* (Crop Product Digest 3), pp. xvi + 435. London: Tropical Products Institute.

Kaye, G. C. 1984. *Wild and Exotic Mushroom Cultivation in North America*, pp. [iii] + 1–32, illus. Cambridge, Mass.: Farlow Herbarium, Harvard University.

Khashagan, S. 1990. A preliminary study on plants used as Mongolian traditional tea. *Acta Bot. Yunnanica* **12**(1): 41–48.

Kia, T. T. and T. S. Kia. 1937. *Plantae Sinicae Cum Illustratanibus*, pp. 1–2 + 1–21 + 1 + 1–12 + 1–4 + 1–14 + 1–1460 + 1–66 + 1–44. Shanghai: Kai Ming Press.

Kingborn, A. D. and D. D. Soejarto. 1985. Sweetening agents of plant origin. *CRC Critical Reviews in Plant Sciences* **4**: 79–120.

Kou, Z. S. 1115. *Elaboration on Chinese Herbs*. Ancient Chinese style, fascicles 20.

Lampe, K. F. and M. A. McCann. 1985. *AMA Handbook of Poisonous and Injurious Plants.* pp xi + 432. Am. Med. Ass.

Lee, B. 1963. *The Easy Way to Chinese Cooking*, pp. x + 1–180, illus. New York: Doubleday.

Lee, C. B. T. and A. E. Lee. 1983. *The Gourmet Chinese Regional Cookbook*, pp. 1–322. Castle.

Liang, J. M and C. W. Chi. 1980. Investigations on the introduction of sweet potato. *Journ. S. China Agric. Coll.* **1**(3): 74–79.

Liou, T. N. 1958–1981. Flora plantarum herbacearum Chinae boreali-orientalis. Vol. I (1958), [v] + I-75, figs. 1–56; Vol. II (1959), [iv] 1 + 1–120, figs. 1-102; Tomus 7 (1981), i + 1–267, figs. 1-126. Beijing: Science Press.

Liu, B. 1974. *True Fungi Used in Chinese Medicine* (中國藥用真菌, *Zhongguo yaoyong zhenjun*), pp. 4 + 2 + 1 + 1–196, figs. 84. Taiyuan: Shanzi People's Press.

Liu, J. C., J. J. Yang and L. Q. Hu. 1979. *Popular Sichuan Dishes* (大眾川菜, *Dazhong chuancai*), pp. 8 + 1–233. Chengdu: Sichuan People's Press.

Liu, T. S. 1942. *List of Economic Plants in Taiwan*, pp. 1-163. Taipei: Cheng Chung Press.

Liu, Y. X. 1985. *Flora in Deserts* (Reipublicae Populorum Sinarum, Tomus I), pp. [iv] + 1–546, figs. 1–191, map.

Makapugay, H. C., N. P. Dhammiki Nanayakkara, D. D. Soejarto, and A. D. Kinqhorn. 1985. High-performance liquid chromatographic analysis of the major sweet principle of Lo Han Kuo Fruits. *Journ. Agr. Food Chem.* **33**: 348–350.

Mathews, R. H. 1931. A Chinese-English Dictionary (1951 rev. Am. ed.), pp. xxii + 1– 1126. Cambridge: Harvard University Press.

McClure, F. A. 1957. *Bamboos of the genus Phyllostachys under cultivation in the United States* (Agr. Handb. 114, USDA), pp. 1–69, figs. 1–53.

Merrill, E. D. 1923–1926. *An Enumeration of Philippine Flowering Plants* **1**(**1**, **2**)(1923): 1– 240; **1**(**3**)(1924): 241–368; **1**(**4**)(1925): 369–463; **2**(1925): 1–530; **3** (1925): 1–628; **4**(1926): 1–515, plates 1–6. Manila: Bureau Printing.

Miller, O. K. *Mushrooms of North America*, pp. 1–353. N.Y.: Dutton.

Mong, S. 684 A.D. *Health Food and Curative Herbs* (食療本草, *Shiliao bencao*), 3 vols, 227 items (a Chinese classics).

Morse, W. J. 1947. The versatile soybean. *Econ. Bot.* **1**: 137–147, figs. 1–10.

Nicholson, B. E., S. G. Harrison, G. B. Masefield, et M. Wallis. 1969. *The Oxford Book of Food Plants*, pp. 1–viii + 1–206. Oxford: Oxford University Press.

Palo, M. A., L. Vidal-Adeva and L. M. Maceda. 1960. A study of Ang-khak and its production. *Philipp. Journ. Sci.* **89**: 1–22, plates 1–6.

Porterfield, W. M., Jr. 1951. The principal Chinese vegetable foods and food plants of Chinatown market. *Econ. Bot.* **5**: 3–37.

Qin, G. F. and P. D. Wang. 1978. *The Chinese Matrimony-vine and Lycium-berry* (枸杞, Gou qi), pp. 5 + 1–132, figs. 1–28. Yinchuan: Ningxia People's Press.

Richands, A. J. and P. D. Sell. 1976. Taraxacum Weber. In Tutin, T. G., et al. *Flora Eumpaea*, Vol. 4, pp. 332–343.

Rosengarten, F. Jr. 1984. *The Book of Edible Nuts*, pp. i–xi + 1–384, illus.

Siegel, R. K. 1979. Ginseng abuse syndrome — Problems with the panacea. *Journ. Am. Med. Ass.* **241**(15): 1614–1615.

———. 1980. Ginseng use among two groups in United States. *Proc. IIIrd Intern. Ginseng Symp*, pp. 229–236. Seoul, Korea.

Shibasaki, K. and C. W. Hesseltine. 1961. Miso I. Preparation of soybeans for fermentation. *Journ. Biochem. Microbiol. Technol.* **3**: 161–174, fig. 1.

———. 1962. Miso II. Fermentation. *Econ. Bot.* **16**: 180–195, figs 1–8.

Simonds, N. 1982. *Classic Chinese Cuisine*, pp. i–xi + 1–353. Boston: Houghon Mifflin.

Simpson, B. B. and M. Conner-Ogorzoly. 1986. *Economic Botany — Plants in Our World*, pp. i–x + 1–640.

Swingle, W. T. 1941. *Monordica growvenori*, sp. nov., the source of the Chinese Lo Han Kuo. *Journ. Arn. Arb.* **22**: 197–203, plates 1–2.

Talekar, N. S. and T. D. Griggs (ed.). 1981. Chinese cabbage. *Proc. First Intern. Symp. Asian Veg.*, pp, i–x + 1–489. Tainan, Taiwan: Center As. Veg. Res. Develop.

Talley, E. A. and I. A. Wolff. 1951. Chemicals from starch and sugar. *Yearbook Agr.*, pp. 136–141.

Tanaka, O. R. Kasai and T. Morita. 1986. Chemistry of ginseng and related plants: recent advances. *Abstr. Chin. Med.* **1** (1): 130–152.

Tseng, C. K. 1983. *Commn Seaweeds of China*, pp. x + 1–316, colored plates 1–149. Beijing: Science Press.

Tsuchiya, H. M. and R. H. Blom. 1951. Mold agents in conversion of starch. *Yearbook Agr.*, pp. 148–158.

Tucker, J. B. 1986. Amaranth: the once and future crop. *Bio Science* **36** (1): 9–13.

Verdcourt, B. 1979. A manual of New Guinea legums. *Bot. Bull.* **11**. Papua New Giuniea: Forests Office.

Wagner, H. 1985. Immunostimulants from medicinal plants. In M. H. Chang, H. W. Yeung, W. W. Tso and A. Koo. *Advances in Chinese Medicinal Material Research*, pp. 159–170.

Wang, P. L. *Recipes for Dietitic Healing* (食譜與治療, *Shipu yu zhiliao*), pp. 1–224. Taipei, Taiwan: Ter-jyp Publ. House.

Wang, X. J. 1630 A.D. *Herbal Monograph* (群芳譜, *Qun fangpu*), fascicles 30 (First record of sun flower as "zhang-ju" 丈菊, Ten-feet Chrysanthemum).

Wang, Y. 1506 A.D. *Food Plants Used in Medicine* (a Chinese classics).

Weis, E. A. 1983. *Oilseed Crops*, pp. x + 1-660. London: Longman.

Went, F. A. F. C. 1895. *Monascus purpureus*, le champignon de l'Ang-guac, une nouvelle Thélébolée. *Ann. Sci. Nat. Bot.* **VIII** (1): 1–18, figs. 1–33.

Winton, A. L. and K. B. Winton. 1932–1935. *The Structure and Composition of Foods*, Vol. 1

(1932), Cereals, starch, oil seeds, nuts, oils, forage plants, PP. xiv + 1–710; Vol. 2 (1932), Vegetables, legumes, fruits, pp. xiv + 1–904. N.Y.: John Wiley.

Wu, C. Y. 1983–1987. *Flora Xizangica* (Series Sci. Exp. Qinghai-Xizang Plateau). Vol. 1 (1983), pp. ix + 1–791, figs. 1–241, photos plates 110–114, photomicrographs 1–12. Vol. 2 (1985), pp. ix + 1–956, figs. 1–305; Vol. 3 (1986), pp. x + I-1047, figs. 1–305; Vol. 4 (1986), pp. vii + I-1021, figs. 1–373; Vol. 5 (1987), pp. ix + 1–955, figs 1–499.

Wu, G. M. 1957. *Chinese Vegetable and Their Culture* (中國蔬菜栽培學, *Zhongguo sucai zaipei xue*), pp. i–xiii + 1–537. Beijing: Science Press.

Wu, P. 220–260 A.D. *Herbal Classics of the Divine Plowman* (神農本草經, *Shennong bencao jing*), 3 fascicles (in Chinese, 799 edition by Sun and Sun).

Wu, T. L. 1985. A preliminary investigation into the origin of ginger (薑的起源初探, Jiang di qiyuan chutan). *Agric. Hist.* (農業考古) **2**: 247–250 (in Chinese).

Xu, L. K. and L. S. Mong. 1981. Nutrient composition of Lo Han Kuo. *Guangzi Plants* **1(2)**: 50–51.

Xu, X. H. and Y. Z. Lai. 1980. *Rare and Useful Plants of South China*, pp. [i]. 1–6 + 1–158, figs. 1–89.

Yamaguchi, M. 1983. *World Vegetables, Principles, Production Nutritive Value*, pp. xv + 1–415, figs. (numbered by chapters). Westport, Conn.: Avi Publ. Co.

Ye, Y. H. 1975. *A Complete Cookbook for Chinese Pastries* (中國點心大全, *Zhongguo dianxin daquan*). Hong Kong: P.O. Box 15635.

Yen, L. T. 1951. *Vegetable Crops of China*, pp. 1–642. Shanghai: Commercial Press.

Ying, J. Z., J. D. Zhao, X. L. Mao, Q. M. Ma, L. W. Xu and Y. C. Zong. 1982. *Edible Mushroom of China* (食用蘑菇, *Shiyong mogu*), pp. i–xiii + 1–255, figs. 1–300. Beijing: Science Press.

Young, E. M. 1931. The morphology and cytology of Monascus ruber. *Am. Journ. Bot.* **18**: 499–517, plates 39–41.

Yu, T. T. 1962. *Common Ornamental Plants in North China* (華北習見觀賞植物), Vol. 2, pp. ix + 1–219, figs. 191. Beijing: Science Press.

———. 1979. *Classification of Chinese Fruit Trees* (中國果樹分類學, *Zhongguo guoshu fenleixue*), pp. i + 1–421. Beijing: Agric. Press.

Yunis, E. J., G. Fernandes and R. A. Good. 1978. Aging and involution of the immunological apparatus. In J. J. Twoney and A. Good. *The Immunopathology of* Lymphoreticular neoplasms, pp. 53–75.

Zhou, L. C., B. Y. Zhong, L. Tan and F. Li. 1981. Surveying the resources of Lo Han-Kuo at the varietal level with suggestions for their uses. *Guangxi Plants.* **1(3)**: 29–33, figs. 1–9.

Zhu X. 1406 A.D. *Famine Food* (救荒本草, *Jiuhuang bencao*), 2 volumes, 414 plants (a Chinese classics).

Abelia engleriana 686
Abelmoschus esculentus 538
Abelmoschus manihot 538
Aberia gardneri 563
Abrus cantoniensis 166, 209, 210, 211
Abrus fruticulosus 211
Abrus precatorius 235
Acanthopanax senticosus 585
Acanthopanax trifoliatus 586
Acer mono 522
Acer pictum 522
Acer saccharum 522
Acer truncatum 522
Achras sapota 621
Achyranthes aspera 235
Achyranthes bidentata 166, 379
Aconitum carmichaeli 42, 166, 389
Acronychia oligophlebia 491
Acronychia pedunculata 235, 491
Actinidia arguta 549
Actinidia callosa var. henryi 550
Actinidia chinensis 7, 9, 180, 550
Actinidia coriacea 551
Actinidia deliciosa 551
Actinidia kolomikta 551
Actinidia latifolia 552
Actinidia polygama 552
Actinidia suberifolia 553
Actinidia tetramera 553
Adansonia digitata 543
Adenophora axilliflora 713
Adenophora capillaris 713
Adenophora khasiana 713
Adenophora perekiifolia 713
Adenophora polyantha 713
Adenophora potaninii 713
Adenophora stricta 189, 713
Adenophora tetraphylla 166, 713
Adenophora verticillata 713
Adenosma glutinosum 235
Aegopodium alpestre 588

Aeschynomene indica 465
Agaricus bisporus 266
Agaricus campestris 266
Agastache rugosa 246
Aglaia odorata 507
Agriophyllum arenarium 375
Akebia quinata 391
Akebia trifoliata var. australis 391
Alaria crassifolia 258
Aletris spicata 310
Aleurites moluccana 509
Alisma plantago-aquatica 246
Allium altaicum 311
Allium ampeloprasum 311
Allium ascalonicum 311
Allium cepa 311
Allium chinense 31, 312
Allium chrysanthum 312
Allium fistulosum 246, 312
Allium hookeri 313
Allium ledebourianum 313
Allium lineare 313
Allium macrostemon 22, 313
Allium mongolicum 314
Allium nipponicum 313
Allium porrum 311
Allium sativum 314
Allium senescens 315
Allium tuberosum 315
Allium victorialis 315
Allospondias lakonensis 514
Alpinia katsumadai 325
Alpinia officinarum 325
Alpinia oxyphylla 325
Alternanthera sessilis 235, 239, 379
Amalocalyx yunnanensis 629
Amaranthus ascendens 380
Amaranthus blitum 380
Amaranthus caudatus 380
Amaranthus inamoenus 380
Amaranthus lividus 380

Amaranthus paniculatus 51, 381
Amaranthus retroflexus 381
Amaranthus tricolor 382
Amaranthus viridis 382
Amelanchier asiatica var. sinica 428
Amomum cardamomum 325
Amomum hongtsaoko 326
Amomum kravanh 326
Amomum maximum 326
Amomum subulatum 327
Amomum tsao-ko 246, 327
Amomum villosum 246
Amorphophallus mairei 305
Amorphophallus rivieri 43, 304
Amygdalus communis 442
Amygdalus kansuensis 444
Amygdalus persica 446
Anacardium occidentale 514
Ananas comosus 308
Anaphalis margaritacea 717
Andrographis paniculata 235
Aneilema angustifolium 308
Aneilema bracteatum 309
Anethum graveolens 589
Angelica anomala 166, 246
Angelica dahurica 166
Angelica grosseserrata 601
Angelica laevigata 595
Angelica miqueliana 601
Angelica omeiensis 589, 590
Angelica polymorpha 590
Angelica polymorpha var. sinensis 591
Angelica pubescens 246
Angelica sinensis 166, 168, 190, 591
Angelica viridiflora 602
Anisogonium esculentum 273
Annona cherimolia 398
Annona muricata 398
Annona squamosa 398
Anredera baselloides 384
Anthriscus aemula 592
Anthriscus cerefolium 591
Anthriscus nemorosa 592
Anthriscus sylvestris var. aemula 592
Antidesma bunius 509
Antiotrema dunnianum 640
Aphananthe aspera 352
Apios fortunei 465
Apium graveolens var. dulce 592
Apocynum venetum 629
Aponogeton natans 282
Arachis hypogaea 465

Aralia chinensis 585
Aralia cordata 585
Arctium lappa 718
Arctostaphylos rubra 610
Arctous ruber 610
Ardisia cornudentata 613
Ardisia humulis 614
Ardisia morrisonensis 613
Ardisia quinquegona 613
Ardisia sieboldii 614
Ardisia solanacea 614
Areca catechu 166, 227, 235, 246, 300
Arenga pinnata 300
Arenga saccharifera 300
Armeniaca ansu 441
Armeniaca mandshurica 441
Armeniaca mume 445
Armillaria mellea 168, 330
Armoracia rusticana 403
Artemisia apiacea 35, 235, 244, 246
Artemisia brachyloba 718
Artemisia capillaries 246
Artemisia lactiflora 719
Artemisia selengensis 719
Artocarpus altilis 355
Artocarpus bicolor 356
Artocarpus heterophyllus 355
Artocarpus hypargyreus 11, 17, 355
Artocarpus integer 355
Artocarpus styracifolius 356
Arytera littoralis 523
Asparagus officinalis 316
Aspergillus oryzae 34, 40, 262
Aspergillus soyae 34, 40, 262
Asplenium nidus 274
Aster ageratoides 720
Aster indicus 738
Aster scaber 720
Aster trinervius 720
Astilboides tabularis 424
Astragalus membranaceus 166, 173, 187, 466
Astragalus mongholicus 187
Athyrium filix-foemina var. angustifrons 273
Athyrium filix-foemina var. multidentatum 273
Athyrium multidentatum 273
Athyrium sinense 273
Atractylodes lancea 246
Atractylodes macrocephala 166, 173, 609
Atriplex hortensis 375
Atriplex rosea 376
Auricularia auricula 266
Auricularia polytricha 266

Avena fatua var. glabrata 284
Avena sativa 284
Averrhoa bilimbi 490
Averrhoa carambola 490
Avicennia marina 642
Azukia angularis 117, 466
Azukia umbellata 472

Baccaurea ramiflora 509
Balanophora involucrata 369
Bambusa beecheyana 284
Bambusa beecheyana var. pubescens 285
Bambusa gibboides 285
Bambusa oldhamii 285
Bambusa spinosa 285
Bambusa stenostachya 285
Bambusa vario-striatus 285
Bambusa vulgaris 286
Bambusa vulgaris var. vittata 286
Basella alba 384
Basella rubra 384
Bauhinia hupehana 467
Bauhinia variegata 467
Begonia fimbristipula 12, 566
Begonia hayatae 566
Benincasa cerifera 691
Benincasa hispida 85, 93, 691
Benincasa hispida var. chiehqua 692
Benincasa hispida var. hispida 691
Berberis heteropoda 394
Berchemia flavescens 527
Berchemia floribunda 528
Berchemia giraldiana 528
Berchemia hypochrysa 527
Berchemia lineata 235, 528
Berchemia racemosa 528
Berchemia yunnanensis 529
Bertholletia excelsa 574
Beta vulgaris 376
Beta vulgaris var. cicla 376
Beta vulgaris var. rapa 376
Bischofia javanica 509
Blechnum orientale 271
Boehmeria nivea 363
Boerhaavia diffusa 383
Boerhaavia repens 383
Boletus edulis 267
Boltonia indica 738
Bombax ceiba 246, 544
Bombax malabaricum 544
Bouea macrophylla 515
Brachybotrys paridiformis 641

Brasenia schreberi 387
Brassica alba 407
Brassica alboglabra 153, 403, 406
Brassica botrytis 410
Brassica campestris 413
Brassica campestris var. oleifera 414
Brassica campestris var. purpuraria 414
Brassica cauliflora 410
Brassica caulorapa 406
Brassica chinensis 50, 406, 407
Brassica chinensis var. parachinensis 411
Brassica hirta 407
Brassica juncea 54, 251, 407
Brassica juncea var. foliosa 28, 29, 408
Brassica juncea var. gracilis 408
Brassica juncea var. integrifolia 408
Brassica juncea var. megarrhiza 28, 408
Brassica juncea var. multiceps 408
Brassica juncea var. multisecta 28, 408
Brassica juncea var. napiformis 28
Brassica juncea var. rugosa 408
Brassica juncea var. tamida 28, 29
Brassica juncea var. tumida 409, 505
Brassica napiformis 408
Brassica napobrassica 409
Brassica narinosa 410
Brassica oleracea 406, 410
Brassica oleracea var. alboglabra 403
Brassica oleracea var. botrytis 410
Brassica oleracea var. capitata 406, 410
Brassica oleracea var. gemmifera 411
Brassica oleracea var. gongylodes 406
Brassica parachinensis 29, 406, 411
Brassica pekinensis 29, 50, 64, 413
Brassica rapa 251, 413
Brassica rapa var. campestris 413
Breea segeta 726
Brugmansia arborea 658
Buchanania latifolia 515
Buchnera cruciata 235
Buckleya graebneriana 367
Buckleya henryi 368
Bupleurum chinense 167, 246
Bupleurum scorzonerifolium 167
Butyrospermum parkii 619

Cadelium radiatum 468
Cajanus cajan 469
Caldariomyces fumago 206
Callicarpa japonica 643
Callicarpa kochiana 643
Callicarpa mollis 643

Callipteris esculenta 273
Callitriche stagnalis 513
Calonyction aculeatum 635
Calonyction muricatum 635
Camellia gigantocarpa 554
Camellia latilimba 554
Camellia oleifera 554
Camellia polyodonta 555
Camellia semiserrata 555
Camellia sinensis 240, 246, 555
Camellia sinensis var. assamica 556
Camellia sinensis var. bohea 556
Camellia sinensis var. cantoniensis 557
Camellia sinensis var. viridis 557
Campanumeoa pilosula 714
Campylaephora hypneoides 260
Canarium album 32, 241, 246, 505
Canarium bengalense 505
Canarium nigrum 506
Canarium pimela 31, 506
Canarium strictum 506
Canarium tonkinense 506
Canavalia ensiformis 469
Canavalia gladiata 469
Canna edulis 329
Cannabis sativa 244, 246
Cansella bursa-pastoris 414
Canthium horridum 677
Capparis cordifolia 423
Capparis spinosa 423
Capsella bursa-pastoris 414
Capsicum annuum 658
Capsicum annuum var. grossum 659
Capsicum annuum var. longum 659
Capsicum frutescens 660
Capsicum frutescens var. conoides 660
Carallia brachiata 574
Cardamine hirsuta 415
Cardamine leucantha 414
Cardamine lyrata 414
Cardamine macrophylla 415
Cardamine regeliana 415
Cardamine urbaniana 415
Cardiocrinum giganteum var. yunnanense 316
Cardiospermum halicacabum 523
Carex kobomugi 299
Carissa carandas 630
Carissa grandiflora 630
Carissa macrocarpa 630
Carlesia sinensis 593
Carpodacus purpureus 55
Carthamus tinctorius 167, 215, 216, 721

Carum carvi 593
Carya cathayensis 336
Carya illinoinensis 336
Carya pecan 336
Caryota mitis 301
Caryota ochlandra 301
Caryota urens 301
Cassia minosoides 469
Cassia nomame 469
Cassia occidentalis 470
Cassia sophora 470
Cassia tora 470
Castanea bungeana 340
Castanea henryi 340
Castanea mollissima 340
Castanea seguinii 341
Castanopsis calathiformis 341
Castanopsis carlesii 341
Castanopsis chinensis 342
Castanopsis concolor 346
Castanopsis delavayi 342
Castanopsis eyrei 342
Castanopsis fargesii 342
Castanopsis fissa 343
Castanopsis fordii 343
Castanopsis formosana 344
Castanopsis hystrix 344
Castanopsis indica 344
Castanopsis jucunda 345
Castanopsis kawakamii 345
Castanopsis lamontii 345
Castanopsis orthacantha 346
Castanopsis platyacantha 346
Castanopsis sclerophylla 346
Castanopsis tibetana 346
Castanopsis tribuloides 347
Catalpa bungei 672
Catalpa ovata 673
Cedrela sinensis 508
Celastrus kusanoi 520
Celosia argentea 382
Celtis biondii 352
Celtis japonica 353
Celtis sinensis 353
Centella asiatica 235, 239, 593
Cephalonoplos segetum 726
Cephalostachyum capitatum 286
Ceramium boydenii 260
Cerastium aquaticum 385
Cerasus avium 442
Cerasus japonica 444
Cerasus pseudocerasus 446

Ceratopteris siliquosa 272
Ceratopteris thalictroides 272
Chaenomeles sinensis 167, 181, 218, 246, 429
Chamaesyce hirta 510
Chenopodium album 14, 42, 377
Chenopodium giganteum 377
Chenopodium hybridum 377
Chimonobambusa metuoensis 286
Chionanthus retusus 626
Chloranthus inconspicuus 334
Chloranthus japonicus 334
Chloranthus spicatus 334
Choerospondias fordii 515
Choerospondias pubinervis 516
Chondrus elatus 31, 260
Chondrus ocellatus 260
Chorda filum 259
Chorispora tenella 416
Chrysanthemum boreale 723, 724
Chrysanthemum coronarium 722
Chrysanthemum coronarium var. spatiosum 57, 722
Chrysanthemum indicum 723
Chrysanthemum lavandulaefolium 723
Chrysanthemum x morifolium 723
Chrysanthemum nankingense 724
Chrysanthemum sinense 723
Chrysanthemum spatiosum 57, 722
Chrysobalanus icaco 429
Chrysophyllum cainito 619
Cibotium barometz 167
Cicer arietinum 471
Cichorium endivia 724
Cichorium intybus 725
Cimicifuga heracleifolia 246
Cinnamomum cassia 167, 399, 400
Cinnamomum chingii 400
Cinnamomum loureirii 399, 400
Cinnamomum wilsonii 400
Cirsium japonicum 725
Cirsium segetum 726
Cistanche deserticola 673
Cistanche salsa 167, 181, 673
Citrullus battich 694
Citrus aurantium 492
Citrus aurantium var. amara 492
Citrus aurantium var. sinensis 498
Citrus deliciosa 496
Citrus erythrosa 496
Citrus grandis 493
Citrus kinokuni 496
Citrus limon 492

Citrus limonia 492, 493
Citrus maxima 493, 495
Citrus medica 494
Citrus medica var. sarcodactylis 495
Citrus microcarpa 495
Citrus mitis 495
Citrus nobilis 496
Citrus nobilis var. ponki 496
Citrus nobilis var. subcompressa 497
Citrus x paradisii 495
Citrus ponki 496
Citrus reticulata 167, 246, 496
Citrus reticulata var. erythrosa 496
Citrus reticulata var. kinokuni 496
Citrus sinensis 246, 495, 498
Citrus sinensis var. brazziliensis 500
Citrus sinensis f. sekkan 499
Citrus sinensis var. tatincheng 499
Citrus suavissima 497
Citrus succosa 497
Citrus suhoiensis 497
Citrus tangerina 498
Citrus tankan 498
Citrus wampi 500
Cladonia alpestris 263
Cladonia macroptera 264
Cladosiphon decipiens 259
Clausena anisum-olens 500
Clausena lansium 500
Clausena wampi 500
Clavaria botrytis 267
Cleidiocarpon cavaleriei 510
Cleistocalyx operculatus 232, 235, 239
Clematis chinensis 247, 390
Clematis dioscoreifolia 389
Clematis hexapetala 389
Clematis paniculata 389
Clerodendron fragrans 643
Coccoloba uvifera 370
Cocos nucifera 302
Cochlearia armoracia 403
Codonopsis lanceolata 713
Codonopsis nervosa 715
Codonopsis pilosula 167, 190, 714
Codonopsis tangshen 167, 173, 190, 715
Codonopsis tsinlingensis 715
Codonopsis tubulosa 715
Codonopsis viridiflora 715
Coffea arabica 678
Coffea arabica var. arabica 679
Coffea arabica var. bourbon 679
Coffea arabica var. typica 679

Coffea canephora 679
Coffea excelsa 680
Coffea liberica 680
Coffea robusta 679
Coix lachryma-jobi 167, 247, 287
Coleus amboinicus 652
Collybia albuminosa 267
Collybia velutipes 267
Colocasia antiquorum 305
Colocasia esculenta 41, 235, 305
Colocasia fallax 307
Commelina benghalensis 308
Commelina communis 308
Commelina nudiflora 235
Convolvulus arvensis 11, 636
Convolvulus batatas 638
Convolvulus chinensis 11, 22, 636
Coptis chinensis 230, 247
Cordia dichotoma 641
Cordyceps sinensis 167, 174, 175, 262
Cordyceps sobolifera 247
Coriandrum sativum 594
Corispermum puberulum 377
Cornus capitata 607
Cornus hongkongensis 607
Cornus kousa 608
Cornus kousa var. angustata 608
Cornus kousa var. chinensis 608
Cornus officinalis 608
Cornus walteri 609
Cornus wilsoniana 609
Cortinellus edodes 268
Corylus avellana 338
Corylus chinensis 338
Corylus chinensis var. fargesii 339
Corylus ferox var. thibetica 340
Corylus heterophylla 339
Corylus heterophylla var. sutchuenensis 339
Corylus mandshurica 339
Corylus sieboldiana var. mandshurica 339
Corylus tibetica 340
Costus speciosus 327
Cotinus coggygria var. glaucophylla 516
Cotoneaster acutifolia 429
Cotoneaster melanocarpus 430
Cotoneaster mongolicus 430
Cotoneaster multiflorus 430
Crassocephalum crepidioides 726
Crataegus cuneata 430
Crataegus dahurica 431
Crataegus hupehensis 430
Crataegus maximowiczii 430

Crataegus pinnatifida 84, 167, 218, 247, 431
Crataegus sanguinea 431
Crataeva nurvala 423
Crataeva religiosa 423
Cratoxylum ligustrinum 232, 235, 557
Cratoxylum prunifolium 558
Cryptotaenia japonica 594
Cucumis acidus 698
Cucumis bisexualis 696, 697
Cucumis callosus 698
Cucumis conomon 697
Cucumis dudaim 698
Cucumis flexuosus 697
Cucumis melo 695, 696
Cucumis melo var. agrestis 698
Cucumis melo var. chinensis 39, 697
Cucumis melo var. conomon 31, 697
Cucumis melo var. dudaim 698
Cucumis melo var. flexuosus 697
Cucumis melo var. inodorus 698
Cucumis melo var. reticulatus 698
Cucumis sativus 698
Cucurbita foetidissima 700
Cucurbita maxima 702
Cucurbita moschata 701
Cucurbita pepo 26, 702
Cudrania cochinchinensis 356
Cudrania tricuspidata 356
Cudrania triloba 356
Cullen corylifolium 167, 471
Cuminum cyminum 595
Curculigo orchioides 167
Curcuma aromatica 327
Curcuma domestica 328
Curcuma longa 328
Curcuma zedoaria 255, 328
Cycas revoluta 275
Cycas siamensis 275
Cydonia oblonga 431
Cydonia sinensis 429
Cymbopogon citratus 287
Cynanchum auriculatum 633
Cynanchum bungei 43, 193, 633
Cynanchum caudatum 633
Cynanchum forrestii 9, 634
Cynanchum otophyllum 634
Cynanchum wilfordii 193, 194, 634
Cynara scolymus 727
Cynomorium songaricum 584
Cyperus esculentus var. sativus 299
Cyphomandra betacea 660
Cystophyllum fusiforme 260

Czernaevia laevigata 595
Czernaevia laevigata f. exalatocarpa 595
Czernaevia laevigata f. latipinna 595

Daucus carota 596
Debregeasia edulis 363
Debregeasia longifolia 363
Decaisnea fargesii 392
Delandia calcaratus 167
Delandia umbellata 117, 244, 247, 472
Dendranthema x grandiflorum 72, 247
Dendranthema lavandulifolium 72
Dendranthema morifolium 723
Dendranthema vestitum 72, 167
Dendrobenthamia angustata 608
Dendrobenthamia capitata 607
Dendrobenthamia hongkongensis 607
Dendrobenthamia japonica 608
Dendrobenthamia japonica var. chinensis 608
Dendrocalamus asper 287
Dendrocalamus latiflorus 288
Dendrocalamus tibeticus 288
Desmodium podocarpum 473
Desmodium styracifolium 235, 239
Desmodium triquetrum 235
Dicliptera chinensis 675
Dictyophora indusiata 267
Dictyophora phalloidea 267
Digitalis purpurea 195
Dillenia indica 548
Dillenia pentagyna 548
Dillenia turbinata 549
Dimocarpus longan 167, 218, 523, 547
Dioscorea alata 144, 321
Dioscorea batata 323
Dioscorea bulbifera 145, 322
Dioscorea esculenta 322
Dioscorea hemsleyi 323
Dioscorea japonica 323
Dioscorea opposita 11, 144, 145, 167, 247, 323
Dioscorea pentaphylla 324
Dioscorea sativa 322
Dioscorea villosa 145
Diospyros kaki 623
Diospyros lotus 624
Diospyros oldhamii 625
Diplazium exculentum 273
Dipsacus asper 167
Dispyros oldhamii var. chartacea 625
Dispyros taitoensis 625
Disporopsis aspersa 317
Disporopsis pernyi 317

Doellingeria scabra 720
Dolichos angularis 466
Dolichos catjang 488
Dolichos lablab 475
Dolichos sinensis 488
Dolichos umbellatus 472
Dolichos unguiculatus 488
Dovyalis hebecarpa 563
Dracontomelon dao 516
Dracontomelon duperreanum 516
Drynaria fortunei 167
Durio zibethinus 544

Eberhardtia aurata 620
Ebulum formosana 687
Ecdysanthera rosea 631
Echinochloa crusgalli var. crusgalli 288, 289
Echinochloa crusgalli var. frumentacea 288
Echinops dahuricus 728
Echinops latifolius 728
Ecklonia bicyclis 259
Ehretia acuminata var. obovata 642
Ehretia thyrsiflora 642
Eichhornia crassipes 310
Elaeagnus angustifolia 12, 568
Elaeagnus bockii 569
Elaeagnus courtoisii 569
Elaeagnus glabra 569
Elaeagnus henryi 570
Elaeagnus lanceolata 570
Elaeagnus loureiri 570
Elaeagnus magna 570
Elaeagnus multiflora 571
Elaeagnus oldhamii 571
Elaeagnus umbellata 571
Elaeocarpus hainanensis 536
Elaeocarpus serratus 537
Elatostema edule 364
Eleocharis dulcis 43, 142, 299
Eleocharis tuberosa 299
Elephantopus mollis 728, 729
Elephantopus scaber 235
Elephantopus tomentosus 728, 729
Elettaria cardamomum 325
Eleusine coracana 289
Eleutherococcus senticosus 585
Eleutherococcus trifoliatus 235, 586
Elsholtzia ciliata 645
Elsholtzia cristata 645
Elsholtzia densa 646
Elsholtzia patrini 645
Elsholtzia penduliflora 646

Elsholtzia splendens 247
Embelia laeta 614
Embelia longifolia 615
Embelia ribes 615
Embelia sessiliflora 616
Embelia subcoriacea 616
Embelia vestita 616
Emilia sonchifolia 729
Empetrum nigrum 513
Enteromorpha compressa 258
Ephedra intermedia 278
Ephedra monosperma 279, 654
Ephedra sinica 279, 654
Epiphyllum oxypetalum 567
Erechtites valerianaefolia 730
Eriobotrya cavaleriei 432
Eriobotrya deflex 432
Eriobotrya japonica 432, 547
Erioglossum edule 524
Erioglossum rubiginosum 524
Eriophyton wallichii 646
Eriosma chinense 473
Eruca sativa 416
Eruca versicaria 416
Eryngium foetidum 596
Erysimum cheiranthoides 416
Erythropalum scandens 369
Eucheuma papulosa 260
Eucheuma spinosa 261
Eucommia ulmoides 42, 167, 205, 206, 428
Eugenia caryophylla 579
Eugenia javanica 581
Eugenia malaccensis 580
Eugenia uniflora 576
Eupatorium rebaudianum 748
Euphorbia hirta 510
Euphorbia humifusa 510, 624
Euphorbia longana 523
Euphorbia longan 523
Euphorbia tashiroi 510
Euryale ferox 106, 167, 387
Euscaphis japonica 520
Eutrema wasabi 416
Evodia lepta 232, 236, 239

Fagara ailanthoides 503
Fagopyrum cymosum 370
Fagopyrum esculentum 370
Fagopyrum tataricum 371
Fallopia nervosa 213
Farfugium japonicum 731
Farfugium japonicum var. formosanum 730

Fargesia melanostachys 289
Fargesia setosa 289
Fargesia spathacea 290
Feijoa sellowiana 577
Ferula tunshanica 597
Ficus auriculata 357
Ficus awkeotsang 357
Ficus beecheyana 357
Ficus carica 358
Ficus chlorocarpa 358
Ficus erecta 357, 358
Ficus erecta var. beeheyana 357
Ficus esquiroliana 11, 358
Ficus foveolata 360
Ficus gibbosa 359
Ficus henryi 359
Ficus hispida 232, 236
Ficus microcarpa 236
Ficus oligodon 359
Ficus pumila 43, 360
Ficus pumila var. awkeotsang 357
Ficus roxburghii 357
Ficus sarmentosa var. henryi 360
Ficus septica 360
Ficus stenophylla 361
Ficus tikoua 361
Ficus variegata var. chlorocarpa 358
Firmiana simplex 545
Flacourtia indica 564
Flacourtia inermis 564
Flacourtia rukam 564
Flammulina velutipes 267
Flemingia philippinensis 478
Foeniculum vulgare 597
Forsythia suspensa 247
Fortunella hindsii 501
Fortunella margarita 501, 547
Fragaria x ananassa 433
Fragaria chiloensis 433
Fragaria chinensis 435
Fragaria gracilis 433
Fragaria moupinensis 433
Fragaria nilgerrensis 434
Fragaria nubicola 434
Fragaria orientalis 431
Fragaria pentaphylla 434
Fragaria vesca 435
Fragaria virginiana 433
Fritillaria camschatcensis 317

Ganoderma japonicum 178, 268
Ganoderma lucidum 167, 177, 178, 268

Garcinia cowa 558
Garcinia mangostana 558
Garcinia multiflora 559
Garcinia oblongifolia 559
Garcinia paucinervis 560
Garcinia xanthochymus 560
Gardenia angusta 680
Gardenia jasminoides 247, 681
Gardenia thunbergii 680
Gastrodia elata 42, 168, 330
Gaultheria veitchiana 11, 611
Gelidium amansii 261
Gelidium latifolium 261
Gelidium rigidum 261
Geum aleppicum 435
Ginkgo biloba 98, 99, 275, 428
Gleditsia sinensis 473
Glehnia littoralis 168, 190, 598
Glossogyne tenuifolia 236
Glycine max 33, 168, 251, 474
Glycine soja 474
Glycosmis citrifolia 501
Glycosmis hainanensis 502
Glycyrrhiza glabra 168, 247, 475
Glycyrrhiza inflata 247, 475
Glycyrrhiza kansuensis 247
Glycyrrhiza uralensis 168, 247, 475
Gmelina arborea 644
Gnaphalium affine 731
Gnaphalium multiceps 731
Gnetum indicum 280
Gossampinus malabarica 544
Gossypium herbaceum 539
Gossypium hirsutum 539
Grateloupia flabellata 261
Gymnadenia conopsea 330
Gynostemma pentaphyllum 703
Gynura aurantiaca 731
Gynura bicolor 732
Gynura divaricata 733
Gypsophila davurica 384
Gypsophila oldhamiana 385

Habenaria conopsea 330
Habenaria delavayi 330
Handeliodendron bodinieri 524
Hedychium coronarium 255, 328
Helianthus annuus 734
Helianthus tuberosus 735
Helicia nilagirica 366
Heliciopsis henryi 366
Heliciopsis terminalis 366

Helicteres angustifolia 232, 236
Helixanthera parasitica 367
Helminthostachys zeylanica 271
Helwingia japonica 610
Hemerocallis citrina 318
Hemerocallis fulva 318
Hemerocallis lilioasphodelus 67, 318
Hemerocallis minor 67, 318
Hemerocallis thunbergii 318
Hemiptelea davidii 353
Hemistepta carthamodes 42, 736
Hemistepta lyrata 736
Hepialus armoricanus 176, 262
Heracleum moellendorffii 598
Hericium erinaceum 268
Hibiscus esculentus 538
Hibiscus manihot 538
Hibiscus sabdariffa 540
Hibiscus schizopetalus 540
Hibiscus syriacus 541
Hibiscus tiliaceus 541
Hibiscus trionum 541
Hippophae rhamnoides 572
Hodgsonia heteroclita 703
Holboellia coriacea 392
Holboellia fargesii 392
Holboellia latifolia 393
Hordeum vulgare 168, 247, 290
Hordeum vulgare var. nudum 290
Hordeum vulgare var. trifurcatum 291
Hordeum vulgare var. vulgare 290
Hosta plantaginea 318
Houttuynia cordata 255, 332
Hovenia dulcis 529
Humulus japonicus 363
Humulus lupulus 362
Humulus scandens 363
Hunteria corymbosa 631
Hunteria zeylanica 631
Hydrocharis asiatica 283
Hydrocharis dubia 283
Hydrocotyle benguetensis 599
Hydrocotyle ranunculifolia 599
Hygrophila lancea 676
Hylocereus undatus 168, 215, 567
Hypericum ascyron 561
Hypericum japonicum 236, 239

Ilex asprella 232, 236, 239
Ilex latifolia 247, 519
Ilex pubescens 236
Ilex rotunda 236, 239

Ilex yunnanensis 520
Illicium verum 369
Impatiens noli-tangere 527
Imperata cylindrica 232, 236, 291
Ipomoea alba 635
Ipomoea aquatica 636
Ipomoea batatas 144, 638
Ipomoea cairica 639
Ipomoea digitata 639
Ipomoea hungaiensis 640
Ipomoea mauritiana 43, 639
Ipomoea muricata 635
Ipomoea repens 636
Ipomoea staphylina 640
Ipomoea wilsonii 640
Ixeris chinensis 737
Ixeris denticulata 737
Ixora chinensis 682

Jasminum sambac 626
Jatropha manihot 511
Juglans cathayensis 337, 338
Juglans mandshurica 337
Juglans regia 168, 337, 338
Juncus alatus 236, 247
Justicia lancea 676

Kaempferia galanga 328
Kalimeris indica 738
Kandelia candel 574
Kochia scoparia 378
Kummerowia striata 475

Lablab purpureus 475
Lactobacillus delbrueckii 40
Lactuca formosana 738
Lactuca sativa 739
Lactuca sativa var. angustana 740
Lactuca sativa var. angustata 740
Lactuca sativa var. capitata 740
Lactuca sativa var. crispa 741
Lactuca sativa var. oblongifolia 741
Lagenaria siceraria 92, 704
Lagenaria siceraria var. cougourda 705
Lagenaria siceraria var. hispida 43, 705
Lagenaria siceraria var. microcarpa 705
Lagenaria siceraria var. siceraria 704
Lagenaria vulgaris 704
Lagenaria vulgaris f. cougourda 705
Laggera alata 236
Laminaria japonica 259
Lamium album 647

Lansium domesticum 507
Lansium parasiticum 507
Lappa edulis 718
Lathyrus davidii 476
Lathyrus japonicus 476
Lathyrus maritimus 476
Lecanora affinis 264
Ledebouriella seseloides 605
Lens culinaris 476
Lens esculentus 476
Lentinus edodes 268
Leonurus artemisia 647
Leonurus heterophyllus 647
Leonurus sibiricus 647, 648
Leonurus tataricus 648
Lepidium sativum 22, 417
Lepisanthes rubiginosa 524
Lespedeza bicolor 476
Lespedeza davurica 477
Lethariella cladonioides 264
Ligularia tussilaginea 730
Ligusticum brachylobum 606
Ligusticum nipponicum 601
Ligusticum sinense 599, 604
Ligusticum wallichii 200, 247, 600
Ligustrum japonicum 626
Ligustrum lucidum 168, 218
Ligustrum ovalifolium 21
Lilium brownii 319
Lilium brownii var. colchesteri 168
Lilium lancifolium 319
Lilium regale 319
Lilium speciosum 319
Lilium tigrinum 319
Limnanthemum cristatum 628
Limnanthemum nymphoides 628
Lindera aggregata 247
Linum usitatissimum 490
Liriope graminifolia 168
Litchi chinensis 525
Lithocarpus cleistocarpa 347
Lithocarpus dealbata 347
Lithocarpus glaber 348
Lithocarpus hancei 348
Lithocarpus polystachyus 348
Lithocarpus truncatus 349
Litsea cubeba 401
Litsea mollifolia 402
Lodoicea maldivica 302
Lonicera confusa 236
Lonicera henryi 686
Lonicera japonica 168, 215, 247, 687

Lophatherum gracile 232, 236, 239, 247
Luffa acutangula 90, 706
Luffa aegyptiaca 90, 707
Luffa cylindrica 90, 707
Lycium barbarum 168, 218, 222, 661
Lycium chinense 168, 218, 222, 663
Lycium halimifolium 661
Lycopersicon esculentum 664
Lycopersicon lycopersicum 664
Lycopus lucidus 648
Lygodium dichotomum 232, 236, 239
Lysimachia fortunei 618
Lythrum salicaria 572

Macadamia ternifolia 366
Maclura cochinchinensis 356
Maclura tricuspidata 356
Macrocarpium officinale 608
Madhuca pasquieri 620
Maesa argentea 617
Maesa doraena 617
Maesa japonica 617
Maesa montana 617
Maesa perlaria 618
Maesa sinensis 618
Magnolia denudata 396
Magnolia heptapeta 396
Magnolia officinalis 168, 247
Magnolia soulangeana 396
Malachium aquaticum 385
Mallotus furetianus 511
Malus asiatica 436
Malus baccata 436, 437
Malus communis 437
Malus domestica 437
Malus hupehensis 436
Malus kansuensis 436
Malus micromalus 437
Malus prunifolia 437
Malus pumila 437
Malus sinensis 429
Malus spectabilis 437
Malva crispa 542, 543
Malva neglecta 13, 61, 542
Malva verticillata 61, 542
Mammea americana 561
Mangifera indica 517
Mangifera persiciforma 517
Mangifera sylvatica 518
Manihot esculenta 511
Manihot utilissima 511
Manilkara achras 621

Manilkara hexandra 621
Manilkara zapota 621
Manilkara zapotilla 621
Mappianthus iodoides 521
Maranta arundinacea 330
Marsilea quadrifolia 274
Mazus japonicus 670
Mazus rugosus 670
Medicago denticulata 477
Medicago hispida 477
Medicago sativa 477
Medinilla radiciflora 581
Mangifera indica 236
Melampyrum lineare 671
Melampyrum roseum 671
Melastoma candidum 236
Melastoma dodecandrum 582
Melastoma normale 582
Melastoma polyanthum 582
Melastoma sanguineum 583
Melicope pteleifolia 236
Meliosma rhoifolia 526
Melissa maxima 652
Melodinus henryi 631
Melodinus khasianus 632
Melodinus suaveolens 632
Melothria heterophylla 711
Melothria indica 712
Mentha aquatica 649
Mentha arvensis 648, 649
Mentha arvensis f. piperascens 649
Mentha haplocalyx 648
Mentha piperita 247
Mentha x piperita 649
Mentha x piperita var. citrata 649
Mentha spicata 649
Merremia hungaiensis 640
Mesona chinensis 43, 650
Metasequoia glyptostroboides 7
Metroxylon rumphii 303
Metroxylon sagus 303
Microcos mala 213
Microcos nervosa 168, 209, 212, 213, 236, 239,
    247, 538
Microcos paniculata 212, 538
Millettia speciosa 478
Mimulus tenellus 671
Mimusops hexandra 621
Moghania philippinensis 478
Momordica charantia 18, 232, 236, 708
Momordica grosvenorii 219, 710
Momordica lanata 694

Monascus purpureus 38, 263
Monochoria hastata 310
Monochoria vaginalis 310
Morchella esculenta 263
Morinda citrifolia 682
Morinda officinalis 168, 683
Moringa oleifera 424
Morus alba 236, 247, 361
Morus australis 361
Morus nigra 362
Mucor stolonifer 262
Mucuna capitata 485
Mucuna sempervirens 479
Murdannia angustifolia 308
Murdannia keisak 309
Murdannia nudiflora 11, 309
Murdannia triquetra 309
Murraya koenigii 502
Musa acuminata 324, 325
Musa balbisiana 325
Musa cavendishii 324
Musa chinensis 324
Musa nana 325
Musa x paradisiaca 325
Musa sapientum 325
Mussaenda pubescens 684
Myosoton aquaticum 385
Myrica adenophora 335
Myrica rubra 335
Myristica fragrans 399
Myrtus communis 577

Nasturtium globosum 421
Nasturtium microspermum 421
Nasturtium officinale 422
Nelumbo lutea 139
Nelumbo nucifera 43, 168, 247, 387
Nemalion helminthoides var. vermiculare 261
Neottopteris nidus 274
Nephelium lappaceum 525
Nephrolepis cordifolia 272
Nervilia fordii 331
Nicandra physaloides 664
Nigella damascena 390
Nigella sativa 390
Nitraria sibirica 490, 584
Nostoc commune 257
Nostoc flagelliforme 257
Notopterygium franchetii 248
Nymphoides hydrophylla 628
Nymphoides indica 628
Nuphar pumilum 388

Nymphaea tetragona 388
Nymphoides cristata 628
Nymphoides peltatum 628
Nypa fruticans 303
Nyssa javanica 575

Ocimum acutum 651
Ocimum basilicum 650
Ocimum crispum 651
Ocimum frutescens 652
Oenanthe javanica 600
Oenanthe stolonifera 600
Olea europaea 627
Opuntia dillenii 567
Opuntia ficus-indica 568
Opuntia megacantha 568
Opuntia stricta var. dillenii 567
Orixa japonica 502
Orobanche coerulescens 674
Oroxylum indicum 232, 236
Orychophragmus violaceus 22, 417
Oryza fatua 292
Oryza minuta 292
Oryza sativa 168, 248, 292
Oryza sativa var. glutinosa 292
Osbeckia chinensis 237
Osmanthus fragrans 76, 627
Osmunda japonica 274
Osteomeles schwerinae 438
Ostericum grosseserratum 601
Ostericum sieboldii 601
Ostericum viridiflorum 602
Ottelia acuminata 283
Ottelia alismoides 283
Ottelia yunnanensis 283
Oxalis corniculata 237
Oxycoccus microcarpus 612
Oxyria digyna 371

Pachira aquatica 545
Pachira macrocarpa 545
Paederia foetida 684
Paederia scandens 684
Paederia scandens var. tomentosa 685
Paederia stenobotrya 685
Paeonia lactiflora 168, 199, 391
Panax ginseng 168, 182, 586
Panax notoginseng 587
Panax pseudo-ginseng var. wangianus 587
Panax quinquefolium 168, 182, 586
Panax schin-seng 586
Panax wangianum 168, 587

Pandanus forceps 281
Pandanus utilis 281
Pandanus tectorius 232, 237, 281
Panicum miliaceum 293
Panicum miliaceum var. effusum 293
Panicum miliaceum var. glutinosa 293
Papaver somniferum 402
Parthenocissus tricuspidata 532
Passiflora edulis 565
Passiflora foetida 565
Pastinaca sativa 602
Pentapanax henryi 588
Penthorum chinense 425
Penthorum sedoides 425
Perilla acuta 651
Perilla crispa 651
Perilla frutescens 248
Perilla frutescens var. acuta 651
Perilla frutescens var. arguta 651
Perilla frutescens var. crispa 651
Perilla maxima 652
Perilla nankinensis 651
Perilla ocymoides 652
Peristrophe roxburghiana 676
Persea americana 402
Petasites japonicus 741
Petroselinum crispum 603
Petroselinum sativum 603
Phaseolus angularis 466
Phaseolus aureus 468
Phaseolus calcaratus 472
Phaseolus coccineus 479
Phaseolus cylindricus 488
Phaseolus limensis 480
Phaseolus lunatus 480
Phaseolus multiflorus 479
Phaseolus mungo 251, 468, 469
Phaseolus radiatus 468
Phaseolus vulgaris 480
Phellopterus littoralis 598
Phoenix dactylifera 303
Phoenix hanceana 304
Phoenix loureiri 304
Photinia serrulata 438
Phragmites karka 237
Phyllanthus cochinchinensis 237
Phyllanthus emblica 512
Phyllostachys angusta 129, 293
Phyllostachys bambusoides 129, 294
Phyllostachys congesta 294
Phyllostachys decora 294
Phyllostachys dulcis 129, 294

Phyllostachys edulis 294, 295
Phyllostachys elegans 294
Phyllostachys flexuosa 129
Phyllostachys nidularia 295
Phyllostachys nigra var. henonis 295
Phyllostachys nuda 129
Phyllostachys pubescens 42, 129, 295
Phyllostachys reticulata 294
Phyllostachys viridis 129
Phyllostachys vivax 129
Physalis alkekengi var. francheti 664
Physalis angulata 665
Physalis ciliata 665
Physalis peruviana 665
Physalis pubescens 666
Phytolacca acinosa 131, 382
Phytolacca americana 41
Phytolacca esculenta 255, 382
Picris hieracioides 742
Picris hieracioides var. japonica 742
Picris japonica 742
Pimpinella arguta 603
Pimpinella brachycarpa 606
Pinus armandii 98, 104, 277
Pinus bungeana 278
Pinus densata 278
Pinus koraiensis 98, 104, 277, 278
Pinus massoniana 269
Pinus tabulaeformis var. densata 278
Pinus wilsonii 278
Piper betle 333
Piper cubeba 401
Piper longum 333
Piper nigrum 333
Pistacia chinensis 518
Pistacia vera 518
Pisum sativum 480
Pisum sativum var. arvense 481
Pisum sativum var. macrocarpa 481
Pisum sativum var. sativum 481
Pittosporopsis kerrii 521
Plantago asiatica 677
Plantago major 237, 677
Platanthera delavayi 331
Platycodon grandiflorus 169, 248, 716
Plectranthus amboinicus 652
Pleurospermum camtschaticum 604
Pleurospermum uralense 604
Pleurotus sajor-caju 268
Plumeria rubra 248
Podocarpus nagi 276
Podophyllum emondii var. chinense 395

Pogonatherum crinitum 237
Pogostemon cablin 248
Polygonatum chinense 169
Polygonatum cyrtonema 169
Polygonatum inflatum 319, 320
Polygonatum involucratum 319, 320
Polygonatum macropodum 319, 320
Polygonatum multiflorum 169, 173
Polygonatum odoratum 169, 197, 319
Polygonatum officinale 319
Polygonatum sibiricum 173, 320
Polygonum bistorta 189
Polygonum chinense 237
Polygonum cuspidatum 371
Polygonum cymosum 370
Polygonum flaccidum 372
Polygonum hydropiper 244, 248, 372
Polygonum multiflorum 169, 192, 194, 372, 633,
    634
Polygonum perfoliatum 237
Polygonum viscosum 373
Poncirus trifoliata 248
Poria cocos 169, 173, 175, 248, 268
Porphyra atropurpurea 261
Porphyra tenera 261
Porphyra yezoensis 261
Portulaca oleracea 384
Potamogeton oxyphyllus 282
Potamogeton pectinatus 282
Potentilla anserina 12, 169, 438
Potentilla chinensis 439
Potentilla discolor 439
Potentilla exaltata 439
Potentilla flagellaris 440
Potentilla fruticosa 440
Poupartia chinensis 514
Poupartia fordii 515
Pouteria annamensis 622
Pouteria grandifolia 622
Prasiola japonica 258
Prinsepia sinensis 229
Prinsepia uniflora 169, 229, 440
Prunella vulgaris 169, 232, 237, 248, 653
Prunus amygdalus 442
Prunus ansu 441
Prunus armeniaca 169, 227, 224, 248, 441
Prunus armeniaca var. ansu 441
Prunus armeniaca var. armeniaca 441, 442
Prunus armeniaca var. mandshurica 441
Prunus avium 442
Prunus cerasoides 442
Prunus communis 443

Prunus domestica 443
Prunus dulcis 227, 442
Prunus hongpiensis 443
Prunus humilis 443
Prunus janonica 444
Prunus kansuensis 444
Prunus mandshurica 441
Prunus mira 445
Prunus mume 445
Prunus persica 446
Prunus persica f. aganopersica 446
Prunus persica f. compressa 446
Prunus persica var. nectarina 446
Prunus persica var. nucipersica 446
Prunus persica var. platycarpa 446
Prunus persica f. scleropersica 446
Prunus pseudocerasus 446
Prunus salicina 447
Prunus simonii 447
Prunus tomentosa 447
Prunus vaniotii 448
Psalliota campestris 266
Pseudostellaria heterophylla 169, 189, 385
Psidium guajava 577
Psidium littorale 578
Psidium lucidum 578
Psophocarpus tetragonolobus 481
Psoralea corylifolia 471
Psychotria rubra 237
Pteridium aquilinum 272
Pueraria edulis 481
Pueraria lobata 248, 482
Pueraria montana 482
Pueraria thomsonii 248, 482
Pueraria thunbergiana 482
Pugionium cornutum 417
Punica granatum 573
Pyracantha fortuneana 448
Pyrularia edulis 368
Pyrularia sinensis 368
Pyrus betulaefolia 448
Pyrus bretschneideri 449
Pyrus communis 450
Pyrus nivalis 450
Pyrus pashia 450
Pyrus pyrifolia 451
Pyrus sativa 450
Pyrus serotina 451
Pyrus ussuriensis 451
Pyrus ussuriensis var. chinensis 449

Quercus acuta 349

Quercus acutidentata 349
Quercus acutissima 350
Quercus glabra 348
Quercus glauca 350
Quercus mongolica 350
Quercus phillyreoides 351
Quercus schottkyana 350
Quercus spinosa 351
Quercus variabilis 352

Raphanus sativus 28, 30, 417, 418
Rehmannia glutinosa 169, 173, 195, 671
Rehsonia sinensis 489
Rhamnus davurica 530
Rhamnus erythroxylon 529
Rhamnus utilis 530
Rhaphiolepis indica 452
Rheum nanum 373
Rheum officinale 248, 373
Rheum palmatum 373
Rheum rhabarbarum 374
Rheum rhaponticum 374
Rheum undulatum 374
Rhizopus nigricans 262
Rhizopus stolonifer 262
Rhododendron oldhamii 611
Rhodomyrtus tomentosa 17, 255, 578
Rhoeo discolor 310
Rhoeo spathacea 310
Rhynchosia volubilis 483
Ribes alpestre 425
Ribes burejense 426
Ribes diacanthum 427
Ribes emodense 426
Ribes fasciculatum var. chinense 426
Ribes grossularis 426
Ribes longiracemosum 427
Ribes manshuricum 427
Ribes pulchellum 427
Ricinus communis 512
Robinia pseudoacacia 80, 483
Rorippa cantoniensis 421
Rorippa globosa 421
Rorippa indica 422
Rorippa microsperma 421
Rorippa montana 422
Rorippa nasturtium-aquaticum 422
Rosa bella 452
Rosa davurica 453
Rosa laevigata 169, 237, 453
Rosa omeiensis 453
Rosa roxburghii 453

Rosa rubus 454
Rosa rugosa 77, 454
Rosa xanthina 454
Rotala indica 573
Rubus amabilis 455
Rubus biflorus 455
Rubus buergeri 455
Rubus calycinoides 455
Rubus chroosepalus 456
Rubus corchorifolius 456
Rubus coreanus 456
Rubus ellipticus 456
Rubus flagelliflorus 457
Rubus ichangensis 457
Rubus idaeus 457
Rubus idaeus var. aculeatissimus 458
Rubus lambertianus 458
Rubus mesogaeus 458
Rubus niveus 459
Rubus palmatus 459
Rubus parvifolius 237, 459
Rubus peltatus 459
Rubus phoenicolasius 460
Rubus pungens 460
Rubus reflexus 460
Rubus rosaefolius 461
Rubus sachalinensis 458
Rubus setchuenensis 461
Rubus simplex 461
Rubus suavissimus 461
Rubus sumatranus 462
Rubus swinhoei 462
Rubus triphyllus 459
Rumex acetosa 374
Rumex patientia 375
Ruta graveolens 503

Sabia schumanniana 526
Saccharomyces cerevisiae 263
Saccharomyces rouxii 40
Saccharum officinarum 296
Saccharum sinense 296
Sageretia thea 237, 530
Sageretia theezans 530
Sagittaria sagittifolia 136, 282
Salix purpurea 334
Salvia miltiorrhiza 189, 217
Sambucus formosana 687
Sambucus williamsii 687
Sandoricum indicum 507
Sandoricum koetjape 507
Sanguisorba officinalis 462

Sanicula chinensis 604, 605
Sanicula rubriflora 605
Sapindus mukorossi 237
Sapium discolor 237
Saposhnikovia divaricata 230, 248, 605
Sargassum enerve 259
Saurauia oldhamii 553
Saurauia tristyla 553
Sauropus androgynus 214, 512
Sauropus changianus 214, 221, 513
Sauropus rostratus 513
Saussurea affinis 736
Saussurea carthamoides 736
Saussurea lappa 169
Scabiosa comosa 690
Scabiosa fischeri 690
Scabiosa tschiliensis 690
Schefflera heptaphylla 248
Schefflera octophylla 237, 243
Schisandra chinensis 169, 218, 224, 225, 226, 396
Schisandra glaucescens 397
Schisandra incarnata 397
Schisandra propinqua var. sinensis 397
Schisandra sphenanthera 224
Schizonepeta tenuifolia 248
Scilla japonica 11, 320
Scilla scilloides 320
Scoparia dulcis 237
Scorzonera austriaca var. glabra 743
Scorzonera glabra 743
Scorzonera ruprechtiana 743
Scrophularia ningpoensis 248
Scutellaria amoena 653
Scutellaria baicalensis 248, 653
Scutellaria barbata 654
Scutellaria scordifolia 655
Scytosiphon lomentarius 259
Secale cereale 296
Sechium edule 709
Senna occidentalis 470
Senna sophora 470
Senna tora 470
Serratula chinensis 744
Serratula coronaria 744
Serratula coronata 744
Sesamum indicum 674
Sesamum orientale 674
Sesbania grandiflora 483
Setaria italica 296
Silene conoidea 386
Silene venosa 386
Siler divaricatum 605

Silybum marianum 745
Sinapis alba 407
Sinocalamus latiflorus 288
Sinofranchetia chinensis 393
Siraitia grosvenorii 153, 169, 181, 218, 710
Sium sisarum 606
Sium suave 606
Sloanea hemsleyana 537
Solanum japonense 666
Solanum melongena var. esculentum 666
Solanum nigrum 669
Solanum nigrum var. pauciflorum 670
Solanum photeinocarpum 670
Solanum torvum 238
Solanum tuberosum 670
Solena amplexicaulis 711
Solena heterophylla 711
Sonchus arvensis 746, 747, 748
Sonchus brachyotus 746
Sonchus oleraceus 747
Sophora japonica 248, 484
Sophora viciifolia 484
Sorbus alnifolia 463
Sorbus folgneri 463
Sorbus pohuashanensis 463
Sorghum bicolor 296
Sorghum saccharatum 296
Sorghum vulgare 296
Sorghum vulgare var. caffrorum 297
Spinacia oleracea 378
Spiraea blumei 464
Spiraea cantoniensis 464
Spiraea reevesiana 464
Spiraea trilobata 464
Spiranthes australis 331
Spiranthes lancea 189, 331
Spiranthes sinensis 331
Spondias axillaris var. pubinervis 516
Spondias pinnata 519
Spuriopimpinella brachycarpa 606
Stachys adulterina 655
Stachys affinis 31, 656
Stachys sieboldii 656
Staphylea bumalda 521
Stauntonia brevipes 392
Stauntonia chinensis 393
Stauntonia duclouxii 394
Stauntonia hexaphylla 394
Stellaria aquatica 385
Stellaria media 387
Sterculia foetida 546
Sterculia lanceolata 546

Sterculia nobilis 546, 547
Sterculia pexa 547
Stevia rebaudiana 748
Stixis suaveolens 424
Stizolobium capitatum 485
Striga asiatica 672
Suaeda glauca 378
Suaeda heterophylla 379
Suaeda maritima 379
Suaeda salsa 379
Symplocos paniculata 625
Syzygium aqueum 578
Syzygium aromaticum 579
Syzygium buxifolium 579
Syzygium cumini 580
Syzygium jambos 580
Syzygium javanicum 581
Syzygium malaccense 580
Syzygium samarrangense 581

Tacca leontopetaloides 321
Tacca pinnatifida 321
Talinum paniculatum 189
Tamarindus indicus 485
Taraxacum mongolicum 748, 751
Taraxacum officinale 748, 750, 751
Taraxacum sinicum 748, 751
Telosma cordata 11, 66, 635
Terminalia bellirica 575
Terminalia catappa 576
Termitomyces albuminosus 267
Tetragonia expansa 383
Tetragonia tetragonoides 383
Tetrapanax papyriferus 248
Thamnolia vermicularis 265
Thea sinensis 555
Theobroma cacao 547
Thespesia lampas 543
Thlaspi arvense 422
Thymus mongolicus 656
Thymus serpyllum 656, 657
Thymus vulgaris 658
Thysanolaena maxima 297
Toona sinensis 508
Torilis japonica 607
Torreya grandis 276
Torreya grandis var. dielsii 277
Torreya grandis var. grandis 277
Torreya grandis var. merrillii 277
Torreya grandis var. non-apiculata 277
Torreya grandis var. sargentii 277
Torreya nucifera 277

Trachycarpus fortunei 304
Tragopogon porrifolius 751
Trapa bicornis 583
Trapa bispinosa 583, 584
Trapa incisa var. quadricaudata 584
Trapa japonica 584
Trapa maximowiczii 584
Trapa natans 584
Trapa quadrispinosa 584
Tremella fuciformis 169, 178, 269
Tremella mesenterica 269
Tribulus terrestris 491
Trichosanthes anguina 712
Trichosanthes kirilowii 217, 248
Trigonella foenum-graecum 485
Triticum aestivum 244, 248, 297
Triticum dococcum 298
Triticum durum 298
Triticum monococcum 298
Triticum turgidum 298
Triumfetta bartramia 238
Trochodendron aralioides 8
Tulipa edulis 321
Turbinaria fusiforme 260
Typha latifolia 280

Ulmus davidiana 353
Ulmus glaucescens 353
Ulmus parvifolia 354
Ulmus propinqua 354
Ulmus pumila 354
Ulva lactuca 258
Ulva pertusa 258
Umbilicaria esculenta 265
Undaria pinnatifida 260
Uraria crinita 238
Urena lobata 238
Urobotrya latisquama 368
Urtica angustifolia 364
Urtica cannabina 364
Urtica dioica var. angustifolia 364
Urtica laetevirens 365
Urtica triangularis 365
Urtica urens 365
Usnea longissima 265
Ustilago esculenta 269, 298
Uvaria rufa 399

Vaccinium bracteatum 611
Vaccinium myrtillus 612
Vaccinium oxycoccos 612
Vaccinium vitis-idaea 612

Valerianella olitoria 690
Vangueria edulis 685
Vangueria madagascariensis 685
Vanilla fragrans 332
Vanilla planifolia 332
Viburnum cordifolium 688
Viburnum dilatatum 688
Viburnum ichangense 689
Viburnum mongolicum 689
Vicia bungei 486
Vicia cracca 27, 486
Vicia faba 7, 33, 486, 645
Vicia hirsuta 487
Vicia pseudo-orobus 487
Vicia sativa 487
Vicia tridentata 486
Vicia unijuga 488
Vigna angularis 466
Vigna calcarata 472
Vigna cylindrica 489
Vigna radiata 468
Vigna sesquipedalis 488
Vigna sinensis 488
Vigna sinensis var. sesquipedalis 488
Vigna umbellata 472
Vigna unguiculata 488, 489
Vigna vexillata 489
Viola acuminata 561
Viola chinensis 562
Viola collina 562
Viola micrantha 561
Viola verecunda 563
Viola yedoensis 563
Vitex cannabifolia 238, 644
Vitex negundo var. cannabifolia 644
Vitex rotundifolia 238
Vitex vestita 645
Vitis amurensis 533
Vitis balanseana 533
Vitis betulifolia 534
Vitis davidii 534
Vitis ficifolia 534

Vitis flexuosa 535
Vitis labrusca 535
Vitis piasezkii 535
Vitis quinquangularis 536
Vitis romanetii 536
Vitis vinifera 536
Volvariella volvacea 270

Wahlenbergia gracilis 716
Wahlenbergia marginata 716
Walsura robusta 508
Wasabia japonica 422
Wedelia chinensis 238
Wisteria sinensis 66, 80, 489
Wolfiporia cocos 268

Xanthium sibiricum 244, 248
Xanthoceras sorbifolia 526
Xantolis stenosepala 623
Xylosma congestum 565

Youngia denticulata 737
Youngia japonica 238, 752

Zanthoxylum ailanthoides 503
Zanthoxylum alatum 503
Zanthoxylum alatum var. planispinum 503
Zanthoxylum armatum 503
Zanthoxylum bungeanum 504
Zanthoxylum planispinum 503
Zanthoxylum schinifolium 504
Zanthoxylum simulans 504
Zanthoxylum trifoliatum 586
Zea mays 298
Zehneria indica 712
Zingiber mioga 255, 329
Zingiber officinale 329
Zizania caduciflora 269, 298
Ziziphus jujuba 87, 169, 218, 531, 532
Ziziphus mauritiana 532
Ziziphus vulgaris 531

A-lin-xi 阿林稀 616
A-wa 阿瓦 288
A-yue-hun-zi 阿月渾子 518
Ai-da-huang 矮大黃 373
Ai-gua 矮瓜 666
Ai-mo-mo 艾饃饃 731
Ai-yu-zi 愛玉子 357
An-li 安梨 451
An-nan-cao 安南草 726
Ao-tou-xian 凹頭莧 380
Ao-zhou-jian-guo 澳洲堅果 366

Ba-dan-xing 八旦杏 227
Ba-dan-xing 巴旦杏 442
Ba-ji-tian 巴戟天 168, 683
Ba-jiao 八角 396
Ba-jiao-hui-xiang 八角茴香 396
Ba-jiao-yu 芭蕉芋 329
Ba-leng-shi 八稜柿 624
Ba-tian-suan-mo 巴天酸模 375
Ba-wang-hua 霸王花 168, 567
Ba-xi-li 巴西栗 574
Ba-yue-gua 八月瓜 391
Ba-zhu-na 巴朱納 373
Bai-bao-hao 白苞蒿 719
Bai-bei-san-qi 白背三七 733
Bai-cai 白菜 50, 55, 204, 407, 413
Bai-cha-ju 白茶菊 723
Bai-chan 白蟬 681
Bai-chuan 白椽 345
Bai-ci 白刺 490
Bai-ci-hua 白刺花 484
Bai-di-li 白地栗 282
Bai-dou 白豆 489
Bai-dou-kou 白豆蔻 326
Bai-fu-pian 白附片 42, 389
Bai-fu-zi 白附子 166
Bai-gua 白瓜 691, 697
Bai-gui-mu 白桂木 355
Bai-guo 白果 99, 100, 102, 104, 275
Bai-he 白合 319

Bai-he 百合 319
Bai-he-qi 百合七 316
Bai-hu-jiao 白胡椒 502
Bai-hua-cai 白花菜 670
Bai-hua-cha 白花茶 617
Bai-hua-cha 百花茶 684
Bai-hua-di-dan-cao 白花地膽草 728
Bai-hua-di-dan-tou 白花地膽頭 728
Bai-hua-li-shu 白花櫟樹 349
Bai-hua-long-xu-teng 白花龍鬚藤 467
Bai-hua-shao-cai 白花苕菜 487
Bai-hua-shi-jie-cai 白花石芥菜 414
Bai-hua-suan-teng-zi 白花酸藤子 615
Bai-hua-zi 白花子 235
Bai-ji-shi-teng 白雞屎藤 685
Bai-jian 白箭 235
Bai-jiao-tong 白腳桐 543
Bai-jie 白芥 407
Bai-la-shu 白蠟樹 616
Bai-la-zhi-tian-jiao 白蠟指天椒 660
Bai-lan 白欖 246, 505
Bai-lan-gua 白蘭瓜 696, 698
Bai-le 白簕 586
Bai-li 白栗 347
Bai-li 白梨 449
Bai-li-xiang 百里香 658
Bai-long-xu 白龍鬚 634
Bai-ma-li 白麻櫟 352
Bai-mo-gu 白蘑菇 266
Bai-mu-tong 白木通 391
Bai-nuan-tiao 白暖條 689
Bai-pi-song 白皮松 278
Bai-ri-sai-cao 百日晒草 308
Bai-rou-rong 白肉榕 359
Bai-shan-yao 白山藥 323
Bai-shao 白芍 168, 197, 199, 391
Bai-shi-li 白石栗 346
Bai-shou-wu 白首烏 193, 194, 633
Bai-si-cai 白絲菜 652
Bai-song 白松 277
Bai-su 白蘇 652

Bai-tan 白檀 625
Bai-tang-zi-shu 白棠子樹 643
Bai-tiao-gou-qi 白條枸杞 662
Bai-yu 白芋 638
Bai-yu 白榆 354
Bai-yu-lan 白玉蘭 396
Bai-yu-gua 白玉瓜 702
Bai-zhi 白芷 166, 246
Bai-zhi-shan 白紙扇 684
Bai-zhu 白朮 166, 171, 173
Bai-zi 稗子 288
Bai-zi-cai 白子菜 733
Ban-guo-teng 斑果藤 424
Ban-jiu-fan 斑鳩飯 684
Ban-li 板栗 340
Ban-zhi-lian 半枝蓮 654
Bang-hua 蚌花 310
Bang-weng-cai 蒡翁菜 718
Bang-zi 棒子 298
Ban-zi-mian 棒子麵 298
Bao 匏 704
Bao-fu-hua 包袱花 716
Bao-gu 包穀 298
Bao-gu-fen 包穀粉 298
Bao-gua 匏瓜 704
Bao-gun 寶棍 687
Bao-li-shu 包栗樹 347
Bao-ta-cai 寶塔菜 31, 656
Bao-tou-li 包頭栗 349
Bao-xin-cai 包心菜 410
Bao-zi-gan-lan 抱子甘藍 411
Bao-zi-jiao 包子椒 659
Bao-zi-rong 包子榕 357
Bei-chai-hu 北柴胡 246
Bei-jiu 北韭 313
Bei-gua 北瓜 702
Bei-jing-ci-gua 北京刺瓜 699
Bei-jing-mi-zao 北京蜜棗 88, 531
Bei-qi 北芪 466
Bei-sha-shen 北沙參 168, 190, 191, 598
Bei-si-shi-zhu 北絲石竹 384
Bei-wu-wei-zi 北五味子 224
Bei-xing-ren 北杏仁 227, 248
Bei-zhuang-kao 杯狀栲 341
Beng-da-wan 崩大碗 235, 593
Bi-bo 畢拔 333
Bi-dang-qie 畢當茄 576
Bi-li 薜荔 360
Bi-li-le 毗黎勒 575
Bi-lu-fan-li-zhi 祕魯番荔枝 398
Bi-luo-chun 碧螺春 6

Bi-ma 蓖麻 512
Bi-qi 荸薺、葧臍 142, 143, 299
Bi-xiao-jiang 閉鞘薑 327
Bian-dou 扁豆 475
Bian-dou-cai 變豆菜 604
Bian-gua 鞭瓜 699
Bian-guo-mu 扁果木 523
Bian-he-tao 扁核桃 228, 442
Bian-pu 扁蒲 705
Bian-tao 扁桃 227, 228, 442, 517
Bian-tao-zi 扁桃子 228
Bian-ye-xiao-zhi-zi 變葉小梔子 681
Bin-lang 檳榔 166, 235, 246, 300
Bin-lang-hua 檳榔花 300
Bin-lang-qing 檳榔青 519
Bin-lang-yu 檳榔芋 306
Bin-mu-huan 濱木患 523
Bing-dou 兵豆 476
Bing-fen 冰粉 664
Bing-fen-shu 冰粉樹 360
Bing-fen-zi 冰粉子 360
Bing-tou-huang-qin 併頭黃芩 655
Bo-bai-da-guo-you-cha 博白大果油茶 554
Bo-cai 菠菜 378
Bo-he 薄荷 648
Bo-ke 薄殼 637
Bo-luo 菠蘿 308
Bo-luo-men-shen 婆羅門參 751
Bo-luo-mi 波羅蜜 355
Bo-luo-zhong 菠蘿種 693
Bo-ye-da-huang 波葉大黃 374
Bu-gu-zhi 補骨脂 167, 471
Bu-zha-ye 布渣葉 168, 209, 211, 212, 213, 236, 247, 538

Cai-dou 菜豆 480
Cai-gua 菜瓜 31, 697
Cai-hu 柴胡 167
Cai-ji-miao 菜雞苗 744
Cai-jue 菜蕨 273
Cai-teng 菜藤 369
Cai-xin 菜心 411
Cai-zi-qi 菜子七 415
Can-dou 蠶豆 486
Cang-er 蒼耳 248
Cang-zhu 蒼朮 246
Cao-dou-kou 草豆蔻 325
Cao-gu 草菇 270
Cao-guo 草果 327
Cao-jue 巢蕨 274
Cao-jue-ming 草決明 471
Cao-ma-huang 草麻黃 279

Cao-mei 草莓 336, 433
Cao-mei-fan-shi-liu 草莓番石榴 578
Cao-mian 草棉 539
Cao-shi-can 草石蠶 656
Cao-teng 巢藤 486
Cao-ye-shu 糙葉樹 352
Cao-yuan-shi-tou-hua 草原石頭花 384
Cha 茶 555
Cha-tian-pao 插田泡 456
Cha-ye 茶葉 246
Cha-ye-shu 茶葉樹 626
Chan-hua 蟬花 245, 247
Chang-chuan-cha-biao 長串茶藨 427
Chang-chun-you-ma-teng 常春油麻藤 479
Chang-jiang-dou 長豇豆 488
Chang-jing-hu-lu 長頸葫蘆 705
Chang-shen 長身 709
Chang-shou-guo 長壽果 438
Chang-shui-gua 長水瓜 708
Chang-si-gua 長絲瓜 707
Chang-tan-guo 長灘果 710
Chang-ye-hu-tui-zi 長葉胡頹子 569
Chang-ye-shui-ma 長葉水麻 363
Chang-ye-suan-teng-guo 長葉酸藤果 615
Chang-ye-wo-ju 長葉萵苣 741
Chao-dou-shu 炒豆樹 645
Che-gen-cai 扯根菜 425
Che-lun-mei 車輪梅 452
Che-qian-cao 車前草 237, 677
Chen-pi 陳皮 167, 246
Chen-pi-mei 陳皮梅 31, 32, 445
Chen-xiang 沉香 197
Cheng 橙 498
Cheng-chui-ye 秤錘葉 397
Chi-bai-shu 遲白薯 321
Chi-cai 赤材 524
Chi-cang-teng 赤蒼藤 369
Chi-che-shi-zhe 赤車使者 364
Chi-gen-cai 赤根菜 379
Chi-hua 遲花 405
Chi-jian 赤箭 330
Chi-jian-peng 翅鹼蓬 379
Chi-nan 赤楠 579
Chi-tou-po 黐頭婆 238
Chi-xiao-dou 赤小豆 167, 247, 472
Chong-cao 蟲草 167, 175, 262
Chong-yang-mu 重陽木 509
Chou-cao 臭草 503
Chou-chang-shan 臭常山 502
Chou-dou-fu-ru 臭豆腐乳 36
Chou-gua 臭瓜 700
Chou-jing-jie 臭荊芥 645

Chou-mo-li 臭茉莉 643
Chou-mu-dan 臭牡丹 643
Chou-shan-yang 臭山羊 502
Chou-teng 臭藤 684
Chou-xiang-ru 臭香薷 646
Chu-nai-mu 出奶木 620
Chuan-dang 川黨 188, 715
Chuan-dang-shen 川黨參 715
Chuan-fang-feng 川防風 606
Chuan-gui 川桂 401
Chuan-guo-teng 串果藤 393
Chuan-jiao 川椒 504
Chuan-li 川梨 450
Chuan-mei 川莓 461
Chuan-mu 川木 246
Chuan-mu-gua 川木瓜 167
Chuan-po-shi 穿破石 356
Chuan-shen 拳參 189
Chuan-wu 川烏 42
Chuan-xin-lian 穿心蓮 235
Chuan-xiong 川芎 168, 200, 201, 247, 600
Chuan-xu-duan 川續斷 167
Chuan-zhen 川榛 339
Chun-bu-lao 春不老 409
Chun-cai 蓴菜，蒓菜 387
Chun-hua 春花 452
Chun-sha 春砂 246
Chun-sun 春筍 129, 295
Chun-yu 春榆 354
Ci-ci-ya 刺刺芽 726
Ci-er-cai 刺兒菜 726
Ci-gu 慈菇，茨菰 136, 137, 282
Ci-hong-hua 刺紅花 721
Ci-huang-bai 刺黃柏 394
Ci-huang-guo 刺黃果 630
Ci-ji-cai 刺薊菜 725
Ci-ji-ya 刺薊芽 726
Ci-kui 刺葵 304
Ci-li 刺李 425, 453
Ci-li 刺梨 426
Ci-li-mu 刺籬木 564
Ci-long-bao 刺龍包 585
Ci-pu-tao 刺葡萄 534
Ci-shan-gan 刺山柑 423
Ci-shao-zi 刺韶子 544
Ci-shou-ma 刺蒴麻 238
Ci-wu-jia 刺五加 585
Ci-xuan-gou-zi 刺縣鉤子 460
Ci-ye-li 刺葉櫟 351
Ci-yu 刺榆 353
Ci-zhu 刺竹 285
Cong 蔥 246, 312

Cu-ke-lang 粗殼榔 352
Cu-liu 醋柳 572
Cui-pi-yu 脆皮魚 31
Cui-sheng-yao 催生藥 582, 583
Cun-jin-fei 寸金桸 277
Cun-mu 楤木 585

Da-ba 打巴 286
Da-bai-cai 大白菜 55, 413
Da-bai-shu 大白薯 322
Da-bo-gen-zi 大脖梗子 424
Da-chao-cai 大巢菜 487
Da-chi-dang-gui 大齒當歸 601
Da-chi-shi-ru 大翅石蕊 264
Da-ci-cha-biao 大刺茶藨 425
Da-cong 大蔥 39, 312
Da-ding-ku-gua 大頂苦瓜 709
Da-dou 大豆 474
Da-fei-niu 大肥牛 733
Da-fei-yang 大飛揚 510
Da-gang-cha 大崗茶 642
Da-gu-qing 大骨青 637
Da-gua 大瓜 694, 702
Da-gua 打瓜 124, 125, 695
Da-guo-rong 大果榕 357
Da-he-guo-shu 大核果樹 622
Da-hong-pao 大紅拋 493
Da-hua-di-lun-tao 大花第倫桃 549
Da-hua-jia-hu-ci 大花假虎刺 630
Da-hua-pi-ba 大花枇杷 432
Da-hua-qiu 大花球 404
Da-hua-wu-ya-guo 大花五椏果 549
Da-huang 大黃 248
Da-huang-gan-gan 大黃竿竿 373
Da-huang-yao 大黃藥 646
Da-huang-ye-bing 大黃葉柄 374
Da-hui-xiang 大茴香 396
Da-huo-cao 大火草 717
Da-ji 大薊 725
Da-ji 大薺 422
Da-ji-bai 大雞白 637
Da-ji-huang 大雞黃 637
Da-jiao 大蕉 325
Da-jie-cai 大芥菜 408
Da-lian-hua 大蓮花 307
Da-liao 大料 148, 151, 152
Da-liao 大蓼 389
Da-li-ka-fei 大粒咖啡 680
Da-li-shu 大力薯 478
Da-lü 大綠 530
Da-lu-mian 大陸棉 539
Da-ma-ye 大麻葉 246

Da-ma-ye-gou-qi 大麻葉枸杞 662
Da-mai 大麥 290, 291
Da-na-hong-ya-yu 大黐紅芽芋 306
Da-na-yu 大黐芋 306
Da-qing 大青 700
Da-qing-pi 大青皮 691
Da-qiu-xin 大球心 412
Da-shu 大薯 321
Da-suan 大蒜 314
Da-sun-zhu 大笋竹 285
Da-teng 大藤 693
Da-tian-cheng 大甜橙 499
Da-tou-cai 大頭菜 10, 28, 408
Da-tou-dian 大頭典 285
Da-tou-pao 大頭泡 665
Da-wu-li-hu-zhi-zi 達烏里胡枝子 477
Da-wu-shu 大無樹 360
Da-ye-bai-tou-weng 大葉白頭翁 717
Da-ye-kao 大葉栲 341
Da-ye-ci-li-mu 大葉刺籬木 564
Da-ye-li 大葉櫟 343
Da-ye-li 大葉藜 377
Da-ye-qin 大葉芹 598, 605
Da-ye-rong 大葉榕 357
Da-ye-suan-teng-zi 大葉酸藤子 616
Da-ye-sui-mi-ji 大葉碎米薺 415
Da-ye-tong-hao 大葉茼蒿 723
Da-ye-wu 大葉烏 288
Da-ye-zhui 大葉錐 343
Da-ye-zhui-li 大葉錐栗 346
Da-ye-zi 大葉子 424
Da-you-cha 大油茶 554
Da-zao 大棗 88
Dai-dai-hua 代代花 492
Dai-si 帶絲 259
Dan-shen 丹參 189, 217
Dan-sheng-mei 單生莓 461
Dan-shu 蛋樹 560
Dan-zhu-ye 淡竹葉 236, 247
Dan-zi-ma-huang 單子麻黃 279
Dang-gui 當歸 157, 166, 181, 186, 187, 188, 190,
    194, 196, 197, 202, 217, 230, 591, 603
Dang-shan-li 碭山梨 449
Dang-shen 黨參 162, 167, 173, 181, 188, 189,
    190, 194, 202, 466, 714, 715
Dao 稻 292
Dao-di-ling 倒地鈴 523
Dao-dou 刀豆 469
Deng-diao-zi 燈吊子 570
Deng-long-cao 燈籠草 665
Deng-long-guo 燈籠果 665
Deng-long-hua 燈籠花 334, 540

Deng-long-jiao 燈籠椒 659
Deng-long-pao 燈籠泡 386
Deng-xin-cao 燈心草 236, 247
Di-bai-zi-cai 地白子菜 733
Di-can-zi 地蠶子 655
Di-cha 地茶 265
Di-di-cai 地地菜 414
Di-ding 地釘 748
Di-er 地耳 257
Di-fu 地膚 378
Di-gu-niu 地牯牛 656
Di-gua 地瓜 361
Di-gua-er-miao 地瓜兒苗 648
Di-hai-shen 地海參 391
Di-hu-lu 地葫蘆 193, 633
Di-huang 地黃 169, 171, 172, 181, 195, 196, 202,
    672
Di-huang-cai 地黃菜 732
Di-huang-pi 地黃皮 272
Di-jiao 地椒 657
Di-jiao-pi 地角皮 257
Di-jiao-ye 地椒葉 657
Di-jin 地錦 510, 533
Di-jin-qi 地錦槭 522
Di-li 地栗 299
Di-li-zi 地梨子 142
Di-nian 地稔 582
Di-ruan 地軟 257
Di-shen 地參 648
Di-sun 地筍 648
Di-yu 地榆 462
Di-zao-zi 地棗子 320
Dian-deng-dan 電燈膽 668
Dian-zhui-li 滇錐栗 342
Diao 椆 348
Diao-deng-hua 吊燈花 540
Diao-si-zhu 吊絲竹 285
Diao-zhong-teng 吊鐘藤 570
Ding-xiang 釘香 579
Ding-xiang-qie 丁香茄 635
Ding-xin-teng 定心藤 521
Dong-bei-niu-fang-feng 東北牛防風 598
Dong-bei-xing 東北杏 441
Dong-bei-yang-jiao-qin 東北羊角芹 589
Dong-cong-xia-cao 冬蟲夏草 175, 262
Dong-dang 東黨 188, 189
Dong-fang-cao-mei 東方草莓 434
Dong-feng-cai 東風菜 720, 732
Dong-gu 冬菇 268
Dong-gua 冬瓜 93, 94, 95, 691
Dong-gua-han 冬瓜漢 711
Dong-han-cai 冬寒菜 542

Dong-lü 凍綠 530
Dong-mang-guo 冬芒果 518
Dong-qiang 東廧 375
Dong-sang-ye 冬桑葉 236
Dong-sun 冬筍 295
Dong-xian-cai 冬莧菜 542
Dong-ye 冬葉 247
Dou-ban-cai 豆瓣菜 422
Dou-ban-jiang 豆瓣醬 7, 35, 36, 149, 150, 262,
    474
Dou-cha-jue-ming 豆茶決明 469
Dou-chi 豆豉 35, 82, 83, 102, 120, 121, 150, 157,
    262, 474
Dou-fu 豆腐 120, 121, 147, 204, 474
Dou-fu-gan 豆腐乾 122, 123, 474
Dou-fu-guo 豆腐果 515
Dou-fu-nao 豆腐腦 474
Dou-fu-ru 豆腐乳 36, 262, 474
Dou-fu-zha 豆腐渣 147
Dou-jiang 豆漿 474
Dou-jiao 豆角 96, 488
Dou-kou 豆蔻 325
Dou-ya 豆芽 121
Dou-you 豆油 474
Du-gen-cao 獨根草 674
Du-huo 獨活 246
Du-huo-gan-gan 獨活桿桿 589
Du-jiao-jin 獨腳金 672
Du-jing-shan 杜荊山 617
Du-li 杜梨 448
Du-maio 獨苗 306
Du-xing-cai 獨行菜 417
Du-xing-qian-li 獨行千里 438
Du-zhong 杜仲 205, 428
Duan-bing-ye-zhi-ma 短柄野芝麻 647
Duan-dou 端豆 665
Duan-geng-suan-teng-guo 短梗酸藤果 616
Duan-guo-hui-qin 短果回芹 606
Duan-shui-gua 短水瓜 708
Duan-sui-yu-wei-kui 短穗魚尾葵 301
Duan-zhi-liu-dao-mu 短枝六道木 686
Dui-jie-ci 對節刺 530
Dui-ye-san-hua 對業散花 689
Dun-tou-fei 鈍頭榧 277
Dun-ye-cha 鈍葉茶 556
Dun-ye-mei 盾葉莓 459
Duo-chi-ti-gai-jue 多齒蹄蓋蕨 273
Duo-hua-mi-hou-tao 多花彌猴桃 552
Duo-hua-ye-mu-dan 多花野牡丹 582
Duo-sui-diao 多穗椆 348
Duo-xian-xuan-gou-zi 多腺懸鈎子 460

E-chang-cai 鵝腸菜 385
E-er-chang 鵝兒腸 387
E-lan-cai 遏藍菜 422
E-li 鱷梨 402
E-mei 鵝莓 426
E-mei-qiang-wei 峨嵋薔薇 453
E-shen 峨參 592
E-shen-mu 鵝腎木 574
E-zhu 莪朮 255, 328
Er-jiao-qin 二角芹 602
Er-se-bo-luo-mi 二色波羅蜜 356
Er-ying-zi 二英子 420

Fa-cai 髮菜 69, 257, 258
Fa-guo-ma-lan-tou 法國馬蘭頭 690
Fan-ba-tuan 飯巴團 397
Fan-bai-cai 翻白菜 439
Fan-bai-cao 翻白草 439
Fan-bao-cao 飯包草 308
Fan-dou 飯豆 489
Fan-gua 番瓜 701
Fan-gua 飯瓜 701
Fan-gui-ning-meng 番鬼檸檬 652
Fan-gui-pu-tao 番鬼蒲桃 578
Fan-li-zhi 番荔枝 398
Fan-lu 繁縷 387
Fan-qie 番茄 664
Fan-shan-di-huang-gua 房山地黃瓜 700
Fan-shi-liu 番石榴 577
Fan-shu 番薯 638
Fan-xing 番杏 383
Fan-ying-tao 番櫻桃 576
Fan-zao 番棗 303, 304
Fan-zhi-xian 反枝莧 381
Fang-dang 防黨 167, 188, 190, 715
Fang-feng 防風 230, 248, 605, 606
Fang-lan 方欖 505
Fang-li 仿栗 537
Fei-cheng-tao 肥城桃 446
Fei-hou 妃后 577
Fei-yang-cao 飛揚草 510
Fei-yong-cao 肺痈草 311
Fei-zi 榧子 277
Fen-bei-huang-lu 粉背黃櫨 516
Fen-ge 粉葛 482
Fen-pi 粉皮 44, 86, 87, 91, 92, 112, 468
Fen-shu 粉薯 330
Fen-si 粉絲 44, 53, 65, 112, 203, 204, 303, 468
Fen-tiao-er-cai 粉條兒菜 310
Fen-zhi-mei 粉枝莓 455
Feng-chuan-ge 風船葛 523
Feng-dou-cai 蜂斗菜 741

Feng-gun-yi 風滾衣 264
Feng-guo 鳳果 559
Feng-li 鳳梨 308
Feng-long-guo 鳳龍果 568
Feng-wei-gu 鳳尾菇 268
Feng-wei-jiao 鳳尾蕉 275
Feng-yan-guo 鳳眼果 546
Fo-du 佛肚 306
Fo-shou-gan 佛手柑 495
Fo-shou-gua 佛手瓜 709
Fo-zhang-gua 佛掌瓜 709
Fu-cai 芣菜 283
Fu-fu-miao 伏伏苗 636
Fu-cai 福橘 498
Fu-ling 茯苓 169, 171, 174, 175, 248, 268, 269
Fu-pen-zi 覆盆子 457
Fu-shen 茯神 269
Fu-shen 福參 601
Fu-ye-pu-tao 複葉葡萄 535
Fu-zhu 腐竹 76, 137

Ga-ge-la 嘎哥拉 327
Ga-li-ye 咖喱葉 502
Gan 柑 496
Gan-cao 甘草 247, 475
Gan-da-li-guo 干達利果 515
Gan-di-huang 乾 (干) 地黃 672
Gan-ge 乾葛 248
Gan-ge-teng 甘葛藤 482
Gan-jiang 乾薑 329
Gan-ju 柑橘 496
Gan-lan 甘藍 410
Gan-lan 橄欖 31, 32, 241, 505
Gan-lu-yi 甘露衣 264
Gan-lu-zi 甘露子 656
Gan-lu-zi 乾露子 654
Gan-shu 甘薯 322, 638
Gan-tai 乾苔 258
Gan-zhe 甘蔗 296
Gang-ma 港麻 541
Gang-mei-gen 崗梅根 236
Gang-ru-shu 檳如樹 514
Gang-ye-wu-zi 剛業霧子 371
Gang-zhi-ma 崗脂麻 236
Gang-zhu 剛竹 294
Gao-jiao-lao-hu-niu 高腳老虎扭 462
Gao-liang 高粱 297, 474
Gao-liang-jiang 高良薑 325
Gao-liang-pao 高粱泡 458
Gao-shan-song 高山松 278
Gao-she-zhu 高舌竹 287
Gao-wu 藁吾 730

Ge-cong 各蔥 315
Ge-gen 葛根 482
Ge-gen-fen 葛根粉 481
Ge-hua 葛花 482
Ge-liu 葛藟 535
Ge-ma-mu 葛麻母 482
Ge-ming-cai 革命菜 726
Ge-shan-jiao 隔山撬 634
Ge-shan-xiao 隔山消 193, 634
Ge-she-shu 割舌樹 508
Ge-xian-mi 葛仙米 257
Ge-ye-mi-hou-tao 革葉獼猴桃 551
Ge-yu-jin 葛郁金 330
Ge-zao 葛棗 552
Ge-zao-mi-hou-tao 葛棗獼猴桃 552
Gong-fen-yang-ti-jia 宮粉羊蹄甲 467
Gong-jia 貢甲 491
Gong-sun-shu 公孫樹 275
Gou-er-cha 鈎兒茶 528
Gou-gan-cai 狗肝菜 675
Gou-ji 枸棘 356
Gou-jiao 狗椒 503
Gou-jin-mai-ping-cao 狗筋麥瓶草 386
Gou-kao 鈎栲 347
Gou-li 鈎栗 347
Gou-pu-tao 狗葡萄 427
Gou-qi 枸杞 62, 63, 663
Gou-qi-tou 枸杞頭 663
Gou-qi-zi 枸杞子 162, 168, 222, 223, 663
Gou-she-cao 狗舌草 640
Gou-suan-jiang 溝酸漿 671
Gou-zao-mi-hou-tao 狗棗獼猴桃 552
Gou-zhua-dou 狗爪豆 485
Gou-zhua-yu 狗爪芋 304, 305, 306
Gu-cong 古蔥 315
Gu-hei-sui-jun 菇黑穗菌 269
Gu-mi 菰米 298
Gu-niu-cai 古鈕菜 670
Gu-sui-bu 骨碎補 167
Gu-ya 穀芽 248
Gu-zi 穀子 292
Gua 瓜 695
Gua-li 瓜栗 545
Gua-zi 瓜籽 124, 694
Gua-zi-cha 瓜子茶 522
Guai-gun-zhu 拐棍竹 290
Guai-zao 拐棗 529
Guai-zi-qin 拐子芹 590
Guan-gui 官桂 399
Guan-li 罐梨 449
Guan-yin-cao 觀音草 676
Guan-yin-lian 觀音蓮 369

Guan-yin-shou-zhi 觀音手指 668
Guan-yin-xian 觀音莧 732
Guang-ban-jin-cai 光瓣堇菜 563
Guang-dong-sheng-ma 廣東升麻 744
Guang-gan 廣柑 498
Guang-he-tao 光核桃 445
Guang-ju 廣橘 498
Guang-lang-mian 桄榔麵 301
Guang-mu-xiang 廣木香 169
Guang-ning-you-cha 廣寧油茶 555
Guang-pi-shu 光皮樹 609
Gui 桂 149
Gui-hua 桂花 76, 180, 627
Gui-jiu 鬼臼 395
Gui-lin-kao 桂林栲 342
Gui-man-tou 鬼饅頭 360
Gui-pi-shu 桂皮樹 401
Gui-xiang-liu 桂香柳 569
Gui-xin 桂心 399
Gui-yuan 桂圓 523
Gui-zhi 桂枝 149, 399
Gui-zhi-tou 鬼指頭 392
Guo-lu-wu-gong 過路蜈蚣 510
Guo-shan-long 過山龍 479

Ha-la-gai 哈拉蓋 364
Ha-mi-gua 哈密瓜 696, 698
Hai-ba-ji 海巴戟 682
Hai-cai 海菜 283
Hai-cai-hua 海菜花 283
Hai-cao-tai 海草苔 283
Hai-dai 海帶 203, 259
Hai-di-ye 海底椰 302
Hai-dou 海豆 642
Hai-hong 海紅 437
Hai-jiao 海椒 148, 658
Hai-jin-sha 海金沙 236
Hai-lan-zi 海欖子 642
Hai-ma 海麻 541
Hai-nan-pu-tao 海南蒲桃 580
Hai-nan-shan-xiao-ji 海南山小桔 502
Hai-nan-shu 海南薯 640
Hai-pu-tao 海葡萄 370
Hai-song 海松 278
Hai-suo-mian 海索麵 261
Hai-tang-guo 海棠果 437
Hai-wo-ju 海萵苣 258
Hai-yun 海蘊 259
Hai-zao 海棗 303, 304
Han-cai 蔊菜 422
Han-mei 寒莓 455
Han-qin 旱芹 594

Han-sheng-you-gua 旱生油瓜 700
Han-yu 旱榆 353
Hang-ju 杭菊 723
Hao-ben 藁本 599
Hao-cai 蒿菜 719, 722
He-bao-qie 荷包茄 668
He-bao-xian-cai 荷苞莧菜 381
He-dong-qiu-qie 鶴洞秋茄 669
He-gua 盒瓜 701
He-hua 荷花 74, 75, 387
He-lan-dou 荷蘭豆 481
He-meng 合萌 465
He-qiu 河楸 673
He-shou-wu 何首烏 169, 191, 192, 193, 194,
    195, 372, 373, 633, 634
He-tang 荷塘 405
He-tao 核桃 168, 337
He-ye 荷葉 158
Hei-bai-he 黑百合 317
Hei-cai 黑菜 259
Hei-dou 黑豆 168
Hei-dou-ban-cai 黑豆瓣菜 526
Hei-duan-dou 黑端豆 669
Hei-ge-lan 黑格蘭 (隔攔) 529
Hei-guo-xiao-bo 黑果小檗 394
Hei-guo-yue-ju 黑果越橘 612
Hei-mai 黑麥 296
Hei-pi-qing 黑皮青 693
Hei-sang 黑桑 362
Hei-sui-jian-zhu 黑穗箭竹 289
Hei-tou-guo 黑頭果 615
Hei-yu 黑榆 353
Hei-zao 黑棗 88, 169, 531, 532
Hei-zhong-cao 黑種草 390
Hong-bei-ji-guo 紅北極果 610
Hong-cai 紅菜 732
Hong-cai-li 紅材栗 349
Hong-cao-kou 紅草蔻 326
Hong-cha 紅茶 555
Hong-dou 紅豆 109, 116, 466
Hong-feng-cai 紅鳳菜 732
Hong-gen-cao 紅根草 618
Hong-geng-cai 紅梗菜 738
Hong-gou-kao 紅鉤栲 345
Hong-gu-niang 紅姑娘 664, 708
Hong-guo-shan-zha 紅果山楂 431
Hong-hou-pi-cai 紅厚皮菜 376
Hong-hua 紅花 167, 216, 721
Hong-hua-ai 紅花艾 647
Hong-hua-cai-dou 紅花菜豆 479
Hong-hua-guan 紅花罐 452
Hong-ji-dan-hua 紅雞蛋花 248

Hong-ju 紅橘 498
Hong-lan-hua 紅藍花 721
Hong-li 紅李 447
Hong-li-meng 紅黎檬 493
Hong-luo-bo 紅蘿蔔 131
Hong-mao-dan 紅毛丹 525
Hong-mei-zi 紅楣子 344
Hong-pao 紅泡 433
Hong-pao-ci-teng 紅泡刺藤 459
Hong-pi-liu 紅皮柳 334
Hong-pi-shu 紅皮薯 322
Hong-ping-xing 紅坪杏 443
Hong-qu 紅麴 38, 263
Hong-rou-shu 紅肉薯 322
Hong-rou-yang-tao 紅肉羊桃 551
Hong-shi 薍柿 623
Hong-si-xian 紅絲線 676
Hong-song 紅松 278
Hong-suan-mo 紅酸模 374
Hong-tang-cha 紅糖茶 611
Hong-tang-guan 紅糖罐 449
Hong-xian-cai 紅莧菜 381
Hong-xian-mi-cao 紅莧米草 380
Hong-xue-cha 紅雪茶 264
Hong-xun-zi 紅栒子 428
Hong-you-cai 紅油菜 414
Hong-yu 紅芋 638
Hong-zao 紅棗 88, 169, 531, 532
Hong-zao 紅糟 37, 38, 149, 155, 262
Hong-zao-wang 紅棗王 87
Hong-zi-su 紅紫蘇 651
Hong-zi-xi-gua 紅籽西瓜 695
Hou-ban-li 猴板栗 340
Hou-guo 猴果 510
Hou-li 猴栗 346, 347
Hou-mian-bao-shu 猴麵包樹 543
Hou-pi-cai 厚皮菜 376
Hou-po 厚朴 168, 247
Hou-tou 猴頭 268
Hou-tui-ti-gai-jue 猴腿蹄蓋蕨 273
Hou-zi-guo 猴子果 632
Hu 胡 314
Hu-bei-shan-zha 湖北山楂 430
Hu-bei-yang-ti-jia 胡北羊蹄甲 467
Hu-cong 胡蔥 313
Hu-die-guo 蝴蝶果 510
Hu-dou 胡豆 486
Hu-dou-ban 胡豆瓣 11, 37, 40
Hu-dou-ban-jiang 胡豆瓣醬 486
Hu-gua 胡瓜 698
Hu-jiao 胡椒 148, 149, 333
Hu-jiao-mian 胡椒麵 333

Hu-li-wei 狐狸尾 238
Hu-lu 葫蘆 92, 93, 704
Hu-lu-ba 葫蘆巴 149, 485
Hu-lu-cao 葫蘆草 716
Hu-lu-cha 葫蘆茶 235
Hu-luo-bo 胡蘿蔔 126, 131, 596
Hu-ma 胡麻 674
Hu-ma-zi 虎麻子 365
Hu-nan-ji-zi 湖南稷子 288
Hu-pi-song 虎皮松 278
Hu-pu-tao 狐葡萄 535
Hu-suan 胡蒜 314
Hu-he-tao 胡核桃 337
Hu-zhang 虎杖 371
Hu-zhi-zi 胡枝子 476
Hu-zi 瓠子 705
Hu-zi-lai 胡子來 446
Hua-bei-lan-pen-hua 華北藍盆花 690
Hua-cai 花菜 410
Hua-cha-biao 華茶藨 426
Hua-dou 花豆 479
Hua-fan-dou 花飯豆 489
Hua-hong 花紅 436
Hua-hong-cha 花紅茶 436
Hua-jiao 花椒 148, 149, 504, 505
Hua-jiao-ye 花椒葉 504
Hua-ku-cai 花苦菜 738
Hua-mei 話梅 31, 445
Hua-nan-he-shi 華南鶴虱 607
Hua-qi-mi-ju 花旗蜜橘 500
Hua-qi-shen 花旗參 586
Hua-qiu-shu 花秋樹 463
Hua-shan-song 華山松 277
Hua-shen 滑身 709
Hua-sheng 花生 465
Hua-tan-li 華檀梨 368
Hua-yao-hong-ya-yu 花腰紅芽芋 307
Hua-ye-hong-pao 花葉紅袍 419
Hua-ye-pu-tao 樺葉萄葡 534
Hua-zhen 華榛 338
Hua-zhong-sui-mi-ji 華中碎米薺 415
Hua-zhu 花竹 294
Huai-dou-zi 槐豆子 484
Huai-hua 槐花 248
Huai-hua-mo 槐花末 484
Huai-shan 懷山 167, 247
Huai-shu 槐樹 484
Huai-xiang 懷香 597
Huan-yang-cao 還陽草 238, 752
Huang-an-cai 黃鵪菜 752
Huang-cao 黃草 744
Huang-ci-mei 黃刺玫 454

Huang-deng 黃登 692
Huang-dou 黃豆 118, 474
Huang-dou-ya 黃豆芽 112, 474
Huang-du 黃獨 322
Huang-er-cha 黃兒茶 518
Huang-gua 黃瓜 698
Huang-guo-gou-qi 黃果枸杞 663
Huang-hai-tang 黃海棠 561
Huang-hua-cai 黃花菜 318, 752
Huang-hua-cao 黃花草 748
Huang-hua-jiu 黃花韭 312
Huang-hua-qiu 黃花楸 673
Huang-hua-tu-gua 黃花土瓜 640
Huang-jiao-ji 黃腳雞 317
Huang-jin 黃槿 541
Huang-jin-cha 黃金茶 653, 654
Huang-jin-guo 黃猄果 519
Huang-jin-jian-bi-yu 黃金間碧玉 296
Huang-jin-qi 黃金七 317
Huang-jing 黃精 169, 320
Huang-lian 黃蓮 247
Huang-lian-ya 黃蓮芽 518
Huang-ling 黃苓 248
Huang-mao-cao-mei 黃毛草莓 434
Huang-mao-jing 黃毛荊 645
Huang-mei 黃楣 344
Huang-mei-kao 黃楣栲 344
Huang-mu-jin 黃木槿 541
Huang-niu-cha 黃牛茶 557
Huang-niu-mu 黃牛木 235, 557
Huang-pao-zi 黃泡子 457
Huang-pi 黃皮 500
Huang-qi 黃耆 162, 166, 171, 173, 181, 187, 188,
  190, 466
Huang-qin 黃芩 653, 654
Huang-shan-teng 黃鱔藤 528
Huang-shu-kui 黃蜀葵 538
Huang-wei-chi 黃尾遲 412
Huang-xi-xin 黃細心 383
Huang-xu-cai 黃鬚菜 379
Huang-ya-bai 黃芽白 413
Huang-ya-guo 黃牙果 559
Huang-yan-mi-ju 黃嚴蜜橘 497
Huang-yang 黃羊 631
Huang-yao 黃藥 646
Huang-yao-zi 黃藥子 322
Huang-zhang-li 黃樟梨 451
Huang-zhi 黃枝 680
Hui-cai-hua 輝菜花 527
Hui-dou 灰斗 692
Hui-hui-cai 灰灰菜 377
Hui-hui-cong 茴茴蔥 313

Hui-hui-su 茴茴蘇 651
Hui-xiang 茴香 597
Hui-xun-zi 灰栒子 429
Hui-yu 灰榆 353
Huo-ba-guo 火把果 448
Huo-cong 火蔥 313
Huo-guo 火果 509
Huo-ji 火棘 448
Huo-long-guo 火龍果 568
Huo-ma 火麻 365
Huo-shi-liu 火石榴 669
Huo-tan-mu 火炭母 237
Huo-xiang 藿香 246, 248
Huo-yan-cai 火焰菜 376

Ji-cai 蕺菜 255, 332
Ji-cai 薺菜 414
Ji-cai 雞菜 733
Ji-chang-feng 雞腸楓 683, 684
Ji-er-chang 雞兒腸 738
Ji-gu-cao 雞骨草 166, 209, 210, 211
Ji-gu-tou 雞骨頭 625
Ji-guan-jun 雞冠蕈 267
Ji-guan-zi-su 雞冠紫蘇 651
Ji-ji-cao 吉吉草 650
Ji-jiao-cai 雞腳菜 260
Ji-ju-zi 雞距子 529
Ji-jun 雞蕈 267
Ji-li 蒺藜 491
Ji-li-kao 蒺藜栲 347
Ji-ling-zi 雞鵋子 436
Ji-ni 薺苨 716
Ji-pi-guo 雞皮果 500
Ji-sang 雞桑 361
Ji-she-xiang 雞舌香 579
Ji-sheng-huang 寄生黃 369
Ji-shi-teng 雞屎藤 684
Ji-tou-shen 雞頭參 320
Ji-tou-zi 雞頭子 387
Ji-tui-cai 雞腿菜 561
Ji-tui-jin-cai 雞腿堇菜 561
Ji-wan-zi 雞杬子 519
Ji-xin-guo 雞心果 623
Ji-xue-cao 積雪草 593
Ji-xue-teng 雞血藤 478
Ji-yan-cao 雞眼草 475
Ji-yan-qing 雞眼青 613
Ji-yan-teng 雞眼藤 683
Ji-yu-dan 鯽魚膽 618
Ji-zao-sin 急早心 412
Ji-zhua-qin 雞爪芹 595
Ji-zhua-shen 雞爪參 330, 331

Ji-zi 稷子 293
Jia-bo-he 家薄荷 649
Jia-deng-long 假燈籠 666
Jia-fang-feng 假防風 593
Jia-hai-tong 假海桐 521
Jia-li-zhi 假荔枝 356
Jia-lü-dou 假綠豆 470
Jia-ma-shu 家麻樹 547
Jia-mi 莢蒾 688
Jia-mu-huan 假木槵 514
Jia-ping-po 假蘋婆 546
Jia-qing-zi 嘉慶子 447
Jia-shan-long-yan 假山龍眼 366
Jia-shan-yao 家山藥 323
Jia-su 家蘇 652
Jia-suan-jiang 假酸漿 664
Jia-suan-zao 假酸棗 514
Jia-teng 加藤 574
Jia-tong-hao 假茼蒿 726
Jia-xiang-wan-dou 假香豌豆 487
Jia-yu 假芋 307
Jian-chi-hui-qin 尖齒回芹 603
Jian-dao-cao 剪刀草 282
Jian-gan-bai 箭杆白 411
Jian-jiao 尖椒 659
Jian-peng 鹼蓬 378
Jian-shen 建參 601
Jian-tou-huang-ye-gou-qi 尖頭黃葉枸杞 662
Jian-tou-yuan-guo-gou-qi 尖頭圓果枸杞 662
Jian-ye-cha 尖葉茶 557
Jian-ye-rong 尖葉榕 359
Jian-ye-tong-cai 劍葉通菜 638
Jian-ye-yu-jiu-hua 箭葉雨久花 310
Jiang 薑 149, 329
Jiang-dou 豇豆 488, 489
Jiang-hua 薑花 328
Jiang-huang 薑黃 149, 328
Jiang-mang 茳芒 470
Jiang-mang-jue-ming 茳芒決明 470
Jiang-you 醬油 262, 474
Jiao-bai 茭白 298
Jiao-cai 角菜 719
Jiao-gan 蕉柑 498
Jiao-gu-lan 絞股藍 703
Jiao-gua 絞瓜 702
Jiao-gua 茭瓜 298
Jiao-sun 茭筍 298
Jiao-yu 蕉芋 329
Jiao-zi 椒子 658
Jie-cai 芥菜 54, 55
Jie-geng 桔梗 169, 248, 716
Jie-gu-mu 接骨木 687

Jie-gua 節瓜 21, 85, 692
Jie-jie-cai 節節菜 573
Jie-lan 芥蘭 153, 404, 406
Jie-mu-huai-tai 借母懷胎 369
Jie-qiu-wo-ju 結球萵苣 740
Jie-tou-gua 截頭瓜 699
Jin-bu-huan 金不換 374
Jin-cai 蓳菜 563
Jin-dou 金豆 501
Jin-er 金耳 269
Jin-gang-teng 金剛藤 529
Jin-gou-ji 金狗脊 167
Jin-gou-li 金鈎梨 529
Jin-gu 金菇 267
Jin-guo-mei 金果梅 429
Jin-jia-dou 金莢豆 480
Jin-jin-cai 蓳蓳菜 562
Jin-ju 金橘 495, 501
Jin-ju-bing 金橘餅 495
Jin-li-zhi 錦荔枝 708
Jin-lu-mei 金露梅 440
Jin-mao-du-juan-hua 金毛杜鵑花 611
Jin-mei 浸梅 445
Jin-qian-cao 金錢草 235
Jin-qiao-mai 金蕎麥 370
Jin-shan-pu-tao 金山蒲桃 581
Jin-shan-wu-wei-zi 金山五味子 397
Jin-si-cao 金絲草 237
Jin-si-li 金絲李 560
Jin-si-qiu 金絲楸 672
Jin-si-shua 金絲刷 264
Jin-sun 金筍 596
Jin-suo-shi 金鎖匙 236
Jin-xian-diao-hu-lu 金線吊葫蘆 716
Jin-yin-hua 金銀花 168, 215, 247, 687
Jin-ying-qiang 金英強 237
Jin-ying-zi 金英子 169, 453
Jin-zan-cao 金簪草 748
Jin-zao-er 錦棗兒 320
Jin-zhen-cai 金針菜 66, 68, 318
Jin-zhu 金竹 295
Jin-zhu-liu 金珠柳 617
Jing-bai-li 京白梨 452
Jing-jie 荊芥 248
Jing-shan-shan-cheng 景山山橙 632
Jiu-bi-ying 救必應 236
Jiu-bing-liang 救兵糧 448
Jiu-cai 韭菜 53, 315
Jiu-cai-hua 韭菜花 315
Jiu-cai-huang 韭菜黃 53, 315
Jiu-cai-tai 韭菜苔 315
Jiu-ceng-ta 九層塔 650

Jiu-ceng-pi 九層皮 547
Jiu-chi-dou-kou 九翅豆蔻 326
Jiu-guang-lang 酒桃榔 302
Jiu-huang-ye-wan-dou 救荒野豌豆 487
Jiu-niang 酒釀 38, 263
Jiu-ping-guo 酒瓶果 582
Jiu-tou-niao 九頭鳥 408
Jiu-ye-zi 酒椰子 301
Ju-hua 菊花 167, 247, 723
Ju-hua-lao 菊花荖 724
Ju-hua-miao 菊花苗 723
Ju-hua-nao 菊花腦 724
Ju-ji 菊薊 727
Ju-ji 菊苣 724
Ju-ju-teng 鋸鋸藤 363
Ju-ruo 蒟蒻 304
Ju-yu 菊芋 735
Ju-yuan 枸櫞 494
Juan-dan 捲丹 319
Juan-xin-cai 捲心菜 413
Juan-ye-gou-qi 捲葉枸杞 663
Jue 蕨 272
Jue-cai 蕨菜 274
Jue-fen 蕨粉 272
Jue-ma 蕨麻 169, 438
Jue-ming 決明 470
Jue-ming-zi 決明子 471
Jue-qi-miao 蕨其苗 272
Jue-tai-cai 蕨苔菜 272
Jun-da-cai 君達菜 376
Jun-qian-zi 君遷子 624

Ka-fei 咖啡 678
Ka-fei-kui 咖啡葵 538
Kai-xin-guo 開心果 518, 519
Kang-lang 康榔 304
Kao 栲 348
Kao-li 栲栗 342
Kao-shu 栲樹 344
Kao-si-li 栲絲栗 341
Ke-zi-ling 柯子嶺 404
Ke-ke 可可 547
Kong-xin-pao 空心袍 461
Kong-zhu 空竹 286
Kou-xiang-tang 口香糖 621
Kou-zi-cao 扣子草 670
Ku-cai 苦菜 408, 747
Ku-di-dan 苦地膽 235
Ku-ding-cha 苦丁茶 8, 247, 519, 558
Ku-dou 苦豆 485
Ku-gua 苦瓜 81, 708
Ku-gua-jin 苦瓜莖 236

Ku-hu-lu 苦壺盧 704
Ku-ju 苦苣 724
Ku-ju-cai 苦苣菜 747
Ku-le-cong 苦芛蔥 235
Ku-li 苦栗 346
Ku-ma-cai 苦媽菜 746
Ku-mai-cai 苦蕒菜 737
Ku-mai-cai 苦麥菜 746
Ku-qiao 苦蕎 371
Ku-qiao-mai 苦蕎麥 371
Ku-tao 苦桃 448
Ku-zhu 苦竹 294
Ku-zhu 苦櫧 346
Kuan-ye-jiu 寬葉韭 313
Kuan-ye-liu-ye-qin 寬葉柳葉芹 595
Kuan-ye-xun-ma 寬葉蕁麻 365
Kui-cai 葵菜 61, 542
Kun-lun-gua 崑崙瓜 667
Kuo-ye-mi-hou-tao 闊葉彌猴桃 552

La-bo-he 辣薄荷 649
La-cai 辣菜 409
La-dou-ban-jiang 辣豆瓣醬 33, 35, 36, 37, 40, 60, 204
La-dou-fu-ru 辣豆腐乳 36
La-jiang-guo 拉江果 711
La-jiao 辣椒 658, 659
La-jiao-jiang 辣椒醬 33, 40, 41
La-la-yang 拉拉秧 363
La-liao 辣蓼 372
La-mu 辣木 424
Lai-fu 萊服 417
Lai-pu-tao 癩葡萄 708
Lai-xiao 蘺蕭 717
Lan-chi 欖豉 506
Lan-hua-cao 藍花草 716
Lan-hua-shen 藍(蘭)花參 716
Lan-pian 蘭片 76
Lan-qi-cai 爛薺菜 670
Lan-ren 欖仁 506
Lan-ren-cai 懶人菜 315
Lan-ren-shu 欖仁樹 576
Lan-sa-guo 蘭撒果 507
Lan-shi 濫柿 623
Lan-xiang 蘭香 650, 651
Lang-gou-li 狼狗脷 583
Lang-yu 榔榆 354
Lao-chou 老抽 34, 156
Lao-dou 嫪豆 474
Lao-hu-ci 老虎刺 237
Lao-hu-pao 老虎泡 458
Lao-jun-dan 老君丹 465

Lao-ke-sao 老可嫂 644
Lao-mu-ji-wo 老母雞窩 377
Lao-niu-piao 老牛瓢 633
Lao-shan-qin 老山芹 598
Lao-shao-nian 老少年 382
Lao-shu-er 老鼠耳 235, 528
Lao-shu-gua 老鼠瓜 712
Lao-shu-he-tao 老鼠核桃 366
Lao-shu-la-dong-gua 老鼠拉冬瓜 712
Lao-shu-shi 老鼠屎 528
Lao-shu-yan 老鼠眼 483
Lao-tou-chou 老頭抽 34, 153
Lao-wu-zui 老烏嘴 743
Lao-ya-guo 老鴉果 438
Lao-zao 糯糟 149, 155, 263
Le-gu-zi 簕古子 281
Le-zhu 簕竹 285
Leng-guo-rong 棱果榕 360
Leng-si-gua 棱絲瓜 706
Leng-zi-qin 棱子芹 604
Li 李 447
Li 藜 377
Li-cha 梨茶 554
Li-cha 黎茶 470
Li-dou 貍豆，藜豆 485
Li-he-tao 離核桃 446
Li-ji 立芨 255, 328
Li-ma-dou 利馬豆 480
Li-meng 黎檬 150, 492
Li-meng-bing 黎檬餅 492
Li-meng-jiang 檸檬薑 493
Li-meng-zi 黎檬子 493
Li-shuo 鱺蒴 343
Li-zhi 荔枝 525
Li-zi 栗子 340
Li-zi-cao 離子草 416
Lian 蓮 138, 387
Lian-hua-bai 蓮花白 410
Lian-qiao 連翹 247
Lian-wu 蓮霧 581
Lian-xing-zao 聯星早 405
Lian-ye 蓮葉 247
Lian-yong 蓮蓉 388
Lian-zi 蓮子 110, 111, 168, 388
Lian-zi-cao 蓮子草 379
Liang-fen-cao 涼粉草 650
Liang-fen-zi 涼粉子 360
Liang-zi 涼子 608
Liang-zi-cai 涼子菜 726
Liao 蓼 248
Liao-shen 遼參 714
Liao-zhen 遼榛 339

Lie-ban-zhu-jin 裂瓣朱槿 540
Lie-dang 列當 674
Lin-di-e-shen 林地峨參 592
Lin-qin 林檎 436
Lin-wei-mu 鱗尾木 368
Ling 菱 583
Ling-dang-hua 鈴鐺花 716
Ling-jiao 菱角 584
Ling-nan-shan-ji-zi 嶺南山桔子 559
Ling-zhi 靈芝 177, 178, 268
Liu-cao 疣草 309
Liu-cheng 柳橙 499
Liu-e-teng 六萼藤 424
Liu-er-ling 六耳苓 236
Liu-jiao-cai 六角菜 598
Liu-lan-xiang 留蘭香 649
Liu-lian 榴槤 544
Liu-su-shu 流蘇樹 626
Liu-ye-qin 柳葉芹 595
Liu-ye-shu-li 柳葉鼠李 529
Liu-ye-zao 柳葉早 404
Long-chuan-hua 龍船花 682
Long-dong-hai-tang 隴東海棠 436
Long-dong-zao-yu 龍洞早芋 306
Long-guo 龍果 622
Long-jing 龍井 220
Long-kui 龍葵 669
Long-li-ye 龍脷葉 209, 214, 513
Long-she-cao 龍舌草 283
Long-tou-zhu 龍頭竹 286, 290
Long-xu-cai 龍鬚菜 272, 316, 598
Long-ya-cao 龍牙草 439
Long-yan 龍眼 167, 523, 524
Long-zhua-cai 龍爪菜 264
Long-zhua-su 龍爪粟 289
Long-zhu-guo 龍珠果 565
Lou-hao 蔞蒿 719
Lou-lu 漏蘆 728
Lü-cha 綠茶 555
Lü-deng 綠燈 666
Lü-dou 綠豆 111, 112, 113, 468
Lü-dou-ya 綠豆芽 112, 114, 116, 468
Lü-hao 驢蒿 719
Lü-hua-shan-qin 綠花山芹 602
Lü-tou-qing 綠頭青 419
Lü-ye 蔞葉 333
Lü-zhu 綠竹 285
Lu-cao 葎草 363
Lu-dang 魯黨 188
Lu-dao-gen 蘆刀根 237
Lu-dao-shu 蘆刀樹 281
Lu-dou-shu 露兜樹 281

Lu-huo 鹿藿 483
Lu-jiao-cai 鹿角菜 31, 260
Lu-shui 滷水 151, 152
Lu-xin-cha 鹿心茶 264
Luo-bi-mei 羅比梅 564
Luo-bo 蘿蔔 131, 132, 133, 134, 135, 417, 418, 421
Luo-bo-ye 蘿蔔葉 417
Luo-bo-zi 蘿蔔子 417
Luo-bu-ma 羅布麻 629
Luo-han-guo 羅漢果 9, 150, 152, 153, 169, 181, 218, 219, 220, 221, 710
Luo-han-shen 羅漢參 465
Luo-hua-sheng 落花生 465
Luo-kui 絡葵 384
Luo-kui-shu 絡葵薯 384
Luo-la 落拉 376
Luo-le 羅勒 650, 651
Luo-mai 裸麥 291
Luo-san-shu 羅傘樹 613
Luo-su 落蘇 667
Luo-wang-zi 羅望子 485

Ma-an-shan-tian-xian-guo 馬鞍山天仙果 358
Ma-bang 罵榜 623
Ma-bao-er 馬㲜兒 712
Ma-cai 麻菜 408
Ma-chang-zi-shu 馬腸子樹 588
Ma-chi-xian 馬齒莧 384
Ma-hua-tou 麻花頭 744
Ma-huang 麻黃 279, 280, 654
Ma-ji-kang 麻雞康 622
Ma-ka-dan-ren 馬卡丹仁 366
Ma-lan 馬蘭 738
Ma-lan-tou 馬蘭頭 738
Ma-li 麻櫟 350
Ma-ling-shu 馬鈴薯 670
Ma-liu-teng 馬騮藤 632
Ma-pao 馬炮 696
Ma-pao-gua 馬泡瓜 698
Ma-pi-cao 麻皮糙 449
Ma-su-li 麻酥梨 451
Ma-ti 馬蹄 142, 300
Ma-ti-jue-ming 馬蹄決明 471
Ma-ti-zhen 馬蹄針 484
Ma-wang 馬旺 322
Ma-wei 馬尾 284
Ma-wei-zao 馬尾藻 259
Ma-ye-xiu-qiu 麻葉繡球 464
Ma-ye-xiu-xian-ju 麻葉繡線菊 464
Ma-ye-xun-ma 痳葉蕁痳 364
Ma-zao 馬藻 282
Mai-fu 麥麩 248

Mai-ga 嘮咖 517
Mai-jie-cai 麥楷菜 417
Mai-men-dong 麥門冬 226
Mai-ping-cao 麥瓶草 386
Mai-ya 麥芽 168, 247, 291
Man-jing-zi 蔓荊子 238
Man-mei-guo 滿美果 561
Man-wei-ling-cai 蔓委陵菜 440
Man-zao 蔓藻 259
Mang-guo 芒果 517
Mang-guo-he 芒果核 236
Mang-ji-shi 莽桔柿 559
Mao-cha 毛茶 511
Mao-che-teng 毛車藤 629
Mao-di-huang 毛地黃 195
Mao-di-li 毛地栗 142
Mao-dong-qing 毛冬青 236
Mao-dou 毛豆 119
Mao-e-cha 毛萼茶 557
Mao-e-mei 毛萼莓 456
Mao-er-shi 貓兒屎 392
Mao-gen 茅根 236, 291
Mao-gua 毛瓜 691, 692
Mao-gua 茅瓜 711
Mao-guo-jin-cai 毛果菫菜 562
Mao-hao-dou 毛蒿豆 612
Mao-hui 茅薈 237
Mao-ji 貓薊 725
Mao-jin-zhu 毛金竹 295
Mao-lai 毛萊 609
Mao-li 茅栗 341
Mao-li-mu 毛栗木 343
Mao-li-zi 毛李子 180, 550
Mao-li-zi-jiu 毛李子酒 550
Mao-lian-cai 毛蓮菜 742
Mao-long-yan 毛龍眼 525
Mao-mei 茅莓 459
Mao-mu-ban-xia 毛木半夏 569
Mao-mu-er 毛木耳 266
Mao-nian 毛棯 582, 583
Mao-nü-er-cai 毛女兒菜 717
Mao-pu-tao 毛葡萄 536
Mao-ren-shen 貓人參 552
Mao-shan-zha 毛山楂 430
Mao-she-xiang 毛麝香 235
Mao-shu 毛薯 322
Mao-suan-jiang 毛酸漿 666
Mao-tao 毛桃 444
Mao-ti-gu 毛提穀 291
Mao-tian-xian-guo 毛天仙果 357
Mao-tuan-zi 毛團子 324
Mao-wu-gua 毛烏瓜 712

Mao-ye-mu-xiang-zi 毛葉木香子 402
Mao-ying-tao 毛櫻桃 447
Mao-zhu 毛竹 129
Mao-zhui-li 毛錐栗 346
Mei 梅 446
Mei-dou 眉豆 475, 489
Mei-gan-cai 霉乾菜 409
Mei-gui 玫瑰 77, 454
Mei-gui-qie 玫瑰茄 540
Mei-guo-fang-feng 美國防風 602
Mei-guo-qi-cheng 美國臍橙 500
Mei-guo-shan-he-tao 美國山核桃 336
Mei-hua-ping 梅花瓶 386
Mei-jiang 梅薑 493
Mei-ju 美橘 500
Mei-mian 美棉 539
Mei-nong 美濃 419
Mei-qiang-wei 美薔薇 452
Mei-wei-niu-gan 美味牛肝 267
Meng-gu-jia-mi 蒙古莢蒾 689
Meng-gu-jiu 蒙古韭 314
Meng-gu-zao 蒙古棗 462
Meng-li 蒙櫟 350
Mi 米 292
Mi-cong 米蔥 313
Mi-guan-ke 蜜罐棵 672
Mi-hai-tai 米海苔 261
Mi-hou-tao 獼猴桃 550
Mi-hua-suan-teng-zi 密花酸藤子 616
Mi-hua-xiang-ru 密花香薷 646
Mi-ju 蜜橘 496
Mi-li 蜜梨 449
Mi-mian-weng 米麵瓮 368
Mi-tang-guo 米湯果 616
Mi-wa-guan 米瓦罐 386
Mi-wu 蘪蕪 200
Mi-ya 米芽 168
Mi-zai-lan 米仔蘭 507
Mi-zai-le 米仔簕 530
Mi-zao 蜜棗 169, 532
Mi-zi-chai 米子柴 341
Mi-zhui 米錐 342
Mian-bao-guo 麵包果 355
Mian-gua 麵瓜 39, 696
Mian-guo-guo 麵果果 431
Mian-jin 麵筋 297
Mian-mao-ping-po 棉毛蘋婆 547
Mian-shen 綿參 646
Mian-tuan-tie-xian-lian 棉團鐵線蓮 389
Mian-weng 麵瓮 367
Mian-xiang-cai 麵香菜 593
Mian-yin-cheng 棉茵陳 246

Mian-you 棉油 539
Ming-ye-qin 明葉芹 606
Mo-dou-fu 墨豆腐 304
Mo-gu 蘑菇 266
Mo-gu-xiao 冇骨消 687
Mo-gui-xi 沒歸息 615
Mo-li 茉莉 73, 626
Mo-tuo-fang-zhu 墨脫方竹 286
Mo-yu 墨芋 304
Mo-yu 魔芋 304
Mu-ban-xia 木半夏 571
Mu-ding-xiang 母釘香 579
Mu-dou 木豆 469
Mu-er 木耳 68, 137, 266
Mu-gua 木瓜 172, 181, 429
Mu-gua-guo 木瓜果 357
Mu-gui-shi 木桂拾 615
Mu-hua-sheng 木花生 620
Mu-huan-gen 木患根 237
Mu-ji-tou 苜雞頭 477
Mu-jiang 母薑 329
Mu-jiang-zi 木姜子 401
Mu-jin 木槿 541
Mu-jing 牡荊 644
Mu-li 木梨 431
Mu-li-zi 木梨子 453
Mu-mei 木莓 462
Mu-mian 木棉 544
Mu-mo-zi 木磨子 542
Mu-shu 木薯 511
Mu-tian-liao 木天蓼 552, 553
Mu-tian-qing 木田青 483
Mu-tong 木通 391
Mu-xu 苜蓿 477
Mu-zhu-guo 母豬果 366

Nai-jiang-cai 奶漿菜 737
Nai-ren 奈仁 229, 440
Nai-ren-rou 奈仁肉 169, 229
Nai-shen 奶參 713
Nan-gua 南瓜 701
Nan-ling-kao 南嶺栲 343
Nan-mu-xu 南苜蓿 477
Nan-ru 南乳 36
Nan-sha-shen 南沙參 166, 713
Nan-she-teng 南蛇藤 520
Nan-suan-zao 南酸棗 515
Nan-wu-wei-zi 南五味子 224
Nan-xing-ren 南杏仁 227
Nan-yang-pu-tao 南洋蒲桃 580
Nan-ye-fen 南椰粉 300
Nan-zao 南棗 88

Nen-ye-ti-gai-jue 嫩葉蹄蓋蕨 273
Ni-hu-cai 泥胡菜 736
Ni-qiu-cai 泥鰍菜 618
Ni-teng 倪籐 280
Nian-he-tao 黏核桃 446
Nian-shan-yao 黏山藥 323
Nian-shu 黏薯 323
Ning-meng 檸檬 492
Ning-meng-liu-lan-xiang 檸檬留蘭香 649
Ning-xia-gou-qi 寧夏枸杞 661
Ning-yang-ci-gua 寧楊刺瓜 699
Niu-bang 牛蒡 718
Niu-bang-geng 牛蒡根 718
Niu-bi-quan 牛鼻圈 528
Niu-da-li 牛大力 478
Niu-er-da-huang 牛耳大黃 374
Niu-er-teng 牛耳藤 369
Niu-er-teng 牛兒藤 527
Niu-jiao-jiao 牛角椒 659
Niu-li-sheng-cai 牛脷生菜 741
Niu-ma 牛麻 289
Niu-ma-teng 牛麻藤 479
Niu-mao-shi-hua-cai 牛毛石花菜 260
Niu-nai-cao 牛奶草 716
Niu-nai-zai 牛奶仔 236
Niu-nai-zi 牛奶子 571
Niu-pang 牛旁 548
Niu-pi 牛脾 692
Niu-pi-xiao 牛皮消 633
Niu-tui-gua 牛腿瓜 701
Niu-xi 牛膝 166, 379
Niu-xin-cha 牛心茶 561
Niu-xin-guo 牛心果 398
Niu-you-shu 牛油樹 619
Niu-zhi-zi 牛枝子 477
Nong-da-hong 農大紅 419
Nü-er-cha 女兒茶 643
Nü-er-teng 女兒藤 526
Nü-luo 女蘿 265
Nü-zhen 女貞 626
Nü-zhen-zi 女貞子 168
Nuo-mi 糯米 292
Nuo-zao 糯糟 37, 38, 39

Ou 藕 137, 138, 139, 387
Ou-fen 藕粉 138, 140, 388
Ou-gan 甌柑 497
Ou-li 歐李 443
Ou-xun-ma 歐蕁麻 365
Ou-zhou-cao-mei 歐洲草莓 435
Ou-zhou-li 歐洲李 443

Pa-lei 帕累 614
Pa-qiang-hu 爬牆虎 533
Pan-er-cao 盤兒草 737
Pan-long-shen 盤龍參 189, 331
Pan-shi 盤柿 624
Pan-tao 蟠桃 446
Pan-zhi-hua 攀枝花 246, 544
Pao-cai 泡菜 29
Pao-shen 泡參 713
Peng-hao 蓬蒿 722
Peng-qi-ju 蟛蜞菊 238
Pi-ba 枇杷 432
Pi-ba-gen 枇杷根 432
Pi-jiu-hua 啤酒花 362
Pi-lan 匹蘭 406
Pi-xian-dou-ban 郫縣豆瓣 40
Pi-xian-dou-ban-jiang 郫縣豆瓣醬 486
Pi-zhen-ye-hu-tui-zi 披針葉胡頹子 570
Ping-guo 蘋果 437
Ping-guo-rong 蘋果榕 359
Ping-peng-cao 萍蓬草 388
Ping-po 蘋婆 546, 547
Ping-po-ye 蘋婆葉 546
Ping-shan-you 坪山柚 494
Po-bu-mu 破布木 641
Po-bu-ye 破布葉 538
Po-bu-zi 破布子 641
Po-gu-feng 破骨風 392
Po-po-ding 婆婆釘 748
Po-qi-liu 簸箕柳 334
Pu 蒲 704
Pu-cai 蒲菜 280
Pu-cao 蒲草 280
Pu-er-cha 普洱茶 556
Pu-gong-ying 蒲公英 748
Pu-gua 埔瓜 711
Pu-gua 蒲瓜 704, 705
Pu-shu 朴樹 353
Pu-tao 葡萄 536
Pu-tao 蒲桃 580
Pu-tao-you 葡萄柚 495
Pu-tong-xi-gua 普通西瓜 695

Qi 萁 272
Qi-cai 溪菜 258
Qi-dun-guo 齊墩果 627
Qi-jie-guo 七姐果 546
Qi-lin-zao 麒麟藻 261
Qi-shu-tang 槭樹糖 522
Qi-shu-tang-jiang 槭樹糖漿 522
Qi-xing-bo-luo-zai 七星菠蘿仔 694
Qi-xing-zai 七星仔 693

Qi-ye-dan 七葉膽 703
Qi-ye-lian 七葉蓮 395
Qi-ye-mei 槭葉莓 459
Qi-zhao-long 七爪龍 639
Qi-zhi-jue 七指蕨 271
Qian-ceng-zhi 千層紙 237
Qian-jin-ba 千斤拔 478
Qian-jin-shu 千觔樹 463
Qian-lang-suo 前郎索 288
Qian-li-xiang 千里香 657
Qian-qu-cai 千屈菜 572
Qian-shi 芡實 106
Qian-sui-gu 千穗穀 381
Qian-xi-cao-mei 縴細草莓 433
Qian-zhang-pi 千張皮 123, 124, 474
Qiang-huo 羌活 248
Qiang-zi-guo 檜子果 615
Qiao-ke-li 巧克力 547
Qiao-mai 蕎麥 370
Qiao-mai-dang-gui 蕎麥當歸 370
Qiao-mai-qi 蕎麥七 370
Qiao-yao 翹搖 487
Qie-dong 茄苳 509
Qie-yi 竊衣 607
Qie-zi 茄子 666, 667
Qin-cai 芹菜 592
Qing-cai 青菜 408
Qing-dou 青豆 119
Qing-gang 青剛 350
Qing-gang-li 青剛櫟 350
Qing-gou-kao 青鈎栲 345
Qing-gua 青瓜 698
Qing-guo 青果 32, 505
Qing-guo-rong 青果榕 358
Qing-hao 青蒿 235, 246
Qing-hua-jiao 青花椒 504
Qing-jia-ye 青莢葉 610
Qing-jia-yu 青莢芋 307
Qing-jiao 青椒 658
Qing-jing-fan 青精飯 611
Qing-ke 青稞 291
Qing-ke-mai 青稞麥 284
Qing-la-zhi-tian-jiao 青蜡指天椒 660
Qing-lü 青蔞 333
Qing-luo-bo 青蘿蔔 131
Qing-mei 青梅 335
Qing-miao 青苗 307
Qing-pi 青皮 246
Qing-pi-guo 青皮果 711
Qing-pi-qing 青皮青 420
Qing-pi-shu 青皮樹 608
Qing-pi-si-gua 青皮絲瓜 706

Qing-sheng-cai 青生菜 740
Qing-si-zhu 青絲竹 286
Qing-su 青蘇 652
Qing-tian-kui 青天葵 331
Qing-tong 青桐 545
Qing-xiang 青葙 382
Qing-yang-shen 青羊參 634
Qiu-cong 球蔥 311
Qiu-guo-han-cai 球果薄菜 421
Qiu-guo-jin-cai 球果菫菜 562
Qiu-jing-gan-lan 球莖甘藍 406
Qiu-jue 球蕨 272
Qiu-ku-mai-cai 秋苦蕒菜 737
Qiu-kui 秋葵 538
Qiu-shu 楸樹 672
Qiu-ya-gan-lan 球芽甘藍 411
Qiu-zi 楸子 437
Qiu-zi-li 秋子梨 451
Quan-tang 泉塘 405
Que-dou 鵲豆 475
Que-mei-teng 雀梅藤 237, 530
Que-shi-rui 雀石蕊 263
Qun-dai-cai 裙帶菜 260

Rang-he 蘘荷 329
Re-su-guo-ba 熱宿戈巴 286
Ren 苒 652
Ren-dong-teng 忍冬藤 215
Ren-mian 人面 516, 517
Ren-shen 人參 168, 181, 188, 586
Ren-shen-guo 人參果 438
Ren-shen-san-qi 人參三七 587
Ren-xian 人莧 51, 381
Ren-xin-guo 人心果 621
Ri-ben-cha 日本茶 557
Ri-ben-chang-shan 日本常山 502
Rou-cao 肉草 309
Rou-cong-rong 肉蓯蓉 167, 202, 673
Rou-dou-kou 肉豆蔻 399
Rou-gui 肉桂 149, 167, 399
Ru-di-long 入地龍 614
Ru-di-wu-gong 入地蜈蚣 271
Ru-ju 乳橘 496
Ru-shu 乳薯 714
Ruan-jiang 軟漿 384
Ruan-mao-chong-shi 軟毛蟲實 377
Ruan-wei-sheng-cai 軟尾生菜 741
Ruan-zao 軟棗 624
Ruan-zao-mi-hou-tao 軟棗獼猴桃 550
Ruan-zao-zi 軟棗子 550
Rui-chi-hu-li 銳齒槲櫟 349
Rui-he 蕤核 440

Sai-bao 洒堡 575
San-ban-jia-wan-shou-zhu 散斑假萬壽竹 317
San-ban-zhu-gen-qi 散斑竹根七 317
San-cha-da-mai 三叉大麥 291
San-chi-ye-wan-dou 三齒野豌豆 486
San-da-jun 三大蕈 267
San-jiao-ye-xun-ma 三角葉蕁麻 365
San-leng-gua 三棱瓜 698
San-lie-xiu-xian-ju 三裂銹線菊 464
San-nai 三奈，三佘 328
San-qi 三七 181, 185, 186, 587
San-ren 三稔 490
San-ya-ku 三椏苦 236
San-ye-ci-wu-jia 三葉刺五加 586
San-ye-lian 三葉蓮 393
San-yue-bai 三月白 411
San-yue-qing 三月青 412
Sang-shen-zi 桑椹子 361
Sang-ye 桑葉 361
Sang-ye-pu-tao 桑葉葡萄 534
Sao-ba-shu 掃把薯 322
Sao-ba-zhu 掃把竹 295
Sao-zhou-cai 掃帚菜 378
Sao-zhou-gu 掃帚菇 267
Sen-lin-cao-mei 森林草莓 435
Sha-cha-jiang 沙茶醬 64, 65, 151, 153
Sha-di-cai 沙地菜 421
Sha-guo 沙果 436
Sha-ji 沙棘 572
Sha-jiang 沙薑 328
Sha-jie 沙芥 417
Sha-li 沙梨 451
Sha-lin-cao 砂林草 377
Sha-long-bai-qie 沙龍白茄 668
Sha-mu-mian 沙木麵 300
Sha-peng 沙蓬 375
Sha-peng-mi 沙蓬米 375
Sha-shen 沙參 181, 189, 190, 221, 229, 383, 713
Sha-tang-guo 沙糖果 438
Sha-tang-ye-zi 沙糖椰子 301
Sha-tian-you 沙田柚 494
Sha-wo-wo 沙窩窩 458
Sha-zan-tai-cao 砂礦薹草 299
Sha-zao 砂棗 568
Shan-ai-cai 山艾菜 730
Shan-bai-cai 山白菜 720
Shan-bai-gan 山白柑 491
Shan-bai-guo 山白果 280, 338
Shan-bai-ju 山白菊 720
Shan-ban-li 山板栗 510
Shan-bo-cai 山波菜 375
Shan-bo-qin 山坡芹 730

Shan-cheng 山橙 632
Shan-ci-gu 山慈菇 321
Shan-ci-mei 山刺玫 453
Shan-cong 山蔥 315
Shan-da-dao 山大刀 237
Shan-dan 山丹 682
Shan-di-du-jin-shan 山地杜莖山 617
Shan-ding-zi 山釘子 436
Shan-dong-mian-li 山東麵梨 452
Shan-dou-zi 山豆子 447
Shan-gan-cao 山甘草 237
Shan-gan-zi 山柑子 423
Shan-ge-lu 山蛤蘆 720
Shan-gua-mian 山掛麵 265
Shan-gui-hua 山桂花 617
Shan-hai-luo 山海螺 713
Shan-hao 山蒿 718
Shan-he-tao 山核桃 336, 337
Shan-he-ye 山荷葉 424
Shan-hu-cai 珊瑚菜 598
Shan-hu-jiao 山胡椒 507
Shan-hu-lu 山葫蘆 366
Shan-hui-xiang 山茴香 593
Shan-ji-zi 山桔子 559
Shan-jiang 山薑 149
Shan-jiang-dou 山豇豆 476
Shan-jiao 山椒 657
Shan-jie-lan 山芥藍 368
Shan-jing-zi 山荊子 436
Shan-jiu 山韭 315
Shan-jiu-cai 山韭菜 309
Shan-jue-mao 山蕨貓 271
Shan-ku-cha 山苦茶 511
Shan-ku-mai 山苦蕒 737
Shan-ku-mai 山苦麥 746
Shan-kui 山葵 422
Shan-li 山梨 451
Shan-li-hong 山裡紅 431
Shan-li-zhi 山荔枝 607, 608
Shan-li-zi 山李子 394
Shan-lian-ou 山蓮藕 478
Shan-liao 山蓼 389
Shan-luo-bo 山蘿蔔 131, 255, 382, 690
Shan-luo-hua 山蘿花 671
Shan-ma-sheng-cai 山馬生菜 385
Shan-ma-zi 山麻子 427
Shan-mang 山芒 273
Shan-mao-yu 山毛榆 353
Shan-mei 山莓 456
Shan-mian-hua 山棉花 389
Shan-niu-pang 山牛旁 549
Shan-pao-zi 山泡子 456

Shan-pi-ba 山枇杷 620
Shan-pu-tao 山葡萄 533
Shan-qie 山茄 540
Shan-qie-zi 山茄子 641
Shan-qin 山芹 591, 601
Shan-qin-cai 山芹菜 588, 601, 604, 605
Shan-shu 山薯 322
Shan-tao 山桃 229
Shan-tao-ye 山桃葉 613
Shan-tu-gua 山土瓜 640
Shan-wu-jiu 山烏桕 237
Shan-wu-rong 山烏茸 267
Shan-xiao-ji 山小桔 501
Shan-xing 山杏 441
Shan-yang-tao 山洋桃 553
Shan-yao 山藥 144, 145, 146, 167, 323
Shan-yao-dou 山藥豆 670
Shan-yin-hua 山銀花 236
Shan-ying-tao 山櫻桃 447
Shan-you-gan 山油柑 235, 491
Shan-yu 山芋 140, 307, 638
Shan-yu-cai 山俞菜 416, 422
Shan-yuan-sui 山芫荽 599
Shan-zao-shen 山棗參 462
Shan-zha 山楂 84, 85, 247, 431
Shan-zha-gao 山楂糕 84, 431
Shan-zha-pian 山楂片 84, 431
Shan-zha-rou 山楂肉 167
Shan-zhu 山竹 290
Shan-zhu-rou 山豬肉 526
Shan-zhu-yu 山茱萸 609
Shang-dang-ren-shen 上黨人參 714, 715
Shang-lu 商陸 255, 382
Shao-cai 紹菜 413, 486
Shao-gua 稍瓜 697
Shao-hua-long-kui 少花龍葵 670
Shao-hua-wu-pao 少花烏泡 457
Shao-yao 芍藥 391
Shao-zi 韶子 525
She-cong 蛇蔥 314
She-gua 蛇瓜 712
She-ma-cao 蛇麻草 362
She-pao-le 蛇泡簕 460
She-tou-cai 蛇頭菜 741
She-xiang-cao 麝香草 658
Shen-liang 神糧 264
Shen-shan-mu-tian-liao 深山木天蓼 551
Shen-shu 參薯 321
Shen-xian-cai 神仙菜 686
Shen-xian-cao 神仙草 309
Shen-xian-mi 神仙米 275
Shen-xian-ye-zi 神仙葉子 685

Shen-ye-shan-liao 腎葉山蓼 371
Shen-zi-ye 腎子葉 467
Sheng-cai 生菜 739
Sheng-chou 生抽 34, 204
Sheng-di 生地 196
Sheng-di-huang 生地黃 672
Sheng-gu-you 省沽油 521
Sheng-gua 生瓜 696, 697
Sheng-ma 升麻 246
Shi-bai 石白 419
Shi-ban-mu 石斑木 452
Shi-bang-zi 石棒子 464
Shi-bing 柿餅 623
Shi-chun 石蓴 258
Shi-dao-bai 石刀柏 316
Shi-di 柿蒂 624
Shi-er 石耳 265
Shi-gen 柿根 624
Shi-hu 柿糊 571
Shi-hua-cai 石花菜 261
Shi-hui-hua-qiu 石灰花楸 463
Shi-hui-tiao 石灰條 463
Shi-li 石栗 509
Shi-li-zi 石栗子 342
Shi-liu 石榴 573
Shi-luo 蒔蘿 589
Shi-nan 石楠 438
Shi-qi 柿漆 624
Shi-tou-fan-jiao 獅頭番椒 659
Shi-tou-shen 獅頭參 714
Shi-tou-shu 石頭樹 348
Shi-yue 石月 394
Shi-yue-xin 十月心 412
Shi-zhu 石竹 293
Shi-zhui-shu 石錐樹 345
Shi-zi 柿子 623
Shi-zi-cao 獅子草 310
Shou-gong-mu 守宮木 512
Shou-shen 手參 330
Shu-di 熟地 196, 197
Shu-di-huang 熟地黃 672
Shu-fan-qie 樹番茄 660
Shu-li 鼠李 530
Shu-qi 樹杞 614
Shu-qu-cao 鼠麴草 731
Shu-shang-cha 樹上茶 367
Shu-su 蜀粟 297
Shu-yu 薯蕷 144
Shu-zi 黍子 293
Shu-zi-cai 樹子菜 512
Shuan-pi-li 栓皮櫟 352
Shuan-ye-mi-hou-tao 栓葉獼猴桃 553

Shuang-hu-die 雙蝴蝶 521
Shuang-shen-teng 雙腎藤 467
Shui-bai-cai 水白菜 283
Shui-bi-zai 水筆仔 574
Shui-bie 水鼈 283
Shui-che-qian 水車前 284
Shui-dong-ge 水東哥 553
Shui-dou-er 水豆兒 282
Shui-fei-ji 水飛薊 745
Shui-fou-lian 水浮蓮 310
Shui-gua 水瓜 707
Shui-hao 水蒿 719
Shui-heng-zhi 水橫枝 680, 681
Shui-hu-luo-bo 水胡蘿蔔 592
Shui-jin-feng 水金鳳 527
Shui-jue 水蕨 272
Shui-la-zhu 水蠟燭 280
Shui-li 水栗 343
Shui-lian 睡蓮 388
Shui-liao 水蓼 372
Shui-liu 水柳 572
Shui-liu-shu 水柳樹 536
Shui-lu-qiang 水蘆強 237
Shui-ma 水麻 363
Shui-ma-chi 水馬齒 513
Shui-man-qing 水蔓青 421
Shui-mu-ji-guo 水母雞果 620
Shui-pi-lian 水皮連 628
Shui-qian-cai 水前菜 732
Shui-qie 水茄 238
Shui-qin 水芹 600
Shui-rong-hua 水榕花 235
Shui-shi-rong 水石榕 536
Shui-suo-yi 水蓑衣 676
Shui-tian-jie 水田芥 422
Shui-tian-sui-mi-ji 水田碎米薺 414
Shui-tong 水桐 673
Shui-yang-liu 水楊柳 425
Shui-yang-mei 水楊梅 435
Shui-ye 水椰 303
Shui-yong 水蕹 282
Shui-yu-hua-qiu 水榆花秋 463
Shui-ze-lan 水澤蘭 (澤藍) 425
Shui-zhu 水竹 294
Shui-zhu-ye 水竹葉 309
Si-chuan-qing-feng-teng 四川清風藤 526
Si-chuan-zha-cai 四川榨菜 409
Si-cong 絲蔥 311
Si-gua 絲瓜 90, 706, 707
Si-hui-gan 四會柑 497
Si-ji-cai 四季菜 719
Si-ji-dou 四季豆 480

Si-ji-ju 四季橘 495
Si-ji-pao 四季拋 494
Si-jiu-xin 四九心 412
Si-leng-dou 四楞豆 481
Si-leng-shi 四稜柿 624
Si-li 絲栗 342, 346
Si-li-kao 絲栗栲 342
Si-li-shu 絲栗樹 342
Si-mao-shan-cheng 思茅山橙 631
Si-rui-mi-hou-tao 四蕊獼猴桃 553
Si-xian-ye 四仙葉 274
Si-ye-cai 四葉菜 716
Si-ying 四英 418
Si-yue-zi 四月子 571
Si-zhao-hua 四照花 608
Song 菘 413
Song-gu 松菇 268
Song-luo 松蘿 265
Song-ren 松仁 104
Song-zi 松子 105, 277
Su 粟 296
Su-tie 蘇鐵 275
Su-ye 蘇葉 248
Su-zhou-qing 蘇州青 411
Su-zi 蘇子 652
Suan-ban 蒜瓣 314
Suan-cai 酸菜 29, 408
Suan-cheng 酸橙 492
Suan-ci 酸刺 572
Suan-ding 酸丁 443
Suan-guo 酸果 517
Suan-ji-teng 酸雞藤 616
Suan-jiang 酸漿 664
Suan-jiao 酸角 485
Suan-jiao-gan 酸腳杆 581
Suan-li 酸梨 451
Suan-ma-zha-cai 酸螞蚱菜 385
Suan-mei-jiang 酸梅醬 33
Suan-miao 蒜苗 60, 314
Suan-ming-guo 酸名果 508
Suan-mo 酸模 374
Suan-nao-shu 蒜腦藷 319
Suan-tai 蒜苔 314
Suan-tai-cai 酸苔菜 614
Suan-teng 酸藤 615, 631
Suan-teng-guo 酸藤果 614
Suan-teng-tou 酸藤頭 614
Suan-wei 酸味 530
Suan-wei-cao 酸味草 237
Suan-ye-po-luo-men-shen 蒜葉婆羅門參 751
Sui-jun-cha 隨軍茶 476
Sui-mi-ji 碎米薺 415

Sui-mi-lan 碎米蘭 507
Sui-ye-qin 碎葉芹 601
Sun 筍 127, 130
Sun-gan 荀乾 288
Sun-gua 荀瓜 702
Suo-yang 鎖陽 584

Ta-di-cai 搨地菜 410
Ta-wo-cai 搨窩菜 410
Tai-bai-cha 太白茶 265
Tai-bai-hua 太白花 263
Tai-cai 苔菜 258
Tai-dong-dou-shi 台東豆柿 625
Tai-shan-he-shou-wu 泰山何首烏 193, 633
Tai-shan-zhu 泰山竹 286
Tai-tiao 苔條 258
Tai-wan-kao 台灣栲 344
Tai-wan-pi-ba 台灣枇杷 432
Tai-wan-shi 台灣柿 625
Tai-zi-shen 太子參 169, 189, 385
Tan-hua 曇花 567
Tan-li 檀梨 368
Tang-cha-biao 糖茶藨 426
Tang-di 唐棣 428
Tang-ge-cai 塘葛菜 422
Tang-hu-lu 糖葫蘆 84, 431
Tang-li 棠梨 448
Tang-li-zi 棠梨子 450
Tang-luo-bo 糖蘿蔔 376
Tang-qiu 糖球 84, 431
Tang-yuan 湯圓 79, 80
Tao 桃 446
Tao-er-qi 桃耳(兒)七 395
Tao-jin-niang 桃金娘 255, 578
Tao-jun-liang 逃軍糧 255, 578
Tao-lan 桃欖 622
Teng-gua 藤瓜 550
Teng-li 藤李 552
Teng-teng-cai 藤藤菜 636
Tian-cai 甜菜 376
Tian-cai-zi 甜菜子 719
Tian-cha 甜茶 348, 461
Tian-cheng 甜橙 498
Tian-ci-gu 甜慈菇 300
Tian-ci-lan 滇刺欖 623
Tian-ci-zao 滇刺棗 532
Tian-cong 天蔥 312
Tian-da-yun 甜大芸 673
Tian-dai-shu 田代薯 321
Tian-e-dan 天鵝蛋 419
Tian-fan-jiao 甜番椒 659
Tian-gao-liang 甜高粱 297

Tian-ge 甜葛 248
Tian-gua 甜瓜 696, 697
Tian-guo-teng 甜果藤 521
Tian-hua-fen 天花粉 248
Tian-huang-qin 滇黃芩 653
Tian-ji-huang 田基黃 236
Tian-jiu 甜酒 38, 263
Tian-ju 甜橘 496
Tian-kui 天葵 331
Tian-lan 滇欖 506
Tian-liao 天蓼 552
Tian-ma 天麻 42, 168, 330
Tian-mian-jiang 甜麵醬 33, 39, 40, 60, 109, 110, 149, 150, 159, 297
Tian-pao-cao 天泡草 665
Tian-peng-cao 天棚草 265
Tian-qi 滇七，田七 587
Tian-qiao 甜蕎 370
Tian-qie 天茄 635
Tian-shi-hao 天使號 658
Tian-shi-zi 滇石梓 644
Tian-shu 甜薯 322
Tian-su-jie 甜粟楷 297
Tian-sun-zhu 甜笋竹 285
Tian-tai-mi-ju 天台蜜橘 497
Tian-tao 天桃 518
Tian-tao-mu 天桃木 517
Tian-xian-cai 天仙菜 257
Tian-xian-guo 天仙果 358
Tian-xiang-lu 天香爐 237
Tian-xing-ren 甜杏仁 227
Tian-xuan-hua 田旋花 636
Tian-ye-ju 甜葉菊 748
Tian-ying-tao 甜櫻桃 442
Tian-zao 甜棗 571
Tian-zhu 田珠 664
Tian-zhu 甜竹 294
Tian-zhu-kao 甜櫧栲 342
Tian-zi-cao 田字草 274
Tiao-qie 條茄 669
Tie-bao-jin 鐵包金 237
Tie-diao 鐵椆 350
Tie-gan-hai-tang 帖梗海棠 429
Tie-gu-san 鐵箍散 397
Tie-li 鐵櫟 350
Tie-shu 鐵樹 275
Tie-xian-geng 鐵線梗 638
Tie-xian-zi 鐵線子 621
Tie-xiang-shu 鐵橡樹 351
Tong-cai 通菜 636
Tong-cao 通草 248
Tong-gan 桶柑 498

Tong-hao 茼蒿 57, 58, 722
Tong-ke-ye 銅殼葉 405
Tong-ma 桐麻 541
Tong-qiao-she-gu 筒鞘蛇菇 369
Tong-quan-cao 通泉草 670
Tong-shan-a-wei 銅山阿魏 597
Tou-chou 頭抽 34
Tu-bai-lian 土白蘞 712
Tu-dan 土蛋 465
Tu-dang-gui 土當歸 585
Tu-dang-shen 土黨參 714
Tu-er-tui 兔兒腿 674
Tu-huang-qi 土黃蓍 478
Tu-lan-tiao 土蘭條 688
Tu-luan-er 土圞兒 465
Tu-mi-zi 荼蘼子 454
Tu-mu-gua 土木瓜 431
Tu-ni-tiao 禿妮條 736
Tu-niu-xi 土牛膝 235, 379
Tu-ren-shen 土人參 189
Tuo-yang-lü-dou 拖秧綠豆 468

Wa-er-cai 娃兒菜 716
Wa-er-cha 娃兒茶 520
Wai-tou-cai 歪頭菜 274, 488
Wan-dou 豌豆 481
Wan-dou-miao 豌豆苗 59
Wan-dou-tou 豌豆頭 481
Wan-jia-guo 萬加果 685
Wan-tian-hong-hua-you-cha 宛田紅花油茶 555
Wang-jiang-nan 望江南 470
Wei 薇 274, 487
Wei-ling-cai 委陵菜 439
Wei-ling-xian 威靈仙 247, 390
Wei-ni-hu-cai 偽泥胡菜 744
Wei-xiang 蔚香 599
Wei-zhi 蕑芝，偎枝 356
Wen-dan-you 文旦柚 494
Wen-guan-guo 文冠果 526
Wen-po 溫桲 431
Wen-wang-yi-zhi-bi 文王一支筆 369
Weng-cai 雍菜 52, 636
Wo-ju 萵苣 739
Wo-sun 萵笋 740
Wo-zhu-sun 萵竹筍 740
Wu-ban-ji-sheng 五瓣寄生 367
Wu-dou 烏豆 474
Wu-fan-shu 烏飯樹 611
Wu-feng-teng 五風藤 393
Wu-gang-li 烏崗櫟 351
Wu-he-hong-zao 無核紅棗 531
Wu-he-zao 無核棗 532

Wu-hua-cha 五花茶 215, 241, 482, 484, 544
Wu-hua-guo 無花果 358
Wu-lan 烏欖 506
Wu-li 烏梨 451
Wu-ling 烏菱 583
Wu-mao-diao-li 無毛稠李 448
Wu-mao-jue 烏毛蕨 271
Wu-mei-kao 烏楣栲 345
Wu-ming-zi 無名子 518
Wu-pi-si-gua 烏皮絲瓜 707
Wu-qi-zhao-yang-cao 五氣朝陽草 435
Wu-qing-gan-lan 蕪菁甘藍 409
Wu-shi-di-lun-tao 五室第倫桃 549
Wu-tong 梧桐 545
Wu-wei-zi 五味子 169, 208, 218, 224, 225, 226, 396
Wu-xiang-fen 五香粉 151
Wu-xin-cai 無心菜 636
Wu-ya-guo 五椏果 548
Wu-yao 烏藥 247
Wu-ye-cao-mei 五葉草莓 434
Wu-ye-gua-teng 五葉瓜藤 392
Wu-ye-shu-yu 五葉薯芋 324
Wu-yi-liu-ye-qin 無翼柳葉芹 595
Wu-ying 五英 418
Wu-yue-cha 五月茶 509
Wu-yue-hua 五月花 682
Wu-zhao-jin-long 五爪金龍 639
Wu-zhao-long 五爪龍 639
Wu-zhu-yu 吳茱萸 609
Wu-zi-song 五子松 277

Xi-dang 西黨 188
Xi-fan-ju 西番菊 735
Xi-fan-lian-guo 西番蓮果 565
Xi-fu-hai-tang 西府海棠 437
Xi-gu-mi 西穀米 511
Xi-gu-zi 西穀子 303
Xi-gua 西瓜 124, 125, 694
Xi-hong-shi 西紅柿 664
Xi-hu-lu 西葫蘆 702
Xi-kang-chi-song 西康赤松 278
Xi-lan-gan-lan 錫蘭橄欖 537
Xi-lun-cu-li 錫倫醋栗 563
Xi-ming 菥蓂 422
Xi-nan-cao-mei 西南草莓 433
Xi-suan-zao 西酸棗 516
Xi-tong 細通 638
Xi-yang-bai-niu-bang 西洋白牛蒡 751
Xi-yang-cai 西洋菜 422
Xi-yang-jun 西洋蕈 266
Xi-yang-shan-yu-cai 西洋山俞菜 403

Xi-yang-shen 西洋參 168, 586
Xi-yang-zhen 西洋榛 338
Xi-ye-jie-you-cai 細葉芥油菜 408
Xi-ye-ren-dong 細葉忍冬 686
Xi-ye-rong 細葉榕 236
Xi-ye-xiang-gui 細葉香桂 400
Xi-yin-xuan-gou-zi 喜陰懸鈎子 458
Xi-zang-cao-mei 西藏草莓 434
Xi-zang-jian-zhu 西藏箭竹 289
Xi-zang-mu-zhu 西藏牡竹 288
Xi-zang-tao 西藏桃 445
Xia-cao 霞草 385
Xia-ku-cao 夏枯草 169, 237, 248, 653
Xia-ye-he-bao-guo 狹葉荷包果 623
Xia-ye-xun-ma 狹葉蕁麻 364
Xia-zhi-jun 夏至蕈 267
Xian-bai-cai 仙白菜 720
Xian-cai 仙菜 260
Xian-cai 莧菜 50, 380, 381
Xian-cao 仙草 260
Xian-cong 線蔥 313
Xian-da-yun 鹹大芸 673
Xian-di-jian-peng 鹽地鹼蓬 379
Xian-luo-mi-ju 暹羅蜜橘 498
Xian-mao 仙茅 167
Xian-mao-xuan-gou-zi 腺毛懸鈎子 462
Xian-qie 線茄 609
Xian-ren-gu 仙人穀 381
Xian-ren-zhang 仙人掌 567
Xian-tao 仙桃 567
Xian-ai-yang 小矮秧 659
Xiang-cai 香菜 594
Xiang-cao 香草 153, 485
Xiang-chun 香椿 508
Xiang-cong 香蔥 311
Xiang-dao-mi 香稻米 292
Xiang-dou 香豆 485
Xiang-dou-kou 香豆蔻 327
Xiang-duan-dou 香端豆 665
Xiang-er-cao 象耳草 718
Xiang-fei 香榧 276
Xiang-gang-si-zhao-hua 香港四照花 607
Xiang-gu 香菇 136, 268
Xiang-gua 香瓜 697, 698
Xiang-guo-lan 香果蘭 332
Xiang-jiao 香蕉 324
Xiang-jiao-shu 香膠樹 399
Xiang-jiao-zi 香椒子 504
Xiang-li-meng 香黎檬 493
Xiang-liao 香蓼 373
Xiang-ling-cao 香鈴草 541
Xiang-mao-cao 香茅草 287

Xiang-pian 香片 555
Xiang-ping-po 香蘋婆 546
Xiang-ri-kui 向日葵 734
Xiang-ru 香茹 247
Xiang-ru 香薷 645
Xiang-si 香絲 651
Xiang-si-cai 香絲菜 597, 651
Xiang-si-zi 相思子 235
Xiang-su 香蘇 652
Xiang-tao-mu 香桃木 577
Xiang-yu 香芋 306
Xiang-yuan 香圓 494
Xiao-bai-cai 小白菜 413
Xiao-bian-dou 小扁豆 476
Xiao-chao-cai 小巢菜 487
Xiao-chong-er-wo-dan 小蟲兒臥單 510
Xiao-cong 小蔥 313
Xiao-fei-jin-cao 小肺筋草 311
Xiao-gen-suan 小根蒜 313
Xiao-guo-hong-mei-tai-zi 小果紅莓笞子 612
Xiao-guo-ye-pu-tao 小果野葡萄 533
Xiao-hong-kao 小紅栲 341
Xiao-hong-li 小紅栗 344
Xiao-hu-lu 小葫蘆 705
Xiao-hua-tang-jie 小花糖芥 416
Xiao-hua-wu-ya-guo 小花五椏果 549
Xiao-huang-gua 小黃瓜 700
Xiao-huang-guo-gou-qi 小黃果枸杞 663
Xiao-hui-xiang 小茴香 589
Xiao-ji 小薊 726
Xiao-jin 肖槿 543
Xiao-ku-mai 小苦蕒 737
Xiao-kun-bu 小昆布 258
Xiao-li-ka-fei 小粒咖啡 678
Xiao-ma-huang 小麻黃 279
Xiao-mai 小麥 297
Xiao-mi 小米 296
Xiao-mi-la 小米辣 660
Xiao-suan 小蒜 313
Xiao-xue-teng 小血籐 397
Xiao-yang-tao 小羊桃 550
Xiao-ye-cha 小葉茶 557
Xiao-ye-cha-biao 小葉茶藨 427
Xiao-ye-qin 小葉芹 588, 601
Xiao-ye-sang 小葉桑 361
Xie 薤 312
Xie-bai 薤白 313
Xie-cong 薤蔥 311
Xie-tou 薤頭 312
Xie-zi-cao 蝎子草 364
Xin-hui-qi-cheng 新會臍橙 500
Xin-hui-tian-cheng 新會甜橙 499

Xin-li-mei 心裡美 420
Xin-ye-jia-mi 心葉莢蒾 688
Xing 杏 441
Xing-cai 莕菜 628
Xing-li 杏李 447
Xing-ping-guo 星蘋果 619
Xing-ren 杏仁 169, 227, 228, 441
Xing-shan-wu-wei-zi 興山五味子 397
Xing-si-cai 荇絲菜 628
Xing-teng 腥藤 369
Xing-su-cai 星宿菜 618
Xing-zi-shu 杏子樹 443
Xiu-li-mei 秀麗莓 455
Xiu-mao-mei 銹毛莓 460
Xiu-mao-suo-zi-guo 銹毛梭子果 620
Xiu-mao-wu-ye-shen 銹毛五葉參 588
Xiu-tie-bang 銹鐵棒 584
Xu-cao 蓿草 477
Xuan-cao 萱草 318
Xuan-gou-zi 懸鉤子 459
Xuan-shen 玄參 248
Xuan-zao 萱藻 259
Xue-bu-tao 血布桃 446
Xue-cha 雪茶 265
Xue-dou 雪豆 481
Xue-feng-teng 雪風籐 265
Xue-gan 雪柑 499
Xue-hua-li 雪花梨 449
Xue-jiao-shu 血膠樹 620
Xue-li 雪梨 450
Xue-li-hong 雪裡紅 28, 55, 408, 410
Xue-li-qing 雪裡青 410
Xue-wei-cai 雪維菜 591
Xue-zhu 血櫧 349
Xun-ma 蕁麻 365

Ya-cong 鴉蔥 743
Ya-er-cai 鴨兒菜 594
Ya-gong-teng 牙公藤 529
Ya-gu 鴉穀 381
Ya-guang-li 鴨廣梨 452
Ya-hu-lu 亞壺盧 705
Ya-jiao 芽蕉 325
Ya-jiao 崖椒 504
Ya-jiao-ai 鴨腳艾 719
Ya-jiao-ban 鴨腳板 524
Ya-jiao-cai 鴨腳菜 719
Ya-jiao-mu 鴨腳木 248
Ya-jiao-mu-pi 鴨腳木皮 237
Ya-li 鴨梨 449
Ya-ma-cai 鴨麻菜 744
Ya-ma-you 亞麻油 490

Ya-mu-gua 崖木瓜 526
Ya-she-cao 鴨舌草 310
Ya-shi-liu 崖石榴 360
Ya-zhang-qin 鴨掌芹 604, 605
Ya-zhua-bai 鴨爪稗 289
Yan-cai 醃菜 29
Yan-gao-lan 岩高蘭 513
Yan-gu 岩菇 265
Yan-hao 岩蒿 718
Yan-lai-hong 雁來紅 382
Yan-mai 燕麥 284
Yan-rong 岩茸 265
Yan-zhi-cai 燕脂菜 384
Yan-zhi-guo 胭脂果 574
Yang-bai-cai 洋白菜 410
Yang-cai 洋菜 261
Yang-cong 洋蔥 311
Yang-du-jun 羊肚菌 263
Yang-duan 漾短 506
Yang-gan-lan 洋橄欖 627
Yang-he 陽荷 255, 329
Yang-huai 洋槐 483
Yang-jiang 洋姜 735
Yang-jiao-fei 羊角榧 277
Yang-jiao-gua 羊角瓜 697
Yang-jiao-jiao 羊角椒 659
Yang-jiao-zi 羊角子 392
Yang-li 洋李 443, 450
Yang-luo-han-guo 洋羅漢果 558
Yang-mei 楊梅 335, 336
Yang-nai-zi 羊奶子 569
Yang-pu-tao 洋蒲桃 581
Yang-rui 漾蕊 506
Yang-shi-shu 羊屎樹 514
Yang-shi-zi 洋柿子 664
Yang-tao 羊桃 550
Yang-tao 洋桃 490
Yang-ti-cao 羊蹄草 729
Yang-ti-jia 羊蹄甲 467
Yang-tou-cai 洋頭菜 409
Yang-xi-cai 羊栖菜 260
Yang-xiang-si 洋香絲 596
Yang-ying-tao 洋櫻桃 442
Yang-yu 洋芋 126, 140, 670
Yang-yuan-sui 洋芫荽 596, 603
Yang-zhua-teng 羊爪藤 394
Yang-zi-jing 洋紫荊 467
Yao-guo 腰果 514
Yao-sang 藥桑 362
Ye-bai-li-xiang 野百里香 657
Ye-bian-dou 野扁豆 470
Ye-cao-mei 野草莓 435

Ye-cha 野茶 629
Ye-er-ba 葉兒粑 731
Ye-fen-tuan-er 野粉團兒 720
Ye-ge 野葛 482
Ye-hai-qie 野海茄 666
Ye-hai-tang 野海棠 436
Ye-he-tao 野核桃 337
Ye-hua-cai 椰花菜 410
Ye-hua-sheng 野花生 471
Ye-huang-dou 野黃豆 473
Ye-ji-guan-hua 野雞冠花 382
Ye-jiang-dou 野豇豆 489
Ye-jiao-teng 夜焦藤 372
Ye-jie-cai 野芥菜 752
Ye-ju 野苣 690
Ye-ku-ju 野苦苣 725
Ye-ku-ma 野苦麻 736
Ye-ku-mai 野苦蕒 746
Ye-kui-cai 野葵菜 542
Ye-la-shu 野辣樹 660
Ye-la-zi 野辣子 660
Ye-lai-xiang 夜來香 635
Ye-li-zhi 野荔枝 608
Ye-ling 野菱 584
Ye-mao-suan 野貓酸 616
Ye-mu-dan 野牡丹 236
Ye-mu-gua 野木瓜 393
Ye-ren-gua 野人瓜 392
Ye-shan-cong 野山蔥 311
Ye-shan-yao 野山藥 323
Ye-shan-zha 野山楂 430
Ye-suan 野蒜 313
Ye-tao 野桃 444
Ye-wan-dou 野豌豆 476
Ye-xi-gua-miao 野西瓜苗 541
Ye-xia-hong 葉下紅 729
Ye-xian-cai 野莧菜 382
Ye-xiang-hua 夜香花 635
Ye-ya-chun 野鴉椿 520
Ye-yan-mai 野燕麥 284
Ye-yu-tou 野芋頭 307
Ye-zhi-ma 野芝麻 527
Ye-zi 椰子 302
Ye-zi-su 野紫蘇 651
Yi-chang-hu-tui-zi 宜昌胡頹子 570
Yi-chang-jia-mi 宜昌莢蒾 689
Yi-dian-hong 一點紅 729
Yi-la-ke-mi-zao 伊拉克蜜棗 303
Yi-mian-qing 一面青 717
Yi-mu-cao 益母草 647
Yi-mu-zi 宜母子 492, 493
Yi-ren 薏仁 108, 109, 167, 247

Yi-wu 宜吾 571
Yi-yi 薏苡 287
Yi-zhi-ren 益智仁 325
Yin-chai-hu 銀柴胡 385
Yin-du-kao 印度栲 344
Yin-du-nan-gua 印度南瓜 702
Yin-er 銀耳 74, 169, 178, 179, 180, 223, 269
Yin-guo-hu-tui-zi 銀果胡頹子 570
Yin-lian-hua 銀蓮花 628
Yin-xian-cao 銀線草 334
Yin-xing 銀杏 100, 275
Yin-ye-du-jin-shan 銀葉杜莖山 617
Ying-pi-si-gua 硬皮絲皮 707
Ying-su 罌粟 402
Ying-tao 櫻桃 447
Ying-tou-shi-li 硬頭石栗 348
Ying-yang-hua 迎陽花 735
Ying-zhua-bai 鷹爪稗 289
Ying-zhua-feng 鷹爪楓 392
Ying-zhui-dou 鷹嘴豆 471
Yong-mao-zhen 絨毛榛 339
You-cai 油菜 414
You-cai-tai 油菜苔 413
You-cha 油茶 554
You-fei 油榧 277
You-gan-zi 油甘子 512
You-hu-lu 油葫蘆 368, 705, 706
You-mai 油麥 284
You-sha-cao 油沙草 299
You-shi 油柿 624
You-tao 油桃 446
You-zha-guo 油渣果 703
You-zi 油子 706
You-zi 柚子 493
Yu 芋 140, 305
Yu-du-nan 魚肚腩 285
Yu-gan-zi 餘甘子 512
Yu-huang-li 玉皇李 447
Yu-jin 郁金 327
Yu-lan 芋蘭 331
Yu-lan-hua-pian 玉蘭花片 75, 76, 396
Yu-lan-pian 玉蘭片 76
Yu-li 郁李 444
Yu-li-ren 郁李仁 444
Yu-lin-mu 魚鱗木 579
Yu-mai 玉麥 291
Yu-mu 魚木 423
Yu-mu 芋母 140
Yu-nai 芋奶，芋艿 305
Yu-qian-zi 榆錢子 354
Yu-rou 萸肉 609
Yu-shan-xuan-gou-zi 玉山懸鈎子 455

Yu-shan-zi-jin-niu 玉山紫金牛 613
Yu-shu-pi 榆樹皮 354
Yu-shu-shu 玉蜀黍 298
Yu-tou 芋頭 140, 141, 305
Yu-tou-gan 芋頭桿 235, 305
Yu-tou-hua 芋頭花 305
Yu-xiang 魚香 151, 152, 597, 645
Yu-xing-cao 魚腥草 332
Yu-ye-jin-hua 玉葉金花 684
Yu-zan 玉簪 318
Yu-zan-hua 玉簪花 318
Yu-zhu 玉竹 169, 181, 191, 197, 319
Yu-zi 芋子 140, 305
Yu-zi-lan 魚子蘭 507
Yuan-bao-shu 元寶樹 522
Yuan-fei 圓榧 277
Yuan-guo-gou-qi 圓果枸杞 662
Yuan-guo-qiu-hai-tang 圓果秋海棠 566
Yuan-jing-qie 遠景茄 668
Yuan-ling-ye-shan-ma-huang 圓菱葉山螞蝗
    473
Yuan-pi 軟皮 706
Yuan-sui 芫荽 594
Yuan-tou-huang-ye-gou-qi 圓頭黃葉枸杞 662
Yuan-wa-mu-li 袁窪木梨 450
Yuan-xi-guo 猿喜果 621
Yuan-ye-rong 圓葉榕 357
Yuan-ye-zhi-zi 圓葉梔子 681
Yuan-you 原油 34
Yue-gua 越瓜 696, 697
Yue-guang-hua 月光花 635
Yue-jiao 越椒 503
Yue-ju 越橘 612
Yue-lan 越欖 506
Yue-si-gua 粵絲瓜 706
Yue-yu 躍榆 354
Yun-dou 雲豆 480
Yun-er 雲耳 266
Yun-nan-da-bai-he 雲南大百合 316
Yun-nan-gou-er-teng 雲南鈎兒藤 529
Yun-nan-ou-li 雲南歐李 442
Yun-nan-shan-zhu-zi 雲南山竹子 558
Yun-nan-su-tie 雲南蘇鐵 275
Yun-nan-ye-mu-gua 雲南野木瓜 394
Yun-nan-zi-shu 雲南紫樹 575
Yun-xiang 芸香 503

Zai-yang-pao 栽秧泡 456
Zang-ci-zhen 藏刺榛 340
Zang-hui-xiang 藏茴香 593
Zang-qing-ke 藏青稞 291
Zao 棗 87, 169, 531

Chinese Index

Zao-bai-shu 早白薯 321
Zao-hong-qie 早紅茄 667
Zao-ji-guan 早雞冠 404
Zao-jia-ren 皂莢仁 473
Zao-ju 早橘 497
Zao-qing-qie 早青茄 668
Zao-yu 早芋 140
Ze-qi-ma 澤漆麻 629
Ze-qin 澤芹 606
Ze-xie 澤瀉 246
Zha-cai 榨菜 28, 29, 409, 505
Zha-gu-ding 榨古丁 356
Zha-mu 扎 (柞) 木 565
Zhai-xiang-cai 塞香菜 650
Zhai-yao-pao 窄腰泡 582
Zhai-ye-tong-hao 窄葉茼蒿 723
Zhan-mao-ye-mu-dan 展毛野牡丹 582
Zhang-ju 丈菊 735
Zhang-li 杖藜 377
Zhang-li 樟梨 402
Zhang-liu-tou 樟柳頭 327
Zhang-ye-mu 掌葉木 524
Zhe-ma 蜇(螫)麻 365
Zhe-ma-zi 螫麻子 364
Zhe-shu 柘樹 356
Zhen 榛 338, 339
Zhen-tou-gua 枕頭瓜 701
Zhen-zhu-cai 珍珠菜 618
Zhen-zhu-cao 珍珠草 331
Zhen-zhu-hua 珍珠花 521
Zhen-zhu-li 珍珠栗 340
Zhen-zhu-mei 珍珠梅 464
Zhen-zhu-mi 珍珠米 511
Zhen-zi 榛子 339
Zheng-xiang-si 正香絲 651
Zhi-gan-cao 炙甘草 168
Zhi-ju 枳棋 529
Zhi-ma 芝麻 674
Zhi-ma 枳麻 674
Zhi-ma 脂麻 490
Zhi-ma-cai 芝麻菜 416
Zhi-ma-jiang 芝麻醬 674
Zhi-que 枳殼 248
Zhi-tian-jiao 指天椒 660
Zhi-xue-cao 止血草 599
Zhi-zhu-xiang 蜘蛛香 606
Zhi-zi 枝子 680
Zhi-zi-hua 梔子花 681
Zhong-chi 中遲 405
Zhong-hua 中花 405
Zhong-hua-ti-gai-jue 中華蹄蓋蕨 273
Zhong-li-ka-fei 中粒咖啡 679

Zhong-ma-huang 中麻黃 278
Zhong-zi-ye 粽子葉 297
Zhou-ye-wo-ju 皺葉萵苣 741
Zhu-bai 竹柏 276
Zhu-du-mu 豬肚木 677
Zhu-er-yan 豬耳眼 383
Zhu-ge-cai 諸葛菜 417
Zhu-gen-jia-wan-shou-zhu 竹根假萬壽竹 317
Zhu-gen-qi 竹根七 317
Zhu-gu-nian 豬姑稔 582
Zhu-guang-xia-qing 珠光香青 717
Zhu-jie-cao 竹節草 235
Zhu-jie-shu 竹節樹 574
Zhu-ju 朱橘 496
Zhu-lan 珠蘭 334
Zhu-lan-cha 珠蘭茶 334
Zhu-li 豬栗 346
Zhu-ma 苧麻 363
Zhu-pi-gu 豬屁股 332
Zhu-sun 竹蓀 267
Zhu-ye-cao 竹葉草 308
Zhu-ye-hua-jiao 竹葉花椒 503
Zhu-ye-jiao 竹葉椒 503
Zhu-ye-rong 竹葉榕 361
Zhu-you-guo 豬油果 510
Zhu-yu 竹芋 330
Zhu-zhe 竹蔗 296
Zhu-zi-li 豬子苓 473
Zhuan-lian 轉蓮 734
Zhuang-jing-ye 壯荊葉 238
Zhui-feng-qi 追風七 435
Zhui-li 錐栗 340, 346
Zhun-ren-gua 隼人瓜 709
Zi 梓 673
Zi-bei-tian-kui 紫背天葵 566, 732
Zi-cai 紫菜 261
Zi-dan-shu 紫彈樹 352
Zi-hua-bian-dou-cai 紫花變豆菜 605
Zi-hua-di-ding 紫花地丁 563
Zi-hua-qin 紫花芹 604, 605
Zi-jiang 子薑，紫薑 329
Zi-lan-shu 子欖樹 631
Zi-li-yu 孖鯉魚 693
Zi-mu-xu 紫苜蓿 477
Zi-ran-qin 孜然芹 595
Zi-rong-teng 紫絨藤 731
Zi-su 紫蘇 651
Zi-teng-hua 紫藤花 80, 489
Zi-wu-lian 子午蓮 388
Zi-ya-qing 紫芽青 420
Zi-zhu 紫珠 643
Zong-lü 棕櫚 304

Zuan-gu-feng 鑽骨風 397
Zuan-yan-jin 礦岩筋 397
Zuo-li 柞櫟 350

A Pen-of-King Wen 369
Abortifacient Herb 582
Abrus 209
Achyranthes 235
Aconite Tuber 166
Acronychia 235
Actinidia Wine 550
Actor Wen's Pomelo 494
Adenophora 713
Adsuki Bean 106, 107, 109, 117, 466
African Baobab 543
African Millet 289
After-taste Sweet 512
Agar-agar 261
Agastache 246
Air Potato 322
Aleppo Avens 435
Alfalfa 477
Alkaline Weed 378
Alligator Pear 402
Almond 45, 125, 224, 228, 441
Alone Traveling a Thousand Miles 438
Alpine Onion 315
Alternanthera 235, 379
Amaranth 50, 51, 57, 380
Amaranth Seed 52
Amelanchier 428
American Cotton 539
American Fangfeng 602
American Ginseng 89, 163, 168, 182, 183, 184,
    185, 221, 223, 226, 586, 587
American Orange 500
American Walnut 336
Amomum 150, 152
Amorphophallus 304
Anaphalis 717
Ancient Button Vegetable 670
Ancient Onion 315
Angelica Shoot 589
Angel's Trumpet 658
Announcing Fountain Herb 670

Aponogeton 282
Apple 81, 98, 132, 437
Apple Fig 359
Apple-of-Peru 664
Apricot 31, 81, 100, 227, 441, 443
Apricot of the Northeast 441
Apricot Plum 447
Apricot Seed 45, 169, 181, 214, 215, 221, 227, 228,
    244, 248, 441
Apricot Tree 443
Apron-ribbon Vegetable 260
Aquatic Morning-glory 52, 57, 636
Arabian Coffee 678
Arabian Obscure Nut 518, 519
Aralia Tree 585
Areca-nut Palm 300
Aroid 46
Aromatic Bell Herb 541
Aromatic Grass 287
Aromatic Limeng 493
Aromatic Madder 247, 645
Aromatic Pea 485
Aromatic Perilla 652
Aromatic Pingpo 546
Aromatic Sliver 651
Aromatic Sliver Vegetable 651
Aromatic Torreya Nut 276
Aromatic Vegetable 594
Aromatic Zanthoxylum 504
Arrowhead Corm 138
Arrow-leaved Monochoria 310
Arrowroot 330
Arrow-shaft White 411
Artemisia Shoot 719
Artemisia Vegetable 719
Asian Cinquefoil 440
Asian Cow-wheat 671
Asian Ginseng 183, 184
Asian Horseradish 416
Asian Pheasant Tree 516
Asiatic Ardisia 613

Asiatic Frog's-bit 283
Asiatic Sweetleaf 625
Asparagus 316, 371
Asparagus Angelica 589
Asparagus Bean 488
Asparagus Lettuce 20, 739, 740
Asparagus Root Tuber 226
Ass Artemisia 719
Assam Tea 556
Astragalus 166, 171, 172, 181, 187, 190, 223, 466, 714
Atractylodes Root 166, 171, 172
August Melon 391
Auricularia 68, 221, 266
Australian Nut 366
Autumn Flower Tree 463
Autumn Mallow 538
Autumn Pear 451
Autumn Squash 702
Avocado 81, 402
Awl Tree 682

Baby Canarium 631
Baby Tea 520
Baby Vegetable 716
Bago 280
Baical Skullcap 248, 653, 654
Bald-headed Maid Shoot 736
Ball Fern 272
Ball Onion 311
Ball-bud Cabbage 411
Ball-fruit Wild Mustard 421
Ball-fruited Violet 562
Balloon Flower 169, 248, 761
Balloon-vine 523
Ball-stem Cabbage 406
Balsam Pear 708
Bamboo 126, 127, 128, 129, 130
Bamboo Flower 267
Bamboo Juniper 276
Bamboo Leaf Vegetable 308, 309
Bamboo Mustard 408
Bamboo Shoot 57, 59, 70, 73, 75, 91, 127, 128, 129, 130, 137, 155, 159, 270, 284
Bamboo Sugar Cane 296
Bamboo Taro 330
Bamboo-joint Tree 574
Bamboo-leaved Fig 361
Bamboo-leaved Sichuan Peppercorn 503
Bamboo-leaved Zanthoxylum 503
Bamboo-root False Disporum 317
Bamboo-root Hematinic 317

Bamboo-shoot Squash 702
Banana 324
Banana Mandarin Orange 498
Banana Taro 329
Bangapur Lime 492
Barbados Ground Cherry 665
Barley 37, 81, 85, 168, 245, 290, 291, 474
Barnyard Grass 288
Barnyard Millet 288
Basella Potato 384
Basil 150, 650
Batoko-plum 564
Beach Hibiscus 541
Beach Wild Pea 476
Bean Curd 36, 55, 56, 64, 69, 71, 82, 83, 120, 121, 122, 123, 135, 147, 153, 204, 262, 474, 485, 515, 681
Bean Curd Bamboo 69, 76, 77, 104, 137
Bean Curd Cheese 33, 36, 262, 263, 474
Bean Curd Residue 147
Bean Horn 488
Bean Sprouts 115
Beancurd Fruit 515
Beautiful Bamboo 294
Beautiful Berry 455
Beautiful Currant 427
Beautiful Rose 452
Bee Bushel Vegetable 741
Beer 37
Beer Flower 362
Beet 376
Beijing Honey Date 32
Beijing Mountain Ash 463
Beijing Prickly Cucumber 699
Bell Pepper 659
Belleric 575
Belleric Myrobalan 575
Belvedere 378
Benefit Mother Herb 647
Bending Head Greens 274
Bengal Canarium 505
Bengal Day Flower 308
Betel Green 519
Betel Leaf 300, 333
Betel Nut 227, 235, 246, 300
Betel Nut Flower 300
Betel Nut Taro 306
Betel Palm 300
Betel-leaf Green 333
Betel-nut 166
Betel-pepper 333
Big Banana 325

Big Fire-herb 717
Big Grain 291
Big Green 530, 700
Big Head Shoot 285
Big Head Vegetable 28, 408
Big Jujube 88
Big Leaf 424
Big Lotus Flower 307
Big Mother Red Bud Taro 306
Big Mother Taro 306
Big Muslim Spice 396
Big Mustard 408
Big Nothing Tree 360
Big Polygonum 389
Big Pumpkin 702
Big Scallion 25, 312
Big Shoot 285
Big Shoot Bamboo 285
Big Shoulder Bitter Melon 709
Big Sweet Orange 499
Big Yam 321
Big-head Cabbage 408
Big-leaved Black 288
Big-leaved Chenopodium 377
Big-leaved Conic Chestnut 343, 346
Big-leaved Oak 343
Big-leaved White-haired Gentleman 717
Big-spine Gooseberry 425
Big-stem Mustard 408
Bilberry 612
Bilimbi 490
Billion-dollar Grass 288
Birch-leaved Grape 534
Bird Cladonia 263
Bird Mei Bush 530
Bird Vetch 486
Bird-nest Fern 274
Birth-leaved Pear 448
Bit Fat Cattle 733
Bitter Bamboo 294
Bitter Buckwheat 371
Bitter Chestnut 346
Bitter Cress 415
Bitter Cucumber 236, 708
Bitter Fellow Tea 519
Bitter Lagenaria 704
Bitter Lettuce 724
Bitter Lettuce Vegetable 747
Bitter Melon 18, 81, 82, 83, 84, 708
Bitter Mother Herb 746
Bitter Mustard 408
Bitter Orange 492

Bitter Pea 485
Bitter Peach 448
Bitter Potherb 737
Bitter Sow-thistle 746
Bitter Tea 558
Bitter Vegetable 747
Bitter Wheat Herb 746
Bitter Wheat-field Herb 746
Bitterweed 742
Black Bean 474
Black Bean Curd 304
Black Canarium 506
Black Chinese Canarium 150, 506
Black Crowberry 513
Black Elm 353
Black Eye Pea 488
Black Grape 536
Black Hair Fern 271, 272
Black Hair Ho 372, 373
Black Head Fruit 615
Black Jujube 88, 89, 169, 531, 532
Black Lily 317
Black Ling 583
Black Lintel Castanopsis 345
Black Locust 45, 483
Black Luxuriant Growth-on-the Mountain 267
Black Mangrove 642
Black Mulberry 362
Black Mushroom 69, 70, 75, 90, 95, 109, 110, 136, 158, 174, 266, 268
Black Nightshade 669
Black Nut 343
Black Pea-berry 669
Black Pear 451
Black Pepper 69, 71, 115, 116, 148, 149, 150, 152, 333, 334
Black Rice Tree 611
Black Ridge Oak 351
Black Rind Green 693
Black Seed Herb 390
Black Separated Block 529
Black Skin Luffa 707
Black Soybean 35, 124, 168, 181, 193, 227, 262
Black Spike Arrow Bamboo 289
Black Split-bean Vegetable 526
Black Vegetable 259
Black-fruited Barberry 394
Black-fruited Bilberry 612
Bladder Campion 386
Blood Gum Tree 620
Blood Lithocarpus 349
Blue Flower Herb 716

Blue Flower Shen 716
Blue Globae Thistle 728
Blue-flowered Orobanche 674
Bone-healing Wood 687
Boring-rock Muscle 397
Boruljian 458
Boston Fern 272
Boston Ivy 532
Bottle Gourd 235, 704
Bower Actinidia 549
Bower Vine 549
Box Squash 701
Bracken 272
Brahman's Vital Root 751
Brake 272
Brazil Chestnut 574
Brazil Nut 574
Breadfruit 355
Break Large Bowl 593
Brinjal 666, 667
Bristly Raspberry 460
Broad Bean 37, 486, 645
Broad Bean Sauce 33, 36, 37, 486
Broad-beaked Mustard 410
Broadleaf Podocarpus 276
Broad-leaved Actinidia 552
Broad-leaved Cat-tail 280
Broad-leaved Holly 247
Broad-leaved Leek 313
Broad-leaved Nettle 365
Broad-leaved Willow Celery 595
Broccoli 65, 130, 410
Broken Bone Vine 392
Broken Rice Capsella 415
Broken Rice Lan 507
Broom Bamboo 295
Broom Mushroom 267
Broom Vegetable 378
Broomcorn 293
Broomrape 167, 181, 202, 203, 205
Brown Rice 109
Brown Rice Gin 173
Brown Sugar Tea 611
Brussels Sprouts 411
Buchnera 235
Bucket Mandarin Orange 498
Buckwheat 370
Buckwheat Angelica 370
Buckwheat Hematinic 370
Buddha Hand 495
Buddha Disciples' Shen 465
Buddha's Belly 306

Buddha's Disciple Fruit 9, 152, 218, 710
Buddha's Hand Gourd 709
Buddha's Palm Gourd 709
Buffalo-gourd 700
Bullet Fruit 615
Bunching Onion 312
Bund Mustard 422
Buni 509
Bur Artichoke 727
Bur Bush 238
Bur Clover 477
Bur Clover Tops 477
Burdock 718
Burdock Root 718
Burning Hemp 364
Burning Nettle 365
Burnut 491
Burnut-bloodwort 462
Bush Vegetable 512
Bushy Melastoma 236
Butter Bean 480
Butter Tree 619
Butterbur 741
Buttercup Squash 702
Butterfly Fruit 510
Butterfly Ginger 328
Button Herb 670

Caa-ehe 748
Cabbage 47, 64, 65, 69, 135, 403, 406, 410
Cacao 547
Cairo Morning-glory 639
Cajan 469
Calabash Gourd 704
Calamondin 495
Caltrop 491
Caltrop Castanopsis 347
Camel's Foot 467
Camphor-willow Shoot 327
Canadian Ginseng 183
Canarium 28, 31, 241, 242, 245, 246, 505, 514, 536
Canarium Kernel 506
Canarium Kernel Tree 576
Candel 574
Candied Ginger 32
Candied Hawthorn 84, 431
Candleberry Tree 509
Candlenut 509
Cane Bamboo 290
Cane Goosefoot 377
Cang Zhu 246

Canna Taro 329
Canton Abrus 166
Canton Ginger 329
Cantonese Bugbane 744
Cantonese Gourd 697
Cantonese Luffa 706
Cantonese Tangerine 498
Cantonese Tea 557
Cape Gooseberry 665
Caper 423
Capillary Wormwood 246
Caraway 593
Cardamon 325
Carombola 490
Carpenter-weed 653
Carp's Egg Mass 618
Carrot 44, 47, 56, 66, 70, 74, 77, 91, 102, 124, 126,
    133, 185, 221, 586, 596
Cart Trail Herb 677
Cashew 514
Cashew Nut 506, 514
Cassava 511
Cassia 148, 149, 150, 151, 152, 399, 400
Cassia Bark 150, 158, 167, 171, 399
Cassia Bark Tree 401
Cassia Branches 399
Cassia Fragrant Willow 569
Cassia Heart 399
Cassia Seed 471
Castanopsis 344
Castanopsis Nut 342
Castor Bean 512
Cat Feces 392
Cat Ginseng 552
Cat Thistle 725
Catalpa 672, 673
Catalpa Crab-apple 437
Catchfly 386
Catjang 488
Cattle Nose-ring 528
Cattle Vine 527
Cauliflower 65, 410
Cayenne Pepper 40, 150, 152, 153, 154, 658
Cedrela 508
Celery 66, 74, 123, 133, 185, 204, 592
Celery Cabbage 29, 55, 64, 72, 73, 91, 115, 116,
    122, 133, 135, 137, 143, 144, 203, 204, 413
Celery Wormwood 235, 246
Celestial Bladder Herb 665
Celestial Eggplant 635
Celestial Fairy Fruit 358
Celestial Malva 331

Celestial Onion 312
Celestial Peach 518
Celestial Peach Tree 517
Centella 593
Central Asian Barberry 394
Central China Anise 603
Central China Bitter Cress 415
Central China Blackberry 457
Central China Castanopsis 342
Central China Honeysuckle 686
Central China Wild Walnut 337
Ceylon Gooseberry 563
Ceylon Olive 537
Chaenomeles 167, 172, 173
Chaff Palm 304
Chard 376
Chase-out-wind Hematinic 435
Chatting Mei 31, 445
Chayote 709
Chea Butter 619
Cheeses 542
Cherimoya 398
Cherokee Rose 169, 237, 453
Cherry 102, 103, 441, 447
Chervil 591
Chestnut 103, 118, 161, 224, 340, 355, 367
Chewing Gum 621, 622
Chick Pea 471
Chicken Bone 625
Chicken Bonewort 209, 210
Chicken Canarium 519
Chicken Droppings Creeper 684
Chicken Foot Shen 330
Chicken Feet Tonic 331
Chicken Feet Vegetable 260
Chicken Head Seed 387
Chicken Head Shen 320
Chicken Heart Fruit 623
Chicken Intestine 738
Chicken Intestine Feng 683, 684
Chicken Leg Vegetable 561
Chicken Leg Violet 561, 562
Chicken Mulberry 361
Chicken Mushroom 267
Chicken Skin Fruit 500
Chicken Vegetable 733
Chicken-blood Vine 478
Chicken-eye Vine 683
Chicken-eye Weed 475
Chicken-eyes Green 613
Chicken-feet Celery 595
Chicken-tongue Spice 579

Chickweed 387
Chicle 621, 622
Chico 621
Chicory 725
Children-of-Yu 140
Chili 658
Chili Pepper 20, 28, 45, 49, 659
China Grass 363
China Pea 488
China Root 108, 169, 171, 172, 173, 174, 248, 268, 319
Chinese Allspice 247
Chinese Amelanchier 428
Chinese Angelica 150, 151, 157, 166, 171, 172, 173, 181, 186, 187, 188, 196, 197, 200, 205, 217, 218, 223, 230, 603
Chinese Angelica Tree 585
Chinese Apple 436, 437
Chinese Artichoke 656
Chinese Athyrium 273
Chinese Banyan Root 236
Chinese Bindweed 636
Chinese Black Currant 427
Chinese Black Elder 687
Chinese Blackberry 462
Chinese Broccoli 403
Chinese Bugbane 246
Chinese Buckthorn 530
Chinese Cabbage 413
Chinese Canarium 241, 505
Chinese Castanopsis 342
Chinese Cherry 446
Chinese Chestnut 340
Chinese Clematis 247
Chinese Cornbind 169, 181, 191
Chinese Cornel 608
Chinese Cudweed 731
Chinese Currant 426
Chinese Date 269, 319, 482, 531
Chinese Ditch-stonecrop 425
Chinese Dogweed 608
Chinese Elm 354
Chinese Ephedra 279
Chinese Eriosma 473
Chinese Field Garlic 313
Chinese Fire-thorn 448
Chinese Foxglove 171, 173, 181, 195, 200, 671, 672
Chinese Ginseng 168, 182, 183, 184, 586
Chinese Globe Thistle 728
Chinese Golden Thread 230, 247
Chinese Gooseberry 550

Chinese Groundsel 733
Chinese Hackberry 353
Chinese Hawthorn 84, 85, 431
Chinese Hazelnut 338, 339
Chinese Herbal Morinda 683
Chinese Hickory 336
Chinese Hobble-bush 688
Chinese Honey-locust 473
Chinese Ixora 682
Chinese Jacinth 320
Chinese Jelly Herb 650
Chinese Jujube 87, 118, 471, 532, 591
Chinese Kale 20, 403
Chinese Lady-fern 273
Chinese Lantern 664
Chinese Laurel 509
Chinese Leek 20, 53, 54, 115, 116, 315
Chinese Licorice 168, 247, 475
Chinese Lovage 599, 600
Chinese Manna 264
Chinese Matrimony-vine 663
Chinese May-apple 395
Chinese Morinda 683
Chinese Olive 32, 131, 241
Chinese Osbeckia 237
Chinese Palm 304
Chinese Parasol Tree 545
Chinese Parsley 594
Chinese Pear 451
Chinese Pillow Squash 701
Chinese Pistachio 518
Chinese Pistacia Shoot 518
Chinese Plum 447
Chinese Privet 168
Chinese Pyrularia 368
Chinese Quince 246, 429
Chinese Radish 28, 30, 122, 126, 131, 132, 133, 148, 151, 417
Chinese Radish Pickle 30
Chinese Red Currant 426
Chinese Red Rice 37
Chinese Rhubarb 373
Chinese Sanicle 604
Chinese Sauropus 513
Chinese Sawwort 744
Chinese Scholar Tree 484
Chinese Shallot 312
Chinese Silverweed 439
Chinese Stargrass 310
Chinese Sweet Cane 296
Chinese Sweet Melon 697
Chinese Tea 246

Chinese Teaberry 611
Chinese Thorowax 246
Chinese Violet 562
Chinese Wedelia 238
Chinese White Pine 277
Chinese Wild Rice 298
Chinese Windmill Palm 304
Chinese Yam 11, 107, 108, 145, 167, 172, 173, 175,
	181, 190, 209, 247, 319, 323, 324, 714
Chinese Yellow Rose 454
Chloranthus Tea 334
Chocolate 547
Chocolate Vine 391
Christophine 709
Chrysanthemum 26, 71, 72, 73, 247, 723
Chrysanthemum Brain 724
Chrysanthemum Lettuce 724
Chrysanthemum Shoot 723, 724
Chrysanthemum Taro 735
Chrysanthemum Thistle 727
Chufa 299
Cicada Flower 245, 247
Cilantro 594
Cinnamon Pear 402
Cinnamon Round 523
Cistache 673
Citron 494
Cliff Artemisia 718
Cliff Floss 265
Cliff Mushroom 265
Cliff Pomegranate 360
Cliff Shrub 452
Cliff Wood Melon 526
Cliff Zanthoxylum 504
Climbing Fern 236
Climbing Fig 360
Climbing Wall Tiger 533
Clingstone Peach 446
Cloth-refuse Leaf 538
Cloud Bean 480
Cloud Ear 266
Clove Eggplant 635
Clover 477
Clover Fodder 477
Clove 150, 151, 152, 579
Club Seed 298
Cluster Fishtail Palm 301
Coarse-shelled Celtis 352
Cock's Comb Mushroom 267
Cock's Comb Perilla 651
Cock's Spur 529
Cocklebur 244, 248

Coco Plum 429
Cocoa 547
Coco-de-mer 302
Coconut 302, 303
Coconut Flake 32
Coconut Palm 302
Codonopsis 171, 172, 181, 188, 466
Coffee 45, 222, 471, 538, 678, 679, 680, 725
Coffee Mallow 538
Coffee Senna 470
Coiled Dragon Tonic 331
Coiling Dragon Shen 189
Cold Berry 455
Cold Jelly Seed 360
Coles 410
Comb Pond Weed 282
Common Apple 437
Common Bamboo 286
Common Cat-tail 280
Common Dandelion 748
Common Field Mushroom 266
Common Fig 358
Common Gardenia 680
Common Lettuce 740
Common Mallow 542
Common Mulberry 361
Common Myrtle 577
Common Nostoc 257
Common Pea 481
Common Pear 450
Common Plantain 237, 677
Common Plum 443
Common Quince 431
Common Rue 20, 503
Common Rye 296
Common Screw-pine 281
Common Smartweed 248, 372
Common Sow-thistle 747
Common Sunflower 734
Common Thyme 658
Common Watermelon 695
Compound-leaved Grape 535
Compressed Peach 517
Compressed Stone Peach 442
Concord Grape 535
Conic Chestnut 340, 346
Connected Vegetable 413
Continental Cotton 539
Cooked Rice 292
Cool Fruit 608
Cooling Vegetable 726
Copper Hemp 541

Coral Hibiscus 540
Coral Vegetable 598
Cordate-leaved Viburnum 688
Cordyceps 175, 176, 262
Coriander 20, 70, 77, 136, 150, 594
Corky Bark Oak 352
Corn 11, 37, 81, 292, 298, 723
Corn Flour 298
Corn Oil 204
Corn-flower Bud 69
Corn-salad 690
Cornel Meat 609
Cornstarch 44, 49, 54, 57, 60, 61, 68, 70, 75, 76,
    82, 83, 84, 85, 94, 95, 97, 115, 120, 123, 138, 143,
    144, 155, 159
Cotton 542, 696
Cotton Oil 539
Cotton-mass Clematis 390
Cotton-seed Oil 539
Country Borage 652
Country Walnut 509
Cow Gourd-ladle 633
Cow Nipple 571
Cow Parsnip 598
Cow Vetch 486
Cowberry 612
Cowpea 108, 488, 489
Crab-apple 436, 437
Crane Cave Autumn Eggplant 669
Crape Ginger 327
Cream Nut 574
Creat 235
Creeping Centipede 510
Creeping Fig 360
Creeping Oxalis 237
Creeping Potentilla 440
Creeping Potherb 636
Creeping Seaweed 259
Creeping Vitex 238
Crinkly-leaved Lettuce 741
Crisp Chinese Pear 449
Crisp-skin Fish 31
Crookneck Squash 701
Cross-the-mountain 634
Crow Grain 381
Crow Onion 743
Crow-berry 438
Crown Daisy 57, 58, 722
Crown Daisy Poterb 722
Crown-sawwort 744
Crow's Beak 743
Cuban Bast 541

Cucumber 20, 21, 82, 159, 589, 698
Cudrania 356
Cudweed Pastry 731
Cultivated Mint 649
Cultivated Pueraria 482
Cumin 595
Cummin 595
Cup-shaped Castanopsis 341
Curly Heart Vegetable 413
Curly-leaved Lycium-berry 663
Curry Leaf 502
Curry-leaf Tree 502
Custard Apple 398
Custard Apple of Peru 398
Cut Head Cucumber 699
Cut-leaved Celery 601
Cut-the-tongue Tree 508
Cylinder Gourd 705

Dahl 469
Dahurian Rose 453
Dandelion 746, 748, 751
Dang-gui 591
Dark Soy Sauce 156
Date 103, 303, 304
Date Palm 303
Date Plum 624
David's Vetchling 476
Davuria Lespedeza 477
Day Flower 308
Daylily 71, 72, 73
Daylily Flower Bud 29, 66, 68, 69, 70, 90, 145
Dead Ripe Persimmon 623
Decisive Eye-sight Promoter 470
Deep Valleys Woody Sky Smartweed 551
Deer Heart Tea 264
Deer Vine 483
Deer-horn Pickle 31
Deer-horn Vegetable 260
Delavay's Castanopsis 342
Delicious Calf-liver Mushroom 267
Dense-flowered Aromatic Madder 646
Dense-flowered Embelia 616
Desert Onion 314
Devil's Fingers 392
Devil's Mortar 395
Devil's Steamed-bread 360
Devil's Tongue 304
Devil's Tuber 304
Diamond Vine 529
Dill 589
Dioscorea 171

Dish Herb 737
Disk Persimmon 624
Ditch Sour Sap 671
Divine Mushroom 268
Dog Feet Taro 306
Dog Grape 427
Dog Liver Herb 675
Dog Paw Taro 304
Dog Zanthoxylum 503
Dogbane 629
Dog-jujube Actinidia 552
Dog-rib Wheat-bottle Herb 386
Dog's Paw Bean 485
Dog's Tongue Herb 640
Domesticated Dioscorea 323
Domesticated Perilla 652
Dookoo 507
Double-nine Tree 509
Dove's Food 684
Downy Ground Cherry 666
Downy Holly 236
Downy Myrtle 578
Dragon Boat Flower 682
Dragon Claw Millet 289
Dragon Crossing the Mountain 479
Dragon Eye 523, 524
Dragon Fruit 622
Dragon Head Bamboo 286
Dragon Pearl Fruit 565
Dragon Well 220
Dragon Whisker Vegetable 272, 316, 598
Dragon-head Bamboo 290
Dragon's Claw Vegetable 264
Dragon's Teeth Herb 439
Dragon's Tongue 214, 221, 513
Dragon's Tongue Herb 283
Dried Bamboo Shoot 288
Dried Fermented Red Rice 38
Dried Fermented Rice 37
Dried Persimmon Cake 623
Dried Rhizome Starch 388
Drop-flower Borne 465
Drought-resistant Elm 353
Dry Bamboo Shoot 288
Dry Bean Curd 465, 474
Dry Di-huang 672
Dry Exposing Seed 654
Dry Ginger 329
Dry Moss 258
Drynaria 167, 172
Duck Pear 450
Duck Sauce 33

Duck Tongue Herb 310
Duck Vegetable 594
Duck-feet Artemisia 719
Duck-feet Celery 604, 605
Duck-feet Tare 289
Duck-feet Tree 524
Duck-feet Vegetable 719
Duck-hemp Vegetable 744
Duku 507
Durian 544
Dutch Pea 481
Dutchman's Pipe Cactus 567
Duzhong 428
Dwarf Banana 325

Eagle Fern 272
Eagle-beak Pea 471
Early Mandarin Orange 497
Early Olive-green Eggplant 668
Early Red Eggplant 667
Early Taro 140
Early Taro of Dragon's Cave 306
Earth Bamboo Shoot 648
Earth Corner Skin 257
Earth Ears 257
Earth Egg 465
Earth Tartary Potherb 736
East Indian Arrowroot 321
Eastern Asian Monkey Flower 671
Eastern Wall 375
East-wind Potherb 720, 732
Edible Banana 324
Edible Burdock 718
Edible Canna 329
Edible Crown Daisy 722
Edible Garland Chrysanthemum 722
Edible Kudzu 481
Edible Tulip 321
Eel Vegetable 618
Egg Tree 560
Eggplant 20, 26, 152, 666, 667
Eight Horns 396
Eight-horned Muslim Spice 396
Electric Light-bulb 668
Elegant Lychee 708
Elegant Plum 444
Elephant-ear Herb 718
Elephants-foot 235
Eleuthero 585
Elm Bark 354
Elm Coin 354
Emblica 512

Emergency Wild Pea 487
Endive 724
English Pea 481
English Walnut 337, 338
Enrich Wisdom Seed 325
Ephedra Tea 280
Ephemeral Flower 567
Erect Hedychium 255, 328
Erect Sword Fern 272
Eriosma 473
Eryngo 150
Escaped Solders' Food 578
Essence-of-the-Earth 648
Eucommia 167, 173, 178, 181, 205, 206, 207, 208, 209, 428
Eucommia Bark 167, 172, 205, 206, 207, 208, 209, 428
European Alliaria 403
European Grape 536
European Hazelnut 338
European Nettle 365
European Olive 241
European Plum 443
European Raspberry 457
European Strawberry 435
European White Burdock 751
Euryale 106, 107, 108, 387
Euryale Seed 106, 157, 164, 167, 171, 181, 207, 227, 269, 319, 387
Euscaphis 520
Evergreen Oil-hemp Vine 479
Everlasting Vegetable 315
Excellent Rejoice Plum 447
Eyebrow Bean 475, 489
Eyre's Castanopsis 342

Fagle-claw Tare 289
Fairy Herb 260
Fairy Peach 567
Fairy Vegetable 260
Fairy's Grain 381
Fairy's Hand 567
Fall Bitter Potherb 737
False Bamboo 236, 245, 247
False Crown Daisy 726
False Fang-feng 593
False Hemistepta 744
False Lantern 666
False Litchi 356
False Mungbean 471
False Pakchoi 411
False Physalis 664

False Pittosporum 521
False Protea 366
False Saffron 721
False Soap-berry 514
False Sour Jujube 514
False Sterculia 546
False Sweet Pea 487
False Taro 307
Far-view Eggplant 668
Fat-craklings Fruit 703
Feicheng Peach 446
Feijua 577
Feminine Net 265
Fennel 20, 150, 151, 152, 597
Fennel-flower 390
Fenugreek 150, 152, 153, 485
Fermented Black Soybean 33, 35, 102, 120, 150, 157
Fermented Broad Bean Sauce 40, 153
Fermented Grain 263
Fermented Soybean 262
Fermented Soybean Paste 35, 40
Fermented Soybean Sauce 133, 204, 474
Fermented-and-Seasoned Black Soybean 474
Fern 272
Fern Frond Vegetable 272
Fern Rhizome 167
Fern Starch 272
Fern Vegetable 274
Fern-hemp 438
Festival Gourd 21, 85, 86, 94, 692
Fetid Gourd 700
Fetus-in-Surrogate Mother 369
Few-flowered Black Berry 457
Few-flowered Nighshade 670
Field Amaranth 20, 381
Field Cucumber-of-Fangshan 700
Field Mint 648
Field Morning Glory 636
Field Mustard 413
Field of Rice 292
Field Pea 481
Field Pearl 664
Field Perilla 652
Field Potato Substitute 321
Field Rat Walnut 366
Field Sow-thistle 746
Filamentous Nostoc 69
Filbert 338
Filiform Nostoc 257
Fine-leaved Schizonepeta 248
Finger Millet 289

Fire Dragon Fruit 568
Fire Fruit 509
Fire Pomegranate 669
Fire Scallion 313
Fire Tongue Vegetable 376
Fire-thorn 448
Fireweed 730
Firm Bean Curd 152
Firm Bean Curd Square 119, 123
Firmiana 545
Fish Belly 285
Fish Spice 151, 152, 597, 645
Fish Tree 423
Fish-egg Lan 507
Fish-scale Bush 579
Fish-smell Herb 332
Fish-smelling Herb 332
Fish-smelling Vine 369
Fishtail Palm 301
Fishtail Palm Flour 301
Fist Shen 189
Fitting-for-mother 492
Five Flowers Tea 215, 241, 245, 482, 484, 544
Five Needle Pine 98, 277
Five Spices Powder 151, 152, 153, 156
Five Vitalities Rejuvenating Herb 435
Five Wind Vine 393
Five-branched Fruit 548
Five-clawed Dragon 639
Five-clawed Golden Dragon 639
Five-flavored Drupelet 224, 396
Five-leaved Akebia 391
Five-leaved Gourd Vine 392
Five-leaved Strawberry 434
Five-leaved Yam 324
Five-petaled Mangrove 574
Five-segmented Parasite 367
Flat Bean 475
Flat Peach 442, 446
Flat-fruited Tree 523
Flat-on-ground Vegetable 410
Flax 490
Fleshy Cistache 673
Floating Heart Vegetable 628
Floating Pond-lily 388
Floating Silk Herb 628
Florist's Chrysanthemum 723
Flour Jar 367
Flower Bamboo 294
Flower Bean 479
Flower Red 436
Flower Red Tea 436

Flowering Almond 444
Flowering Bitter Vegetable 738
Flowering Plum 31
Flowering Stalk 315
Flowering Umbels 315
Flowerless Fruit 358
Flower-of-an-Hour 541
Flying-spreading Herb 510
Following Army Tea 476
Food Bag Herb 308
Food Bean 489
Foreign Apricot 383
Foreign Cabbage 410
Foreign Canarium 627
Foreign Cherry 442, 576
Foreign Coriander 596, 603
Foreign Eggplant 664
Foreign Ginger 735
Foreign Gourd 701
Foreign Head Vegetable 409
Foreign Jujube 303, 304
Foreign Lychee 398
Foreign Onion 311
Foreign Peach 490
Foreign Pear 450
Foreign Persimmon 664
Foreign Plum 443
Foreign Pogoda Tree 483
Foreign Pomegranate 577
Foreign Red Bud 467
Foreign Rose-apple 581
Foreign Silk-spice 596
Foreign Taro 670
Foreign Yam 638
Foreign-devil's Rose-apple 578
Foreigner's Lemon 652
Forest Delight 436
Forest Strawberry 435
Forsythia 247
Fortress Sweet Herb 650
Fortune Herb 650
Four Sages Leaf 274
Four Times Flowering 494
Four-angled Bean 481
Four-angled Water-lily 388
Four-leaved Vegetable 716
Four-season Bean 480
Four-season Orange 495
Four-season Vegetable 719
Four-shining Flower 608
Four-stamened Actinidia 553
Fourth Month Fruit 571

Fowl's Nest 377
Fox Grape 535
Fox's Tail Pea 238
Foxtail Millet 296
Fragrant Ailanthus 508
Fragrant Angelica 166
Fragrant Cardamon 327
Fragrant Herb 153
Fragrant Glorybower 643
Fragrant Hemp 674
Fragrant Litsea 401
Fragrant Melon 697
Fragrant Mushroom 268
Fragrant Round 494
Fragrant Smartweed 373
Fragrant Tea 555, 556
Frangipani 245, 248
Freestone Peach 446
French Horse Aster 690
Fresh Young Ginger 329
Fringe Tree 626
Frog's Bit 283
Froth Shen 713
Frozen Green 530
Fujian Tagerine 498
Fujian Tea 556
Fuling 268
Furry Longan 525
Fusty Iron-club 584

Galangal 152
Gandaria 515
Ganoderma 174, 177, 178, 268
Gansu Crabapple 436
Garden Amaranth 380
Garden Asparagus 316
Garden Beet 376
Garden Cress 417
Garden Euphorbia 510
Garden Gardenia 681
Garden Green 60
Garden Lettuce 739
Garden Orach 375
Garden Pea 59, 153, 481
Garden Pea Shoot 59, 65, 75, 154
Garden Sorrel 374
Garden Strawberry 433
Garden Thyme 658
Gardenia 215, 247
Gardenia Fruit 153, 680
Garlic 20, 45, 48, 49, 51, 54, 60, 64, 65, 68, 69, 81,
  82, 83, 84, 91, 93, 97, 104, 116, 120, 121, 123,
  124, 130, 133, 145, 148, 150, 154, 157, 185, 200,
  204, 208, 264, 314, 354, 716
Garlic Clove 29, 73, 93, 96, 123, 133, 153, 314
Garlic Head Bulb 319
Garlic Scape 60, 314
Garlic Seedling 314
Garlic-leaved Brahman's Vital Root 751
Gastrodia 330
Ge Onion 315
Generation Flower 492
Germinated Barley 247
Giant Bean 474
Giant Chickweed 385
Giant Panda Bamboo 290
Giant Timber Bamboo 294
Gin 152, 153, 154, 173, 194, 196, 199, 202, 208,
  209, 217, 589
Ginger 28, 48, 49, 52, 59, 60, 61, 62, 64, 65, 66, 68,
  69, 71, 72, 73, 74, 75, 82, 83, 84, 92, 93, 94, 95,
  97, 98, 104, 109, 110, 114, 115, 116, 120, 121, 123,
  124, 130, 133, 134, 141, 144, 145, 148, 149, 150,
  151, 152, 154, 155, 156, 157, 158, 176, 179, 181,
  185, 186, 187, 188, 197, 200, 204, 213, 224, 243,
  264, 265, 329, 471, 492
Ginger Flower 328
Ginger Lily 328
Ginkgo 45, 69, 98, 99, 100, 101, 102, 103, 104,
  106, 175, 275
Ginkgo Nut 275
Ginseng 44, 45, 89, 108, 162, 163, 164, 168, 169,
  170, 172, 173, 181, 182, 183, 184, 185, 187, 188,
  189, 194, 202, 218, 221, 223, 226, 386, 473, 586,
  587, 601, 603, 609
Ginseng Berry 438
Ginseng of Shangdang 714, 715
Ginseng Sanchi 587
Ginseng Yam 321
Girosole 735
Glabrous Petal Violet 563
Glabrous Plum 448
Glabrous Tanoak 348
Glandular Nodding Hooked Berry 462
Glandular Raspberry 460
Glaucous Smoke Tree 516
Glaucous-branched Berry 455
Glehnia Root 191
Globe Artichoke 727
Glorious Herb Flower 527
Glutinous Millet 293
Glutinous Rice 37, 38, 39, 73, 102, 103, 105, 109,
  138, 191, 198, 263, 292, 297
Glutinous Rice Flour 79, 136

Glutinous Rice Zao 37, 38
Glutinous Yam 323
Goat Droppings Tree 514
Goat Floating Weed 260
Goat Horns 392
Goat Nipple 569
Goat Stomach Mushroom 263
Goat-feet Herb 729
Goat-hoof Vine 394
Goat-horn Gourd 697
Goat-horn Pepper 659
Goat-horn Torreya Nut 277
Goat's Foot 467
Goat's Peach 550
Goddess-of-Mercy Fingers 668
Goddess-of-Mercy Herb 676
Goddess-of-Mercy Lotus 369
Goddess-of-Mercy's Amaranth 732
God's Food 264
Gold Hardback 440
Gold Orange 501
Gold-and-green Jade 286
Golden Bamboo 295
Golden Buckwheat 370
Golden Button 238
Golden Coin Herb 235
Golden Dew Mei 440
Golden Hair Grass 237
Golden Hair Rhododendron 611
Golden Hairpin Herb 748
Golden Hematinic 317
Golden Hips 453
Golden Hook Pear 529
Golden Key 236
Golden Mushroom 267
Golden Needle Vegetable 26, 68, 318
Golden Orange 495, 501
Golden Orange Cake 495
Golden Pea 501
Golden Pearl Willow 617
Golden Plum 429
Golden Pod Bean 480
Golden Shoot 596
Golden Tea 653, 654
Golden Thread Catalpa 672
Golden Thread Plum 560
Golden Tremella 269
Golden Wire Brush 264
Golden-hairy Strawberry 434
Golden-thread Holding Hulu 716
Gold-silver Flower 215, 687
Gombo 538

Good-for-me 571
Goose Intestine 387
Goose Intestine Potherb 385
Gooseberry 426
Goosefoot 377
Goose-intestine Potherb 385
Goose-kidney Tree 574
Gourd 695
Gourd from South China 696
Gourd-from-South 701
Gourd-of-the-Kunlun Mountain 667
Goutweed 588
Governor's Plum 564
Grain Conic Chestnut 342
Grain Cooking Fuel 341
Grain-of-Paradise 246
Grains of Immortal Ge 257
Grandfather-and-Grandson Tree 275
Grandmother's Nail 748
Grape Fruit 495
Grape 536, 689
Grassland Berry 459
Grassland Raspberry 459
Gravel Forest Herb 377
Gray Bushel 692
Gray Cotoneaster 429
Gray Elm 353
Gray Vegetable 377
Great Aromatic Herb 599
Great Burdock 718
Great Green Skin 691
Great Sterculia 546
Great Strength Potato 478
Greater Flying-spreading 510
Greater Oil Tea 554
Greater Red Bean 107, 109, 117, 118, 466
Greater Shepherd's Purse 422
Greater Thistle 725
Greater Vetch 487
Greater White Vegetable 413
Greater Yellow Herb 646
Green Bamboo 285
Green Bark Tree 608
Green Bean 112, 117, 119, 468
Green Buds 307
Green Canarium 143
Green Celestial Malva 331
Green Celosia 382
Green Fruit 32, 505
Green Gourd 698
Green Hard Oak 350
Green Herbage 307

Green Hook Castanopsis 345
Green Immature Soybean 119
Green Lantern 666
Green Mugwort 35
Green Myrica 335
Green Pea 481
Green Pepper 20, 114, 658
Green Perilla 652
Green Pod-on-leaf 610
Green Radish 143
Green Rind Luo-han-guo 711
Green Salad 740
Green Skin Luffa 706
Green Stem 291
Green Tea 216, 220, 555, 556
Green Vegetable 408
Green Wheat 284
Green Zanthoxylum 504
Green-barked Tong 545
Green-flowered Hillside Celery 602
Green-fruited Fig 358
Green-hook Castanopsis 345
Green-in-Snow 410
Green-silk Bamboo 286
Green-steel 350
Ground Berry 582
Ground Bottle-gourd 633
Ground Brocade 510
Ground Cherry 666
Ground Chestnut 299
Ground Clausena 272
Ground Cover 378
Ground Elm 462
Ground Embroidery Maple 522
Ground Jujube 320
Ground Little Pear 140
Ground Melon 361
Ground Melon Top 648
Ground Nail 748
Ground Nut 465
Ground Sea Cucumber 391
Ground Silkworm 655
Ground Slug 656
Ground Soft 257
Ground Spice 657
Ground Spice Leaf 657
Ground Tapestry 533
Ground Tea 265
Ground Vegetable 414
Ground White-seeded Vegetable 733
Ground Yellow Vegetable 732
Guangdong Mandarin Orange 498

Guangdong Mangosteen 559
Guava 577
Guiana Chestnut 545
Guihua 627
Gui-lin Castanopsis 342
Gulf Weed 259
Gumbo 538
Gynostemma 703

Hainan Glycosmis 502
Hainan Rose-apple 580
Hainan Yam 640
Hair Vegetable 69, 257
Hairy Ball 324
Hairy Bamboo 295
Hairy Bean 119
Hairy Berry 582
Hairy Black Gourd 712
Hairy Cart Vine 629
Hairy Cherry 447
Hairy Chestnut 343
Hairy Chestnut Tree 343
Hairy Conic Chestnuts 346
Hairy Dogwood 609
Hairy Elm 353
Hairy Fruit 562
Hairy Girl Vegetable 717
Hairy Golden Bamboo 295
Hairy Gourd 691, 692
Hairy Grape 536
Hairy Ground Chestnut 140
Hairy Hawthorn 430
Hairy Immortal Fruit 357
Hairy Lotus Vegetable 742
Hairy Oleaster 569
Hairy Peach 444
Hairy Pingpo 547
Hairy Plum 180, 550
Hairy Portia Tree 543
Hairy Sepal Tea 557
Hairy Sour Sap 666
Hairy Tea 511
Hairy Wood Ear 266
Hairy Wormwood Pea 612
Hairy Yam 322
Hairy-fruited Violet 562
Hairy-leaved Fragrant Litsea 402
Hairy-sepaled Berry 456
Half-branch Flower 654
Hami Melon 696, 698
Hand Ginseng 330
Hand-bell Flower 716

Hanging Bell Vine 570
Hanging Lantern Berry 570
Hanging Lantern Flower 540
Hanging Purse Eggplant 668
Hanging Red Silk 285
Hangzhou Chrysanthemum 723
Harbar Hemp 541
Hard Rind Luffa 707
Hard-cupped Lithocarpus 348
Hare's Ear 167
Hargan Chai 464
Haul-the-Root Herb 425
Hawk's Talon Vine 392
Hawthorn 84, 85, 178, 245, 247, 431
Hawthorn Cake 431
Hawthorn Flake 431
Hazel 338
Hazelnut 338, 339
Head Cabbage 410
Head Lettuce 740
Heal-all 653
Heaven Hemp 330
Heaven Immortals Vegetable 257
Hedge Sageretia 237, 530
Hedgehogs 268
Hedge-parsley 607
Helwingia 610
Hemp Leaf 244, 246
Hemp Mustard 408
Hemp Oak 350
Hemp Palm 304
Hemp-flower Shoot 744
Hemp-leaved Embroidered Ball 464
Hemp-leaved Lycium-berry 662
Hemp-leaved Nettle 364
Hemp-leaved Spiraea 464
Hemp-leaved Vitex 238
Herb Cassia 471
Herbaceous Cotton 539
Herbaceous Rock Silkworm 656
Herbal Hematinic 415
Herb-like Ephedra 279
Herb-of-the-Goddess 20, 309
Hercules Club 705
He-shou-wu of Tai Mountain 633
High Bush Tiger Twist 462
High Ligule Bamboo 287
High Montane Pine 278
High Ridge Berry 458
High Rock Lan 513
High Shen 592
Highland Celery 594

Hill Blackberry 459
Hill Celery 605
Hill Gooseberry 578
Hill Mallow 422
Hillside Actinidia 553
Hillside Bamboo 290
Hillside Bitter Potherb 737
Hillside Bitter Tea 511
Hillside Bitter Wheat 746
Hillside Cabbage 720
Hillside Celery 589, 601, 730
Hillside Celery Vegetable 601
Hillside Cotton 389
Hillside Eggplant 543, 641
Hillside Fennel 593
Hillside Ginkgo 338
Hillside Hairy Elm 353
Hillside Horse Salad 385
Hillside Jujube Shen 462
Hillside Lotus Rhizome 478
Hillside Lotus-leaf 424
Hillside Nail-fruit 436
Hillside Net Flower 671
Hillside Niupang 549
Hillside Oil Tangerine 491
Hillside Osmanthus 617
Hillside Pea 447
Hillside Peach-leaf 613
Hillside Pepper 507
Hillside Plum 394
Hillside Polygonum 389
Hillside Screw-pine 281
Hillside Sesame 236
Hillside Tallow-tree 237
Hillside Taro 307, 638
Hillside Thorn-fruit 436
Hillside Walnut 337
Hillside White Chrysanthemum 720
Hillside Yam 322
Himalayan Costus 169
Himalayan Currant 426
Hinggan Red Pine 278
Hispid Berry 583
Hodgsonia 703
Hog Chestnut 346
Hog Fruit 366
Hog Stomach Bush 677
Hog-plum 519
Holding Baby Cabbage 411
Hold-on-branch Flower 544
Hollow Bamboo 286
Hollow Heart Berry 461

Hollow Potherb 636
Holy Thistle 745
Home-made Pickle 29
Homestead Hemptree 547
Honey Bottle Plant 672
Honey Jujube 532
Honey Mandarin Orange 496
Honey Mandarin Orange of Tiantai 497
Honeydew 698
Honey-locust Kernel 473
Honeysuckle Flower 72, 168, 215, 216, 245, 247
Hong Kong Castanopsis 343
Hong Kong Hawthorn 452
Hong Kong Orchid Tree 467
Hong Kong Wild Fig 358
Hongkong Dogwood 607
Hongping Apricot 443
Hook Castanopsis 347
Hook Chestnut 347
Hook Tea 528
Hooked Spines 356
Hook-vine of Yunnan 529
Hop 362
Horn Vegetable 719
Horse Alga 282
Horse Aster 738
Horse Aster Green 738
Horse Bean 486
Horse Bubble Gourd 698
Horse Bullet 696
Horse Tail 284
Horse-bell Potato 670
Horse-hoof 142, 300
Horse-hoof Needles 484
Horse-intestine Tree 588
Horseradish 403
Horse-radish Tree 423
Horseshoe Cassia 471
Horse-teeth Amaranth 384
Horseweed Mugwort 722
Hosta 318
Hosta Flower 318
Hot Bean Sauce 33
Hot Cayenne Pepper 148
Hot Peanut Sauce 64, 151, 153
Hot Pepper 121, 148, 150, 153, 658, 659
Hot Pepper Sauce 33, 40, 41, 64
Hot Rotten Bean Curd Cheese 36
Hot Soybean Sauce 64, 65, 81, 97, 98, 133, 135, 145
Huangchy 466
Huangyan Honey Mandarin Orange 497

Hua-shan Pine 277
Hubei Artichoke 655
Hubei Bauhinia 467
Hubei Hawthorn 430
Hulu 20, 92, 93, 704
Hunan Millet 288
Hundred Day Herb 308
Hundred Union 319
Hundred-flowered Tea 684
Hundred-mile Fragrant 658
Husk Tomato 665
Huzi 705
Hyacinth Bean 475

Icaco 429
Ice Jelly 664
Ice Jelly Seed 360
Ichang Elaeagnus 570
Ichang Viburnum 689
Ill-smell Herb 503
Immortal Cabbage 720
Immortal Rice 275
Immortal's Leaf 686
Imperata Root 291
Imported Buddha Disciple Fruit 558
Inch-of-gold Torreya Nut 277
Indian Almond 546, 576
Indian Castanopsis 344
Indian Corn 298
Indian Fig 568
Indian Jujube 532
Indian Long Pepper 333
Indian Mulberry 682
Indian Mustard 407
Indian Pumpkin 702
Indian Walnut 509
Ink Tuber 304
Intermediate Ephedra 278
Intermediate Grain Coffee 679
Inverted Basin Berry 457
Iraqi Honey Jujube 303, 304
Irgai 429
Irish Potato 126, 140, 670
Iron Hoop Scattered 397
Iron Oak 350, 351
Iron Revival 275
Iron Tree 275
Iron Wrapping Gold 237
Iron-thread Seed 621
Italian Bean 486

Jackfruit 355

Jade Bamboo 169, 171, 181, 191, 197, 198, 319
Jade Barley 291
Jade Millet of Sichuan 298
Jade Mountain Ardisia 613
Jade-leaf Gold-flower 684
Jamaica Sorrel 540
Jambolan 580
Japanese Apricot 31, 445
Japanese Barnyard Millet 288
Japanese Bush Cherry 444
Japanese Changshan 502
Japanese Cornelian Cherry 608
Japanese Hazelnut 339
Japanese Honeysuckle 687
Japanese Hop 363
Japanese Horseradish Wasabi 422
Japanese Ivy 532
Japanese Privet 626
Japanese Raisin Tree 529
Japanese Tea 557
Japanese Thistle 725
Japanese Yam 323
Jasmine 26, 73, 74, 77, 150, 215, 220, 555, 556,
 626
Java Apple 581
Java Plum 580
Java Sauropus 512
Jelly Herb 650
Jelly Tree 360
Jerusalem Artichoke 655, 735
Jiang-mang Positive Bright 470
Jing-shan Melodinus 632
Jinshan Schisandra 397
Jira 595
Job's Tears 106, 107, 108, 109, 111, 117, 118, 145,
 157, 164, 167, 171, 173, 174, 175, 181, 185, 187,
 198, 207, 227, 229, 245, 247, 269, 287, 319, 387
Joint Potherb 573
Juda's Ears 266
Jug Pear 449
Jujube 26, 87, 88, 89, 90, 103, 107, 110, 157, 164,
 169, 181, 187, 209, 211, 215, 221, 228, 269, 531,
 547, 704, 723
Jute-leaved Raspberry 456

Kaa-hee 748
Kaempferia 149, 150, 151, 152, 328
Kaki 623
Kales 406, 410
Kamchatka Lily 317
Kaoliang 296
Karalaka 650, 651

Karanda 630
Katydid Cage Hulu 705
Kechapi 507
Kelp 259
Kelupang 546
Kidney Fruit 514
Kidney-leaf 467
Kidney-leaved Sorrel 371
Kiryat 235
Kiwi Fruit 7, 180, 550, 551
Kobomugi Sedge 299
Kohlrabi 406
Korean Blackberry 456
Korean Ginseng 168, 182, 183, 184, 586
Korean Pine 278
Kousa 608
Kudzu 245, 482
Kumquat 32, 501, 547
Kundang 515

Lablab Bean 475
Lace-bark Pine 278
Lagenaria 704
Lagenaria Herb 716
Lageniform Hulu 705
Lamb's Quarter 377
Lambs-lettuce 690
Laminaria 69, 203, 259, 591
Lamont's Castanopsis 346
Lamp-wick Rush 236, 245, 247
Lanceolate-leaved Elaeagnus 570
Langsat 507
Lantern Bubble 386
Lantern Flower 334, 540
Lantern Fruit 665
Lantern Herb 665
Lantern Pepper 659
Lard Nut 510
Large Fruit Fig 357
Large Grain Coffee 680
Large Hemp-leaved Lycium-berry 662
Large Slope Tea 642
Large Stone Fruit 622
Large Vine 693
Large Wing Rock-flower 264
Large-flowered Crown-of-thorns 630
Large-flowered Dillenia 549
Large-flowered Five-branched Fruit 549
Large-flowered Loquat 432
Large-fruited Bobai Oil Tea 554
Large-leaved Bayan 357
Large-leaved Bitter Cress 415

Large-leaved Castanopsis 341
Large-leaved Celery 598, 605
Large-leaved Crown Daisy 723
Large-leaved Embelia 616
Large-leaved Spiny Hedge 564
Large-rooted Mustard 408
Large-toothed Dang-gui 601
Lazy Fellow's Vegetable 315
Leaf Wrapper 731
Leaf-mustard 407
Leaking Hut 728
Leather-leaved Actinidia 551
Leaves Red Beneath 729
Leechee 525
Leek 311
Lemandarin 492
Lemandarin Cake 492
Lemon 28, 90, 216, 492
Lemon Grass 287
Lemon Mint 649
Lemon Thyme 657
Lentil 476
Leopard Palm 304
Lesser Blood Vine 397
Lesser Galangal 150, 325
Lesser Ixeris 737
Lesser Kelp 258
Lesser Lung-muscle Grass 311
Lesser Muslim Spice 589
Lesser Red Bean 117, 167, 175, 227, 472
Lesser Red Berry 612
Lesser Red Castanopsis 341
Lesser Red Chestnut 344
Lesser Thistle 726
Lesser Vetch 487
Lesser Water Snowflake 628
Lesser White Vegetable 413
Lesser-flowered Dillenia 549
Lettuce 106, 120, 121, 739
Lettuce Shoot 740
Levant Cotton 539
Liberian Coffee 680
Licorice 35, 152, 168, 171, 172, 173, 218, 230, 245,
    445
Li-ethnic Tea 470
Light-colored Soy Sauce 133, 144, 145, 155, 159
Like-chestnut 537
Lily Bulb 107, 164, 168, 171, 174, 181, 191, 269,
    387
Lily Hematinic 316
Lima Bean 480
Lime Mountain Ash 463

Lime Twig 463
Limeng 492
Limeng Ginger 493
Ling 583
Ling Horn 584
Lingberry 612
Lingchih 268
Lingnan Wild Mangosteen 559
Linseed Oil 490
Lion Grass 311
Lion-head Foreign Pepper 659
Lion-head Tangshen 714
Liquid Preserved Mei-fruit 445
Litchi 525
Lithocarpus 348
Little Ephedra 279
Little Hulu 705
Little Millstone 542
Little Persimmon 571
Little Red Bean 472
Little Yellow-fruit Lycium-berry 663
Lizard Reed 720
Lobi Plum 564
Lobi-lobi 564
Local Angelica 585
Local Astragalus Root 478
Local Bai-lian 712
Local Ginseng 189
Local Ox-knees 379
Local Wood Melon 431
Lock-up Yang 584
Lofty Elm 354
Lonely Vegetable 417
Long Luffa 707
Long Body 709
Long Pepper 150, 333
Long Vigna 488
Long Water Gourd 708
Longan 11, 26, 164, 171, 173, 181, 221, 547
Longevity Berry 438
Long-leaved Elaeagnus 569
Long-leaved Embelia 615
Long-leaved Lettuce 741
Long-leaved Water Hemp 363
Long-necked Hulu 705
Long-string Currant 427
Looking South Beyond the Yangtze River 470
Loose Seed Herb 416
Loose-leaved Cabbage 413
Lophatherum 245
Loquat 432, 547
Lotus Flower 75, 387

Lotus Flower White 410
Lotus Leaf 158, 159, 388
Lotus Plant 387
Lotus Rhizome 32, 69, 137, 138, 139, 221, 387
Lotus Seed 32, 95, 106, 110, 111, 118, 139, 145, 164, 170, 171, 174, 181, 198, 223, 227, 269, 387, 388
Lotus Seed Herb 379
Lotus Starch 140
Love Apple 664
Love Jade Seed 357
Love Pea 235
Love-in-a-Mist 390, 565
Love-lies-bleeding 380
Low Fishtail Palm 301
Luffa 90, 91
Lump-of-Cooked Rice 397
Lung-cancer Grass 311
Luobo 417
Luohanguo 710
Luo-han-guo of Lajiang 711
Luo-han-guo of Long Sandy Bank 710
Luxurious Jujube 320
Lychee 11, 26, 525
Lycium Green-shoot 663
Lycium Tea 661
Lycium-berry 164, 168, 181, 185, 187, 188, 190, 197, 198, 218, 222, 223, 466, 591, 661, 663, 714, 723

Macadamia Kernal 366
Macadamia Nut 366, 367
Macaron Potherb 636
Madagascar Plum 564
Madeira Vine 384
Magenta Rose 454
Magenta-colored Rose 454
Magnolia Flower 75, 76
Magnolia Petals 76, 396
Magpie Bean 475
Ma-huang 279
Maiden Tea 643
Maiden Vine 526
Maidenhair Tree 275
Maize 297, 298, 406
Major Spice 148, 151
Malabar Plum 580
Malabar Spinach 384
Malaysian Rose-apple 580
Mallow Green 61, 542
Malt 290, 291
Mamey 561

Mamey Fruit 561
Mammee 561
Mammee Apple 561
Man Amaranth 51, 381
Man Face 516, 517
Manchu Rose 454
Manchurian Walnut 337
Mandarin 496
Mandarin Lime 492
Mandarin Orange 496
Mandarin Orange Peel 167, 246
Manggis 558
Mango 517
Mango Stone 236
Mangosteen 558, 559
Mangrove Palm 303
Man-heart Fruit 621
Manico 511
Many-flowered Actinidia 552
Many-flowered Melastoma 582
Many-spiked Lithocarpus 348
Many-toothed Athyrium 273
Maolizi 550
Maple Sugar 522
Maple Syrup 522
Maple-leaved Berry 459
Maple-leaved Goosefoot 377
Marrow-leaved Nettle 364
Marsh-varnish Hemp 629
Mashed-cooked Lotus Seed 388
Matrimony-vine 62, 63, 65, 162, 164, 168, 172, 663
May Flower 682
May Raspberry 237
May Tea 509
Meadow Mushroom 266
Meadow Rock Flower 384
Meal Squash 701
Mealy Fruit 431
Mealy Gourd 696
Mealy Melon 39
Mealy Pueraria 482
Meat Herb 309
Meaty Cardamon 399
Medicinal Magnolia Bark 247
Medicine Mulberry 362
Mei 31, 32, 441, 445, 446
Mei Flower Vase 386
Mei-flavored Ginger 493
Mei-hua 445
Meikuei 454
Melon 695

Melon Seed 694
Melon Seed Tea 522
Melon-from-the-west 694
Melt Across-the-mountain 634
Mesetor 558
Messenger-of-the-red-cart 364
Mice Melon 712
Mice Plum 530
Mice Pulling Winter Melon 712
Mice Yeast Herb 731
Microcos 213, 236, 247
Midnight-noon Water-lily 388
Milk Herb 716
Milk Orange 496
Milk Potato 714
Milk Shen 713
Milk Thistle 745
Milk Vegetable 737
Milky Tree 620
Millet 81, 111, 293, 296, 701
Millet Hot Pepper 660
Millet Lan 507
Millet Thorn 530
Mimetic Mallow 543
Mint 20, 157, 216, 648
Mock Lime 507
Mock Pakchoi 411
Mold Dried Vegetable 409
Moli 626
Moneywort 235, 593
Mongolia Jujube 462
Mongolia Lespedeza Tea 477
Mongolia Sagebrush 718
Mongolia Tea 279
Mongolian Leek 314
Mongolian Oak 350
Mongolian Peach 446
Mongolian Tea Clematis 389
Mongolian Viburnum 689
Mongolian Whiterod 689
Monkey Bread Tree 543
Monkey Chestnut 340, 346, 347
Monkey Creeper 632
Monkey Fruit 632
Monkey Head 268
Monkey Nut 510
Monkey Peach 550
Monkey-legged Athyrium 273
Monkey's Delight 621
Monk's Rhubarb 375
Montane Maesa 617
Moonlight Flower 635

Moon-on-rock 394
Morel 263
Morinda 168, 172, 173, 682
Moses-in-a-Boat 310
Moslem Perilla 651
Moslem Scallion 313
Moss Ribbon 258
Moss Vegetable 258
Mother Ginger 329
Mother Herb 318
Mother of Clove 579
Mother Seaweed 259
Mother-of-Gin 38
Mother-of-Kudzu 482
Mother-of-Kumquat 501
Mother-of-Wine 263
Mother-of-Yu 140
Mother's Delight 493
Mottled Bamboo-root Hematinic 317
Mottled False Disporum 317
Mottled-fruit Vine 424
Mountain Artemisia 718
Mountain Berry 456
Mountain Caper 423
Mountain Celery 591
Mountain Chestnut 510
Mountain Cinnabar 682
Mountain Coriander 599
Mountain Cranberry 612
Mountain Fern Cat 271
Mountain Ginkgo 280
Mountain Grape 533
Mountain Ground Gourd 640
Mountain Hemp 427
Mountain Langenaria 366
Mountain Leek 315
Mountain Loquat 620
Mountain Lychee 607, 608
Mountain Mustard Blue 368
Mountain Orange 632
Mountain Pear 451
Mountain Periwinkle 713
Mountain Pork 526
Mountain Radish 382, 416, 690
Mountain Sagittaria 321
Mountain Sorrel 371
Mountain Spinach 375
Mountain Thorny Rose 453
Mountain Vermicelli 265
Mountain Walnut 336, 337
Mt. Morrison Ardisia 613
Mt. Morrison Raspberry 455

Mt. Omei Rose 453
Muku Tree 352
Mulberry 142, 236, 361
Mulberry Leaf 35, 247, 361
Mulberry-leaf Grape 534
Mungbean 44, 106, 111, 112, 113, 114, 117, 191, 224, 468
Mungbean Skin 468
Mungbean Silk 53, 54, 64, 65, 66, 69, 112, 115, 116, 203, 303, 468
Mungbean Sprout 112, 114, 116, 468
Mushroom 91, 95, 110, 136, 177, 266, 267, 270
Musk Adenosma 235
Musk Herb 658
Muslim Spice 597
Mustard 150, 153, 154, 403, 406
Mustard Blue 404
Mustard Greens 54, 407
Myrobalan 512

Nagami Kumquat 501
Nail Spice 579
Naked Barley 291
Nanjing Pickle 29
Nanking Chrysanthemum Shoot 724
Narrow Waist Berry 582
Narrow-leaved Chinese Dogwood 608
Narrow-leaved Crown Daisy 723
Narrow-leaved Pouch Fruit 623
Naseberry 621
Natal Plum 630
Native Hobble Bush 688
Nectarine 446
Nepal Cardamon 327
Nest Vegetable 486
Nest-fern 274
Net Look Seed 485
Netted Stinkhorn 267
Netting Hemp 629
Nettle 365
New Zealand Spinach 383
Niggertoes 574
Night Fragrant 635
Night Fragrant Flower 635
Night Shrivel Vine 372
Night-blooming Cereus 168, 215, 567
Nightshade 669
Nine-headed Bird 409
Nine-layered Bark 547
Nine-story Pagoda 650
Nine-winged Cardamon 326
Ningpo Figwort 248

Ningpo Tangerine 497
Ningxia Lycium-berry 661
Ningxia Matrimony Vine 661
No Bone Melt 687
No Exchange-for-Gold 374
Noble Bottle Tree 546
Nodding Hooked Berry 459
No-name Seed 518
North China Baby's Breath 384
North China Blue Pot Flower 690
North China Hazel 339
North China Scabiosa 690
North China Violet 563
North-China Catalpa 672
Northeastern Cow Parsnip 598
Northeastern Goat-horn Celery 589
Northeastern Hazelnut 339
Northern Apricot Seed 227
Northern Astragalus 466
Northern China Motherwort 647
Northern China Peppercorn 504
Northern Gypsophila 384
Northern India Castanopsis 344
Northern Leek 313
Northern Sand Shen 190, 191, 198, 229, 598
Northern Schisandra Fruit 224
Northern Squash 702
Notopterygium 248
Nutmeg 150, 399, 471
Nypa Palm 303

Oat 284
Obtuse-leaved Tea 556
Ocean Jujube 303
Ocean Red 437
Octagonal Persimmon 624
Official Cassia Bark 399
Official Silk Umbrella Tree 613
Oil Bottle Gourd 368
Oil Hemp 490
Oil Nut-grass 299
Oil Peach 446
Oil Persimmon 624
Oil Rape 414
Oil Rape Shoot 413
Oil Sweet Seed 512
Oil Tea 554
Oil Tea of Guangning 555
Oil Torreya Nut 277
Oil Wheat 284
Okra 538
Old Man Extract 34

Old Mountain Celery 599
Old World Arrowhead 136, 282
Oldham Bamboo 285
Oldham's Azalea 611
Oleaster 568, 571
Olive 32, 241, 242, 537, 627
Olive Oil 627
One-foot Gold 672
One-seeded Ephedra 279
One-side Green 717
Onion 28, 45, 47, 48, 51, 55, 56, 57, 58, 59, 60, 62,
    63, 64, 65, 66, 68, 69, 70, 71, 72, 73, 74, 75, 82,
    83, 84, 86, 87, 91, 92, 93, 94, 95, 97, 98, 102, 104,
    109, 110, 114, 115, 120, 121, 122, 123, 124, 130,
    131, 133, 135, 136, 137, 139, 140, 141, 144, 145,
    148, 150, 152, 154, 155, 157, 159, 176, 179, 185,
    186, 187, 188, 200, 202, 203, 204, 213, 221, 223,
    241, 242, 245, 311, 354, 474, 604
Open Sack Tree 281
Open-heart Nut 518, 519
Opium Poppy 402, 403
Opposite Thorn 530
Opposite-leaved Fig 236
Opposite-leaved Loose Flower 689
Orach 375
Orange 496, 498, 499
Orange Peel 150, 246, 445
Orange Peel Mei 31, 445
Orchid Spice 650
Oriental Pickling Melon 697
Oriental Strawberry 434
Orobanche 674
Osmanthus 76, 77, 110, 150, 180, 627
Osmunda 274
Oster Plant 310
Oulana 443
Oval Kumquat 501
Over-water Branches 680
Ox Bush-clover 477
Ox Ear Vine 369
Ox Flank 718
Ox Spleen 692
Ox Tongue 742
Ox Yokewood 235, 557, 558
Ox-ear Rhubard 374
Ox-hair Gelidium 260
Ox-heart Fruit 398
Ox-heart Tea 561
Ox-hemp Vine 479
Ox-hide Melt 633
Ox-horn Pepper 659
Ox-knees 379

Ox-leg Squash 701
Ox-strength 478
Ox-tongue Lettuce 741
Oyster Plant 751

Paddy Mustard 422
Paddy St. John's Wort 236
Paddy Water-cress 414
Paederia 684
Pagoda Pickle 31
Pagoda Tree 248, 484
Paired Butterflies 521
Paired Carps 693
Paired Skullcap 655
Pakchoi 407
Palace-security Shrub 512
Palm Flour 301
Palm Tree 304
Palmate-leaved Tree 524
Palmate-leaved Yam 324
Palo De Tomate 660
Panacea Holly 236
Pandanus Root 237
Papaya 81
Parsley 603
Parsnip 585
Partially Cured Di-huang 672
Passion Fruit 565
Patchouli 248
Patience Dock 375
Pea 98, 161, 220, 480, 501
Pea Shoot 59, 481
Pea-berry 665
Peach 81, 441, 446
Peach Canarium 622
Peach Kernel 172, 173, 218
Peach Hematinic 395
Peach-for-Goddess 446
Peanut 88, 101, 102, 107, 125, 128, 153, 161, 224,
    465, 734, 735
Peanut Butter 674
Peanut Oil 77
Pearl Sago 303
Pear Tea 554
Pearl Chestnut 340
Pearl Flower 521
Pearl Grain 511
Pearl Herb 331
Pearl Mei 464
Pearl Orchid 334
Pearl Vegetable 618
Pearly Anaphalis 717

Pearly Everlasting 717
Pecan 336
Pegion Pea 469
Peking Cherry 443
Peking Honey Jujube 88, 531
Peltate-leaved Raspberry 459
Penetrating Bone Rheumatism 397
Penetrating Rock 356
Pennycress 422
Peony 199, 391
Peony Root 168, 171, 173, 181, 199, 200
Pepper 66, 82, 83, 84, 91, 92, 104, 105, 116, 156, 658
Pepper-across-the-ocean 658
Peppercorn 64, 66, 504
Peppermint 247, 649
Perilla 150
Perilla Nutlet 652
Perilla Shoot 35
Persian Melon 698
Persian Walnut 337, 338
Persimmon 26, 220, 623, 625
Persimmon Calyx 624
Persimmon Tannin 624
Philosopher Laotzu's Elixir 465
Phoenix Dragon Fruit 568
Phoenix Eye Fruit 546
Phoenix Fruit 559
Phoenix Mushroom 268
Phoenix Pear 308
Phoenix Tail Cycas 275
Phoenix Tail Mushoom 268
Photinia 438
Piba 432
Piba Root 432
Piece Cabbage 406
Piercing Tree 565
Piercing Tree Oil 565
Pig Ear-opening 383
Pig Rump 332
Piglet Basket 473
Pigweed 377
Pimento 659
Pine Mushroom 268
Pine Net 265
Pine Nut 277
Pine Seed 104, 105, 106
Pineapple 308
Pineapple Guava 577
Pineapple Type 693
Pingpo 546, 547
Pingpo Leaf 546

Pink Baby's Breath 385
Pink Botton 266
Pink Goat's Foot 467
Pinon 277
Pinshan Pomelo 494
Pinto Bean 489
Pistachio 518, 519
Pistachio Nut 518
Pitanga 576, 577
Pitted Jujube 532
Pitted Sugared Jujube 531
Pixian Bean Sauce 486
Plain Melon 711
Plantain 178, 325
Plantain-lily 318
Plateau Thyme 656
Pleasant Flavor Taro 306
Pleasant Raspberry 455
Plum 31, 81, 441, 447
Plum-leaved Apple 437
Plum-leaved Holly 236
Pogoda Tree 245
Pointed Leaf Tea 557
Pointed Pepper 659
Pointed-head Round Fruit Lycium-berry 662
Pointed-head Yellow-leaved Lycium-berry 662
Pointed-leaved Banyan 359
Pokeweed 382
Police Dog Tongue 583
Polished Rice 292
Polygonatum 171
Polynesian Arrowroot 321
Pomegranate 81, 573
Pomelo 493
Pommelo 493
Poppy Millet 402
Potato 126, 133, 670
Potato Bean 465
Potentilla 439
Powdered Pepper 333
Precious Pogoda Vegetable 656
Precious Stick 687
Premier Zhu-ge's Vegetable 417
Preserved Black Canarium 506
Pressed Old Nail 356
Pressed Vegetable 28, 29, 409
Preventing Infection 605
Prickle Cucumber from Ningyang of Shandong 699
Prickley Pear 426
Prickly Dragon 585
Prickly Elegant Fruit 544

Prickly Ipomoea 635
Prickly Pear 453, 567
Prickly Potherb 726
Prickly Red Berry Vine 459
Prickly Red Flower 721
Prickly Thistle Vegetable 725
Prickly Vine 363
Prickly Water-lily 387
Prime Sauce 34
Prince Ginseng 169, 385
Prince Shen 189, 385, 386
Prinsepia 440
Prinsepia Seed 169, 227, 229, 230, 440
Proper Restoration 591
Prune 90, 441
Pseudo-melon 697
Psychotria 237
Pu 704
Pu Herb 280
Pueraria 482
Pueraria Flower 482
Pueraria Root 482
Pueraria Root Starch 481
Pugua 705
Pulling Pussy Grain 291
Pull-off-skin 432
Pumpkin 701, 702
Puncture Vine 491
Pungent Mint 649
Pungent Smartweed 372
Pungent Tree 424
Pungent Vegetable 409
Purple Back Tea Begonia 566
Purple Bullet Tree 352
Purple Celestial Mallow 732
Purple Clover 477
Purple Ginger 329
Purple Granadilla 565
Purple Ground-nail 563
Purple Osier 334
Purple Passion 731
Purple Pearl 643
Purple Perilla 245, 651
Purple Perilla Leaf 248
Purple Strawberry Guava 578
Purple Vegetable 261
Purple Velvet Vine 731
Purple-backed Begonia 566
Purple-flowered Celery 604, 605
Purple-flowered Sanicle 605
Purse Amaranthus 381
Purslane 384

Queen-of-the-Night 567
Queensland Arrowroot 329
Queensland Nut 366
Quince 431

Rabbit Leg 674
Radish 47, 126, 131, 132, 133, 135, 136, 403, 417, 420, 421
Radish Leaf 417
Radish Seed 417
Rag Leaf 211, 538
Rag Seed 641
Rag Tree 641
Ragi 289
Raisin 103, 117
Rambutan 525
Ramie 363
Raspberry 460
Rat Droppings 528
Rat Ear 528
Rat-eyes 483
Raw Gourd 696, 697
Recurved Red Lily 319
Red Amaranth 381
Red Amaranthus Weed 380
Red Amomum 326
Red Arrow 330
Red Bark Osier 334
Red Bean 106, 111, 116, 137, 164, 466
Red Berry 433
Red Brewer's Yeast 263
Red Cardamon 326
Red Chilli Pepper 26
Red Daylily 318
Red Drop 729
Red Fermented Rice 37, 150, 152, 153, 154, 155, 156, 263
Red Fermented Grain 263
Red Flower 721
Red Flower Potherb 721
Red Flowered Bean 479
Red Gram 469
Red Groundsel 732
Red Hawthorn 431
Red Hook Castanopsis 345
Red Hot Pepper 40
Red Jujube 88, 89, 102, 109, 110, 111, 169, 175, 190, 191, 195, 198, 207, 214, 229, 531, 532
Red Jujube King 87
Red Limeng 493
Red Lintel 344
Red Maiden 664, 708

Red Meat Goat's Peach 551
Red North Pole Berry 610
Red Oil Rape 414
Red Pepper 91, 114, 116, 120, 658, 716
Red Perilla 651
Red Phoebe 579
Red Phoenix Vegetable 732
Red Pine 278
Red Plum 447
Red Raspberry 457
Red Root Herb 618
Red Root Vegetable 378
Red Shadbush 428
Red Silk-cotton Tree 544
Red Silk-thread 676
Red Snow Tea 264
Red Sorrel 374, 540
Red Stalk Vegetable 738
Red Swiss Chard 376
Red Tangerine 498
Red Taro 638
Red Tea 150, 158, 555
Red Timber 524
Red Vegetable 732
Red Zao 37
Red-flowered Artemisia 647
Red-flowered Indigo 721
Red-flowered Wan-tian Oil Tea 555
Red-green Vine 369
Red-in-Snow 28, 55, 56, 57, 408
Red-in-the-mountains 431
Red-seeded Watermelon 695
Red-wood Oak 349
Reeves Spiraea 464
Refined Seaweed 259
Rehmannia 171, 172
Repairing Bone Grease 471
Repairing Material 161
Return Cooking Pork 314
Returning-to-Youth 382
Retuse-leaved Amaranth 380
Reverse Bell 523
Reverse Branch Amaranth 381
Revitalizing Herb 238, 752
Revolution Potherb 726
Rhubarb 248, 374
Rhubarb Stalk 373
Ribbon Eggplant 669
Rice 37, 85, 105, 107, 108, 109, 111, 114, 117, 118,
145, 149, 158, 164, 175, 198, 201, 219, 245, 248,
250, 262, 292, 293, 297, 387, 466, 536, 546, 584,
647, 652, 701, 723

Rice Bean 117, 167, 175, 244, 247, 472
Rice Flour Jar 368
Rice Gin 151, 196, 199, 627
Rice Paper 248
Rice Planting Berry 456
Rice Seaweed 261
Rice Soup Fruit 616
Rice Tamale Leaf 297
Rice Urn 386
Rice-planting Blackberry 456
Ridged-fruit Banyan 360
Ridged-fruited Celery 604
Ridgy Luffa 706
Ridgy Silk Gourd 706
Rising Shaking 487
River Catalpa 673
Roasted Bean Tree 645
Robusta Coffee 679
Rock Bamboo 293
Rock Chestnut 342, 509
Rock Ear 265
Rock Stick 464
Rock Tree 348
Rocket 416
Rock-flower Vegetable 261
Romaine Lettuce 741
Rose 77
Rose Eggplant 540
Rose Mallow 238
Rose Petal 150
Rose-apple 580
Rose-leaved Raspberry 461
Roselle 540
Rose-of-Omei Mountain 453
Rose-of-Sharon 541
Rosy Herb 385
Rotten Bean Curd Cheese 36
Rotten Shepherd's Purse 670
Rouge Fruit 574
Rouge Vegetable 384
Rough Aster 720
Rough Chestnut 341
Rough Leaf Tree 352
Round Cardamon 325
Round Fruit Lycium-berry 662
Round Head Torreya Nut 277
Round Torreya Nut 277
Rounded-rhomboid-leaved Tick-clover 473
Round-fruited Begonia 566
Round-headed Yellow-leaved Lycium-berry 662
Round-leaved Bayan 357
Round-leaved Gardenia 681

Roxburgh's Rose 453
Rue 150, 503
Rugose Rose 454
Running Pop 565
Russet Pear 448
Russian Olive 568
Rust-hair Berry 460
Rusty-hairy Five-leaved Shen 588
Rusty-hairy Shuttle Fruit 620
Rusty-hairy Vitex 645
Rutabaga 409
Rye 81, 296

Sacred Lotus 74, 75, 106, 107, 108, 110, 126, 138, 168, 194, 387
Sacred Lotus Leaf 247
Sacred Lotus Rhizome 138
Saddle Mountain Fig 358
Safflower 167, 172, 173, 202, 215, 216, 217, 218, 721
Sagittaria 126, 136, 137, 138, 221, 282
Sago Palm 275, 303
Sago-grain 303
Saigon Cinnamon 149, 399, 400
Salad 739
Salep 321
Saline Soil Suaeda 379
Salsify 751
Salt Goosefoot 379
Saltbush 376
Salty Big Cistache 673
Samarang Rose Apple 581
San Francisco Rose-apple 581
Sanchi 587
Sand Dragon White Eggplant 668
Sand Ginger 328
Sand Jujube 568
Sand Mustard 417
Sand Pear 451
Sand Raspberry 458
Sand Shen 189, 383
Sand Thorny Bush 572
Sand Weed 375
Sand Weed Grain 375
Sandalwood Pear 368
Sandbar Sedge 299
Sandy Land Potherb 421
Sang 586
Santol 507, 508
Sapodilla 621
Sappan Wood 217
Satsuma Orange 496

Saving Soldier Food 448
Scabby Grape 708
Scallion 39, 49, 96, 113, 115, 124, 133, 134, 144, 150, 152, 153, 154, 155, 156, 185, 187, 204, 207, 208, 224, 312
Scaly-tailed Shrub 368
Scarlet Mandarin 496
Scarlet Runner-bean 479
Scarlet Shen 189
Scarlet Sterculia 546
Scarlet Wisteria Tree 483
Scented Peach Tree 577
Schefflera 248
Schefflera Bark 237
Schisandra 169, 224, 226, 396
Schizopetalous Red Hibiscus 540
Scholar Hat Fruit 526
Scissors Herb 282
Scoparia 237
Scorpion Plant 364
Scurfy Pea 167, 208, 471
Sea Blite 378, 379
Sea Cabbage 258
Sea Canarium 642
Sea Grape 370
Sea Lettuce 258
Sea Pea 642
Sea Pine 278
Sea Vegetable 261
Sea Vermicelli 261
Sea-belt 259
Sea-buckthorn 572
Seashore Morinda 682
Seashore Soapberry 523
Seaside Hemp 541
Seed Watermelon 125
Selfheal 169, 237, 248
Sensitive Joint Vetch 465
Sentol 507
Serpent Root 743
Sesame 43, 52, 78, 79, 131, 136, 146, 674
Sesame Butter 42, 64, 65, 674
Sesame Oil 43, 44, 53, 54, 56, 57, 59, 61, 62, 64, 65, 69, 71, 73, 81, 83, 84, 87, 91, 92, 93, 94, 96, 97, 98, 115, 116, 120, 122, 123, 124, 130, 133, 134, 135, 137, 140, 144, 145, 148, 150, 154, 155, 157, 158, 185, 186, 187, 200, 204, 224 264, 354, 539, 674
Sesame Paste 153
Sesame Vegetable 416
Sessile-fruited Crab Apple 429
Setting-heart Vine 521
Seven Finger Fern 271

Seven Little Stars 693
Seven Sisters Fruit 546
Seven-clawed Dragon 639
Seven-leaved Bile 703
Seven-leaved Lotus 395
Seven-star Little Pineapple 694
Seville Orange 492
Shadbush 428
Shade-loving Blackberry 458
Shaji 572
Shallot 311, 312
Shallot Bulb 312
Shallot White 313
Shandong Silk Oak 350
Shantung Maple 522
Sharp-toothed Anise 603
Sharp-toothed Oak 349
Shashen 166, 713
Shatian Pomelo 494
Shea Tree 619
Sheathy Snake Mushroom 369
Shed Gourd 711
Shepherd's Purse 151, 414
Shield-leaved Berry 459
Shiny-leaved Celery 606
Shiny-leaved Privet Fruit 172
Short Gourd 666
Short Rhubarb 373
Short Water Gourd 708
Short-branched Abelia 686
Short-fruited Anise 606
Short-spiked Fishtail Palm 301
Short-stalked Sour Vine Fruit 616
Short-stalked Wild Sesame 647
Showy Millettia 478
Shrubby Cinquefoil 440
Shrubby Hot Pepper 660
Siam Cardamon 326
Siamese Cycad 275
Siamese Honey Mandarin Orange 498
Siberian Crab-apple 436
Siberian Elm 354
Siberian Nitraria 490
Sichuan Aconite 42, 181, 389
Sichuan Angelica 246
Sichuan Berry 461
Sichuan Blackberry 461
Sichuan Cinnamon 401
Sichuan Hazelnut 339
Sichuan Ligusticum 247
Sichuan Lovage 168, 171, 172, 173, 181, 197, 200,
    201, 202, 218, 600

Sichuan Pear 450
Sichuan Pepper 504
Sichuan Peppercorn 49, 60, 61, 86, 87, 91, 102,
    120, 122, 135, 137, 139, 144, 504, 505
Sichuan Preserved Vegetable 409
Sichuan Pressed Vegetable 409
Sichuan Red Oil Sauce 154
Sichuan Sabia 526
Sichuan Tangshen 715
Sichuan Teasel 167, 172, 209
Sichuan Zanthoxylum 504
Sicklepod 470
Sieva Bean 480
Sihui Mandarin Orange 497
Silk Castanopsis 341
Silk Chestnut 342, 346
Silk Chestnut Castanopsis 342
Silk Chestnut Tree 342
Silk Gourd 706, 707
Silk Scallion 313
Silkworm Bean 486
Silver Apricot 100, 275
Silver Ear 74, 162, 169, 174, 178, 269
Silver Lotus 628
Silver Thread Herb 334
Silver-ingot Tree 522
Silver-vine 552
Silverweed 169
Silvery Fruit Elaeagnus 570
Silvery Hare's Ear 385
Silvery-leaved Maesa 617
Si-mao Melodinus 631
Simon's Plum 447
Simpoh 548
Singapore Almond 576
Single-rooted Herb 674
Six Horn Vegetable 598
Six Sepals Vine 424
Six-ears Angled 236
Skirret 606
Skirt Belt Vegetable 260
Sky Smartweed 552
Sky Tent Herb 265
Skyward-pointing Pepper 659
Slanting-head Vegetable 488
Sleeping Water-lily 388
Slender Artemisia 719
Slender Strawberry 433
Sliced Laminaria 259
Slippery Body 709
Sliver Aloeswood 197
Small Bulb Garlic 313

Small Cranberry 612
Small Cucumber 700
Small Flat Pea 476
Small Garlic 313
Small Goat Peach 550
Small Grain 296
Small Insects' Bedding-sheet 510
Small Lima Bean 480
Small Rhubarb 373
Small Scallion 313
Small Variegated Gardenia 681
Small Wide Calamondin 501
Small-flowered Sugar Mustard 416
Small-fruited Wild Grape 533
Small-grain Coffee 678
Small-leaved Aromatic Cassia 400
Small-leaved Celery 588, 601
Small-leaved Curran 427
Small-leaved Honeysuckle 686
Small-leaved Mulberry 362
Small-leaved Mustard 408
Small-leaved Tea 557
Smartweed 244
Smashed Melon 694
Smoke Tree 516
Smooth Bark Tree 609
Smooth Stone Peach 445
Smooth Supple-jack 528
Snake Blackberry 460
Snake Gourd 712
Snake Hemp 362
Snake Melon 697
Snake Palm 304
Snake-head Herb 741
Snap Bean 480
Snapweed 527
Snow Orange 499
Snow Pea 481
Snow Pear 450
Snow Tea 265
Snow Wind Vine 265
Soap Berry Root 237
Soft Date 624
Soft Jujube 550
Soft Jujube Monkey Peach 550
Soft Shen 646
Soft Skin 706
Soft Soybean Curd 474
Soft-hairy Bugseed 377
Soft-juice 384
Soft-tailed Lettuce 741
Soldier Pea 476

Solitary Berry 461
Solitary Leaf-set 306
Solitary Rose 452
Solomon's Seal 169, 172, 173
Sophora Flower-bud 484
Sophora Seed 484
Sorghum 29, 37, 39, 88, 290, 291, 296, 297, 298, 636
Sorrel 374
Sour Ankle 581
Sour Bean 485
Sour Cherry 446
Sour Chicken Vine 616
Sour Creeper 631
Sour Embelia 614
Sour Fellow 443
Sour Fruit 517
Sour Grasshopper Vegetable 385
Sour Jujube 515
Sour Mangosteen 560
Sour Mint 652
Sour Mustard 29
Sour Orange 492
Sour Pear 451
Sour Sap 664
Sour Taste 530
Sour Thorn 572
Sour Vegetable 408
Sour Vine 615, 631
Sour Vine Berry 614
Sour Vine Shoot 614
Sour-shoots Vegetable 614
Soursop 398
South China Castanopsis 343
South China Crane-lice 607
South China Dayflower 235
South China Eleuthero 235
South China Mangosteen 559
South China Motherwort 647
South China Oleaster 570
South China Sour Gum 575
Southern Apricot Seed 227
Southern Astragalus 478
Southern Bean Curd Cheese 36
Southern Clover 477
Southern Jujube 88
Southern Mulberry 361
Southern Palm Starch 300
Southern Schisandra Fruit 224
Southern Shashen 713
Southern Snake Vine 520
Southern Sour Jujube 515

Southwestern China Strawberry 433
Soy Sauce 33, 34, 49, 53, 54, 56, 57, 60, 61, 64, 65,
    68, 71, 73, 77, 80, 82, 83, 84, 86, 87, 93, 96, 97,
    104, 109, 110, 114, 115, 116, 120, 121, 123, 124,
    130, 133, 134, 141, 144, 145, 148, 150, 152, 153,
    154, 155, 156, 158, 159, 185, 186, 187, 200, 204,
    208, 224, 262, 264, 474
Soybean 36, 37, 56, 106, 109, 118, 119, 120, 124,
    127, 147, 149, 150, 262, 290, 291, 474, 542, 594,
    665, 685, 696
Soybean Curd 119
Soybean Milk 474
Soybean Oil 474
Soybean Sauce 33, 604
Soybean Sheet 119
Soybean Skin 119
Soybean Sprout 112, 119, 121, 122, 474
Spaghetti Potherb 310
Spanish Onion 312
Spearmint 649
Spice Herb 485
Spice Silk Vegetable 597
Spiced Soybean Preserve 262
Spice-on-the Mountain 657
Spider Spice 606
Spider Tree 423
Spiked Loosestrife 572
Spinach 14, 64, 204, 273, 378, 543
Spine-leaved Oak 351
Spiny Bamboo 285
Spiny Caper 423
Spiny Grape 534
Spiny Hedge Plant 564
Spiny Hedge Plum 564
Spiny Knotweed 237
Spiny Liquid 151, 152
Spiny Palm 304
Spiny Phellodendron 394
Spiny Plum 425
Spiny Raspberry 460
Spiny Straps 281
Spiny Tender Shoot 726
Spiny Wild Pear 450
Spiny Yellow Fruit 630
Split Cabbage 406
Split-bean Vegetable 422
Sponge Gourd 20, 21, 707
Spread-hairy Melastoma 582
Spreading Vegetable 410
Spring Elm 354
Spring Flower 452
Spring Shoot 129, 295

Spring Tender Green 409
Spring Vetch 487
Sprouted Rice 168
Square Bamboo in Mei-tuo 286
Square Canarium 505
Square Herb 274
St. John's-wort 561
Staff Celery 590
Staphyllea Oil 521
Star Anise 77, 150, 151, 152, 158, 396
Star Apple 619, 620
Star Vegetable 618
Starch Tuber 330
Stars And Stripes Flag Honey Orange 500
Star-stripes Flag Sang 586
Stealthy-on-cloth 607
Steel Leaves Fog Seed 371
Steelyard Weight Leaf 397
Sticky Yam 323
Sting Nettle 365
Stinging Hemp 364
Stinging Nettle 365
Stinking Aromatic Madder 646
Stinking Catnip 645
Stinking Common-in-Mountain 502
Stinking Creeper 684
Stinking Goat 502
Stinking Jasmine 643, 644
Stinking Peony 643
Stinking Weed 470
Stone Brasenia 258
Stone Conical Chestnut Tree 345
Stop Bleeding Herb 599
Stop Blue Herb 422
Straw Mushroom 69, 270
Strawberry 81, 336, 433
Strawberry Guava 578
Stream Vegetable 258
Strike Forehead Bubble 665
Strike Melon 695
String Bean 480
String-of-fruits Vine 393
Stripe Bamboo 286
Suberose-leaved Actinidia 553
Succhini Squash 702
Sugar Apple 398
Sugar Ball 431
Sugar Beet 376
Sugar Cane 296
Sugar Coconut 301
Sugar Fruit 438
Sugared Jujube 169

Sugar Maple 522
Sugar Palm 300, 301
Sugarplum 428
Sugar-tea Berry 426
Summer Cypress 378
Summer Solstice Mushroom 267
Summer Squash 20, 702
Summer Withered Herb 653
Sun-facing Mallow 734
Sunflower Seed 81, 88, 125, 126
Sun-lotus 255, 329
Supple-jack 235
Supreme Deity Plum 447
Surinam Cherry 576
Surpass Elm 354
Suzhou Green 411
Swamp Celery 606
Swamp Potato 282
Swamp Sword-tree 281
Swamp-loosestrife 618
Swatow Mustard 28, 55, 65, 407, 408
Sweet Almond 227, 228, 442
Sweet Apricot Seed 227
Sweet Bamboo 285, 294
Sweet Basil 20, 650
Sweet Big Cistache 673
Sweet Blue 410
Sweet Buckwheat 370
Sweet Castanopsis 342
Sweet Cherry 442
Sweet Coltsfoot 741
Sweet Dew Lichen 264
Sweet Flour Sauce 33, 39, 60, 297
Sweet Foreign Pepper 659
Sweet Fruit Orchid 332
Sweet Gourd 696
Sweet Gum Tree 399
Sweet Herb 475
Sweet Jujube 571
Sweet Kudzu 248, 482
Sweet Liquor 263
Sweet Mandarin Orange 496
Sweet Melon 696, 697
Sweet Olive 627
Sweet Orange 498
Sweet Pea-berry 665
Sweet Pepper 659
Sweet Potato 11, 26, 114, 126, 140, 144, 638
Sweet Potherb 719
Sweet Sagittaria 300
Sweet Shoot Bamboo 285, 294
Sweet Smelling Gourd 698

Sweet Sop 398
Sweet Sorghum 297
Sweet Tall Cereal 297
Sweet Tea 348, 461
Sweet Vegetable 376
Sweet Wild Grape 534
Sweet Yam 332, 638
Sweet-as-dew 656
Sweet-fruit Vine 521
Sweet-kernel Apricot 442
Sweet-leaved Chrysanthemum 748
Swiss Chard 376
Sword Bean 469

Tai-bai Peak Flower 263
Taidong Pea Persimmon 625
Tai-shan Bamboo 286
Taiwan Castanopsis 344
Taiwan Elder 687
Taiwan Farfugium 730
Taiwan Loquat 432
Taiwan Orange 499
Taiwan Persimmon 625
Taiwan Wild Persimmon 625
Tall Cereal 297
Tamarillos 660
Tamarind 485
Tamarind of the Indies 685
Tangerine 496
Tangerine Peel 150, 152, 167, 547
Tangled Herb 387
Tangled Vine 363
Tangshen 714
Tangshen from Fangxian in Western Hubei 715
Tangshen from Liaoning 714
Tangshen of Sichuan 715
Taoist Goat Shen 634
Tapioca 138, 511
Tare 288, 487
Taro 126, 140, 141, 305
Taro Head 305
Taro Orchid 331
Taro Petiole 235, 305
Taro Scapes 305
Taro Son 305
Tartar Garlic 314
Tartar Scallion 313
Tartary Buckwheat 371
Tartary Bush-clover 476
Tartary Gourd 698
Tartary Hemp 674
Tartary Oat 284

Tartary Pepper 333
Tartatian Bean 486
Tassel Flower 380, 729
Tassel Tree 626
Tea 555
Tea Cassia 469
Tea Crab-apple 436
Tea of Tai-bai Shan 265
Tea Spiraea 464
Tea-leaf Tree 626
Tea-on-Tree 367
Teeth Banana 325
Teeth-man Vine 529
Tender Lady-fern 273
Tender Shoot Bamboo 285
Tender-leaved Athyrium 273
Ten-feet Chrysanthemum 735
Tetragonal Persimmon 624
The Virgin 626
Thick Neck 424
Thick Skin Vegetable 376
Thimbleberry 459
Third Month White 411
Thistle Shoot 726
Thorny Bamboo 285
Thorny Bramble 677
Thorny Elm 353
Thorny Hemp 674
Thoroughly Cured Di-huang 672
Thousand Bending Vegetable 572
Thousand Ears Grain 381
Thousand Ribs Tree 463
Thousand Sheet 474
Thousand-fold Paper 237
Thousand-leaves 408
Thousand-mile Fragrant 657
Thread Eggplant 669
Thread Onion 313
Three Big Mushroom 267
Three Prosperity 490
Three-and-Seven 185, 587
Threefold Bitter 236
Three-leaved Eleuthero 586
Three-leaved Lian 393
Three-toothed Wild Pea 486
Through Wood 391
Tibetan Arrow Bamboo 289
Tibetan Barley 291
Tibetan Fennel 593
Tibetan Male Bamboo 288
Tibetan Peach 445
Tibetan Spiny Hazelnut 340

Tibetan Square Bamboo 286
Tibetan Strawberry 434
Tiger Berry 458
Tiger Lily 319
Tiger Nettle 365
Tiger Skin Pine 278
Tiger Stick 371
Timber Bamboo 294
Toddy Palm 302
Tomato 45, 81, 133, 664
Tomentose Hazelnut 339
Tong Shan Asafetida 597
Tonkin Canarium 506
Toog 509
Tooth-cup 573
Torch Fruit 448
Torreya Nut 276, 277
Trailing Smartweed 237
Trapa 583
Treated Persimmon 623
Tree Cotton 245, 246, 544
Tree Ear 266
Tree Ginger Seed 401
Tree Hibiscus 541
Tree Lycium 614
Tree Melon 357
Tree Myrica 335
Tree Pea 469
Tree Peanut 620
Tree Potato 511
Tree River Hemp 483
Tree Sorrel 490
Tree Strawberry 335, 336
Tree Tomato 660
Tremella 74, 163, 179, 223, 269
Triangular-leaved Nettle 365
Tricosanthes Root 248
Trifoliate Orange 248
Trifurcate Barley 291
Trigonous Gourd 698
Trilobate Spiraea 464
True Aromatic Sliver 651
True Ginger 329
Tsaoko 150, 152, 246, 327
Tumi 454
Turban Oak 349
Turkestan Rose 454
Turmeric 149, 150, 217, 327
Turning Gourd 702
Turning Lotus 734
Turunj 494
Twin-kidney Vine 467

Twisting Vine 703
Two-colored Jackfruit 356
Two-flowered Raspberry 455
Two-horned Celery 602
Typha Vegetable 280

Ulcer Herb 309
Underground Centipede 271
Underground Dragon 614
Underground Lagenaria 193
Unicorn Seaweed 261
United Germinating 465
Upland Cotton 539
Uprooting Thousand Pounds 478
Usurper's Flower 567

Vanilla 332
Variant Bean Vegetable 604, 605
Variegated Loins Red Buds Taro 307
Vegetable Bean 480
Vegetable Fern 273
Vegetable Gourd 697
Vegetable Heart 20, 411
Vegetable Humming-bird 483
Vegetable Marrow 702
Vegetable Oil 51, 52, 53, 54, 55, 57, 58, 59, 61, 63,
    65, 69, 70, 73, 74, 75, 82, 83, 86, 91, 92, 93, 97,
    102, 114, 115, 120, 122, 123, 126, 130, 134, 135,
    136, 137, 139, 141, 144, 146, 152, 154, 155, 156,
    157, 187, 465
Vegetable Oyster 751
Vegetable Sponge 707
Vegetable Vine 369
Vegetable-chicken Shoot 744
Vegetable-of-the-Immortal 686
Velutinous Blackberry 460
Velvet Foot 267
Velvet Plant 731
Vetch 486, 487
Viburnum 688
Vietnam Canarium 506
Vietnam Pepper 503
Vietnam Weed 726
Vigna 488, 489
Vine Gourd 550
Vine Herb 636
Vine Jujube 552
Vine Malva 384
Vine Plum 552
Vine Turmeric 330
Vinegar Willow 572
Vine-jujube Actinidia 552

Violet Green 562
Violet Vegetable 563
Violet-flowered Earth Nail 563
Vitex Bush 644
Voa-vanga 685

Wall Litchi 360
Walnut 88, 105, 168, 208, 337, 506
Walnut Catalpa 337
Wampi 500
Wasabi 416, 422
Washington Navel Orange 500
Water Apple 581
Water Artemisia 719
Water Bamboo 294
Water Bamboo Leaf 309
Water Banyan 235
Water Bean 282
Water Block-up 282
Water Cabbage 284
Water Caltrop 584
Water Candle 280, 281
Water Carrot 592
Water Celery 600
Water Chestnut 32, 69, 73, 77, 126, 140, 142, 143,
    144, 194, 299, 343
Water Chestnut Starch 138
Water Chickweed 513
Water Coconut 303
Water Creeping Green 421
Water Cress 403, 422
Water East Brother 553
Water Eupatorium 425
Water Fiddlehead 272
Water Fly Thistle 745
Water Front Vegetable 732
Water Golden Phenix 527
Water Gourd 707
Water Hemp 363
Water Hen Fruit 620
Water Hyacinth 310
Water Myrica 435
Water Paulownia 673
Water Pen 574
Water Plant Scape 283
Water Plantain 246, 284
Water Poppy 628
Water Portulaca 513
Water Rock Banyan 536
Water Rose-apple 578
Water Shield 387
Water Smartweed 35, 372

Water Spinach 636
Water Starwort 513
Water Straw-cloak 676
Water Surface Lotus 628
Water Turtle 283
Water Vegetable 283
Water Vegetable Flower 283
Water Willow 425, 536, 572
Water Yam 321
Water-elm Mountain Ash 463
Waterfern 272
Waterhen Egg 436
Watermelon 81, 124, 125, 694
Watermelon Seed 106, 124, 125, 126
Waternut 584
Wavy-leaved Rhubarb 374
Wax Bean 480
Wax-leaf Privet 626
Welcome-the-sun Flower 735
Welsh Onion 20, 25, 39, 246, 312
West Court Crab-apple 437
Western European Common Foxglove 195
Western Foreign Chrysanthemum 735
Western Ginseng 586
Western Lagenaria 702
Western Mushroom 266
Western Red Persimmon 664
Western Rice 511
Western Sour Jujube 516
Western Vegetable 422
Wet Fermented Red Rice 38
Wet Fermented Rice 37
Wheat 11, 37, 81, 150, 248, 262, 291, 292, 297,
   298, 299, 406, 636, 696, 723
Wheat Bottle Herb 386
Wheat Flour 40, 149, 150, 244, 297, 474
Wheat Gluten 44, 69, 70, 152, 297
Wheat Scallion 313
Wheat Straw Vegetable 417
Wheel Mei 452
Whip Cucumber 699
White Aconite Slice 389
White Acronychia 491
White Akebia 391
White Aster 720
White Beam 345
White Bean 489
White Branch Lycium-berry 662
White Canarium 505
White Cardamon 326
White Chicken Droppings Creeper 685
White Chrysanthemum 65, 245

White Cicada 681
White Dragon Whisker 634
White Elm 354
White Fermented Rice 37, 38, 150
White Flesh Banyan 359
White Fruit 100, 275
White Ginger 328
White Gourd 691, 697
White Jade Pumpkin 702
White Lanzhou Melon 696, 698
White Laurel Tree 355
White Lily 319
White Magnolia 396
White Mugwort 719
White Mulberry 361
White Mushroom 266
White Mustard 407
White Oak 352
White Paederia 685
White Pear 449
White Peony 171, 391
White Pepper 333, 334, 502
White Perilla 652
White Pine 277
White Potato 670
White Prickly Stem 586
White Radish 135
White Rock Chestnut 346
White Sliver Vegetable 652
White Spine Flower 484
White Tan 625
White Taro 638
White Tea Chrysanthemum 723
White Thorn 490
White Vegetable 407, 413
White Warm Rod 689
White Water Chestnut 282
White Wax Tree 616
White Wild Pear 643
White Wild Rose 454
White Yam 145, 321, 323
White-back Hematonics 733
White-bark Pine 278
White-beneath Herb 439
White-beneath Vegetable 439
White-bracted Artemisia 719
White-flowered Deadnettle 647
White-flowered Dragon Whisker 467
White-flowered Elephant's Foot 728
White-flowered Elephant's Foot Shoot 728
White-flowered Elephant's Foot Weed 728
White-flowered Embelia 615

White-flowered Gourd 92, 704
White-flowered Mugwort 20
White-flowered Oak 349
White-flowered Rock Mustard 414
White-flowered Tea 617
White-flowered Vegetable 670
White-flowered Vetch 487
White-foot Tong 543
White-hair Black 633
White-headed Crow 193
White-leaved Bramble 535
White-paper Fan 684
White-seeded Vegetable 733
Whortleberry 612
Wild Amaranth 382
Wild Apricot 441
Wild Artemisia Vegetable 730
Wild Berry 456
Wild Bitter Hemp 736
Wild Bitter-lettuce 746
Wild Cabbage 410
Wild Cat Bean 485
Wild Cat Sour 616
Wild Celery 604
Wild Cherry 447
Wild Cock's Comb 382
Wild Cocoa Tree 545
Wild Cornel 609
Wild Cowpea 476
Wild Crab-apple 436
Wild Crow Ailanthus 520
Wild Dioscorea 323
Wild Endive 725
Wild Fern 273
Wild Flat Bean 470
Wild Garlic 313
Wild Geese Return Red 382
Wild Ginger 329
Wild Hawthorn 430
Wild Hot Pepper 660
Wild Hot Pepper Tree 660
Wild Hundred-mile Fragrant 657
Wild Kudzu 248, 482
Wild Leek 308
Wild Lettuce 690
Wild Lychee 608
Wild Mallow Vegetable 542
Wild Mangosteen 559
Wild Man's Melon 392
Wild Mustard 422, 752
Wild Oat 284
Wild Oceanic Nightshade 666

Wild Onion 311, 312, 315
Wild Pea 476
Wild Peach 229, 444
Wild Peanut 471
Wild Pear 448
Wild Perilla 651
Wild Persimmon 624
Wild Piece-of-dough 720
Wild Sesame 527
Wild Soybean 473, 474
Wild Spice 657
Wild Strawberry 433, 435
Wild Tangshen 714
Wild Taro 307
Wild Tea 629
Wild Thyme 657
Wild Trapa 584
Wild Vigna 489
Wild Walnut 337
Wild Waterlemon 565
Wild Watermelon Seedling 541
Wild Wood-gourd 393
Wild Zanthoxylum 504
Willow Orange 499
Willow-leaved Celery 595
Willow-leaved Mice Plum 529
Wind Rolled Lichen 264
Windmill Palm 304
Wind-sail Vine 523
Wine Coconut 301
Wine Grape 536
Wine Palm 302
Wine-bottle Berry 582
Wing Bean 481
Winged Suaeda 379
Wingless Willow Celery 595
Winnowing Fan Willow 334
Winter Amaranth 542
Winter Cold Vegetable 542
Winter Mallow 542
Winter Mango 518
Winter Melon 93, 94, 95, 96, 691
Winter Melon Fellow 711
Winter Mushroom 268
Winter Shoot 295
Winter Tolerant Vine 215
Winter-worm Summer-herb 174, 175, 262
Wire Eggplant 669
Wisteria 45, 80, 81, 489
Witchweed 672
Wolf-teeth Spines 484
Wood Ear 68, 69, 71, 73, 90, 137, 145, 155, 174,

221, 266
Wood Melon 429
Wood Pear 431, 453
Wood-ear of Yunnan 266
Woodland Chervil 592
Woodland High Shen 592
Woody Berry 462
Woody Sky Smartweed 552, 553
Woolly Grass 291
Woolly Pingpo 547
Worm-herb 262
Wormseed Mustard 416
Wrap-heart Vegetable 410
Wrapped Grain 298
Wrapped Heart Mustard 408
Wrapped Oak 347
Wrapped-seed Bayan 357
Wrapper Flower 716
Wuni 509
Wu-wei-zi 396

Xerophytic Oil-gourd 700
Xikang Red Pine 278
Xing-shan Schisandra 397
Xinhui Navel Orange 500
Xinhui Sweet Orange 499
Xi-suan-zao 516

Yam 126, 144, 145, 146, 171, 174, 482
Yam Pea 670
Yangmei 335
Yard-long Bean 21, 26, 96, 97, 98, 122, 488
Yellow Alone 322
Yellow Asarum 383
Yellow Bean 474
Yellow Bud White 413
Yellow Climbing-up 692
Yellow Crab Apple 561
Yellow Daylily 318
Yellow Deer Fruit 519
Yellow Drug 322
Yellow Eel Vine 528
Yellow Elixir 320
Yellow Floating Heart 628
Yellow Flower Vegetable 318, 752
Yellow Gardenia 680
Yellow Goat 631
Yellow Gourd 698
Yellow Herb 646, 744
Yellow Hibiscus 541
Yellow Jiu-cai 53, 315
Yellow Lintel 344

Yellow Lintel Castanopsis 344
Yellow Locust 483
Yellow Nut-grass 299
Yellow Ox Tea 557, 558
Yellow Parasite 369
Yellow Pond-lily 388
Yellow Quail Vegetable 752
Yellow Raspberry 456, 457
Yellow Rose-of-Sharon 541
Yellow Sichuan Mallow 538
Yellow Skin 500
Yellow Spiny Rose 454
Yellow Tea 518
Yellow Teeth Fruit 559, 560
Yellow Tremella 269
Yellow Whisker Vegetable 379
Yellow-feet Chicken 317
Yellow-flower Herb 748
Yellow-flowered Catalpa 673
Yellow-flowered Ground Gourd 640
Yellow-flowered Leek 312
Yellow-fruit Lycium-berry 663
Yellow-rice Wine 432
Yichang Raspberry 457
Young Fern Frond 272
Young Pig Berry 582
Young Radish Plant 417
Youthful Strength Rice 611
Yu Shan Raspberry 455
Yunnan Canarium 506
Yunnan Cardamon 327
Yunnan Cardiocrinum 316
Yunnan Conic Chestnut 342
Yunnan Cycad 275
Yunnan Gmelina 644
Yunnan Holly 520
Yunnan Mangosteen 558
Yunnan Nyssa 575
Yunnan Ottelia 283
Yunnan Plum 442
Yunnan Prickly Canarium 623
Yunnan Prickly Jujube 532
Yunnan Sanchi 587
Yunnan Skullcap 653
Yunnan Supple-jack 529
Yunnan White Oak 347
Yunnan Wild Wood-melo 394

Zanthoxylum 20, 26, 29, 40, 49, 60, 61, 62, 66, 74,
    86, 87, 91, 102, 114, 116, 120, 121, 122, 135, 137,
    139, 140, 144, 148, 149, 150, 151, 152, 154, 158,
    504, 505, 716